MARTIN FROBISHER, Śc.D.

Formerly Instructor in Bacteriology and Pathology, Johns Hopkins University Medical School; Associate Professor of Bacteriology, Johns Hopkins University; Associate Professor of Bacteriology, Emory University Medical School, Atlanta, Georgia; Special Member, International Health Division, Rockefeller Foundation; Chief, Bacteriology Section, Communicable Disease Center, United States Public Health Service, Atlanta, Georgia; Professor and Head, Department of Bacteriology, University of Georgia, Athens

RONALD D. HINSDILL, Ph.D.

Associate Professor of Bacteriology, University of Wisconsin, Madison

KOBY T. CRABTREE, Ph.D.

Professor of Microbiology; Chairman, Department of Biological Sciences, University of Wisconsin Center System, University of Wisconsin, Wausau

CLYDE R. GOODHEART, M.D.

Visiting Professor, Department of Microbiology, Rush-Presbyterian-St. Luke's Medical Center, Chicago; President, BioLabs, Inc., Northbrook, Illinois

NINTH EDITION

FUNDAMENTALS of MICROBIOLOGY

FROBISHER · HINSDILL · CRABTREE · GOODHEART

W.B. SAUNDERS COMPANY · Philadelphia · London · Toronto

W. B. Saunders Company: West Washington Square
Philadelphia, Pa. 19105

1 St. Anne's Road
Eastbourne, East Sussex BN21 3UN, England

833 Oxford Street
Toronto, M8Z 5T9, Canada

The Cover: A scanning electron micrograph showing a typical bacterial population in the digestive system of the rat. (Courtesy of S. E. Erlandsen and G. Wendelschafer.)

Fundamentals of Microbiology

ISBN 0-7216-3922-4

Last digit is the print number: 9 8 7 6 5

preface
to the ninth edition

This, the 9th edition of *Fundamentals of Microbiology,* unlike all previous
editions, represents the combined efforts of four authors, three—Koby T.
Crabtree, Clyde R. Goodheart, and Ronald D. Hinsdill—being new to this
volume. The field of microbiology has become so broad, and so much new
knowledge has accrued so rapidly, that it has seemed advisable to the
hitherto sole author to seek synergy with others in presenting a revision of a
book at this level. Nonetheless, this book retains its basic character and
approach: it is intended to be an elementary survey of microbiology for
college students, as complete and broad as possible. The new authors of this
edition include men highly knowledgeable about modern advances in the
field and in various specialties, as well as competent in writing. All four
authors are experienced in undergraduate and/or graduate teaching. The
original author congratulates himself in having had the collaboration
of these scientists.

As a result of the joint efforts of the group, the old (8th) edition has been
largely rearranged, extensively rewritten, brought up to date, and put as
much in line with current knowledge as possible. One of several new
features in the ninth edition is the considerable expansion of illustrative
material and the inclusion of numerous carefully selected color plates. Other
additions include a glossary and two new chapters: one on the role of
microorganisms in our daily lives and one on the ecology of natural waters.
Supplementary reading now consists less of lists of specific technical articles
and more of general reviews, monographs, and books which the beginning
student of microbiology can read and comprehend but which may also
possibly be of some use to instructors.

The authors fully intended that this text would contain more information
than can be covered by most instructors teaching a one-semester course. This
will allow students to probe more deeply into subject areas that are given
only cursory treatment in lectures because of time limitations. The division
of each chapter into numbered sections should greatly simplify the assignment of both optional and required reading.

The 9th edition should also prove useful as a general reference text for
students taking microbiology as a major. It is one of the few to consider the

various groups of bacteria in some detail and one of the very few to conform, as far as presently possible, to the new (8th) edition of *Bergey's Manual of Determinative Bacteriology.* The chapters on classification and systematic study, with integration of the taxonomy of the 7th (1957) and 8th (1974) editions of *Bergey's Manual,* together with new materials on GC ratios of the various groups and other data, were a special project and should be of great help in this difficult field. Many students and teachers will also appreciate the updated coverage of the various aspects of applied microbiology.

Because of continued demand, chapters on the chemical basis for microbial life have been included again in the 9th edition. Even students with adequate prerequisites for microbiology may find these chapters helpful as "quick refreshers," while students with little chemistry and no biochemistry may find them indispensable to an understanding of modern microbial physiology and genetics.

Thanks are due, as always, to the expert staff of W. B. Saunders Co., especially to Mr. Richard H. Lampert, Biology Editor, for very effective co-ordination of the work as a whole and painstaking assistance with illustrations, and to Mrs. Jean Fraley, Manuscript Editor, for expert handling of the numerous problems that inevitably arise in the preparation of a book of this size and complexity.

A valuable new collateral feature is a class-tested companion book for laboratory teaching, prepared for general use, by two of the authors of this textbook: K. T. Crabtree and R. D. Hinsdill.

Martin Frobisher

Ronald D. Hinsdill

Koby T. Crabtree

Clyde R. Goodheart

contents

FUNDAMENTALS of MICROBIOLOGY

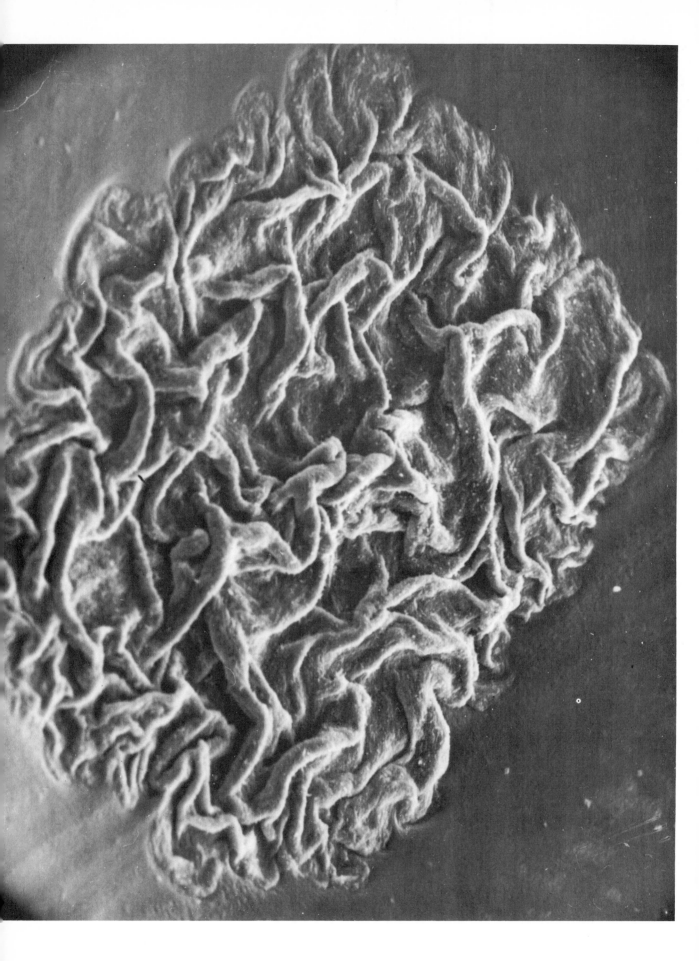

SECTION ONE

THE ORIGINS OF MICROBIOLOGY AND THE CHEMICAL BASIS OF MICROBIAL LIFE

If monkeys are people, as once averred, they are the queerest. One of the most amusing queernesses of monkeys is their intense curiosity. But among all of the primates *Homo sapiens* is perhaps the queerest if this attribute be measured by the intensity of the desire to know, to probe, to investigate, to find out the why, and how, about everything everywhere: natural phenomena, the midnight glories of the celestial firmament, and so on, including the nature and origin of life and the possibility of its existence elsewhere than upon this planet.

So intense and urgent is the human desire to uncover the secrets of the universe, both animate and inanimate, and even the mysteries beyond life, that much of the time, energy, and resources of many intellectually endowed individuals have, since time immemorial, been expended in what is now called scientific investigation: the search for experimentally demonstrable truth.

The accumulation of fascinating, profitable, and useful exact knowledge, through science, has each year attracted thousands of new workers, each highly motivated by one or more of many different considerations: pure curiosity, the wish to serve humanity and ameliorate the lot of mankind, desire for fame and/or fortune, and so on through the entire gamut of human desires and ambitions.

The segment of scientific investigation that is now called **biology** has, like all living things, grown enormously over the millennia of human existence. A little over three centuries ago, fathered by the invention of powerful magnifying lenses, the mother science, biology, produced an offspring of tremendous vigor that has now grown to great size. Nurtured since then by constant improvements in high-power, compound microscopes and, within the last quarter century, by the development of practical electronic microscopy, the offspring has concerned itself entirely with living things too

Facing page, Courtesy of Ivan L. Roth and Cheryl Weinmeister.

3

small to be seen without the use of microscopes and has consequently been named **microbiology.** It is the purpose of this book to introduce, to interested persons, this relative newcomer in the ancient and honorable community of the sciences.

The first three chapters of the book present some details of the origin and growth of microbiology, with some useful instructions for beginners to aid those who intend to develop the acquaintance further. Ideas concerning the chemical basis of microbial life are given in the next four chapters, while uses of such knowledge in microbiological methodology comprise the two remaining chapters of this section. Thus our first section of 10 chapters prepares the reader technologically to follow intelligently the synoptic descriptions of the fantastic inhabitants in the unseen, living cosmos that are the subject of chapters in Section Two.

CHAPTER 1 · MICROBIOLOGY — A RELEVANT SCIENCE

1.1
MICROORGANISMS AND MICROSCOPY

We may define microbiology as the study of organisms too small to be seen clearly, as individuals, by the unaided human eye. Such organisms are commonly spoken of as "microbes" or, more exactly, as **microorganisms** (hence *microbiology*). Simple magnifying glasses ("hand lenses") capable of enlargements up to about 10 diameters (×) can make clearly visible only comparatively enormous, multicellular organisms such as "chigger" fleas, larval mites, and plant structures like the filaments of some molds; properly speaking these are neither microbes nor microorganisms. The microorganisms of the microbiologist bear about the same size ratio to chiggers and mites as fleas bear to the dogs they torment. The true microorganisms are absolutely invisible at magnifications of 10× or even 100×. They are unimaginably minute.

To visualize clearly any single living cell (except certain eggs and a few species of large algae), a microscope magnifying at least 300× is at best a minimal requirement. The advanced and professional microbiologist of today customarily uses compound microscopes giving magnifications of from about 400× to 1,200×. For the greater magnifications now in common use he must rely on electron microscopes giving magnifications of from around 2,500× to a million or more. The virologist (student of viruses) is restricted to electron microscopy to visualize his **virions** (individual virus particles). The diameters of virions are measured in millionths of an inch, or in **nanometers** (1×10^{-9} meter) on the metric scale. Clearly, familiarity with the use of powerful magnifying instruments (Chapt. 3) is an indispensable qualification for admission into the mysterious realms of microscopic life.

1.2
PROTISTS, MONERA, AND PROCARYOTAE

Most of the microorganisms discussed in this book are single, independently living, wholly autonomous cells like protozoans and bacteria. They have been, at various times, by various experts, assigned to various so-called kingdoms: plant, animal, and others discussed below. There are also billions of microscopic cells constituting all "higher" animals and plants. Each cell in such animals or plants is a part—interconnected with and dependent on others—of an extraneously governed system, each system having a highly specialized organic function, such as liver, muscle, xylem, cambium, and so on. Strictly speaking these cells also are microorganisms; they are alive, they grow and multiply, they are visible only with powerful microscopes, they have many anatomical and physiological properties in common with all other cells, and many types may be artificially cultivated like bacteria or protozoans as separate, independent individuals in test tubes.

However, they are easily identifiable by the microbiologist as being, or having been, parts of higher plants or animals. As such they are of special interest to virologists, plant and animal histologists, pathologists, and so on. As individuals these cells are not assigned to any of the **kingdoms,** and they are given only the necessary incidental consideration in this book, especially in the chapters on virology.

Kingdom Protista. Although some of the wholly autonomous and free-living microorganisms mentioned in the opening sentence of the paragraph above are obviously animals or animal-like (e.g., protozoans and some bacteria) and some are plants or plant-like (e.g., various algae and some bacteria), many others are neither one nor the other but possess characteristics of both plant and animal kingdoms, ranging, in their differing combinations of various structural and physiological properties, like a spectrum, from one extreme to the other. To avoid confusion arising from attempts to define and classify organisms with such intermingled and overlapping characters Haeckel, a German zoologist, in 1866 proposed grouping all microscopic, unicellular, free and autonomously living, photosynthetic and nonphotosynthetic, motile and nonmotile bacteria, protozoans, fungi, and algae in a third kingdom, for which he coined the term Protista. Viruses were not discovered until 1882 (by Iwanowski) and so were unknown to Haeckel. Since then, however, although neither plant nor animal, cellular nor even alive, as these characteristics are currently defined, viruses have been included by a number of workers as protists "by courtesy" (and for convenience). With the passage of time and the accumulation of knowledge, the kingdom Protista, like so many other kingdoms in recent years, has undergone modification. Some biologists would now exclude from the kingdom all but the protozoans. Others view it in other ways. Here the term **protist(s)** is used in a collective sense to mean all naturally free-living, autonomous microorganisms and viruses. This book is an introduction to these microorganisms.

Two Kinds of Cells. As will be discussed more fully in Chapter 11 and elsewhere, there are certain fundamental likenesses between all cells: plant, animal, or otherwise. Nevertheless, there are two absolutely distinct types of cells: (a) **eucaryotic** (Gr. *eu* = true or real; *karyon* = nucleus) and (b) **procaryotic** (Gr. *pro* = primitive).

The nucleus of eucaryotic cells is segregated from the cytoplasm by an enclosing nuclear membrane having numerous perforations in it (Fig. 1–1). Genetic recombination typically involves the familiar merging of nuclei, commonly of well-differentiated sex cells or gametes. The eucaryotic cell is the structural unit of all "modern" (higher) organisms, including ourselves. It is generally thought that the eucaryotic cell represents a much more "advanced" stage in evolution than the procaryotic cell. Eucaryotic protists are often called "higher protists."

In procaryons the nuclear material (DNA) is not enclosed within a nuclear membrane but is distributed, often as apparently discrete masses, throughout the cytoplasm (Fig. 1–2). Although genetic recombination occurs as readily in procaryons as in eucaryons, sexual phenomena of the familiar type seen in eucaryons are absent. The procaryons are commonly called "lower protists." Because of their supposed ancient origin and primitive characteristics procaryons are sometimes called "living fossils."

The only known procaryotic organisms are the bacteria (Schizomycetes) and the blue-green algae (Schizophyceae, Cyanophyceae, or Cyanobacteria). In 1969 these two groups were placed

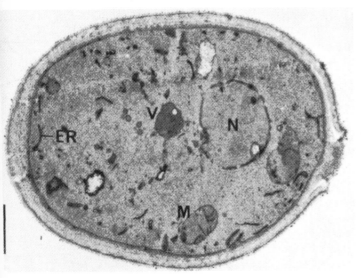

Figure 1–1. Electron micrograph of a thin section through a cell of the common bakers' yeast, *Saccharomyces cerevisiae,* showing many functional structures typical of any eucaryotic protist, including the nucleus (*N*), completely enclosed in a perforated nuclear membrane. (Other structures: *M,* mitochondrion; *ER,* endoplasmic reticulum; *V,* vacuole. A "budding scar" in the cell wall is seen at right; millions of ribosomes form the dark background.) Compare with the procaryotic nucleus in Figure 1–2. Further details of the structure and life cycles of eucaryotic fungi are given in Chapter 12. The vertical bar represents 1 μm (0.00003937, or 1/25,400, inch).

Figure 1-2. Lengthwise thin section of a common sausage-shaped bacterium (*Escherichia coli*), showing a typical form of procaryotic nuclear structure. The nuclear material is the light gray, finely fibrillar masses distributed randomly throughout the cytoplasm and ribosomes (black dots), seen without an enclosing membrane, and occupying about half the area of the picture. (Approximately 20,000×.)

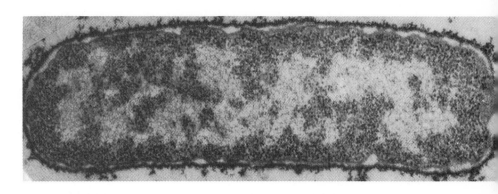

together by Whitaker in a separate kingdom, **Monera.** More recently (1973), the kingdom **Procaryotae** has been proposed for the procaryotic microorganisms. *All other cells are eucaryotic.*

More detailed discussions of eucaryons and procaryons are given in Chapters 11 through 14.

1.3
BACTERIA AS LIVING MODELS

If anyone were to ask why we study microorganisms we would reply that, except for a few highly evolved functions like nervous reactions and mental activities, the microorganisms not only are important in every phase of human life but furnish unparalleled models for the study of the underlying and really intimate physiological and biochemical mechanisms and genetic anatomy of all living organisms in all kingdoms.

Although all living cells—plant, animal, or protist—are the legitimate province of the microbiologist, many microbiologists work especially with bacteria. This is because bacteria can be housed and manipulated conveniently, inexpensively, and, with the exception of a few dangerous pathogens, safely in test tubes, bottles, and the like, and because they can be nourished and maintained in easily prepared nutrient solutions (culture media) occupying relatively small space. Many species may be investigated as individual cells or as "populations" that can multiply in a few hours from a dozen or so cells, or even a single cell, to thousands or millions within a test tube or on a square inch or less of solid, jelly-like surface. Each bacterial cell is a single, discrete self-multiplying, physiologically complete organism. Yet of all microorganisms that are actively living (this excludes viruses) bacteria are seemingly the smallest, structurally the simplest, and physiologically the most primitive. Nevertheless, the forms, arrangements,

and anatomy of bacteria, now fully revealed by electron microscopy, are not only amazingly complex but highly diversified. Except for a few specialized functions, such as cerebration, each tiny cell is, in point of vital activities and physiological functions, a prototype of the human body.

Because bacteria of all species are insensate we can subject them to experimental procedures that would torture higher animals: total deprivation of oxygen or prolonged low oxygen tension (some bacteria *prefer* little oxygen; some are poisoned or killed by it; others must have it freely at all times), or exposure to slow boiling or baking heat (a number of species thrive in nature at almost boiling temperatures, while others can survive for hours in the oven at bread-baking temperatures). We may subject many to seeming starvation on diets of dilute salt solutions, without nutrient sources of energy or vitamins (some can get their energy from sunlight and many can independently synthesize all the vitamins they require). We can desiccate most species completely, freeze them in solid CO_2 at $-76C$ or liquid nitrogen at $-198C$, and confine them in a total vacuum in the dark. All but a few species can survive such treatment unchanged and unharmed for decades. We can immerse them in carbon monoxide (to us a deadly poison but for some of them an excellent source of food energy), expose them to radiations that cause cancer in man and animals but merely produce interesting and sometimes useful genetic mutations in bacteria, and observe and make motion pictures of their reproductive and other processes under many experimentally designed conditions. We can cause them to grow together in various combinations and types of culture media and environmental conditions, thus using them as extremely flexible and adjustable models for industrial and ecological experiments. We can infect them with viruses and study their disease processes. In short, using the methods and materials of microbiology, in cooperation with related sciences, man has advanced his knowledge of life phenomena more in the last few decades than in all of the thousands of previous years of human existence.

1.4
NEWER "PUBLIC IMAGE" OF MICROORGANISMS

Early studies of microorganisms dealt very largely with **pathogenic** (disease-producing) bacteria, protozoans, fungi, and other minute forms of life, and the benefits of such studies were widely and justly acclaimed and publicized. Partly because of this all microorganisms were thought of collectively and indiscriminately by the laity as dangerous and mischievous "germs," invisible and mysterious

Figure 1–3. The old "public image" of microorganisms, shown in this woodcut of a plague victim attended by a physician and his helper. (Courtesy of the Bettman Archive, Inc., New York.)

(Fig. 1–3). Development of higher-powered microscopes and the unceasing invention of more precise and incisive methodologies have now revealed, even to the general public, that the term "microbe" includes only a few dozen really harmful kinds of microorganisms and that there are thousands of species that are perfectly harmless, many of them of enormous value industrially, medically, economically, and in many other ways. Some are highly specialized types on whose activities the entire pageant of life on this planet has been absolutely dependent.

1.5
CELL REPRODUCTION

Except gametes, *all* actively growing cells —plant, animal, or protist, procaryon or eucaryon—can multiply indefinitely in the entire absence of sex. This includes all of the **somatic** cells of our own bodies—muscle, liver, and so on—and of plants—leaf, stem, and root. Protists typically multiply asexually. In addition to asexual multiplication, some eucaryotic protist species (e.g., protozoans, green algae) also exhibit well-differentiated gametes and sexual phenomena.

A few species of bacteria and fungi (molds) exhibit genetic interchanges, though these are of a very primitive sort, suggestive of the earliest beginnings of sexual differentiation; the "sexes" are to some extent interchangeable. The "sex life" of protists is detailed in Chapters 11, 14, and 19.

1.6
ORIGINS OF CELLS

The origins of protists, and of life itself, although experimentally investigated by biochemists and speculated upon by astrobiologists, chemical evolutionists, and other savants, are still lost in the mists of geological antiquity. Yet, as detailed in Chapter 2, it seems evident to many microbiologists that the entire brilliant panoply of life as we see it around us today has evolved, during the billions of years since the earliest beginnings of the earth, from obscure combinations of mere atoms and simple molecules that are known to exist in the remotenesses of interstellar space. It now appears certain that man's earliest ancestors were neither shrewish nor primate, but protist.

1.7
MICROORGANISMS AS ENVIRONMENTAL FACTORS

Marine and Aquatic Microbiology

All forms of protists abound in the waters of the earth: bacteria, diatoms and other algae (Fig. 1–4), fungi, protozoans in thousands of forms, nuisance organisms such as those that cause "red tides" (dinoflagellates), and so on through a long list. Although microorganisms are found in abundance in even the most remote ocean waters (arctic or antarctic), microbial life of all forms is naturally most abundant in the littoral zones near land masses where runoff provides great quantities of food, dissolved and particulate.

A familiar area of special interest is pollution of fresh water by domestic and industrial sewage. Improperly treated sewage involves public health and recreational problems. Many toxic industrial wastes, discharged via rivers into the sea, have both local and worldwide effects on all kinds of marine and aquatic life (Chapt. 44). Microorganisms of both land and seas are largely responsible for decomposition of both solid and dissolved wastes of plant, animal, or industrial origin. Certain alga-like bacteria (e.g., *Beggiatoa*) reduce pollution due to toxic hydrogen sulfide by oxidizing it as a source of food energy. Other bacteria attack and destroy petroleum wastes, and so on.

At stake in the fight against pollution of all sorts is the preservation of photosynthesis by marine and aquatic plants, without which the world's supply of oxygen would diminish drastically and, with continued destruction of our green fields and forests, even fatally for the human race. It has been reported that minute quantities of biologically recalcitrant polychlorinated hydrocarbons inhibit photosynthetic activities of marine algae. Microorganisms are at the very foundation of the marine and aquatic food chains on which humanity ultimately depends for much of its nourishment. The microorganisms, as well as macroorganisms, convert many dissolved and solid materials, both organic and inorganic, into organic food substances on which man ultimately depends.

We are indebted to, and to some extent dependent on, many species of marine protists for such valuable items as chalk, limestones, infusorial earth, flint, some building materials, and probably petroleums and natural gas. Full exploitation of the microbiological resources of

Figure 1–4. Some marine algae (phytoplankton), illuminated by electronic flash and magnified 110×.

fresh and marine waters awaits the attention of well-educated and ingenious marine microbiologists, limnologists, biochemists, engineers, and related professionals.

On the minus side, numerous species of marine organisms cause fouling and destruction of ship bottoms, pilings, cordage, asphalt and concrete sea walls and docks, wood, and other materials constantly exposed to the sea and weather, especially in tidal areas. Some species of microorganisms cause catastrophically destructive epizootics and epiphytotics among marine animals and plants that are valued in the millions of dollars as food, fertilizer, and other products for human uses. Control of all such destructive organisms is a challenge to the student of marine microbiology.

Agricultural Microbiology

This subject has many ramifications that invite the attention of the specialist in microbiology. Among them are plant pathology, including studies and control of microorganisms that cause numerous highly destructive and ruinous diseases of crop plants, forest trees, plants of horticultural and landscaping value, and so on. Similar studies related to livestock, while a province of veterinary medicine, are within the broad realm of agricultural microbiology.

Microbiology of the Soil

All soils, whether desert, fertile loam, or newly emerged lava, contain soluble substances that are available as foods to at least some microorganisms as well as to certain higher plants. Microorganisms are the prime geochemical agents in the formation of soils. Growth depends on the nature of the soil and environmental conditions, such as temperature and moisture. Many microorganisms of the soil are identical with, or closely similar to, those found in the waters of the earth. Some microorganisms appear to be restricted to life in moist, fertile, well aerated soils. Others thrive in muds of swamps in the absence of air, or in the desiccated soils of deserts. The upper few feet or inches of soil in the differing areas of the earth provide innumerable kinds of environmental niches or microenvironments to which one or more of thousands of different species of minute animals, plants, and protists have become adapted: protozoans, algae, molds, bacteria of many biochemical and physiological types, nematodes, insects, annelids, amphibians, rodents, green plants, and so on, all living together commensally, antagonistically, or otherwise. The microbial ecology of the different kinds of soil is even more complex than that of the seas (Chapt. 43).

Useful Species

Antibiotic-Producers. Among the most valuable bacteria of the soil are the mold-like *Streptomyces*, producers of antibiotics such as streptomycin, chlortetracycline, erythromycin, and so on (Chapt. 23). The blue mold *Penicillium notatum*, commonly seen on decaying citrus fruits, produces the antibiotic penicillin, as discovered in 1929 by Sir Alexander Fleming (Fig. 1–5), one of the many microbiologists who have won the Nobel Prize.

Nitrogen Fixation. Other useful, and indeed absolutely indispensable, microorganisms indigenous to the soil are those spoken of as **nitrogen-fixers.** Nitrogen, constituting about 80

per cent of the earth's atmosphere, is essential to all life. Yet, in its elemental, molecular form (N_2) it is completely beyond use by any living thing, with the exception of a few species of very primitive bacteria, some blue-green algae, and some molds. These microorganisms are the only known forms of life capable of utilizing atmospheric nitrogen in the synthesis of cellular materials, a process known as **biological nitrogen fixation** (Chapt. 43). Without the activities of the nitrogen-fixers, green plants (e.g., grasses, trees, ferns), unable to fix N_2 for themselves, could not have evolved. The entire animal kingdom, therefore—herbivores, carnivores, and eventually *Homo sapiens* himself—would never have existed at all. The preparation and sale of preinoculated seed or packaged cultures of nitrogen-fixing microorganisms (*Rhizobium* spp.) is an industrial activity of some value.

Microbial Insecticides. A most interesting and challenging combination of microbiology with entomology is the development of "biological insecticides." For example, certain bacteria (e.g., *Bacillus thuringiensis*) produce poisons (Chapt. 36) that are very effective against certain insects but harmless to other forms of life. The **milky-white disease** used to control Japanese beetles is an actual infection of the larvae with sporeforming bacteria (*Bacillus popilliae*). Destructive "cabbage looper" worms and a number of other pests are very susceptible to certain specific viruses (Chapt. 16). In view of rising objections to chemical insecticides like DDT and other chlorinated hydrocarbons, and biologically recalcitrant chemicals in general, the exploitation of microbial insecticides offers a wide field for the ecologist and the environmentalist.

A converse aspect of applied entomology concerns development of means for control of microbial diseases of valuable insects like honey bees (European or American **foulbrood disease**). Pasteur's studies of viral diseases of silkworms, already mentioned, were pioneering steps in this direction.

1.8
ECOLOGICAL IMPLICATIONS

Of vital importance in the ecological cycles of the earth are hundreds of common species of protists, including scores of kinds of bacteria, in both the soils and seas, that cause decomposition (**mineralization**) of the organic materials in the wastes and dead bodies of animals and plants, and their conversion into simple soluble compounds or elements that can be assimilated by higher plants. All such organisms thus maintain a natural scavenger service coupled with recycling of wastes into valuable fertilizers and these into foods. Means of exploiting microorganisms as agents in the prevention of pollution, reclaiming of waste lands, and related environmental control activities (e.g., improved methods of composting) are under constant investigation. The ingenuity and know-how of the microbiologist and the expertise of the engineer, ecologist, and chemist are daily under increasing pressure from the environmentalist to keep the earth from eventually becoming an uninhabitable planet for the human species.

A most interesting and extremely important phase of research in soil microbiology concerns one aspect of nitrogen fixation. Among the nitrogen-fixing bacteria previously mentioned is the genus *Rhizobium*. These fix nitrogen only when growing in fleshy nodules (Fig. 1–6) which they produce on the roots of the class of plants called Leguminosae (beans, peas, etc.). Because fixed nitrogen is enormously expensive for the farmer, the nitrogen-fixing bacteria of the soil are great aids to agriculture. It would be of tremendous importance if some species of microorganisms could be induced, by mutation or other means, to fix nitrogen in roots of nonleguminous plants like wheat, cotton, or timothy grass! It would change the course of history, and it is already under investigation.

Sanitary Microbiology

In addition to the general considerations of aquatic microbiology and its ecological aspects mentioned in the foregoing paragraphs, very specialized fields revolve around cooperation of the microbiologist with the sanitary engineer in the provision of potable water for human consumption and the disposal of domestic industrial sewage.

Under modern conditions of human life, almost no major water supply continues long without pollution. We depend on microorganisms almost entirely to decompose this offensive organic matter and reduce it to soluble and inoffensive or useful materials. One such useful material is methane, valuable as fuel. The engineering and microbiological exploitation of microorganisms in sewage disposal plants has been highly successful for decades, but new methods are in constant demand to meet problems arising from tremendous increases in urban populations. Similarly, the combined

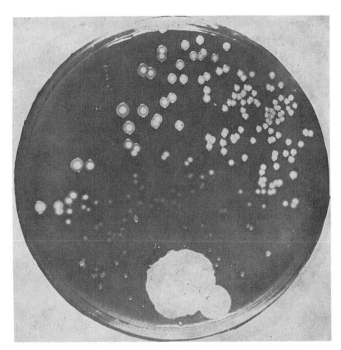

Figure 1–5. Fleming's original culture of staphylococci contaminated with *Penicillium notatum*. The remains of colonies near the mold can barely be seen, but healthy colonies exist comfortably in the areas that are relatively distant from the mold.

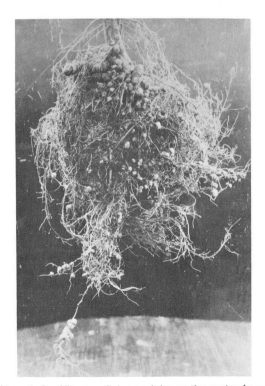

Figure 1–6. Nitrogen-fixing nodules on the roots of a soybean plant. (Photograph courtesy of the U.S. Department of Agriculture.)

efforts of microbiologists, engineers, and chemists are constantly needed to provide water that is **potable,** i.e., clean, clear, odorless, virtually tasteless, and free from pathogenic microorganisms and deleterious chemicals, for drinking and other domestic purposes. Hundreds of microbiologists specialize in the fields of water purification and waste water disposal. These matters are discussed more fully in Chapter 45.

Microorganisms in Foods

Many kinds of microorganisms multiply very well in most of the organic substances used as food by humans. Viruses cannot multiply in foods but can readily be transmitted from person to person in or on them. Most organisms commonly present in or on foods are harmless, and we ingest millions of these gastronomic hitchhikers daily without harm. However, spoilage by these ubiquitous nuisances is a common calamity, both in various food and dairy industries and in the domestic kitchen.

Several kinds of pathogenic microorganisms, principally bacteria, can grow vigorously in many moist, warm foods, such as milk, puddings, creamed macaroni and similar dishes, salads and sandwiches containing mayonnaise, meats or eggs, cream gravies, and the like. Pathogens generally gain entrance to foods from soil (especially the cause of botulism or food poisoning, *Clostridium botulinum*), from infected handlers of the foods (e.g., persons with diarrhea, colds, sore throats, boils, etc.), or from infected animals. Foods from infected animals—e.g., milk from cattle with tuberculosis, brucellosis, or mastitis; eggs from hens with salmonellosis; meats or poultry from any infected stock, including pork from trichinous swine and flesh from wild animals with tularemia—are obviously dangerous vectors of animal diseases transmissible to man (i.e., **zoonoses**).

Prevention of food-borne diseases requires constant veterinary inspection of livestock and of meats in the slaughter houses; periodic medical and microbiological examination of persons handling foods, food-animals and their products, and premises where foods are prepared for market; and enforcement of sanitary regulations (rigid asepsis) and pure food laws by federal, state, and local authorities. Many hundreds of specialists in the microbiological aspects of food preparation and marketing are constantly employed in microbiological laboratories maintained for the purpose.

Preservation of Foods. Among other important measures used to prevent spoilage and disease transmission by foods are: pasteurization of milk; freezing and refrigeration of foods; expert heat-processing of canned foods; and proper processing of frozen foods, whether cooked, partially cooked, or raw (meat pies, TV dinners, vegetables, juices, and so on). In the home, such processes as canning, pickling, cooking, drying, freezing, and refrigeration not only can save many dollars worth of so-called "leftovers" but, by following certain approved practices in cleanliness and sanitation, can prevent serious food infection and food poisoning.

Foods Made with Microorganisms. Some species of microorganisms in foods are neither infections nor causes of spoilage. A number of such species are of immense value and are carefully selected, nurtured, and depended on for the commercial production of a variety of food products: butter, many different cheeses, yogurt, buttermilk, baked goods, pickles, sauerkraut, candies, alcoholic beverages, and others. More details are given in Chapters 47 and 48. All require highly trained microbiologists, generally at very good salaries.

Microorganisms in the Atmosphere

To speak of the microbial flora of the air is to talk of nothing; neither micro- nor macroorganisms are indigenous to the air. Neither water as liquid nor any form of nutriment is available to aerial indigens. Rain, snow, and gravitation tend to remove them from the air. Nevertheless, the outdoor air has a suspended population of transients whose true habitat is mainly the soil, plants, animals, and sea. The organisms found in any sample of outdoor air will depend on local and distant wind conditions, including velocity and humidity; moisture of the area over which the wind blows; temperature; rain- and snowfall; sunshine; toxic gaseous pollution, both natural and industrial; nearness of dusty agricultural and other human activities; and so on. The atmosphere in motion is a powerful carrier of both micro- and macroorganisms: seeds and plants; insects; pathogenic and harmless microorganisms, including bacteria, yeasts, protozoans, etc.; small birds; rodents; and the like. It is said that the southeastern United States is contaminated with dust from wind storms in the Sahara Desert some 5,000 miles distant. Darwin noted microorganisms in dust from Africa over 1,000 miles at sea on the *Beagle*.

Microbial Pollution of the Air. The air in occupied buildings, buses, and other enclosed spaces often carries considerable numbers of pathogenic microorganisms, mainly bacteria and viruses of the respiratory tract. Talking, coughing, and sneezing can produce sprays of saliva and mucus laden with etiological agents of such diseases as influenza, "common colds," septic sore throat ("strep throat"), pneumonia, diphtheria, and whooping cough. Many apparently healthy persons carry these and other organisms, pathogenic and otherwise, in their throats, noses, and mouths, intermittently or constantly.

Respiratory organisms, protectively coated with mucus and saliva, often become dried as dust particles and remain suspended in the air as **droplet nuclei** for considerable periods. Studies of the functions of air as a vector of disease by means of collecting samples of air from mountain tops, the stratosphere, mines, hospital rooms, and so on, often involve very specialized techniques and equipment. Studies of the microorganisms themselves, their enumeration, and identification are other aspects of aerobiology. Still others concern methods for controlling airborne microorganisms and means for disinfecting potentially infectious air by physical and chemical means (Chapt. 46). The various parameters of aerobiology occupy many microbiological and environmental specialists as well as engineers, chemists, and others.

1.9
APPLIED AND INDUSTRIAL MICROBIOLOGY

For centuries foods and many other substances produced partly or wholly by microorganisms have been used by mankind: fermented milk products like yogurt, kefir, and buttermilk; innumerable varieties of cheese; leathers; alcoholic beverages; sauerkraut; ensilage; linens; coffee; olives; breads; pickles; etc. Knowledge of the role of microorganisms in these processes, and improvements and modifications of the products, had to await the advent of the microbiological specialist. Pasteur was one of the first industrial microbiologists. Some of his earliest researches concerned diseases of silkworms in France that were decimating the French silk industry. His investigations into what he called diseases of French wines and beers (sourings, off-flavors, putrefactions, etc.) revealed the basic fact that a different specific microorganism was responsible for each specific disease. From his studies of wines and beers (Fig. 1–7) evolved the process of preventing industrial spoilages by the process now called **pasteurization.** His views on specificity of industrial diseases led him to the concept of a different, specific microorganism as the cause of each specific infectious disease of man. Since Pasteur's times the roles of microorganisms and the microbiologist in many of the most important human activities have enlarged enormously.

Figure 1–7. A view of the French wine industry, a chief beneficiary of Pasteur's research, in a 17th century engraving. (Courtesy of the Bettmann Archive, Inc., New York.)

Applied and industrial microbiology are today the basis for the discovery, production, standardization, and clinical trial of scores of drugs like the antibiotics, cortisone, and related steroids; manufacture of a wide variety of dairy products; production of foods and food supplements for man and his livestock from waste materials like garbage and sawdust through the activities of yeasts, molds, and bacteria; production and use of bacterial and fungal enzymes in various manufacturing processes; the production of alcoholic beverages, industrial solvents, vitamins, and so on through a long list (Chapt. 49). All of these processes had their origins in the ingenuity and discoveries of innumerable dedicated, ambitious, and hard-working microbiologists, with the collaboration of physicists, chemists, engineers, physicians, veterinarians, and other specialists.

Greater industrial utilizations of microorganisms possible in the future include development of new types of hydrocarbon-decomposing species capable of rapidly dissipating "oil-spills" in the sea; use of particular species of moldlike bacteria as sensitive detectors of hydrocarbon vapors in the soil when prospecting for oil-bearing soils or shales; use of petroleum wastes and a variety of industrial refuses as foods for certain microorganisms, especially bacteria and yeasts. Certain species can metabolize these substances, synthesizing masses of valuable "single-cell" (i.e., single-species of cell) proteins, lipids, vitamins, and other cellular substances in forms that can serve as feeds for livestock or for other purposes still to be devised. Some of these processes are already in the course of development and improvement; others will be contrived as need for them arises.

1.10
MEDICAL MICROBIOLOGY

Though the role of microorganisms as causes of diseases of man, plants, and animals was shrewdly suspected by many ancient and medieval philosophers, accurate knowledge has evolved only since the advent of the microbiologist and his works.

Growth of the science of microbiology and knowledge of infectious diseases have developed concomitantly and interdependently since the beginnings of pure culture methods and modern improvements on Leeuwenhoek's microscopes. Although work in the field of medical microbiology requires knowledge of pathology, biochemistry, and related preclinical disci-

plines, as well as epidemiology, immunology, and so on, the basic principles of medical microbiology are the same as those that underlie all microbiology, though some of the techniques used in medical microbiology are specialized for clinical applications. The medical microbiologist must, of course, have a good knowledge of the major groups of pathogenic microorganisms: protozoans, fungi, viruses, and bacteria; their habitats, modes of transmission, and mechanisms of pathogenesis; and a few common-sense rules for preventing their spread to himself or others. However, a medical microbiologist usually specializes as, for example, a protozoologist, a virologist, a mycologist, a bacteriologist, or an immunologist (Fig. 1–8).

Many students of microbiology use their knowledge in their home and community relationships and/or in other professions: nursing, pharmacy, dentistry, teaching, anthropology, sanitary engineering, and so on. Knowledge of pathogenic microorganisms is especially valuable for tourists, explorers, missionaries, and other travelers to areas outside of North America where such ancient scourges of the human race as cholera, dysentery, smallpox, yellow fever, dengue fever, hookworm, amebiasis, bubonic plague, typhus, leprosy, malaria, and the like are still present in endemic or epidemic form. A knowledge of means of immunization against (Chapt. 28), and avoidance of (Chapt. 25), these

Figure 1–8. Edward Jenner, the pioneer in immunization, treating a child to resist smallpox. Neither the general principles of immunization nor the response of the patients has changed greatly since Jenner's demonstration of the method in 1798.

and other diseases, including diseases of childhood—diphtheria, measles, German measles, poliomyelitis, whooping cough, and so on—to say nothing of venereal disease, is of value to every person, medical or nòt. Many of the more common diseases, their etiologic agents, pathogenesis, and prevention are discussed in later chapters.

1.11
TEACHING MICROBIOLOGY

The training and education of microbiologists is a full-time or part-time occupation for many experienced professionals. The demand for good teachers and for their assistants and associates is great. Good teachers are at the very roots of the supply of competent microbiological personnel needed to supply our ever-growing national needs. While not necessarily the most lucrative or easiest of occupations, it is nevertheless amply rewarding and, in many ways, one of the most satisfying and pleasant of careers in microbiology. Really good teachers of micro-

biology are as much in demand as really good microbiologists. Each professor or instructor must have a full knowledge of his subject. The teacher needs also a personal aptitude for, and a liking of, imparting knowledge to his or her pupils.

As new methods of exploiting useful microorganisms and of controlling noxious species develop, and older methods are improved to meet massive demands of increasing populations, so new job openings for both teachers and pupils will occur. Those best trained and educated in the field will be most in demand, will have the most responsible positions, and will command the highest salaries.

Of all the members of the American Society for Microbiology, one of many affiliated groups of professional microbiologists (the total numbers running into the tens of thousands), about half have a Ph.D or Sc.D degree, some 20 per cent have master's degrees, others bachelor's or M.D. degrees. At any level, the field is excellent for anyone interested in the science of Life. A considerable proportion of microbiologists are women.

CHAPTER 1
SUPPLEMENTARY READING

Brock, T. D.: Milestones in Microbiology. Prentice-Hall, Inc., Englewood Cliffs, N.J. 1961.

Clark, P. F.: Pioneer Microbiologists of America. University of Wisconsin Press, Madison. 1961.

Demain, A. L.: Application of the Microbe to the Benefit of Mankind—Challenges and Opportunities. Am. Soc. Microbiol. News, 38:237, 1972.

Doetsch, R. N. (Ed): Microbiology: Historical Contributions from 1776 to 1908. Rutgers University Press, New Brunswick, N.J. 1960.

Dobell, C.: Antony van Leeuwenhoek and His "Little Animals." Dover Publications, Inc., New York. 1960.

Dubos, R.: Louis Pasteur: Free Lance of Science. Little, Brown & Co., Boston. 1950.

Dubos, R.: The Unseen World. Rockefeller University Press, New York. 1960.

LaRiviere, J. W. M.: International Implications of Microbiology. Am. Soc. Microbiol. News, 38:237, 1972.

Microbiology Training Committee, Ralph B. DeMoss, Chairman, National Institutes of Health: Microbiology for the future. Am. Soc. Microbiol. News, 38:33, 1972.

Monogr. Am. Soc. Microbiol.: Microbiology in Your Future. Washington, D.C. 1972.

Porter, J. R.: The environment—ours or the microbes'? Am. Soc. Microbiol. News, 39:173, 1973.

ORIGINS OF MICROBIOLOGY

• **CHAPTER 2**

2.1
THE DAWN OF SCIENCE

Before the dawn of civilization in the Mesopotamian regions and farther east some 7,000 to 8,000 years ago there was little exact knowledge of either the causes or nature of natural phenomena. They were feared by most of the populace as supernatural and beyond human understanding. However, scholarly thinkers and their works were not wholly lacking. By the time writing and written history had evolved (ca. 3100 B.C.) in Sumer (cuneiform), Egypt (hieroglyphic), Syria, India (sanskrit [ca. 600 B.C.]), and adjacent regions following Phoenician influences (ca. 1500 B.C.) many keen and ambitious minds in the ancient priesthoods, secular upper classes, and royal families had learned of the medicinal and poisonous properties of certain minerals and plants and of the venoms of certain snakes and insects. They knew how to exploit superstitious dread of thunder and lightning for political and other purposes. For thousands of years after the beginnings of civilization magic, incantation, abracadabra, and witchcraft passed for science and usually also for religion. Even as recently as the Middle Ages (ca. A.D. 500–1400) and later in the European Renaissance (ca. A.D. 1400–1700) astrology (aided by imaginative charlatans, with weird grimaces and impressive costumes) passed for astronomy; alchemy (strongly flavored with wizardry) masqueraded as chemistry; the most outrageous quackery was accepted, even by royalty, as medicine.

As always, however, honest, imaginative,

and inquisitive men here and there were still capable of analytical and creative thought and the proposing of working hypotheses to be tested experimentally. They were sometimes reviled, persecuted, and tortured for their supposed dealings with "The Evil One." Century after century these pioneer scientists (seekers after experimentally demonstrable truth) began to establish a system of knowledge based on accurate, purposeful observation; logical inference; imaginative hypothesis; and ingenious experiments designed to establish indisputable fact or destroy fallacy.

Because of great difficulties in travel and communications, ancient scholars shared little of one another's learning. As the centuries passed, exploration began and travel became more common, populations increased, and vast interminglings of peoples occurred because of wars and trade. Scientific information thus began to spread from country to country and, more recently, from continent to continent. Instead of a few great scholars who were thought (even by themselves!) to know everything, men began to realize that there were boundless deserts and plains and illimitable dark forests of ignorance only awaiting the axe and plow of the devoted researcher to yield rich crops of wonderful, golden knowledge. Men also realized the awesome truth that knowledge is power—to create or to destroy utterly. Eventually scientific thought, experimentation, and communications became permissible and even respectable. They also became incalculably profitable and frightening.

Scientists interested in the mysteries of life

collected, over the centuries, a considerable mass of increasingly accurate information about such living things as could be seen with the naked eye, and even with "magnifying glasses" (magnifications up to about 10 diameters). By 350 B.C. Aristotle and his students had drawn up a systematic, though limited and (as we now know) often erroneous classification of hundreds of plants and animals. Accumulating knowledge of living organisms slowly became arranged into a more or less orderly system and the study of life was eventually dignified with a given name: **biology** (Gr. *bios* = life; *logos* = study or description). Most of biology was at first largely descriptive of outward form, color, motion when present, habitat, and other macroscopic details, including anatomy. These descriptions became the basis of classifications and taxonomic systems, major preoccupations of many early botanists and zoologists. Until the 18th and 19th centuries chemistry and physics were almost completely separate fields of study and were little used in biology. Life and living substance were commonly thought of as mysterious and beyond physical and chemical analysis.

2.2
BEGINNINGS OF MICROSCOPY

The very best human eyes, unaided, fail to see objects less than about 100 μm (1/254 inch) in diameter (Table 3–1). Nor can the eye clearly perceive as separate objects (i.e., resolve) particles separated by distances less than this. Even "large" objects, such as some insects with diameters of about 150 μm—e.g., larvae of "jigger" fleas (*Tunga penetrans*) or the "itch mites" causing scabies (*Sarcoptes scabei*)—are seen with difficulty or not at all. Microorganisms are enormously smaller. They range downward in diameter from the ca. 50-μm diameters of animal tissue cells to the diameters of bacteria (1 to 5 μm) and those of viruses (ca. 0.25 μm). They are as far beyond the vision of persons untrained in high-power microscopy as Phobos and Deimos* are from persons without powerful telescopes.

From the remotest antiquity all human knowledge of plants and animals was limited to what could be seen with the naked (or very feebly assisted) human eye and dependent on such experiments and inferences as could be based on such knowledge. Man's visual limita-

---

*Satellites of the planet Mars.

tions are due to the relatively long wavelength of visible light, the relatively coarse structure of the optic nerve endings in the retina, and the nature of the eye lens and cornea. Before the invention of high-power microscopes these shortcomings of the human eye had always stood, like an impenetrable curtain, between mankind and the glittering cosmos of the microscopic universe.

The First Microscopes. Simple magnifying glasses ("hand lenses") giving enlargements of 2 to 10× had been known and used for many years prior to this and are familiar to everyone today. Technically, they are microscopes since they can make visible some objects that the eye does not see. But high-power microscopy awaited the enormous and exciting discovery made by a Dutch spectacle maker, reputedly Zaccharias Janssen who found, about 1590, that by using a second lens he could greatly enlarge the image formed by a primary hand lens. He produced what were, at that time, really spectacular magnifications, probably of the order of 50 to 100×. This is the basic principle of the modern compound microscope that is used by every microbiologist today. In 1610 Galileo invented an "improved" microscope, though its resolving power was very limited compared to later instruments.

Robert Hooke (1635–1703) made and used a compound microscope in the 1660's and described his fascinating explorations of the newly discovered universe of the microscopic in his classic "Micrographia" (1665), published at the request of the Royal Society in London (Fig. 2–1). Although Hooke's highest magnifications were possibly enough (200×?) to reveal bacteria, he apparently made no observations of them, probably because he studied mainly opaque objects in the dry state by reflected light, conditions that, as will be explained, are not optimal for observation of microorganisms. However, his pictures of a "mould" and a wide variety of other objects are very informative and accurate (Fig. 2–2).

A contemporary of Hooke, and the first man to reveal the hitherto unknown and unseen world of unicellular, independently living microorganisms (Protista) did not use a compound microscope. He was the Dutch investigator, Antonj van Leeuwenhoek (1632–1723), a linen merchant by trade and a man of public and commercial affairs in the city of Delft (Fig. 2–3). He was not a trained scientist but was self-educated, and amused himself by means of his skill and craftsmanship in glass blowing and fine metal work. He lived in relatively easy circumstances

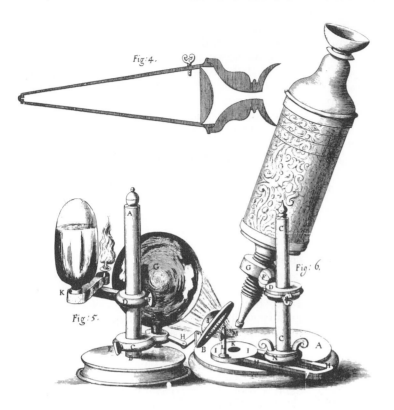

Figure 2–1. Hooke's compound microscope, drawn by himself. Note that the object is seen by light reflected from above, unlike Leeuwenhoek's "animalcules," which were seen by light transmitted through them.

Figure 2–2. Drawing of a mold by Robert Hooke. This is probably a species of *Mucor*, which Hooke listed as a "blue mould" but described as "a small white spot of hairy mould" with "round and white knobs" (sporangia), growing on a leather book.

Figure 2–3. Antonj van Leeuwenhoek. A fanciful delineation based on a famous portrait. The picture shows accurately the size and shape of the first microscopes, the manner in which they were used, and the simple laboratory apparatus of the "Father of Bacteriology." (Courtesy of Lambert Pharmacal Co.)

with leisure time for his avocation of making minute, simple but powerful lenses. These are known to have given magnifications up to about $300\times$ — sufficient to visualize bacteria, protozoa, and numerous other microorganisms. With such lenses Leeuwenhoek delighted in examining a great variety of objects: saliva, pepper infusions, cork, the leaves of plants, circulating blood in the capillary vessels in the tail of a salamander, seminal fluid, urine, cow dung, scrapings from his own teeth, and so on. In many of these he saw living things, some of which we now know were protozoa and bacteria but all of which he called "animalcules."

Leeuwenhoek's invention and use of his high-power lenses was like turning on a 500-watt lamp in a pitch-dark curiosity shop. It gave men the power to see, for the first time in human history, a universe of incredibly fantastic and beautiful objects, living and inanimate, so minute that their existence had never before even been suspected. The discovery was as momentous as Columbus's first landfall in the New World or man's first footsteps on the moon.

...he showed rare ingenuity and expert craftsmanship in the grinding and mounting of his simple lenses, a skill which he zealously kept to himself; and in spite of the requests of his learned friends, he refused to disclose the secret of his success.

...Leeuwenhoek's instruments are not true microscopes at all in the sense in which we think of microscopes, but rather simple magnifying glasses generally consisting of a small, single, biconvex, almost spherical, lens. The object, and not the lens, was moved into focus by means of screws [Fig. 2–4].

To adjust the lens to the object was so long and tedious a task that it is not surprising that Leeuwenhoek used an individual lens for each object.... The magnification varied and at best did not exceed two hundred to three hundred diameters.... The size of objects which Leeuwenhoek examined was determined by comparison. For this purpose he used at various times a grain of sand, the seed of millet or

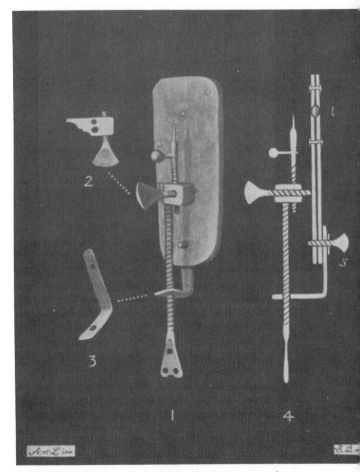

Figure 2–4. One of Leeuwenhoek's microscopes: front and side views. The minute biconvex lens (*l, 4*) was mounted between two thin oblong plates of brass (or other metal). When the apertures had been made to coincide exactly, the lens was clamped between the plates, in the concavities (*4*), and secured by four rivets forming the corners of a square (*1*).

Leeuwenhoek left 26 of his lenses, mounted in silver, to his daughter Maria. To the Royal Society he left many others. In all, he made some 419 lenses, some mounted in gold, most in brass. Only one of his microscopes is known to be extant, in the University of Utrecht.

mustard, the eye of a louse, a vinegar eel, and still later hair or blood corpuscles. In this way he secured fairly accurate measurements of a great variety of objects . . . he was forced to admit that the sand grain was more than one million times the size of one of the animalcules.*

Leeuwenhoek was so interested in the things he observed that, like Hooke, he wrote minutely detailed reports about them to the Royal Society in London, beginning in 1674. He was later elected a fellow of the Royal Society. Some of his observations are at once quaint and epoch-making. For example, after examining material which he scraped from between his teeth, he said:

Though my teeth are kept usually very clean, nevertheless when I view them in a Magnifying Glass, I find growing between them a little white matter as thick as wetted flour; in this substance, though I could not perceive any motion, I judged there might probably be living Creatures.

I therefore took some of this flour and mixt it either with pure rain water wherein were no Animals; or else with some of my Spittle (having no Air bubbles to cause a motion in it) and then to my great surprise perceived that the aforesaid matter contained very small living animals, which moved themselves very extravagantly. The biggest sort had the shape of A. Their motion was strong and nimble, and they darted themselves through the water or spittle, as a Jack or Pike does through the water. These were generally not many in number. The second sort had the shape of B. These spun about like a top, or took a course sometimes on one side, as is shown at C and D. They were more in number than the first. In the third sort I could not well distinguish the Figure, for sometimes it seem'd to be an Oval, and other times a Circle. These were so small they seem'd no bigger than E and therewithal so swift, that I can compare them to nothing better than a swarm of Flies or Gnats, flying and turning among one another in a small space† [Fig. 2–5].

Note that, unlike Hooke, Leeuwenhoek made many of his observations by light transmitted *through* the object and that the microorganisms were suspended in various fluids, not immobilized or otherwise altered by drying.

For over 150 years after the discoveries of Janssen, Galileo, Leeuwenhoek and Hooke, knowledge of microscopy and microorganisms developed but slowly. There were relatively few microbiologists, and great technical problems in the science of optics and instrumental mechanics had to be solved. Many workers contributed over the years: use of two convex

*Fred, E. B.: Antonj van Leeuwenhoek. J. Bact., 25:1, 1933.
†Ibid.

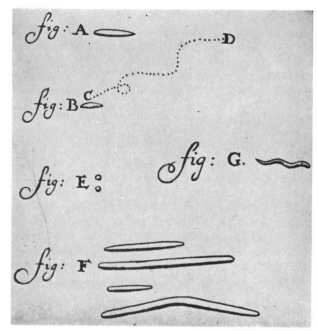

Figure 2–5. Leeuwenhoek's drawings of bacteria. Here may be seen cocci, bacilli, and (probably) a spirochete. The motion of one of the bacilli is clearly indicated. Today such observations are commonplace. But Leeuwenhoek was seeing them for the first time in the history of the human race! It was as momentous a discovery as that of Columbus —a new world!

lenses, J. Kepler (1611); development of the two-lens eyepiece, C. Huygens (ca. 1684); apochromatic objectives and substage condenser, E. Abbé (1840–1905) and many others. Microscopes which were forerunners of those with which we are familiar today came into general professional use only after about 1820. All use visible light as a source of illumination and are called light or optical microscopes. None, even the most modern, can give accurate images at magnifications greater than about 1,000× (Fig. 2–6). Instruments using beams of electrons, x-rays, and similar forms of illumination, some giving magnifications of a million or more diameters, are late 20th century developments. Forms of electron microscopy are discussed more fully in Chapter 3.

2.3
MICROORGANISMS AND THE ORIGIN OF LIFE

The ancients knew nothing of microorganisms, of evolution, or of the doctrine enunciated by Redi (1626–1697) "omne vivum ex vivo" (all living things are produced by other living things). The ancients believed in spontaneous generation, i.e., that creatures like frogs, mice, bees, and other animals sprang fully formed from fertile mud, decaying carcasses, warm rain or fog, and the like. Van Helmont (1577–1644) devised a method for manufacturing mice. He recommended putting some wheat grains with soiled linen and cheese into an appropriate receptacle and leaving it undisturbed for a time in an attic or stable. Mice would then appear. This observation may still be experimentally confirmed but the conclusions drawn from the results differ today.

Belief in spontaneous generation, now called **abiogenesis**, lived on for years, as it had for centuries. For example, an elderly lady of the writer's early acquaintance complained bitterly that she had been cheated by a merchant who sold her a woolen coat which was of such a quality that it turned entirely into moths when left undisturbed in a closet for some months!

In the earlier years, in the absence of exact

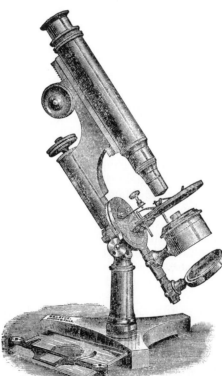

Figure 2–6. A microscope, as advertised in February, 1883.

knowledge of microorganisms or chemistry, there had arisen much skepticism and bitter feeling over the question of the origin of life. One "scientist" who still held to the ancient ideas says of the views of another who doubted,

So may we doubt whether, in cheese and timber, worms are generated, or if beetles and wasps in cow dung, or if butterflies, locusts, shell-fish, snails, eels, and such life be procreated of putrefied matter which is to receive the forms of that creature to which it is by formative power disposed. To question this is to question reason, sense and experience. If he doubts this let him go to Egypt and there he will find the fields swarming with mice begot of the mud of Nylus, to the great calamity of the inhabitants.

There was a great deal of such acrid discussion by wordy savants of the times, who tried to settle everything by argument. Experimentation was regarded as rather undignified and even smacking of relations with the devil.

Francesco Redi (1626–1679). The experimental method was, however, being invoked here and there by true scientists, i.e., seekers after experimentally demonstrable truths. For example, it had always been supposed that the maggots in decaying meat were derived spontaneously from transformations of the putrid meat itself. Francesco Redi, a physician of Arezzo, questioned this hypothesis. He placed meat and fish in jars covered with very fine gauze and saw flies approach the jars and crawl on the gauze. He saw the eggs of the flies caught on the gauze and observed that the meat then putrefied without maggot formation. Maggots developed only when the flies' eggs were deposited on the meat itself. Obviously the meat itself did not turn into maggots. Redi's many experiments of this kind absolutely defeated the doctrine of abiogenesis, at least with respect to forms of life then known, but were not widely noted.

Louis Joblot (1645–1723). After Leeuwenhoek's discovery of microorganisms it was thought by many who believed in the Aristotelian doctrine of abiogenesis (ca. 384 B.C.) that animal or vegetable matter contained a "vital or vegetative force," capable of converting such matter into new and different forms of life. A popular notion was that geese and lambs could grow from certain kinds of trees. Leeuwenhoek's animalcules were hailed by many as proof of this, though Leeuwenhoek himself rejected the idea. In 1710, Louis Joblot observed that hay, when infused in water and allowed to stand for some days, gave rise to countless animalcules, or **infusoria** (bacteria and protozoa). The hay was thought by some to change into animalcules; anyone today can observe the development of these for himself. Joblot, however, boiled hay infusion and divided it into two portions, placing one in a carefully baked (sterilized) and closed vessel. The other portion was not heated and was kept in an open vessel. The infusion in the open vessel teemed with microorganisms in a few days. In the closed vessel no life appeared as long as it remained closed, thus showing that the infusion alone, once freed of life by heat, was incapable of generating new life spontaneously.

John Needham (1713–1781). Similar experiments with mutton gravy and other infusions carried out by an English scientist, John Needham, gave conflicting results. Life developed in Needham's heated closed vessels as well as in the open unheated ones. He therefore believed in abiogenesis. This result was due to insufficient heating which failed to kill extremely heat-resistant forms of bacteria called **endospores.** But nothing was known about spores until the work of Tyndall, Cohn, and others after about 1876.

Lazzaro Spallanzani (1729–1799). Spallanzani, an Italian naturalist, published the results of a whole series of the same type of experiments which disagreed with those of Needham. He showed that if heating was prolonged sufficiently and the vessels kept closed to exclude dust and air, no animalcules developed in hay infusions or in any other kind of organic matter, such as urine and beef broth. Needham, in reply, said that the prolonged heating destroyed the "vegetative force" of the organic matter which, he said, was necessary for the spontaneous generation of life. Spallanzani answered Needham's objections by showing that the heated infusions in the closed flasks could still develop animalcules when exposed to air (i.e., when microorganisms were introduced with dust).

In 1775 Lavoisier discovered oxygen and the relation between air and life. This renewed the controversy about abiogenesis, the objection to Spallanzani's results now being that it was the exclusion of air (oxygen) from the flasks which prevented the development of life.

Schulze and Schwann. A half century later new experiments were performed in which unheated air was admitted freely to previously heated infusions of meat or hay in flasks, but only after passing through sulfuric acid or potassium hydroxide solutions (Schulze, 1836) (Fig. 2–7) or through very hot glass tubes (Schwann, 1837) (Fig. 2–8), their idea being that untreated air introduced the germs of life into the in-

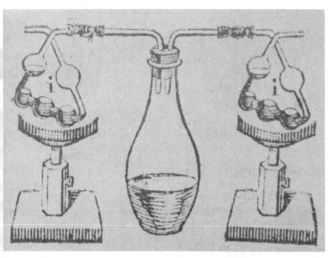

Figure 2–7. Apparatus of Schulze (1836) for treating air before admitting it to flasks of putrescible material.

fusions. When the infusions exposed to air so treated failed to develop any life it was claimed by others that this was not due to a destruction of germs of life in the air by acid, alkali, or hot glass, but that the "life-giving" power of the air had been destroyed, thus preventing abiogenesis. That heat had not destroyed any life-giving property of the air was demonstrated by maintaining contented frogs in previously heated air.

Schroeder and von Dusch. A generation later, objections to the experiments of Schulze and Schwann, noted above, were overcome by Schroeder and von Dusch (1854–1861), who per-

formed similar experiments in which the air was not heated or passed through acid or alkali but merely filtered through cotton wool (Fig. 2–9). This method prevented the appearance of animalcules in the heated broth or infusions until the vessels were opened. It was therefore apparent that the method of treatment of the air had nothing to do with the development of animalcules and that these did not develop spontaneously, but that there were particles of living matter floating on dust in the air which not only were killed by heat, acids, and alkalis but which could be caught and withheld by the cotton wool alone. The presence of the microorganisms in the cotton wool was later proved by Pasteur. The experiments of Schroeder and von Dusch were the origin of our present-day use of cotton plugs for bacteriological culture tubes and flasks.

In spite of these demonstrations, long and bitter controversies still raged. Schroeder and von Dusch were not convinced by their own experiments and admitted the possibility that abiogenesis might occur under natural conditions.

Louis Pasteur (1822–1895). Pasteur, one of the most famous French scientists, was born in Dôle, son of a moderately prosperous tanner who had fought for and been decorated on the battlefield by Napoleon. Pasteur had a great admiration for his father's soldierly accomplishments, and some of his best scientific achievements were motivated by patriotic zeal. In his boyhood an indifferent student, he later became an enthusiastic scholar, especially of chemistry. His discovery of the stereoisomeric forms of tartrate crystals revealed a whole new series of possibilities in physical chemistry and was the basis of a brilliant career and enduring fame (Fig. 2–10). He might have had a career in another field, that of art. He was a skillful and well-known portrait painter.

"Diseases" of Wines and Beers. For patriotic reasons Pasteur turned his attention to the phenomenon of fermentation, which was then believed to be a chemical process. An Englishman had written to him:

People are astonished in France that the sale of French wines should not have become more extended here [in England] since the Commercial Treaties. The reason is simple enough. At first we eagerly welcomed those [French] wines, but we soon had the sad experience that there was too much loss occasioned by the diseases [souring] to which they are subject.*

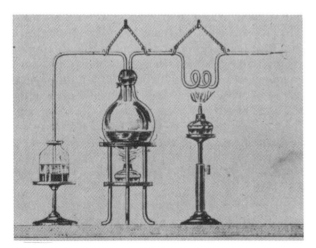

Figure 2–8. Apparatus of Schwann (1837) for treating air before admitting it to flasks of putrescible material.

*From The Life of Pasteur, by René Vallery-Radot; reprinted with permission from Doubleday Company, Inc.

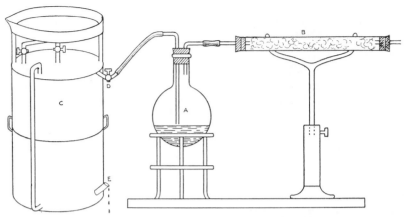

Figure 2–9. Apparatus of Schroeder and von Dusch (1854), drawn from their description. Flask *A* contained beef broth, diluted egg white, etc. Tube *B*, containing cotton, was heated at 100 C for some hours before beginning the experiment. All junctions were made tight with "hot wax." Air was drawn through the cotton by suction as water drained from the aspirator *C* at tube *E*. The broth was heated at 100C before aspirating air through the apparatus. It *usually* remained sterile for days afterward. When preheated in sealed tubes at 160C (350F) before the experiment it *always* remained sterile. This was nearly 20 years before spores were discovered by Koch, Cohn, Tyndall, and others about 1875.

Figure 2–10. Louis Pasteur (1822–1895) from the painting by Martin Frobisher, Sr. The original now hangs in the School of Hygiene, The Johns Hopkins University.

Germany was at that time making a much better beer than France, and Pasteur undertook to make France a successful rival in that respect. He made an intensive study of beer manufacture and of the cause of souring and spoilage ("diseases") of beer and wines. As a result of these studies he arrived at far-reaching conclusions. He said at the Académie des Sciences in January, 1864:

Might not the diseases of wines be caused by organized ferments, microscopic vegetations [yeasts, molds, bacteria], of which the germs would develop when certain circumstances of temperature, of atmospheric variations, of *exposure to air* [author's italics], would favour their evolution or their introduction into wines?... I have indeed reached this result that the alterations ["diseases"] of wines are co-existent with the presence and multiplication of microscopic vegetations.*

Pasteur had found that acid wines, "ropy" (slimy) wines, bitter wines, sour beer, and so on were caused by the growth in them of undesirable contaminating organisms which produced these so-called diseases.

The solution of the problem, as later proved by Pasteur, lay in preventing the growth of foreign organisms, "wild" yeasts, and bacteria, which caused the undesirable conditions. After considerable experimentation along these lines he discovered that the wine did not spoil in transit if it were held for some minutes at a temperature between 50 and 60C. He said,

---

*Ibid.

I have...ascertained that wine was never altered by that preliminary operation [heating], and as nothing prevents it afterwards from undergoing...improvement with age — it is evident that this process [heating] offers every advantage.

His experiments were so successful that a practical test of the efficacy of his methods was made. He wrote to a friend:

...experiments on the heating of wines will be made by the Minister of the Navy. Great quantities of heated and of non-heated wine are to be sent to Gabon so as to test the process; at present our colonial crews have to drink mere vinegar.

Pasteur laid down three great principles:

1. Every alteration, either of beer or of wine, depends on the development in it of microorganisms which are ferments or "diseases" of the beer or wine.
2. These "germs or ferments" are brought by the air, by the ingredients, or by the apparatus used in breweries.
3. Whenever beer or wine contains no **living** microorganisms it remains unchanged.

Pasteurization. In the same way that wines could be preserved by heating from various causes of alteration, bottled beer could escape the development of disease ferments by being brought to a temperature of 50 to 55C. The application of this process soon gave rise to the new word, **"pasteurized"** beer, which became current in technical language. Today, pasteurization of milk (heating at 63C for 30 minutes or at 72C for 15 seconds) is routine. The heating kills pathogenic microorganisms but does not sterilize the milk.

Pasteur on Specificity of Disease. Pasteur foresaw the consequences of his studies, and wrote in his book on beer:

When we see beer and wine subjected to deep alterations because they have given refuge to microorganisms invisibly introduced and now swarming within them, it is impossible not to be pursued by the thought that similar changes may, *must*, take place in animals and in man.*

It was obvious from Pasteur's studies that each kind of fermentation or disease of beer or wine was the result of the growth and activity in it of a distinct form of yeast or other microorganism, depending on the type of fermentation or

*Ibid.

disease under investigation. This furthered an idea, already old, of the **specificity** of biological action, and supported the view that animal and human diseases also, like different sorts of putrefaction and fermentation, were each caused by a single, specific type of microorganism.

Pasteur on Abiogenesis. After Pasteur's views on the nature of fermentation had been made public he became involved in the bitter quarrel over the apparently spontaneous appearance of "germs" in fermentable or putrescible liquids like urine and broths. Pasteur, who like Redi and Leeuwenhoek rejected the doctrine of abiogenesis, carried out many ingenious experiments to show that the animalcules in spoiled beer and wine were merely descendants of microorganisms that had gained access to the fluids from dust in the air and that, by their growth and metabolism, caused fermentation and putrefaction.

First, he redemonstrated, with cotton air filters, that living microorganisms float in the air attached to dust particles. Then he showed, as Schroeder and von Dusch had done, that when dust could be mechanically excluded from various substances, such as sterilized broth and urine, these did not ferment or putrefy. By using flasks with long open necks with vertical bends in them ("swan-necked" flasks; Fig. 2–11), he showed that although unheated, untreated, and unfiltered air communicated freely with the interior, the dust was caught by gravity in the bends of the neck and no life appeared in the infusions. Not until the flask was tilted so that the fluid came into contact with this dust and was allowed to run back into the flask, or until the neck of the flask was cut off close to the body, did growth occur in the fluids. Some of

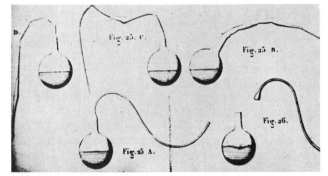

Figure 2–11. Pasteur's open flasks in which boiled broth remained free from microorganisms because dust particles carrying them were trapped in the bends of the necks.

Pasteur's flasks which were sterile in 1864 have been preserved, still sterile, in the Institute Pasteur, Paris.

John Tyndall (1820–1893). Note that in the experiments of Joblot, Spallanzani, Needham, Schulze, and Schwann, heat was depended on to prevent development of animalcules in their broths and infusions. It succeeded if boiling was sufficiently prolonged, especially if the broth or infusion had been previously heated, or, like fresh meat or urine (often used), contained relatively little dust and few spores. Spores of some bacteria are killed quickly by boiling; others are very resistant. Needham's results were probably due to insufficient heating or other technical errors. In contrast to the above, Schroeder, von Dusch, and Pasteur used only mechanical means (filtration and gravity) to sterilize the air admitted to their flasks. Though Pasteur described spores of bacteria and molds in the cotton air filters that he used, both he and all of his predecessors were ignorant of the true nature of bacterial endospores.

In 1876 John Tyndall, an English physicist investigating light, had studied the "motes" that are seen floating in the air traversed by a sunbeam in a partly darkened room. By heating air with a burner he could free it from these floating dust particles. He described such air as "optically empty." Being a contemporary of Pasteur and an admirer of the latter's works and ideas on abiogenesis, Tyndall devised an experimental "culture chamber" consisting of a wooden box with a door in back and a glass front. A small glass window was inserted in each of two opposite sides. Through a pipette piercing a thin sheet of rubber tightly closing a small hole in the top center of the box, fluids (mutton broth, etc.) could be introduced into culture tubes suspended through holes in the floor of the chamber. The entire interior of the box was coated with glycerine to trap motes. Air was admitted through tubes in the top but was made to pass through several vertical bends as suggested by Pasteur's devices (Fig. 2–12). After standing several days a beam of light passing through the side windows revealed no motes; the air in the chamber was said to be optically empty. Various nutrient fluids were then introduced through the pipette into the culture tubes which were afterwards immersed for long periods in boiling brine (well over 100C, i.e., sporicidal heating). After cooling to room temperature the broths in the tubes remained sterile indefinitely.

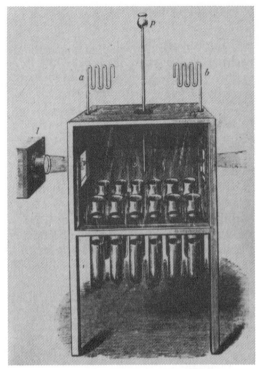

Figure 2–12. Tyndall's cabinet for experiments on spontaneous generation.

After many studies on thermoresistance of bacteria Tyndall concluded that certain bacteria exist in two phases: **thermolabile** (which we now call active or vegetative) and **thermostable** (which we now call sporulated or dormant). Tyndall observed that if he allowed time for the heat-resistant stage to become active or germinate, he could easily kill the vegetative forms. This led to his ingenious method of fractional sterilization, now often called **Tyndallization.** Tyndall's work gave the coup de grace to the doctrine of abiogenesis as then understood (1876).

The actual phenomena of sporulation, and the germination of bacterial endospores, their thermoresistance, and their appearance (Fig. 2–13) were first accurately described by the German botanist Ferdinand Cohn and (independently) by Koch in the same year (Fig. 2–14).

The work and conclusions of Pasteur and other anti-abiogenists were vitriolically attacked by numerous opponents, notably Bennett, Pouchet, and Bastian. Their experiments were technically faulty but upon them they based voluminous dialectical disquisitions: a form of paralogism indispensable to metaphysicians and other intellectual mountebanks who make fallacies appear as truth to ignorant, uncritical, or prejudiced people. Pouchet, in 1859, called the microorganisms that grew in his heated (and by him supposedly sterile) flasks, ". . . the offspring of death and the embodiment of life." The "offspring of death" were all outgrowths from heat-resistant spores or contaminants. It is now clear that, in the pioneering years of microbiology, prior to the time of Koch, Cohn, Tyndall, and Pasteur, ignorance concerning such resistant microorganisms, their ubiquity, numbers, and variety, was a prolific source of tragic diagnostic errors, discouraging frustrations, bitter disputes, and even unseemly polemics among biological and other scientists.

Even as late as the early decades of this century some microbiologists undertook to explain, like Pouchet, Bastian, and Bennett before them, the presence of bacteria that appeared unexpectedly in certain cultures thought to contain no visible or cultivable forms of bacteria at all.

It was notable that the experimental manipulations required in such experiments were of such complexity and nature that they simply invited contamination by airborne bacteria and spores.

Today the microbiologist knows that one must constantly be aware of the hordes of invisible organisms that swarm on one's fingers, all over the body as well as inside it, on the objects that are handled, in the food that is eaten, and in dust in the circumambient air; yea, one's very breath, like exhalations of the dragons of old, can bring forth contamination and, at times (as will be explained later), contagion, disease, and death. The ability of microorganisms to insinuate themselves into even the most carefully managed laboratory procedures may become surprisingly evident to the student in this course, and sometimes embarrassingly so even to experts.

Modern methods of eliminating all sources of contamination and various kinds of mechanical and technical error are based on thorough knowledge of the ubiquity of contaminants, improved techniques, and **pure culture methods,** the last developed originally by Koch and his contemporaries. These methods, pillars of modern microbiology, are more fully detailed in Chapter 10.

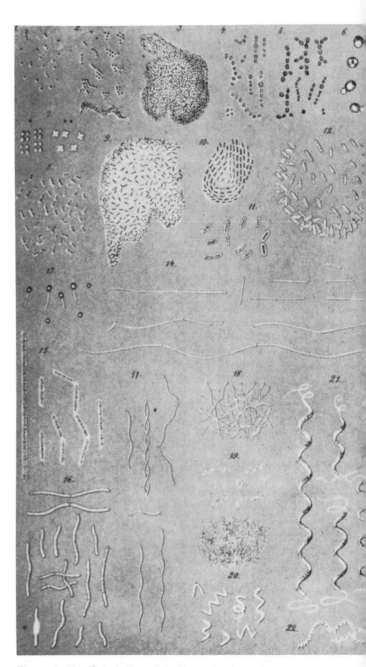

Figure 2–13. Cohn's first plate illustrating bacteria. Note especially the sporeforming bacilli (*13*).

Figure 2-14. Ferdinand Cohn (1828–1898).

2.4
ABIOGENESIS—MODERN STYLE

In 1864 Pasteur received a prize from the French Academy of Science for his studies that conclusively disproved the ancient doctrine of spontaneous generation. Since that time, studies in synthetic chemistry have produced a mass of data that have actively revived the doctrine of abiogenesis in a modern form called **chemical evolution.** There is now good reason to believe that terrestrial life evolved during many millions or billions of years from combinations of a few elements (chiefly carbon, hydrogen, oxygen, nitrogen, sulfur, and potassium) by a series of reactions that were compatible with physical laws and were actually inevitable under what are called "primitive earth" or prebiotic conditions. These are conditions that are thought, on the basis of astute inference from a variety of data, to have existed during the early periods of the formation of the earth some 4,000,000,000 years ago (Fig. 2–15), eons before the Cambrian period. The latter began as recently as 600,000,000 years ago—practically yesterday, geologically speaking (Fig. 2–16). The prebiotic terrestrial atmosphere is thought to have been strongly reducing, i.e., virtually devoid of oxygen.

Chemical Evolution

Prior to 1828 the formation of organic compounds was believed to be absolutely restricted to living organisms (hence **organic**). In 1828 Friedrich Wöhler, a German chemist, produced urea, $O{=}C{=}(NH_2)_2$, an organic substance common in the wastes of many animals, by the simple process of evaporating an aqueous solution of ammonium cyanate, $O{=}C{=}N(NH_4)$, an inorganic substance.* The old-time distinction between **organic** and **inorganic** evaporated with the water from Wöhler's solution; Wöhler had demonstrated that "spontaneous generation" of organic substances from inorganic ones could occur. Everyone was naturally excited with the implications of Wöhler's discovery for the possibility of abiogenesis. If urea could be made to appear "spontaneously," what about other organic substances? Could living substance, or anything like it, be made to appear artificially?

Innumerable subsequent studies of these problems have been made and are still being carried on. To avoid excessive details we may say that many fairly complex compounds that occur in living organisms have been synthesized artificially in the laboratory from a few elements and simple, naturally occurring substances like NH_4, CH_4, H_2S, CO_2, and CO, using conditions and energy sources (electrical discharges, ultraviolet light, solar light, intense sonic and super sonic vibrations, compressed

---

*The term organic chemistry is commonly used to mean the chemistry of the compounds of carbon. Some exclude certain simple carbon compounds, e.g., oxides, sulfides, and metallic carbonates, cyanides, and cyanates. Here, for convenience, CO_2 and metallic and ammonium carbonates and cyanates are treated as inorganic.

Geologic Eras and Epochs		EVOLUTIONARY SUCCESSION OF LIFE FORMS (A few representative types)
Duration of Eras (in years)	Years Ago Epochs Began	
Cenozoic Era (70,000,000)		
"Space Age"	20	
"Stone Age"	500,000	
Pleistocene	1,000,000	
Pliocene (First "men"?)	11,000,000	
Miocene	25,000,000	
Oligocene	40,000,000	
Paleocene	70,000,000	
Mesozoic Era (145,000,000)		
Cretaceous	135,000,000	
Jurassic	180,000,000	
Triassic	225,000,000	
Paleozoic Era (375,000,000)		
Permian	270,000,000	
Carboniferous	310,000,000	
Devonian	375,000,000	
Silurian	425,000,000	
Ordovician	475,000,000	
Cambrian	600,000,000(?)	
Precambrian or Archeozoic (??? billion)		
Entirely	1,000,000,000(??)	
Speculative	2,000,000,000(??)	
Oldest dated* rocks	4,500,000,000	

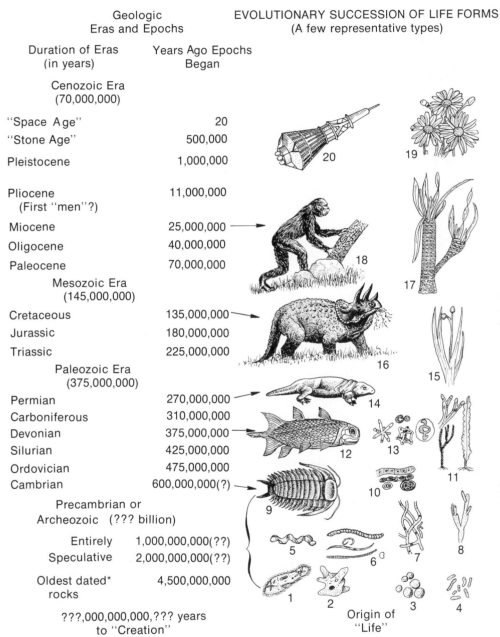

???,000,000,000,??? years to "Creation"

Origin of "Life"

Figure 2–15. Geologic time scale (left) showing ancient origin of bacteria and other microorganisms in pre-Cambrian times (entirely speculative) at foot of evolutionary scale. Pictures *1, 2,* protozoa; *3, 4,* bacteria; *5,* spirochete (protozoa-like bacterium); *6,* marine worms; *7,* mold-like bacteria; *8,* aquatic fungus (*Saprolegnia*); *9,* trilobite (fossil marine arthropod); *10, 13,* bacteria-like algae (Cyanophyceae, desmids); *11,* higher algae; *12,* fossil fish; *14,* cotylosaur (fossil reptile); *15, Psilopsida,* first vascular land plants (Silurian); *16, Triceratops,* a dinosaur; *17,* cycad tree (Jurassic); *18,* fossil man-like ape; *19,* hybrid (1973) daisies; *20,* man in space. *Oldest rocks dated by modern measurements of radioactivity.

Figure 2–16. Reconstruction of a Middle Cambrian sea floor about 600,000,000 years ago. The fauna include siliceous sponges (the upright cones), jellyfish, and two genera of trilobites (*Paradoxides,* the large form, and *Ellipsocephalus,* the small form). Higher forms of life had not yet evolved. Bacteria had probably already been in existence for millions of years. (Drawings by Z. Burian under the supervision of Prof. J. Augusta.)

steam, etc.) that undoubtedly were present on the earth under prebiotic (abiotic) conditions.

Among the numerous substances or their chemical precursors found in living cells, the pentoses ribose and deoxyribose have been artificially synthesized from such compounds as acetaldehyde or formaldehyde, materials that could have formed spontaneously under the influence of ultraviolet and γ-rays. These sugars, as will be detailed later, are parts of the ribonucleic and deoxyribonucleic acids (RNA and DNA, respectively) that constitute the genes of all living cells and viruses. All of the nucleic acid bases, including thymine, which are essential parts of RNA and DNA, are also reported to have appeared under simulated abiotic or prebiotic conditions. Other substances that have been produced under such conditions are thiourea, urea phosphate, thioacetamide, alcohols, aldehydes, propioaldehyde, the vitamin nicotinic acid, and so on. Most importantly, many amino acids, the very "building stones" of which all cellular proteins (enzymes, muscle, antibodies, etc.) are made, have been artificially synthesized by heating together formaldehyde and ammonia, both of which exist free in interstellar space. More recently methionine, a sulfur amino acid, has been made from mixtures of CH_4, N_2, H_2O, NH_3, and CH_3SH. The existence in lunar dust and terrestrial lava of molecules precursor to amino acids has also been reported.

Polymerization of such component molecules into macromolecules, the next step toward formation of living cells, requires little energy input and has been demonstrated to occur under simulated prebiotic conditions. Mononucleotides have been caused to polymerize into oligodeoxyribonucleotides, forerunners of "genes"; amino acids, under certain conditions resembling prebiotic, readily polymerize into oligopeptides and globules spoken of as "proteinoids," precursors of proteins; aldol condensations (additions) can occur spontaneously, producing forerunners of polysaccharides and other complex substances; and so on.

Beutner, the American philosopher whose published works on abiogenesis appeared in 1938 independently of the British scientist Haldane and the Russian Oparin, said:

> . . . we find no evidence in nature of anything like an aim or purpose in creating life. We may therefore state that the appearance of life on the earth is nothing more than a cosmic event. In other words: just as mountains, rivers and oceans are formed by the play of the forces of nature, so also arise living structures on the surface of the earth.

Beutner envisaged the earliest, quasi-living forms as unstable, enzyme-like structures which later became "self-regenerating enzymes." These were strongly suggestive of Oparin's "coazervates" and of more recent proteinoid microsystems. Of Oparin's work Beutner said:

Obviously the essential feature of Oparin's hypothesis is that structural units with a remote resemblance to living organisms were *first* formed from organic matter of the early ocean and that subsequently enzymes formed in them. . . . In this manner some of the coazervates eventually developed into living organisms. Our opinion, on the contrary, was that life-producing enzymes were the first to appear, through the action of electric discharges, without any structure around them, as self-regenerating enzymes. Later only, a structure was built up around them as we assumed. The entire difference between the opposing views is therefore only concerned with the order of the essential events which preceded the appearance of life.

On careful consideration of both views one wonders whether the self-regenerating enzymes were not, in fact, forerunners of coazervates. In either case the primitive enzymes may have been relatively simple protoproteins or other molecules having catalytic properties.

After all, some very active commercial catalysts (e.g., lead, platinum) are simple metals. Like Leduc (1911), Beutner produced some strikingly plant-like and animal-like structures from wholly inorganic materials (Fig. 2–17), while budding and proliferating cell-like proteinoid microspheres have been more recently prepared (Fig. 2–18). In 1973 Merek, like Beutner, produced marvelously microorganism-like structures from solutions of Fe_2SO_4, $FeCl_2$, Na_2S, and silicate. These experiments seem to answer in the affirmative the crucial question as

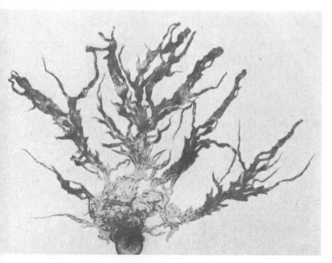

Figure 2–17. An artificial structure grown from a "seed" which is composed of sugar and copper salt. The "seed" is placed in a dilute solution of prussiate, where it sprouts, owing to the osmotic pressure of the sugar as it goes into solution.

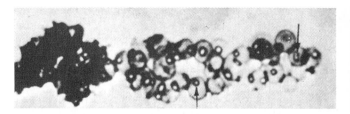

Figure 2–18. Communication between proteinoid microsystems. Endoparticles of informational proteinoid transfer through hollow junctions which arise with high efficiency.

to whether nonvital, inorganic structures, resembling living microorganisms, can occur.

2.5 COSMOBIOSIS

Over the years a number of savants have suggested that terrestrial life could have originated extraterrestrially. Undoubtedly conditions suitable for some form of life could exist on other celestial bodies. As pointed out by Cohn in 1872, "In certain meteorites carbon, and certain combinations containing carbon, have been found, which points to organic combination."

Theories of Cosmozoa and Panspermia. Among early speculations about life's extraterrestrial origin were those of **cosmozoa** and **panspermia**. The idea of cosmozoa received strong support from the studies of Tyndall and Cohn. The theory of panspermia was supported by many persons, notably the Swedish physicist Arrhenius (Nobel Prize winner) as recently as 1938. Both theories attributed life on earth to "germ cells" or "cosmic spores" scattered through the universe. Cohn in 1876 had said,

It is perhaps not impossible that an ascending particle of spore-carrying dust, which has floated for a long time in space, may reach the atmosphere of another world, and if it finds there circumstances favorable to life, it multiplies. . . . it is possible to think that a germ of *Bacterium* or any other exceedingly small and simple form, from some other life-nourishing world, may have been moving about in space, and that such a germ, finally reaching our atmosphere, settled to the earth.

Beutner dismissed the problem with these words, "If it is true that life first arose outside the earth, the question of its origin presents the same problem; the scene is merely shifted." The question is not so much where, as *how*, life originated.

Observations since about 1960 have focused new interest in the possibility of extraterrestrial life, its nature, and its relation to the origin of terrestrial life. Examinations of lunar rocks and soil have so far revealed no convincing evidence that anything like life as we know it ever existed on the moon, though amino acids and their molecular precursors have been reported in lunar dust. Lunar conditions appear to have been always unfavorable to life, due in part to total absence of water. Nevertheless it is not impossible that some primitive biotic or prebiotic process may now exist, or have existed, on other extraterrestrial bodies. It could have developed in cosmic pebbles, or other astral condensations, from simple compounds like CO, CH_4, H_2O, H_2S, CO_2, HCN, $HCOH$, CH_3OH, CS, OCS, and several others, possibly with metallic elements as catalysts. All of these substances are now believed to occur in cosmic clouds of gases and dust. The probability is strongly supported by renewed findings of organic matter in carbonaceous chondrites or meteorites. These include a considerable variety of amino acids, some of which are rare on earth and never found in earth proteins. Various aromatic hydrocarbons also occur in some meteorites, all presumably of biotic or prebiotic origin. That amino acids and other organic matter may form under actual prebiotic earth conditions (i.e., not modern laboratory simulations) is indicated by the report that such materials have been found in Precambrian terrestrial rocks possibly 5 billion years old and therefore formed under truly prebiotic conditions. Electron microscopic examinations of rocks over 3.5 billion years old (Barghoorn, 1971) have revealed outlines of what appear to be fossil bacteria, one of which has been named *Eobacterium isolatum* (Gr. *eo* = early or dawn). Possible fossils of cell-like proteinoid microsystems from the earliest beginnings of subliving structures have been hypothesized.

Unfortunately, the spectre of possible terrestrial contamination of meteorites and rock samples and dusts has haunted some of the evidence from studies of extraterrestrial materials. Entirely conclusive evidence must await results of continuing investigations into the time, manner, and place of the origin of life, both terrestrial and/or extraterrestrial.

2.6
ORIGINS OF IMMUNOLOGY

By 1880 it had been observed for many years that some diseases are restricted to certain species: pebrine to silkworms; leprosy and gonorrhea to man. Other diseases, e.g., smallpox, bubonic plague, measles, and so on, were seen to occur only in persons who had not previously had those diseases. Survivors of such diseases were well known to be resistant or **immune** to them. However, nothing was known of the mechanisms underlying infection or immunity until investigations were made into bodily reactions to infectious diseases, researches that have become known as the science of **immunology**.

In 1884 the Russian scientist Metchnikoff (Fig. 2–19) (Nobel Prize winner) described the phenomenon which he called "phagocytosis" (Chapt. 24). He pointed out its importance as a prime defense mechanism against invading microorganisms, thus establishing what has been called the "cellular school" of immunologists. Another of Koch's students, the English investigator Nuttall (see Figure 3–10) demonstrated the existence of bactericidal properties in normal, cell-free serum. These properties were later explained by the Belgian Bordet (Nobel Prize winner) as due, in part, to two substances in the serum: one now called **specific antibodies** but which he called "sensitizer," and one now called **complement** but which he called "alexin." Their combined

Figure 2–19. Elie Metchnikoff (1845–1916).

action, then known as Bordet's phenomenon, was the basis of what is now called the **complement-fixation reaction.** In 1906 this reaction was adapted by Wassermann and others as a test for syphilis. In 1930 it was adapted by Frobisher and by Davis to the diagnosis of yellow fever. Complement-fixation tests using different infectious agents are widely used today, especially in the study of viruses.

The pioneer investigations into the antibacterial properties of serum **(serology)** established what has been called the "humoral (fluid) school" of immunologists. Each school at first claimed to be the most important. Today we know that both cellular and humoral (serumborne) elements are absolutely essential to resistance or immunity to infection (Chapts. 24 to 28).

Active Immunization. Pasteur was of neither the humoral nor the cellular school of immunologists. He was more immediately concerned with the prevention of communicable diseases than with the mechanisms of immunity. He has sometimes been given the title "Father of Immunology" albeit a more accurate title (if any be needed) would be "Father of Active Immunization" (Chapt. 26). Probably even more appropriately, this title should go to Edward Jenner, the English country doctor who, about 1798, had developed the practice of vaccination against smallpox by actually infecting persons with the virus of the mild disease cowpox. Cowpox may be, or may have been, smallpox virus much weakened or **attenuated** by passage from man to cow and then from cow to cow (Chapt. 27).

About 1882 Pasteur noticed that old cultures of the bacterium of fowl cholera (*Pasteurella multocida*), ordinarily highly virulent for chickens and a scourge to poultry farmers, sometimes lost much of their infectivity and became attenuated when maintained in the laboratory. Aware of Jenner's success with attentuated poxvirus, he inoculated fowls with attenuated *P. multocida* and found the birds fully immune to virulent cultures. The basic principle of **active immunization** with live or active but attenuated infectious agents was thus revealed by Jenner and Pasteur. Pasteur, in a series of brilliant and dramatic experiments, extended the application to the prevention of anthrax, rabies, and swine erysipelas, using various methods of attenuation of the specific causative agents.

In 1888 a potent poison (diphtheria toxin) was found in **cell-free filtrates** of cultures of diphtheria bacilli by Roux and Yersin. In 1890 "lockjaw" (tetanus) toxin was found in culture filtrates of tetanus bacilli by von Behring (Nobel Prize winner) and Kitasato. These scientists soon found that animals which had recovered from a sublethal dose of tetanus toxin were immune to otherwise lethal doses of the toxin given about two weeks later. Perhaps more importantly it was found that the serum of the immune animals contained a substance (now called tetanus antitoxin) that "neutralized" tetanus toxin and protected normal animals from large doses of it. On December 11 of the same year (1890) von Behring published identical discoveries regarding diphtheria toxin. These epoch-marking discoveries opened the door to active immunization with microbial **toxoids** and to the use of serum that contains specific antibodies (antitoxins, etc.) for serum therapy (**passive immunization;** Chapt. 28). Both procedures are now in daily use.

Within the next half century, and continuing today, methods of active immunization, both with attenuated infectious agents of disease and with the same agents killed or inactivated, as well as passive immunization with specific antibodies contained in the serum of animals or persons, were developed by scores of workers against many other diseases: whooping cough, epidemic hepatitis, poliomyelitis, yellow fever, measles and German measles, bubonic plague, and so on. Some of the immunizing agents are much less effective than others, and these are still under investigation. Openings exist in these fields for many new investigators, offering opportunities for all who seek fame and/or fortune in the age-old conflict of humanity against disease.

CHAPTER 2
SUPPLEMENTARY READING

Barghoorn, E. S.: The oldest fossils. Sci. Am., *224*:30, 1971.

Bernal, J. D.: Origin of Life. Wiedenfeld & Nicholson, Ltd., London. 1967.

Beutner, R.: Life's Beginning on the Earth. The Williams & Wilkins Co., Baltimore. 1938.

Bulloch, W.: History of Bacteriology, *in*: A System of Bacteriology in Relation to Medicine, Vol. 1. His Majesty's Stationery Office, London. 1930.

Calvin, M.: Chemical Evolution. Oxford University Press, New York. 1969.

Cohn, F. (C. S. Dolley, Tr.): Bacteria: The Smallest of Living Things. The Johns Hopkins University Press, Baltimore. 1939.

Dobell, C.: Antony van Leeuwenhoek and His "Little Animals." Dover Publications, Inc., New York. 1960.

Engel, A. E. J.: Time and the earth. Am. Sci., 57:458, 1969.

Haggard, H. W.: Devils, Drugs and Doctors. Garden City Publishing Co., Garden City, New York. 1929.

Johnson, F. M. (Ed.): Interstellar molecules and cosmochemistry. Ann. N.Y. Acad. Sci., 194:1–98, 1972.

Keosian, J.: The Origin of Life, 2nd ed. Reinhold Book Co., New York. 1968.

Merek, E. L.: Imaging and life detection. Bioscience, 23:153, 1973.

Pasteur, L.: Sesquecentennial commemorative symposium. Am. Soc. Microbiol. News, 39:1, 1973.

Swain, F. M.: Paleomicrobiology. Ann. Rev. Microbiol., 23:455, 1969.

PRINCIPLES OF MICROSCOPY

The basic principle of the modern compound microscope is the same as that discovered by a Dutch spectacle maker, reputedly one Zaccharias Jensen, about 1590: the use of a lens (now a **lens system**), called an **eyepiece** or **ocular**, to magnify an already enlarged image produced by a first lens, called an **objective**. The total magnification produced by modern objective and ocular together is the product of the two magnifications. Microscopes in which the final magnified image of the object, illuminated by **visible** light, is seen through glass lenses are called **optical** or **light** microscopes. The use of visible light distinguishes these instruments from x-ray or electron microscopes. In the latter the final image is formed on a fluorescent screen or photographic film or plate by **invisible** electrons focused by circular magnets instead of glass lenses.

3.1
THE OPTICAL OR LIGHT MICROSCOPE

The compound optical microscope commonly used in advanced microbiology today (Fig. 3–1) consists of a strong metal stand with a broad base or foot, from which rises a short, stout pillar supporting an upright, curved arm. Situated in the base is a strong light source—either an adjustable, built-in electric lamp or a mirror. Attached to the pillar above the light source is a system of one or more horizontal iris or condenser diaphragms, or **stops**, which regulate the passage of light and eliminate undesirable peripheral rays from the light source. Also

attached to the pillar, above the diaphragm(s), is a one- or two-lens vertically adjustable **substage condenser**, which concentrates the light rays on the object (Fig. 3–2). Above the condenser is the horizontal working platform or **stage** of the instrument, some 3 to 4 inches square or circular, with an opening in the center to admit light from the condenser below. Objects to be examined, e.g., bacteria, are placed on the upper surface of glass slides, which are centered on the stage over the condenser.

Positioning, centering, and alignment of all working parts of the microscope, including lenses and light source, are of critical importance to obtain clear images. The mountings of the oculars and objectives and all other parts of the microscope must therefore at all times be handled gently and never subjected to jars, pressures, impact against hard objects, excessive heat or cold, improper immersion oils (see farther on), or oil solvents.

Attached to the curved upright arm of the microscope is the vertical **barrel** or body tube. In most modern instruments this is double (binocular) and contains a system of prisms and reflectors that permit tilting of the barrel for ease in viewing—luxuries denied to early microscopists. The barrel is mounted on a rack-and-pinion mechanism for vertical coarse and fine adjustment or focusing. At the lower end of the barrel is the **objective lens system**, commonly used with immersion oil and giving a magnification of 90 to 100×. Most high-power instruments also have two to four other objectives of lower magnifying powers: 45×, 16×, etc.

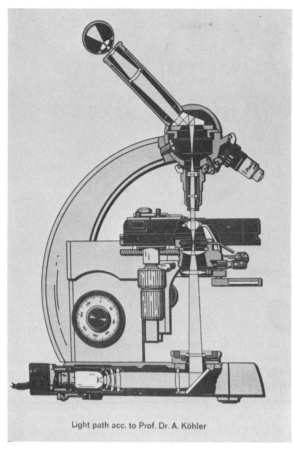

Light path acc. to Prof. Dr. A. Köhler

Figure 3–1. Vertical section of a modern form of compound microscope suitable for student use. Focused light rays from an electric lamp in the base (left) are reflected upward from a mirror (base, right) through a lamp field stop placed immediately above the mirror for light control. The rays are focused, by a multilens condenser with aperture diaphragm on a vertical-motion rack-and-pinion mount, onto the object or specimen slide held in place on the stage by a spring clip. The light rays pass through the specimen slide to a prism and mirror complex where they are reflected at an angle through the ocular system to the eye. There they produce an enlarged image on the retina. (Courtesy of Carl Zeiss, Oberkochen, Württemberg.)

These are mounted on a turret or "nose-piece" on which they may be rotated into position under the barrel, correctly aligned. The lower-power objectives are commonly used without immersion oil and are spoken of as "high-dry" objectives (Fig. 3–3). The objective lens produces a **real image** within the instrument.

At the top of the barrel is the **ocular lens system** or eyepiece, usually containing two or three lenses. These magnify the real image, which then appears as a greatly enlarged **virtual image**, seeming to be projected to a position just above the light source and below the condenser diaphragm. In the compound microscope, light rays from below the condenser or iris diaphragm are refracted through the condenser and emerge from the top surface of the slide, at the plane of the object, as a cone of light with the apex downward (Fig. 3–4).

3.2
RESOLVING POWER

The true measure of a microscope is its **resolving power**. This is the ability to make clearly visible objects less than about 200 nm in diameter, i.e., not visible to the human eye, or to reveal space between lines or objects so closely adjacent (<200 nm apart) as not to appear separate to the unaided eye. Because of its neuromuscular anatomy and its curious optical properties, the resolving power of the unaided human eye is pitiably limited. A simple demonstration of this sad fact may be made by observing the faces in Figure 3–10 with a 2× hand lens. The imperceptible photoengraver's "screen" at once becomes readily visible as well-separated **(resolved)** dots.

The resolving power of any microscope depends on two factors. One is the length of the electromagnetic waves used for illumination.

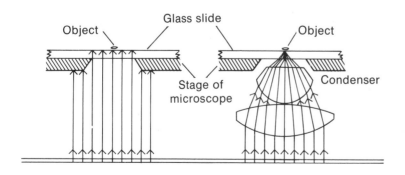

Object — Glass slide — Object
Stage of microscope — Condenser

Figure 3–2. Illumination of microscopic object with and without substage condenser. The condenser focuses all of the light from the light source onto the object. The angular aperture of the objective is thereby greatly increased.

TABLE 3–1. METRIC UNITS AND SYMBOLS COMMONLY USED IN MICROBIOLOGY

Units and Symbols →	kilometer (km)	hektometer (hm)	dekameter (dam)	meter (m)	decimeter (dm)	centimeter (cm)	millimeter (mm)	micrometer (μm)			nanometer (nm)	Ångstrom (Å)		picometer (pm)
	1,000	100	10	1.	0	0	0,	0	0	0,	0	0	0,	0	0	0
Log$_{10}$ of meters →	3	2	1	0	−1	−2	−3	−4	−5	−6	−7	−8	−9	−10	−11	−12

↓ (at −6) Formerly micron (μ) ↓ (at −9) Formerly milli-micron (mμ) ↓ (at −12) Formerly micro-micron (μμ)

Units of weight: substitute gram (or g) for meter (or m) in the above scale, e.g., nanogram (ng). (Gram is often abbreviated as gm for clarity.)

Units of volume: substitute liter (or l) for meter (or m) in the above scale, e.g., nanoliter (nl).

Useful equivalents:

1 meter = 39.37 in	1 gram = 0.035 oz	1 liter = 1.0571 qt
1 inch = 2.54 cm	1 ounce = 28.35 gm	1 quart = 0.946 l
1 mm = ~ 1/25 in		

In optical microscopes this is the wavelength(s) of visible light: about 400 to 750 nm. **Resolution** is *inversely proportional* to **wavelength.** The second factor in resolving power is the maximum light-admitting power of the objective lens. This is determined by its numerical aperture, or N.A., which is related to its **angular aperture.**

Angular Aperture. The objective lens most commonly used in high-power optical microscopes is plano-convex, with an almost hemispherical convexity and a very short focal length; i.e., it is a very "short-sighted" lens. Because of mechanical and other exigencies of manufacture such lenses are very small and have a relatively narrow visual angle or **angular aperture.** The angular aperture of an objective lens is the angle between the most divergent rays of the inverted cone of light emerging from the condenser that can enter the objective

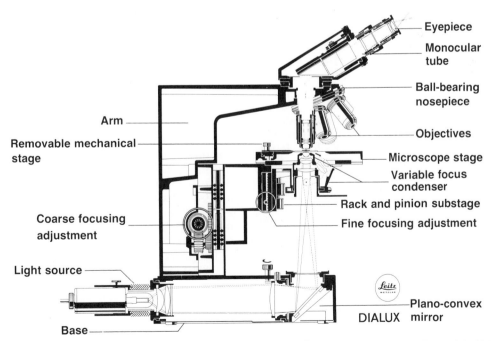

Figure 3–3. Typical working parts of the optical microscope. (Courtesy of E. Leitz, Inc., Rockleigh, N.J.)

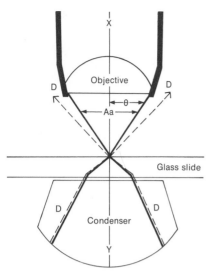

Figure 3–4. Angular aperture. The angular aperture of the lens is shown at *Aa*. The angle θ is that between the optical axis (*X–Y*) and the most divergent rays that can enter the objective. Aberrant rays *D* and *D* might be captured with immersion oil, in effect increasing the angular aperture (see Figure 3–5).

lens. Rays cannot enter the lens if their divergence from the normal ray or **optical axis** is greater than half the angular aperture (Θ, Fig. 3–4). Devices are accordingly used to cause as much as possible of the cone of light from the condenser to enter the objective lens.

Numerical Aperture and Immersion Oil. The extreme range within which a lens of a certain angular aperture can be made to admit divergent rays is expressed as its working aperture or **numerical aperture** (N.A.). N.A. depends primarily on angular aperture and secondarily on the refractive index of the medium between object and objective lens. If the medium is air, which has a refractive index of only 1, rays whose divergence from the optical axis is greater than half the angular aperture obviously cannot enter the objective and are lost. If air is replaced by immersion oil or other specially designed fluid which has a refractive index of, say, 1.52, many of the divergent peripheral rays, and rays lost by reflection and refraction at the surfaces of condenser, slide and objective lens, are refracted within the angular aperture; the N.A. of the objective is thus increased (Fig. 3–5, *1*). The lens can be brought much closer to the object, admitting more of the divergent rays, with a resulting "working distance" of about 0.2 mm instead of 1.8 mm (Fig. 3–5, *2*). The resolving power is greatly increased. The relations between angular aperture, medium

between object and lens, and N.A., are shown in the equation:

N.A. = refractive index of medium ×
$$\text{sine } \frac{\text{angular aperture}}{2}$$

Oil-immersion lenses are specially designed for use with immersion fluids having a refractive index like that of glass, or otherwise specified by the manufacturer. The immersion fluid is placed on the slide and should be removed from the lens after use or before the instrument is stored. In cleaning the lens, avoid using fluid solvents that can act on cement, which may be used to hold the lens in place. Use only lens paper to wipe the lens, as almost any other material may be abrasive enough to scratch the lens.

The diameter of the smallest object, or the least distance between objects which, under optimal conditions of lighting, may be resolved by a lens system (i.e., its resolving power, R.P.), may be calculated by appropriate substitutions of values in the equation:

$$\text{R.P.} = \frac{\text{wavelength}}{2 \text{ N.A.}}$$

e.g.,

$$\frac{600 \text{ nm}^*}{2 \times 1.4} = 213 \text{ nm} = 0.21 \ \mu\text{m}$$

Spherical and Chromatic Aberration. In the early microscopes difficulties arose from inequalities of refraction and focus by peripheral portions of the objective lens. These caused **spherical aberration** and resulted in fuzzy images. Another difficulty, called **chromatic aberration** (color rings), was due to the prism-like effect of the outermost peripheral portion of the lens (Fig. 3–6). In modern microscopes spherical aberration is overcome partly by use of the iris diaphragm to eliminate excess peripheral rays and partly by ingenious combinations of lenses of various curvatures and lens materials such as flint glass, crown glass, fluorite, etc. (Fig. 3–7). Lens systems corrected for chromatic aberration in the red and blue ranges are said to be **achromatic**; those corrected in red, blue and other ranges are called **apochromatic** (Abbé, ca. 1886).

---

*600 nm = average wavelength of visible light; 1.4 = maximum N.A. of oil-immersion lens.

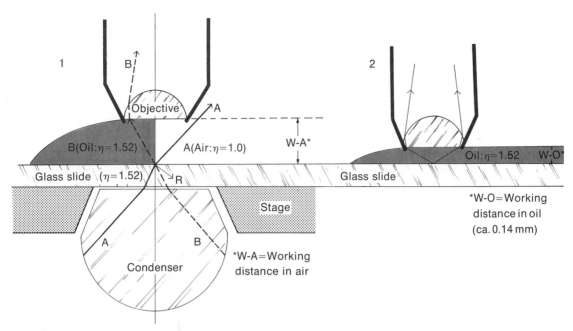

Figure 3–5. Influence of immersion oil having the same refraction index as glass ($\eta = 1.52$). In diagram *1* peripheral ray *A,* on emerging from the slide, is refracted through the air ($\eta = 1.0$) beyond the scope of the angular aperture of the objective lens (see Figure 3–4). The oil captures such rays (*B*) and brings them within the angular aperture of the objective lens. The oil also eliminates loss of light by reflection (*R*) from the upper surface of the slide and the face of the objective lens.

In diagram *1* the working distance (distance between slide and objective lens) is diagrammed for illustrative purposes, as it would be without oil; actually, the lens would not function without oil. The working distance is much decreased (to about 0.14 mm) in oil, as shown in diagram *2*.

The whole effect of immersion oil is to collect aberrant rays (thus widening the N.A.), to eliminate loss of light due to reflection, to eliminate distortion due to diffraction (bending of rays in passing around particles), and to increase resolution with decreasing working distance.

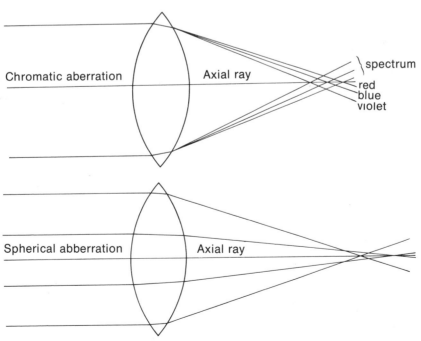

Figure 3–6. Common forms of aberration in lenses.

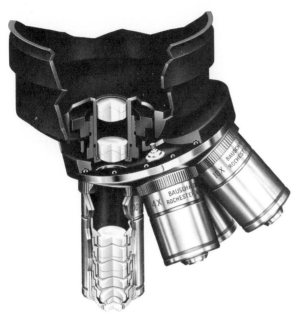

Figure 3–7. Assembly of corrective lenses in modern oil-immersion objective giving 100× magnification without distortion. (Courtesy of Bausch & Lomb, Rochester, N.Y.)

3.3
METHODS IN LIGHT MICROSCOPY

Hanging-Drop Preparations. As Leeuwenhoek discovered, many microorganisms, especially bacteria in a natural living state, are best viewed when suspended in a clear fluid of some sort, usually water, saline solution, or broth. The cells are transparent, colorless, refractile, and so tiny that they are often difficult to find and even to identify in the drop of fluid. This is especially true of the spherical types (cocci), which may be only 1 or 2 μm in diameter.

To observe bacteria and similar microorganisms in fluid media it is necessary to put a loopful* of the fluid in which they are suspended in the center of a thin cover slip. On each of the four corners place a tiny droplet of mineral oil. Hold a hollow-ground slide, depression side down, over the slip and bring the two into contact. Invert the slide quickly so that the drop cannot run off to one side. A tiny additional drop of clear mineral oil may now be

---

*By a **loop** is meant the space included by a tiny ring or loop made at the end of a thin wire. The wire is fixed into some sort of handle so that it may be sterilized in a flame. Loops are usually about 2 mm in diameter; a loopful is a small drop, about 1/50 ml.

run under the edges of the cover slip at each corner if needed. This spreads under the cover slip and prevents drying of the drop (Fig. 3–8).

In the hanging drop we may see the size, shape, and arrangement of microorganisms and their motion if they are motile. Sometimes bright refractile granules and spores may be seen within the living cells. Since cells of yeasts, molds, protozoa, and bacteria are generally colorless and transparent, it is often extremely difficult to study them because this lack of density prevents details from showing clearly. Further, when suspended in fluid they move about, because of either their own motility, currents in the fluid, or brownian movement. They often have almost the same refractive index as the fluid in which they are suspended, which makes them nearly invisible, like a splinter of ice immersed in water. The observation of microorganisms in hanging-drop preparations, therefore, yields limited though valuable information.

Wet Mounts. These are among the simplest procedures in microscopy. A drop of fluid containing cells to be examined is placed on a slide. A thin coverslip is placed over the fluid. The organisms remain alive. The slide is placed on the stage, and usually one of the high-dry objectives is focused on it, though the oil-immersion objective may be used if care is taken to hold the coverslip in place. It is of interest to note that strongly aerobic organisms (those requiring oxygen to live) (Chapt. 8) often migrate toward the margins of the coverslip

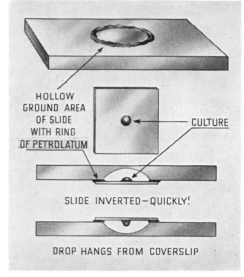

Figure 3–8. Hanging-drop preparation.

but not if the margins are sealed with petroleum jelly to prevent drying as is sometimes done.

The organisms are best observed in such preparations if the light intensity is diminished or, better, if darkfield or phase-contrast illumination is used. These methods are described farther on in this chapter.

3.4
DEVELOPMENT OF STAINING METHODS

Robert Koch (1843–1910). Early in his career (ca. 1870) Robert Koch was provincial health officer in Wollstein. He had occasion in an official capacity to investigate anthrax (a disease of animals and man caused by *Bacillus anthracis*, a species of cylindrically shaped bacteria). He decided to study the disease in his laboratory during his spare time.

At first Koch and other microbiologists examined all their specimens of microorganisms in the living state, usually in drops of fluid mounted on a bit of glass. They thus became familiar with motility, when present, and they observed refractile granules inside various microorganisms. But the transparency and constant motion of some of the bacteria (either that purposeless oscillation due to molecular impact, known as **brownian movement**, or real progressive motion due to gliding or the action of flagella) made accurate and prolonged study difficult. Koch realized that it would be much better for his drawings and especially for his photographs (both of which, by the way, were excellent; Fig. 3–9) if the bacteria could be made to remain still. He tried spreading out the drops of fluid to be examined in thin films (commonly called "smears") on glass slides and allowing them to dry. By heating the smear to about 70C for 10 to 20 seconds the cells were "fixed," i.e., they were dried, solidified, and adhered to the slide when afterward immersed in staining solutions (see below). The cells apparently had not shriveled or changed in any visible way. However, the bacteria were nearly transparent and colorless. They offered little contrast with surrounding materials or the colorless glass slide. It was very difficult to observe the fine details of their structure and equally difficult to photograph them. He obtained ideas from other workers, a procedure commended to all investigators.

Weigert, a German scientist contemporary with Koch, had observed the use by Cohn and others of various dyes to make clear the details of cell structures in histological preparations

(histology deals with microscopic anatomy). The natural dyes carmine (from the cochineal insect, *Dactylopius coccus*) and hematoxylin (from the logwood tree, *Hematoxylon campechianum*) were, and still are, widely used, especially by histologists. Paul Ehrlich, a renowned chemist and Nobel Prize winner, had recently improved methods discovered by W. H. Perkin, Jr. (1838–1907), a brilliant British chemist, of preparing very fine dyes from coal-tar distillates. Perkin's first dye was called **mauve**, and its popularity and wide use for ladies' dresses, draperies, etc., gave rise to the term "mauve decade" for the 1880's. These were the first coal-tar or aniline dyes.

Weigert, the bacteriologist, tried the methods of the histologists with the dyes invented by the chemists. His first success was in 1875 when he found that the dye methyl violet could be used to reveal bacteria in histological preparations. This method of making bacteria easy to find and study, when before they had

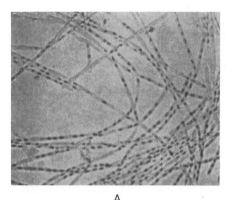

A

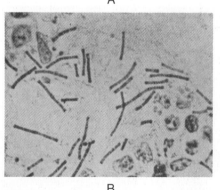

B

Figure 3–9. Micrographs of *Bacillus anthracis* made in 1877 by Robert Koch. (About 1000×.) The bacilli in *A* are unstained; the dark bodies are spores which appear dark in the photograph because they are highly refractile. The bacilli in *B* are stained. The spores are unstained and appear as tiny transparent "holes" in the stained rods. The background material contains tissue cells and dried fluid.

been visible only with difficulty, was adopted by Koch and soon came into wide use. Koch became one of the foremost bacteriologists and teachers of his day. His discoveries attracted scholars from all over the world (Fig. 3–10).

Nature of Dyes. Strictly speaking, a dye may be any colored substance that, when combined with a second substance, imparts a color to that substance. However, most of the dyes used in microbiology are organic compounds, chiefly benzene derivatives, especially those with azo ($-N=N-$), nitroso ($-N=O$), azoxy ($-N=^+N-$),

thiocarbonyl ($>C=S-$), and triphenylmethane groups (Fig. 3–11).

Most microbiological dyes may be thought of as salts of two kinds: (1) acidic, those in which the color-bearing ion, the **chromophore**, is the anion, e.g., sodium$^+$ eosinate$^-$ (the dye, **eosin**), and (2) basic, those in which the chromophore is the cation, e.g., methylene blue$^+$ chloride$^-$ (the dye, **methylene blue**). The first type of dye is acidic in the sense that as an acid the chromophore combines with a base (e.g., NaOH) to form the dye salt; the second is basic because the chromophore acts like a base, combining with an acid (e.g., HCl) to form the dye salt.

In general, acidic dyes combine more strongly with cytoplasmic (basic) elements of cells; the basic dyes combine best with nucleic (acidic) elements of cells.

Factors in the environment that can affect the staining reaction of a cell are: ionic concentration, temperature, composition, and pH of the surrounding fluid.

Dyes as pH Indicators. With respect to pH, some dyes are weak acids or bases that ionize in aqueous fluids, the ionized and nonionized parts having different colors. The color of the mixture varies over a range depending on the kind of dye and on the relative proportions of its ionized and nonionized parts. These proportions depend, in turn, on the ionization constant of the dye and on the pH of the fluid. Such dyes are commonly used as **pH indicators**; e.g., bromcresol purple (range, pH 5.1 = yellow to pH 6.7 = purple); phenol red (range, 6.9 = yellow to 8.5 = magenta red) (Chapt. 30).

Some dyes depend neither on forming salts nor on chemical combinations with the stained material. Some merely coat the surface of the cell by adsorption; others dissolve, or precipitate, in some special part of the cell. Some have little or no affinity for the cell or its components (e.g., cell wall, flagella, capsules), staining only through the intermediation of a **mordant**, i.e., a substance that, though not itself a dye, causes the dye to adhere to, or combine with, the substrate in some way. Probably different physical and chemical reactions occur during different staining processes, and probably many different interactions are involved in the staining of different species, or of different parts or sub-

Figure 3–10. One of the groups of famous scientists who studied microbiology under Koch. Standing, left to right: Alphonse Laveran (1845–1922), discoverer of the malarial parasite; Emile Roux (1853–1933), co-discoverer of diphtheria toxin; Edmund Étienne Nocard (1850–1903), French veterinarian and mycologist; George H. F. Nuttall (1862–1937), British microbiologist. Sitting, left to right: Robert A. Koch (1843–1910), discoverer of the tubercle bacillus and pioneer microbiologist; Karl Joseph Eberth (?) (1835–1926), discoverer of the typhoid bacillus; Elie Metchnikoff (1845–1916), Russian zoologist and discoverer of phagocytes and phagocytosis. (Courtesy of Wiley A. Penn, Director of Laboratories, Department of Health, Savannah, Georgia.)

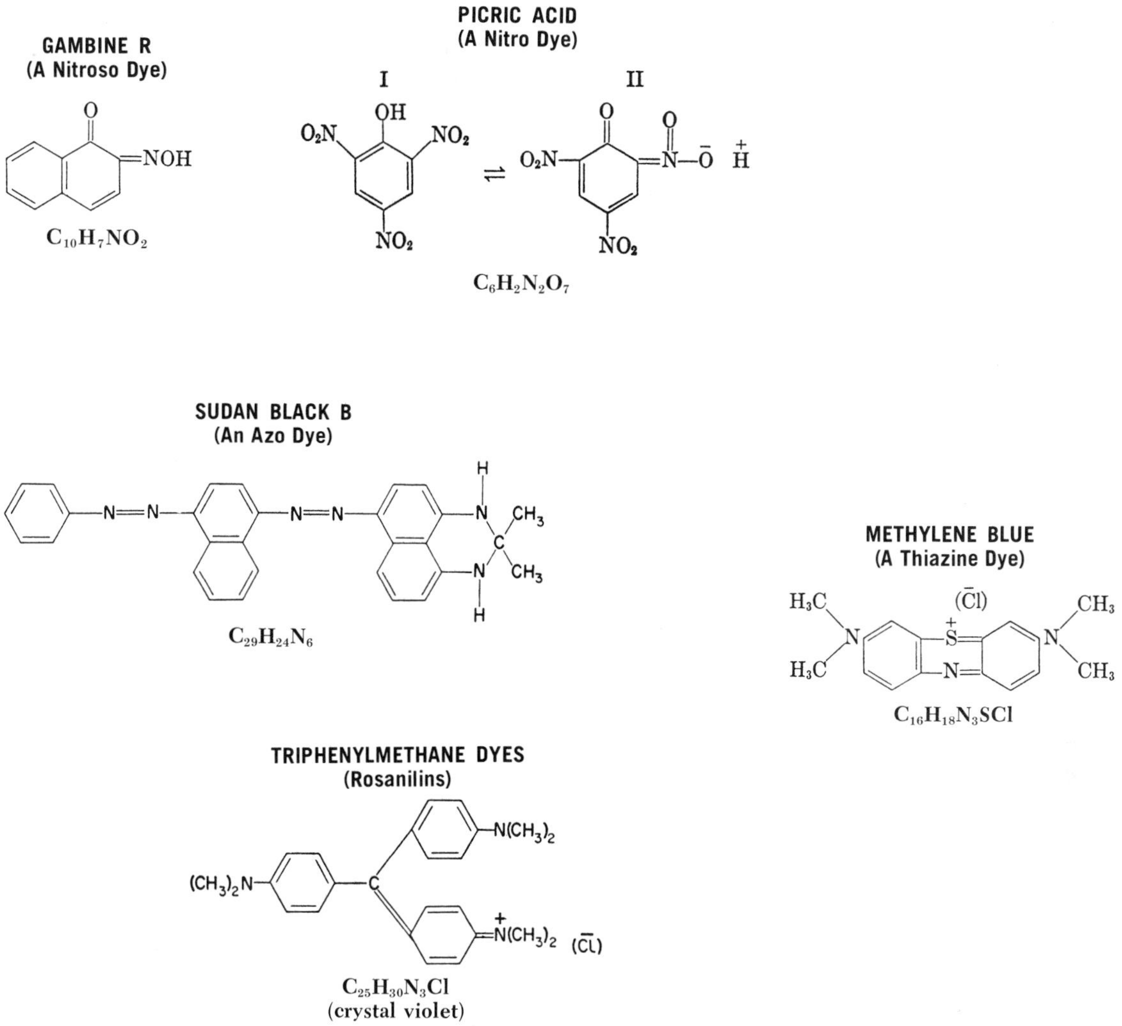

GAMBINE R
(A Nitroso Dye)

$C_{10}H_7NO_2$

PICRIC ACID
(A Nitro Dye)

I ⇌ II

$C_6H_2N_2O_7$

SUDAN BLACK B
(An Azo Dye)

$C_{29}H_{24}N_6$

METHYLENE BLUE
(A Thiazine Dye)

$C_{16}H_{18}N_3SCl$

TRIPHENYLMETHANE DYES
(Rosanilins)

$C_{25}H_{30}N_3Cl$
(crystal violet)

Figure 3–11. Some representative dye formulas.

stances in the cell. Few biological staining reactions are fully understood.

Simple Stains. Because bacteria generally react toward stains as though the cells are composed principally of acidic components (nucleic acids, acidic polysaccharides, proteins, etc.), basic dyes like methylene blue, crystal violet, basic fuchsin, eosin Y, and safranin O are widely used. All belong to the group of aniline (coaltar) dyes. Any simple aniline dye solution may be applied by flooding the smear with it. Löffler's methylene blue solution is very widely used and reveals many details of form and structure. The dye is allowed to remain in contact with the smear for about one minute and is then washed off with a gentle stream of cool water. The slide is then blotted (not rubbed) between two pieces of filter or blotting paper. When dry

it is ready for examination with the oil-immersion lens as previously described.

Gram's Stain. A simple stain such as Löffler's is of great value for many purposes. But another staining procedure, devised by the Danish scholar Christian Gram in 1884, is more valuable because it enables us to differentiate between kinds of bacteria that are morphologically indistinguishable yet of different species. It is therefore called a **differential stain.**

To the smear which, for purposes of discussion, we shall assume to contain a variety of bacteria (e.g., saliva), crystal violet solution is applied for 30 seconds. This is gently rinsed off and an iodine solution is applied for 30 seconds. This, in turn, is rinsed off. All of the cells are deeply stained (Fig. 3–12, *A* and *B*). Ninety-five per cent ethyl alcohol is applied and renewed

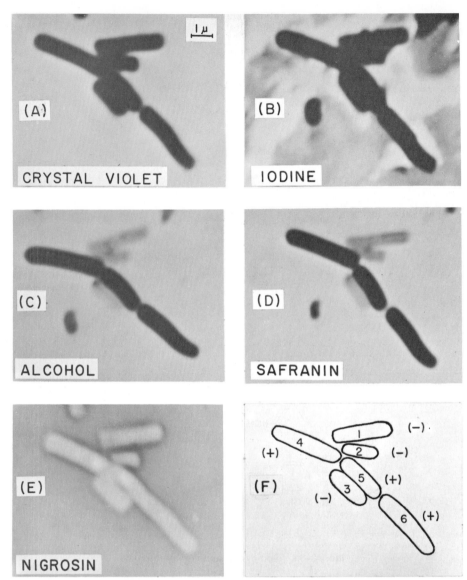

Figure 3–12. Appearance of gram-positive and gram-negative bacteria following each step of the Gram stain. The gram-positive and gram-negative cells are identified by + and – signs in diagram in *F*. After applying crystal violet and iodine all cells are deeply stained (dark) (*A* and *B*). After decolorizing with alcohol (*C*) the gram-negative cells are very light; the gram-positive cells retain the stain (dark). Counterstaining with the transparent red dye, safranin, colors the gram-negative cells but they remain relatively light and translucent (*D*). The nigrosin preparation at *E* illustrates negative staining (see text).

until all but the thickest parts of the smear have ceased to give off dye. (This usually takes from 20 seconds to one minute.)

The differential feature of the method is now apparent. Examination with the microscope will reveal the fact that, as Gram found, while many bacteria retain the violet-iodine combination, others will have yielded it largely to the alcohol and are almost as nearly invisible as before (Fig. 3–12, *C*). Those species of bacteria which retain the stain are called **gram-positive**. Those which yield it to the alcohol are called **gram-negative**.

But the staining process is not yet complete. There is still the important final step of applying the **counterstain**—a dye of some contrasting color, usually eosin (red), safranin (red), brilliant green, or Bismarck brown. Any one of these dyes colors the gram-negative species; they become as visible as the gram-positive ones but are readily differentiated by their color (Fig. 3–12, *D*).

Thus, by applying Gram's stain, which takes but five or six minutes, we can learn a great deal about any bacterium. We make visible not only form, size, and certain other structural details, but we can at once assign the organisms to one of two great artificial groups of bacteria: gram-negative or gram-positive.

Some organisms are "borderline" cases in respect to Gram's stain, sometimes being positive, other times negative; and sometimes, both positive and negative cells of the same organism are seen in the same culture. As a rule, repeated tests will reveal the true nature of the bacterium. Slight variations in cultural conditions or staining technique can affect the result. For example, many bacteria are not definitely gram-positive unless cultivated in the presence of at least 5 per cent blood or serum. The acidity or alkalinity (pH) of the fluid in which the bacteria are suspended, as well as the age of the microorganisms, will also markedly affect their reaction to the Gram stain. In acid media, gram negativeness is the rule.

Mechanism of the Gram Stain. Although still the subject of some debate, it now appears that the property of gram positiveness depends on the presence and physicochemical nature of the cell wall of gram-positive organisms. For example, after physical disruption of gram-positive cells no part of the disrupted cells retains gram positiveness. Even mere removal of certain parts of the cell wall (e.g., lipids by means of solvents; ribonucleic acid or RNA by means of the enzyme ribonuclease) results in gram negativeness. Removal of the entire cell wall (i.e., production of protoplasts) likewise results in gram negativeness. Protoplasts are plant cells, including bacteria, without cell walls (Chapt. 14).

All evidence on the subject indicates that the iodine and violet dye form, inside the cell wall, a compound or precipitate that cannot readily pass outward in the presence of the decolorizing agent. The cell wall of gram-negative organisms, quite different in physicochemical structure (Chapt. 14), cannot retain the dye-iodine complex in the presence of the decolorizing agent.

Correlation of Gram Reaction with Other Properties. Whatever the explanation of the Gram reaction it is important to note that there are characteristic differences between most gram-positive and gram-negative bacteria. Several of the most obvious of these are shown in Table 3–2. It is evident that the property of gram positiveness is related to very fundamental physiological properties of the cell.

Ziehl-Neelsen (Acid-Fast) Stain. Another differential stain is that of Ziehl-Neelsen. It is used especially for staining tuberculosis bacilli (*Mycobacterium tuberculosis*) and related organisms (genus *Mycobacterium*) having an abundance of particular acid-fast waxy materials (mycolic acids) in the cell. Such organisms are gram-positive, but the Gram stain does not give as useful information about them as the Ziehl-Neelsen, or **acid-fast**, stain.

In using this stain a smear of the material to be examined (e.g., tuberculous sputum) is made as usual, dried, and fixed by heat. The smear is then flooded with a special solution of carbolfuchsin and heated to 90C over a steam bath for four minutes. This softens the wax and the dye supposedly penetrates. After washing off the excess dye, the smear is treated for five

TABLE 3–2. SOME DIFFERENCES BETWEEN GRAM-POSITIVE AND GRAM-NEGATIVE BACTERIA

Property	Gram-Positive	Gram-Negative
Susceptibility to sulfonamide drugs and penicillin	Marked	Much less
Inhibition by triphenyl-methane dyes, e.g., crystal violet	Marked	Much less
Susceptibility to acriflavine	Marked	Much less
Susceptibility to low surface tension	Marked	Much less
Susceptibility to anionic detergents	Marked	Much less
Susceptibility to lysis by complement	Slight	Marked
Digestion by trypsin or pepsin (dead cells)	Resistant	Susceptible
Cell wall digestible by lysozyme	Many species	Requires pre-treatment of the cell wall
Dissolved by 1% NaOH	Resistant	Susceptible
Ratio of RNA to DNA in the cell (approx.)	8:1	Almost equal
Aromatic and S-containing amino acids in cell wall	None	Numerous
Fat-like substance (lipid) in cell wall	Little (ca. 3%)	Much (ca. 20%)
Resistance to sodium azide	Marked	Much less
Mesosomes	Present	Rare or absent
Ornithine transaminase	Present	Absent

minutes with *cold* 95 per cent alcohol containing 5 to 10 per cent hydrochloric acid. The organisms retain the red dye in spite of the acid-alcohol, which removes the color from everything else. Organisms retaining the red stain are said to be acid-fast. If methylene blue or brilliant green is now applied as a counterstain, the acid-fast bacilli stand out as bright red objects in a blue or green field. The Ziehl-Neelsen stain is a **differential** stain because it differentiates acid-fast organisms from non-acid-fast ones. Sometimes acid-fast bacilli appear to contain intracellular "beads." Extensive studies of these indicate that they are usually, but not invariably, staining artifacts.

Mechanism of the Acid-Fast Stain. It has been suggested that acid fastness is a matter of relative solubilities. The red dye, fuchsin, is more soluble in phenol than in water or acid-alcohol. Phenol, in turn, is more soluble in lipids or waxes, such as are present in tubercle bacilli, than in water. In the acid-fast staining procedure the phenol, with red fuchsin in it, enters the cell lipids and remains because it is there more soluble than in the decolorizing agent (acid-alcohol). The intact cell coatings prevent the red-stained lipids from leaving the cell. If the cell coatings are broken, the lipids leave the cell and the acid-fast property disappears. The role of the waxy mycolic acids in acid fastness is indicated by the fact that washing with hot (90C) water removes the waxes, and with them, acid fastness.

Negative Staining. This is not really a method of staining bacteria, but of staining the background a solid black, usually with nigrosin. The dye fails to penetrate the microorganisms at all, and leaves them unstained to appear as light areas in the darkened field (Figs. 3–12, E, 3–13). Only the outlines of the organism are made apparent by this method. Consequently, it has a limited application, being used mainly for species that, like spirochetes, cannot be satisfactorily stained by ordinary methods.

A source of error in this method is shrinkage and distortion of the cells of many microorganisms during the drying process.

Special Stains. In addition to the stains described, there are others designed to bring out special details such as spores, capsules, and flagella. Description of each of these methods will be reserved to the discussion of the structural features to which they apply.

3.5
DARKFIELD MICROSCOPY

The ordinary compound microscope may easily be equipped for darkfield illumination by substituting a darkfield condenser for the ordinarily used Abbé condenser, or by using a centrally placed **stop** with the Abbé condenser.

A high-aperture darkfield condenser is designed to prevent the entrance of central rays of light straight upward into the tube of the microscope. All peripheral rays are reflected obliquely to the center of the upper surface of the microscope slide. They emerge from the upper surface of the slide as a **hollow** cone of light, apex down and centered on the object. Intensification of the light source is sometimes required. The oblique rays forming this inverted cone do not reach the eye unless some object is present to reflect or diffract them upward (Fig. 3–14). The empty field, therefore, appears dark. When a fluid containing any particles such as dust or microorganisms is placed on the slide at the focal point of these oblique rays, each particle becomes visible as a brightly illuminated speck because of the light reflected upward from its surface into the barrel of the microscope (Fig. 3–15). The remainder of the field appears dark, hence the term **darkfield**. Unlike the process of negative staining, the darkfield shows not only outward form but motility, since the organisms, covered with a cover slip, are examined in a moist, living state. Perfect alignment of all parts of the system and total elimination of undiffracted or unreflected rays are essential. For accurate, high-power work, immersion oil is necessary between the slide, objective lens, and condenser lens.

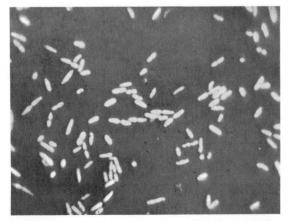

Figure 3–13. Negative staining or relief demonstration. The background is darkened by nigrosin. The bacilli (agriculturally important *Azotobacter chroococcum*) are unstained and transparent. (978×.) (Courtesy of Dr. Robert L. Starkey, College of Agriculture, Rutgers University, New Brunswick, N.J.)

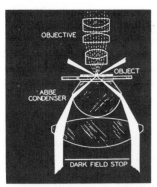

Abbé Condenser with
Darkfield Stop

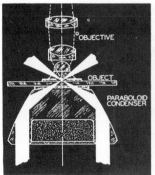

Paraboloid
Condenser

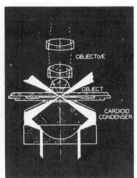

Cardioid
Condenser

Figure 3–14. Various forms of condensers for oblique illumination of the darkfield. In the upper picture some object on the slide is reflecting light up through the objective lens. In each arrangement note that only peripheral light rays pass the condenser. (Courtesy of Bausch & Lomb, Rochester, N.Y.)

One adaptation of fluorescence microscopy has been in the study of tuberculosis. The fluorescent dye, auramine, has a strong affinity for wax-like substances in tubercle bacilli. The hard-to-find bacilli are stained with auramine. All the dye is washed from everything else. The slide is then examined in the dark by ultraviolet light. The tubercle bacilli fluoresce with a brilliant yellow glow and a diagnosis can be quickly made (Fig. 3–16). More widely used applications of fluorescence in microscopy are described in Chapter 27 (fluorescent antibody staining) and shown on Color Plate III, *A*.

3.7
PHASE-CONTRAST MICROSCOPY

If one examines the smaller microorganisms such as bacteria in their living state suspended in a hanging drop of fluid, not only is it difficult to see the organisms but it is almost impossible to discern clearly any of the internal structures. This is because there is almost no **contrast** or difference in refractive index or density between the object (e.g., bacterial cell), its internal structures, and their surrounding fluids. The situation is almost like trying to see a tiny fragment of glass or ice in a bowl of water. However, in biological materials there *are* slight differences in refractive index and density. By means of special optical devices, these differences can be greatly enhanced so that readily perceptible contrast (**phase contrast**) is produced between these objects and their surroundings.

3.6
FLUORESCENCE MICROSCOPY

By fluorescence is meant the property of emitting rays having a wavelength different from that of the incident rays. For example, objects invisible by ultraviolet light may become brilliantly luminous if coated with a fluorescent substance such as quinine sulfate, which fluoresces violet (wavelength, 400 nm) in ultraviolet light (wavelength, 150 to 350 nm), or the dye auramine, which fluoresces yellow (wavelength, 600 nm) in ultraviolet rays. Bacteria stained with a fluorescent dye and then observed through an ordinary microscope, using ultraviolet light, appear as luminous objects readily seen and differentiated from nonfluorescent objects. A special, ultraviolet-opaque filter must be placed in the microscope tube to protect the eyes from ultraviolet rays.

Figure 3–15. Darkfield preparation. A group of spirochetes and other bacteria from a lung abscess. Note that the thicker bacteria are seen only as luminous outlines. This is because they are visible only by means of light reflected from their outer surfaces. (900×).

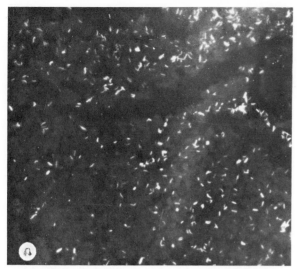

Figure 3–16. Fluorescence microscopy of tubercle bacilli in lung tissue treated with auramine. (Courtesy of American Society for Microbiology.)

To produce phase-contrast an optical microscope is equipped with a special annular diaphragm at the lower focal plane of the condenser (Fig. 3–17). This permits only a ring of light to pass upward through the condenser and objective. Inside the objective mounting, a transparent disk (**phase-shifting plate**) is placed at the rear focal plane of the objective lens system. This disk has upon it a ring of optical dielectric material on which the ring of light from the annular diaphragm is focused. The ring has the property of retarding or advancing (depending on the material used on the ring) the phase of the light waves traversing it.

For each transparent or translucent particle in the object, consider a single ray of incident light. From this two rays result. One, the direct or undiffracted ray (Fig. 3–17, A), comes through the annular diaphragm, passes through the object, and is focused on the phase-shifting ring which either retards or advances the ray one-quarter wavelength with respect to the second ray. This second ray is also derived from the incident ray but is modified by being scattered and diffracted in passing around the margin of the object. This ray does not pass through the phase-shifting ring but traverses the other areas of the transparent disk: its wavelength is neither advanced nor retarded. There is thus an optical difference of one-quarter wavelength between the **undiffracted** waves passing through the phase-shifting ring and those scattered (**diffracted**) in passing around it; i.e., the waves are one-quarter wavelength **out of phase** (Fig. 3–18).

Waves of any kind (sound, water, light) are **out of phase** when troughs and peaks of neighboring waves are asynchronous. At certain points they mutually reinforce, producing **amplification** and, in the case of light waves, increased brightness (**bright-phase**); at other points the waves mutually interfere, producing darkness or less brightness (**dark-phase**).

In conventional microscopy the unaided eye fails to detect these differences. However, the phase-shifting plate greatly enhances the optical effect of the difference, and the object appears dimmer or brighter than the surrounding material. The contrast between the two, called **phase-contrast,** is thus made readily visible to the eye, enabling the observer to see objects otherwise not visible to him (Fig. 3–19). Phase-contrast microscopy is especially valuable in the examination of wet-mounts and hanging-drops. It is clearly a great advance in the visualization of cellular structure. For this contribution Fritz Zernike was awarded the Nobel Prize in Physics in 1953.

3.8
TRANSMISSION ELECTRON MICROSCOPY

As pointed out in a foregoing paragraph the resolving power of optical microscopes is limited by the human retina and its physiological restriction to the wavelengths of visible light: between about 400 nm (violet) and 750 nm (red)—relatively very long! Particles with a diameter less than one-half the shortest wavelength of visible light (200 nm), or lines separated by a distance of less than about 200 nm, cannot be resolved by the eye because light waves of the order of 400 to 750 nm can entirely by-pass such minute, individual particles or cannot pass between particles less than 200 nm apart, which thus fail to be visualized clearly, if at all (Fig. 3–20).

By 1930 it had long been known that electromagnetic radiations such as electrons at energy values of 60 kV or more have wavelengths of around 0.5 nm (their wavelength being partly determined by the accelerating voltage), and that, because of their short wavelength, they would yield resolutions over 1,000 times that available with optical microscopes. However, their use in optical microscopes was formerly impossible because glass lenses are opaque to electrons.

The discovery that electrons are deflected from their line of propagation by magnetic fields, and that circular magnetic fields could

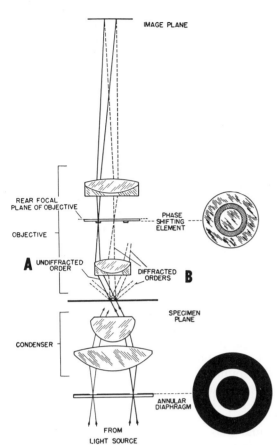

Figure 3–17. Image formation by phase contrast. An **annular aperture** in the diaphragm, placed in the focal plane of the substage condenser, controls the illumination on the object. The aperture is imaged by the condenser and the objective at the **rear focal plane** of the objective. In this plane the **phase-shifting disk** (diffraction or phase plate) is placed.

With the particular plate shown, light waves, A (solid lines: undiffracted order), are **transmitted** through the object and pass through the phase-altering ring on the phase plate. At this point they acquire a one-quarter-wave-length advance over light waves, B (broken lines: diffracted order), which do not pass through the object but are partly **diffracted** around it. Waves (B) do not pass through the phase-altering ring on the phase plate. The resultant **interference** or **resonance** effects of the two portions of light form the final image. Altered **phase** relations in the illuminating rays, induced by otherwise invisible elements in the specimen, are translated into brightness differences (**contrast**) by the phase-altering plate; hence, phase-contrast. (Courtesy of Dr. J. R. Benford, Bausch & Lomb Scientific Bureau, Rochester, N. Y.)

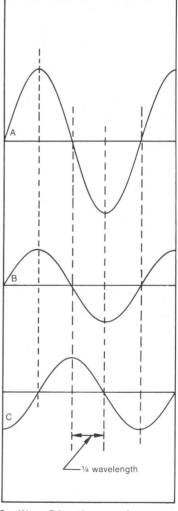

Figure 3–18. Wave B has the same **frequency** (vibrations per unit of time) as A (troughs and crests coincide; i.e., the two waves are **in phase**), though B has less **amplitude** (less depth of troughs and height of crests) then A. Though amplitude and frequency of wave C are the same as B, the crests and troughs of C are one-quarter wavelength behind (or ahead of) crests and troughs of B; i.e., C is one-quarter wavelength **out of phase** with B.

focus electron beams, thus forming electronic images, made possible the use of electron beams as a source of illumination in microscopy. After years of intensive research the first practical electron microscopes were constructed by von Borries and Ruska in Berlin in 1938 and by Hillier and Vance, of RCA in the United States. Commercial instruments first came into general use around 1940. The name of V. K. Zworykin is prominent in all phases of the development of electron microscopes and television scanning tubes (see section on scanning electron microscopy).

From these basic pioneer works the modern electron microscope has evolved (Fig. 3–21). The units of this instrument are analogous to optical units in an inverted compound optical microscope (Fig. 3–22) but deal with electron beams instead of visible light rays.

Operation. The electron source is commonly a tungsten filament at 30 to 150 kV potential. The electron beams pass through the center of the ring-like magnetic condenser (analogous to an ordinary microscope condenser) and are converged on the specimen. After being *transmitted through the specimen,*

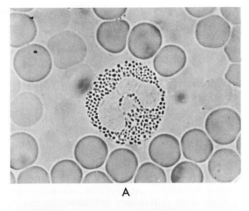

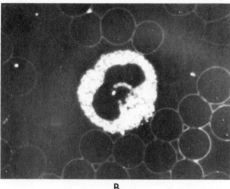

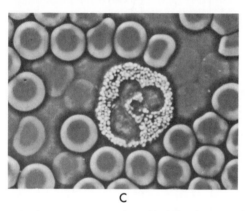

Figure 3–19. Appearance of a white blood corpuscle or leucocyte (eosinophil) (large, granular, central object) and red blood corpuscles or erythrocytes, unstained and magnified about 1,500 times, as seen by three different methods of microscopy: *A,* visible transmitted light (ordinary type of microscopy); *B,* darkfield; *C,* phase-contrast microscopy. (From Scope; courtesy of the Upjohn Co., Kalamazoo, Mich.)

(hence **transmission electron microscopy [TEM]**; cf. **scanning electron microscopy [SEM]**, Section 3.9) the magnetic objective-lens coil focuses the electrons into a first (real) image of the object enlarged; in modern instruments this may be up to 2,000 times. The magnetic intermediate-image projector (analogous to an optical eyepiece or ocular lens) then magnifies a portion of the first image, producing magnifications up to 250,000 or more. The final enlarged image can be viewed by causing it to strike a fluorescent screen which makes it visible. The image can also be thrown upon a photographic plate for permanent record. Portions of the photographs may be enlarged four to six times without undue loss of detail, thus giving pictures in the range of two million times as large as the object (Fig. 3–23). This degree of magnification is inconceivably tremendous. The lower-case letter "o" on this page, say 2 mm in diameter, if magnified 2 million times would be an area 4 million mm (over 2.5 miles) in diameter—large enough for several football fields, complete with stadiums. Direct magnifications up to several millions are available by amplification with an ordinary TV monitor.

Because the motion of electrons is impeded by air, the interior of the electron microscope must be maintained at a vacuum by means of suitable pumps. This necessitates air locks for the insertion and removal of objects and photographic plates. The operator can look into the main tube by means of portholes or magnifying binocular glasses and can scan the images on the fluorescent screen, manipulate the object, and make suitable adjustments of alignment and of field strength of the focusing magnets. Some of the control devices concerned with these details are seen in Figures 3–21 and 3–22.

Under usual conditions of transmission electron microscopy (drying, vacuum, etc.) living organisms cannot survive and physiological processes in live cells cannot be studied because of the destructive action of the stream of electrons. However, in 1972 special apparatus became available for high-energy (100+ kV) transmission electron microscopes that permits observation of cells, viruses, etc., in natural, moist conditions.

Specimens for Transmission Electron Microscopy. Objects like bacteria and viruses may be prepared for examination with the transmission electron microscope by spraying or sedimentation of fluids containing the objects on very thin films of electron-transparent organic materials like collodion. These are supported on small, electroplated wire grids with extremely fine meshes. The films must not only be thin but be free from electron-scattering or -absorbing materials (especially impurities of

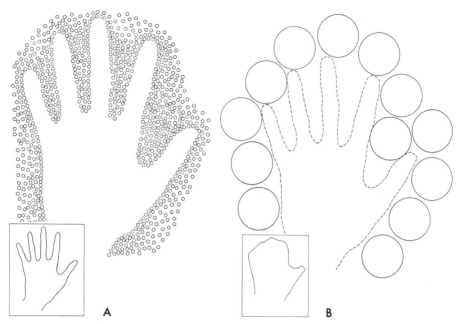

Figure 3–20. Effect of wavelength on resolution. In each sketch the circles represent wavelengths: in *A*, electrons; in *B*, visible light. The difference is actually enormously greater than shown. In *A* the projected image reveals all of the details of the form of the object. Each finger is clearly resolved. In *B* the image lacks detail; the long waves yield poor resolution.

heavy atoms like iron or lead), sufficiently strong to stand necessary manipulation, and stable enough not to volatilize under intense electron bombardment in vacuo.

Staining for Transmission Electron Microscopy. Unlike the images of microorganisms that are produced in optical microscopes by transmitted visible light, electron images are "shadows" produced mainly by the scattering of electrons passing through the specimen. Scattering is produced when electrons encounter atoms. More scattering (denser shadows) is produced by heavy atoms than by light atoms. Some materials may also absorb electrons and thus appear as shadows in electron pictures. Objects that entirely scatter the incident electrons are "opaque" to electrons or "electron dense." They are not necessarily opaque to light, and vice versa.

As in unstained bacteria, **contrast** in electron images of the organic components of cells is very slight. This is because these substances are composed chiefly of atoms of low atomic weight—C, 12; N, 14; O, 16; etc.—which scatter electrons very little; i.e., they are almost transparent to electrons. Contrast may be greatly increased by "positive" staining; i.e., combination of the organic matters in the cell with metals of high atomic weight: Pb, 207; U, 238; Os, 192, etc. These strongly scatter electrons, i.e., they are electron "opaque." Because they have different weights and combine differently with different organic compounds their use permits a helpful degree of selective or differential staining and even tentative identification of certain substances.

Negative Staining. For the electron microscope this is achieved in a manner analogous to that used for optical microscopes. The background is stained with an electron-opaque material, commonly phosphotungstic acid (W; at. wt., 184), which does not penetrate the cells themselves but darkens the background.

Ultramicrotomy. For the study of animal and plant tissues with optical microscopes the tissues are first soaked in appropriate **fixing** (coagulating) solutions, then in water, and afterward in dehydrating fluids and paraffin solvents. They are then impregnated with paraffin and cooled in small "blocks." The blocks, with the embedded tissues, are cut into thin slices (0.05 to 0.2 mm) with a special steel blade in an instrument called a **microtome**. Each thin section is mounted on a glass slide, paraffin is re-

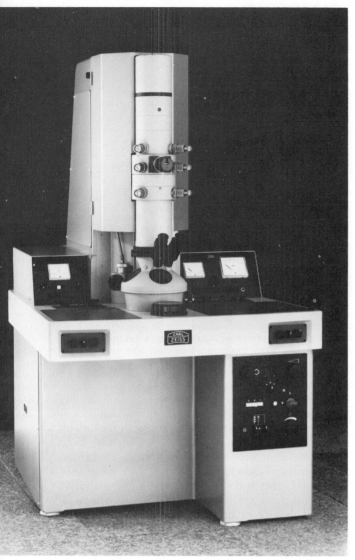

Figure 3–21. A modern electron microscope. (Courtesy of Carl Zeiss, Inc., Oberkochen, Württemberg.)

microtome for cutting such materials the knives are usually the edge of a fragment of broken glass or of diamond.

The ultramicrotome must be specially mounted to eliminate irregularities due to vibrations of the building, slippage on films of lubricating oils in the instrument, variations due to a few degrees of change in temperature, and so on.

Shadowing. Topographical details of surfaces of objects, or the objects themselves, can be accurately reproduced by various replica techniques. Dilute collodion (e.g., Formvar) is poured over the film to be examined and allowed to dry. The topographical details are accurately molded in the plastic film or **replica.** The film is removed and cleaned, and a thin coherent film of carbon vaporized in a vacuum is deposited on it. The replica is then placed on a supporting grid and cleaned with an appropriate solvent, and vaporized metal (e.g., gold) is deposited on the surface in a vacuum at an oblique angle (Fig. 3–25). Faces of raised portions of the object that are toward the source of vaporized metal will be more heavily coated with the electron-dense deposit than areas facing away from the source of vapor and hence will appear as highlights in the final picture; the less heavily coated areas will appear as dark "shadows." The length, form, and locations of the shadows permit three-dimensional measurements of the objects to be made (Fig. 3–26).

Operational Problems. Magnetic lenses suffer from difficulties similar to those found in glass lenses, though of electrical rather than of refractive origin; for example, **spherical** and **chromatic aberration** caused by irregularities in "refraction" (i.e., axial and peripheral electrons come to focus at different distances from the lens); differences in "wavelength" (i.e., electron velocities and energies); and distortions like astigmatism resulting from irregularities in field strengths of the magnets and various extraneous magnetic influences. A constant difficulty is **contamination** of the specimen by volatilized impurities such as oils, solvents, and other materials from various sources within the instrument itself. This is caused by the intense bombardment by electrons within a high vacuum. These contaminations form blurring and obscuring films over the object within a matter of seconds or minutes. Electron bombardment is also destructive to the specimen.

Modern instruments have devices to simplify maintenance of vacuum, align the electron beam and adjust its brightness, eliminate contamination and lessen or eliminate many of the difficulties with older instruments.

moved with solvents, and the tissues are then stained and examined with a microscope.

An analogous procedure is used for making thin sections of bacteria and other organisms for transmission electron microscopy. But since even large bacteria are at most only 3 to 5 μm in diameter and since electrons have so little penetrating power, the sections must be exceedingly thin: of the order of 0.02 μm. For this an **ultramicrotome** is used (Fig. 3–24). The embedding material, instead of paraffin, is at first a liquid plastic (e.g., methacrylate). After mixing with the bacteria it is caused to solidify (polymerize) by warming to about 40C. In an ultra-

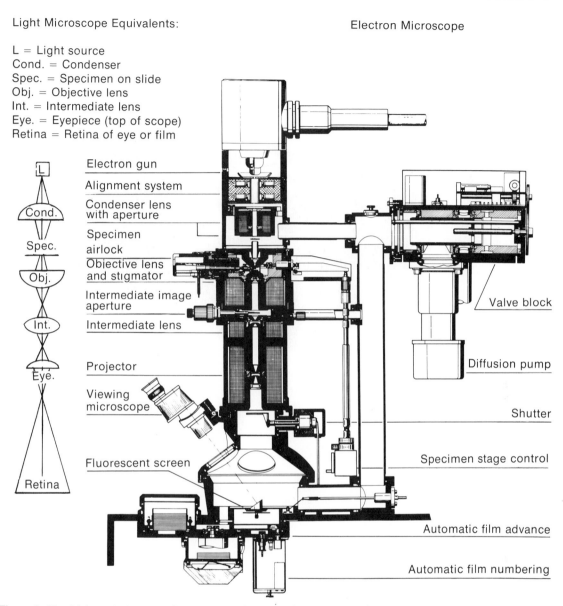

Light Microscope Equivalents:

L = Light source
Cond. = Condenser
Spec. = Specimen on slide
Obj. = Objective lens
Int. = Intermediate lens
Eye. = Eyepiece (top of scope)
Retina = Retina of eye or film

Electron Microscope

Electron gun
Alignment system
Condenser lens with aperture
Specimen airlock
Objective lens and stigmator
Intermediate image aperture
Intermediate lens
Projector
Viewing microscope
Fluorescent screen

Valve block
Diffusion pump
Shutter
Specimen stage control
Automatic film advance
Automatic film numbering

Figure 3–22. Light and electron microscope equivalents. (Courtesy of Carl Zeiss, Inc., Oberkochen, Württemberg.)

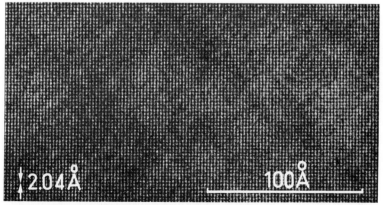

Figure 3–23. Lattice image of single gold crystal. In this astonishing picture details of the molecular lattice only 0.204 nm (about 1/127,000,000 inch) apart (opposing arrows) are clearly resolved. The white line at lower right shows the extent of 10 nm (1/2,540,000 inch) at this enormous magnification. (5,500,000×) (Courtesy of JEOL USA, Medford, Mass.)

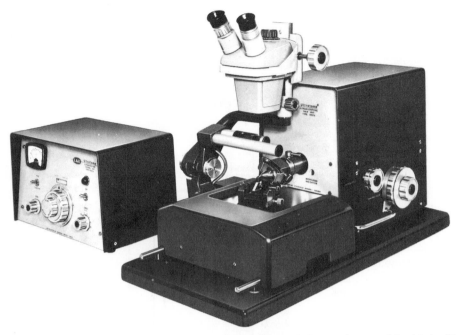

Figure 3–24. This ultramicrotome is designed with systems to protect both the specimen and the blade. (Courtesy of LKB Instruments, Inc., Rockville, Md.)

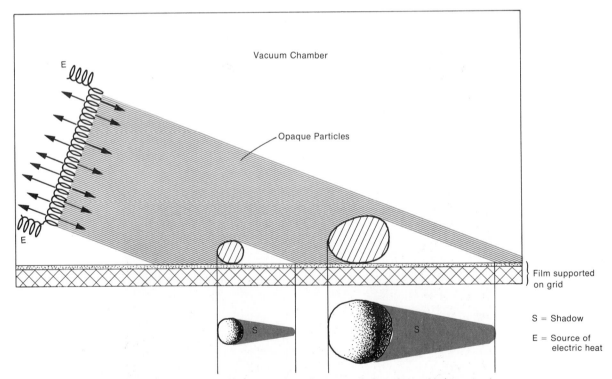

Figure 3–25. Shadowing technique for transmission electron microscopy.

3.9
SCANNING ELECTRON MICROSCOPY (SEM)

Scanning electron microscopes combine the mechanisms of electron microscopy and television. SEM became commercially available only in the early 1960's after over three decades of intensive researches by Knoll, von Ardenne, Zworykin, Oatley, Hayes, Pease, and many others. SEM differs from TEM in that, in SEM, the electrons are not transmitted through the very thin specimen from below but impinge on its surface from above. The specimen may be opaque or electron dense and may be of any manageable thickness and size. If the specimen is an electron conductor it need only be held on an appropriate support by a conducting adhesive. If a nonconductor, as is true of most biological specimens, it is allowed to dry or, if moist, is freeze-dried in liquid nitrogen at −198C. The water is sublimated at approximately −70C and the specimen then coated with metal vapor (e.g., gold) in vacuo. Thickness of the coating (5.0 nm to about 30 nm) determines completeness of coverage. This affects detail and resolution because only the coated areas can yield a final image. For some purposes, specimens may be fixed and stained by conventional techniques before coating.

The electrons originate at high energy (20,000 V) from a hot tungsten or lanthanum hexaboride cathode "gun." These electrons are sharply focused, adjusted, and narrowed or demagnified (not demagnetized!) by an arrangement of magnetic fields and lenses functionally analogous to the condenser of an optical microscope (Fig. 3–27). These, like other lenses, require arrangements for correction of spherical and chromatic aberrations and exclusion of disturbing peripheral rays. Instead of forming a broad, inverted cone of rays illuminating a wide field, as in optical microscopes or transmission electron microscopes, the electrons are made to form a needle-sharp **probe**, about 5.0 nm to 10 nm in diameter. This primary beam or probe acts only as an exciter of image-forming secondary electrons that emerge from the surface of the specimen; no focusing is necessary except to sharpen the probe.

The probe scans the specimen in a raster pattern like that on a blank TV screen. It can impinge on depths and heights with equal speed and accuracy, giving great depth of field and producing images with three-dimensional or perspective effect. Because of its fineness and the short wavelength of electrons, the probe can enter deep, narrow openings and can resolve details with dimensions of 10 to 20 nm, far beyond the resolving power of optical microscopes and over 500 times the depth of field of TEM, whose depth of field is limited by the shallowness of the very thin specimens necessarily used for it. Images (electronic signals) are elicited from wherever the probe strikes the metal-coated areas of the specimen. Magnification is the ratio of size of final image to the diameter of the area scanned. Since the probe "sees" an area only about 5 to 10 nm in diameter at any moment of its scanning, it cannot, of itself, produce any broad view of the specimen as a whole. The final image is derived from a succession of secondary electrons as follows.

Impact of the high-energy probe beam on the specimen causes some of the probe electrons to enter the surface to depths of the order of about 0.5 nm to 20 or more nm, the depth depending on the nature of the specimen and the energy of the electrons. Of those that enter, some directly activate secondary, relatively low-energy, electrons. Other entering electrons are "backscattered" or, as it were, reflected, especially by heavy atoms, with energy only slightly diminished. These can activate additional secondary electrons within the specimen.

Figure 3–26. Electron micrographs of a virus (bacteriophage). *A*, Without metal shadowing. *B*, With metal shadowing. (Courtesy of Dr. D. Gordon Sharp, Duke University.)

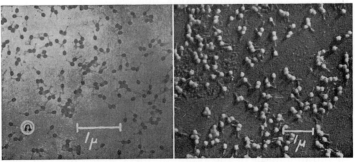

A   *B*

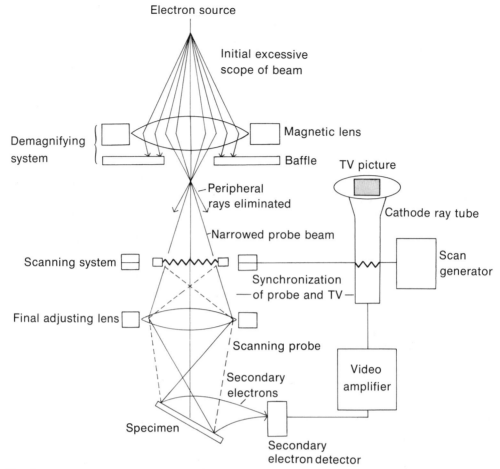

Figure 3–27. Scanning electron microscope. Electrons from the cathode source are accelerated through a grid and grounded anode, the whole assembly constituting the "electron gun." The electron beam, too broad at first, is reduced, or demagnified, by magnetic lens systems. Passing through the deflector yoke of a scanning coil system, and a final lens system, the now-much-narrowed and adjusted beam scans the sample in a raster pattern. Secondary electrons from the sample are collected by the detector. They are transmitted to a video amplifier and into the TV cathode ray tube, the scanning rate of which is synchronized with that of the probe of the microscope. Thus any given point on the TV screen, at any moment, represents a point on the specimen. The final image is really a series of pictures of different points on the sample, each about 10 nm in diameter or the diameter of the probe, seen in such rapid succession as to provide, for the eye, a unified view of the entire surface of the sample.

Any of the secondary electrons with sufficient energy, and originating not too deeply within the specimen (about 0.5 to 20 nm), can emerge from the surface. Those that emerge not too far from the point of impact of the probe can be used to form an image. Widely dispersed (peripheral) secondary electrons can cause blurring of the image and are screened out. The useful secondary electrons are magnetically deflected to a **collector** or **detector.** Here they produce a beam or signal that represents, at any single moment, only the 5- to 10-nm area or spot of impingement of the probe on the specimen.

The **successive** signals from the collector are amplified and transmitted to a cathode ray (TV) tube. The scanning beam and the TV tube beam are synchronized (Fig. 3–28). The image seen by the eye on the TV screen is thus a sequence of signals representing, in a raster pattern, the successive areas traversed by the primary probe beam (Fig. 3–29). Usually the rate of oscillation of the probe is slowed to allow time for adequate concentrations of electrons to develop an image of the specimen at each point. Exposure times may range from a few seconds to one-half hour or more. The TV image may be photographed, video-taped or processed in motion on a computer.

Freeze-Etching. Microorganisms prepared for ultramicrotome sectioning by chemical treatment and embedding in plastic may show distortions and unnatural changes due to the preparatory processes. These artifacts can be eliminated to a great extent by freezing the specimen instantaneously in ice at temperatures of −198C or lower, and then cutting. The ice is then removed from the specimen by sublimation in a vacuum, leaving the cut cell surfaces with structural details revealed in depth by the removal of the water. Shadowed carbon-plastic replicas of such cut surfaces prepared by methods mentioned above and examined with a scanning electron microscope, yield three-dimensional views of the interiors of cells.

Character of Image. The character of SEM pictures varies, depending on the nature of the specimen and the manipulation of the instrument. Commonly, things are what they seem, but certain types of work require some experience to make correct interpretations of the images.

Contrast in SEM images represents, in part, uneven generation of secondary electrons when surfaces are sloped or curved at angles of 30 to 60° from the axis of the probe; more secondary electrons are engendered by sloping or curved surfaces than by flat surfaces. Backscattered, high-energy, electrons produce excessive contrast, obscuring details. Other variable sources of contrast are: temperatures, angle of the probe, alterations in the probe beam by closely adjacent or overlying objects, and so on.

3.10
OTHER METHODS OF MICROSCOPY

For research and various advanced applied purposes, other methods of microscopy are available. These need not be detailed here but two will be mentioned, with references to literature for those interested in extending the range of human vision "far back among the atoms and electrons."

X-ray Microscopes. X-rays provide a source of even shorter wavelengths than electrons and may be used for various optical purposes. They also have a much greater penetrating power than electrons and thus can serve to make observations which are impossible with

Figure 3–28. A scanning electron microscope, with the electron column, sample port, and vacuum equipment in the left console; controls and cathode ray tubes for viewing in the center console; and a digital counter for various specialized applications to the right. (Courtesy of JEOL USA, Medford, Mass.)

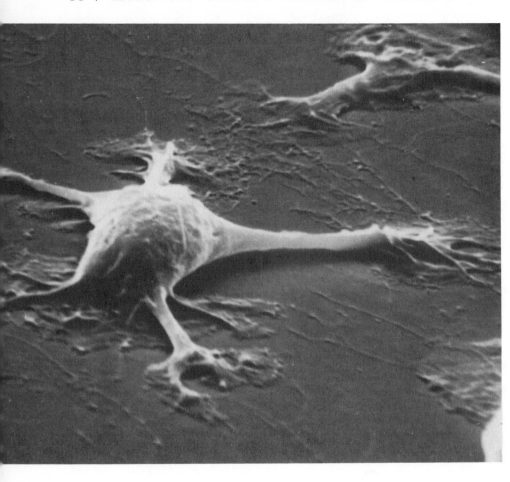

Figure 3–29. This scanning electron micrograph of a macrophage from the peritoneum of a mouse demonstrates the three-dimensional effect of SEM images. (17,000×.) (Courtesy of R. Albrecht.)

electrons. Construction and operation of satisfactory x-ray microscopes for general use are still in the pioneering stage.

Interference Microscopes. An additional means of studying transparent objects like living cells is the interference microscope. This combines a double-beam interferometer and a polarizing microscope. The object is seen as in phase contrast but the polarization and wave interference produce greater contrast—and color. It is also possible to measure optical path differences between different parts of the object.

CHAPTER 3
SUPPLEMENTARY READING

Everhart, T. E., and Hayes, T. L.: The scanning electron microscope. Sci. Am., 226:55, 1972.

Hayat, M. A.: Principles and Techniques of Electron Microscopy, 2 vols. Van Nostrand Reinhold Co., Cincinnati. 1972.

Hearle, J. W. S., Sparrow, J. T., and Cross, P. M.: The Use of the Scanning Electron Microscope. Pergamon Press, Inc., New York. 1972.

Hooke, R.: Micrographia. 1665. Dover Publications, Inc., New York. 1961.

Lillie, R. D., et al.: H. J. Conn's Biological Stains, 8th ed. The Williams & Wilkins Co., Baltimore. 1969.

Martin, L. C.: The Theory of the Microscope. Elsevier Publishing Co., Inc., New York. 1966.

Seeger, R. J.: Galileo Galilei, His Life and His Works. Pergamon Press, Inc., New York. 1966.

Wren, L. A., and Corrington, J. D.: Understanding and Using the Phase Microscope. Unitron Instrument Co., Newton Highlands, Mass. 1963.

CHAPTER 4 · CHEMICAL BASIS OF MICROBIAL LIFE

Atoms, Ions, Isotopes

All cells, however they may differ in physical and physiological characteristics, are composed of essentially identical elements combined in similar or identical compounds which, with small modifications from species to species, constitute similar or homologous cellular structures. Chemical reactions of the same general (often identical) types underlie the processes by which foodstuffs are utilized in different species of cells as sources of energy and of cell substance. And, finally, it is now known that all of the variations in structure and physiology, including enzymic activities, that characterize different species can be ascribed to relatively small variations in the structure of certain complex molecular groups that are common to all cells; i.e., the heredity-determining nucleic acids. Thus virtually all of the phenomena of life may be explained in terms of molecular biology and referred ultimately to the molecular structure of nucleic acids: deoxyribonucleic acid (DNA) and ribonucleic acid (RNA).

4.1
STRUCTURE OF ATOMS

A knowledge of molecular structure is essential to a modern appreciation of biology, especially microbiology. Therefore we shall, for the convenience of the reader, recapitulate some basic concepts concerning atoms, molecules, their structure and their roles in the phenomenon that we call life. It is assumed that the reader has at least a speaking acquaintance with physics and chemistry.

Molecules, as the reader will doubtless remember, are made up of atoms. Each atom, in turn, consists of a heavy, central nucleus containing a distinctive number of positively charged **protons** and a variable number of electrically neutral particles called **neutrons.** The nucleus is surrounded by one or more negatively charged particles called **electrons** that encircle the nucleus in various inner and outer orbits.

Under usual conditions the negative charges on the electrons of an atom are balanced by the positive charges of the protons in the nucleus of that atom: i.e., the numbers of protons and of electrons are alike. Such atoms are electrically neutral. The number of electrons (or of protons, since their numbers are equal) in any given type of atom is called the **atomic number** of that atom. The number of protons plus the number of neutrons is the **atomic weight** of that atom, since protons and neutrons each weigh one unit. Because the electrons weigh about 1/1,800 as much as protons or neutrons, they contribute little to the atomic weight of an atom.

Atomic weight is a relative quantity, usually expressed as the number of times heavier a given atom is than an atom of some other element taken as a standard. This is now commonly carbon (6 protons + 6 neutrons = at. wt. 12; 6 electrons = at. no. 6). The hydrogen atom is unique in having in its nucleus only one proton and no neutron; atomic weight = 1.00+, expressed as 1 **dalton** (for John Dalton, 1766–1844, English physicist, originator of Dalton's law of partial pressures of gases in mixtures).

Electrons occur in concentric shells or orbitals with different energy levels surrounding the nucleus. Each shell contains, at most, a certain specified number. The configuration of electrons in the shells, especially in the unfilled shells, accounts for the chemical properties of a given atom. The most energetic electrons available for the formation of bonds with other atoms or radicals are called **valence electrons** and their number is the valence number (**valence**) of that atom or element. Because the electronic configurations recur in atoms of increasing weight, different elements can be arranged in a **periodic table**. The elements thus are grouped in families with similar chemical properties.

4.2
IONIZATION

Substances whose aqueous solutions readily conduct electricity are called **electrolytes.** Very active ones in this respect are said to be strong electrolytes. Generally these are inorganic compounds: salts like sodium chloride, acids like sulfuric acid, bases like potassium hydroxide. Except carbon monoxide, carbon dioxide, carbonates, and cyanides, **inorganic** compounds characteristically lack carbon, carbon-to-carbon, or carbon-to-hydrogen bonds. These bonds are always found in **organic** compounds, typically *not* strong electrolytes.

Substances whose aqueous solutions do not readily conduct electricity are called **nonelectrolytes.** They are generally organic compounds. Organic colloids like protein and fat emulsions are virtually nonelectrolytes. Organic acids like citric (of lemons), acetic (of vinegar), and lactic (of sour milk) are weak electrolytes; that is, their aqueous solutions conduct electricity to a slight degree.

Electrolytes have several other properties that distinguish them from nonelectrolytes. Compared with nonelectrolytes, electrolytes markedly affect the colligative properties of water: i.e., they (a) lower the freezing point,

(b) raise the boiling point, (c) increase the osmotic pressure, and (d) lower the vapor pressure of their aqueous solutions.

Dissociation. Electrolytes are electrically conductive because their molecules in aqueous solutions split (**dissociate**) into positively and negatively charged particles called **ions.** They dissociate in this way because: (a) pure water has a high dielectric constant; i.e., it has practically no conductance for electricity and (if pure!) is therefore a very good insulating material. (Actually it has a very slight though measurable conductance that is extremely important. See under pH.) (b) Water molecules are **polar** because in each molecule the oxygen has attracted the electrons of the hydrogen unequally, giving it unbalanced negativity. These seemingly unrelated facts are easily integrated as follows.

If an electrolyte, for example HCl, ($H^+ Cl^-$) is dissolved in water $\left(\begin{array}{c} H^+ \\ \\ H^+ \end{array} \!\!\! \diagdown\diagup \; O^{2-} \right)$ the positively charged hydrogen ion (H^+) (a **cation**) attracts around itself swarms of water molecules, each with its negative pole (oxygen) nearest the H^+ (Fig. 4–1). In a like manner, on electrolytic dissociation the negatively charged chlorine ion (Cl^-) (an **anion**) attracts other water molecules with their positive poles (hydrogen) nearest the Cl^-. The two ions (H^+ and Cl^-) are thus well insulated from each other by the dielectric action of the water molecules. The H^+ and Cl^- move about freely in the water. If two oppositely charged, chemically inert electrodes from a battery (e.g., platinum electrodes) are immersed in the solution at opposite sides of the vessel, any cations in the solution migrate to the negative electrode (the **cathode**), accepting electrons therefrom; any anions in the solution migrate to the positive electrode (the **anode**), yielding electrons thereto and thus demonstrating the conductivity of the solution. Each pair of H^+ ions, accepting electrons, becomes a hydrogen molecule, and passes off in hydrogen bubbles at the cathode. Each pair of Cl^- ions, yielding electrons, becomes a chlorine molecule, and passes off in bubbles at the anode: an example of **electrolysis.** Various other reactions may occur at the electrodes, depending on the electrolyte in solution, the nature of the electrodes, the solvent, the voltage applied, and other factors. The water itself is concomitantly decomposed into $2H_2$ and O_2. If the forces of attraction between ions in aqueous solution are great enough to overcome the insulating effect of the water mole-

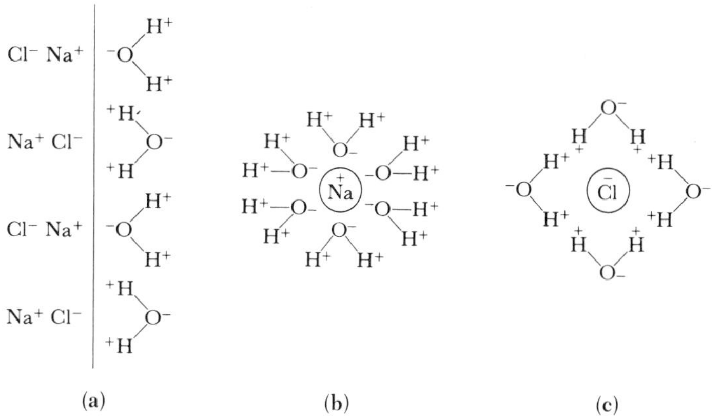

Figure 4–1. Dissociation of sodium chloride in water.

(a)   (b)   (c)

cules, then those ions will combine or react together.

Radicals (Groups). A **radical** is a group of atoms held together by strong bonds but containing an excess (or deficiency) of electrons. The group commonly acts as an ion, the valence number of which depends on the net charge of the group; e.g., ammonium, valence $= +1$ (NH_4^+); hydroxyl, valence $= -1$ (OH^-); nitrate, valence $= -1$ (NO_3^-); sulfate, valence $= -2$ (SO_4^{2-}); phosphate, valence $= -3$(PO_4^{3-}).

Hydrogen and Hydroxyl Ions

Hydrogen and hydroxyl ions (H^+ and OH^-) are extremely important in biology because they are the constituents of water and the basis of the properties of acidity and alkalinity. These influence all aspects of cell life. Degree of **acidity** is commonly expressed as **pH,** with a number representing the concentration of hydrogen ions, since it is ionized hydrogen that determines the immediate acidic activity of any solution. **Alkalinity** of solutions may be expressed as concentration of hydroxyl ions (**pOH**). However, since concentrations of hydrogen ions and of hydroxyl ions are reciprocally related, as will be shown, degrees of either are commonly given in terms of pH only.

pH. Since the concentration* of H ions determines acidity, acids or alkalies may be strong or weak, depending on their degree of dissociation. This is always a fixed value for any given electrolyte and is generally expressed as the **dissociation** or **ionization constant** (K_a).

---

*Physical chemists prefer to relate pH to the chemical activity of H ions; at practical concentrations, however, activity is essentially equal to concentration.

Strong acids are those which, when dissolved in water, dissociate largely into positively charged hydrogen ions and negatively charged ions. For example, sulfuric acid dissociates into two hydrogen ions and a sulfate ion. Weak acids like acetic or citric also dissociate, but to a lesser degree. The acidic activity of any acid solution depends upon the concentration of ions of hydrogen, and this is obviously dependent upon the ability of the acid to give them off into the solution or to dissociate. Thus, two acid solutions may be of the same concentration with respect to the total amounts of hydrogen available, yet have widely differing activities due to differences in the amount of active or ionized or dissociated hydrogen. Here we deal with a capacity effect, i.e., total available (dissociated plus undissociated) acid, as contrasted with an intensity or activity effect (dissociated acid or hydrogen ions alone).

As an example, let us compare acetic acid and hydrochloric acid. A liter of a normal solution of each contains exactly 1 gm of total available hydrogen, yet the activity of the N/1 acetic acid is slight, while that of the N/1 hydrochloric acid is great. Of the gram of available hydrogen in the acetic acid solution only 1.36 per cent is in an ionized state, so that there is, in the liter of solution, only 0.0136 gm of hydrogen ions. The gram of hydrogen in the liter of N/1 HCl solution is 91.4 per cent ionized, giving 0.914 gm of hydrogen ions per liter. The N/1 HCl, therefore, is about 67 times as active or "strong" as the N/1 acetic acid.

If one were to titrate the solutions, i.e., add N/1 NaOH solution until each became neutral, the total amount of alkali required would be the same in each case. A measurement of hydrogen ion concentration differs from such a titration, in that the former determines the actual concentration of **ionized** hydrogen at the moment, without calling out any of the reserve, **undissociated** acid.

In acidimetry the term "normal" refers to the presence of 1 gm of **total available** hydrogen per liter (dissociated* plus undissociated). By contrast, a solution normal only with respect to **ionized** hydrogen contains 1 gm of hydrogen ions per liter. This implies the presence of 1 gm-equivalent of a completely (100 per cent) dissociated acid; a N/10 solution would contain 0.1 gm-equivalent of a completely dissociated acid, and so on. (See Table 4–1.)

As shown in Table 4–1, if we were to express hydrogen ion concentrations or normality in terms of grams of hydrogen ions per liter we should have to deal with long words and long rows of zeros, a confusing and laborious system of nomenclature. In 1909 Sorensen devised a simpler system based on the fact that water is itself a very weak electrolyte. As noted previously, the extent of dissociation of any electrolyte is a physical constant (K_a) for that electrolyte under standard conditions. A liter of pure, neutral water at 20C always contains 0.0000001 gm (1×10^{-7} moles) of H^+ and, reciprocally, since water is actually $H^+ + OH^-$, 0.0000017 gm (1×10^{-7} moles) of OH^-. In Sorensen's system, the term "grams of hydrogen ions per liter" is replaced by the symbol

---

*The concept of hydronium (hydrated hydrogen) ions (H_3O^+) in the dissociation of water, while indispensable in advanced studies, is omitted here to simplify discussion.

pH, while the number of moles of H^+ per liter (1×10^{-7} in the neutral water under discussion) is expressed as the logarithm of the reciprocal of the fraction, i.e., the positive number 7. The reaction of neutral water, and of any neutral solution, is therefore expressed as pH 7.

Now the product of the concentration of H^+ and of OH^- in neutral water is always 10^{-14} ($H^+ \times OH^- = K_w = 10^{-7} \times 10^{-7} = 10^{-14}$). Since the product of the two is always the same (i.e., since the two are reciprocally related), the term pH is commonly used to express either. For example, the pH of a solution containing 1 gm of H^+ per liter (i.e., normal [N/l] with respect to hydrogen ions or 1 gm-equivalent of a completely dissociated acid) is 0 (log 1 = 0). Reciprocally, this is also pOH 14, the smallest fraction of a gram of OH^- per liter possible on the Sorensen scale. Similarly, pH 6 implies pOH 8; pH 2 implies pOH 12, and so on.

Since the number representing pH is derived from a fraction, the larger the fraction the smaller the pH number. Therefore, pH numbers between 7 and 0 represent increasing degrees of acidity, and numbers between 7 and 14, increasing degrees of alkalinity (see Table 4–1). Unless one is familiar with the numbers, they can at first be misleading. For example, a change in pH from 7 to 6 represents a 10-fold increase in concentration of hydrogen ions since the 7 and 6 are logarithms; a change from pH 7.0 to 7.3 represents a 50 per cent decrease in the concentration of hydrogen ions ($1/2 \times 10^{-7} = \log 2 + 7 = 0.3 + 7 = $ pH 7.3).

TABLE 4–1. RELATIONSHIPS OF HYDROGEN ION CONCENTRATIONS EXPRESSED IN VARIOUS WAYS

Reaction	Fraction of Normality*	Hydrogen Ions per Liter (grams)	Logarithms of H Ion Concentrations	Expressed as pH
Acid	N/1	1.0	− 0	0.0
Acid	N/10	0.1	− 1	1.0
Acid	N/100	0.01	− 2	2.0
Acid	N/1,000	0.001	− 3	3.0
Acid	N/10,000	0.000,1	− 4	4.0
Acid	N/100,000	0.000,01	− 5	5.0
Acid	N/1,000,000	0.000,001	− 6	6.0
Neutral	Pure water	0.000,000,1	− 7	7.0
Alkaline	N/1,000,000	0.000,000,01	− 8	8.0
Alkaline	N/100,000	0.000,000,001	− 9	9.0
Alkaline	N/10,000	0.000,000,000,1	−10	10.0
Alkaline	N/1,000	0.000,000,000,01	−11	11.0
Alkaline	N/100	0.000,000,000,001	−12	12.0
Alkaline	N/10	0.000,000,000,000,1	−13	13.0
Alkaline	N/1	0.000,000,000,000,01	−14	14.0

*With respect to hydrogen or hydroxyl **ions**.

Buffers

The majority of chemical reactions in living systems involve addition or deletion of H^+ and OH^- ions. Therefore, pH control within cells is vitally important to assure a suitable environment for the necessary reactions. Evolutionary processes have selected 7.4 as the physiological pH. Usually, extracellular pH is also optimal for growth of microorganisms when maintained at about 7.4. Even if an organism is growing in a medium with a pH different from neutrality, the cell maintains intracellular pH at about 7.4.

Both microbiologists and microorganisms use **buffer systems** to achieve pH control. Solutions whose pH's change little upon addition of strong acid or strong base are said to be buffered. Buffer action ordinarily requires that the solution contain two different components, one that reacts with H^+ ions, the other with OH^- ions. The microbiologist usually uses a weak acid and the salt of that acid in the medium as the components of his buffer system. The acid chosen is one with a pK_a ($pK_a = -\log K_a$) close to the desired pH. Phosphate buffers are commonly used, the two components being Na_2HPO_4 and KH_2PO_4. Another buffer system contains carbon dioxide (which forms carbonic acid when dissolved) and sodium bicarbonate. Both buffer systems are useful in the physiologic range.

The ability of a buffer system to maintain pH control when a large amount of either acid or base is added to the system depends on the **buffer capacity,** which in turn depends on the concentration of the components in the buffer system. Phosphate and bicarbonate buffers are rarely used at concentrations exceeding 0.1 molar and are usually much less concentrated. These systems are useful for maintaining pH control in the face of small additions of acid or base, but their capacity can be exceeded easily.

Microorganisms, like culture media devised by microbiologists, contain inorganic salts and acids and obtain some pH control from these inorganic buffer systems. Most control of intracellular pH, however, comes from certain organic molecules, such as amino acids and proteins, that have different regions on a single molecule, some of which can react with acid and some with base. Thus, a single protein molecule, which contains a large number of amino acids, can have a considerable buffering capacity. Buffering capacity in a cell is high because of the high intracellular concentration of proteins, amino acids, and other organic molecules.

4.3 ISOTOPES

All atoms of a given element have the same number of electrons and the same number of protons to balance those electrons. Atoms thus remain electrically neutral. But not all atoms of a given element necessarily have the same number of neutrons; i.e., different atoms of a given element may have different numbers of neutrons and therefore different atomic weights. Such atoms are called **isotopes.**

In any given sample of an element there is usually a mixture of isotopes. The international atomic weight of any element is listed as an average of the weights of all the various isotopes normally (commonly) present. For example, the weight of a single atom of normal (the most common isotope of) chlorine is 35. But any considerable quantity of the element also contains about 25 per cent of heavy chlorine, ^{37}Cl. The international atomic weight of chlorine as an element is therefore given as 35.457. Similarly, the atomic weight of normal hydrogen is 1, but the international atomic weight is 1.00814. Three isotopes of hydrogen are known: ^1H, mass (at. wt.), 1.00814; deuterium (^2H), mass, 2.01474; tritium (^3H), mass, 3.01701. Tritium is radioactive.

There are isotopes of nearly every element; not all are radioactive. In the periodic table all of the isotopes of an element are placed together, since the position of each element in the table depends only on the number of its protons (at. no.), not its nuclear structure (at. wt.). Isotopes of an element are **chemically** similar since it is mainly the electrons that determine the chemical properties. However, since isotopes have different numbers of neutrons, they may differ markedly in **physical** properties because these are affected by their atomic weights. For example, of the four principal carbon isotopes ^{11}C, ^{12}C, ^{13}C, and ^{14}C one (^{11}C) is lighter and two are heavier (^{13}C and ^{14}C) than the "normal" ^{12}C carbon atom.

Radioactivity. Nuclei of atoms of certain isotopes are unstable and tend to decay or disintegrate spontaneously by releasing energy (γ-rays) or part of the nucleus (α- or β-particles). A γ-ray is a form of energy resembling x-rays; α-particles are part of the nucleus, composed of 2 protons and 2 neutrons and thus identical to helium nuclei; β-particles are electrons released from the nucleus with high energy and therefore high speed.

The nucleus of an atom of a radioactive isotope has a certain probability, characteristic for that isotope, that it will decay at any instant. The probability remains constant, independent of chemical reactions the atom may undergo, and independent of the temperature or other physical conditions of the atom. Because the probability is constant for all atoms of a given isotope, in a specified period of time a constant fraction of the nuclei will have decayed. The **half-life** of an isotope is that period of time during which half the radioactive atoms originally present will have decayed.

^{14}C is commonly used in biological experiments (see below). When ^{14}C decays, β-particles are released. The half-life of ^{14}C is 5,770 years, plenty of time for even a slow microbiologist to do his experiments. ^{15}C is not used in biology: it releases a β-particle, but its half-life is only 2.25 seconds, not allowing much time for an experiment. Tritium (^3H) is useful also; it emits a β-particle upon decay, and its half-life is 12.3 years. Biochemists and microbiologists have also used many other radioactive isotopes (^{32}P, ^{15}N, ^{131}I, etc.) in their studies of metabolic pathways.

Uses of Isotopes. Isotopes are of special interest to the biologist because he can use them to study many biological phenomena, such as determining exactly what a living cell does with any given element or compound that it takes in with its food; for example, carbon assimilation. By the use of carbon dioxide made with ^{14}C ($^{14}CO_2$) it has been shown that, in photosynthesis, carbon dioxide is not combined directly with any organic compound in the cell. It is first reduced by transfer of electrons of hydrogen split from water (Chapt. 32). The carbon (^{14}C) is then found in a number of compounds formed during the synthetic process; e.g., phosphoglyceric acid:

$$^{14}CO_2 \xrightarrow[\text{synthesis}]{\text{Photo-}} \begin{array}{c} H_2C\!-\!O\!-\!PO_3H_2 \\ | \\ HC\!-\!OH \\ | \\ ^{14}COOH \end{array}$$

3-phosphoglyceric acid

If $H_2{}^{18}O$ is provided in photosynthesis, all of the free oxygen given off is $^{18}O_2$. No ^{18}O is found in the synthesized products. In many species of chemosynthetic bacteria carbon dioxide combines directly with already formed organic molecules. For example, Werkman, Wood, and their colleagues have "fed" carbon dioxide made with ^{14}C to propionic acid bacteria. The radioactive carbon can be followed, like a tracer bullet, through its course in the synthetic processes of the bacterial cell. For each molecule of ^{14}C one molecule of succinic acid or propionic acid is formed, probably from pyruvic acid via oxalacetic acid, or from glycerol:

$$\begin{array}{c} H_2COH \\ | \\ HCOH \\ | \\ H_2COH \end{array} + {}^{14}CO_2 \rightleftharpoons \begin{array}{c} COOH \\ | \\ CH_2 \\ | \\ CH_2 \\ | \\ {}^{14}COOH \end{array} + H_2O$$

glycerol   succinic acid

A common soil saprophyte, *Aerobacter indologenes*, produces acetic, lactic, and succinic acids as waste products of glucose utilization. When $^{14}CO_2$ is added, ^{14}C is found in these acids. *Proteus vulgaris*, another common saprophyte of soil and water, produces lactic and succinic acids. The $^{14}CO_2$ is found mainly in the carboxyl group of the acids. The pathway to the formation of these acids is via pyruvic acid according to the **Wood-Werkman reaction:**

$$\begin{array}{c} CH_3 \\ | \\ C\!=\!O \\ | \\ COOH \end{array} + {}^{14}CO_2 \rightleftharpoons \begin{array}{c} {}^{14}COOH \\ | \\ CH_2 \\ | \\ C\!=\!O \\ | \\ COOH \end{array}$$

pyruvic acid   oxalacetic acid

CHAPTER 4
SUPPLEMENTARY READING

Bigeleisen, J.: Chemistry of isotopes. Science, *147*:463, 1965.

Kasting, R., and McGinnis, A. J.: Radioisotopes and the determination of nutrient requirements. Ann. N.Y. Acad. Sci., *139*(Art. 1): 98, 1966.

Wang, C. H., and Willis, D. L.: Radiotracer Methodology in Biological Science. Prentice-Hall, Inc., Englewood Cliffs, N.J. 1965.

CHAPTER 5 · CHEMICAL BASIS
OF MICROBIAL
LIFE

Structure of Compounds

Compounds are substances consisting of atoms of two or more different elements in fixed proportions and steric relations, held together by chemical bonds. Molecules may be combinations of atoms of the same element (e.g., H_2, Cl_2, O_2) or of different elements (e.g., CH_4). A molecule is that minimal amount of a substance which, if further subdivided, ceases to be that substance.

5.1
CHEMICAL BONDS IN COMPOUNDS

The manner in which atoms are bonded together is of critical importance in all biochemical reactions. Three kinds of bond may be briefly considered here: electrovalent, covalent, and hydrogen.

Inorganic Compounds: Electrovalence. An electrovalent bond is formed by the **transfer** (not sharing) of one or more electrons from the outer shell of one element, say sodium, to another, say chlorine. Sodium has a single outer electron; it lacks seven of the eight electrons needed for stability of the outer shell. Chlorine has seven outermost electrons; its outermost

shell is one electron short of complete stability. Each atom would gain greater stability by combining with the other. The sodium tends to give up its single electron; the chlorine tends to saturate its outer electron shell by accepting the electron. Each becomes ionized thereby. The two ions therefore combine, the sodium transferring its electron to the chlorine, to form sodium chloride (NaCl), common table salt. The bond holding the two ions together is electrostatic and is therefore called an **electrovalent** or **ionic bond.** Such bonds are not very strong. Electrovalent bonds are typical of **inorganic** compounds. Since electrovalence results in aggregations of ions, the compounds thus formed are actually **ion aggregates,** not true molecules. However, the term molecule is commonly used for them. Unlike "true" molecules (see following material on covalence) ions held together by electrovalence are readily separated; i.e., such compounds readily ionize or dissociate in aqueous solutions.

Organic Compounds: Covalence. By **covalence** is meant the merging or linking together by **sharing** unpaired outermost electrons of two atoms so that the shared electrons are attracted by each nucleus. **Paired** inner electrons such as the two in the k (inner) shell of carbon are very strongly held by the nucleus and are not shared. It is the **unpaired** outer (valence) electrons that tend to be shared. In the carbon atom there are four unpaired outer (valence) electrons; in hydrogen, one; in nitrogen, three; in oxygen, two. When electron pairs are shared (not transferred), the shared electrons are strongly held by each nucleus. Each pair of shared electrons is called a **covalent bond** or **electron-pair bond:** a strong and stable type of bond. Covalent bonds are most common in organic compounds. If two or three pairs of electrons are shared between two atoms, double and triple covalent bonds, respectively, are formed (see previous discussion of carbon). Instead of being stronger, however, double and triple bonds are progressively less stable.

Among the most important covalent bonds are those formed by the removal of the elements of water from two adjacent molecules: a hydrogen ion from one, a hydroxyl ion from the other. The two ions combine to form water. The residues of the two altered molecules combine by a covalent bond in a **condensation** reaction.

For example:

$$H_3C-\overset{H_2}{C}-O\boxed{H+HO}\overset{\overset{\displaystyle O}{\|}}{-C}-CH_3 \rightleftharpoons$$

ethyl alcohol   acetic acid

$$CH_3\overset{H_2}{C}-O\underset{\uparrow}{-}\overset{\overset{\displaystyle O}{\|}}{C}-CH_3$$

Ester bond

ethyl acetate

Condensation reactions are extremely important in the process of **polymerization,** the mechanism by which complex **macromolecules** (proteins, polysaccharides, nucleic acids) containing thousands of smaller subunits (e.g., glucose, amino acids, nucleotides), are put together by living cells. Condensation reactions are important also because they are readily and usually (in biological systems) formed by action of specific enzymes. Condensation reactions, polymers, and their functions in living cells are more fully discussed farther on.

Condensation Reaction Bonds. Certain bonds formed by condensation reactions, because they are important and frequent in biological systems, are illustrated here.

ESTER BONDS. Between the carboxyl

$$R-\overset{\overset{\displaystyle O}{\|}}{C}-OH$$

group of an organic acid and the hydroxyl group (R—OH) of an alcohol, forming an ester (organic salt): these bonds occur in fats, oils and waxes, and related compounds:

$$R\overset{*}{}-\overset{\overset{\displaystyle O}{\|}}{C}-\boxed{OH+H}O-\overset{\overset{\displaystyle H_2}{}}{C}-R \rightleftharpoons$$

$$R-\overset{\overset{\displaystyle O}{\|}}{C}\underset{\uparrow}{-}O-\overset{\overset{\displaystyle H_2}{}}{C}-R$$

Ester bond

PHOSPHATE-ESTER BONDS. Between phosphoric acid and a hydroxyl group of a sugar, forming a phosphorylated sugar: these bonds are important in polymerization and the storage and transfer of energy in the cell:

$$R-\overset{\overset{\displaystyle H_2}{}}{C}-\boxed{OH+H}O\cdot P\cdot O_3H_2 \rightleftharpoons$$

$$R-\overset{\overset{\displaystyle H_2}{}}{C}\underset{\uparrow}{-}O\cdot P\cdot O_3H_2$$

Phosphate-ester bond

Such bonds, linking phosphoric acid with the nucleoside adenosine, as adenosine monophosphate (AMP), adenosine diphosphate (ADP), and adenosine triphosphate (ATP), and with certain other compounds, are exceedingly important in accepting, storing, and transferring energy from various sources to polymerization and other endothermic reactions in the cell (see Chapters 6 and 8).

THIOESTER BONDS. Between the carboxyl group of an organic acid and a sulfhydryl (—SH) group in a sulfhydryl-containing compound:

---

*R symbolizes any **radical** to which the particular group mentioned (e.g., carboxyl or hydroxyl) is attached. R may range in complexity from CH_3 to an enormously complicated structure.

the thioester bond is important in the storage and transfer of energy in the cell:

$$R-\overset{\overset{\textstyle O}{\|}}{C}-\boxed{OH + H}\,S-R \overset{H_2O}{\rightleftharpoons} R-\overset{\overset{\textstyle O}{\|}}{C}-S-R$$

Thioester bond

PEPTIDE BONDS. Between the carboxyl group of one amino acid and the amino ($R-NH_2$) group of another, forming peptides, polypeptides and proteins: these are basic to the synthesis of living matter:

$$R-\overset{\overset{\textstyle O}{\|}}{C}-\boxed{OH + H}\overset{\overset{\textstyle H}{|}}{-N}-R \overset{H_2O}{\rightleftharpoons}$$

$$R-\overset{\overset{\textstyle O}{\|}}{C}-\overset{\overset{\textstyle H}{|}}{N}-R$$

Peptide bond

GLYCOSIDE BONDS. Between hydroxyl or related groups in simple sugar molecules (e.g., glucose) to form various disaccharides, and polysaccharides (polymers) like starch and cellulose: the glycoside bond is basic to the synthesis of many structural materials and energy-storing compounds in the cell:

$$\rightleftharpoons$$

Glycoside bond (intersaccharidic)

Note that all five reactions are indicated as being reversible. This is because most enzymic reactions are reversible unless too much energy is released by the reaction in one direction (see Chapters 7 and 8).

Hydrolysis. When the synthetic process is reversed, the elements of water are reinstated, and the constituent residues or subunits are separated and released in their original form. This process, called **hydrolysis,** is exceedingly important in biology because it is the basic reaction of virtually all processes of digestion of proteins, fats, polysaccharides (e.g., starches), and many other compounds.

Hydrogen Bonds. As previously noted, hydrogen is unique in having only one electron, one proton, and no neutron. The single proton exerts a strong, electrostatic, attractive force that extends beyond its single electron. Hydrogen therefore tends to exert an unbalanced electropositive effect in certain molecules or groups containing it. Such molecules or groups tend to be polarized. Strongly polarized groups or molecules (dipoles), like magnets, attract other polarized groups or molecules. When sufficiently strong the attraction results in the formation of a hydrogen bond through the sharing of a hydrogen atom (not an electron) between adjacent polarized groups or molecules, especially those containing oxygen or nitrogen in polarized groupings. For example, water molecules as well as ammonia molecules, all polarized groupings, become **associated** through hydrogen bonding:

Similarly, ammonia and hydrogen chloride form ammonium chloride through hydrogen bonding:

$$+\overset{+}{\underset{+}{\diagdown}}H_3N^- + + HCl \rightarrow +\overset{+}{\underset{+}{\diagdown}}H_3N^-$$

$$\cdots +HCl^- \text{ or } NH_4Cl$$

(In the above diagrams hydrogen bonds are indicated by $\cdots$.) In many combinations the proton of hydrogen is actually transferred to the other atom and the bonds are called proton bonds. Hydrogen bonds, although relatively weak, are essential in the phenomenon of inheritance and the reproduction of living cells because they help to support the structure of DNA (Chapter 6).

5.2
STRUCTURE OF ORGANIC MOLECULES

Most of the organic molecules or compounds found in living cells are more or less complex derivatives of one large family of compounds, the alkanes or saturated hydrocarbons (hydrogen + carbon). These are the basis of the great series of **aliphatic** (Gr. *aleiphatos* = from fats or oils, e.g., petroleum or hydrocarbon) compounds. The simplest saturated hydrocarbon is methane (CH_4), principal constituent of marsh and sewer gas. It is said to be saturated because all of its four valences are linked to hydrogen atoms. Hydrogen-carbon and carbon-carbon bonds are characteristic of organic compounds and are rare in inorganic compounds. Except methane, saturated hydrocarbons are straight chains of from two to scores of carbon atoms linked together by covalent bonds, each with its remaining valences saturated by a hydrogen atom:

$$H-\overset{\overset{\displaystyle H}{|}}{\underset{\underset{\displaystyle H}{|}}{C}}-\overset{\overset{\displaystyle H}{|}}{\underset{\underset{\displaystyle H}{|}}{C}}-\overset{\overset{\displaystyle H}{|}}{\underset{\underset{\displaystyle H}{|}}{C}}-\overset{\overset{\displaystyle H}{|}}{\underset{\underset{\displaystyle H}{|}}{C}}-\overset{\overset{\displaystyle H}{|}}{\underset{\underset{\displaystyle H}{|}}{C}}-\overset{\overset{\displaystyle H}{|}}{\underset{\underset{\displaystyle H}{|}}{C}}-H = \text{hexane} = C_6H_{14}$$

Compounds with this type of structure are termed **normal** hydrocarbons. They have the general formula C_nH_{2n+2}.

Simple modifications of this form are seen in branched chains or **isomers** (Gr. *isos* = like; *meros* = part):

$$H-\overset{\overset{\displaystyle H}{|}}{\underset{\underset{\displaystyle H}{|}}{C}}-\overset{\overset{\displaystyle H}{|}}{\underset{\underset{\displaystyle |}{C}}{C}}-\overset{\overset{\displaystyle H}{|}}{\underset{\underset{\displaystyle H}{|}}{C}}-\overset{\overset{\displaystyle H}{|}}{\underset{\underset{\displaystyle H}{|}}{C}}-\overset{\overset{\displaystyle H}{|}}{\underset{\underset{\displaystyle H}{|}}{C}}-H = \text{isohexane} = C_6H_{14}$$

$$H-\overset{\overset{\displaystyle H}{|}}{\underset{\underset{\displaystyle H}{|}}{C}}-H$$

Isomers are compounds having the same empirical formula but different chemical or physical properties. Saturated hydrocarbon isomers also have the general formula C_nH_{2n+2}.

A second modification of the straight chain is the cyclic form in which the two ends of a chain are joined covalently:

$$= \text{cyclohexane} = C_6H_{12}$$

Two of the hydrogens are not needed in such molecules, and thus the general formula of the cyclic form is C_nH_{2n}. This molecule is suggestive of the six-sided ring compound, benzene:

$$= \text{benzene} = C_6H_6$$

Benzene, however, is the basis of the enormous family of **aromatic** compounds, e.g., phenols. For convenience, benzene may be thought of as resembling cyclohexane in form but unsaturated, with three double bonds alternating with single bonds between the carbon atoms. Double and triple bonds between carbon atoms are also found in many hydrocarbons. Such hydrocarbons are said to be **unsaturated**. The double and triple bonds are increasingly unstable and easily broken. Hydrogen atoms of hydrocarbons may be replaced—for example, with hydroxyl or with halogens: i.e., **substitution compounds.**

Functional Groups. The organic compounds that occur in protoplasm generally contain certain groups of atoms that are always linked together and are attached, as a group, by covalent bonds to the molecule of which they are a part. Such groups function as distinct entities. They are spoken of as radicals or functional groups, and they have an electric charge, positive or negative. Their charge is derived from the fact that when the covalent bonds that hold such a group to its molecule are broken, paired, shared electrons are separated. Both the molecular residue and the separated group then have unshared and unpaired electrons and are "free" radicals or ions. These, like electrons, under normal conditions exist free only theoretically or under special conditions. They tend to combine immediately with something else.

The combining valence of a group, radical or ion, depends on the net charge of the group. Examples from inorganic chemistry are the ammonium radical (NH_4^+), valence 1; the sulfate radical (SO_4^{2-}), valence 2; the phosphate radical (PO_4^{3-}), valence 3. A free organic radical would be the methyl group: CH_3^-.

In organic chemistry different substances often have similar chemical potentialities because they contain the same functional groups. For example, glucose, a sugar, has at one end of the molecule a **hydroxyl** (OH^-) group, an arrangement that is characteristic of alcohols or organic **bases**:

$$CH_3 \overset{H_2}{\underset{\|}{C}} OH$$

ethyl alcohol

At the other end is a carbonyl (=C=O) group, an arrangement that is distinctive of **aldehydes**:

$$CH_3 \overset{H}{\underset{|}{C}} = O$$

acetaldehyde

Glucose, therefore, with its curious structure, has the combining properties both of an alcohol and of an aldehyde because its molecule contains those functional groups. Glucose is sometimes called an **aldose** (aldehyde-sugar). In a glucose isomer, fructose (fruit sugar), the carbonyl oxygen is attached to a carbon atom *within* the carbon chain rather than at the end of it:

	D-glucose	D-fructose
1	H\C=O	H₂COH
2	HCOH	C=O
3	HOCH	HOCH
4	HCOH	HCOH
5	HCOH	HCOH
6	H₂COH	H₂COH

When arranged in this way the group is called a **keto group** ($R \overset{O}{\underset{\|}{C}} R$), and sugars containing it are called ketoses. Note that the carbon atoms in the glucose chain are numbered, for reference, from 1 through 6.

If we oxidize the aldehyde group of glucose it changes to the carboxyl group ($R\overset{O}{\underset{\|}{C}} OH$). This is characteristic of all organic acids. These ionize to yield a hydrogen ion. By thus replacing the aldehyde group in glucose with the carboxyl group, we form the **monocarboxylic sugar acid** gluconic acid. If we also oxidize the alcoholic group at the other end of the glucose molecule we form a **dicarboxylic sugar acid** (a saccharic or glycaric acid).

Any basic alcoholic hydroxyl group usually reacts readily with any acidic carboxyl group to form an ester or organic salt, plus water. The formation of an ester is analogous to the reaction between the hydroxyl ion of an alkali (e.g., Na^+OH^-) and the hydrogen ion of an acid (e.g., H^+Cl^-) to form the salt NaCl and H_2O. Although alcoholic OH^- generally acts as a group and acidic $R \overset{O}{\underset{\|}{C}} OH$ generally ionizes to give H^+ as above, in forming an ester it is the hydroxyl (not the H ion) of the $\overset{O}{\underset{\|}{C}} OH$ group that combines with the alcoholic hydrogen:

$$CH_3 \cdot \overset{O}{\underset{\|}{C}} \boxed{OH + H}O - CH_2 \cdot CH_3 \longrightarrow$$

$$CH_3 \cdot \overset{O}{\underset{\|}{C}} O - CH_2 \cdot CH_3 + H_2O$$

This illustrates a difference between the organic hydroxyl group and the inorganic hydroxyl ion. It also illustrates the principle that the way an atom or group acts depends to a great extent on what other atoms it is combined with.

Using the glucose molecule again as our model, we may replace the hydroxyl group of carbon atom 2 with an amino group (—NH$_2$), producing **glucosamine:**

CH$_2$OH

glucosamine

CH$_2$OH

D-glucose

(Haworth cyclic formula)*

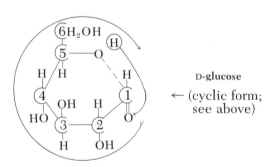

D-glucose

← (cyclic form; see above)

Glucosamine, an **amino sugar,** is found in the polysaccharide **chitin,** the rigid, structural substance in the skeletons of insects and Crustacea and in the cell walls of fungi, Protozoa, and

---

*The straight-chain formula of the aldose is bent, as shown by long, outer, circular arrow, into a closed hexagon by bringing the OH⁻ group of carbon atom 5 into combination with the aldehyde group of carbon number 1 (shorter arrow):

some other protists. The large chemical family of amines, of which glucosamine is only one example, may be thought of as ammonia molecules (NH$_3$) with one or more of the hydrogen atoms replaced by organic groups (e.g., methyl amine: H$_3$C—NH$_2$). The **amino group** (—NH$_2$) is the ammonia residue. If the amino group is attached to the alpha carbon atom (first carbon atom in the chain after the carboxyl carbon atom) of a carboxylic acid, such as propionic acid, we have an **alpha amino acid,** alanine:

$$CH_3\overset{\overset{\displaystyle NH_2}{|}}{C}HCOOH.$$

In alanine and, indeed, in all of the amino acids that occur in protoplasm, the amino group is attached to the **alpha carbon atom,** i.e., next to the carboxyl carbon. Thus the general formula for amino acids that occur in nature is $R-\overset{\overset{\displaystyle NH_2}{|}}{C}H \cdot COOH$. The R may be very simple or very complex and may contain additional amino and carboxylic groups and cyclic and other groups. The amino group is basic and, with the acidic carboxyl group, makes amino acids **amphoteric** (able to react with acids or bases), and thus effective buffers.

Since all amino acids have at least one basic amino group and at least one acidic carboxylic group, they can form bonds between the acidic and basic groups of adjacent amino acid molecules, producing long chains or polymers called **polypeptides,** the basis of protein structure. The bonds between the amino groups and the carboxylic groups are called peptide bonds.

Other molecules with both acidic and basic groups may react similarly. For example, the molecules of a fatty acid like β-hydroxybutyric acid, having acidic carboxyl groups and basic hydroxyl groups, can join, forming ester bonds between the carboxylic and hydroxyl groups of adjacent molecules. Long chains of the polymer poly-β-hydroxybutyric acid (PHB) are formed:

β-hydroxybutyric acid

poly-β-hydroxybutyric acid

Another important functional group in protoplasmic substances is the sulfhydryl or mercapto group (—SH). It may be thought of as analogous to a hydroxyl group in which the oxygen has been replaced by sulfur. The —SH group is found in the amino acids cystine, cysteine, and methionine, all of which are ex-tremely important in the formation of enzymes and other essential structures in the cell. Other important functional groups will be mentioned in the appropriate place.

CHAPTER 5
SUPPLEMENTARY READING

Gray, H.: Electrons and Chemical Bonding. W. A. Benjamin, New York. 1964.

Mulliken, R. S.: Spectroscopy, molecular orbitals, and chemical bonding. Science, *157*:13, 1967.

Pauling, L., and Hayward, R.: The Architecture of Molecules. W. H. Freeman & Co., San Francisco. 1964.

Steiner, R. F., and Edelhoch, H.: Molecules and Life. D. Van Nostrand Co., Inc., Princeton, N.J. 1967.

CHEMICAL BASIS · CHAPTER 6
OF MICROBIAL
LIFE

Macromolecules

Most of the essential substances of living cells are built up of macromolecules. These are made of smaller molecules that are combined into macromolecules mainly by the process of **polymerization** (Gr. *polys* = many; *meros* = parts). By polymerization is meant the formation of compounds of high molecular weight by the combination of many small molecules that have identical or similar structures and functional groups that permit interlinkages. Each of the small molecules entering into any polymerization is called a **monomer** (Gr. *monos* = single). When two such molecules are joined chemically, the result is a **dimer;** when three are joined, they constitute a **trimer.** A combination of more than three molecules in this way is called a **polymer**—a macromolecule.

6.1
POLYMERS

Polymerization is one of the commonest synthetic reactions occurring in living cells, and living cells consist largely of various kinds of polymers. If all of the units in a polymer are identical, the polymer is called a **homopolymer** (Gr. *homos* = the same). Examples of homopolymers are starch, made of α-glucose units,

and cellulose, made of β-glucose units—two very different substances, both made of glucose isomers that differ slightly (Fig. 6–1). If the units in a polymer are related but not identical substances, for example the various amino acids in proteins (see Protein Synthesis, this chapter), the substance is said to be a **heteropolymer** (Gr. *hetero* = different). This illustrates the important point that in any given polymer all the **linkages** between the units are identical, though the **units** may be very different (as in a heteropolymer), provided they have the appropriate functional groups to form the linkage characteristic of that kind of polymer, e.g., polysaccharide or polypeptide.

Polymerization. There are two general types of polymerization: (a) **addition polymerization** or **polymerization by association** and (b) **polymerization by condensation.**

Addition Polymerization. Addition polymerization commonly occurs spontaneously or under the activating (catalytic) action of oxygen or a few ions of hydrogen or hydroxyl or some metal. Addition polymerization is one of the important processes in chemical evolution. Nothing is added to or removed from any one of the combining monomers. Thus, the empirical formula of an addition polymer (if a homopolymer) is theoretically a multiple of the em-

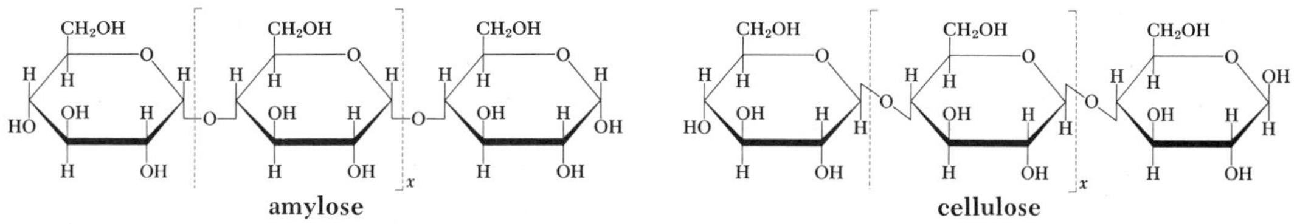

Figure 6–1. Starch and cellulose formulas, showing marked similarities.

pirical formula of the monomer; actually, the terminal groups may differ slightly.

A simple example of addition polymerization is **aldol addition** or aldol condensation. (Do not confuse aldol condensation with condensation polymerization.) Molecules of aldehydes like formaldehyde and acetaldehyde (with alpha hydrogens), in the presence of H^+ or OH^-, can link together to form dimers, trimers, and long polymers with high molecular weights (Fig. 6–2). By addition polymerization, formaldehyde forms solid **paraformaldehyde;** molecules of acetaldehyde may polymerize to form cyclic **metaldehyde:**

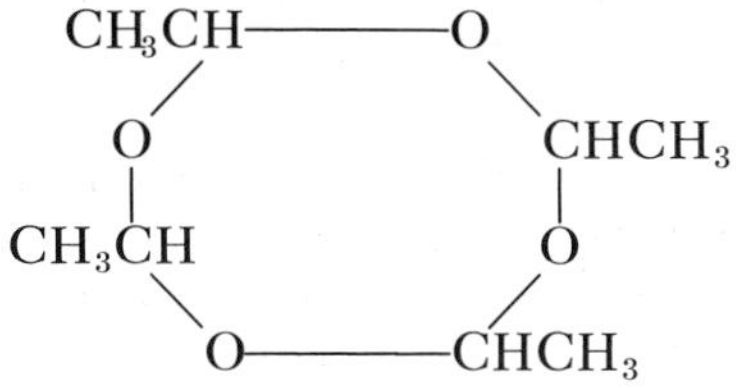

Ketones, alkenes, and alkynes can also undergo addition polymerizations. It is evident that many complex and varied substances can be formed by simple addition reactions between relatively simple substances and under **prebiotic** conditions (spontaneously). Most addition polymers are easily depolymerized by heat.

Condensation Polymerization. In a condensation polymerization, each monomer is combined with the next by removing H^+ from one and OH^- from the other, forming water and causing the two residues to combine (polymerize). Unlike addition polymers, the empirical formula of the polymer formed by condensation is not a simple multiple of that of the monomer, since x number of molecules of water

have been withdrawn in the process of polymerization.

Polymerization by condensation generally involves at least two kinds of enzymes (Chapt. 7): one or more to activate the units and one to withdraw the H^+ and OH^- and cause the residues to combine. The process also requires energy. This is derived from elsewhere in the cell via the ADP–ATP and similar energy-transfer systems (Chapt. 8) and is used in forming the bonds between the units of the polymer.

Energy Mechanisms in Polymerization. Condensation polymerization, whether of glucose, amino acids, or other units, requires an input of energy; i.e., polymerization is an endothermic process. The energy is derived ultimately from biooxidations of foodstuffs (Chapt. 8).

Energy derived from foodstuffs is transferred first into the phosphate bonds of ATP and similar compounds and stored there. These phosphate bonds are said to be energy-rich. When hydrolysis of the ATP occurs during the process of polymerization, the energy is released.

In polymerization the energy released by breaking the energy-rich phosphate bonds of the ATP is not lost. For example, in the polymerization of glucose molecules, each hexose molecule (or other monomer) is first activated or energized by combination with one energy-rich phosphate group. The transfer of the phosphate to the monomer is achieved via several intermediate steps involving transphosphorylases and phosphokinases (Gr. *kinema* = activity) or **synthetases.** In the case of starch formation glucose-1-phosphate is formed by uridine diphosphate (UDP) via uridine triphosphate

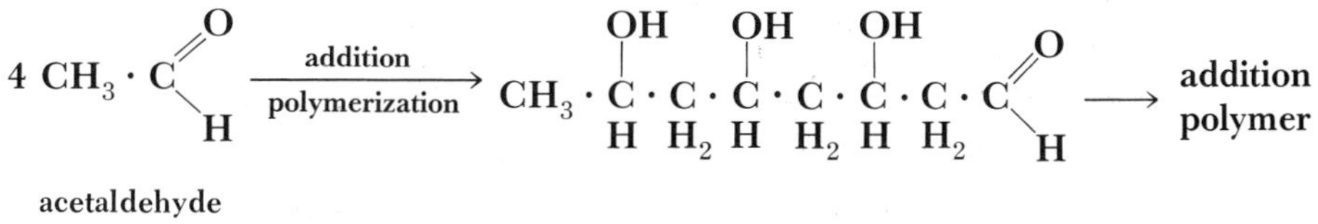

acetaldehyde

Figure 6–2. In addition polymerization, molecules of acetaldehyde link together to form long polymers.

(UTP). UDP and UTP constitute a nucleotide energy-transferring system working via, and analogous to, ADP and ATP. As each glucose-1-phosphate unit is added to the polymer chain, UDP is liberated and the energy of the phosphate bond goes into the making of the glycoside linkage between the glucosyl groups. UDP then receives energy-rich phosphate from ATP, becoming UTP; ATP becomes ADP. The ADP then receives new phosphate from substrate-level and oxidative-chain phosphorylations (and from photophosphorylations in photosynthetic cells) and is re-energized as ATP.

The role of ATP-ADP and UTP-UDP in the formation of starch furnishes an illustration of one common way in which energy from foodstuffs is transferred, by means of energy-rich bonds, into polymers. With some modifications to be discussed, the starch model holds also for the synthesis of proteins, RNA, DNA, and lipids as described in the following paragraphs.

6.2
MONO-, DI-, TRI-, AND POLYSACCHARIDES

A sugar consisting of single molecules (e.g., glucose) is called a **monosaccharide.** Some sugars consist of two sugar molecules joined by glycosidic linkages. They are called **disaccharides.** Examples of familiar disaccharides are **lactose** (a molecule of glucose linked with a molecule of galactose), **sucrose** or ordinary cane sugar (one molecule of glucose and one of fructose), and **maltose,** the sugar in malted barley (two molecules of glucose) (Fig. 6–3). Three-unit sugars are called **trisaccharides.**

In plants "higher" than fungi the carbohydrate polymers starch and cellulose are among the most important structural and nutritional substances. Starch consists of masses of long chains of alpha glucosyl residues. By releasing glucose on hydrolysis, starch functions as a reserve food, furnishing both energy and carbon. In the formation of cellulose, beta glucosyl residues are used instead of alpha residues. The strength of wood depends greatly on the tensile strength of the covalent bonds in the chains of glucosyl units. These cellulose fibrils are bundled together lengthwise with a tough "glue," a complex substance called **lignin.**

In all of the "higher" plants the cell walls are cellulose. Masses of these, with lignin, constitute wood. In many bacteria the cell walls are heteropolymers made up of several amino acids and amino sugars. They are more fully discussed in the section on cell walls (Chapt. 14). Glycogen (animal starch), found mainly in animal livers and only rarely in plants, is also a homopolymer of glucose, but the polymer chains are branched.

Several polysaccharides of biological importance are heteropolymers. Some of these, conjugated with lipids (lipopolysaccharides) and other substances, are major constituents of the cell walls of many species of bacteria. Most important of all, perhaps, are the polymers made up of phosphated molecules of the **pentoses** (sugars with five linked carbon atoms) **ribose** and **deoxyribose.** Ribose and deoxyribose phosphate polymers form the very backbone of the chains of nucleotides that make up the nucleic acids RNA and DNA described farther on in this chapter. These control and direct all heredity and all of the enzymic activities and resulting properties of every living cell.

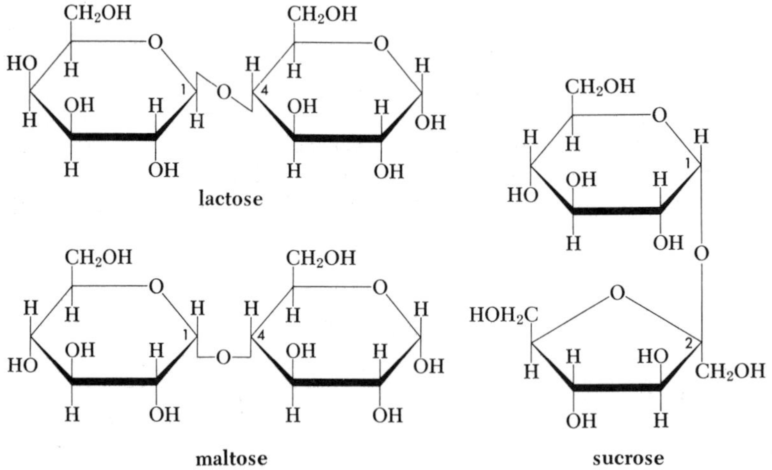

Figure 6–3. Three disaccharides. These are common and important components in and sources of energy for living cells.

6.3
LIPIDS

This term includes fats, oils, waxes (not petroleum waxes), some fatty acids, and polymerized fatty acids. Nearly all are complex esters or similar compounds of the higher fatty acids. Lipids are characteristically insoluble in water but soluble in organic solvents like ether, acetone, alcohol, and chloroform. They are important constituents of cells, especially of cell walls and membranes.

Most natural fats are mixtures of triglycerides. A glyceride is a molecule of the tribasic (three replaceable —OH groups) alcohol, glycerol, combined by ester linkages with three molecules of fatty acid: commonly palmitic, stearic (both saturated), and oleic or linoleic (both unsaturated). Usually there are two or three different fatty acids per molecule (Fig. 6-4). Melting points of fats are determined by the fatty acids. The longer-chain fatty acids, e.g., carnaubic acid (24 carbon atoms), produce tallow-like and harder fats and waxes. Fats that are fluid at about 22C are commonly called "oils," though the term oil has no exact chemical meaning.

Esters, including lipids, are produced by reversible acid-base reactions; i.e., the withdrawing (or restoring) of the elements of water (hydroxyl from the acid, hydrogen from the alcohol) and combination of the acid and alcohol residues as an ester. When fats are hydrolyzed in the presence of alkali (e.g., potassium hydroxide) **saponification** occurs (Fig. 6-5). The glyceryl receives the hydroxyl group; the fatty acid forms a salt with the potassium and becomes a soluble soap (e.g., toilet soap). Soaps of calcium (e.g., formed in hard water) and of most heavy metals form insoluble curds. In living cells fats serve mainly as reserve fuel globules. (See also poly-β-hydroxybutyric acid.) In higher animals fats in large deposits also serve as thermal insulation.

Some waxes that are constituents of living cells are fatty acid esters of complex, mono-

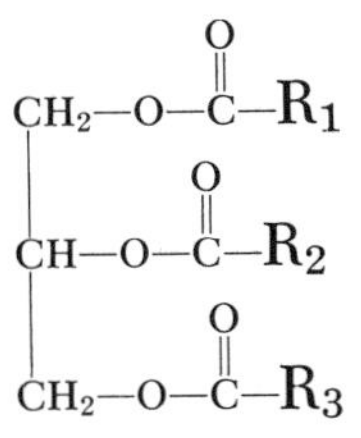

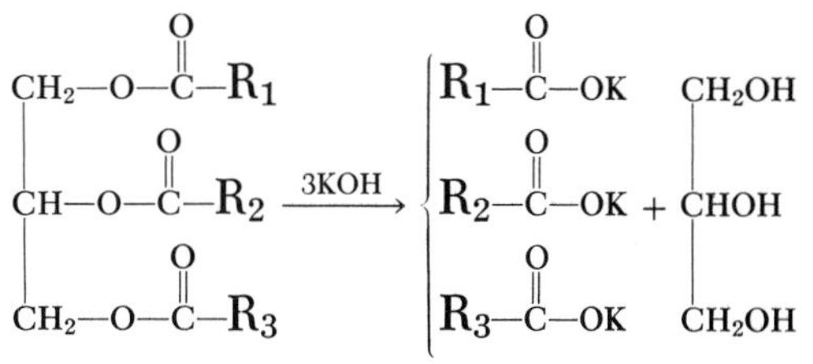

Figure 6–5. Saponification of a triglyceride. Note that glycerol is liberated by saponification of any triglyceride.

hydric (one —OH group) alcohols other than glycerol. Some of these waxes, such as **cholesterol esters,** are important in the physiology of higher animals; e.g., humans sometimes worry about cholesterol in the blood and arterial walls. In microorganisms, however, waxes are of lesser (known!) importance. In one group, *Mycobacterium*, which includes the tubercle bacillus, waxes are present in considerable amounts, though their exact physiological function has not yet been fully clarified. Waxes are important in methods of medical diagnosis (see acid-fast stain) and many confer some pathogenicity on acid-fast organisms like the tubercle bacillus.

Lecithins. Many triglycerides are not true fats but **lecithins.** Unlike fats, lecithins have a phosphate group attached to glycerol, with two fatty acid groups. Thus, lecithins belong to the group of **phospholipids.** Various additional groups may be attached to the phosphate group: inositol (the **phosphoinositides**), ethanolamine (the **cephalins**) and so on. Many bacteria produce enzymes (**lecithinases**) that destroy lecithins. Several important diagnostic tests are based on this property.

Lipoproteins. A large number of compounds called **lipoproteins,** and others called **proteolipids,** that occur in living cells are lipids combined with proteins by linkages that are very little understood. Lipids are also often combined with carbohydrate complexes or polysaccharides (**lipopolysaccharides**). Lipoproteins are important constituents of the cell membranes of many types of cells, including important species of bacteria. They also occur in the outer coverings of some viruses that infect animals. In viruses the lipoproteins confer special and distinctive properties that will be mentioned in the section on viruses. In certain bacteria such as typhoid bacilli the lipopolysaccharides of the cell walls, complexed with polypeptides, have important pathogenic properties to be mentioned later (see endotoxins).

One of the most important fatty-acid com-

Figure 6–4. Formula of a triglyceride.

pounds is poly-β-hydroxybutyric acid (PHB), a polymer of β-hydroxybutyric acid:

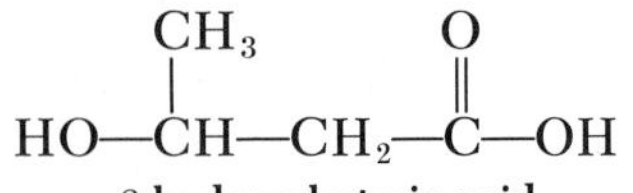

$$
\underset{\beta\text{-hydroxybutyric acid}}{HO-\overset{\displaystyle \overset{CH_3}{|}}{CH}-CH_2-\overset{\displaystyle \overset{O}{\|}}{C}-OH}
$$

The units are joined by ester linkages between the alcoholic hydroxyl group of one molecule and the carboxyl group of the next, a molecule of water being eliminated.

Poly-β-hydroxybutyric acid is found as granules and globules in the cells of a number of important and common species of bacteria. It serves a purpose like that of fat or starch, or inorganic metaphosphates (**volutin**), i.e., as a reserve food. Like starch, fat, and volutin, it is insoluble and compactly stored as a polymer but readily hydrolyzed to release soluble, readily metabolizable, individual units: β-hydroxybutyric acid. Similarly, soluble glucose is released from insoluble starch, soluble phosphorous compounds from insoluble volutin, fatty acids and glycerol from water-insoluble fats and oils.

In the synthesis of lipid polymers, energy for the linkages is derived from ATP as in polysaccharide synthesis, though not by direct formation of energy-rich phosphate bonds with the monomers. In lipid polymers the energy appears in an energy-rich acyl ($R-\overset{\displaystyle \overset{O}{\|}}{C}-R$) bond introduced into the fatty acid by combination with coenzyme A (CoA) (see Chapter 8, material on the tricarboxylic acid cycle). The acyl group is then transferred from the fatty acid, leaving the acyl-bond energy. The energy

is used in forming the ester linkage between the activated fatty acid and the alcohol (or alcoholic group) with which it combines.

6.4
NUCLEIC ACIDS

Unlike carbohydrates, lipids and proteins, nucleic acids (NA's) contribute little as energy sources or structural members of cells. Nucleic acids are like the pilothouse of a ship, contributing nothing as fuel or cargo, hull or superstructure, yet they guide the entire vessel, its contents and crew. As the name implies, NA's were formerly thought to occur only in the nucleus of cells. As will be explained, some NA's are strictly intranuclear, some are not. In the nucleus (or the procaryotic nucleoid) NA's are loosely combined (conjugated) with proteins as **nucleoproteins**. The NA's are easily separated from their protein conjugates by mild hydrolytic processes. Structurally, NA's are long-chain heteropolymers. Unlike polymers made up of simple molecules like glucose or amino acids, NA's are made up of complex units called **nucleotides**: i.e., they are **polynucleotides.** Each nucleotide in NA is made up of three components: one molecule of phosphoric acid, one of the pentose ribose or deoxyribose and either a purine base (**adenine or guanine**) or a pyrimidine base (**cytosine, thymine, or uracil**) (Fig. 6–6). Alternate forms of cytosine (**methyl cytosine or hydroxymethyl cytosine**) may occur in some forms of nucleotides (see farther on). The nucleotide units are joined by phosphate diester bonds between alternating phosphate and sugar groups: sugar-phosphate-sugar-phosphate-sugar and so on. To each sugar group a purine or pyrimidine

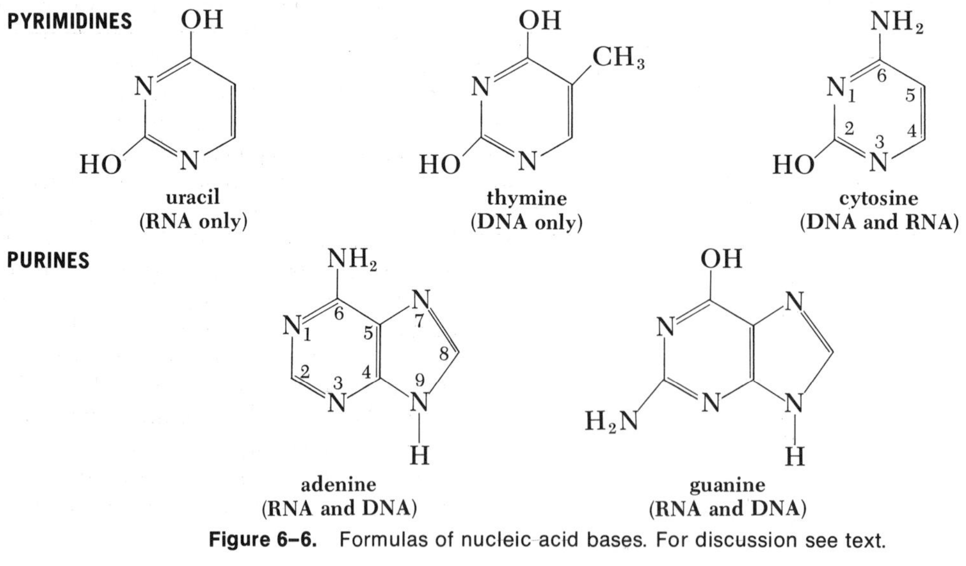

Figure 6–6. Formulas of nucleic acid bases. For discussion see text.

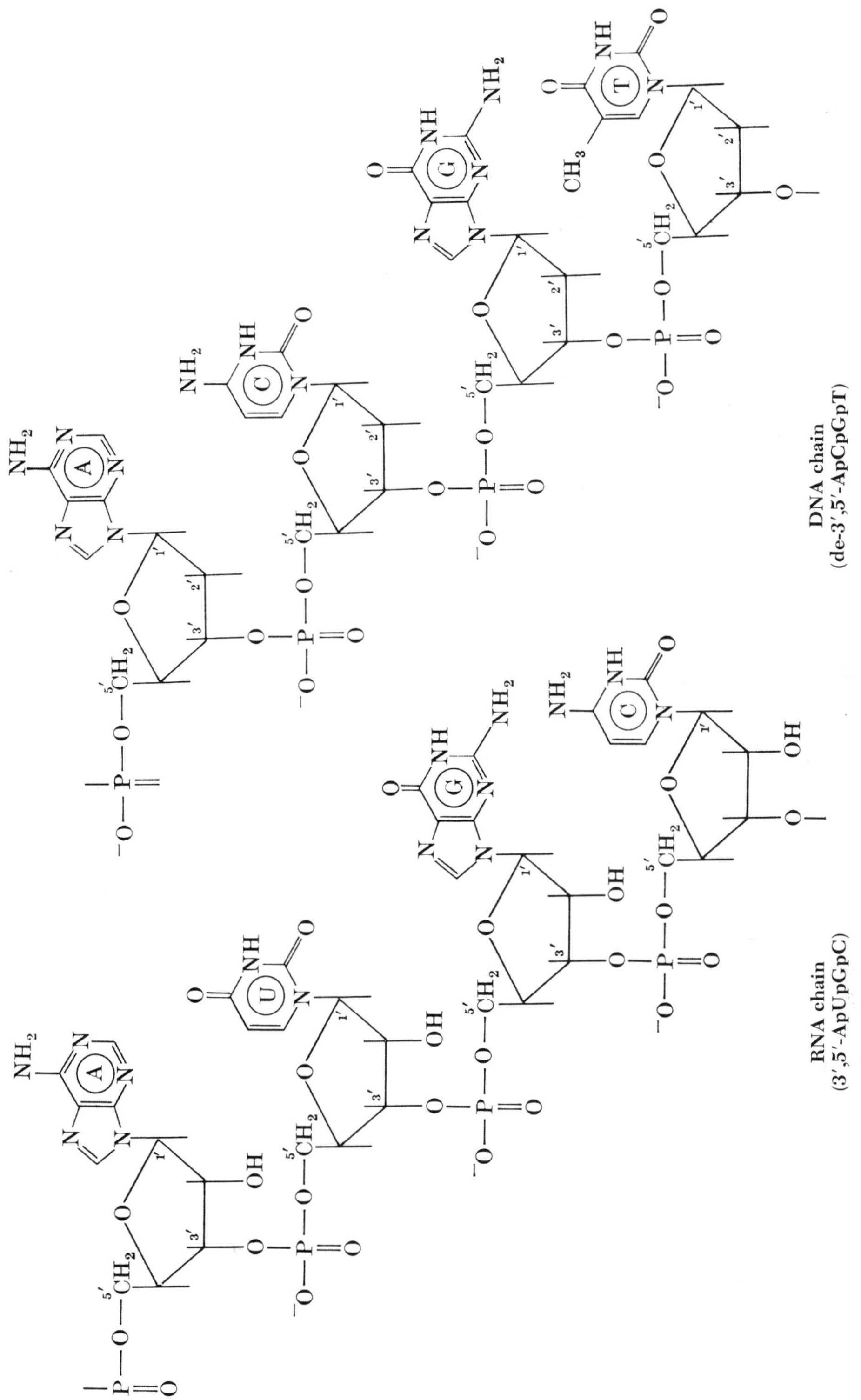

Figure 6-7. Structure of sections of RNA and DNA chains. (A) = adenine; (T) = thymine; (G) = guanine; (C) = cytosine; (U) = uracil.

base is attached by covalent C to N bonds (Fig. 6–7). The chains of nucleotides in NA's are immensely long. In mammalian cells the number of nucleotides per chain is thought to be of the order of $10^9 = 1$ billion.

DNA and RNA. There are two kinds of NA. One, ribonucleic acid (RNA), is so called because its sugar moiety is the pentose **ribose.** In the other, the sugar moiety is very similar but has one less oxygen atom than ribose and is called **deoxyribose.** This NA is called DNA. In RNA the purine bases may be adenine or guanine and the pyrimidine bases, cytosine or uracil. The bases in DNA are the same except that thymine occurs in place of uracil.

The nucleotides (base + pentose + phosphate) in RNA are adenylic acid (or adenosine-5'-monophosphate, familiar as AMP and in NAD), guanylic acid, cytidylic acid and uridylic acid; in DNA: deoxyadenylic, deoxycytidylic, deoxyguanylic and deoxythymidylic acids. All are **nucleoside phosphates.** While discussing nucleotides it is of interest to note that nucleotides other than adenylic acid also serve as coenzymes. Among these are uridylic acid or uridine-5'-monophosphate or UDP (see starch synthesis); guanylic acid or guanosine-5'-monophosphate or GDP (see Chapter 8, material on the tricarboxylic acid cycle). Adenylic acid is also found in FAD and CoA (see Chapter 8, material on biooxidation and the tricarboxylic acid cycle).

The full story of synthesis of the NA macromolecules logically begins with synthesis of the nucleotides and of the ribose, purines, and pyrimidines of which they are composed. These details are not essential to the present discussion; we stipulate the ready-formed nucleotides. The formation of the diester bonds linking the pentose-phosphate groups of the various nucleotides into a long chain, the backbone of NA, occurs by the familiar condensation reaction. The free nucleotides occur as triphosphates. In the presence of a NA polymerase enzyme system they join, each giving off pyrophosphate, leaving the energy of the phosphate bond in the polymer chain.

DNA, as is generally known, is the substance of genes, chromosomes, and inheritance. In all "higher" cells DNA is confined to the chromosomal structures inside the nuclear membrane. In all procaryotic cells DNA is confined to the chromosomal **equivalent,** the fibrillar nucleoid material which is *not* enclosed in a nuclear membrane. In both "higher" as well as in procaryotic cells RNA appears in both nuclear and cytoplasmic regions. As will be explained, RNA does not determine heredity, but it acts as an intermediary between the DNA of the nucleus and all synthetic functions in the cell, which are carried on mainly in the cytoplasm. In any true virus, there is only one kind of NA, either RNA or DNA, never both. (See Chapters 15, 16, and 17.)

The importance of DNA or RNA in all forms of life can hardly be overstated. In all forms of life it is the DNA (or RNA) of the genes and chromosomes that undergoes the alterations called **mutations.** Mutations, with natural selection, are the basis of the great pageant of evolution of all forms of life on this earth. An understanding of the structure, replication, and alterations in NA's is the key to the *real* "facts of life."

Structure of DNA. The DNA macromolecules that constitute genetic material are intertwined **pairs** of very long chains (strands) of nucleotides. The two nucleotide polymer chains are connected by weak hydrogen bonds between opposite or complementary purine and pyrimidine bases, much as the two uprights of a ladder are connected by rungs. In the famous Watson-Crick model of DNA, the ladder-like structure is twisted into a double helix (Figs. 6–8, 6–9).

A most important feature of DNA structure is the **pairing** of the purine and pyrimidine bases to each other. Adenine (A) of one chain is always connected by two hydrogen bonds with thymine (T) of the complementary strand; guanine (G) always is connected by three bonds with cytosine (C) (Fig. 6–9). The pairings occur in this way because this diminishes the energy associated with the structure to the greatest extent; hence, they are thermodynamically directed. The two chains of nucleotides in the double helix of DNA "run" in opposite directions: i.e., they have opposite polarity; the deoxyribose residues are oppositely oriented. The pairs of bases, while always A to T and G to C, are highly irregular as to numbers and sequences. Any number or sequence is possible. Since the chains of DNA may contain many thousands of nucleotide units, it is evident that the number of possible combinations and arrangements of nucleotide pairs is virtually unlimited; we may have a sequence of pairs such as G—C, A—T, A—T, C—G, G—C, T—A, with many other sequences, almost ad infinitum. The helical form is maintained by electrostatic attractions between adjacent nucleotides.

Genetic Significance of DNA. The DNA of every species of cell has its own distinctive and specific arrangement of groups of nucleotides. A **genetically functional group** of few to many, even thousands, of nucleotides consti-

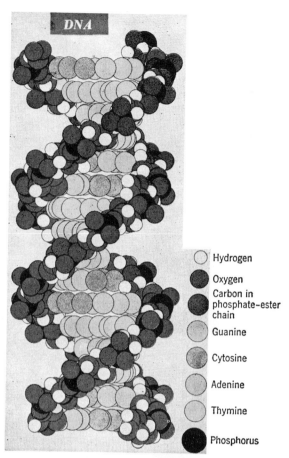

○	Hydrogen
●	Oxygen
●	Carbon in phosphate-ester chain
○	Guanine
◐	Cytosine
○	Adenine
○	Thymine
●	Phosphorus

Figure 6–8. A drawing of the DNA molecule, using solid circles to illustrate atoms. It can be seen that there are two helical grooves of unequal size on the outside of the DNA molecule.

the process of duplication of the DNA double helix begins, the H-bonds between the A—T and G—C base pairs are broken enzymically at a particular point in the double helix; the two individual helices separate at that point and are held apart by a special protein. This leaves unattached the H-bonding sites of the purine or pyrimidine bases in each separate chain, A, T, G, or C. Therefore, each nucleotide residue in each separate strand of the unraveled helix immediately attracts, and combines with, a ready-formed nucleotide from the "pool" of nucleotides in solution in the fluid matrix of the cell; each A combines with a new T nucleotide; each G combines with a new C nucleotide; and so on.

The new nucleotides, thus selected and arranged side by side according to the existing arrangement and sequence of bases in the strand to which they are attached, then are joined enzymically to form a nucleotide polymer by formation of diester bonds between the deoxyribophosphate residues. This process continues along the length of the double helix, probably one short segment at a time, from one end to the other. Thus each strand of the original double helix builds up a new complementary strand attached to itself by H-bonds as before. The nucleotide pairs of the new strand are necessarily in the same order and arrangement as before, since the order and arrangement of nucleotides in the old half of the new double helix was conserved and A could combine only with T, and G only with C. Thus the original

tutes a **gene.** Within a gene, one of several types of changes in one or more nucleotides can occur, resulting in a **mutation.** There may be several possible arrangements of nucleotides in a given gene that are consistent with life and therefore have persisted in a species through evolution. A given individual of a species may have a different set of genes from another individual of the same species and therefore may have different eye and hair colors, for example. In brief, the helix of DNA is the basis of heredity; indeed, there is no (cellular) life without DNA. It has been called "the thread of life" (Kendrew). Not the least remarkable feature of this unique vital machine is the fact that it is self-replicating both inside the cell and outside the cell under laboratory conditions. So far as we are aware, without the replication of DNA life soon ceases.

   Replication of DNA. Very briefly, when

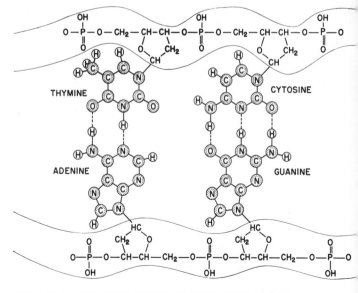

Figure 6–9. Molecular drawing of components of DNA.

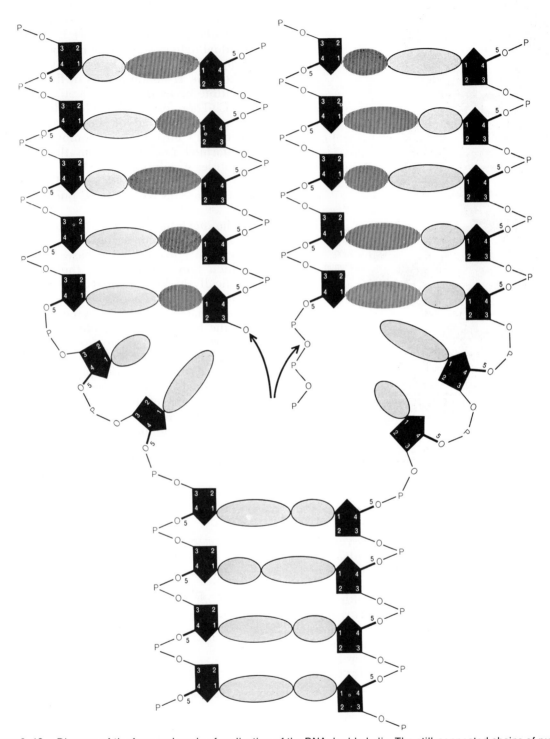

Figure 6–10. Diagram of the form and mode of replication of the DNA double helix. The still-connected chains of nucleotides of the old double helix are at the lower part of the diagram; their separation is at the center. New nucleotides (bases shaded) are joining the nucleotides of the separated right and left parental (outermost) old chains as fast as the old chains separate. Note that pyrophosphate is split from each new nucleotide triphosphate as it joins the nucleotide chain. Note also the opposite polarities of the nucleotide arrangements in the complementary chains.

double helix has now produced two complete, double helices, each an exact replica of the original. Each of the two new double helices consists of one chain of the old double helix and a new complementary chain (Fig. 6–10). This mode of replication of DNA is spoken of as semiconservative.

Chromosome Replication. The mechanism of replication of DNA as a substance, described above in a simplified and superficial way, probably also accounts for the replication of a bacterial chromosome. The bacterial chromosome is a long thread of DNA formed into a closed circle (Fig. 6–11), consisting of thousands of nucleotides. The opened circle of thread is estimated to be about 1 mm (1,000 μm) long. It is packed into a cell only about 2 by 8 μm, and it constitutes a single, doubly helical macromolecule with a molecular weight of the order of two billion daltons.

RNA commonly differs structurally from DNA in being single-stranded. The exact mode of synthesis of RNA is still under investigation. However, double-stranded RNA and single-stranded DNA occur in some viruses (Chapter 15). The single-stranded RNA in cellular organisms is of vital importance in protein synthesis and therefore in inherited characteristics, as explained in the following paragraphs.

In some bacterial viruses the replication of RNA involves the formation, at least transitorily, of two double-stranded forms of RNA in addition to the single-stranded form. One is spoken of as a replicative form (RF), the other as replicative intermediate (RI). The significance of these findings is clearly of fundamental importance though complete elucidation awaits further investigation.

6.5 PROTEINS

Amino Acids; Peptides. Amino acids are organic acids that have an amino group (—NH$_2$) attached to the alpha carbon atom, i.e., the carbon atom next to the carboxyl (—COOH) group (e.g., alanine):

$$\begin{array}{c} CH_3 \\ | \\ HC-NH_2 \\ | \\ COOH \end{array}$$

D-alanine

The alpha carbon atom (except that of glycine)

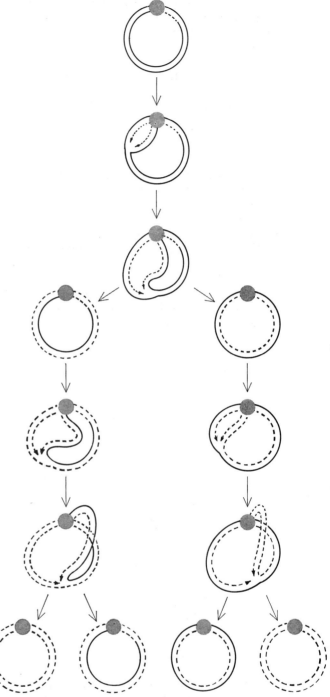

Figure 6–11. Replication of bacterial DNA molecule. The two chains of the molecule, represented by concentric circles, are joined at a "swivel" (spot). DNA labeled with radioactive (isotope-containing) thymine is shown by dotted lines; part of one chain of the parent molecule is labeled by having taken up the isotopic thymine, as are two generations of newly synthesized DNA. Duplication starts at the swivel and proceeds counterclockwise. The arrowheads mark the points at which the two parental strands are being separated and new DNA is being synthesized in each chromosome. Two complete replications are represented.

is asymmetrical. Thus, as among carbohydrates, D and L steric forms of amino acids exist. The steric orientations are referred to D- and L-serine:

$$
\begin{array}{cc}
\text{COOH} & \text{COOH} \\
| & | \\
\text{NH}_2\text{CH} & \text{HCNH}_2 \\
| & | \\
\text{CH}_2\text{OH} & \text{CH}_2\text{OH} \\
\text{L-serine} & \text{D-serine}
\end{array}
$$

instead of to D- and L-glyceraldehyde, as are glucose and other carbohydrates.

It is of interest that the amino acids in natural proteins are nearly all of the L form. Some D amino acids, once regarded as "unnatural" because they had never been found in natural substances, occur only in bacteria and molds, notably in antibiotics produced by those organisms, and in cell walls of bacteria. Not all of the amino acids that occur in cells are as simple as alanine: e.g., tryptophan and thyroxine:

$$\text{H}_2\text{N}-\text{CH} \cdot \text{COOH}$$

tryptophan

$$\text{CH}_2 \cdot \text{CH(NH}_2) \cdot \text{COOH}$$

thyroxine

Like various monosaccharides, amino acids can join together by condensation polymerization to form **peptides**. A combination of two amino acids constitutes a **dipeptide**; three make a **tripeptide**; many make a **polypeptide**. Proteins are enormously long-chain polypeptides. When protein macromolecules are broken, as by hydrolysis, into large groups of amino acids these are called **peptones** and **proteoses**. Complete hydrolysis of protein yields only amino acids.

The linkage between the amino acid residues in peptides including all proteins is between the carboxyl group of one and the amino group of the other—a peptide bond or linkage:

$$
\text{R}-\overset{\overset{\text{O}}{\|}}{\text{C}}-\text{N}-\overset{\text{H}_2}{\text{C}}-\text{R}
$$

Peptide bond

This is exemplified in the formation of glycylglycine, a dipeptide:

glycine   glycine

$$
\begin{array}{cc}
\text{COOH} & \text{CH}_2\text{NH}_2 \\
| & \text{H} \quad | \\
\text{H}_2\text{C}-\text{N}-\text{C}=\text{O}
\end{array}
$$

glycylglycine

Formation of the very long peptide chains that constitute proteins is a more complex process than the formation of a glucose homopolymer like starch, because all proteins are complex heteropolymers. Each of the score or more of the amino acids that go to form proteins is structurally different.

Protein Structure. The different amino acids that are joined to form proteins occur in innumerable sequences. The amino-acid sequence of a protein is said to constitute its **primary structure** (Fig. 6–12). Secondly, the peptide chains may be held parallel to each other by H-bonds between adjacent R—NH and O=C—R groups (R—NH $\cdots$ O=C—R) and by disulfide (R—S—S—R) bonds between adjacent sulfur-containing amino acids. Many parallel fibers thus joined can form protein sheets that may be flat or pleated. In another type of arrangement the polypeptide chains are twisted into helices such as the alpha helix. The helices are held in their coiled form chiefly by H-bonds between adjacent coils. A protein is said to be **denatured** when these bonds are broken, as by heating: the coil "unfolds" or straightens out. Many biologically extremely important substances (like keratins of hair, horn, and fingernails) consist of protein complexes with such helical forms.

These parallelings in sheets, twistings into helices and similar such arrangements of poly-

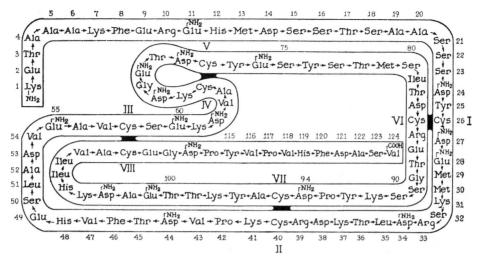

Figure 6–12. The complete amino acid sequence of the enzyme ribonuclease. Standard three-letter abbreviations are used to indicate individual amino acid residues. Four disulfide bonds are seen, symbolizing a secondary structure.

peptide chains constitute their **secondary structure** (Fig. 6–13). A **tertiary structure** results when the sheets, helices or bundles are folded and twisted on themselves like the twisted yarn in a skein of wool or a bundled-up towel. The forms of the skeins or sheets are maintained largely by hydrogen and sulfide bonds (Fig. 6–14). Because of tertiary structures we find proteins in the forms of globules, ropes, coiled fibers and many other arrangements. A **quaternary structure** results from aggregation of units that are in tertiary form. The subunits in quaternary structures are held together by disulfide bonds.

Specificity. The primary, secondary, tertiary, and quaternary structural arrangements of protein molecules are not hit-or-miss. They depend on certain amino-acid sequences. Therefore, they are genetically determined and relatively constant. Each protein has its own peculiar, distinctive, and specific inherited structure; i.e., it possesses the property of **specificity** (each protein differs from all others).

The number of different proteins possible from different arrangements of some 24 amino acids in the primary, secondary, tertiary, and quaternary structures is obviously astronomical.

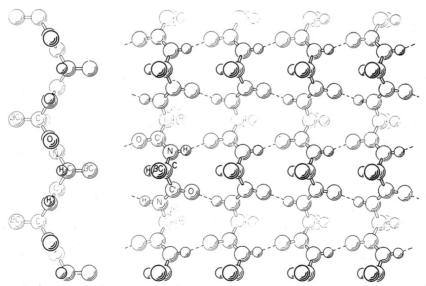

Figure 6–13. The parallel-chain pleated sheet structure.

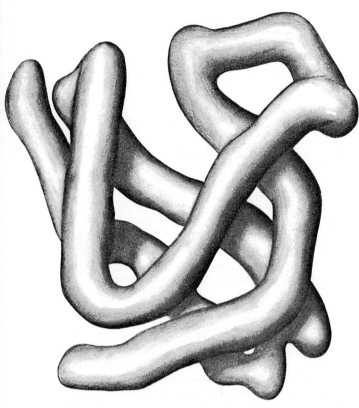

Figure 6–14. Typical conformation of peptide chain in a globular protein, after myoglobin structure of Kendrew et al. and hemoglobin structure of Perutz et al.

When proteins are conjugated with other substances, e.g., metals, lipids, and polysaccharides, the number of possible permutations and variations is virtually infinite. The property of specificity is of the greatest importance in all enzymic functions, in resistance to disease, in an individual's identity or in inheritance. Specificity will be discussed again in later sections.

Protein Synthesis

All biochemical reactions in the cell are catalyzed by **specific** enzymes, the specificity of which resides in the primary, secondary, tertiary, and quaternary structures of their protein moieties (apoenzymes). In the preceding discussion it was stated that the structure of all proteins in the living cell, including enzymes, is determined by the nucleotide units in the DNA of each cell. The obvious question is, how? The answer, drawn from nature only after years of the most exquisitely incisive, inductively imaginative, and toilsome researches by hundreds of highly ingenious biochemists,

molecular biologists, microbial geneticists, and associated scientists, lies in: (a) the sequence of nucleotides (called the **genetic code**) that make up the DNA chains and (b) the mechanism by which this **code** or **genetic information** or **message** is transferred from the DNA polynucleotide in the nucleus to the protein-synthesizing mechanisms in the cytoplasm.

These protein-synthesizing mechanisms are ultramicroscopic granules called **ribosomes** that consist of ribonucleoprotein (40 per cent protein, 60 per cent RNA). Ribosomes make up a great part of the finely granular, apparently diffuse substance of cytoplasm that is seen in specially prepared electron micrographs of cross sections of cells. In "higher" cells, ribosomes are attached to membranous structures, the endoplasmic reticulum; in bacteria they seem to be diffusely scattered or attached to the cytoplasmic membrane and mesosomes (Chapt. 14).

rRNA and tRNA. The RNA of the ribosomes, **ribosomal RNA** (rRNA), plays a basic role in the synthesis of proteins by providing a place for the specific **selection, arrangement,** and **joining** of the amino acids. Another form of RNA, called **transfer RNA** (tRNA), consists of relatively small molecules (70 to 100 nucleotides). It exists in solution in the fluid matrix of the cell. Each molecule of tRNA has a specific terminal nucleotide sequence that enables it to accept a single molecule of an **activated** amino acid and transfer it to a ribosome.

mRNA and the Genetic Code. Each ribosome has attached to it a third form of RNA that appears to be formed only by an enzyme, RNA-polymerase, under the direct influence of DNA. This class of RNA molecules is called **messenger RNA** (mRNA), because it contains the same sequence of nucleotides as the strand of DNA from which it was copied, but complementary to it (i.e., a T in the DNA sequence is represented by an A in the RNA strand, C by G, A by U, and G by C), and thus embodies the genetic message encoded in the DNA.

Transcription of the Genetic Code. By the use of these three forms of RNA (tRNA, mRNA, and rRNA), the construction of a polypeptide chain proceeds (in much simplified form) as follows:

Since each amino acid ($H_2N \cdot R$) that goes into a polypeptide (protein) chain has a different structure, a different specific enzyme is required to activate each. The energy for activation is derived from ATP bonds by the formation of the complex: Enz + ATP + $H_2N \cdot R$. Pyrophosphate is liberated from the ATP (yielding

AMP), and the bond energy is left in the complex.

The enzyme then catalyzes combination between the activated amino acid and a molecule of tRNA that has a sequence of three nucleotides (a **nucleotide triplet**) that is specific for that amino acid. Each molecule of tRNA, which is shaped somewhat like a cloverleaf, has a distinctive triplet of nucleotides at the end of the nucleotide chain (at the base of the "stem" of the cloverleaf) that enables it to combine with one, single, specific amino acid.

The activated amino acid is then transferred by its specific tRNA molecule, with energy derived from the high-energy compound guanidine triphosphate (GTP, analogous in function to ATP), to a ribosome where the actual synthesis of the protein occurs. Probably many ribosomes act together in groups called **polysomes** or **polyribosomes**. These bodies are visible only with electron microscopes (Chapt. 14).

The rRNA itself appears to be nonspecific, but it acts as a working place for the tRNA and the mRNA. As mentioned before, mRNA is a long, single chain of nucleotides whose sequence is complementary to that of the nucleotide sequence (genetic code) in the DNA. Each activated amino acid brought by tRNA to the mRNA becomes attached to the strand of mRNA at a place in which the sequence of the three nucleotides in the tRNA corresponds exactly with that of a triplet of nucleotides in the mRNA. The tRNA triplet is thus oriented opposite the mRNA nucleotides as though the tRNA triplet were part of a complementary strand of DNA.

For convenience in visualizing the relationships we may imagine that the tRNA is a three-nucleotide fragment of a single DNA strand while the mRNA is the complementary strand. This complementary relationship between the tRNA and the mRNA results in the pairing of the nucleotide triplets in the tRNA and mRNA according to the DNA code of which mRNA is a single-stranded copy.

Each molecule of tRNA, with its attached amino acid, recognizes its place on the mRNA strand (which carries the DNA code). Thus the sequence of amino acids that are attached to the strand of mRNA by the tRNA is determined by the DNA code.

Final Synthesis. Once the activated amino acids are properly aligned in the coded order along the strand of mRNA, the polypeptide bonds are formed enzymically to join the amino acids. The now complete polypeptide chain is released from the ribosome or poly-

some as a full-fledged protein of highly specific structure (Fig. 6–15). The tRNA and the activating enzyme are liberated and repeat the process. The mRNA appears to remain attached to the ribosomes to act repeatedly as a template for the ordering of more polypeptide chains when the tRNA's bring the activated amino acids to it (see §7.6, Enzyme Control).

The Codon. Because of the almost infinite number of possible arrangements of the four kinds of nucleotides (letters or digits: A—T, C—G) in the DNA, and considering that one DNA macromolecule contains perhaps 100,000 nucleotides, an almost infinite number of three-letter "words" or nucleotide triplets, each representing an amino acid, can be arranged. Each word thus constructed of three nucleotides in DNA is called a **codon**.

Each protein thus bears an inherited structure and, because of its enzymic or other activities, confers an inherited characteristic on the cell of which it is a part.

Universality of the Genetic Code. A thought-provoking aspect of the genetic code as a biological mechanism is that DNA is constructed on the same general form throughout the living universe. Genes, codons, triplets, amino acids, and nucleotide "words" are all involved in the same way and in the same process, though the nucleotide sequences occur in endless variations. The code, i.e., the language of heredity, is said to be universal. The tRNA and mRNA from a bacterial cell will serve the same function and direct the synthesis of the same protein from the same amino acids in an artificial enzyme mixture derived from a mammalian cell.

Does this universality of the genetic language and mechanism point toward DNA as the common, primeval starting point of all life? There is evidence indicating that various amino acids, the purine and pyrimidine bases, and other constituents of RNA and DNA could have been formed by lightning flashes under prebiotic conditions in vapors of water, ammonia, and methane. Thus a single, primeval DNA helix, spontaneously formed, could, in the course of thousands of millions of years, by mutations, genetic recombinations, and natural selection, account for ourselves and all of the living things that now surround us. It is interesting to realize that each helix of DNA in each person was derived from ancestral DNA and that these ancestral strands of DNA have successively produced each other by helix replication without interruption over eons of time. The DNA in each person has been "immortal," al-

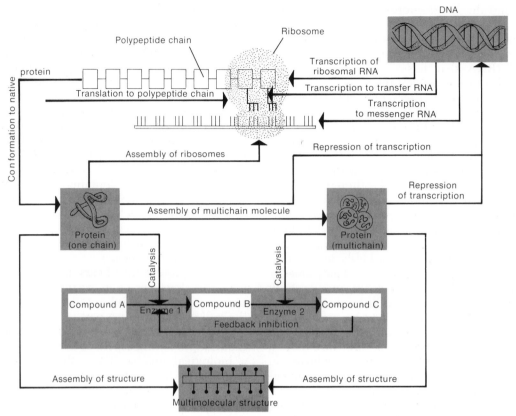

Figure 6–15. Overview of the process by which biological information is transferred from DNA via RNA to specific polypeptides. The peptide subunits are then assembled into multichain proteins.

though the bodies of the individuals carrying the ceaselessly replicating strands from one generation to another have in time disintegrated. We may draw an analogy with a lighted torch carried for billions of miles and years since life began by successive relay runners who, having handed the torch on to the next runner, themselves drop panting or lifeless by the wayside.

Deciphering the Genetic Code. One of the most magnificent pieces of decoding in human history, including Champollion and the Rosetta stone, was not done by international spies or supersleuths in Washington but by Watson and Crick (Nobel Prize winners), Benzer, and others when they discovered in 1961 that each DNA code word consists of a sequence of three nucleotides (nucleotide triplets). These have been called codons. Each codon occurring in tRNA specifies one amino acid.

Also in 1961 Nirenberg (Nobel Prize winner) and Matthaei mixed, in vitro, ribosomes (separated from sonically disrupted cells by high-speed differential centrifugations) with a suspension of cytoplasmic materials from the same sort of cells. The mixture contained, among other materials, enzymes, amino acids, tRNA, ribosomes, and energy-transfer coenzymes such as ATP and GTP. Here were found all, except one, of the many elements necessary for the synthesis of polypeptides. The one missing ingredient was mRNA.

Now Nirenberg and Matthaei had synthesized a simple form of mRNA in which the entire chain was made up of only one kind of nucleotide, uridylic acid (U). This historic polynucleotide is now often called "poly U." Be it noted that, although in RNA uracil replaces thymine, uracil and thymine are equivalent in respect to pairing (Fig. 6–6). When U was added to the cellular "purée" just described, the cellular elements in the purée cooperated as they do in a living cell to form a polypeptide chain. But since there was only one kind of triplet or codon, UUU in the synthetic mRNA, the polypeptide formed consisted of only one kind of amino acid, phenylalanine.

$$NH_2$$

$$CH_2CHCOOH$$

phenylalanine

It was thus revealed that the codon UUU meant phenylalanine; the first genetic code word ever to be deciphered! Of course, many eager researchers were at work deciphering other code words.

The huge mass of data resulting from these studies fills many long shelves in libraries and carries the names of scores of brilliant scientists. It is sufficient to say here that, in the same or similar ways, codons have been found since then for all the amino acids that occur in natural proteins. AAA means lysine, AGA means arginine, UUU means phenylalanine, UUA means leucine, and so on. It was found also that several amino acids have more than one codon; for example, leucine is specified by CUC, CUU, UUA, and UUG. The code is therefore said to be **degenerate** in that each amino acid is not restricted to only one codon. This is true in vitro, but the code appears not to act degenerately in vivo.

Genetic Punctuation. In reading the entire sequence of triplets in a long chain of DNA (about 10^5 units), scientists found it difficult to determine where one triplet ended and another began, i.e., to know how the genetic message was punctuated. One might have a series of nucleotides like ACTGATGAGCAT. If we mark off triplets from the left, we find that they are ACT, GAT, GAG, and CAT. They make "sense" words (in English). But if we overlook the first A we find CTG, ATG, AGC, and so on, i.e., nonsense words. English words are used as an example, but of course the triplet letters could be arranged in any sequence in a polypeptide. (After all, when genetic words were forming, there wasn't any English language.) In the above example using English words, a single error at the beginning would clearly result in making nonsense words such as CTG of many or all of the succeeding triplets and, in any case, in producing a faulty message (polypeptide) or none at all.

Not only is it necessary to have a starting point but the end of the genetic message must be indicated. In between, the structure of the tRNA molecule (see Fig. 6–15) provides for reading the nucleotides as triplets, i.e., as three-letter words, so that no further separation between words is necessary. The message will be correct if the reading begins correctly and if no nucleotides have been added or deleted.

Two codons, AUG and GUG, are now believed to be the initiation codons, with AUG being more efficient. The chain is terminated by UAA, UAG, or UGA, which have been called "nonsense" codons since they do not correspond to any amino acid.

Thus, the genetic language has only three-letter words and it provides for indicating a starting point and the end of a genetic message, as do periods, spacing, and capital letters of the sentences in this book.

CHAPTER 6
SUPPLEMENTARY READING

Bodanszky, M., and Bodanszky, A. A.: From peptide synthesis to protein synthesis. Am. Sci., 55:185, 1967.

Goren, M. B.: Mycobacterial Lipids: Selected topics. Bact. Rev., 36:33, 1972.

Horecker, B. L.: The biosynthesis of bacterial polysaccharides. Ann. Rev. Microbiol., 20:253, 1966.

Kendrew, J. C.: The Thread of Life. Harvard University Press, Cambridge, Mass. 1967.

Meselson, M., and Stahl, F. W.: Demonstration of the semiconservative mode of DNA duplication, in Cairns, J., Stent, G. S., and Watson, J. D. (Eds.): Phage and the Origins of Molecular Biology, p. 246. Cold Spring Harbor Laboratory of Quantitative Biology, Cold Spring Harbor, N.Y. 1966.

Nomura, M.: Bacterial ribosome. Bact. Rev., 34:228, 1970.

O'Leary, W. M.: The fatty acids of bacteria. Bact. Rev., 26:421, 1962.

Shaw, N.: Bacterial glycolipids. Bact. Rev., 34:365, 1970.

Stent, G.: Molecular Genetics. W. H. Freeman & Co., San Francisco. 1971.

Symposium: The Genetic Code. Cold Spring Harbor Symp. Quant. Biol., 31, 1966.

Watson, J. D.: Molecular Biology of the Gene, 2nd ed. W. A. Benjamin, New York. 1970.

ENZYMES • CHAPTER 7

The macromolecules described in the preceding chapter comprise a wonderful system for passing genetic information from generation to generation and for permitting "read-out" of the information in the form of specific proteins, which are themselves macromolecules. They are the effectors of the system—the molecules that provide the structure of the cell and that carry out the work of the cell. Specialized proteins called **enzymes** do the majority of the work of the cell and are the subject of this chapter.

7.1
CATALYSIS

Catalysis is the speeding up of chemical reactions. A catalyst can accelerate only reactions that are energetically feasible—but without the assistance of the catalyst the reaction may be exceedingly slow. Catalytic agents may be organic or inorganic. Many inorganic catalysts consist of sheets or "sponges" of various inert metals (platinum, lead). They are widely used in industry.

Catalyzed reactions typically occur at the surface of the catalytic agent. Therefore, the greater the surface area or state of subdivision of the catalyst, the more the reaction can occur.

For example, a cube of catalyst 1 cm on each edge has a surface area of 6 sq cm. Cut into two parts the catalyst has a surface area of 8 sq cm. Cut into 100 slices each 0.1 mm thick it presents 204 sq cm. Divided into millions of colloidal particles it presents a surface area measuring many hundreds of square centimeters.

Substances are in the colloidal state when they are in the form of ultramicroscopically minute particles stably suspended in a fluid (gas or liquid). For example, smoke is a colloidal suspension of minute particles of carbon, tars, and other substances in air; milk is a colloidal suspension of casein and fat in whey. Some enzymes are colloidal proteins.

A common industrial inorganic catalyst is finely divided or colloidal platinum. Among its catalytic potentialities is the oxidation of ethyl alcohol. Alcohol and oxygen at room temperature do not combine to a readily perceptible degree. In the presence of finely divided platinum they are greatly concentrated on its surfaces by **ad**sorption (Chapt. 21) (not **ab**sorption). A reaction then occurs between alcohol and oxygen, which is facilitated and controlled by the nature and extent of the catalyst, by temperature, and by moisture. The alcohol is rapidly oxidized to acetic acid. The platinum does not enter into the reaction but remains to adsorb more oxygen and more alcohol on its surface. It continues the process of oxidizing the alcohol, first to acetic acid and then to water and carbon dioxide, as long as the products of the reaction, or extraneous side products, are continuously removed and do not remain to block or "poison" the surfaces of the catalyst.

In a simple system of this kind we can predict, from a knowledge of the substances and surfaces involved, what the result of a given combination will be. Stable, inorganic catalysts such as platinum are simple and constant, and are extensively used in industry. In living systems the situation is more complex: numerous physically and chemically complex organic catalysts act simultaneously or in rapid succession.

Enzymes are organic catalysts and act

mainly by forming transitory chemical combinations with the **substrates** (substances altered by the enzyme). Following the reaction catalyzed between the substrates, the enzyme separates from them. Theoretically it remains unchanged; actually an enzyme can "wear out" (chemically deteriorate) after prolonged activity.

Enzymes usually act exceedingly rapidly and efficiently. Properly concentrated and in contact with optimum amounts of substrates under suitable conditions (temperature, pH), a very small quantity of enzyme, probably of the order of a few molecules, can bring about a relatively large amount of catalyzed reaction in a comparatively short time. For example, 5 ml of an aqueous extract of pig's pancreas contains perhaps 1 mg of the enzyme **trypsin** that hydrolyzes (digests) protein in the intestinal tract. This can decompose 5 lb of beef (protein) within about five hours at 37C at a pH of about 7.5. The ratio of specific substrate (in this case beef protein) to pure enzyme (trypsin) probably exceeds one million to one.

7.2
DISCOVERY OF ENZYMES

Before the time of Pasteur the nature of the fermentations that produce beers and wines was virtually unknown. Since it is the basis of very large industries the process of fermentation has been the subject of much study. Fermentation was thought by Liebig and many other brilliant chemists to be a spontaneous chemical change entirely independent of life. However, after many ingenious experiments and demonstrations it was made clear by Pasteur around 1860 that fermentation does not occur spontaneously but is wholly dependent on living microorganisms, notably brewers' yeast. Microorganisms were often called "living ferments." Pasteur also showed that true fermentation occurs only in the **absence** of free oxygen. In Pasteur's words, "La fermentation est la vie sans l'air." He called life without air **anaerobiosis.**

It was soon realized that beer and wine fermentations were not caused by the yeast cells themselves, but by some active principle associated with them. The active principle was thought to be inside the cells and was first called an **enzyme** (Greek: *en* = in; *zyme* = yeast or leaven) by Kuhne in 1878. Buchner (Nobel Prize winner), in 1897, found that filtered, cell-free juice of crushed yeast cells would cause sugar to ferment. Thus the fermentative enzyme of yeast, and later a great variety of other sorts of enzymes from many other kinds of living cells, were found to be distinct, nonliving entities mechanically separable from the cells that produced them. We now know that, in a sense, Liebig was right, since fermentation can occur in the absence of living cells. However, the necessary enzymes are produced only by living cells.

7.3
STRUCTURE OF ENZYMES

Coenzymes

In 1905 Buchner and others dialyzed cell-free yeast juice. Dialysis is carried out by enclosing the fluid to be dialyzed in a sac of **selectively permeable** material (i.e., material permeable to some substances but not to others), such as cellophane or animal membrane. The sac is then suspended in water. Ions and small molecules that are soluble in water (salts, glucose, amino acids) pass out of the sac, through the ultramicroscopic pores in the membrane, into the water surrounding the sac. Large molecules, such as those of proteins and complex polysaccharides, cannot pass through the membrane but remain inside the sac. In Buchner's experiments with yeast juice it was found that neither the **dialysate** (i.e., the material that passed out of the sac) nor the **residue** (i.e., the material that remained inside the sac) could alone produce fermentation. When mixed together, however, they produced normal fermentation. Obviously each contained something essential to the fermentation.

It was soon demonstrated that the essential substance in the residue was a nondialyzable, colloidal protein, readily destroyed by heat (100C): i.e., it was thermolabile, as are virtually all proteins. Such a protein moiety of an enzyme is now called an **apoenzyme** (Gr. *apo* = part of).

The essential material in the dialysate was found to be nonprotein, noncolloidal, of small molecular weight, and thermostable. This part of an enzyme, easily separable from the protein part, is now called either a coenzyme or a cofactor, depending on whether it is organic or inorganic. Apoenzyme plus the coenzyme constitutes the complete active enzyme, called a **holoenzyme** (Gr. *holos* = entire). The term enzyme is generally used to mean holoenzyme.

The molecular structure of many coenzymes is now well known, and some have been synthesized in vitro. Coenzyme molecules generally carry the distinctive, reactive portion of an enzyme. Acting with the apoenzyme, the co-enzyme brings about the specific substrate reaction that is characteristic of that particular enzyme. In many coenzymes the distinctive, reactive portion is a familiar vitamin: nicotinic acid or its popular derivative "niacin"; others are vitamins of the B complex, such as thiamine (vitamin B_1) or riboflavin (vitamin B_2). In fact, all vitamins that function physiologically have been found to act as the reactive group of one or another coenzyme.

Cofactors

Some enzymes are first produced by the cell in an incomplete or inactive form. They are called **zymogens** or **pre-enzymes**. These must then be activated or completed by contact with another agent called a **kinase**, an **activator**, or a **cofactor**. Cofactors may be hydrogen ions or ions of iron, magnesium, copper, molybdenum, cobalt, or zinc; they also may be coenzymes, vitamins, or other enzymes, depending on the particular enzyme involved. Trypsin, for exam-ple, exists in the pancreas as inactive trypsino-gen, which becomes activated in the intestine when in contact with a substance called en-terokinase. Phosphatase, an enzyme that hy-drolyzes organic phosphates, must be activated by magnesium ions. The exact function of some activators, especially vitamins and certain me-tallic ions, is well known; of others it is not so clear. The seemingly curious and often very specific requirements of many living cells, in-cluding our own body cells, for minute quanti-ties of certain metals or vitamins (i.e., micro-nutrients) are evidence of the fact that these are absolutely essential parts of various coenzymes or prosthetic groups.

Prosthetic Groups

In some enzymes there is a nonprotein, specifically reactive portion, which is covalently bound and not readily dissociated from the re-mainder of the molecule. The nonprotein por-tion is then called a **prosthetic group**; the combi-nation of the protein portion and the prosthetic group is called a **conjugated protein**. A familiar example of a conjugated protein is **hemoglobin**. This is a combination of the red, iron-bearing, porphyrin pigment **heme** (the prosthetic group) (Fig. 7–1) with the globular protein, **globin**. Hemoglobin is the oxygen-carrying pigment of

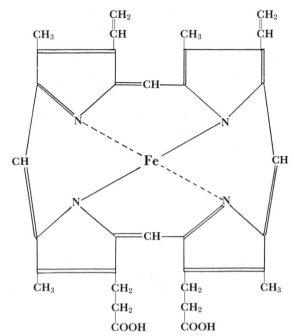

Figure 7–1. The heme molecule. Note the position of the iron in this ring-like or claw-like iron-porphyrin molecule. Heme is part of the red coloring matter of the blood. Chelat-ing agents, such as the porphyrins, by combining with (chelating) harmful metallic ions, perform a function analogous to that served by buffers in combining with H^+ to maintain a favorable pH. Some pH buffers also act as metal chelating agents.

vertebrate red blood cells. The iron of the prosthetic heme group of hemoglobin is readily oxidized and reduced. It combines with oxygen in the air via the lungs, releases oxygen to the body tissues, and returns to the lungs for more oxygen.

Although not generally classed as an en-zyme, hemoglobin closely resembles some en-zymes in both structure and function. There are several metal-bearing, enzyme-like proteins (**metalloproteins**) other than hemoglobin. Gen-erally the active metal is carried in a porphyrin residue much like heme. The metal in the green, sunlight-utilizing pigment **chlorophyll** in higher plants is magnesium instead of iron (Fig. 32–1). In the Crustacea the active metal is copper, though it is not carried in porphyrin.

Many common bacteria use atmospheric oxygen, which is combined directly with yellow hemoproteins (closely analogous to hemoglobin) called **cytochromes** (Fig. 7–2; see also Chapt. 8). The combination is mediated by an oxidizing enzyme called **cytochrome oxidase**. Some details of these reactions will be given later. All such metalloprotein pigments that are involved in

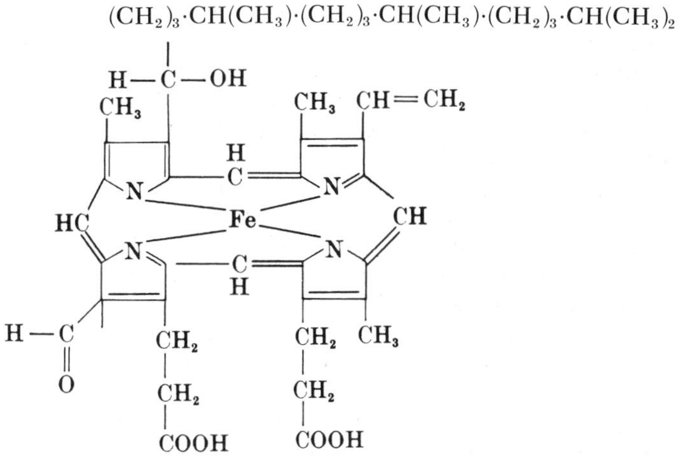

(CH2)3·CH(CH3)·(CH2)3·CH(CH3)·(CH2)3·CH(CH3)2
H—C—OH
CH3
CH3 CH=CH2
H
C
N
N
HC
Fe
CH
N
N
C
H
H—C
O
CH2
CH2
COOH
CH2 CH3
CH2
COOH

Figure 7–2. Iron-porphyrin of cytochrome *a.*

biological oxidations, or respirations, are often called **respiratory pigments.**

7.4 SPECIFICITY OF ENZYMES

The protein moiety (apoenzyme) of each enzyme is characterized by a property called **specificity** that is typical of proteins in general. Specificity of a protein molecule depends upon the physicochemical configuration of its **surface.** This, in turn, is determined by the number, kind, and sequence of amino acids that constitute the peptide chains making up the protein (i.e., its **primary structure**), by the distinctive coiling or helical structure of the peptide chains (**secondary structure**), and by the manner of folding of the long peptide chains of proteins upon themselves (**tertiary structure**) (Chapt. 6). Now, since an almost infinite number of permutations and combinations of the approximately 22 amino acids that make up the protein chains (i.e., the primary structure of the proteins) is possible, and since the chains can be twisted (secondary structure) and then folded like long skeins of wool yarn (tertiary structure) in billions of different arrangements, it is evident that there can be an astronomical number of different enzyme proteins (apoenzymes). Each protein is a macromolecule that may consist of hundreds of the 22 to 24 amino acid residues linearly linked in different arrangements. Each is different from all others in respect to molecular configuration; i.e., each enzyme is unique or **specific.** Each enzyme can react only with certain particular substrates that have a corresponding stereochemical structure (Fig. 7–3). This correspondence between enzyme and substrate is of a reciprocal nature, such, for example, as the correspondence of a plaster cast to its mold. Exact details of enzyme-substrate interactions remain to be elucidated. Be it noted that the correspondence between enzymes and substrates extends beyond mere physical form of the molecules involved, and includes the correspondence of mutually attractive electrostatic forces, hydrogen and sulfur bondings, van der Waals forces, and so on.

Specificity of enzymes varies greatly in degree. For example, one enzyme that catalyzes the oxidation of L-amino acids cannot oxidize the corresponding D-amino acids. A certain enzyme that destroys the carboxyl group of (**decarboxylates**) pyruvic and related keto acids will not decarboxylate fatty acids like acetic acid. Some enzymes (e.g., trypsin of the intestine) are

SUBSTRATE

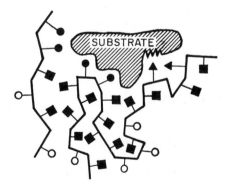

Figure 7–3. A schematic representation of an active site of enzyme-substrate interaction. ●, amino acid residues whose fit with substrate determines specificity; ▲, catalytic residues acting on substrate bond, indicated by a jagged line; **O**, nonessential residues on the surface; ■, residues whose interaction maintains three-dimensional structure of the enzyme protein.

more broadly specific and hydrolyze many different proteins, because these enzymes attack peptide bonds between certain linked amino acids. These are common to all proteins. Such enzymes do not attack carbohydrates, fats, or other classes of substrates not having peptide bonds. Some enzymes, however, can attack only peptide bonds at the end of a peptide chain. Other enzymes attack only carbohydrates. For example, amylase, an enzyme in saliva, attacks glycosidic bonds and splits starch into simpler sugars: dextrins, disaccharides, and monosaccharides. Such enzymes do not act on proteins and fats.

It is worth noting at this point that many biological phenomena other than enzyme actions involve specific proteins. **Specificity** therefore characterizes many nonenzymic protein functions. Other examples of specificity will be seen in the sections on antigens and antibodies, microbial poisons (toxins), virus infections, and certain industrial processes.

Isoenzymes (Isozymes)

Studies of enzymes by physicochemical methods have shown that numerous enzymes occur in several forms, even in the same tissue or cell. These variant forms are called **isoenzymes** or **isozymes.** They appear to represent different structural arrangements of the same protein subunits, since all forms of a single enzyme have the same molecular weight. All appear also to have the same specificity as to substrate, though they may act in slightly different ways depending on the arrangement of the subunits in the apoenzyme. These structural variations may explain certain hitherto puzzling irregularities in enzyme action and may relate also to obscure immunological discrepancies or actions.

7.5
HOW ENZYMES ACT

Mechanism of Enzyme Action

In most enzymic catalyses, the specific apoenzyme involved appears first to attach to the substrate at certain specific sites. These sites represent reciprocally corresponding physical structure and anionic and cationic groups in the molecule of the substrate (Fig. 7–3). This preliminary combination appears to place certain

bonds in the substrate under stress. The coenzyme, because of its appropriate molecular structure, then combines with (**accepts**) a part of the substrate: for example, a hydrogen ion or glycosyl or amino group. This ion or group is then either passed by the coenzyme to another coenzyme, or to a different substrate molecule, or it may be liberated as waste into the surrounding fluid. The final result depends on the nature of the reaction being catalyzed. The enzyme, freed of the altered substrate residue, is then ready to combine with another substrate molecule and repeat the process. If any energy is released by the reaction, it is partly lost as heat and in part taken up into the cell substance by the formation of "energy-rich" compounds that contain certain types of bonds (e.g., organic phosphate bonds, thioester bonds, and some others) in certain coenzymes (see Chapt. 8). The energy stored in such energy-rich compounds is later used by the cell in motility and cell synthesis.

Enzyme Induction

Each species of living cells has a genetically determined (inherited) natural endowment with certain functioning enzymes. These are constantly present in, and distinctive of, all of the cells in that species of cell. Such inherited enzymes, part of the normal constitution of the cell, are called **constitutive enzymes.** In addition many cells possess genetic determinants (genes) for the synthesis of numerous enzymes that, curiously, do not ordinarily appear. The mechanisms for the synthesis of such enzymes are genetically repressed. (This is more fully explained in Chapt. 19.) Each such repressed genetic determinant finds expression only when a corresponding or inducing substrate (or related substance; see next paragraph) enters the cell. In the presence of the inducer the repressor of the appropriate synthetic mechanism is removed and the enzyme specific for the inducer substrate is synthesized. Such enzymes are said to be **inducible.** They were formerly called "adaptive" enzymes. Thus, although genes (i.e., genetically functional units) determine the full enzymic potentialities of a cell, environmental factors in the form of inducers, for example, determine just which of the latent enzymic potentialities of a cell shall appear under any given circumstances. The cell is evidently not under the necessity of synthesizing all its potential enzymes all the time but only such as may be needed from time to time. This is an important economy of the food and energy resources of the cell, since enzyme synthesis requires energy and food substance.

In a number of cases an enzyme may be induced in a cell by any one of several substances that are not themselves substrates but are merely chemically related to a specific metabolizable substrate. A much-studied example is the induction of the enzyme β-galactosidase in a common bacterium, *Escherichia coli*. Synthesis of this enzyme is inducible by not only its normal substrate, lactose (milk sugar; a β-galactoside readily metabolized by *E. coli*) but also by melibiose (an α-galactoside *not* metabolizable by *E. coli*) and also by several thio-β-galactosides. In another bacterium, *Proteus vulgaris*, the enzyme **leucine decarboxylase** is induced by the amino acids alanine (*not* a substrate for *P. vulgaris*) and valine, chemically similar to leucine.

Enzyme Equilibria and Reversibility

The action of many enzymes is demonstrably reversible. For this reason the symbols used in equations involving enzyme action often indicate reversibility. For example, if we place in a solution of amyl butyrate (an organic salt or **ester**) a little **esterase** (an enzyme from the pancreas that hydrolyzes amyl butyrate under the proper conditions of temperature and pH), the amyl butyrate is hydrolyzed to its constituents, butyric acid and amyl alcohol. The decomposition automatically ceases when a certain concentration of the acid and alcohol has been reached, i.e., at a definite **equilibrium point**:

$$C_9H_{18}O_2 + H_2O \xrightleftharpoons{\text{Enzyme}} C_4H_8O_2 + C_5H_{12}O$$

| amyl butyrate | | butyric acid | amyl alcohol |

Conversely, if we put the acid and alcohol together in a beaker with esterase, amyl butyrate and water are re-formed, ceasing at a definite concentration of end products. This is an excellent example of two of the most important catalyzed reactions in all living forms: **hydrolysis** (separating with water) and **condensation** (joining by withdrawing water). (See also Chapter 6.) This also illustrates the important point that, in general, accumulation of end products inhibits the action of any enzyme in either direction, i.e., at the same equilibrium point. Under any set of constant conditions, the equilibrium point for an enzyme-catalyzed reaction is constant.

There is clearly a constant relationship between concentration of enzyme and concentration of substrate. Up to the point of saturation, the rate (not the ultimate extent or quantity) of reaction increases with increase of ratio of one component to the other. With a constant amount of enzyme, increase of substrate increases rate of reaction until every molecule of enzyme is fully occupied (saturated) with substrate. Further additions of substrate cannot increase the rate of reaction (Fig. 7–4). Conversely, with a fixed amount of substrate, rate of reaction increases with additions of enzyme until all molecules of substrate are in contact with enzyme. Further additions of enzyme do not affect the rate of reaction (Fig. 7–5).

In many instances enzyme-catalyzed reactions appear to proceed in only one direction because the equilibrium point is very far in that direction. In other cases one or more of the end products may be removed constantly by some mechanism so that equilibrium is never reached. Under normal conditions in the living cell, enzyme reactions are constantly pushed in this manner toward the one or the other side of the reactions.

When markedly different **energy levels** are involved, theoretically reversible reactions cannot actually reverse. In the hydrolysis $\rightleftharpoons$ synthesis reaction of amyl butyrate just given, very little energy is lost in the hydrolysis or required to complete the resynthesis. The reaction proceeds in either direction because both are at very nearly the same energy level. When a great deal of energy is released, as in the complete

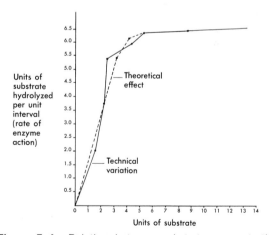

Figure 7–4. Relation between substrate concentration and rate of enzyme action with a fixed enzyme concentration. Beyond a certain point (in this illustration about 4 units of substrate) a fixed amount of enzyme becomes saturated with substrate and will not act any faster no matter how much substrate is added. The activity of the enzyme may be inhibited.

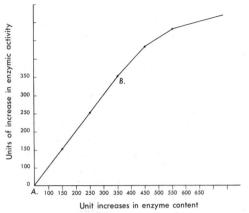

Figure 7–5. Relation between enzyme action and enzyme concentration with fixed amount of substrate. The relationship is linear within limits: i.e., between points A and B. Beyond B, all of the substrate eventually comes into contact with enzyme, and no increase in enzyme action occurs regardless of how much enzyme is added.

enzymic oxidative decomposition of glucose (688,500 calories), resynthesis cannot be brought about by the same enzymes because they cannot restore the lost energy. To reverse the reaction requires that "work" be done by *other* systems of enzymes that capture new energy derived from solar or other radiant sources by green plants or the biooxidation of foodstuffs by nonphotosynthetic species.

As an analogy, compare the stepwise, theoretically reversible process of raising and lowering a car with a jack. The jack may be thought of as an enzyme system. Lowering the car is like the enzymic decomposition of glucose: potential energy is liberated. Raising the car is like resynthesizing the glucose: new energy is required. The two processes take place at totally different energy levels. To complete the analogy, the new energy for raising the car is also derived (ultimately) from biooxidation of the glucose in the muscles of the man (or woman) operating the jack.

7.6 ENZYME CONTROL

"Feedback" Controls. Inside a living cell most enzymes do not act individually but as parts of well-organized, coordinated, and sequentially operating systems. Whatever affects one portion of the intracellular enzymic system has some effect on all parts, like the parts of a spider web. As pointed out before, the activity of an enzyme is inhibited by accumulation of the end products of the catalyzed reaction. In a stepwise sequence of cooperating enzymes (a "biological production line") excessive accumulation of a reaction product at the end of the line may inhibit the action not only of the enzyme at the end of the line but of all the enzymes in that sequence, all the way back to the beginning of the line. This is an important form of automatic control called **feedback inhibition.**

Eventually, in the presence of excessive amounts of end products, not only is enzyme activity inhibited but the actual synthesis of the enzymes themselves may be repressed. For example, if a cell normally synthesizing a certain substance, say the amino acid alanine, is artificially abundantly supplied with that substance (the end product of the enzyme) from an extraneous source, not only is the enzyme inhibited, but synthesis of some or all of the enzymes in the production line for that substance is repressed until the enzymes are needed again. This is called **feedback repression.** Note that it is necessary to differentiate between (a) inhibition of the **action** of enzymes by their end products (feedback inhibition) and (b) feedback repression of the **synthesis** of the enzymes themselves by the accumulation of end products, especially in enzyme series.

Contrary to its repressive action, the end product of each enzyme in a series or production line can be the **inducer** of the next enzyme in the series and the inhibitor or repressor of the preceding enzyme, thus carrying forward the work of the enzyme machine. Various such start-stop, induce-repress-inhibit mechanisms of enzyme control result in amazingly complex and efficient "automation."

Energy Controls. An important aspect of enzyme control is that related to the liberation of energy from foodstuffs. Clearly, in any regulated mechanism using energy, including living cells, if the energy supply is uninhibited and uncontrolled it can soon become injurious and, in cells, fatal. Conversely, failure to provide energy at a sufficient rate would prevent the normal functioning of the cell. In the cell, as in the machine shop, there must be a "stop-go" mechanism to control the energy supply.

In the machine shop, energy for the power source is supplied by the burning (i.e., the **oxidation**) of fuels or the use of water power. The energy from the power source is transmitted to the machines by belts, shafts, and gears. In living machines (e.g., the living cell) energy is also derived from the burning (i.e., the enzymic **biooxidation**) of foods. In the cell the energy liberated by biooxidation is transmitted not by gears and shafts but via high-energy compounds as noted elsewhere, notably certain organic phosphates: e.g., phosphoenol pyruvic acid, acetyl phosphate, and, most importantly, adenosine triphosphate (ATP). The energy of these compounds appears to relate to the special phosphate bonds. Now, ATP is derived during oxidative phosphorylation by addition of a phosphate group + energy (Chapt. 8) to adenosine diphosphate (ADP). ADP is derived from adenosine monophosphate (AMP):

$$AMP + OPO_3H_2^- \rightarrow ADP + OPO_3H_2^- \rightarrow ATP$$

ATP is a high-energy compound. The exact manner in which the energy is transferred from the low-energy level oxidized foods into the high-energy ATP is not yet fully clarified.

From the standpoint of energy control in the cell, ADP is an essential factor because it is needed to accept the energy of foodstuffs and become phosphorylated to high-energy ATP in the process. But if the energy stored in the ATP is not used up as fast as it is derived from the biooxidation of foodstuffs, ATP accumulates to excess. None is broken down to replace the supply of ADP that was used in its formation. The supply of ADP is thus depleted. The depletion of the ADP supply prevents its further acceptance of high-energy phosphate. This, in turn, inhibits the entire succession of oxidative enzymes that "collect" the energy from the foodstuffs. The supply of energy is thus automatically cut off until utilization of the energy stored in the excess ATP changes it back to ADP, which at once begins to accept more phosphate and energy. The whole effect is analogous to the backing up of a long line of traffic behind a stoppage far ahead.

The kind of energy control described might be considered a form of feedback inhibition of an enzyme series due to concentration of an end product, in this case ATP. However, the mechanism differs from feedback inhibition since it is not the concentration of ATP that blocks the enzyme system, but deprivation of the ATP-forming mechanism of the necessary ADP.

Location of Enzymes in the Cell

Exoenzymes. Probably many enzymes exist free in colloidal suspension in the fluid matrix of the cell. Some of these are secreted to the exterior. They are called **exoenzymes.** Exoenzymes are mainly digestive in function. By hydrolysis they decompose complex organic matter in the outer world, such as proteins, cellulose, and fats, to simple, soluble molecules of amino acids (from proteins), glucose (from polysaccharides), and glycerol and fatty acids (from fats). These relatively small molecules can pass through the cell membranes of many microorganisms, there to be utilized as food.

Endoenzymes. Foodstuffs, once they get inside the cell, are acted upon by whole systems of enzymes that act only inside the cell. These are **endoenzymes.** Endoenzymes of many kinds cooperate in two general types of processes inside the cell: (a) **synthesis** of cell components and food reserves and (b) **bioenergetics**— i.e., the release of energy from foodstuffs. The energy is either stored in reserve nutrients like starch, fats, or β-hydroxybutyric acid inside the cell or is immediately used for any of the active processes of the cell. It is evident that synthetic and energizing mechanisms are closely knit (**coupled**) into a definite and very efficient organization that carries on the complex chemistry of life but whose most intimate interreactions still escape us.

Endoenzymes concerned in the synthesis of proteins are, as noted in Chapter 6, intimately associated with **ribosomes.** In the cells of animals and higher plants endoenzymes concerned in energy production are organized into granular and membranous structures inside organelles called **mitochondria** and, in green plants, are closely associated with the light-absorbing organelles called **chloroplasts.** In bacterial cells the energy-mediating mechanisms appear to be closely associated with the cell membrane from which the more highly evolved **endoplasmic reticulum** of cells of animals and higher plants is derived. Energizing endoenzymes may also possibly be associated with **mesosomes** in bacterial cells that show these structures (Chapt. 14). The appearance of mesosomes in bacteria suggests primitive stages in the evolution of endoplasmic reticula.

The term exoenzyme must be used with care, because enzymes that are found in the medium surrounding cells may actually be endoenzymes that have been liberated by rupture of the cells.

7.7
FACTORS THAT AFFECT ENZYMES

Since enzymes are protein complexes, they are sensitive to all the various precipitating and coagulating (denaturing) factors that affect proteins in general: i.e., temperatures over about

80C, excessive concentrations of ions of heavy metals or of hydrogen (H^+) or hydroxyl (OH^-) ions. Active chemicals like chlorine and corrosive agents like strong alkalies quickly destroy all types of proteins. Any substance or physical agent that destroys protein can act as a **disinfectant**. (See Chapter 21.) At temperatures below about 75C (boiling point = 100C; human body temperature = 37C) each enzyme has an **optimal** temperature at which it functions best. This varies considerably with different enzymes and different organisms. Extreme limits range from about −2 to 85C. Most cells thrive from about 20 to 40C, their optimal range (Fig. 7–6). Temperatures below optimal are usually not destructive; they merely slow or inhibit enzyme action. Temperatures much above 75C destroy even the most resistant enzymes, except enzymes in bacterial endospores (Chapt. 36).

In addition to being sensitive to deviations in temperature, enzymes are very sensitive to variations in pH or pOH. Different enzymes and different organisms have different requirements but, with a few striking exceptions, optima generally range from about pH 4.5 to pH 8.5 (Fig. 7–7). The optimal pH for most cells is near 7.0 (neutrality). Optimal temperatures and

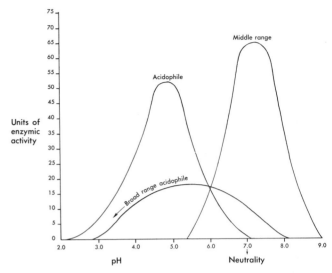

Figure 7–7. Relation between pH and activity of a certain type of proteolytic enzyme. In microorganisms that grow best in an acid medium (pH 5.5 to 4.5 or lower; acidophiles) the enzyme has maximum activity around pH 4.8. In certain species that grow best in a near-neutral or middle range of pH (mesophiles as to pH) it has its maximum activity at about pH 7.4. In species capable of growth over a broader range of pH its activity is not so sharply restricted by pH. Probably the enzyme differs chemically in the different species but has the same specific function in each.

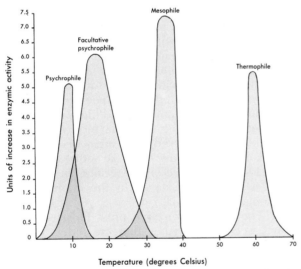

Figure 7–6. Relation between temperature and activity of a certain hydrolytic enzyme from different species of microorganisms. If obtained from species growing best in a cold environment (psychrophiles), the enzyme has its maximum activity at around 7C. In species capable of growing over a wider range of low temperatures (facultative psychrophiles) it acts well over a wider range but has its maximum activity at around 20C. In species preferring a middle range (mesophiles), the enzyme is most effective at around 37C. Note that it is rather sharply limited above 37C. In species growing only at temperatures above 50C (thermophiles) it has a rather narrow range with maximum at about 60C.

pH's for various species will be cited later. Some are very narrowly restricted.

Other factors of importance in enzyme functioning are hydrostatic and osmotic pressures and ultraviolet light and other radiations. Some enzymes in order to function require high concentrations of salt. Such factors will be discussed farther on (Chapt. 20). In general, whatever affects enzymes also affects microorganisms, including the individual cells of our own bodies, since the normal activity of all life depends on unhampered action of enzymes. The enzyme equipment of any cell is a major expression of its **genetic constitution** (DNA structure) (Chapt. 6).

Enzyme Inhibitors. In addition to chemicals that totally destroy all enzymes, there are many substances (disinfectants and sterilizing agents) that can combine chemically with certain particular enzymes or classes of enzymes, or with their coenzymes, and thus suppress their activity without destroying them or affecting others nearby. Some of these inhibitors combine with, or attack, the sulfhydryl group (—SH) that occurs in most proteins, regardless of whether the protein is enzymic or not.

Other enzyme inhibitors are more specific in that they affect only certain particular coenzymes, or even certain parts of certain coenzymes. For example, each molecule of hemoglobin, an enzyme-like metalloprotein, contains an atom of iron that is the essential part of the prosthetic group heme. It is this atom that carries oxygen from the lungs to the tissues. The iron in hemoglobin can combine more readily with the deadly poisons carbon monoxide (CO) and cyanide (HCN) than it can with oxygen. The combining site of the iron in the hemoglobin, normally occupied by oxygen, is pre-empted by the poisons. The oxygen is thus excluded; it is said to be **antagonized.** In this state the hemoglobin can no longer function as an oxygen carrier. The animal so poisoned dies, essentially of anoxia. A parallel example is poisoning of cytochrome oxidase of bacteria by HCN, CO, and azides (Na · CO · N$_3$). As shown in a foregoing paragraph, cytochrome is a heme-like oxygen carrier vitally important in many aerobic cells. In either case, hemoglobin or cytochrome, the iron of the prosthetic group is said to have been poisoned.

7.8
METABOLITE ANTAGONISM

The functioning of some enzymes may be inhibited by certain nonmetabolizable substances whose molecular structure is like (or analogous to) that of a true, metabolizable substrate (**metabolite**). Such metabolite-like inhibitory agents are called **metabolite antagonists** or **metabolite analogs.** A metabolite analog or antagonist, because of its molecular structure, can pre-empt the specific combining site on a particular enzyme protein (or in its coenzyme) to the exclusion of the true metabolite. The enzyme or coenzyme is not necessarily destroyed, but it can then no longer function. The cell involved may soon die or may remain for days or years in a state of suspended animation called **microbiostasis.** It is a static microbe! The phenomenon is called **metabolite antagonism.** The metabolite antagonist prevents functioning of the enzyme but remains attached to it, a "monkey wrench in the machinery."

For example, the dehydrogenase enzyme of lactic acid is inhibited by such compounds as hydroxy malonic acid and oxalic acid.

$$
\begin{array}{ccc}
\text{CH}_3 & \text{COOH} & \\
| & | & \\
\text{HCOH} & \text{HCOH} & \text{O}{=}\text{C}{-}\text{OH} \\
| & | & | \\
\text{COOH} & \text{COOH} & \text{O}{=}\text{C}{-}\text{OH} \\
\text{lactic acid} & \text{hydroxy malonic acid} & \text{oxalic acid}
\end{array}
$$

These antagonize the lactic acid because they possess, in their structure, the combination

$$
\begin{array}{c}
| \\
{-}\text{C}{-}\text{OH} \\
| \\
\text{COOH}
\end{array}
$$

Lactic acid is not antagnoized by malonic acid or methyl malonic acid:

$$
\begin{array}{cc}
\text{COOH} & \text{COOH} \\
| & | \\
\text{CH}_2 & \text{HC}{-}\text{CH}_3 \\
| & | \\
\text{COOH} & \text{COOH} \\
\text{malonic acid} & \text{methyl malonic acid}
\end{array}
$$

These nonantagonists, while somewhat resembling lactic acid in structure, lack the specific molecular combination requisite for antagonistic attachment to the enzyme. They have instead

$$
\begin{array}{c}
| \\
{-}\text{C}{-}\text{H} \\
| \\
\text{COOH}
\end{array}
$$

Be it noted, however, that oxalic and malonic acids are readily metabolized by some other organisms, while lactic acid is not; the metabolite antagonist of one may be the food of another.

Metabolite antagonism is often reversible when there is an excess of the true metabolite. The antagonist and the true metabolite are said to compete for the specific combining site on the enzyme.

It is worth making a special note of these basic principles of metabolite antagonism and of enzyme inhibition by specific poisons because they underlie the action of numerous antimicrobial drugs, such as sulfonamides and antibiotics.

7.9
CLASSIFICATION AND NOMENCLATURE OF ENZYMES

Early students of plant and animal physiology who discovered new enzymes often gave them descriptive names without consideration of any systematic scheme of nomenclature. Such names as **pepsin,** the protein-digesting enzyme of the stomach; **trypsin,** a proteolytic enzyme from the pancreas; and **ptyalin,** the starch-digesting enzyme in human saliva, are time-honored examples of the early method of naming enzymes. Later it became necessary to systematize nomenclature. For ordinary purposes of discussion it is convenient to name an enzyme by adding the suffix "ase" to the name of the substance acted upon (the substrate) or to the name of the activity of the enzyme. This simple scheme is used in the following outline, in which it can be seen that most enzymes can be gathered into six main groups on the basis of their overall functions—catalyzing oxidation and reduction reactions, catalyzing group transfer reactions, catalyzing hydrolytic reactions, catalyzing addition of groups to double bonds, catalyzing isomerization, and catalyzing condensation reactions.

The list contains only a few representative examples of each type of enzyme. Hundreds more are known; probably there are thousands yet to be discovered. Numerous enzymes are not readily classifiable, and the exact position and nomenclature of several in this listing are debatable.

Oxidizing and Reducing Enzymes (Oxidoreductases).
These enzymes catalyze the transfer of electrons, oxygen, or hydrogen. They are basically electron transferases.

Electron-Transfer Oxidases (formerly aerobic dehydrogenases or oxidases).

OXYGEN-OBLIGATIVE OXIDASES. These enzymes remove hydrogen atoms, and concomitantly their electrons (H^+e^-), from the substrate, thus oxidizing it (see Chapter 8). The hydrogen is transferred to an intermediate coenzyme or carrier and then to oxygen. These enzymes are restricted to systems using free oxygen as final acceptor for the substrate H^+e^-, hence are said to be oxygen-obligative.

$$O_2 + (4e^- + 4H^+) \rightarrow 2H_2O$$

OXYGEN-FACULTATIVE OXIDASES (or aerobic dehydrogenases). These enzymes remove substrate hydrogen like the foregoing but are not restricted to free oxygen as (H^+e^-) acceptor. The reduced carrier coenzyme can react with either free oxygen to form H_2O_2 or with other (H^+e^-) acceptors:

$$O_2 + (2e^- + 2H^+) \rightarrow H_2O_2$$

The H_2O_2 is commonly decomposed immediately by catalase (see Hydroperoxidases).

OXYGENASES (or oxygen transferases). These enzymes catalyze transfer of free oxygen directly to the substrate: $O_2 + 2Subs \rightarrow 2SubsO$. (Subs = substrate, commonly inorganic.)

HYDROXYLASES (or mixed-function oxidases). These catalyze a direct oxidation of the substrate with $\frac{1}{2}O_2$ instead of with O_2. Of a molecule of free oxygen (O_2), one atom is combined with an organic substrate while the other is reduced to H_2O by a separate or "auxiliary" (H^+e^-) donor, commonly an adjacent reduced coenzyme (Coenz H):

$$\text{Subs} + \begin{cases} \frac{1}{2}O_2 \longrightarrow \text{SubsO} \\ 2\text{CoenzH} + \frac{1}{2}O_2 \longrightarrow 2\text{Coenz} + H_2O \end{cases}$$

DEHYDROGENASES (or anaerobic dehydrogenases). Unlike oxidases, the coenzymes of neither dehydrogenases nor of the first carrier to which they transfer hydrogen can be directly reoxidized by free oxygen, hence the term anaerobic formerly used for these dehydrogenases. To be reoxidized these dehydrogenase coenzymes must pass the substrate (H^+e^-) to a second coenzyme and, in some cells, to a series of others and to the cytochrome system (see Chapter 8).

In some bacteria (facultative anaerobes) the substrate (H^+e^-) may, in the absence of free oxygen (i.e., under anaerobic conditions), be combined with oxygen from some readily reduced substance like $NaNO_3$, Na_2SO_4, or Na_2CO_3 to form $NaNO_2$, H_2S, or CH_4. All the dehydrogenases can operate in the presence of free oxygen but they cannot use it as an (H^+e^-) acceptor.

In this connection differentiate carefully between bacteria whose dehydrogenases may act in the presence of free oxygen or in its total absence (i.e., under entirely anaerobic conditions; e.g., facultative bacteria) and bacteria that, while they may contain similar dehydrogenases, are **strictly anaerobic** in the sense that they are poisoned by free oxygen and die in its presence (see Chapter 36).

Dehydrogenases are divided on the basis of their coenzymes: **NADP-linked** are those linked to the cytochrome system by a pyridine nucleotide, nicotinamide-adenine-dinucleotide phosphate (NADP), and **FAD-linked** are those linked to the cytochrome system by a flavin coenzyme, flavin-adenine dinucleotide (FAD), or alloxazine adenine dinucleotide.

Hydroperoxidases. As we have noted, oxygen-facultative, electron-transfer oxidases usually produce hydrogen peroxide as an end product. Many cells are very sensitive to hydrogen peroxide, including many medically and industrially important bacteria. Many peroxide-sensitive cells produce a hydroperoxidase called **catalase,** the coenzyme of which is a heme-like

TABLE 7-1. SOME TRANSFERRING ENZYMES

Types	Group Transferred	Coenzyme
Transaminases Transphosphorylases Transpeptidases	Amino ($-NH_2$) Phosphate ($H_2PO_4^-$) Entire peptide units:	Pyridoxal phosphate
Transglycosylases (formerly phosphorylases)	Entire glycosidic units:	In synthetic processes, uridine diphosphate
Transacylases	Acetyl $\left(\overset{O}{\underset{\|}{-C}}-CH_3\right)$ or other acyl groups	CoA

molecule. This decomposes H_2O_2 to oxygen and water:

$$2H_2O_2 \longrightarrow 2H_2O + O_2$$

Catalase activity of human salivary gland cells is easily demonstrated by mixing a few drops of drugstore hydrogen peroxide with saliva.

Transferring Enzymes (Transferases). The transferases catalyze transfer, from one molecule to another, of various groups that are not in the free state during the transfer (Table 7-1).

Hydrolyzing Enzymes (Hydrolases). The hydrolases catalyze reactions involving hydrolysis as shown in Table 7-2.

Lyases. This group of enzymes catalyzes the addition of groups to, or removal of groups from, double bonds. Threonine aldolase, for ex-

ample, catalyzes the formation of glycine from L-threonine by removal of acetaldehyde.

Isomerases. These enzymes catalyze isomerization by intramolecular rearrangements of H^+ and OH^-:

$$\text{glucose-6-phosphate} \xrightarrow{\text{phosphohexose isomerase}}$$

$$\text{fructose-6-phosphate}$$

Ligases. These enzymes do not catalyze hydrolysis, oxidation-reduction, or transfer of chemical groups from one molecule to another. Groups are *added to* molecules from the free state in the surrounding medium or *released from* molecules to the surrounding medium in the free state.

For example, carboxylases catalyze addition

TABLE 7-2. SOME HYDROLYZING ENZYMES

Substrate Types	Substrates	Kind of Linkage Attacked
Carboxylesterases	Esters of carboxylic acids	Simple ester
Lipases	Fats (triglycerides)	Lipid ester
Phosphatases	Esters of phosphoric acid	Phosphate ester
"Nucleases" (phosphodiesterases)	Nucleic acids	Phosphate diester
Peptidases	Proteins, polypeptides	Peptide
Glycosidases	Polysaccharides, oligosaccharides	Glycosidic

of CO_2 to organic acids to form carboxyl groups (require preliminary phosphorylation):

$$CH_3 \cdot CO \cdot COOH + CO_2 \longrightarrow$$

pyruvic acid

$$COOH \cdot CH_2 - CO \cdot COOH$$

oxalacetic acid

In addition to the enzymes in this listing there are numerous others not readily classifiable. Among these are the permeases.

Transporting Enzymes (Permeases). Cell membranes are said to have a **selective permeability,** i.e., they are highly selective and discriminatory in regard to the substances that may pass through them, inward or outward. Selective permeability is related to molecular structure of both membrane and substance passing through it. Many of the seemingly great physiological differences between various species of cells exist merely because of differences between (1) the molecular structures of their cell membranes or (2) certain enzymes. The structure and function of membranes will be discussed further in Chapter 14. Here, we wish only to point out that various substances pass through cell membranes by at least two mechanisms: passive transport and active transport.

Passive Transport. The substance (or substrate), especially if it is an electrolyte or a relatively small molecule, may diffuse passively through the membrane (a) by diffusion (osmosis), (b) because of solubility in certain components, especially lipids, of the cell membrane, or (c) because the substrate is in higher concentration outside the cell membrane than inside. The substrate tends to move with the concentration gradient, i.e., from higher to lower concentration. When the inner and outer concentrations are in equilibrium or when the inner concentration is physiologically ideal for that particular species of cell, inward diffusion either ceases or is balanced by equal outward diffusion. This solubility-diffusion type of mechanism is called **passive transport.** The ionic and solubility relationships are extremely complex.

Active Transport. Substrates may also be transported through cell membranes **counter** to concentration gradient. Since this requires added energy, the process is called **active transport.** Active transport may be accomplished by one or both of two systems: enzymic and nonenzymic.

ENZYMIC SYSTEM. The substrates, including large molecules such as proteins that could not passively diffuse through the membrane, are transported through by the action of one or more enzymes or enzyme transport systems sometimes called **permeases.** The permeases, like other enzymes, are specific for the substrate involved, they are inducible, and they are therefore genetically controlled. The exact mechanisms of permease action are still under investigation.

NONENZYMIC SYSTEM. Not all active transport is necessarily enzymic. For example, some seaweeds (e.g., kelp) concentrate such large quantities of iodine (as potassium iodide) from the minute quantities in sea water that these algae have served as valuable commercial sources of potassium and iodine.

CHAPTER 7
SUPPLEMENTARY READING

Atkinson, D. E.: Regulation of enzyme activity. Ann. Rev. Biochem., *35*:85, 1966.
Commission of Editors of Biochemical Journals: Enzyme nomenclature. Science, *150*:719, 1965.
Neidhardt, F. C.: Roles of amino acid activating enzymes in cellular physiology. Bact. Rev., *30*:701, 1966.
Philips, D. C.: The three-dimensional structure of an enzyme molecule. Sci. Am., *215*:78, 1966.

CHAPTER 8 · BIOENERGETICS

Chemical reactions that yield heat or energy are said to be **exothermic** or **exergonic** (Gr. *ex* = out from; *therme* = heat; *ergon* = work or energy). In microbiology, the most important exergonic reactions are concerned with the oxidation of nutrients as sources of energy. Such reactions, if they occur in the presence of available oxygen, are often called **respiration**. However, the term respiration seems to imply to many the presence of lungs, gills or other complex breathing mechanisms. In dealing with unicellular organisms the term biological oxidation or biooxidation is preferable, because it implies only the type of chemical reaction and not the machinery. Perhaps a still more accurate term would be "biological electron transfer" or, for brevity, "bioelectronics" since, as will be explained, all processes of oxidation, bio- or otherwise, and whether or not available oxygen is present, are fundamentally transfers of electrons from the oxidized food substance or **substrate.**

8.1
OXIDATION AND REDUCTION

Physiologically, the term biooxidation is associated with an old idea that free or atmospheric oxygen is necessary to life. This view was eclipsed when Pasteur showed that there are numerous species of microorganisms, e.g., yeasts, that can thrive without free oxygen— a phenomenon that Pasteur called **anaerobiosis:** "life without air." As now defined, any biological oxidation is fundamentally the removal of electrons from various substrates and does not necessarily involve oxygen.

Electrons cannot, in the living cell, remain in a free state. If removed from a substrate (an **electron donor** in this context) which is thereby oxidized, there must be something (an **electron acceptor**) immediately present to which the electrons can be transferred. The electron acceptor is thereby reduced. Thus when we speak of the oxidation of something we imply the concomitant reduction of something else. Oxidations are always, therefore, coupled reactions and are actually oxidation-reduction or **redox** reactions. Oxidation is loss of electrons; reduction is gain of electrons. An electron acceptor is an oxidizing agent; an electron donor is a reducing agent.

In the "traditional" form of oxidation, i.e., addition of oxygen to the substrate, electron transfer is not obvious but is based on the fact that the oxygen accepts two electrons from the substrate: e.g., $Cu + O = Cu^{2+}O^{2-}$, an ionic crystal. There are several species of soil bacteria that obtain energy from the enzymic oxidation of such strange nutrients as molecular hydrogen, sulfur, and carbon monoxide. In such oxidations free oxygen is combined with the substrate:

$$H_2 + \frac{1}{2}O_2 \longrightarrow H_2O + Energy$$
$$CO + \frac{1}{2}O_2 \longrightarrow CO_2 + Energy$$
$$2S + 2H_2O + 3O_2 \longrightarrow 2H_2SO_4 + Energy$$
$$NaNO_2 + \frac{1}{2}O_2 \longrightarrow NaNO_3 + Energy$$

Oxidation can also occur in the total absence of oxygen. Electron transfer is involved in any increase in positive valence (i.e., loss of an electron), as when ferrous iron is oxidized to ferric iron: $Fe^{2+} - (e^-) + (e^- \ acceptor) \longleftrightarrow Fe^{3+} + (acceptor \cdot e^-)$. This is an important type of reaction that occurs in the iron of iron-containing respiratory pigments, such as the cytochrome pigments of most aerobic bacteria (to be discussed).

Dehydrogenation

In most cells oxidation of organic substrates is accomplished by the enzymic removal of hydrogen, or **dehydrogenation**. The enzymes that catalyze dehydrogenations are called **dehydrogenases** or **oxidoreductases**. When hydrogen is thus removed from a substrate molecule, an electron accompanies it; i.e., a hydrogen atom may be thought of as a hydrogen ion (H^+) with an electron (e^-) attached: $H^+ + e^- = H$ or (H^+e^-) or hydrogen(e^-). The H^+ may be considered to go into solution, the electron being taken into an enzymic electron-transfer system or "oxidative chain" that is described farther on. The enzyme systems, including the initial dehydrogenases, involved in hydrogen(e^-) transfers are often called **electron-transfer systems**. An example of oxidation of an organic substrate by dehydrogenation is seen in the enzymic oxidation of alcohol to aldehyde, with free oxygen as hydrogen(e^-) acceptor:

$$H-\underset{\underset{H}{|}}{\overset{\overset{CH_3}{|}}{C}}-OH - 2(H^+e^-) + \tfrac{1}{2}O_2 \longrightarrow$$

ethyl alcohol

$$H-\underset{}{\overset{\overset{CH_3}{|}}{C}}=O + H_2O$$

acetaldehyde

Oxidation of an inorganic substrate by dehydrogenation is seen in the enzymic oxidation of hydrogen sulfide to sulfur:

$$H_2S - 2(H^+e^-) + \tfrac{1}{2}O_2 \longrightarrow S + H_2O$$

This reaction occurs in several species of soil bacteria (Chapt. 43).

Hydrogen Transport. Dehydrogenases, like other enzymes, are highly specific for their respective substrates. Their coenzymes are less restricted; one coenzyme may act in turn with several different dehydrogenase apoenzymes or, conversely, several different dehydrogenase apoenzymes may use the same coenzyme. The molecular structure and functions of the dehydrogenase coenzymes are of particular interest.

NAD. Dehydrogenation in many cells is initiated by the coenzyme **nicotinamide-adenine-dinucleotide** (NAD) or by **NAD phosphate** (NADP). These coenzymes were long known as diphospho-pyridine nucleotide (DPN) and triphosphopyridine nucleotide (TPN), respectively, and are found in all cells. These coenzymes act with many different apoenzymes that determine specificity of the enzyme for its substrate. The essential portion of NAD is in the nicotinamide group: the vitamin **niacin** (Fig. 8–1). The pyridine ring of this group accepts substrate hydrogen, being reduced to $NADH_2$. The $NADH_2$ yields the hydrogen to a second hydrogen acceptor, thus being reoxidized to NAD. Dehydrogenases that transfer substrate hydrogen by way of NAD or NADP are said to be **NAD-** or **NADP-linked**. In aerobic cells NAD and NADP commonly transfer hydrogen(e^-) to a second hydrogen(e^-) carrier at a lower energy level, e.g., FAD.

FAD. FAD is a coenzyme common to several dehydrogenases. It is a yellow-colored nucleotide commonly called **flavin-adenine dinucleotide** (FAD). Dehydrogenases that transfer substrate hydrogen(e^-) via FAD are said to be **flavin-linked**. As a point of interest it may be noted, as seen in Figure 8–2, that FAD contains the vitamin **riboflavin (vitamin B_2)**. So far as is known all vitamins are analogous parts of coenzymes.

When reduced by substrate hydrogen to $FADH_2$, the coenzyme may be reoxidized by enzymic transfer of the hydrogen(e^-) directly to oxygen, forming H_2O_2. Very commonly $FADH_2$ is reoxidized to FAD by transfer of the hydrogen(e^-) to an iron-bearing respiratory pigment of the **cytochrome system**.

The Cytochrome System

The **cytochromes** (Gr. *cyto* = cell; *chroma* = color) are a group of yellow pigments (cytochromes a, a_3, b, c, and others) commonly found in aerobic or facultative cells—cells that use free oxygen or an alternative substance as final hydrogen(e^-) acceptor, e.g., $NaNO_3 + H_2 \longrightarrow NaNO_2 + H_2O$. Like the **heme** (red matter) in our red blood cells, cytochrome pigments contain iron in organic combination (**porphyrin**, Fig. 7–2). This iron can be readily reduced or oxidized by changing its valence as previously explained. The energy level of each successive cytochrome pigment in the series is progressively lower. Cytochromes occur only rarely in cells that cannot use free oxygen or an alternative as their final hydrogen(e^-) acceptor.

In the operation of the cytochrome system,

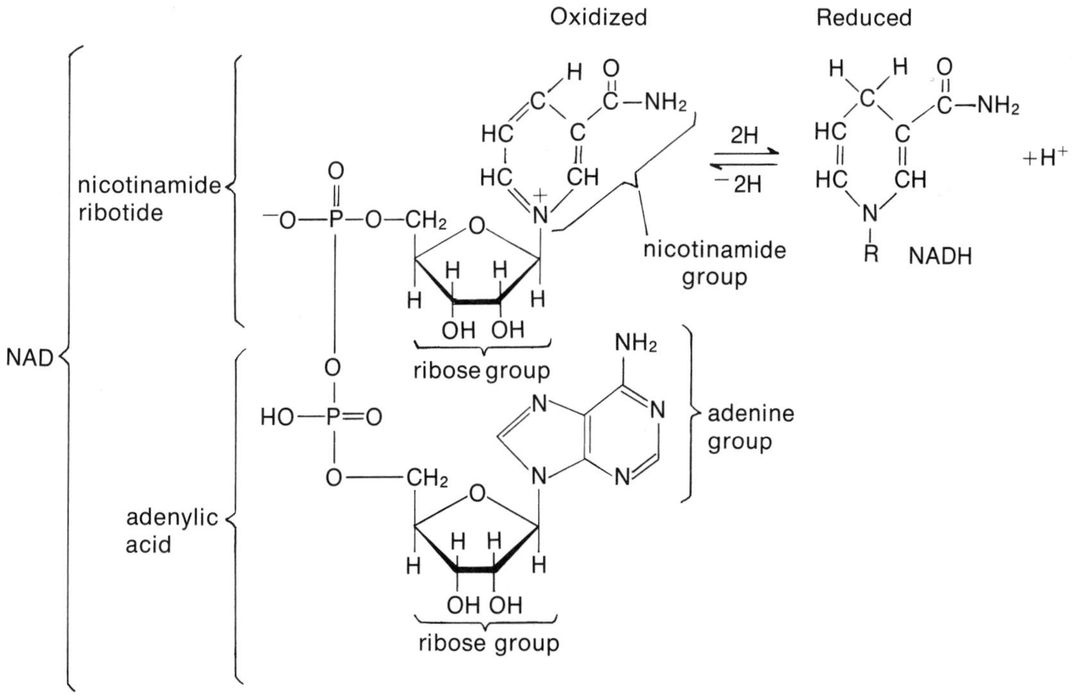

Figure 8-1. Diagram of the structure of the molecule of nicotinamide adenine dinucleotide (NAD).

substrate hydrogen is ionized, the H^+ going into solution. The electrons are transferred to the cytochrome system. The cytochromes mediate the transfer of the e^-, not of the H^+, which remains in solution. The electrons descend the cytochrome "stairway," being transferred from one energy level to the next lower, yielding energy at each step. The final cytochrome of the series is relieved of its electrons by **cytochrome oxidase** (cytochrome a_3), the enzyme that catalyzes final combination of the hydrogen ion, the electron, and free oxygen—or an alternative hydrogen(e^-) acceptor, such as $NaNO_3$, H_2SO_4, or CO_2 (Fig. 8-3).

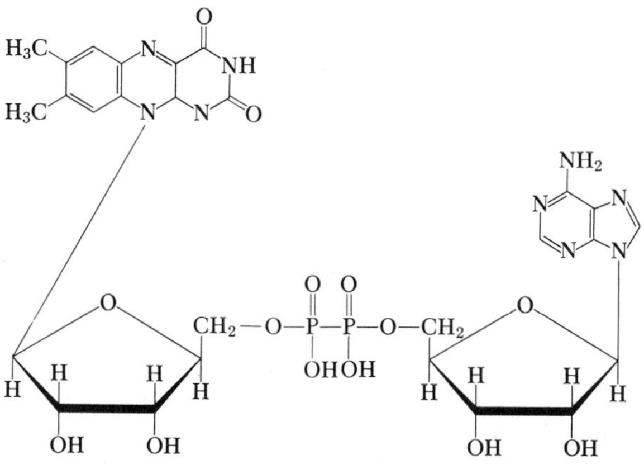

Use of Combined Oxygen

Numerous species, among them certain bacteria and fungi, preferentially utilize uncombined (free) oxygen as hydrogen acceptor by means of the cytochrome system; when free oxygen is scarce or entirely absent, they are able to use oxygen in combined form as in $NaNO_3$ (see reaction on page 102). Such organisms are said to be **facultative** with regard to oxygen. The process involves the activity of an inducible enzyme called **nitratase** that activates the combined oxygen. Mo^{2+} is an essential cofactor. The activated oxygen appears then to be passed to the cytochrome oxidase of the cytochrome system. Presence of free oxygen competitively interferes with nitrate reduction. Diagnostic tests for nitrate reduction should therefore be made under anaerobic conditions. Some organisms also reduce nitrates to nitrites via **nitritase** and then to nitrogen. Some cells use part of the nitrogen for synthesis of amino acids. The exact

Figure 8-2. Flavin-adenine dinucleotide (FAD). The adenine group is shown at upper right (compare Figure 8-1) and a ribose group below it. The two are connected through a phosphate bond to the riboflavin 5'-phosphate group at the left.

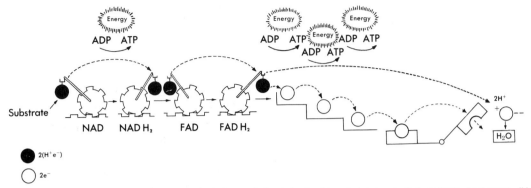

Figure 8–3. A typical hydrogen (e⁻) transport chain or "physiological bucket brigade." Substrate hydrogen (H^+e^-) is transferred to the dehydrogenase coenzyme NAD which thus becomes $NADH_2$. $NADH_2$ yields its $2(H^+e^-)$ to the dehydrogenase coenzyme FAD, which thus becomes $FADH_2$, with the $NADH_2$ being thereby reoxidized to NAD. Energy is released and stored in ATP. The $2(H^+e^-)$ then go into solution and the $2e^-$ are transferred to the cytochrome system, where further energy is released to ATP. Final combination of $2H^+$, $2e^-$, and oxygen of the air outside the cell is catalyzed by the enzyme cytochrome oxidase.

chemical pathways differ in species and remain to be fully clarified.

Several important bacteria can use combined oxygen in the form of sulfates or carbonates (CO_3^{2-}). These organisms generally have little or no cytochrome system and are therefore restricted to anaerobic conditions of growth. They are said to be **strict anaerobes.** Sulfates may be reduced to H_2S; CO_2, to CH_4. The cells may also derive their S and C from such materials (see also Chapter 43).

Energy-Rich Bonds

Energy released by biooxidation of a substrate is immediately taken up into a particular kind of hydrolyzable bond spoken of as a high energy or **energy-rich** (e-r) bond, for which ~ is the conventional symbol. Actually, the bond itself is not solely concerned; the energy is distributed in the molecule itself. For present purposes, however, the concept of the bond is a convenient usage if its broader implication is kept in mind. A more precise term might be "compound with high energy level or potential." Several compounds in living cells have such e-r bonds: thioesters like

$$\overset{\text{O}}{\overset{\|}{}}$$

acetyl~SCoA: $CH_3C~SCoA$ and phosphates like phosphoenol pyruvic acid:

$$\begin{array}{c} \text{COOH} \\ | \\ H_2C{=}C{-}O{\sim}PO_3H_2 \end{array}$$

Functions of some of these will be discussed farther on.

Among the most important phosphates in energy transfer and storage for the present discussion (i.e., in bacteria) are those of **adenosine diphosphate** (ADP) and **adenosine triphosphate** (ATP). These, like NAD and FAD, are **nucleotide coenzymes** (Fig. 8–4). Unlike NAD and FAD, ADP and ATP accept and transfer phosphate groups instead of hydrogen(e⁻). Generation of e-r bonds, especially those of ATP, is one of the most important immediate results of biooxidation.

In nonphotosynthetic cells, the formation of e-r phosphate bonds of ATP may occur at several stages, or energy levels, of the stepwise process of oxidation of an organic substrate such as glucose. Glucose is a source of energy for almost all living cells, though utilization may involve varied metabolic pathways.

8.2 PHOSPHORYLATION

The Embden-Meyerhof Pathway. In the dissimilation of glucose via the series of steps called the **Embden-Meyerhof pathway** or **glycolytic pathway** (Table 8–1), one derivative of the glucose molecule, i.e., 3-phosphoglyceraldehyde, is dehydrogenated via NAD and concurrently combined with inorganic phosphate (P_i) to form $NADH_2$ and 1,3-diphosphoglyceric acid. The added phosphate of the 1,3-diphosphoglyceric acid, with energy derived from the simultaneous dehydrogenation, is taken over by

ADP, forming e-r ATP. The 3-phosphoglyceric acid residue then undergoes a molecular rearrangement to form 2-phosphoglyceric acid. This is dehydrated, providing an e-r bond in the resulting phosphoenol pyruvic acid. This e-r phosphate is then transferred to ADP to form more ATP, leaving pyruvic acid as a final residue at this stage in the dissimilation of the glucose molecule. The formation of new e-r phosphate bonds in derived portions of the substrate itself is generally called **substrate-level phosphorylation,** and is best exemplified in the Embden-Meyerhof pathway after the formation of the trioses (step 4, Table 8–1).

Oxidative-Chain Phosphorylation. In the presence of available oxygen, when the cytochrome system is involved in substrate oxidations, at least three additional phosphorylations occur during the progress of the substrate electrons from FAD, down the "steps" of the oxidative chain, to lower and lower energy levels in the electron-transporting cytochrome system. At each "descent" the electrons give up some of their energy. In the presence of P_i, ADP, and several enzymes, more e-r ATP is formed. These **oxidative-chain,** or **oxidative, phosphorylations,** as they are called, yield much additional energy (Fig. 8–4). As previously mentioned, the use of free oxygen as a final hydrogen(e^-) acceptor is often called respiration. If, in the absence of free oxygen, an alternative **inorganic** hydrogen (e^-) acceptor is used, e.g., $NaNO_3$, the biooxidation is sometimes called **anaerobic respiration.**

Photophosphorylation. In addition to the phosphorylations just described, a third type of phosphorylation, termed **photophosphorylation,** occurs in photosynthetic species of cells. The process derives its energy from the effect of light on electrons in chlorophyll. The net result is the same as in substrate-level or oxidative-chain phosphorylations: i.e., generation of ATP. However, the reaction is light-dependent. It is discussed more fully in Chapter 32.

Phosphate-Bond Energy

The energy of the phosphate bonds in ATP formed during all biooxidations, including photosynthesis, is released when the phosphate bonds are hydrolyzed. ATP is thereby degraded to ADP; the energy released is used in synthesis (Chapt. 6) and other vital functions of the cell. The ADP is then available to accept a new e-r phosphate group. Thus ADP and ATP act together as an energy transfer and storage system; this is one of the most important mechanisms in biooxidation (Fig. 8–4).

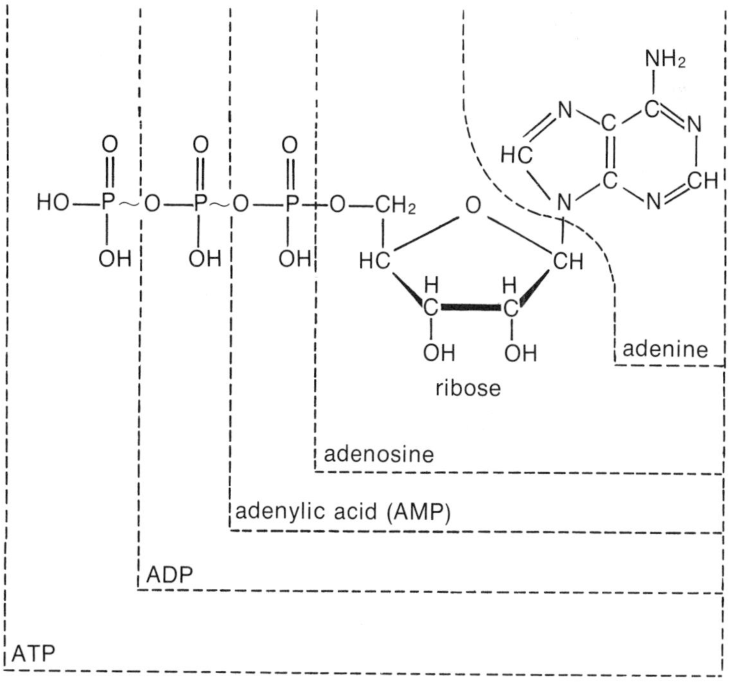

Figure 8–4. Molecular structure of ADP and ATP.

TABLE 8-1. THE EMBDEN-MEYERHOF (GLYCOLYTIC) PATHWAY FOR GLUCOSE DISSIMILATION

Step No.	Substrate	Enzyme	Products	Result of Reaction
1.	$C_6H_{12}O_6 + AT \approx P$ glucose	$\xrightarrow{\text{hexokinase}}$	$C_6H_{11}O_6 \cdot PO_3H_2 + AD \sim P$ **glucose-6-phosphate**	*Energy* is yielded by breaking one energy-rich phosphate bond ($\sim$) of $AT \approx P$, leaving $AD \sim P$.
2.	Glucose-6-phosphate	$\xrightarrow{\text{phosphohexose isomerase}}$	$C_6H_{11}O_6 \cdot PO_3H_2$ **fructose-6-phosphate**	Molecule is rearranged; no yield of energy.
3.	Fructose-6-phosphate $+ AT \approx P$	$\xrightarrow{\text{phosphohexokinase}}$	$C_6H_{10}O_6 \cdot (PO_3H_2)_2 + AD \sim P$ **fructose-1,6-diphosphate**	Molecule takes phosphate from $AT \approx P$, yielding *energy* and leaving $AD \sim P$.
4.	Fructose-1,6-diphosphate	$\xrightarrow{\text{zymohexase}}$ **(aldolase)**	$C_3H_4O_2 \cdot OH \cdot PO_3H_2 + CHO \cdot CHOH \cdot CH_2O \cdot PO_3H_2$ dihydroxyacetone phosphate (a form of triose) glyceraldehyde-3-phosphate (a form of triose)	Molecule is split into two interchangeable triose molecules. No yield of energy.
5.	Glyceraldehyde-3-phosphate $+ P_i + NAD$	$\xrightarrow{\text{phosphoglyceraldehyde dehydrogenase}}$	$C_3H_4O_4 \cdot (PO_3H_2)_2 + NADH_2$ **1,3-diphosphoglyceric acid**	Triose molecule takes up phosphate. H_2 is yielded to NAD, forming $NADH_2$ and yielding *energy* to be stored (see step 6).
6.	1,3-Diphosphoglyceric acid $+ AD \sim P$	$\xrightarrow{\text{diphosphoglyceric dephosphorylase}}$	$C_3H_5O_4 \cdot PO_3H_2 + AT \approx P$ **3-phosphoglyceric acid**	*Energy* yielded in step 5 is stored as $\sim$ by changing $AD \sim P$ to $AT \approx P$. Molecule loses phosphate.
7.	3-Phosphoglyceric acid	$\xrightarrow{\text{phosphoglyceromutase}}$ **(triose mutase)**	$C_3H_5O_4 \cdot PO_3H_2$ **2-phosphoglyceric acid**	Molecular rearrangement. No yield of energy.
8.	2-Phosphoglyceric acid	$\xrightarrow{\text{enolase}}$	$C_3H_3O_3 \cdot PO_3H_2$ **phosphoenol-pyruvic acid**	2-phosphoglyceric acid yields H and OH. *Energy* is stored in step 9.
9.	Phosphoenol-pyruvic acid $+ AD \sim P$	$\xrightarrow{\text{phosphopyruvate dephosphorylase}}$	$C_3H_4O_3 \cdot \left\{ \begin{array}{l} CH_3 \\ C{=}O \\ COOH \end{array} \right\} + AT \approx P$ **pyruvic acid**	Molecule is decomposed to pyruvic acid; phosphate and $\sim$ being transferred to $AD \sim P$ which thus becomes $AT \approx P$.

The energy of the phosphate bonds may also be readily transferred into some other types of e-r bonds (e.g., thioesters) at about the same energy level, and vice versa. In some reactions both of the phosphate bonds of ATP are hydrolyzed, forming ADP and AMP successively. In other reactions pyrophosphate is split off from ATP, leaving AMP. AMP is restored to ATP by successive phosphorylations or by combination of AMP with pyrophosphate. Similar nucleotides of other nucleosides, e.g., guanidine, uridine, and cytidine, carry on similar functions in other energy-yielding reactions and energy-transfer systems (see material on the Krebs* cycle, this chapter; protein synthesis, Chapter 6; etc.) in various cells.

8.3
COMPLETE AND INCOMPLETE BIOOXIDATIONS

Biooxidation in the Presence of Available Oxygen. In most aerobic and some facultative bacteria, biooxidation of a substrate is complete. If the substrate is an organic compound such as glucose, the end products are CO_2 and H_2O. This is illustrated by the aerobic utilization of glucose by bakers' yeast (*Saccharomyces cerevisiae*):

$$C_6H_{12}O_6 + 6O_2 \longrightarrow 6CO_2 + 6H_2O + 688{,}500 \text{ cal}$$

If the substrate is inorganic, the oxidation may be equally complete, even though the end products differ, as in the oxidation of sulfur to sulfuric acid (see page 119). Such oxidations theoretically release all the energy available from the substrate. They are characteristic of aerobic metabolism. Actually, parts of the substrate are usually utilized in cell synthesis.

However, even in the presence of free oxygen, biooxidation may be incomplete. Much depends on the species, the enzymic equipment of the cell, and the availability of an alternate hydrogen(e⁻) acceptor. An example of aerobic but incomplete oxidation is the formation of vinegar (acetic acid) from wine (alcohol) by *Acetobacter*. These aerobic bacteria, long familiar as "mother of vinegar," grow as a scum on the surface of wine and convert it to vinegar:

*Named for H. A. Krebs, biochemist, Nobel Prize winner with F. A. Lipmann.

$$C_2H_5OH + O_2 \longrightarrow CH_3COOH + H_2O + 118{,}000 \text{ cal}$$

Much of the available energy of the alcohol is left in the acetic acid. Many aerobic fungi can utilize this energy by oxidizing the acetic acid to H_2O and CO_2:

$$CH_3COOH + 2O_2 \longrightarrow 2CO_2 + 2H_2O$$

Biooxidation in the Absence of Available Oxygen. Biooxidations may occur in the total absence of available oxygen; in several genera of strict anaerobes, biooxidations can occur *only* in the total absence of free oxygen. Biooxidations in the total absence of free or available oxygen, when the hydrogen(e⁻) acceptor is an **organic** substance, are called **fermentations**.

8.4
FERMENTATION

Elsewhere (Chapts. 7 and 36) we have indicated that some organisms, called **strict aerobes,** are enzymically equipped to use *only* free oxygen as final hydrogen(e⁻) acceptor; that others, called **facultative aerobes,** are equipped to use, as final hydrogen(e⁻) acceptor, either free oxygen or some alternative, reducible, **inorganic** substrate, commonly a nitrate. Here we shall consider biooxidations that occur in the absence of available inorganic oxygen and in which both hydrogen donors and acceptors are different parts of the same organic substrate molecule. Such biooxidations are classed as **fermentations;** the substrates are commonly, but not necessarily, carbohydrates. In fermentations usually only NAD or NADP functions as hydrogen(e⁻) carrier. FAD and the cytochrome system are not required since the final hydrogen(e⁻) acceptor is not oxygen but an organic substance, commonly pyruvic acid.

The term fermentation is sometimes erroneously used for certain strictly aerobic processes, especially industrial processes, e.g., the aerobic oxidation of alcohol to acetic acid by *Acetobacter* sp., in "vinegar fermentation."

Fermentation may be caused by facultative organisms under anaerobic conditions, e.g., brewer's yeast (*Saccharomyces cerevisiae*); by **strictly anaerobic** organisms, e.g., bacteria of the genus *Clostridium;* or even in the presence of free oxygen by "indifferent" organisms that do not ordinarily utilize free oxygen, e.g., species of *Lactobacillus.*

Products of Fermentation. Depending on the conditions of growth, the substrate, and the organisms involved, the end products of fermentations vary greatly. *Clostridium* species commonly ferment glucose to yield butyl and other alcohols and certain acids; *Lactobacillus lactis* yields almost entirely lactic acid, while *L. brevis* yields lactic and acetic acids, ethyl alcohol, and carbon dioxide. Other microorganisms produce numerous other valuable end products of fermentation (Fig. 8–5; see also Chapter 49).

Pyruvic Acid (Pyruvate). Pyruvate is physiologically the most important first intermediate product in most biooxidations of glucose, aerobic or anaerobic. It is commonly produced from glucose by stages, as shown in Table 8–1. However, several other analogous series of reactions occur in many bacteria and other microorganisms, e.g., the Entner-Doudoroff cycle in bacteria of the large and important genus *Pseudomonas* (Chapt. 33). In this cycle the pathway to pyruvate progresses via glucose-6-phosphate, 6-phosphogluconic acid, 2-keto-3-deoxy-6-phosphogluconic acid, to pyruvate plus 3-phosphoglyceraldehyde.

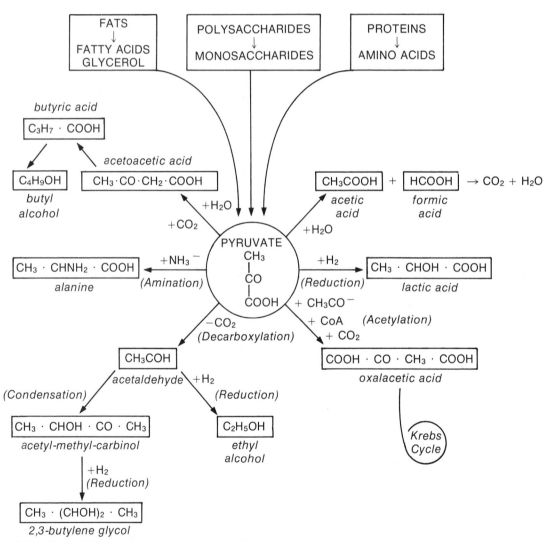

Figure 8–5. Some of the products of aerobic-anaerobic dissimilation of pyruvate by a common facultative bacterium, *Enterobacter aerogenes.* Note that fats, carbohydrates and proteins can undergo preliminary digestive hydrolysis to yield products that may yield pyruvate. The derivatives of pyruvate shown here are only a few of the many possibilities.

Pyruvate may also be reached via the metabolism of sugars other than glucose, and by pathways other than those cited via the metabolism of fatty acids and of amino acids. The monosaccharides, fatty acids and amino acids are, of course, derived mainly from the preliminary digestive hydrolysis of starch, glycogen, cellulose, fats, poly-β-hydroxybutyric acid, chitin, or proteins depending on the species of organism. Thus, many kinds of organic nutrient may contribute to the pyruvate content of the cell.

Pyruvate is a sort of chemical "Grand Central Station" in that it is the point of arrival and departure for a wide variety of metabolic substrates and products (Fig. 8–5). Pyruvate serves as the electron acceptor in most fermentations, being reduced in many types of cell to lactic acid. It may also be decarboxylated and the product reduced to ethyl alcohol. Conversely, it may serve as a **source** of amino acids, fatty acids, aldehydes, or other important cell-building materials.

For example, a series of reactions by which many amino acids are changed to pyruvate is summarized in the oxidative deamination of alanine:

$$
\begin{array}{ccc}
\text{CH}_3 & & \text{CH}_3 \\
| & & | \\
\text{H}-\text{C}-\text{NH}_2 + \tfrac{1}{2}\,\text{O}_2 \longrightarrow & \text{C}=\text{O} & + \text{NH}_3 \\
| & & | \\
\text{O}=\text{C}-\text{OH} & \text{O}=\text{C}-\text{OH} & \\
\text{alanine} & \text{pyruvic acid} &
\end{array}
$$

Long-chain fatty acids undergo somewhat more complex changes, but the residues of the changes often become pyruvate. For example, propionic acid may be oxidized to pyruvate:

$$
\begin{array}{ccc}
\text{CH}_3 & & \text{CH}_3 \\
| & & | \\
\text{C}=\text{H}_2 + \text{O}_2 \longrightarrow & \text{C}=\text{O} & + \text{H}_2\text{O} \\
| & & | \\
\text{O}=\text{C}-\text{OH} & \text{O}=\text{C}-\text{OH} & \\
\text{propionic acid} & \text{pyruvic acid} &
\end{array}
$$

A common source of pyruvate, especially in fermentations, is lactic acid:

$$
\begin{array}{ccc}
\text{CH}_3 & & \text{CH}_3 \\
| & & | \\
\text{H}-\text{C}-\text{OH} + \tfrac{1}{2}\,\text{O}_2 \longrightarrow & \text{C}=\text{O} & + \text{H}_2\text{O} \\
| & & | \\
\text{O}=\text{C}-\text{OH} & \text{O}=\text{C}-\text{OH} & \\
\text{lactic acid} & \text{pyruvic acid} &
\end{array}
$$

If a substance, e.g., acetic acid, CH_3COOH, lacks a carbon atom to form pyruvate, it can be changed (via a series of enzymic reactions, with energy from CoA~S) to a three-carbon acid by addition of —C=O; if it has too many carbon atoms, it may be split enzymically; pyruvate is the biochemical procrustean bed which many molecules are made to fit.

A wide variety of end products of incomplete decomposition of glucose may be derived under a common condition of growth (e.g., in a test tube of glucose-infusion broth which is partly aerobic and partly anaerobic). For example, from a common saprophyte, the gram-negative, facultative bacillus *Enterobacter aerogenes*, the following end products have been shown to occur: hydrogen, carbon dioxide, ethyl alcohol, acetic acid, acetyl-methyl-carbinol, 2,3-butylene glycol, lactic acid, glycerol, succinic acid. There are undoubtedly others in minute amounts (Fig. 8–5).

Some of the end products of metabolism, e.g., acetyl-methyl-carbinol, H_2, and CO_2, are easily tested for in a culture tube and are often distinctive of certain species. Distinctive products of metabolism of various substances, glucose among others, are extremely useful in identifying certain organisms. Also, as already indicated, some of these products, e.g., lactic acid and butyl alcohol, are of great industrial value (Chapt. 49).

8.5
THE CITRIC ACID CYCLE
(Tricarboxylic Acid or Krebs Cycle)

From a physiological standpoint one of the most important derivatives of pyruvate is **oxalacetic acid (oxalacetate)**. This is a "stepping-stone" into the cyclical series of reactions that are involved in the terminal oxidation process in many aerobic biooxidations, i.e., the **citric acid** or **Krebs cycle.**

In the presence of available inorganic oxygen and in cells of aerobes and facultative anaerobes with necessary enzymes, pyruvate undergoes complete oxidation to CO_2 and H_2O through a cyclical series of interdependent reactions. The pyruvate is first dehydrogenated by pyruvic dehydrogenase, with liberation of energy that is taken up in ATP. In this reaction, as in many other enzymic catalyses, various ions, e.g., Mg^{2+}, NH_3^+, K^+, are essential as cofactors.

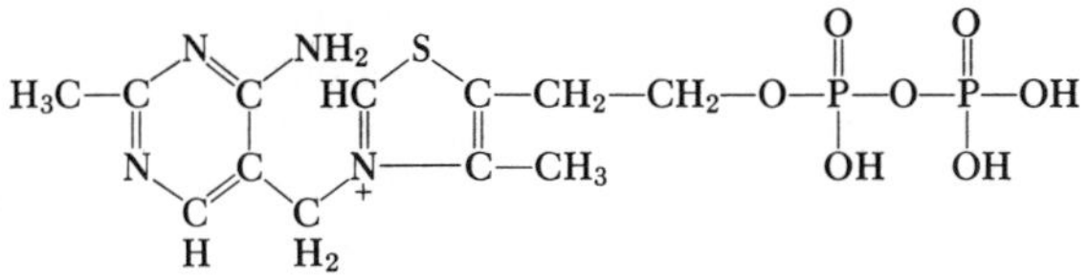

Figure 8-6. Thiamine pyrophosphate (cocarboxylase).

The pyruvate is also decarboxylated by the keto-acid-decarboxylating coenzyme called **cocarboxylase** or **thiamine pyrophosphate** (Fig. 8-6). Note the presence in this coenzyme of the antiberiberi vitamin, **thiamine**, or **vitamin B$_1$**. The acetyl residue (CH_3CO—) is transferred to the acetyl-transferring coenzyme, coenzyme A (CoA), carrying an e-r S-bond ($CH_3CO \cdot S \sim CoA$; Fig. 8-7):

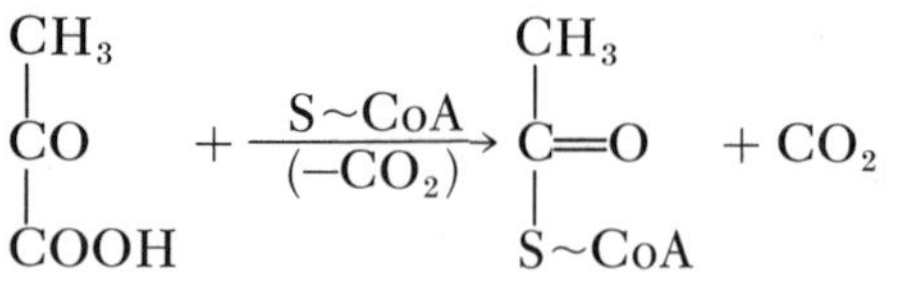

Note the presence, in CoA, of **pantothenic acid, vitamin B$_6$**.

At the same time, oxalacetate is derived from pyruvate by carboxylation (combination with CO_2) through a series of steps, the overall reaction being:

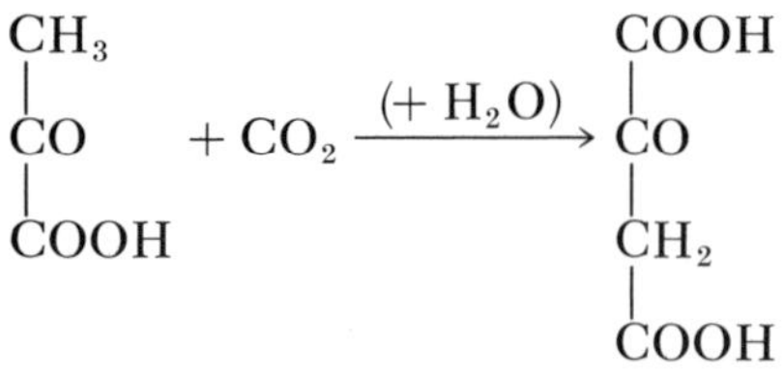

pyruvate   oxalacetate

The $CH_3 \cdot CO \cdot S \sim CoA$ then combines with oxalacetate to form citric acid (citrate) by means of a condensing enzyme, liberating $HS \cdot CoA$:

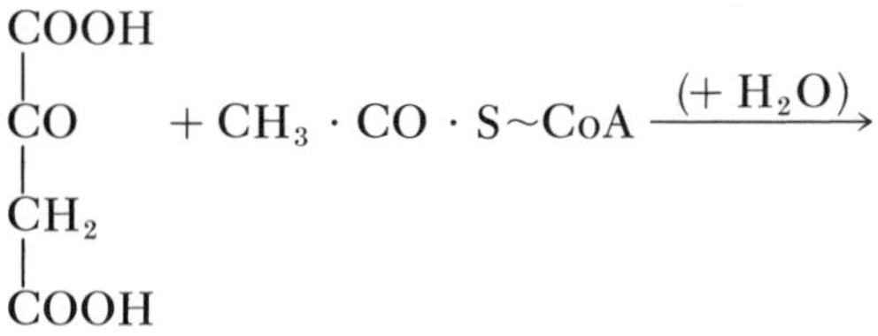

oxaloacetic acid

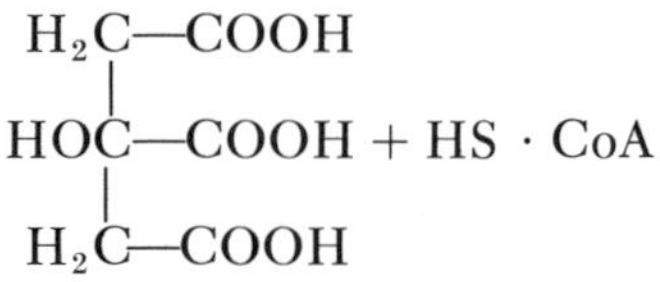

citric acid

Citrate then undergoes the series of changes called the **citric acid, tricarboxylic acid,** or **Krebs cycle** (Table 8-2; Fig. 8-8). The overall reaction shows the complete oxidation of pyruvate to CO_2 and H_2O:

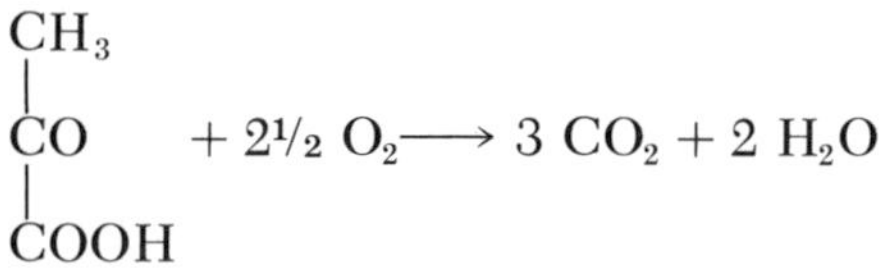

pyruvic acid

Several mechanisms of biooxidation are shown in Figure 8-9.

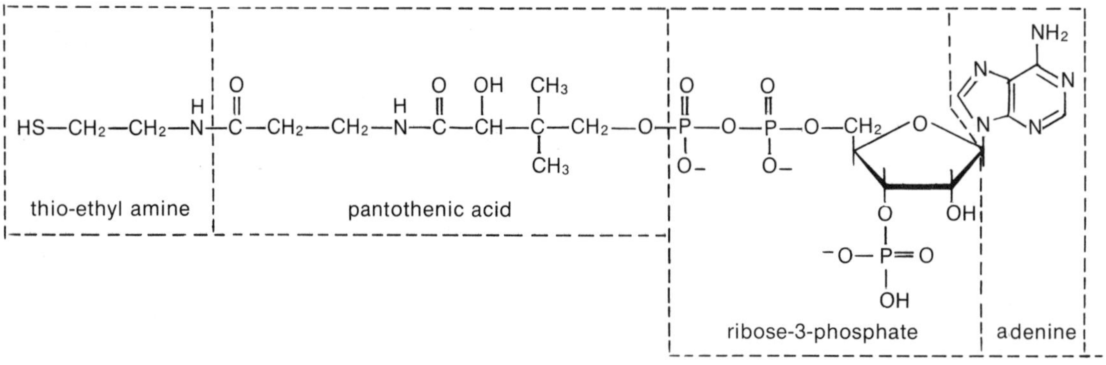

Figure 8-7. Structure of coenzyme A. Compare with structure of DNA (Fig. 6-7), noting the presence in coenzyme A (CoA) of various groups found in DNA and in other coenzymes.

TABLE 8-2. THE KREBS (OR CITRIC ACID OR TRICARBOXYLIC ACID) CYCLE°

Step No.	Substrate	Enzyme	Products of the Reaction	Nature of the Reaction
1.	CH_3 $C{=}O$ + (CoA) + NAD + AD ~ P + P_i‡ COOH **pyruvic acid**	pyruvic dehydrogenase pyruvic decarboxylase $\rightarrow$	Acetyl CoA‡ + CO_2 + NADH + AT ≈ P	Pyruvic acid is decarboxylated, the acetyl portion ($CH_3 \cdot CO$) combining with CoA to form acetyl CoA. H is transferred to NAD, forming NADH and yielding *energy* to AT ≈ P.
2.	Acetyl SCoA ("active acetyl")	condensation $\rightarrow$	$C_6H_8O_7$ + SCoA **citric acid** (a tricarboxylic acid)	"Active acetyl" combines with oxalacetic acid to form citric acid. SCoA is liberated to combine with more $CH_3 \cdot CO{-}$. No AT ≈ P is formed.
3.	Citric acid	aconitase $\rightarrow$	$C_6H_6O_6$ + H_2O **cis-aconitic acid**	Citric acid is dehydrated to form *cis*-aconitic acid. No AT ≈ P is formed.
4.	cis-Aconitic acid + H_2O	aconitase $\rightarrow$	$C_6H_8O_7$ **isocitric acid**	*cis*-Aconitic acid is rehydrated to form isocitric acid. No AT ≈ P is formed.
5.	Isocitric acid + NADP + AD ~ P + P_i	isocitric dehydrogenase decarboxylase $\rightarrow$	$C_6H_6O_7$ + H_2O + NADPH₂ + AT ≈ P **oxalosuccinic acid**	Isocitric acid is dehydrogenated to form oxalosuccinic acid. H is transferred to NADP, forming NADPH₂ and yielding *energy* to form AT ≈ P. CO_2 is lost.
6.	Oxalosuccinic acid	oxalosuccinic decarboxylase $\rightarrow$	$C_5H_6O_5$ + CO_2 + H_2O **α-ketoglutaric acid**	Oxalosuccinic acid is decarboxylated to yield α-ketoglutaric acid and CO_2. No AT ≈ P is formed.
7.	α-Ketoglutaric acid + NAD + AD ~ P + P_i + H_2O + SCoA	α-keto-glutaric dehydrogenase decarboxylase $\rightarrow$	Succinyl SCoA + CO_2 + NADH₂ + H_2O + AT ≈ P	α-Ketoglutaric acid is decarboxylated and dehydrogenated. Hydrogen is transferred to NAD, yielding NADH₂. The *energy*, with phosphate, is transferred to AD ~ P, forming AT ≈ P.
8.	Succinyl SCoA + H_2O + AD ~ P + P_i	succinyl thiokinase $\rightarrow$	Succinic acid + AT ≈ P + SCoA	The succinyl group of α-ketoglutaric acid combines with SCoA. This is converted to succinic acid, yielding SCoA and energy to form AT ≈ P.
9.	Succinic acid + FAD + P_i + AD ~ P	succinic dehydrogenase $\rightarrow$	$C_4H_4O_4$ + FADH₂ + H_2O + AT ≈ P **fumaric acid**	Succinic acid is dehydrogenated to fumaric acid, H_2 being transferred to FAD to form FADH₂ and yielding *energy* to AT ≈ P.
10.	Fumaric acid + H_2O	fumarase $\rightarrow$	$C_4H_6O_5$ **malic acid**	Fumaric acid is hydrated to form malic acid. No energy is released.
11.	Malic acid + NAD + P_i + AD ~ P	malic dehydrogenase $\rightarrow$	Oxalacetic acid + H_2O + NADH₂ + AT ≈ P	Malic acid is dehydrogenated to form oxalacetic acid. This returns to step 2 to combine with active acetyl and continue the cycle. Hydrogen from malic acid is transferred to NAD, forming NADH₂ and yielding *energy* to AT ≈ P.

°See also Figure 8-8.
†Inorganic phosphate.
‡See Figure 8-7.

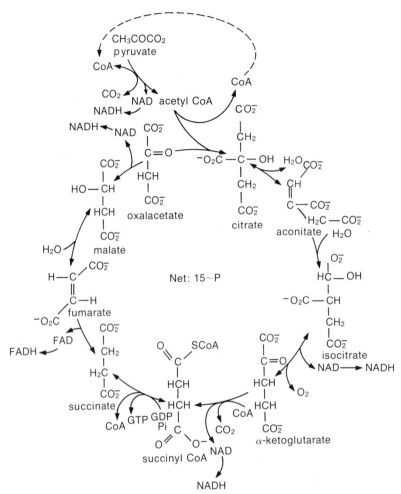

Figure 8–8. Diagram of the Krebs (citric acid or tricarboxylic acid) cycle. For explanation see text and Table 8–2. Hydrogen taken up by NAD and by FAD is transferred to the cytochrome system and so to oxygen.

(a) Organic Substrate $2(H^+e^-) \longrightarrow$ NAD $\longrightarrow$ (FAD) $\longrightarrow$ Cytochrome system $+ O_2 \longrightarrow H_2O$ — Obligate aerobes

(b) Inorganic Substrate $\begin{cases} CO \\ S \\ NH_3 \\ H_2 \end{cases} + n/O_2 \xrightarrow{\text{Oxidases}} \begin{cases} CO_2 \\ SO_4^= \\ NO_3^- \\ H_2O \end{cases}$ — Obligate aerobes

(c) Organic Substrate $2(H^+e^-) \longrightarrow$ NAD $\longrightarrow$ organic (H^+e^-) acceptor; e.g., pyruvate (fermentation)   Obligate anaerobes

(d) Organic Substrate $2(H^+e^-) \longrightarrow$ NAD $\longrightarrow$

Pyruvate $\longrightarrow$ lactic acid, etc. (fermentation)

$\begin{cases} NO_3^- \longrightarrow N_2 \\ SO_4^= \longrightarrow H_2S \\ CO_3^= \longrightarrow CH_4 \end{cases}$ anaerobic

Pyruvate + oxalacetate $\longrightarrow$ Krebs cycle $\longrightarrow CO_2 + H_2O$ (aerobic)

Facultative species

Figure 8–9. Some representative types of mechanisms of biooxidation. Shown are mechanisms of oxidation of (a) organic and (b) inorganic substrates by strict aerobes; (c) oxidation of organic substrates by strict anaerobes using organic H(e⁻) acceptors (fermentation); (d) the anaerobic use of either **organic** H(e⁻) acceptors, e.g., lactic acid (fermentation), or **inorganic** H(e⁻) acceptors (depending on species) or terminal oxidation via the Krebs cycle, by facultative species.

CHAPTER 8
SUPPLEMENTARY READING

Harold, F. M.: Conservation and transformation of energy by bacterial membranes. Bact. Rev., 36:172, 1972.

Kalckar, H. M.: High energy phosphate bonds: optional or obligatory? in Cairns, J., Stent, G. S., and Watson, J. D. (Eds.): Phage and the Origins of Molecular Biology, p. 43. Cold Spring Harbor Laboratory of Quantitative Biology, Cold Spring Harbor, N.Y. 1966.

Kaplan, N. O., and Kennedy, E. P. (Eds.): Current Aspects of Biochemical Energetics. Academic Press, New York. 1966.

Lieberman, M., and Baker, J. E.: Respiratory electron transport. Ann. Rev. Plant Physiol., 16:343, 1965.

NUTRITION OF MICROORGANISMS

9.1
INTRODUCTION

Napoleon Bonaparte is reputed to have said, "An army marches on its stomach."* In establishing this military aphorism he missed the opportunity to become a biological immortal. He could have stated the obvious fact that all life depends on food. To Bonaparte, food meant army rations. To the biologist, food means any substrate that can be metabolized to provide building material and/or energy for the cell. In order to be metabolized, food must enter the cell.

Osmotrophic Cells. Entrance of food substances into virtually all plant cells, and most animal cells, is by passage through the cell membrane (and cell wall if present) of nutrients in aqueous solution by the processes of **diffusion** and **osmosis.** This type of nutrition is said to be **osmotrophic.** In many cells passage of solutes through the cell membrane is assisted by certain **permease enzymes** (Chapt. 7).

Plant cells, including bacteria and fungi, are typically osmotrophic, partly because they have rigid or tough cell walls outside of the cell membrane that offer strong mechanical barriers to inward passage of solid particles.†

Phagotrophic Cells. Animal cells are typically without cell walls. Two large groups of animal cells have the ability to ingest solid par-

ticles of food by drawing them into the cell through the cell membrane by the process called **phagocytosis** (Chapt. 24). This term was devised in 1883 by Metchnikoff (Nobel Prize winner) who made fundamental studies of the phenomenon. Phagocytic cells are said to have a **phagotrophic** type of nutrition. Not all animal cells are phagocytic.

One of the two groups of phagocytic cells referred to above comprises the enormous phylum (over 250,000 species) of Protozoa: typically free-living, unicellular, microscopic animals, motile and widely distributed in moist, fertile soils, sewage, seas, and fresh waters around the world wherever organic food particles exist. Protozoa are represented by organisms of superclass Sarcodina, especially the genus *Amoeba,* and subphylum Ciliophora, genus *Paramecium,* though there are two or three other classes. As usual in biology, there are exceptions: some few species of parasitic protozoa, e.g., malaria parasites and trypanosomes, which live in the blood of vertebrates, have no cell walls but are nevertheless entirely osmotrophic in their nutrition.

Other phagocytic cells are component parts of many kinds of multicellular animals. One sort is exemplified in our own bodies by the ameba-like, polymorphonuclear, neutrophile leucocytes or "white blood corpuscles" or phagocytes that circulate in our blood stream. Phagocytic cells of another kind found in most vertebrates are said to be "fixed" (i.e., not freely moving about like the polymorphonuclear cells). They constitute what is called the **reticu-**

---

*He did not mean that armies are gastropods!

†An exception among the bacteria is the order Mycoplasmatales: bacteria without cell walls (Chapt. 42).

loendothelial system and are part of the lining of blood vessels in most organs of the animal body; some also occur in loose connective tissues in the body. As pointed out by Metchnikoff, phagocytic cells in any animal organism (he observed them in very minute Crustacea) are extremely important as mechanisms of defense against infection (Chapt. 24). In some lower animals (e.g., Porifera and Coelenterata) phagocytic cells also constitute part of the food-gathering and nutritional system.

Pinocytosis. Many kinds of animal cells, though lacking cell walls, are not phagocytic. However, they can engulf fluids, and possibly pass minute particles inward through the cell membrane by a process called **pinocytosis.** The term is derived from Greek words meaning the process of drinking or swallowing by the cell. In the membranes of cells showing this phenomenon there appear myriads of tiny invaginations or "gullets" that seem to act like minute mouths. They take in fluids, and probably suspended particles that are molecular or colloidal in size (i.e., too small for phagocytosis but too large for osmosis or diffusion), and engulf, close upon, and introduce them into the cell contents through the cell membrane.

Pinocytosis is probably not common in plant cells (or typical bacteria) since these have cell walls, though certain bacteria of the soil may ingest colloidal sulfur through the cell membrane by a process like pinocytosis. Certain viruses that infect animal cells appear to gain entrance to the cell by passage through the cell membrane possibly by a mechanism like pinocytosis (Chapt. 17). By a similar process called **endocytosis,** large, complex molecules, such as proteins, nucleic acids, some viruses, and possibly colloids like sulfur, are taken into the mammalian cell by minute invaginations of the cell membrane.

9.2 DIGESTION

As mentioned in Chapter 7, physiological digestion is commonly catalyzed by hydrolytic enzymes. Hydrolysis involves the enzymic introduction of one molecule of water between each two attached molecules of substrate or foodstuff (Fig. 9–1).

The processes of digestion and of assimila-

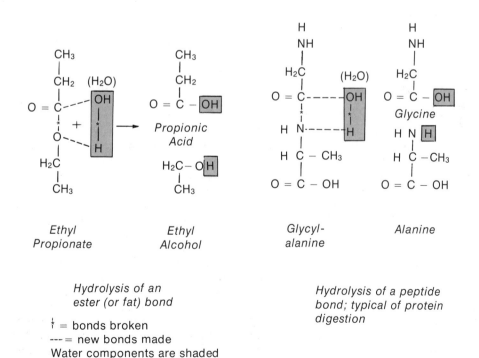

Ethyl Propionate   Ethyl Alcohol

Glycyl-alanine   Alanine

Hydrolysis of an ester (or fat) bond

Hydrolysis of a peptide bond; typical of protein digestion

$\overset{+}{\dagger}$ = bonds broken
--- = new bonds made
Water components are shaded

Figure 9–1. Typical hydrolytic reactions. The water molecule in each is split, giving H^+ to one part of the hydrolyzed substrate, OH^- to the other.

tion (utilizing for energy and/or cell synthesis the food substances taken into the cell) are astonishingly alike in all living cells: protist, plant, or animal. Like the various mechanical devices in different makes of automobile, the biochemical processes and mechanisms of cell metabolism are all obviously modifications of the same fundamental plan. However, each species of cell differs genetically and enzymically and structurally to some degree from all others. These differences are the basis of the sciences of classification and taxonomy.

Extracellular Digestion. The osmotrophic mode of nutrition might seem to impose severe dietary limitations on microorganisms restricted to it. However, many microorganisms, though entirely osmotrophic, can use the same sorts of solid foods as those that nourished Napoleon's (and other) soldiers, their horses, and other animals. In fact, numerous strange foods that are wholly unavailable to humans and other vertebrates are readily used by various species of bacteria, fungi, and so on. This is accomplished by a process called "extracellular digestion," i.e., digestion outside the cell. Basically it differs little from other kinds of digestion.

Digestion, whether intracellular or extracellular, results in the decomposition of complex and often insoluble food materials, such as polysaccharides, fats, and proteins into their constituent molecules of various soluble substances. These may be amino acids (from proteins), alcohols, glycerol and fatty acids (from fats), and various monosaccharides such as glucose (from starch, cellulose, or other polysaccharides). Unlike the original solid food masses, the molecules produced by extracellular digestion are small enough to pass through the cell membrane (or the vacuole membrane in some protozoa) into the living cell substance (Fig. 9–2). There they are further changed by other enzyme systems and are used as sources of energy, cell substance, or both.

In the higher animals, digestion occurs in the gastrointestinal tract. The soluble, molecular digestion products are taken from the lumen of the gut into the cells lining the intestine through the cell membranes and thence to the blood and lymph. Among the lower fungi, blue-green algae, and bacteria restricted to extracellular digestion, digestive enzymes are to a large extent excreted aimlessly into the surrounding fluid, where they may or may not come into contact with food. They may be wholly dissipated by dilution, convection currents, or other factors.

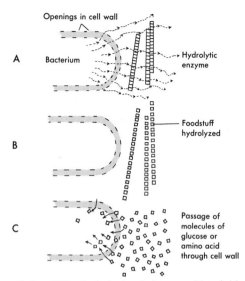

Figure 9–2. Utilization of grossly large and insoluble food particles by extracellular digestion. Sketch *A* shows the bacterium secreting hydrolytic enzyme which attacks the food particle (e.g., cellulose, protein, fat). In *B* the foodstuff is hydrolyzed and its constituent molecules (glucose, amino acids fatty acids, etc.) are separated from one another. In *C* the molecules derived by hydrolysis of the food pass readily (or are actively passed by permease enzymes) through the bacterial cell wall and cell membrane into the cytoplasm to be further metabolized.

9.3
FOODS AVAILABLE TO MICROORGANISMS

Regardless of their unorganized and hit-or-miss digestive systems, an interesting advantage possessed by Protista over sensate animals is their indifference in matters of taste. For example, certain species of bacteria and molds, given time and numbers (and the proper enzymic equipment), may use as food, with equal avidity, railroad ties, crab shells, animal hoofs and horns, feces, paper, leather, sulfur, petroleum greases, sawdust, and old rubber tires. Few if any bacteria or molds are actually so very versatile as to use all these foods, but some approach this degree of enzymic capability. All the substances mentioned, plus hundreds of others equally distasteful or poisonous to us (such as carbon monoxide, carbolic acid, paraffin, soap, molecular hydrogen, aviation gasoline, hydrogen sulfide, house paint), can serve as at least part of the food for one or another species of microorganisms or, more often perhaps, for combinations of bacteria acting together and with molds (see §43.2, Syntrophism in the Soil). In contrast, some microorganisms are highly restricted and fastidious in the matter of food and can thrive only with the aid of certain particular compounds, such as various derivatives of human blood or living tissue components.

Nutrition of Viruses. The nutrition of viruses differs basically from that of all cellular organisms (viruses are not cellular) and is discussed with the viruses (Chapt. 15).

Kinds of Food Requirements. We know that all living microorganisms—plant, animal, or protist—consist of very complex organic compounds—proteins, carbohydrates, lipids, DNA, RNA, vitamins, and so on—and many varied combinations of these: lipoproteins, mucopolysaccharides, chitin, starch, cellulose, and the like. Viruses are an exception; they consist of about three kinds of protein, either RNA or DNA, and nothing else (unless enveloped in a cell-derived envelope). However, the molecular structures of cellular organisms differ according to the genetic make-up of the various species of cells. Each kind of cell must synthesize its own complex and characteristic constituents from simpler substances found in its environment. Actively multiplying cells of any species are necessarily in contact with nutrients suitable for them. Most commonly these nutrients are in the form of a suspension of particulate matter and/or aqueous solution in sea, river, lake, or ground water, animal blood or lymph, sewage or other decomposing organic matter, and so on; the physical and chemical nature of the habitat determines the kind of organisms it will support. The nutritive and environmental requirements of the thousands of species of microorganisms vary widely.

Elements in Nutrition

Aside from water, seven elements—carbon, oxygen, nitrogen, hydrogen, phosphorus, sulfur, and potassium—are the *major* components of *all* living matter, their proportions per cell usually being in the order listed. They comprise about 95 per cent of the dry weight of the cell. Their presence in the same sorts of compounds (proteins, fats, carbohydrates, DNA, RNA, etc.) in *all* cells, and also in viruses, attests the evolutionary relationship of *all* organisms.

In addition to the elements just mentioned, about five others are required in lesser amounts: sodium, calcium, and chlorine, quantities and chemical form differing with species. Iron is required in electron transport by aerobic (oxygen-using) organisms, including humans; Mg is essential to all photosynthetic organisms, since it is the active, energy-transporting metal in chlorophylls. Diatoms require silicon for their "valves"; most vertebrates require iodine for the manufacture of thyroxin, as well as fluorine, silicon, tin, vanadium, and copper in extremely minute amounts.

In general, all species of cells—plant, animal, and protist—can assimilate the needed metallic elements in the form of soluble salts, chiefly chlorides or sulfates. Soluble inorganic phosphates (e.g., KH_2PO_4) commonly serve as sources of phosphorus (P). Some of these elements can also be used as they occur in organic compounds, uses depending on species. Carbon in the form of CO_2 can be used by many species of bacteria; others *must* have their carbon in organic form. It is in respect to the nonmetallic elements—carbon, nitrogen, oxygen, and sulfur—that the really spectacular differences between species are manifest. So distinctive are these differences that certain large groups of organisms are distinguished and named on this basis alone. Some of these are described farther on in this chapter, others in succeeding chapters.

Nitrogen and Sulfur. These *must* be in reduced form for organic combinations, e.g., nitrogen as in amino acids ($R—NH_2$) or sulfur as in sulfhydryl ($R—SH$) compounds. Certain species of true fungi, most blue-green algae, and a few genera of bacteria can enzymically (enzyme **nitrogenase**) reduce atmospheric nitrogen for the synthesis of organic compounds in the cell (Chapt. 43), a process called **biological nitrogen fixation.** Organisms that lack the enzymes necessary to catalyze the reductions of N and S must receive these elements in already-reduced forms: e.g., N as ammonium salts or N-containing organic compounds such as amino acids; S as H_2S or other sulfides, or S-containing organic substances such as mercapto ($R—SH$) compounds.

Hydrogen and Oxygen. These elements are commonly used as water for such purposes as solution, hydrolysis, ionization, osmosis, etc. In molecular form they have other functions.

Molecular hydrogen is used by some species of bacteria (e.g., genus *Hydrogenomonas*) as a source of energy. These remarkable organisms can grow in an entirely mineral solution, using CO_2 from the air as a source of carbon and oxidizing H_2 to H_2O as a source of energy.

Molecular oxygen is of major importance in cellular energy-yielding mechanisms and has other relationships to cell life. Sources and uses of oxygen differ with different species of bacteria.

Oxygen Sources

At least five different types of relationships of various organisms to molecular oxygen may be listed:

1. **Strict or obligate aerobes** must have free access to O_2 as final electron acceptor in biological oxidation or **aerobic respiration.**
2. **Strict or obligate anaerobes** not only do not require free oxygen but, on exposure to it, are inhibited or killed. Unable to use oxygen in their energy-yielding metabolic processes, they obtain their energy in another manner, i.e., by using one portion of a molecule of a substance (e.g., glucose) as donor of electrons and transferring them to another part of the same molecule, which thus serves as electron acceptor, with a net yield of energy to the cell. This process is called **fermentation.**
3. **Facultative aerobes** (or **facultative anaerobes**) can use O_2 as electron acceptor or, alternatively, the oxygen of salts like $NaNO_3$, Na_2SO_4, or carbonates. These alternative uses are sometimes called **anaerobic respiration.**
4. **Microaerophilic organisms** are inhibited or killed by full atmospheric oxygen tension. They grow best, or only, in the presence of a limited concentration of oxygen.
5. **Indifferent organisms** are those which neither require free oxygen nor are inhibited or killed by it except under certain special conditions.

Examples of each type of oxygen relationship are discussed in several following chapters.

Sources of Carbon

Two types of organisms have long been differentiated on the basis of whether (a) they can thrive, like typical green plants, on an entirely inorganic or "mineral" diet, using CO_2 or carbonates as a sole source of carbon, or whether (b) they cannot use CO_2 as a sole source of carbon but require, in addition to minerals, one or more organic substances (e.g., glucose or amino acids) as sources of carbon. The former (a) have been designated as **autotrophs** (Gr. *autos*=self; *trophe*=nourishing); the latter (b), as **heterotrophs** (Gr. *hetero*=other), since it was once commonly held that carbon could be combined in organic form only by *other* living organisms. This was disproven by Wöhler in 1828.

Autotrophs typically possess the enzymic equipment to catalyze direct combination of carbon atoms, such as the combination of carbon from carbon dioxide with carbon in an organic complex already existing in the cell. Typical autotrophs are represented by familiar green plants, algae, and a number of species of bacteria important in agriculture and industry.

Heterotrophs cannot utilize CO_2 as a sole source of carbon but require organically combined carbon (e.g., glucose, amino acids). Heterotrophs are the commonest and most widely distributed nutritional types. They are represented by *Homo sapiens* and all other animals, and by most species of true fungi (Eumycophyta) and bacteria.

With advances in knowledge, the traditional differentiation between autotrophs and heterotrophs lost much of its usefulness because various overlappings, exceptions, and borderline cases came to light, blurring the lines of demarcation. For example, some soil-inhabiting bacteria (genus *Hydrogenomonas*) and others of the family Athiorhodaceae are capable of living either autotrophically or heterotrophically. Many species living on an otherwise completely inorganic diet must have an organic source of carbon and energy (e.g., glucose) or certain vitamin-like (i.e., organic) substances. Certain common intestinal bacilli can grow well in a simple and completely inorganic solution if provided with sodium citrate as a source of carbon.

Another question arose when it was found that many supposedly typical heterotrophs require carbon dioxide in addition to organic carbon. Although it is not surprising that in the course of billions of years of organic evolution some metabolic aberrations have occurred, it makes classification difficult. Matters were more satisfactorily resolved when it became clear that sources and uses of carbon were closely related to sources of energy.

Sources of Energy

According to some authorities the primeval earth was completely dark because of heavy clouds of water vapor in the skies. It may (or may not—you may argue with equal profit on either side) be as a result of their evolution from progenitors which thrived in the dark that most species of bacteria and higher fungi, and all typical animal cells (including our own), are independent of sunlight as their source of energy and can thrive in complete darkness. Some of these **scotobiotic** (darkness-living; Gr. *skotos*= darkness) cellular types, such as large, outdoor fungi, have become adapted to withstand sunlight, commonly by means of protective pigments. However, most scotobiotic cells are injured by direct sunlight or artificial ultraviolet light.

CHEMOTROPHS

Cells capable of thriving in complete darkness obtain the energy for their activities and self-synthesis from chemical reactions (biological oxidations) that can occur in the dark. Such organisms are said to be **chemosynthetic** or **chemotrophic.**

Chemoautotrophs. Species of chemotrophs differ with respect to the chief (or only) sources from which they derive their carbon. For many species of chemotrophs the main source of carbon is CO_2. These are called **chemoautotrophs.** Some of these species are wholly restricted to CO_2 as their carbon source and also to the oxidation of only certain specific inorganic substances as sources of energy:

$$H + \tfrac{1}{2}O_2 \longrightarrow H_2O + \text{energy (56 kcal)}$$
$$NaNO_2 + \tfrac{1}{2}O_2 \longrightarrow NaNO_3 + \text{energy (17 kcal)}$$
$$H_2S + 2O_2 \longrightarrow H_2SO_4 + \text{energy}$$
$$CO + \tfrac{1}{2}O_2 \longrightarrow CO_2 + \text{energy (74 kcal)}$$

These are sometimes called **chemolithotrophs** (Gr. *lithos*=stone or mineral). Many bacteria of both these types are of great ecologic and economic importance, being active in various industrial decompositions and in the fertility of soils.

Chemoheterotrophs. Chemotrophs whose chief sources of carbon are organic compounds (e.g., glucose, amino acids) are called **chemoorganotrophs** or **chemoheterotrophs.** For most of these species the source of carbon (e.g., glucose) is also the source of energy. Many must also have carbon as CO_2. In the latter form it is used chiefly in intracellular syntheses. Chemoheterotrophs are the commonest of all nutritional types, including all animals and most species of bacteria and fungi.

PHOTOTROPHS

Although photosynthesis by green plants and algae is widespread and represents a tremendous tonnage, as well as being basic to the very existence of all forms of animal life,

among the bacteria it occurs in only three relatively small groups. These are like all other typical photosynthetic organisms in being able to use radiant energy, but they differ in that they do not release free O_2. Like chemotrophs, they, too, differ in respect to their sources of carbon. All are gram-negative; flagella, if present, are polar. Most are strict anaerobes.

Photoautotrophs. With few exceptions, green plants, algae, and two small divisions of the photosynthetic bacteria assimilate CO_2 as their sole or main source of carbon. They use the CO_2 in photosynthesis and are therefore called **photoautotrophs** or **photolithotrophs.**

Photoheterotrophs. These are represented by a third group of bacteria (Athiorhodaceae, or purple nonsulfur bacteria) that are chiefly aquatic or marine. These generally require organic compounds as sources of carbon for photosynthesis; hence, **photoheterotrophs** or **photoorganotrophs.** The group is a relatively small one, though of considerable economic and ecologic importance.

Hydrogen Donors in Photosynthesis. The phototrophic organisms may be divided into three groups on the basis of the donor of hydrogen that they typically use in the reduction of CO_2 for photosynthesis.

Hydrogen Donor H_2O. All green plants and all but a few species of algae, as mentioned above, *but no bacteria,* use water as the hydrogen donor in the photosynthetic reduction of CO_2 for the synthesis of their cellular organic compounds (Chapt. 32).

$$CO_2 + 2H_2O \xrightarrow{\text{light}} (CH_2O) + H_2O + O_2$$

The O_2 is released to the atmosphere.

Main Hydrogen Donor H_2S. Two of the groups of phototrophic bacteria, the green bacteria and the purple sulfur bacteria, because their photosynthetic pigments are **bacteriochlorophyll** instead of chlorophyll, *cannot use water as hydrogen donor* in photosynthesis but can use H_2S:

$$CO_2 + 2H_2S \xrightarrow{\text{light}} (CH_2O) + H_2O + 2S$$

Note that *no oxygen is released* but that free sulfur is produced and collects either intracellularly or extracellularly (Chapt. 34). Some species can use other reduced inorganic compounds.

Main Hydrogen Donor Organic Compounds. One relatively small group of photosynthetic bacteria (Athiorhodaceae, or purple nonsulfur bacteria; see Photoheterotrophs, above) can use neither H_2O nor H_2S as hydro-

TABLE 9-1. METABOLIC TYPES OF LIVING ORGANISMS

| | Electron Donor | |
Energy Source	INORGANIC SUBSTANCE (LITHOTROPH)	ORGANIC SUBSTANCE (ORGANOTROPH)
I Oxidizable Substance (Chemotrophs)	*Chemolithotrophs* H_2 bacteria Colorless sulfur bacteria Nitrifying bacteria Iron bacteria	*Chemoorganotrophs* Most bacteria Fungi Protozoa Animals, including *Homo sapiens*
II Light (Phototrophs)	*Photolithotrophs* Green plants Algae Purple sulfur bacteria Green sulfur bacteria	*Photoorganotrophs* Purple nonsulfur bacteria

gen donor in photosynthesis but depend on organic compounds for this function:

$$CO_2 + 4H{-}R \xrightarrow{\text{light}} (CH_2O) + H_2O + 4R$$

Hydrogen Donor Molecular Hydrogen. Most of the phototrophic bacteria mentioned

above can also use molecular hydrogen to reduce CO_2 in photosynthesis:

$$CO_2 + 2H_2 \xrightarrow{\text{light}} (CH_2O) + H_2O$$

For convenience in review, we may list the various nutritional types of microorganisms as shown in Tables 9–1 and 9–2.

TABLE 9-2. SUMMARY OF ENERGY METABOLISM

| | Phototrophic Organisms | | | Chemotrophic Organisms | | |
Green Plants, Algae, Blue-Green Algae	GREEN SULFUR (Chloro-bacteriaceae)	PURPLE SULFUR (Thiorhodaceae)	PURPLE NONSULFUR (Athiorhodaceae)	RESPIRATION	ANAEROBIC RESPIRATION	FERMENTATION
Chlorophyll *a*, carotenoids (and other chlorophylls)	⊢—— Various bacteriochlorophylls and carotenoids ——⊣			Glycolytic pathway Citric acid cycle	Glycolytic pathway Citric acid cycle	Glycolytic Oxidation incomplete
Cyclic phosphorylation	⊢——— Cyclic phosphorylation ———⊣			Oxidative PO_4	Oxidative PO_4	Stops at intermediate— e.g., lactic acid, ETOH, etc.
Noncyclic phosphorylation	⊢——— No noncyclic phosphorylation ———⊣			O_2 final e^- acceptor	O_2 not final e^- acceptor°	
Produce O_2	⊢——— Do not produce O_2 ———⊣					
Electron donor or reducing power [H] from H_2O	From H_2S, H_2, thiosulfate	From H_2S, H_2, thiosulfate	From organic compound	Glucose→$CO_2 + H_2O$ Net 38 ATP, efficiency of 55%	Glucose→$CO_2 + H_2O$ Net 38 ATP, efficiency of 55%	Net 2 ATP, efficiency of 35%
	Obligate anaerobe	Obligate anaerobe	Facultative: Dark—chemo-organotroph Light—photo-organotroph			
	Sulfur deposition extracellular	Sulfur deposition intracellular	No sulfur deposition			
	Usual habitat is anaerobic aquatic environment—e.g., ponds, ditches, sewage lagoons, mud at bottom of lake or river, and muddy soil.					

°Final e^- acceptors other than O_2: $SO_4 \xrightarrow{e^-} H_2S$, $CO_2 \xrightarrow{e^-} CH_4$, $NO_3 \xrightarrow{e^-} NO_2$

9.4
SHORT DEFINITION OF BACTERIA

It is notable that the cells of all bacteria and blue-green algae are distinguished from the cells of all other organisms in having primitive (non-membrane-enclosed) nuclei; bacteria and blue-green algae are therefore said to be **procaryotic** (Gr. *pro*=primitive; *karyon*=nucleus). Among the procaryons, bacteria differ from blue-green algae, and indeed from *all other* living organisms, in *not* containing the photosynthetic pigment **chlorophyll *a*** (though some species of bacteria contain **bacteriochlorophyll *a***). These two characters, procaryotic nuclei and lack of chlorophyll *a*, distinguish all bacteria from all other living organisms and form the basis of a sophisticated, four-word definition of bacteria as: **procaryons without chlorophyll *a*.** In using the definition, the student must be prepared to define procaryons, blue-green algae, bacteriochlorophyll *a*, and chlorophyll *a*!

9.5
MICRONUTRIENTS

Substances that do not yield energy to the cell or contribute materially to its bulk, but which are absolutely necessary to growth and functioning (analogous to the tiny distributor points in an automobile engine), are variously called **essential nutrilites, growth factors,** or **micronutrients.**

Inorganic Micronutrients. A few of the heavier elements (Co, Mo, Cu, Mn, Zn) are indispensable to cell life, though they are used in extremely small amounts. An exact list of such elemental requirements is hard to formulate because some metals are present in such minute amounts that it is very difficult to detect them or to be sure that, when found, they are functional and not merely impurities. Those known to be **essential** (see above) are commonly called **trace elements** or **inorganic micronutrients.** They usually function as key molecules in certain enzymes and vitamins: e.g., Co in vitamin B_{12}, Mo in nitratase, and so on (Chapts. 7, 8).

Organic Micronutrients. Vitamins are good examples of organic micronutrients. Many bacteria require none, while others require almost as many as people do. For humans, any "balanced diet," or (at unnecessary expense) a single small capsule containing mere micrograms of vitamins and inorganic micronutrients ("minerals"), supplies all, or most, of both organic and inorganic micronutrients of the nor-

mal person. In our food requirements, both organic and inorganic, we are remarkably like many common species of bacteria, though our digestive systems and "tastes" differ considerably.

Certain amino acids, e.g., tryptophan, have the status of organic micronutrients. Without this amino acid, though only a minute amount is needed, the typhoid bacillus (*Salmonella typhi*), the lockjaw (tetanus) organism (*Clostridium tetani*), the diphtheria bacillus (*Corynebacterium diphtheriae*), and a number of other species cannot grow, even though the nutrient solution (culture medium) in which they are suspended is complete in all other respects. In contrast, there are many species of microorganisms that can synthesize their own tryptophan and therefore do not need to have it fed to them.

Of peculiar interest is a group of iron-chelating agents called **mycobactins**, obtainable solely from bacteria of the genus *Mycobacterium*, to which belong the causative agents of tuberculosis (*Mycobacterium tuberculosis*), leprosy (*M. leprae*), and Johne's disease of cattle (*M. paratuberculosis*). Whereas mycobactins are synthesized by most species of mycobacteria, *M. paratuberculosis* is absolutely dependent on it for growth in culture medium (Fig. 9–3). Mycobactins are related to ferrioxamines in cellular iron transport systems.

Other substances acting as organic micronutrients for many species are purines and pyrimidines, components of DNA and RNA.

One role of organic micronutrients in metabolism is exemplified by the vitamin pantothenic acid, as described by Fildes. Assume that an organism (a bacterium or you or I) requires pantothenic acid for growth. This vitamin is essential to the synthesis of coenzyme A, which in turn is essential in the energy-yielding metabolism of the cell (Chapt. 8). Assume that the organism is able to carry out all but one of the synthetic steps necessary to the formation of

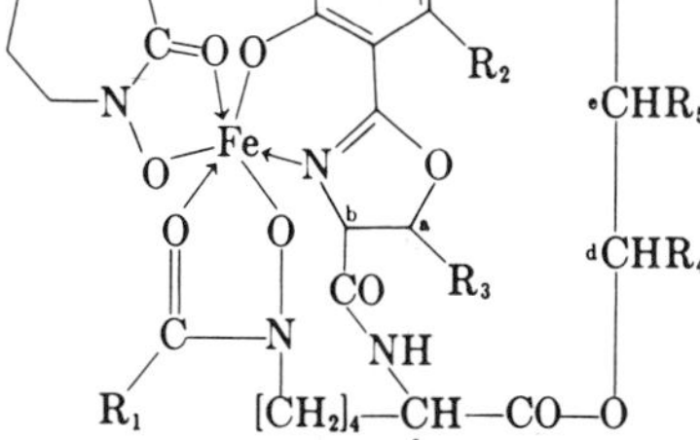

Figure 9–3. Type of structure of ferric mycobactins. R = alkyl groups with numbers of C atoms indicated.

pantothenic acid. Beginning with ammonia, it synthesizes β-alanine, but it lacks the enzymic equipment to synthesize pantoic acid, which is absolutely essential to the complex pantothenic acid molecule (Fig. 9–4). Given only a few micrograms of pantoic acid, completion of the pantothenic acid (commonly sold to humans at high prices as calcium pantothenate) is provided for and growth occurs. Another species of organism, able to synthesize pantoic acid as well as the other components of the pantothenic acid molecule, encounters no such difficulty, and we say that it does not require pantoic acid or pantothenic acid. Both organisms may actually *require* pantothenic acid, but one can manufacture it internally; the other cannot. Humans are woefully delinquent in this respect. They must eat foods (available in any well-stocked supermarket) containing some 8 to 12 vitamins or suffer from such dietary deficiency diseases as scurvy (vitamin C), beri-beri (vitamin B_1), pellagra (niacin), pernicious anemia (vitamin B_{12}), and others. Several species of bacteria can synthesize their entire complex of cell materials, vitamins, and all other structures to obtain their life energy and thrive, in an aqueous mineral solution containing only the essential elements, meanwhile obtaining their carbon from the CO_2 of the atmosphere (see farther on).

9.6
ECOLOGIC RELATIONSHIPS

Saprophytes. Presumably most of the earliest heterotrophic fungi, including bacteria, obtained their carbon from inert organic compounds—either those occurring spontaneously in the fluids around them, or those available from the wastes and dead remains of other organisms. Whatever may have been their evolutionary origin, such heterotrophic forms are commonplace today among both microorganisms and macroorganisms. Many are exceedingly important as scavengers. Among the microorganisms, most "molds" and bacteria are scavengers. They are spoken of as *saprophytes* because they are involved in the decomposition or decay ("recycling") of lifeless organic matter. (Gr. *sapros* = rotten, decaying).

Parasites. Probably still later than saprophytes there appeared, either through progressive or regressive evolution or both, types of heterotrophic microorganisms, both plants and animals, that could metabolize not only dead and waste organic matter but also enter and damage the substances inside other living cells or tissues. They caused disturbances of the delicate physical and chemical equilibria of the organisms they poisoned or in which they lived.

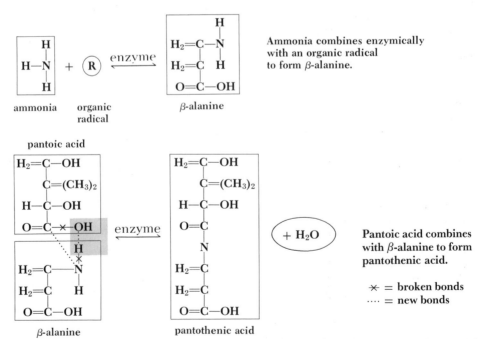

Figure 9–4. Role of an essential metabolite (in this case β-alanine) in metabolism. β-Alanine is essential to the formation of pantothenic acid, which is essential to coenzyme A (Fig. 8–7), which is essential to the Krebs cycle (Fig. 8–8) which is essential to life! (For explanation see text.) Note that β-alanine and pantoic acid are joined by a reaction that is the reverse of hydrolysis (Fig. 9–1).

If their victims were multicellular, their tissues were affected. This was **disease** and often resulted in the death of the injured organisms. Disease-producing organisms, microscopic and macroscopic, are said to be **parasitic** and **pathogenic**. A woefully long list of pathogenic microorganisms may be cited, including the amebae of dysentery, the bacilli of whooping cough, bubonic plague, brucellosis, and leprosy, and so on and on.

Obligate Parasites. In the course of organic evolution some of the parasitic organisms presumably became so fully adapted to a parasitic existence that they became partly or wholly dependent on this mode of life and on the organisms that they parasitized. They apparently lost the ability to live saprophytically and could not multiply in the outer world. Since they are obliged to live a wholly or partly parasitic existence they are said to be **obligate parasites.** Shining examples of obligate parasites are all of the viruses; rickettsias such as those that cause Rocky Mountain spotted fever; animals such as hook- and tapeworms, malaria parasites, spirochetes of syphilis, and bacilli that cause leprosy.

CHAPTER 9
SUPPLEMENTARY READING

Bodily, H. L., Updyke, E. L., and Mason, J. O.: (Eds.): Diagnostic Procedures for Bacterial, Mycotic and Parasitic Infections, 5th ed. American Public Health Association, New York. 1970.

Bovallius, A., and Zacharias, B.: Variations in the metal content of some commercial media and their effect on microbial growth. Appl. Microbiol., 22:260, 1971.

Butlin, K. R., and Postgate, J. R.: The economic importance of autotrophic micro-organisms, in: Fourth Symposium, p. 271. Society for General Microbiology. Cambridge University Press, London. 1954.

Frieden, E.: The biochemistry of copper. Sci. Am., 218:103, 1968.

Frieden, E.: The chemical elements of life. Sci. Am., 227:52, 1972.

Heine, J. W., and Schnaitman, C. A.: Entry of vesicular stomatitis virus into L cells. J. Virol., 8:786, 1971.

Hutner, S. H.: Inorganic nutrition. Ann. Rev. Microbiol., 26:313, 1972.

Kelly, D. P.: Autotrophy: Concepts of lithotrophic bacteria and their organic metabolism. Ann. Rev. Microbiol., 25:177, 1971.

Taras, M. J., Greenberg, A. E., Hoak, R. D., and Rand, M. C. (Eds.): Standard Methods for the Examination of Water and Wastewater, 13th ed. American Public Health Association, New York. 1971.

CULTIVATION AND GROWTH OF BACTERIA

By the cultivation of microorganisms is meant the process of inducing them to grow. For most purposes of microbiology they are cultivated **in vitro,** i.e., in glass (L. *vitro* = glass) flasks, test tubes and other vessels. Today the term in vitro includes containers made of plastic or steel, as in the huge tanks (see Figure 23–2) used for commercial purposes. Cultivation in vitro necessitates the preparation of substances that the microorganisms can use as food. Such nutrient preparations are called **culture media.**

10.1
CULTURE MEDIA

There are three main types of culture media: natural or empirical, synthetic or "defined," and living. They vary widely in form and composition, depending on the species of organism to be cultivated and the purposes of the cultivation. Here we shall describe only a few representative types. Certain specialized methods and media are necessary for cultivating bacteria of the order Mycoplasmatales (Chapt. 42). Viruses, chlamydias, and rickettsias must be cultivated inside other living cells ("living media"). Special methods are described in the discussions of those forms requiring them.

Empirical Culture Media

Empirical media are those used on the basis of experience and not on the basis of exact knowledge of their composition and action. In the early days of microbiology natural, empirical culture media were widely used: milk, urine, diluted blood, vegetable juices, and the like. Some of these are still widely used. They contain all of the required elements and a rich assortment of soluble organic and inorganic compounds that satisfy the requirements of many, though not all, microorganisms, especially bacteria and fungi. Such media are convenient and inexpensive. However, their exact composition is not known and is variable, and they are unsuitable for the cultivation of numerous important species.

Peptone. Most empirical media in common use today contain as a major, or only, ingredient 1 to 2 per cent of peptone; i.e., trypsinized or hydrolyzed protein from animal (e.g., meat or casein) or vegetable (e.g., soy beans, cottonseed, etc.) sources. In composition, peptones are mixtures containing, in only partly known concentrations and identity, a variety of peptides and polypeptides, proteoses, amino acids, carbohydrates, etc., including inorganic, and many organic, micronutrients. Peptones provide, in soluble and assimilable or otherwise available forms, all of the phosphorus, sulfur, and essential mineral content of living material as well as the organic carbon and nitrogen sources noted above. For most purposes aqueous extracts or hydrolysates of yeast in 1 per cent concentration are entirely satisfactory substitutes for peptone.

Meat Extracts and Infusions. Beef tea (beef extract) and an aqueous **meat infusion**

made by soaking (infusing) fresh, ground meat in water are common ingredients of media useful for many species of microorganisms (as well as for people). Beef infusions, especially, are rich in minerals, organic micronutrients, proteins, protein derivatives, and carbohydrates. They are often supplemented with 1 per cent peptone or yeast extract. The acidity (or alkalinity) is "adjusted" to near neutrality for most organisms (pH 7.0 to 7.6). (pH and pOH are symbols used to express degrees of acidity and alkalinity. They are factors of critical importance in cell life [Chapts. 4, 20].) The fluid is then filtered to clarify it, dispensed in culture tubes or flasks which are then plugged with cotton, and all are autoclaved (sterilized with superheated steam) (Chapt. 22). Culture fluids made from beef infusion are commonly called **infusion broth;** those made from beef extract are called **extract broth.** All such empirical media, including simple solutions of peptones or yeast extracts, are sometimes loosely included in the general terms **nutrient broth** or **nutrient solution.**

Meat is not the only useful source of organic matter. Fresh extracts of vegetables of various kinds are often used. Canned tomato and orange juices make excellent media for numerous bacteria, yeasts, and molds. Some workers use the flesh or juices of shellfish, while those interested in the microbiology of milk sometimes use whey or skim milk. Coagulated eggs are often used also, especially for tubercle and diphtheria bacilli.

Some media are made by adding bits of kidney, spleen, or other tissues freshly removed from dead animals under aseptic precautions (i.e., in the absence of contaminating microorganisms) to tubes of meat-infusion broth. Extracts of yeast are often included. Because the composition of peptones, extracts of meat, yeast, and the like is known only approximately and is variable, such media are included as empirical.

Adsorption of Nutrients at Surfaces. The incorporation in bacteriological culture media of small amounts of some solid substance such as ground meat or even sand is often advantageous, as many microorganisms seem to grow best in the crevices of, or in contact with, the surfaces of such matter, forming little "nests" or **niduses** there (Fig. 10–1).

Special Media. To any of the media we have mentioned, many different substances may be added for numerous purposes. Blood, a wide variety of carbohydrates, esters, alcohols, glucosides, and many other compounds may be added to media for experimental or diagnostic purposes. The medium is then referred to by

the name of the special added substance, e.g., blood-infusion broth, lactose broth, tomato juice agar, and so on.

Synthetic Culture Media

These media consist wholly of dilute, reproducible solutions of chemically pure, known inorganic and/or organic compounds. They have special uses in research and industry. Artificial media of exactly known, reproducible composition are called **synthetic media** or **chemically defined media.**

The formulation and use of specific synthetic media necessitate exact knowledge of the nutritional requirements of the organism to be cultivated. As a result of extensive investigations, such knowledge is now available for a number of medically or commercially or otherwise important microorganisms. However, synthetic media are not so commonly used for routine purposes as empirical media because they are often expensive and time-consuming to prepare; the exact nutrient requirements of numerous organisms are still incompletely

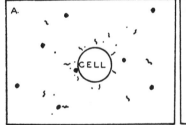

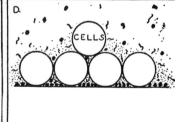

Figure 10–1. *A,* A free-floating bacterial cell surrounded by a few suspended particles of food (dark circles) which must be hydrolyzed by the exoenzyme (wavy lines) before the resulting hydrolysates (dots) can be assimilated. *B,* Particles of food concentrated in a monomolecular layer on a solid surface. *C,* Food particles are more available to the cell on the solid surface where the interstices at the tangent of the bacterial cell and the solid surface retard the diffusion of exoenzymes and hydrolysates away from the cell. *D,* Multiple cells form additional interstitial spaces.

known; the organism to be cultivated is, itself, often unknown; for example, in an attempt to identify a bacterium that is causing a particular industrial spoilage problem or an infection of the blood or spinal fluid.

Inorganic Synthetic Media. These are among the simplest of media. A medium used for a common, sulfur-oxidizing, chemolithotrophic species of bacteria found in the soil (*Thiobacillus thiooxidans*) is as follows:

$(NH_4)_2SO_4$	0.2	gm
$MgSO_4 \cdot 7H_2O$	0.5	gm
KH_2PO_4	3.0	gm
$CaCl_2$	0.25	gm
Powdered sulfur	10.0	gm
Distilled H_2O to make	1,000.0	ml

This provides all needed elements and a source of energy; the phosphate serves as a buffer (i.e., it helps maintain a suitable pH) and also as a source of phosphorus and potassium. Inorganic micronutrients (Cu, Fe, Mg, Mo, Na, and a few others) are sometimes present as impurities in the water or other ingredients. If deionized water is used, they may be specifically added as inorganic salts in concentrations of about 0.25 mg each per liter. The powdered sulfur is a source of energy. It is oxidized to H_2SO_4 (see Chapter 43). Carbon is obtained as CO_2 from the atmosphere. Nitrogen is provided in an ammonium salt. This is a typical **synthetic** or **chemically defined** inorganic medium. From this simple solution *T. thiooxidans* is able to synthesize, in the form of a living organism, proteins, carbohydrates, lipids, DNA, RNA, and so on, completely beyond the powers of even the most advanced human chemist. This medium would serve for many nonsulfur oxidizers if the sulfur were replaced by glucose as a source of energy and carbon.

Organic Synthetic Media. These sometimes seem complicated but are not nearly so complex as a solution of peptone with its scores of various and often unknown components. An organic synthetic medium for a rather fastidious pathogenic bacterium, *Corynebacterium diphtheriae* (cause of diphtheria), contains some 21 chemically pure ingredients in accurately weighed amounts:

    Eight amino acids
    Three vitamins
    Salts of Ca, Mg, Cu, K, P, S
    Several carbohydrates and esters as sources of
        carbon and energy

As noted above, an advantage of synthetic media is that they are exactly reproducible. In some instances they may be less expensive and troublesome to prepare than media made with meat infusions or peptones (especially at current meat prices!). Also, they can be prepared, if necessary, without proteins and therefore without **antigenic** or **allergenic** properties when injected into man or animals in vaccines or for experimental purposes. By virtue of these properties they lend themselves well to exact experimental research and to medical and commercial uses. (See also sections on vitamin assay, Chapter 49.) A difficulty is that a solution suitable for one species is often not suitable for another, and it is frequently difficult to determine the exact requirements for a given species.

Dehydrated Media

In the modern laboratory commercially available, portable, dehydrated, powdered, bottled mixtures of a great variety of media or their ingredients may be used simply by adding weighed portions to the required amount of water, dispensing, and sterilizing. These save time, expense, and shelf space and promote uniformity of composition, pH, freshness, and so on.

Such powdered bulk mixtures are now replaceable by compressed sterile tablets of media or ingredients. These require merely aseptic addition to a stated volume of sterile deionized or distilled water. Serum, blood, and other substances may also be added aseptically.

10.2
PURE CULTURE METHODS

If one studies the properties of some substance—say, ferric chloride—and over a period of months performs a series of laborious, time-consuming and expensive tests, carefully noting all the properties of $FeCl_3$ revealed by the tests, it is discouraging to find afterward that a little copper sulfate had been inadvertently mixed with the iron salt, completely vitiating the whole set of data. So also one may spend time, energy, and money determining the exact and absolute properties of, say, *Saccharomyces cerevisiae*, the brewers' yeast. The findings must be thrown out the window if it is afterward found that the culture of yeast cells was **contaminated** (inadvertently mixed) with some bacteriological weed like *Bacillus subtilis*, a common dust bacterium found, like dandelions, almost everywhere. When studying $FeCl_3$ we must have the pure substance; when studying a microorganism we must have a **pure culture,** i.e., material containing a single species of microorganism.

Difficulties with Contaminated Cultures. As outlined in Chapter 2, early microbiologists had to contend with the difficulty that, as Leeuwenhoek had observed, the natural habitats (market milk, feces, sewage, the skin, soil, river water, air, dust, and so on) of microorganisms generally contain many different species living together. The occurrence of any single species in natural materials, unmixed with any other microorganisms, i.e., as a **pure culture,** is rare, though it does happen under certain circumstances such as "blood poisoning" (infection of the blood stream or **septicemia**). But as late as the 1860's no good method of separation or **isolation** of organisms in pure culture was known.

Joseph Lister (1827–1912), physicist, physician, and pioneer in aseptic surgery, described a method of isolating bacteria from aqueous suspensions by successive dilutions of the suspensions in sterile fluid to the point that one tube among the highest dilutions would contain only one of the desired organisms, a point extremely difficult to determine. He succeeded but rarely, and perhaps only three or four of his contemporaries were ever able to succeed with the dilution method.

Origin of Pure Culture Technique. In 1872 a microbiologist named Schroeter had observed the growth of different sorts of bacteria in isolated masses (**colonies**) of various colors on slices of decaying potato. With his microscope he saw that all the microorganisms in any one colony were always exactly the same. It was obvious that by cultivating microorganisms on **solid** nutrient surfaces it was possible to obtain isolated colonies of any single kind—each a pure culture!

Use of Gelatin. Extending this principle, Koch in 1880 used a 5 to 10 per cent gelatin solution to prepare a transparent, solid jelly with a moist, sticky nutrient surface on flat pieces of glass. In addition, various nutrient solutions and test substances could be added to the gelatin before it was allowed to set. Here was a very important advance—a revolutionary advance, one that has become the basis of all our present-day bacteriology. Modified in technical details, it is currently used in the pure-culture study of viruses, of cells of cancer, and of humans, plants, and insects.

In summer, and when held in body-temperature incubators, the gelatin melted. Being a protein it was often digested and liquefied by the metabolic processes of the microorganisms, especially molds. Besides that, particles of dust settled on it with various microorganisms from the air or soil, which contaminated it, obscuring and confusing the results as badly as ever.

First Use of Agar-Agar. Many students flocked to Koch's laboratory from all over the world to learn his methods. One of these was W. Hesse. To the wife of this man the science of microbiology is indebted for suggesting, in 1881, agar-agar (commonly called **agar**) as a substitute for gelatin. This gum is a polysaccharide derived from seaweed (*Gelidium* sp.) and was used at that time in making jellies. Today agar is often used as "bulk" in laxatives, also as "vegetable gum," a common commercial food additive. For most bacteria, and humans, it is wholly indigestible and has no nutritive value. Agar is transparent and colorless, melts in water only at boiling temperature and, once melted, does not "set" again until approximately body temperature (39C). Agar has not been improved upon as a solidifying agent for culture media and is in general use for this purpose today. It is added in 1.5 to 3.0 per cent concentrations to nutrient solutions, forming fairly stiff gels.

Origin of the Petri Plate. In order to prevent contamination of the pure cultures by dust, another student in Koch's laboratory, R. J. Petri, suggested the simple expedient of pouring the melted nutrient agar into circular, shallow dishes and immediately covering them with a glass cover. This permitted prolonged examination of the cultures but excluded dust. Such dishes are widely used today and are called **Petri plates.**

Materials from which pure cultures are to be obtained are spread on the surface of agar or other solid media in Petri-type plates in such a manner that single, isolated colonies will be formed by growth where single (or supposedly single) cells are deposited on the agar by the streaking process (Figs. 10–2, 10–3). Plastic, disposable dishes of the Petri type, and many other vessels and instruments, such as flasks, syringes, and pipettes, are now available and solve problems of dishwashing and sterilization in the preparation rooms.

After 1883 the preparation and study of pure cultures proceeded rapidly in microbiological laboratories around the world (Fig. 10–4).

Silica Gels. Soon after the first uses of agar as a solidifying agent it was discovered that several important species of chemolithotrophic microorganisms of the soil are "poisoned" or

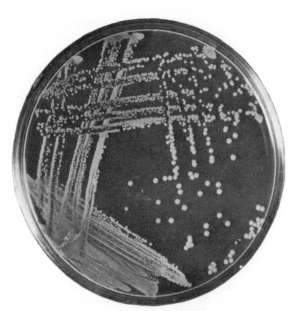

Figure 10–2. The surface of nutrient agar in a Petri plate has been streaked with material containing bacteria capable of growing on the nutrients provided and then incubated at 37C overnight. Note the method of graded cross-streaking used in order to obtain isolated colonies. (Courtesy of Naval Biological Laboratory.)

inhibited by the presence of an organic substance such as commercial agar (or impurities in it). Further, a number of species of heterotrophic microorganisms of the sea and soil readily digest and liquefy agar; hence there arose a need for an inorganic solidifying agent. Silica (SiO_2) was found to be an acceptable substitute for agar when made to assume a jelly-like state (silica gel). Preparation of silica gel is not difficult but requires meticulous attention to details such as pH, temperature, concentration of silica and the presence of various ions. Further details of materials and mode of preparation can be found in the literature.

Other Solidifying Agents. Several substances other than agar, gelatin, or silica are widely used to prepare solid microbiological media. Serum, blood, and whole mixed eggs, which coagulate readily on heating to 80C or above, are often used either alone or as the basis for mixtures. In addition, slices of potato, bread, carrot, and the like, as well as pieces of meat, can serve as solid media for numerous species of microorganisms.

Selective Cultivation. As previously mentioned, microorganisms living under natural conditions in the soil, sea, lakes, saliva, feces, decaying organic matter, and the like, seldom exist in pure culture. Several means of selec-

tively cultivating (Figs. 10–5, 10–6) specific organisms from such mixtures are available. The basic principle of all is the provision of a type of physical environment and a medium in, or on, which the desired organism will thrive—ideally to the exclusion of all others. The medium may be solid or liquid. Further details of selective cultivation are given in Chapter 30.

10.3
GROWTH AND MULTIPLICATION

Growth means increase and may refer to size, numbers, weight, mass, and numerous other parameters of things animate and inanimate. While increases in size or mass of single, individual cells usually occurs during maturation, such changes are generally temporary and secondary to the main process of multiplication of those cells.

Multiplication and Fission. Under ordinary conditions of growth, and with only a few exceptions including coenocytic organisms (e.g., filamentous molds), all actively growing (vegetative) cells (plant, animal, and protist) multiply by the asexual process of cell fission. This does not include viruses; these are not cellular as currently defined. Fission results in

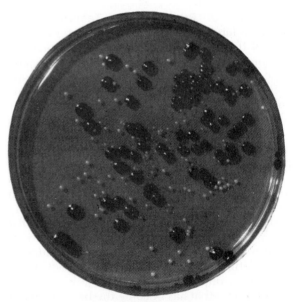

Figure 10–3. Petri plate containing nutrient agar streaked with a mixture containing two harmless bacteria: *Sarcina lutea* (small colorless colonies) and *Serratia marcescens* (large, dark red colonies), showing differentiation by colonial characteristics. (Courtesy of Naval Biological Laboratory.)

division of the cell into two or more vegetative cells whose progeny, similarly produced, can continue this asexual reproductive process indefinitely, provided food and energy are available and environmental conditions (pH, temperature, absence of pollution by toxic wastes, and so on) remain favorable.

Among multicellular plants and animals, fission of body (somatic) cells results only in increase in size of the individual plant or animal (Fig. 10–7, *B*), not in increase of numbers of individuals. In contrast, among the autonomously living, unicellular microorganisms (protists) fission *is* multiplication. Each fission leads to two (or more in multiple-budding yeasts and molds) independent, complete individuals. Most bacteria multiply by **transverse, binary** fission, i.e., division into two *equal* cells (Fig. 10–7, *A*). Not all of the factors that initiate and control cell fission are known, but several are. These may be summarized as follows:

Cell Wall, Cell Membrane, and Cell Division. Typically, in bacterial fission the nuclear material (in bacteria a single chromosome) undergoes replication and two daughter chromosomes (or masses of identical chromosomes) separate. The cell begins to elongate by synthesis of new cell membrane, cell wall, and cytoplasmic elements.

Generally this is accompanied by centripetal growth of new cell wall and cell membrane material forming a narrow ring around the cell, usually about midway between its poles and at a right angle to the long axis of the cell. The cytoplasmic membrane usually first forms a double-layered partition between the two cells-to-be. Two layers of cell wall material are concomitantly, or later, synthesized between the two layers of cell membrane (Figs. 10–8, 10–9).

Mesosomes and Fission. Mesosomes are saccular invaginations of the cytoplasmic membrane. They appear to have a special relationship to bacterial fission. Studies of exponentially dividing cells show that mesosomes during this period are sacs of membranes apparently derived from the cytoplasmic membrane and in direct contact with the nucleoid (nuclear material). In the early stages of cross-wall or septum formation mesosomes are small and attached by a stalk to the cytoplasmic membrane portion of the ingrowing septum (Figs. 10–8, 10–9). At this stage mesosomes seem either to be closely associated with, or actually to initiate, formation of the septum and synthesis of new proteins and replication of the DNA (nuclear material).

The mesosome increases in size and com-

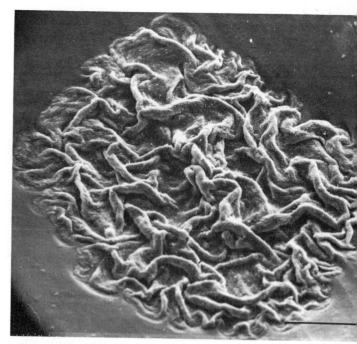

Figure 10–4. This scanning electron micrograph of a colony of *Mycobacterium phlei* demonstrates the complex morphology found in many bacterial colonies. Bar equals 100 μm. (Original magnification 450×; reduced to 45% of original size.) (Courtesy of Ivan L. Roth and Cheryl Weinmeister, Department of Microbiology, The University of Georgia, Athens.)

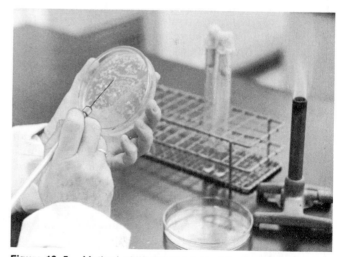

Figure 10–5. Method of "fishing" (picking) from a colony of bacteria on the surface of agar medium in a Petri plate. The bacteriologist's "needle" (wire) has been heated to redness in the Bunsen flame at the right and then cooled. The tip of the needle is carefully brought into contact with a selected colony; the adhering cells are then immediately transferred to tubes of sterile medium in the nearby wire rack.

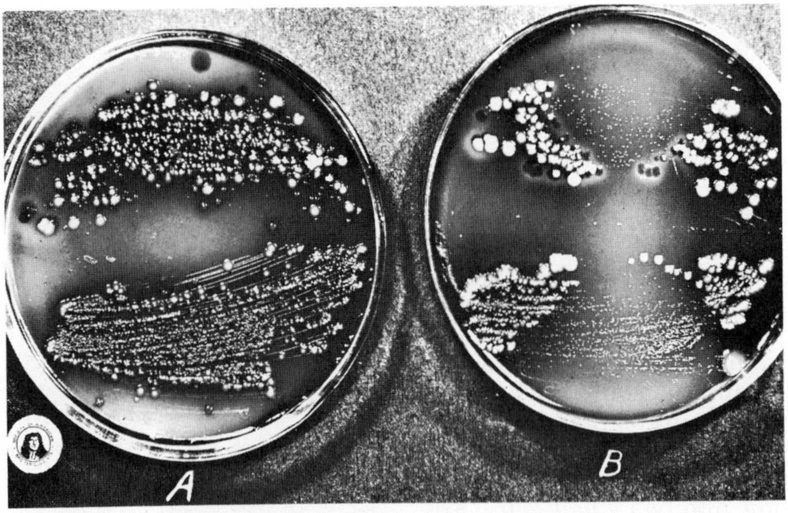

Figure 10–6. Bacterial colonies on blood-agar in Petri plates (about one-half actual size). This picture also illustrates a method of **selective cultivation** of a particular species of bacteria; in this case the organisms of whooping cough, or pertussis. A nasopharyngeal swab with mucus from the patient was passed over both plates, and the mucus was spread over the agar with a sterile loop. At the same time a drop of penicillin solution was placed on two sides of plate B. The plates were then incubated at 37C for three days. Plate A shows the customary heavy growth of common bacteria usually found in the nasopharynx. The pertussis organisms (*Bordetella pertussis*) are completely overgrown and obscured. Plate B shows two areas where the common organisms have been completely inhibited by the penicillin, permitting B. pertussis (tiny white colonies) to grow unhampered by competition of the other organisms. The use of penicillin to aid isolation of pertussis bacilli was devised by Fleming, discoverer of penicillin. (Courtesy of Dr. William L. Bradford. From the collection of the American Society for Microbiology.)

plexity of its membranes during the process of septum formation and DNA synthesis, until the septum is nearly complete and the daughter cells about to separate. In a manner roughly suggestive of centriole and spindle-fiber action in mitosis of eucaryotic cells, the mesosome appears to draw apart the two replicates of the chromosome so that one goes into each daughter cell. The mesosome then separates from the nearly completed septum and apparently later disintegrates, possibly into molecular precursors to be reassembled for a new cycle of fission. Mesosomes are more fully discussed in Chapter 14. Final separation of the cells results from a split between the two new cell walls suggestive of abscission of leaves from higher plants, especially in yeasts in which a "bud scar" is left (see Figure 1–1).

Form, Size, and Fission. Because the volume of any cell increases more rapidly than its surface area the size and shape of all living cells bear important relationships to the nutrition and reproduction of each cell.

As the spherical or ovoid cell grows there arrives a critical point in the ratio (Table 10–1) of volume (cell content which must be fed) to surface (cell membrane, which, alone, supplies the foods and removes the wastes). Obviously,

when the volume exceeds the ability of the surface (cell membrane) to provide food and remove wastes, the favorable, initial relationship between surface and content must be renewed. This (among other important physiological functions) is accomplished by fission into two small daughter cells. When fission proceeds more rapidly than cell volume, the ratio of volume to surface diminishes to a fixed, optimal minimum as the numbers of cells increase. Young, rapidly growing (e.g., exponential phase) cells therefore tend to be smaller than old or dormant cells, or cells just before fission.

The evolution of **cylindrical** or **filamentous** cells with minimum **diameters** may be regarded as a step toward overcoming the difficulties imposed by the volume-surface relationship of spheroid cells. For example, in a cylindrical bacterium or mold filament, the axial center is never much more than a micrometer or so from the cell surface (i.e., from the source of food supply and the means of waste disposal). Length and branching can extend indefinitely, with no increased difficulty in feeding the cell contents or in removing wastes.

Whether, in filamentous cells, as contrasted with spherical cells, cell fission is initiated by **volume–surface ratio,** or whether (as seems more

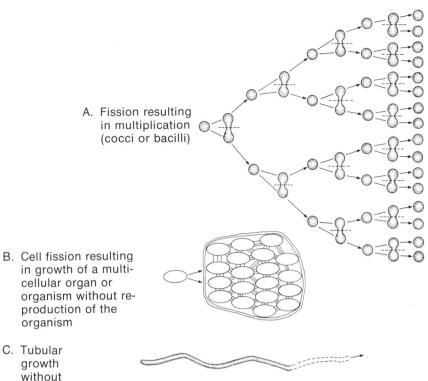

A. Fission resulting
in multiplication
(cocci or bacilli)

B. Cell fission resulting
in growth of a multi-
cellular organ or
organism without re-
production of the
organism

C. Tubular
growth
without
multiplication
(coenocytic mold)

D. Tubular growth
with septation
and fragmentation
resulting in reproduction
(septate mold)

Figure 10–7. Various relationships in growth and reproduction. In *A,* growth results in fission, which in turn results in reproduction of many new individual microorganisms. In *B,* fission results in reproduction of the initial cell, but only an increase in size of the multicellular organism of which the cell is a part. In *C,* growth of a coenocytic tube appears only in length of the tube, not in numbers of new organisms as shown by fragmentation of a septate filament in *D.*

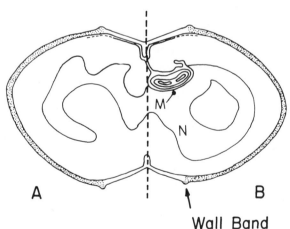

Wall Band

Figure 10–8. Diagram of a central longitudinal section of a cell unit taken from an exponentially dividing culture of *Streptococcus faecalis.* The protuberances on the external portions of the cell wall of this particular organism are called wall bands and appear to act as lines of demarcation between old cell wall and new cell wall from which the septum is derived. In some bacteria they are not so prominent. A membranous sac or mesosome (*M*) is shown attached to the cytoplasmic membrane of the developing septum and closely associated with the nucleoid (*N*), which is undergoing division.

Figure 10–9. Electron micrograph of a dividing cell of *Streptococcus faecalis* showing septum formation, with mesosome and other features diagrammed in Figure 10–8. The bar represents 100 nm.

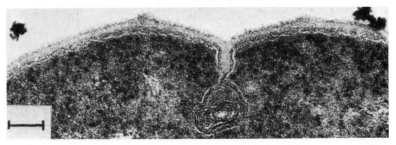

TABLE 10–1. RELATION OF FORM TO AREA AND VOLUME *

Shape of Cell	Area	Volume
Sphere	$4 \pi r^2$	$\dfrac{4 \pi r^3}{3}$
Cylinder	$d \pi h + 2 \pi r^2$	$\pi r^2 h$

*Abbreviations: r = radius; d = diameter; h = height or length.

likely) wastes, nutritive requirements, reproductive cycle and other factors initiate cell fission is not fully known. Probably all of these factors play a combined role.

Eucaryotic cells overcome the volume–surface difficulty to a large extent by the formation of extensively branched and ramifying invaginations of the cell membrane deep into the cell substance, the whole branched system constituting the **endoplasmic reticulum** (Fig. 11–2). Procaryotic cells have no obvious endoplasmic reticulum, though there are sometimes fairly extensive invaginations of the cell membrane called **mesosomes.**

Growth without Fission. Growth of microorganisms does not necessarily involve fission. Many species of normally rod-shaped bacteria, under the influence of numerous extraneous influences, may fail to undergo fission, although nuclear fission and growth of the cell wall, membrane, and contents may continue. As a result, instead of numerous individual cells, long nonseptate filaments are formed (Fig. 10–7, C). Some fission-inhibiting agents are soap and bile salts; ultraviolet irradiations; certain antibiotics, nutritional defects, and mutations. These factors inhibit **septum formation** but not growth. The exact mechanisms are not yet entirely clear.

10.4
MULTIPLICATION AND SEX

Among many of the "higher" microorganisms (including the eucaryotic protists; Chapt. 11), both sexual and asexual reproductive processes are commonly demonstrable under proper conditions of growth and observation. In these microorganisms a major role of the reproductive process in addition to multiplication is the transfer and recombination of genes (genetic recombination) between male and female cells. Genetic recombination, as biology students know, is one of the main factors in organic evolution.

Among bacteria, sex has no role whatever in multiplication; multiplication is entirely asexual. Similarly, genetic recombination is brought about by several mechanisms that do not involve sex at all. A simplified, primitive form of conjugation is observed in some species, in which the "sexes" are to some extent interchangeable. Genetic recombination by such means is commonly only partial. These genetic phenomena are more fully discussed in Chapter 19.

Despite the antiquated character of microorganismal sex life as outlined in the foregoing paragraph, the protists have no difficulty in multiplying. Under optimal conditions they can increase in the entire absence of sex and with extraordinary speed. For example, a single, active young cell of a common bacterium (*Escherichia coli*), if left undisturbed and uninhibited in an ideal environment could, within 24 hours or so, produce a **population*** running into the billions. It has been calculated that the total mass could weigh thousands of times as much as this planet. It is unlikely that this will ever occur because of many natural restrictions of overgrowth, such as competition between species, and numerous others which will occur to the imaginative student.

10.5
ENUMERATION OF MICROORGANISMS

The numbers of microorganisms present in various natural materials such as soil or river water vary constantly because of changing conditions. In an ideal situation such as a pure laboratory culture under uniform, optimal physical and chemical conditions the numbers of organisms vary in a perfectly regular and predictable way in accord with general biological laws.

To illustrate phenomena of growth, multiplication, and enumeration of microorganisms we shall use bacteria as our model because they are convenient to work with, the methods are well developed, and the principles they illustrate are fairly generally applicable in microbiology. The basic laws hold even in the "special cases" of viral growth, multiplication by conidiospores, and certain types of sexual reproduction, in all of which there are sudden releases of large numbers of new cells or growth units.

*The term **population** (L. *populus* = people) is generally used to mean all the microbial cells living in the same material or environment; perhaps "*microbulation*" would be more exact.

We generally measure growth (multiplication) of bacteria and other unicellular microorganisms by measuring increases in numbers in relation to time.

As an illustration we may select a culture flask containing 50 ml of sterile infusion broth at 35C.

Let us introduce by means of a sterile pipette a drop of fluid containing about 10 cells of the common, harmless bacterium of the intestinal tract—*Escherichia coli*. Let us assume that these 10 cells are from an inactive or **dormant** stock culture on an agar slant held for weeks in the refrigerator. The newly inoculated broth culture is held at 35C. The problem before us is to measure the population at regular intervals by counting the numbers of cells present in an aliquot (commonly 1 ml) of the broth. The numbers present at the different periods are then plotted in relation to time, and a **growth curve** is obtained. Enumeration may be carried out by several methods which may be direct or indirect. The results obtained by any method are, at best, estimates having mainly statistical value. They are approximately correct only within certain limits of cell numbers but are sufficiently accurate to be useful for most purposes of microbiology.

Direct Methods
(Estimates of Cell Numbers)

In any of these methods accuracy declines with increases in concentrations of cells that result in crowding and excessive, uncountable numbers; accuracy decreases also with such small numbers of cells as to rob the count of statistical validity, as well as with irregularities of distribution, e.g., clumping.

Specimens that contain large numbers of bacteria (upward of about 10^4 per ml) are customarily diluted from about 1:10 to $1:10^5$ or more, depending on sample and counting method, to make the numbers more manageable and simplify the counting.

Cell Counts. In the hemacytometer or counting-chamber method a minute drop of the fluid is placed in a tiny, shallow, rectangular glass vessel called a **hemacytometer** (because it was devised originally for counting hemocytes or blood corpuscles). The counting chamber is partitioned off by ridges into regular, cubical chambers of exactly known volume (Fig. 10–10). By counting the individual cells in each chamber under a microscope and adding them, the numbers of organisms per milliliter may be computed. This is a total count of live and dead organisms. The method is applicable to any suspension of microscopic particles. Bacteria are so small that numbers of the order of at least 1×10^7 per ml of suspension are requisite for statistical validity of the count.

Smear Counts. **Direct.** Another procedure is to smear an exact volume of the culture over an exact area on a slide, stain with methylene blue or other appropriate dye and count the organisms in a known portion of the total area. Knowing the diameter of the microscopic field from previous measurements (by means of

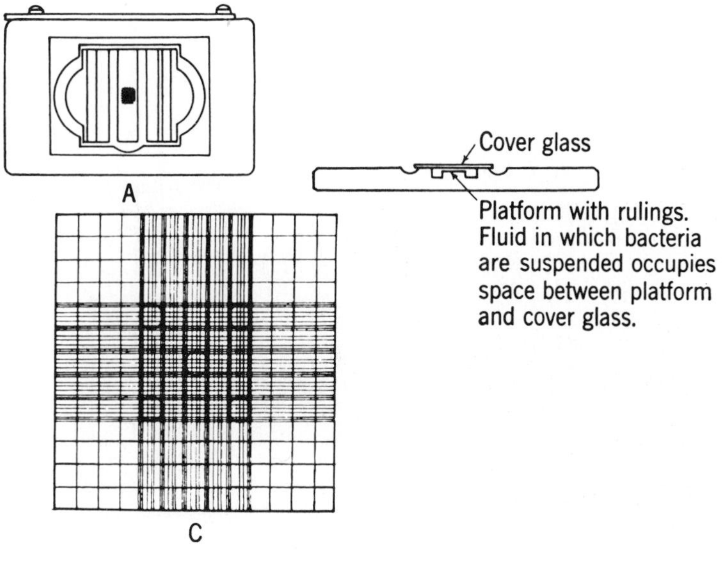

Cover glass

Platform with rulings. Fluid in which bacteria are suspended occupies space between platform and cover glass.

Figure 10–10. A hemacytometer (Petroff-Hausser type) adapted for counting bacteria and other microorganisms. *A*, Plan view, showing the area (dark central square) covered by the ruled chambers which are seen enlarged at *C*. *B*, A vertical section, about two-thirds actual size, with cover glass in place. It is customary to count only the cells in the representative areas encircled in *C*, though all may be counted for greater accuracy. (Courtesy of Arthur H. Thomas Co., Philadelphia, Pa.)

a stage micrometer), one can calculate the numbers of organisms per milliliter of culture. This is a total count also, since no distinction is readily made between living and dead organisms. (See Direct Counts of Milk, Chapter 47.) This method is applicable to any suspension of microscopic particles that can be visualized under the microscope.

Comparative. If 1 ml of male human blood and 1 ml of culture are well mixed and a stained smear of the mixture prepared, an estimate of the numbers of bacteria may be obtained by counting both blood and bacterial cells in a certain number of fields and noting their relative proportions. Since we know that male human blood contains about five million erythrocytes per cubic millimeter, an estimate of the numbers of bacteria is merely a matter of arithmetic. This is a total estimate.

Membrane Filter Counts. Measured samples of fluid may be passed through sterile, porous-membrane filters and the microorganisms on the filter then counted directly (see Figure 21–9). The organisms must not be too numerous and must be uniformly distributed. They are first stained in situ on the membrane and then counted in calibrated fields. Before counting, the filter is made transparent by saturating it with immersion oil. This is a total count of dead and live organisms.

Electronic Counters. These instruments are capable of accurately counting thousands of cells, alive and dead, in a few seconds. Most are based on the principle of electronic gating, which is roughly analogous to the "electronic eye" that operates the familiar automatic door in a supermarket. Basically they depend on interruptions of an electronic beam that traverses a space between two closely adjacent electrodes. Each particle, as it passes between the electrodes, causes an interference with the electron beam due to different conductivities of cells and fluid. The interruption is taken up by instruments and recorded electrically (Fig. 10–11).

Other instruments with complex circuitry are based on high-speed scanning beams and can count enormous numbers of colonies on large numbers of plates in seconds (Fig. 10–12).

Indirect Methods (Cell Mass Determinations)

Determination of Total Volume. One standard method is to place a standard volume — say, 10 ml — of the culture in a kind of test tube, called a Hopkins tube, having a narrow, hollow, cylindrical column projecting from the bottom and graduated in millimeters. The organisms are packed into the column by centrifugation at a standard speed and for an exactly measured time, and their total volume is read on the graduated scale. From a knowledge of the average volume of the individual cells an estimation of numbers is possible. This is a total estimate. In a modified form it is commonly used in medical diagnostic studies to measure the total volume of blood corpuscles in a **hematocrit determination**

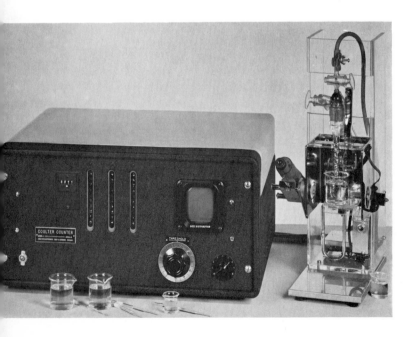

Figure 10–11. Estimation of numbers of microorganisms by an electronic "gating" method. For explanation see text. (Courtesy of Coulter Electronics, Chicago, Ill.)

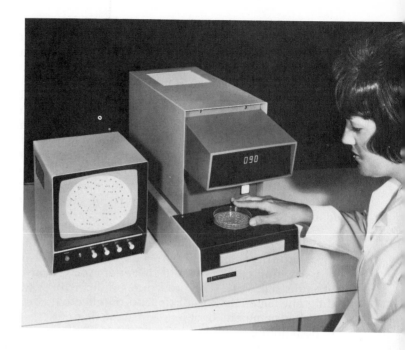

Figure 10–12. With automatic bacterial colony counters such as this, the technician can determine at a glance the number of colonies growing in a Petri dish. The colonies are scanned, the count is registered digitally, and the colonies are marked and shown on a vidicon screen. (Courtesy of New Brunswick Scientific Co., Inc., New Brunswick, N.J.)

(expressed as mm of erythrocytes per 100 mm of column height in a Wintrobe tube; normals: men, 40 to 50 mm; women, 35 to 45 mm).

Turbidometric Methods. A widely used technique measures turbidity or scattering of light in the culture due to accumulation of evenly dispersed cells suspended in it. A measured volume of the culture is placed in a special, clear glass tube of known diameter. This is interposed between a unit source of light and a photoelectric unit, which is attached to a galvanometer. The reading on the galvanometer depends on the passage of light through the culture from the unit source. Of the total light from the unit source, the percentage transmitted through the tube will be diminished in proportion to the turbidity (Fig. 10–13). The method is subject to errors due to variation in size and shape and clumping of cells, as well as to different degrees of translucency of various species and other materials in cultures. However, the method is one of the quickest and simplest and is reasonably accurate. Note that turbidity data are not numbers of bacteria and cannot correctly be used as such in calculations based on exponential expressions of cell numbers. Turbidity readings may be standardized in terms of numbers of cells by hemacytometer counts or electronically, with numerically standardized suspensions of bacteria.

Chemical Methods. Quantitative determinations of substances that are always present in fairly constant amounts in living cells are some-times used as equivalents of total cell growth, with or without fission. Such methods are hardly applicable to bacteria but are much used industrially in measuring heavy growths of filamen-

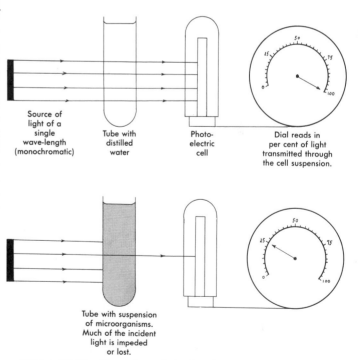

Figure 10–13. Use of a photoelectric turbidimeter to estimate numbers of microorganisms. Virtually 100 per cent of the light from the source at left passes through the tube with distilled water (*top*); less than 25 per cent through the tube turbid with suspended microorganisms (*bottom*).

tous microorganisms. One is total nitrogen determination by the Kjeldahl method. Since nitrogen is always present in protein in the proportion of about 16 per cent, Kjeldahl values can easily be converted to protein values. Protein may also be measured indirectly and approximately by means of the **Folin reagent,** which gives a color reaction with tyrosine and tryptophan, two amino acids always present in protein in relatively constant amounts. Other chemical methods involve determinations of free amino groups (Van Slyke), nucleic acids (DNA and RNA) or the phosphorus of these acids, and so on.

Dry Weight Measurements. These, like the above chemical methods, are not much used for measuring bacteria but are a useful method of measuring growth of molds in certain phases of industrial work. The procedures differ with different materials but all depend on effective washing; complete, or at least constant, degrees of dehydration; and accurate weighing. The increase in weight represents biological synthesis and, with data on cell volume available, can be used to calculate cell numbers.

Dilution Methods

Serial Dilutions. The method of serial dilutions is widely used to estimate numbers of viable bacteria in various fluids: water, milk, cultures, etc. In a typical procedure 1-ml quantities of the sample (of, let us say, milk) diluted in decimal, four-fold, two-fold or other convenient series are placed in tubes of nutrient broth. After incubation of the tubes of broth, presence or absence of growth is recorded. For example, in a 10-fold dilution series suppose there is growth in the tube that received the 1:1,000 dilution but no growth in the tube receiving the 1:10,000 dilution. Then there were (theoretically) between 1,000 and 10,000 organisms (viable in that kind of broth and under those environmental conditions) per milliliter of the sample of milk tested.

Indicated Number. This number of organisms per milliliter is spoken of as the **indicated number** (the reciprocal of the highest dilution showing growth). But it is not a very exact estimate. It ignores sampling errors due to unequal distributions of the cells, especially in the highest dilutions. For example, the cell that produced growth in the tube of broth receiving the 1:1,000 dilution not infrequently, by chance of distribution, appears in the next tube (1:10,000), leaving the 1:1,000 tube sterile.

Most Probable Number. In the example just cited, *theoretically* there should be 1,000 organisms, but there may be, *theoretically,* any number up to 9,999 per milliliter in the sample. What is the true number? This cannot be stated. However, mathematicians have shown that the number most probably present may be calculated if the results from duplicate or triplicate simultaneous determinations are known. Tables are available showing the most probable number calculable from all possible combinations of results in such series. These tables are much used in examination of water. Tables are found in *Standard Methods for the Examination of Water and Wastewater** with directions for use. It is to be borne in mind that these are most probable numbers, not exact numbers.

This method can be made qualitative as well as quantitative by adding special indicator substances to the broth. For example, it is standard practice to add lactose and to note the highest dilution of sample in which **lactose-fermenters** are found, as indicated by production of acid and gas in the broth. **Growth** may occur in the 1:100,000 tube but **fermentation** of lactose only in the 1:100 and lower dilution tubes. (See Chapter 45.)

Colony Counts. The colony count is widely used for determining approximate numbers of microorganisms in milk, water, and many other materials. It is applicable to any microorganisms that will grow as colonies on solid laboratory media. These media must, of course, provide good nutrition and the requisite environmental conditions for the microorganisms under investigation.

Continuing the examination of our flask of culture inoculated with *Escherichia coli*, mentioned earlier, we may, at any desired moment during incubation, withdraw exactly 1 ml of the culture from the flask with a sterile measuring pipette and transfer it to a sterile Petri dish. Immediately afterward, about 15 ml of nutrient agar, previously melted (and cooled to about 40C so as not to kill the bacteria), is poured into the dish. The culture is thoroughly mixed with the still-fluid agar by a gentle horizontal rotation of the dish. In a few minutes the agar will have solidified. This plate culture is held in an incubator at about 35C for 24 hours and is then examined for the presence of colonies distributed throughout the agar.

With continued incubation of the broth culture, the number of bacteria per ml increases

---

* 13th ed., American Public Health Association, 1971.

rapidly toward several thousands or hundreds of thousands. The 1 ml of material removed for the plate count is diluted so that plates are obtained which show only about 50 to 300 colonies (Fig. 10–14). These numbers are optimal for colony counts. Appropriate dilutions are easily estimated after a little experience. Lethal crowding of the colonies is thus avoided and the colonies are separated so that counting is more accurate. The number of colonies, multiplied by the reciprocal of the dilution, gives the indicated number of organisms per ml.

Colony Counting. The counting of colonies in agar is greatly facilitated by the use of a 2× or 3× stereoscopic microscope with both direct and indirect illumination. Electrically lighted colony counters are used in examining relatively small numbers of plates in which each colony is marked by hand with an electric needle and is recorded automatically (Fig. 10–15). Colonies on large numbers of plates are often counted by means of scanning TV tubes (Fig. 10–12).

Roller Tube. Samples or dilutions of fluid specimens may be mixed with melted nutrient agar at a temperature of about 42C (and of composition appropriate to the organisms expected

to grow in it) in cylindrical vials called **roller tubes,** instead of in Petri dishes. While the agar is still fluid these vials are rotated rapidly in an electric **spinner** until the agar solidifies in a thin film evenly distributed over the inside surface of the tube. After incubation, colonies are easily counted in the film of agar (Fig. 10–16). The method is especially convenient on field trips, since several sealed roller tubes with warm, fluid, sterile agar in them can be carried along and inoculated immediately after collection of specimens.

Whatever the method used for obtaining colonies, each represents, theoretically, the progeny of a single cell that was in the original inoculum and that was imprisoned in or on the agar at that point. Actually, several organisms, if stuck together in a clump, will give rise to only a single colony. The colony count, therefore, does not give a wholly accurate enumeration of the live **individual** cells present in the material under investigation. However, the errors in colony-count methods are fairly well known, and, within limitations, such counts are among our most useful means of enumerating microorganisms. The basic principle is widely used and should be fully understood at this point. It

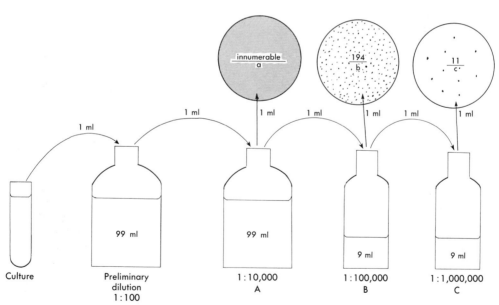

Figure 10–14. A colony count of a young culture of *Escherichia coli*. One milliliter of culture was transferred to 99 ml of buffered diluting fluid in the preliminary dilution. After gentle mixing, 1 ml of this was transferred to 99 ml of fluid in bottle *A*. One milliliter of this dilution, after mixing, was placed in plate *a* and 1 ml was transferred to bottle *B* and so on through plate *c*. After adding agar to the plates and incubating (see text) the plates were examined. Plate *a* contained too many colonies to count; they were confluent and crowded each other out. Plate *b* contained 194 well-distributed, well-separated and easily counted colonies. Plate *c* contained only 11 of a theoretical 19 or 20. The percentage error in each colony in such a high dilution makes such a count unreliable, and such plates are disregarded. It is customary to make colony counts in duplicate and triplicate to minimize such error. This culture contained approximately 20,000,000 viable cells of *E. coli* per milliliter.

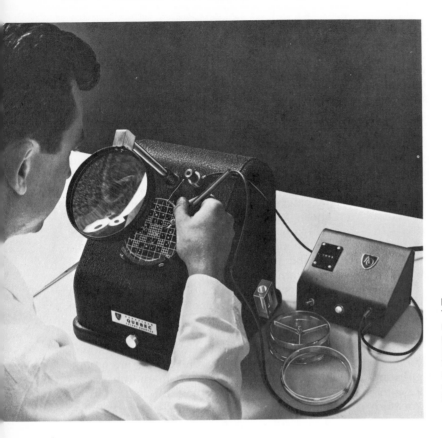

Figure 10-15. Electronic colony counter. The probe (actually an electrode) is used to mark the location of each counted colony. A probe-sterilizing socket is seen on the lower right side of the plate holder. Each colony touched by the probe is automatically recorded on the counter. (Courtesy of American Optical Corp., Scientific Instrument Div., Buffalo, N.Y.)

measures *only organisms viable under the conditions of growth* (medium, temperature, etc.) provided.

10.6
GROWTH CURVES

Let us suppose that during incubation we make colony counts of our culture of *Escherichia coli* every two hours at first, and plot the numbers, and logarithms of the numbers, of colonies (roughly, live organisms per ml) against time. If we were to continue to plot actual numbers on an arithmetical graph instead of logarithms on semi-log paper, we would need a sheet of paper several miles long because, within 24 hours, the numbers can run into millions per ml (Fig. 10–17). We may continue making counts until no further significant changes in numbers occur. At the end of this time a curve will have been obtained which will look somewhat like that seen in Figure 10–18.

A totally different type of curve would be obtained if we were to count the bacteria in the fluid by means of one of the **total count** methods we have described. This is because many of the bacteria die in the culture in increasing num-

bers during the period of incubation. While appearing in the total count, they cannot produce growth in the dilution-tube series or colonies in Petri plates since these enumerate only organ-

Figure 10-16. Colonies in agar in one Astell tube after incubation. (Courtesy of Consolidated Laboratories, Inc., Chicago Heights, Ill.)

isms viable under the growth conditions provided. A curve showing total counts as compared with viable counts is seen in Figure 10–19.

Phases of the Growth Curve

The curve shown in Figure 10–18 has several portions that deserve discussion. These are shown by brackets and labels in the figure.

Initial Phases. Portion A, usually called the **latent** or **initial stationary** (or **lag**) **phase,** represents a period during which the dormant organisms used as inoculum are probably imbibing water, restoring RNA (chiefly ribosomal) essential to synthesis of new cell proteins, possibly producing inducible enzymes (Chapt. 7) to cope with new nutrient substances, swelling, and otherwise becoming adjusted to the new environment, much as might occur when a dormant tree is set out in the spring.

There is growth in **size** of cells but no immediate increase in numbers. The dotted line indicates that some few of the cells may actually die off during this period, only the more vigorous going on to multiplication.

Initiation of Growth. When **dormant** cells are used as inoculum, factors of critical importance in initiating growth are pH, temperature, the presence of suitably high or low oxygen concentrations (oxidation-reduction potential), and favorable concentrations of carbon dioxide.

If the new medium contains nutrients that are not assimilable by most of the cells in the inoculum but are utilizable by perhaps one or two **mutant** cells, then the unmutated cells will die off. Perceptible growth will finally appear, but only after an unusually long lag. Sometimes the production of needed inducible enzymes in unmutated cells also takes a very long time.

Phase of Accelerated Growth. Once growth begins it is soon manifested in the rising inflection of portion B, which is commonly called the **phase of accelerated growth.** The first two phases together are often called the **lag phase.** During this early period, when fission is slow, the size of the cells is large: near the maximum for the species.

During the phase of accelerated growth the time required for each cell to divide gradually decreases, cell size diminishes, and fission rate reaches a maximum determined by the species of microorganism and growth conditions.

Exponential or Logarithmic Phase. As growth continues, the cells reach their maximum rate of fission. Numbers increase in linear relationship to time. Fission may become so rapid that the number of organisms doubles

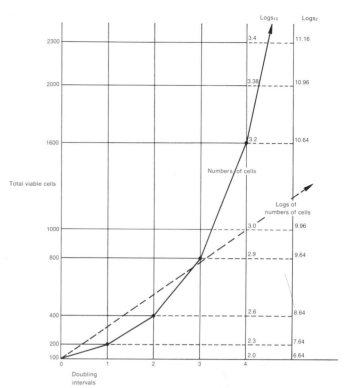

Figure 10–17. Arithmetic (———) and logarithmic (– – –) curves of growth in a flask of *Escherichia coli* culture inoculated with 100 viable cells from another culture in the logarithmic phase of growth and incubated at 35C. The numbers of cells double every time unit, assuming that none dies. Note that the arithmetic curve for numbers of viable cells, plotted against time units, goes up like a rocket! The logarithms of these numbers, plotted against time units, proceed in a straight line. The faster the growth, the steeper the inflection of the curves. Thus the rate, and changes in rate, of growth can be calculated from the slope of the curve. Note also that, of numbers that represent successive doublings of the original 100-cell inoculum, the $\log_2$ increase by 1.0 from the original $\log_2$ of 100: 6.64.

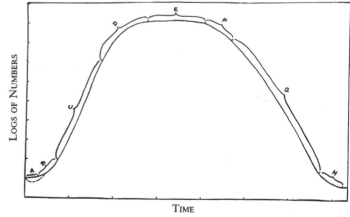

Figure 10–18. Growth curve of unicellular organisms under optimal conditions of growth. Phases shown here are: A, lag; B, accelerated growth; C, logarithmic or exponential; D, negative growth acceleration; E, maximum stationary; F, accelerated death; G, logarithmic death; H, readjustment.

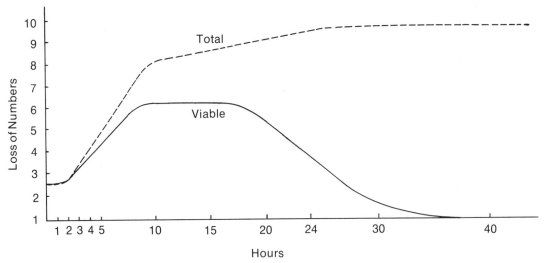

Figure 10–19. Relation of total to viable counts of bacteria in a pure culture under optimal conditions of growth. For explanation see text.

within each 10 minutes. During this phase counts of viable cells almost equal counts of total cells because relatively few of the cells die; it is a "youthful" population. The average size of the cells is at its minimum for the species during this time. It is also conceivable that cell membranes or walls are thinnest and metabolic activities at their highest rate during this period; hence, in part, the commonly observed vulnerability of young cells to numerous deleterious influences that do not affect mature, less active cells. Fission rate varies greatly with different species and under different conditions of growth. Tubercle bacilli, for example, probably divide only about once a day at the highest rate of growth on solid media but every hour or less in certain fluid media.

During this period of most active multiplication (C) the logarithms of the numbers of live organisms counted at short intervals, plotted against time, produce a straight line as shown in Figure 10–18. This period is spoken of as the **logarithmic phase** of multiplication or as the **phase of exponential increase.** Were this to continue uninterrupted, the culture would theoretically become a solid mass of bacteria in a few hours. Actually, numerous factors soon interfere.

During this phase most of the cells are physiologically young and biologically active. If a subculture is made from the flask to a new flask of the same sort of warm, sterile broth, growth continues at the logarithmic rate; there is no lag or dormant phase. The lag and dormant phases become evident to some degree in subcultures made during any other phase of the growth cycle. Biochemical and physiological properties that are commonly used for identification of organisms are usually most manifest during the logarithmic period.

Growth Rate and Generation Time. The time for a single cell to undergo fission is called the **generation time** of that cell. This varies with species of microorganism, nutrients, environmental conditions, and growth phase.

By making one count (C^1) at a specified time and a second count (C^2) after an interval during the logarithmic phase of growth and knowing the number of elapsed time units (tu) (and assuming that none of the daughter cells dies), we can calculate the total number of organisms produced (C^3), the total number of fissions (f) or doublings of numbers, and both growth rate (gr), and generation time (gt).

$C^2 - C^1$ gives the total number of new cells (C^3). Since each fission produces two organisms or (theoretically) a doubling of total numbers in the culture, the exponent of cell increase per unit of time is 2. In expressing numbers of cells, logarithms to the base 10 ($\log_{10}$) are commonly used for convenience in calculating and plotting. Where counts and numbers double per time unit (tu), it is especially convenient to use logs to the base 2 ($\log_2$). One is easily converted to the other:

$$\log_2 N = \frac{\log_{10} N}{0.301}$$

At any moment the number of organisms in

our culture can be expressed as the log to the base 2 ($\log_2$) of that number. The difference between the initial count ($\log_2 C^2$) and the second count ($\log_2 C^1$) gives the total number of generations or doublings or fissions (f) during the total number of time units (tu):

$$f = \log_2 C^2 - \log_2 C^1$$

By substituting common logs ($\log_{10}$) for logs to the base 2, and using 0.301 as the value of log 2:

$$f = \frac{\log_{10}C^2 - \log_{10}C^1}{0.301}$$

The rate at which fissions occur per time unit (growth rate, or gr) can be calculated as:

$$gr = \frac{f}{tu}$$

The time required for each generation to occur (generation time, or gt) is found as:

$$gt = \frac{ltu}{f}$$

Thus, if $C^1 = 50$ and $C^2 = 838{,}860{,}800$ (see Table 10–2):

$$f = \frac{8.92369 - 1.69897}{0.301} = 24 \text{ doublings (approximately)}$$

In this example let tu equal one hour. In 18 hours:

$$gr = \frac{24.00}{18} = 1.33 \text{ doublings } (f) \text{ per hour}$$

Then:

$$gt = \frac{1}{1.33} = 0.75 \text{ hour (45 min) per generation}$$

TABLE 10-2. BASES FOR CALCULATION OF GROWTH DATA

f	C^3	$Log_{10}C^3$	Log_2C^3
0	50	1.69897	5.64
1	100	2.00000	6.64
2	200	2.30103	7.64
3	400	2.60206	8.64
8	12,800	4.10721	13.31
24	838,860,800	8.92369	29.65

Phase of Negative Growth Acceleration. Within a few hours (or days) after the commencement of the logarithmic phase, the organisms begin to encounter difficulties. Food begins to run out, poisonous waste products accumulate, pH changes, hydrogen acceptors are used up, energy transfers are diminished, and the cells interfere with each other. The rate of fission begins to decline and the organisms die in increasing numbers, so that the increase in number of live cells slows, as shown in the portion of the curve labeled D of Figure 10–18. This is spoken of as the **phase of negative growth acceleration.**

A number of workers have studied the development of this phase by additions of fresh sterile medium but without removal of wastes or dead cells. The population increases somewhat with each addition of food, but the overall form of the growth curve develops as usual; exponential growth soon ceases.

Final Phases. Eventually (the time depending on the temperature, the size of the flask and volume of fluid, the composition of the medium, and numerous other factors) the number of cells dying balances the rate of increase, and the **total viable population** remains unchanged for a time. The **total count** continues to increase, but not as rapidly as at first. This phase, the **maximum stationary phase,** is shown at E, Figure 10–18.

As conditions become more and more inimical to the microorganisms, the cells reproduce more slowly, and death overtakes them in ever-increasing numbers, as shown at F. This is the phase of accelerated decrease or **accelerated death phase.**

This progresses into the **logarithmic death phase** (G), during which decrease in number occurs at a regular, unchanging rate.

Finally, conditions begin to reach an equilibrium such that both rate of death and rate of increase tend to balance each other again at a very low population level, and the **phase of readjustment** (H) and a **final dormant phase** are attained. Complete sterility of the culture may quickly ensue or be delayed for weeks or months, depending on the kind of organism, the number of viable cells remaining, whether or not the culture is very acid, and so forth.

Factors Affecting Growth Phases

The form of the growth curve may be affected by many factors. For example, if the culture is suddenly plunged into ice water, the curve at once ceases its upward trend, remains flat for a time, and then begins to decline. If the culture is held at 22C instead of 35C (for *Escherichia coli*), the rise in the positive phases is much less abrupt and much more extended. Other factors such as pH, concentration of food, and so on have their effects. (See also Chapter 19.)

Colony Growth

Cells forming a colony on the surface of a solid medium like nutrient agar encounter an environment very different from that of free cells bathed in a fluid medium. Unless the surface of the solid medium is very moist or the organisms are very actively motile, the colony is restricted to growth in one limited area (Figure 10–20).

The obvious limitations to colony expansions are the following: (a) Nutrient solution can diffuse from the agar to the uppermost cells in the colony to only a limited extent. (b) The available nutrient and moisture in the agar in the immediate area are soon exhausted. (c) Wastes do not readily diffuse away and therefore accumulate in the colony and in the agar beneath. The cells at the top of the heap are obviously at a disadvantage, and soon the upper and central portions of the colony undergo the effects of aging and senescence and other ef-

fects of unfavorable nutrition and nonremoval of wastes. (d) Colonies that are too crowded compete with and overgrow one another.

10.7
CONTINUOUS CULTIVATION

Exponential growth (but not **synchrony;** see §10.8, following) can be prolonged greatly by various means, e.g., a **chemostat.** Apparatus is arranged so that there is slow, but adjustable, admixture of new, sterile, nutrient fluid containing a **limiting** concentration of some growth-controlling micronutrient. In the presence of the limiting micronutrient exponential growth proceeds; as concentration of the limiting micronutrient approaches exhaustion, growth slows. By a concomitant and equal removal of old medium, with its accumulation of toxic metabolic products and older and dead cells, a constant volume of culture, concentration of micronutrient, and numbers of exponentially growing cells is maintained in the main culture vessel: a **steady-state** culture (see Figure 49–2).

Exponential growth may also be stabilized by means of a **turbidostat.** In this, the steady state of the culture is maintained by constant electronic monitoring of changes in turbidity due to growth or removal of cells in the main culture vessel. Changes in turbidity retard (or increase) passage of light through the culture. These

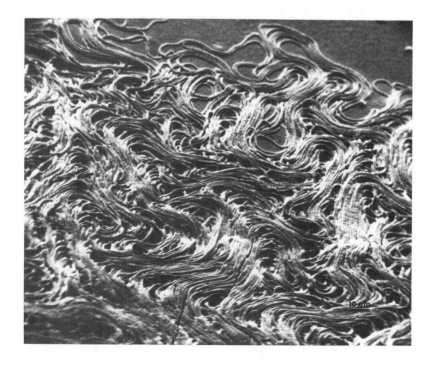

Figure 10–20. This scanning electron micrograph of the edge of a colony of *Bacillus cereus* on trypticase soy agar demonstrates the decreasing density of organisms near the edge. (Original magnification 1,600×; reduced to 45% of original size.) (Courtesy of H. Farzadegan and Ivan L. Roth, Department of Microbiology, University of Georgia.)

changes activate mechanisms that control the flow of nutrient into, and flow of waste out of, the main culture vessel.

Practical possibilities of continuous cultivation combined with selective enrichment are numerous. For example, through the prolonged use of a continuous-flow chemostat containing a medium entirely mineral except for a man-made (synthetic) poison (pentachlorophenol) as a sole source of organic carbon and energy, it has been possible to obtain, in pure or **axenic** culture, from polluted industrial waste, a species of bacterium capable of metabolizing this substance formerly considered to be beyond the metabolic capabilities of microorganisms. Such organisms, and the means of finding them, have obvious ecological potentialities. A highly active microbial digester of cellulose (waste paper), if found, could be very useful.

10.8
SYNCHRONOUS GROWTH

If a single, vegetative bacterial cell in the exponential phase of growth is placed in a new container of fresh, warm medium of the same composition as that in which it was growing, it promptly undergoes maturation and fission into two *almost* equal cells. Being nearly equal, these cells mature and undergo fission again in virtual (but *not absolute*) synchrony. After a few divisions, accumulating inequalities in rate of fission result in a randomly dividing population.

Almost perfect synchrony can be maintained for a time if a randomly dividing culture in the exponential phase of growth at 35C, with a generation time of, say, 15 minutes, is held for approximately 30 minutes at a lower, fission-retarding temperature (say, 20C). During this interval all cells mature to the point of fission, including those that had not previously done so. However, at 20C none divides. On sudden return of the culture to 35C all cells, being now at the point of fission, divide synchronously. By repeating the alternations of temperature, synchrony can be maintained in the culture for several generations. However, randomness finally returns.

Lowering the temperature to a degree at which all of the cells mature to the point of fission *but remain undivided* (poised, as it were, until the temperature is suddenly raised to the optimum) may be thought of as analogous to bringing racetrack contestants to a starting line, poised for the sound of the starter's signal. If nearly evenly matched, they are at first, and for some successive paces or "laps," close together or synchronized. Later, their different speeds randomize their relative positions on the track.

In another procedure, a culture in the exponential phase of growth is passed through a specially graded membrane filter. The smallest cells, i.e., all those that have most recently divided, pass into the filtrate together. Their synchrony of fission continues for a few generations.

Synchronized cultures have special uses in research.

10.9
MICROORGANISMS IN NATURAL AND SIMULATED NATURAL ENVIRONMENTS

Most of our knowledge of bacteria and related microorganisms, and most modern microbiological methods—systematic (for classification), morphological, diagnostic, and so on—have been derived from studies of pure cultures artificially prepared in the laboratory. Yet, as previously mentioned, microorganisms in nature (e.g., fertile soil, sewage, sea water) are nearly always subject to influences, favorable or unfavorable, of millions of cells of perhaps hundreds of different species of neighboring protists, as well as varying chemical and physical environmental factors.

To create artificially, in the laboratory, an exact replica of even one complex natural environment is usually impossible. It is possible, however, to provide, within natural environments, artificial areas or segregated niches to which the investigator has periodic access with little or no disturbance of the microcommunity. Examples are glass slides placed on a lake bottom on which adherent organisms may localize and form slimy films or matted growths; or flat, capillary tubes immersed in any desired natural milieu (e.g., tidal muds), giving protected sites for colony growth, and so on. Such methods, with numerous ingenious modifications, are much used and invaluable in studies of microbial ecology, effects of industrial and sewage pollution on aquatic and marine flora and fauna, and numerous other parameters of microbial ecology and the role of microbial life in human activities.

Dialysis Culture. Dialysis culture presents a means of subjecting a segregated pure culture (or a mixed culture if desired) to the influences of an artificial environment resembling a natural one, or to the influences of another pure culture, or to a great variety of mixed cultures or fluids whose compositions and variations are subject only to the ingenuity and imagination of the investigator.

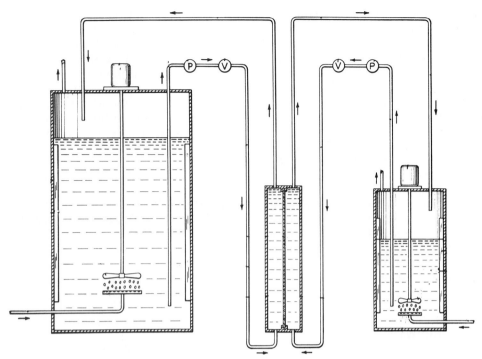

Figure 10–21. One arrangement of a dialysis culture. A sterile medium reservoir is on the left, with aerator and agitator (optional); a pure culture (or mixed culture, e.g., sewage) is in the reservoir on the right. The two environments are separated only by a semipermeable membrane in the central chamber.

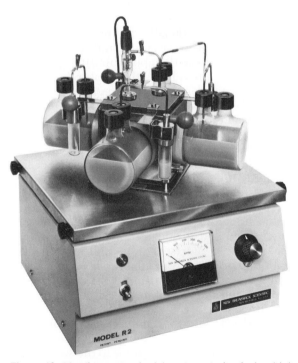

Figure 10–22. Apparatus for laboratory study of microbial ecological relationships (Ecologen®). The central chamber may contain sterile water, broth medium, or other liquid. Each flask is separated from the central chamber by a semipermeable membrane of any desired composition and porosity. (Courtesy of New Brunswick Scientific Co., Inc., New Brunswick, N.J.)

In dialysis culture, pure cultures and complex environments are in close contact, separated only by semipermeable membranes (Fig. 10–21), or even merely by interfaces, as between two immiscible fluids. In early experiments (e.g., Metchnikoff, 1888; Frost, 1904; and numerous others) membranes for dialysis culture were commonly made of collodion. Today films may be of cellophane, various plastics, regenerated cellulose, parchment, filter membranes (see Figure 45–12), and the like, offering a variety of physical properties and pore sizes. Exchange dialysis of many diffusible and dialyzable substances may occur across the membrane, from one environment to the other. Very simple or extremely complex and interesting systems and interrelationships can be arranged by the ingenious and well-informed researcher.

The general method of dialysis culture may be adapted to small laboratory containers (Fig. 10–22), to large industrial vats with machine operation, or to tiny membrane sacs or envelopes implanted surgically in living animals, following Metchnikoff, who thus implanted sacs containing cholera organisms to see if they excreted a diffusible toxin. The many aspects and possible variations of dialysis culture are of great and increasing interest and importance in every field of microbiology.

CHAPTER 10
SUPPLEMENTARY READING

Alexander, M.: Biochemical ecology of microorganisms. Ann. Rev. Microbiol., 25:361, 1971.

Berkley, C.: Potentials for automatic control methods in defined media studies. Ann. N.Y. Acad. Sci., 139:39, 1966.

Brock, T. D.: Microbial growth rates in nature. Bact. Rev., 35:39, 1972.

Hartman, P. A.: Miniaturized Microbiological Methods. Academic Press, New York. 1968.

Lwoff, A.: From Protozoa to bacteria and viruses. Fifty years with microbes. Ann. Rev. Microbiol., 25:1, 1971.

Pato, M. L.: Regulation of chromosome replication and the bacterial cell cycle. Ann. Rev. Microbiol., 26:347, 1972.

Payne, W. J.: Energy yields and growth of heterotrophs. Ann. Rev. Microbiol., 24:17, 1970.

Schultz, J. S., and Gerhardt, P.: Dialysis culture. Bact. Rev., 33:1, 1969.

Swoager, W. C., and Lindstrom, E. S.: Isolation and counting of Athiorhodaceae with membrane filters. Appl. Microbiol., 22:683, 1971.

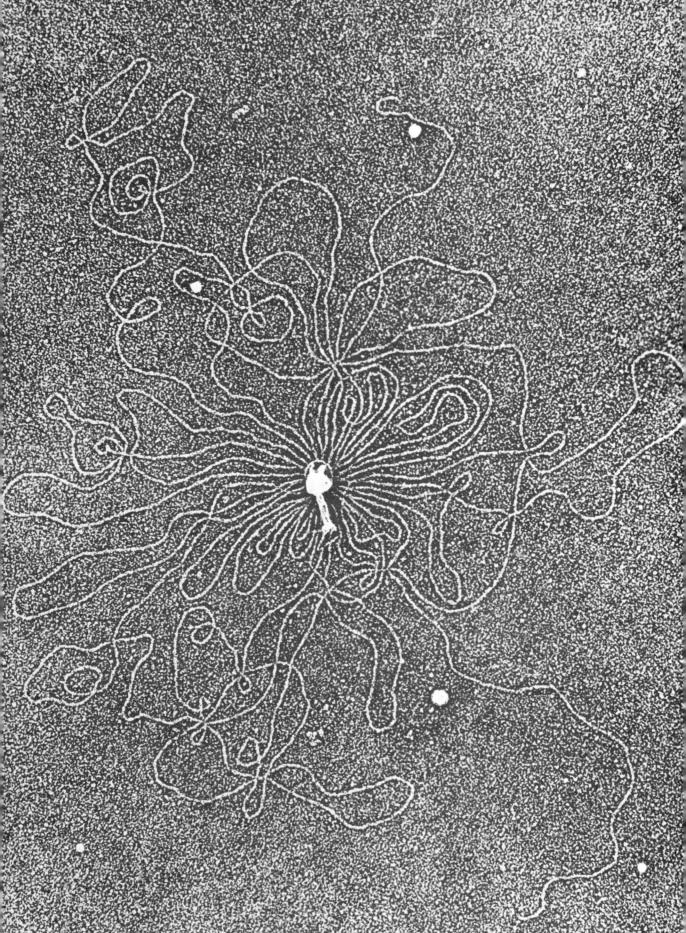

SECTION TWO

THE PROTISTS: EUCARYOTIC AND PROCARYOTIC; THE VIRUSES

Section One of this volume has provided the student with some knowledge of the origin, nature, methods and principles of microbiology and of the underlying mechanisms of all cellular life. In Section Two we now describe some microorganisms that exemplify the workings of those principles. Here we delve somewhat more informatively into the structural and physiological properties of the various types of protists: their similarities, differences, and relations to other forms of life.

Certain representative fungi, algae, and protozoans are described in detail as models of eucaryotic or higher protists; previews are given of some common bacteria as exemplars of the procaryotic or lower protists that are described in greater detail in later sections of the book.

Usefully detailed general descriptions of viruses are included in this section. They are regarded as essential to the reader because, throughout the book, it is evident that not only do viruses play a major role in the life and/or death of all plants, animals, and protists with which the viruses make effective contact, but the structures and activities of viruses provide some of our most informative models of intracellular machinery in general. In these chapters on viruses it becomes clear that knowledge of viruses is indispensable to obtaining a truly "close-up" view of those absolutely basic "facts of life"—sex, heredity, and genetics—at the most intimate level possible: the molecular level. The fields of molecular genetics and molecular biology are rapidly growing daughters of the new science of microbiology: granddaughters of Mother Biology, as it were.

Facing page, Courtesy of A. K. Kleinschmidt.

CHAPTER 11 · THE PROTISTS: EUCARYON AND PROCARYON

Robert Hooke, mentioned in Chapter 2 as a pioneer English microscopist, was also an inventor, artist, physician, philosopher, and enthusiastic investigator of almost everything with the microscope. In 1665 he applied the term **cell** (L. *cella* = a small enclosure) to the microscopic "boxes" or spaces that are readily seen with the microscope in the honeycomb-like structure of dried cork. Hooke, however, had no knowledge of the living matter that the cork "cells" had once contained. To him a cell in cork was a hollow shell composed of cork substance (cellulose). Likewise, although living microorganisms were first described by Leeuwenhoek in 1674, the Dutch microscopist had no knowledge of the true nature and structure of microscopic organisms.

11.1
CELLS AND PROTOPLASM

Accurate knowledge of the sizes, forms, habitats, functions, and anatomy of microorganisms (including the tissue cells of higher plants and animals) accumulated slowly until about 1820, when microscopes were perfected that yielded magnifications as high as 600×. In 1824 Dutrochet, a French scientist, called attention to the fact that all organized tissues were aggregations of billions of microscopic units which he called cells, though their true nature was still far from fully realized. The fact that the internal substance of such cells was not entirely homogeneous was revealed in 1831, when Robert Brown described an often-seen though unidentified intracellular body now recognized as a nucleus. Its structure and function were not fully clarified for many years after Brown's description and, in some respects, are still enigmatic.

Microbiologists of the 18th and early 19th centuries thought of the entire contents (except obviously inert inner granules of substances such as starch or fat) of microorganisms as a homogeneous, clear, viscous, colorless substance that had the unique and mysterious property of being alive. In 1839 the physiologist Purkinje gave this substance the name **protoplasm** (Gr. *protos* = original or primitive; *plasma* = fluid substance). Protoplasm was recognized as **living substance.** Max Schultz, in Germany about 1860, called it the substance of life and Thomas H. Huxley, in England in 1865, called it "the physical basis of life." Because we now know it to be a complex mixture of many chemically different substances, and to contain many distinct parts with a variety of structures and physiological functions, the term protoplasm is no longer used to refer to a single substance. The word is still sometimes used, however, as a convenient synonym for "*total living cell contents,*" and the term will occasionally be so used here.

The first formulation of the modern cell theory in 1838 is generally ascribed to two German scientists: Schleiden, a botanist, and Schwann, a zoologist. Today the cell in biology is defined as the smallest living unit capable of autonomous growth and reproduction using food substances chemically different from itself. This definition of a cell excludes: (a) inorganic crystals that grow only by accretion of the same substance as themselves and (b) viruses. Viruses, as will be explained later, are not cellular in structure, according to the currently accepted definition of a cell, and do not have any autonomous metabolism or, indeed, any life as we know it. Viruses are "alive" only when inside other living cells and while being replicated at their expense. Otherwise, viruses are inert.

11.2
PLANTAE AND ANIMALIA

Before the general use of the high-powered microscope, biologists had relatively little difficulty in classifying the various forms of life visible to their unaided eyes as animals (the animal kingdom, **Animalia**) or as plants (the plant kingdom, **Plantae**). No other kingdom was known. Differentiation was based mainly on such readily visible characteristics as motility, green color, and the presence of leaves, flowers, and stems as contrasted with eyes, teeth, and legs. Plants were regarded as typically not motile and relatively simple; animals as typically mobile and obviously more complex. Similarly, early microscopists readily differentiated between the unicellular or microscopic **green** (**chlorophyll**-containing) algae on the one hand and the **colorless** (non-chlorophyll-containing) unicellular animals, yeasts, protozoans, molds, and bacteria on the other. All of the microscopic algae, fungi, and bacteria were assigned to the plant kingdom because of their relative immobility, their supposedly simple, plant-like structure, and their inability to catch and eat solid foods like animals. On the other hand, Protozoa were readily classed as animals because of their more complex structure, their conspicuous motility, and their ability to catch and eat solid foods such as other Protozoa, bacteria, and cell fragments. Table 11–1 illustrates a simplistic classification scheme which attempts to place organisms of interest to the microbiologist into two kingdoms and to show the division between procaryotic and eucaryotic cells (see §11.4 for further details).

As generally happens, this comfortable arrangement soon encountered difficulties. With advances in knowledge and improvements in scientific techniques, including microscopy, it became evident that, among the small, more primitive ("lower" or less highly evolved) organisms, and especially among microorganisms, classification as plant or animal was becoming difficult and often impossible. Many puzzling overlappings and inconsistencies were discovered. For example, it was found that plant cells generally have cell walls of cellulose (the basic material of all wood, stems, and leaves), and that animal cells do not have any true cell wall at all but only thin, flexible cell membranes which are normally quite fragile. In some animal cells, this membrane, condensed and thickened, serves as a protective and retaining integument. Many species of animal cell synthesize supporting structures of chitin: a characteristically animal substance (insect and crustacean skeletal material). Yet many fungi also have cell walls of chitin, whereas certain primitive animals, the group of chordates called sea squirts (tunicates), have an outer covering (cell wall) of cellulose.

Plants contain the conspicuous green pigment chlorophyll that enables them to use sunlight as a source of energy. Typical animal cells and fungi do not contain chlorophyll and cannot use radiations as a source of energy. They must obtain their energy from oxidations of their foodstuffs. Yet one group of flagellated protozoans, *Euglena*, contains chlorophyll arranged in chloroplasts like those of higher plants and can use sunlight as a source of energy.

By contrast, some species of *Euglena* not only lack chlorophyll, in this resembling fungal and animal cells, but are nonmotile (in this resembling typical plant cells). Several species of nonmotile protozoans are known also. Some algae (obviously [?] plants) produce motile, flagellate, reproductive cells somewhat resembling protozoans. Finally, there are some types of bacteria which contain special chlorophylls and are photosynthetic. There are numerous other illustrations of confusing combinations and overlappings of properties of organisms that were once classified in supposedly distinct and mutually exclusive groups.

11.3
KINGDOM PROTISTA

To avoid the confusion arising from growing knowledge of microorganisms, a third kingdom, the **Protista** (Gr. *protistos* = primitive or first), was proposed for *all microorganisms* in 1866 by Ernst Haeckel, one of Darwin's stu-

TABLE 11-1.   SIMPLISTIC CLASSIFICATION SCHEME BASED ON TWO KINGDOMS (PLANT AND ANIMAL) AND SHOWING THE DIVISION BETWEEN PROCARYONS AND EUCARYONS IN THE MICROBIAL WORLD[*]

Procaryotic	Kingdom: PLANTAE Division I. Protophyta (fission plants) Class 1. Schizophyceae (blue-green algae) 2. Schizomycetes (bacteria and related forms) 3. Microtatobiotes (rickettsias and viruses)
Eucaryotic	Division II. Thallophyta (thallus plants, i.e., undifferentiated into leaves, stems, roots) Subdivision I. Algae (seaweeds) Class 1. Chlorophyceae (green algae) 2. Phaeophyceae (brown algae) 3. Rhodophyceae (red algae) 4. Euglenophyceae (euglenoids) 5. Chrysophyceae (yellow-green and gold-brown algae, diatoms) 6. Pyrrophyceae (dinoflagellates) Subdivision II. Eumycotina (true fungi, including molds and yeasts) Class 1. Ascomycetes (sac fungi) 2. Basidiomycetes (club fungi) 3. Phycomycetes (primitive fungi) 4. Deuteromycetes (Fungi Imperfecti) Subdivision III. Myxomycotina (slime molds) Division III. Bryophyta (mosses and liverworts) Division IV. Pteridophyta (ferns) Division V. Spermatophyta (seed plants) Kingdom: ANIMALIA Division I. Protozoa Class 1. Mastigophora (flagellates) 2. Sarcodina (ameboids) 3. Sporozoa (sporozoans) 4. Ciliata (ciliates) Division II. Metazoa

[*]Adapted in part from Breed, Murray, and Smith (Eds.): Bergey's Manual of Determinative Bacteriology, 7th ed., The Williams & Wilkins Co., 1957.

dents. In Haeckel's time microscopes still lacked the high resolving power of modern instruments, and many species of microorganisms, including the rickettsias, the viruses, and several other curious types that will be mentioned later, were then unknown. These have since been included in the kingdom Protista. Except for viruses and certain bacteria which are obligate parasites, all members of the Protista kingdom (**protists**) are distinguished from all members of the plant and animal kingdoms by the single fact that protists exist as autonomously synthetic, **unicellular** organisms. Some protists, such as unicellular blue-green algae and true bacteria (formerly order Eubacteriales), are obviously plant-like; others, e.g., Protozoa, are obviously animal-like. Many others are intermediate: not closely resembling either plants or animals. Viruses, being noncellular, may (or may not!) be included as protists "by courtesy"; the exact position and status of viruses in the living world is still debatable. In fact, some authorities prefer not even to recognize the kingdom Protista, while others would prefer to

have a scheme based on four or five kingdoms. Figure 11-1, though highly abbreviated, gives some idea of what form a five-kingdom scheme might take. Note that the major changes are to separate the bacteria and blue-green algae into a separate kingdom and to raise the fungi (Eumycotina) to full kingdom status. Viruses are not considered in this scheme. Whether or not a classification system encompassing five kingdoms will eventually win the approval of the majority of biologists and microbiologists is not clear. The term protist is commonly used to mean unicellular microorganism as previously defined and will be frequently so used here.

The group of protists may for convenience be divided into six main groups, as follows:

1. Protozoa (Gr. *protos* = primitive or original; *zoion* = animal): unicellular animals.
2. Algae (L. *alga* = seaweed): eucaryotic, unicellular, photosynthetic plants.
3. Eumycotina (Eumycetes) or true fungi (Gr. *eu* = true; *mykes* = fungus; *tina* = ending assigned to subdivisions in

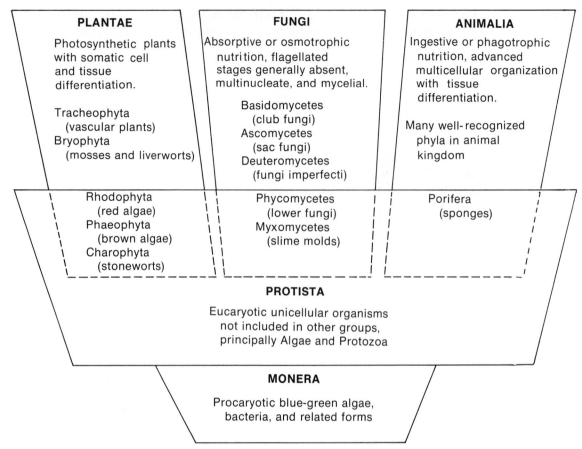

Figure 11–1. Simplified five-kingdom scheme.

botanical code): true fungi, including yeasts and molds.

4. Schizophyceae or blue-green algae (Gr. *schizo* = fission; *phyco* = seaweed): procaryotic, unicellular microorganisms with chlorophyll *a*.

5. Schizomycetes or fission fungi (Gr. *schiza* = fission or separation): bacteria of many divergent types, including the rickettsias and chlamydias.

6. Viruses (L. *virus* = noxious, slimy fluid).

Each of these six groups will be discussed in greater detail in the chapters that follow.

11.4
EUCARYOTIC AND PROCARYOTIC CELLS

As indicated in Chapter 1, all available evidence strongly suggests that the protists of today were derived from still simpler cellular forms of life billions of years ago by evolutionary processes. These simpler cells had in turn presumably developed during still more remote ages from complex but not actually living organic molecules, formed spontaneously in the rich organic milieu of primeval tropical seas by a process now called **chemical evolution,** as outlined in Chapter 2. See also the list of Supplementary Reading at the end of this chapter.

Eucaryons. At some time during the untold millennia of the early development of cellular forms there seem to have occurred hereditary changes (mutations) and natural selections (evolutionary processes) that resulted in greatly increased complexity of structure and function. The protists that underwent these presumed evolutionary upgradings and became animal-like are probably represented today by the Protozoa. Their modern, presumed evolutionary derivatives include all of the "higher" (more highly evolved) animals including man: the animal kingdom.

The "improved model" protists that became plant-like are probably represented today by the higher fungi, molds, and yeasts and the unicellular green (*not* blue-green) algae. Their modern, presumed evolutionary derivatives (or collaterals) include all of the higher plants: the plant kingdom. (There are various other ideas on evolutionary relationships, hence the use of words such as "presumed" and "apparently.")

Whatever their evolutionary history may be, the cells of all of these "upgraded" or higher plant and animal forms, including large animals, man, trees, and so on, as well as the higher protists (Protozoa, molds, yeasts, and higher fungi and green algae) are said to be **eucaryotic** (Gr. *eu* = true; *karyon* = nucleus) because the nucleus of each cell is enclosed within a well-defined **nuclear membrane** and thus is segregated from the cytoplasm; it is a **true nucleus**. In addition, with a few exceptions among the simplest forms of molds, typical eucaryotic protists exhibit well-defined sexual or mating phenomena and often well-differentiated sexual cells (gametes).

By the late 1950's, students of cytology had shown that all eucaryotic cells typically* consist of several morphologically and physiologically distinct parts: (1) a definite **membrane-enclosed nucleus** (L. *nucis* = kernel) containing a complex of deeply staining (**chromatinic**) fibrillar material which is the basis of **chromosomes**, plus a spherical, stainable region called the **nucleolus**. (2) They observed a mass of clear, semifluid matter surrounding the nucleus, the **cytoplasm**, having a colloidal (continuous phase + discontinuous phase) structure. The cytoplasm was seen to contain several kinds of definite, **membrane-enclosed,** subcellular bodies called plastids or organelles: **mitochondria** (the seat of energy-yielding enzymic activities); **chloroplasts** (the seat of photosynthesis in green plants); also granules (**blepharoplasts**), associated with motility; **centrosomes** (involved in mitosis); **Golgi complex** or **apparatus** (apparently involved in vacuole formation and transport); **kinetoplasts** or **kinetonuclei** and basal bodies, associated with motion and flagella or cilia; a **centriole** (in certain plants and animals only; important in fission); other organelles whose function and structure were then obscure. (3) Surrounding the cytoplasm a thin, pliable and relatively fragile **cell**

---

*The use of the words "typically" and "typical" here and elsewhere implies that there are, or may be, "atypical" specimens that do not conform to the general description.

(or **cytoplasmic**) **membrane** was demonstrable. It was seen to function as an enclosing and retaining sac for the whole cell, segregating the cell contents from the outer world (Fig. 11–2, *A*).

Typical eucaryotic plant cells (*not* typical animal cells) were seen to have a relatively thick, strong, rigid, retaining and protective **cell wall** outside the cell membrane. In the higher plants this consists generally of cellulose (the principal component of paper and wood) (Fig. 11–2, *B*). In many eucaryotic fungi, the cell wall contains chitin, which is the skeletal substance of insects and crustacea. The electron microscope reveals all the above details (Fig. 11–3) and others as well. The major groups of eucaryotic protists, i.e., fungi, algae, and protozoans, will be considered in greater detail in the next two chapters.

Procaryons. It has been proposed that those protists that did not participate in the evolutionary upgrading that produced the eucaryotic form of cell, as previously described, continued with their primitive nuclear structure. Presumably, they are with us today as the blue-green algae and the bacteria and have been considered by some to be "living fossils." It should be pointed out, however, that the evolutionary relationship between eucaryons and procaryons is still an enigma. One fascinating theory which is receiving considerable attention is that certain organelles of eucaryotic organisms were derived from free-living procaryons which entered the eucaryons and established a symbiotic relationship. After a period of time, the intracellular procaryons lost their ability to lead an independent existence. Lending credibility to this theory is the finding that both chloroplasts and mitochondria have their own complement of nucleic acids and the fact that mitochondrial ribosomes more closely resemble bacterial ribosomes than eucaryotic ribosomes in sedimentation value. Sophisticated techniques are presently being employed to compare the nucleic acids and protein-synthesizing ability of eucaryotic organelles with those of their host cells and possibly related procaryotic cells. The results of these investigations may provide new insight into the evolutionary pathways of these two types of cells.

An alternate theory is that the mitochondria and other organelles so characteristic of eucaryotic cells, instead of having originated from separate cellular organisms that invaded or were ingested by the eucaryotic cells, evolved gradually from the procaryons themselves as a result of many genetic mutations over long periods

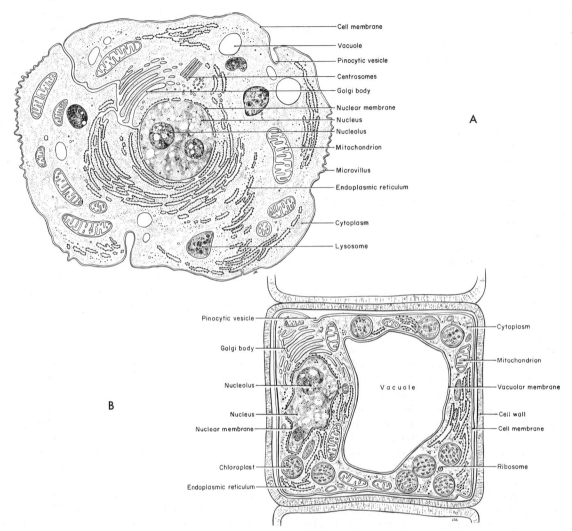

Labels for A (top diagram):
Cell membrane
Vacuole
Pinocytic vesicle
Centrosomes
Golgi body
Nuclear membrane
Nucleus
Nucleolus
Mitochondrion
Microvillus
Endoplasmic reticulum
Cytoplasm
Lysosome

A

Labels for B (bottom diagram):
Pinocytic vesicle
Golgi body
Nucleolus
Nucleus
Nuclear membrane
Chloroplast
Endoplasmic reticulum
Cytoplasm
Mitochondrion
Vacuolar membrane
Cell wall
Cell membrane
Ribosome
Vacuole

B

Figure 11–2. Eucaryotic cells: *A*, Composite diagram of a ''typical'' animal cell based on various cytological studies and use of the electron microscope. The **nucleus** controls hereditary properties and all other vital activities of the cell. Both nucleus and **nucleolus** have functions in the synthesis of cell material. The well-defined **nuclear membrane** appears to have pores or openings for communication with the cytoplasmic structures and direction of their synthetic and energy-yielding activities. The **cytoplasm** contains immense numbers of granules called **ribosomes**, concentrated especially along the periphery of a cell-wide labyrinth of connected, narrow sacs called the **endoplasmic reticulum**. These granules are involved in the continuous reactions which synthesize cell materials under direction from the nucleus. The **mitochondria** are involved in another set of enzymic reactions, called **biological oxidation** (Chapt. 8), which yield the energy for all the cell activities. **Lysosomes** appear to liberate the enzymes which digest part of the food of the cell. The gullet-like **pinocytic mouth** or invagination is a means of ingesting fluids or extremely minute food particles (see Chapter 9). The **centrosomes** are especially active in mitosis during reproduction of the cell by fission. The Golgi body (or complex or apparatus) is the site of vesicle formation, which may serve as an intracellular transport mechanism. There are other structures whose nature and function await elucidation by future cytological studies. Compare this complex cell with pictures of bacterial cells shown in this chapter. *B*, Diagram of typical plant cell. Note the thick cell wall.

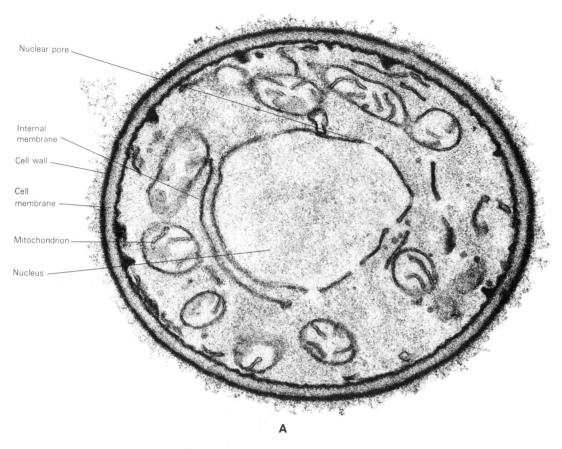

Nuclear pore

Internal
membrane

Cell wall

Cell
membrane

Mitochondrion

Nucleus

A

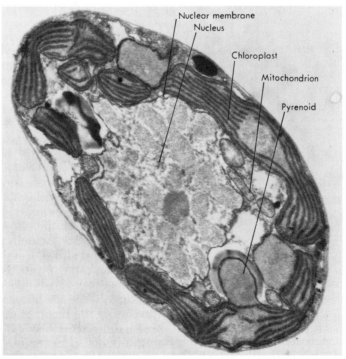

Nuclear membrane
Nucleus

Chloroplast

Mitochondrion

Pyrenoid

B

Figure 11–3. Electron micrographs of thin sections of typical eucaryotic cells, showing many of the structures illustrated in Figure 11–2. *A,* Yeast cell. (39,000×.) *B,* Small dinoflagellate, a member of the algae. (14,600×.)

of time, mutations that produced various intra-cellular metamorphoses of already-existing parts of the procaryons, especially the cyto-plasmic membrane.

At least one species of microorganism, the ameba *Pelomyxa palustris,* seems to have evolved only one of the evolutionary steps toward being eucaryotic: acquisition of a membrane-enclosed nucleus. In all other respects it is procaryotic: lacking mitochondria, endoplasmic reticulum, Golgi-like bodies, and complex cilia. Most striking is its failure to exhibit mitotic figures during fission: a microscopic "connecting link."

The nucleus of procaryotic cells is not enclosed in a nuclear membrane and is not sharply segregated from the cytoplasm, though most of their fibrillar, chromatinic nuclear material is commonly aggregated into an irregular mass called a **nucleoid.** Hence this type of cell is said to be procaryotic (Gr. *protos* = primitive; *karyon* = nucleus). There are numerous other differences: (1) The procaryotic nucleus exhibits none of the complex phenomena of mitosis and meiosis that are distinctive of typical eucaryotic cells. (2) In their manner of reproduction, procaryons are also more primitive than eucaryons, being without typical gametes and exhibiting, though only in some species, a primitive form of conjugation. (3) Procaryotic cells contain none of the complex, membranous, membrane-enclosed organelles which occur in typical eucaryotic cells. (4) The cell walls of eucaryotic plant cells (animal cells do not have cell walls) differ markedly in chemical composition from those of procaryons. The cell walls of eucaryotic green plants are of cellulose, those of eucaryotic fungi are principally of chitin. Most procaryons do have a cell wall (Fig. 11–4) and a principal component of the cell wall is always a polysac-charide–amino acid heteropolymer called **peptidoglycan** (or any of several other accepted names: mucopeptide, mucocomplex, murein, etc.). Other substances, e.g., polysaccharides, lipids, and lipoproteins, may be attached to or complexed with the peptidoglycan, but it is the latter substance which provides most of the strength and allows the procaryotic cell wall to serve the same protective and retaining purposes as the eucaryotic cell wall. (5) Another striking difference between eucaryons and procaryons is the fact that the cytoplasm of typical eucaryons constantly exhibits active streaming movements. This results in, among other effects, motility of many protists, e.g., amebas. The cytoplasm of procaryotic cells, on the contrary, exhibits no apparent streaming. Some species of both types of cells, however, show **gliding mo-**

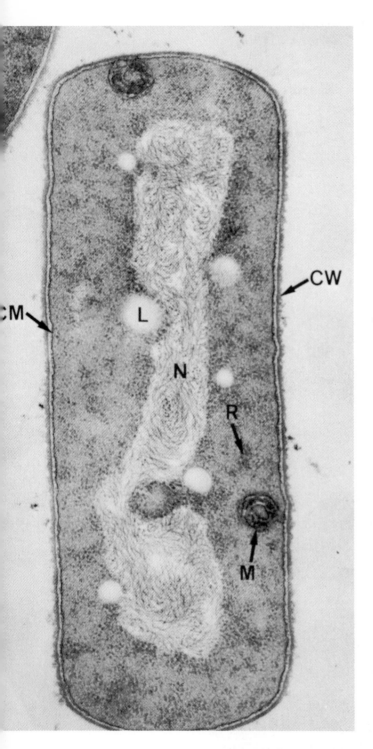

Figure 11–4. Electron micrograph of a thin section of a typical gram-positive procaryotic cell (*Bacillus cereus*). Note that the numerous organelles typical of eucaryotic cells are absent. Only the cell wall (*CW*), cytoplasmic membrane (*CM*), ribosomes (*R*) and the nucleoid (*N*) are regularly seen. Inclusions, e.g., lipid droplets (*L*), and meso-somes (*M*) may or may not be present. These structures are described further in Chapter 14. (64,000×.)

TABLE 11–2. MAJOR DIFFERENCES BETWEEN CELL TYPES

	Procaryotic[*]	Eucaryotic
I. NUCLEAR AREA		
1. Nuclear membrane	−	+
2. Mitotic division	−	+
3. Chromosome number	1	>1
II. CYTOPLASM		
1. Mitochondria	−	+
2. Chloroplasts	−	+ or −
3. Golgi apparatus (dictyosome)	−	+
4. Cytoplasmic ribosomes	70S	80S
5. Organelle ribosomes	−	70S
6. Lysosomes	−	+
7. Membrane-enclosed vacuoles	−	+
8. Phagocytosis and ameboid movement	−	+ or −
III. CELL WALL (when present)		
1. Muramic acid	+	−
2. Diaminopimelic acid (DAP)	+ or −	−
IV. FLAGELLA (when present)	Simple	Complex

[*]Symbols: + = present; − = absent.

tility in contact with solid objects, which is not yet fully understood. Table 11–2 summarizes the differences between eucaryons and procaryons. Additional information on procaryotic cells will be found in Chapter 14. The systematic study of bacteria is discussed in Chapter 30, and detailed information on classification will be found in Chapter 31.

CHAPTER 11
SUPPLEMENTARY READING

Barghoorn, E. S., and Schopf, J. W.: Microorganisms from the late Precambrian of central Australia. Science, 150:337, 1965.

Brachet, J.: The Living Cell. W. H. Freeman & Co., San Francisco. 1966.

Cohen, S. S.: Are/were mitochondria and chloroplasts microorganisms? Am. Sci., 58:281, 1970.

Goodenough, U. W., and Levine, R. P.: The genetic activity of mitochondria and chloroplasts. Sci. Am., 223:22, 1970.

Jeon, K. W. (Ed.): The Biology of Amoeba. Academic Press, Inc., New York. 1973.

Margulis, L.: The origin of plant and animal cells. Am. Sci., 59:231, 1971.

Margulis, L.: Symbiosis and evolution. Sci. Am., 225:48, 1971.

Margulis, L.: Whittaker's five kingdoms of organisms: minor revisions suggested by considerations of the origin of mitosis. Evolution, 25:242, 1971.

Neutra, M., and Leblond, C. P.: The Golgi apparatus. Sci. Am., 220:100, 1969.

Oro, J., and Tornabene, T.: Bacterial contamination of some carbonaceous meteorites. Science, 150:1046, 1965.

Raff, R. A., and Mahler, H. R.: The nonsymbiotic origin of mitochondria. Science, 177:575, 1972.

Schopf, J. W., Barghoorn, E. S., Maser, M. D., and Gordon, R. O.: Electron microscopy of fossil bacteria two billion years old. Science, 149:1365, 1965.

Whittaker, R. H.: New concepts of kingdoms of organisms. Science, 163:150, 1969.

THE EUCARYOTIC PROTISTS · CHAPTER 12

Fungi

There are three major groups of protists that have eucaryotic cell structure: (1) Algae (higher algae, not Cyanophyceae); (2) the Protozoa; (3) the true fungi (Eumycotina, also called Eumycetes). Differentiation between typical (i.e., most familiar) species of the three major groups of eucaryotic protists is easy; for example, a green, nonmotile alga like *Chlorella* is obviously unlike a colorless, actively motile and complexly structured ciliate like *Paramecium* or the large, cottony mycelium of a mold like *Rhizopus* (bread mold). However, on closer examination it is found that numerous other, less typical species in each group have important properties in common and that lines of demarcation are not always so clear. It seems certain that some species in each group are merely evolved forms (i.e., selected mutants) of species in other groups—a reasonable, but not necessarily correct, inference being that all may have originated from a common source billions of years ago. The origins of many are obscure: lost in the mists of antiquity.

12.1
GENERAL CHARACTERISTICS

The Greek word *mykes* (anglicized as *myces*) means fungus, and in current usage includes those microorganisms which have certain plant-like features but that (a) do not form embryos (as in seeds); (b) are without physiologically differentiated or functional roots, stems, leaves, or flowers; (c) consist of only one cell or of characteristic aggregations of many undifferentiated (or very slightly differentiated) cells; (d) are *not* photosynthetic. In the older classification schemes (see Table 11–1) both the Eumycotina (fungi) and Myxomycotina (slime molds) were subdivisions under the Thallophyta in the plant kingdom. A newer system of classification used by Alexopoulos (1962) in his basic text on the subject, most of which appears to have been widely accepted, gives the fungi division status and places them in the kingdom Protista (Table 12–1). Note also that the class Phycomycetes has been split into six new subclasses. Some authorities prefer to eliminate Phycomycetes as a class and elevate the six subclasses to the higher status. When this is done, the endings should be changed to conform with the rules on botanical nomenclature, e.g., *Zygomycetidae* becomes *Zygomycetes*. Despite the divergence among the Phycomycetes, they do differ from the "higher" fungi in several important aspects (to be discussed), and thus the collective term Phycomycetes would appear to be worth retaining.

Except for some aquatic Phycomycetes, most fungi can grow with little free water, thriving on such as may be absorbed from damp atmospheres. They can take water from materials that have very high osmotic pressures,

TABLE 12-1. PARTIAL CLASSIFICATION OF FUNGI*

Kingdom: PROTISTA
  Division: Mycota (Fungi)
    Subdivision 1. Eumycotina (true fungi, Eumycetes)
      Class 1. Phycomycetes (primitive fungi)
        Subclass 1. Chytridiomycetidae (water molds)
            Order: Chytridiales
                Blastocladiales
                Monoblepharidiales
          2. Hyphochytriodiomycetidae (water molds)
            Order: Hyphochytriales
          3. Oömycetidae (water molds, blights, downy mildews)
            Order: Saprolegniales
                Leptomitales
                Leginidales
                Peronosporales
          4. Plasmodiophoromycetidae (club-root organisms)
            Order: Plasmodiophorales
          5. Zygomycetidae (bread molds, pin molds)
            Order: Mucorales
                Entomophthorales
                Zoopagales
          6. Trichomycetidae (commensals with arthropods)
      Class 2. Ascomycetes (sac fungi)
        Subclass 1. Hemiascomycetidae
          2. Euascomycetidae
          3. Loculoascomycetidae
      Class 3. Basidiomycetes (club fungi)
        Subclass 1. Heterobasidiomycetidae
          2. Homobasidiomycetidae
      Class 4. Deuteromycetes (Fungi Imperfecti)
            Order: Sphaeropsidales
                Melanconiales
                Moniliales
                Mycelia Sterilia
    Subdivision 2. Myxomycotina (slime fungi)
      Class 1. Myxomycetes
        2. Acrasiomycetes

*Adapted from scheme by Alexopoulos (1962).

e.g., jams, jellies, syrups, pickling brines, wood, and bread. Fungi are typically aerobic. As a result of their various distinctive properties, many species of fungi are found in damp, dark places where organic matter and oxygen occur. Like many bacteria, fungi can utilize solid food materials by secreting extracellular hydrolytic enzymes. Many species of fungi are adapted to marine and freshwater habitats.

Typical Eumycotina differ from all other eucaryotic plants in having chitinous instead of cellulosic cell walls and, excepting yeasts and torulas, in the characteristically filamentous form of their basic structural units. Fungi resemble animal cells in that all are chemoorganotrophic. Fungi differ from typical animal cells in: (a) having a chemically and morphologically differentiated, rigid cell wall and (b) frequently exhibiting a vegetative manner of growth (i.e., continuous, without regard to size and without the formation of specialized reproductive [fruiting] structures).

Activities of Fungi

Most fungi are saprophytes and active producers of various hydrolytic enzymes. Consequently they are of great value as scavengers and as promoters of soil fertility.

Many fungi decompose cellulose and lignin, and therefore ruin paper and wood products not protected from them. Some filamentous fungi grow in and under paint on walls and cause flaking and deterioration of the paint. Some can grow on the surfaces of lenses in binoculars in the tropics, diminishing clarity of vision. Some fungi grow well on rubber, including rubber tires and insulation, thereby ruining it. They also grow on the surface of electrical insulators, causing them to transmit electricity. Fungi, especially molds, are merry jokers of the microscopic world! Several species cause diseases, often serious, of man, lower animals, and valuable crop plants.

The sea also has a most interesting indigenous flora of fungi. Marine fungi participate in the destruction of ropes and timbers exposed at water level. Some molds have done enormous damage by infecting and killing commercially valuable fish and shellfish, and animal and fish foods such as eel grass upon which many edible marine forms live.

On the other hand, a number of species of fungi, especially yeasts and molds, are of great commercial value in the production of various organic compounds which are used as foods, flavors or drugs. From carbohydrates, various species of fungi produce hundreds of valuable substances that cannot easily be made by artificial processes. Among these products are penicillin, acetone, butanol, sorbitol, and takadiastase. Yeasts are the familiar servants of brewers and bakers.

Some fungi also have very special relationships with plant roots (see Mycorrhizia in Chapter 43).

Structure

Fungi typically exhibit two phases of growth: an asexually growing vegetative or thallus-plant phase and a "fruiting" or sexually reproducing phase. **Spores**, "seed-like" reproductive forms, are produced in both phases, and students should be careful to distinguish between the spores produced by the asexual and sexual processes. Figures 12–1 and 12–2 illustrate some common types of spores.

During the vegetative phase virtually all filamentous fungi are seen to consist of tubular, widely branching filaments called **hyphae**, with rather uniform diameters of 10 to 50 μm in different species. An entire mass of growth (thallus) consisting of such filaments is called a **mycelium**. A new mycelium may be formed by fragmentation of a hypha or by outgrowth from a sexually or asexually produced spore. In the more primitive fungi, the Phycomycetes, the hyphae are nonseptate, i.e., have no cross-walls or septa, and the entire thallus can be considered a single multinucleate (coenocytic) cell. Active growth occurs at the tips of the hyphae, and nutrients are transported to the sites by protoplasmic streaming. The hyphae of higher fungi are generally divided into uninuclear or multinuclear compartments by the formation of septa. At least in the actively growing portions of the mycelium, these septa contain a central hole or pore (Fig. 12–3) which allows the transfer of protoplasm, organelles, and nuclei between compartments. An interesting bit of speculation is that the development of septa represents one of nature's early experiments in the transformation of unicellular to multicellular organisms.

Not all fungi are of the woolly, hairy, or cobwebby sort, popularly called **molds**. Mushrooms, puffballs, brain fungi, shelf fungi, and the like are not filamentous as viewed with the naked eye. On microscopic examination, however, the fleshy portions are seen to consist of compact masses of the branching hyphae that are characteristic of Eumycotina in general. On the other hand, some species of fungi, the yeasts and torulas, exist primarily as single, spheroid or ovoid cells or small aggregations of such cells. Some fungi, especially certain pathogenic spe-

A. **Ascospores:** Spores produced in a sac or **ascus.**

B. **Basidiospores:** Spores produced at the surface of a club-shaped structure, the **basidium.**

C. **Zygospores:** Spores produced by the fusion of similar-appearing gametes formed at the tips of hyphae (limited to the Phycomycetes).

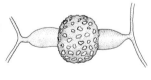

D. **Oöspores:** Spores resulting from the mating of two unlike gametes (limited to the Phycomycetes).

Figure 12–1. Major types of sexual spores resulting from the fusion of nuclei or mating of gametes to produce a diploid cell.

A. **Sporangiospores:** Spores inside swollen fertile structure called **sporangium** (limited to Phycomycetes).

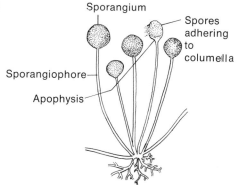

Sporangium

Spores adhering to columella

Sporangiophore

Apophysis

B. **Conidiospores (conidia):** Spores supported by a specialized fertile structure, the **conidiophore.**

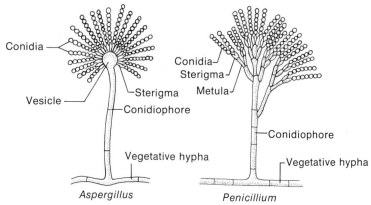

Conidia

Vesicle

Sterigma

Conidiophore

Vegetative hypha

Aspergillus

Conidia
Sterigma

Metula

Conidiophore

Vegetative hypha

Penicillium

C. **Thallospores:** Spores resulting from changes in the vegetative hyphae or **thallus.**

1. **Arthrospores (oidia):** Hyphae fragment into small spores with thickened cell walls.

2. **Chlamydospores:** Hyphae divide into spore-like cells with large food reserve and resistance to unfavorable environment.

3. **Blastospores:** Produced by budding.

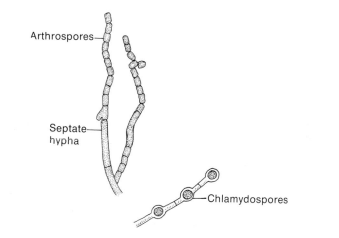

Arthrospores

Septate hypha

Chlamydospores

Blastospores (buds)

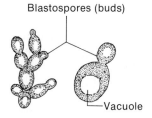

Vacuole

Figure 12–2. Major types of asexual spores.

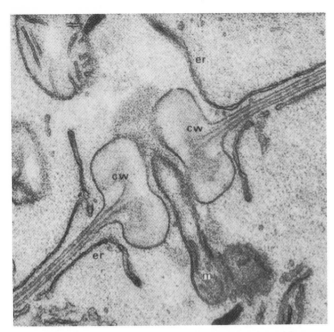

Figure 12–3. Electron micrograph of a septal pore in the cross-wall or septum of a hyphal filament. *CW,* Enlarged ends of the cell wall forming the pore; *M,* mitochondrion passing through pore. (36,000×.) (Photograph by C. E. Bracker and E. E. Butler.)

cies, may grow in either the yeast-like form or the filamentous form; they are said to be **biphasic** or **dimorphic.**

Four major groups of Eumycotina can be differentiated on the basis of sexual reproductive mechanisms. Further subdivisions are based on structure of the mycelial filaments and on forms of asexual reproductive bodies. A convenient arrangement is as follows:

I. Sexual spores are **free zygotes:** Phycomycetes.

Asexual spores are endogenous, i.e., enclosed in a sac-like structure called the **sporangium;** mycelium is nonseptate (no cross-walls) except at reproductive sites. Examples: *Mucor* sp., *Rhizopus* sp.

II. Sexual spores are enclosed in sacs or **asci:** Ascomycetes.

Asexual spores are exogenous, i.e., formed at ends of special hyphae; mature mycelium is septate. Examples: *Aspergillus* sp., *Penicillium* sp., *Saccharomyces cerevisiae.*

III. Sexual spores are borne on **basidia:** Basidiomycetes.

Asexual spore production very rare; mature mycelium is septate. Examples: mushroom, puffball.

IV. Sexual stages not seen: Deuteromycetes (Fungi Imperfecti).

Asexual spores and mycelia mainly resemble those of Ascomycetes, though some are like Phycomycetes and Basidiomycetes. Examples: *Candida albicans, Alternaria.*

The total number of species of Eumycotina is not known, but there are probably over 80,-000. Here we shall confine our discussion to a survey of a few representative species of the four major groups.

12.2
THE PHYCOMYCETES

The Phycomycetes are probably the most primitive of the Eumycotina. Although the name refers to an algal habitat (Gr. *phyco* = seaweed; *myces* = fungi), not all species are aquatic, many being terrestrial and familiar in the household as bread mold and similar nuisances. Unlike all other fungi (but like some algae), the aquatic species produce motile, flagellate, asexual spores, **zoospores** or swarm spores, that can thus disseminate themselves in fluid media.

The most primitive forms of Phycomycetes are the aquatic Chytridiales or **chytrids.** Their life cycle is simple (Fig. 12–4). An anteriorly uniflagellate, ameboid, unicellular zoospore attaches itself to a site favorable for growth, loses its flagellum and initiates the growth of a rootlike (**rhizoid**) mycelium into the substrate. The zoospore enlarges to form a sac, which by repeated divisions of the nucleus becomes a sporangium filled with motile zoospores, nourishment being drawn from the substrate via the rhizoids. Rupture of the sporangium wall liberates the zoospores. Zoospores may become gametes and conjugate, forming dormant, thickwalled, free zygospores. Some species of chytrids are pathogenic for plants of agricultural value.

The more mold-like type of aquatic Phycomycetes is exemplified by the order Saprolegniales or water molds. They are common in ponds and pools, growing on dead organic matter. The mycelia of water molds are often extensive. The **asexual zoospores** are biflagellate, and the zoosporangia form on the tips of special hyphae and are commonly cylindrical or pyriform. The sporangium is separated from the supporting hypha by a septum. Rupture of the sporangium liberates the zoospores.

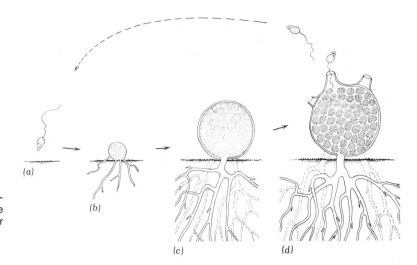

Figure 12–4. The life cycle of an aquatic chytrid, one of the most primitive forms of the Phycomycetes (see text for explanation).

(a)

(b)

(c)

(d)

Sexual Reproduction of Saprolegnia. When conditions no longer favor asexual reproduction the saprolegnias can also reproduce sexually (Fig. 12–5). The sex structures are morphologically distinct (compare with chytrids), being an **antheridium** ($\male$) and an **oogonium** ($\female$).

The sexual process is said to be **heterogamous** since the gametes are morphologically distinct. In species in which morphologically indistinguishable gametes are involved, the process is said to be **homogamous**. Both male and female gametes may occur on one mycelium or thallus;

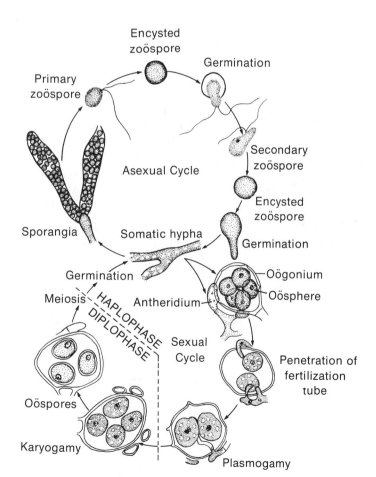

Figure 12–5. The asexual and sexual life cycles of *Saprolegnia* sp., a typical mold-like aquatic Phycomycete (see text for explanation).

such species are said to be **homothallic.** Species in which the male and female elements are on separate mycelia are said to be **heterothallic.**

In *Saprolegnia* the oogonium is a bulbous outgrowth from a hyphal wall; within it many nuclear fissions occur. Most of the nuclei thus formed degenerate, but several grow into uninucleate **oospheres** (egg cells) ready for fertilization. The antheridia are formed at the tips of hyphae and penetrate into the eggs, with resulting fusion of male and female nuclei. The mature, fertile oospore forms a thick wall and becomes dormant for some months. Meiosis occurs and the oospore sends out a short hypha that forms an asexual zoosporangium at its tip.

Some species of *Saprolegnia*, e.g., *S. parasitica*, are familiar to those who work in fish hatcheries or aquaria as white, cottony growths on fish, sometimes on "domestic" goldfish.

Terrestrial Phycomycetes (Order Mucorales). This group is well exemplified by the familiar genera *Rhizopus* and *Mucor*. *Rhizopus nigricans* is the common, black, bread mold familiar to all who have seen bread after it has stood in a humid place for some days during the summer. Species of *Rhizopus* spread rapidly because they produce enormous numbers of asexual conidiospores and because they send out **stolons** or "runners" like Bermuda grass or strawberry plants. These runners take hold of the substrate by means of rhizoids or **holdfasts.** At each such holdfast several erect, unbranched sporangiophores* (short hyphae) are produced. The tip of each enlarges to form a **columella.** This varies in form according to species. The sporangiospores are formed in a mass about the columella by many nuclear divisions inside the **sporangial membrane** (Fig. 12–2, *A*). The spores are liberated by rupture of the sporangial membrane. Not all Mucorales send out stolons like *Rhizopus* species. Zygospores, which are sexual spores, are formed as depicted in Figure 12–6.

---

*The suffix *phore* is from the Greek word *phorein*, meaning carrier or bearer. Do not confuse sporangio*phore* with sporangio*spore*; similarly, conidio*phore* with conidio*spore* (Gr. *conidio* = dust).

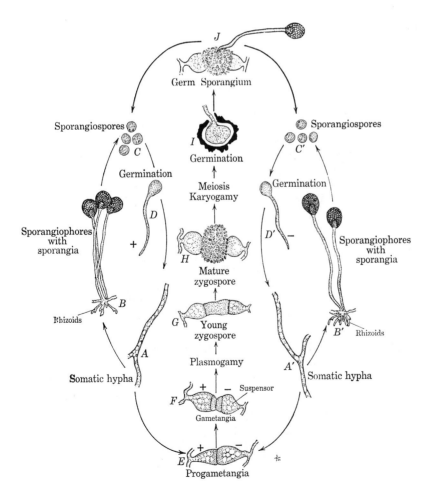

Figure 12–6. Life cycle of *Rhizopus nigricans* showing the production of both asexual and sexual (zygospore) types of spores.

Molds of the order Mucorales, like *Rhizopus* and *Mucor* species, are important commodity- and food-destroyers as well as valuable scavengers. Occasionally certain species of *Rhizopus* and *Mucor* may cause serious infection in man, especially if large numbers of the conidiospores are inhaled.

12.3
THE ASCOMYCETES

The Ascomycetes consist of three major groups, of which we will consider only two: (a) Euascomycetes, typically filamentous molds that form **ascocarps,** and (b) Hemiascomycetes, typically nonfilamentous yeasts that form their asci in separate cells or small clusters of such cells not enclosed in an ascocarp. (An ascocarp is a complete fruiting body including ascospores and surrounding, supporting structures.)

Filamentous Ascomycetes (Euascomycetes) ("Molds")

There are about 30,000 species of the filamentous Ascomycetes alone. They exhibit a bewildering variety of forms and mechanisms of sexual and asexual reproduction, habitat and activities. The unifying character of all, both filamentous molds and yeasts, is the formation of **sexual** spores inside sacs (**ascospores**).

Asexual Reproduction. In addition to the sexual spores there are various types of asexual spores. Some species of filamentous Ascomycetes form, in the mycelium, a number of closely spaced divisions, which develop into a number of short, more or less ovoid cells. These cells are sometimes called **oidia** (Gr. *oion* = egg; i.e., an ovoid body) or **arthrospores.** They tend to leave the parent filament by **fragmentation** (Fig. 12–2, *C*). They then continue **vegetative** growth, each starting a new plant.

Some filamentous Ascomycetes also form yeast-like buds or **blastospores** as outgrowths along the hyphae. Blastospores develop much as do yeasts. Some fungi grow readily in either yeast or filamentous form, or both simultaneously (e.g., *Candida*), the form of growth depending on such factors as temperature and concentration of oxygen.

During maturation of many protists, including Ascomycetes, one or more cells may acquire thick walls and become dehydrated and filled with granular reserve material. In this form they can remain dormant and resist drying and sunlight for long periods. They are called **chlamydospores.**

Conidiospores, or Conidia. Conidiospores are formed by filamentous Ascomycetes and by many Fungi Imperfecti at the free ends of branching special hyphae called **conidiophores** (Fig. 12–2, *B*).

Each conidiophore arises as a branch of a cell in the mycelium called a **foot cell.** Each develops as a stem consisting of several cells end to end. From these, in various distinctive types of arrangements, numerous short stems or **sterigmata** (singular, **sterigma**) arise. The conidial chains are produced at the tips of the sterigmata. Instead of being enclosed in sporangia, as are spores of all Phycomycetes, the conidia of Ascomycetes are free, sometimes being produced in long chains like strings of beads. Since they are not formed inside an enclosing structure, they are said to be **exogenous,** or **exoconidia.** The size, form, and arrangement of the branched conidiophores, the form of the chains, and the color of the conidia are distinctive of the different genera and species.

Sexual Reproduction. In the formation of sexual spores (ascospores) by filamentous Ascomycetes, the tip cells of certain fertile hyphae become multinucleate, reproductive structures of + and − "sexes." When they meet, the cell walls dissolve at the points of contact. The nucleus of each + hypha passes into a − hypha. Pairs of + and − nuclei then fuse, forming **diploid zygotes.** Meiosis then occurs, followed by mitotic divisions, forming four or more **haploid nuclei.** In the process, each nucleus acquires some of the cytoplasm of the original fertilized − hypha, surrounds itself with a thick wall, and becomes dormant; i.e., it becomes an **endogenous spore**— in this case an ascospore since all of the spores are enclosed in the same sac or ascus. The ascospores are then liberated by rupture of the sac wall. Commonly, asci are formed in clusters within a protective and enclosing, distinctively shaped (bowl-like or cup-shaped) mass of hyphal cells. The whole "fruiting" structure is called an **ascocarp.** Depending on arrangement, the surrounding structure is sometimes called a **perithecium,** an **apothecium,** or a **cleistothecium.**

Among the more familiar filamentous Ascomycetes are the common genera, *Aspergillus* and *Penicillium*.

Genus Aspergillus. The fruiting hyphae of aspergilli have enlarged globular tips (**vesicles**). From the surface of these, numerous **sterigmata** radiate in all directions. On the tips of these the conidiospores are borne in long chains (Fig. 12–2, *B*).

There are some pathogenic species of *Aspergillus*. For example, a pulmonary infection of birds due to *Aspergillus fumigatus* is not uncommon, and infection of man by aspergilli (aspergillosis) is not rare. Infection generally occurs by inhalation of large numbers (clouds) of conidiospores. Pulmonary aspergillosis is often fatal.

Aspergillus flavus, a common saprophyte, is now known to produce several dangerous poisons (toxins), among them a group called **aflatoxins,** when growing in certain foods, notably stock feeds like ground peanuts. Aflatoxins, when eaten, cause illness and sometimes death. Their role in human disease remains to be determined. Aflatoxin appears to inhibit enzymes involved in the synthesis of DNA and hence in the synthesis of RNA and proteins in certain cells of susceptible animals and also in some bacteria. In 1973 outbreaks in humans were reported due to eating moldy corn containing a mycotoxin called T–2, which causes a highly fatal disease resembling leukemia. Another mycotoxin in moldy corn is called zearalenane.

Genus Penicillium. The penicillia are widely distributed and contribute to the spoilage of various objects and materials composed of organic matter, especially ripe fruits. The conidiophores are composed of hyphae which branch at the tip into finger-like clusters of sterigmata, the whole roughly suggestive of the bone structure of the hand. The spores extend in chains from the ends of the sterigmata. This arrangement gives the whole conidiophore, with its chains of conidiospores, a form suggestive of a tiny paint brush, from which the generic name is derived (L. *penicillus* = paint brush or pencil) (Fig. 12–2, *B*). As in other groups of molds, the color and form of the fruiting body are of value in classification.

Some species of *Penicillium* are differentiated chiefly by their habitat. The blue-green molds found in Roquefort cheese (*P. roqueforti*), Camembert cheese (*P. camemberti*), and other cheeses of the same nature are distinguished chiefly by their occurrence there. The molds grow in or on the cheese (which must be perforated to admit air since the molds are aerobic), producing various enzymatic changes in the fat, carbohydrate, and protein of the cheese, which result in characteristic aromas, flavors, and textures. (See Chapter 47.)

There are many other species of penicillia. They are frequently seen on old bread, cheese, lemons, and other fruits. They may usually be recognized as members of the genus by their sky-blue or green color. *Penicillium notatum* and *P. chrysogenum*, very similar species, have come into great prominence as sources of penicillin (Chapt. 23).

Nonfilamentous Ascomycetes (Hemiascomycetes) ("Yeasts")

The term yeast, like the term mold, is one of convenience only; the term as commonly used, and as used here, refers to those Ascomycetes that are typically not filamentous but unicellular, ovoid, or spheroid. However, some species of Basidiomycetes and Fungi Imperfecti are also yeast-like in form.

In size, yeasts average somewhat larger than bacterial cells, about 5 μm in diameter by 8 μm in length. Some oval yeast cells have a volume hundreds of times that of *Staphylococcus* cells. The yeast cell wall consists of two or probably three layers and contains chitin, like other eucaryotic fungi. Within the cytoplasm are a well-differentiated nucleus, numerous vacuoles containing food or waste substances and granules of glycogen and of **volutin** (polymerized phosphates). Yeast cells sometimes contain large quantities of fat, which may be used commercially.

Asexual Reproduction. **Budding.** Budding is the commonest and most distinctive method of asexual reproduction in yeasts (Fig. 12–7). In budding, large, mature cells divide, each giving rise to one or more daughter cells or **buds** that are at first much smaller than the mother cell and that characteristically cling to the parent cell, often even after the daughter cell has divided. A cell may form several buds at different sites simultaneously. Clumps and chains of cells, sometimes called **rudimentary filaments,** or **pseudomycelia,** are thus formed. The manner of forming buds is used to differentiate between species of yeasts.

Fission. Some species of yeast, in the genus *Schizosaccharomyces*, divide by transverse, binary fission, much as do the bacteria.

Sexual Reproduction. **Ascospores.** The ascospores of yeasts are dormant, thick-walled and resistant to heat, drought, and other unfavorable environmental conditions. They are not so thermoresistant as bacterial spores, being killed by a temperature of 60C in a short time. Since ascospores of yeasts are generally produced in groups of four or more per cell, they represent a process of multiplication as well as preservation, thus differing from bacterial endospores, of which, as a rule, only one is produced by each cell.

In some species ascospores are formed fol-

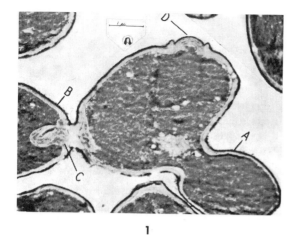

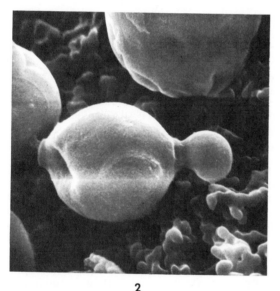

Figure 12–7. *1*, Electron micrograph of a longitudinal section through a budding yeast cell (*Saccharomyces cerevisiae*). *A*, A young bud with its cytoplasm still continuous with that of the mother cell. *B*, A mature bud with the developing cross-wall between mother and daughter cell. *C*, The extension of newly formed cell wall material into the cytoplasm, a phenomenon which appears to be characteristic of the later stages of the budding process. *D*. A bud scar, the surface of which is always convex. (Degree of magnification indicated by the scale line showing 1 μm at top, center.)

2, Scanning electron micrograph of a budding yeast cell. (Original magnification 16,000×; reduced to 55% of original size.) (Courtesy of R. Albrecht and A. MacKenzie.)

lowing **sexual fusion** of two vegetative, haploid yeast cells. In this process the cell walls between two adjoining haploid cells disappear at the point of contact, and the nuclei unite. The diploid nucleus thus formed then usually undergoes meiosis and two or more divisions, resulting in four or eight haploid **ascospores**

within the original cell wall, which is now an **ascus** (Fig. 12–8). These ascospores grow out into asexually budding, haploid yeasts as before.

Ascospores may also be formed without preliminary fusion. The nucleus of a cell undergoes meiosis and then two or more divisions occur within the cell wall. Each portion of each haploid nucleus receives some of the cytoplasm of the parent cell and develops a new cell wall, becoming an ascospore. These spores fuse sexually, each pair forming a diploid **zygote** which goes on reproducing asexually by budding as before. Thus, some yeasts exist most of the time in the diploid state, and some in the haploid state. Ascospores are usually of a very distinctive form that is used in identification of species.

The Torulas. These constitute a large group, Torulopsidaceae, containing several genera. These organisms resemble the yeasts, but sexual stages have not been observed; i.e., they are Fungi Imperfecti.

Habitat and Activities of Yeasts

Yeasts are widely distributed in nature. They commonly occur on grapes and on other fruits and vegetables. The spores pass the winter in the soil. The kind of wine made from grapes depends to some extent on the varieties of yeasts occurring upon them naturally. Yeasts, molds, and torulas may also be found in dust, dung, soil, water, and milk, and are not infrequently observed in cultures made with swabbings from the normal throat. Many species are found as contaminants in brewers' and picklers' vats, and many appear to live on the human skin and in the nectar of flowers.

The common bakers' and brewers' yeasts (species of *Saccharomyces*, chiefly varieties of *S. cerevisiae*) are facultative with respect to oxygen, producing alcohol and carbon dioxide by **fermentation** of sugar under **anaerobic** conditions of growth. Their alcohol-forming power is used in the manufacture of wine, beer, and industrial alcohol.

The ability of fermenting yeasts to form carbon dioxide is important in beer and wine making. It is also important in baking because the carbon dioxide produced by the yeast cells growing in newly made dough causes it to rise. This gives the finished bread its light, porous texture. Yeasts synthesize proteins and also several vitamins, especially those of the B complex, and therefore have great nutritive value. For this reason they are sometimes cultivated **aerobically** in vats. There they produce carbon dioxide and water instead of alcohol. They use

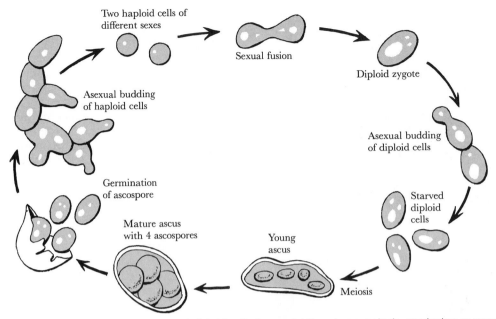

Figure 12–8. The life cycle of a yeast. Starved diploid cells (lower right) undergo meiosis, producing an ascus with four haploid ascospores which eventually mature, germinate, and leave the ascus (lower left). These haploid cells may multiply for a time by asexual budding of the haploid cells (upper left). Sooner or later sexually different haploid cells conjugate (upper center) and form a diploid zygote (upper right) which multiplies for a time in the diploid state (right).

all of the nutrient substrate to produce masses of nutritive yeast substance that can be fed to livestock (Chapter 48).

Yeasts are also important as nuisances. Various kinds of yeasts, and also torulas, cause "disease," or off-flavors, unpleasant odors, and sliminess of various foods. Torulas, like most fungi, are common in soil, water and dust.

One group of torulas, called Rhodotorulaceae, produces reddish pigments. These torulas cause various reddish discolorations and spoilage of fresh foods in such places as butcher shops and seafood establishments. They are not pathogenic. Most torulas have little fermentative ability and consequently are of little commercial value.

12.4
THE BASIDIOMYCETES

These are familiar to everyone as mushrooms, toadstools, rusts, smuts, and shelf or bracket fungi. Some are highly poisonous. The distinctive structure of all Basidiomycetes is the **basidium,** a swollen, club-like, sexually reproductive cell somewhat like the multinucleate reproductive structures (asci) of Ascomycetes. The basidium is produced by a binucleate cell at a hyphal tip. The two nuclei of this cell fuse,

forming a diploid zygote that, as in Ascomycetes, divides meiotically to form four haploid nuclei. Unlike spores of the Ascomycetes, these haploid nuclei are not held inside any structure like an ascus but are extruded to the tips of four tiny projections, each called a **sterigma,** at the surface of one end of the basidium. In the process each haploid nucleus acquires cytoplasm and forms a thick spore wall, thus becoming a **basidiospore.**

The mycelial structure of most Basidiomycetes is commonly hidden underground or in a porous substrate like a rotting log, and is often many feet in extent. The large, fleshy parts of these fungi (e.g., mushrooms) are the fruiting bodies. The basidia and basidiospores are formed along the undersides of the "gills" of these plants. The spores are disseminated in wind and rain. In a suitable environment each germinates and produces a new thallophyte or mycelium (Fig. 12–9).

Asexual spores of Basidiomycetes are rare.

12.5
LICHENS

These lowly plants are striking examples of highly successful "togetherness" (conjunctive symbiosis or mutualism). (See also Symbiotic

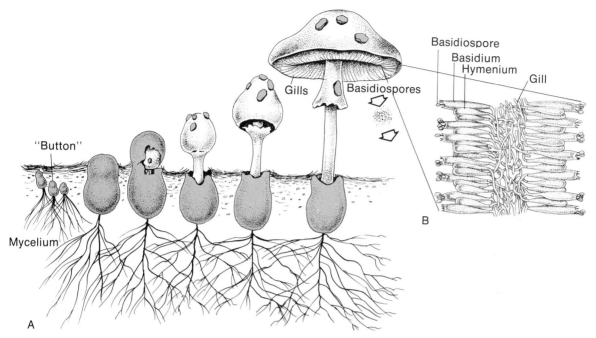

Figure 12–9. *A,* Stages in the development of a mushroom from the mycelium, the mass of white, branching threads found underground. A compact "button" appears and grows into the fruiting body or mushroom. On the undersurface of the fruiting body are "gills," thin perpendicular plates extending radially from the stem. Basidia develop on the surface of these gills and produce basidiospores, which are shed, and, if they reach a suitable environment, give rise to new mycelia.

B, Section of a gill from the underside of a mushroom cap, magnified 250 times, to show the basidia and their basidiospores.

Nitrogen Fixation, Chapter 43.) Each lichen plant (there are over 12,000 species) consists of a fungus and an alga (Fig. 12–10). The fungus is generally an Ascomycete or Basidiomycete that has become adapted to intimate life with an alga that may be eucaryotic or procaryotic. The alga, being photosynthetic, provides organic nutrients for the fungus which, in turn, provides mechanical support and protection as well as minerals for the alga. Each plant can, however,

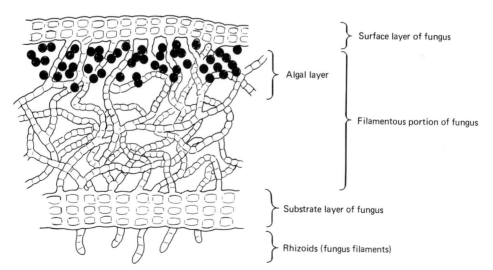

Figure 12–10. Diagram of a cross section of a lichen showing the relationship between the algal cells and fungus. In some species, the algal cells are not confined to a layer but are scattered throughout the central portion of the lichen.

multiply independently. Lichens are familiar as flat, leafy, crusty, or flaky growths, or moss-like plants on such inhospitable and barren places as rocks and the bark of trees as well as on rotting wood and the soil, from pole to pole. They have a variety of commercial uses and are important sources of food for a variety of animals, e.g., "reindeer moss." (Lichens are illustrated in Color Plate II, C and D.)

12.6
FUNGI IMPERFECTI (DEUTEROMYCETES)

This group of about 20,000 species is entirely artificial and is heterogeneous. It contains species of fungi that cannot be assigned to any other group because no sexual process, the primary basis of classification of fungi, has been observed. This is because many of the filamentous fungi are heterothallic, and in many of these only one of the "sexes" has been discovered. When both are found and observed to combine sexually, then the species are removed from the Fungi Imperfecti to their proper group: usually the Ascomycetes. It is possible that some of the fungi may exist only in the form of one of the combining types. A threadbare comment about Fungi Imperfecti is that the imperfection usually lies not in the fungi but in our knowledge of them.

Carnivorous Fungi. Some of the Fungi Imperfecti are among the most useful and interesting organisms because of their carnivorous habits. They are found in fertile, moist soils. They entrap and digest microscopic animals of the soil such as amebae, crustaceans, rotifers, and also the much larger (though still microscopic) nematodes (round worms, or eel worms).

An interesting example of these soil-dwelling fungi is *Dactylaria gracilis.* Under some conditions, such as the abundance of certain crops (turnips, for instance), enormous numbers of root-eating nematodes appear and cause many dollars' worth of damage. The fungus sends out short hyphae that are extremely sticky, and the nematodes are caught in the adhesive. Each hypha is tipped with a doughnut-shaped loop of three cells. When a nematode, struggling in the sticky secretion, thrusts its head or tail inside the loop, the cells of the loop instantly swell up, obliterating the opening in the loop and constricting powerfully on the nematode (Fig. 12–11). They hold it in an inexorable grip in which it soon dies despite its

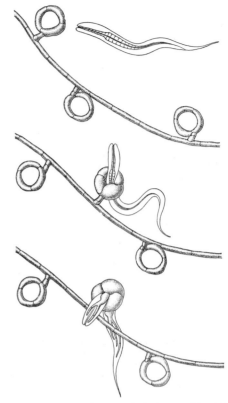

Figure 12–11. Fate of a nematode trapped by a carnivorous fungus. At the top the worm nears the living snare, which is part of a filament of the fungus. Once inside the ring the worm is immediately caught by expansion of the cells of the ring. As seen at the bottom, these cells quickly send out hyphae which penetrate the worm and secrete enzymes which digest it and adsorb the nourishment thus produced.

struggles. The cells of the loop of fungus then send vegetative hyphae into the body of the worm; these excrete enzymes, so the worm is digested and absorbed by the fungus. Even if the worm tears the loop from the mycelium, the ring still holds on and sends in its fatal vegetative hyphae. It grows and forms a new mycelium at the expense of the luckless nematode.

An interesting aspect of this phenomenon is that the ring constricts only when the worm touches the inner surface of the loop. No constriction occurs if the outer surface is touched. The sensitivity and exceedingly rapid action of these cells suggest some sort of nervous and muscular activity, yet none of this can be seen.

Other species of similar molds kill worms which infect sheep. Practical use of these molds has been made by inoculating sheep pens with them; the worms soon disappear.

12.7
PATHOGENIC FUNGI

Although the vast majority of fungi are saprophytic, some 50 species are **opportunists,** i.e., they can cause disease in man when introduced into the body under the right conditions. Frequently, the more serious fungal infections take place in a patient who is already debilitated in some fashion, e.g., undergoing cancer therapy, lungs damaged, normal body ecology disturbed by antibiotic therapy, etc. The pathogenic fungi are of two general types: (a) those causing generalized or systemic infections of deep tissues, often fatal, and (b) those affecting only superficial tissues: i.e., skin, nails, and hair, and causing annoying but not dangerous diseases. The fungi of this second group are collectively called **dermatophytes,** i.e., skin fungi. Infections by any fungi are collectively called **mycoses.** Table 12-2 lists some of the more important ones.

Fungi Causing Systemic Mycoses. Nearly all of these fungi normally live as saprophytes in the soil. Infection of man usually occurs only accidentally through contamination of cuts and abrasions by soil, or by inhalation of dust containing spores or conidia. These diseases are rarely transmitted from person to person. Infection and poisoning by Aspergillus was described in an earlier section; only three representatives of the Fungi Imperfecti will be described here.

Candida Albicans. This organism is commonly found on the oral or vaginal mucous membranes or in feces of normal individuals, and rarely on the skin. Unlike most of the systemic fungi it rarely occurs normally in the soil. Growth on Sabouraud's agar (widely useful in mycology) is best at 37C. The colonies are creamy and have a yeasty odor. The growth consist mainly of small (2 to 4 μm), yeast-like cells. Mycelial elements may occur, with blastospores. The ability of this fungus to form clusters of budding cells and distinctive, round chlamydospores on cornmeal agar or other suitable agar is useful in identification. *C. albicans* can cause a number of serious infections (*candidiasis*) in man and animals, especially of mucous membranes (thrush and vulvovaginitis), the skin (cutaneous candidiasis, Fig. 12-12) and lungs (pulmonary candidiasis). Microscopic examination of scrapings from lesions usually reveals the organisms, but diagnosis should be made only by a skilled mycologist.

Coccidioides Immitis. *C. immitis* is the cause of a disease called coccidioidomycosis, recognized originally in the San Joaquin Valley in California and known to occur in arid regions elsewhere. In nature the organisms live in the soil, and their resistant cells (arthrospores) are blown about with dust and inhaled. Many of these infections pass unnoticed or result in a febrile disease in association with bronchitis, rheumatism, or pneumonia. The disease is often confused with tuberculosis in x-ray and clinical examinations. When cultivated in dilute media, pH 6.0 to 8.0, in the dark, at 37C at reduced surface and oxygen tensions, or when invading the tissues of the body, *C. immitis* forms yeast-like cells, but these never form buds. On the contrary, the contents of these yeast-like cells divide into many smaller cells within the cell wall, forming a **spherule** or sporangium-like

TABLE 12-2. EXAMPLES OF HUMAN MYCOSES

Disease	Foci of Infection	Causative Fungus
I. Systemic Mycoses:		
Candidiasis (moniliasis)	Lungs, mucous membrane, skin, intestinal tract	*Candida albicans*
Coccidioidomycosis	Lungs	*Coccidioides immitis*
Histoplasmosis	Lungs	*Histoplasma capsulatum*
Blastomycosis	Lungs	*Blastomyces dermatitidis*
Aspergillosis	Lungs	*Aspergillus fumigatus*
Cryptococcosis	Lungs, meninges	*Cryptococcus neoformans*
II. Dermal Mycoses:		
Ringworm (tinea)	Scalp, face	*Microsporum audouini* and others
Athlete's foot (tinea pedis)	Toe webs, feet	*Epidermophyton floccosum* and others, especially *Trichophyton*
Barber's itch (tinea barbae)	Bearded area of face and neck	*Trichophyton* sp. and *Microsporum* sp.

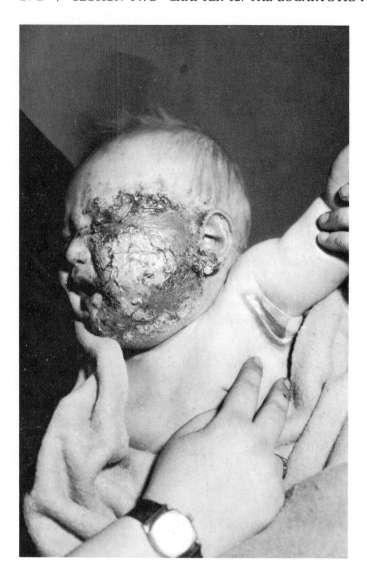

Figure 12–12. Candidiasis of face and axilla. (Photograph courtesy of Dr. John H. Stokes.)

body. The sporangium or spherule wall ruptures, liberating a large number of small cells. These are then distributed by the blood throughout the body and repeat the cycle. When the fungus is grown aerobically on agar media or in the soil, the mycelial filaments are generally formed. Thus it is a dimorphic (or diphasic) fungus.

Histoplasma Capsulatum. This organism resembles *Coccidioides immitis* in several respects. The spores are airborne and may be derived from soil, especially if polluted with dung of birds or animals. It causes infections (**histoplasmosis**) in man which commonly pass unnoticed yet cause fatal generalized infections in certain individuals. *H. capsulatum,* like *C. immitis,* is dimorphic. It may be cultivated on 10 per cent blood-infusion agar at 37C. Un-

der such conditions the cells are yeast-like. In infected tissues, as in tissue cultures, only the yeast-like form is seen. When cultivated at room temperatures on such media as Sabouraud's or glucose agar, a cottony, white, filamentous growth appears.

The Dermatophytes. These fungi invade only the superficial skin in contrast to those that invade the subcutaneous tissues, lymphatics, and deeper tissues. Mycotic infections of the skin (generally called **dermal mycoses**) are common; in fact, the dermal mycoses such as ringworm (**tinea**), **athlete's foot** (**tinea pedis**), and various other forms of dermal mycosis of the hands and feet are among the commonest of infectious diseases. The species of Fungi Imperfecti most commonly involved are listed in Table 12–2.

12.8
MYXOMYCOTINA (SLIME MOLDS)

These interesting microorganisms are generally studied with the fungi because their fruiting structures have traditionally been studied by mycologists. In truth, they do not closely resemble the fungi except for the manner in which they produce spores, and they are considered by many to be modified protozoans.

A complete description of the various kinds of slime molds and their classification is beyond the scope of this chapter, but there are two major types: (1) In the **acellular** slime molds, a spore germinates to form a haploid ameboid type of cell without any cell wall. During the life cycle (Fig. 12–13), the ameboid cells fuse to form eventually a diploid, multinucleate plasmodium in which the individual cells have lost their identity. (2) With the **cellular** slime molds a pseudoplasmodium is formed in which the haploid ameboid cells form a coherent mass and move together. Fusion of the individual cells

does not occur. The life cycle of one of the most thoroughly studied cellular slime molds, *Dictyostelium discoideum*, is shown in Figure 12–14.

In nature, the slime molds are frequently found on moist or decaying vegetation where other types of microorganisms are likely to be abundant. If such material is brought into the laboratory and kept moist, various colorful fruiting bodies soon appear and can be readily examined with a dissecting microscope. Their food consists mainly of bacteria, spores of fungi, and small pieces of organic matter, all of which they ingest by phagocytosis in a typical protozoan fashion. Many biologists, including some cancer researchers, are interested in the slime molds because their cells show migration with protoplasmic streaming, aggregation, and a primitive type of cell differentiation. Identifying the biochemical mediators controlling these activities may give scientists some idea of how to modify similar activities in higher plants and animals. Several species are also known to cause plant disease.

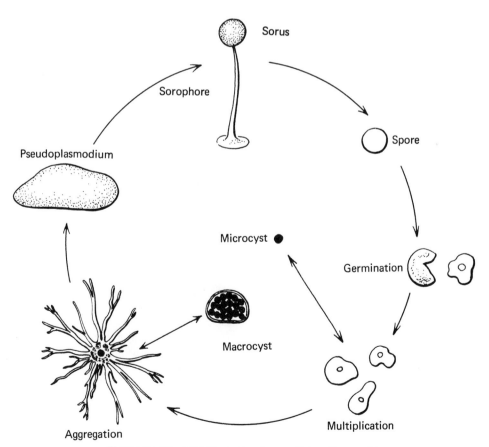

Figure 12–13. The life cycle of an **acellular** slime mold.

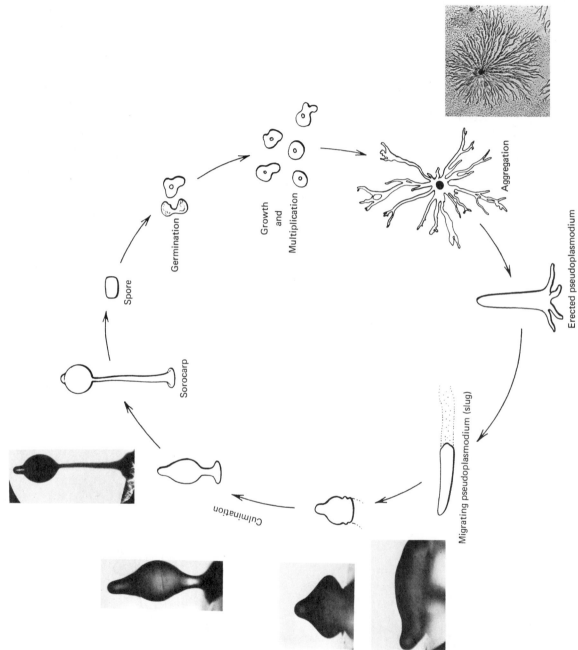

Figure 12–14. The life cycle of a cellular slime mold, *Dictyostelium discoideum*. (Photographs courtesy of Dr. K. B. Raper.)

CHAPTER 12

SUPPLEMENTARY READING

Ahmadjian, V.: Lichens. Ann. Rev. Microbiol., *19*:1, 1965.

Ainsworth, G. C., and Sussman, A. S. (Eds.): The Fungi: An Advanced Treatise. 3 vols. Academic Press, New York. 1965–67.

Alexopoulos, C. J.: Introductory Mycology, 2nd ed. John Wiley & Sons, Inc., New York. 1962.

Alexopoulos, C. J., and Bold, H. C.: Algae and Fungi. The Macmillan Co., New York. 1967.

Bartnicki-Garcia, S.: Cell wall chemistry, morphogenesis and taxonomy of fungi. Ann. Rev. Microbiol., *22*:87, 1968.

Bonner, J. T.: The Cellular Slime Molds, 2nd ed. Princeton University Press, Princeton, N.J. 1967.

Burnett, J. H.: Fundamentals of Mycology. St. Martin's Press, New York. 1968.

Conant, N. F., Smith, D. T., Baker, R. D., and Callaway, J. L.: Manual of Clinical Mycology, 3rd ed. W. B. Saunders Co., Philadelphia. 1971.

Duddington, C. L.: Beginner's Guide to the Fungi. Drake Publishers, Inc., New York. 1972.

Gray, W. D., and Alexopoulos, C. J.: Biology of the Myxomycetes. The Ronald Press Co., New York. 1968.

Kreger-van Rij, N. J. W., and Veenhuis, M.: Electron microscopy of some special cell contacts in yeasts. J. Bact., *113*:350, 1973.

Kurtzman, C. P., Smiley, M. J., and Baker, F. L.: Scanning microscopy of ascospores of *Schwanniomyces*. J. Bact., *112*:1380, 1972.

Lechevalier, H. A., and Pramer, D.: The Microbes. J. B. Lippincott Co., Philadelphia. 1971.

Oujezdsky, K. B., Grove, S. N., and Szaniszlo, P. J.: Morphological and structural changes during yeast-to-mold conversion in *Phialophora dermatitidis*. J. Bact., *113*:468, 1973.

Poindexter, J. S.: Microbiology, An Introduction to Protists. The Macmillan Co., New York. 1971.

Rose, A. H., and Harrison, J. S. (Eds.): The Yeasts. 3 vols. Academic Press, New York. 1969–70.

Taber, W. A., and Taber, R. A.: The Impact of Fungi on Man. Rand McNally & Co., Chicago. 1967.

Talens, L. T., Miranda, M., and Miller, M. W.: Electron microscopy of bud formation in *Metschnikowia krissi*. J. Bact., *114*:413, 1973.

Webster, J.: Introduction to Fungi. Cambridge University Press, Cambridge. 1970.

THE EUCARYOTIC PROTISTS
Algae and Protozoa

13.1
THE ALGAE

The eucaryotic algae are typical, photosynthetic, thallus plants (Gr. *thallos* = plant growth). They may be multicellular, "leafy," branching, and attached to solid objects by root-like **holdfasts.** They are, nevertheless, without vascular or differentiated tissues, flowers, seeds, physiologically differentiated true roots, stems, or leaves. The cells are entirely undifferentiated; i.e., any single cell could asexually reconstitute the entire structure. There are scores of thousands of species of algae. Some (e.g., Phaeophyceae or brown algae) form very large plants (seaweeds), such as the kelps of the Pacific Coast (Fig. 13–1). These are often hundreds of feet in length and may weigh many tons. They contain **fucoxanthin,** a brown pigment that masks the green chlorophylls. The red algae (Rhodophyceae) are also multicellular, branching seaweeds (Fig. 13–2) containing the red pigment **phycoerythrin** in addition to chlorophyll. This pigment absorbs the longer wavelengths (blue–violet) of light. These are able to penetrate water and thus allow the algae to grow at depths of several hundred feet where other algae cannot. Several red algae are edible, e.g., **Irish moss** and **dulse.** Some, as *Gelidium* species, furnish the jelly-forming polysaccharide **agar** that is well known to the microbiologist, the pharmacist, and the housewife. There are also numerous multicellular species of green algae (Chlorophyceae), e.g., sea lettuce. The other groups of algae are mostly unicellular and microscopic. Table 13–1 lists some of the prominent features of the various types of algae.

The origin of the animal kingdom is as obscure as that of the plant kingdom. In view of their eucaryotic structure, absence of photosynthetic pigments and cell walls, animal cells might have developed as loss mutants[*] from a eucaryotic type of plant cell; no procaryotic animal cells are known. Another possibility is that animal cells could have evolved, before the appearance of sunlight and photosynthesis, from a cell wall–less, nonphotosynthetic, procaryotic progenitor like the mycoplasmas. There is evidence, as will be explained, that at least some of the protozoans did evolve from flagellate algae. Animals that contain chitin may be related, distantly, to the fungi whose cell walls also contain chitin. Students should always bear in mind, however, that family trees among the protists are frustratingly delusory!

There are many species of algal protists. The **desmids,** freshwater green algae, are distinguished by their elaborate forms and delicate, symmetrical beauty (Fig. 13–3). One large group of the class Chrysophyceae (golden brown

---

[*]Mutants that result from loss of a functional genetic unit such as ability to synthesize cell walls. (See Chapter 19.)

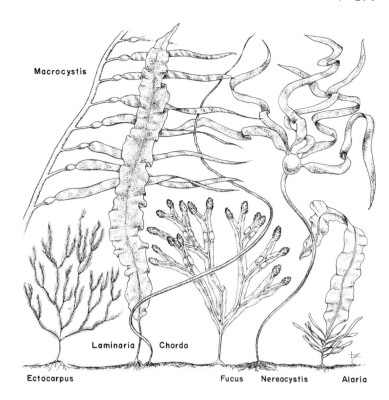

Figure 13-1. Some of the kinds of brown algae or kelps, all of which are multicellular marine plants. The sketches are not drawn to the same scale.

Macrocystis

Laminaria Chorda

Ectocarpus Fucus Nereocystis Alaria

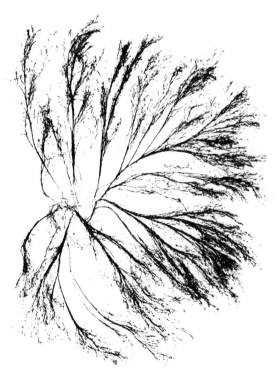

Figure 13-2. A species of red algae. Note the lacy, delicately branched body. (Courtesy of the New York Botanical Garden.)

algae) comprises the **diatoms** (Fig. 13–4). Like the kelps, these contain fucoxanthin. They are widely distributed as marine and fresh water plankton.* Of special interest are their formation of silica-containing cell walls (compare Radiolaria among the Protozoa) that consist of two symmetrically formed and "decorated" parts, like a cut-glass Petri dish and its cover. They exhibit a sedate, gliding motion. They are also distinguished by intracellular storage of food as oil instead of as starch, a supposed origin of at least some petroleum deposits. Like the protozoan Radiolaria, diatoms, through eons of time, have formed enormously thick deposits of their silica shells on the sea bottom. When raised above the sea by geologic movements, these deposits are available as **diatomaceous earth,** which has many industrial uses as an abrasive and filtering agent.

Many algal protists are nonmotile, ovoid cells like the familiar green *Chlorella:* the guinea pig of the space and submarine scientists because it grows well under simple laboratory conditions and can be cultivated in space-

---

*Plankton (Gr. *planktos* = wandering) includes minute plants and animals and also protists that spend their lives floating passively in the sea or fresh water.

TABLE 13-1. CHARACTERISTICS OF MAJOR GROUPS OF ALGAE ARRANGED IN ORDER OF DECREASING SIZE AND COMPLEXITY *

Group or Division	Habitat	Size and Structure	Reproduction	Additional Pigments †	Food Reserve ‡	Cell Wall
Phaeophyta (brown algae: seaweeds, kelp, etc.)	Marine (warm salt water)	Multicellular, large (up to 150 feet in length)	Sexual by gametes; asexual by zoospores, also by fragmentation	Chlorophyll *c*; xanthophylls predominate and mask chlorophyll	Laminarin (P), mannitol, fat	Cellulose and pectin
Rhodophyta (red algae: seaweeds)	Mostly marine (cooler salt waters)	Mostly multicellular and macroscopic (up to 4 feet)	Sexual by gametes; asexual by spores	Chlorophyll *d*; phycocyanin and phycoerythrin mask chlorophyll	Starch	Cellulose and pectin
Chlorophyta (green algae)	Mostly fresh water, soil, vegetation	Unicellular to multicellular, microscopic to macroscopic	Asexual fission, zoospores, primitive sexual fusion	Chlorophyll *b*	Starch	Cellulose and pectin
Chrysophyta (golden algae, diatoms, etc.)	Fresh and salt water, soil, vegetation	Mostly unicellular, microscopic	Mostly asexual	Chlorophyll *c*; xanthophylls and carotenes may mask chlorophyll	Chrysolaminarin (P) and oil	Silicon impregnated
Pyrrophyta (fire algae, dinoflagellates)	Fresh and salt water	Unicellular, microscopic	Asexual fission	Chlorophyll *c*; xanthophylls may mask chlorophyll	Starch, oil	Cellulose or may lack typical cell wall
Euglenophyta (euglenoids, e.g., *Euglena*)	Mostly fresh water	Unicellular, microscopic	Asexual fission (longitudinal)	Chlorophyll *b*	Paramylum (P)	Lacks typical cell wall

*Some authorities prefer to recognize nine divisions. The blue-green algae are discussed along with bacteria in Chapter 14.
†All contain chlorophyll *a* and β-carotene.
‡(P) indicates that the substance is a polysaccharide.

ships and undersea colonies with ultraviolet lamps. It serves man in such situations by using organic waste materials as nutrients, removing carbon dioxide from the air, and providing a constant source of oxygen by its photosynthetic activities. It synthesizes proteins and starch and can therefore also provide food, though it has not, by anyone's account, achieved the status of a gourmet dish.

13.2
FLAGELLATE ALGAE

Chlamydomonas, another green algal protist, is of special interest because it typifies uni-cellular plants that are animal-like in being actively motile by means of flagella. Although *Chlamydomonas* resembles flagellate protozoans, it differs in having a large chloroplast and a thick cellulose cell wall. The ovoid cell can reproduce asexually by repeated nuclear fissions, forming two to eight biflagellate **zoospores.** These are liberated by rupture of the cell wall, and then repeat the process. It may also produce morphologically indistinguishable male and female, biflagellate, haploid **gametes.** Fusion of a pair of the gametes produces a quadriflagellate, diploid zygote that becomes an unflagellated, dormant cyst or zygospore. On germination of the zygospores meiosis occurs, and the resulting

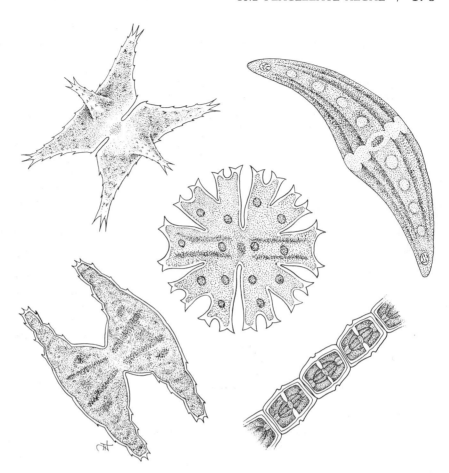

Figure 13–3. Several different species of desmids, unicellular green algae, highly magnified, showing the symmetry of the cells.

four progeny are biflagellate, haploid, potential gametes.

The group of photosynthetic algal flagellates includes the biologically important dinoflagellates. The cells of most dinoflagellates have two unequal flagella and a complex outer wall or shell (armor) of cellulose. The armor is made up of segments or interlocking plates, arranged with transverse grooves or channels

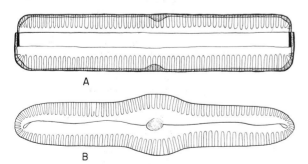

Figure 13–4. A side view (*A*) and a top view (*B*) of a typical diatom, highly magnified. Note the characteristic fine lines on the shells and the way the upper and lower shells fit together.

that give the cells a very distinctive appearance (Fig. 13–5). Commonly, one flagellum lies in the transverse channel (annulus) and through undulating movement causes the cell to rotate. The second flagellum originates within the transverse channel or within a longitudinal channel (sulcus) which intersects the annulus. This flagellum is responsible for the forward motion of the cell. Multiplication may be by asexual cell fission or, depending on species, by intracellular division of the nucleus and the formation of zoospores analogous to *Chlamydomonas*. Sexual reproduction has also been described in some species; it is analogous to reproduction of *Chlamydomonas* though differing in details. A curious feature of dinoflagellates is their ingestion of solid food particles, e.g., other algae, bacteria, and protozoa, by phagocytosis, a distinctly protozoan-like (phagotrophic) process.

Most dinoflagellates live as marine plankton in enormous numbers, their total weight and capacity for photosynthesis far exceeding those of all the terrestrial forests. Sometimes, under conditions especially favorable to their multiplication, certain dinoflagellates become so

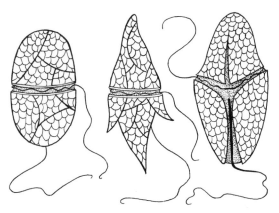

Glenodinium   Peridinium   Gymnodinium

Figure 13–5. Three species of dinoflagellates. Note the plates which encase the single-celled body and the characteristic two flagella, one of which is located in a transverse groove.

numerous as to impart a distinct color to the sea. Some, e.g., *Peridinium* and *Gymnodinium*, produce the dreaded "red tide" that kills tons of fish. Others are taken into mussels as food and are very poisonous to any human eating those mussels.

Thus we see, in the protozoan-like dinoflagellates, actively motile plants, some species of which exhibit phagocytosis.

Another group of photosynthetic flagellates, the euglenoids, is claimed by the botanists as plants (phylum Euglenophyta) and by the zoologists as animals (order Euglenoidina). Of the euglenoids *Euglena gracilis* is a widely studied curiosity, consisting of free-living cells with one or more chloroplasts with typical eucaryotic plant chlorophylls, accessory light-trapping pigments, and alga-like **pyrenoids** that function in starch synthesis. Nutrition may be osmotrophic, or phototrophic (with vitamins). Some colorless euglenoids are phagotrophic. Although the canal and reservoir (Fig. 13–6) may resemble the mouth-like cytostome and the cytopharynx, respectively, of the *Paramecium*, some authorities believe that these organelles are not associated with food gathering (at least in the majority of species studied). Those species which are phagotrophic are believed to possess a separate cytostome and ingestion apparatus that are distinct from the canal and reservoir.

Euglena cells contain contractile vacuoles: a distinctly protozoan feature (see Protozoa, this chapter). These organisms are Protozoa-like also in having no rigid or differentiated cell wall of cellulose or peptidoglycan but instead, like protozoan cells in general, a thickened and

more or less plastic pellicle. A red-pigmented **stigma** (eyespot) occurs in *Euglena* (Fig. 13–6) and other flagellate algae such as *Chlamydomonas*. This is light-sensitive and serves to orient the flagellate to (or from?) the light source. The red pigment is rare and, outside of flagellates, occurs only in an animal group, the Crustacea. Are such light-sensitive organelles the evolutionary forerunners of eyes?

In some protists the ability to synthesize chloroplasts or chromatophores appears to be an unstable property, subject to complete loss. When the chloroplast is lost by a eucaryotic, flagellate, green alga, e.g., *Euglena,* the result is, to all appearances, a flagellate protozoan; it becomes a "naturalized" member of the animal kingdom. When the chromatophore is lost by a procaryotic, filamentous blue-green alga like *Oscillatoria,* it automatically becomes a procaryotic, filamentous bacterium like *Beggiatoa.* Loss of the photosynthetic system in either type of organism apparently can occur as a result of spontaneous mutation. In the case of the flagellate, *Euglena,* it can also be brought about in the laboratory by exposure to ultraviolet light and by other mutagenic methods. Thus we can ex-

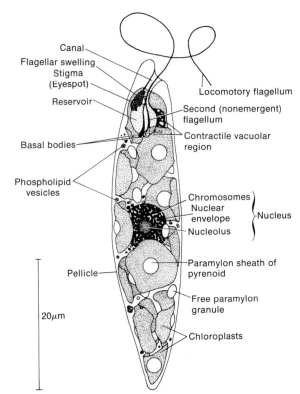

Figure 13–6. Diagram of *Euglena gracilis* showing the location of various structures mentioned in the text.

perimentally change a plant into an animal. Data appear to be lacking on the experimental transformability of blue-green algae into morphologically similar nonphotosynthetic bacteria, though it doubtless can be done.

In this connection it is interesting to note that many algae possess dual energy systems. As photolithotrophs they can utilize radiant energy by virtue of their photosynthetic systems; as chemoorganotrophs they can grow very well in the dark like fungi and protozoa. The interesting point at present is that, at the protist level, the demarcations between major groups often become very hazy or disappear. Many modern Protozoa are not so obviously like the algae, having undoubtedly become altered in the course of millions of years of evolution. On close comparative study, however, their origin from algae seems likely.

The group of procaryotic algae, the blue-green algae or class Schizophyceae (Cyanophyceae), are discussed in the next chapter along with the bacteria.

13.3
SIGNIFICANCE OF ALGAE

The algae constitute a large segment of the microbial world, and their contribution to life on this planet, as we know it, is of the greatest significance. The following is a summary of the more important aspects.

Evolution. For those interested in phylogeny, the algae hold special interest. Their photosynthetic mechanism, storage materials, cell organization, etc., link them strongly with the higher plants. On the other hand, their method of reproduction, and in some cases their nutrition, often resembles that of members of the animal kingdom. The Euglenophyta certainly form a special link between the two kingdoms. Considerable excitement was generated recently when a dinoflagellate, *Gyrodinium cohnii*, was found to exhibit chromosomal separation during mitosis by a mechanism quite different from the usual mitotic spindle apparatus. The nuclear membrane is apparently active in the separation of the chromosomes, thus mimicking one of the functions of the cytoplasmic membrane in procaryons undergoing fission. In-depth studies of the algae may turn up other examples of microorganisms which clearly bridge the two fundamental levels of cell organization.

Food Chains. The algae are considered primary producers of food in the oceans and seas because, through photosynthesis, they convert chemically simple nutrients into complex organic matter and thus serve as the first edible substance in an incredibly long food chain involving most of the higher marine life. In some countries, certain types of seaweed are harvested as feed for animals or food for human consumption. In view of the current deficit in protein for humans in certain parts of the world and the specter of global starvation, algae are being given serious consideration as a possible large-scale food supplement.

Ecology. Algae are often abundant in moist soil, sometimes exceeding 10^4 organisms per gram of soil. Their dead cells contribute to the formation of the complex organic material called **humus** (see Chapter 43) which provides food for other microorganisms. The blue-green algae, considered in the next chapter, are of particular interest because of their ability to convert ("fix") gaseous nitrogen into organic nitrogen. An ecology-minded scientist will pay special attention to the types of algae present in a body of water suspected of being polluted. Badly polluted waters with reducing conditions and much organic matter will contain no algae at all since they require oxygen. As the organic load is decreased and the oxygen levels increase, certain species of algae begin to appear.

Ironically, algae can themselves sometimes be a cause of pollution. This typically takes place during the warm summer months in a lake where there is excessive **eutrophication** (nutritional enrichment) due to sewage effluent, storm run-off from surrounding farms, etc. Under these conditions, there is very rapid algal growth ("bloom") followed later by a death phase with the attendant stench, water discoloration, and fouling of the beaches.

Commercial Value. Besides being used for food, some kinds of algae can be harvested for fertilizer. In fact, ecologists have forced many communities to discontinue the use of chemicals to control algae growth in lakes and streams. High-speed underwater cutters and harvesters are being developed to control growth, with the idea that the harvested seaweed will have commercial value as a fertilizer or feed supplement. **Agar,** familiar to all microbiologists as a solidifying agent in the preparation of media, is obtained from the washed and dried seaweed *Gelidium*, a red alga. Two other polysaccharides, carrageenin and alginic acid, are also obtained from seaweeds and are of great importance in the food and pharmaceutical industries as emulsifiers and thickening agents. Many tons of diatomaceous earth, or kieselguhr, are used yearly by industry for filtration and the manufacture of insulation, paints, and paper products. This material originates from the silicious cell walls of diatoms deposited along certain coastal waters over long periods of time.

13.4
THE PROTOZOANS

Like other major divisions of protists, the phylum Protozoa is a very heterogeneous group. There are probably well over 30,000 species of Protozoa. One or two examples of each major type may be described to show their distinguishing properties and also some of the characteristics common to most protozoans. In general, protozoans are the most highly specialized, and their cell structures and modes of life and reproduction the most complex, of all the protists. Unless one includes the flagellate algae, the phylum Protozoa is the only animal group in the entire kingdom Protista. (See also Color Plate II, *B*.)

Examine with a microscope at a magnification of between 400× and 1,000× a drop of sewage or of water in which a bit of dried grass has been allowed to soak at room temperature for two or three days. Not only are many kinds of bacteria seen, but also many fantastic and elegant creatures that resemble bacteria in some respects but that can, as a rule, be readily differentiated by their relatively larger size, their elliptical or ovoid form, complex internal structure, and other distinctive features. These are protozoans—microscopic animals, each consisting of a single cell or groups of cells to form a colony.

One or more species of Protozoa may be found in almost every habitable situation on earth: stagnant water, pond mud, surface waters, feces, the soil, dust, the ocean. They live, in part, upon other minute living things, including other protozoans and bacteria. All of the protozoans seem to have evolved in the direction of complexity of physiological function without differentiation of specialized organs such as heart, liver, or brain.

Classification. All protozoans are motile during at least one stage of their existence. Differences in the mechanism of locomotion were adopted by early workers as one of the major criteria for separating the Protozoa into five groups:

Sarcodina (**amebas**):
  Adults move by protoplasmic flow with the formation of pseudopodia.

Ciliata (**ciliates**):
  Movement by cilia.
Suctoria (**suctoreans**):
  Young stages are ciliated; adult stages are sessile and provided with tentacles.
Mastigophora (**flagellates**):
  Movement by flagella.
Sporozoa (**sporozoans**):
  Move by pseudopodia only in immature stages; the male gamete is flagellate.

While the above division is still a convenient one for discussing the Protozoa at an elementary level (and will be used in this chapter), it should be pointed out that modern classification systems are based on locomotion, morphology, mode of reproduction, and nutrition. The placement of the above groups in a widely adopted system of classification is shown in Table 13–2. A discussion of all the groups of Protozoa is beyond the scope of this book; only certain aspects and selected groups can be described here.

Structure. Protozoa vary greatly in size according to species and physiological state. Most of them are hundreds of times as large as most bacteria, but some are not much larger. A commonly studied ciliate, *Paramecium*, is roughly elliptical in shape and has dimensions of about 200 μm by 40 μm (see Figure 13–10).

As previously noted (Chapt. 11), eucaryotic plant cells typically have a chemically and morphologically differentiated cell wall of cellulose or, in many fungi, chitin. In contrast, animal cells, such as protozoans and cells of the human body, have a typically eucaryotic structure but no cell wall.

TABLE 13–2. ABRIDGED CLASSIFICATION OF THE PROTOZOA*

PHYLUM: PROTOZOA
  Subphylum: Sarcomastigophora (flagellates and amebas)
    Superclass: Mastigophora (flagellates)
      Class: Phytomastigophorea
             Zoomastigophorea
    Superclass: Opalinata
               Sarcodina (amebas)
  Subphylum: Sporozoa (sporozoans)
    Class: Telosporea
           Toxoplasmea
           Haplosporea
  Subphylum: Cnidospora (sporozoans)
    Class: Myxosporidea
           Microsporea
  Subphylum: Ciliophora (ciliates and suctoreans)
    Class: Ciliatea

*Adapted from Honigberg et al.: J. Protozool., *11*:7, 1964.

The outer covering of protozoans consists of a **pellicle** or **periplast** which is composed of lipoprotein and takes the place of a cell wall. In some groups, such as the amebas, the pellicle is either absent or present as a thin, elastic covering that does not interfere with ameboid movement and phagocytosis, both of which are motivated by **cytoplasmic streaming.** In other types, especially the large and active flagellated or ciliated species, the pellicle is thicker and less flexible (giving a definite shape to the organism) and may be ridged or grooved or show other surface patterns. In addition to the pellicle, some species also possess a chitinous, cellulosic or silicious exoskeleton for additional protection.

Located just underneath the pellicle is the **plasma membrane** which encloses the cytoplasm. The inner, very fluid zone of cytoplasm is called **endoplasm** and contains many of the common microbial organelles (food vacuoles, contractile vacuoles, nucleus, mitochondria, etc.). The outer, less fluid, peripheral zone of cytoplasm is called **ectoplasm.** This layer gives rise to a number of specialized structures used for locomotion, defense, and feeding (flagella, cilia, trichocysts, cytostome, cytopharynx (gullet), anal pore, and contractile fibers).

The use of scanning electron microscopy and new fixation techniques have allowed biologists to observe the delicate, beautiful outer structures of many protozoans not easily seen by other methods (Fig. 13–7). Such work continues to make substantial contributions to our knowledge of the structure and organelles of these and other organisms.

Contractile vacuoles are conspicuous in species of Protozoa that live in fresh water but are also found in some marine and parasitic species. The vacuoles appear to be regulators of intracellular water and hence of intracellular osmotic pressure. Because of the low osmotic tension of fresh water and the much higher osmotic tension of intracellular fluids, cells in fresh water tend to take in water. Like bacterial protoplasts or PPLO, the cells would burst unless the excess water taken in could be excreted.

Flagella and Cilia. As is true of the flagella of motile green algae, protozoan flagella act as swimming appendages. Cilia are much like flagella in structure and function but are shorter. They tend to move in a coordinated and rhythmical fashion but apparently not because of any primitive neuromuscular control mechanism, as was proposed earlier. Cilia, like flagella, produce movement by sending out undulations and also appear to originate in basal granules in the ectoplasm. As will be shown in Chapter 14,

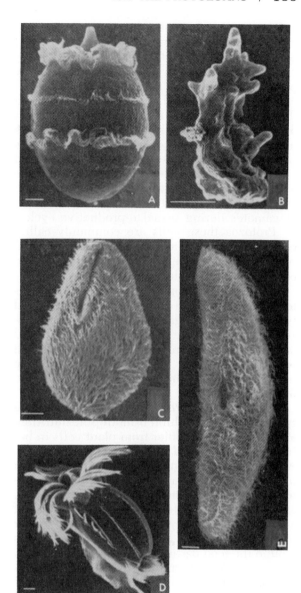

Figure 13–7. Scanning electron micrographs of fixed, frozen, and dried Protozoa. *A, Didinium nasutum* (scale, 10 μm). *B, Amoeba proteus* (scale, 100 μm). *C, Nyctotherus ovalis* (scale, 10 μm). *D, Uronychia* sp. (scale, 10 μm). *E, Paramecium multimicronucleatum* (scale, 10 μm).

eucaryotic flagella and cilia have a complex inner multifibrillar structure not seen in flagella of procaryons.

Reproduction. As in most protists, asexual reproduction by cell fission is the most efficient and commonly seen means of increasing numbers of cells or new protozoan individuals. In many species, no other kind of reproductive process is known. In other species genetic recombination and multiplication, usually resulting in rejuvenation of the asexual processes, is commonly brought about by **conjugation.** In some species this may be a quite complicated process as will be explained farther on.

Encystment. During the life cycle of many protozoans some cells may produce a thick cell wall, lose water, and become dormant. During this stage metabolism is reduced to a minimum or ceases entirely, and the dormant cell resists unfavorable environmental conditions such as prolonged drought, summer heat, increased salinity, or unfavorable pH. Such a stage may be formed by mature, growing cells during asexual cycles of development or just after conjugation of gametes during sexual reproductive cycles. In Protozoa these cells are commonly called **cysts.** Cells going into the cyst stage are said to be **encysting.** In some species, encysting may be coupled with reproduction.

After a physiologically and genetically determined period or, depending on species and type of cyst, when growth conditions again become favorable, the cell inside the cyst wall takes in water, resumes activity, bursts the cell wall, and emerges in the actively growing **trophozoite** stage. The encysted cell is said to have **excysted.**

Nutrition of Protozoa. Although some types of Protozoa exhibit osmotrophic nutrition, Protozoa typically differ from plant cells in being able to take *solid* food particles into the cell. All Protozoa are chemoorganotrophic. Ingestion of foods by Protozoa is accomplished by three methods: by **phagocytosis,** by means of **cytostome,** and by **pinocytosis.** In phagocytosis, typical of amebas and leucocytes of the blood, two or more pseudopodia are extruded like fingers around the food particle. The fingers merge into one, with the particle trapped within. The section of plasma membrane surrounding the food particle is pinched off and becomes the food vacuole within which the food undergoes digestion. In other protozoans, most typically ciliates, food particles are wafted by cilia into a deep pouch or invagination of the cell coating called a **cytopharynx.** The food particle passes through the cell coating at the inner end of the cytopharynx into a cytoplasmic digestive food vacuole. Pinocytosis is a somewhat similar process for ingesting fluid rather than particulate matter, though not dependent on cilia; the pinocytic vesicles are numerous and very tiny (see Chapter 9). Protozoa commonly eat bacteria, other minute organisms, and—with a voracity which must surely tax their digestive systems—each other (Fig. 13–8).

Responsiveness of Protozoa. These little animals appear to have a certain intelligence, possibly of the type referred to by Immanuel Kant as "transcendental knowledge"; i.e., inherited or inherent. Protozoa accept or reject food particles with discrimination; e.g., a *Didinium* such as shown in Figure 13–8 will back away from a dead *Paramecium* but will readily attack and ingest a live one. They also have quite a delicate sense of touch (**tactile sense**), so that they recoil on contact with hard objects, turning aside quite as though they were highly sensate and responsive creatures. Some (e.g., *Euglena*) have eyespots and can distinguish light of different wavelengths. Many are quite sensitive and responsive to such stimuli as heat, chemicals, gravity, and electricity.

13.5
SOME REPRESENTATIVE PROTOZOANS

The Amebas

There are many species of amebas, most of them harmless to man and higher animals. They are common in pools and ponds in which there are bacteria and other minutely particulate matter and dissolved substances on which they can feed.

Form. Amebas differ from most other Protozoa in that while alive they have no particular form. Cells of *Amoeba proteus*, typical of the genus, continually change from round or oval to very irregular shapes with temporary protrusions and finger-like processes sticking out from, or being retracted into, various portions (Fig. 13–9). They move by means of these finger- or foot-like processes called **pseudopodia** (Gr. *pseudo* = false; *pod* = foot). The ameba thrusts out a pseudopodium and moves by flowing into the projected portion. This sort of motility is called **ameboid movement.** The pellicle, if present, is very thin and flexible. As in all typical eucaryotic cells, the protoplasm maintains a constant intracellular streaming motion (**cytoplasmic streaming**), thus performing the function of a circulatory system. Waste products are excreted through the plasma membrane and pellicle to the outside. Ameboid movement appears to be dependent on protoplasmic streaming. Such movement cannot occur in species with rigid pellicles.

The cells of some genera of the superclass Sarcodina, unlike the genus *Amoeba*, surround themselves with a protective and supporting shell or skeletal structure. In *Difflugia* the cells secrete a gummy cement substance that binds together a supporting and protective coating of sand grains. *Arcella* secretes a covering of chitin.

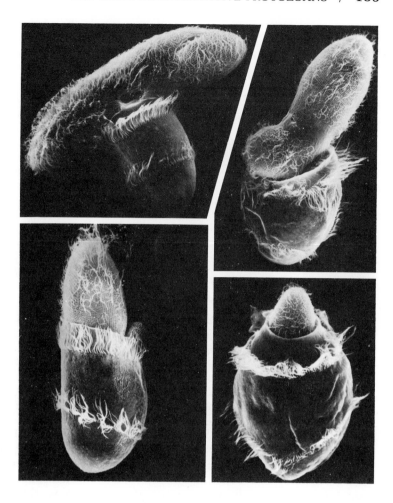

Figure 13–8. The capture and ingestion of *Parmecium* by another protozoan, *Didinium*.

Members of the group called Foraminifera, a common and numerous marine type, characteristically form rather complex, chambered shells of calcium carbonate. The white cliffs of Dover are the accumulations of the shells of Foraminifera (chalk) deposited on the ocean floor during millions of years and later raised above the sea. A related marine group called Radiolaria secrete skeletons of silica (SiO_2), some of which are exquisitely delicate and radially symmetrical. Accumulated on the ocean floors and compressed during millions of years, these have become flint or flint-like. In all of the species just mentioned the cell itself is ameba-like; it thrusts its pseudopodia through openings in the shell or skeletal structure for locomotion or food gathering.

Reproduction. Amebas, like all other protists, possess the primitive power of **asexual** or **vegetative** multiplication. In this process the cell divides itself into two cells, a phenomenon called **cell fission.** Cell fission is said to be binary when the cell divides into two equal parts, each part having (theoretically) all of the physio-logical and genetic potential of the parent cell. During binary fission amebas divide by a constriction somewhere near the physiological middle of the cell, i.e., by **transverse fission.** The nucleus undergoes mitosis, typical of eucaryons.

SOME PATHOGENIC AMEBAS

Amebiasis. Infection by any species of ameba is properly spoken of as **amebiasis.** The most harmful and best known of the species pathogenic for man bears the name *Entamoeba histolytica. E. histolytica* attacks the walls of the intestine. These amebas feed, characteristically but not exclusively, upon red blood cells. They penetrate the intestinal lining and cause intense inflammation and ulcers **(amebic dysentery).** By means of proteolytic enzymes they can burrow through the lining and penetrate deep into the intestinal wall so that, occasionally, rupture of the intestine occurs. The patient may then die of peritonitis caused by escape of the bacteria of the feces into the abdominal cavity.

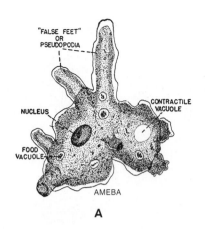

A

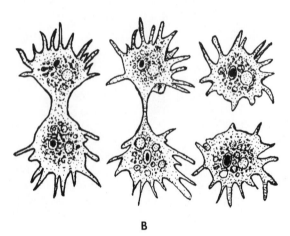

B

Figure 13–9. A species of harmless ameba from stagnant water (trophozoite forms). *A,* Note the numerous pseudopodia, the nucleus and the vacuoles inside the cell. This organism, itself microscopic in size, feeds upon bacteria and is thousands of times larger than they are. *B,* Ameba dividing.

Entamoeba histolytica may also get into the lymph and blood vessels and then be carried to the liver, lungs, brain, and other organs where they become localized and cause the formation of large abscesses.

Transmission of Amebiasis. *Entamoeba histolytica* is passed in the feces, most commonly in the dormant, thick-walled, encysted form. Amebiasis is often chronic and may be present with little definite symptomatology for a long time. Thus amebic infection, especially of the intestine, like many other infections is often unknowingly disseminated widely by mild cases or **carriers.** Carriers who handle food may transmit cysts to food via their soiled hands. These cysts, after being swallowed, rapidly excyst in the intestine. Feces-soiled hands and flies appear to be major vectors of most intestinal pathogens: protozoan, bacterial, and viral.

Anything recently contaminated with feces from a chronic case or carrier of amebiasis may transmit the cysts. Transmission on fruits and vegetables can occur when human sewage and feces are used for fertilizer. Under such circumstances fresh vegetables (lettuce and celery, for instance) may have viable *E. histolytica* cysts upon them when eaten.

Amebiasis is common in all regions in which sanitation of sewage is neglected or absent. Some serious outbreaks of amebic dysentery in the United States have been caused by sewage-polluted water supplies.

The Ciliates

This group of Protozoa takes its name from the cilia that are distributed over the surfaces of the cells. Ciliates are commonly ovoid or pear-shaped (pyriform). A much studied species, *Paramecium caudatum,* has been called the "slipper animalcule" because of its slipper-like outline (Fig. 13–10). *P. caudatum* is commonly found in pond water. It is a representative of one of the most complex unicellular microorganisms. It maintains its form by means of a flexible but tough external pellicle that has protruding through it symmetrically arranged rows of cilia. The cilia move in a highly coordinated rhythmic manner which permits the organism to swim in a spiral manner as gracefully as a seal. Near one end of the cell is a depression (sometimes called the oral groove or **peristome**) which leads to the mouth or cytostome. Food (bacteria, smaller protozoans) is wafted by cilia in this area through the cytostome into the cytopharynx. The food is then taken into food vacuoles which form at the end of the cytopharynx. Digestion and absorption occur, and indigestible material is extruded through a posterior **anal opening** or **cytopyge. Contractile vacuoles** in the cytoplasm help to maintain ion and water balance.

Paramecium can presumably defend or position itself by means of **trichocysts.** These are small organelles beneath the pellicle. When activated, the trichocysts secrete a substance which (outside the pellicle) forms long, sticky filaments. The major function of trichocysts is still obscure; defense, anchoring, and ion regulation have been suggested.

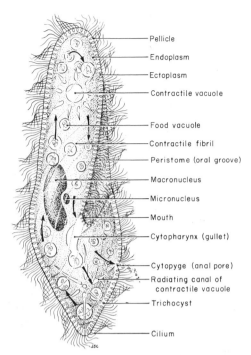

Figure 13-10 labels:
- Pellicle
- Endoplasm
- Ectoplasm
- Contractile vacuole
- Food vacuole
- Contractile fibril
- Peristome (oral groove)
- Macronucleus
- Micronucleus
- Mouth
- Cytopharynx (gullet)
- Cytopyge (anal pore)
- Radiating canal of contractile vacuole
- Trichocyst
- Cilium

Figure 13-10. Diagram of *Paramecium caudatum*, showing the location of structures described in the text.

Paramecium multiplies asexually by binary fission and sexually by conjugation. The process of conjugation in ciliates is well exemplified by *Tetrahymena pyriformis,* a related species of ciliate.

Reproduction. In most vegetative cells of ciliates such as *Tetrahymena* there are two differing nuclei: a large **macronucleus** or "vegetative," polyploid nucleus and a minute, diploid **micronucleus.** Some cells contain only the macronucleus; these cells can multiply only vegetatively, i.e., by binary fission. Only cells containing micronuclei can conjugate. (Compare F^+ bacteria, Chapter 18.) Most conjugating cells are morphologically indistinguishable as to sex.

Typically, the cells fuse and each diploid micronucleus undergoes meiotic divisions, producing four haploid nuclei in each cell. Of these, three degenerate; one divides mitotically. Thus there are two haploid (gamete-like) nuclei. One haploid nucleus from each cell of the conjugal pair is exchanged with the counterpart in the other cell. The residual and the exchange haploid nuclei in each cell then combine, thus forming a diploid nucleus and achieving genetic recombination.

In each cell the new nucleus, now a diploid zygote, twice divides by mitosis, yielding four diploid nuclei, two of which become macronuclei, the other two micronuclei. The old macronucleus disintegrates. The conjugants now separate, each with two macro- and two micronuclei. In each cell one of the micronuclei disintegrates; the remaining one divides by mitosis. Each of the conjugants now again has two of each kind of nucleus. Each of these **tetranucleate** cells divides by binary fission. Each daughter cell normally has one macronucleus and one micronucleus. In some instances a micronucleus may be "lost in the shuffle"; that cell has lost its sex factor and multiplies only asexually.

The Flagellates

There are numerous species of the superclass Mastigophora with ovoid, spindle-like or pear-shaped cells that resemble *Paramecium,* but that move actively about by means of one or more flagella commonly attached at one end of the animal. Many varied types can be found in sewage or infusions of hay. Some species contain chlorophyll and are regarded by many workers as flagellate algae, as previously described. Some species inhabit the human body and are pathogenic (Fig. 13-11, 13-12).

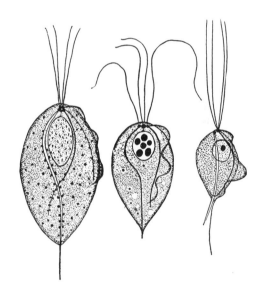

Figure 13-11. Three species of *Trichomonas* found in humans: (left to right) *T. vaginalis, T. buccalis* (or *T. tenax?*), *T. hominis.* The size differences shown are not very constant. *T. buccalis* is not known to be pathogenic, but its continuous presence in the mouth in considerable numbers indicates bad oral hygiene. (2,000×.) (Courtesy of Dr. S. J. Powell, Institute for Parasitology, Durban.)

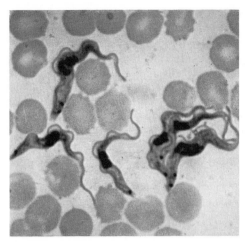

Figure 13–12. *Trypanosoma gambiense* in a droplet of blood (the large, round objects are erythrocytes). This is one of the species of trypanosomes causing African trypanosomiasis (African "sleeping sickness"). Note the prominent flagellum along the edge of the wavy, keel-like membrane on each trypanosome. (1,525×.) (Courtesy of Dr. A. Packchanian, The Medical School, University of Texas.)

Sporozoans

The most important (to us!) representatives of the sporozoans are parasites. Among these, probably the best known and most deadly are the malaria parasites. Since every advanced nation is now cooperating with the World Health Organization in a huge effort to eliminate these deadly organisms from the earth, we will consider them in some detail.

MALARIA

Malaria is one of the most important of the arthropod-borne diseases (Gr. *arthron* = joint; *pod* = foot). An arthropod is an animal with jointed feet or legs, the most familiar of which are insects and Crustacea (scorpions, lobsters, crabs, shrimp).

In the United States, malaria as an epidemic or endemic disease has been eliminated. Persons entering the country from malarious areas occasionally reintroduce it, but because of constant surveillance and control measures, it does not spread. However, it still remains a widespread and death-dealing scourge in many areas outside the United States, especially in tropical and subtropical zones. Hundreds of thousands of persons die annually from this disease and millions are made chronically ill by it.

The Malarial Parasite. Malaria in man is caused by a protozoan parasite of the genus *Plasmodium*. Its life history involves two stages

of development: one, asexual, passed in the human body; the other, sexual, passed in the female of various species of mosquitoes of the genus *Anopheles*.

Cycle in Man. An infected mosquito introduces the parasites into the blood of its victim with its saliva when it bites (Fig. 13–13). The female mosquito bites only to obtain blood proteins for egg production. The males live inoffensively on plant juices. The parasites undergo a short period of multiplication in certain tissue cells, particularly in the liver. This is called the **exoerythrocytic** (or **pre-erythrocytic**) cycle. Very soon their asexual progeny enter red cells and grow within them. This stage of the parasite is the **trophozoite** stage. The parasite multiplies asexually within the red cells, forming a number of small bodies or segments. Finally the affected erythrocyte breaks up, and the segments escape into the circulating blood. Each segment is a new, active parasite called a **merozoite.** It no sooner gets out of one red cell than it attacks another erythrocyte and in turn multiplies. In this way the blood is soon teeming with the parasites and the infection becomes clinically manifest; the **intrinsic incubation period** (bite to first symptom) is completed. The patient becomes anemic and weakened by the loss of so many red cells, and possibly also suffers from poisonous products formed by the parasites.

The parasites appear in the blood in successive generations, all the individuals of which divide and burst out of the erythrocytes at about the same time. Each such process causes the chills and fever so characteristic of malaria. A chill indicates that a fresh crop of parasites has matured and entered the circulation.

After passing through several cycles of asexual development as just described, round, distinctive gametocytes being to appear in the blood of the patient. These are larger than the asexual forms and are easily recognized by trained persons in the smears of the patient's blood examined under the microscope. Gametocytes undergo no further development in human erythrocytes. They die if not taken up by a mosquito.

Cycle in the Mosquito. When an *Anopheles* mosquito bites a person who has mature malarial gametocytes in his blood, the sexual stage of the parasite begins.

After fertilization of the female by the male gamete in the stomach of the mosquito, the motile **zygotes** invade the cells lining the mosquito's stomach and multiply there, forming a sac (**oöcyst**) wherein the parasites undergo further development by fission. The sac ruptures,

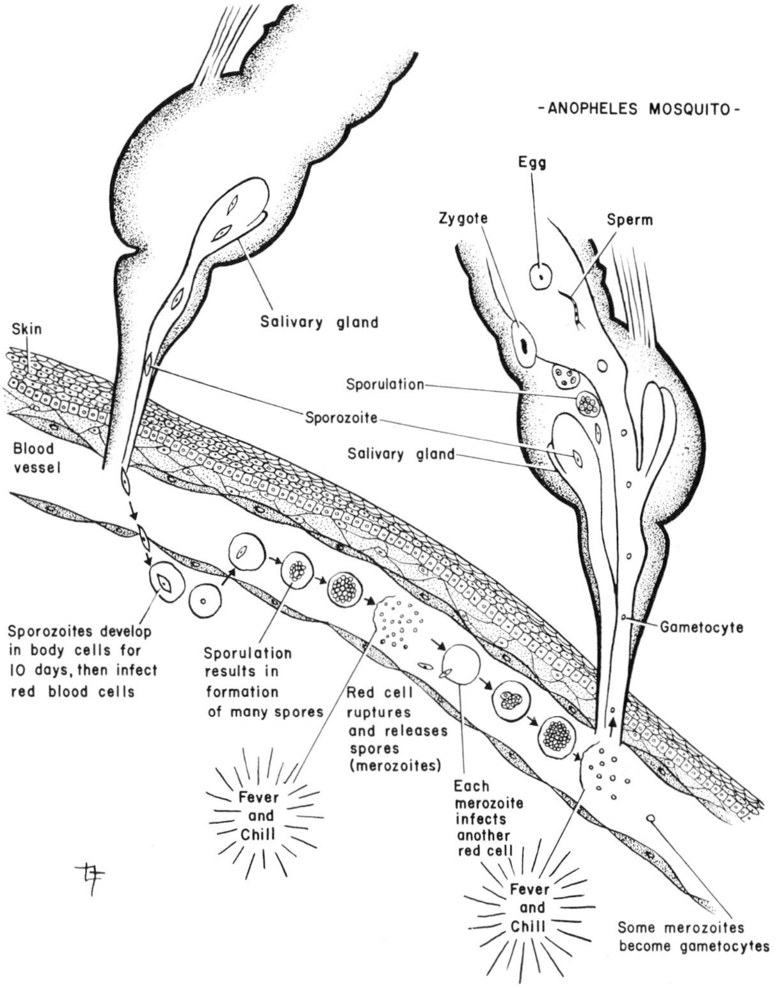

Figure 13–13. A diagram of the life cycle of the malaria parasite, *Plasmodium*. An infected mosquito (*left*) bites a man and injects some *Plasmodium* sporozoites into his bloodstream. These reproduce asexually by sporulation within the red blood cells of the host. The infected red cells rupture, and the new crop of merozoites released then infects other red cells. The bursting of the red cells releases toxic substances which cause the periodic fever and chill. In time some merozoites become gametocytes which can infect a mosquito if one bites a man. The gametocytes develop into eggs and sperm (*right*) and undergo sexual reproduction in the mosquito, and the zygote, by sporulation, produces sporozoites which migrate to the salivary glands.

liberating numerous new young parasites. After moving about for some days inside the mosquito, these reach the mosquito's salivary glands and from there are injected into man when the insect bites. The life cycle is thus complete. Because of the necessary period of sexual reproduction of the parasite, a mosquito that has bitten a malaria patient cannot transmit the disease to another person until the end of about 12 days — the **extrinsic incubation period.** A mosquito, once infected, remains so for the rest of its life, which may be two months or more.

Species. There are four species of the human malarial parasite. One of the most widely distributed in temperate zones is called *Plasmodium vivax* (from the *vivacious* activity of its trophozoite stage). It requires about 48 hours to complete its development within the red cells. The chills, therefore, commonly occur at intervals of 48 hours, or every third day, although there is considerable variation. This type of malaria is called **tertian (third) fever.** A second species, called *Plasmodium malariae*, requires about 72 hours for development, and groups of parasites mature approximately every

fourth day. This species causes **quartan (fourth) fever.** A third form, *Plasmodium falciparum* (the word *falciparum* is derived from the curved or sickle-shaped—**falciform**—gametes) requires from 24 to 48 hours or more for development. This type of malaria is called **estivo-autumnal fever** because in temperate climates it typically occurs in the late summer and autumn. It is most prevalent in tropical zones. It is more severe than the other forms of malaria and is less easily controlled by antimalarial drugs. *Plasmodium ovale*, a fourth species, resembling *P. vivax*, causes a disease much like tertian malaria, but milder.

Mosquitoes, like most other two-winged insects (*Diptera*), pass through four stages of development: the **egg,** the **larva** (wiggler), the **pupa,** and the fully developed insect (**imago**). The first three stages develop in water.

The female *Anopheles* mosquito may be recognized by her stance as she bites: a "head-on" position, with hind legs in the air. She usually has spots of silver or gray on her wings and often gray bands on her legs. The non-malaria-bearing varieties are usually brownish or brown-gray and bite with body nearly parallel to the skin.

Diagnosis of Malaria. The laboratory diagnosis of malaria is commonly made by spreading a small drop of the patient's blood on a slide and either examining it in the fresh state or staining it with any special stain used for blood smears. The parasites can be seen in (or occasionally upon) the blood cells and may have various appearances, depending on the species and the stage of their development in the red cells.

Control of Malaria. As in other arthropod-borne diseases, control is directed primarily against the vector arthropod. Aircraft from areas where any foreign arthropod-borne diseases exist are generally sprayed with aerosols before landing. The control of mosquitoes in swamps and airplanes is an engineering problem. Mosquitoes may be kept out of homes and away from sleepers by screens and insect repellents. The repeated use of DDT, dieldrin, or other residual sprays has been most effective since it kills the infected mosquitoes in the house. Unfortunately, insecticide-resistant mutants soon appear in numbers.

Another very important measure is to eliminate the parasites in infected persons by treating them with antimalarial drugs like quinacrine, primaquine, chloroquine, and amodiaquine. Also unfortunately, drug-resistant mutants often soon predominate. Some drugs are suppressive, but do not always cure. Quinine, known as a valuable febrifuge extracted from bark of the cinchona tree by pre-Columbian Peruvians, is less used than formerly. However, in those situations in which the malaria parasites are resistant to other drugs, the use of quinine is increasing.

Unfortunately, some species of *Plasmodium* formerly thought to be restricted to monkeys, e.g., *P. cynomolgi, P. brasilianum,* and *P. knowlesi,* are also transmissible to man by *Anopheles* mosquitoes in the jungles under natural conditions. This discovery, by Eyles, Coatney, and others, greatly complicates plans for control and complete elimination of malaria. The situation is somewhat similar to that which developed after the discovery of jungle yellow fever in various species of monkeys.

CHAPTER 13
SUPPLEMENTARY READING

Ahmadjian, V.: Lichens. Ann. Rev. Microbiol., *19*:1, 1965.

Alexopoulos, C. J., and Bold, H. C.: Algae and Fungi. The Macmillan Co., New York. 1967.

Chapman, V. J.: The Algae. The Macmillan Co., New York. 1968.

Corliss, J. O.: The Ciliated Protozoa. Pergamon Press, Inc., New York. 1961.

Grimstone, A. V.: Structure and function in protozoa. Ann. Rev. Microbiol., *20*:131, 1966.

Hawking, F.: The clock of the malaria parasite. Sci. Am., *222*:123, 1970.

Hunter, G. W., Frye, W. W., and Swartzwelder, J. C.: A Manual of Tropical Medicine, 4th ed. W. B. Saunders Co., Philadelphia. 1966.

Jackson, D. F. (Ed.): Algae and Man. Plenum Press, New York. 1963.

Jackson, D. F. (Ed.): Algae, Man and the Environment. Syracuse University Press, Syracuse, N.Y. 1968.

Kudo, R. R.: Protozoology, 5th ed. Charles C Thomas, Springfield, Ill. 1966.

Palmer, C. M.: Key for the identification of algae, *in*: Standard Methods for the Examination of Water and Wastewater, 12th ed. American Public Health Association, New York. 1965.

Poindexter, J. S.: Microbiology, An Introduction to Protists. The Macmillan Co., New York. 1971.

Prescott, G. W.: The Algae: A Review. Houghton Mifflin Co., Boston. 1968.

Public Health Service, U.S. Department of Health, Education, and Welfare: Algae in Water Supplies. U.S. Government Printing Office, Washington, D.C. 1959.

Round, P. E.: The Biology of Algae. E. Arnold, Ltd., London. 1965.

Smith, G. M.: The Fresh Water Algae of the United States. McGraw-Hill Book Co., Inc., New York. 1950.

Villee, C. A.: Biology, 6th ed. W. B. Saunders Co., Philadelphia. 1972.

PROCARYOTIC CELL STRUCTURE AND FUNCTION

Among the procaryons certain species of bacteria have been more thoroughly studied than others, and for this reason we shall use them to exemplify procaryotic cell structure in general. It must be borne in mind, however, that in biology, as in many other fields, one always treads on thin ice when he attempts to extrapolate data concerning one species, individual, or single cell, to others; what is true of one or two is not necessarily true of all.

The principal anatomic features of the bacteria that have been investigated, proceeding from the outermost inward, are listed in Table 14–1 along with their prime function(s) and composition. The location of these structures is shown in Figure 14–1. Not all of these structures are present in all species, and some have been investigated in only a few species. Spores are discussed in Chapter 36. The reader may find it profitable to review quickly the principal differences between eucaryotic and procaryotic cells (Chapt. 11) before starting this chapter. The systematic study and classification of bacteria are covered in Chapters 30 and 31, respectively.

14.1
MOTILITY AND FLAGELLA

Among the many protists at least four types of translatory motility may be distinguished: ameboid; gliding; progress caused by the rotatory motion of a helically coiled cell (see Spirochaetales); and motion caused by the oar-like or sculling action of flagella or cilia.

Ameboid Movement. Ameboid movement is distinctive of the protozoan class Sarcodina, which includes the amebas. Details of the motion are given elsewhere (page 184). The formation of pseudopodia in these organisms is dependent on protoplasmic streaming which, with the possible exception of gliding blue-green algae and certain alga-like bacteria, is not found in procaryotic cells. Even if protoplasmic streaming occurred in procaryotic cells, they could not exhibit ameboid motion because, with a few exceptions (*Mycoplasma*), all have semi-rigid or thick cell walls.

Gliding Motion. This form of motion is found in numerous species of blue-green algae, in some alga-like bacteria (order Cytophagales [Chapt. 41] and order Rhodospirillales [Chapts. 32, 34]), and in the Protozoa-like slime bacteria, order Myxobacterales (Chapt. 41). The motion consists of a slow, steady, to-and-fro or steadily progressive, gliding of the cell in contact with a solid surface. The mechanism remains obscure.

Rotatory Movement. The rotatory movement of the helically coiled cells of spirochetes is also not fully understood. Two axial filaments arising from opposite ends of the cell meet or join at the center of the cell. The filaments appear structurally more complex than bacterial flagella and are believed to consist of an inner core filament, an inner coat, and a helically wound outer coat. Just how these filaments

TABLE 14-1. CHARACTERISTICS OF TYPICAL BACTERIAL STRUCTURES*

Structure	Function(s)	Predominant Chemical Composition
Flagella	Locomotion.	Protein (flagellin).
Pili (fimbriae)	Conjugation, possibly attachment to certain surfaces.	Protein (pilin).
Capsule (or slime layer)	Protection against drying, food reserve, waste disposal; virulence factor.	Frequently polysaccharide, sometimes protein or polypeptide.
Cell wall	Confers shape through corset-like restraining of intracellular components.	Peptidoglycan frequently complexed with other substances (e.g., teichoic acid, lipid, polysaccharides).
Cytoplasmic membrane	Permeability, transport.	Protein and lipid.
Mesosomes	In some, may involve separation of nuclear material during cell division; other roles less clear.	Protein and lipid.
Ribosomes	Synthesis of proteins.	Mostly RNA and protein.
Inclusion vacuoles	Storage products, food reserve.	Highly variable, frequently carbohydrate.
Chromatophore	Site of photosynthesis.	Protein, lipids, bacteriochlorophylls.
Nucleoid (nuclear region)	Genome of cell.	Mostly DNA.

*Composite list; no single species expected to exhibit *all* of these structures.

bring about both rotary and translational movement is not known.

Flagella. Flagella (and cilia) are common in the plant, animal, and protist kingdoms. Among eucaryotic protists, two large classes of Protozoa (Mastigophora and Ciliata) are classified on the possession of flagella or cilia, respectively. Among the procaryons the blue-green algae are typically without flagella (**aflagellate**). On the other hand, many species of bacteria move actively by means of flagella. The flagellum has proved to be a very useful swimming appendage in both plant and animal kingdoms, in eucaryons and procaryons. When we

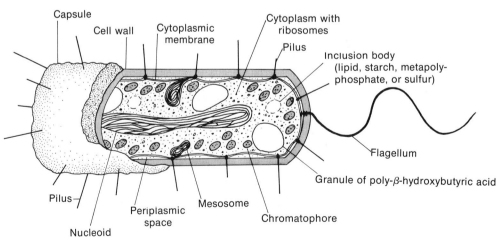

Figure 14-1. Composite drawing showing various bacterial structures, some of which are present only in certain groups of bacteria.

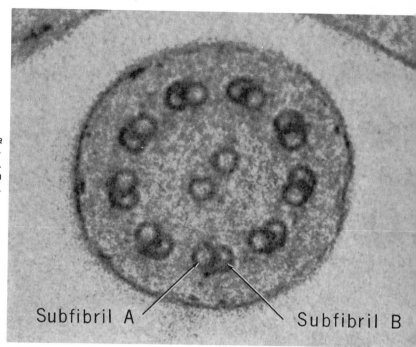

Figure 14–2. Cross section of flagellum of *Giardia muris* (300,000×) showing the symmetric arrangement of 10 pairs of longitudinal fibers (**microtubules**). The nine outer pairs differ from the central pair in being doublets composed of subfibrils (A and B). (Micrograph courtesy of Dr. Daniel Friend.)

examine the fine structure of flagella, however, we find one of the most marked differences between procaryons and eucaryons.

The fine structures of the flagella (and cilia) of all **eucaryotic** cells are remarkably alike. They are also astonishingly complex in view of their extremely small diameters of approximately 0.2 μm. Flagella may be as long as 200 μm.

The typical **eucaryotic flagellum** (Fig. 14–2) is enclosed in an outer membrane that is continuous with the cell membrane. The structures of the flagellum penetrate entirely through the cell wall and cell membrane, and the capsule when one is present. Inside the membrane of the flagellum is a symmetrical arrangement of longitudinal fibers (microtubules), one axial pair and nine peripheral pairs or doublets that are evenly spaced in a circle around the central pair. Just within the cell membrane all of the fibrils of a flagellum are connected together in a distinct basal granule, the **kinetoplast** (or **parabasal body**). In representative species the central pair of fibrils arise in the kinetoplast. The peripheral pairs seem to go beyond the kinetoplast and to be rooted deep inside the cell, apparently very near to, or in contact with, a centriole, or a polar cap in plant cells without centrioles. The whole structure—flagellum with kinetoplast and "roots"—acts as a type of autonomous, self-replicating plastid, coordinate with mitochondria and chloroplasts. Plastid replication is generally independent of nuclear fission.

The exact mechanism of motion of eucaryotic flagella is not yet entirely clarified, but it appears to result from rhythmical, coordinated contractions of the peripheral fibers. These consist of coiled elastic strands of filament protein (**tubulin**) that resembles the elastic protein, **myosin**, of muscles. Flagella also require ATP for activity and possess ATP-ase activity; hence they resemble muscles. It is of interest to note in passing that the basic structure of eucaryotic flagella is used in various adaptations wholly unconnected with motion: e.g., in the olfactory hairs in the noses of mammals (rabbits) and in the rods and cones in the retina of the human eye.

The **procaryotic flagellum** (Fig. 14–3; see also Fig. 14–8), unlike eucaryotic flagella, has no definite membrane and consists of a single (apparently hollow) very small filament made up of three or more parallel or intertwined longitudinal fibers of flagellin-type protein (Fig. 14–4). Each fibril is a polypeptide chain in α-helix form. At least some, and perhaps all, procaryotic flagella appear to be attached to a structure called the **hook**, which in turn is attached to the **basal body**. Figure 14–5 shows the basal body of a gram-negative bacterium and a drawing depicting the attachment of the rings of the basal body to the cell wall and membrane. Gram-positive bacteria lack a distinct lipopolysaccharide outer layer in their cell wall, and the basal bodies of their flagella, when present, are correspondingly simpler (only two instead of

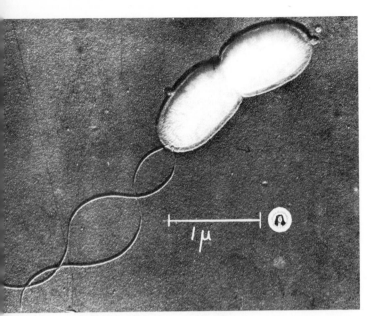

Figure 14–3. Dividing bacterial cell (*Pseudomonas fluorescens*) with two polar flagella. (Shadowed electron photomicrograph courtesy of A. L. Houwink and W. van Iterson; A.S.M. LS-275.)

four rings). Precisely how flagella propel procaryotic cells is still a mystery. Certainly there is an expenditure of energy, most likely by an enzyme triggering the release of energy from one of the high-energy compounds like ATP. However, all attempts to demonstrate ATP-ase activity in isolated bacterial flagella have failed. It may be that a contraction taking place at the basal-hook region is simply transmitted by the flagellum, the latter having no contractile ability. More recently, however, evidence has been presented to show that bacterial flagella rotate individually; each flagellum thus represents a biological rotary motor which causes the cell to move due to lateral displacement of fluid as the flagellum turns.

Removal of wall or capsule does not interfere with the flagella except that, when the wall is absent (see protoplasts) motility is diminished. Flagella may be broken off without injury to the cell and can then be regenerated. Oddly enough, regeneration takes place at the tip rather than at the base. Isolated filaments may be dissociated by several chemical and physical means into protein monomers of **flagellin** with a molecular weight ranging from 20,000 to 40,000, depending on the procedure used and species examined. Under suitable conditions, these protein monomers show the remarkable ability to repolymerize into helically intertwined filaments resembling the original flagellum. This is one of several bacterial structural components which have the ability to self-assemble.

Although all procaryotic flagellar proteins belong to the same general chemical class of flagellins, the amino acid composition of the flagellin of each species is sufficiently different from those of other species to confer **immunologic specificity** on the flagella of each bacterial species, subspecies, or type. This is of importance in methods of identifying various species, as in medical diagnosis. (See Flagellar [H] Antigens, Chapter 27.)

The speed of flagellar motility, average being perhaps 30 to 50 μm/sec (approximately 0.0001 mph), is either very slow or very fast depending on your point of view. Some bacteria can cover a distance equal to scores of times their length in a second. A 6-foot man would have to run at nearly 200 mph to obtain the same distance/size relationship as the small *Vibrio cholerae* moving at 200 μm/sec. Still other microorganisms move very slowly and sedately, e.g., the syphilis spirochete.

Demonstration of Motility. Although bacterial flagella are not ordinarily visible with light microscopes unless specially stained, motility caused by movement of cilia or flagella is easily seen by direct microscopic observation

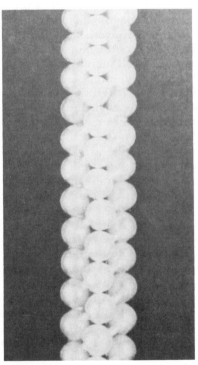

Figure 14–4. Model depicting a bacterial flagellum composed of three helically wound strands.

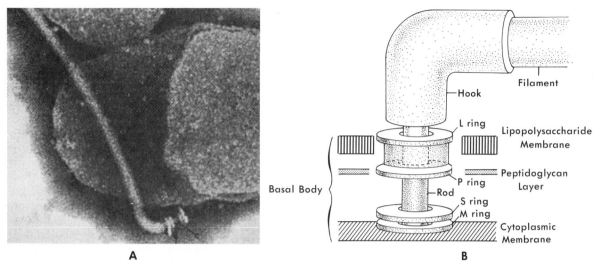

Figure 14–5. Attachment site on a bacterial flagellum. *A*, Electron photomicrograph of a flagellum from *Rhodospirillum molischianum*, showing the hook and basal body with four rings. (180,000×.)
B, Interpretive drawing, showing how the rings are attached to the cell wall and cytoplasmic membrane in gram-negative cells.

of motile organisms in a droplet (**hanging drop**) of the fluid in which they are living. In any culture of bacteria, especially old cultures, motile cells may be difficult to find among thousands of dead or senescent cells. Young cultures should always be used. In cultures acidified, as by fermentation, bacteria lose their motility; strict anaerobes lose motility on contact with air.

If a drop of broth culture of motile bacteria is placed on a slide under a cover slip (a wet mount) and observed with a microscope, strict aerobes will be seen to migrate to the periphery of the slip, near the air. Strict anaerobes migrate to the center, and microaerophils to an intermediate position. This phenomenon is called **aerotaxis** (Gr. *aer* = air; *taktikos* = ordered arrangement).

Another method of detecting motile bacterial cells is to thrust a wire, previously dipped into a culture of the bacteria to be tested, deep into a column of semisolid (0.5 per cent agar) medium. After 12 hours to six days of incubation motile organisms are seen to have migrated away from the line of initial stab for several millimeters, into the agar, forming a peripheral cloud of growth.

Brownian Movement. It is necessary to distinguish carefully between true motility and brownian movement. Truly motile bacteria progress definitely and continuously in a given direction. Brownian movement is a purposeless, undirected oscillation of any minute particles suspended in aqueous fluid, within a very limited area.

Among bacteria the power to move by means of flagella appears to have evolved mainly with elongation of form. Very few spherical bacteria have flagella.

Demonstration of Flagella. Bacterial flagella, though up to 70 μm long, are generally only 10 to 20 nm (0.02 μm) in diameter. Thus, they can be observed readily in the electron microscope but are below the resolution of light microscopes. In order to make bacterial flagella visible with the light microscope it is necessary to increase their apparent diameter by first coating them with a mordant (a substance like tannic acid, which will adhere to them and hold a dye) and then applying one of the special stains (e.g., Leifson's stain) developed for flagella.

Forms and Arrangements of Flagella. There are several forms of bacterial flagella: coiled, curly, normal, or wavy. Several forms may occur on one cell and even on one flagellum. The wavelengths and amplitudes of these coils (or curves or waves) appear to be fairly constant and to bear some significant relationship to each other and possibly to species. For example, the wavelength of a curly flagellum may be one-half the wavelength of a wavy or normal flagellum on the same cell. This may be distinctive of the species (Fig. 14–6).

Flagella may be arranged in various ways on bacterial cells (Fig. 14–7). The flagellation is said to be **monotrichous** if only one flagellum protrudes from one end, or pole, of the cell; **lophotrichous** if several or numerous flagella protrude from one pole; **amphitrichous** if at least

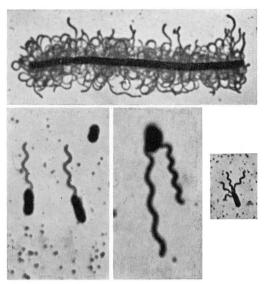

Figure 14–6. Bacterial flagella stained by Leifson's method. Note that in the bacillus with two flagella, the wavelength in one flagellum is twice that in the other. In two bacilli the wave forms are identical. Knowledge of the arrangements and wavelengths of flagella is an important factor in identification of bacterial species. (About 3,000×.) (Courtesy of Dr. Einar Leifson, Strich School of Medicine, Loyola University.)

one flagellum is at each end; and **peritrichous** if the flagella protrude from all portions of the bacterial surface. The type of flagellation has often been used as a major taxonomic feature.

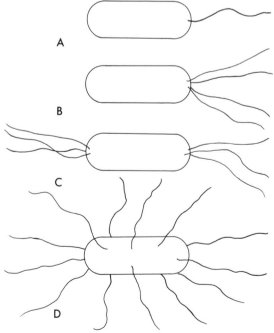

Figure 14–7. Types of flagellation: *A,* monotrichous; *B,* lophotrichous; *C,* amphitrichous; *D,* peritrichous.

14.2
PILI (FIMBRIAE)

Pili are extremely fine, filamentous appendages extending outward from the surfaces of bacteria. With a few possible exceptions, they have been observed only on gram-negative rods. Their size ranges from 3 to 25 nm in diameter and 0.5 to 20 μm in length, depending on the type of pilus (see Table 14–2). Thus they are visible only with electron microscopes (Fig. 14–8). Like flagella, pili appear to originate in basal bodies and to pierce cell wall and capsule. They are straight or nearly so, and apparently rigid and immobile. Like flagella and capsules, they can be removed mechanically without affecting growth or viability of the cell.

Chemical studies show that pili are composed entirely of a protein, **pilin,** made up of subunits or monomers (mol. wt., approximately 17,000 for Type I pili) arranged helically to form a single rigid filament with a hollow core. Since they are protein and vary in structure, they confer antigenic specificity on the cell in the same manner as flagella (see Flagellar [H] Antigens, Chapter 27). The presence or absence of pili is genetically controlled by structural and regulatory genes, and is therefore subject to mutation. Like R and S mutations (Chapt. 18), mutation rates for pili at ordinary temperatures are high—around 1:100. Thus piliate and apiliate cells of the same species are common. Environmental conditions such as pH, temperature, and oxygen tension strongly affect the presence or absence of pili.

The function of pili is not understood for the most part. It is known that bacteria with Type I pili have a strong tendency to adhere to each other and cause **agglutination** (clumping together) of themselves and various other microscopic particles, such as plant or animal cells, yeast cells, red blood cells, especially erythrocytes of sheep. In broth cultures the agglutinative tendency of piliate bacteria is manifested by the formation of thin films or pellicles of adherent cells near the surface (and closer to the oxygen supply).

More recently, Type I pili have been described as sex pili, and evidence has been presented to show that they promote the transfer of genetic material (resistance transfer factor; see Chapter 19), much like the F pili, and can also serve as a phage receptor site. With more study, similar functions may be found for the other so-called "common" pili.

The F pili are believed to play a role in conjugation, perhaps by acting as conjugal tubes through which DNA is transferred from donor to

TABLE 14-2. PROPERTIES OF BACTERIAL PILI°

Morpho-logical Type	Bacterial Strain where Originally Observed	Diameter (nm)	Typical Length (μm)	Typical Number on Cell	Distribution on Cell	Function
I	*E. Coli*	7	0.5–2	100–200	Uniform	Surface adhesion, sexual, conjugation, bacteriophage receptor
II	*Klebsiella pneumoniae*	4.8	0.5–2	100–200	Uniform	Unknown
III	*Proteus*	3–4	2–6	200–500	Uniform	Unknown
IV	*Proteus*	About 7 (helical)	1–2	100–200	Uniform	Unknown
V	*E. coli*	About 25	1–2	1–2	Uniform	Unknown
VI	*Pseudomonas*	About 5	2–10	1–5	Polar	Unknown
VII	*Pseudomonas*	10–20	0.2–0.5	10–20	Polar	Adhesiveness
F	*E. coli*	8.5	1–20	1–4	Uniform	Surface adhesion, sexual conjugation, bacteriophage receptor

°Adapted from C. C. Brinton, Jr., *in*: Davis, B. D. and Warren, L. (Eds.): The Specificity of Cell Surfaces. Prentice-Hall, Inc., Englewood Cliffs, N.J. 1967.

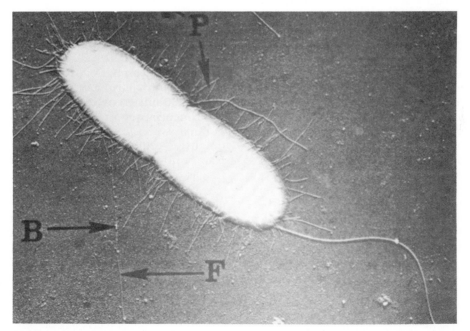

Figure 14-8. Electron micrograph of a dividing *Escherichia coli* showing a single polar flagellum and two types of pili. Type I pili (*P*) are numerous, while only a few of the longer F pili (*F*) are present. The latter can be identified easily by mixing a small RNA bacteriophage with the culture prior to examination. Here the phage (*B*) are seen as small dots along the F pilus, which serves as the receptor site for the phage. (Photomicrograph by Dr. D. Kay, Sir William Dunn School of Pathology, Oxford.)

recipient (see Chapter 19). The formation of sex pili, unlike the other pili, seems to be under the control of an episome. The F pili can be distinguished clearly from others on the same cell since they act as the receptor sites for both a small round RNA bacteriophage (Fig. 14–8; see also Figure 35–1) and a filamentous DNA phage.

14.3
CAPSULES

The composition of the outer surfaces of all cells, especially cells of protists, is of great importance, because it is here that the organism is in contact with all of the influences of the environment. The outermost structure of many procaryotic cells is the capsule: a slimy, gummy, or mucilaginous coating presumably synthesized at the cell membrane (cell walls have no known enzymatic activity) and extruded to the exterior surface through the meshes of the cell wall. Bacterial capsules vary in thickness from a fraction of a micrometer (see Microcapsules, this section) to 10 μm or more. Distinction between cell wall and capsule is not always clear. Sometimes wall and capsule appear to merge into each other. Generally, however, wall and capsule are distinct and separate structures, and the latter can be demonstrated by simple means (Fig. 14–9). In most species the outer boundaries of capsules are sharply defined; in others the outer limits are vague and hazy as though the outer portion were dissolving in the surrounding fluid.

Capsules are not essential to the life of a cell. They may be removed artificially (by enzyme action or washing) or naturally (by mutation) without affecting the **viability** (ability to live and grow) of the cells.

The presence or absence of capsules is genetically controlled. Encapsulated and nonencapsulated mutants of many species of bacteria occur frequently. As will be discussed in Chapter 18, encapsulated mutants generally produce smooth (S) colonies and are immunologically S-type cells. Absence of capsules is associated with rough (R) colonies and immunologically R-type cells. Encapsulated mutants are also associated with virulence of pathogenic species. The capsules protect these bacteria from the defensive mechanisms of infected animals, including ingestion by the ameba-like white blood cells of animals. It should be mentioned that not all bacteria giving rise to smooth (S) colonies are encapsulated. Many of the gram-negative enteric bacteria do not make a capsule

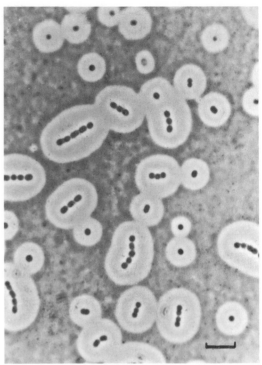

Figure 14–9. Photomicrograph of wet mount of an encapsulated bacterium in India ink. The scale line shows the size of 10 μm at this magnification.

but are S-type cells because a rather large proportion of their cell wall is lipopolysaccharide. Again, rough variants of pathogenic species tend to possess low virulence or be avirulent, while smooth strains are virulent (see Chapter 24).

Since capsular material consists of about 98 per cent water, capsules probably also serve as defensive buffers against too rapid influx of water into the cell and also against dehydration, i.e., as osmotic barriers.

Development, or enhancement of size of capsules or amount of capsule-like materials, is genetically controlled but is also subject to environmental modification (see Modifications, Chapter 18). In many species large, easily seen capsules are evoked by special conditions of nutrition. For example, some species of spherical bacteria (*Leuconostoc dextranicus,* Chapter 38) form large, gummy masses consisting of billions of the cocci surrounded by **glucan** (dextran) or **fructan** (levan) gums. *Leuconostoc* is a nuisance bacterium when it grows in sugar-refining vats or dairies, since in both these locations appropriate carbohydrate is available from which to synthesize their capsules.

Composition of Capsules. Capsules are

generally made up of polymers of various subunits, the compositions varying with species.

Because of their distinct and individual molecular structures, most capsular substances have the property of **immunological specificity** (see Haptens, Chapter 27). This permits experimental distinction between closely similar species of bacteria that could not otherwise be differentiated, an important point in identification of unknown species for medical diagnosis, industrial processes, and other purposes. Even with a given species, immunological **subspecies** or **types** may often be distinguished because of slight differences in the composition of their capsular substance. For example, there are about 75 distinct immunological types of the single species *Streptococcus (Diplococcus) pneumoniae* (a cause of pneumonia). The capsule of type III, for example, is largely a polymer of β-1,4-glucuronosidoglucose units; type VI capsule is composed of galactose, rhamnose, glucose, and ribitol-5-phosphate. The immunological mechanisms and methodology are described in Chapter 27.

In the blue-green algae, capsules or slimy matrices are gummy in nature and consist of **pectin** (a homopolymer of D-galacturonic acid) or of various pectin-like (i.e., gummy) polymers of glucose (glucans).

Among the simplest of the bacteria there are striking differences in capsular substance between species. For example, a common species, *Acetobacter xylinum* (Chapt. 49), familiar as a thick surface growth in vinegar ("mother of vinegar") produces a crudely tangled matwork of long fibers of cellulose (unique among bacteria) as an outermost secretion. Some might argue, however, that the secreted substance is not sufficiently organized to be called a capsule. Capsules of many other bacteria consist of polymers of glucose or other sugars; some also contain amino (nitrogen-containing) sugars and sugar acids. Some capsules consist of polypeptides. Capsular substances generally combine with water to form viscous, gelatinous slimes.

In the animal body viscous polymers of amino sugars are found in synovial (joint) fluids, in mucus, and as a "cement" (**hyaluronic acid**) between tissue cells. Curiously, among bacteria the dangerous streptococci (*Streptococcus pyogenes*) that cause "blood poisoning" and scarlet fever have capsules of hyaluronic acid. Some dysentery bacilli have very thin capsules (see Microcapsules, this chapter) made up of polysaccharide, phospholipid, and polypeptide; capsules of some *Bacillus* species (Chapter 36) are polypeptides of D-glutamic acid units.

Demonstration of Capsules. Most capsules do not have a marked affinity for the dyes commonly used to stain bacteria. While unstained capsules are often large enough to be dimly visible with light microscopes, they are commonly colorless and have low refractive indices. Special means are therefore generally used to demonstrate them. Methods differ with capsules of different compositions. Figure 14–9 shows a method employing India ink (negative staining) which has the advantage that the bacteria may be examined in a moist, living state. The capsule is not stained in negative staining but is visible because of the space it occupies around the bacterial cell.

Microcapsules. Some organisms on which a definite capsule is not microscopically demonstrable may nevertheless possess a similar or analogous, though very thin, layer of distinctive molecular structure on the cell surface. These very thin layers have been called **microcapsules, sheaths,** or **envelopes.** They determine the physicochemical and immunological reactivity of the cell to various surface-acting agents in much the same manner as a capsule. Like that of capsules, the physicochemical composition of microcapsules is distinctive for each species.

14.4
THE CELL WALL

Chemically and structurally distinct cell walls are typically lacking in the animal kingdom. Some of the Protozoa, like *Paramecium* and some flagellates, have a relatively thick, tough, protective outer integument that performs several functions of a cell wall, including that of maintaining the form of the protozoan. Other Protozoa (Radiolaria, Foraminifera) secrete protective exoskeletons of silica or calcium carbonate. Except for these and possibly a few other primitive or specialized and parasitic forms, animal cells are typically pliable when encountering another object.

In the plant kingdom, on the contrary, nearly all species are characterized by the synthesis of chemically and structurally distinct cell walls. Cell walls of cellulose in cork were first accurately drawn and described by Hooke in 1665 (Chapter 3). Cell walls may be flexible or rigid; all are relatively tough. They are not ruptured readily by osmotic pressure from within: i.e., they are osmotically protective. With a few rare exceptions, the cell walls of photosynthetic eucaryotic plants are made wholly or mainly of cellulose. The cell walls of eucaryotic fungi derive their rigidity and tensile strength from fibers of **chitin,** the glucosamine polymer of which the tough exoskeletons of Crustaceae are composed.

In typical bacteria and blue-green algae, (except *Mycoplasma,* which have no cell walls), the rigidity and strength of cell walls are due mainly to strong fibers composed of heteropolymers generally called **peptidoglycans** or **mucopeptides,** but also referred to as glycopeptide, muropeptide, glycosamino-peptide, mucocomplex, murein, etc. These fibers form a relatively coarse (at the molecular level), three-dimensional, tough meshwork rather than a solid structure. Such a corset-like structure offers no obstacle to the inward passage of water, food substances such as minerals, glucose, amino acids, and even larger organic molecules. All wastes of the cell can pass outward unobstructed.

The peptidoglycans consist of alternating units of N-acetylglucosamine and N-acetylmuramic acid with β-1,4 linkages (Fig. 14–10). This "backbone" structure appears to be the same regardless of the procaryon examined. Linked to the muramic acid is a short peptide which varies in composition but always contains a minimum of three amino acids: alanine, glutamic acid, and either diaminopimelic acid (DAP) or the structurally related amino acid lysine (Fig. 14–10). Note that the glutamic acid and one of the alanines of peptidoglycans are in D-isomeric form rather than the L configuration which is characteristic of amino acids found in proteins.

Structural rigidity of the peptidoglycan is achieved by cross-linking the polymers, as shown in Figure 14–11. The type and extent of cross linkage varies considerably among different species. In some bacteria, a short peptide, e.g., a pentaglycine, is used to link the tetrapeptide side chains extending from the muramic acid units, while in other species, the terminal D-alanine of one tetrapeptide may be covalently linked to the DAP of an adjacent tetrapeptide. The peptidoglycans of gram-positive bacteria are more extensively cross-linked than those found in gram-negative bacteria.

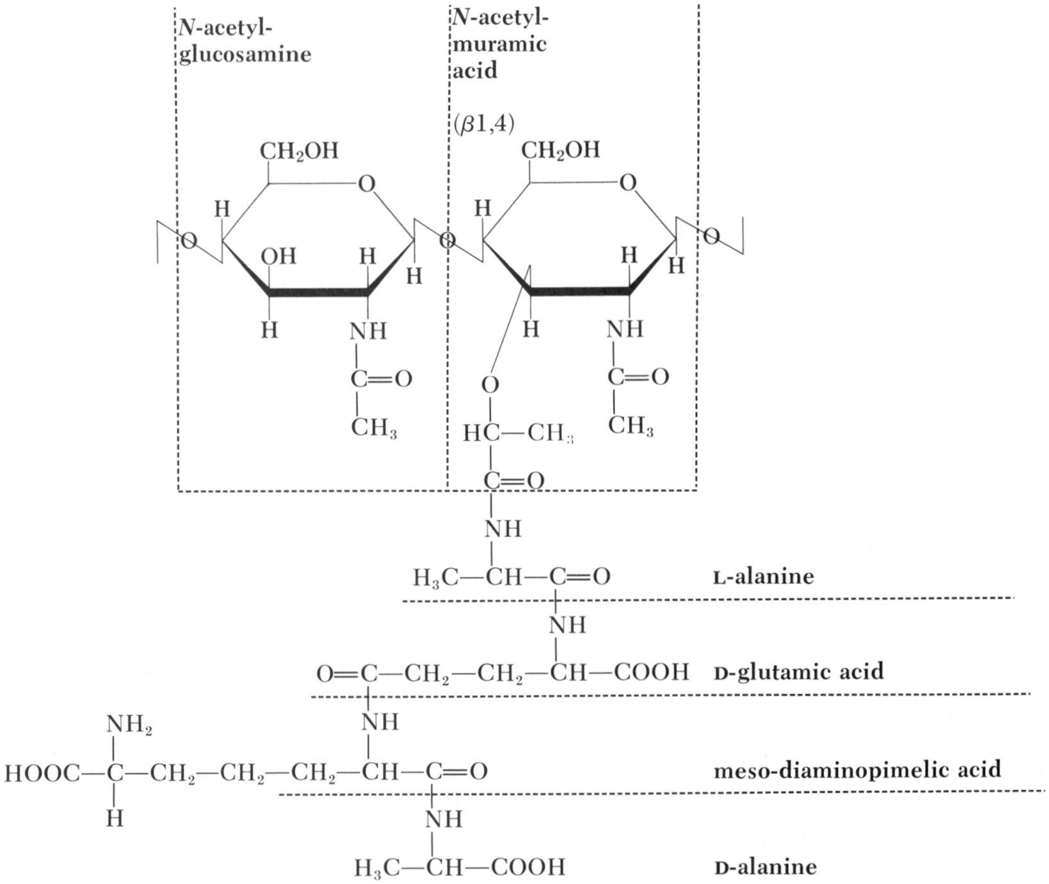

Figure 14–10. The structure of the basic repeating unit in the peptidoglycan (mucopeptide) component of the cell walls of procaryotic cells. The amino acids listed are those found in *Escherichia coli* and many other gram-negative bacteria. See text for further details.

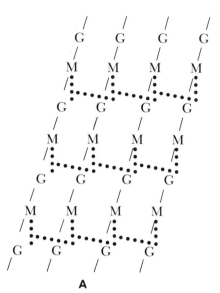

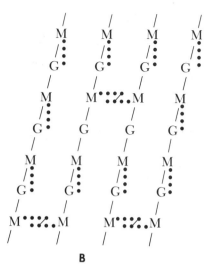

A

B

Figure 14–11. Schematic representation of the cross-linkage between repeating units of peptidoglycan. Alternating molecules of *N*-acetylglucosamine (*G*) and *N*-acetylmuramic acid (*M*), with a tetrapeptide attached to the latter (vertical rows of four dots) constitute the basic heteropolymer. The polymers may be extensively cross-linked to form a tight, fairly rigid network as in *A*, typical of gram-positive bacteria; or a loose network as shown in *B*, typical of gram-negative bacteria. Note that in example *A*, a pentapeptide (horizontal row of dots) cross-links two tetrapeptides, while in example *B*, the latter are linked directly to each other. See text for further details.

Muramic acid, DAP, and several D-amino acids are found *only** in association with the procaryons, a fact which indicates to many investigators a strong evolutionary relationship within this group.

The peptidoglycan layer, through found in most procaryons and responsible for the overall shape of the organism, does not constitute the cell wall by itself. In fact, the cell walls are structurally and chemically very complex. Electron micrographs of thin sections of typical gram-negative bacteria often show a multi-layered outer covering (Fig. 14–12). One interpretation of such micrographs is shown in Figure 14–13, where five layers of different composition are envisioned. They appear separated from the cytoplasmic membrane by an electron-transparent zone called the periplasmic space (Fig. 14–1; also see §14.5, The Cytoplasmic Membrane). In general, the cell walls of gram-negative bacteria are thin (10 to 15 nm thick and making up 10 to 20 per cent of the dry weight of the cell) and have the following composition: 5 to 15 per cent peptidoglycan, 35 per cent phospholipid, 15 per cent protein, and 50 per cent lipopolysaccharide. Gram-positive cell walls tend to be thicker (25 to 30 nm and 20 to 40 per cent of total dry weight) and contain 20 to 80 per cent peptidoglycan plus various other substances, e.g., proteins, polysaccharides, and teichoic acids. Teichoic acids are polymers made up of either ribitol or glycerol phosphate and amino acids, glucose, and *N*-acetylglucosamine. The gram-positive cell walls appear for the most part to be more homogeneous, in that

*There may be some rare exceptions elsewhere in the living cosmos.

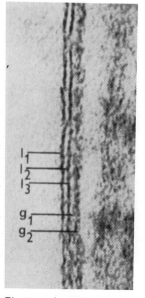

Figure 14–12. Electron photomicrograph of a thin section of the cell wall of *Escherichia coli*, showing the various layers. (98,000×.) See Figure 14–13 and text for further details.

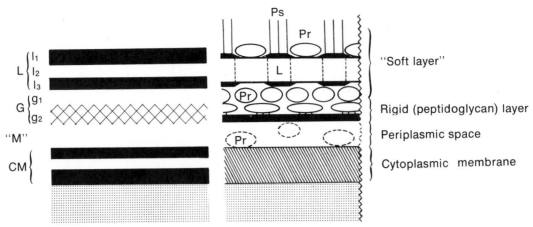

Figure 14–13. Interpretive drawing of layers seen in gram-negative cell walls. l_1 is the outermost layer composed of protein (*Pr*) and polysaccharide (*Ps*) with "whisker-like" projections extending out from the core structure; l_2 is chiefly lipid; l_3 is possibly lipopolysaccharide; g_1 may be protein; g_2 represents the rigid peptidoglycan layer. As discussed elsewhere, the periplasmic space contains proteins such as degradative enzymes.

distinct layers are not readily apparent and the peptidoglycan is distributed throughout the cell wall more uniformly in a three-dimensional network.

Protoplasts and Spheroplasts. Some microorganisms, both eucaryotic and procaryotic (e.g., yeasts, molds, bacteria, and some higher plants), that normally have a functionally and chemically distinct cell wall can at times exist without a cell wall, in what might be called a wall-less state. The cell membrane and its intact contents are then called a **protoplast** (Fig. 14–14). Unless the solute concentration of the suspending fluid is **osmotically protective,** i.e., high enough (e.g., 3 to 20 per cent glucose; 2 to 5 per cent sodium chloride; 10 to 20 per cent serum)

to balance the intracellular osmotic pressure, protoplasts, no longer retained by their thick, strong cell walls, usually burst. Protoplasts are therefore said to be **osmotically fragile.**

Protoplasts and Penicillin. An important property of bacterial cell walls is sensitivity to the antibiotic penicillin. Penicillin acts by inhibiting the cross-linking of polymers of peptidoglycan of cell walls. This synthesis occurs only in actively growing cells. Young, actively growing, gram-positive bacteria are, therefore, typically susceptible to penicillin. Without their strong, retaining cell walls they are protoplasts and therefore osmotically fragile. They promptly burst unless the surrounding fluid is osmotically protective. Most gram-negative bacteria, on the contrary, are usually partly or wholly resistant to the action of penicillin because, with little peptidoglycan in their cell walls, they are not wholly dependent on the peptidoglycan for their cellular integrity. Other components of their cell walls prevent rupture. Penicillin susceptibility is thus largely a function of peptidoglycan content and not of staining properties, although gram positiveness and penicillin sensitivity are commonly parallel. Animal cells, including human cells, are unsusceptible to penicillin. (Why?)

The lipopolysaccharides and lipoproteins of gram-negative bacteria, besides making them less susceptible to the effects of penicillin, also protect them to a great extent from the action of a group of enzymes typified by **lysozyme,** an enzyme discovered by Fleming (who also discovered penicillin). Lysozyme is found in egg white, secretions of skin and mucous mem-

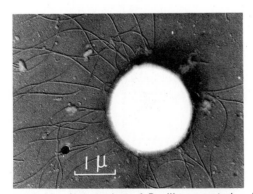

Figure 14–14. A protoplast of *Bacillus megaterium* in a medium made osmotically protective with 15 per cent sucrose. Ordinarily rigid and distinctly rod-shaped, it is now a limp spherical sac devoid of supporting cell wall. Note that the flagella are retained, being rooted within the protoplast membrane.

branes, tears, and elsewhere. It has the property of attacking specifically the glycosidic bonds in the polysaccharide backbone of the peptidoglycan of bacterial cell walls. When lysozyme is applied to many species of gram-positive bacteria, the organisms are rapidly denuded of their cell walls and become naked protoplasts. In contrast, although the cell walls of gram-negative bacteria contain some peptidoglycan, these fibers are protected or supplemented to a great extent by the outer layers of lipocomplexes. To make gram-negative bacteria vulnerable to lysozyme it is necessary first to remove the cell wall lipids with lipid solvents such as NaOH or ethylenediaminetetraacetate (EDTA). Even then the wall is not completely removed and the result is osmotically fragile cells still retaining some remnants of cell wall. Such cells are called **spheroplasts.**

In osmotically protective media, protoplasts are able to carry on all of the essential functions of the intact cell, including growth and fission. However, in the absence of the rigid wall, protoplasts often assume bizarre ring forms or ameboid forms with pseudopodium-like (though nonmotile) protuberances, or long, branching filaments containing many granular structures, like strings of beads. Protoplasts may undergo subdivisions by a process superficially resembling budding. Some of the bud-like extensions may be extremely minute, yet capable of multiplying and reverting to the forms of the cell from which they originated. Some of these forms are the smallest cells capable of independent growth. (Most viruses are smaller, but are neither cells nor capable of independent growth.)

Some bacterium-like organisms (*Mycoplasma;* see Chapter 42) exist normally in a wall-less state resembling protoplasts. Also, many bacteria and some other microorganisms may assume the protoplast form spontaneously as by mutation, or they may be made to assume the protoplast form experimentally. Bacteria temporarily in the protoplast form are referred to as **L-forms.** The relationships are complex. Protoplasts, L-forms, and *Mycoplasma* are discussed further in Chapter 42.

The physical nature of the protoplast membrane and the protective role of the cell wall are seen when protoplasts are derived from long, cylindrical bacilli or from branching mold filaments. These protoplasts promptly assume a spherical but easily distorted shape. It is evident that the cell wall is a semi-rigid, retaining structure, whereas the protoplast membrane is fragile and easily ruptured.

14.5
THE CYTOPLASMIC MEMBRANE

The cytoplasmic membrane of procaryotic cells is a distinct and separable structure. By gentle enzymic removal of the bacterial cell wall, followed by osmotic rupture of the resulting protoplast in a hypotonic solution, the contents of the cytoplasmic membrane can be removed and the resulting sac, sometimes called a "ghost," cleaned by gentle washing and centrifugation. Cell membranes can also be readily seen in cross sections of intact bacterial cells properly prepared and stained for observation under the electron microscope.

Studies of such materials show that bacterial cell membranes typically consist of three distinct layers (Fig. 14–15). The two outer layers

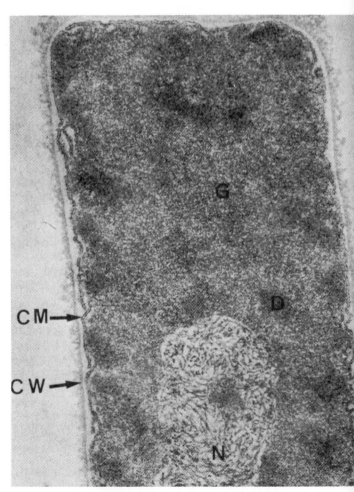

Figure 14–15. Electron photomicrograph showing the triple-layered cell membrane (*CM*) proximal to the cell wall (*CW*) in *Bacillus cereus.* Also clearly evident is a portion of the nucleoid (*N*). (87,500×.)

are electron-dense (i.e., opaque to electrons) and are each about 2.5 nm thick. The middle layer, about 5 nm thick, is much less electron-dense and appears to consist of a bimolecular layer of lipid oriented so that hydrophilic poles are outermost, hydrophobic groups opposed within. The chemical composition of bacterial membranes (approximately 60 per cent protein, 30 per cent lipid, 10 per cent carbohydrate) is similar to that of mammalian membranes but contains no sterols. There is no peptidoglycan present in the bacterial cell membrane, since the membrane is not dissolved by lysozyme. Many of the proteins found in the membrane are enzymes or have enzyme function.

The cell membrane is of extreme importance to the cell and has three major functions. The first is to maintain a favorable intracellular osmotic pressure for the cell. Thus, it behaves as an **osmotic barrier** and is impermeable to ionized substances and to nonionized substances with molecules larger than those of glycerol. Then, you may ask, how do extracellular enzymes, cell wall precursors, etc., exit through the cell membrane, and how do essential salts, amino acids, and sugars enter? This involves a second function of the cell membrane, the housing of a number of highly specific **active transport systems,** frequently called **permeases,** albeit an oversimplification. Each transport system is specific for a given compound or for a group of structurally related compounds, e.g., a cell may transport fructose, but not maltose.

Active transport systems require an energy input to function and are capable of establishing a concentration gradient, i.e., the concentration of a substance within the cell will exceed the concentration of the substance in the environment. Thus, transport systems do not involve merely selective diffusion. Under some conditions where energy input is not available, some formerly active transport systems can continue to shuttle their specific "substrates" across the cell membrane, but at a reduced rate. Such transport has been called "passive" transport or "facilitated diffusion." Interestingly, a number of passive transport systems have been described for various types of eucaryotic cells, e.g., yeasts and red blood cells, but such cells typically inhabit a highly nutritious environment.

Just outside the cytoplasmic membrane is the periplasmic space (see Figures 14–1, 14–12, and 14–13) where, in some bacteria, degradative enzymes are present. Thus, large molecules passing through the cell wall can be broken down at this site into simple sugars, amino acids, etc., and transported across the cell membrane by transport systems (permeases).

A third function of the cell membrane is to provide a site for many of the key enzymic reactions involved in energy metabolism. Examinations of isolated bacterial cell membranes have revealed the presence of short-stalked, small particles adhering to the membrane. These particles closely resemble the particles found within mitochondria of eucaryotic cells and shown to possess ATP-ase activity.

14.6
MESOSOMES

Microbiologists are currently very interested in the curious, intracellular, membranous structures called **mesosomes** (Gr. *mesos* = middle; *soma* = body). Although mesosomes have been observed in some gram-negative bacteria, they are most frequently seen in gram-positive cells. There appear to be many variations in form, but the vesicular type and the whorl type are seen most frequently (Fig. 14–16). Mesosomes appear to be highly convoluted invaginations of the cell membrane. Electron micrographs show the cell membrane to be continuous with at least a portion of the mesosomes, and they disappear if the cells are treated so as to form protoplasts, i.e., if the cell membrane is allowed to stretch out. Chemically, the mesosomes are identical with the cell membrane.

The function of mesosomes is the subject of active research and of considerable debate. There is evidence that mesosome formation may precede and coordinate septum or cross-wall formation prior to cell division and to forespore formation (Chapt. 36). In addition, mesosomes may adhere to nuclear material and act as a primitive mitotic apparatus to insure that each daughter cell receives the appropriate share of DNA (Fig. 14–17). Speculation abounds concerning the evolutionary significance of mesosomes, and in this light it is interesting to note that a form of cell division "intermediate" between typical procaryons and eucaryons has been reported. In certain dinoflagellates, separation of chromosomes appears to be accomplished through the attachment and movement of the nuclear membrane rather than the conventional spindle apparatus.

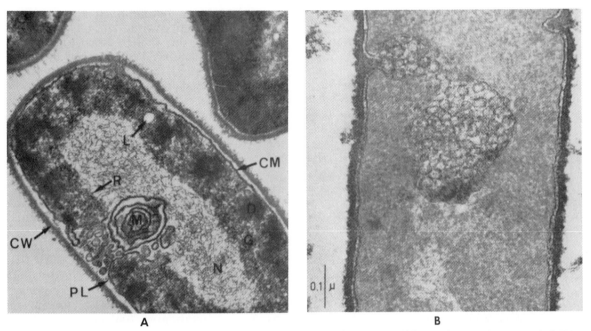

Figure 14–16. Electron photomicrographs showing the two forms of mesosomes (*M*) most frequently observed. *A,* Whorl type, as seen in *Bacillus cereus.* (51,000×.) *B,* Vesicular type, as seen in *B. subtilis.* (114,000×.)

14.7

CYTOPLASM

The term **cytoplasm** (Gr. *kytos* = cell; *plasma* = a substance), like the term protoplasm, is an inclusive one; i.e., cytoplasm is not a homogeneous substance. Except for the nuclear material, the cytoplasm comprises all the different substances and structures inside the cell membrane: a variety of **microsomes** (Gr. *mikro* = microscopic; *soma* = body) or subcellular particles that are mainly proteins and nucleoproteins, with some lipoproteins and other materials. All the particulate matter in the cytoplasm is surrounded by, and suspended in, an aqueous fluid or semifluid ground substance or matrix. The matrix is a complex mixture containing in solution a variety of ions (H^+, PO_4^{3-}, Na^+, Cl^-), amino acids, some proteins, lipocomplexes, peptides, purines, pyrimidines, glucose, ribose, vitamins, nucleotides, coenzymes, disaccharides, and so on. Functionally these are: (a) precursors ("prefabricated" molecules) and other building materials to be used in cell synthesis; (b) sources of energy (e.g., glucose or other oxidizable material); (c) waste products of the cell to be excreted to the exterior. The matrix may also contain tRNA and complete, active enzymes in solution. Also present within the cytoplasmic substance are granules or globules of inert stored food substances, the composition of which depends on the species of cell and the nutritive conditions surrounding it: starch and similar polysaccharides, lipids, poly-β-hydroxybutyric acid, sulfur.

Figure 14–17. Schematic representation of how mesosomes might be involved in the separation of the nucleoid material into equal shares for two daughter cells. Since mesosomes do not appear to be present in all dividing cells, it must be conceded that mesosomes are not indispensable for cell division.

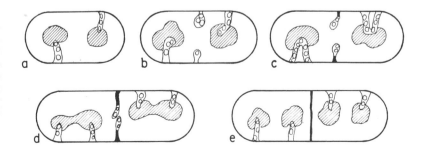

Cytoplasmic Ultrastructures

Ribosomes. The minutely granular appearance of the major portion of the cytoplasm in all cells, eucaryotic and procaryotic, seen in electron micrographs is due to the presence of enormous numbers of minute, diffusely scattered particles. They are called **ribosomes** (Fig. 14–18; also see Chapter 6). Depending on their origin (mitochondria, eucaryotic or procaryotic cytoplasm), ribosomes differ with respect to size and density. Hence, they are found in different layers (density gradients) in differential centrifugation, and are designated according to their sedimentation coefficients (S*) as 70S or 80S.

Each ribosome consists of a small (about 30S) subunit and a larger (about 50S) subunit. Ribosomes tend to form aggregates of varying sizes. These are called **polyribosomes** or **polysomes.** Some polysomes appear to consist of chains of ribosomes strung along a thread of connecting substance which may be messenger RNA. Indeed, ribosomes consist largely of ribosomal RNA (rRNA) with some protein (ribonucleoprotein). At least part of the RNA of ribosomes is messenger RNA (mRNA). Ribosomes are thus responsible for the synthesis of **specific** proteins, including the proteins of all enzymes. (See Chapter 6.)

---

*S = ratio of sedimentation velocity to gravitational force (G) under given conditions of viscosity.

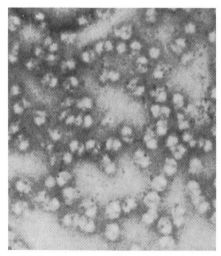

Figure 14–18. Electron photomicrograph of isolated ribosomes from *Escherichia coli.* Certain of the particles clearly demonstrate that each 70S ribosome is composed of a larger (50S) and smaller (30S) portion. (200,000×.)

Intracytoplasmic Membranes. In eucaryons, ribosomes are concentrated at, and probably connected to, the inner surfaces of the cell membrane and along the surfaces of a very complex and all-pervasive membranous structure called the **endoplasmic reticulum.** This is formed by extensive, branching invaginations of the cytoplasmic membrane. The dendritic ramifications of these membranous invaginations in eucaryons form tubules and channels everywhere deep in the cytoplasm: a sort of molecular vascular system. The endoplasmic reticulum brings both interior and exterior surfaces of the cell membrane close to all of the innermost reaches of the cytoplasm—an obvious advantage since the life of the entire cell is dependent on its communications with the outer world via the cell membrane.

Procaryons display no conspicuous endoplasmic reticulum. Their ribosomes appear to have no obvious arrangement or organization on, or in, membranes, beyond a tendency to accumulate close to the cell membrane. However, ribosomes in some procaryons appear to be clustered together in polysomes like those described for eucaryotic cells. Attachment of ribosomes to some sort of net-like (not membranous) structure has also been described in some species of bacteria.

Mesosomes (previously described) may, among other things, constitute a procaryotic, surface-extending organelle in lieu of the more extensive eucaryotic endoplasmic reticulum. As pointed out earlier, species of procaryons appear to vary considerably in the extent and form of these intracytoplasmic invaginations of the cell membrane. These variations may represent evolutionary stages between very primitive and minute cells with no inward extensions of the cell membrane, to the endoplasmic reticulum and self-replicating **mitochondria** of larger eucaryotic cells familiar to students of elementary biology.

Inert Cytoplasmic Inclusions. Many species of microorganisms, procaryotic and eucaryotic, store up reserve food substances in intracellular granules or globules. When appropriate extracellular food is abundant such granules may be very conspicuous and occupy a considerable portion of the intracellular space, only to diminish or disappear when food is scarce. The chemical nature of most of these inclusions is well known because of their distinctive staining or other properties.

Lipids. Many bacteria, blue-green algae, and eucaryons like yeasts, higher fungi, and human tissue cells accumulate intracellular globules of fat. These are readily recognized because they are (a) easily extractable with organic solvents; (b) readily stained by fat-soluble lipid stains: blue by naphthol blue, black by Sudan black; (c) microscopically structureless.

In bacteria, many of the inclusions formerly regarded as fat are actually a highly polymerized form of a fatty acid, β-hydroxybutyric acid (BHA). Butyric acid is one of the constituent molecules of many fats, a familiar one being butter. The β-hydroxybutyric acid units of poly-β-hydroxybutyric acid (PHB) are joined by ester bonds linking the basic hydroxyl group of one molecule of butyric acid with the acidic carboxyl group of another, with elimination of water.

Poly-β-hydroxybutyric acid granules or globules (Fig. 14–19) can be stained by fat-soluble dyes in the same way as fats. When needed by the cell, the PHB is hydrolyzed to soluble, oxidizable BHA or its dimers, depending on species.

Polysaccharides. Many species of cells, both eucaryotic and procaryotic, synthesize and store up excess soluble carbohydrate food substances in the form of insoluble polysaccharides. These are commonly polymers of glucose, though polymers of other monosaccharides or oligosaccharides are not uncommon. In green plants these food reserves usually appear as **starch,** and in animal cells, as **glycogen** (so-called **animal starch**). When cells containing them are treated with iodine, starch granules become conspicuously blue, while glycogen granules may be brown or purple.

Volutin. This substance was originally described in a bacterium called *Spirillum volutans* and was named for that species. It has since been observed in a wide variety of bacteria and blue-green algae and in some eucaryotic species. Volutin is especially rich in organic phosphates. It consists largely of polymerized metaphosphates (**polymetaphosphates**) in insoluble form associated with nucleic acids and lipids in various linkages and proportions. These differ in various species (Fig. 14–20).

Volutin has a marked affinity for basic dyes and is also **metachromatic;** i.e., it often appears to be of a color different from that of the applied dye. For example, stained with methylene or toluidine blue, volutin granules often appear ruby red, probably due to red impurities in the

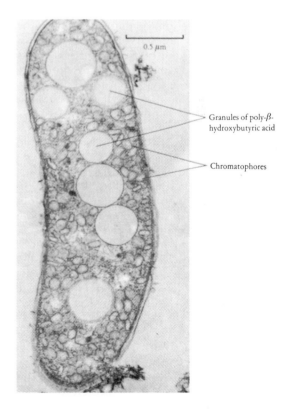

Granules of poly-β-hydroxybutyric acid

Chromatophores

Figure 14–19. Thin section of *Rhodospirillum rubrum* showing the size and location of chromatophores and granules of poly-β-hydroxybutyric acid.

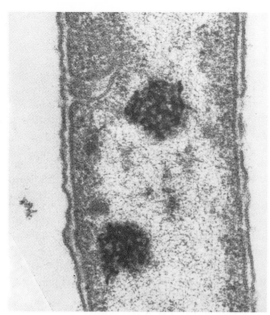

Figure 14–20. Polymetaphosphate granules (volutin) in *Myxococcus xanthus.* (110,000×.) (Electron micrograph by H. G. Voelz.)

dye. Volutin therefore often appears in large, conspicuous granules called **metachromatic granules,** and is usually abundant under conditions of good nutrition and slowed metabolism, especially in media containing glycerol or carbohydrates. It appears to accumulate when phosphates are not being used by the cell in its energy metabolism. Because of their marked affinity for basic dyes, volutin granules were formerly often confused with bacterial nuclei, which also are basophilic.

Other Stored Substances. In some species of aquatic and soil-inhabiting bacteria, globules of pure, elemental sulfur accumulate in the cell as unused food. This sulfur is often derived from the intracellular dehydrogenation (oxidation) of hydrogen sulfide or other inorganic reduced forms of sulfur. The elemental sulfur in the globules can be extracted from the cells with carbon disulfide from which the sulfur is deposited as distinctive crystals.

Food Storage. The accumulation of food substances in the cell as polymers such as PHB, starch, glycogen and volutin has at least three advantages for the cell: (a) polymers are a sort of molecular *multum in parvo* in that many molecules, with accompanying energy, are condensed into a small space; (b) the substance and energy in the insoluble polymers are quickly made available for use by the cell by the simple process of hydrolysis into the soluble units; (c) being insoluble, the polymers do not affect the intracellular osmotic pressure and other colligative properties* of the cell contents as would an equivalent amount of the unpolymerized materials in soluble form.

14.8
PHOTOSYNTHETIC ORGANELLES

Eucaryotic cells are more "advanced" than procaryotic cells in respect to photosynthetic mechanisms.

Chloroplasts. These organelles, familiar to all students of elementary biology, contain the green chlorophyll pigments that are the basis of the utilization of radiant energy, e.g., ultraviolet or sunlight, for photosynthesis by all eucaryotic green plants. Like mitochondria, chloroplasts are self-duplicating.

The intimate structures of chloroplasts are complex. Each chloroplast is discrete and is enclosed within a unit-type membrane. Within the membrane are elaborately and systematically ordered lamellar cross-membranes or layered disks. These structures are so arranged as to provide maximum surfaces, with many reactive molecular points, at which the biochemical activities of the chloroplast can occur to the fullest extent. The layered disks and membranes also expose a maximum surface to the light.

Chromatophores and Thylakoids. In the procaryons, chloroplasts like those of eucaryotic plants are lacking. Nevertheless, all of the blue-green algae and some alga-like species of bacteria characteristically exhibit active photosynthesis. The photosynthetic functions of procaryotic cells are centered in pigment-bearing structures called, in bacteria, **chromatophores** (Fig. 14–19) and, in blue-green algae, photosynthetic lamellae or thylakoids. These bodies,

---

*Colligative properties of a fluid are those that are affected only by the *number* of particles in suspension or solution, not their chemical nature. For example, 1 gm of sodium chloride dissolved in 500 ml of water will yield billions of tiny ions in the solution, and is easily shown to have a marked effect on all of the colligative properties. Osmotic pressure, vapor pressure, and boiling points are raised; freezing points are lowered. One gram of sugar or protein or bacteria, all of which introduce relatively few, comparatively large, molecules (or objects) into the solution or suspension, have lesser effects in the order named. One gram of birdshot would have no measurable effect.

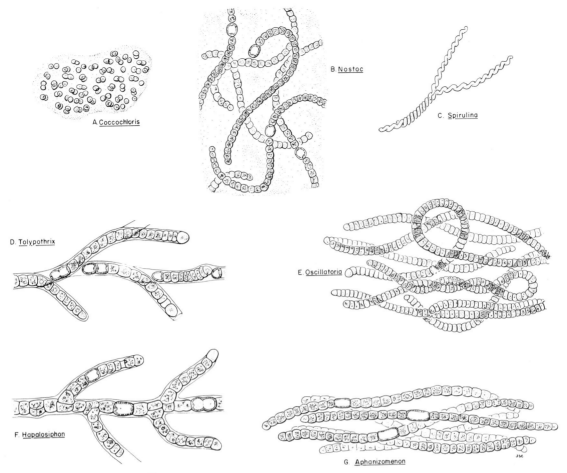

Figure 14–21. Some common species of blue-green algae.

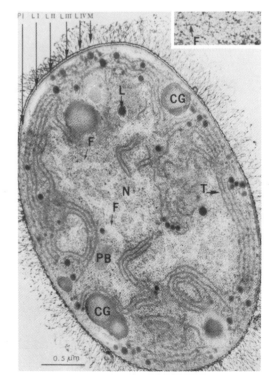

Figure 14–22. Electron micrograph of a thin section of a blue-green alga (*Anabaena variabilis*). The cell wall appears to consist of four layers (*LI* to *LIV*) and is coated with mucilage (*M*). Just beneath the cell wall is the cytoplasmic membrane or plasmalemma (*PL*). Note how the photosynthetic lamellae or thylakoids (*T*) are distributed around the periphery of the cell. Other structures which are apparent include the fibrils (*F*) in the nuclear area (*N*), pigmented (cyanophycin) granules (*CG*), lipid inclusion (*L*) and polyhedral body (*PB*) of unknown composition.

209

TABLE 14–3. DISTINCTIVE PROPERTIES OF EUCARYOTIC AND PROCARYOTIC PROTISTS AND VIRUSES*

Properties	Eucaryons	Procaryons	Viruses
NUCLEUS			
Membrane	Distinct; unit type	None	None
Nucleolus	Distinct	None	None
Nucleic acids	DNA and RNA	DNA and RNA	RNA or DNA; not both
Mitosis	Distinct	None	None
Chromosomes	Typically more than one; rods in mitosis	One circular, never rod-shaped; plus episomes and plasmids in some	One replicon
CYTOPLASM			
Centrioles	Distinct†	None	None
Streaming	Distinct (not all species)	None seen	None
Ameboid movement	Distinct in *Amoeba*	None seen	None
Ribosomes	Polysome groupings; membrane-oriented	Polysome groupings; membrane-oriented	None
Endoplasmic reticulum	Extensive throughout cytoplasm	Suggestive membranous structures in some	None
Mesosomes	Not evident as such; (pinocytic invaginations?)	Sac-like invaginations of cell membrane in numerous species	None
Autonomous metabolic enzyme systems	Complex; active exergonic and synthetic	Complex; active exergonic and synthetic	None
Mitochondria	Numerous; distinct	None	None
Chloroplasts	Distinct in green plants	None	None
Thylakoids and chromatophores	None	In blue-green algae and some bacteria	None
Chlorophyll *a*	Present in chloroplasts	In blue-green algae only	Absent
Lysosomes	Numerous; distinct	None	None
Microtubules, microfilaments	Distinct	Absent	None
FLAGELLA‡	Sheath containing 10 pairs of fibrils	Single naked filament of flagellin	None
CELL MEMBRANE	"Unit type" (protein-bimolecular lipid leaflet-protein [trilaminate] structure)	Unit membrane	None

*In all groups there are some species for which data are lacking or incomplete or which are exceptions or variations from the "typical," the "usual," or "selected examples."
†Animal cells only; in many plants replaced by polar caps.
‡When present.
§Morphologically and chemically differentiated.
‖Established only in some species of bacteria; data on blue-green algae, rickettsias, and chlamydias incomplete or absent.

Table 14–3 continued on opposite page

TABLE 14-3. DISTINCTIVE PROPERTIES OF EUCARYOTIC AND PROCARYOTIC PROTISTS AND VIRUSES*
(Continued)

Properties	Eucaryons	Procaryons	Viruses
CELL WALL (principal strengthening material)	Fungi: chitin§ Green plants: cellulose§ Animals: none morphologically or chemically differentiated; plasmalemma	Peptidoglycan; morphologically and chemically differentiated; absent in *Mycoplasma*	Protein coating (capsid); envelope of host origin containing lipopolysaccharides in some animal viruses
CELL VOLUME (μm^3)	Generally > 20	Generally < 5.0	< 0.02
SEXUALITY			
Gametes	Distinct (not all species)	Not morphologically differentiated; obscure	None
Recombination	Generally only by conjugation in which entire genomes are combined; true, diploid zygotes are formed	Transformation; transduction; conjugation in which variable segments of (rarely complete) genomes are transferred, sometimes with episomes; episomes may be transferred independently of genome; merozygotes are formed ‖	Can occur in vegetative state; mingling of NA strands may occur in multiple infection of cells
MULTIPLICATION			
In inanimate media	Common	Common, with some notable exceptions	None
Basic mechanism	Self-synthesis of entire cell followed by conjugation, fission, budding, etc.	As in eucaryons except conjugation not common; see recombination above.	Synthesized largely by host enzyme systems, host-derived materials; no processes of fission, budding or independent growth.

* In all groups there are some species for which data are lacking or incomplete or which are exceptions or variations from the "typical," the "usual," or "selected examples."
† Animal cells only; in many plants replaced by polar caps.
‡ When present.
§ Morphologically and chemically differentiated.
‖ Established only in some species of bacteria; data on blue-green algae, rickettsias, and chlamydias incomplete or absent.

while membranous, are not membrane-bounded, self-duplicating organelles like chloroplasts. The thylakoids of blue-green algae have a finely layered and complex laminated structure (see Section 14.10; also Fig. 14–22), but they typically lack the elaborate internal, membranous structure of the chloroplasts of eucaryotic plants. The chlorophyll-bearing structures of photosynthetic bacteria are of various morphologies. They are more fully discussed in Chapter 32.

14.9
THE NUCLEUS

As mentioned elsewhere the procaryotic nucleoid or nuclear area differs most obviously from the eucaryotic nucleus in that the latter is enclosed within a definite nuclear membrane and exhibits mitotic cycles and, in forming gametes, meiosis. The procaryotic nucleoid is not membrane-enclosed and exhibits no mitotic or meiotic phenomena. There are other, less obvious, differences. Structurally, the procaryotic nucleoid, as seen in ultrathin cross sections of blue-green algae, bacteria, rickettsias, chlamydias, and *Mycoplasma,* appears as an amorphous, lobular mass of fibrillar, intensely chromatinic material. It occupies from one-half to two-thirds of the intracellular space. Histones, the basic protein associated with eucaryotic chromosomes, are absent. Chromosomes, such as those seen during prophase and metaphase in mitosis of eucaryotic cells, are never seen. The procaryotic nuclear material (Figs. 14–15, 14–16) appears to remain in a state resembling the interphase of the eucaryotic chromosomal matter.

The fibrils seen in bacterial nucleoids with the electron microscope appear to represent a long (about 1,400 μm), thin (about 3 nm), flexible, circular (i.e., no free ends) filament of DNA. The arrangement of this structure in the cell may be visualized as being like a 6- to 10-foot-long filament of thin, two-stranded, twisted cotton thread, ends fastened together, the whole collected into an irregular, tightly packed bundle in the hand. Sometimes in actively multiplying cells several such strands, identical replicas of each other (see Chapter 19), are present in one cell because chromosome replications precede cell division. The circular filament of DNA just described is generally spoken of as the **bacterial chromosome.**

In bacteria (and in other procaryons?) extrachromosomal DNA, in the form of small rings, may replicate autonomously (not in synchrony with the chromosome) and may also act as genetic determinants. These structures are discussed in Chapter 19 under **episomes** and **plasmids.**

14.10
THE PROCARYOTIC ALGAE

Of all the procaryons, the blue-green algae are the largest and most plant-like. About 2,000 species are known, some of which live, like bacteria, as individual cells of microscopic size. The blue-green algae, like bacteria and many fungi, inhabit a wide range of environments from pole to pole, in arctic and antarctic seas and in thermal springs as well as in soils and fresh waters. Most are harmless and some, in mass growth, give bright colors to seas and lakes; many provide food for plankton and, collected in mass, fertilizer for land crops. Their food requirements are of the simplest and they are often found as the first forms of life on cooled outflows of new lava.

Most blue-green algae resemble bacteria in several respects but are much larger, although still microscopic. With a few exceptions these organisms contain three pigments: the green pigment, photosynthetic **chlorophyll a;** a blue pigment, **phycocyanin,** formed in no other plants; and other pigments, e.g., **phycoerythrin, carotene,** and **xanthophyll.** The blue-green algae are the smallest and most bacteria-like of the sunlight-dependent plants. Examples are *Oscillatoria, Nostoc, Coccochloris,* and *Spirulina* (Fig. 14–21).

Some eucaryotic algae possess flagella and move actively like protozoa of the class Mastagophora. The blue-green algae, however, are nonflagellate and not actively motile, although some species, like certain bacteria, can bend and glide by unknown mechanisms. Nutritionally, typical blue-green algae are photolithotrophs and aerobic.

Like bacteria, the blue-green algae multiply by binary fission and have no distinct sexual cycle as do some protozoa. The nucleus is without nuclear membrane, as would be expected of a procaryotic organism. The chlorophyll of blue-green algae is not confined to chloroplasts, as it is in eucaryotic plants, but rather is contained in thin membrane-like, folded lamellae within the cell (Fig. 14–22).

The properties of procaryotic and eucaryotic cells, contrasted with those of viruses, are summarized in Table 14–3.

CHAPTER 14
SUPPLEMENTARY READING

Berg, H. C., and Anderson, R. A.: Bacteria swim by rotating their flagellar filaments. Nature (Lond.), 245:380, 1973.

Brinton, C. C., Jr.: The properties of sex pili, the viral nature of "conjugal" genetic transfer systems, and some possible approaches to the control of bacterial drug resistance. CRC Crit. Rev. Microbiol., 1:105, 1971.

Brinton, C. C., Jr.: The structure, function, synthesis and genetic control of bacterial pili and a molecular model for DNA and RNA transport in gram-negative bacteria. Trans. N.Y. Acad. Sci., 27:1003, 1965.

Buckmire, F. L. A.: The physical structure of the cell wall as a differential character. Int. J. Sys. Bact., 20:345, 1970.

Cruden, D. L., and Stanier, R. Y.: The characterization of chlorobium vesicles and membranes isolated from green bacteria. Arch. Mikrobiol., 72:115, 1970.

Doetsch, R. N.: Functional aspects of bacterial flagellar motility. CRC Crit. Rev. Microbiol., 1:73, 1971.

Echlin, P.: The photosynthetic apparatus in prokaryotes and eukaryotes, in: Organization and Control in Prokaryotic and Eukaryotic Cells. Twentieth Symposium, Society for General Microbiology. Cambridge University Press, London. 1970.

Fox, C. F., and Keith, A. D. (Eds.).: Membrane Molecular Biology. Sinauer Associates, Inc., Stamford, Conn. 1972.

Fuhs, G. W.: Fine structure and replication of bacterial nucleoids. Bact. Rev., 29:277, 1965.

Ghuysen, J. M.: Use of bacteriolytic enzymes in determination of wall structure and their role in cell metabolism. Bact. Rev., 32:425, 1968.

Glauert, A. M., and Thornley, M. J.: The topography of the bacterial cell wall. Ann. Rev. Microbiol., 23:159, 1969.

Hayflick, L. (Ed.): The Mycoplasmatales and the L-Phase of Bacteria. Appleton-Century-Crofts, New York. 1969.

Hendler, R. W.: Biological membrane ultrastructure. Physiol. Rev., 51:66, 1971.

Higgins, M. L., and Shockman, G. D.: Procaryotic cell division with respect to wall and membranes. CRC Crit. Rev. Microbiol., 1:29, 1971.

Jahn, T. L., and Bovee, E. C.: Movement and locomotion of microorganisms. Ann. Rev. Microbiol., 19:21, 1965.

Knox, K. W., and Wicken, A. J.: Immunological properties of teichoic acids. Bact. Rev., 37:215, 1973.

Luderitz, O., Staub, A. M., and Westphal, O.: Immunochemistry of O and R antigens of *Salmonella* and related Enterobacteriaceae. Bact. Rev., 30:192, 1966.

Murrell, W. G., Ohye, D. F., and Gordon, R. A.: Cytological and chemical structure of the spore, in Campbell, L. L. (Ed.): Spores, Vol. IV. American Society for Microbiology, Bethesda, Md. 1969.

Nomura, M.: Bacterial ribosomes. Bact. Rev., 34:228, 1970.

Pollock, M. R., and Richmond, M. H. (Eds.): Function and structure in Microorganisms. Fifteenth Symposium, Society for General Microbiology. Cambridge University Press, London. 1965.

Rogers, H. J.: Bacterial growth and the cell envelope. Bact. Rev., 34:194, 1970.

Rogers, H. J., and Perkins, H. R.: Cell Walls and Membranes. E. & F. N. Spon Ltd., London. 1968.

Rothfield, L. I. (Ed.): Structure and Function of Biological Membranes. Academic Press, Inc., New York. 1971.

Ryter, A.: Structure and functions of mesosomes of gram-positive bacteria. Curr. Top. Microbiol. Immunol., 49:151, 1969.

Salton, M. R. J.: The Bacterial Cell Wall. American Elsevier Publishing Co., Inc., New York. 1964.

Salton, M. R. J.: Bacterial membranes. CRC Crit. Rev. Microbiol., *1*:161, 1971.

Schleifer, K. H., and Kandler, O.: Peptidoglycan types of bacterial cell walls and their taxonomic implications. Bact. Rev., *36*:407, 1972.

Schlessinger, D.: Ribosomes: development of some current ideas. Bact. Rev., *33*:445, 1969.

Sharon, N.: The bacterial cell wall. Sci. Am., *220*:92, 1969.

Tipper, D. J.: Structure and function of peptidoglycans. Int. J. Sys. Bact., *20*:361, 1970.

van Iterson, W.: Bacterial cytoplasm. Bact. Rev., *29*:299, 1965.

Wolk, P.: Physiology and cytological chemistry of blue-green algae. Bact. Rev., *37*:32, 1973.

CHAPTER 15 • THE VIRUSES

Discovery, Structure, Properties, Classification

In 1891, bacteria were viewed as the boundary between the living and the inanimate. Investigators of that time felt that they had probed the depths of the mystery of life and had discovered its extreme limit with respect to minuteness of size and simplicity of structure. Yet many clear, colorless, and seemingly sterile fluids that they examined with their most powerful optical microscopes teemed with billions of complex, often deadly "living" particles that escaped their vision and their knowledge.

The first viral disease of plants was discovered in 1892, when Iwanowski demonstrated that a common disease of the tobacco plant called **tobacco mosaic** could be transmitted to healthy plants in the sap from diseased plants, even though the sap had been passed through filters fine enough to remove all bacteria. Furthermore, the disease could be transmitted with sap filtrates from plant to plant in indefinite series. Yet no living thing capable of producing tobacco mosaic grew from the filtered sap of diseased plants on any culture medium in the laboratory, and nothing could be seen in the crystal-clear fluid with any microscope then available. Beijerinck, a famous microbiologist of the time, found that the filtrable, invisible, and noncultivable infectious principle would diffuse through an agar gel like a fluid. He thought the fluid itself alive, and called it **contagium**

vivum fluidum—a living infectious fluid! This concept is embodied in the very word **virus,** which is derived from a Latin root meaning a slimy, noxious liquid, a sort of living snake venom!

We now know that the sap from the diseased tobacco plant contained billions of particles of the virus of tobacco mosaic, the first known virus. Iwanowski had opened the door to the world of the ultramicroscopic, much as Hooke and Leeuwenhoek had opened the door to the world of the microscopic. In 1935, Stanley (Nobel Prize winner) showed that the tobacco mosaic virus (TMV) can be crystallized but that the virus crystals, instead of being inorganic matter, were aggregations of thousands of submicroscopic nucleoprotein complexes; i.e., individual particles (**virions**) of tobacco mosaic virus. Scores of viral diseases of other plants are now known.

Viral diseases of vertebrates were well known by 1892, since Pasteur had been studying canine rabies for some time. He spoke of the causative agent as a virus, although during his early studies he was apparently not aware of its true nature. The term virus was then commonly used for a variety of infectious agents, including bacteria. In 1898, the foot-and-mouth disease of cattle was shown by Loeffler and Frosch to be caused by an agent that, like TMV, passed through bacteria-retaining filters, and was neither visible with the microscope nor cultivable on inanimate media. These three properties were, at that time, the basic identifying properties of a virus. Foot-and-mouth disease was the first known viral disease of lower vertebrates. In 1900 Walter Reed and his associates discovered the virus of yellow fever, the first known viral disease of man. Today many viral diseases of vertebrates are well known.

Viral diseases of insects were studied by Pasteur during his investigations of diseases of commercial silkworms. The infectious agents of those diseases were first recognized as viruses by Wahl, by von Prowazek and by Escherich. Since then many viral diseases of arthropods have been recognized, important among them being sacbrood of honey bees and a viral disease of value in the control of destructive cabbage-looper worms.

Viruses that attack bacteria were first described in 1915 by the British scientist, Twort, and were independently observed and more fully studied about 1917 by the French investigator, d'Herelle, who named these viruses bacteriophage (Gr. *phagein* = to eat). The bacteriophage was originally (and erroneously) thought of as eating the bacteria from within. The shorter term, **phage,** is commonly used for bacteriophage and will be so used here.

Subsequent studies of many animals, bacteria and higher plants revealed the existence of hundreds of other viruses, and new ones are being discovered almost daily as techniques for their detection and identification are improved.

It is now clear that the three properties once thought to be unique and absolutely distinctive of viruses are no longer valid. Originally described as ultramicroscopic or invisible (with optical microscopes), viruses and their intimate structural details are now readily made visible by means of high-power electron microscopes (Chapter 3). For many years described as **filtrable** because they passed through unglazed porcelain, diatomaceous-earth (kieselguhr) and similar bacteria-retaining filters, they are now easily retainable (i.e., **nonfiltrable**) on specially prepared molecular filters made of very fine-pore collodion or plastic materials (Chapter 21). For decades thought to be **noncultivable** (i.e., on inanimate media), they are now easily propagated in **living culture media,** as will be explained.

Historically, an unknown infectious agent was considered to be a virus by exclusion, i.e., if the unknown agent was *not* a bacterium or other identifiable organism, it must be a virus. This resulted in erroneously including certain agents, like *Chlamydia,* among the viruses, because they grew in living cells and were exceedingly small. Viruses, however, now are defined more clearly, and an unknown agent can be determined to be a virus if it meets the definition, rather than if it is shown to be nothing else!

Viruses contain only a single kind of nucleic acid—either RNA or DNA, but never both. In almost all viruses, the nucleic acid exists as a single molecule, and it can be either single- or double-stranded. The nucleic acid is usually enclosed in a protein shell (**capsid**) that serves to protect the nucleic acid and to aid in infection. The capsid can be simple or complex, and the protein molecules of which it is composed confer antigenic specificity. The virion contains no metabolic enzymes or protein synthetic machinery of its own (sometimes proteins or other components of the cell are accidentally enclosed in the capsid when the virion is assembled), although its nucleic acid can contain the **information** to direct the cell to make these components. Viruses have a method of replication that is different from any other organism: they do not grow. Instead, the information encoded in the nucleic acid directs the cell to make the various parts of the virus and then to assemble these parts into complete, infectious progeny virions.

These unique qualities of viruses will be discussed more thoroughly in the sections to follow.

15.1
STRUCTURE AND MORPHOLOGY

As indicated in the preceding definition, viruses are not cellular and therefore do not have a nucleus, cytoplasm, or cell membrane. Instead, the genetic information that permits new generations is contained in a (usually) single molecule of either RNA or DNA. The possession of only one kind of nucleic acid distinguishes viruses from all cellular forms, eucaryotic or procaryotic, because all cellular organisms contain both RNA and DNA. In some viruses this nucleic acid molecule has a demonstrated structural form, i.e., it may be wound into a ball or is characteristically in some other form; it is then called a **nucleoid** or **core.**

Surrounding the nucleic acid core is the second major portion of the virion: the previously mentioned protein shell or coating called a **capsid.** Since it is a protein, it has antigenic specificity (Chapt. 27). The capsid is made

Figure 15–1. Model of a reovirus virion. The 92 hollow protein capsomers on the outer surface of the nucleocapsid are shown. Note that some of the capsomers are pentagonally faced, others hexagonally faced.

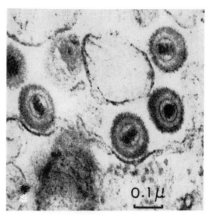

Figure 15–2. Canine herpes virions after release from an infected cell. Note the bar-shaped helix at the core, surrounded by a thin, dark membrane and a thicker outer mantle or envelope; also the spike-like projections from the outer surface (cf. Figure 15–4). This preparation was not stained to show the 162 capsomers demonstrable by negative staining.

up of many identical structural units called **capsomers** (Gr. *meros* = part); their composition, numbers, and forms vary with the kinds of viruses (Fig. 15–1). The capsid is physiologically inert and is believed to serve only as a protective shell. The core, with its capsid, is called the **nucleocapsid** of the virus. Many mammalian viruses also have, outside the capsid, an **envelope** or **limiting membrane** or mantle. As will be explained farther on, this appears to be derived largely or wholly (depending on kind of virus and cell) from the nuclear or cytoplasmic membrane of the infected cell (Fig. 15–2). It is made up of structural units sometimes called **peplomers** and usually contains lipids or lipoproteins. These lipid components confer on the viruses that have them the distinctive property of sensitivity to lipid solvents such as ether and

chloroform, or to emulsifying agents such as bile salts and detergents. Such viruses are sometimes called **lipoviruses.** Many viruses have no envelope; they are naked virions.

In outward form virions differ widely. They may be elongate, like a piece of insulated electric cable, or rounded, polyhedral, or cuboidal. Several appear to be pleomorphic. In elongate forms the nucleic acid strand of the core is usually coiled like a helical spring. The capsid is closely coiled about it or is arranged as a succession of rings. Such a virion is said to have **helical symmetry** (Fig. 15–3). In some helical

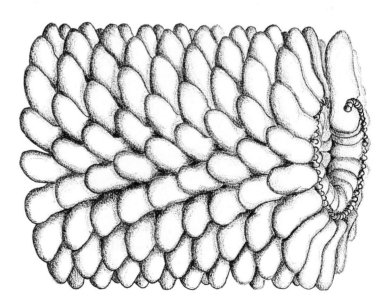

Figure 15–3. Diagram of the tobacco mosaic virus virion showing the protein subunits and the helical coil of RNA.

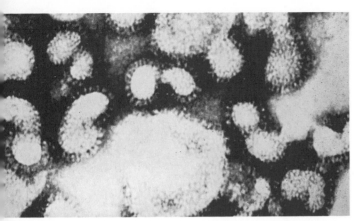

Figure 15–4. Influenza virus PR8 after one hour of incubation at 37C. Projecting spikes or rods resemble bacterial fimbriae.

viruses the whole nucleocapsid is sufficiently flexible to be secondarily coiled on itself into a spheroidal form. Helical viruses of mammals generally have a lipid-containing envelope. Many are covered externally by thin, radially projecting spikes or rods suggestive of bacterial fimbriae (Chapt. 14) (Fig. 15–4).

A common viral form is based on **cubic symmetry.** The cores of such viruses are surrounded by well-defined capsomers usually arranged in **icosahedral symmetry** (Fig. 15–5).

An icosahedron is a polyhedron with 20 identical triangular facets (Fig. 15–6). A distinguishing character of an icosahedron is that it may present any of several different symmetrical appearances, depending on which of three axes it rotates about. Thus, when rotated about axis *a* it presents five identical appearances, about axis *b* it presents three appearances, and about axis *c*, two appearances. It is said to have 5-3-2 (icosahedral) symmetry.

The numbers of capsomers of virions that show icosahedral symmetry are determinable and are distinctive of certain virus types. Capsomer numbers vary from 12 to 812 or more.

The symmetry and form of numerous viruses, especially the large poxviruses (Fig. 15–7) are complex and often vague or indeterminable. Capsomers of some helical viruses with envelopes are not sufficiently distinct to be accurately counted. Instead, the diameter of the helical nucleocapsid inside the envelope is measured as a distinguishing character (Fig. 15–8). Some viruses are bullet-shaped (Fig. 15–9).

Bacterial viruses are known in many shapes and sizes, some with double-stranded DNA, some with single-stranded DNA, and some with single-stranded RNA. The morphology of phages varies from exceedingly simple, such as the small, single-stranded DNA phage, ϕX-174, to phages with a complex, tadpole-like shape (Figs. 15–10, 16–6). The latter phages also are subdivided on the basis of the relative size of

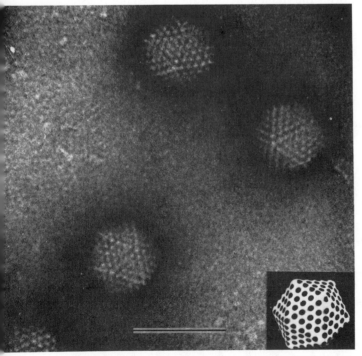

Figure 15–5. Purified adenovirus type 2 particles. (Bar is 100 nm.) The model in the right corner shows the icosahedral form of the virions.

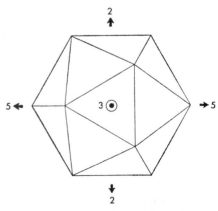

Figure 15–6. Diagram of an icosahedron showing axes of rotation that reveal 2-, 3-, and 5-face symmetry.

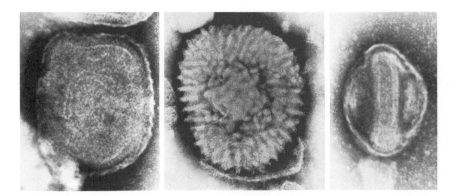

Figure 15–7. Three particles of vaccinia virus. All three forms are seen in electron micrographs. The one on the left may be artifactual or defective. The other two forms are usually seen. (Bar is 100 nm.)

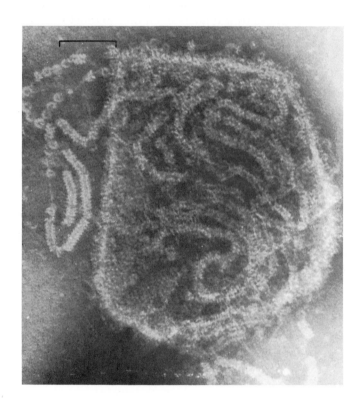

Figure 15–8. A mumps virion showing the internal component. Note its similarity to that of tobacco mosaic virus (Fig. 15–3). The envelope is severely disrupted. (Bar is 100 nm.)

Figure 15–9. Three bullet-shaped viruses: *A* shows the rabies virus; *B*, the virus-like particle of plantain; and *C*, vesicular stomatitis virus. The striations and indentation from the flat end are clearly shown, as well as surface projections similar to those of influenza virus (Fig. 15–4).

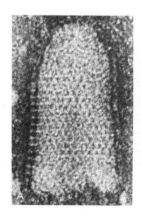

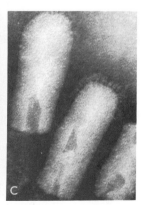

the various components. The many varieties of phages are diagrammed in Table 15–1.

15.2
SIZES OF VIRUSES

Size and Visibility. Except for certain "large" viruses (e.g., smallpox, vaccinia), a major distinguishing property of viruses is their invisibility under even the best optical microscopes (magnifications up to 2,000×). However, electron microscopes, which can give clear resolutions at magnifications of several hundred thousand times, make even the smallest viruses and their innermost structures readily visible.

An idea of the degree of magnification implied by 200,000× may be gained by imagining the letter O on this page to be so magnified. Approximately 2.5 mm in diameter, the letter would appear as an oval 200,000 × 2.5 = 500,000 mm = 500 meters or nearly one-half mile in diameter: a large race track! Magnified 2,000× by an optical microscope the O would appear only about 16 feet in diameter. The nucleocapsid of a

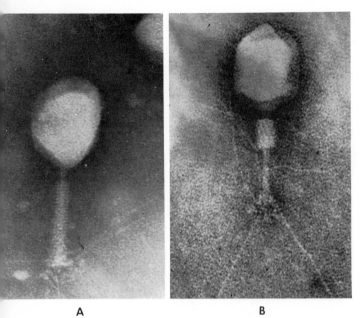

A   B

Figure 15–10. Two virions of T2 phage. In *A* the sheath surrounding the tail as rings or a helix is extended. In *B* the phage has been "triggered" with contraction of the sheath. Some of the six fibers attached to the tip of the tail are visible. The tail plate and spikes are not well shown. In *B* the contracted tail sheath has revealed the tail core.

TABLE 15–1. MAJOR PROPERTIES OF THE BASIC MORPHOLOGIC TYPES OF BACTERIOPHAGES*

Bradley Type	Examples (Coliphages)	Basic Morphology	Description	Nucleic Acid Type
A	T-even		Angular head, contractile tail	2-DNA
B	T1 T5		Angular head, long non-contractile tail	2-DNA
C	T3 T7		Angular head, short non-contractile tail	2-DNA
D	ϕX174		Angular virion, no tail	1-DNA
F	fd M13 f1		Flexible filament	1-DNA
E	f2 R17 fr		Angular virion, no tail	1-RNA

*From Bradley: Bact. Rev., *31*:230, 1967.

herpesvirion, which measures 100 nm in diameter, would have an image size of 2 cm at 200,000× magnification.

Size and Filtrability. Iwanowski, Walter Reed, and later many others demonstrated what was thought, until about 1930, to be a uniquely distinctive property of viruses: the ability when suspended in fluids to pass through porcelain filters of such fine porosity as to retain all known bacteria. However, it is now known that under some experimental conditions bacteria can readily be made to pass through the same porcelain filters by special means, since size is not the only factor that determines filtrability through silicious filters. Surface electrical charge, which can be altered experimentally, is also very important. The "filtrable" viruses, however, cannot pass through filters made from collodion or cellulose acetate (called **membrane filters**) if the filters are made with sufficiently fine pores. Thus the term filtrable, once widely used to distinguish viruses, is no longer applicable without appropriate qualifications.

Certain details of structure as well as sizes of viruses are often inferred from the results of indirect methods. Graded ultrafiltration is sometimes used. It consists of passage of an aqueous suspension of virus particles through successive collodion or plastic membranes of known, graded, decreasing pore sizes, the effect being like that of a gravel sorter. An important source of error is electrostatic adsorption of the virions to opposite charges on the membrane surfaces. Measurement by direct observation of virus particles with electron microscopes is valuable but subject to error resulting from dehydration of the specimen and distortion by the electron beam. Done carefully, however, measurement of virions by electron microscopy yields the most accurate size determinations available.

Virion size determinations are usefully accurate and are important in primary differentiations of virus types and classifications of viruses.

There is a considerable range in size of viruses (Fig. 15–11). Some, such as those of yellow fever, foot-and-mouth disease, and poliomyelitis, are "small," with diameters about 25 nm (nm = nanometer, or one billionth of a meter). Others, like the virus of smallpox (vaccinia), are "large," about 250 nm in size. These are just within the range of resolution by the best optical microscopes. There are many intermediate sizes.

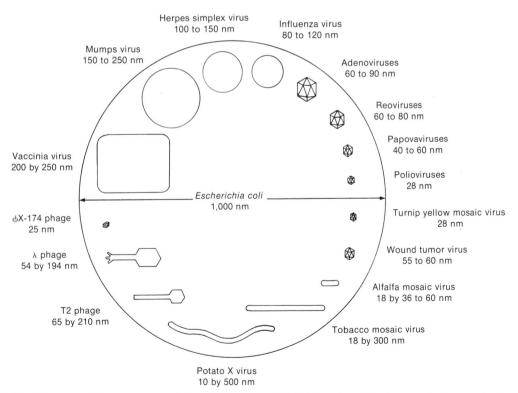

Figure 15–11. Diagrammatic comparison of sizes of viruses and related structures. The largest circle, enclosing the whole, represents the diameter of *Escherichia coli,* a small cylindrical bacterium about 1 μm (1,000 nm) in diameter. The other organisms are drawn to approximately the same scale.

15.3
STABILITY OF VIRUSES

Phage and plant viruses such as tobacco mosaic virus (TMV) are in general much more stable chemically and physically than viruses of warm-blooded vertebrates. Many animal viruses tend to become inactivated on exposure outside the animal or cell in which they multiply. There are wide variations, however; TMV will last for years in infected dried plant tissues, and poliovirus can survive for days or weeks in sewage. Yellow fever virus and many similar viruses, on the contrary, are inactivated merely by storage overnight in blood in a refrigerator, or in a few minutes by dilution of infectious blood in saline solution.

In general, mammalian viruses are inactivated in a few minutes by temperatures like that for pasteurization (63C for 30 minutes) (Chapt. 20); some phages are inactivated in 10 minutes at temperatures as low as 56C. There are exceptions, however, an important one being the virus of serum hepatitis, which can withstand boiling for some minutes.

Most viruses are highly resistant to intense cold and remain infective at −76C for a year or longer. They are commonly stored in carbon dioxide ice. They are well preserved by rapid desiccation in vacuo (Chapt. 20) after rapid freezing (freeze-drying, or **lyophilization**). Ordinary bactericidal strengths of disinfectants such as phenol, cresol, β-propiolactone, formaldehyde, and halogens are not always effective in dealing with viruses. Certain disinfectants may inactivate some viruses but not all. Surface-active agents, such as soap, some detergents, and bile salts (Chapts. 20, 21), can inactivate some viruses readily in vitro, especially those having lipids in the envelope. Ultraviolet light is rapidly destructive to most viruses, since nucleic acids are very sensitive to irradiation. (See Mutagenic Agents, Chapt. 18; Environment, Chapt. 20.) Differences in resistance to pH, heat, and lipid solvents have been used as bases for classification secondary to type of nucleic acid.

15.4
CLASSIFICATION OF VIRUSES

Beginning with the discovery of tobacco mosaic virus by Iwanowski in 1892, continuing discovery of large numbers of viruses and virus-like organisms with one or more obvious characteristics in common has demanded, in the face of chaos, some system of classification. An ordered arrangement (classification) into groups or **taxa** of similar viruses and a ruled system of descriptive nomenclature of the groups and species (**taxonomy**) has gradually evolved. The history of the classification and taxonomy of viruses is similar to that of bacteria, enzymes and other biological groups: one of trials, casual classifications and names, rearrangements and successive improvements with increasing knowledge. Some of these casual classifications and names, based on host, vector or diseases, are outlined in the following paragraphs.

For convenience, viruses may be subdivided primarily on the basis of the **hosts** in which they multiply in nature. In microbiology a host is a plant or animal or single cell that is parasitized by another organism, generally smaller. For example, certain beetles are infected by bacteria; the bacteria are, in turn, infected by virus (bacteriophage). The beetle is host to the bacteria; the bacterium is host to the virus. One or more representatives of most (all?) major divisions of both the animal and plant kingdoms are subject to invasion by one or more viruses; a few examples follow.

Viruses that infect:
  I. Plants only
     A. Green plants: tobacco mosaic virus and many others
     B. Blue-green algae
     C. Fungi
        1. Bacteria: bacteriophages
        2. Other fungi, e.g., edible mushrooms
 II. Arthropods only
     Insects: numerous "polyhedral" diseases such as cabbage-looper (*Trichoplusia ni* Hbn) disease; a viral disease of fruit flies (*Drosophila melanogaster*)
III. Dual-host viruses
     Arthropods and warm-blooded vertebrates: viruses causing diseases in birds and mammals, transmitted by and multiplying in arthropods, e.g., Eastern and Western Equine Encephalitis (EEE and WEE) occurring in man, equines, and birds and transmitted by mosquitoes (*Culiseta melanura* and others)
 IV. Warm-blooded vertebrates only
     Measles, influenza, fowlpox, chicken sarcoma, rabies, foot-and-mouth disease and many others

V. Cold-blooded vertebrates
    Pancreatic necrosis of brook trout,
    Lucké virus of frogs, and others

A convenient classification of the viruses of animal disease may also be based on the types of tissues or organs principally affected. Another sort of classification could be based on mode of transmission and another on type of disease caused.

Classifications such as those just discussed, although convenient for purposes of discussion, are arbitrary and incomplete and cannot take into consideration the fact that viruses often become modified and cause atypical or entirely different types of disease. Further, the tissue affinities of some viruses can change completely. Also, one is at a loss where to place some viruses, such as that of mumps and fowl plague. These groupings were useful when little was known of viruses except that they caused disease.

A Modern Classification. Current methods of classifying viruses are based to only a minor extent on type of disease caused (e.g., polioviruses, poxviruses, herpesviruses). Recent groupings depend largely on detailed and accurate knowledge of physical and chemical properties of viruses. These data have accumulated only during the last decade and continue to accumulate daily.

The entire group of viruses, plant and animal, is currently divided primarily into two types on the basis of the kind of nucleic acid in the core: DNA or RNA. Each of these two groups is then subdivided on the basis of other characteristics.

In a recent system all of the viruses—bacterial, plant, and animal—are grouped in the phylum Vira. This is divided into subphyla, classes, orders, suborders, and families on the basis of four major features called "essential integrants."

1. Type of nucleic acid: DNA or RNA.
2. Symmetry: helical, cuboidal, or binal (binal refers to possession of head and tail by viruses like bacteriophage).
3. Presence or absence of envelope around the nucleocapsid.
4. Diameter of helical nucleocapsids; number of capsomers in cuboidal viruses.

The families established on these four integrants are named according to rules adapted to viruses by a single, generally recognized authority, The Provisional Committee on Nomenclature of Viruses (P.C.N.V.) established by the executive committee of the International Association of Microbiological Societies. The P.C.N.V., headed by Sir Christopher Andrews, provisionally adopted in slightly modified form a system drawn up by Lwoff, Horne, and Tournier. Neither perfection nor immutability is claimed for the system but, as the first of its kind, it marks a milestone in the science of virology. A grossly abbreviated version follows.

PHYLUM: VIRA

Subphylum: Deoxyvira (DNA viruses)
  Class: Deoxyhelica (helical symmetry)
    Order: Chitovirales (Gr. *chiton* = tunic or envelope)
      Family: Poxviridae (poxviruses) ("large viruses")
  Class: Deoxycubica (cubical symmetry)
    Order: Haplovirales (Gr. *haploos* = simple: i.e., without envelope)
      Family: Microviridae   12 caps° (phage φX174)
            Parvoviridae   32 caps (a rat virus)
            Papilloviridae   72 caps (papovaviruses)
            Adenoviridae   252 caps (adenoviruses)
            Iridoviridae   812 caps (insect viruses)
    Order: Peplovirales (mantle viruses)
      Family: Herpesviridae 162 caps (herpesviruses)
  Class: Deoxybinala (viruses with head and tail)
    Order: Urovirales
      Family: Phagoviridae (bacteriophages)

Subphylum: Ribovira (RNA viruses)
  Class: Ribohelica (helical symmetry)
    Order: Rhabdovirales (Gr. *rhabdos* = rodlike)
      Suborder: Rigidovirales
        Family: Dolichoviridae   12–13 nm ⎫
             Protoviridae   15 nm ⎬ Plant viruses
             Pachyviridae   20 nm ⎭
      Suborder: Flexiviridales
        Family: Leptoviridae   10–11 nm ⎫
             Mesoviridae   12–13 nm ⎬ Plant viruses
             Adroviridae   15 nm ⎭
    Order: Sagovirales (L. *sagum* = mantle)
      Family: Myxoviridae   9 nm
            Paramyxoviridae 18 nm
            Stomatoviridae
  Class: Ribocubica (cuboidal symmetry)
    Order: Gymnovirales
      Family: Napoviridae 32 caps ⎰ A. Plant viruses
                      ⎱ B. Picornaviruses
         Reoviridae   92 caps (reoviruses)
    Order: Togavirales (L. *toga* = Roman mantle)
      Family: Arboviridae (arboviruses)

Properties that determine genera within families include (for example):

1. NA: base sequence, relative numbers of bases, number of nucleotides.
2. Capsomers: structure, antigenic properties, molecular weight.
3. Capsid: number of capsomers, antigenic properties, reaction to heat, pH, other physical and chemical agents.

° "Caps" refer to numbers of capsomers (cubical and cuboidal viruses).

Additional features for generic and specific differentiations include data on envelope or mantle, enzymes, mode of development, sensitivity to interferon, specificity for host, virulence, clinical effect.

15.5
VIRUSES IN INDUSTRY

There are several important industrial aspects of viruses. For example, the preparation of viral vaccines, both inactivated and active but weakened or **attenuated** in virulence, is a multimillion-dollar industry. Some viruses are cultivated on a very large scale in enormous tissue-cell culture vats for the preparation of inactivated vaccines such as the Salk polio vaccine, as well as the active Sabin oral vaccine. Some viruses, such as that of rabies, are propagated on a large scale in living avian embryos. Various factory-sized means of cultivation and propagation are used to prepare other viral vaccines of veterinary importance. Some mass-produced viruses of insects are used to combat arthropods of medical and economic importance.

A contrary role of viruses in industry is their destructive activity. For example, a bacterial virus called **actinophage** is virulent for the mold-like bacterium *Streptomyces griseus*, from which the antibiotic streptomycin is obtained. Unless carefully excluded from the culture vats, the actinophage can be an expensive nuisance.

Another industrially important phage is one active against *Streptococcus lactis* and related species of bacteria that are used to cause souring of milk, i.e., to produce lactic acid and "buttery" flavor in the milk and cream used for dairy products. These phage particles are widely distributed, and when such a phage gets into the creamery vats, whole batches of valuable culture-soured dairy products are spoiled. The control of such phages is an important industrial problem (see Chapter 47). One method of control consists in selecting strains of milk-souring streptococci that are genetically resistant to the phage.

Many bacteriophages have been found, each active against a specific bacterium. Bacteria for which phages have been found include tubercle bacilli, diphtheria bacilli, and certain very valuable nitrogen-fixing bacteria (*Azotobacter*) of the soil. The latter phages can constitute a serious problem to agriculture. Other phages infect sporeforming bacilli of the genus *Bacillus*; and bacilli of the genus *Brucella* (brucellaphage), active against the cause of brucellosis (undulant fever).

A virus active against the blue-green alga *Plectonema boryanum* was discovered in 1963. Others have been found since then. These viruses resemble bacteriophages. Algal viruses are significant because blue-green algae are important agents in sewage disposal (Chapt. 45), and their destruction by viruses constitutes a problem of some magnitude.

CHAPTER 15
SUPPLEMENTARY READING

Goodheart, C. R.: An Introduction to Virology. W. B. Saunders Co., Philadelphia. 1969.

Lwoff, A. and Tournier, P.: The classification of viruses. Ann. Rev. Microbiol. *20*:45, 1966.

Wildy, P.: Classification and Nomenclature of Viruses. S. Karger, Basel. 1971.

CHAPTER 16

THE VIRUSES

Cultivation,
Enumeration,
Replication

The major, unique characteristic of viruses that distinguishes them from all other organisms is their method of replication, mentioned briefly in Chapter 15. Because they contain no metabolic enzymes, viruses are unable to use environmental nutrients. Instead, they must have available nutrients and enzymes contributed by living cells. To do this, they actually become a part of the host cell during their replicative cycle—when not within a suitable cell, a virus is completely inanimate. Whether viruses are alive has therefore been a subject of considerable debate and ultimately requires definition of the word "life." How viruses subvert a cell to the production of more viruses is the subject of this chapter.

16.1
CULTIVATION OF VIRUSES

The requirement for living cells for viral replication can be met in any of several ways. The cells, for example, can be part of an intact plant or animal—the cells of our nasal mucosa are all too often host to viruses of the common cold. Or, the cells can be grown individually in culture, as with bacterial or mammalian cells.

Bacterial cultures will not be discussed in

detail in this chapter, since they are described in detail elsewhere in this book. It is sufficient to say at this point that phages are conveniently grown in broth cultures or in cells growing on agar. The latter method is most frequently used to enumerate infective virions by the plaque method, as will be discussed farther on. Methods for animal viruses are somewhat more complicated.

Cultivation in Chick Embryos. Chick embryos are live animals. For microbiological purposes they have as good a scientific standing as monkeys, rabbits, or mice, and many obvious advantages. For example, they are conveniently packaged in their shells and have considerable natural resistance to growth of contaminating bacteria. They are incubated as for hatching until the embryo is well developed (from ten to fourteen days).

One procedure for introducing microorganisms is to inoculate the chorioallantoic membrane (one of the vascular membranes surrounding the embryo). Modifications of this method include injection through small, appropriately located cuts in the shell by a needle directly into the amniotic and chorionic cavities, yolk sac, and embryonic tissue (Fig. 16–1).

Despite considerable natural resistance of embryonated eggs to growth of contaminating organisms, many saprophytes, especially yeasts,

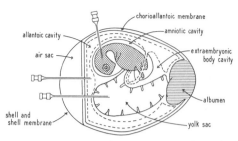

Figure 16–1. Diagrammatic representation in sagittal section of the embryonated hen's egg 10 to 12 days old. The hypodermic needles show the routes of inoculation of the yolk sac, allantoic cavity, and embryo (head). The chorioallantoic membrane is inoculated after it has been dropped by removing the air from the air sac.

molds, and bacteria that may enter as contaminants, are able to multiply as well in the fluids of the chick embryo as in a culture tube. The embryo is usually killed and destroyed by such contaminants. Extraneous contamination with such organisms from the air, from shell dust, from implements, or from contaminated inocula such as feces or saliva is therefore an ever-present source of error in the use of chick embryos for the study of viruses and rickettsias. As in handling tissue cultures, rigid precautions against such contaminations and considerable technical skill are required (Fig. 16–2). Antimicrobial drugs that are known not to affect viruses (e.g., antibiotics) are usually mixed

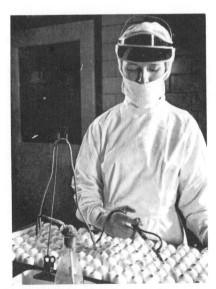

Figure 16–2. Carefully dressed technician using a pressure-fed syringe to inoculate fertile eggs with suspension of living rickettsiae. The eggs, after suitable incubation, will be used in the preparation of vaccines or of antigens for diagnostic purposes. (Photograph courtesy of E. R. Squibb & Sons, Princeton, N.J.)

with materials such as feces or saliva before using them as inocula.

Embryonated eggs, once used extensively for the propagation of many different kinds of viruses, are now used only for production of certain vaccines and other special purposes. Mammalian, avian, and other vertebrate cells grown in culture have virtually replaced embryonated eggs for experimental purposes with viruses, and most vaccines are now prepared commercially in vertebrate cell cultures.

Vertebrate Cell Culture. In vitro propagation of viruses eliminates the use of living animals or plants, but it does not eliminate the necessity for living *cells* of animals or plants. It therefore makes necessary the cultivation of these cells in vitro. A brief outline of methods used in cultivating cells of vertebrates will illustrate the various techniques. For invertebrate and plant cells appropriate modifications are made in details, such as nutrients supplied and temperature of incubation. All cell-culture work requires care and skill to keep out contaminating bacteria, molds, yeasts, and mycoplasmas (Chapt. 42). Antibiotics, such as kanamycin, penicillin, and gentamicin, often are used to suppress contamination.

A subtle and unfortunately common source of error is the unsuspected presence of extraneous viruses which are already infecting the supposedly sterile tissues when they are removed from the animal. Several viruses commonly remain latent in animal tissues for long periods. When penicillin is used in cell cultures, bacterial contaminants accidentally introduced can become L-forms or protoplasts and have been mistaken for mycoplasmas, which have no cell walls.

Primary Cultures. Cultures made directly from live tissues are **primary cell cultures**. As indicated, they have the disadvantage that, in addition to being mixtures of different kinds of cells (e.g., muscle, blood vessel, and fibroblasts, present in the tissue), they are limited in size. However, primary cultures in larger volume are now feasible for bulk cultures, used in the commercial preparation of large amounts of vaccines such as polio vaccine, and for other purposes. Although they are initiated from the tissues of animals immediately after they are killed, the cells may be contaminated with microorganisms that were infecting the living animal. However, these difficulties have been recognized and are being overcome to a great extent. A full and detailed description of the procedures is beyond the scope of the present discussion. A condensed outline of the principal steps in a representative procedure for preparing a primary culture follows. There are many modifications of this procedure, especially in regard to the growth medium used. New improvements are constantly being made. The first step, after preparation of the desired tissue fragments, is **dispersion** of the cells from the tissue. This is commonly done by digesting the intercellular cement substance with trypsin. Other dispersing agents are collagenase and Versene or ethylenediaminetetraacetic acid (EDTA).

Trypsinization. Suppose that a primary culture is to be made of rabbit kidney. The animal is killed by an overdose of anesthesia, the kidneys are immediately removed aseptically and stripped of their capsule, and the cortex is minced and washed repeatedly with a sterile saline solution. The minced, washed tissue is then placed in a flask with sterile trypsin solution at room temperature for several hours. Trypsin, a proteolytic enzyme, digests the material that binds the cells together, and they are thus dispersed in the fluid as individual cells; i.e., each cell is now an independent microorganism!

Enzyme solutions should be free of Ca^{2+} and Mg^{2+}. These ions tend to interfere with the desired dispersion of the cells by making the intracellular "cement substance" or matrix more resistant to the dispersing agent.

Washing. The cell suspension is then passed through a wire screen filter to remove coarse particles and to provide a smooth suspension of separated cells. The cells are washed free of trypsin by centrifuging gently and resuspending the pelleted cells in growth medium.

Cultivation. After appropriate adjustment of the number of cells per unit of volume, the cells are diluted in **growth medium** and dispensed in tubes, flat-sided flasks, or Petri dishes.

Growth media of many formulations are available. Choice of the medium to be used is based on the type of cell to be grown, the purpose for making the culture, and the type of incubation to be used. Media contain salts that simulate physiologic concentrations, buffers, vitamins, amino acids, and usually a source of protein that is obtained by adding animal serum at a concentration of up to 20 per cent. pH is adjusted to about 7.4, although it can be varied, depending on the type of cell to be grown.

Monolayers. When prepared as just described and freely suspended in a glass or special plastic vessel, cells quickly settle out of suspension and attach firmly to the bottom. If undisturbed for five to seven days at 36C, they grow and spread over the surface in a sheet of

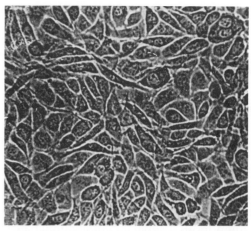

Figure 16–3. Monolayer of L cells. These are normal cells. Note the distinctive, relatively uniformly polygonal, sharply angular appearance of the cells, all of which are joined together by intercellular cement substance into a **flat** layer. Cells that have been transformed to malignancy often pile up in heaps in uninhibited growth.

one-cell thickness called a **monolayer** (Fig. 16–3). After renewal of the growth medium, this monolayer culture may be inoculated with viruses for a variety of purposes.

Secondary Cultures. For commercial and other purposes requiring large volumes, secondary cultures can be made from primary monolayers. The growth or maintenance medium is removed and the monolayer is washed with saline solution devoid of calcium and magnesium ions. The cells are then dispersed by adding a few ml of 1/3,000 Versene (ethylenediaminetetraacetate or EDTA) to **chelate** the calcium and magnesium ions in the intracellular matrix that holds the cells together and to the glass surface. After a few minutes at 36C, the loosened cells are shaken to separate them and resuspended in growth medium. The suspension is dispensed in fresh culture vessels, where the cells again settle to the bottom, attach, and grow into a new monolayer.

Necessity of Carbon Dioxide. In all cell cultures the atmosphere must contain about 5 per cent carbon dioxide, similar to requirements of some bacteria (*Brucella, Neisseria,* and autotrophs). Special buffer systems or special incubators may be used to maintain carbon dioxide-rich atmospheres.

Established Cell Lines. As previously mentioned, a culture derived directly from an animal (or plant) which is not transferred or subcultured is referred to as a **primary culture** and usually consists of several different kinds of cells. Subcultures from the primary culture are called **diploid cell strains.** These cell strains cannot be immediately differentiated from the primary culture, and both may contain more than one kind of cell. However, repeated serial subculturing of a cell strain may result in final dominance of the fastest-growing cell type in the original culture, in mutations suggesting malignancy or in other undesirable changes in the cells. More important, they can become contaminated with extraneous, possibly oncogenic, viruses that can remain occult for long periods.

Clonal Cell Lines. Although primary cultures and diploid cell strains may contain several different kinds of cells, it is possible to isolate single cells from these cultures. Each cell then forms a separate colony, much as bacteria do. They can then be propagated separately as pure cultures, thus establishing **clones.** A clone is a population in which all individual cells are descendants of a **single cell.**

Many established clonal cell lines (lung, kidney, various cancers such as HeLa and L strains) are available commercially.

A new *Registry of Animal Cell Lines,* second edition, of the American Type Culture Collection, Rockville, Maryland, gives the complete life history and description of cell lines available to industry and researchers.

Infection of Cell Cultures. Once vigorous, uncontaminated cultures are obtained they may be infected (or inoculated) with any desired microorganism. Bacteria and molds grow well in cell cultures, but it is not customary to cultivate them in such expensive media since most of them grow well in simple nutrient fluids. However, very interesting observations have been made on the effects of various bacteria and molds, and their toxins, on living cells in vitro.

When a cell culture is inoculated with a virus, the virus is replicated inside the multiplying cells. The infected cells commonly, but not always, disintegrate. If the virus is **virulent,** new virus particles are liberated into the surrounding fluid. Most virus-infected cell cultures show visible evidence of cell damage caused by the virus. If cells in the culture are examined daily under the microscope, the progress of the infection, and disintegration if it occurs, can readily be followed. Viruses which thus damage cells are said to be **cytopathic.** Their effects are referred to as **CPE** (cytopathic effects). CPE are sometimes quite distinctive in appearance (Fig. 16–4).

16.2
ENUMERATION OF VIRUSES

The simplest method of enumeration is by direct visual count. In the case of viruses this involves the use of the electron microscope. By this means virus particles in a suspension may be counted, relative to a known number of minute, opaque particles of polystyrene included in the suspension. If the diameter of the opaque plastic particles is known, the size of the virions may be measured at the same time. Another method of enumeration is by plaque assay.

Plaque Assay. The plaque assay procedure consists in distributing a suspension of the virions to be counted, in contact with susceptible host cells, in (or on) a medium appropriate for the host cells, over a flat surface such as a Petri dish or flat-sided flask. Each virulent virion infects a single susceptible cell. During the ensuing incubation period each virion, with its progeny, manifests itself by forming, in the immediately surrounding area, a typically clear, transparent, readily visible circular area of lysed cells called a **plaque** (Fig. 16–5; see also

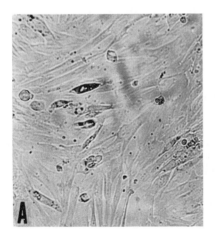
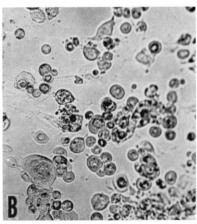

Figure 16–4. The cytopathic effect of a virus on human chorion cells in culture. (The chorion is one of the membranes protecting the embryo in the uterus.) *A,* Normal cells: culture five days old showing a fairly regular, organized network of uniformly elongated, spindle-shaped cells of rather distinctive appearance. *B,* The same sort of cells 72 hours after being inoculated with the virus of herpes simplex ("fever blister"). The cells have lost their distinctive form and arrangement and have become mere shapeless, disconnected, disorganized blobs of dead protoplasm. (1,500×.)

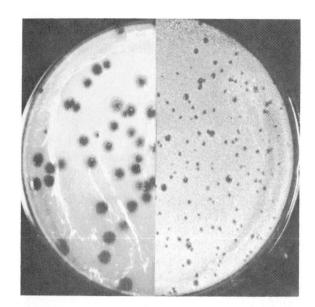

Figure 16–5. Bacteriophage plaques. On the surface of agar of appropriate composition was spread a culture of tubercle bacilli (*Mycobacterium tuberculosis*) mixed with bacteriophage specific for this organism. After incubation, the generalized growth of tubercle bacilli is seen as a whitish film. (Such growth of bacteria on agar is often called a "lawn.") The plaques of phage are seen as dark, circular holes in the "lawn." Each plaque is, in effect, a colony of phage that has grown at the expense of the tubercle bacilli. Two different types of phage are shown here: large-plaque and small-plaque. Phages specific for other organisms show these and other colonial peculiarities.

Color Plate I, *C*). The number of virulent virions or **plaque-forming units** (PFU) in the original suspension is determined by counting the plaques and applying appropriate arithmetic.

The procedures differ somewhat, depending on the type of virus dealt with. If phage virions are being assayed, the suspension of virus may be mixed with the suspension of bacteria immediately before distributing it over the agar, or the virions may be applied in a measured amount to the agar surface shortly after applying the bacterial suspension.

If it is necessary to assay an animal virus, susceptible host cells are allowed to form a monolayer in a suitable growth medium in a Petri dish or flat-sided flask. A measured volume, say 0.2 ml, of a suspension of virions is then spread over the monolayer of host cells. The infected monolayer is then covered with a thin layer of nutrient agar (the **agar overlay**) to hold the cells in place and decrease diffusion of virions.

In either procedure, incubation is at a temperature suitable for the host cells and continues until the plaques are well formed, usually within about 24 hours for bacteria, 48 to 72 hours or even up to two weeks for animal cells. Each plaque contains the progeny of the originating virion, and these may be "fished," like bacteria in a colony, to a new culture as a (theoretically) pure culture of the virus.

In order to avoid confluence of neighboring plaques and resultant complete lysis of the entire **lawn, sheet,** or **carpet** of bacteria, or the entire monolayer of tissue cells, by an overwhelmingly high ratio of virions to cells (i.e., **multiplicity** of virions), the original suspension of virulent virions is diluted. The required dilution may be determined beforehand if necessary. If the number of plaques per Petri dish (or equivalent area) is much less than 50 or much over 350 the count becomes accordingly less accurate.

Titration by Dilution. Like bacteria, viruses may be enumerated by preparing serial dilutions and determining the highest dilution capable of initiating infection of cells in culture medium. Multiplication of viruses may be determined in several ways. For example, the presence or absence of viral growth can be checked by microscopic examinations for cytopathic effects (Fig. 16–4).

Let us suppose that the contents of tubes 1 through 3 (dilutions 10^{-2} through 10^{-4}) show viral growth and the cultures in tubes 4 through 7 (dilutions 10^{-5} through 10^{-8}) do not, indicating that no active virions were introduced into those tubes. Theoretically, the undiluted sample contained at least 10^4 infective virions but less than 10^5 infective virions per ml. In this case the **indicated number** of virions would be 10^4 per ml, a result that is useful for some purposes though far from precise. Greater accuracy can be obtained by setting up the series of dilutions in duplicate, triplicate, or even pentuplicate and combining the results. Tables are available that show calculated numbers for various combinations (Chapt. 10). Such a result is called a "**most probable number**" (MPN)— somewhat nearer the actual number but not the actual number. Theoretically, each tube in our series contains a known fraction of the number of virions in the undiluted suspension. For example, if there are 1,000 virions per ml in an undiluted suspension, then in 1 ml of a 1:500 dilution there would be (ideally) two virions; one virion in a 1:1,000 dilution (the theoretical **100 per cent end point**); and none in a greater dilution.

However, it is well known that, by the laws governing chance variations of distribution, any given dilution tube might contain many more or less organisms than the theoretical or "ideal" number. In the foregoing example the 1:500 tube might contain none or 78, the 1:1,000 dilution might contain several, and so on. Such variations limit the accuracy of such titrations. Accuracy of titrations can be increased in several ways, one of which is the 50 per cent end point method.

The 50 Per Cent End Point Method. This procedure yields a much closer approach to the true number of virions in a solution than the MPN method just described. In the 50 per cent end point process we prepare several tubes of each dilution, say five of each. We select the dilution (A) that infected *more* than 50 per cent (but less than 100 per cent) of the tissue cultures inoculated with that dilution and (B) the next higher dilution. We then mathematically interpolate between these two dilutions a dilution that should give an **exact** number of virions. Such a number is the reciprocal of the highest dilution between the two that produces death of the cells in exactly 50 per cent of the inoculated cultures. From this number is derived the **tissue culture dose$_{50}$** (TCD$_{50}$) of the undiluted suspension.

For example, suppose that our virus suspension, diluted 10^{-4}, killed the cells in 75 per cent (instead of exactly 50 per cent) of the tubes of tissue culture inoculated with that dilution, while the 10^{-5} dilution produced cell death in 27 per cent of the tubes inoculated with that dilution. What dilution would have killed the cells in exactly 50 per cent (the **50 per cent end point**) of the culture tubes? Obviously a dilution between 10^{-4} and 10^{-5}. This dilution can be calculated by means of a formula devised by Reed and Munch in 1938:

$$\frac{A - 50}{A - B} = C$$

in which:

A = per cent of tubes with CPE (i.e., cells killed) at the dilution (in this case 10^{-4}) next *lower* than that which should kill all cells in 50 per cent of the tubes (*over* 50 per cent of the tubes are infected by the 10^{-4} dilution).

B = per cent of tubes with CPE at dilution (in this case 10^{-5}) next *higher* than that which should kill all cells in 50 per cent of the tubes (*less* than 50 per cent of the tubes are infected by the 10^{-5} dilution).

C = the negative log of the difference between 10^{-4} and the dilution that should give the 50 per cent point.

In the example given:

$$\frac{75 - 50}{75 - 27} = \frac{25}{48} = 0.52$$

C (−0.52) is added to the log (−4) of the lower dilution (10^{-4}) to give the exponent of the dilution required to give the 50 per cent end point: i.e., $-4 + (-0.52) = -4.52$. Thus the required dilution is $10^{-4.52}$.

If the series of dilutions is not 10-fold (dilution factor = 10; log = 1) but some other, say fourfold, then the amount to be added to the negative exponent of the lower dilution (in this case −0.52) is multiplied, not by the log of the 10-fold factor, but by the log of the fourfold dilution factor, e.g., $-0.52 \times \log 4$ or $0.6 \times -0.52 = -0.312$.

It may be noted incidentally that, by using other indicators, other 50 per cent doses or end points may be determined: e.g., half of the number of animals inoculated with a particular substance may die – the LD$_{50}$ (lethal dose$_{50}$) of that substance; exactly half of the chicks inoculated with a virus dilution may become infected – the CID$_{50}$ (chick infective dose$_{50}$), and so on.

The 50 per cent end point is based on statistical considerations and is widely used in measuring potency of poisons, drugs, and organisms and in many other biological applications.

16.3
MULTIPLICATION OF VIRUSES

The Lytic Cycle

Because viruses multiply only intracellularly and because the in vitro cultivation of cells of animals and eucaryotic plant tissues is time-consuming and expensive, many virologists have turned to bacterial cells and bacterial viruses (bacteriophages) for the study of viruses. Bacterial viruses are therefore commonly used as the model for viruses in general and will be so used here. It must be remembered, however, that we deal here with a procaryotic cell-virus system and that data obtained with such a system are not completely or necessarily applicable to eucaryotic cell-virus systems such as the multiplication of measles virus in human embryonic tissue cells. Nevertheless, the two types

PLATE I

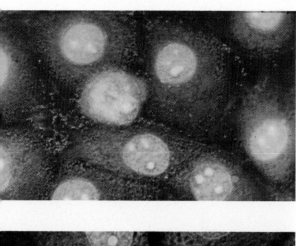

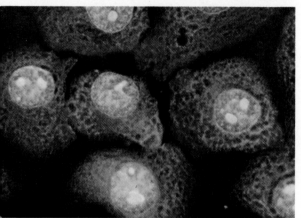

A

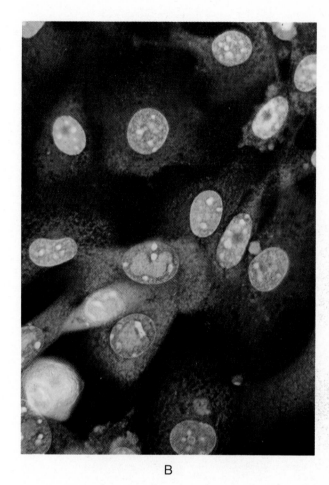

B

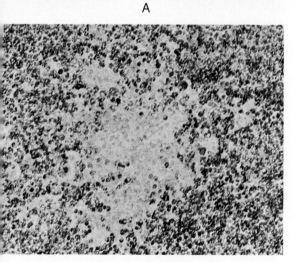

C

A, The top view shows a culture of monkey kidney cells (MK_2) contaminated with mycoplasmas. The mycoplasmas are the small granular bodies, chiefly around the cell membranes. The bottom view shows the same culture after treatment with the antibiotic lincocin.

B, This shows a colony of cells, some of which (particularly the one in the center) are infected with herpesvirus. The intense green body at the center of the nucleus is an inclusion body caused by the infection. The small yellow bodies are nucleoli.

C, The appearance of a typical plaque. The clear area contains cells killed by the virus; the surrounding cells are alive and stained with neutral red, a vital dye.

PLATE II

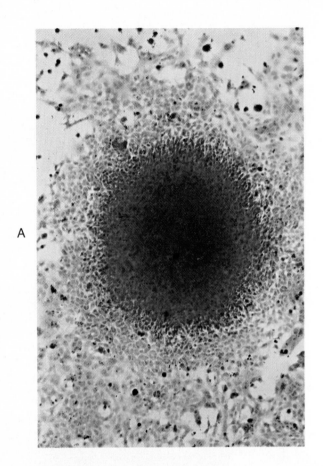

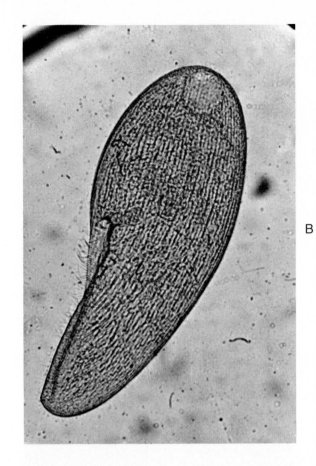

A

B

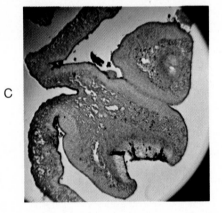

C

A, A growth of malignant cells, induced by infection with an adenovirus. Because of an evident loss of contact inhibition, the cells are overcrowded at the center of the field. Giemsa stain. (About 40×.) (Photograph by C. R. Goodheart.)

B, Blepharisma is a pigmented protozoan, an example of a eucaryotic protist. (Courtesy of Dr. A. C. Giese.)

C, A lichen is a symbiotic association between two eucaryotic protists. This picture of *Lichen apothecium* has an original magnification of 50×. (Photograph by K. T. Crabtree.)

D, A foliose lichen on a fallen birch. (Photograph by K. T. Crabtree.)

D

of system have been shown to be remarkably parallel in most fundamental respects. Many types of phage are known, each capable of parasitizing one or more different species of bacteria. A bacterium much used for virological studies is called *Escherichia* (for a German scientist, Escherich) *coli* (because the organism commonly occurs only in the colon). *E. coli* is a gram-negative, harmless (usually!), nonspore-forming, facultative, motile, rod-shaped bacillus (Chapt. 35). It grows vigorously and rapidly in simple nutrient solutions such as 1 per cent peptone, or in saline solutions with a little glucose. Under optimal growth conditions new generations (fissions) of *E. coli* occur every 20 to 30 minutes.

Isolation and Demonstration of Phage. Various types of phage are commonly present in sewage or in the intestines of coprophagous insects, or in other situations in which they find bacteria on which to feed. Bacteriophages may be isolated easily and their bacteriolytic action demonstrated as follows. Pass about 30 ml of raw sewage, or of 1 per cent peptone solution in which a dozen flies or cockroaches have been thoroughly macerated, through a sterile bacteria-retaining and virus-passing filter. Add 1 or 2 ml of the bacteria-free, phage-containing filtrate to about 10 ml of a young broth culture of bacteria (e.g., *Escherichia coli* or *Shigella sonnei*) that is just beginning to become turbid in its early, logarithmic phase of growth. Incubate the culture at 37C, observing it every two to three hours. Usually the culture will at first become more turbid and then, within two to three hours or less, much less turbid and sometimes crystal clear, because of lysis of the bacterial cells by the phage.

If the lysed broth culture is filtered, lytic phage will appear in the filtrate and can be demonstrated by plaque formation as previously described and can be transferred in high dilutions to new broth cultures. The process can be repeated indefinitely without loss, and usually with a gain, in lytic potency of the phage in the filtrates, showing that the phage multiplies.

"T" Phages. Several types of *Escherichia coli* phages (**coliphages**) are known, designated respectively as T1 (T for type), T2, T3, . . . T7. Phage lambda and some others are known also. The discussion here will deal mainly with the *E. coli*-T2 phage system, since this has been widely used as a model. The **T-even** (2, 4, 6) phages differ in several respects (e.g., form, size, chemical composition) from the **T-uneven** or **T-odd** (3, 5, 7) phages.

Although there are some tail-less phages,

most known types, including "T" phages, have a tadpole-like or sperm-like form with a polyhedral head or nucleocapsid about 50 nm in diameter and a narrow tail, whose length is two to six times the diameter of the head. They vary considerably in form (Fig. 16–6). The head contains all the active material of the phage virion, i.e., a single, linear molecule of nucleic acid. In most known phages this is double-stranded DNA. The phage tail is a hollow tube of protein attached to the protein coating of the head. The tube is enclosed within a retractable sheath. A system of fibers and an enzyme at the tip of the tail play a critical role in attaching the virion to the cell and penetrating the cell wall and membrane. Bacteriophages commonly destroy (i.e., cause lysis of) the bacterial cell that they infect.

Lysis of Bacteria by Phage. Analysis of the process of bacteriolysis by phage has resolved it into several steps, as described in the following paragraphs. The information thus gained has been of inestimable value in studying the mechanisms by which other viruses attack and destroy plant and animal cells, in the study of genetics and in the elucidation of many other biological phenomena. It is important, therefore, to have a good understanding of the mechanisms of **phage lysis.**

Adsorption and Cell Receptors. The bacterial cell wall, as previously described, is made up of molecular aggregations having various functional groups (—OH, —COOH, —NH₂) with definite physicochemical structures and distributions of electrical charges. These are sites with physicochemical **specificity** and are called **receptor sites.** Let us suppose that a phage particle comes into contact with a certain kind of bacterial cell. If the physicochemical structures in the tip of the phage tail correspond exactly and reciprocally to a specific receptor site in the bacterial cell wall, the phage tail is irreversibly adsorbed to the bacterial receptor site. The tail fibers thus play a critical role in the adsorption process.

It is important to note that bacteriophages appear unique in their mode of attachment to, and entry into, the host cell via a tail. Other viruses without such obvious structures, e.g., animal viruses, exhibit the same degree of specificity and the phenomena of attachment of the virion and entrance of nucleic acid. However, the exact mechanisms are not yet fully clarified for all viruses. Some animal viruses appear to be taken into the cell by pinocytosis (Chapt. 9).

Penetration. Within a few seconds a lysozyme-like enzymic mechanism in the tip of

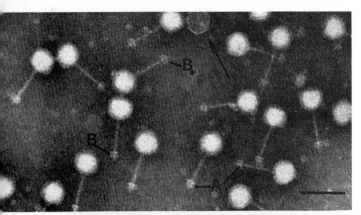

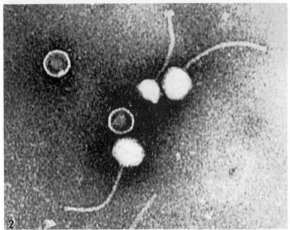

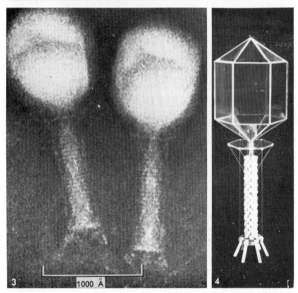

Figure 16-6. Electron micrographs of bacteriophages. Phosphotungstate preparations. *1,* Field of marine bacteriophages, showing properties of bacteriophages in general, i.e., heads and tails. An empty head displaying the hexagonal shape is seen at arrow. Base plates (*A*) reveal a triangular configuration. Base plates (*B*) present a rectangular appearance. (150,000×.) *2, Bacillus cereus* bacteriophage, phage B. (120,000×.) *3,* Phage 66t⁻. (333,000×.) *4,* Model of phage T4. (300,000×.)

the tail of the phage makes an opening through the bacterial cell wall and possibly the membrane. Through this opening the nucleic acid from inside the head of the phage is forced through the tail and into the bacterial cell (Fig. 16–7). The sheath around the tail is retracted at this time (Fig. 16–8). The protein coating of the head and tail of the phage, its mission accomplished, remains as an inert shell or "ghost" on the outside of the cell wall (Fig. 16–9).

Eclipse. Once inside the cell, the nucleic acid of the phage is no longer infective, since it has emerged from its means of entering a new cell. If the cell containing it is experimentally ruptured at this point, no infectious phage can be demonstrated. The phage is said to be in **eclipse.**

Formation of New Phage. The immediate effects of entry of the viral nucleic acid are: **immunity,** i.e., formation of a specific enzyme repressor that prevents multiplication of any other phage of that type (but not of other types); and **suppression** of all further synthesis of messenger RNA (mRNA) and of ribosomal RNA (rRNA) by the cell and, consequently, suppression of synthesis of cell enzymes. Existing mRNA of the cell ceases to function. However, cell enzymes and rRNA already present at the moment of infection are not destroyed, but are made to serve the purposes of the phage, as will be outlined.

A third important effect of entry of the phage nucleic acid is formation by the phage, from the amino acid "pool" of the cell matrix, of new enzymes foreign to the cell. These enzymes are sometimes called **early proteins.** These do not become part of the phage but (1) seal the orifice of entry of the viral nucleic acid, (2) **depolymerize** the cell DNA, and (3) synthesize the **viral nucleic acid** from some of the nucleotides thus freed from the cell DNA, and from newly synthesized nucleotides. Thus the viral genome is replicated many times. It directs synthesis of new mRNA bearing the code of the virus. Using the cell rRNA and tRNA, the viral mRNA then directs synthesis of the protein portions of the phage virion—head or tail, according to the code of the virus. These first appear early in the eclipse period as separate, "prefabricated" parts. They are then assembled, with the already-formed phage chromosomes, during a brief period of **maturation.** The result is dozens or hundreds of new, complete, mature, **infective** virions. Excess, unused, unassembled heads, tails, or nucleic acid fragments often appear as surplus "incomplete" or "defective" units

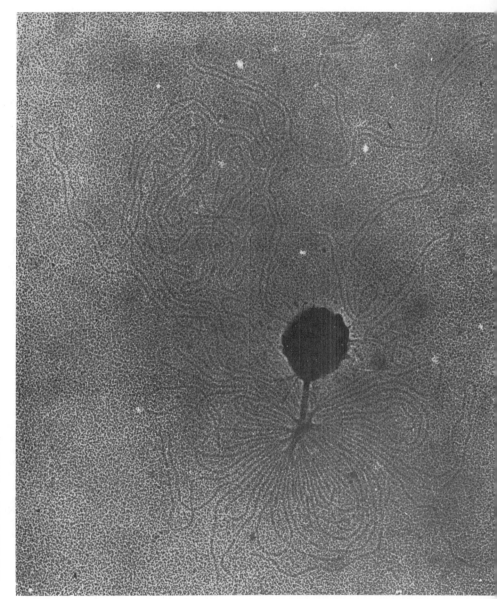

Figure 16–7. Fowlpox virus particle releasing DNA after exposure to sodium lauryl sulfate. Although this is an animal virus the DNA was frequently extruded in a configuration closely resembling a bacteriophage tail. Sample was dialyzed against phosphate buffer and prepared by the Kleinschmidt technique. Note that, except for two ends, the molecule of DNA is one continuous thread. (85,000×.)

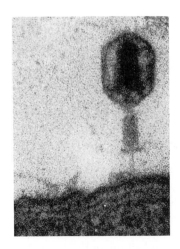

Figure 16–8. Bacteriophage T4r$^+$ infecting *E. coli.* The retracted tail sheath, base plate, and fibers are evident. (210,000×.)

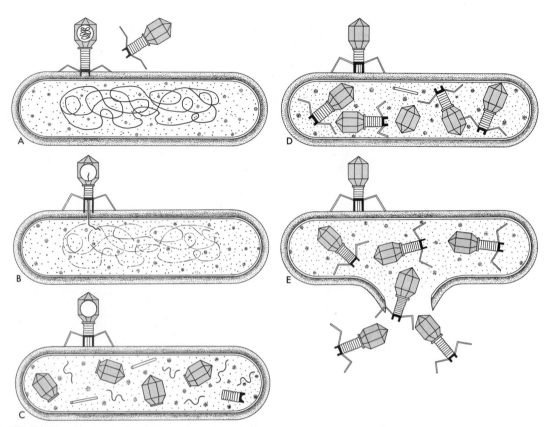

Figure 16–9. Development of a virulent phage in a susceptible bacterium. At *A* the polyhedral head of the phage (here shown in section to reveal the phage DNA) is seen attached to the outer surface of the bacterium by its tail and tail fibers. At *B*, within a few seconds or minutes the bacterial cell wall and membrane have been perforated by lytic enzymes in the tail, and the DNA of the phage has entered the bacillus through the phage tail. Note the retraction of the tail sheath. The opening through the cell wall is resealed. The inert protein coat of the phage remains on the outside of the cell. Disintegration of the cell DNA begins at once. The phage, as such, is no longer demonstrable. (What is this stage called?) During the next 12 minutes phage protein heads, tails, and phage DNA replicons are being synthesized, as seen at *C*. At *D*, about 12 minutes later, following a brief period of **maturation** during which some of the heads, tails, and DNA molecules have been assembled, infective phage virions are first demonstrable by artificial rupture of the cell. (This is the end of what period?) At *E*, about 12 minutes later, the assembly of parts into phage virions is complete and the now "eviscerated" cell, an inert sac, ruptures by enzymic lysis of the cell wall, liberating many new phage particles. Their number is characteristic of the **burst size** of the particular phage involved; in this case, six. The whole process, from *A* to *E*, occupies about 30 to 40 minutes in this particular virus-cell system. Note that in this diagram the phage virions have been drawn about three times as large proportionately in order to show more clearly certain structural details.

(Fig. 16–10). If the cell is ruptured at this time, the new infective units can be demonstrated, also the phage fragments and the incomplete or defective units.

The formation of mature, infective virions inside the cell signalizes the end of the eclipse period. The **eclipse period** is thus the interval between injection of viral nucleic acid through the cell wall and first appearance of infective virions *inside* the cell. This is often called the **vegetative phase** of the life cycle of a virus. The duration of the eclipse period in the *E. coli*-T2 phage system is about 12 minutes under optimal conditions of growth. It varies for other virus-cell systems. For example, the eclipse period of the influenza virus in tissue culture cells is about seven hours.

The Latent Period. The assembly of new virions from the prefabricated parts continues inside the cell for some time after the end of the eclipse period. In the *E. coli*-T2 phage system the cell wall ruptures about 18 minutes after the end of the eclipse period, and the new phage units are then set free to begin the cycle anew. The cell is said to have undergone **lysis from within.** This signalizes the end of the **latent period.** The latent period is the time from adsorption up to, but not including, cell rupture. It includes the eclipse period. Do not confuse latent period of viral growth with **latency** (see farther on).

The Lytic Cycle. The period from adsorption to cell lysis comprises one **lytic cycle** of phage activity (Fig. 16–11). As will be described, there are other cycles of viral activity.

Burst Size. The average number of mature virions released per cell by rupture of the cell wall at the end of the latent period is more or less constant in any given cell-virus system. It may range from about 20 to 200 or more. This number is referred to as the **burst size** of the cell at the end of the latent period. In the *E. coli*-T2 phage system just described the burst size is about 200.

Cause of Cell Lysis. Cell lysis is apparently not caused by expansive pressure from within. Under the influence of the phage the cell appears to form, or to cease to inhibit, enzymic agents capable of destroying the cell wall. This causes **lysis from within.** Similar lytic agents have been found in **phage lysates** (suspensions of cells lysed by phage) of numerous bacterial species. In some respects these lytic agents resemble an enzyme called **lysozyme** that hydrolyzes the mucocomplex of cell walls (Chapt. 14). Some of the enzymes found in phage lysates can depolymerize capsular polysaccharides; others digest mucopeptides of cell walls and cause **lysis from without.** Whether such lytic agents are involved in lysis of virus-infected cells of animals or higher plants is not yet clear.

Animal cells, of course, have no cell walls. This affects the mechanisms of attachment of viruses and penetration of animal cells by animal viruses. As will be explained, certain types of animal viruses continue to be matured slowly on or in the cell membrane during a period of days or weeks. In passing out of the cell, each viral nucleoid acquires a part of the nuclear or cell membrane as its outer envelope (Fig. 16–12).

The growth of phage during the latent period has been cleverly investigated. As shown by providing nutrients containing radiophosphorus (^{32}P) and radioisotopes of sulfur, carbon, and nitrogen, the virus seems to derive some of its phosphorus and nitrogen from compounds in the medium in which the host cell is growing, but about one-fifth of each of these elements in the virus is derived from molecules of the host cell itself. Improperly nourished cells may fail completely to support multiplication of a virus to which they are ordinarily fully susceptible.

Host Specificity. An important factor in multiplication of viruses is **host specificity.** In the case of phages, host specificity, as mentioned previously, depends on the presence of reciprocally adaptive adsorption sites on cell wall and phage tail. In the case of viruses of animal cells

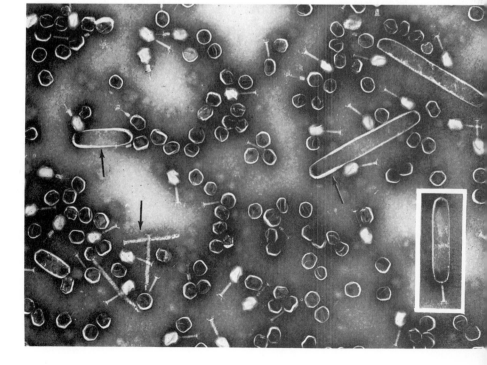

Figure 16–10. Abnormal forms and faulty maturation of T-even bacteriophage as induced by various substances; in this case by an amino acid analogue, L-canavanine. Note the numerous long ovals which are several connected heads (polyheads), some with tails attached (insert), tail-less heads, headless tails, empty heads, polytail sheaths, polytail tubes and small heads.

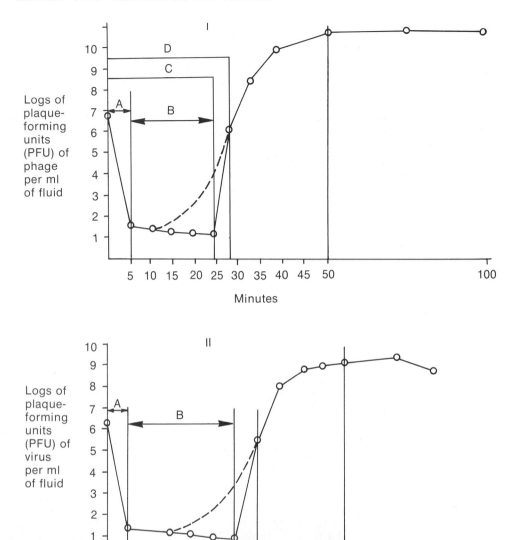

Figure 16-11. Representative growth curves of viruses in cell cultures: (*I*) bacteriophage; (*II*) one type of animal virus, e.g., poliovirus. Each young cell culture was inoculated with 10⁶ to 10⁷ active virions. In about 5 minutes (**period of adsorption and penetration,** *A*) most virions had infected susceptible cells. A few, unadsorbed, remained demonstrable in declining numbers in the suspending fluid during the **eclipse period** (*B*). During this period viral NA in each infected cell induced synthesis of many new, unassembled, phage parts: heads, tails, NA, etc. (broken line; exact form hypothetical). At the end of the eclipse period, intracellular assembly of these parts began (**period of maturation**). During this period (about 4 minutes in *I*, 3 hours in *II*) complete, infective virions were demonstrable **intracellularly** in increasing numbers by rupturing cells artificially at successive intervals. Immediately following maturation of all virions the cell walls and membranes ruptured by **lysis from within** (end of **latent period,** *D*, and of the **lytic cycle**). Many new virions were thus released from each infected and lysed cell almost simultaneously, causing a sudden rise in numbers of **free** virions. Total lysis of all infected cells in each culture occupied about 22 minutes in *I*, 13 hours in *II*.

While growth curves for some animal virus-cell systems resemble *II* (poliovirus) as shown here, others differ markedly owing to different rates and mechanisms of release of new virions; e.g., see Figure 16-12.

and cells of higher plants, the phenomenon of host specificity is evident to the same degree, but its exact mechanism is not so clear.

Under natural (as distinguished from ex-perimental) conditions, all viruses, including phages, are more or less restricted as to host. For example, plant and fish viruses cannot directly infect mammalian cells (so far as we know at this moment!). However, plant viruses commonly infect their insect vectors (e.g., leaf-hoppers), and some animal viruses infect the insects (e.g., mosquitoes) that transmit them. Smallpox, measles, and polioviruses normally infect only humans. On the other hand, rabies virus will infect virtually any mammal but not any plant or insect. Yellow fever virus in nature is restricted to man, certain monkeys, and a few lower animals; however, under experimental conditions it will infect avian embryos, mice, and guinea pigs' brains if introduced with a hypodermic syringe. Among bacterial viruses a given phage is usually restricted to a single species of bacteria or even to a single special type (called a **phage** type) of that species. Some animal viruses will infect only cells from certain specific tissues of certain specific animals; others can infect cells of a wide range of species.

Phage Typing of Bacteria

If we continuously propagate a virus in a single type of cell, the virus often becomes highly adapted to that single type of cell and will not infect cells of any other type. In a practical application of this fact, adaptation of phage may be carried to such a degree that a given **strain*** of phage becomes so selective with respect to a single species of bacterium, or even to a certain type or subdivision of that species, that it will not infect any other. Thus, it will distinguish, by its lytic effect, between apparently identical bacteria that are indistinguishable by any other means.

*Any designated specimen or culture, or progeny of same. (See Chapter 31.)

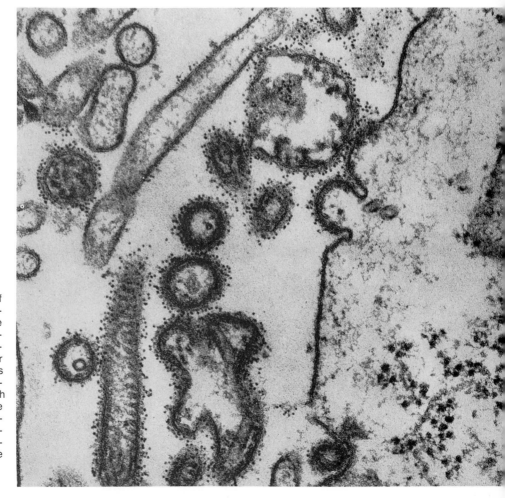

Figure 16–12. Maturation of parainfluenza virions at the surface of a human tissue cell. The proteins (antigens) of the virions have combined with ferritin-labeled antibody (see Chapter 27) and are seen as numerous minute black specks surrounding the virions, some of which are rounded (left, center), some apparently misshapen. The release of two new virions is exemplified by the rounded, bud-like bodies (right, center) at the cell margin. (70,000×.)

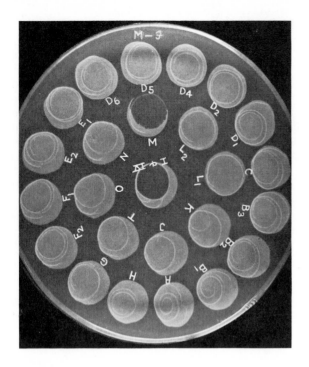

Figure 16–13. Use of highly specific bacteriophages to type typhoid bacilli. Separate drops of a culture of *Salmonella typhi* of unknown type have each been mixed with a different type of *Salmonella typhi* bacteriophage (indicated by letters) on the plate. Lysis (circular dark area in the white growth) occurs only where phage type and bacillus type correspond—in this case, type M. The lysis in the center of the plate is a control test. Closely parallel processes are used in the phage typing of staphylococci, *Shigella,* and other bacteria. (Photograph courtesy of U.S. Public Health Service, Communicable Disease Center, Atlanta, Ga.)

For example, when we propagate a given phage on a certain selected strain of *Salmonella typhi* (typhoid bacillus), the phage becomes so specific for that particular strain of *S. typhi* that when appropriately diluted* it will not act on any other strain of the same species. The first

highly specific typhoid phage of this sort was designated as **typhoid phage A**, and the corresponding susceptible strain of typhoid bacilli as **phage-type A** of *S. typhi*. By a similar process several other phage types of *S. typhi* were discovered and designated by the letters A, B, C, D, E, and so forth. Similar bacteriophage typing systems have been developed for several other species of bacteria, notably *Staphylococcus* (see Chapter 37). Phage typing is widely used in medical diagnosis and industry (Fig. 16–13).

---

*If not diluted, the selective specificity is masked by an overwhelming action on nearly all forms of typhoid bacilli. The highly specific dilution used for typing is called the **critical dilution.**

CHAPTER 16
SUPPLEMENTARY READING

Goodheart, C. R.: An Introduction to Virology. W. B. Saunders Co., Philadelphia. 1969.

CHAPTER 17 · THE VIRUSES

Intracellular and Genetic Phenomena, Lysogeny, Transduction

How viruses perpetuate their kind is quite different from the reproductive process of any other organism. All cellular life—even to the most degenerate and dependent form of bacteria—propagates by an increase in size of an individual organism to some critical size, followed by reproduction of the organism, frequently by simple fission, to yield new individuals.

But viruses, not being cellular, do not grow in this sense. They merely insinuate a new set of genetic instructions into a susceptible host cell that unwittingly proceeds to do what these new instructions require—to make virus parts and finally to assemble these parts into new virions like the infecting "parental" virus. During its sojourn in the infected cell, the virus, or rather its nucleic acid, actually becomes a part of the cell and is expressed in many ways in addition to directing the cell to make new virions. Interactions between virus and cell, to be discussed next, have been some of the most fascinating and informative aspects of the study of viruses.

17.1
FUNCTION OF VIRAL NUCLEIC ACID

When a phage infects a bacterial cell, only the viral nucleic acid enters the cell. The viral protein coat is left outside as an empty "ghost" (Chapt. 16). But because the progeny virions are complete and include newly synthesized viral coats, the information for the coat must have been encoded in the viral nucleic acid.

Animal viruses, including the coat, are completely engulfed by the infected cell, making less obvious the distinct role of the nucleic acid in perpetuating the infection. However, various experimental means have been developed to remove the protein artificially from the nucleic acid. The nucleic acid can then be purified, and under certain conditions can be introduced into susceptible cells. The cells thus infected with viral nucleic acid alone still produce complete virions with the appropriate protein capsid.

Such naked nucleic acids lose some of the specificity for host cell type possessed by intact virions because the specificity for attaching to a cell resides in the capsid. For example, intact poliovirus cannot attach to chick cells and therefore does not replicate in that cell. The RNA extracted from poliovirus, however, can be made to penetrate into the interior of chick cells, whereupon the cells synthesize complete infectious virus.

After being set free in the cells, the viral nucleic acid begins to function. With the aid of cellular enzymes, some of the viral genes are transcribed into complementary copies of mRNA. These virus-specific mRNA molecules substitute for the host cell's mRNA in the ribosomal-protein synthetic machinery (see Chapter 6).

The first genes to be transcribed are those concerned with replication of the viral nucleic acid. With DNA viruses, for example, enzymes for synthesizing DNA precursors and DNA polymerase are the main enzymes synthesized early after infection. In some cases, enzymes or other proteins also are synthesized that either degrade the cell's DNA or render it nonfunctional.

Collectively, those functions that occur soon after infection, prior to replication of the viral nucleic acid, are referred to as "early" functions.

Next, the cell begins to replicate the viral nucleic acid, making use of the new enzymes and new precursors that the cell has synthesized so cooperatively according to the viral instructions. As the new viral nucleic acid molecules are set free, they too begin to function. However, other genes are transcribed from the new molecules in addition to the "early" genes. Now, genes that are transcribed have the information for synthesizing proteins of the viral capsid and other proteins necessary to complete the assembly and release of infectious progeny virions. These functions are referred to as "late" functions. The mechanisms controlling the switch from "early" to "late" functions are under intense investigation with many viruses. It is believed that when this problem is understood, scientists will have a much better understanding of genetic function and control in all species.

The eclipse period, described as part of the replicative cycle of **lytic** or **virulent** viruses in Chapter 16, therefore is seen to be a period of great activity in the infected cell. Although the virus is not identifiable as such during the eclipse period—artificial disruption of the cell releases no infective virus—the viral nucleic acid is anything but inactive.

During this period, many antiviral chemicals can block viral synthesis, viral-specific proteins are detectable serologically and in some cases by their biological activity, and the viral nucleic acid makes changes in the cell that may be visible microscopically.

17.2
SOME ANTIVIRAL SUBSTANCES: CHEMOTHERAPY

Chemotherapy is the treatment of infections with specific chemical substances. These may be antibiotics, sulfonamides, or, in preantibiotic days, empirical treatments such as malaria with quinine and syphilis with mercury compounds or arsenicals (salvarsan or 606 of Paul Ehrlich, Nobel Prize winner). Typical modern chemotherapeutic agents are effective in amounts that are harmless to the infected host but that inhibit or destroy the infecting agent. Typically, the inhibition or destruction of the parasite is achieved through chemical combination of the chemotherapeutic agent with a specific enzyme, coenzyme, vitamin, or other essential metabolic mechanism of the parasite. In their highly specific action, chemotherapeutic agents differ markedly from the wholesale, generalized, destructive action of disinfectants like the cresols, bichloride of mercury, and the halogens. These, in effective amounts inside the body, would be fatal to the host as well as to the parasite (Chapts. 21 to 23).

The sulfonamide drugs and the antibiotics derived from bacteria and molds, e.g., streptomycin and penicillin, in general act on cellular parasites such as bacteria or Protozoa by interference with certain specific enzyme systems, most of which are not present in viruses. Viruses are not vulnerable to most such enzyme-inhibiting chemotherapeutic agents. However, viral nucleic acid must be synthesized if viruses are to multiply. An agent that interferes with synthesis of viral nucleic acid, or that causes the formation of faulty or distorted viral nucleic acid, could be an antiviral chemotherapeutic agent if not toxic for the host cells in the amounts needed to inhibit the virus. A number of substances have been found that are more or less effective **experimental** antiviral agents though, so far, only a very few have practical clinical usefulness. The field is under active investigation, however, and more will certainly appear. (See Mutagenic Agents, Chapter 18.)

Analogues in Chemotherapy. Antiviral action may be brought about by chemical analogues of the purine or pyrimidine bases that occur naturally in nucleic acid. Analogues of normal nucleosides of nucleic acid can be prepared by halogenation of the bases and coupling them to ribose or deoxyribose. These altered or "ersatz" nucleosides can occupy the place of the normal substances but cannot perform the normal functions. They are chemical "impostors." Some of them, e.g., 5-fluoro-2′-deoxyuridine, inhibit the enzyme system thymidylate synthetase that synthesizes the nucleoside deoxythymidine, essential to the synthesis of viral DNA. Others, e.g., 5-bromo-2′-deoxyuridine, are built into the viral DNA. The resulting "erroneous" DNA containing the analogue cannot form perfect virions. In other examples, 5-iodo-2'-deoxyuridine (IUD or IUDR) and isatin-β-thiosemicarbazone (IBT) and its methyl derivative (MIBT) (Fig. 17–1) appear to act by preventing final maturation of some DNA viruses; synthesis of DNA and other parts of the virion occurs but they remain unassembled. These drugs have preventive or curative value in certain cases of herpes simplex and smallpox; both are caused by DNA viruses.

RNA viruses (e.g., several picornaviruses) are selectively inhibited by 2-(α-hydroxybenzyl) benzimidazole (HBB) and guanidine. These drugs appear to inhibit formation of virus-induced RNA polymerase. As a result viral RNA and, secondarily, capsid proteins are not produced. The drugs have little or no effect on host cells or on DNA viruses. Most of these mechanisms are still under investigation.

In another example of analogue effectiveness 5-fluorouracil "fed" to infected cells can be taken into virus-coded mRNA in place of normal uracil. It prevents replication of the viral DNA though not necessarily of other parts of virions. Thus, in herpesvirus so treated, the "virions" that are produced have no DNA cores.

As will be seen in Chapter 22, chemical analogues of many sorts, and their stoppage or distortion of various enzymic functions, are basic factors in antimicrobial chemotherapy.

Interferon. Unlike artificial inhibitors of viral replication, interferon is a natural substance produced by cells infected by a virus. Actually, several interferons exist.

The nucleic acid of many animal viruses, on entering a mature (not embryonic) animal cell, induces the cell to produce a new type of protein that prevents replication of the infecting virus and also of many others. The first such protein was described by Isaacs and Lindenmann in 1957. Because it was found to **interfere** with **intracellular** viral multiplication, it was called **interferon.**

The inhibitory effect of any interferon appears to be exerted early in the eclipse phase of the viral growth. Probably the interferon inhibits the viral nucleic acid synthetase system, although its exact mode of action, and the mode of its induction, are still under intensive investigation. It does not interfere with normal cell metabolism. Its action is entirely intracellular; it does not act in any way upon the intact virion outside the cell—i.e., it does not prevent infection. Interferon production may be induced in cells growing in culture (in vitro) as well as in cells in living animals (in vivo). It is now known that there are several proteins of slightly differ-

ent chemical and physical properties that have the characteristics of interferon, i.e., there are several different forms of interferons. All are produced only by cells, not by viruses.

An interferon is liberated by an infected cell into the surrounding fluid. If taken into adjacent cells of the same species, it interferes with intracellular viral replication in them also. Interferons are thus probably important factors in recovery from (not prevention of) many viral infections. Their value as therapeutic agents is still under investigation.

A practical difficulty in therapeutic use of interferons is production in sufficient quantities. However, interferons may also be induced in cells in the absence of viral infection by various bacteria, bacterial cell walls (i.e., endotoxins; see Chapters 26 and 28), animal nucleic acids, yeasts, etc. Thus, therapeutic quantities of interferons may become available without the necessity of infecting cells with viruses.

Interferons taken into cells appear to act indirectly against viruses by inducing the cell to synthesize (or by removing a repressor substance that prevents synthesis?) a substance that is the actual antiviral agent. The exact nature of this substance is not yet clear. The cell must be genetically capable of such synthesis. The antiviral agent appears to act by inhibiting synthesis of viral mRNA and hence of viral nucleic acid.

In physicochemical properties interferons are neither enzymes, coenzymes nor antibodies as defined in Chapter 27. Interferons have no immunological relationship to the viruses that induce them. Interferons are relatively small, yet nondialyzable protein molecules ranging in molecular weight from about 20,000 to 100,000, depending on source and inducer. They are readily hydrolyzed by proteolytic enzymes, e.g., trypsin. Unlike enzymes, they are relatively stable to heat (60C for 1 hour) and remarkably stable to a wide range of pH (2.0 to 10.0). An exact definition of interferon(s) in terms of physicochemical properties remains to be formulated.

A striking property of any interferon is its relative species specificity. Although it can interfere with replication of many different kinds of viruses, it does so only in cells of the species that produced it or in phylogenetically closely related species. For example, interferon produced by chick cells will not interfere with virus replication in cells of mice, and vice versa. However, the species specificity of interferons is not absolute, since it may extend, in a limited degree of effectiveness, across boundaries between closely related species. Thus, monkey in-

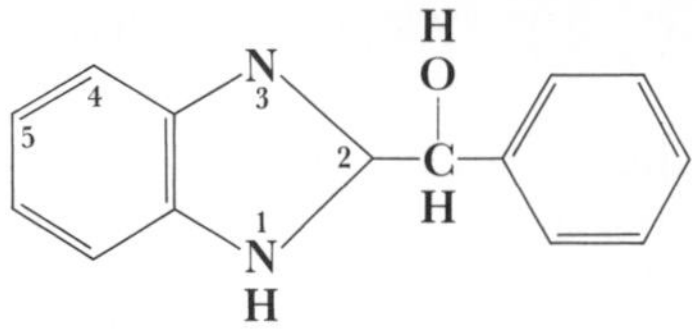

HBB 2-(α-hydroxybenzyl)-benzimidazole

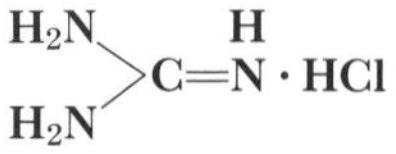

Guanidine HCl

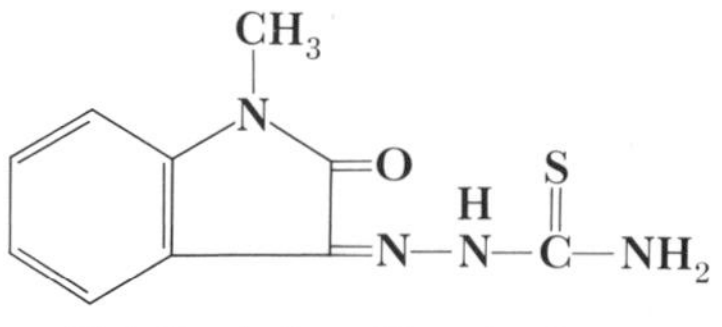

Methylisatin beta-thiosemicarbazone
(MIBT)

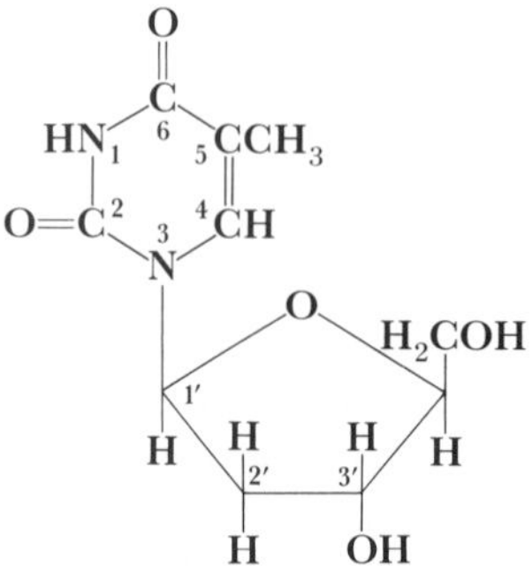

Thymidine

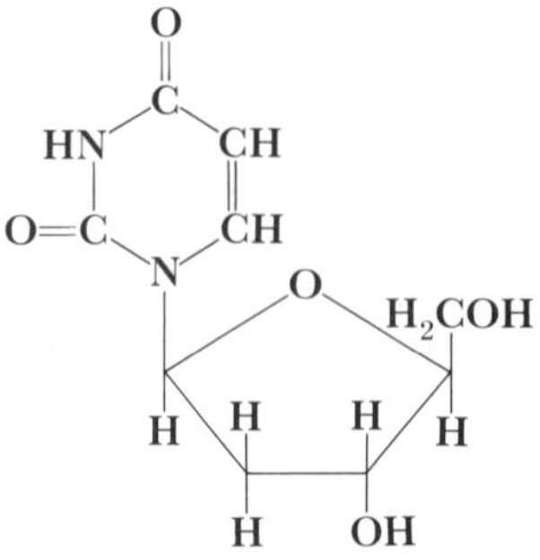

2'-deoxyuridine

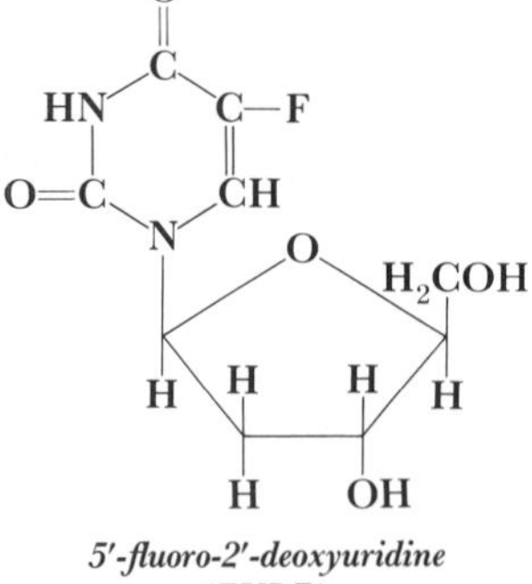

5'-fluoro-2'-deoxyuridine
(FUDR)

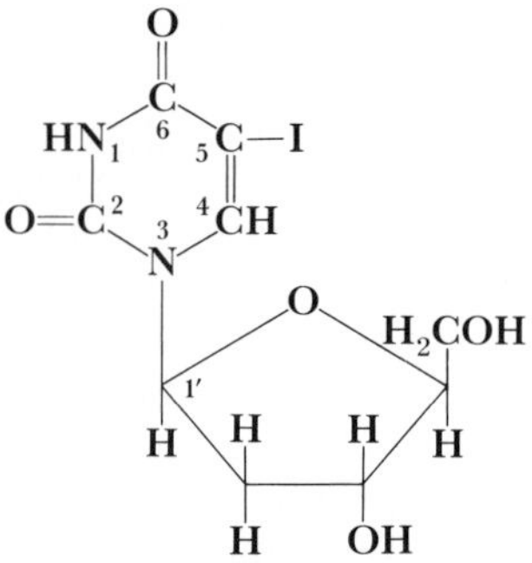

5-iodo-2'-deoxyuridine
(IUDR)

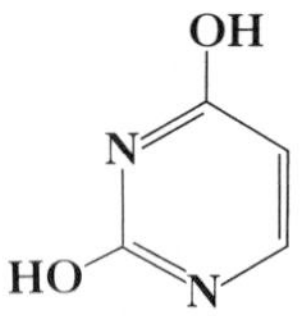

Uracil

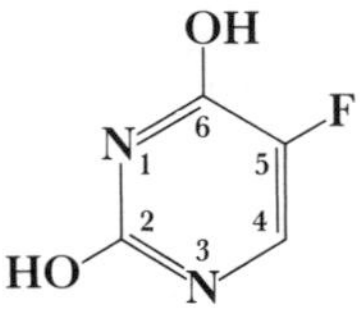

5-fluorouracil

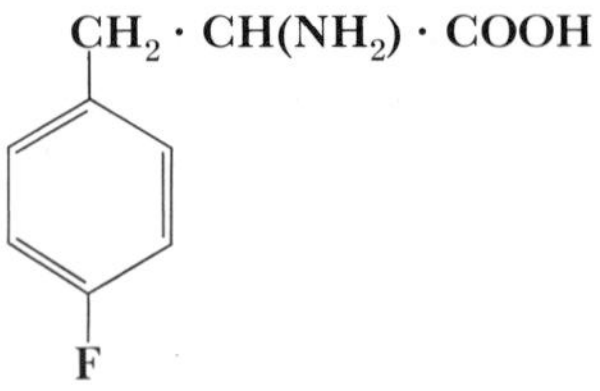

$CH_2 \cdot CH(NH_2) \cdot COOH$

p - fluorophenylalanine

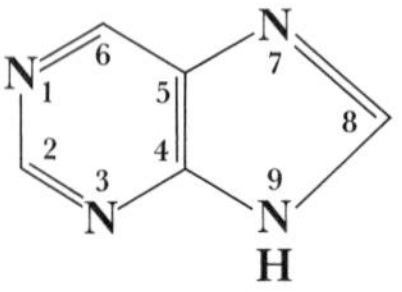

Purine

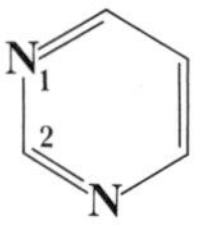

Pyrimidine

Figure 17-1. Formulas of purine, pyrimidine, halogenated derivatives and nucleosides, and some other antiviral drugs. Note similarities between normal and derived substances (analogues).

terferon will have a small but definite effect in human cells, and mouse interferon will show some effect in cells of related rodents such as rats and hamsters. Specificity sometimes extends also to particular types of cells in animal species.

There are a number of substances that resemble interferons in physicochemical and biological properties. These substances occur inside cells and in association with them, but their source and nature are not yet understood. They are collectively called "interfering substances." Some may actually be interferon normally present in cells, and some may be mixtures of other antiviral substances. Some are found free in blood and in other biological substances in the absence of viral infection. Most may be differentiated from true interferon(s) on the basis that they are normal, preformed components of cells in which they occur and not newly synthesized proteins like interferon. The antibiotic actinomycin D, which blocks synthesis of DNA-dependent mRNA, also prevents synthesis of true interferons but not of the "interferon-like" substances.

17.3
SOME INTRACELLULAR EFFECTS

Intracellular Inclusions. Within many types of cells infected by a virus, various abnormal granules are visible with optical microscopes. These abnormalities are called **intracellular inclusions.** They may occur in the nucleus (**intranuclear inclusions**) or cytoplasm (**cytoplasmic inclusions**) or both (Fig. 17–2). Some appear to be dead and altered cellular debris, and others may be aggregations of virus particles. (See also Color Plate I, *B*.)

Depending on location and staining properties, intranuclear inclusions have been classified as Type A (a single, **eosinophilic** particle filling most of the central nuclear area) and Type B (variable numbers and sizes of **basophilic** intranuclear granules). The location and type of intracellular inclusions are highly distinctive of certain viruses and can be used diagnostically. For example, cells in certain areas of the brain of rabid animals contain characteristic and diagnostic inclusion bodies called Negri bodies. Intranuclear inclusion bodies formed by human adenovirus type 1 are shown in Figure 17–3 and by human cytomegalovirus in Figure 17–4.

Virus Crystals. These are aggregations of virus particles arranged in infected cells as orderly arrays. If the viruses are spherical or

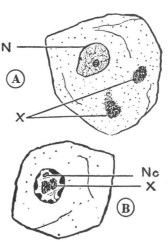

Figure 17–2. *A,* One type of cytoplasmic viral inclusion body is shown at *X.* The nucleus is shown at *N. B,* Intranuclear inclusions in liver cell of victim of yellow fever. *Nc,* nucleolus; *X,* inclusion bodies. These are eosinophilic da Rocha-Lima bodies and probably of the A type mentioned in the text. Note the characteristic basophilic lobulations at the nuclear membrane resembling inclusions of the B type.

polyhedral, they often appear in electron micrographs like many marbles or buckshot placed as close together as possible in ordered rows (Fig. 17–5). These arrangements are typically three-

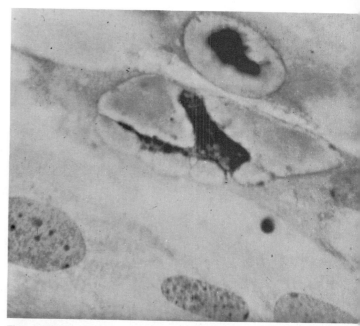

Figure 17–3. Human embryonic fibroblasts infected with human adenovirus type 1. Intranuclear inclusions are visible in the upper two cells as the darkly staining, irregularly shaped bodies. Several nuclei of normal cells are seen in the lower part of the picture. The cytoplasm is faintly stained in these cells. (1,200×.)

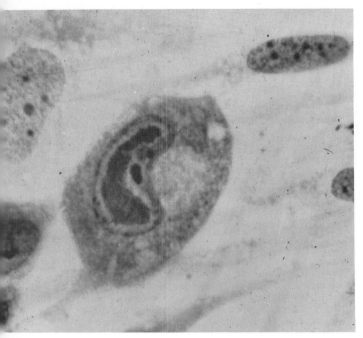

Figure 17–4. A human fetal fibroblast infected with the human cytomegalovirus. The infected cell in the center shows several features characteristic of infection with this virus: The cell seems enlarged and rounded; the cytoplasm is more darkly stained than in the uninfected neighboring cells; and the nucleus is somewhat kidney-shaped, with marginated chromatin separated from the intranuclear inclusion by a pale-staining halo. The dark bodies lying in the halo are the nucleoli and the cytoplasmic inclusion appears as a pale-staining area adjacent to the nucleus. (1,000×.)

Resistance and Interference.

In preceding sections viral infection of a susceptible cell was described as resulting in lysis of the cell or a lytic cycle of the virus. A virus that, on entering a cell, promptly becomes **vegetative** in the cell and destroys it, liberating many new mature, infective virions, is said to be **virulent**. As indicated, cell lysis is not the only or inevitable result of contact between a cell and a virus particle. The phenomena of **resistance** and **interference** may be exhibited or, as detailed in the next section, **lysogeny** may result.

A cell may be wholly **resistant** to a virus because adsorption cannot occur. Failure of adsorption may be caused by the absence of specific receptor sites for that specific kind of virus on the cell surface because of species peculiarities of that cell. A cell species, ordinarily susceptible, may undergo a genetic mutation altering its receptors for a specific virus. Further, receptor sites present on a cell may be destroyed or altered by some nongenetic environmental factor such as heat, pH, enzymes from other cells or viruses, or presence or absence of certain metallic ions.

In **interference,** adsorption of an active virion by a fully susceptible cell may fail because all the specific receptors of the cell have been preempted or saturated by the same sort of specific virions which have been previously inactivated by heat, ultraviolet light or other agents. Adsorption is said to be **interfered with.** Receptors may also be blocked by defective or incomplete virions. Some fully active viruses can interfere with secondary infection by a closely related active virus, presumably by prior pre-emption of receptors specific to both forms of virus. For

dimensional. In this arrangement they constitute **crystals** of viruses (Fig. 17–6). The crystal form is characteristic of the virus-cell system involved.

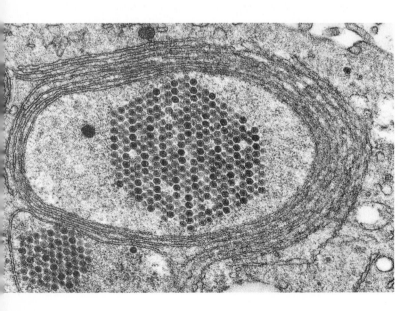

Figure 17–5. An electron micrograph of a thin section of an adenovirus-infected cell showing a crystalline array of virions surrounded by multiple layers of nuclear membrane. (45,000×.) (Courtesy of Dr. C. Morgan.)

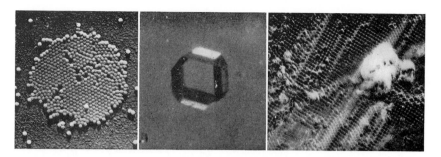

Figure 17–6. Poliovirus particles in a flat array (*left*) and in a three-dimensional crystal (*center*). A cut surface of a crystal (*right*) shows it to consist of virus particles in orderly arrays.

example, if monkeys are infected with a neurotropic (i.e., adapted specifically to nerve tissue) variant of yellow fever virus they are immediately resistant to the normal or viscerotropic (i.e., adapted to the viscera) yellow fever virus; if rabbits are infected with fibromavirus (which causes a disease of rabbits) they are immediately wholly resistant to the closely related myxomavirus. Irradiated virus of influenza or Newcastle disease of chickens will interfere with infection by an active virus of the same type. The immunizing effect of the Sabin oral polio vaccine, an active virus that immunizes by initiating a mild intestinal infection, may be interfered with by prior presence of several related common intestinal viruses (enteroviruses). (See Classification of Viruses, Chapter 15.) Interference of the types just described must not be confused with the action of **interferon**, which is synthesized by tissue cells *after* they become infected, and which prevents replication of viruses in other cells of the same type and species (see Interferon, this chapter).

17.4 LYSOGENY

A virus may contact a susceptible cell and the viral nucleic acid may enter that cell without causing immediate lysis. In a phage infection, lysis may fail to occur or be delayed because the phage is not fully **virulent** for that cell; i.e., the virus nucleic acid is not able to initiate the immediate, unrestricted, vegetative replication of phage that is necessary to the lytic cycle. The phage is said to have undergone a change (**reduction**) to what is designated as the **temperate** state. When the phage becomes temperate it becomes closely associated with, or actually a segment of, the single, circular, bacterial chromosome.

Lambda (λ) is one of the best-studied temperate phages and usually serves as the model system, with *E. coli* as its host cell. A strain of

E. coli that carries λ is said to be **lysogenic.** This means that every cell of a lysogenic culture has the potential, transmitted genetically to daughter cells, for producing λ. A healthy, growing lysogenic culture continually produces small amounts of λ when the virus in a certain small fraction of individual cells in the culture becomes vegetative, replicates, and causes lysis of those few cells with release of virus.

Infection of a nonlysogenic *E. coli* cell with λ can result in either a complete replicative cycle as described in Chapter 16 or lysogenization. If the replicative cycle takes place, all the steps of the growth cycle occur, including the "early" and "late" functions, replication of the viral DNA, synthesis of viral parts, and assembly and release of infective virions.

In **lysogenization,** however, the replicative cycle begins with certain early functions, but instead of continuing, the viral DNA attaches to a specific site on the bacterial chromosome where there is a region of base sequence homology between viral and cellular DNA. The bacterial DNA breaks in that region, the phage DNA is inserted, and the bacterial chromosome reassumes its complete, circular structure. A set of enzymes actually forms covalent bonds between the viral and cellular DNA, so the viral DNA becomes a part of the cellular DNA in every sense and even replicates with it. Lysogenization is illustrated diagrammatically in Figure 17–7.

The following model has been proposed to explain whether lysis or lysogeny takes place: After infection of a nonlysogenic cell with a temperate phage, various phage genes begin to function. The "early" genes, concerned with synthesis of enzymes, start to transcribe messenger RNA to direct the synthesis of a **repressor** substance. A postulated "race" ensues between the synthesis of repressor substance and early proteins. If the enzymes become sufficiently concentrated, new viral DNA is synthesized, and the cell produces more phage particles and lyses. If, however, the repressor substance

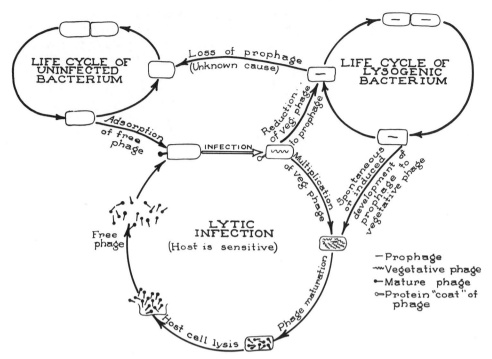

Figure 17–7. Phage-host life cycles. For explanation see text.

reaches a sufficient concentration in the cytoplasm, it represses the expression of other phage genes, and lysogenization results.

Once it has become a part of the bacterial DNA, the viral genome is called a **prophage.** The repressor substance is produced continually; as long as it is present in the cell at a certain minimum concentration, the prophage remains repressed and a part of the cellular DNA.

Induction. If a lysogenic cell is subjected to certain influences such as hydrogen peroxide, ultraviolet irradiation, or certain chemicals, the concentration of repressor decreases, the prophage is excised from the bacterial DNA, and the lytic cycle begins. It is said to have been **induced.** Artificial induction of a culture can result in simultaneous lysis of the majority of cells.

Induction can also occur spontaneously, as indicated above, and does so in a small proportion of cells. Sometimes spontaneous induction is due to an accidental decrease in concentration of the repressor below the critical level in an individual cell. It can also occur when bacteria mate. As a bacterial chromosome containing a prophage is transferred from a cell with high concentration of repressor into a cell with low concentration of repressor, the prophage is no longer repressed and enters the lytic cycle. This is called **zygotic** induction.

Lysogenic Immunity. A cell containing a prophage (a lysogenic cell) is immune to superinfection by any other virion of that specific kind of virus even though an infective virion may be adsorbed and its DNA penetrate into the cell. Immunity thus conferred on a cell by viral infection appears to be caused in part by the repressor described above that not only keeps the prophage already present from entering the lytic cycle, but also represses vegetative replication of any secondarily entering genome of the same specific kind of virus. Further, if the attachment site is already occupied, a second genome of the same type of virus cannot attach. However, one or more viruses of some other type(s) may enter the cell and either destroy it at once or be severally reduced to the temperate state if there are specific, unoccupied loci in the cell genome for their attachment.

The immunity of lysogenic cells is called **lysogenic immunity.** Lysogenic immunity to a given virus persists only so long as that prophage is present in the cell. If, as sometimes happens, the cell is "cured" of its lysogenicity; i.e., if the prophage is inactivated or for unknown reasons fails to be transmitted to a daughter cell, lysogenic immunity of that cell and of its progeny to that type of virus disappears. Note that lysogenic **immunity** and **resistance** of a cell to viral infection are two distinct phenomena dependent on totally different mechanisms. (Explain. Compare also with interferon.)

Many strains of lysogenic bacteria are well known and are widely used in virological investigations. They are designated by adding to their specific name, parenthetically, as a suffix, the designation of the prophage that they carry; thus the lambda (λ) phage in the lysogenic strain of *Escherichia coli* known as K-12 is indicated as *E. coli* K-12 (λ).

Lysogeny can also be a problem in industry. A strain of bacteria used in a fermentation process, for example, may be lysogenic and this fact can be undetected for a long period. Suddenly, all cultures may lyse due to induction, much to the embarrassment of the microbiologists in charge.

Lysogenic Conversion. The establishment of a prophage in the genome of a bacterial cell results in the addition of new genetic units. This commonly alters the heritable characters of the cell, often so profoundly that the bacterium becomes essentially a new species, except for the fact that the prophage is apt to leave at any time, taking its genetic characters with it. A genetic change caused by addition of genetic characters by infection with a virus is sometimes called **infective heredity.** It is also known as **lysogenic conversion** or **phage conversion.** For example, *Corynebacterium diphtheriae* is the bacillus which causes diphtheria (see Chapter 39). The principal effects of diphtheria on the patient are those caused by diphtheria **toxin,** a poisonous protein waste product which the toxigenic bacteria secrete whenever they grow in a suitable medium in a laboratory, or in a patient's throat. Strains of *C. diphtheriae* are often found which are **atoxigenic** although typical in all other respects; that is, they appear never to have had (or to have lost?) the power to produce toxin. In 1951 it was first observed by Freeman that a certain strain of temperate bacteriophage from toxigenic *Corynebacterium diphtheriae,* when propagated in completely nontoxigenic strains of *C. diphtheriae,* caused lysogenic conversion to toxigenicity in the formerly nontoxigenic strains. The newly altered strains remained toxigenic only as long as the prophage was present. Cells that lost the phage lost their toxigenicity. The toxigenicity of some other bacteria is likewise based on lysogenic conversion.

Transduction. Genetic material may be transferred from one cell to another by conjugation and transformation (Chapt. 19). **Transduction,** another mechanism of genetic transfer, is the transmission of a portion of the chromosome of one bacterial cell (say, *a*), into another bacterial cell (*b*), by a matured, temperate bacteriophage that lysogenizes bacterial cell *b*. The phenomenon of transduction was first observed by Zinder and Lederberg (Nobel Prize winner with Beadle and Tatum) in 1952.

The prophage nucleic acid, initially an integrated part of the genome of bacterium *a*, has been induced or activated and has separated from the cell chromosome. In so doing it has exchanged a bit of itself for a bit of the genome of bacterium *a*. This is an abnormal and faulty separation and occurs relatively rarely. The phage now infects cell *b*. The bit of bacterial *a* genome (sometimes called an **exogenote**) carried into bacterium *b* is said to have been **transduced** from *a* to *b* by the phage. The genome of the recipient cell *b* is called an **endogenote.** The genome of *b* (endogenote), after the addition and integration of the exogenote, is called a **heterogenote** (Fig. 17–8). Since both the phage nucleic acid *and* the *a* exogenote can add new genetic characters to a recipient cell, it is evident that both viral conversion and transduction can occur simultaneously.

If the exogenote fails to be integrated into the endogenote, it may manifest its phenotypic character in that single generation, but it is not replicated and disappears from the cell. **Abortive transduction** is said to have occurred. It is the most common event in transduction.

Some phages can transduce any of a variety of genetic units of the donor cell. This is spoken of as general or **nonspecific transduction.** For example, in the genus *Salmonella*,* some phages have transduced various special nutrient requirements (auxotrophies), enzymic functions, resistance or susceptibility to various antibiotics, motility, and the chemical makeup of the flagella. Some phages carry only the specific genetic unit to which they are most closely linked in the genome of the donor cell.

It is worth noting here some points of differentiation between the types of genetic transfer among bacteria. In **transformation** the transforming nucleic acid exists in a free, soluble, naked condition and is completely vulnerable to the action of pH, electrolytes or specific enzymes (DNase or RNase). This is not true in **conjugation** or **transduction.** In these the transferred nucleic acid either remains intracellular or is inside a virion and so is not accessible to

---

*The genus of gram-negative rods containing the bacillus of typhoid and paratyphoid fevers and named for the American bacteriologist, Salmon.

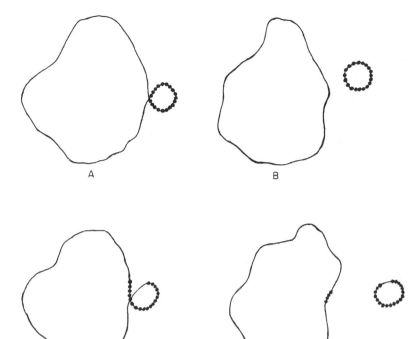

Figure 17–8. *A* and *B*, Normal separation of prophage DNA from its association with the bacterial chromosome. *C* and *D*, Aberrant separation resulting in the incorporation of a piece of chromosomal DNA into the phage DNA, and vice versa.

any extracellular enzymes or other deleterious influences.

Viral conversion is differentiated from transduction by the fact that in viral conversion the altered characters of the cell are caused by the **viral nucleic acid** and, as in lysogenic immunity, depend on the continued presence of prophage in the converted cell. The characters leave when prophage leaves (or is lost by failure to be transmitted to a daughter cell). In transduction, on the contrary, the altered characters of the transductant are due to **bacterial** (not viral) **nucleic acid.** The exogenote remains as part of the heterogenote whether the prophage stays or leaves; i.e., a permanent genetic mutation has occurred.

17.5
COLICINS, C FACTOR, AND PROPHAGE

Colicins (often called bacteriocins; see last paragraph of this section) are proteins or polypeptides released into the surrounding medium by certain strains of *Escherichia coli.* Colicins are extremely lethal to other strains of bacteria of the same species or to closely related species. A single molecule of colicin, once adsorbed to a susceptible cell, kills that cell. Colicins are often compared with antibiotics but colicins are proteins and thus differ markedly from the substances (e.g., penicillin; see Chapter 23) commonly thought of as antibiotics. Colicins are specific as to the strains for which they are lethal; i.e., they will adsorb only to certain strains (Fig. 17–9). Adsorption of colicin appears to occur, like adsorption of phage, at specific receptor sites on the outer cell surface. It is notable that some of these receptors serve also for the adsorption of certain bacteriophages. The exact mode of action and the biological significance of colicins remain to be fully elucidated.

Various colicins are lettered and numbered primarily according to the letter-number (if any) of the producing strain. Some strains of *E. coli* may produce more than one type of colicin; these are often given letters such as H, K, or E. For example, a colicin produced by a strain of *E. coli* designated as R450 might be identified as colicin R450-K. Classification and nomenclature of colicins have not been fully systematized.

The ability to produce colicin constitutes the property of **colicinogenicity.** Colicinogenicity appears to be a transferable property of bacteria dependent on the presence of an episome like

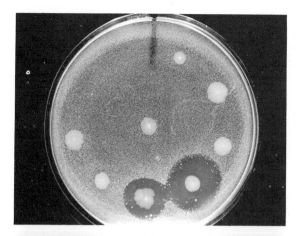

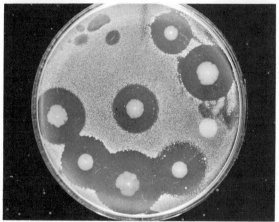

Figure 17–9. Action of colicins (bacteriocins). The two Petri plates contain nutrient agar, the entire surface of which was inoculated with two different types of dysentery bacilli (*Shigella boydii* 2). Growth of these bacteria is seen as smooth, grey, overall pebbling. Bacteriocin-producing bacteria were "spot" inoculated at center and several points around the periphery corresponding in the two plates. Action of bacteriocin is seen as inhibition of growth in dark circular zones around "spot" colonies of the colicin-producers. The two types of *S. boydii* 2 (both plates) differed in their susceptibility to the various bacteriocins.

the F factor. The episome of colicinogenicity is called the **C factor,** or **Cf.** Unlike F, Cf does not appear to integrate with the cell chromosome. It is transferred by conjugation (cell contact) but not in association with the cell chromosome or F. It may also be transferred by transduction. Presence of Cf confers immunity to the colicin produced, though the colicin may be adsorbed on the producing cell. This immunity to colicin suggests the immunity to superinfection conferred on a cell by the presence of a prophage, though the mechanisms of the immunity seem to differ.

The presence of Cf does not necessarily result in production of colicin. However, like prophage, Cf can be **induced** by ultraviolet light and other agents that induce prophages to enter a lytic cycle. Cf then becomes active and causes colicin production. Thus, there are similarities (Table 17–1) between prophage, F, and Cf, though they are clearly not identical. There is evidence that imperfect or defective prophages are associated with the presence of Cf. Whether Cf is, or is influenced by, incomplete phage or prophage is currently under investigation. Cf's, and the colicins they induce, differ considerably among themselves in physiological and morphological characters.

In the foregoing paragraphs colicins (so called because originally derived from *Escherichia coli*) have been used as a model to describe what has since been found to be a whole class of these curious microbicidal agents. Similar lethal agents are produced by numerous other species of bacteria. The more inclusive term **bacteriocins** is now commonly used to include the entire group. Some bacteriocins are named according to the organisms that produce them, e.g., pesticins (from *Yersinia pestis*),

TABLE 17–1. PROPERTIES AND EFFECTS OF EPISOMES

Factor	Effect of Induction	Attachment to Chromosome	Replication	Mode of Transfer	Distinctive Effect of Presence
F	. . .	Attaches to chromosome at any site; may remain unattached	Replicates synchronously with chromosome whether attached or not	Transferred by conjugation	Confers mating ability (F⁺); sometimes Hfr state
Cf	Induces colicin production	Rarely if ever attaches to chromosome or F	Replicates synchronously with chromosome	Transferred by conjugation and by transduction	Confers ability to synthesize colicin (bacteriocin)
Prophage	Induces separation of prophage from chromosome and replication as virulent phage	Attaches at a specific site of chromosome	Replicates synchronously with chromosome only when attached (integrated); very rapidly when free	Transferred as part of cell chromosome when attached; otherwise independent as mature, free phage	Confers new genetic properties (phage conversion) and property of lysogenicity

megacins (from *Bacillus megaterium*), and staphylococcins. Perhaps the symbol for the bacteriocinogenic episome should be Bf (bacteriocin factor) instead of Cf.

17.6
DEFECTIVENESS

Defective Phages. A phage not only contributes new genetic properties to a lysogenized cell but is itself subject to genetic mutation, as is any genetic material. For example, mutation in a prophage may destroy its power to mature. Even though it may confer immunity to superinfection on induction, as by ultraviolet irradiation, it reveals its defective state by failing to develop into a vegetative virus. Although the infected cell may die, no mature, infective virus particles are released. The provirus is said to be **defective**. It produces **incomplete** or defective virus particles, perhaps minus their protein coating or unassembled fragments of virions. Some may have multiple tails or no tails. Sometimes these incomplete virus particles may actually cause the cell to lyse, but the incomplete virus particles thus released cannot infect other cells (Fig. 16–10). Such defective prophages may act as bacteriocinogenic factors, as previously mentioned.

Complementation. Some mutant viruses are defective in such a way that they cannot multiply normally or at all. However, they may be helped to multiply if their defect can be compensated for by a second virus which may be mutated in another locus. The two **complement** each other. Neither alone can cause cytolysis but together they can do so. This relationship is suggestive of syntrophism or satellitism as observed among bacteria, though the mechanisms are apparently quite different.

Helper Viruses. Several instances are known of viruses that are unable to replicate in a cell unless another virus has infected the same cell. For example, a virus called adeno-associated virus (AAV) can replicate only if an adenovirus—the helper—has also infected the host cell. The AAV and helper adenovirus are quite unrelated, being different sizes and having different DNA molecules. Other instances are known in which a virus can replicate in one cell type without a helper, but in some other cell type a helper is required. It is believed that the helper virus codes for some enzyme essential to virus replication that the dependent virus does not code for. These are examples in which one virus parasitizes another virus!

Phenotypic Mixing. Although infection of a bacterial cell by phage precludes superinfection with an identical phage, mixed, simultaneous infections by two or even three related, but not identical, phages can occur. In such mixed infections virions of two or more types of phages may mature at the same time. In the process of assembly, interchange can occur with some of the "prefabricated" parts such as heads, tails, DNA, or proteins. The resulting mature phages (**phenotypes**) exhibit mixtures of the characters of the various phages involved. The process is spoken of as **phenotypic mixing**. Note that the molecular structures of the respective genomes have not changed; i.e., neither genetic mutation nor genetic recombination has occurred. The change is not heritable. On later reproduction the proper capsid protein is produced.

17.7
VIRUSES AND TUMORS

A tumor is a swelling (L. *tumere* = to swell). In pathology, plant or animal, the word tumor generally means a mass of abnormal, independently growing tissue that is without physiological function or that exhibits abnormal activity. A mass of such new growth is also called a **neoplasm** (Gr. *neos* = new; *plasm* = formation or substance).

Neoplasms may be **benign**, i.e., they may grow slowly, apparently under at least some physiological restraint, remain **in situ** and neither vigorously invade adjacent normal tissue nor give off numerous rapidly growing tumor cells that migrate and localize elsewhere in the body. In contrast, some neoplasms consist of cells that are markedly different from normal cells. They grow rapidly and are unimpeded by contact with other cells. They tend to damage and invade neighboring tissues and give off many cells that, distributed in body fluids (e.g., lymph), localize elsewhere (**metastasize**) in the body and rapidly extend the process. Such tumors are said to be **malignant**. Depending on the type of cell of origin (in animals), malignant neoplasms may be **carcinomas** (Gr. *karkinos* = crab) or **epitheliomas**, both of which are epithelial in origin, or **sarcomas** (originate from connective and supporting tissues or striated muscle) or **leukemias** (involve cells of the lymphatic and circulatory systems). The term **cancer** (the Latin word for **crab**, from the fancied resemblance of the shape of some skin cancers to crabs) is widely used to mean any or all malignant neoplasms in general. Neoplasms, especially if malignant, do damage by absorbing nutrients necessary to the host, by swelling and blockage of body channels, by pressures, by killing normal cells, and possibly by the effects of toxic waste products of their abnormal metabolism. Neoplasms are not inheritable, although a tendency to develop certain types of neoplasm seems to be associated with certain families.

A basic cause of malignant neoplasms appears to be a genetic mutation in a body-tissue cell (a **somatic** mutation). As is true of other cells, mutations in body-tissue cells may be caused by a variety of agents: ionizing radiations, ultraviolet light, various chemical substances such as methylcholanthrene, substances in smoke from wood, coal, and vegetable leaves and in automobile exhaust fumes, cresols, and mechanical irritants. These agents, physical and chemical, are collectively called **mutagens**. Any mutagen is a potential producer of neoplasms or tumors. Tumor-producing agents are often said to be **oncogenic** (Gr. *onkos* = mass or swelling). Virus-induced changes in normal cells that result in malignancy are called **malignant transformations** (do not confuse with transformation of bacteria by **free DNA**).

Numerous malignant tumors of lower animals are well known to be caused by viruses. One of the first-discovered viral causes of malignant tumors was the virus of chicken sarcoma (P. Rous, 1911; Nobel Prize winner, 1966), the "Rous sarcoma." An oncogenic virus that causes a wart-like disease (papilloma) of rabbits was discovered by Shope in 1933; a leukemia virus of mice was discovered in 1951. A virus of mice producing a wide variety of malignant neoplasms in mice, rats, and hamsters and called **polyoma virus** was discovered in 1957.

Many groups of viruses, both RNA and DNA, have representatives that induce malignancy in one species or another. Most of the viruses that induce leukemias of mammals and birds are RNA viruses resembling myxoviruses. Other closely related RNA viruses induce sarcomas (Rous sarcoma virus, for example). Somewhat similar viruses induce breast cancer in mice and perhaps monkeys, and morphological-ly similar virus-like bodies have been found in human breast cancers and human milk (but simply finding such particles does not mean that they caused the tumors).

Among the DNA viruses, polyoma virus and simian virus 40 (SV40), both members of the papova group, induce solid tumors in hamsters and mice. More than half of the 33 known serotypes of human adenoviruses and of the 20 or so serotypes of monkey adenoviruses induce malignancies in hamsters and rats. (See Color Plate II, *A*.) The herpesvirus group has several members that induce leukemias and lymphomas in cottontail rabbits, monkeys, chickens, and guinea pigs. One herpesvirus is consistently isolated from certain tumors that occur in children in Africa, called Burkitt's lymphoma after the man who described the tumor and showed the relationship. (The same virus is believed also to cause infectious mononucleosis.) Another herpesvirus, herpes simplex type 2, causes genital infections and is transmitted venereally; this virus is under intense investigation as a prime suspect in causing cervical carcinoma, which is one of the most common tumors of women.

Many of the viruses described above can cause malignant change of cells of various species growing in culture. It is believed that the mechanism of transformation is the same whether a tumor is induced in an animal or in cell culture.

Because the presence of the viral genome can be demonstrated in the malignant cells, and because new, virus-specific **tumor antigens** (the T antigen) are present in the transformed cells, it is believed that the viral genome becomes integrated into the cell's genetic material, very much like lysogenization in bacteria. The viral genome then changes the cell to malignancy, and the code for the change is perpetuated to progeny cells.

Cancers are one of the leading causes of death and illness. Therefore, viruses as possible causes of human cancers are under intensive investigation.

CHAPTER 17
SUPPLEMENTARY READING

Baron, S., and Levy, H. B.: Interferon. Ann. Rev. Microbiol., *20*:291, 1966.

Goodheart, C. R.: An Introduction to Virology. W. B. Saunders Co., Philadelphia. 1969.

Nomura, M.: Colicins and related bacteriocins. Ann. Rev. Microbiol., *21*:257, 1967.

Rous, P.: The challenge to man of the neoplastic cell. Science, *157*:24, 1967.

Stent, G.: Molecular Genetics. W. H. Freeman & Co., San Francisco. 1971.

MUTATION AND MODIFICATION

The genetic material DNA is the same chemically in all types of cells, differing only in type, quantity, and arrangement of the various nucleotides. The hereditary traits transferred from generation to generation—feathers of a certain color, a Hapsburg jaw, or red pigment in a certain bacterial type—are remarkably stable. A bluejay, for example, is always blue; he is seldom seen to have red wings. The stability of genetic traits in any species depends on the normally unchanging sequence of nucleotides that make up the DNA of that particular species. If the nature, number, or sequence of nucleotides in any genetic DNA is altered so that one or more heritable characters are lost, gained, or altered, a genetic change or *mutation* is said to have occurred. A mutation may therefore be defined as a sudden change in heritable characters that is not a result of transfer, segregation or recombination (Chapt. 19) of normal genes but of an alteration of the numbers, molecular structure or sequence of nucleotides that constitute the genes themselves.

18.1
MUTATION

The Gene. The term **gene**, attributed to Johanssen in 1911, is derived from the classical science of heredity founded about 1870 by the Austrian monk Gregor Mendel (1822–1884). By 1940, through the work of Morgan, Muller, Sutton (Nobel Prize winners), and many others, the gene had been recognized as the genetic unit: the basis of mutation, recombination, the transmission of hereditary traits and the phe-

nomenon of dominance and recessiveness. However, the true chemical and physical structures of the gene were clarified only within the current decade by Watson, Crick (Nobel Prize winners), Benzer, and many others.

The term gene is now recognized as a sequence of nucleotides that functions genetically as a unit. A gene may consist of from one nucleotide pair to thousands of pairs linked together, and is often made up of several distinct and separable segments of the DNA filament. (Although described here in terms of DNA, it must be remembered that RNA is the genetic material of certain viruses, such as poliovirus and RNA phages, so that the definition of a gene applies as well to RNA.) An entire, double-helix DNA macromolecule with its genes constitutes a **chromosome** (chromosomes of eucaryotic cells also contain certain proteins, such as basic histones and structural proteins), a cytological unit familiar to all students of elementary biology. A gene is characterized by its ability to transmit heritable characteristics. It is subject to various kinds of alterations that cause progeny to have heritable characteristics different from their ancestral cells, i.e., to be **mutants.**

Mutations in either eucaryons or procaryons under familiar conditions of life are rare events. Commonly, only one cell in many thousands or many millions may be a **mutant**. Most mutations are ineffective or trifling and have no significant effect on the cell one way or another. Others are occult; i.e., they are present in the **genome** (the genome is the entire genetic complement of a cell or virus) but not manifest in the **phenotype** (the phenotype is the detect-

252

able expression of a gene's function, such as a bacterial cell that makes a particular pigment; the genotype is the genetic composition, whether or not the genes are expressed). They remain unrecognized or become evident only under special, selective environmental conditions (see Auxotrophic Mutants, this chapter). Many mutations are lethal or so detrimental that the cell dies or fails to reproduce.

The Muton. Although a mutation commonly involves more than one pair (perhaps thousands) of nucleotides, the **minimal genetic unit** capable of mutation is a single nucleotide: a **muton.**

Any change, however slight, in even one of a single pair of complementary nucleotides in any **recon** (a minimal genetic unit transmissible intact from one chromosome to another; see material on recombination, §19.2) will produce a changed or misspelled nucleotide word that results in no enzyme or in a faulty enzyme. The error is automatically repeated in subsequent DNA replications, just as the error of a misspelled or omitted word on a printing press will be repeated endlessly until corrected.

In **multicellular, sexually reproducing** organisms a mutation in one of the body (somatic) cells is inherited only by the progeny of that cell. The cell may immediately die, the mutation may pass unnoticed, it may express itself as an overgrowth of the tissues locally (a benign tumor), or it may become manifest as a malignant cancer. The somatic cell mutation is not transmitted to the progeny of the multicellular individual. Only when a mutation occurs in the gamete is it transmitted to the progeny of the plant or animal in which the gamete mutation occurs. In protists that multiply both sexually and asexually, mutation in either the haploid (equivalent to gamete) or diploid (equivalent to somatic) cell may produce mutant progeny since the single cell is the entire individual.

Chromosomal Mutations. These generally result from: **deletions** (loss) of or failure to replicate, a segment of a chromosome consisting of numerous genes; **addition** or duplication of chromosome segments; **translocation** or transfer of one chromosome segment to a nonhomologous chromosome in a new position; **inversion** or end-for-end reversal of a segment of a chromosome; and in eucaryons, **nondisjunction** or failure of chromosomes to separate at the second meiotic division so that one daughter cell lacks a chromosome while the other has it in duplicate. Many of these changes involve breakage of chromosomes.

Point Mutations. These involve smaller groups of genetic units than chromosomal mutations; often only a single muton. They may result from: chemical change in one or more mutons; deletion or gain of one or more mutons; or change in sequence in one or more pairs of nucleotides by reversal of nucleotide pairs (e.g., A—T to T—A), or inversion of a segment of the DNA chain involving one or many contiguous nucleotides.

Mutation Rates. Under natural conditions mutations occur spontaneously—i.e., the cause of the mutation is not identified. However, most recognized mutations occur as a result of exposure of growing cells to recognized mutagenic agents such as cosmic rays, sunlight, or certain substances in smoke. Probably many "spontaneous" mutations result from familiar causes. Some probably cause some types of cancer.

Spontaneous mutation occurs among microorganisms at widely varying rates. For example, consider mutation to phage-resistance. In an actively growing culture of a phage-**sensitive** bacterium, a mutant cell **resistant** to that particular phage may occur as rarely as one cell in 10^5. This is a fairly high mutation rate. In another culture only one mutant (resistant) cell may be found among 10^{10} cells or more. On the other hand, one culture was reported in which 1.3 to 15 per cent of all live cells were mutants. Spontaneous mutation rates of various characteristics are commonly about 1 in 10^8 cells. Further, the rate of mutation in a given property may vary greatly. One reason for this is that in any given gene a mutation in any one of the mutons in that gene may produce the same effect; also, different mutons may vary greatly in their degree of mutability.

It is evident that unless very large numbers of individuals, human or bacterial, are examined, mutants are not likely to be found. In a small culture of bacteria (5 ml), the billions of cells (e.g., 5×10^9) that could be present might contain only 500 mutant cells of a given type. The detection and isolation of one of these 500 mutant cells among 5 billion are not easy without special techniques. These are described on page 261 et seq.

Mutagenic Agents. A mutagenic agent may be defined as any substance (e.g., certain chemicals) or any physical influence (e.g., ultraviolet irradiation) that causes a change in the number, kind, sequence or structure of nucleotides (i.e., mutons) in the genetic materials in a cell such that alteration in heritable characteristics of that cell results.

By exposing microorganisms to mutagenic agents, the mutation rate is greatly increased; detection and isolation of mutants are thus facilitated. The changes thus caused are called **induced mutations**.

Chemical Mutagens. Among the most interesting chemical mutagens are the **base analogues** (cf. metabolite analogues). These are substances that closely resemble in molecular structure (i.e., are **analogous** to) the purine or pyrimidine bases in DNA. Because of the similarity in structure these analogues can preempt positions of the correct bases in the nucleotides of the DNA and by so doing interfere with, or modify or destroy, the action of the nucleotide of which they become a part. They may alter the nucleotide sequence of the entire succeeding DNA chain. Very complex effects may result.

A simple example is seen in the effect of 5-bromouracil (BU), analogue of the pyrimidine base thymine. A glance at Figure 18–1 will show the similarity between bromouracil, uracil (U) and thymine (T). BU can occupy the place of T in a strand of DNA. In the next replication process each BU may attract guanine (G) instead of adenine (A), as the displaced T would normally do. Thus a BU—G pair appears in place of each normal T—A pair. Since G has been drawn into the positions usually occupied by A, at the next replication each G attracts C. Thus in two DNA replications one or more A—T pairs have been replaced by a number of G—C pairs. This type of mutation by base analogues and many other substances that can usurp a variety of key positions in DNA has, as may be supposed, been the subject of a great deal of research in molecular genetics.

Among other substances that can cause point mutations are ethyl methane sulfonate (EMS) and ethyl ethane sulfonate. EMS acts by ethylating the 7 position of guanine (Fig. 18–2) or adenine. The bond between the purine and the deoxyribose in the nucleotide is hydrolyzed and the purine base is lost, leaving an unoccupied space; a **deletion mutation**. On replication of the DNA, if the purine base is replaced by the correct base no mutation occurs. But the purine may be replaced by other bases, resulting in a change in base sequence. The end result can be a mutation or a complete failure to reproduce. Other chemical mutagens include nitrous acid, hydroxylamine, amino purines, proflavine, the nitrogen and sulfur mustards, formaldehyde, manganese chloride, methylcholanthrene, arsenic, chromium, urethane, creosote, tars, and organic peroxides (see also Ionization, this chapter). Not all these mutagens act in the same manner; the effects of some are not fully elucidated. Some mutagens, both physical and chemical, act directly by combinations and changes in the nucleotides of the DNA. Others act indirectly by affecting the mechanisms (e.g., synthetases) or materials (e.g., purine and pyrimidine bases in solution) involved in DNA synthesis. The bases themselves may change by tautomeric alterations. In addition, various faults of nutrition, such as starvation of the necessary purine or pyrimidine bases, can result in the omission of nucleotide pairs or the incorporation of spurious substitutes for correct nucleotides or their constituent residues.

Physical Mutagens. DNA is susceptible to several physical agents. For example, temperatures above 80C may cause denaturation by "melting" (breaking H-bonds). Slightly lower temperatures cause lesser changes, notably liberation of purine bases from DNA, with resulting deletions of nucleotide pairs or replacement by "wrong" bases in the next replication. Data on melting points of DNA are used in classification (Chapt. 31). DNA may also be destroyed or mutations may be induced by various forms of radiant energy. One effect of ultraviolet radiation (absorbed especially by bases of nucleic acids) is to cause a junction between adjacent thymine groups in DNA, with the formation of thymine dimers. This distorts the double helix and can be lethal to that cell.

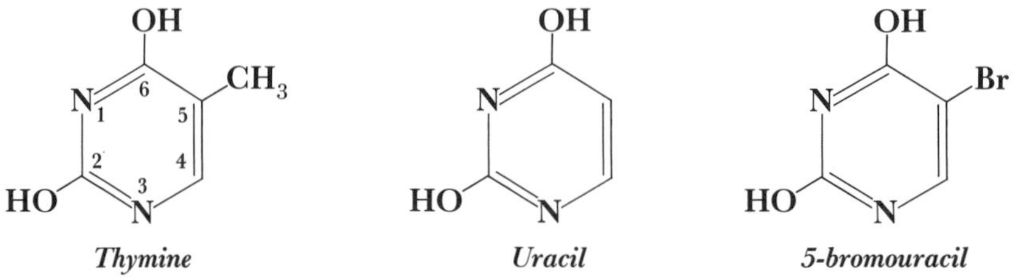

Thymine *Uracil* *5-bromouracil*

Figure 18–1.

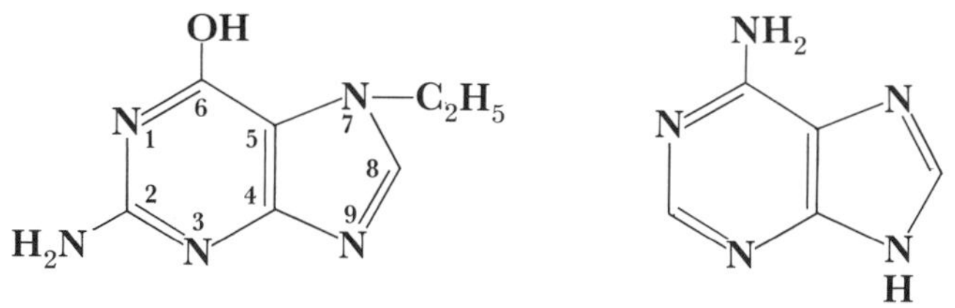

7-ethyl-guanine   *Adenine*

Figure 18–2.

18.2
RADIANT ENERGY

Radiant energy is energy which travels through space as a wave motion. Electromagnetic activity liberates radiant energy in the form of rapidly vibrating waves—the electromagnetic spectrum (Chapt. 20).

The Electromagnetic Spectrum. Electromagnetic radiations may be arranged according to wavelength in a spectrum which includes radio waves, x-rays, ultraviolet, and the visible solar spectrum (red to violet), the latter being only a small part of the whole system of electromagnetic waves. The physiological effects of these various radiations differ greatly, ranging from warmth and vision to mutations and death. The study of effects of radiant energy on cells is often called **radiobiology**.

Radiobiology. The longest waves having known physiological effect are (1) the infrared or invisible "heating waves." Next are (2) the visible red rays, which affect the retina of the eye and bacteriochlorophyll (photosynthetic pigment of bacteria; see Chapt. 32). Then come (3), successively, yellow, blue, and violet, which affect the retina and have photodynamic (see page 256) and possibly other biophysical actions. Then come (4) the invisible ultraviolet rays, which have marked physiological and biochemical activities (page 256). Shorter waves are (5) x-rays (relatively long x-rays, called "soft," and shorter x-rays, called "hard"); (6) α-, β-, and γ-rays, and finally (7) the little-understood cosmic waves coming to the earth from outer space.

Biological Effects of Irradiations

In radiobiology we consider two types of radiations: (1) long waves, specifically ultraviolet radiations (wavelengths 200 to 295 nm), which are much used in radiobiology (these waves have little power of penetration and do not cause ionization); and (2) short waves,

especially x-rays (wavelengths 0.6 to 100 nm) and γ-rays (wavelengths of 0.1 to 14 nm), which have high energy content and great powers of penetration and cause ionization. Both types of waves can kill living cells. The lethal effect is of great value in disinfection and treatment of cancer. Nonlethal changes, however, are in many ways more important and useful.

Ionization. When short-wave radiant energy (e.g., x-rays) is passed through a cell, certain atoms absorb a quantum (unit amount) of energy. This removes an electron from each affected atom, which becomes a **positively** charged ion. The free electron immediately attaches to a neutral atom, which becomes **negatively** charged. Thus an **ion pair** is formed. Since changing the electron structure of atoms changes chemical bonds, molecular structures of cells containing those atoms undergo alterations. If the dosage of radiant energy is sufficient, the effect is lethal. If not lethal, then other changes are produced, most conspicuously in the **genetic** mechanisms of the cell. These changes are of tremendous importance to man, his offspring, his industries, his diseases, and indeed his whole way of life. Some examples will be mentioned later. Atomic bombs and radioactive fallout produce a considerable amount of ionizing radiation and radioactive substances.

Target vs. Free Radical. Two theories concerning the effects of **ionizing radiations** are as follows:

THE TARGET-HIT THEORY. This theory holds that the effect is due to a direct hit of a quantum of radiant energy (photon) upon a **target** (e.g., an enzyme molecule, a gene molecule, nucleotide or muton). Undoubtedly, many such targets exist in a cell, and the greater the dosage of energy the more that are hit; a logarithmic relationship exists between dosage and effect. While explaining many of the observed phenomena, this theory does not explain all of the biological effects of ionizing radiation.

THE FREE-RADICAL THEORY. This theory is not the whole explanation either. It is based on the idea that water or some other nonspecific and generalized material in the cell (as contrasted with specific materials such as genes and enzymes) is broken down, liberating various **free radicals,** such as HO_2. Liberated throughout the cell in a very active condition, many free radicals form organic peroxides (in themselves mutagenic). These set up chains of destructive oxidation-reduction reactions. It is of great interest, in relation to these oxidative effects, that the presence of reducing agents or absence of oxygen tends to offset them. Catalase, a peroxide-destroying enzyme, has a similar effect. Probably both target-hits and free radicals are involved in the effects of ionizing radiations; undoubtedly other as yet unknown actions also play a role.

Some Genetic Effects of Ionizing Radiations. The nuclear structures, especially nucleic acids (DNA, RNA), appear to be most sensitive to ionizing radiations, although cytoplasmic structures (e.g., enzymes) are also affected. Ionizing radiation greatly slows the process of mitosis. Division of most cells, regardless of species, is usually much retarded. Chromosomes are often broken and may be partly or wholly destroyed. Genetic mutations usually result. These mutations are characteristically manifested only in later generations of cells because the alterations in DNA in the parent cell are not effective until the next one or two replications of the DNA. A **mutational lag** is observed. The genetic effects are usually permanent and irreversible.

Effects of Environmental Conditions. Several factors, such as temperature, pH and concentration of oxygen, affect the action of ionizing radiations. It is of interest, in view of the danger from radioactive fallout, to find agents to offset the effects of ionizing radiations. Although this extremely important subject is beyond the scope of the present discussion, it might be said that since the presence of oxygen increases the effects of ionizing radiations, and since oxidations appear to be involved in their destructive effects, certain chemicals which take up oxygen (cysteine, glutathione) seem to offer promise in attempts to eliminate or ameliorate the effects of radiations. Other defensive reactions involving sulfur compounds and heat are also under investigation.

Ultraviolet (Nonionizing) Radiation. Ultraviolet radiation, though not possessing the penetrating and ionizing properties of x-rays, is able to penetrate into exposed cells and to set up—*not* ion pairs or free radicals—but **excitations** or **activations,** which have a variety of effects. Most of these effects are similar to those of ionizing radiations, although probably caused by a different mechanism. Sunlight contains much ultraviolet light.

As a rule, actively multiplying cells are most susceptible to irradiation, both ionizing and ultraviolet. In these cells DNA is in an actively replicating condition. Such are usually the cells most susceptible to other deleterious agents also: phage, antibiotics, various bacteriostatic agents and heat. Bacteria in a dormant condition or in a mature, slowly growing stage are markedly resistant compared with cells in an actively multiplying phase of growth.

Reactivation. In contrast to the irreversible injury caused by ionizing radiations, injury resulting from ultraviolet radiation may in large part be cured or reversed if not too extensive. This is done in one way by **postirradiation** exposure to waves longer than the ultraviolet waves. This is often called **photoreversal** or **photoreactivation.** Generally, visible light (wavelengths 420 to 540 nm) is used in photoreactivation. Exposure to the reactivating waves must be within 30 minutes after the inactivating irradiation. Curiously, among the various microorganisms, bacteria often fail to be reactivated. Some types of injury caused by ultraviolet rays may be prevented by **pre-irradiation** exposure to visible light.

Ultraviolet-irradiated cells may be reactivated in one way by incubation of the irradiated cells in a favorable growth medium at temperatures 5 to 30C lower than optimal. Another method is treatment of the cells with various chemical agents: indoacetate, pyruvate, and a long list of others. These phenomena are not yet fully explained but much interesting and valuable research is being done in this field, especially in relation to radioactive fallout and nuclear warfare. In dealing with bacteria many variables affect reactivation: pH, temperature, age, and composition of the culture.

Photodynamic Sensitization. Certain substances, combining with vital target portions of microorganisms, although not necessarily harmful per se, can render the cells very sensitive to visible light. This is usually spoken of as **photosensitization** or **photodynamic** action. For example, many bacteria can grow well in contact with low concentrations of certain dyes such as eosin and fuchsin provided no light reaches them. The combined effect of light and dye for only a few minutes usually kills the organisms. Photodynamic sensitization may also induce mutations. Interestingly, methylene blue, in the light of an ordinary electric lamp, rapidly kills certain gram-positive bacteria that withstand much higher concentrations of the dye in the dark. Gram-negative bacteria are not so affected. One might infer that such differences are mediated by composition of the cell wall. Other organisms, such as viruses, and enzymes and proteins, are also affected. The exact mechanism of photodynamic sensitization is not clear but may represent weakening of bonds by the dye and their rupture by the radiant energy in light. It has been shown that oxygen is essential to photodynamic action, since the effects are not observed if the microorganisms are placed in a vacuum or in atmospheres devoid of oxygen. (Compare with relation of oxygen to the effect of x-rays, page 256.) The action increases in acid media and in undiluted media.

18.3
SOME TYPES OF MUTANTS

Auxotrophic Mutants. Mutations often result in the production of **auxotrophs** (Gr. *auxein* = to add to, *trophe* = food), which require additional food substances. Auxotrophic mutants are caused by changes in, or destruction of, one or more mutons that are responsible directly or indirectly for the synthesis of certain substances that are essential for cell growth and multiplication. In the auxotrophs these essential nutrients (**growth factors**)* fail to be produced by the cell. Such mutants absolutely require that the missing growth factor(s) be supplied preformed in their diet.

As a single example, a common species of intestinal bacteria (*Escherichia coli*), in its normal state, is able to thrive and to synthesize all its cell components in a solution containing several minerals and glucose as a source of energy and carbon. While these bacteria can grow in a much more complex medium, this simple solution amply supplies all of their minimal requirements; such a medium is called a **minimal medium.** (Minimal media may vary in composition according to the minimal nutritional requirements of the species under investigation.) If a single ingredient is omitted, they

cannot grow. After contact of the bacteria with a mutagenic agent, say UV radiation, mutant progeny may appear that have lost the ability of the ancestral (wild-type) **prototroph** (Gr. *protos* = original or primitive) to synthesize the amino acid tryptophan, a substance that is absolutely essential to their growth and reproduction. To these mutants tryptophan has become a growth factor. It must be added preformed, as such, to the minimal medium to make the medium (for them) a **complete medium.** As an additional example, humans are nutritional auxotrophs in respect to many substances, notably, vitamin C (ascorbic acid) and "niacin" (nicotinic acid). They cannot synthesize these substances for themselves. Ample vitamins are present in any normal, balanced diet. Humans share their auxotrophism in respect to vitamin C with guinea pigs.

Auxotrophic bacterial mutants are commonly designated by the name or a symbol of the growth factor that they cannot synthesize, e.g., "tryptophanless" (or Trp⁻) mutants. Other bacterial mutants may fail to produce certain enzymes that ferment a sugar, say lactose. They are designated Lac⁻, and their wild-type ancestors, Lac⁺. Auxotrophic mutants are particularly common after irradiation with ultraviolet light or x-rays but they also not uncommonly appear spontaneously. Some mutagens appear to affect particular mutons: e.g., particular nucleotide bases in certain genes. They are said to be specific for that muton.

Back-Mutations. Any given mutation, especially point mutations, may be "repaired," i.e., either reversed or suppressed, resulting in a reversion to the parent type of cells. For example, a certain species of bacteria (*Proteus mirabilis*) that is normally susceptible to the antibiotic drug streptomycin (ss) produced mutants that, instead of being susceptible to streptomycin, could not grow and reproduce without it; i.e., they were **streptomycin-dependent** (sd). Mutations to dependence on various drugs and other substances are not at all rare. The sd mutants in this example produced other mutants that were no longer sd but had reverted to the prototype that was ss. Such repaired mutants are sometimes called **revertants.**

In a second type of reversion from the mutant sd cells back to the prototype, not only was the prototype ss state restored but the back-mutating cells had mutated to dependence on the amino acid valine (dv). In this case, two mutations could have occurred simultaneously: from sd "back" to ss; or "forward" to dv. It was also thought that the mutation to the dv state

---

*Because the term metabolite is used sometimes to mean a waste substance and sometimes to mean a necessary nutrient, we here use growth factor to mean a specific, pure substance: e.g., a vitamin or an amino acid that is a necessary nutrient, and metabolite to mean a specific waste substance (e.g., antibiotics).

could have suppressed the effect of the sd muton. Experimental studies showed that the effect of the sd muton had indeed been suppressed by the dv muton.

These examples suffice to show that very complex genetic interactions and problems can arise in such situations. It is not always possible to be sure that an apparent back-mutation is not a suppression or some other effect. An apparent revertant may superficially resemble the prototype but carry new, occult mutant genes. A phenotype may represent the net result of several interacting simultaneous mutations.

Occult Mutations. Not all mutations result in readily visible changes such as presence or absence of pigments or spores. A mutation that results in a change in the molecular structure of a certain part of the cell that is not visibly expressed may be called **occult** in that the change in molecules is not in itself visible but is made manifest only indirectly. Some examples of occult mutants among bacteria are auxotrophic mutants, drug or amino acid-dependent mutants and mutants with different enzymic activities or inducible enzyme potentialities. These are often called **biochemical mutants.**

Mutations That Affect Cell Surfaces. Another type of mutation is seen in changes that affect the cell surface structure. In bacteria the nature of the cell surface is closely related to a variety of important properties, four of which are: **ability to conjugate** (in the presence of F factor); **virulence** (if the species of bacterium is pathogenic for plants or animals); **immunological and antigenic properties** that, in turn, reflect specific molecular composition of proteins, enzymes and polysaccharides of the cell; and **colony form.** All these properties are subject to mutational change. Conjugation, virulence, immunology, and antigenicity are discussed elsewhere.

Mutations that result in changes in colony type are relatively frequent and appear to undergo back-mutation or progressive mutation quite often. Here we may briefly describe some of the colony forms that are often associated with immunological and pathogenic (virulence) properties.

Types of Colonies. Many microorganisms, among them yeasts, molds, and bacteria, produce several different forms of colonies. Common among these are the smooth (S) and rough (R) colony types (Fig. 18–3).

S COLONIES. These colonies on nutrient agar media are commonly about 2 to 4 mm in diameter, circular, with regular margins, convex, translucent, homogeneous, smooth, moist, and glistening. In broth media S-type cells produce an even turbidity; i.e., a cloudy or milky suspension of individual cells. S-type cells are frequently encapsulated. The presence of capsular material markedly affects their surface properties: electrical charge, susceptibility to various unfavorable influences, motility, agglutinability by metallic ions, and serum proteins (antibodies). Frequently S forms of cells seem to represent a defensive reaction to unfavorable environmental influences. For example, contact with disinfectants and certain salts may stimulate the appearance of S forms, though sometimes these appearances are modifications (see page 260), not mutations.

Among bacteria capable of infecting animal tissues, the mutational changes necessary to produce the S type of cell and its subsequent rapid multiplication are furnished by contact with the tissues of the infected host which tend to resist the infection. Virulence is thus commonly associated with the S form and this, in turn, is associated with the formation of a capsule or other protective surface modification. For example, capsulated cells more easily escape phagocytic white blood cells as well as antibodies of the infected animal.

Surface modifications of any sort usually affect the immunological and antigenic properties of cells, since these are largely surface phenomena (Chapt. 26). Whole systems of results, e.g., identification of species, medical diagnosis, and value of vaccines, may depend on whether S or R forms of organisms are being used.

When S forms of bacteria are cultivated on artificial media in pure culture, under wholly uniform, benign, and favorable conditions, removed from competition with other species

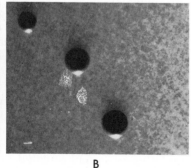

Figure 18–3. *A,* Rough (R) colonies of the bacillus of diphtheria (about three times actual size). *B,* Smooth (S) colonies of the bacillus of diphtheria (about three times actual size).

A   B

of microorganisms that commonly occur in natural environments such as soil, feces, or throat, R mutants are often selected and the populations tend to lose their virulence and their capsules or other protective surface properties.

R COLONIES. The R type of colony on nutrient agar is not glistening, like S, but is dull, granular or matte, and opaque or less translucent than S colonies; it has rough or wrinkled surfaces and crenated or irregular edges. It is also dry and crumbly (friable) in consistency. On the surface of solid media, rod-forms of bacteria (bacilli) in rough colonies commonly form long, tangled filaments of cells as contrasted with cells in S colonies, which tend to occur singly. The long filaments of rough-type cells suggest a delayed or absent formation of septa or, if septa are formed, some alteration in the ends of the cells that cause them to cling together in strands instead of separating as in S cells. R cells are generally not encapsulated. In broth the R-type growth is usually granular or flaky and irregular, commonly forming a thick scum or pellicle on the surface or settling in flakes or lumps to the bottom.

MUCOID COLONIES. Many species of bacteria produce what are called mucoid (M) colonies (Fig. 18–4). These are large, slimy, and viscous, often almost watery. The mucoid material is like an exaggerated capsule or slime layer. Like the slime coating on a garden slug or the mucous secretions of the respiratory tract, and possibly the capsule of S cells, bacterial mucoid material has protective properties. In some species of bacteria the slime contains excess genetic DNA (see Chapter 19).

DWARF COLONIES. These are very minute, often just at the limit of unaided vision. Cultures derived from dwarf (D) colonies often have properties such as virulence and certain metabolic characters that differ markedly from corresponding properties of cultures from R, S, M, or other type colonies. As is true of other colonial types, D mutants tend to change to M, R, or S or others. Cells in D colonies often consist of smaller-than-normal cells and cells with certain metabolic deficiencies or peculiarities.

It is to be remembered that R, S, M, D, and other colonial types are not mutations in themselves. Colony type is only one evidence of profound enzymic and other mutational changes within the cell. These mutations, while often associated, are not necessarily linked. Like many genetic characters, they can change independently. Thus, although R, S, M, and D forms are described as if they are distinct types, hun-

dreds of intermediate overlapping forms occur. A wide range of various combinations of properties can occur. For example, a cell may have some surface properties that cause the formation of a rough colony yet have some of the virulence or immunological properties of S-type cells.

So important are the antigenic and virulence characters of cells that the letters R, S, etc., are sometimes used to designate the antigenic types commonly but not necessarily associated with those colonial types. The words rough, smooth, etc., are reserved for the gross, visible forms of colonies regardless of their antigenic and other occult properties. Change in gross colony form, although seemingly great, may in itself represent a very small change in composition of the cell wall of cells that are basically of another type.

Acriflavine Test. That surface structure of cells reflects more accurately their fundamental antigenic (enzymic, protein, polysaccharide) composition than does gross colony form is shown by studies of the reaction of various cells to the orange, fluorescent dye acriflavine. Cells from a given colony are suspended in a 0.1 per cent solution of the dye in 0.85 per cent saline. Cells that are unaffected by the dye are found to behave antigenically like S cells, even though they may have come from a rough or other type colony. Cells that are antigenically R when in contact with acriflavine gather together in floc-

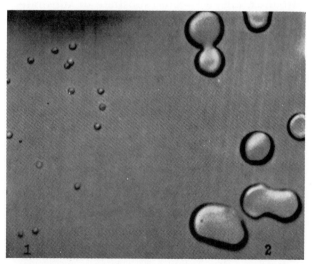

Figure 18–4. *1*, Colonies of R variant of Type II pneumococci plated on blood agar from a culture grown in serum broth in the **absence** of the transforming substance. (3.5×.) *2*, Colonies of the same cells after transformation during growth in the **presence** of active transforming principle isolated from Type III pneumococci. The smooth, glistening, mucoid colonies shown are characteristic of Type III pneumocci. (Photograph by J. B. Haulenbeck.)

cules or clumps though they may be S in colony form. Cells that are antigenically M-type form slimy curds in the presence of the dye.

Modifications. Some seeming mutations are merely temporary effects of factors in the environment. They do not involve alterations in genes. Such nongenetic changes are spoken of as **modifications** or fluctuations. Typically, modifications cease when the environmental stimulus that evoked them is withdrawn. Sometimes the reasons for modifications, like some causes of spontaneous mutations, are quite obscure.

Many different properties of organisms may undergo modification. For example, certain bacteria when cultivated on ordinary nutrient agar produce markedly rough-type colonies. If the same organisms are transferred to a new plate of the same agar, modified by the addition of a little phenol or other irritating substance, the growing cells produce slimy or mucoid-type colonies. On transferring the mucoid growth again to plain (phenol-free) medium the rough-type growth is produced promptly. The sudden and temporary nature of modifications, their dependence on an environmental influence and the generalized effect of the influence on all of the cells at once tend to distinguish modifications from mutations.

Modifications often produce changes, visible or occult, that closely resemble mutations. For example, induced enzyme production is a result of, and dependent on, an **environmental** stimulus, e.g., a specific substrate or substrate analogue (Chapts. 7, 19). The enzyme induction is not a genetic change but merely an evoked expression of an existing, but repressed, genetic property.

Some very complex situations can arise when mutation accompanies or follows modification. For example, the cells in the phenol-treated population described above produce apparently identical mucoid cells on phenol agar yet, on calcium chloride agar, both mucoid and smooth colonies may be formed. Obviously there are cells of two genotypes: on phenol agar the two phenotypes are seemingly identical; on calcium chloride agar their differences become manifest.

Mutations and Population Changes. Growth of a microbial population in a tube of culture medium generally changes environmental conditions. For example, all available essential growth factors or hydrogen acceptors (Chapt. 8) may be used up, or acids, alcohols, or other noxious waste substances may be produced. If, as is nearly always the case, there are mutant cells in the population that are favored by the new conditions, those mutants may grow and eventually partly or entirely replace and supersede the prototype. For example, in aging cultures of *Brucella* (cause of undulant fever) that were originated with S-type cells, some agent or condition (or both) develops that is inimical to S cells of *Brucella*. The S cells cease growth. However, R-type *Brucella* cells are not injured by these changed conditions and may be favored by them. The R form eventually replaces the S form.

Factors responsible for such replacement of one mutant form by another of the same species vary. In the case of *Brucella* cultures the change in population may occur partly because of the accumulation in the broth culture of the specific waste product or metabolite, D- or DL-alanine, which suppresses the S form. Other organisms are affected by other metabolites. Oligodeoxynucleotides from DNA digestion, added to cultures of various bacteria, cause increased growth of S cells and suppression of R cells. Equally important for various species may be merely the depletion of oxygen and other hydrogen acceptors in the culture. Probably still other factors alter populations in other ways. An increase in acidity, for example, a common result of growth, could result in selection of an aciduric mutant and thus change the population from an acid-sensitive one to an aciduric one. Very complex interrelationships can occur, involving, for example, multiple mutations and changing environmental factors.

Whatever the mechanisms, the important biological fact is illustrated that the activities of a population, microbial or human, may so alter their environment as to replace the original type with a new type capable of coping with the new order of things. On the other hand, the environment may change independently of the population (reduction in temperature or decrease in water). Then only those capable of reproduction under the new conditions will survive (natural selection).

Mutations in Colonies. On a solid agar surface a bacterial colony grows radially. The oldest (and senescent) growth is at the center. The newest (and most active) growth is around the periphery. In the development of a colony the single cell initiating the colony may produce billions of progeny. If a visible type of mutation (for example, production of a bright red pigment) occurs in a single cell during the radial growth of the colony, a roughly triangular and visible sector of red cells appears. Typically the apex of the triangle is toward the center of the colony and is the point at which mutation occurred. The base of the sector is at the periphery of the colony (Fig. 18–5). Another visible type of sector in a colony is produced by loss (or gain) of the property of spore formation. If, in a colony of cells producing spores as they grow, sporeless mutation occurs, the progeny are sporeless and they appear as a translucent sector in the colony.

Secondary Colonies. The formation of small excrescences, papillae or outgrowths from ordinary colonies of many species of bacteria after the first growth is mature and begins to age is a common phenomenon. The outgrowths are called secondary colonies, or **daughter colonies** (Fig. 18–6). They may appear on the surface, develop from within, or grow out from the edges. They vary in size, form, numbers, and appearance. The cells in secondary colonies differ genetically from the original in many properties, both morphological and physiological, and are able to grow under the conditions of the **aged** colony.

Detection and Isolation of Mutants

Detection of obvious mutants such as sporeforming or nonsporeforming, or production of different colors, in cultures of molds, yeasts, and bacteria, is simple. For example, anyone who knows about culture methods and the resistance of spores to heat can devise a simple procedure to detect the presence of sporeforming mutants in a culture. Also, detection of mutants resistant

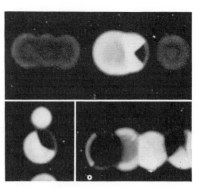

Figure 18–5. Colonies of one species of bacterium growing on the same agar plate, originating from the same inoculum. Variation between white pigmentation (light) and nonpigmentation or transparent (dark) is clearly evident. Clearly shown also is the appearance of mutants during the development of each type of colony, with resulting pigmented (or nonpigmented), triangular sector in the colony.

to various unfavorable factors is obviously possible by submitting a culture containing a billion or so actively growing cells to graded concentrations of the unfavorable factor, such as heat, an antibiotic or a disinfectant. Survivors at each level of concentration obviously must be resistant to that concentration. An adaptation of this principle, especially for detection and isolation of mutants of graded resistance to antibiotics or other substances, consists of the gradient plate.

The Gradient Plate. In this method one pours a shallow layer of nutrient agar (10 ml) into a Petri plate and allows it to solidify in a

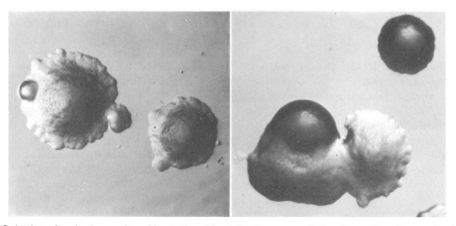

Figure 18–6. Colonies of a single species of bacterium (*Acetobacter rancens*) showing various forms of colony, daughter colonies, and mutant outgrowths.

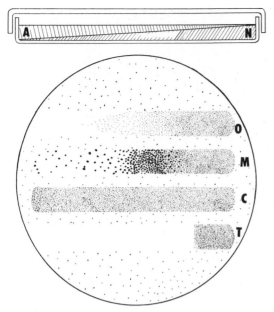

Figure 18–7. Use of the gradient plate to detect drug-resistant mutants and to measure their degree of resistance. In the Petri plate above, the bottom layer (*N*) consists of ordinary nutrient agar allowed to solidify in a slanting position. The upper layer (*A*), poured and allowed to solidify with the plate in a level position, consists of the same sort of agar but with a drug (say, Aureomycin or chlortetracycline) added in measured concentration. The antibiotic diffuses from the upper into the lower layer, leaving a gradient of concentrations of the drug at the surface; greatest at the left of the plate shown above, least at the right. Broth cultures of four different species of bacteria are streaked across the plate, with a sterile brush, parallel with the slope of the agar (in this case, left to right). After incubation it is seen that all organisms grow well at the very lowest concentrations of the drug but react differently in the increasingly greater concentrations. Organism *O* shows moderate resistance of most colonies; a few more than others, but none thriving much more than the majority. Species *M* shows some resistance by most cells but contains some that are wholly resistant and even appear to be stimulated by the drug. Organism *C* shows complete indifference to the drug. Species *T* contains many slightly resistant cells, but all are completely inhibited by exactly the same concentration of the drug.

sloping position (Fig. 18–7). When solid, a second 10 ml of agar containing an appropriate concentration of the test substance, e.g., an antibiotic, is poured over the first layer and the plate is held horizontal until the agar is solid. Owing to the reciprocating gradations in the thickness of the layers of agar, the concentrations of antibiotic on the top surface will be graded from high at one side of the plate to low at the other.

A culture is introduced at the area of low concentration and spread thinly toward the area of high concentration. Of the cells that are deposited on the areas of higher concentration, only those of higher resistance will grow and form colonies. They are thus revealed as resistant mutants. Their degree of resistance is indicated by their position on the gradient plate. They are transferred to other media for further study in pure culture.

Replica Plating. This is a clever adaptation of the printer's art to bacteriological purposes by J. Lederberg (Nobel Prize winner) and his wife E. M. Lederberg. One prepares a wooden disk with a flat surface, in diameter about 9 cm (1 cm less than that of a common Petri plate). A piece of velvet about 12 cm square is placed smoothly over the surface of the disk, drawn firmly down over the sides and fastened in place with a band. A handle, like that of a rubber stamp, is fastened firmly on the back of the disk (Fig. 18–8). Many modifications of the velvet disk have been devised. The whole is sterilized.

Let us now select a Petri plate containing nutrient agar, on the surface of which are several hundred colonies. How do we determine which colony consists of (a) streptomycin-resistant mutants and which of (b) "threonineless" **auxotrophs**? Their appearance in no wise distinguishes them. We could pick a portion of each of the several hundred colonies on the plate with a needle and transfer it to (a) a medium containing streptomycin, (b) a medium containing threonine and (c) one without threonine. If inoculum from any colony grows on the streptomycin-containing medium it is obviously streptomycin-resistant. If it grows on the medium with threonine and not on the same medium minus threonine, it is a threonine-deficient auxotroph. But this entails at least 300 (usually thousands) of inoculations (requiring 28 hours of work per day!) in order to find the one or two mutant colonies which may (or may not!) be among the hundreds of colonies on the plate.

Time and labor are saved if we press the sterile velvet disk gently down on the agar plate. Each colony leaves a small spot of cells where it touches the velvet. If, now, we press the velvet gently on the surfaces of three sterile plates successively, we have in three motions "printed" all of the colonies on the original plate in three **replica plates**: (a) one plate containing medium with streptomycin, (b) a plate with agar containing threonine, and (c) a plate containing agar without threonine. The three replica plates are then incubated and colonies develop. Each colony is in its own, readily determined location on the surface of each of

PLATE III

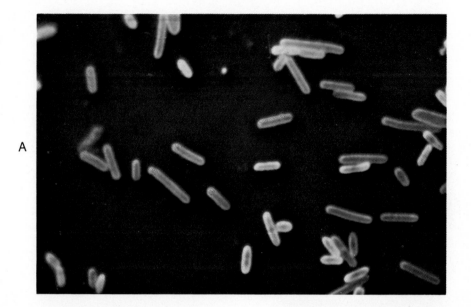

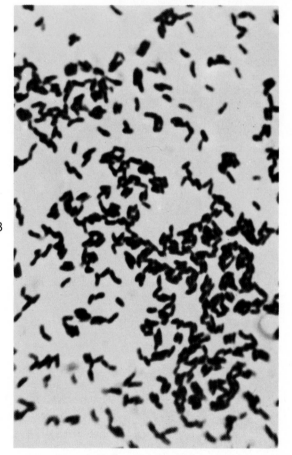
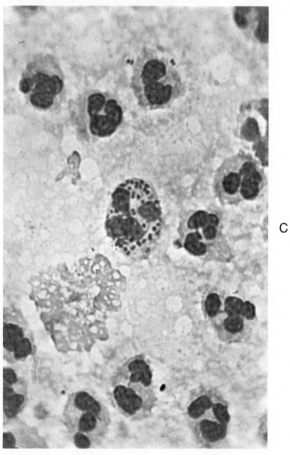

A, A mixed smear of *Clostridium chauvoei* and *Clostridium septicum,* stained with Lissamine Rhodamine B200–labeled antiserum to *C. chauvoei* and fluorescein isothiocyanate–labeled antiserum to *C. septicum.* This is a specific application of the fluorescent antibody–staining technique, described in more detail in Figure 27–25. (Courtesy of Dr. Irene Batty and Dr. Max Moody, Burroughs Wellcome Co., Research Triangle Park, N.C.)

B, An example of a gram-positive organism, *Corynebacterium pyogenes,* showing the characteristic color of Gram's stain.

C, The gram-negative organism *Neisseria gonorrhoeae* shows only the color of the counterstain.

PLATE IV

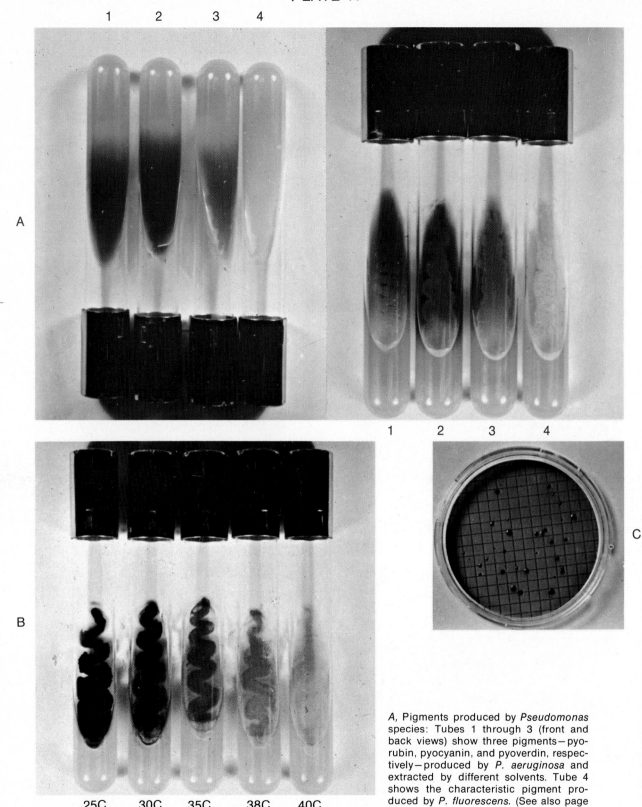

A, Pigments produced by *Pseudomonas* species: Tubes 1 through 3 (front and back views) show three pigments—pyorubin, pyocyanin, and pyoverdin, respectively—produced by *P. aeruginosa* and extracted by different solvents. Tube 4 shows the characteristic pigment produced by *P. fluorescens*. (See also page 470.)

B, *Serratia marcescens,* an enteric species, produces different pigments in culture, depending on temperature. Here, from left to right, we see the pigments produced at 25C, 30C, 35C, 38C, and 40C, respectively.

C, An example of the standard test for total coliform on a membrane filter. This culture was incubated in m Endo broth (Difco) for 24 hours at 35C. (See also Chapter 45, and compare with Color Plate VIII, *A* and *B.*) (Courtesy of Millipore Filter Corporation, Bedford, Mass.)

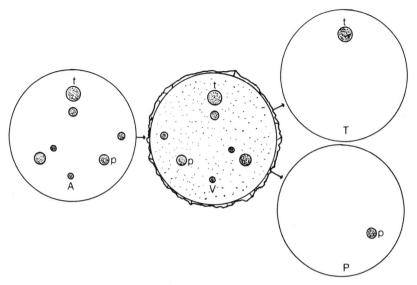

Figure 18–8. Use of the **replica plate** for detection of colonies of mutants. Plate *A* contains agar complete in all essential nutrients and free of penicillin. It was inoculated with a culture of a certain species of bacillus. It was then incubated. It is desired to determine which of the colonies (if any) consists (1) of mutants able to grow in the absence of the amino acid, threonine, and (2) of mutants resistant to penicillin. The sterile velvet disk *V* is lightly pressed down upon plate *A*, so that it acquires spots of live organisms in positions corresponding to (i.e., in a mirror image of) the positions of the colonies on plate *A*. Immediately thereafter, plate *T*, containing sterile agar like that in plate *A*, but *devoid of threonine,* is inoculated by the velvet disk's being pressed on it. Plate *P*, containing agar in all respects like that in plate *A* but *with penicillin added* in measured concentration, is likewise inoculated by the velvet disk. After incubation, plate *T* has only one colony (*t*), obviously consisting of mutants able to synthesize their own threonine (or able to grow without it). Plate *P* has a colony (*p*) resistant to penicillin, in the concentration, at least, in which it exists in plate *P* or perhaps wholly resistant.

the selective replica plates (a, b, and c). Only the colony of the streptomycin-resistant mutant will grow on the streptomycin-containing plate (a); the threonine-deficient mutant will grow on the threonine plate (b), but cannot grow on the plate without threonine (c). The mutant colonies are thus easily identified by their growth and locations on the replica plates.

Isolation by Penicillin. This method, devised by Lederberg and also by Davis, depends on the fact, as previously explained, that organisms in the logarithmic (most actively multiplying) phase of growth are most susceptible to numerous deleterious influences, including the action of penicillin.

Keeping this in mind let us irradiate a culture of bacteria with ultraviolet light. Suppose purineless auxotrophs are to be looked for. They are not immediately apparent. At least two generations of growth must be allowed before the mutational lag is overcome. After the proper, postirradiation incubation, all the cells are removed from the culture medium by sedimentation (centrifugation). They are then resuspended in an inert saline solution in which they will go into a resting state. To separate the auxo-

trophs from the prototrophs, penicillin is added to the suspension along with just enough **purine-free** medium to permit the purine-synthesizing prototrophs to grow but not the purineless auxotrophic mutants.

As soon as the prototrophs start to grow the penicillin kills them. The still-dormant auxotrophs are not affected. They are now removed to a penicillin-free, purine-containing (complete) medium where they grow vigorously.

A little scientific joke is sometimes played by the prototrophs in the final stage of the foregoing procedure, when the unsuspecting prototrophs are being done to death by encouraging them to grow in the presence of penicillin. As they start to grow they may synthesize just enough purine to start the auxotrophs growing also, and so the auxotrophs, too, are killed by the penicillin. Nevertheless the penicillin method is a very useful one.

A number of very ingenious modifications have been devised. In one the culture is irradiated on a microscopically porous membrane filter (Chapt. 21) that is lying on moist agar containing a minimal medium. The membrane is then transferred to minimal agar with penicillin.

The prototrophs grow and are killed by the penicillin. The filter is transferred then to plain minimal agar to absorb and remove the penicillin. Then it is placed on agar with complete medium, i.e., in this case containing threonine. These penicillin methods obviously cannot be used with organisms that are resistant to penicillin.

CHAPTER 18
SUPPLEMENTARY READING

Hayes, W.: The Genetics of Bacteria and Their Viruses, 2nd ed., Chapter 13. John Wiley & Sons, Inc., New York. 1968.

Stent, G.: Molecular Genetics. W. H. Freeman & Co., San Francisco. 1971.

Watson, J. D.: Molecular Biology of the Gene, 2nd ed., Chapter 8. W. A. Benjamin, Inc., New York. 1970.

GENETIC EXCHANGE: RECOMBINATION IN PROCARYONS; GENE FUNCTION

Bacteria utilize several mechanisms to exchange genetic material between cells. Transduction (Chapt. 17) depends on a phage to carry a piece of DNA from one cell to another. Other mechanisms—transformation and conjugation—will be described in this chapter. Exchange of all or part of the genetic material, DNA, from one cell to another, however, is only one step. In becoming functional in the new cell, the DNA is introduced into and becomes a part of the genome of the new cell. This process is called **recombination**.

19.1
GENETIC EXCHANGE

Transformation. Before 1930 it had been observed that if one species of bacterium were cultivated in the presence of a different but closely related species the one would acquire distinctive properties of the other. Apparently something carrying an inheritable character passed from one to the other. This was called **entrainement** by Burnet in 1925. Later, a species of harmless streptococci from cream cheese was shown by Frobisher and Brown to acquire the property of forming scarlet fever toxin when grown in contact with scarlet fever streptococci. It was suspected at that time that a "scarlet fever virus" passed from one species of cell to the other, an idea subsequently abandoned. The mechanism of these alterations was not known at that time, nor is it now.

In later studies by Griffith and others it was shown that cell-free extracts of **dead**, encapsulated pneumococci of Type III (see material on immunological specificity of capsular substance, Chapter 14) would cause cells of live, nonencapsulated (and therefore not type-specific) pneumococci derived from Type II, to produce Type III capsular substance. By similar methods a variety of genetic changes have been produced in bacteria of several other species. Genetic changes induced in this way are called **transformations**.

The Transforming Substance. Subsequent studies by Avery, McCleod and McCarty, and many others revealed that the substance in the extracts of encapsulated Type III pneumococci responsible for the change of nonencapsulated Type II pneumococci to encapsulated Type III was Type III DNA. In all other cases of transformation the transforming principle has likewise been found to be DNA. Transforming DNA is derived from supernumerary replications of DNA. In some species it accumulates on the outside of the cell in a capsule or slime layer. It may be obtained from such cells by washing them. It can also be obtained by rupture of the cell.

The chromosomal fibril of DNA is fragile. Consequently, when DNA is removed from a donor cell by washing or by rupture the DNA macromolecule (bacteria have only one true chromosome) can be broken into segments. Some, perhaps 10 per cent, of the fragments may consist of only one, less commonly two or three, closely linked complete genes. Some may be large (entire?) segments of chromosome. In any case, if they can pass through the recipient cell wall and membrane, one or more new genes may thus be integrated into the chromosome of the recipient cell. However, the DNA segments, large or small, before they enter the new cell are likely first to be destroyed by DNase, an extracellular enzyme produced by many cells, or the DNA segments may be destroyed by other unfavorable factors such as pH or salinity. The recipient cell must also be in a state of **competence.**

Competence. Cells are "competent" for transformation if they do not secrete DNase, do not have thick, interfering capsules and have other, still obscure, qualifications. Into a small (5 to 10?) percentage of these cells the donor DNA will quickly (in about 10 seconds) penetrate. Having once penetrated into the recipient, several genes of the DNA segment may or may not be integrated ("incorporated" or "fixed") into the chromosome of the recipient cell by recombination. When and if this occurs it may take from 10 minutes to many hours or generations. Once integrated, the transforming DNA is a permanent part of the recipient genome and is replicated as such. The appearance of the effects of transformation in the progeny (**transformant**) cells awaits transcription of the new genetic message into mRNA, synthesis by the ribosomes of the necessary new enzymes and one or two fissions. The donor DNA that is integrated by recombination with recipient DNA appears to replace a part of the recipient DNA. Replaced recipient DNA, and donor DNA not integrated into the recipient genome, do not replicate and are eliminated.

Cross Transformation. Cross transformations between different genera, or *distantly* related species, are extremely rare. For successful transformations the respective DNA's must have similar base compositions, i.e., similar ratios of total number of G—C pairs to A—T pairs in their DNA. Even with similar base compositions, transformations between even quite closely related species may fail because of total dissimilarity between nucleotide sequences or because of slight chemical differences in DNA (e.g., attachment of a methyl group.) As to whether transformation occurs in nature there are experimental data indicating that it can occur.

Since we now know how to transform bacteria artificially, may we ever transfer DNA between cells of higher plants and animals, or add artificially synthesized DNA to nucleotide sequences of appropriately prepared gametes for the production of "better" animals? It would seem not much more difficult than a trip to the moon (and much less expensive!). Shall we, in the future, genetically manufacture men with specially coded characteristics, perhaps especially suited to weightlessness, prolonged space travel, and life on distant planets? Stranger things have happened.

Conjugation. The term conjugation as applied to bacteria means physical contact between two genetically different cells of the same (or closely related) species and the establishment of a bridge or "conjugation tube" between them. This is usually followed by passage of genetic material from one (donor or "male") to the other (recipient or "female"); there is no **exchange** of material between the conjugants. Conjugation involves certain properties of the cell wall, to be discussed.

As is true of transformation, conjugation in procaryons is a variable and uncertain means of gene transference. As previously mentioned, procaryons, as represented by bacteria, have only a single chromosome. Bacteria are therefore normally in a **haploid** state. As a self-replicating unit the bacterial chromosome is often called a **chromosomal replicon.** Replication of the donor chromosome occurs before the time of conjugation.

Episomes. In several species of bacteria (possibly also in other procaryons) there are, in addition to a chromosomal replicon, smaller, self-replicating DNA units sometimes called **extrachromosomal replicons, plasmids, episomes,** or **transfer factors.** They are small, closed circles of DNA entirely independent of the major chromosome, but replicating synchronously with it; at times, as will be explained, they become integrated with the chromosomal replicon. They are subject to the same agencies that affect the chromosome itself. Episomes could be viewed, speculatively, as multiple chromosomes, e.g., as primitive forerunners of the multiple chromosomes of eucaryons. Their presence confers on the bacterial cell containing them fertility or the ability to **conjugate**. At least three types of transfer factors or extrachromosomal genetic determinants are known: the F or fertility factor, the Cf or colicinogenic factor, and the R or resistance transfer factor.

The F Factor. Donor cells normally have one F factor or replicon for each chromosome in the cell. Cells with at least one F replicon are designated F$^+$, and cells lacking the F factor are designated F$^-$. The F factor is about one-tenth the size of the chromosomal replicon. The F factor may fail to replicate or it may be destroyed through the action of various agents, among them a dye called acridine orange. The F factor never *appears* spontaneously; it is acquired by a recipient (F$^-$) cell only by conjugation with an F$^+$ cell. Once the F factor is transferred to an F$^-$ cell the F$^-$ cell becomes F$^+$ and the F factor is rapidly replicated in both donor and recipient cells. The F factor confers on F$^+$ cells the genetic property of synthesizing a distinctive polysaccharide at the cell surface that is necessary for conjugation. F$^-$ cells do not synthesize this surface material and cannot initiate conjugation. The surface of the F$^+$ cell is altered *antigenically* (Chapt. 26) and in other ways.

In conjugation between bacteria an F$^+$ cell by chance contacts an F$^-$ cell and a conjugation tube opens between them. The F factor is readily transferred to the F$^-$ cell, but the chromosome of the F$^+$ cell rarely enters the F$^-$ cell. The result is that although fertility (F factor) is spread rapidly by conjugation in a population of F$^-$ cells, genetic recombination (integration of chromosomal material) rarely occurs in such matings. F pili may serve as conjugation tubes, but this has not been proven.

Hfr Cells. In some F$^+$ cells the F factor becomes attached to the chromosomal replicon. Both circles break open, and one end of the F factor attaches to one end of the chromosome. The F factor is then a part of the chromosome and the two replicate as one.

Association of the F factor with the chromosome facilitates transfer of the chromosome during conjugation. As a result of mating between such cells and F$^-$ cells **recombination** occurs with high frequency; hence such F$^+$ cells (F factor associated with chromosome) are designated as Hfr (high frequency of recombination) cells.

In Hfr × F$^-$ matings, recombination is frequent but transmission of the complete chromosome is not. The recipient cell is rarely in a fully diploid state; it is not a complete zygote. It is referred to as a **merozygote** (Gr. *meros* = partial).

Curiously, the merozygotes produced by the Hfr × F$^-$ matings are usually F$^-$ (unlike F$^+$ × F$^-$ merozygotes, which are commonly F$^+$). This is explained by experimental data indicating that in Hfr cells the F factor, or a major part of it, attaches to the "rear" end of the chromosome, i.e., farthest from the end of the chromosome that first enters the F$^-$ cell. All (or part of?) the F factor may be thought of as a locomotive at the rear end of a train of cars, i.e., with the chromosome ahead of it. Because of the fragility of the DNA thread, the filament of DNA usually breaks before the termination of the transfer process, leaving part of the chromosome and usually all of the F factor in the donor cell. As in transformation, unintegrated donor material is eliminated from the recipient and the recipient is soon restored to the haploid state.

When, as rarely happens, the entire Hfr chromosome, including the F factor at the final tip of the chromosome, is transferred to the F$^-$ cell, then, and only then, the recipient becomes an HfrF$^+$ cell.

F-duction. In Hfr cells the F replicon may become disassociated from the chromosomal replicon, re-establishing the F$^+$ state. The point of break between the two replicons may occur slightly to one side or the other of the true junction. One replicon thus acquires a small portion of the other by exchange (or donation or crossing over). The exact mechanism remains to be fully clarified, though plausible hypotheses doubtless closely approximate the truth (Fig. 19–1). F$^+$ cells in which the F factor thus carries some of the chromosomal material are said to be F′. The F factor, plus its bit of chromosome, is called an F-genote. F′ cells act like F$^+$ but transfer their acquired chromosomal

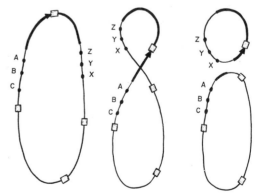

Figure 19–1. Formation of an F-genote by imperfect relooping of the F-chromosome–integrated replicon. Heavy lines denote F-DNA; light lines, chromosomal DNA.

material to F⁻ cells as a unit along with the F factor. This transfer of chromosomal material from one cell to another with the sex (F) factor is spoken of as F-duction or sexduction.

The R (Resistance Transfer) Factor. This factor is, in all essential respects, similar to the F factor but transfers, in addition to fertility, genetic determinants of resistance to several antibiotics. These transfers occur mainly between gram-negative bacteria closely related to the enteric species: *Escherichia coli, Salmonella typhi,* the cause of typhoid fever, and *Shigella dysenteriae,* the cause of epidemic bacterial dysentery. Antibiotic resistance, probably caused in large part by the spread of the R factor among bacteria, is becoming increasingly prevalent in several epidemic areas and presents a grave problem in the treatment and control of enteric bacterial diseases. As mentioned in another chapter, antibiotic resistance may also develop as a result of **point mutations.**

The Colicin Factor (Cf). This is an extra-chromosomal replicon similar to the F and the R replicons. It transfers the property of producing **colicins.** These are antibiotic-like substances lethal for closely related strains or species of the bacteria producing them. They were first observed in *E. coli,* hence their name. Similar substances are produced by other species of bacteria and are given the more general name of **bacteriocins.** Since all of these substances are suggestively similar to bacterial viruses (prophages), they are discussed more fully in Chapter 17.

Chromosome Mapping. In Hfr cells the F element has a regular point of attachment to the chromosome. It is at this point that the chromosome ring appears commonly to open for transference to the F⁻ cell during the conjugation. In any given cell type, the genes on the chromosome thread therefore usually enter the recipient in a fixed and predictable order, like knots on a string being pulled through a hole in a door separating two rooms, the F replicon last. As shown by many workers, led by Jacob, Lwoff, Monod (Nobel Prize winners), and Wollman in the early 1960's, it is possible to determine the exact **locus** (location in the chromosome) of any gene, and the order of loci of all genes on the chromosomal thread, by the order in which they are transferred from the Hfr cell to the F⁻ cell during mating. This location of genes on the chromosome is spoken of as **chromosome mapping.**

For example, a culture of Hfr cells having certain genetic markers (e.g., various auxotrophies, enzymic characters, or resistance to

various antibiotics) is mixed under suitable conditions in a culture vessel with F⁻ cells. A sample of the mixture taken about eight minutes after mixing shows no cells with recombination of characters (markers). Two or three minutes later, and at each of several short intervals thereafter, a sample is taken and violently shaken in a Waring blendor. It is known that this is sufficient to break all DNA filaments and therefore to separate all conjugating cells. The shaken samples are immediately plated on agar media that are selective for the various markers to be identified. By observing (in thousands of experiments!) which markers appear in recombinants removed from the conjugation culture after each successive interval, it is possible to determine that certain markers always appear first, followed by the others in their regular and predictable order at predictable intervals. Having established the order of arrangement of markers on a chromosome, it became possible to draw up circular **linkage maps** of the bacterial chromosome (Fig. 19–2).

The point of breakage, and consequently the point of attachment of the F factor to the opened chromosome of the Hfr donor cell, may differ in different species or in particular cultures (strains) of the same species. Thus, while the same sequence of genes may characterize all the cells, the first markers to be transferred may differ. For example, we may have a circular linkage l-m-n-o-p-q-r. In one strain of this species, breakage and attachment of the F factor may occur between l and r, the l and r being adjacent in the circle. The order of transfer then is l-m-n-o-p-q-r-F; markers l and m enter the F⁻ cell first, F last, if at all. If, in another strain, breakage occurs between markers n and o, then the first marker to enter the F⁻ cell may be n or o, depending on whether F is associated with n or o.

19.2
GENETIC RECOMBINATION IN PROCARYONS

In all cells, during the nondividing or resting stage (**interphase** in eucaryons) that exists prior to the beginning of cell **fission,** each chromosome, whether single as in procaryons, or multiple as in eucaryons, **replicates** itself. In bacteria this is from end to end by the process described in Chapter 6. Each bacterial chromosome thus forms two, usually identical, double-helix chains of nucleotides (analogous to **chromatids** in eucaryons). The two identical chromosomes are so closely approximated that they

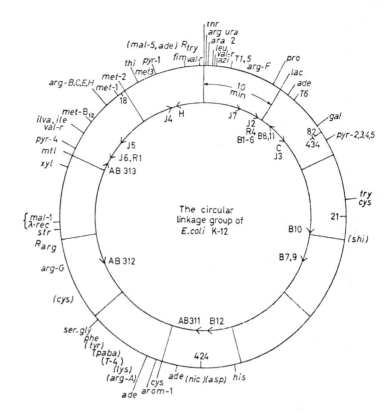

Figure 19–2. Circular genetic linkage map of *E. coli* K-12. The double circle is divided into 11 sections, each representing a 10-minute transfer interval as determined from interrupted mating experiments. Around the outer circle are located various genetic markers: e.g., *gal* = galactose fermentation; *str* = streptomycin resistance. Prophage loci are indicated inside the outer circle. The various Hfr substrains of *E. coli* K-12 are shown inside the inner circle at the arrowheads (*J4, H, J7*, etc.), which mark the leading end and direction of the transfer of markers. (Prophages are discussed in Chapter 17.)

appear to be one. During interphase the chromosome threads seem to be in the form of amorphous granular material or of tangled masses of long fibrils of chromatinic material mingled with cytoplasmic matter. When the cell undergoes fission one of each pair of identical DNA chains goes to each of the two progeny. Each daughter bacterial cell therefore receives an entire haploid set of all of the genetic determinants (a **genome**) of the parent cell.

It is at this point, i.e., the manner of distribution of chromosomes to progeny, that one of the major differences between eucaryotic and procaryotic cells becomes manifest: eucaryotic cells exhibit the phenomena of mitosis and meiosis; procaryotic cells do not. It is assumed that the student has already made the acquaintance of mitosis and meiosis as they occur in eucaryotic cells in elementary biology. Therefore, we may turn our attention to hereditary mechanisms of procaryons as represented by bacteria.

When the procaryotic cell undergoes fission, the genetic material shows no evidence of systematic organization into chromosomes of the eucaryotic type or any of the phenomena associated with mitosis or meiosis. The procaryotic analogy of mitosis appears to be a gathering together of the interphasic tangle of DNA strands, i.e., the **nucleoid** into an amorphous skein, which then appears to divide merely by a constriction near the middle, like dividing a hank of yarn into two equal parts.

Although processes of nuclear fission in procaryons may seem haphazard, the replication and division of the DNA in procaryotic cells is as exact and as effective a hereditary mechanism as in eucaryons.

In preceding sections, mechanisms have been described by which DNA is transferred between bacterial cells. But transfer is not sufficient—the DNA must become integrated into a replicating unit if it is to be passed on to daughter cells and if it is to function in the new cell on a long-term basis.

If a bacterium is simultaneously infected with two phages, each having different identifiable genetic characters (such as plaque morphology, host range, etc.), some of the progeny phage are found to possess genetic characters of both parental phages. A mutation that has occurred in one phage is thus given the opportunity to be associated with genetic properties of another phage, resulting in improved chances of coping with a changing environment.

From the standpoint of evolution, recombination gives a new mutant many different combinations of other genes, the best combination, of course, being the one that is most likely to succeed. The vast majority of mutations are "harmful;" that is, they decrease the chance of survival. Since mutations are usually rare events, it would take a very long time for a given "beneficial" mutation to arise in sufficient combination with other genes to find the most successful. Likewise, the "harmful" mutations are quickly weeded out. Recombination is therefore a tremendous help in evolution and is seen at all levels of biological organization.

Study of the mechanism of recombination in bacteria and their viruses has led to a much deeper understanding of molecular genetics. During recombination, there occurs physical exchange of genetic material between the two "parental" organisms. The two molecules of DNA are actually broken and subsequently rejoined by an elaborate enzyme system; the same system also appears to be involved in repair of DNA damaged by chemicals, physical agents like ultraviolet light, and other noxious agents.

Recombination exemplifies the essence of sex in the simplest and most fundamental molecular sense. Exchange of DNA between two phage genomes is called a mating event and, biologically speaking, does not differ in principle from sex in higher organisms. The sex life of viruses, however, is so rudimentary that it is of primary interest only to other viruses and certain virologists!

19.3
GENETIC FUNCTION

One of the ultimate goals of genetic research is to determine how genes work: How they control a cell's morphology and how they control the various enzyme systems involved in a cell's metabolism. Some of these findings will be described briefly here.

The Cistron. During intensive studies of the location of genes on chromosomes (see Chromosome Mapping, this chapter), it was discovered by Benzer and others about 1957 that if two similar genes, each having suffered different minor deletions due to some agent like x-irradiation, were on chromosomes of mating cells, one defective gene of each chromosome could, by recombinational crossing over or equivalent means, compensate, within the limits of that gene, for the defect in the

other. For example, imagine a gene having a nucleotide sequence (No. 1) l, m, n, o, p, q, r, made defective in that it lacks o: l, m, n, ..., p, q, r. Imagine the sequence on the corresponding gene of the mating chromosome (No. 2) to lack q: l, m, n, o, p, ..., r. If the two sequences overlap, i.e., are both on the same side of a given point (in the *cis* [= this side] position), they can complement each other. **Complementation** is said to have occurred. If they are on opposite sides of the point at which they overlap (in the *trans* [= other side] position), then they do not complement each other. The limits of a nucleotide sequence can thus be established. The gene whose nucleotide sequence is thus established by complementation as a functional unit in its relation to a complementary gene is termed a **cistron**, from "cis-trans." A cistron usually consists of hundreds or thousands of recons.

The Operon. Not all genes or cistrons are responsible solely for polypeptide synthesis. Those that are involved in the synthesis and specific structure of proteins, as previously described, are called **structural genes**. Other genes or cistrons are part of the "stop-go" mechanisms (Enzyme Control, Chapter 7). They are called **regulatory genes**.

Regulatory genes may be thought of as a sort of on-off switch. For example, the enzyme β-galactosidase is inducible in several species of bacteria (Chapt. 7); it is an enzyme involved in fermentation of lactose. Inducibility consists in the fact that there is a genetic mechanism for β-galactosidase production that is repressed by a removable repressor. The inducing agent (lactose, for example) in contact with the cell removes the repressor, allowing the β-galactosidase mechanism to become functional.

The genetic mechanism for β-galactosidase formation consists of several factors: (1) two structural genes for synthesis of β-galactosidase (z in Fig. 19–3) and another for synthesis of the permease enzyme (y in Fig. 19–3) to allow entrance of the substrate or inducer into the cell; (2) a regulatory gene that produces the repressor substance (i in Fig. 19–3); and (3) a receptor site or unit called an **operator** (o in Fig. 19–3) that is affected by the repressor (R in Fig. 19–3) and inhibits production of the messenger RNA that is necessary to synthesis of the structural genes. The whole mechanism, including several cooperating structural genes, and several ancillary initiatory units (e.g., a promotor; p in Fig. 19–3), the whole influenced by one operator, is called an **operon**. Several operons may be interconnected and influence each

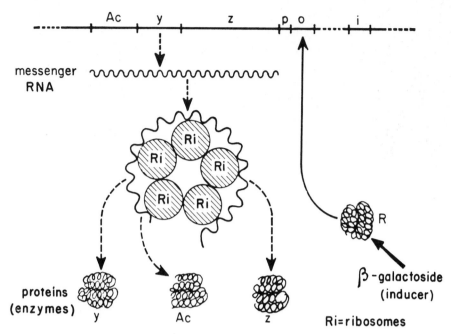

Figure 19–3. Diagram of an operon, in this case the genetic mechanism of β-galactosidase formation in *Escherichia coli.* At the top an enlargement of the lactose region is shown. *i,* Regulatory gene; *o,* operator; *p,* promotor; *z,* structural gene for β-galactosidase; *y,* structural gene for β-galactosidase permease enzyme; *Ac,* structural gene for β-galactosidase transacetylene; *Ri,* ribosome. See text for discussion.

other. Some genes are said to be **modifiers** because they increase or diminish the activity of other genes. Complete details may be found in books and articles on genetics. (See list of Supplementary Reading below.)

It is clear that genes are far from being autonomous agents, and that all are closely interadjusted and interdependent and tuned to the activities of each other and of the cell as a whole.

CHAPTER 19
SUPPLEMENTARY READING

Boyes, B. C.: The impact of Mendel. Bioscience, *16*:85, 1966.

Braun, W.: Bacterial Genetics, 2nd ed. W. B. Saunders Co., Philadelphia. 1965.

Clowes, R. C.: The molecule of infectious drug resistance. Sci. Am., *228*:19, 1973.

Hanawalt, P. C., and Haynes, R. H.: The repair of DNA. Sci. Am., *216*:36, 1967.

Stern, C., and Sherwood, E. R. (Eds.): The Origin of Genetics. A Mendel Source Book. W. H. Freeman & Co., San Francisco. 1966.

Watson, J. D.: The Molecular Biology of the Gene, 2nd ed., Chapter 10. W. A. Benjamin, Inc., New York. 1970.

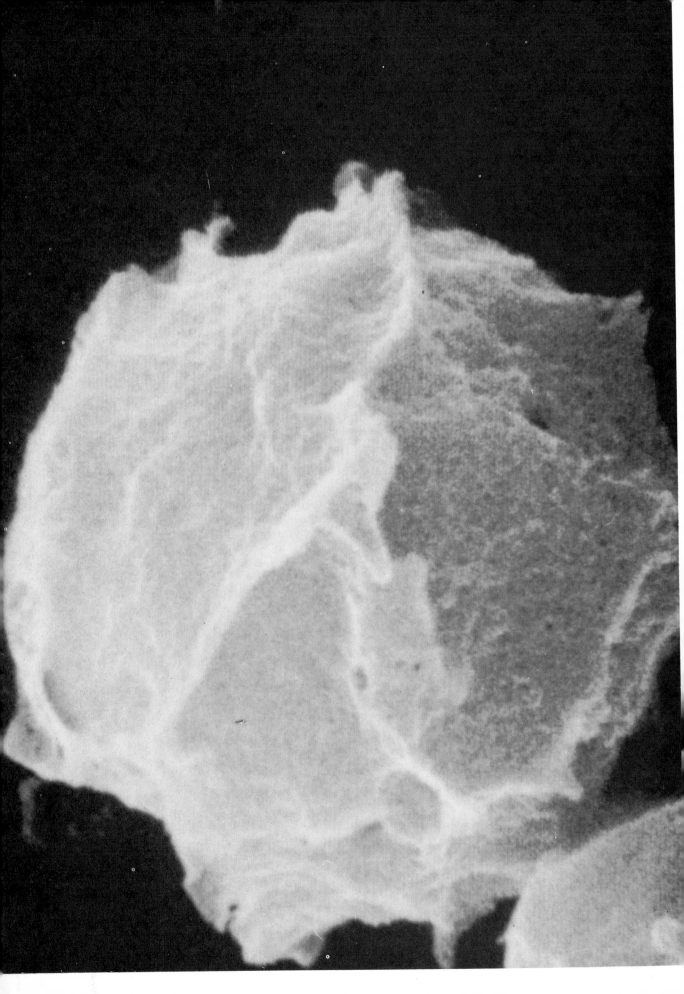

SECTION THREE

MICROBIOLOGY AS AN ECLECTIC SYSTEM

Unlike such pure scientific disciplines as mathematics, physics, celestial mechanics, and optics, modern microbiology is a complex structure, embracing selected parts of many other disciplines; a mélange or pasticcio, as it were, of segments of knowledge drawn from many other fields of study, experimentation, invention, and discovery.

This section of *Fundamentals of Microbiology* presents, first, a series of chapters describing the effects of various physical and chemical agents on microorganisms. Parts of such knowledge form the basis of methods for controlling microorganisms by such means as heating; producing high osmotic pressures and unfavorable levels of pH; freezing; desiccation or dehydration; radiations both lethal and mutagenic; application of a variety of chemically active, indiscriminately destructive poisons like chlorine or carbolic acid (phenol); use of highly specific antibiotics like penicillin, tetracycline, and other cytologically selective chemotherapeutic substances, both bacteriostatic and bactericidal; and many others. All of this information is absolutely essential to the practicing microbiologist in any field: agriculture, human or veterinary medicine, domestic economy, and a wide variety of industries ranging from the making of cottage cheese to the production of cortisone, whiskey, or poliomyelitis vaccine.

The last-mentioned activity requires knowledge not only of virology, as outlined in previous chapters, but also of the basic principles of biological methods of preventing infectious disease (immunology); both of these areas of knowledge are still other segments of the patchwork quilt of microbiology. Familiarity with the basics of immunology, presented in the closing chapters of this section, is essential to a full understanding of almost every facet of microbiology, including systematics and diagnostics and the role of microorganisms in disease, in the histories of nations, in human ecology, and in almost every other aspect of human life.

Facing page, Courtesy of R. Albrecht.

MICROORGANISMS · CHAPTER 20
AND THEIR
ENVIRONMENT

We can only guess at the environmental conditions under which the most primitive microorganisms made their first appearance on this planet. At present their descendants live all over the earth under a wide variety of environmental conditions. Since some of their present habitats, such as the living tissues of mammals and plants, presumably did not exist in Archeozoic ages, it may be assumed that evolution selected many microorganisms capable of adapting to conditions vastly different from those which their predecessors first encountered. Microorganisms exist today that are capable of utilizing what seem to us the most indigestible of foods (pure sulfur, kerosene, naphthalene, carbolic acid) and of thriving in the most remote, dismal, and uninhabitable places, such as the depths of ocean ooze, subterranean slimes, petroleum wells, hot sulfur springs, and inside nuclear reactors, as well as inside other living cells. So complete and highly specialized are some adaptations that many forms of microorganisms perish when transferred suddenly to other environments.

In this chapter we shall discuss some of the environmental factors governing microbial life. Table 20–1 represents a minimal list of factors which must be appreciated and understood by the beginning student if microorganisms, particularly bacteria, are to be properly cultivated and maintained. Many of these factors are discussed further in Chapters 43 and 44.

20.1
TEMPERATURE

Temperature and Growth. Temperature is one of the most important factors influencing growth, propagation, and survival of all living organisms. Low temperatures generally slow down cellular metabolism, while higher tem-

TABLE 20–1. SOME IMPORTANT PHYSICAL FACTORS OF THE ENVIRONMENT

Environmental Factors	Microbial Activity Principally Influenced	Cell Component(s) Likely to Be Most Affected
Temperature	Growth (and survival)	Enzymes, structural proteins
Gases	Growth (and survival)	Enzymes
Ionic effects	Growth (and survival)	Enzymes, plasmoptysis, plasmolysis
Desiccation	Survival	Enzymes, structural components
Radiant energy	Survival	Nucleic acids, enzymes
Mechanical stress	Survival	Cell envelope
Surface tension	Survival	Cell envelope

peratures increase the rate of cell activities. However, every organism has certain lower and higher temperature limits beyond which growth ceases, as well as an optimum temperature range for growth and reproduction. These three temperature ranges are called **cardinal temperatures** or **points** (Fig. 20–1).

Minimum Growth Temperature. The minimum is the lowest temperature at which the organism grows. Many microorganisms and almost all bacteria will survive for varying lengths of time below this temperature but will show negligible growth.

Maximum Growth Temperature. The maximum is the highest temperature at which growth occurs. Temperatures only slightly above this point frequently kill the microorganism by inactivating critical enzymes.

Optimum Growth Temperature. The optimum is the temperature at which the most rapid rate of multiplication occurs. For most organisms, optimum growth occurs over a temperature range rather than at a fixed temperature, and the upper limit of this range is generally only a few degrees below maximum growth temperature.

These **cardinal temperatures** of microorganisms differ widely, with optima ranging from as low as 5 to 10C to a high of 70 to 75C. Some microorganisms have a minimum growth temperature below freezing (−12C) where the freezing point of the environment or medium has been depressed by a high concentration of solutes. Others, especially those found near hot springs, may grow at temperatures in excess of 90C. However, many of the microorganisms so frequently encountered in water, soil, or decaying matter, as well as most pathogens, have cardinal temperatures ranging from 10 to 45C. Bacteria are frequently classified into three or

TABLE 20–2. CLASSIFICATION OF BACTERIA BASED ON GROWTH AT VARIOUS TEMPERATURE RANGES

Group	Growth Temperature (C)		
	MINIMUM	OPTIMUM	MAXIMUM
Mesophiles	10–15	30–45	35–47
Psychrophiles			
Facultative	5 and below	25–30	30–35
Obligate	5 and below	15–18	19–22
Thermophiles	40–45	55–75	60–85

four groups according to their temperature characteristics (Table 20–2). These groups are not sharply defined, as the temperature ranges frequently overlap. Although the placement of a particular organism in one of these groups could be described as arbitrary in certain instances, microbiologists still find this sort of classification useful in describing the collective properties of groups of microorganisms adapted to life in certain environments. In the following discussion of temperature effects it is to be understood that the organisms are suspended in an aqueous fluid medium unless otherwise noted.

Mesophiles. Many species indigenous to the soil, waters, and the vertebrate body can grow at temperatures from 10 to 47C (human body temperature is 37C) (Fig. 20–2 and Table 20–3). Their **optimum** growth temperatures, however, are about 30 to 45C and vary with species. Species growing best at such temperatures are called **mesophilic** (Gr. *meso* = middle or medium; *philic* = prefer).

Psychrophiles. Psychrophiles are commonly defined as microorganisms capable of growth at 0C. Other microorganisms that are adapted to life in the sea or soil grow best at temperatures below or near the freezing point (10 to −2C). These also are referred to as **psychrophilic** (Gr. *psychros* = cold). Many psychrophiles will also grow well at temperatures in the lower mesophilic range. They might be called **facultative psychrophiles.** On the other hand, certain marine bacteria, adapted to life at about 4C, the temperature at profound depths, die if held at about 30C for more than a few minutes. These could be called **obligate psychrophiles,** since they are limited to low temperature as a condition of life. Psychrophilic microorganisms occur all over the earth.

The importance of psychrophilic microorganisms is easily realized when we consider the enormous volumes of organic matter to be decomposed continually in the oceans and soil.

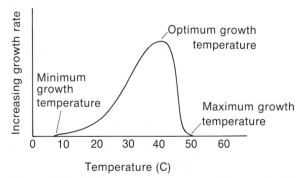

Figure 20–1. Influence of temperature on the growth rate of a typical mesophile. Note that optimum growth occurs closer to the upper temperature extreme. Also see Figure 7–6.

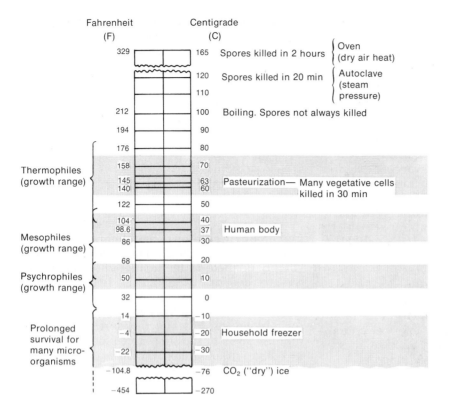

Figure 20–2. Some temperatures of significance in microbiology. Note the slow killing action of dry air heat and the relatively rapid killing action of steam (moist heat) at lower temperature. Boiling kills some but not all spores. Note overlapping growth ranges of thermophilic, mesophilic and other forms.

However, psychrophilic molds, yeasts, and bacteria can cause spoilage of foods and many other materials that are stored at refrigeration temperatures. They are thus important in the huge frozen food industry. Few species of psychrophilic microorganisms are pathogenic; most are aerobic or facultative.

Thermophiles. Microorganisms that grow at temperatures above 45 to 50C are frequently referred to as **thermophilic** (Gr. *therme* = heat). However, thermophilic species that thrive only at high temperatures (50C and above) are called **obligate thermophiles.** Such organisms are frequently found in hot sulfur springs, like those in Yellowstone National Park, Wyoming, but also in other heat-producing environments, such as manure, compost, or silage. Thermophiles have also been isolated from milk, soil, and sea water. Thermophilic bacteria may grow at temperatures above 90C, or perhaps slightly higher. Some have been cultured with an optimum growth temperature of 80C.

Recent investigations have shown that certain types of microorganisms are better suited for life at high temperatures than others. For example, a natural thermal gradient is set up when water flows from a hot spring into an adjacent stream. By examining the kinds of micro-organisms that are growing in the various parts of the run-off with successively lower temperatures, the upper temperature limits for different kinds of organisms can be ascertained. Thus it seems, at least in the natural situations studied, that procaryotic organisms can grow at higher temperatures than eucaryotic organisms, non-photosynthetic organisms grow at higher temperatures than photosynthetic organisms and, finally, that organisms less complex structurally can grow at higher temperatures than more complex organisms.

Endurance Compared with Growth. A distinction should be made in all cases between ability to **endure** a given temperature and ability to **grow** well under the same conditions. Many microorganisms, including fungi, mammalian tissue cells, most bacteria, and viruses, can live for months or years frozen in "dry ice" (carbon dioxide ice) at −76C or liquid nitrogen (−195C), yet they do not grow at all. They might be called **psychroduric. Thermoduric** (heat-enduring) organisms survive well at temperatures above 50C, but only **thermophilic** species grow well under such conditions. Some microbiologists would draw the line between mesophiles and thermophiles at somewhere between 44 and 52C.

TABLE 20-3. EQUIVALENTS OF CENTIGRADE AND FAHRENHEIT THERMOMETRIC SCALES °

C	F	C	F	C	F
−40	−40.0	9	48.2	57	134.6
−39	−38.2	10	50.0	58	136.4
−38	−36.4	11	51.8	59	138.2
−37	−34.6	12	53.6	60	140.0
−36	−32.8	13	55.4	61	141.8
−35	−31.0	14	57.2	62	143.6
−34	−29.2	15	59.0	63	145.4
−33	−27.4	16	60.8	64	147.2
−32	−25.6	17	62.6	65	149.0
−31	−23.8	18	64.4	66	150.8
−30	−22.0	19	66.2	67	152.6
−29	−20.2	20	68.0	68	154.4
−28	−18.4	21	69.8	69	156.2
−27	−16.6	22	71.6	70	158.0
−26	−14.8	23	73.4	71	159.8
−25	−13.0	24	75.2	72	161.6
−24	−11.2	25	77.0	73	163.4
−23	−9.4	26	78.8	74	165.2
−22	−7.6	27	80.6	75	167.0
−21	−5.8	28	82.4	76	168.8
−20	−4.0	29	84.2	77	170.6
−19	−2.2	30	86.0	78	172.4
−18	−0.4	31	87.8	79	174.2
−17	+1.4	32	89.6	80	176.0
−16	3.2	33	91.4	81	177.8
−15	5.0	34	93.2	82	179.6
−14	6.8	35	95.0	83	181.4
−13	8.6	36	96.8	84	183.2
−12	10.4	37	98.6	85	185.0
−11	12.2	38	100.4	86	186.8
−10	14.0	39	102.2	87	188.6
−9	15.8	40	104.0	88	190.4
−8	17.6	41	105.8	89	192.2
−7	19.4	42	107.6	90	194.0
−6	21.2	43	109.4	91	195.8
−5	23.0	44	111.2	92	197.6
−4	24.8	45	113.0	93	199.4
−3	26.6	46	114.8	94	201.2
−2	28.4	47	116.6	95	203.0
−1	30.2	48	118.4	96	204.8
0	32.0	49	120.2	97	206.6
+1	33.8	50	122.0	98	208.4
2	35.6	51	123.8	99	210.2
3	37.4	52	125.6	100	212.0
4	39.2	53	127.4	101	213.8
5	41.0	54	129.2	102	215.6
6	42.8	55	131.0	103	217.4
7	44.6	56	132.8	104	219.2
8	46.4				

°To convert degrees F to C, subtract 32 from the F measurement and take ⁵/₉ of the remainder. To convert degrees C to F, take ⁹/₅ of the C figure and add 32.

It is worth noting that the optimal temperatures for growth of nearly all microorganisms are near the upper maximal limits of their range and that lethal temperatures are only a little above optimal. Most microorganisms have a wide tolerance to temperatures far below optimal.

Growth Temperatures and Enzymes. The optimal and limiting growing temperatures for microorganisms and, indeed, for all living cells, are in general the optimal and limiting temperatures of their enzymes. As pointed out in Chapter 7, enzymes have minimal, optimal and maximal reaction temperatures.

At temperatures well below the optimum, most enzymes function more slowly or not at all. This is in part because low temperatures generally decrease chemical reaction rates and cause increases in viscosity of fluids and hardening of lipids. Growth is retarded or inhibited, but low temperatures are not highly destructive as are high temperatures. High temperatures cause rupture of hydrogen bonds in proteins and in DNA, resulting in denaturation of proteins and "melting" (separation of strands) of DNA.

Thermal Resistance. Most microorganisms in an actively growing (vegetative) state are killed by exposures to temperatures of about 70C for one to five minutes. Some are killed in 10 minutes at temperatures as low as 54C. Commercial **pasteurization** of milk (63C for 30 minutes or 72C for 15 seconds) kills all vegetative **pathogens** (disease-producers) in milk, including tubercle bacilli (*Mycobacterium tuberculosis*) and the Q fever–producing rickettsia, *Coxiella burneti*. However, numerous thermoduric **saprophytes** (Gr. *sapros* = decay) in milk survive pasteurization, thus pasteurized milk is safe but not sterile. (Saprophytes, involved in decay of dead organic matter, are generally nonpathogenic, though there are some notable exceptions—see Chapters 24 and 36.)

Thermophiles are quite resistant. Some vegetative thermophilic cells can survive 80 to 90C for as long as 10 minutes. These obviously can survive pasteurization temperatures and hence certain species are great nuisances in dairies. Boiling (212F or 100C) kills all vegetative microorganisms (certain viruses excepted) within 10 minutes.

Endospores of bacteria of the family Bacillaceae are the most heat-resistant of all living things. Some of these spores can survive boiling or higher temperatures for many hours. (Killing of spores is more fully discussed in Chapters 21 and 22 on disinfection.) Spores of eucaryotic fungi and of some mold-like bacteria (family Streptomycetaceae) are resistant to drought, unfavorable pH, and salinity, but have only slightly greater thermal resistance than their vegetative cells.

Thermal Death Point. The lowest temperature at which all the bacteria of a given species in a given culture are killed within 10 minutes is called the **thermal death point** of that species. This is an inexact expression because at a given temperature, say 70C, the bacteria in a given culture or situation do not all die simultaneously and suddenly just as the clock registers the expiration of 10 minutes. Some are more heat resistant than others, as are humans. Unless the temperature is catastrophic, as when a culture is dropped into a furnace, thermal death point merely tells us when the last survivor of all has expired. Furthermore, the thermal death point is influenced greatly by the numbers of bacteria originally present, their physiological state (growing or resting), their age, the acidity of the suspending fluid, its osmotic pressure, and so forth. The term **thermal death point,** therefore, can be correctly used only if the exact conditions of an experiment are known. When the exact conditions are carefully controlled, information on thermal death points can be of some limited use, in canning and in the commercial preservation of milk and other products.

Thermal Death Time. The time required to kill all the bacteria of a certain species in a given substance at a stated temperature is called the **thermal death time.** The thermal death time of various species of microorganisms is influenced by the same factors that affect thermal death point, but under known conditions knowledge of thermal death time is of practical use, especially in the canning industry.

Rate of Death (Thermal Death Rate). As we have noted, death of all the bacteria of any given species in a given material does not occur simultaneously, but in a definite relationship to time, the rate being determined by factors such as number, age and kind of cells, temperature, moisture, and acidity.

Mature (early stationary phase) cells are more resistant than young ones. Moisture and acidity greatly increase the vulnerability of cells to heat. Cells inside solid material such as mucus or canned meats may escape heat or the action of disinfectants longer than the same cells suspended naked in distilled water or broth, because the heat or disinfectant does not penetrate immediately into the center of solid masses.

If we know the number of cells initially present, we can determine the number and percentage of survivors of organisms present in a disinfection or heat-resistance test. If these are determined at various intervals and plotted against time, regular curves are formed (Fig. 20–3). If the logarithms of the numbers are similarly plotted, a straight line is formed (Fig. 20–4). Under certain influences such as pH, presence or absence of certain cations (Mg^{2+}, Ca^{2+}), or osmotic conditions, these theoretical straight lines may be much distorted: rapid at first, slow toward the end, and vice versa (see Chapter 21).

20.2
GASEOUS ENVIRONMENT

The types and concentration of gases present in the environment profoundly influence

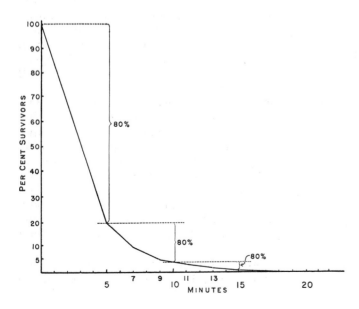

Figure 20–3. Idealized death curve plotted arithmetically. Relation between time and per cent survivors in a situation in which the lethal agent acts at a constant rate under uniform conditions.

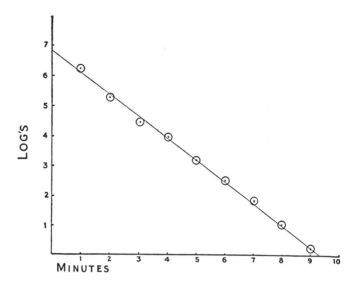

Figure 20–4. Idealized death curve plotted log-arithmically. Relation between logarithms of numbers of surviving organisms and time in a 1 per cent aqueous solution of phenol at fixed temperature and pH. The straight line is characteristic of the relation between time and survivors under fixed adverse conditions when no growth occurs.

growth, and often survival, of microorganisms. Paramount in consideration are the oxygen tension and the oxidation-reduction potential. This topic is discussed at length in Chapter 36 in the section on anaerobiosis. A comparison of metabolic pathways of aerobes and anaerobes was made in Chapter 8.

Other gases of notable importance are: carbon dioxide, an important by-product of organic carbon degradation and a carbon source for photosynthetic microorganisms and certain other bacteria; nitrogen and ammonia, essential constituents of the nitrogen cycle; and H_2S, which plays a major role in the cycling of sulfur. Detailed information on the cycling of elements, including those in gaseous forms, and the principal microorganisms involved will be found in Chapter 43. While the diversity and complexity of microbial metabolism make generalizations difficult, as a rule the accumulation of gaseous products in the microbial environment much beyond a level considered optimum will have a deleterious effect on growth and survival. Finally, as will be described in Chapter 22, certain gases are known to be highly toxic and can be used as microbicidal agents.

20.3
OSMOTIC PRESSURE

If living cells are immersed in fluids with abnormally high osmotic pressures, water will be drawn out of the cells until the cell collapses (**plasmolysis**). If cells are immersed in a fluid of abnormally low osmotic pressure, water will be drawn into the cells until they rupture (**plasmop-**

tysis). Ordinarily, osmotic pressure within the cell is sufficient to keep it slightly distended or in a **turgid** condition. When the osmotic pressure of the surrounding fluid is in equilibrium with that of the cell contents, the surrounding fluid is said to be **isotonic** with that of the cell. If the osmotic pressure of the surrounding fluid is less than that of the cell contents the fluid is **hypotonic** with respect to the cell. If the osmotic pressure of the surrounding fluid is greater than that of the cell contents, the surrounding fluid is **hypertonic** with respect to the cell.

Extremely hypertonic solutions like pickling brines and concentrated sugar syrups have a "preservative" value because they withdraw water from cells, which has a microbiostatic effect on many organisms. A method of preserving numerous species of bacteria in salt solutions has been described. The action of strongly hypertonic solutions is comparable with desiccation, which removes water entirely, and with freezing, which immobilizes water in the cell. Only organisms that can imbibe water from very hypertonic solutions, e.g., many molds and some bacteria, can grow under such conditions.

Probably because of their minute volume, relatively strong cell wall, and thin cytoplasmic membrane which permits rapid adjustment of osmotic equilibria, most bacteria are not highly sensitive to variations in salt concentrations between about 0.5 and 3 per cent. Concentrations much above this may adversely affect some of the more sensitive strains. Some marine bacteria adapted to the salinity of ocean water (about 3.5 per cent) are quite sensitive to lower or higher salinities and will not grow if the salinity is less than about 2 per cent or over about 15 per cent.

Halophilic Organisms. There are bacteria that have become adapted to the high salinity (about 29 per cent) of various salt waters such as the Dead Sea and the Great Salt Lake of Utah. These organisms cannot grow in lake water diluted to a salinity of less than about 13 per cent. Such organisms are spoken of as **halophilic** (salt-loving). There are halophilic and also **haloduric** bacteria that grow in commercial pickling brines with salt concentrations up to 30 per cent. Some are a cause of spoilage of various commodities preserved with salt, such as fish, meat, hides, and pickle stock (see Chapters 48 and 49). Certain of these halophiles produce brilliant red pigment when growing on spoiled salted herring. This has been said to be the origin of the common expression "a red herring."

Alterations in Membrane Permeability. Even though the favorable osmotic pressure of a given culture medium may continue unchanged, there often appear in it, as a result of aging and metabolic activities, substances that alter the permeability of the cell membrane so that excessive water diffuses inward and the cells become swollen and distorted. Such alterative substances may be various waste products (e.g., acids or alcohol) of cells that have grown in the culture. Such distorted forms of bacteria are often called **involution forms.** Cell lysis finally results.

20.4
DESICCATION

Many species of microorganisms can survive complete drying or **desiccation** for long periods, although they do not grow under such conditions. Many substances, such as hay, fruits, fish, and meat, "preserved" by drying, contain large numbers of living microorganisms that are dormant but that soon grow and cause spoilage if the dried product becomes moist. On the other hand, some microorganisms, especially marine and aquatic species and delicate pathogenic species, are quickly killed by drying. Spores, conidia, arthrospores, chlamydospores, and cysts of protozoans, of course, are dormant cells specially adapted to withstand drying for long periods.

Desiccation and Vacuum. Not only do many kinds of microorganisms in a vegetative state withstand desiccation, but they may be simultaneously subjected to the highest possible vacuum without harm. In fact, some organisms appear to be more thermoduric in high

vacua and therefore possibly capable of travel through space on or in spacecraft. The possibility of transporting terrestrial microorganisms to the moon and other celestial bodies is a major concern of space scientists.

A simple method for preserving cultures in the laboratory is first to suspend the organisms in a harmless or protective fluid like broth or blood. Then, about ¼ ml of this suspension is placed on absorbent material like bits of filter paper, unglazed ceramic beads, or sand previously sterilized in a small, cotton-plugged vial. Many such vials may be placed in a single jar that contains a desiccant, such as calcium sulfate, and that can be evacuated rapidly and sealed while evacuated. Water in the suspending fluid and in the microorganisms is rapidly transferred to the desiccant. Desiccation must be rapid (hence the vacuum) to avoid prolonged exposure of the cells to deleterious osmotic and other effects of slowly increasing concentrations of solutes in the suspending fluid. A more widely used procedure is called **freeze-drying,** or **lyophilization.** The organisms are first suspended in broth, skim milk, or other suitable fluid and approximately 0.5 to 1.0 ml is dispensed into small glass vials or ampules. A very cold (−76C) slurry is prepared by adding small pieces of dry ice (solid carbon dioxide) to 2-methoxyethanol (Methyl Cellosolve) or a similar solvent. The vials are then rotated in this slurry to produce almost instantaneous freezing of the bacterial suspension in a thin layer within the vials. The more rapid the freezing, the smaller the ice crystals formed with a concomitant decrease in the amount of structural damage to the cells. The vials are connected to a freeze-dry apparatus which consists of a trap or condenser (also cooled by a slurry of dry ice and Methyl Cellosolve) and a high-vacuum pump. While under vacuum, the water passes directly from the solid form to the vapor state without melting (sublimation). When all the vapor has been removed, the necks of the vials are sealed quickly with the needle flame of an oxygen-gas torch. Freeze-drying is widely used in industry and medicine to preserve antibiotics, antisera, and many other biologicals (see §20.5.)

To reactivate the desiccated microorganisms it is only necessary to transfer them to a suitable fluid medium and then incubate. Survival of microorganisms under these conditions has both practical and philosophical implications.

Practical Implications. The survival of bacteria when desiccated in a vacuum is of great practical importance. Once preserved, a large number of cultures may be stored in a small space, and many cultures will remain viable for decades. When cultures are maintained in this state, contamination and mutation are avoided, and both time and money are saved as compared with frequent renewal of active cultures. These considerations are of great importance in science and industry. While the percentage of viable cells slowly decreases in these preserved cultures, the "half-life" is frequently measured in years. Cultures of various pathogenic species of streptococci kept for reference and research, for example, have survived for 25 years, while diphtheria bacilli have survived for 15 years and tubercle bacilli for 17 years. Many can survive much longer under these conditions. However, though many species of bacteria tolerate freezing, vacuum, and drying, this is not true of all. Some sensitive species will be discussed farther on.

Philosophical Implications. In a cell in a state of virtually complete desiccation and in a high vacuum, there must be an unimaginably small vital activity. The metabolic processes must stop almost completely, since these depend on osmosis, diffusion, ionization, and the colloidal state, all of which are dependent on hydration. How, then, can microorganisms exist desiccated in a vacuum? How can living beings survive in the entire absence of vital activity? The condition of microorganisms desiccated in a vacuum must be as near an approach to suspended animation as can be imagined. Interestingly, the spores of some microorganisms can achieve a similar state of suspended growth without any manipulation by man (see Chapter 36). The use of desiccation in vacuo with or without freezing to preserve microorganisms is one means of producing the condition called **microbiostasis.** The organisms, though in a vegetative state to begin with, are reduced to a completely static condition.

20.5
EXTREME COLD

Many microorganisms are highly resistant to extremes of cold even when in the vegetative state. It is possible that certain species of organisms have survived glacial epochs of the earth's history because of this capacity. Some species can survive (not grow) while frozen in ice for weeks. The fragile syphilis spirochetes and numerous viruses are routinely maintained frozen in carbon dioxide ice at -76 C (-104.8F) or liquid nitrogen (-198C) for years with only small loss of infectivity. Many species of bac-

teria and animal cells will grow, apparently unaffected, even after subjection to the temperature of liquid hydrogen (-252C or -421 F).

The microbiostatic effect of freezing in some respects resembles that of desiccation. In both, water is made unavailable for physiological purposes: freezing immobilizes it; desiccation removes it.

Uses of Extreme Cold. The preservation of microorganisms by temperatures ranging from -76C to that of liquid nitrogen is of great importance in industry and medicine. For example, sperm of valuable breeding animals are kept for artificial insemination; living cells of various cancers and normal human and animal tissues are kept for preparation of vaccines, and for use in cancer research and in the diagnosis of disease; and strains of bacteria, molds, and commercial yeasts are kept for industrial fermentations, antibiotic production, and enzyme manufacture.

One of the most fascinating problems in cryobiology (Gr. *kryos* = freezing cold) is whether microbiostasis can be extended to cell masses of higher forms of life. Would it be possible, for example, to have an inventory of preserved human organs ready for transplant when needed? Will the time come when distant space probes are manned by astronauts in a state approximating microbiostasis? Certainly there are some formidable problems to be dealt with, but research is currently being conducted to improve present methods and discover new techniques of freezing, dehydration, and rehydration. We know that certain insects will withstand freezing and the high vacuum of a scanning electron microscope prior to reactivation. While this is a feeble beginning, the future of cryobiology promises to be both fascinating and highly controversial.

Lethal Mechanisms of Freezing. For practical reasons the factors favoring either survival or death during freezing are the subject of much study.

One cause of death of frozen cells is believed to be the formation of intracellular ice crystals with consequent mechanical destruction of the cells. Very rapid freezing (1 to 10 seconds) has therefore been used as a means of preventing excessive intracellular crystallization and favoring **vitrefication** (formation of a solid, uncrystallized, glass-like mass of ice). Other researchers hold that slow freezing (20 to 30 minutes) is a preferable method, if the cells are suspended in a 5 to 50 per cent glycerol solution or in solutions that (a) withdraw water from the cell to avoid crystallization, (b) act as "antifreeze" inside the cell, or (c) allow time for osmotic and other adjustments between the interior of the cell and the environmental fluid, or a combination of these factors. Some regard **excessive** drying (above about 85 per cent) as harmful.

Autolysis. In the preservation of microorganisms by freezing, desiccation, or similar means, the preserving agent must act quickly in order to prevent deleterious effects of autolysis. Autolysis is self-digestion by intracellular enzymes which are released or cease to be inhibited when the cell dies or becomes almost wholly inert (Chapt. 48). This may occur if cells undergoing preservation are held at the threshold between active and inactive existence too long.

20.6
IONIC EFFECTS

Hydrogen Ion Concentration. A factor profoundly affecting all microorganisms, including each individual cell of all plant and animal tissues, is the acidity or alkalinity of the fluid by which they are surrounded. You will recall that the pH of an aqueous fluid depends entirely on the hydrogen ion (H⁺) concentration (Chapt. 4). Figure 20–5 shows the pH of some familiar substances.

It is important for the microbiologist to remember that, as a rule, increases in temperature increase dissociation of acids. Thus, a solution which is neutral or slightly alkaline, and therefore favorable to growth of most microorganisms at room temperature (about 22C), may become definitely acid and lethal if incubated at a commonly used incubator temperature (37C). If a nutrient solution is prepared at a definite pH while near the boiling point, it will be more alkaline when cool.

Unfavorable influences of many sorts are enhanced in acid fluids. For example, coagulation of protein by heat occurs more readily in acid solutions. Thus, milk that is only very slightly sour (not sour as to odor or taste) may curdle on being warmed.

Enzymes are even more sensitive to alterations in pH than they are to temperature. They have definite minimal, optimal and maximal zones and limits in respect to pH just as they do to temperature. What is true of enzymes in these respects is consequently largely true of living cells. The pH of microbiological culture media must, therefore, be very carefully adjusted. The pH to be selected depends upon the organisms to be cultivated. This is usually about pH 7.0, but commonly extends from pH 6.5 to pH 8.0 and, for some species, considerably beyond (Fig. 20–5). For example, the vibrio that causes cholera is **alkaliphilic,** i.e., it prefers an alkaline environment (about pH 9.0). Some soil bacteria (*Agrobacterium* sp.) grow well at about pH 12. Many yeasts and molds and certain bacteria prefer acid media of about pH 5.0. Still others, said to be **acidophilic,** can grow at a pH as low as 2.0.

Buffers and Buffer Action. It is often necessary in microbiology to change the pH of culture fluids (i.e., to "adjust the pH"). With media such as meat-infusion broth it is found that, unlike the titration of aqueous solutions of acid or alkali, no sharp end point is reached at which a single drop of 0.1 N acid or alkali added to a liter of solution can change the pH 10 million–fold. In a fluid such as meat broth the change from an acid to an alkaline reaction or vice versa is very gradual, requiring the continuous addition of relatively large amounts (10 to 100 ml per liter) of acid or alkali. In other words, even at or near the neutral point, the solution being titrated shows a marked tendency to resist any change in its hydrogen ion concentration. This resistance is caused by the **buffer action** of certain constituents of the broth. Important among these constituents from a biological standpoint are amino acids and their polymers. These have both amino and carboxyl groups and will combine with either acid or alkali: they are said to be **amphoteric.** For example, the amino acid glycine combines with HCl or NaOH (Fig. 20–6). Sulfhydryl (—SH), imidazolium, and

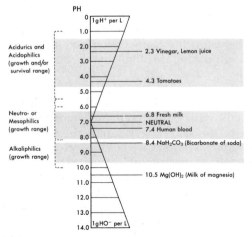

Figure 20–5. Some pH's of significance in microbiology. The slanting lines starting at pH 7.0 indicate increasing concentration of H⁺ or OH⁻ to the maximum of 1 gram per liter.

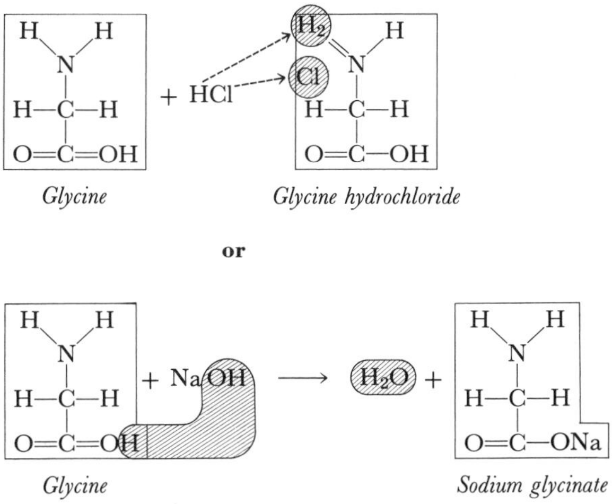

Glycine   *Glycine hydrochloride*

or

Glycine   *Sodium glycinate*

Figure 20–6. Glycine as buffer.

other groups that occur in proteins also serve as buffers (see also page 63).

A commonly used buffer in biological laboratories is a mixture of the monobasic and dibasic phosphates, KH_2PO_4 and K_2HPO_4. These dissociate relatively little in aqueous solution, i.e., they are neither strongly acid nor basic. When acid is added the dibasic salt absorbs H^+, changing into the monobasic and forming a potassium salt with the acid. When alkali is added, the monobasic salt releases H^+ to form H_2O with the ^-OH, and the incomplete salt combines with the cation of the alkali:

$$KH_2PO_4 + NaOH \rightleftharpoons KNaHPO_4 + H_2O$$
$$KNaHPO_4 + NaOH \rightleftharpoons KNa_2PO_4 + H_2O$$
$$K_2HPO_4 + HCl \rightleftharpoons KH_2PO_4 + KCl$$

Other important buffer constituents of organic media are carbonates and bicarbonates:

$$Na_2CO_3 + 2HCl \longrightarrow 2NaCl + H_2O + CO_2 \uparrow \text{gas}$$

or

$$NaHCO_3 + NaOH \longrightarrow Na_2CO_3 + H_2O$$

Mixtures of acids and their salts also act as buffer systems. If hydrogen ions are added they suppress ionization of the acid and combine with anions of the salt. If hydroxyl ions are added they combine with the acid to form water and a salt. Buffer mixtures may be prepared at any desired pH.

Other Ions: Cations. Ions of heavy metals such as gold, silver, lead, copper, and mercury, except in minimal amounts, are usually more

toxic to microorganisms than are ions of light metals such as sodium, potassium, and calcium. We may think of heavy metal ions as acting in at least three different ways, two inimical and one essential:

In relatively large concentrations (e.g., 1.0 per cent or more), they act by causing denaturation of proteins (e.g., enzymes); this is as lethal to cells as a blow with a hammer is to a mosquito! Consequently, certain heavy metal salts such as copper sulfate and bichloride of mercury are often used as general disinfectants or antiseptics. Copper, zinc, and iron compounds are commonly used in garden sprays to control fungal and bacterial diseases of plants. Silver nitrate is effective in decontaminating burns but has been replaced by antibiotics for such uses.

Some metals (e.g., Hg^{2+}, Ag^+, Pb^{2+}, and Zn^{2+}) are poisonous even in very low concentrations (of the order of 0.0001 per cent) because they combine with and inactivate certain essential functional groups in the cell. For example, Hg^{2+} combines with the —SH group of enzymes, inhibiting their action.

In minute, barely detectable (**trace**) amounts, several heavy metal ions, including some of those just mentioned, are essential for cell growth because these ions are parts of coenzyme molecules. Examples are iron in the red respiratory pigment hemoglobin in our own red blood cells and in the cytochrome system of respiratory pigments of aerobic bacteria. Molybdenum is essential in nitrogen fixation by certain soil organisms (Chapt. 43); others will be mentioned farther on.

Toxicity of metallic ions depends greatly on the species of microorganism being dealt with, the presence of chelating* agents and other substances that combine with the metal ions, and the pH of the surrounding medium. Some metals are poisonous for one species but not at all for others. As a rule, not only are light metals such as sodium, potassium, lithium, strontium, magnesium, calcium, and nonmetal ammonium (usually present as chlorides) harmless but most are essential to bacteria as salts in concentrations of from 0.05 to about 1.5 per cent.

The less toxic metal ions act favorably in a great variety of ways, sometimes by effectively suppressing the ionization of unfavorable sub-

---

Chelating agents are organic compounds which form a complex compound with a heavy metal, enclosing it in a ring-like molecular structure as though held in a claw (Gr. *chēlē* = claw). The position of iron in the molecule of heme, part of the respiratory pigment of our red blood cells, is a typical example (Fig. 7–1).

stances, or by reacting with these substances to prevent them from affecting bacteria unfavorably. For example, the relatively benign calcium chloride protects against noxious sodium oxalate by forming an insoluble precipitate of calcium oxalate, thus removing the toxic oxalate radical from the solution.

Ion Antagonisms. Some cations are biologically antagonistic to each other. For example, Li^+ and Zn^{2+} are definitely toxic to certain bacteria of importance in the dairy industry (*Lactobacillus* and *Leuconostoc*) and to others as well. These ions appear to drive H^+ from its normal position and to take its place in certain enzymes. This results in a stoppage of the action of the enzymes. The ions are said to **antagonize** the H^+. By increasing the concentration of H^+, i.e., lowering the pH, this toxicity of Li^+ and Zn^{2+} is reduced or eliminated. The antagonism is said to be reversed. Complex organic molecules can act to block and unblock enzymes in much the same way.

As will be seen later, knowledge of this sort of antagonism, its reversal, and competition between ions and molecules in living cells is of fundamental importance in the science of chemotherapy, as well as in cell physiology, industry, genetics, oncology, and other activities. It would be well to keep the basic idea of ionic and molecular antagonisms in mind.

Undissociated Molecules. The biological activity of most ionizing compounds, inorganic as well as organic, depends not only on anion and cation but also on the undissociated molecule. For example, the toxic effects of benzoic, acetic, and sorbic acids are much greater than would be expected from the pH of their solutions. Thus, hydrochloric acid and sulfuric acid, although "strong" acids, are much less poisonous to most bacteria at a given pH than benzoic, sorbic, or acetic acid at the same pH. About 7.5 to 7.7 parts per million of hydrochloric or sulfuric acid are required to produce the same toxic effect as 0.1 and 1.2 parts per million, respectively, of benzoic or acetic acid. Benzoic and acetic acids, also sorbic acid, are common food preservatives because of the high toxicity of the undissociated molecules for microorganisms (and relatively low toxicity for you and me!).

20.7
RADIANT ENERGY

The nature of radiant energy and its effects on living cells, especially its genetic or mutagenic effects, are discussed in Chapter 18. We may here mention some practical uses made of lethal effects of radiations.

Infrared Rays. To be effective radiations must penetrate into the irradiated substance, and their energy must be liberated within it. They are ineffective if they are completely reflected from the surface of the substance, or if they pass completely through it. The energy content of each quantum of radiant energy is an inverse function of the wavelength. Waves longer than about 1,000 to 1,200 nm (invisible infrared) contain less energy than invisible ultraviolet and x-rays and are reflected by surfaces that are opaque to visible light. When absorbed by nonreflecting materials their relatively low energy is given off as heat ("heating rays"). Such long waves are neither ionizing nor "exciting," although their heat may be lethal. They are sometimes used in quick cooking.

Ultraviolet Light. Ultraviolet light is a component of sunlight and produces suntan and sunburn. It is also mutagenic and carcinogenic (Chapt. 18) and activates green-plant chlorophyll. Ultraviolet light has very slight powers of penetration and is therefore effective mainly at surfaces or in thin films. It is **non-ionizing** but produces lethal photochemical changes in enzymes and other cell constituents if exposure is sufficient. Ultraviolet light consists of a spectrum of different wavelengths from about 400 to about 14 nm (see Figure 20–7). Those between 280 and 230 nm, especially 254 nm, are most useful biologically, being strongly microbicidal yet readily produced and managed.

It is of interest to note that the microbicidal effects of ultraviolet radiation differ from those of heat, in that very thermostable endospores are easily killed with ultraviolet light. Ultraviolet light is as effective in the absence of oxygen as in its presence; short-wave **ionizing** radiations, like x-rays, on the contrary, are more effective in oxygen.

X-rays. Unlike ultraviolet and longer waves, x-rays (about 50 nm for "soft" x-rays to about 0.1 nm for "hard" x-rays) have marked powers of penetration. X-rays cause abnormal making and breaking of hydrogen bonds of DNA and disturbance of secondary molecular structures. Short exposures are mutagenic and carcinogenic. Longer exposures are lethal, but the use of x-rays for routine sterilization is costly and could be dangerous.

Sunlight. The microbicidal or disinfectant ("purifying") action of sunlight has been known for centuries. We now know that it is caused mainly by the ultraviolet rays (400 to 295 nm) in solar light. The heating and drying effects of sunlight also have a bactericidal effect.

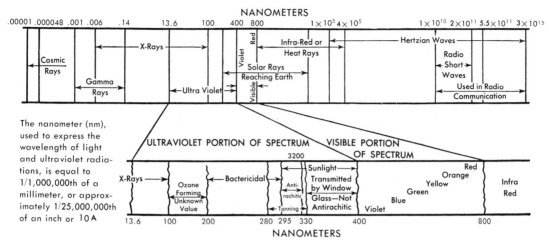

Figure 20-7. Spectrum charts.

Photodynamic Sensitization. See page 256 for the discussion of photodynamic sensitization.

20.8
HYDROSTATIC PRESSURES

If living microorganisms in aqueous fluid are placed in a steel cylinder and a piston is pressed down upon the suspension with great force, the organisms may or may not be affected, depending on the source and species of organism, the pH and salinity and other properties of the medium and the temperature. Some species of bacteria found in the deepest known valleys in the floor of the Pacific and Indian oceans obviously thrive under pressures of over 16,000 lb per square inch. Deep-sea bacteria and others from deep oil wells appear to be favorably influenced by such pressures (see Chapter 44); they are said to be **barophilic** (Gr. *baros* = weight; *philic* = prefers). In the range above 9,000 lb per square inch, death of most familiar shallow-living species occurs within one hour.

Pressures much above 15,000 lb per square inch denature ordinary proteins and inactivate ordinary enzymes. Sublethal high pressures also increase rates of some chemical and enzymic reactions, and cause diminution in volume of organic colloids, enzymes, and molecules, increases in viscosity of many fluids, and increased electrolytic dissociation. Probably all of these changes are involved in the biological effects of high pressures, but exact details of the mechanisms are still under investigation.

Pressure and Temperature. In general there is a compensatory relation between pressure and temperature. Unduly high incubation temperatures may be thought of as causing deleterious **expansion** of enzymes (colloids), perhaps with weakening of hydrogen bonds of primary and secondary structures, while increased pressures tend to prevent the expansion and so lessen this unfavorable effect of high temperatures.

Cooling may be thought of as causing undue **contraction,** and hence diminution or loss of activity of enzyme colloids. High pressures enhance the retarding effect of low temperatures, while relief from pressure tends to overcome the contraction. For example, organisms normally growing best at 20 to 30C were completely inhibited at 20C by 4,500 lb. per square inch pressure but they grew well at 40C under the same pressure. Experimental data may vary somewhat under different experimental conditions.

20.9
MECHANICAL IMPACT; VIBRATION

In order to investigate the inner substances of cells, liberation of the cell contents by mechanical rupture of the cell walls and membranes is often preferable to the use of chemical extractants, enzymes, or solvents.

Crushing. Most unicellular microorganisms (not viruses) are easily crushed by the impingement upon them of solid particles such as steel or glass balls. Violent shaking in a vessel half filled with minute glass beads or sand is often used for the mechanical disruption of bacteria (Fig. 20–8). They shatter especially readily if made brittle by freezing. Cells may also be ruptured by grinding them with fine abrasives or by repeated freezing and thawing. Machines for such mechanical breaking of cells are available commercially.

Rapid Vibrations. Bacteria, because of their small masses and their relatively tough cell walls, are not readily disrupted by slow vibrations. Supersonic vibrations (generally around 20,000 cycles per second) for microbiological purposes are generally produced in fluids by electrically vibrated disks or rods of nickel, titanium, stainless steel, or quartz (Fig. 20–9). Immersed in a fluid that is subjected to vibrations of appropriate rate and intensity, bacteria are torn, the protoplasm disrupted, and the cells killed (Fig. 20–10). Much of the damage is believed to be caused by **cavitation**: the **intracellular** formation of a foam of minute bubbles of the gas that is ordinarily in solution in the protoplasm, or extracellularly in the fluids surrounding the cell. Bacteria differ greatly in their susceptibility to mechanical rupture by any means. Some are much tougher than others.

Osmotic Shock. Another method of disrupting cells without using chemicals is the immersion of prepared protoplasts (see page 202) in distilled water (**osmotic shock**). They immediately burst.

Figure 20–9. Ultrasonic probe. See text for discussion. (Courtesy of Bronwill Scientific, a division of Will Scientific, Inc., Rochester, N.Y.)

20.10
SURFACE FORCES

For purposes of the present discussion a **surface** is the boundary between a **solid** and a **fluid** (L. *fluidum* = substance that flows; liquid or gas) or between two immiscible fluids. Commonly encountered interfaces are (1) between a cell and its surrounding medium; (2) between a colloidal particle and the fluid matrix of the cell; (3) between a bubble of gas and the fluid surrounding it; (4) between a globule of oil and the aqueous fluid surrounding it, or, conversely, between a droplet of aqueous fluid and the oil in which it is suspended.

In any cell, most of the chemical and physical changes on which life depends take place at surfaces. Biologically effective surfaces are found at the outer surfaces of cells and at the surfaces of the intracellular colloidal particles, enzymes, lipids, membranes, and ribosomes.

Surface Tension. One of the most important surface forces is called **surface tension.** Surface tension results from the universal attraction, or cohesive force, between molecules. As shown in Figure 20–11, molecules (*a, á*) in the depths of a fluid are attracted equally from all sides. At the air-liquid interface the attraction between the liquid molecules (*b*) is along the interface and from below. These unbalanced attractions produce surface tension. The surface tension of any fluid is a physical constant under any stated condition and, for liquids, is expressed in terms of dynes per cm of surface (a dyne is the force required to accelerate one

Figure 20–8. In mechanical cell homogenizer, cells are crushed in collisions between thousands of tiny glass beads in the violently agitated cylinder. (Courtesy of Bronwill Scientific, a division of Will Scientific, Inc., Rochester, N.Y.)

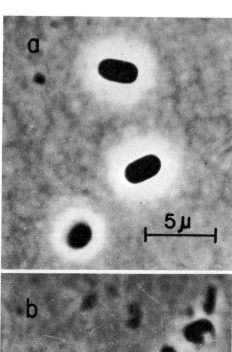

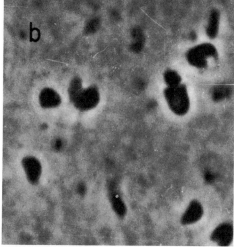

Figure 20–10. Effect of intense sonic vibration on a common bacterium of the soil, *Azotobacter vinelandii. a,* Before vibration; *b,* after vibration for one minute. Note the shattering effect on the bacilli.

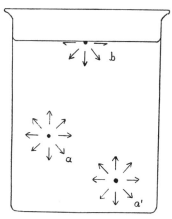

Figure 20–11. Diagram showing how surface tension acts. The molecules *a* and *a'* are attracted equally from all directions and are in a state of equilibrium. Molecules at the surface, like *b,* are under a greater tension from below than above and the entire surface therefore tends to pull inward much as though the surface film were a rubber membrane.

is allowed to flow slowly from a given orifice (e.g., the tip of a pipette with an inside diameter of about one mm). It will be seen that the lower the surface tension, the smaller and more numerous and lighter are the drops (Fig. 20–12). Some factors that affect the measure-

gram a distance of 1 cm per second per second). Increases in temperature usually lower surface tension. The lower the surface tension of a liquid, the more easily molecules escape from the surface, i.e., the more volatile it is.

Surface Tension Reducers. Many organic substances, such as alcohols, organic acids, polypeptides, bile, and certain substances (e.g., soaps and household detergents), can weaken this intermolecular pull and thus lower surface tension. Such substances are called **surface tension reducers** or, commonly, **surfactants.**

One simple means of illustrating and measuring surface tension (roughly for comparative purposes) is by noting the number and/or weight of drops of a liquid that form if the liquid

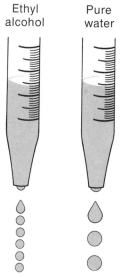

Ethyl alcohol   Pure water

Figure 20–12. The surface tension of alcohol being "weak" (28 dynes per cm of surface), drops fall from the supporting film at the tip of the pipette before they attain much weight, even though the specific gravity of alcohol is small. There are many drops per ml. Water not only is heavier but has more than double the surface tension of alcohol (77 dynes per cm of surface); hence fewer and larger drops are formed per ml. Addition of a little household detergent would have what effect on the number of drops per ml of water? Could this be used as a "screening" test for domestic sewage pollution of drinking water?

ments are: temperature and humidity of the air, diameter and wall-thickness of the tube, and density, viscosity, and rate of flow of the fluid.

While an explanation of just how various surfactants work—which involves considerable physical chemistry—is beyond the scope of this chapter, it is useful to note that the presence of almost any surfactant in a culture medium will influence the growth of microorganisms. The effects will vary from highly beneficial to extremely detrimental, depending on the concentration and type of surfactant, the other medium constituents present, and the type of microorganism involved. Beyond that, generalizations are difficult to make; some specific examples will be found in the chapters that follow.

CHAPTER 20
SUPPLEMENTARY READING

Alexander, M.: Microbial Ecology. John Wiley & Sons, Inc., New York. 1971.

Brock, T. D.: Principles of Microbial Ecology. Prentice-Hall, Inc., Englewood Cliffs, N.J. 1966.

Farrel, J., and Rose, A.: Temperature effects on microorganisms. Ann. Rev. Microbiol. 21:101, 1967.

Rose, A. H. (Ed.): Thermobiology. Academic Press, Inc., New York. 1967.

Williams, R. E. O., and Spicer, C. C. (Eds.): Microbial Ecology. 7th Symposium, Soc. Gen. Microbiol. Cambridge University Press, London. 1957.

CHAPTER 21 • STERILIZATION AND DISINFECTION

Basic Principles

An important phase of microbiology is knowledge of methods for killing, for removing and for inhibiting (preventing growth of) microorganisms. Species of microorganisms vary in the ease with which they may be destroyed, removed, or inhibited, and the situations in which they may occur differ greatly (e.g., blood, foods, water, sewage, soil). Therefore, no one or two methods are generally applicable. Each situation is a problem in itself, and the methods employed must depend on the knowledge, ingenuity, and purposes of the operator. There are basic facts, however, which guide the procedure in any given situation.

There are four main reasons for killing, removing or inhibiting microorganisms. They are: (1) to prevent infection of man, his animals, and plants; (2) to prevent spoilage of food and other commodities; (3) to prevent interference by contaminating microorganisms in various industrial processes that depend on pure cultures; (4) to prevent contamination of materials used in pure-culture work in laboratories (diagnosis, research, industry) so that studies of the growth of one kind of organism in a particular medium or infected animal will not be confused by the presence and growth of others at the same time. Some microbicidal and microbiostatic agents are listed in Table 21–1. In this chapter we shall discuss some basic principles underlying com-

TABLE 21–1. SOME MICROBICIDAL AND MICROBIOSTATIC AGENTS

I. **Destruction** by
   A. Heat (boilers, ovens)
   B. Chemical agents (disinfectants)
   C. Radiations (x-rays, ultraviolet)
   D. Mechanical agents (crushing, shattering by ultrasonic vibrations)

II. **Removal** (especially bacteria) by
   A. Filtering
   B. High-speed centrifugation

III. **Inhibition** by
   A. Low temperatures (refrigeration, "dry ice")
   B. Desiccation (drying processes)
   C. Combinations of A and B (e.g., freeze-drying)
   D. High osmotic pressures (syrups, brines)
   E. Chemicals and drugs
      1. Certain dyes such as eosin and methylene blue, crystal violet; bile (deoxycholate)
      2. Chemotherapeutic drugs such as sulfonamides and antibiotics

mon methods of killing or removing microorganisms.

21.1 DEFINITION OF TERMS

Several new terms used in this chapter may be explained as follows:

289

Sterilization. In microbiology sterilization means the freeing of any object or substance from *all life of any kind.* For microbiological purposes microorganisms may be killed in situ by: heat; gases such as formaldehyde, ethylene oxide, or β-propiolactone; solutions of various chemicals; ultraviolet or gamma irradiation. Organisms may be removed mechanically by very high speed centrifugation or by filtration (see farther on, this chapter).

Disinfection. Disinfection means the killing or removal of organisms capable of causing infection. Disinfection does not necessarily include sterilization, although some processes of disinfection accomplish sterilization. Disinfection is usually accomplished by chemicals such as phenol (carbolic acid), formaldehyde, chlorine, iodine, or bichloride of mercury. In the case of milk, disinfection—but not sterilization —is brought about by **pasteurization,** a heating process described elsewhere. Disinfection is generally thought of as killing the more sensitive vegetative cells but not heat-resistant spores.

A **disinfectant** is an agent accomplishing disinfection. The term is often used synonymously with **antiseptic.** However, one ordinarily thinks of disinfection and disinfectants as applicable mainly to inanimate objects: floors, dishes, laundry, and bedding. The terms **sanitizer** and **sanitization** are often used with this meaning.

Antiseptic. Antiseptic is an ill-defined term, closely allied to disinfectant. Antiseptics are generally considered to be substances that kill or inhibit microorganisms, especially in contact with the body, without causing extensive damage to flesh. Most disinfectants are too destructive of tissues to be useful as antiseptics.

Bactericide. Any substance or agent that kills bacteria is a bactericide or bactericidal agent. The suffix, **-cide,** indicates **killer** and is used with germ (**germicide**) and virus (**virucide**).

Sepsis. **Sepsis** is the toxic or diseased state resulting from the growth of harmful microorganisms in contact with living tissue.

Asepsis. In a strict sense, **asepsis** is the absence of infectious microorganisms in living tissue, i.e., the absence of sepsis. However, the term is usually applied to any technique designed to keep all unwanted microorganisms out of any field of work or observation. Gnotobiotics (see Chapter 24) is asepsis developed to its utmost extent. The work of microbiologists and surgeons involves aseptic technique. The surgeon and his assistants have sterile instruments, handle them with sterilized gloves, cover the patient with sterilized sheets, and wear sterilized caps, gowns, and masks to prevent infected hair, dust, droplets of saliva, perspiration, or sputum from entering the sterile field and possibly infecting the patient. The patient's skin cannot be absolutely sterilized without injury because microorganisms live deep in sweat and sebaceous glands, but the site of the operation is disinfected as thoroughly as possible by applications of some suitable antiseptic or dilute disinfectant.

The microbiologist works in a "germfree" cabinet or uses sterilized culture media and sterilized glassware kept sterile until the moment of use by coverings of paper and cotton plugs and by **aseptic technique** (i.e., avoidance of touching sterile materials with hands or unsterile objects and the exclusion of dust).

Microbiostasis. **Stasis** is a Greek word meaning to stand still. **Microbiostatic** (bacteriostatic, fungistatic) agents are substances or conditions that do not immediately kill microorganisms but that inhibit metabolism, so that, after an initial decline in numbers, the microorganisms die over a period of hours, days, or many years without significant multiplication. Important microbiostatic agents are desiccation, very low temperatures, antibiotics, sulfonamide drugs, extremely hypertonic solutions, and certain dyes, such as crystal violet.

Microbiostasis is more fully discussed in Chapter 22. Here it is necessary to note only that numerous disinfectants are chiefly microbiostatic; i.e., they appear to kill microorganisms whereas actually the microorganisms are merely temporarily inhibited. Under certain conditions, such as introduction into the body, they are reactivated.

21.2
TYPES OF CELL DAMAGE

Denaturation and Coagulation. When solutions of protein are heated at temperatures above 80C, or when acids or certain divalent or trivalent cations are added to them, a white precipitate appears. The process by which this change from liquid to solid form occurs is termed **coagulation.** Coagulation is preceded by a partial dissolution or **denaturation** of the protein by breakage of hydrogen (or disulfide) bonds in the secondary and tertiary structures. The resulting separated and extended protein fibers become less soluble and the solution becomes more viscous and turbid. Functional properties such as enzymic activity are destroyed, and the protein is more readily hydrolyzed. The protein becomes flocculent and the floccules coalesce into the solid mass typical of coagulation (a hard-boiled egg). Some agents that cause the preliminary denaturation may or may not also cause coagulation. They may cause coagulation if in the presence of strong, colloid-precipitating agents such as H^+, heavy metal salts, and especially heat.

Colloidal protein particles may be thrown out of suspension if their surface charges are neutralized. They are then said to be at their **iso-electric point.** For example, proteins are precipitated by certain cations: H^+, Cu^{2+}, Zn^{2+}, Fe^{3+}. Because of this, several heavy metals are commonly used as disinfectants: $CuSO_4$, $AgNO_3$, $HgCl_2$, ZnO.

Certain organic substances are also important as denaturants or coagulants, among them alcohols, phenols, formaldehyde, and many related and derived substances.

The general rule may be stated that any agent, physical or chemical, that induces denaturation or coagulation is lethal to living cells.

Nonspecific Chemical Combinations. Various chemically active substances will combine indiscriminately with any and all proteins and related and derived compounds. Chlorine is such a substance, iodine another; cresols, carbolic acid (phenol), and formaldehyde are others. Lye (strong alkali) and strong acids are destructive of nearly all organic matter. Such substances are entirely nonspecific in their action; that is, they will combine as readily with body tissues, casein, feces, mucus, blood, wood, or leather, as with the protoplasm of microorganisms.

Specific Chemical Combinations. There are several classes of substances of relatively low molecular weight that can enter certain cells and, even in very low concentrations, interfere with or completely stop the action of one or more **specific molecular groups** (e.g., coenzymes). This may result promptly in death or, more commonly, it may produce microbiostasis, depending on type of cell and agent. Therapeutically useful substances of this nature are represented by sulfonamide drugs and antibiotics.

Cyanide and carbon monoxide act similarly by "poisoning" hemoglobin and cytochrome oxidase, combining with the enzyme to the exclusion of oxygen, and thus inhibit cell respiration. However, they are far too poisonous for common or therapeutic use. Sodium fluoride interferes with carbohydrate metabolism; arsenicals with respiration, but these, too, are very poisonous for humans and must be used

cautiously if at all. These phenomena are discussed in Chapter 22.

Action on Surfaces. As discussed in Chapter 20, surfactants are substances which reduce surface tension. Their chemical nature is such that they tend to accumulate at interfaces, e.g., between the cell and its surrounding medium (called adsorption). Disinfectants that are also surfactants would be expected to have added effectiveness, because they tend to concentrate and form coatings on microbial and enzymic surfaces. Indeed, some surfactants appear to act largely by coating the microbial surfaces and the effective surfaces of enzymes. This interferes with contact between enzymes and their substrates, with permeability of the cell wall and also with permease and other enzyme systems, thus preventing entrance and utilization of food substances.

Many surfactants also appear to injure or destroy the cell wall, causing inimical alterations of permeability or immediate **lysis.** These effects are probably caused, at least in part, by the dissolving or emulsifying effects of several surfactants on lipids of cell membranes and cell walls. Some important antibiotics (e.g., polymyxin) are surfactants of this nature.

Many disinfectant substances probably act by more than one of the mechanisms just mentioned, and by others that are only partly understood.

21.3
FACTORS THAT AFFECT STERILIZATION AND DISINFECTION

Hydration. The role of **hydration** in denaturation or coagulation by heat is shown by experimental data listed in Table 21–2. Obviously, coagulation proceeds best when protein is well hydrated. The same principle holds true, within

TABLE 21–2. EFFECT OF HYDRATION AND HEAT ON EGG ALBUMEN *

Water Content (Per Cent)	Approximate Coagulation Temperature (C)
50	56
25	76
15	96 †
5	149
0	165 ‡

*Albumen is one form of protein.
†Boiling water = 100C.
‡165C = Oven temperature.

limits, in coagulation by chemicals. The resistance of bacterial **endospores** to heat is probably caused in part by their almost completely dehydrated condition. In this discussion, unless otherwise noted, all statements on disinfection refer to vegetative (nonsporulating) cells in aqueous suspension.

An example of the resistance of dehydrated protein to coagulation is seen in the difficulty of making dried-egg powder coagulate. Heated in a test tube, it will turn brown or char, but it will not coagulate unless a considerable quantity of water has been added. Of course, mixing dry egg with dry chemical disinfectants such as powdered bichloride of mercury produces no coagulation at all until moisture is added.

Aside from purely coagulative effects, the chemical reactions necessary to the actions of a variety of disinfectants are facilitated by the presence of water. Solution or **ionization** is essential.

Time. No disinfectant as ordinarily used acts instantly. Sufficient time for contact and for whatever chemical and physical reactions will occur must be allowed. The time required will depend on: nature of the disinfectant; concentration; pH; temperature; nature of the organisms; and existence in the bacterial population of cells having varying susceptibilities to the disinfectant.

Rate of Disinfectant Action. Under uniform conditions of temperature, concentration, and pH, the **rate** at which death of organisms in a pure culture occurs in contact with a disinfectant (including heat) is theoretically a function of time only.

If all the cells in a given culture are absolutely equal in vulnerability to the lethal agent, the line obtained by plotting logs of numbers of surviving organisms against units of time should be **straight** and **vertical**; i.e., the cells should all die at the same instant. Experience shows, however, that such curves are rarely vertical but are instead **slanting**, indicating that some of the cells are more vulnerable than others and thus that some cells die early, others after various intervals (Fig. 21–1). If the slope is steep the death rate is high; if it slopes gently, the death rate is low. Further, the theoretically straight, sloping line is often a curve. The line may be concave (steep at first, less steep in the middle segment and almost level toward the end) indicating that a large proportion of the cells in the culture are more rapidly killed (more vulnerable) than those that survive to the end of the experiment. For example, in an old culture many cells may become almost dormant, with their enzymic and reproductive functions slowed or stopped. Passage of microbicidal substances inward is diminished; cell walls are possibly thickened; some of the cells may even be encapsulated. Such matured cells would probably be less vulnerable to heat or chemical disinfectants than young cells. Thus, the proportion of old to young cells, as well as other factors such as species of organism and number

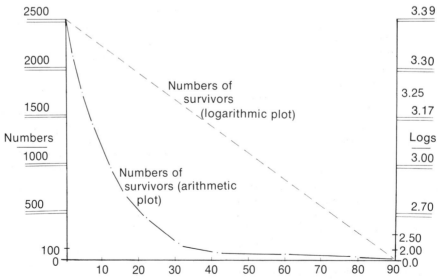

Figure 21–1. Relation between time and bactericidal action of heat (90C.) in saline solution buffered at pH 7.0. Both logarithmic and arithmetic curves (idealized) are shown. Actual curves are not always so ideally symmetrical.

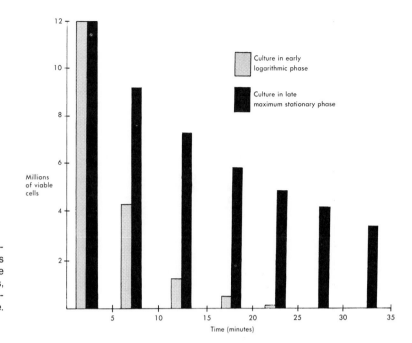

Figure 21–2. Comparison between logarithmic phase and stationary phase cells as to susceptibility to a lethal agent. The former are all killed within 25 minutes, while a considerable part of the more resistant, stationary phase cells still survive.

of cells initially present, can affect both the slope of the death curve (rate of death) and its form (differences in vulnerability) (Fig. 21–2).

Of course, if very high concentrations of disinfectants or very high temperatures are applied, all of these gradual depopulations and measurable fluctuations are minimized or entirely masked and lost in one instantaneous, catastrophic stroke. Note that this effect was achieved in raising the temperature from 53C to 57C in the experiment shown in Figure 21–3.

Temperature. With respect to microbicidal action of heat, temperature is inversely related to time. The lower the temperature, as a rule, the longer the time required to kill the organisms (Fig. 21–4).

In the case of chemical microbicides, as a rule, the warmer a disinfectant, the more effective it is. This is based partly on the principle that chemical reactions in general are speeded up by raising the temperature. Usually, within the range of growing temperatures for microorganisms, a rise in temperature of 10C increases reaction rates two to eight times. However, since many disinfectant actions are partly physical in character, the laws governing chemical reactions do not apply exclusively. Higher temperatures generally reduce surface tension, increase acidity, decrease viscosity, and diminish adsorption. The first three increase, and the last diminishes, the effectiveness of a disinfectant.

Concentrations. Within narrow limits, the more concentrated a disinfectant, the more rapid and certain is its action.

Effectiveness is generally related to concentration exponentially, not linearly. For example, doubling a 0.5 per cent concentration of phenol in aqueous solution does not merely

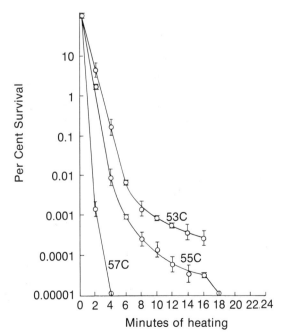

Figure 21–3. Heat destruction of *Staphylococcus aureus* in neutral buffer.

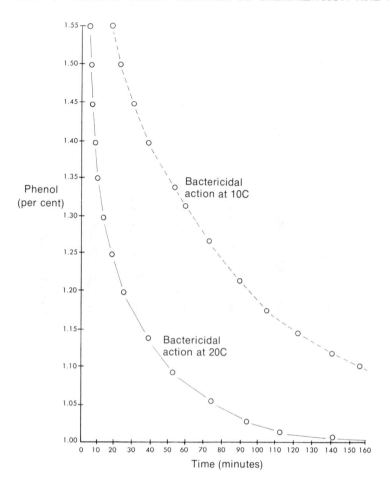

Figure 21–4. Effect of temperature and of concentration of a disinfectant on killing time for a nonsporeforming bacterium in broth at pH 7.0. Note that: (1) at a given temperature a slight increase in concentration of phenol produces a relatively marked decrease in killing time; (2) a 10C increase in temperature at a given concentration greatly enhances the rate of death. For example, a 1.1 per cent solution at 20C sterilizes in 50 minutes while at the same temperature a 1.29 per cent solution sterilizes in 14 minutes; in 64 minutes at 10C.

double the killing rate for bacteria but may increase it by 500 or 900 per cent. Doubling the concentration again may increase the effect by only a negligible amount. There is clearly an optimal concentration of phenol at about 1 per cent. Thus, a concentration of a disinfectant beyond a certain point accomplishes increasingly less, and is wasteful (Fig. 21–4). Microorganisms vary considerably in their susceptibility to phenol. Some saprophytic species (pseudomonads and *Arthrobacter*; see Chapters 33 and 40) can even utilize phenol and phenolic derivatives as nutrients.

When a disinfectant is in a colloidal state (**emulsion**), the material in each minute globule is highly concentrated. The bacterial cell in contact with these colloidal globules is, therefore, in contact with a high concentration of disinfectant. Many commercial preparations of cresols and chemically related substances are in the colloidal state.

Oligodynamic Action. Several substances that are toxic in relatively high concentrations

stimulate growth at low concentrations. To illustrate, a bright piece of a heavy metal, such as copper, silver, or gold, is placed on a plate of nutrient agar which has previously been heavily inoculated with an organism such as *Staphylococcus aureus* (a cause of boils). The plate is incubated. Small quantities of the metal diffuse into the agar and inhibit the growth of the organisms in a zone around the metal piece. This phenomenon is ascribed to what is called **oligodynamic action** (Gr. *oligo* = little; *dynamis* = power; hence, acting in small amounts) of the metal (Fig. 21–5).

At the periphery of this barren zone one might suppose that, as the concentration of metal ions diminishes with distance from the metal piece, growth would gradually increase and finally reach a density equal to normal in all areas beyond the zone where the metal ions had not yet migrated. On the contrary the sterile zone is sharply demarcated. At the periphery of the sterile zone, where concentration of the toxic metal ions is minimal, a narrow but dis-

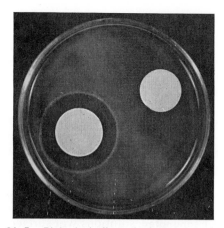

Figure 21–5. Biological effect of minute amounts of certain metals (oligodynamic action). The agar medium was heavily inoculated with micrococci while warm, well mixed, and poured into the plate. When it was solid, the metal disks (coins) were placed on the agar surface and the plate was incubated. The bacteria grew where they could, producing a greyish granular appearance. Note that the silver disk (quarter) is surrounded by a clear zone where no growth occurred, while the nickel disk shows little or no zone of inhibition. Note also the increased density of the growth at the outer margin of the clear zone around the quarter. This may be due to: (1) less competition for food at the edge of the sterile area, (2) stimulation by a critically small concentration of the metallic ions, or (3) both. The difference in action between nickel and silver is not necessarily a general one. With another species of test organism the situation could be reversed. Further, the coins are not pure silver or nickel. (Courtesy of Dr. W. C. Burkhart, Department of Bacteriology, University of Georgia, Athens, Ga.)

tinct opaque ring of **extra dense** growth is often clearly evident. This may be caused by absence of competition for food in the nearby barren zone or to stimulation of growth by an optimal low concentration of metal ions. The same phenomenon is seen in similar experiments with some other antimicrobial substances.

pH. As a general rule, the lethal or toxic action of harmful agents, both physical and chemical, is increased by increased concentrations of H^+ (or of OH^-). The synergistic relationship between heat and H^+ has already been discussed. Heat tends to increase acid effects in part by causing greater dissociation of acids. The close relation of pH and temperature in lethal processes is illustrated in Table 21–3. In this table it is seen that OH^- also is an adjuvant of heat but to a lesser degree than H^+. Both acidity and alkalinity increase denaturation and coagulation by heat.

The increase in effectiveness of benzoates, sorbates and salts of some other organic acids in acid solution is an excellent example of one effect of H^+ on disinfectant action. In the case of these salts of weak organic acids, the **undissociated** molecules, and not the sorbate or ben-

zoate anions per se, are the active agents. The effect of lowered pH (**increased** acidity) of their solutions is to suppress ionization of these weak acid salts, thus increasing the concentration of the toxic undissociated molecules.

Extraneous Organic Matter. Most common disinfectants, such as compounds of phenol, bichloride of mercury, strong acids, and the halogens, are quite general in their affinity for protoplasm and many other organic materials, whether part of a living cell or not. The presence of considerable quantities of extraneous organic materials such as blood serum, plant or animal tissues, mucus, or feces in any material being disinfected will therefore protect the organisms to a great extent, since any of these materials will combine with and inactivate the disinfectant before it reaches the organisms.

Osmotic Pressure. Fluids of high osmotic pressure (e.g., food-preserving syrups and brines) tend to dehydrate the cell contents and so increase resistance of microbial cells to heat and chemical disinfectants.

Chemical Antagonisms. Many antimicrobial agents are inactivated, or their action is reversed (**antagonized**), by certain specific substances. For example, bichloride of mercury exerts a **specific toxic** (not coagulative) effect in high dilutions (e.g., 1:100,000 or more). The **toxic** action of **dilute** bichloride of mercury is an effect of its combining specifically with the sulfhydryl (—SH) group, which, as explained elsewhere, is a very important, functioning part of the glutathione or cysteine molecule of many proteins and enzymes (Fig. 21–6). When one enzyme is stopped or blocked by any agent, a whole series of dependent enzyme reactions may also stop, both above and below the blocked enzyme, as in an assembly line in a factory. Bichloride of mercury may thus be viewed as a poison of specific enzymes and proteins, as well as a coagulative agent. However, the toxic action of bichloride of mercury may be completely antagonized or reversed by putting —SH com-

TABLE 21–3. RELATION OF pH AND TEMPERATURE TO SURVIVAL OF TETANUS SPORES

	Survival (minutes) at			
pH	105C	100C	95C	90C
1.2	4	5	6	6
4.1	6	11	14	23
6.1	9	14	38	54
7.2	11	29	53	65
10.2	5	11	21	24

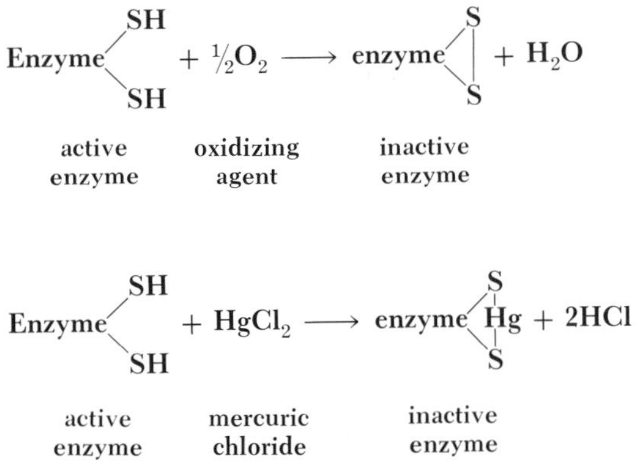

$$\text{Enzyme}\begin{array}{c} \text{SH} \\ \\ \text{SH} \end{array} + \tfrac{1}{2}O_2 \longrightarrow \text{enzyme}\begin{array}{c} \text{S} \\ | \\ \text{S} \end{array} + H_2O$$

active   oxidizing   inactive
enzyme   agent   enzyme

$$\text{Enzyme}\begin{array}{c} \text{SH} \\ \\ \text{SH} \end{array} + HgCl_2 \longrightarrow \text{enzyme}\begin{array}{c} \text{S} \\ | \\ \text{S} \end{array}Hg + 2HCl$$

active   mercuric   inactive
enzyme   chloride   enzyme

Figure 21–6. Inactivation of an enzyme by combination of its sulfhydryl groups (SH—) with an oxidizing agent (above) and with bichloride of mercury (below). The effect of $HgCl_2$ is bacteriostatic; i.e., it is reversible by H_2S, etc.

pounds such as glutathione and cysteine into the suspending fluid. These combine with the bichloride of mercury, eliminating blockage of —SH groups of enzymes so that they are enabled to function again. In another example, the lethal action of dilute phenol on micrococci and on *Salmonella typhi* may be stopped and the apparently dead organisms "revived" by removal of the phenol with activated charcoal or ferric chloride.

In Chapter 19, it is explained how microorganisms apparently killed by ultraviolet irradiation can be reactivated by visible light and some other agents. Everyone knows how men pronounced dead have been restored by cardiac massage. Thus, all is not dead that seems so. The terrible, dark, and secret realms of death shrink daily before the onslaughts of scientific research!

Surface Tension. Surface tension is of basic importance in disinfection (see Chapter 20). In dealing with aqueous solutions of disinfectants, there are two aspects of this factor: **adsorption** of surfactant disinfectants or interfering substances on the surfaces of cells, and the effect of surfactants on the wetting and spreading properties of the solution. Both affect contact between disinfectant and microorganisms.

Adsorption. For example, when phenol is added to an aqueous suspension of bacteria, contact between disinfectant and bacteria is immediate. The bacteria float naked, as it were, and are reached by the disinfectant in effective concentration without delay, partly because phenol lowers surface tension and is therefore **adsorbed** and consequently concentrated upon their surfaces. A nonsurfactant like bichloride of mercury is not concentrated in this way. Once the surface-active substance is in contact with the organism, further action depends on such factors as toxicity of the agent, its coagulative action, presence of interfering substances, pH, temperature, species of organism, and whether or not the disinfectant penetrates readily inside the cell.

Wetness. Surface tension reducers enhance the effectiveness of disinfectants by increasing their ability to spread and to come into intimate contact with surfaces, i.e., to increase their **wetness.**

Soap, although a relatively weak disinfectant in most situations, is a good surfactant; soapy, or **saponated,** solutions wet surfaces thoroughly. Phenol and related compounds such as cresols also lower surface tension and are powerful germicides besides. A combination of soap, with "carbolic acid" (crude phenol) or cresols, would therefore seem to have exceptional possibilities as a disinfectant. Indeed, mixtures containing these substances in effective proportions (see next paragraph) are widely used in hospitals and laboratories. Products of this type (for example, Lysol) are available on the market or can be made up as Liquor Cresolis Compositus from the U.S. Pharmacopeia. Solutions of iodine with low surface tension (i.e., **iodophors,** or surface tension reducers combined with iodine) are now available commercially (e.g., Wescodyne and Ioclide).

Competitive Adsorption. An excess of soap, however, can interfere with the adsorption of some disinfectants. If more than minimal amounts of a bactericidally ineffective but very surface-active substance like soap are added to surfactant disinfectant solutions, the relatively inert soap is adsorbed on the bacterial surfaces to the exclusion of the disinfectant. Such displacement of one surfactant by another is called **competitive adsorption.**

If organisms are coated with waxy material, as is the case with tubercle bacilli, a disinfectant solution affects them with difficulty unless it contains a good surface tension reducer which allows it to wet the wax. Certain surfactant disinfectants (**quaternaries**; see page 310) are very effective against tubercle bacilli if dissolved in alcohol, since alcohol is both a wax solvent and a potent surface tension reducer.

21.4
EVALUATION OF DISINFECTANTS

Disinfectants or microbicidal agents are usually described as "strong," "weak," or "mild." These terms are inexact and convey different meanings to different people. To one the term "strong" may mean a disinfectant odor; to another, pain on application to a scratch; to still another, corrosive action; to another, a bright red or brown or other color. Rarely does the untrained person think of disinfectants in terms of **microbicidal** activity, or **toxicity** for human beings, plants, or animals. Actually, the value of any substance as a disinfectant depends on a number of factors. The ideal disinfectant should be:

1. Highly effective against a wide variety of microorganisms in concentrations so low as to be economical to use and nontoxic for animals (or plants).
2. Noninjurious and nonstaining to materials like fabrics, furniture, or metal wares, and nonoffensive to odor or taste.
3. As specific as possible for microorganisms, i.e., not inactivated by extraneous materials.
4. A good surface tension reducer (have good wetting and penetrating properties).
5. Stable in storage.
6. Readily available and not expensive.
7. Easily applied under household or other practical conditions of use.
8. And very important: completely microbicidal within a few minutes or an hour at most, and not inducing a state of microbiostasis, leading to a false sense of security.

No single disinfectant has all of these ideal properties. Some agents may be ideal under some conditions but not under others; e.g., strong lye or cresol may be ideal for stable floors or sanitary plumbing but harmful in infants' eyes. Not all disinfectants are equally effective, and some of them are more effective against certain bacteria than against others. Some are effective in pure cultures in the test tube but not in contact with organic matter such as blood, feces, or dead tissues; some are effective in the vapor phase but not as liquids.

The microbicidal effectiveness of disinfectants may be estimated by mixing them with cultures of whatever microorganisms are to be destroyed, and then measuring the time required for the substance to kill the organisms. If this is done under carefully standardized conditions (i.e., using a constant quantity of culture of a stated composition, a fixed temperature and a suspension of measured numbers of microorganisms of known and constant resistance) the results will be accurate and reproducible.

The Phenol Coefficient Determination.

The effectiveness of a water-soluble disinfectant that is chemically similar to phenol can be tested by determining its **phenol coefficient**. The phenol coefficient is based on the effectiveness of the test disinfectant as compared with that of pure phenol under the carefully standardized conditions just noted.

In a commonly used and representative procedure, 5-ml amounts of a series of dilutions of a disinfectant to be tested (here called X) are placed in a row of tubes of standard size. A similar series of dilutions of pure phenol (here called P) of 1:80, 1:90, and 1:100 is prepared. The temperature of all is brought to 20C in a water bath. To each tube is added 0.5 ml of a standard, young, broth culture of the test microorganism (selected strains of *Salmonella typhi, Staphylococcus aureus,* or *Pseudomonas aeruginosa*). At intervals of 5, 10, and 15 minutes, a standard loopful (4-mm loop of No. 23 B and S wire) is transferred from each tube in succession to a corresponding tube containing 10 ml of sterile broth of standard composition.

After 48 hours of incubation at 37C, growth in the broth tubes is recorded. If all of the organisms in all of the disinfectant X tubes were killed, **no growth** should appear in any of the corresponding subcultures. The test must then be repeated with higher dilutions of the disinfectant X such that not all of the organisms are killed in at least some of the tubes and some of the corresponding broth subcultures show growth.

The phenol coefficient is then calculated as the ratio of the highest dilution of X not killing the organisms in five minutes (evidenced by growth in the corresponding broth tube), but killing in 10 minutes (evidenced by no growth in the broth), to the corresponding dilutions of P. The values obtained are shown in Table 21–4. In this experiment the 1:90 dilution of phenol (P) failed to kill in five minutes but killed all the *S. typhi* cells in 10 minutes. This is compared with the "unknown" (X) which did the same in a dilution of 1:450. The ratio of X to P is $\frac{450}{90}$ or 5, the FDA* phenol coefficient of X. A slightly modified procedure is prescribed by the AOAC.† If desired, dilution plate counts of the numbers of organisms remaining alive in the disinfectant tubes after various intervals of time

---

*United States Food and Drug Administration.
†Association of Official Agricultural Chemists. There are many modifications of the procedure, such as the Rideal-Walker and Chick-Martin methods.

TABLE 21-4. TYPICAL DATA FROM A PHENOL COEFFICIENT DETERMINATION

Disinfectants	Dilutions	5-Minute Sub-cultures	10-Minute Sub-cultures	15-Minute Sub-cultures
Phenol	80	−°	−	−
	90	+°	−	−
	100	+	+	+
"Unknown"	350	−	−	−
	400	+	−	−
	450	+	−	−
	500	+	+	−
	550	+	+	−
	600	+	+	+
	650	+	+	+

° + = growth; − = no growth.

can be plotted on graph paper, thus establishing a "death curve."

The significance of a phenol coefficient has definite limitations. For example, a disinfectant dissolved in distilled water may have a phenol coefficient as high as 50. However, it may be wholly ineffective if applied in the presence of blood, or used in contact with organic matter such as pus, saliva, feces, or milk, as these may combine with the disinfectant and remove it from the bacteria. Further, it may have a high coefficient when tested against *Salmonella,* but only a coefficient of 2 or 3 when it is tested against some other organisms, such as *Staphylococcus aureus.* When a substance is said to have a certain phenol coefficient, the limitations of the method must be kept in mind.

The Use-Dilution Test. Once a phenol coefficient has been determined for a disinfectant, it has been customary to use the disinfectant in an arbitrary concentration 20 times the phenol coefficient (i.e., if the phenol coefficient is 5 the **use dilution** of the disinfectant is 1:100). The arbitrary use dilution has been found an often unreliable guide to disinfection. A **use-dilution test** (UDT) is therefore often applied as a check.

Small steel cylinders are wet with broth culture of one or more test organisms and dried. They are then placed for a standard interval in the disinfectant diluted for use. On removal to separate tubes of sterile broth no growth should occur in any tube if the recommended use dilution is to be certified as effective.

In another test, whole milk is added to the use dilution of a disinfectant to measure its effectiveness. This simulates the presence of extraneous proteins in actual use. Sometimes blood and serum are used in a similar way. A variety of test organisms often gives a still better idea of the effectiveness of a disinfectant under general conditions. Obviously, no one test method can give an accurate evaluation of all disinfectants under all conditions.

As a result, many different modifications have been suggested by various workers. For example, some workers with the UDT logically substitute, for the steel cylinders, bits of the actual materials on which the disinfectant is to be used, e.g., asphalt tile, linoleum, or wood flooring, and use a large variety of test organisms. In testing effects of disinfectants on viruses, cell cultures may be used instead of tubes of the bacteriological culture media for the testing of viability of the virus after contact with the disinfectant dilutions.

Inactivation. It is important to note that many disinfectant substances are also highly microbiostatic. If, in the phenol coefficient or other test, a microbiostatic substance clings to the test organisms on removal from the disinfectant-dilution tubes to the broth subcultures, no growth of the organisms occurs even though they are still alive. To eliminate this error it is sometimes sufficient merely to **dilute out** the microbiostatic agent by making two or three successive transfers of test objects from tube to tube of broth, or to use large volumes of broth in the subcultures. However, some disinfectants, such as bichloride of mercury, sulfonamide drugs, and quaternary ammonium compounds (see farther on), appear to be adsorbed upon or inside the cells and cannot be removed or stopped by dilution. A definite antagonist or inactivator is then added to the broth subculture tubes. For sulfonamides, *p*-aminobenzoic acid is used (see paragraphs on microbiostasis). Bichloride of mercury is fully inactivated by sodium thioglycollate (an organic compound of sulfur) in the broth culture medium. For quaternaries, Luramin sodium combined with sorbitan mono-oleate (Tween 80) and Azolectin, and other substances acting as competitive adsorbents have been found useful.

Gaseous Disinfectants. Methods of testing disinfectants that are applied in the gaseous state, of which several are in fairly wide use (page 312), obviously cannot be based on procedures like the phenol coefficient determination. Final methods for the testing of germicidal gases are still to be perfected. One method recently under experimental trial is designed to catch a standardized number of microorganisms on a bacteria-retaining membrane filter. The filter is then placed in a specially designed exposure chamber to which carefully metered gas is admitted at fixed temperature and humidity for a measured time. The organisms are then washed from the filter membrane into broth and incubated. The measure of effectiveness is not complete killing, but **injury** and **percentage kill** as indicated by slowed rate of development of turbidity in the broth as compared with an untreated **control**. Other methods of testing are under investigation.

Toxicity of Disinfectants. One of the great problems in the selection of disinfectants or antiseptics to be used in contact with living tissues is excessive **toxic effects.** One may determine the toxic dose of a disinfectant in various ways—for example, by observing the smallest quantity necessary to stop completely the action of leucocytes (white blood cells) in a test tube in a given time. Other methods measure the inhibitory or lethal effect of the tested substance on various tissues, while still others measure the respiratory quotient (i.e., the ratio of carbon dioxide given off to oxygen consumed) of tissue cells in contact with the germicides. One method is based on observing the survival time of chick embryos into which the tested substances are injected. A time-honored, simple, and effective method for many substances is the direct trial on living animals. The results are sometimes expressed as **toxicity index** (i.e., ratio of minimal toxic dose to minimal germicidal dose).

Probably no single test gives a generally applicable result, and species of microorganisms as well as tissue cells vary greatly in their susceptibility to different substances. For any disinfectant, therefore, the toxicity index will vary for each type of tissue cell and microorganism.

21.5

BACTERIOLOGICAL FILTERS

Special filters have been used in microbiology for many years to remove bacteria from fluids that cannot be subjected to heat or chemicals without destroying their usefulness. Bacteriological filters may not always sterilize fluids, in a strict sense, since viruses and bacteriophages, if present, may pass through the filters. However, for most purposes microbiologists wish only to remove any microorganisms that would grow and interfere with the planned usage of the fluid. This goal is often achieved by a single passage of the fluid through any one of the commonly used filters. Where exclusion of virus particles is demanded, special filters, to be discussed, are available with extremely small pore sizes.

Many materials in various shapes and forms have been used to make bacteriological filters. Three types which were very widely used, but now mostly replaced by the newer membrane filters, are shown in Figure 21-7. The Seitz filter employed a compressed filter pad made of asbestos fibers. The hollow tube or "candle" filters were made of diatomaceous earth (Berkefeld filter) or of unglazed porcelain (Chamberland-Pasteur filter). The sintered glass filters employed a disc of fused powdered glass. An interesting point is that all these filters contain passages larger than the bacteria they remove. Their effectiveness depends largely on adsorption of the bacteria (which are more negatively charged than the filter material) to the more

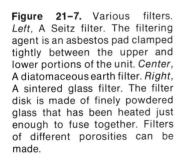

Figure 21-7. Various filters. *Left,* A Seitz filter. The filtering agent is an asbestos pad clamped tightly between the upper and lower portions of the unit. *Center,* A diatomaceous earth filter. *Right,* A sintered glass filter. The filter disk is made of finely powdered glass that has been heated just enough to fuse together. Filters of different porosities can be made.

positively charged surfaces of filter passages. Some industrial processes still use filters similar to those described above but much larger in size and designed for heavy and continuous work loads.

A particularly useful and versatile type of filter is the **membrane filter** made of cellulose acetate or similar substance. A common form is a paper-like disk about 50 mm in diameter and 0.1 mm in thickness (Fig. 21–8). It has myriads of very fine tubular openings from upper to lower surface, as shown by the scanning electron microscope (Fig. 45–12). The diameter of these tubulations may be varied by the manu-

facturer from over 1.0 μm to less than 0.005 μm, the last sufficiently fine to strain out even the smaller viruses from fluid passing through the filter membrane.

Membrane filters, unlike the filters discussed previously, depend for their effectiveness to a greater extent on the sieve action of their perforate structure and on the fineness of the pores. Adsorption and van der Waals forces play a lesser but still significant role.

A particular advantage of the membrane filter is that all of the microorganisms in a relatively large volume of fluid (water, milk, urine or diluted blood) may be collected on one small

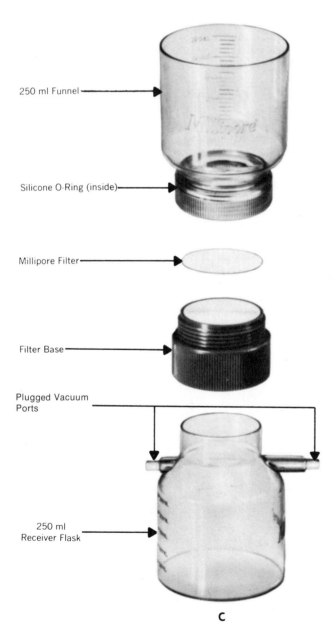

250 ml Funnel

Silicone O-Ring (inside)

Millipore Filter

Filter Base

Plugged Vacuum Ports

250 ml Receiver Flask

C

Figure 21–8. Sterifil® aseptic filtration system disassembled. For use, the entire unit is sterilized. Fluid to be filtered is poured into the funnel. Suction is applied via one of the side arms of the 250-ml receiving flask after capping the opposite side arm. (Courtesy of Millipore Filter Corporation, Bedford, Mass.)

Figure 21–9. Plastic Petri-type dish containing a broth selective for bacteria (*Escherichia coli*) which indicate sewage pollution of water. The membrane filter through which the water was filtered was removed from an apparatus like that shown in Figure 21–8 and laid in the dish on a sterile pad soaked with broth, which the membrane absorbed. After incubation for 18 hours at 37C the distinctively colored, glistening colonies of *E. coli* had developed. (Approximately actual size.) (Courtesy of Millipore Filter Corporation, Bedford, Mass.)

disk, where they may be observed directly or cultivated in situ (Fig. 21–9). There is no need to handle hundreds of tubes or flasks with large-volume cultures. Some special uses of these filters are mentioned in discussing the bacteriological examination of drinking water supplies (Chapter 45).

A more recent development is the Nuclepore (General Electric Co.) filter, a very thin (10 µm thick) polycarbonate film in which minute holes have been etched after bombarding the film with charged particles. The filters can be obtained with pore sizes ranging from 0.1 to 0.8 µm. Variation (or tolerance) from the stated pore size is remarkably small. The surface of such a filter is shown in Figure 45–12. High flow rates, low toxicity, chemical inertness, and resistance to damage by most biological fluids and chemicals are a few of the desirable features of these filters.

CHAPTER 21
SUPPLEMENTARY READING

(See Supplementary Reading at end of Chapter 22.)

STERILIZATION · CHAPTER 22
AND
DISINFECTION

Practical Applications

22.1
STERILIZATION BY HEAT

Heat may be applied for sterilization in three ways: by steam or hot water (moist heat), by prolonged baking in the oven (dry heat), and by complete incineration. The last needs no comment beyond pointing out that common sense will direct what may be burned and that care must be taken to see that such material is completely destroyed and not carried out in the hot gases or ash to pollute the environment.

Moist Heat

Boiling in Water. The role of moisture in the lethal effects of heat was described in the foregoing chapter. The use of boiling for preserving foods and for disinfection is very simple. It is necessary to remember only that bacterial endospores (family Bacillaceae) may remain alive even after hours of boiling. For ordinary household purposes of **disinfection** (not **sterilization**), five minutes of boiling is usually sufficient, provided that the hot water actually comes into contact with the microorganisms and not merely with the outside of lumps of food or packets of instruments or other objects contain-

ing the microorganisms. Boiling in water can never be depended on for sterilization, especially at high elevations above sea level, where the temperature at the boiling point is lower (Table 22–1).

Free-Flowing (Live) Steam. **Live,** or free, steam is usually applied in a loosely covered container that will hold steam without pressure. Boiling water and free steam never reach a temperature above 100C (212F). Free steam is sometimes used to accomplish **fractional sterilization,** or **tyndallization.**

Tyndallization. As described in Chapter 2, John Tyndall devised a process of sterilization by steaming for a few minutes at 100C on three or four successive occasions, separated by 24-

TABLE 22–1. BOILING POINT OF WATER AT VARIOUS ALTITUDES

Location	Altitude°	Boiling Point†
New York City	0	100
Chicago, Ill.	589	98.9
Denver, Colo.	5,280	94.3
Fort Laramie, Wyo.	7,380	92.2
Tahoe, Nev.	10,000	89.1

° Feet above sea level.
† Degrees C.

hour intervals at room temperature. The intervals permit the dormant, resistant spores to become active, vulnerable vegetative cells, readily killed by 100C. This process renders an infusion sterile, whereas one single continuous boiling for one hour may not, since many spores can remain in their dormant and resistant state during this time. An advantage of the method is that it requires no special apparatus. A disadvantage is that it is time-consuming and in some fluids, such as water, spores may not grow out promptly. Also, if the material is freely exposed to air, anaerobic spores may not germinate and may survive the process. If not freely exposed to air, aerobic spores will not grow out.

Compressed Steam: Autoclaving. Anyone familiar with the operation of a home pressure cooker is familiar with the principle of an **autoclave**, because the cooker is a simple form of autoclave. In the autoclave (Fig. 22–1), be it a small and simple home pressure cooker or an apparatus large enough to fill a room and fitted with various gauges, pipes, valves, clocks, and wheels, the object of both is alike: to heat the articles to be sterilized by means of steam under considerable pressure.

Steam under pressure is hotter than boiling water or free-flowing steam such as is used in tyndallization. The higher the steam pressure, the higher the temperature will be.

The relation of steam pressure to temperature is shown in Figure 22–2.

It must be remembered that it is the compressed steam (moisture, hydration) that sterilizes and not compressed air (dry and usually not as hot as steam).

For example, pure steam at 15 pounds of pressure has a temperature of 121C. If the steam is mixed with an equal amount of air, at the same pressure the temperature is only 110C, while if the mixture is two thirds air the temperature is only 109C.

Steam **hydrates** and thus promotes coagulation; air does neither. Steam, being water vapor, also produces **hydrolysis** at autoclave temperatures. Dry air cannot do this at any pressure or temperature. In autoclaving, therefore, as in using a pressure cooker, a valve is left open for the escape of *all air* before the steam pressure is allowed to rise.

The actual amount of water present as steam in an autoclave is small, and articles soon dry off after removal, especially if removed from the autoclave while hot.

The **thermometer** on the autoclave is the important guide to the process of autoclaving, not the pressure gauge. However, the latter, as

well as a steam-escape or safety valve, is essential to safety.

The common practice in autoclaving fluids or freely exposed surfaces such as those of dishes and instruments is to apply 115 to 125C (achieved at 10 to 20 pounds of pressure) for 20

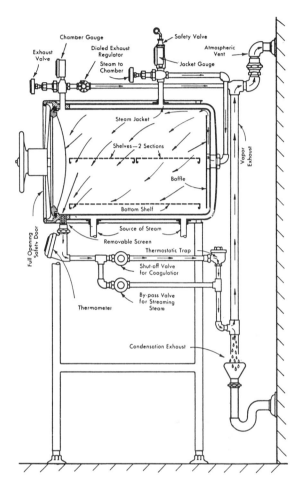

Figure 22–1. Diagrammatic illustration of steam jacketed autoclave. Steam enters the **jacket,** a double-walled shell, at two places just beneath the cylinder. It passes out the top through a pipe to which are attached: a wheel valve admitting the steam to the inner **chamber,** a **safety valve,** a gauge showing **pressure in the jacket.** The steam enters the inner chamber at the right of the diagram, filling the upper portion. Its pressure registers on the **chamber gauge.** It may be allowed to escape rapidly by the **exhaust valve.** If this is closed, steam pushes the cooler air in the lower portion out at the bottom (left), where the **thermometer** registers proper temperature only when the air is gone and is followed by the hot steam. The escaping steam may be allowed to flow out without building up any pressure if the **by-pass valve** is fully opened. If the by-pass valve is closed and the **shut-off valve** is opened, steam passes through the **thermostatic trap** where the heat shuts off all but a pinhole opening. This causes pressure to build up in the chamber, yet prevents stagnation by permitting a constant minute flow of steam through the apparatus. (Courtesy of American Sterilizer Co., Erie, Pa.)

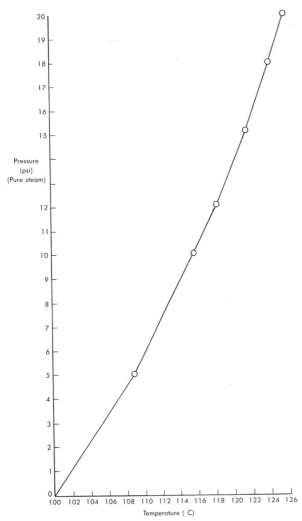

Figure 22–2. Relation between temperature and pressure of pure steam.

temperatures for sterilization than low-acid foods (pH, 6.5 to 7), such as milk, corn, and meats. This is because of the synergistic action between pH and temperature. (See Chapter 21.)

Dry Heat

Dry heat is used in oven sterilization (Fig. 22–3). It is necessary to bear in mind that significant coagulation does not occur when moisture is not present.

Articles in ovens are very dry and, therefore, in order to be freed of live spores, must reach a very high temperature (165 to 170C; 329 to 338F). It is customary to apply 165C for a period of two hours or more. This accomplishes no coagulation but, what is more effective, slight charring.

A home oven can easily be used for sterilization. A "moderate" temperature (330F) is satisfactory, and the heating should be allowed to proceed for two hours *after* reaching that temperature. Paper wrappings should be slightly browned but not brittle; muslin or string should be faintly yellow from the heat.

Only dry articles not injured by baking (glassware, bandages, instruments, mineral oils, petrolatum, talcum powder, and the like) may be thus sterilized. Solutions containing water, alcohol, or other volatile substances will, of course, boil away and be ruined.

22.2 TIME AND CONTROLS

Necessity of Thorough Heating. In any process of disinfection or sterilization by heat it is absolutely essential that the object be heated through and that the center of the object be held at a killing temperature long enough to destroy the microorganisms.

Thus, in home canning, a quart jar of spinach may be held in free steam (100C) for 5 to 10 minutes and when grasped with the hand will feel very hot, yet the center may be well below a microbicidal temperature. Large masses of nonfluid materials like quart jars of canned vegetables and roast meat, in which the contents cannot circulate, require a long time (one and one-half to two hours), even in the autoclave, to be heated so thoroughly that the center reaches a sporicidal temperature. Pieces of meat or vegetables to be sterilized in jars by heating should be loosely packed, allowing space for circulation of the fluid in the jar. Air

minutes. The pressure must be allowed to subside *slowly* after the heating is completed or the **superheated** fluids in open (cotton-plugged) vessels will boil over. *Tightly sealed vessels may explode.* Any large, solid masses must be heated a longer time to allow for heat penetration. Packages must be spaced so as to allow free circulation of steam. Substances such as mineral oil or petrolatum, sand, or any dry objects in tight jars, or substances that are impervious to moisture, cannot be satisfactorily sterilized in the autoclave. The temperature may rise as high as 125C, but in the absence of moisture this may be ineffective. Such materials are more effectively sterilized by dry heat in an oven.

Acid materials, such as canned tomatoes, acid fruits, pickles, or sauerkraut (pH, more acid than 4.5), require much shorter periods or lower

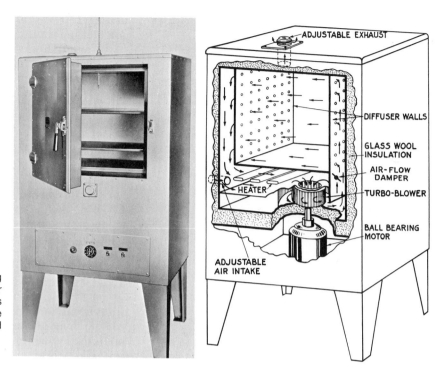

Figure 22–3. A type of sterilizing oven for microbiology. The motor at the bottom of the oven forces streams of hot dry air through the chamber of the oven, as indicated by the arrows.

pockets in the jars should be carefully removed. Penetration of the heat is facilitated if the cans or jars or pieces of roasting meat are small and not packed too closely together, promoting free circulation of steam or hot air around and between them.

Sterility Checks. For many purposes — industrial, hospital, research — it is important to have some absolute means of determining after a heat sterilizing process that even the most heat-resistant spores have been killed. Two general types of check are commonly used. One consists of pieces of paper, gauze, or thread to which are applied spots of some thermolabile dye. Inserted in the materials being sterilized, the dye turns a distinctive color if the contents

of the sterilizer have been subjected to a sterilizing temperature for a sufficient time (Fig. 22–4).

A more direct (though more time-consuming) check consists of strips of filter paper having on them the very heat-resistant spores of *Bacillus subtilis* (Fig. 22–5). This is a harmless aerobic bacterium. The spore-impregnated strips are inserted (enclosed in sterile envelopes) inside materials being sterilized. Afterward the strips are removed to tubes of culture medium and incubated. If growth occurs in the tubes within seven days then it is evident that the sterilization process in which they were involved was inadequate to kill the spores. In a modification of this process, sealed ampules

Figure 22–4. One type of device for indicating proper autoclaving. Before heating the word "NOT" appears clearly as on the upper indicator. After partial heating the leftmost black bar appears as seen on the lower indicator. Not until autoclaving is complete both as to time and temperature is the word "NOT" obliterated by a second dark bar on the lower indicator. (Courtesy of Sterilometer Laboratories, Inc., N. Hollywood, Calif.)

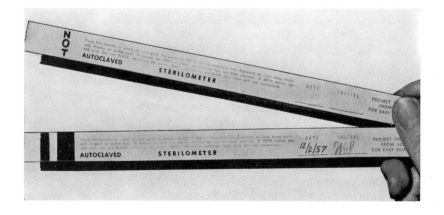

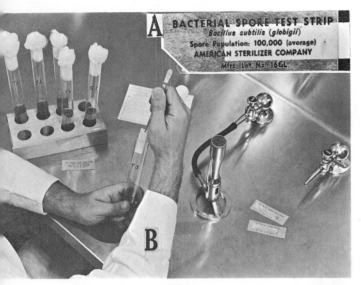

Figure 22–5. One method of checking effectiveness of heat sterilization is by testing the sterility of a strip of paper (A) containing 100,000 very heat-resistant bacterial spores (*Bacillus subtilis*). The strip of paper is placed inside the material being sterilized before the application of heat. After the heating process the strip of paper, with its spores, is placed by means of sterile forceps in a culture tube of sterile broth (B). If any of the spores grow during seven days of incubation, the heating process did not achieve complete sterilization. (Courtesy of American Sterilizer Co., Erie, Pa.)

(commercially available) containing culture medium, with dormant spores of harmless *Bacillus stearothermophilus* already in it, are inserted in the materials being sterilized. After sterilization the complete ampules are incubated at 55C (since *B. stearothermophilus* is a strict thermophile). If no growth occurs in seven days it is proof of effective sterilization, since spores of *B. stearothermophilus* are among the most heat-resistant of known living things.

22.3
SOME USEFUL DISINFECTANTS

There are hundreds of disinfectants on the market, and it would obviously be impossible to discuss even a small part of them here. A few commonly used representatives of several classes will illustrate general principles and usages. Table 22–2 lists the properties of a number of widely used disinfectants, while a more detailed account of the various classes of disinfectants follows in the text. A point worth emphasizing again is that disinfectants and antiseptics generally do not have a "specific target," as do some of the antibiotics; i.e., their action is indiscriminate. Cells of human tissue can be

damaged by almost all of these agents, depending on the duration of contact and concentration of the agent. For convenience we may group the disinfectants as follows: halogens and halogen compounds, compounds of heavy metals, phenol and its derivatives, alcohols, detergent disinfectants, microbicidal gases, and radiations.

Halogens and Halogen Compounds

Iodine and chlorine, both strong oxidizing agents, are the most widely used of the halogens. Chlorine gas is used to disinfect filtered water at all municipal water-purification plants and many sewage-disposal plants. It is usually handled in tanks or tubes like oxygen. It is applied to drinking water in a final concentration of about 1.0 part per million.

A more convenient form of chlorine for the individual and household user is **calcium** or **sodium hypochlorite**. Solutions (5 per cent) of sodium hypochlorite are purchasable in all grocery stores under various names (e.g., **Clorox**, household bleach). They have a multitude of household and sanitary uses. They depend for their effectiveness probably on their liberation of free chlorine and the subsequent oxidation of essential enzymes and proteins.* Directions for use of these solutions are on each bottle. Four points need to be remembered in their use: the chlorine tends to evaporate from the solution; they give a bad odor to the hands; chlorine is poisonous; and it is readily inactivated by all organic matter.

Chloride of Lime. Calcium hypochlorite (chloride of lime, 0.5 to 5 per cent aqueous solution) is excellent for similar purposes and is inexpensive. The ordinary chloride of lime of commerce is unstable and soon loses most of its free chlorine. Many organic chlorine compounds which liberate their chlorine more slowly are very effective. **Azochloramid** is one of these; **dichloramine toluol** is another. The odor of chlorine may be objectionable.

In addition to the above substances, there are many new proprietary chlorine compounds on the market based on principles given above.

Iodine. The most actively antimicrobial of the halogens is iodine. However, it is not com-

---

*Various theories ascribe the antimicrobial action of hypochlorites to HOCl, HCl, **nascent** oxygen and free chlorine. Thus: $2\ CaOCl_2 + 2\ HCl \longrightarrow 2\ CaCl_2 + 2\ HOCl$; $HOCl + HCl \longrightarrow H_2O + Cl_2$; or $2\ HOCl \longrightarrow 2\ HCl + O_2$; or $2\ Cl + H_2O \longrightarrow 2\ HCl + \frac{1}{2}O_2$; or $Cl_2 + H_2O \longrightarrow HCl + HOCl$.

TABLE 22-2. EXAMPLES OF CHEMICAL AGENTS USED FOR THE CONTROL OF MICROORGANISMS

Agent	Major Chemical Action	Practical Uses	Disadvantages
HALOGENS			
1. Chlorine and compounds of chlorine (e.g., Clorox, halazone, dichloramine T).	Oxidation of essential enzymes or proteins.	General sanitizer and disinfectant useful against most microorganisms.	Inactivated by organic matter, also corrosive.
2. Iodine and iodophors (e.g., Tr. Iodine, Isodyne, Wescodyne, Betadine).	Inactivation of proteins through iodination.	Antiseptic on skin, relatively nontoxic, useful against most microorganisms.	Inactivated by organic matter; alcoholic solutions volatile and irritating. Toxic if taken internally.
HEAVY METALS			
1. Mercuric chloride; organic mercurials (e.g., Mercurochrome, Merthiolate, Metaphen).	Inactivation of enzymes or proteins through coupling to sulfhydryl groups.	Disinfectant for bench tops, inanimate objects. Organic mercurials useful as antiseptics on skin.	Inactivated by organic matter; mostly bacteriostatic, not effective against spores and certain pathogens.
2. Silver nitrate (AgNO$_3$, Argyrol, Protargol).	Denatures essential enzymes and proteins.	Nonirritating antiseptic eyedrops.	Inactivated by organic matter. Not effective against many microorganisms.
PHENOLIC COMPOUNDS (e.g., Lysol, pHisohex, cresols).	Surface active—disrupts cell membrane, inactivates enzymes and proteins.	Broadly germicidal, long lasting. Not inactivated by organic matter.	Can be irritating and toxic. Not generally effective against endospores. Some have objectionable odor.
ALCOHOLS (Ethyl and isopropyl).	Lipid solvent, denatures proteins, inactivates enzymes.	Mostly skin antiseptic. Good penetration, low toxicity.	Volatile, irritating, inactivated by organic matter. Some pathogens and all endospores resistant.
QUATERNARY AMMONIUM COMPOUNDS (e.g., Zephiran, Roccal, Cetrimide).	Surface active—disrupts cell membrane, denatures proteins, inactivates essential enzymes.	Skin antiseptic, disinfection of utensils. Relatively nontoxic, odorless.	Neutralized by soap. Not effective against spores nor many viruses. May be only microbiostatic.
FORMALDEHYDE (HCHO).	Inactivates enzymes, strong reducing agent.	Penetrating disinfectant. Largely replaced by other gases.	Irritating, produces film on surfaces. May be only microbiostatic. Needs high humidity to be effective.
ETHYLENE OXIDE (CH$_2$OCH$_3$ Carboxide, Oxyfume, Cryoxcide).	Inactivates enzymes and various other organic compounds through alkylation.	Sterilization of heat-labile materials. Leaves no residue.	In pure state, it is toxic and explosive. Comparatively slow-acting, temperature and humidity must be controlled for maximum effectiveness.

monly used for the same large-scale purposes as chlorine because of its physical properties and its cost. Formerly used widely as a local antiseptic in a 7 per cent tincture (alcoholic solution), this strength is far too caustic and poisonous for local application and does more harm than good. The 2 per cent tincture is much better and is the most generally useful for small cuts and abrasions. An aqueous solution containing approximately 1.85 per cent of iodine and 2.2 per cent of potassium iodide does not sting as much when placed on a small wound, but the solution has a high surface tension and thus does not penetrate as well.

Iodophors. These are nonirritating, nonstaining, virtually odorless, and very effective solutions of organic compounds of iodine. The iodine is loosely combined with a surfactant organic **iodophor** (Gr. *phor* = bearer or carrier). The iodophor releases the iodine slowly and also lowers the surface tension of the solution.

Examples of such iodine products are **Wescodyne, Ioclide,** and **Betadine.** In such forms iodine may be used for a variety of purposes, from clinical uses to sanitization of dishes, with aqueous solutions containing 200 parts per million. Under certain conditions of concentration and temperature such solutions may be sporicidal. They appear to kill bacteria, molds and some viruses quickly. Iodine is a much more effective germicide than chlorine.

Iodine combines indiscriminately with organic matter and the iodination of proteins leads to their inactivation. Its solutions are unstable and, like chlorine, it is poisonous.

Compounds of Heavy Metals

The first scientifically designed, specifically acting drug was based on the heavy metal arsenic. Arsenic was compounded with an organic group to diminish its poisonous properties. Prior to 1907, Paul Ehrlich and his co-workers had synthesized and tried many such compounds in the treatment of spirochetal diseases, notably syphilis, without success. The 606th compound ("SOS" or "606" or Salvarsan or **arsphenamine**), discovered in 1907, was very effective and, with various modifications such as neoarsphenamine, was used for years in the treatment of syphilis and other spirochetoses. Because it has marked toxic properties, it fell into disuse after it was found that penicillin is much more active against the same organisms and almost nontoxic. The antimicrobial effects of arsenicals presumably depend upon their

combination with sulfhydryl groups of key proteins within the cell.

Mercury. Long used as a general disinfectant and as an antiseptic, mercuric chloride is now obsolete. It is very toxic to tissues, very poisonous internally, and corrosive to materials, especially metals (hence its older name of **corrosive sublimate**). In many situations where high concentrations are used, mercuric chloride fails because it forms a thick coating of coagulum on the outer surface of particles of mucus, tissues, and feces, and thus protects the microorganisms within. In dilutions of 1:10,000 or more, its action is essentially bacteriostatic and may be reversed by sulfhydryl compounds if they are also present.

Organic Mercurials. Attempts have been made, following Ehrlich's lead with arsenic, to decrease the toxic, corrosive, and irritating qualities of mercuric disinfectants by incorporating mercury in complex organic molecules. This has yielded a number of products that are less toxic and irritating than mercuric chloride, among them **Mercurochrome, Merthiolate,** and **Metaphen.** In the presence of organic matter they are more effective than bichloride of mercury. Mercurochrome is often used for local, superficial application for disinfecting cuts, wounds and skin. Some of these organic mercury compounds are used as preservatives in biological materials. Other useful organic mercurials are phenylmercuric nitrate, very effective and of sufficiently low toxicity for external use, and ammoniated mercury, long used as a 10 per cent ointment for external wounds and fungal infections of the skin. These mercurials all mainly act microbiostatically, and their action, like that of mercuric chloride, is reversed by compounds with sulfhydryl groups. All mercury compounds are poisonous internally to some degree, and mercury is absorbed through the skin from ointments, etc.

Silver Nitrate. Silver nitrate (1 per cent aqueous solution) is used principally for application to infants' eyes at birth to prevent infection. Silver is also used as a local disinfectant in the form of organic colloidal preparations such as **Argyrol** or **Protargol,** in which form it is nonirritating and may be used in adult eyes at 20 per cent concentration. It is not for use in infants' eyes in place of silver nitrate. The use of silver nitrate or an equivalent, approved antibacterial in the eyes of infants at birth is required by law in most states as a safeguard against blindness of the newborn **(ophthalmia neonatorum).**

Copper Sulfate. Copper sulfate is used

chiefly to control the growth of algae in open water reservoirs and as a fungicide in garden sprays (**Bordeaux mixture**). Organic compounds of iron are also used as fungicides in garden sprays.

Phenol and Its Derivatives

Phenol (C_6H_5OH, carbolic acid) in the pure state is not commonly used as a disinfectant because it is expensive and because there are more than a dozen derivatives that are not only less costly but more effective. All act similarly to phenol (i.e., disrupting cell membrane, causing lysis, and inactivating essential enzymes and proteins). Unlike the halogens, which are indiscriminately effective against a wide range of microorganisms, susceptibility to phenolic disinfectants varies greatly.

Surface (Residual) Disinfection. Phenol and many of its derivatives are surface tension reducers and are commonly used in solutions containing additional surfactants to improve wetting properties. Since they are surfactants themselves, the phenolic disinfectants tend to be adsorbed in thin, and more or less durable, films on inert surfaces to which they are applied, resulting in what is termed **surface,** or **residual,** disinfection. As compared with the volatile halogens or alcohols their action is thus prolonged, often for many hours. This is an especial value of certain preparations of phenol derivatives for surgical and hospital hand washes.

Types of Phenolic Disinfectants. Phenolic disinfectants are of many types, common among which are: the **cresols,** the molecules of which are like phenol but have methyl groups attached; and the **bisphenols,** the molecules of which consist of two phenolic groups joined directly or through some other radical (Fig. 22–6).

Crude Cresols. Crude cresols form colloidal (milky) suspensions in water and are therefore especially effective as disinfectants, because each colloidal droplet consists of **concentrated** cresol. This is true of many colloidal disinfectants. There are many excellent cresol preparations on the market. Some have a cling-ing odor. Most of them contain a surface tension reducer in addition to the disinfectant. They are commonly used in 1 to 5 per cent concentrations for disinfecting floors, furniture, barns, and stables. Some may also be used in temporary contact with human tissues in 0.5 to 1.0 per cent strengths. Good examples are Lysol and saponated cresol solution, U.S. Pharmacopeia. The preparation of such solutions is an important industry.

Bisphenols. Like the cresols, bisphenols have surfactant properties. In order to increase their solubility in water they are commonly used in solutions with other surfactants or detergents.* In general they are less toxic than the cresols. Among well-known bisphenols is one combined with chlorine (one of the class of **halogenated bisphenols**), called hexachlorophene. Because it is a surfactant, it provides good residual surface disinfection of the skin, and it is not irritating if used properly. It is incorporated especially in surgical soaps and hand washes (Gamophen, pHisohex and Hexosan).

A recent development is that hexachlorophene, once widely used in deodorants, deodorant soaps, shampoos, and numerous hygiene products, has now been virtually banned from such products. Hospital personnel are being warned not to use cleansing agents containing hexachlorophene *unless* there is a clear need for doing so, since recent studies have indicated that hexachlorophene may be absorbed (slowly with prolonged use or rapidly in the case of burn patients) and reach toxic levels in the central nervous system. While there is no sound evidence that the widespread use of hexachlorophene was a health hazard, except in certain clinical applications, health authorities regard the indiscriminate use of this compound to be unwise. No doubt the excellent health enjoyed by people in the civilized parts of the world is attributable in large measure to cleanliness and good sanitation. However, we have to make sure that we are not poisoning ourselves

---

*Detergency (L. *detergere* = to cleanse) does not necessarily confer disinfectant properties. Virtually all household detergents are strong surface tension reducers but are not therefore necessarily good disinfectants.

Figure 22–6. Structures of phenol, representative cresols, and one bisphenol (hexachlorophene).

  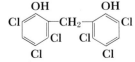

phenol   *o*-cresol   *p*-cresol   hexachlorophene (a bisphenol) as well as the bacteria with the use of our disinfectants, sanitizers, soaps, preservatives, etc.

Orthophenylphenol. A compound similar to hexachlorophene, orthophenylphenol, is used in mixtures such as **O-syl.** Two other halogenated bisphenols, **chlorothymol** and **chlorohexidine,** are similar in general properties. All are broadly germicidal, and each has distinctive properties that adapt it to particular situations. Some can be used effectively as irrigations in surgical wounds in 1:1,000 to 1:10,000 concentrations.

The activity of all phenolics is markedly depressed in the presence of extraneous organic matter (pus, blood, feces, etc.), and they are highly microbiostatic. Phenolic compounds or mixtures should never be left long in contact with skin or tissues.

Alcohols

There are many substances that chemically are alcohols: many of these are effective microbicides. Like various organic disinfectants, their microbicidal activity is directly related to increased length of their carbon chain and molecular weight; beyond their critical point (different for each type) microbicidal activity decreases. The two alcohols most commonly used for disinfection or sanitization are ethyl ("grain") alcohol and isopropyl ("rubbing") alcohol. They are mild disinfectants and are nontoxic on external application. Methyl or "wood" alcohol and its vapor are extremely poisonous and should never be used in any bodily contact whatever.

Ethyl alcohol is a surface tension reducer, a lipid solvent, and a coagulating agent, properties that should contribute to microbicidal potency. However, it is also a potent dehydrating agent, a property that interferes greatly with its coagulating power and probably also with its other antimicrobial properties. Because of its dehydrating power it is best used as a 70 per cent aqueous solution. Stronger concentrations (95 or 100 per cent "absolute") are much less effective. As a coagulating agent it tends to combine nonspecifically with extraneous organic matter and can produce a thick, protective coating of coagulum around microorganisms. As a lipid solvent it is useful as a **cleansing** agent ("disinfectant"?) for skin prior to hypodermic injections. It adversely affects viruses and cells with envelope lipids. As a surface tension reducer it is a potent adjuvant of other disinfectants such as phenolics and quaternaries. Similar considerations apply to isopropyl alcohol.

Detergent Disinfectants

As just mentioned, not all detergents are disinfectants. However, several compounds combine the properties of detergent, surface tension reducer, and disinfectant.

Soaps. Among these versatile compounds are the common soaps: potassium or sodium esters of higher fatty acids (stearic or glyceric). They are actively lethal for certain species of microorganisms: *Treponema pallidum,* the syphilis spirochete; *Streptococcus (Diplococcus) pneumoniae,* a cause of pneumonia, and some other streptococci; and a few other organisms. Unfortunately, with these few exceptions, soaps are relatively ineffective as general disinfectants. However, soaps are potent surface tension reducers and consequently are good emulsifying and foaming or "sudsing" agents. Foaming results from the fact that, since the surface tension is greatly weakened, the surface can be extended almost indefinitely as foam or as droplets in an emulsion; compare with crude cresols, page 297. When soapy water is applied to skin or instruments, microorganisms are carried into the emulsified grease droplets and removed by rinsing. Detergency is the chief virtue of most soaps. Soluble soaps are precipitated by acids, by ions of heavy metals, and by Ca^{2+}.

Synthetic Detergents. As generally understood, this term comprises a group of potent surface tension reducers that are not typical soaps. They are synthesized commercially and much used for household and commercial cleansing and laundry purposes. They are available in "large, economy" packages in all grocery stores. Most of them are long-chain, alkylbenzylsulfonates (ABS) (see Chapter 45). Few are markedly microbicidal. They are **anionic** compounds (see following paragraphs).

Another group of synthetic compounds, classed as detergents because they are strong surface tension reducers, are much more important as disinfectants and sanitizing agents. The first was discovered in 1936 by Domagk (Nobel Prize winner), who described the properties of benzalkonium chloride, now familiar as **Roccal** or **Zephiran** (Fig. 22–7, A). This is a quaternary alkyl ammonium salt, and is representative of a number of similar compounds that have since been developed commercially as disinfectants or detergents, although they are not effective against most viruses. As a group they are commonly called **quaternaries.** Basically, they are ammonium halides (for example, ammonium chloride) in which the hydrogen atoms have been replaced by various alkyl groups (Fig. 22–7, B).

Chemical Structure and Activity. In general, the more effective of these detergent-disinfectant compounds are those in which at least one of the alkyl groups contains a chain of 12 to 16 carbon atoms, whereas the less effective are those of lower or higher molecular weight. This represents a common observation previously mentioned; i.e., the germicidal efficacy of many organic compounds is greatly affected by chemical structure, especially length of carbon chains.

Cationics, Anionics, and Nonionics. There are three types of microbicidal detergents: (1) those in which the surfactant portion is the **cation** (e.g., quaternary alkyl ammonium halides) (Fig. 22–7, A); (2) those in which the surfactant portion is the **anion** (e.g., complex soaps like sodium lauryl sulfate) (Fig. 22–7, C); and (3) those that do not ionize. These compounds are classified as **cationic, anionic,** and **nonionic** detergents, respectively. The cationic compounds appear to be the most generally effective as disinfectants and are what is generally meant by the term "quaternary." Soaps are anionic detergents. The nonionic compounds are relatively weak disinfectants. They are sometimes used with anionic compounds in household detergents as adjuvants.

Mechanism of Action. Since detergents like the quaternaries and sodium lauryl sulfate

$$R-N^+_{\ |}\!\!\begin{array}{c} R \\ | \\ -R \end{array}\ Cl^- (^-Br,\ \text{etc.})$$

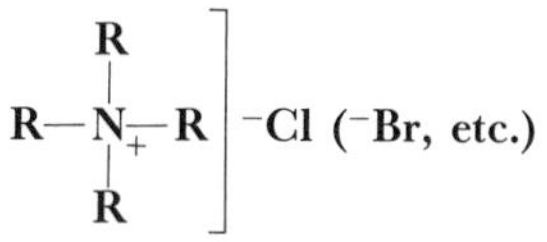

A. General formula of alkyl ammonium halides

$$CH_3-(CH_2)_{11}-\underset{CH_3}{\overset{CH_3}{\underset{|}{\overset{|}{C}}}}-N^+-CH_2\ \bigg]\ ^-Cl$$

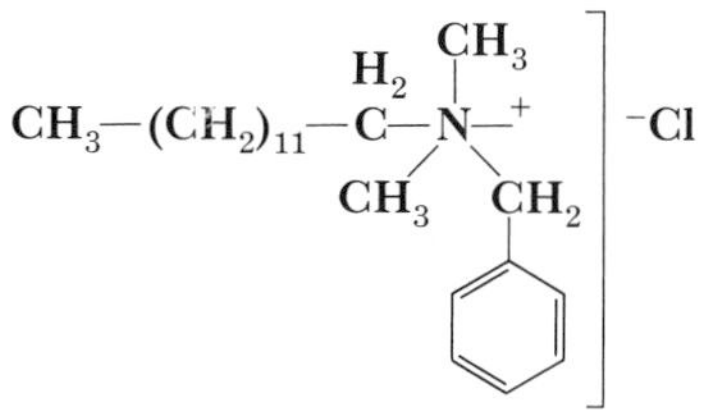

B. Alkyl-dimethyl-benzyl-ammonium chloride; a cationic, microbicidal detergent

$$CH_3-(CH_2)_{11}-O-\underset{O}{\overset{O}{\underset{\|}{\overset{\|}{S}}}}-O^-\ \bigg]\ ^+Na$$

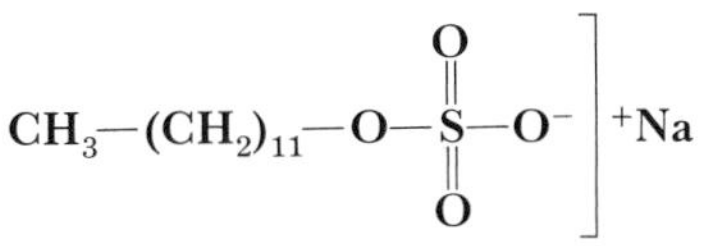

C. Sodium lauryl sulfate; an anionic microbicidal detergent.

Figure 22–7. Forms of microbicidal detergents.

are generally alkyl compounds of chlorine or sodium, they are bipolar, i.e., they exhibit both fat-soluble (lipophilic) and water-soluble (hydrophilic) sides. When dissolved in aqueous culture media, the longer alkyl chains tend to be oriented toward, and to dissolve in, the cell lipids. The charged residue (chlorine or sodium ion) is oriented away from the cell and is drawn toward the surrounding aqueous fluid. The surface tension reducer molecules are systematically arranged side by side at the cell—aqueous medium interface, as closely placed as possible (Fig. 22–8). Since the reducer molecules at any surface have a lower mutual attraction than the water molecules in which they are dissolved, the surface tension of their solution is lowered. Lipids upon which detergents in aqueous solution become oriented tend to become divided into smaller and smaller particles, each coated with the surface tension reducer (i.e., to be **emulsified**); detergents thus characteristically "cut" (remove) grease.

The lipophilic property of quaternaries and alkyl sulfates is thought to disrupt the lipids, and therefore the whole structures, of bacterial cell walls and membranes, especially of those rich in lipids like many gram-negative bacteria. Disruption of the cell walls and membranes would permit outward passage of vital intracellular components and passage inward of the surface tension reducer. Electron micrographs show the destruction of bacterial cell walls and inner structures by antimicrobial substances which reduce surface tension. Probably these detergents also act in part by occluding the bacterial surfaces by being adsorbed there. Also, many proteins are denatured in the presence of these agents.

In use, soaps and other anionic compounds should *not* be mixed with cationic compounds, because the oppositely charged molecular complexes neutralize each other and the agents are precipitated or inactivated. For the same reasons hard, acid, and iron-rich waters tend to inactivate them. Unlike many other disinfectants, quaternaries retain much of their activity in the presence of considerable organic matter. The quaternaries tend to be microbiostatic or slowly lethal rather than to kill instantly.

When the effectiveness of quaternaries as disinfectants is being measured, their marked microbiostatic action must be offset by the use of special inactivators to allow the microorganisms to grow in the subculture (see §21.4). **Tinctures** (alcoholic solutions) of quaternaries appear to be very effective indeed as disinfectants against tubercle bacilli, and some are effec-

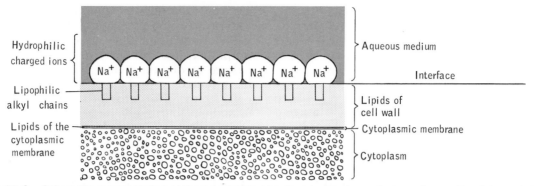

Figure 22–8. Schematic representation of the orientation of polar groups of an **anionic,** surface-active detergent, in this case sodium lauryl sulfate [$Na^+(C_{12}H_{25}OSO_3)^-$], in relation to the surrounding aqueous medium and a cell having lipids in its cell wall or cytoplasmic membrane. The hydrophilic group (here represented by Na^+) orients at the fluid-cell interface; the hydrophobic group (here represented by the lauryl sulfate group$^-$) is oriented in the lipids of the cell wall or membrane.

Conversely, at any air-water interface, the hydrophilic groups orient themselves within the surface water while the hydrophobic alkyl groups are at the air-water interface, weakening the intermolecular attraction between water molecules at the surface; i.e., **lowering the surface tension** of the water.

tive against molds, but they are of limited value against certain viruses. Since they are emulsifying agents for lipids, quaternaries should be effective against viruses that have lipid-rich components, e.g., envelopes, and there are data in support of this view. Quaternaries are not sporicidal. They are much used for disinfection of skin, both for purposes of sanitization in restaurants and hospitals and in some types of throat lozenges and mouthwashes (though here, their effectiveness has been questioned).

Microbicidal Gases

Attempts to cure and prevent disease with vapors and smokes date from the earliest human records, when "aromatics," smudges, perfumes, and the like were endowed by popular supposition with almost miraculous, and generally quite factitious, powers to drive away evil spirits and "noxious humors."

More objective studies in the last few decades have yielded several exceedingly valuable results in the form of definitely sporicidal, and therefore sterilizing, gases: formaldehyde, ethylene oxide, and β-propiolactone. Chlorine and sulfur dioxide, formerly used, are too destructive and poisonous for general purposes.

As sterilizing agents, gases have the advantages that destructive heat and aqueous solutions are not needed and that they can be applied in large volumes to disinfect all or part of a building. On the other hand they affect only exposed surfaces, except objects made of porous or permeable materials, and they must be applied under conditions of controlled temperature and humidity. Such sterilizing agents are often referred to as **sterilants.**

Formaldehyde. This very irritating gas (formula, HCHO) is generated by heating paraformaldehyde or concentrated solutions of formaldehyde. Unless very carefully applied at raised temperatures, formaldehyde tends to polymerize as paraformaldehyde or **paraldehyde** in a thin white film over all surfaces, and hence its use is limited to places not containing valuable furniture or equipment. Formaldehyde is a strong reducing agent and inactivates essential enzymes and other organic components of the cell. Its action is to some extent microbiostatic and reversible with sulfite ions. In some solutions it is definitely sporicidal and therefore sterilizing but, like mercuric chloride, it tends to form protective coatings of coagulum around microorganisms if extraneous organic matter is present. For best results a relative humidity of about 70 per cent and a temperature of about 22C are requisite. Formaldehyde vapor has low penetrability into such materials as blankets. It is much less used now than formerly.

Ethylene Oxide. This substance (formula, C_2H_4O) is liquid at temperatures below 10.8C (51.4F), its boiling point. It is relatively inexpensive and is handled in metal tubes or bottles like other compressed gases. In the pure state it is very toxic, irritating, and explosive. For use it is mixed with other substances: ethylene oxide, 10 per cent, and carbon dioxide, 90 per cent (sold as Carboxide); ethylene oxide,

20 per cent, and carbon dioxide, 80 per cent (sold as Oxyfume); ethylene oxide, 11 per cent, and halogenated petroleum, 89 per cent (sold as Cryoxcide). Each preparation has special advantages in certain situations.

The desired mixture is generally applied in a modified or specially built autoclave (Fig. 22–9), or other closed container where humidity, temperature, and pressure may be controlled. The concentration of the gas is generally expressed in milligrams of pure ethylene oxide per liter of space. A typical use situation is 500

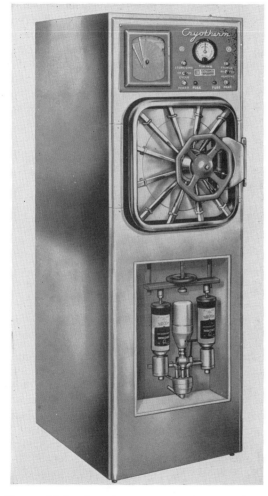

Figure 22–9. Autoclave arranged for sterilization by sporicidal ethylene oxide gas (Cryoxcide). Below the steel door are seen the containers of gas and a water inlet so arranged that the vapors can be mixed and humidified (much as gasoline, air and water vapor are mixed in a modern automobile carburetor) before admission of the sterilizing vapor mixture to the chamber of the autoclave. Above the door are the dials controlling sterilizing temperature and time, vacuum pump, gas and humidity inlet, and a maximum and minimum recording thermometer. Operation is almost completely automatic. (Courtesy of American Sterilizer Co., Erie, Pa.)

mg of gas per liter at about 58C (130F) and a relative humidity of 40 per cent for four hours. If one factor is varied the others must be varied accordingly. For example, if the concentration is doubled, the time may be halved. At a relative humidity of 30 per cent the action of the gas is about 10 times as fast as at 95 per cent relative humidity.

Ethylene oxide is a very effective general microbicide and kills bacterial endospores. It is therefore a sterilizing agent. It acts by alkylation or introduction of an organic group in place of a hydrogen atom, in various compounds in protoplasm. It does not injure even delicate organic materials that are injured by heat and aqueous solutions. It penetrates well, does not condense on objects, and soon dissipates from all materials.

It can also be used in the liquid state in 1 per cent concentration to sterilize unstable organic fluids, such as bacteriological culture media, by mixing fluid ethylene oxide with them at 10C. After a few hours the ethylene oxide is driven off by a short incubation at 37C.

β-Propiolactone (BPL). In pure form at 20C this is a liquid with a sweetish but very irritating odor (formula, $C_3H_4O_2$). Unlike ethylene oxide, it is not explosive. It may be stored at 4C, but is more stable in sealed glass containers at −25C. Its aqueous solutions are very unstable. It kills most microorganisms, including bacterial spores, and is therefore a sterilizing agent. It probably acts by alkylation, as does ethylene oxide. In fluid form it has been used to sterilize vaccines, tissues used for grafting and other delicate biological materials that could not be sterilized in any other way. Like ethylene oxide, it must be used in closed chambers where temperature and humidity can be accurately controlled. A relative humidity of 70 to 80 per cent and a temperature of about 25C are optimal, with a final concentration of β-propiolactone of about 2 to 4 mg per liter of air and a sterilizing time of two to three hours (Fig. 22–10).

When properly applied, β-propiolactone is much more active than either formaldehyde or ethylene oxide. It is noncorrosive, and does not condense on surfaces as does formaldehyde. However, β-propiolactone lacks the penetrating power of ethylene oxide and is more like formaldehyde in this respect. It also requires higher relative humidities. It is generally recommended to replace formaldehyde for use in sterilizing rooms.

Toxic and Destructive Properties. Experience shows that prolonged and repeated ex-

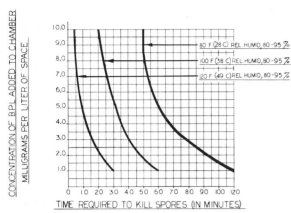

Figure 22–10. Relation between concentration of β-pro-piolactone and time required to kill spores of *Bacillus globigii* on filter paper strips. Curves are shown for three different temperatures at a relative humidity of 80 to 95 per cent. (Courtesy of Wilmot Castle Co., Rochester, N.Y.)

posure to any chemically active disinfectant, or any sterilant, may cause tissue irritation or destruction, and may damage articles containing certain unstable organic and inorganic substances such as magnesium (instrument housings) and some plastics. Users must always be alert to this possibility.

They must also be alert to toxic properties, especially dangerous potentialities of newly developed disinfectants. For example, with continued study of β-propiolactone, data have appeared suggesting that it may be carcinogenic as well as irritating. The toxic or carcinogenic properties of a disinfectant do not necessarily militate against its use; otherwise, such substances such as magnesium (instrument hous-chloride, and phenol, and radiations such as ultraviolet, x-rays, and gamma rays would be banned. However, proper precautions must be taken in handling such agents to prevent ingestion or absorption through the skin or undue exposure or contact.

Radiations

Ultraviolet Light. As described in Chapters 20, 45, and 46, ultraviolet light is actively microbicidal when applied properly. It has very slight powers of penetration. Exposure must be direct, intense, and sufficiently prolonged. Eyes must be protected, as they can be severely damaged. It is sometimes used in hospitals, especially in operating rooms, reportedly with good results, to kill microorganisms on instrument tables, in the air and on floors and walls. Ultra-

violet light is used on a large scale to prevent growth of molds in meatpacking houses, bakeries, and other commercial and industrial establishments.

Another application is in the treatment of biological fluids, such as blood plasma and vaccines, to kill contaminating viruses. The rays are directed on fluid that is made to flow in a micro-thin film that ultraviolet rays can penetrate. The wavelength commonly used in disinfectant applications of ultraviolet light is 254 nm, because it is most practical. Longer waves, such as those that penetrate through dust, smoke, and clouds in sunlight (around 350 nm), are proportionally less effective. Shorter waves, such as 185 nm, are more rapidly lethal but are difficult to control and apply. Ultraviolet light damages nucleic acids especially, notably by producing thymine dimers in DNA, which suppress DNA replication and can cause mutations. Ultraviolet light kills spores less readily than vegetative cells.

THYMINE DIMER

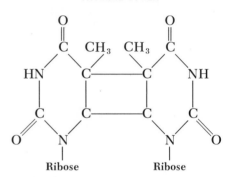

Cathode Rays. A recent development in radiation sterilization, cathode rays (electron beams) are expensive to apply, but are being developed on a commercial scale. They can be made to penetrate thin metal, paper or plastic sheets and have been used successfully to sterilize sealed packages for surgical use, packaged meats, vegetables, and other foods. (See also Chapter 48.)

22.4
INHIBITION OF MICROORGANISMS (MICROBIOSTASIS)

As previously indicated, microbiostasis denotes a condition of microorganisms in which, although they are alive (or **viable**), they do not multiply. Their enzymic or other vital functions are in a static condition.

In one sense the dormant ascospores of yeasts, conidia, or molds, the cysts of some protozoans, and the endospores of bacteria represent microbiostasis as a part of their normal life cycles. However, microbiostasis is usually thought of as affecting vegetative cells and as being produced artificially. Microbiostasis is exceedingly important in all aspects of microbiology. There is even published discussion of preserving "important" people in "deep-freeze" ("homostasis"?).

Physical Methods. As mentioned elsewhere, there are several physical methods of inducing microbiostasis. Most of them are fundamentally methods of depriving the cell of liquid water: desiccation; immobilizing intracellular water by changing it into ice; drawing most of the water out of the cell by immersion in a fluid of high osmotic pressure; and combinations of these.

In addition, mild irradiations with ultraviolet light and x-rays are often microbiostatic instead of lethal. Their microbiostatic action is not clearly understood but it is not related to dehydration. Apparently, oxidations are among the most important reactions that occur (Chapter 20).

Reversibility. The action of microbiostatic agents is **characteristically reversible;** that is, the organisms can be reactivated if the inhibiting action has not been so prolonged that the cells have gradually died off. We have seen in foregoing chapters how the microbiostatic effect of irradiations may be reversed by visible light and certain chemicals, and how poisoning by mercuric chloride may be "cured" by sulfhydryl-containing compounds. Tubercle bacilli dried in a vacuum were found fully viable and virulent after 17 years; diphtheria bacilli and scarlet fever streptococci, after 25 years. Microbial spores (natural microbiostasis) will survive for decades. Viruses are commonly held unaltered for months or years frozen at −76C. Cattle breeders and others preserve sperm by freezing. In all cases, a good percentage of the inactive cells can be reactivated.

Chemical Microbiostasis

The distinction between **chemical microbiostasis** and disinfection is interesting but entirely arbitrary, a matter of semantics. The differentiation depends on what one means by killing "quickly" (i.e., disinfection) and what one may mean by killing "slowly" (i.e., microbiostasis). Even under the most effective conditions of microbiostasis, some of the weaker cells die fairly soon; presumably the others are not actually immortal and die eventually, even though only after days, weeks, or much longer. We may note three general types of chemical microbiostatic agents: certain aniline dyes; sulfonamide and similarly acting drugs; and antibiotics.

Aniline Dyes. We have already referred to selectively bacteriostatic culture media containing dyes (Chapter 10). A representative dye used in such media is crystal violet, commonly used in the Gram stain. Since it is a basic dye, it has a marked affinity for acidic constituents of the bacterial cell, notably nucleic acids. With these the dye forms unstable salts that (in the absence of iodine) soon dissociate, and are either disposed of by the cell or can be removed by washing the cell with a harmless solvent such as water. Cells stained in this way with crystal violet, especially gram-positive cells, are in a static condition. They will not grow in media containing such dyes. They can be quickly reactivated by washing.

On the other hand, gram-negative cells in contact with crystal violet are not inhibited in the same way and generally grow well on media containing dyes like crystal violet, basic fuchsin, or methylene blue–eosin. Such dye-containing media are therefore commonly used for **selective** bacteriostasis, especially to isolate gram-negative bacteria from materials contaminated with gram-positive bacteria.

Sulfonamide Drugs. The sulfonamides are excellent examples of bacteriostatic agents. They were discovered by Domagk (Nobel Prize winner) about 1936. Sulfonamide drugs are derived from the same basic material (coal tar) as aniline dyes. All are derivatives of sulfanilamide (Fig. 22–11). Like basic dyes, they are effective mainly against gram-positive bacteria (with a few notable exceptions: *Neisseria, Shigella, Proteus, Klebsiella,* and *Escherichia*).

Sulfonamide Drugs and Metabolite Antagonism. The true sulfonamide drugs are those that readily release the sulfanilamide residue in the living cell. The structure of sulfanilamide so closely resembles that of *p*-aminobenzoic acid (PABA) that the drug can displace PABA from the absolutely essential vitamin folic acid (Fig. 22–12). This important vitamin is necessary to a coenzyme (CoA; Chapter 8) involved in synthesis of amino acids and hence of proteins. Such absolutely essential specific substances as PABA, necessary to the completion of a particular functional part of the cell, but not used as sources of energy or bulk, are some-

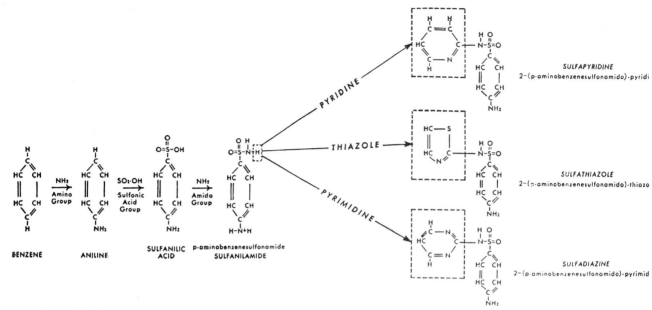

Figure 22–11. How the sulfonamide drugs are derived from benzene. First, an amino group is added, yielding aniline (a base from which dyes are synthesized). The addition of a sulfonic acid group to aniline yields sulfanilic acid. The addition of another amino group yields sulfanilamide. The addition to this, in place of one hydrogen atom, of various other chemical groups yields other sulfanilamide drugs, three of which are shown in the diagram.

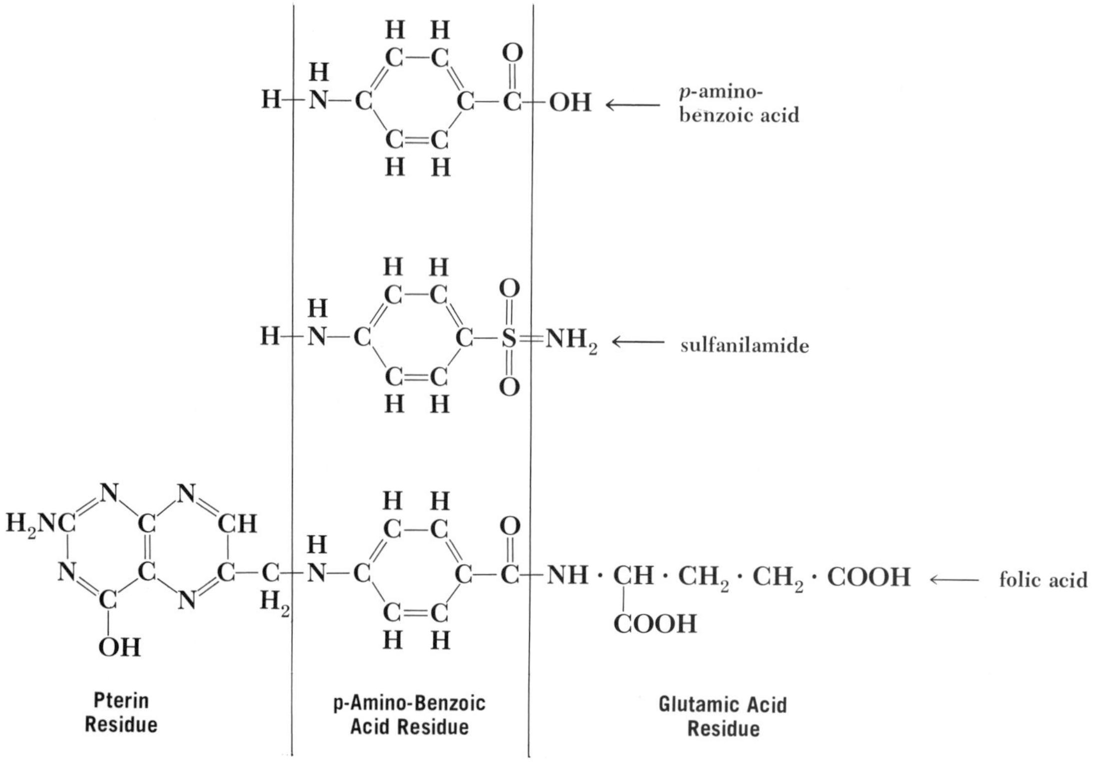

Figure 22–12. Structural resemblance between molecules of *p*-aminobenzoic acid (PABA) and sulfanilamide (*center*), and the antagonism by sulfanilamide of PABA in the folic acid molecule (*below*).

times called **essential metabolites** or **micronutrients.** In the present discussion PABA is considered to be an essential metabolite.

When we compare the diagrams of the molecules of sulfanilamide and PABA (Fig. 22-12), the resemblance between the major parts of the two is made clear. As shown in the diagram, the folic acid molecule consists of three functional groups: pterin, PABA, and glutamic acid. If any form of a sulfonamide drug is present in a sufficient amount, part of it goes into the folic acid complex in place of PABA; i.e., it is in competition with PABA for a position in the folic acid molecule. The sulfonamide is therefore often called a **competitive inhibitor.** Because the structure of the sulfonamide is analogous to that of PABA, the drug is also called a **metabolite analogue,** or **antimetabolite.** But no form of sulfonamide is capable of carrying on the functions of PABA. The sulfonamide is, in effect, a molecular impostor. The enzymic functions of the microbial cell depending on folic acid are therefore promptly stopped by sulfonamides. Since all of the functions of the cell are closely interdependent, the whole cell mechanism stops also. The cell is then in a state of microbiostasis. The production of microbiostasis in this way by chemicals or drugs is the basis of modern **chemotherapy** (see §22.6).

There are many substances besides sulfonamides (e.g., amino acids, certain ions [see ion antagonism], organic acids and esters, purines, pyrimidines, and sugar derivatives) that, because of similarity of structure, can act as metabolite analogues. They antagonize or displace ions or molecules of vital cell structures such as proteins, enzymes, and coenzymes or nucleic acids. In the nucleic acids nucleotide analogues can produce mutations (Chapt. 18).

Reversal of Metabolite Antagonism. It is important to note that, like most other microbiostatic reactions, the enzyme blockage produced by sulfonamide is not fatal if not too pro-

longed and may be reversed in several ways— for example, by adding enough PABA to displace and exclude the sulfonamide; by supplying new folic acid ready-made in the medium; by supplying, in the culture medium, the product that would have been formed by the blocked enzyme (or any subsequent enzyme in the chain); or by supplying the finished product of the entire enzymic series, in this case certain amino acids.

22.5
DRUG RESISTANCE

The cell itself may be capable of bypassing or overcoming the effects of metabolite antagonists. For example, let us suppose that in a parallel series of enzyme reactions (series I and II, Fig. 22-13) enzyme D is blocked by some microbiostatic drug. However, let us assume that some of the substance usually made by the blocked enzyme D is produced as a side reaction or intermediate stage in the function of another enzyme, H. This product becomes available to E and thus the blockage at D is bypassed. Such a cell is more or less **drug-resistant,** depending on how much of the product needed by E is available from H.

Such enzymic rearrangements and bypassings (**alternate metabolic pathways**) are quite common. From the clinical standpoint, sulfonamide resistance of an infecting organism means that the patient must receive some other drug.

Resistant Mutants. In Chapter 18 it was shown that among any large number of microbial cells in a given pure culture, mutant forms may be found. Among the large numbers of bacterial cells of a given species infecting a patient there are, therefore, likely to be mutants with enzymic arrangements that will be unaffected by a microbiostatic drug such as sulfonamide. In the presence of the drug these

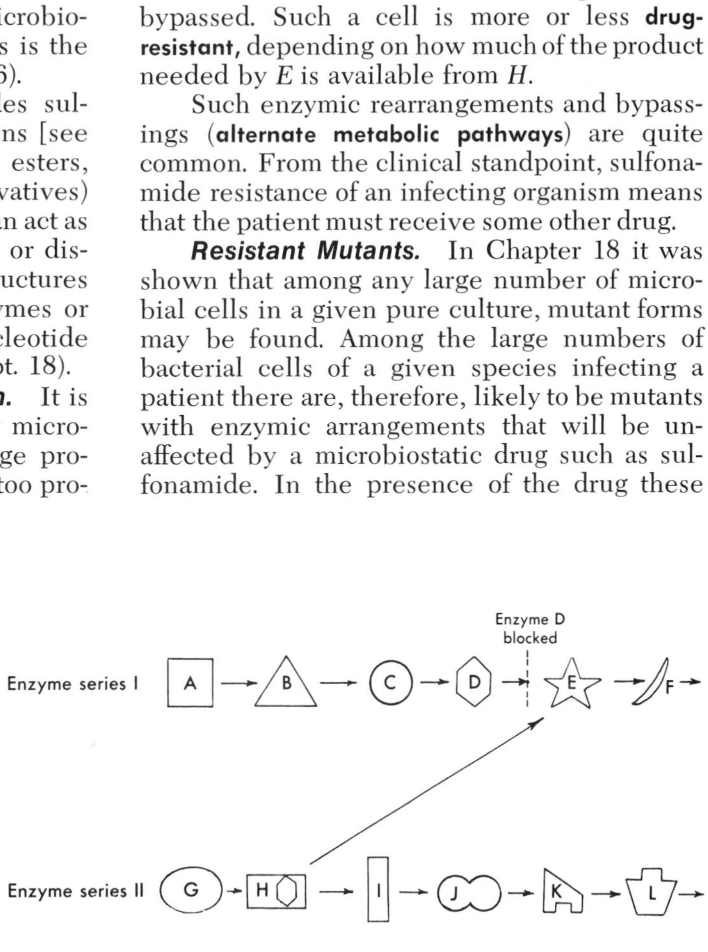

Figure 22-13. Two parallel series of enzymes carrying on different synthetic functions. If enzyme D in series I becomes blocked, as by an antimetabolite, the entire series of reactions in series I stops. However, enzyme H in series II may, because of similar structure of one of its parts, produce a small amount of the substance produced by D and required by E (or the substance required by E may be derived from some other source in the organism). In either case series I continues to function, though possibly at a reduced rate. The different geometrical figures are intended merely to indicate different specificities of the different enzymes.

mutants will thrive, while all the other microbial cells are killed off. We thus have a new population of sulfonamide-resistant microorganisms, a condition which is dangerous to the patient and to those who contract the resistant organism from him.

In some microorganisms that are ordinarily susceptible to penicillin, notably *Staphylococcus*, penicillin resistance can depend on production of an extracellular enzyme (**penicillinase**) which destroys the drug. Mutants that produce such enzymes are able to grow unhampered in the presence of the drug, to the eventual exclusion of the non-enzyme-producing sensitive mutants.

Induced Drug Resistance. In addition to the occurrence of drug-destroying enzymes that appear as a result of mutation, such enzymes may appear also as **induced** or **adaptive enzymes** resulting from contact of the cell with the drug. The mechanisms of drug resistance, not only in microorganisms but in insect pests (e.g., resistance of mosquitoes and flies to insecticides), is a major problem in chemotherapy and pest control, and an excellent field for profitable research.

Drug Dependence. The appearance of strains that are not merely resistant to drugs but wholly **dependent** on them seems as strange as the emergence of resistant strains. The two are probably not related except that they are two manifestations of the phenomenon of mutation. Drug-dependent mutants obviously could not survive and develop significantly except in an environment containing the drug on which they are dependent. It is only because we have tried to cultivate microorganisms in contact with antimicrobial drugs that these curious mutant forms have been revealed.

22.6 CHEMOTHERAPY

Chemotherapy, broadly speaking, is therapy by means of chemicals. The term is generally restricted, however, to the use of drugs that act by metabolite antagonism or enzyme blockage. Ideally, chemotherapeutic drugs should act on the parasite in concentrations that have no effect (or at least no serious "side" effects) on the host. Some chemotherapeutic drugs are toxic to the patient as well as to the parasite and so must be used with great care and under medical supervision. Such drugs may be of great value when the microorganisms are resistant to the drugs normally used. Often, toxic manifestations may be ameliorated by other drugs or careful adjustment of dosage.

Chemotherapy and Cancer. A very interesting aspect of chemotherapy concerns the possibility of selectively poisoning the cells of **neoplasms** (Gr. *neo* = new; *plasm* = form) (e.g., cancers). Neoplasms are thought to have some metabolic mechanisms different from those of normal cells. The obvious possibility presents itself of finding some metabolite antagonist that will poison an enzymic mechanism peculiar to the neoplasm cell and not present in the normal cell.

CHAPTER 22
SUPPLEMENTARY READING

Ehrlich, R.: Application of membrane filters, *in* Umbreit, W. W. (Ed.): Advances in Applied Microbiology, Vol. 2. Academic Press, Inc., New York. 1960.

Hedgecock, L. W.: Antimicrobial Agents. Lea & Febiger, Philadelphia. 1967.

Lawrence, C. A., and Block, S. S.: Disinfection, Sterilization and Preservation. Lea & Febiger, Philadelphia. 1968.

Perkins, J. J.: Principles and Methods of Sterilization in Health Sciences. 2nd ed. Charles C Thomas, Springfield, Ill. 1969.

Runkle, R. S., and Phillips, G. B. (Eds.): Microbial Contamination Control Facilities. Van Nostrand Reinhold Company, New York. 1969.

Witkin, E. M.: Ultraviolet induced mutation and DNA repair. Ann. Rev. Microbiol., 23:487, 1969.

CHAPTER 23 · ANTIBIOTICS

The term antibiotic was introduced by S. A. Waksman (Nobel Prize winner) in 1945, as the name of a class of substances of **biological** origin that are antagonistic to microorganisms. Strictly speaking, antibiotic means "against life." Since antibiotics, as commonly used, are inimical only to microorganisms they might logically be called "antimicrobiotics." Although substances of biological origin with antimicrobial properties were known before 1900, antibiotics as now defined first came into worldwide prominence about 1940. There followed a widespread search for new antibiotics. It was found that antibiotic substances are produced by a wide variety of living organisms, ranging from man to microbes. The most important and widely used antibiotics at present are derived from microorganisms. A few come from common species of molds (e.g., *Penicillium*), but most of them are obtained from various species of the mold-like bacteria in the genus *Streptomyces*. Very few useful antibiotics have been isolated from "true" bacteria, with the exception of those from several species of *Bacillus*. Table 23–1 shows some of the antibiotics commonly used to treat infections in man and animals. Antibiotics vary considerably in their chemistry, mode of action, and spectrum of activity, and while it would be impossible to cover this large subject area in any depth in a single chapter, a great deal can be learned by considering a few antibiotics produced by each of the three taxonomic groups listed above.

23.1
PENICILLIN

Since the earliest studies with pure cultures, microbiologists had known that when cer-tain airborne, saprophytic microorganisms contaminated their cultures they suppressed the growth of the desired species. The phenomenon was so commonplace and attention was so fixed on other problems that the antagonistic action of such contaminants was pushed aside as merely an inevitable nuisance. The true significance of the "nuisance" was largely overlooked until 1929, when Alexander Fleming (afterwards Sir Alexander Fleming, Nobel Prize winner with Florey and Chain) realized it, acted upon the basis of his idea, and discovered, named, and described penicillin.

The antagonistic airborne saprophytic organism that first attracted Fleming's attention was a colony of the common mold, *Penicillium notatum*. When it contaminated one of his cultures in a Petri plate (Fig. 23–1), Fleming experimented with its antimicrobial action. He passed broth cultures of the mold through filters, removed the mold filaments and was thus able to study the activity of the soluble growth products alone as they occurred in the clear broth. He found that the clarified, sterile broth contained a highly potent antimicrobial principle, the activity of which was readily demonstrated in contact with sensitive microorganisms, especially gram-positive bacteria. He called this principle **penicillin.** In a much purified and refined form it is still one of the forms of penicillin used therapeutically today.

Fleming realized the practical possibility of his discovery and made use of it in his laboratory to eliminate gram-positive contaminants from cultures but was not, at that time, in a position to develop it more fully. For some years the value of such antibiotic phenomena remained relatively unknown.

TABLE 23–1. PARTIAL LIST OF COMMON ANTIBIOTICS*

| Antibiotics | | | |
Common Name	Trade Name	Source	Major Site of Action
			Antimicrobial Spectrum†
Amphotericin B	Fungizone	*Streptomyces nodosus*	*Candida* sp.	Cell membrane
Bacitracin		*Bacillus subtilis*	Like penicillin	Cell wall
Carbomycin‡	Magnamycin	*Streptomyces halstedii*	Like erythromycin	Protein synthesis
Chloramphenicol	Chloromycetin	*Streptomyces venezuelae*	Broad-spectrum	Protein synthesis
Chlortetracycline	Aureomycin	*Streptomyces aureofaciens*	Broad-spectrum	Protein synthesis
Colistin (Polymyxin E)	Coly-mycin	*Bacillus colistinus*	Gram-negative bacteria	Cell membrane
Cycloheximide	Actidione	*Streptomyces griseus*	Saprophytic fungi	Protein synthesis
Cycloserine	Seromycin	*Streptomyces orchidaceus*	Resistant staphylococci; *Mycobacterium*	Cell wall
Demethylchlortetracycline	Declomycin	Synthetic; also from *Streptomyces aureofaciens*	Broad-spectrum	Protein synthesis
Dihydrostreptomycin		*Streptomyces*; also some species of *Streptomyces*	Like streptomycin	Protein synthesis
Erythromycin	Ilotycin, Erythrocin	*Streptomyces erythreus*	Broad-spectrum (not Enterobacteriaceae)	Protein synthesis
Griseofulvin	Grifulvin	*Streptomyces griseus*	Pathogenic fungi	DNA function
Kanamycin	Kantrex	*Streptomyces kanamyceticus*	Broad-spectrum	Protein synthesis
Lincomycin	Lincocin	*Streptomyces lincolnensis*	Staphylococci	Protein synthesis
Neomycin B‡	Flavomycin	*Streptomyces fradiae*	Mycobacteria	Protein synthesis
Nystatin		*Streptomyces noursei*	Pathogenic fungi	Cell membrane
Oleandomycin	Matromycin	*Streptomyces antibioticus*	Broad-spectrum	Protein synthesis
Oligomycin		*Streptomyces diastatochromogenes*	Fungi of plants	Energy metabolism
Oxytetracycline	Terramycin	*Streptomyces rimosus*	Broad-spectrum	Protein synthesis
Paromomycin	Humatin	*Streptomyces rimosus*	*E. histolytica*	Protein synthesis
Penicillin‡		*Penicillium notatum*	Gram-positive bacteria; *Treponema*, *Neisseria*	Cell wall
Polymyxin B‡		*Bacillus polymyxa*	Gram-positive bacteria	Cell membrane
Streptomycin		*Streptomyces griseus*	*Mycobacterium tuberculosis*; Gram-negative bacteria	Protein synthesis
Tetracycline	Achromycin	Chlortetracycline	Broad-spectrum	Protein synthesis
Vancomycin	Vancocin	*Streptomyces orientalis*	Resistant staphylococci	Cell wall

*Several not listed here are valuable commercially, agriculturally, and horticulturally.

†Not necessarily the only activity.

‡Several of these antibiotics are in reality mixtures consisting of related compounds such as the penicillins, polymyxin A, B, C, D, carbomycin A and B, and so on.

Figure 23–1. Sir Alexander Fleming, penicillin discoverer and Nobel Prize winner, points with his inoculating needle to a giant colony of *Penicillium notatum,* the organism that produces penicillin, on agar in a Petri dish. (Courtesy of Chas. Pfizer & Co., Inc., Brooklyn, N.Y.)

Interest was reawakened when, realizing that many nonsporeforming pathogenic organisms in the soil are rapidly destroyed, presumably by antimicrobial substances produced by saprophytic microorganisms, Waksman suggested that the search for antibiotic-producing microorganisms be carried to the soil. In 1939 Dubos found, in bogs, organisms (*Bacillus brevis*) that produced two valuable antibacterial substances (**gramicidin** and **tyrocidin**). These antibiotics, while of enormous value therapeutically for surface application and of great interest scientifically, were too toxic for internal use. Dubos's discoveries stimulated new interest in Fleming's observations on *Penicillium notatum.* Work on penicillin was begun in England on a large scale by Florey, Chain, Abraham, and others (the Oxford Group) in 1940, just after the outbreak of World War II. Because of the exigencies of the war, work on mass production methods was transferred largely to the United States.

Since that time penicillin has played a role in the greatest war in human history; in the most far-reaching piece of cooperative research ever organized prior to atomic research; in a great industrial development; in the most complete control over a variety of diseases ever achieved by man in a short time; and, because of its im-

pact on venereal diseases, in a tremendous new social and moral trend, the possibilities of which are still only partly realized. The military importance of penicillin in World War II, then just beginning, can hardly be estimated. The development of American facilities for mass production of penicillin, not then available in Europe, was soon of great benefit to all concerned.

Production. Production of penicillin and other antibiotics is now a billion-dollar industry in the United States alone. In 1943 an ounce of penicillin cost around twenty thousand dollars. Now it costs about three dollars because of mass production methods.

The manufacturing process involves cultivation of *Penicillium notatum* under conditions most favorable to growth and penicillin production (i.e., at about 24C and at a pH between 7 and 8). As the mold is strictly aerobic, exposure to air is essential.

Commercially, a process using submerged growth in closed but vigorously aerated tanks holding thousands of gallons of medium (Fig. 23–2) is most widely used today.

In all methods, suitable medium is first inoculated with suspensions of the conidia of *Penicillium notatum* or the closely related *P. chrysogenum,* which also produces penicillin. The entire procedure is a **pure-culture** process and necessitates costly aseptic technique at all stages.

During the 7 to 14 days of incubation in the large vat of aerated fluid medium the mold excretes at least three waste substances that are of importance: the yellow pigment **chrysogenin,** which must be removed by adsorption with charcoal or similar means; **penicillin;** and **notatin,** or **penicillin B.** The last occurs especially if the acidity of the medium is too great. It is removed during the final purification process. Other substances are also produced, some of them related to penicillin. In general, if a microorganism produces one antibiotic it produces others of the same nature, some valuable, others of little significance.

Medium. Different formulas are doubtless in use by various manufacturers, but basically they are similar and call for the addition of several essential minerals and salts to a liquid rich in easily digestible carbohydrate and protein, e.g., corn-steep liquor* (see Chapter 49). After

---

*Corn-steep liquor is a by-product of the distilling industry, being the water used to soak (steep) corn prior to fermentation. It contains various growth factors (vitamins), proteins, and carbohydrates, and is one of the best sources of nutrients for antibiotic production.

Figure 23–2. Like a steel igloo, the top of a 9,000-gallon aerated-growth tank rises above the floor level of the penicillin plant at the Lederle Laboratories. Like an iceberg, nine-tenths of the huge fermentation tank is out of sight below the floor. The small tank at the right contains chemicals to prevent excessive foaming of the liquid containing the mold. (Courtesy of Peter Winkler, Lederle Laboratories, a division of American Cyanamid Co., Pearl River, N.Y.)

quickly destroyed by acid, so that in the unmodified form it is ineffective if taken orally. In the crystalline form it is quite stable, even for several days at 100C or for months refrigerated in the dark; however, in aqueous solution it is very unstable and quickly decomposes.

Penicillin acts mainly by inhibiting the enzymes concerned with the cross-linkage of peptide side chains of the polymers making up the peptidoglycan of the cell wall. Since synthesis of cell wall is carried on only by young, actively multiplying cells, the antibacterial action is manifested only during active growth. Because septum formation appears particularly vulnerable to penicillin, rod-shaped bacteria tend to form aseptate filamentous forms like mycelia in the presence of nonlethal concentrations of penicillin.

The clinical uses of penicillin are beyond the scope of this chapter, but some of the common infectious organisms against which natural penicillins are more or less effective are listed in Table 23–2. Note that most of the susceptible species are gram-positive. The reason generally given for this is that the cell wall of many gram-positive bacteria is made up largely of peptidoglycan, the formation of which is inhibited by penicillin. Gram-negative bacteria, on the other hand, have a more complex cell wall with a smaller percentage of peptidoglycan and thus are not as susceptible to the effects of penicillin at the concentrations generally employed. The above is not a hard and fast rule, however, since several types of gram-negative bacteria are known to be very sensitive to penicillin, e.g., *Treponema pallidum*, the cause of syphilis, and *Neisseria gonorrhoeae*, the cause of gonorrhea. Several of the biosynthesized and semisynthetic penicillins (see under following section) have a broader spectrum of activity than do the natural penicillins and are more effective on a variety of gram-negative bacteria.

Among the most valuable properties of penicillin is its relatively low toxicity for man and animals, even in large doses, coupled with the fact that its antibacterial potency against susceptible microorganisms is such that a concentration of 0.000001 gm per ml will exert marked bactericidal effects. It may be administered subcutaneously, intravenously or locally. A particular disadvantage of penicillin is that it is a potent allergen in some persons and can cause fatal allergic reactions (Chapt. 29).

Chemistry of Penicillins. The naturally occurring penicillins are compounds of the strong, monobasic acid, **penicillanic acid**. The penicillins usually occur as salts of sodium and

about seven days, growth is complete, pH rises to 8.0 or above, and penicillin production ceases.

When no more penicillin is being formed, the masses of mold growth are separated from the culture fluid by centrifugation and filtration. The complex process of extracting the penicillin from the clear fluid then begins. The method involves various extractions with organic solvents and recrystallizations. These are chemical engineering problems and do not concern us at present.

Properties and Uses. The material extracted from the culture medium is in reality a mixture of six chemically related penicillins called X, G, two forms of F, dihydro F and K. (These are discussed on page 324.) The most important is penicillin G (benzyl penicillin), which is usually meant when the term penicillin is used. Penicillin is readily soluble in water and alcohol, but the latter inactivates it. It is

TABLE 23–2. ANTIMICROBIAL DRUGS FREQUENTLY CHOSEN TO TREAT HUMAN INFECTIONS

Organism	Penicillins	Tetracyclines	Erythromycin	Cephalosporin	Chloramphenicol	Others
GRAM-POSITIVE BACTERIA						
Streptococcus pyogenes A (Rheumatic fever and others)	Pref.*		Alt.*	Alt.		
Streptococcus (Diplococcus) pneumoniae (Pneumonia)	Pref.		Alt.	Alt.		Lincomycin and others – Alt.
Staphylococcus aureus (Abscesses)	Pref.		Alt.			
Bacillus anthracis (Anthrax)	Pref.	Alt.	Alt.			
Bacillus perfringens (Gas gangrene)	Pref.		Alt.			
Clostridium tetani (Tetanus)	Pref.	Alt.		Alt.		
Corynebacterium diphtheriae (Diphtheria)	Alt.		Pref.	Alt.		
GRAM-NEGATIVE BACTERIA						
Neisseria gonorrhoeae (Gonorrhea)	Pref.†	Alt.	Alt.			
Salmonella sp. (Salmonellosis)	Alt.					Chloramphenicol – Pref.
Shigella sp. (Shigellosis)	Pref.	Alt.				
Escherichia coli (Gastroenteritis)	Alt.	Alt.				Kanamycin – Pref. if enteritis
Pseudomonas aeruginosa (Various infections)						Polymyxin – Pref.
Brucella sp. (Brucellosis)		Pref.			Alt.	
Vibrio cholerae (Cholera)		Pref.			Alt.	
MISCELLANEOUS						
Treponema pallidum (Syphilis)	Pref.		Alt.			
Mycobacterium tuberculosis (Tuberculosis)						Isoniazid in combination with others – Pref.
Rickettsia sp. (Various diseases)		Pref.			Alt.	

*Pref. = Usual drug of choice; Alt. = Alternate antibiotics which are effective.
†Ampicillin is the penicillin usually chosen to treat infections caused by gram-negative bacteria.

potassium. The general chemical structure assigned to them is shown in Figure 23-3. They are unstable to heat and to strong acids and alkalies. Laboratory synthesis of penicillin was announced shortly after the formula was known. The substance artificially synthesized was penicillin G.

Biosynthesis of Penicillins. It has been found possible to induce *Penicillium* cultures to synthesize new forms of penicillin. The method consists of "feeding," to cultures of *Penicillium notatum,* synthetic substitutes for the normal radicals, or "R's," (i.e., the distinctive side chains) of the penicillin molecule. These substitutes, "prefabricated" molecules, are referred to as **precursors** of the new "R." In the presence of such abnormal precursors, the mold unwittingly conjugates them "as is" with penicillanic acid, thus forming new and more valuable **biosynthetic penicillins** (synthesized partly by man, partly by the mold). An important product is **phenoxymethyl penicillin (penicillin V)**,

produced by feeding the precursor phenoxyacetic acid to the mold. Penicillin V is fairly stable to gastric acidity and is less quickly excreted from the blood than other penicillins. It is widely used for oral administration.

Semisynthetic Penicillins. When it was discovered, about 1958, that 6-aminopenicillanic acid, the molecular "nucleus" of all penicillins, could be produced in commercial quantities, it became possible to attach various artificially synthesized acyl groups to 6-aminopenicillanic acid on a large scale. Hundreds of such semisynthetic penicillins have been made and tried. Some are extremely valuable—e.g., phenethycillin, which is quickly absorbed into the blood and is as effective as penicillin G; methicillin, cloxacillin, and nafcillin, which are not readily destroyed by gastric acidity and which are resistant to penicillinase (β-lactamase); and ampicillin, which is effective against a wider range of bacteria, both gram-positive and gram-negative, than other penicil-

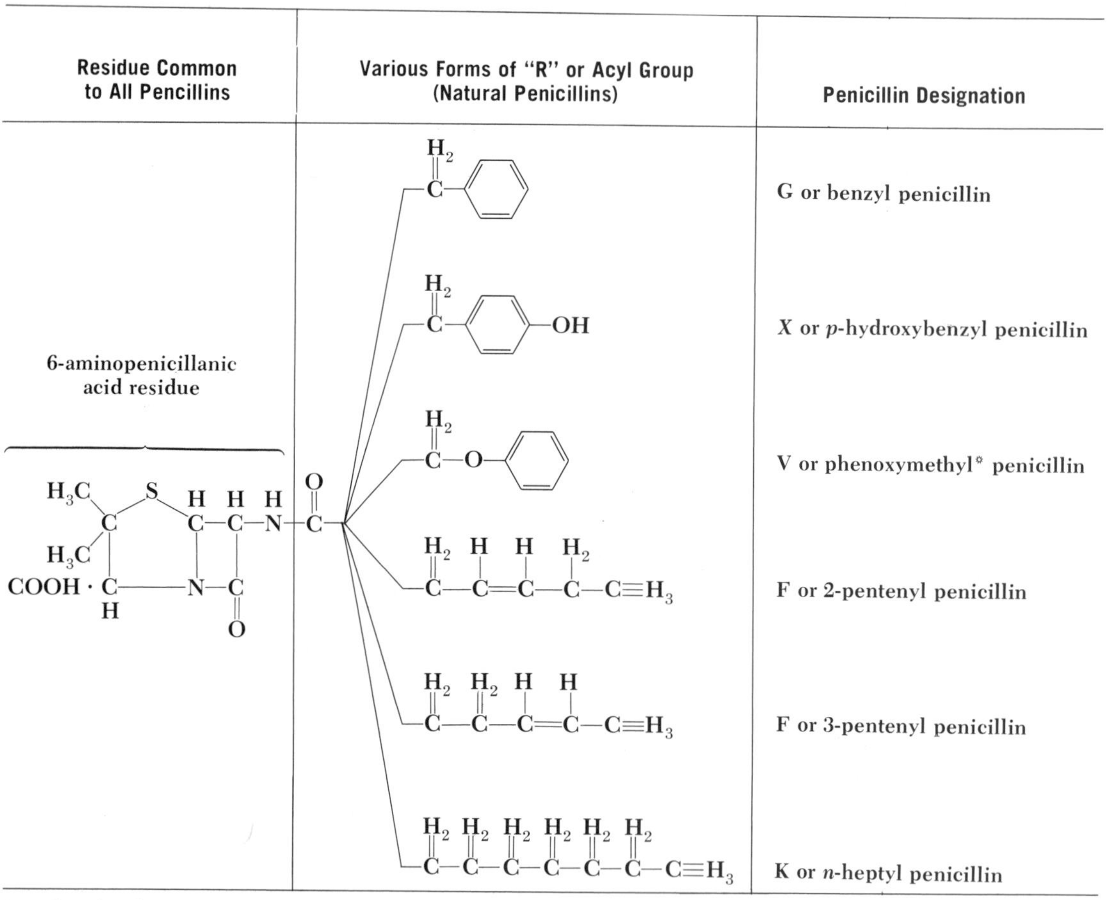

Residue Common to All Penicillins	Various Forms of "R" or Acyl Group (Natural Penicillins)	Penicillin Designation
6-aminopenicillanic acid residue	(structure)	*G* or benzyl penicillin
	(structure with OH)	*X* or *p*-hydroxybenzyl penicillin
	(structure with O-phenyl)	*V* or phenoxymethyl° penicillin
	(2-pentenyl structure)	*F* or 2-pentenyl penicillin
	(3-pentenyl structure)	*F* or 3-pentenyl penicillin
	(n-heptyl structure)	*K* or *n*-heptyl penicillin

°Produced only if phenoxyacetic acid is present in the medium as a precursor.

Figure 23-3. Various forms of penicillin. They differ only in the form of radical (R) in the general formula. The different forms of R are shown in the second column.

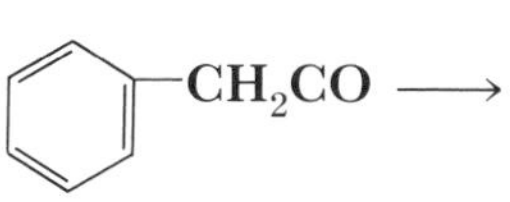

CH₂CO →

benzyl penicillin (G)

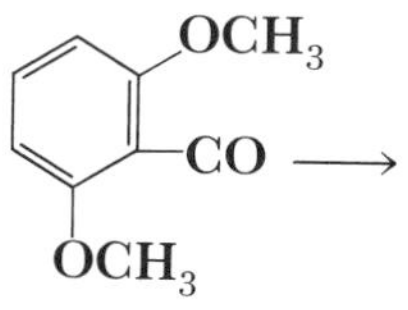

Methicillin
(dimethoxyphenyl penicillin)

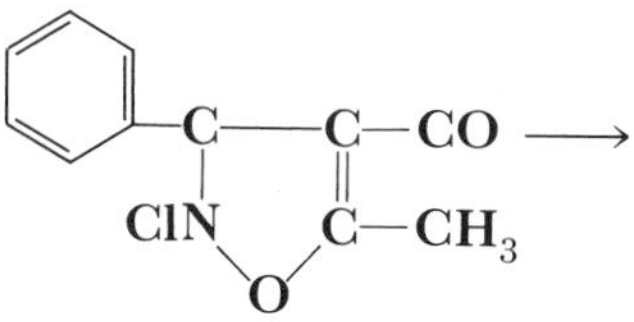

Cloxacillin
(5-methyl-3-phenyl-4-isoxyzolyl penicillin)

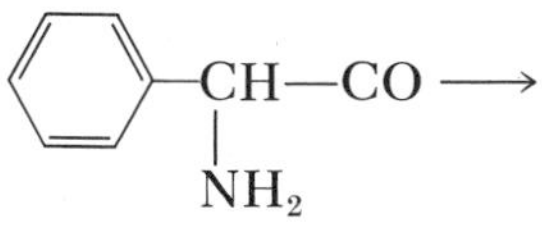

Ampicillin
(α-aminobenzyl penicillin)

Figure 23–4. R groups of some representative semisynthetic penicillins compared with that of a natural penicillin, benzyl penicillin (G).

lins, and is effective in oral administration as well. The acyl groups of some of these are shown in Figure 23–4 for comparison with natural penicillins.

Penicillinases. Several hydrolytic enzymes can attack and destroy penicillins. One, commonly called **penicillinase,** is produced by a number of dangerous pathogenic bacteria, notably *Staphylococcus* and several species of gram-negative intestinal bacteria. By virtue of this enzyme these and other bacteria are penicillin-resistant and so defeat therapeutic uses of natural penicillins in many cases. These bacterial penicillinases attack the β-lactam bond of penicillins G, X, and, to a lesser extent, V, and are often referred to as β-lactamases (Fig. 23–5).

The ability of some bacteria to produce penicillinase is constitutive, while in other strains it may be inducible. For example, certain strains may be sensitive to penicillin because they are not producing penicillinase. However, contact with nonlethal concentrations of penicillin may induce the formation of penicillinase and from then on, the bacteria can withstand higher concentrations of penicillin.

Another important point to remember is that a penicillinase-producing strain of bacteria will usually be killed by penicillin if there is only a small number of bacteria involved, i.e., if the concentration of penicillin is large compared to the amount of penicillinase being produced. On the other hand, dense cultures or clumps of bacteria, such as would be found in tissue infections, will generally be able to produce sufficient penicillinase to inactivate the penicillin present. From this, we can understand why it is considered good practice to achieve effective levels in blood and body fluids as quickly as possible when treating patients with penicillin and to keep the blood level high by repeated doses at specified intervals.

While the widespread, and sometimes careless, use of penicillin can be blamed in part for selecting penicillin-resistant strains of bacteria and furthering their spread among patients and hospital staff, there is ample evidence that penicillinase-producing strains existed before the advent of antibiotic therapy. You may wonder, then, what possible ecological or survival advantage penicillinase production might have had in earlier times. Recent work has suggested that penicillinase is simply a peptidase capable of acting on a variety of peptides, some of which could be expected to be present in a natural environment. Hence, penicillinase production may well have had survivor value before the era of antibiotic therapy.

Another kind of penicillinase is produced by many common genera of fungi, including *Penicillium, Aspergillus,* and *Mucor.* The penicillinases produced by these organisms differ from bacterial β-lactamase in that they attack the peptide bond between the 6-aminopenicillanic acid residue and the "R's". These enzymes are sometimes called **penicillin amidases.** They attack penicillins V and K most rapidly, G more slowly (Fig. 23–5).

The Cephalosporins. These are a group of antibiotics, the first described type of which, cephalosporin C, is produced by the mold,

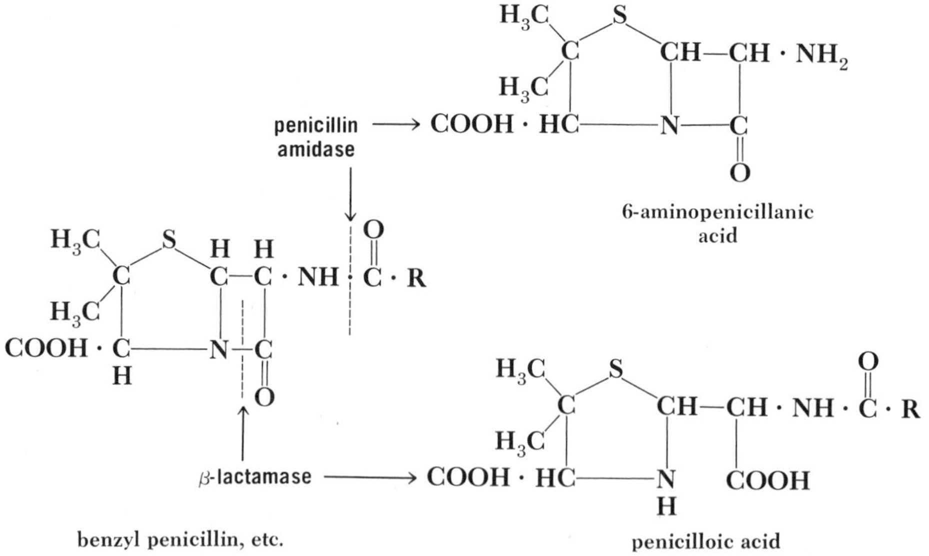

Figure 23–5. Action of two major types of penicillinase: penicillin amidase (penicillin acylase) and β-lactamase (commonly called penicillinase). The dotted lines and arrows show which bonds are broken.

Cephalosporium acremonium. Although a relatively ineffective antibiotic, cephalosporin C is of great interest because it has a formula very similar to that of penicillin (Fig. 23–6). Like penicillin, cephalosporin C lends itself to production of a large number of synthetic derivatives. Knowledge of these processes is based on the work of Woodward (Nobel Prize winner) and many other skillful chemists and biologists. The central residue of the cephalosporins is indicated in Figure 23–6. Various acyl groups may be attached to this residue, with the production of much more valuable cephalosporins than cephalosporin C; e.g., **cephalothin** (Fig. 23–6).

Penicillin and L Bodies. Numerous species of bacteria cultivated in the presence of penicillin in a hypertonic medium (i.e., one of increased osmotic pressure) grow but produce no cell wall; that is, they produce protoplasts or spheroplasts. In media of ordinary osmotic pressure these immediately burst because the strong, retaining cell wall is lacking. This is the basis of the lethal effect of penicillin; i.e., removal of the cell wall followed by osmotic rupture. In fact, only actively growing bacteria which are synthesizing cell walls are susceptible to penicillin. Animal cells (e.g., our own body cells), viruses, and *Mycoplasma* (Chapt. 42) are completely resistant to penicillin because they do not synthesize cell walls. In general, the gram-positive bacteria are most susceptible to penicillin because their cell walls consist almost entirely of peptidoglycan.

Cells which have been grown in the presence of penicillin in a hypertonic medium will generally revert to normal growth when the penicillin is removed, i.e., they will begin to form cell walls and assume a normal appearance. Occasionally, however, certain cells will appear to have lost the ability to regenerate cell walls and will remain in the wall-less (L-form) state. Multiplication continues so long as the medium remains hypertonic. There is considerable evidence to indicate that certain pathogens may revert to L-forms during antibiotic treatment and then persist intracellularly or in deep tissues of the patient as a latent, penicillin-resistant infection (also see Chapter 42). In view of this possibility, it becomes even more imperative, especially in cases of infection in which the causative agent is known, to initiate antibiotic therapy as early as possible and to use a dosage schedule that will surely maintain bactericidal levels if this is clinically feasible.

Various Strains of Penicillium Notatum. Fleming's strain of *P. notatum* for a time was thought to be unique. It was later found that some "wild" strains of this organism were better producers of penicillin and that other closely related organisms, especially *P. chrysogenum*, also produce penicillin. These findings led to a very extensive search for better strains of *P. notatum* and for penicillin-producing strains of other organisms. Many thousands of cultures were tested, and it was found that the penicillin-producing property is widely distributed in the *P. notatum-chrysogenum* group and that some strains are better than others, especially for the various methods of production (i.e., some were good in tanks, not so good on surfaces, and vice versa). Irradiation with ultraviolet light has furnished many thousands of mutant strains for testing, among them some very potent penicillin-producers.

The same principle of hunting for new and more effective mutants for a variety of purposes can be applied in most fields of industrial microbiology, using yeasts, molds, bacteria, and viruses.

Laboratory Uses of Penicillin. As mentioned, the value of penicillin as a selective bacteriostatic agent in laboratory bacteriology had been immediately recognized by Fleming and was adopted by others. It is a convenient aid to the isolation of penicillin-resistant organisms (such as the gram-negative whooping cough and influenza bacilli) from throat cultures, since it suppresses the growth of unwanted penicillin-sensitive (mainly gram-positive) organisms. It is also used for suppressing bacterial growth in animal cell cultures used in studies of viruses, since viruses and animal cells are wholly resistant to most antibiotics.

23.2
ANTIBIOTICS FROM STREPTOMYCES

Streptomyces (Chapt. 40) is a genus of bacteria that has some of the physiological and gross morphological characteristics of molds. *Streptomyces* species are, however, minute in size as compared with true molds and are entirely procaryotic in cellular structure.

A number of species of *Streptomyces* produce antibiotics of medical and commercial value. More are constantly being found. Some of the antibiotics produced by *Streptomyces* are said to be active against chlamydias (Chapt. 42), others against certain viral neoplasms and some against certain Protozoa. Some of the more common *Streptomyces*-derived antibiotics are listed in Table 23–1. Only several of these will be discussed because of space limitations, but this should suffice to introduce the student to the various types of antibiotics available and their different modes of action.

Streptomycin. The discovery of streptomycin dates from 1944, and followed naturally in a long series of researches by Waksman and his collaborators, Schatz and Bugie, into the numbers and kinds of antagonistic microorganisms in the soil. The source organism, *Streptomyces griseus*, produces mold-like conidia and aerial mycelia suggestive of some of the filamentous Ascomycetes (e.g., *Aspergillus*) but is one of the species listed under the bacterial order of Actinomycetales. The principles underlying the production, purification, assay, and standardization of streptomycin are analogous to those of penicillin, differing mainly in respect to cultural details and technical procedures appropriate to the organism and antibiotic involved.

The molecular formula of streptomycin is $C_{21}H_{39}O_{12}N_7$. It is structurally much more complex than penicillin. Numerous salts and derivatives have been prepared, some of them of great therapeutic value. Dihydrostreptomycin, in which two hydrogen atoms are added, was widely used at one time, but proved more toxic than the parent compound. Both streptomycin and dihydrostreptomycin are much more stable than penicillin, both in dry form and in solution.

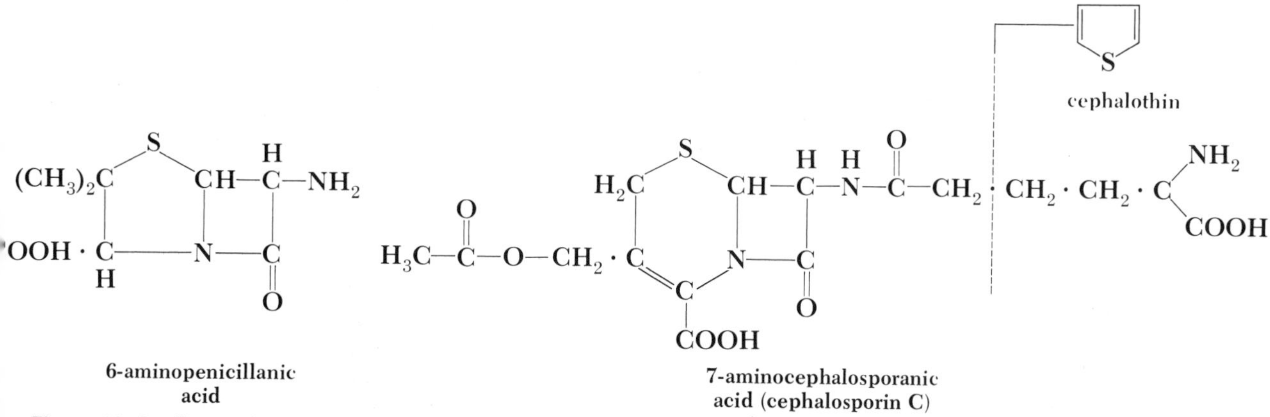

Figure 23–6. Comparison of formulas of 6-aminopenicillanic acid, 7-aminocephalosporanic acid, and cephalothin as a derivative of cephalosporin.

In the refrigerator the solutions retain potency for months; at 37C for about two weeks. Both drugs, in vitro, are bactericidal in commonly used concentrations.

Scope of Action. The therapeutic range of streptomycin is broader than that of penicillin; it is active against a variety of gram-negative and several gram-positive organisms, some of which are listed in Table 23–3. It has been used in tuberculosis with striking results and in many infections with species of gram-negative rods that are not susceptible to penicillin or sulfonamides. Streptomycin is not so rapidly destroyed in, or excreted from, the body as penicillin. Some toxic effects (vertigo, deafness) have been described and it is no longer the drug of choice for many of the pathogens (as can be seen from Table 23–2). Streptomycin is now used mostly in combination with other antibiotics to treat infections that are not responding to the antibiotic of first choice.

Mode of Action. One of the most pronounced effects of streptomycin on growing bacteria is interference with protein synthesis. This apparently occurs because streptomycin is able to attach to the 30S portion of the bacterial ribosomes, and in doing so, causes a misreading of messenger RNA. Thus, faulty proteins are synthesized and are incapable of sustaining vital cell functions. In a short time the cell is killed, but without lysis such as is seen with penicillin. Streptomycin is not alone in its ability to interfere with protein synthesis by combining with ribosomes. Neomycin and kanamycin also attach to the 30S subunit, but apparently in a slightly different way. Cells resistant to streptomycin, for example, may still be susceptible to neomycin. Two other antibiotics, chloramphenicol and erythromycin, are known to combine with the 50S portion of the bacterial ribosomes and inhibit protein synthesis.

No streptomycin-destroying enzyme has been reported, but resistant organisms appear and, unfortunately, often occur in patients during treatment. Strains of tubercle bacilli wholly dependent on streptomycin have also been found in treated patients.

Chloramphenicol. The organism producing chloramphenicol is much like other species of *Streptomyces*. It was found in 1947 by Paul R. Burkholder in soil collected near Caracas, Venezuela, and is called *S. venezuelae*. Like most other natural antibiotics, the drug is a by-product of growth. It may be produced in a manner analogous to the production of penicillin and streptomycin, but is now produced mainly synthetically. In fact, it is one of the few antibiotics that can be made synthetically on a profitable basis.

It affects many species of both gram-positive and gram-negative bacteria in high dilutions and may be given either orally or intravenously. Unlike either penicillin or streptomycin, it is also effective against the chlamydias and rickettsias. It is the first antibiotic discovered to have definite rickettsicidal properties. It has proved of inestimable value in the treatment of many infectious diseases, including especially typhoid fever, but must be used with extreme care because of its toxicity (blood dyscrasia, "gray" syndrome in infants, etc.). Because of its wide range of activity, it is classed as a **broad-spectrum** antibiotic. As mentioned above, chloramphenicol inhibits protein synthesis by combining with the 50S portion of the bacterial ribosome.

The Tetracyclines. This group includes four closely related antibiotics: tetracycline (**Achromycin**), chlortetracycline (**Aureomycin**), oxytetracycline (**Terramycin**), and demethylchlortetracycline (**Declomycin**). They differ from one another only by the nature of the substitutions on the basic structure shown in Figure 23–7. For tetracycline, $R_1 = H$, $R_2 = CH_3$, $R_3 = H$; for chlortetracycline, $R_1 = Cl$, $R_2 = CH_3$, $R_3 = H$; for oxytetracycline, $R_1 = H$, $R_2 = CH_3$, $R_3 = OH$; and for demethylchlortetracycline, $R_1 = Cl$, $R_2 = H$, $R_3 = H$.

Unlike penicillin and streptomycin, tetracyclines are bacteriostatic in clinical concentrations. They are effective against many gram-negative and gram-positive species of bacteria, chlamydias and some rickettsias. For this reason, they are also classed as broad-spectrum antibiotics and have similar ranges of therapeutic activity. All are effective orally or parenterally.

TABLE 23–3. SOME BACTERIAL SPECIES SENSITIVE TO STREPTOMYCIN

GRAM-POSITIVE
    Mycobacterium tuberculosis
    Staphylococcus aureus
    Streptococcus (Diplococcus) pneumoniae
    Streptococcus species (others)
    Bacillus subtilis and related species

GRAM-NEGATIVE
    Salmonella typhi
    Francisella (Pasteurella) tularensis
    Klebsiella species
    Brucella abortus
    Proteus vulgaris
    Salmonella paratyphi B
    Bordetella pertussis
    Haemophilus influenzae
    Pseudomonas aeruginosa
    Escherichia species

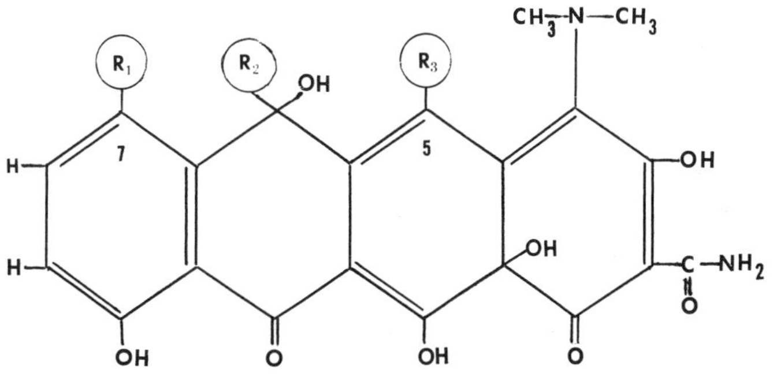

Figure 23-7. Composite formula of the tetra-cycline group of antibiotics. See text for details.

The tetracyclines act on the 30S ribosomal units to block protein synthesis in susceptible bacteria. Drug resistance can occur and is believed to be due to altered membrane permeability or to enzymatic inactivation of the antibiotic. The side effects most commonly encountered through the use of the tetracyclines include diarrhea, bone lesions, and staining of the teeth in young children.

Erythromycin. This antibiotic, one of a group including erythromycins A and B, is representative of several valuable and widely used antibiotics, including **carbomycin, oleandomycin** and **spiramycin.** All are substances of a basic nature derived from *Streptomyces* species and having relatively high molecular weights. Because of their large, complex molecules, the group name **macrolides** has been given to them. They have approximately the same spectrum of activity as penicillin and act principally against gram-positive bacteria, but they also attack several species of gram-negative bacteria, as well as other types of pathogens. They are therefore generally included with the broad-spectrum antibiotics. Their chief effect is to inhibit synthesis of proteins by blocking the function of the 50S ribosomal unit. Erythromycin is the most potent of the macrolides.

Antibiotic resistance to erythromycin is observed frequently, and ribosomes isolated from resistant strains of bacteria show a lower capacity to bind erythromycin. Side effects from the administration of erythromycin are usually absent or mild, although occasionally, severe allergic reactions are encountered.

23.3
ANTIBIOTICS FROM BACILLUS

Several useful antibiotics have been derived from gram-positive, aerobic, sporeforming bacteria closely related to, or identical with, *Bacillus subtilis.* The first to be described, **tyrothricin** or **gramicidin,** from *B. brevis,* was discovered by Dubos, as previously mentioned.

Another is **bacitracin,** discovered by F. L. Meleney.

Bacitracin. Bacitracin was first described in 1945. The source organism is a particular strain of *Bacillus subtilis,* a common and usually harmless organism widely distributed in dust. It was found contaminating a wound in a patient named Tracy, hence baci*tracin.* Bacitracin is quite toxic if given internally and is therefore used only for topical applications. It acts mostly on gram-positive bacteria by preventing an essential step in cell wall synthesis.

Other Polypeptide Antibiotics. Other polypeptide antibiotics derived from species of *Bacillus* include: the polymyxins, A, B, C, D, E (*B. polymyxa*); subtilin and bacitracin (*B. subtilis*); the gramicidins and tyrocidin (*B. brevis*); biocerin (*B. cereus*); and circulin (*B. circulans*). Commercial preparations of these antibiotics are undoubtedly mixtures of related substances.

These are curious chemically in that they contain rare types of peptide linkages and amino acids of the D-series which have not been found elsewhere in nature except in bacterial cell walls (Chapt. 14). They are alike in being surface tension reducers, or surfactants.

Mode of Action. Antibiotics such as the polymyxins resemble other surfactants, such as the quaternaries and soap, in quickly destroying susceptible bacteria by attacking the cell membrane and cell wall, possibly by acting as an emulsifying agent on the lipids in the membranes and walls. They affect many gram-negative species, the cell walls of which are rich in lipids. They act so quickly as to suggest that, unlike most other antibiotics, interference with enzymes is not involved. Unlike penicillin, these antibiotics leave the cell wall morphologically intact.

23.4
NONMEDICAL USES OF ANTIBIOTICS

Many antibiotics are extremely effective as antimicrobials but are too toxic for unlimited use. For example, gentamicin, an aminoglycoside and the first known to be derived from a species of *Micromonospora (M. purpurea),* is valuable for suppressing growth of a wide range of bacterial contaminants in cultures of animal cells. Such cultures are much used experimentally and commercially to propagate viruses for vaccines and other purposes, and gentamicin has little or no deleterious effect on the animal cells or viruses. Gentamicin is the drug of choice in certain infections of man, but, like streptomycin, another aminoglycoside, it must be used with care because of toxicity. The industrial, agricultural, and other uses of such antibiotics have grown enormously. Over three million pounds of antibiotics are used for such purposes annually. They have been used to prevent spoilage of various products; to supplement stock feeds, thereby greatly increasing growth; and to control bacterial diseases of plants. Further information on nonmedical uses is given in Chapters 40, 47, and 48.

23.5
STANDARDIZATION OF ANTIBIOTICS

Standards of potency for all antibiotics are established by international agreement as to what shall constitute an **International Standard Sample** and an **International Unit.** Generally, the International Standard Sample of an antibiotic consists of a certain highly purified and tested lot of that antibiotic. For example, the 1960 International Standard Sample of the antibiotic tetracycline was a 500-gm lot of highly purified tetracycline chloride furnished by an American manufacturer. The International Unit of tetracycline was defined as the antibiotic activity contained in 1.01 μg of that particular standard sample; 1.01 μg was equivalent in activity to 1 μg of pure, crystalline tetracycline chloride.

Assay and Sensitivity Testing. Several types of procedures are in use for assaying the potency of antibiotic preparations for therapeutic purposes. This is commonly done in manufacturers' control laboratories under the supervision of the U.S. Food and Drug Administration. These methods, conversely modified, are also used for measuring sensitivity (**sensitivity testing**) of "unknown" organisms to antibiotics (Figs. 23–8, 23–9). This is commonly done in hospital laboratories and is essential to rational chemotherapy, although not always feasible. There are sources of error in all of these methods, and so they must be used and interpreted by experienced personnel. (See Chapter 49.)

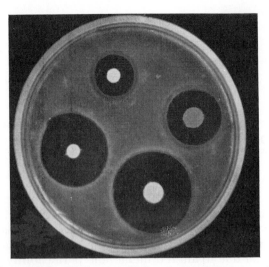

Figure 23–8. Testing sensitivity of a bacterium to antibiotics or other chemotherapeutic agents by the "disk method." The entire surface of agar medium in a Petri plate is inoculated with the organism to be tested. Paper disks of uniform thickness containing graded amounts of the agent to be tested (or the same amount of different agents if a comparison is desired) are then placed on the surface of the agar. The agent diffuses into the agar and prevents growth of the bacterium in a zone around the disk. The width of the zone indicates, roughly, the sensitivity of the organism to the agent or agents being tested, though the **presence** or **absence** of a zone is of greater significance. (About ½ actual size.) (Courtesy of Drs. R. W. Fairbrother and A. Rao, Department of Clinical Pathology, Manchester Royal Infirmary, Manchester, England.)

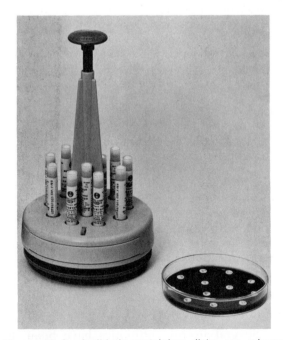

Figure 23–9. Antibiotic-containing disks are released automatically onto an inoculated plate when plunger of Sensi-Disc dispenser is pushed. (Courtesy of Baltimore Biological Laboratories, Division of Becton-Dickinson Laboratories, Baltimore, Md.)

CHAPTER 23
SUPPLEMENTARY READING

Anderson, E. S.: The ecology of transferable drug resistance in the enterobacteria. Ann. Rev. Microbiol., 22:131, 1968.

Demain, L., and Inamine, E.: Biochemistry and regulation of streptomycin and mannosidostreptomycinase (α-D-mannosidase) formation. Bact. Rev., 34:1, 1970.

Dulaney, E. L., and Laskin, A. I. (Eds.): The problems of drug-resistant pathogenic bacteria. Ann. N.Y. Acad. Sci., 182, 1971.

Falconer, M. W., Norman, M. R., Patterson, H. R., and Gustafson, E.: The Drug, the Nurse, the Patient, 4th ed. W. B. Saunders Co., Philadelphia. 1970.

Falconer, M. W., Patterson, H. R., and Gustafson, E.: Current Drug Handbook 1972–1974. W. B. Saunders Co., Philadelphia. 1972.

Garrod, L. P., and O'Grady, F.: Antibiotics and Chemotherapy, 3rd ed. The Williams & Wilkins Co., Baltimore. 1971.

Gause, G. F.: The Search for New Antibiotics—Problems and Perspectives. Yale University Press, New Haven. 1960.

Goodman, L. S., and Gilman, A. (Eds.): The Pharmacological Basis of Medical Therapeutics, 4th ed. The Macmillan Co., New York. 1970.

Hedgecock, L. W.: Antimicrobial Agents. Lea & Febiger, Philadelphia. 1967.

Hobby, G. L. (Ed.): Antimicrobial Agents and Chemotherapy. Eleventh Proceedings, Interscience Conference on Antimicrobial Agents and Chemotherapy. American Society for Microbiology. The Williams & Wilkins Co., Baltimore. 1972.

Lester, W.: Rifampin: A semisynthetic derivative of rifamycin—A prototype for the future. Ann. Rev. Microbiol., 26:85, 1972.

Mitsuhashi, S. (Ed.): Drug Action and Drug Resistance in Bacteria. 2 vols. University Park Press, Baltimore. 1972.

Newton, B. A.: Mechanisms of antibiotic action. Ann. Rev. Microbiol., 19:205, 1965.

Newton, B. A., and Reynolds, P. E. (Eds.): Sixteenth Symposium, Biochemical Studies of Antimicrobial Drugs. Society for General Microbiology, London. 1966.

Riva, S., and Silvestri, L. G.: Rifamycins: a general review. Ann. Rev. Microbiol., 26:199, 1972.

Watanabe, I.: Infections and drug resistance. Sci. Am., 217:19, 1967.

Wehrli, W., and Staehelin, M.: Actions of rifamycins. Bact. Rev., 35:290, 1971.

Zähner, H., and Maas, W. K.: Biology of Antibiotics. Springer-Verlag, New York. 1972.

CHAPTER 24 • MICROORGANISMS AND INFECTION

In biology the term **host** means any organism that is used by another organism for shelter or food. The second organism is called a **parasite.** The relationship of each plant and animal to each of the many other organisms with which it is in contact under natural conditions varies from complete mutual indifference or **neutralism** to lethal, mutual **antagonism** or antibiosis. A number of terms, familiar to the student of elementary biology, are used to describe the different types of relationships between these extremes: **commensalism,** in which one species benefits from the association, the other being unaffected; **mutualism** or **symbiosis,** in which both are benefited and often wholly interdependent; **antagonism** or **amensalism,** in which one species is injured, the other unaffected, by the association; **predation** (usually of large animals), as when hawks eat field mice or trouts eat minnows. In **parasitism** the **parasite** lives in or on the parasitized plant or animal (the **host**) at the expense of the host. Parasitism is not necessarily harmful to the host and often approaches commensalism. The term parasite is usually applied to small species such as lice, hookworms, and microorganisms such as yeasts, molds, bacteria, viruses, and amebas.

24.1
HOST-PARASITE RELATIONSHIPS

The first requisite of continued successful parasitism by microorganisms is the ability to live in or on the host without stirring up a de-

fensive reaction on the part of the host with which the parasite cannot cope. In the absence of significant damage to the host this could be viewed as a kind of commensalism and is the commonest form of relationship of man to most microorganisms. If a violent defensive reaction is elicited by the parasite, there are three possible outcomes: the parasite may be killed or cast out; the host may be killed; or the invasive and pathogenic properties of the parasite and the defensive mechanisms of the host may reach an equilibrium, during which the two live together in "peaceful coexistence" or "an armed truce." Either is a potential aggressor against the other if the equilibrium (the "balance of power") is disturbed.

Infection exists when parasites are enabled to penetrate or bypass the defensive barriers of, and to live inside, the host. Infection may or may not result in disease. When the host is visibly or sensibly injured by the parasite, **disease** exists and the parasite is said to be **pathogenic** (Gr. *pathos* = sadness or pain). Most microorganisms lack the physiological properties that might enable them to multiply within the tissues of (i.e., to **infect**) a plant or animal host. They are not infectious. An organism may, however, be highly infectious for one host and harmless for another or, in the same host, pathogenic at one time, commensal at another.

A parasite may be called a **primary pathogen** if it can, unaided, penetrate or evade the normal defensive mechanisms of the healthy, susceptible (i.e., not specifically immunized) body and establish an infection. It is highly **aggressive.**

Examples would be measles virus, gonococci, or rickettsias of typhus fever. An **opportunist** or **secondary pathogen** is one that cannot, of its own powers, pass the normal defensive mechanisms of the healthy, susceptible host but that can infect and cause disease if defense mechanisms have been broken, as by wounds, prolonged disease, old age, or poisons. A good example would be the staphylococci which are normally found in the nares and on the skin of healthy people. If these organisms make their way to the blood or deep tissues, serious infections may result.

24.2
GERM-FREE (AXENIC) LIFE

As we know, microorganisms are found almost everywhere in our environment as well as on all parts of the body surface, the entire alimentary canal (Fig. 24–1), the upper respiratory system, the eyes, the ears, and the urogenital openings. The newborn animal is contaminated at birth and remains an involuntary and often unknowing host to billions of various species of microscopic parasites or commensals throughout his life. Some of them help digest his food and furnish him with vitamins. Even the interior of the living cell is not immune from parasitic invasion by viruses, rickettsias, and some bacteria. These relationships have existed since Archeozoic periods, and animal life is well adapted to existence in contact with most microbial life. Only a few species of microorganisms cause much disturbance in modern man; among them are pathogens like typhoid bacilli, hookworms, malaria parasites, viruses, certain fungi, pneumococci, gonococci, and tubercle bacilli. It would be of great interest and extremely useful to know how this ageslong contact between host and parasite has affected the hereditary, physiologic, biochemical, anatomic, and other properties of each of the participants in the relationship.

Interrelations of Organisms. What effect does the enormous number of supposedly harmless bacteria in the intestine have on our nutrition? Do they synthesize vitamins for us? Do they produce antibiotics which are taken into our systems? Do they produce poisons which damage us subtly and shorten our lives, or do they exert favorable or even absolutely necessary influences? Do they immunize us to infection? What is the role of each of the species?

We begin to realize the importance of such

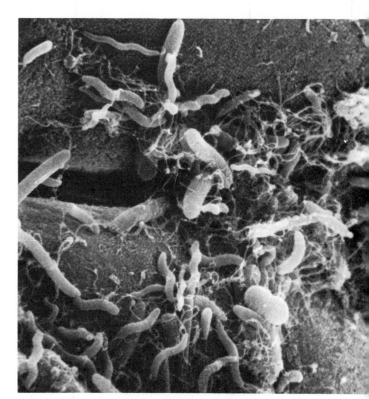

Figure 24–1. A scanning electron micrograph taken at the microvillous border of an intestinal villus in the rat ileum shows a typical population of microorganisms found within an animal's digestive system. (Original magnification 10,400×; reduced to 56% of original size.) (Courtesy of S. E. Erlandsen and G. Wendelschafer.)

interrelationships when it is found that suppression of certain groups of bacteria which normally inhabit the gastrointestinal tract (as sometimes occurs during presurgical antibiotic therapy) permits growth of other groups which are normally held in abeyance by the suppressed microorganisms. This often has evil and sometimes fatal results. Could we answer some of the above questions by producing an animal entirely free from microorganisms and then contaminating it with single species of microorganism at a time?

To do such a thing it would be necessary, by the use of fantastically rigorous aseptic technique, to separate an animal from all demonstrable microorganisms at the beginning of life, to maintain it for months or years free from any demonstrable live microorganisms, and then to observe how it fares in life without its usual living mates, the microorganisms. Such an animal might be said to represent "germ-free" life, or **axenic** life (Gr. *a* = without; *xenos* = foreign material).

Experimental Approaches. Pasteur realized the value of such investigations as early as 1885 and actually reared some germ-free or axenic chicks. Because of the great difficulties involved, only four workers prior to 1928, including Pasteur, had successfully reared any germ-free animals. In 1928 Reyniers started work on chickens and, with his coworkers, succeeded in raising numerous germ-free animals. Completely axenic insects, fish, chickens, rodents, dogs, pigs, and monkeys have been born and held axenic for many generations.

The original apparatus for such work consisted of large steel cylinders that could be steam-sterilized, with closed glass observation ports; hand holes fitted with airtight, arm-length, seamless rubber gloves; and airtight systems of outer chambers or locks through which sterilized food, water, and equipment could be passed into the chamber while waste materials were passed out. Air was passed through sterilizing filters and conditioning apparatus. Such equipment and its operation necessitated solution of some difficult engineering problems. In the original apparatus, one chamber was big enough for an attendant to enter, dressed in a diving suit, through a deep tank of disinfectant.

As an improvement over these ponderous steel containers, vinyl-plastic, tent-like "flexible film isolaters" have been introduced. These can be quickly and easily sterilized by filling the interior with germicidal gas. They are made airtight and have the advantages of being completely transparent, flexible, easily portable, relatively inexpensive, and convenient. They come in a variety of sizes (Fig. 24–2).

Some of the technical difficulties are great, as for example, feeding young rats delivered aseptically inside the germ-free compartment at cesarean section (to say nothing of doing a completely germ-free cesarean section!). They require milk every hour, 24 hours a day, for weeks on end. It took much research to synthesize a satisfactory substitute for mother rats' milk!

Careful control is necessary at all times, and all animals, their feces and bodies, and the dust, feed and water in the germ-free compartments are examined bacteriologically at short intervals to detect any contamination. The first significant studies on germ-free life were carried on at the Laboratories of Bacteriology at the University of Notre Dame (Lobund). Within the last few decades germ-free life has taken an important role in many aspects of biology. Axenic animals are reared in a number of places and shipped in special containers all over the world. Who will produce the first axenic human family, and *must* it be on the moon?

Germ-free animals thrive. Axenic animals in general live longer and seem healthier than ordinary (conventional) animals. However, they are often susceptible to fatal infections with usually harmless bacteria. The living cells of axenic rats and chickens seem to remain "younger" than those of conventional animals. Is this because they are germ-free? If antibiotics are fed to farm stock, this possibly reduces their bacterial burden. Young stock fed with antibiotics certainly grow much faster. If antibiotics are fed to **axenic** poultry there is no such improvement. This suggests that unless the birds are contaminated the antibiotic is without effect.

Gnotobiosis and Its Effects. The study of axenic animals that have been experimentally infected with *known* microorganisms is spoken of as gnotobiology (Gr. *gnos* = known; *bios* = life). The animals are said to be **gnotobiotic.** Studies of gnotobiotic animals have yielded interesting data.

An important observation was made when it was found that axenic guinea pigs did not develop dysentery or amebic abscesses when infected with the ameba *Entamoeba histolytica*, which usually causes a particularly severe form of dysentery in guinea pigs. Now, it is known that *E. histolytica* feeds on bacteria. In axenic

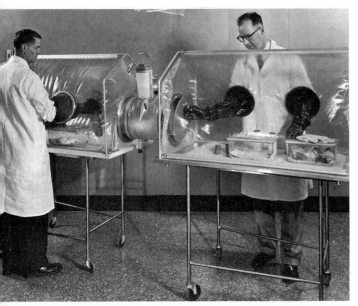

Figure 24–2. One type of plastic film apparatus for rearing animals under germ-free (axenic or gnotobiotic) conditions. (Courtesy of American Sterilizer Co., Erie, Pa.)

guinea pigs there were no bacteria. The ameba-infected axenic guinea pig, free from disease, quickly developed dysentery (or amebic abscesses in the liver) and died when fed ordinary "harmless" intestinal bacteria! It is suggested that certain antibiotics are effective in curing amebic dysentery, not because they affect amebae, but because they deprive the amebae of their food, the bacteria. The problem requires further study.

Studies of the blood of axenic animals show that they have less defense against infection (fewer antibodies) than conventional animals. Numerous other important differences between axenic and conventional organisms have been noted. A serious problem has arisen from the finding that apparently axenic animals harbor intracellular viruses (see Chapters 40 to 42) in latent form and transmit them from generation to generation in utero or even in ovum.

24.3
PARASITISM AND PATHOGENICITY

It is sometimes difficult to distinguish between parasitism and pathogenicity. A parasite may or may not be significantly pathogenic. For example, a mosquito might be regarded as a harmless parasite. But if its bite causes a severe allergic reaction it could be considered a pathogen.

Pathogenic Saprophytes. On the other hand, pathogenicity does not always involve parasitism. For example, certain organisms, such as the bacilli causing tetanus (*Clostridium tetani*), cannot invade or live in normal tissues and are in no sense parasites. They can live only in dead (necrotic) material, such as might occur in a crushed foot; they are true saprophytes. They can cause fatal disease, however, because they produce a toxin that diffuses from the site of their growth and is absorbed by the body.

Similarly, the organism *C. botulinum* cannot as a rule multiply in the tissues or on the body, but it is able to produce a potent toxin if it grows in inadequately processed food. When such food is eaten, the preformed toxin produces a highly fatal disease: food poisoning, or **botulism.** Such organisms may be considered **pathogenic saprophytes.**

Pathogenicity Is Fortuitous. Those not familiar with microbial disease sometimes consider microorganisms to be purposeful predatory creatures that invade the tissues and produce poisons for the sole purpose of causing harm, as though that were to their benefit in the sense that a tiger, snake, or spider benefits from killing its prey. Microorganisms are sometimes thought to be endowed with special powers for injury. Like dandelions in the lawn, microbial growth occurs only where (and because) conditions permit. Dandelions, although highly irritating to people who cherish all-blue-grass lawns, have no evil intent. Their only interest is propagation of the dandelion; so also the microbe. If the lawn owner (or host) reacts against them with 2,4-D (or antibiotic), it is only because he is not adapted to them. Some people like the dandelion flowers or make delicious salads out of their leaves. However, excessive irritation to the host by plant or microbe usually is unfavorable to the parasite. The greatest advantage to the parasite results from *unnoticed* parasitism.

Mutual Adaptation. The process of adaptation between host and parasite must be thought of as a mutual one, requiring many generations of natural selection of the most compatible. Microbial generations are sometimes very short, minutes or hours; human generations are long, about 25 years. Thus, microorganisms can become adapted to growth in contact with human tissues before man acquires species or racial resistance to invasive microorganisms. Over a sufficient period of time, however, and without interference from life-saving drugs and antibiotics, selection for resistant populations will become evident. Some specific examples will be considered in Chapter 26.

The Carrier State. In still other instances, an individual human host and dangerous pathogens, such as the typhoid bacillus, hemolytic streptococci, pneumococci, and meningococci, rapidly become mutually adapted, with or without perceptible disease. They may live together for years with no evidence of disease. This is a very common situation and is referred to as the **carrier state.** The person or animal who thus carries infectious organisms (i.e., a **carrier**) although remaining well himself can often transmit the pathogens to others who may become very ill as a result.

Frequently carrier relationships represent in reality only a temporary "armistice" or a microbiological "cold war." If one side weakens, the other automatically takes advantage of the situation and serious infection of the carrier can occur. The important point here is that we are constantly in contact with microorganisms capable of making us very ill or of killing us. That they do not do so is because we have ample defensive mechanisms. Let these become weakened and disease, severe or mild, results.

Hence, one of the most fundamental concepts in microbiology is that the interaction between a pathogen and its host is never a static affair but rather a very dynamic contest to see who will win. The severity of the disease, then, will depend on (1) what mechanisms the pathogen can employ to invade and multiply in the host, and (2) what types of defensive tactics the host can rally to oust the intruder. The latter will be dealt with in some detail in Chapters 26 through 28.

24.4
FACTORS IN THE OCCURRENCE OF DISEASE

Besides the weakening of defensive mechanisms by various extraneous factors such as exposure, exhaustion, alcoholism, and drug addiction or too-frequent child-bearing, several other factors are important in the occurrence of infection: the means by which the parasite enters the body (**portal of entry**); parasite **virulence;** and **dosage,** or numbers, of infecting organisms.

Portal of Entry. The portal by which an organism enters the body is important in determining the occurrence and kind of disease. If the skin is kept intact, no ordinary microorganism can get through it. (Some few microorganisms are said to be able to penetrate the intact skin; probably they get into hair follicles and sweat glands.) But if any slight cut or scratch exists, then microorganisms can get into the tissues. The thin mucous membranes about the eye (**conjunctivae**), in the nose, throat, and lungs, and in the genitourinary tract are less able to withstand invasion of some microorganisms than the tough outer skin. Numerous infections readily begin in such situations.

Certain microorganisms under ordinary circumstances can gain a foothold in the body *only* when they come into contact with the respiratory tract, others only through contact with the intestinal lining, and so on. For example, dysentery bacilli (*Shigella*) rubbed over the hands or even into a wound would ordinarily cause no infection, while if swallowed they might produce a fatal disease. On the other hand, *Neisseria gonorrhoeae* (cause of gonorrhea) might be swallowed without harmful effect, but if rubbed into the eye or genitalia it could cause gonorrheal infection of the mucous membranes.

Virulence. Beginning students of microbiology sometimes have difficulty in distinguishing between the terms **pathogenic** and **virulent.** A species is generally designated pathogenic if experience tells us that the microorganism is likely to cause disease in one host or another. We know, however, that various strains of a pathogenic species can differ tremendously in their ability to cause disease. Those organisms able to produce serious disease or able to initiate an infection with very few cells (sometimes as low as 1 to 10) are said to be highly virulent. Those producing only mild disease or requiring large numbers to establish an infection are said to possess low virulence. Thus, many microbiologists use the term pathogenic in a qualitative sense and virulent in a quantitative sense.

Two other terms which appear frequently in the literature and which sometimes cause confusion are **avirulent** and **attenuated.** An avirulent strain is generally one which cannot infect. The line between *total* avirulence and *very low* virulence is often difficult to determine. An attenuated strain is one which was originally virulent but is now of lowered virulence for a particular host. It may elicit antibodies yet cause no perceptible disease. Such strains frequently, but not always, show full virulence if put into a different host.

While it is easy to define virulence, it is difficult to explain the basis for it. Table 24–1 lists some of the specific bacterial products which may contribute to the virulence of certain pathogens. Virulence may also be looked upon as the net effect of three components: (1) **Infectiousness,** or the ability to initiate and maintain an infection in a host; (2) **invasiveness,** or the power to progress farther into the host from the initial infection; and (3) **pathogenicity,** or the ability to injure a host once an infection is established (Fig. 24–3). Pathogens that pass from one host to another (e.g., "animal passage") seem to acquire enhanced virulence. The mechanism is not clear in many cases but may result from the development of protective capsules, toxins, or other properties.

Infectiousness. Infectiousness depends on a complex of properties, several still obscure, that enable the parasite to establish an initial beachhead in the host by evading or overcoming local defensive measures such as antibodies (protective serum proteins, to be discussed later) and phagocytes, which are able to engulf and destroy many types of pathogens.

Infectiousness need not involve great pathogenicity. Infection and even invasion frequently occur without any perceptible disease. Indeed, very fortunately, this is the general rule. For example, yellow fever and polio-

TABLE 24-1. FACTORS WHICH MAY CONTRIBUTE TO MICROBIAL VIRULENCE

Factor	Important Characteristics
Capsule	Protects cells from serum components and makes phagocytosis difficult. Some capsules are toxic for phagocytic cells.
Intracellular growth	Certain pathogens are not killed after phagocytosis but survive and may even multiply intracellularly and be transported throughout the body.
Leucocidin	Enzyme-like exotoxin which can kill white cells (leucocytes), including phagocytic cells.
Protopectinase	Enzyme produced by certain plant pathogens which digests the intercellular cementing substance of plant tissues (pectin) and allows the spread of the bacteria in the tissues.
Hyaluronidase	Enzyme which digests the intercellular cementing substance of animal tissues (hyaluronic acid) and facilitates the spread of toxins and organisms through host tissues.
Lecithinase	Causes lysis of red blood cells and results in local anoxia and anemia.
Exotoxins	Large variety of highly specific, potent toxins released by growing bacteria. Many act directly on a particular type of cell or nerve ending.
Endotoxin	Frequently present during infection with gram-negative bacteria. Causes symptoms of generalized shock.

myelitis have been so widespread in some communities that virtually all of the children under 10 have become infected and have developed demonstrable antibodies, yet few have shown any perceptible evidence of disease at all. Before the advent of highly effective preventive measures that are commonplace today (i.e., prior to about 1940), most adult urban residents were immune to poliomyelitis and tuberculosis yet had never had a *recognized* case. Obviously, there are many mild, immunizing infections yet little disease.

Invasiveness. Invasiveness depends on characteristics that enable the parasite to leave the initial site of infection and grow in other tissues. Invasion of blood and tissues may or may not result in perceptible disease, depending on the character of the organisms and the resistance of the host.

Mutations undoubtedly occur that confer infective properties on an otherwise harmless commensal. Additional mutations can occur during the ensuing invasion, such that the invaders are selectively favored by conditions in the host. For example, such mutants often possess capsules that protect them from phagocytosis and antibodies. They can then grow into the lymph spaces, invading widely through the tissues. Others may invade and grow in the blood; then we have the condition called **bac-**

teremia, viremia, or **rickettsemia. Septicemia,** or "blood poisoning," is the term used to indicate that the microorganisms and/or their toxic products are circulating in the blood. Once microorganisms invade the blood they may be carried to various other points in the body (e.g., liver, spleen, bone marrow, or lymph nodules of the intestine). Secondary abscesses (**secondary foci** of infection, or **metastatic infections**) may then result.

Invasiveness is enhanced by the production of certain exoenzymes that attack defensive measures. For example, several common bacterial pathogens produce enzyme-like substances called **leucocidins** that kill leucocytes. Many pathogens of plants produce the enzyme **protopectinase** that digests the supporting structures (middle lamella) of plants, causing soft-rot diseases. Streptococci such as those causing scarlet fever, erysipelas, and septicemia produce an enzyme-activator called **streptokinase** or **fibrinolysin.** This helps to digest the fibrin of blood clots that surround and retain sites of injury and infection. Destruction of the retaining net of fibrin presumably enables the organisms to invade distant tissues.

Hyaluronidase is an enzyme that destroys hyaluronic acid, a clear, gummy, intercellular cementing substance of animal tissues which normally opposes progress of microorganisms through the tissues. For this reason hyaluronidase has been spoken of as a **spreading factor.** Some organisms produce lipolytic enzymes such as **lecithinase.** This causes destruction of erythrocytes (**hemolysis**) with resulting anemia and anoxia. Such organisms are said to be **hemolytic.**

Production of these enzymes and others may be, in some cases, **constitutive** properties of

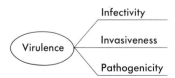

Figure 24-3. Factors in virulence. For explanation see text.

the organisms involved; in some microorganisms the enzymes may appear as a result of **mutation;** others may represent **induction** by the substrates in the body tissues. Other organisms are known to produce poisonous products of normal tissues by means of their exoenzymes. The production of **catalase,** an enzyme that decomposes H_2O_2 to H_2O and $\frac{1}{2}O_2$, appears to be definitely associated with disease production by several species of bacteria pathogenic for man, though the exact mechanism is not yet clear.

Pathogenicity. Like invasiveness, pathogenicity is a complex of factors, some known, others still to be elucidated. A principal cause of pathogenicity is the production of poisons called toxins, as well as of exoenzymes as just described. Microbial toxins may be secreted or released into the surrounding fluids as **exotoxins,** or remain attached to the producing cell and are then called **endotoxins.**

EXOTOXINS. Exotoxins are proteins and, like typical proteins, are sensitive to temperatures above 70C, 50 per cent alcohol, formaldehyde, and dilute acids. When subjected to mild denaturation, exotoxins lose their toxicity but retain most of their chemical structure and thus become **toxoids.** When injected into animals, toxoids elicit the formation of antitoxins (antibodies), which can be harvested as serum and shown to neutralize the original toxin. A number of antitoxins are available commercially and can be used clinically to prevent injury and death if the toxin has not already done irreversible damage. Some exotoxins, such as those of *Clostridium botulinum* and certain staphylococci (causes of food poisoning), are harmful only when swallowed (except under unusual conditions, such as experimental injection). Others, such as diphtheria and tetanus toxins, can be taken by mouth with impunity, but if injected or adsorbed into the blood from infected lesions, even very tiny doses may cause death: additional evidence of the importance of portal of entry. Most bacterial exotoxins appear to have an affinity for nervous tissue and often for heart muscle, kidney and certain other specialized tissues. They cause damage principally to those tissues. Most of these toxins are much more potent than cobra venom. Indeed, botulinal toxin is the most poisonous biological substance known (for mammals).

Various toxins and toxin-like substances of specific organisms are discussed with the specific organisms that produce them.

ENDOTOXINS. The term **endotoxin** in a broad sense includes toxic substances derived from the structural components of the microorganism, i.e., **somatic toxins** as contrasted with toxins secreted or released into the surrounding medium, e.g., exotoxins. However, the term endotoxin as currently used is synonymous with the lipopolysaccharide portion of the cell walls of gram-negative bacteria and is sometimes referred to as the somatic or O antigens of the Enterobacteriaceae. Unlike the proteinaceous exotoxins and exoenzymes previously discussed, the endotoxins are lipid-polysaccharide complexes (often coupled with protein when in crude form) which are resistant to heat, alcohol, and dilute acids. They do not form toxoids. Endotoxins are antigenic, i.e., will elicit the formation of antibody, but neutralizing antibody is more difficult to obtain than with exotoxins. Incomplete neutralization may be attained when the endotoxin is mixed with the homologous antibody.

The toxic effects of endotoxins have been widely investigated. Endotoxins are markedly **emetic** and **pyrogenic** (cause vomiting and elevations of temperature) and at first may temporarily stimulate an increase of activity of phagocytic cells, both of the reticuloendothelial system and the leucocytes. However, the reticuloendothelial system removes endotoxins from the blood rapidly and the cells are injured thereby. Thus, phagocytic activity is temporarily suppressed; so also is antibody formation. Repeated doses of endotoxins soon induce a tolerance that seems unrelated to antibody formation. (Phagocytic cells and antibody are covered in detail in Chapter 26.)

Endotoxins produce several of the effects of gastrointestinal infection with pathogens of the group of Enterobacteriaceae: weakness, nausea, diarrhea, lowered arterial blood pressure, and intestinal hemorrhage. Infection of pregnant animals can induce abortion. Curiously, there seems to be little fixed relationship between pathogenicity and endotoxin content of an organism.

It is of interest that, in tumor-bearing animals, injection of endotoxin causes hemorrhage and necrosis of the *tumor.* Therapeutic use of the endotoxin is limited by its toxicity and the rapid appearance of tolerance.

Endotoxins are very important as **potentiators** of other infections (i.e., they greatly weaken the phagocytic and other defenses against pathogens). For example, infections with certain endotoxin-containing gram-negative rods of relatively little virulence (swine influenza bacilli) can cause an otherwise mild viral disease (swine influenza) to become a devastating scourge in a herd. Long-standing chronic bacterial infections (e.g., tuberculosis) may be

TABLE 24-2. GENERAL CHARACTERISTICS OF EXOTOXINS AND ENDOTOXINS

Feature	Exotoxin	Endotoxin
Source	Predominately, but not exclusively, excreted by certain gram-positive bacteria.	Principally released from gram-negative cell walls.
Chemical nature	Proteins.	Lipopolysaccharides.
Heat sensitivity	Easily inactivated at 60 to 80C.	Resistant, will withstand autoclaving.
Immunological features	Toxin easily converted to toxoid; readily neutralized by antitoxin.	Toxoids not formed; neutralization absent or more difficult to achieve with antitoxin.
Lethal dose	Small; among most potent toxins known.	Usually larger than most exotoxins.
Pharmacologic action	Generally specific for a particular type of cell or nerve ending.	Various effects, mostly symptoms of generalized shock.

suddenly converted to acute, quickly fatal infections by injecting minute amounts of endotoxins. Table 24-2 compares some of the features of endotoxins with those of exotoxins.

Dosage. This factor in establishing infection is a simple quantitative one, yet it also involves other factors. As a generality we may say that under ordinary circumstances, the larger the dose of infective microorganisms, the greater the chance that an infection will result. However, certain qualifications are necessary. For example, very large numbers of some organisms may be present in certain situations without causing any difficulty at all. The intestine contains thousands of billions of deadly bacteria at all times, yet if only a dozen or so of some of these are placed in the peritoneal cavity or injected into the brain, they can quickly set up a fatal infection (an example of the importance of portal of entry). Similarly, one might swallow three or four typhoid bacilli with impunity, yet a dosage of several hundred might overcome local resistance and cause typhoid fever. With some organisms a single cell or particle is invariably sufficient to infect. Obviously, much depends on the virulence of the particular organism involved and on the resistance of the tissues that it contacts, as well as on dosage alone. It is conceivable that among a thousand rough (R) unencapsulated, readily phagocytized bacilli or cocci, there might be one encapsulated smooth (S) mutant that by itself could initiate an infection.

Viruses. Many of the factors that influence the occurrence of disease caused by bacteria also apply to viruses, but obviously some do not. Certainly, the relationship between virus and host is just as dynamic and the outcome perhaps even more variable (as in the case of tumors). The principles of virulence—i.e., infectiousness, invasiveness, and pathogenicity—still apply, but the mechanisms used by the parasite represent a substantial departure. This is not to say that viruses never produce toxins. Large numbers of mumps, vaccinia, or influenza virions are sufficiently toxic to kill experimental animals without viral replication, but the relevance of this to actual infections is not entirely clear. The reader is referred to Chapters 16 and 17 for a more detailed account of virus-host interactions.

24.5
KOCH'S AND RIVERS' POSTULATES

It is not always possible to be certain that the microorganisms isolated from a given disease lesion or from pus, blood or feces are the cause of the observed disease condition. Many harmless microorganisms are found growing in feces, sputum, and ulcerating wounds. Some would not grow there unless the diseased condition existed first. Such adventitious organisms are called **secondary invaders.** To prove that a certain microorganism is the primary and unassisted cause of a given disease often requires careful study.

Koch's Postulates. The question of the etiological relationship of various bacteria to specific diseases was a very live one long before the time of Koch, and there was much loose discussion and profitless argument regarding many bacteria and their relation to disease, owing to unrecognized contamination of cultures. When Koch established the pure-culture technique it became possible to apply exact methods to the study of the etiology of disease. He was very conservative in stating the relationships of any given organism to any particular disease.

His ideas on the subject were crystallized largely by his studies of the relationship of tubercle bacilli (*Mycobacterium tuberculosis*) to tuberculosis. Koch, like others before him, observed the bacilli in the lesions of persons and animals dead of the disease. But he was not too ready to believe that he had discovered the cause of tuberculosis just because he found certain organisms present in the lesions of tuberculosis. Might not this bacillus appear in the tissues merely by accident because the host, being so ill, is too weak to resist its invasion? Might it not be merely a relatively harmless opportunist? Might it not represent contamination with a common saprophyte capable of living in the necrotic tissue? Koch, involved in a discussion of the problem, finally stated what he believed to be the evidence necessary to prove an organism to be the cause of a disease. The evidence consists of four postulates, generally called **Koch's postulates** today, and they are essentially as follows:

1. The organism must be associated with all cases of a given disease and in logical pathological relationship to the disease and its symptoms and lesions.
2. It must be isolated from victims of the disease in pure culture.
3. When the pure culture is inoculated into susceptible animals or man, it must reproduce the disease or engender specific antibodies. (Many such inoculations into man have been made on courageous volunteers. In others, accidental infections have occurred which have provided long-wanted evidence. The value of animal experimentation is here very evident.)

4. It must be isolated in pure culture from such experimental infections.

Even today the etiological relationship of some bacteria to diseases that they are thought to cause has not been established on the basis of Koch's postulates, e.g., leprosy. However, it has been established recently that human leprosy bacilli can cause leprous infection in mice.

Rivers' Postulates in Viral Diseases. Viruses were unknown at the time of most of Koch's major works, so he failed to take these invisible, noncultivable agents of disease into consideration when he stated the criteria by which the causal relationship of a pathogen to a disease might be determined. Rivers, in 1937, outlined criteria similar to Koch's postulates, which might apply in the cases of viruses. Essentially these are as follows:

1. The virus must be present in the host cells showing the specific lesions or in the blood or other body fluids at the time of the disease.
2. Filtrates of the infectious material (blood or tissue triturates) *shown not to contain bacteria or other visible or cultivable* * *organisms* must produce the disease or its counterpart, specific antibodies, in appropriate animals. (In response to infections, plants produce antimicrobial substances, though these are not true antibodies or proteins.)
3. Similar filtrates from such animals or plants must transmit the disease.

---

*In inanimate media.

CHAPTER 24
SUPPLEMENTARY READING

Agrios, G. N.: Plant Pathology. Academic Press, Inc., New York. 1969.

Ajl, S. J., Kadis, S., and Montie, T. C. (Eds.): Microbial Toxins, Vol. 1. Academic Press, Inc., New York. 1970.

Burrows, W.: Textbook of Microbiology, 20th ed. W. B. Saunders Co. Philadelphia. 1973.

Buxton, E. W.: Speculations on plant pathogen-host relations, p. 145, *in* Fourteenth Symposium, Microbial Behaviour in vivo and in vitro. Society for General Microbiology, London. 1964.

Davis, B. D., Dulbecco, R., Eisen, H. N., Ginsberg, H. S., and Wood, W. B.: Microbiology, 2nd ed. Harper & Row, New York. 1973.

Deverall, B. J.: Substances produced by pathogenic organisms that induce symptoms of disease in higher plants, p. 165, *in* Fourteenth Symposium, Microbial Behaviour in vivo and in vitro. Society for General Microbiology, London. 1964.

Fucillo, D. A., and Sever, J. L.: Viral teratology. Bact. Rev., 37:19, 1973.

Glynn, A. A.: Bacterial factors inhibiting host defense mechanisms, *in* Smith, H., and Pearce, J. H. (Eds.): Twenty-Second Symposium, Microbial Pathogenicity in Man and Animals. Society for General Microbiology, London. 1972.

Gordon, H. A., and Pesti, L.: The gnotobiotic animal as a tool in the study of host microbial relationships. Bact. Rev., 35:390, 1971.

Horsfall, F. L., and Tamm, I.: Viral and Rickettsial Infections of Man, 4th ed. J. B. Lippincott Co., Philadelphia. 1965.

Hungate, R. E.: The Rumen and Its Microbes. Academic Press, Inc., New York. 1966.

Joklik, W. K., and Smith, D. T.: Zinsser, Microbiology, 15th ed. Appleton-Century-Crofts, New York. 1972.

Lev, M.: Studies on bacterial associations in germ-free animals and animals with defined floras, p. 325, *in* Thirteenth Symposium, Symbiotic Associations. Society for General Microbiology, London. 1963.

Luckey, T. D.: Germ-free Life and Gnotobiology. Academic Press, Inc., New York. 1963.

Mudd, S. (Ed.): Infectious Agents and Host Reactions. W. B. Saunders Co., Philadelphia. 1970.

Nowotny, A.: Molecular aspects of endotoxic reactions. Bact. Rev., 33:72, 1969.

Pearce, J. H., and Lowrie, D. B.: Tissue and host specificity in bacterial infection, *in* Smith, H., and Pearce, J. H. (Eds.): Twenty-Second Symposium, Microbial Pathogenicity in Man and Animals. Society for General Microbiology, London. 1972.

Savage, D. C.: Survival on mucosal epithelia, epithelial penetration and growth in tissues of pathogenic bacteria, *in* Smith, H., and Pearce, J. H. (Eds.): Twenty-Second Symposium, Microbial Pathogenicity in Man and Animals. Society for General Microbiology, London. 1972.

Smith, H.: Biochemical challenge of microbial pathogenicity. Bact. Rev., 32:164, 1968.

Smith, H.: The little-known determinants of microbial pathogenicity, *in* Smith, H., and Pearce, J. H. (Eds.): Twenty-Second Symposium, Microbial Pathogenicity in Man and Animals. Society for General Microbiology, London. 1972.

CHAPTER 25 • SOURCES AND VECTORS OF INFECTION

In ancient times diseases of man and animals were thought to be due to or transmitted by odors, vapors, and miasmas from rotting organic matter or from swamps. The disease malaria is named for the Italian words *mala-aria* (for "bad air" or "air-disease"). Occult and mysterious "influences" were also blamed for disease. *Influenza* is the Italian word for influence. Today we know virtually all the sources of infection, the causative agents, and their means of transmission.

Sources of Infection. The primary **sources** of infection of man and other animals are: (a) infected man and other animals and (b) the soil.

Most infectious agents of man and other animals may be transmitted from one to another by various body fluids, which thus serve as **original** or **prime** or **direct** vectors. The principal sources of infectious body fluids, spoken of as **portals of exit,** are: the oral and respiratory tracts, including eyes and ears, which are directly connected to the respiratory tract; the intestinal tract; the genitourinary tract; open lesions anywhere on the body; tissues, e.g., stillborn infected animals, fluids, and placentas; eggs of birds; blood and blood derivatives; and intravenous fluids. Table 25–1 lists the usual portals of exit used by pathogens frequently causing human disease.

TABLE 25–1.   USUAL PORTAL OF EXIT FOR VARIOUS HUMAN PATHOGENS °

I. Oral and respiratory tracts:
    A. BACTERIA:
       Bordetella pertussis (whooping cough)
       Chlamydia sp. (psittacosis)
       Corynebacterium diphtheriae (diphtheria)
       Mycobacterium tuberculosis (tuberculosis)
       Neisseria meningitidis (meningitis)
       Streptococcus (Diplococcus) pneumoniae (pneumonia)
       Streptococcus pyogenes (scarlet fever)
    B. VIRUSES:
       Adenoviruses
       Chickenpox
       Measles
       Mumps
       Myxoviruses
       Rabies
       Rhinovirus
    C. FUNGI:
       See those listed under VI.

°Adapted from Frobisher and Fuerst: Microbiology in Health and Disease, 13th ed., W. B. Saunders Co., 1973.

(*Table 25–1 continued on opposite page.*)

TABLE 25-1. USUAL PORTAL OF EXIT FOR VARIOUS HUMAN PATHOGENS * *(Continued)*

II. Intestinal and/or urinary tract:
  A. BACTERIA:
    Brucella sp. (brucellosis)
    Clostridium sp. (tetanus, gas gangrene)
    Leptospira sp. (leptospirosis)
    Salmonella sp. (salmonellosis, typhoid fever)
    Shigella sp. (dysentery)
  B. VIRUSES:
    Coxsackie
    ECHO
    Hepatitis (epidemic hepatitis)
  C. PROTOZOA:
    Entamoeba histolytica (amoebic dysentery)
    Giardia lamblia (enteritis)
    Trichomonas hominis (enteritis)
  D. HELMINTHS:
    Many types—tapeworms, pinworms, *Ascaris*, etc.

III. Genital tract:
  A. BACTERIA:
    Neisseria gonorrhoeae (gonorrhea)
    Treponema pallidum (syphilis)
  B. PROTOZOA:
    Trichomonas vaginalis (vaginitis)

IV. Blood via clinical procedures (injections, transfusions, etc.):
  A. BACTERIA:
    Almost any type that can cause a bacteremia, e.g., *Brucella, Staphylococcus, Streptococcus,* etc.
  B. VIRUSES:
    Almost any type that can cause a viremia, but of great concern are those causing serum hepatitis.

V. Blood via arthropods:
  A. BACTERIA:
    Yersinia (Pasteurella) pestis (bubonic plague)
    Francisella (Pasteurella) tularensis (tularemia)
  B. RICKETTSIAS:
    Most of those of medical importance, e.g., the cause of typhus, Rocky Mountain spotted fever, etc.
  C. VIRUSES:
    Mostly the arboviruses, the cause of such diseases as eastern equine encephalitis, yellow fever, etc.
  D. PROTOZOA:
    Plasmodium sp. (malaria)
    Leishmania sp. (leishmaniasis)
    Trypanosoma sp. (trypanosomiasis)

VI. Various portals of exit, but soil is principal reservoir of infection:
  A. BACTERIA:
    Bacillus anthracis (anthrax)
    Clostridium sp. (gas gangrene)
    Clostridium botulinum (botulism; toxin only—no infection)
    Clostridium tetani (tetanus)
  B. FUNGI:
    Blastomyces dermatitidis (blastomycosis)
    Coccidioides immitis (coccidioidomycosis)
    Histoplasma capsulatum (histoplasmosis)
    Sporotrichum schenckii (sporotrichosis)
  C. HELMINTHS:
    Same as group II. D.

*Adapted from Frobisher and Fuerst: Microbiology in Health and Disease, 13th ed., W. B. Saunders Co., 1973.

The student should remember that under certain circumstances, many of these pathogens can, and often do, use alternate routes of transmission.

Soil is an important primary source of several dangerous agents of disease. Several species of pathogenic and invasive fungi are indigenous to the soil (see Chapter 12). Several pathogenic but *not invasive* bacteria are also commonly found in soil: *Clostridium tetani*, the cause of tetanus or lockjaw; *C. botulinum*, the cause of food poisoning; *C. perfringens* and several similar organisms associated in causing gas gangrene of wounds. These soil organisms are more fully discussed under each specific disease.

Transmission of Infection. Among microorganisms, self-mobility is generally limited to distances of a few microns, inches or feet. Microorganisms cannot travel or swim long distances or fly or climb of their own volition. Unless they are transferred from one animal or plant to another by **direct contact,** they must depend on indirect transmission in (or on) extraneous **vectors,** which may be various **substances** (foods, water, milk), **objects** (hands, bedding, toys, eating utensils, cutting instruments), or certain **arthropods** contaminated with, or containing, the infectious agent.

Even though many pathogens are fairly durable in the outer world, especially those that form spores, conidia, or cysts, a great difficulty in travel for many pathogens, especially those of mammals, is that conditions in the world outside the body are too harsh. This is a penalty of extreme adaptation. Drying is fatal to some—meningococci, gonococci, and syphilis spirochetes, for instance. Exposure to sunlight quickly kills many mammalian pathogens, such as tubercle bacilli. Others cannot live long in natural bodies of water, in soil or in feces. Others can travel only inside insects or animals.

Furthermore, not only are many infectious agents much restricted in the modes of travel available to them, but, if they are to infect, they must find a suitable **portal of entry** into a host at the end of their journey. The host must in addition be a **susceptible** subject, plant or animal. Nevertheless, in spite of their difficulties, like the lowly bedbug,* microorganisms get there just the same.

contact). However, the term **contagious** is often used interchangeably with **transmissible, communicable,** and **infectious.** Transmission by direct contact is easy to guard against, especially if one avoids transferring the contagium to its special portal of entry. One does not voluntarily come into physical contact with feces, sputum, or the visible sores of pustules of infected persons or the flesh or fluids of infected animals. If, in the course of professional or home-nursing duties, this cannot be avoided, one should wear rubber gloves or wash and disinfect the hands immediately afterward without touching anything first. However, an innocent kiss may transmit various respiratory diseases and all too often does, sometimes with tragic results to infants, young children, and very ill or old persons. Common examples of diseases transmitted by direct contact among adults are several venereal diseases, including syphilis and gonorrhea, both spread by coitus; syphilis is spread by kissing, also, if there are open labial or oral lesions (Fig. 25–1).

Hands. The practice of shaking hands doubtless transmits many pathogenic intestinal and respiratory organisms, notably poliomyelitis, bacillary dysentery, and respiratory diseases. If hands are to be held they should be clean! (Fig. 25–2).

Milk supplies and the food in any kitchen may become infected from the hands of careless milkers, dairymen, or cooks who are carriers of respiratory or intestinal infectious microorganisms.

Washing the hands after defecation, urination, or blowing the nose is a partial safeguard against transmission of intestinal and respiratory diseases, but careless and ignorant persons are often very lax in this respect.

Thorough cooking of foods followed by prompt eating or prompt refrigeration is another

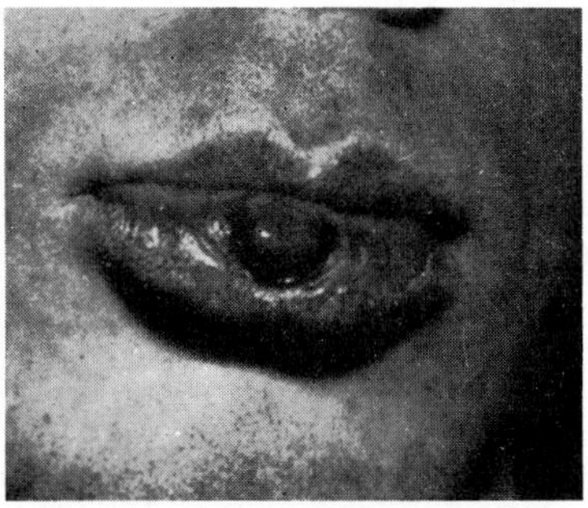

25.1
VECTORS OF INFECTION: AGENTS AND MECHANISMS

Direct Contact. Obviously, if one rubs against infectious material he runs a risk of infection. Diseases so transmitted are properly said to be **contagious** (L. *contagio* = touch or

---

*The moth has wings of velvet;
  The butterfly, wings of flame.
  The bedbug has no wings at all,
  But he gets there just the same! (*Old rhyme*)

Figure 25–1. Primary syphilitic lesion (chancre) of the lip; typically swollen and firm or hard; contains *T. pallidum* and is highly infectious.

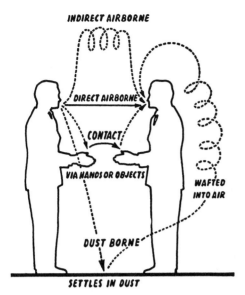

Figure 25–2. Transmission of infectious microorganisms of oral and respiratory tracts from one person to another. Absent from the picture but active, nevertheless, are contaminated foods, water, and milk. (Courtesy of American Sterilizer Co., Erie, Pa.)

safeguard. Persons who handle foods for restaurants or institutions, as well as dairy workers, should be required by law to pass bacteriological examinations and are required to do so in many communities; however, enforcement is difficult and expensive. There is danger from enteric and respiratory infections in food handlers, but little from venereal disease or tuberculosis unless present in an open, acute, or active stage.

Saliva and Nasal Secretion. The mucous secretions of the nose, throat, mouth, and lungs, all combined to some extent with saliva, constitute one of the most formidable vectors of disease. Pneumococci, streptococci, meningococci, diphtheria bacilli, and tubercle bacilli, as well as influenza virus, poliomyelitis virus, measles and mumps viruses, and other organisms of respiratory disease, are thus transmitted. These organisms are frequently carried in the upper respiratory tract by normal persons (Fig. 25–3).

We are all very careless in our habits in regard to mucus and saliva, far more so than we care to realize. The case has been stated vividly by a famous physician (Chapin):

If infection by contact is of such very great importance in the fecal-borne diseases, how much more important must it be in diseases in which the infective agent is found in the secretions of the nose and mouth, as is the case with diphtheria, scarlet fever, smallpox, mumps, measles, whooping cough, tuberculosis, influenza, and cerebrospinal meningitis.

Everyone avoids feces and urine, but it is only the very few who have any objection to saliva.

Not only is the saliva made use of for a great variety of purposes, and numberless articles are for one reason or another placed in the mouth, but for no reason whatever, and all unconsciously, the fingers are with great frequency raised to the lips or to the nose. Who can doubt that if the salivary glands secreted indigo the fingers would continually be stained a deep blue.

Droplet Infection. Droplets of saliva are presumably responsible for much disease transmission. Sneezing or coughing in public without a handkerchief is, like exceeding the speed limit, reprehensible but commonplace, and can have fatal results. Every cough or sneeze inevitably results in a microbe-laden spray. The smallest spray droplets remain suspended for some time in the air and may be carried many feet by drafts. The bacteria in such droplets may easily be demonstrated on an agar plate held near the sneeze and then incubated (Fig. 25–4). They land on food, lips, hands, furniture. After the microbe-laden droplets become dry the mucus-coated and protected bacteria and viruses that they contain then constitute what are called **droplet nuclei.** These may float about through the air for hours like very fine dust particles.

Air Disinfection. The possibilities of disinfecting air in public places have been the subject of intensive and large-scale investigations. The two methods giving most promise are irradiation with ultraviolet light and the use of bactericidal vapors, sometimes called **aerosols.** Both methods are strongly bactericidal but neither is of significant value for practical purposes, except in special situations like operating rooms and other closed areas. These are discussed in more detail in Chapter 46.

Dust. From what has been said, little imagination is needed to understand how disease may be transmitted by household or intramural dust. If not exposed to excessive heat or sunlight or other unfavorable influences, the organisms in droplet nuclei may survive in dust for considerable periods. When the dust is stirred up, persons inhaling it or getting it into operative or accidental wounds may suffer an attack of disease. One of the major problems of today in hospitals is the prevention of transmission of *Staphylococcus aureus* (the "golden killer" of popular newsprint) in air, in dust, and on fomites. Probably respiratory diseases like tuberculosis, pneumonia, diphtheria, and scarlet fever are often transmitted by such means, since the organisms involved resist drying and

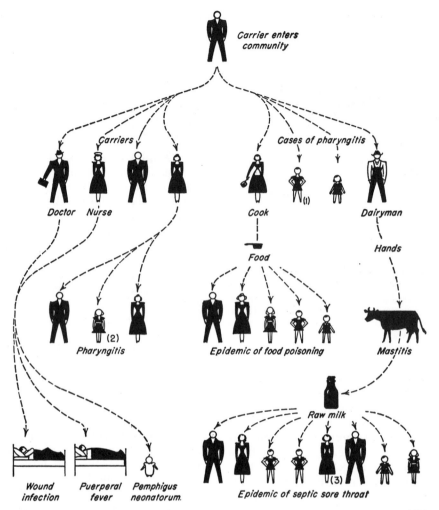

Figure 25-3. Spread of streptococcal infection from a single carrier. Many persons became additional carriers and others developed various diseases. A child (1) with a case of pharyngitis also developed middle ear infection. A child (2) also developed the dreaded rheumatic fever. A woman (3) with a case of septic sore throat also developed subacute bacterial endocarditis, a frequently fatal heart infection.

Figure 25-4. Unstifled sneeze explodes a cloud of highly atomized, bacteria-laden droplets. Some droplets travel at such high speed that they are streaks even at 1/30,000 of a second. (Courtesy of M. W. Jennison, Department of Plant Sciences, Syracuse University.)

exposure to diffuse daylight. Good examples of outdoor dustborne fungal diseases are coccidioidomycosis and histoplasmosis (Chapt. 12). Infections transmitted directly or indirectly from person to person in institutions like hospitals are called **nosocomial** infections (L. *nosocomium* = hospital).

The dust in places where **psittacine birds** (e.g., parrots) are raised and sold can be a source of serious and often fatal infection with the organisms (see chlamydias) of **parrot fever (psittacosis)**. The organisms occur in feces and nasal secretions of infected birds. These dry and are scattered as dust about the building. Strict federal laws of quarantine and control are now in effect.

In barracks and hospital wards dust and lint from clothing and bedding can be important means of disease transmission, especially of respiratory infections. One method of controlling this in certain situations is to impregnate bedding with imperceptible oils that tend to keep the dust from flying about. Floors and sweepings are also oiled. The oil merely controls dust; it does not kill microorganisms. Bactericidal sweeping compounds (oiled, disinfected sand or sawdust) are often used.

Fomites. Fomites are any inanimate objects or substances that serve to transfer infectious microorganisms from one host to another: man, animal or plant. Thus, soiled bed linen or clothing, eating utensils, toys, pencils, and similar objects are dangerous after having been in contact with infected hosts harboring microorganisms that can be transmitted by such means, especially those of respiratory and intestinal diseases. Plant pathogens are transmitted by pruning instruments, gardeners' gloves, and the like.

Eating Utensils. Sanitization of public eating places has been developed on the basis of scientific study. Most restaurants have standardized dishwashing equipment that cleans and disinfects mechanically (Fig. 25–5). Proprietors of smaller restaurants, who have the well-being of their patrons in mind, either carefully scald all dishes after washing them or, after thoroughly washing them in *hot* water with a good detergent, rinse them in clean cool water containing at least 100 parts per million of available chlorine and dry by drainage. The odor of chlorine around a lunch counter is a favorable sign.

Other disinfectants, without taste or odor, are also widely used.

Examination of Utensils. Methods for measuring and controlling the amount of bacterial contamination of dishes are not yet exact. Most of the present methods for bacteriological examination of eating utensils revolve around some modification of the **swab-rinse technique.** In a simple procedure, a swab made of cotton or of a soluble material, such as calcium alginate, is moistened in a bland collecting fluid (water, broth, or buffered saline solution). It is then used to wipe a certain prescribed area of the utensils shortly after they have been washed and dried. The swab is then shaken thoroughly in a known volume (10 ml) of sterile saline solution or, better, broth, in a vial.

After shaking the swab in the collecting fluid, dilution-plate counts are made of the bacteria in the fluid. From the numbers of colonies obtained an estimate is made of the degree of contamination on the dishes. Commonly, a minimum standard of not more than 100 organisms per utensil is recommended.

A standardized method taking these and other factors into consideration has been outlined by a committee of the American Public Health Association. This group specifies formulae for media, solutions, area swabbed, method of swabbing and cultural details. There are also other, direct methods (Fig. 25–6).

Unfortunately, there is no exact method of measuring the amount of disease spread by dirty dishes and eating utensils in unhygienic

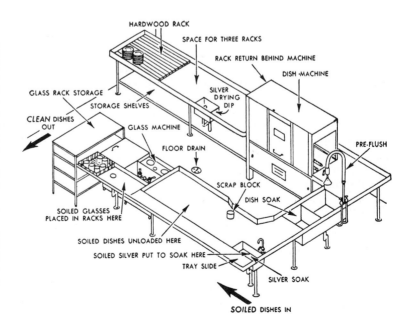

Figure 25–5. One form of modern, sanitary dishwashing equipment. The working bench is of stainless steel. Soiled dishes are piled on the bench in the foreground. They are sorted and scraped, the larger scraps of food dropping into a barrel beneath the counter. Glasses are rinsed over rotating brushes, dipped in disinfectant, and placed in trays in a rack (at left). Silverware soaks in a pan of special detergent solution (right foreground). The dishes, arranged in baskets, are soaked and then given a preliminary rinse with **hard** streams of **hot** water (right). They then pass through a machine dishwasher (center, background). The silver, after soaking, passes through the same process as the dishes and is self-dried after a dip into a drying agent. Afterward all utensils are stacked and stored in dustproof cabinets. Eating utensils handled in this way are virtually sterile.

Figure 25–6. Direct bacteriological examination of eating utensils. Warm sterile fluid agar was poured into the plate. The fork was placed in the agar and moved about to dislodge contaminating material. The dish was then covered, the agar allowed to solidify, and the dish incubated. The bacteria from the fork developed into colonies which are clearly visible. Among these hundreds of colonies there are undoubtedly pathogenic bacteria.

restaurants. From an esthetic viewpoint alone, one does not like to feel that a little saliva from previous patrons is being included, gratis, with his meal.

Paper Dishes. The use of paper cups, dishes, and eating utensils is an effective step toward eliminating the sanitary evils of public glass and chinaware and metal spoons and forks. Not only is expensive dishwashing equipment with its noise, sloppiness, and heat eliminated, but labor and fuel costs are reduced, breakage costs are trifling, and esthetic and sanitary standards enormously improved.* Bacteriological studies of paper used for containers and tableware show negligible content of microorganisms, of which none is pathogenic. Paper dishes are used with success in many snack bars and smaller restaurants but have not yet been perfected to the point where the public accepts them in place of china and glass at formal meals.

Foods. Foods that are moist and not very acid, e.g., soups, puddings, meat stew, or pie (not pickles [acid] or bread [dry]), are excellent

*Possible disadvantages would include increasing waste material and using more wood pulp. But these ecological factors must be balanced against all other factors, such as the large amounts of waste water, detergents, and disinfectants used in some larger operations.

culture media for many microorganisms. The great majority of microorganisms in foods are harmless molds and saprophytic species of bacteria. They may cause the food to become sour, putrid, or otherwise "spoiled" but do not cause infection. However, carriers of respiratory or intestinal pathogens may infect foods with the typhoid bacillus (*Salmonella typhi*) and dysentery bacilli (*Shigella* species) (Chapt. 35); toxin-producing *Staphylococcus aureus* (Chapt. 37); the hemolytic streptococci (*Streptococcus pyogenes*) which cause scarlet fever and septic sore throat (Chapt. 37); the diphtheria bacillus (*Corynebacterium diphtheriae*) (Chapt. 39); and others. The infection of foods with respiratory or enteric pathogens is often the result of carelessness by food handlers who are unrecognized carriers of the organisms and who sneeze or cough over foods or manipulate them with hands unwashed after toilet or after nose-blowing. Raw meats from infected animals are notorious sources of infection: cattle, bovine tuberculosis; swine and poultry, salmonellosis; wild rabbits, tularemia; cattle, swine, and goats, undulant fever. (See the specific diseases for details.)

Foods may or may not be sterilized by cooking. The center of large masses of food, such as a deep pan of bread pudding, macaroni with cheese, or hash is not always raised to a bactericidal temperature by baking or boiling. Further, if the cooked food is infected during handling *after* it has cooled, and is left standing (actually, **incubating**) for hours in a warm kitchen, the persons who eat it might (in some instances) just as well drink a culture in the laboratory. In case of doubt, discard the food or, second best, recook it. This is *second* best because the toxins of staphylococcal food poisoning (one of the commonest) are not destroyed by cooking (see following paragraph). Always keep perishable foods **covered** (avoiding contamination) and **refrigerated** (avoiding incubation).

Food Poisoning. Two other bacterial diseases associated with improperly handled foods are not infections like typhoid fever, scarlet fever, and diphtheria but are **poisonings** caused by toxins preformed in the foods. One form of food poisoning, called **botulism,** is caused by *Clostridium botulinum* (Chapt. 36). The other type of food poisoning, **staphylococcal food poisoning,** is caused by certain strains of staphylococci (Chapt. 37). Botulism is highly lethal; staphylococcal food poisoning is not usually fatal but is highly unpleasant and prostrating.

Milk as a Disease Vector. Sterilized milk is often used in the laboratory as a culture medium, and it is a good one. It is thus clear why milk, incubated for hours in the sun on a loading platform, was, in days before general sanitation and pasteurization, the vector of scores of epidemics of diphtheria, typhoid fever, dysentery, scarlet fever, and sore throat ("strep throat") derived from infected cattle and dairy workers. There are two principal methods by which milk may become infectious for man and cause epidemics:

The milk, as it is drawn from the udder, may contain pathogenic microorganisms which are infecting the udder of the cow. The organisms of most importance in this respect are: *Mycobacterium tuberculosis* var. *bovis; Brucella* (cause of undulant fever); *Coxiella burneti* (the rickettsias of Q fever); *Streptococcus pyogenes* (cause of scarlet fever and septic sore throat), introduced into the udder by a milker carrying the organisms; and *Staphylococcus aureus,* cause of boils, deep tissue infections, and mastitis.

Milk freshly drawn from a *normal* udder usually contains a few harmless contaminants but is free from pathogens. Pathogens of *human* origin may be introduced into the milk *after it is drawn,* by infected persons, utensils, washwater or other vectors. The more important organisms of this sort are: *Salmonella* and *Shigella* (causes of typhoid fever and dysentery); *Streptococcus pyogenes; Corynebacterium diphtheriae;* and *Staphylococcus aureus. S. aureus* usually does not **infect** but if allowed to grow extensively in the milk can make it very poisonous by excreting staphylococcal **enterotoxin** (page 556). The milk may also contain the toxin if secreted by a cow with mastitis caused by *S. aureus.*

Pasteurization. As mentioned previously, pasteurization consists in holding the milk in tanks at 145F (63C) for 30 minutes and immediately refrigerating. Pasteurization is accomplished in many dairies by heating the milk rapidly in a coiled tube or in thin layers between metal plates to 71.6 to 80C and holding at that temperature for 15 to 30 seconds, then cooling. These high-temperature-short-time (**HTST** or **flash**) methods save time and money and are effective so far as sanitation of milk is concerned.

Pasteurization does not sterilize milk. Many bacteria, especially sporeformers, survive 63C for 30 minutes or 80C for 15 seconds. These survivors will cause pasteurized milk to spoil if it is not properly refrigerated. However, pasteurization eliminates all of the **pathogens** referred to above. Staphylococcal enterotoxin is very ther-

mostable and if already present in the milk, as in milk from an udder infected with staphylococci, is not inactivated by pasteurization. Never drink unpasteurized milk or **uncertified** raw milk. (See Chapter 47.)

Blood and Blood Derivatives. Blood not infrequently contains pathogenic microorganisms. In certain infectious diseases the etiologic agents circulate in the blood for varying periods. Typhoid bacilli are readily found in the blood during the first week of the disease. Meningococci not infrequently occur in the blood, even in the absence of meningitis. Rickettsias are present in the blood during typhus and Rocky Mountain spotted fevers. Many viruses (e.g., yellow fever, dengue, encephalitis) and protozoa (malaria parasites or trypanosomes of **sleeping sickness**) also circulate in the blood.

Any organisms circulating in the blood may be transmitted by improperly sterilized cutting or piercing instruments. Rickettsias and many viruses and some protozoa are also transmissible in blood via certain blood-sucking arthropods, their natural vectors (see following pages). A notorious means of transmitting infectious blood is the use of an unsterilized needle by groups of persons to inject drugs intravenously.

Serum Hepatitis. Two closely similar viruses may occur over long periods in the blood of apparently healthy persons: (a) the virus of infectious hepatitis (also called hepatitis virus A); (b) the virus of serum hepatitis (also called hepatitis virus B).

Hepatitis due to virus A has a shorter incubation period and is rarely fatal. Virus B hepatitis has a longer incubation period and may cause 20 per cent fatality.

The latter virus is often carried in blood of donors, serum, plasma, blood-bank blood, and by syringes, needles or instruments not properly sterilized. Very rigid precautions must be taken in handling any human blood, tissues, derivatives thereof, or blood-contaminated instruments to avoid transmission of this virus. Virus B has the property, rare if not unique among viruses, of being able to withstand boiling for some minutes (virus A may withstand near-pasteurization temperatures: 56C for 30 minutes, or more).

It is worth noting that hepatitis virus A is transmitted by contaminated feces, urine, sewage, and food, including shellfish, and causes serious epidemics; hence the term epidemic hepatitis. Virus B is not transmitted in feces unless blood is present.

Blood-Bank Blood. Human blood may temporarily have many organisms in it immediately after any severe injury, after some tooth extractions or surgery, or even in the absence of any injury at all. Such organisms are normally quickly removed by phagocytes and natural antibacterial factors in the blood. Blood drawn at such times for blood-bank purposes may, if not properly refrigerated, contain large numbers of bacteria, because the few bacteria that may have been initially present soon multiply to thousands. Even refrigerated blood may support growth of some psychrophilic organisms. Sometimes blood is contaminated by bacteria introduced by the hypodermic needle from the surface of the skin. However, under proper conditions of collection, storage, and use, danger of infection with bacteria from blood-bank blood is relatively remote.

25.2
INFECTED ARTHROPODS

Arthropod Bites. As noted in the section on viruses, many plant diseases are transmitted by bites of leafhoppers and other insects. There are also many diseases of man and lower animals, which, in nature, are transmitted only by the bites of arthropods. In 1878 a mosquito (*Culex fatigans*) was shown to transmit the worm *Filaria bancrofti*, agent of one form of **filariasis** (a notorious symptom of which is **elephantiasis**). The classical observations of Smith and Kilborne in 1893 on transmission of Texas fever of cattle by the cattle tick (*Boophilus annulatus*) were the first on tick transmission of protozoan disease. Certain mosquitoes (*Anopheles*) were later found to transmit malaria. Usually, but not always, each disease has its own **specific insect vector.** Several arthropods of importance as vectors of pathogenic microorganisms are shown in Figure 25–7.

Several arthropod-borne pathogenic agents are listed in Table 25–2 describing zoonotic diseases.

Arthropod Feces. Cockroaches were shown as early as 1914 to transfer *Vibrio cholerae* in their intestines for at least 48 hours after feeding on human cholera feces. Ants transmit cholera and probably other enteric diseases in the same manner. Flies have long been under indictment for the same crimes.

Many bloodsucking insects deposit feces on the skin when they feed. The feces of lice infected with typhus rickettsias will infect if scratched into the skin. Feces of infected fleas from rats, prairie dogs, and similar rodents contain plague bacilli and may contaminate small wounds or scratches. Indeed, many bloodsucking insects may pass infective agents in the feces and may also cause infection by being crushed on the skin near or in an abrasion or wound. Engorged ticks on dogs are especially dangerous in this respect because they contain a relatively large volume of blood that can transmit the rickettsias of Rocky Mountain spotted fever.

Bodies of Arthropods. Arthropods that fly or crawl from unsanitary, unscreened, and undisinfected privies to hospitals or to dwellings may mechanically transmit intestinal and other disease organisms on their feet and bodies. In areas where flies abound, especially rural or city slum areas, if there is access to infectious sewage or feces, enteric fevers will usually be more prevalent during the summer months when flies are numerous. In places where city sewerage systems are not available, flyborne disease can be avoided to a large extent by the construction of screened and deep-pit or other sanitary types of privies or, better still, by the installation of sanitary plumbing and septic tanks. Plans and specifications for such structures can be obtained from state health departments.

25.3
ZOONOSES

Animals constitute an enormous and everpresent reservoir of agents infectious for human beings. Diseases primarily of animals but transmissible to man are called **zoonoses.** A partial list of zoonoses is shown in Table 25–2.

Animal Bites. Any animal (or human) bite will introduce a mixture of the microorganisms present in the saliva and on the teeth. There is always crushing of tissues and violation of the defensive barriers. The so-called "normal flora" of the mouth is introduced in large numbers. Several ordinarily "harmless" organisms, introduced under such circumstances, can become dangerous, invasive pathogens. Such bites are always infectious and should immediately be opened, cleaned, disinfected, and covered with sterile gauze.

The most notorious pathogen transmitted by animal bites is the virus of **rabies,** or hydrophobia. All mammals are susceptible to rabies and can transmit it. Cats, dogs, foxes, and wolves are particularly dangerous in this respect. Three varieties of bats—vampire, insectivorous, and fruit-eating—have been shown to harbor and transmit rabies among themselves and to cattle, other animals, and man.

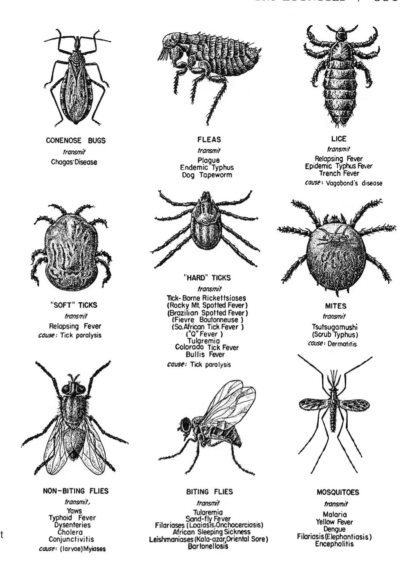

CONENOSE BUGS
transmit
Chagas'Disease

FLEAS
transmit
Plague
Endemic Typhus
Dog Tapeworm

LICE
transmit
Relapsing Fever
Epidemic Typhus Fever
Trench Fever
cause: Vagabond's disease

"SOFT" TICKS
transmit
Relapsing Fever
cause: Tick paralysis

"HARD" TICKS
transmit
Tick-Borne Rickettsioses
(Rocky Mt. Spotted Fever)
(Brazilian Spotted Fever)
(Fievre Boutonneuse)
(So.African Tick Fever)
("Q" Fever)
Tularemia
Colorado Tick Fever
Bullis Fever
cause: Tick paralysis

MITES
transmit
Tsutsugamushi
(Scrub Typhus)
cause: Dermatitis

NON-BITING FLIES
transmit,
Yaws
Typhoid Fever
Dysenteries
Cholera
Conjunctivitis
cause: (larvae)Myiases

BITING FLIES
transmit
Tularemia
Sand-fly Fever
Filariases (Loaiasis.Onchocerciasis)
African Sleeping Sickness
Leishmaniases (Kala-azar,Oriental Sore)
Bartonellosis

MOSQUITOES
transmit
Malaria
Yellow Fever
Dengue
Filariasis (Elephantiasis)
Encephalitis

Figure 25–7. Types of insects that transmit disease.

Eggs. Even the fragile and inscrutable egg is guilty as a disease vector. Poultry are frequently carriers of *Salmonella* (paratyphoid and food-infection organisms). Numerous large outbreaks of food infection (diarrhea) due to *Salmonella* species have been traced to foods made with raw eggs (mayonnaise). Eggs often contain infectious organisms when laid by an already infected hen (Fig. 25–8).

Rats. Rats are well known as vectors of disease to man. Their feces transmit food infection (salmonellosis); their urine, leptospirosis or hemorrhagic jaundice. The fleas of rats transmit bubonic ("black") plague and murine typhus. Their bites, especially in the Orient, introduce the agent of sodoku or rat-bite fever.

Rats should be eliminated by poisoning and trapping and deprivation of food and breeding places through cleanliness and rat-proof construction. Their fleas may be temporarily exterminated by dusting runways and places of refuge with DDT.

Domestic Environments and Diseases. Infectious diseases, both of man and animals, are usually much more frequent and widespread in crowded, unsanitary living quarters than in clean spacious dwellings. This is well illustrated in the case of insect-borne diseases of man such as typhus (body lice) and plague (rat fleas), which are notoriously associated with low-grade living conditions, often as a result of wars. It is equally true of respiratory diseases and enteric infections. Microorganisms spread by oral and nasal secretions, as in sneezing and coughing, and by soiled hands can much more readily be transmitted from person to person in

TABLE 25–2. SOME REPRESENTATIVE ZOONOSES

Diseases	Causative Organism	Animals Principally Involved	Transmitting Agent
VIRAL:			
Eastern equine encephalitis	Eastern equine virus	Birds, equines	Mosquitoes
Japanese B encephalitis	Japanese B virus	Birds, horses, swine	Mosquitoes
St. Louis encephalitis	St. Louis virus	Birds	Mosquitoes
"Jungle" yellow fever	Yellow fever virus	Monkeys	Mosquitoes
Rabies	Rabies virus	Canines and other mammals	Infected saliva
RICKETTSIAL:			
Murine typhus	*Rickettsia mooseri*	Rats	Rat fleas
Q fever	*Coxiella burneti*	Domestic animals (cattle, sheep)	Infected milk, dust, ticks
Scrub typhus (tsutsugamushi)	*Rickettsia tsutsugamushi*	Rodents	Larval mites
Rocky Mountain spotted fever	*Rickettsia rickettsi*	Rodents, dogs	Ticks
CHLAMYDIAL:			
Psittacosis Ornithosis	Chlamydias	Psittacine birds, poultry, pigeons	Oronasal secretions, feces, infected dust
BACTERIAL:			
Anthrax	*Bacillus anthracis*	Domestic livestock	Infected tissues, fluids, dust, hair
Brucellosis	*Brucella* species	Domestic livestock	Animal fluids and tissues, milk
Bubonic plague	*Yersinia (Pasteurella) pestis*	Rodents	Fleas, ticks
Leptospirosis	*Leptospira* species	Rodents, dogs, swine, wild mammals	Animal tissues, fluids, urine
Relapsing fevers	*Borrelia* species	Cave rodents	Ticks, lice
Salmonellosis	*Salmonella* species	Rats, poultry, dogs	Excreta, flesh, eggs
Tuberculosis	*Mycobacterium tuberculosis*	Domestic livestock	Milk, flesh
Tularemia	*Francisella (Pasteurella) tularensis*	Wild rabbits	Deer flies, various ticks, flesh
FUNGAL:			
Ringworm (tinea) favus	*Microsporum* species *Trichophyton* species	Various domestic animals	Contact, hair, dander
PROTOZOAL:			
Leishmaniasis (various forms)	*Leishmania* species	Dogs, cats, rodents	Sand fly
Toxoplasmosis	*Toxoplasma gondii*	Domestic cats	Cat feces
Trypanosomiasis	*Trypanosoma* species	Man, wild game	Tsetse flies (Africa); "kissing bugs" (South America)
Malaria	Simian species of *Plasmodium*	Monkeys	Mosquitoes

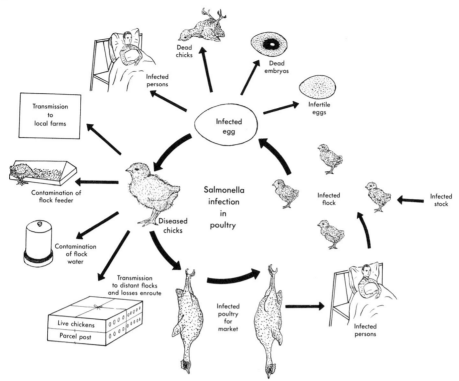

Figure 25–8. Transmission of salmonellosis to and from poultry and man.

close, crowded, cold, and damp rooms than in spacious, well-ventilated, warm, and dry quarters.

Infection by enteric viruses (polio, hepatitis) and other microorganisms of the intestinal tract (dysentery, typhoid and related bacilli, intestinal worms, and protozoa) are obviously transmitted by feces-soiled hands, clothing, soil, water, or food. It is very significant that a direct correlation has been shown to exist between many of these diseases and the **availability of ample clean water for domestic purposes,** especially for washing of hands, and **installation of sanitary plumbing.**

CHAPTER 25
SUPPLEMENTARY READING

Benenson, Abram S. (Ed.): Control of Communicable Diseases in Man, 11th ed. American Public Health Association, New York. 1970.

Betts, A. O., and York, C. J.: Viral and Rickettsial Diseases of Animals, Vol. 1. Academic Press, Inc., New York. 1967.

Hirschhorn, N., and Greenough, W. B. Cholera. Sci. Am., 225:15, 1971.

Isolation Techniques for Use in Hospitals. U.S. Department of Health, Education, and Welfare, Center for Disease Control, Atlanta, Ga. 1970.

Kalter, S. S., and Heberling, R. L.: Comparative virology of primates. Bact. Rev., 35:310, 1971.

Matumoto, M.: Mechanism of perpetuation of animal viruses in nature. Bact. Rev., 33:404, 1969.

Reimann, H. (Ed.): Food-borne Infections and Intoxications. Academic Press, Inc., New York. 1969.

Rosebury, T.: Microorganisms Indigenous to Man. McGraw-Hill Book Co., Inc., New York. 1962.

Top, F. H., and Wehrle, P. F. (Eds.): Communicable and Infectious Diseases, 7th ed. The C. V. Mosby Co., St. Louis. 1972.

CHAPTER 26 · HOST DEFENSE MECHANISMS

Having reviewed how pathogenic microorganisms infect and damage host tissues and how microorganisms are spread, we might now consider the way in which the host deals with this continuous assault upon its integrity. The study of host defenses is generally considered part of, or closely allied to, the rapidly growing field called **immunology.**

Immunology derives its name from the fact that 18th and 19th century microbiologists, many of whom were physicians, were deeply concerned with the causes and prevention of communicable diseases and the reasons for the high degree of resistance or solid **immunity** to the then rampant scourges of typhoid fever, diphtheria, smallpox, scarlet fever, and so on. Many of the basic discoveries of immunology were made during those times. It was only in the first decades of the 20th century that it became generally recognized that the mechanisms of immunity to disease were only one aspect of a very broad and general biological phenomenon.

A modern definition of immunology might state that it is the study of reactions of vertebrates, or certain cells of vertebrates, to proteins, protein complexes, and certain polysaccharide complexes foreign to the reacting cells, whether the foreign substances are of microbial or any other origin. Modern methods and phenomena of immunology are widely used as laboratory tools in many fields of work such as biochemistry, diagnostic and preventive medicine, genetics, ethnology, microbiology, and criminology.

In Chapters 27 and 28 we will consider in greater detail some of the principles, implica-tions, and applications of immunology, but here we wish to examine, in a broad, general fashion, those factors which the host has at its disposal to combat invasions by pathogenic microorganisms.

26.1
BLOOD

Because blood and its constituents play a very important role in host defense and immunological phenomena in general, it is advisable to review some facts concerning blood. For the purposes of this discussion blood may be considered to have seven important constituent parts: plasma, fibrin components, platelets, serum, lymph, erythrocytes, and leucocytes.

Plasma. First, there is the **plasma**—the yellowish, transparent, fluid part of the unclotted, circulating blood. It consists of about 92 per cent water and 7 per cent proteins, and is a solution of salts, buffers (to maintain a constant reaction of pH 7.4) and other soluble substances including cell foods (such as amino acids and glucose) and cell wastes. The plasma also has in solution the components of **fibrin,** which are essential in the clotting of blood and also of plasma. Essentially, plasma is blood minus all of its cellular elements.

Fibrin Components. Fibrin, as such, does not normally occur in circulating blood. Fibrin components ordinarily combine only after the blood leaves the blood vessels. Fibrin is a protein which forms an elastic network of microscopically fine fibrils. As a result of the formation of fibrin, blood clots. The elastic fibrin meshwork soon shrinks to about half the original volume of blood, squeezing out of its meshes the fluid part of the blood, which is now called **serum.** Serum is equivalent to plasma minus the fibrin components. Most of the blood cells (and bacteria if any are present) are caught and held in the fibrin clot (Fig. 26–1).

Platelets. Associated with fibrin production are small bodies called **platelets,** or **thrombocytes** (Gr. *thrombus* = clot). These are deeply staining, amorphous or stellate particles, variable in size but smaller than red blood cells (Fig. 26–2). They are fragments of marrow cells. There are normally about 350,000 platelets per cubic millimeter of blood. Their role in the coagulation of blood is important, though not entirely clear. Platelets appear to facilitate clotting by liberating one of the components of fibrin called **thromboplastin.** They also have other functions in the clotting process. The ability of blood to clot appears to be directly related to the number of platelets, calcium ions, and ample vitamin K in the blood.

Serum. Since fibrin usually enmeshes the blood cells as it forms, serum is yellowish and transparent, like plasma. Serum contains most of the soluble substances, especially **gamma globulins (antibodies),** in which the immunologist is interested. After blood has clotted in a test tube or another type of container, the serum may be withdrawn in pipettes, centrifuged to remove stray blood cells, and stored in the refrigerator. It must be handled with every aseptic precaution to keep it sterile,

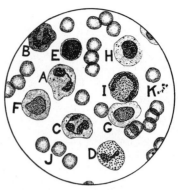

Figure 26–2. Drawing of a smear of blood stained with Jenner's stain, showing common forms of blood cells. *A, B,* and *C,* Polymorphonuclear leucocytes with two-, three-, and four-lobed nuclei, respectively. *D,* Eosinophil showing lobular nucleus and prominent, eosinophilic (red-staining) granules. *E,* Lymphocyte. *F, G,* and *H,* Various forms of monocytes. *I,* Lymphocyte with horseshoe-shaped nucleus ("transitional cell"). *J,* Erythrocytes (red blood cells); note biconcave-disk shape, thin at center. *K,* Platelets. (About 1,000×.)

since microorganisms gaining access to it would soon cause it to decompose. Immunologists often add minute quantities of preservatives.

Lymph. Lymph is very much like blood that has been deprived of red cells and about half its protein content by passing through the intercellular spaces in the thin walls of the smaller blood vessels as through a fine filter (i.e., lymph closely resembles plasma). The lymph thus seeping out of the vessels travels slowly in the fine spaces surrounding the blood vessels and between the tissue cells and various organs. It contains **leucocytes** or white blood cells that are derived from lymphatic tissues. Like plasma, it can clot. Lymph is in contact with, or close to, all active tissue cells. Lymph is the nutrient fluid or "culture medium" in which tissue cells grow in the animal body. It also collects the cell wastes. It eventually collects from all parts of the body in large drainage vessels (the **lymphatic ducts**) and is returned to the blood. Wastes carried from cells by serum and lymph are excreted in the kidneys.

Erythrocytes. In the plasma, before clotting, are suspended the red cells or **erythrocytes** (Gr. *erythro* = red; *cyte* = cell). These are the non-nucleated* cells that give blood its opacity and red color and that carry oxygen from the lungs to the tissues. The color of

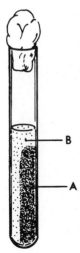

Figure 26–1. Tube with clot (*A*) which has shrunk, enmeshing virtually all of the blood cells and exuding clear serum (*B*).

---

*Mammalian erythrocytes are non-nucleated; erythrocytes of birds, amphibians, and reptiles are nucleated.

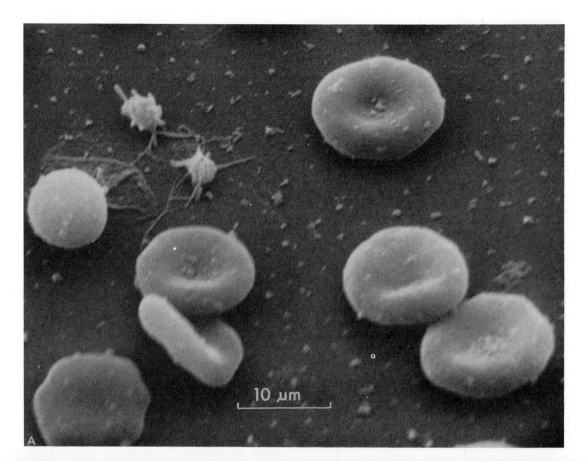

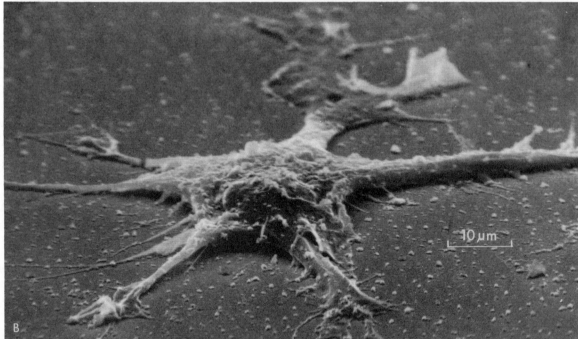

Figure 26–3. Scanning electron micrographs of formed elements in the blood of the mouse. *A,* Erythrocytes along with one lymphocyte (*center left*) and two platelets (*upper left.*) *B,* A macrophage.

Figure 26–3 continued on opposite page.

erythrocytes is caused by the red, oxygen-carrying substance, **hemoglobin** (respiratory pigment), which they contain.

Erythrocytes are about 8 μm in diameter and number four to five million per cubic millimeter of blood. When they are ruptured by plasmoptysis or other means, the hemoglobin is released and the blood becomes transparent, like red ink. The released hemoglobin appears in the urine. The cells are said to have been "laked," or **hemolyzed.** Certain bacteria are very active in producing **hemolysis** by means of toxins. Some snake venoms are very hemolytic.

26.2
THE PHAGOCYTIC CELLS

Leucocytes. In addition to erythrocytes, blood contains several kinds of colorless cells called, collectively, **leucocytes** (Gr. *leukos* = white). Leucocytes are relatively large, 10 to 20 μm in diameter. Unlike mammalian erythrocytes, leucocytes have a definite nucleus. Normally there are seven to eight thousand leucocytes of all types per cubic millimeter of blood (Fig. 26-2). Recent photomicrographs of various mammalian leucocytes taken with the scanning electron microscope reveal their striking features much more vividly than in conventionally stained smears (Fig. 26-3). In many infectious processes, such as appendicitis, the leucocytes increase greatly in numbers (up to 15,000 or more), and the patient is said to have a **leucocytosis.** In some infections they diminish in numbers to 3,000 or fewer, a **leucopenia.**

The several kinds and sizes of leucocytes may be listed as follows:

    I. Granulocytes (or Myelocytes)
       A. Polymorphonuclear leucocytes (12 to 14 μm)
         1. Neutrophils
         2. Basophils
         3. Eosinophils
   II. Agranulocytes
       A. Monocytes (16 to 22 μm)
       B. Lymphocytes
         1. Large (15 to 20 μm)
         2. Small (10 to 14 μm)

The **polymorphonuclear leucocytes** are so called because their nuclei are divided into two to five distinct lobes connected by thin threads. They are classed as **granulocytes** because they contain numerous conspicuous granules. They originate in bone marrow where erythrocytes are formed (myeloid tissue), and are therefore

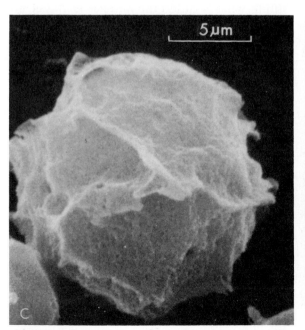
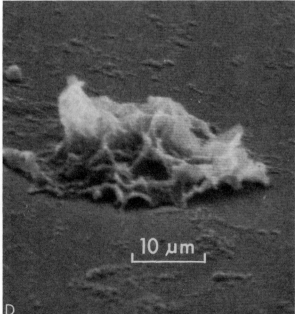

Figure 26-3 *Continued.* *C,* A splenic lymphocyte. *D,* A neutrophil, one variety of polymorphonuclear leucocyte. (Courtesy of R. Albrecht.)

also classed as **myelocytes**. Three types are differentiated by the nature of their granules. Granules of the **neutrophilic** polymorphonuclear leucocytes have no particular affinity for either acidic or basic dyes. Granules of the **basophils** stain darkly with basic dyes like crystal violet, while granules of the **acidophilic** type (**eosinophils**) stain bright red with acid dyes like eosin. The functions of the granules are not clear.

The **monocytes** are the largest of the leucocytes and have a large, kidney-shaped or lobular nucleus. No prominent granules are seen in the cytoplasm of these cells, and they are therefore classed as **agranulocytes**. They are long-lived cells which develop from lymphocytes or lymphoid stem cells. Unlike erythrocytes and granulocytes, they have the capacity to multiply. When monocytes leave the blood and enter the tissues, they become indistinguishable from macrophages and histiocytes (see following section). The neutrophils (sometimes called microphages), monocytes, and macrophages are the cells primarily responsible for phagocytosis of microorganisms and of dead or injured body cells.

The **large** and **small lymphocytes** have few if any conspicuous granules and are classed as agranulocytes. They are produced mainly in the lymphatic tissues; i.e., germinal centers in the spleen, lymph nodes, adenoid tissues, and, until the onset of puberty, in the thymus. Their nuclei are large and rounded. The lymphoid cells that are the sites of the continued production of small lymphocytes in spleen and adenoid tissues (outside the thymus) probably have their origin in the thymus in fetal and neonatal life.

A very important property of the polymorphonuclear neutrophilic granulocytes ("polymorphs") is that they can move about in the tissues and body fluids by means of pseudopodia, remarkably like amebas. Also like amebas, they can ingest small solid particles such as bacteria, cellular detritus, and other foreign particulate matter (Figs. 26–4, 26–5). Thus they serve as scavengers and also as "policemen" in the blood. By means of intracellular enzymes they injure or kill most types of invading microorganisms that they ingest and then digest the organic particles. Because these cells can "eat" cells and other small particles they are called **phagocytes** (Gr. *phagein* = to eat). Because the neutrophils are the most numerous and active of the granulocytes, the term leucocyte is often used as though it were synonymous with polymorphonuclear neutrophilic leucocyte.

In normal human blood the ratios (per cent or "**differential count**") of different types of leucocytes vary greatly but commonly are:

	Per Cent
Neutrophils	60–70
Basophils	0–2
Acidophils	0–4
Lymphocytes	25–30
Monocytes	2–8

In acute infections the differential count shows a relative increase in neutrophils or lymphocytes.

The lymphocytes show little if any phagocytic activity but, as the source of **antibodies**, play an extremely important role in the defensive mechanisms that are discussed farther on.

Macrophages and Histiocytes. The phagocytic cells we have described so far float freely in the blood and lymph or wander in the tissue spaces. They are sometimes called free or "wandering" phagocytes. There is another group of phagocytic cells called **fixed phagocytes** or, because they are relatively large, **macrophages** or, because they are fixed portions of tissues, **histiocytes** (Gr. *histion* = tissue). Fixed phagocytes are principally cells that occur in reticular connective tissue; some are special endothelial cells of sinuses of liver (Kupffer cells; Fig. 26–6), spleen, lymphatic tissue, and bone marrow. Some are monocytes that have become localized in tissues. The macrophages and histiocytes phagocytize foreign particles from the blood or lymph as it flows past them. Together they constitute what is called the **reticuloendothelial** system of phagocytes; it is one of the most important defensive mechanisms of the body.

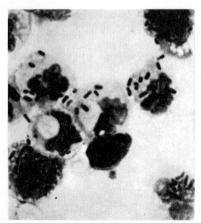

Figure 26–4. Stained smear of pus from lung of mouse inoculated with a species of pathogenic bacilli. The bacilli are seen to have been engulfed by the leucocytes (phagocytized) in large numbers. This is an excellent illustration of one of the most important defensive measures.

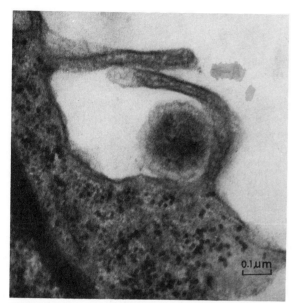

Figure 26–5. A guinea pig polymorphonuclear leucocyte engulfing a single cell of *Brucella abortus*. (Courtesy of Dr. Marilyn Zanardi Tufte, University of Wisconsin, Platteville, Wis.)

26.3
CELLULAR VS. HUMORAL IMMUNITY

As mentioned in Chapter 2, the relation of phagocytosis to defense against infection (im-

munity) was first pointed out by Elie Metchnikoff (Nobel Prize winner) about 1882 and was the origin of the doctrine of **cellular immunity**. In 1890 certain proteins, now known to be **gamma globulins** and called **specific antibodies** or **immunoglobulins,** were discovered by von Behring (Nobel Prize winner), Buchner, and others when they examined the serum of persons and animals who had been vaccinated against, or who had recovered from, certain specific infectious diseases. These specific antibodies were found to be defensive only against the specific microorganisms, or their toxins, that evoked those particular antibodies. For example, diphtheria antibodies (**antitoxin**) "neutralize" diphtheria toxin, but not the toxin of tetanus or any other kind. Specific antibodies in the blood were therefore hailed as the real basis of immunity to disease and gave rise to the doctrine of **humoral** (fluid) **immunity.** A long, and often acrimonious, controversy arose between the two schools of thought.

Antibodies and Phagocytes. Long and intensive studies proved that both phagocytic cells and antibodies are important in immunity. Phagocytes can act without the aid of antibodies, but specific antibodies, with one or two possible exceptions, are important **adjuvants** to phagocytosis (see discussion of **opsonins**).

A great deal has been learned through cross-

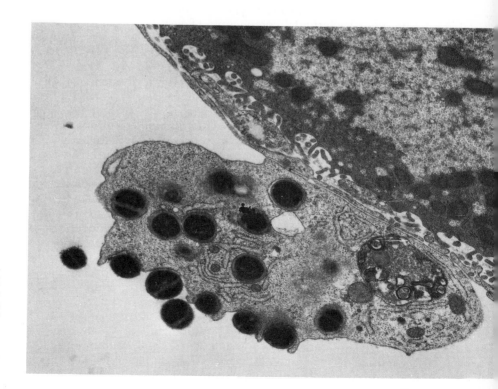

Figure 26–6. Electron micrograph showing a rabbit Kupffer cell (fixed phagocytic cell in liver) ingesting cocci (*Staphylococcus aureus*). (14,000×.) (Courtesy of Drs. R. G. Horn and E. D. Collins, Department of Pathology, Vanderbilt University School of Medicine.)

protection studies in experimental animals. For example, if *Streptococcus (Diplococcus) pneumoniae,* an encapsulated coccus, is injected into a mouse, generally the mouse will die. However, if specific antibody against the pneumococci or their capsular substance is injected into the mouse prior to challenge with the virulent microorganism, the mouse will be protected and will not die. The explanation for this observation is that without specific antibody present, the phagocytes are unable to engulf and destroy the parasite. Specific antibody, when present, alters the surface of the parasite and promotes phagocytosis and killing (see discussion of opsonization in the following section). Thus antibody can be shown to play a vital role in a number of bacterial infections and to be a useful adjunct in still other infections, including rickettsial and viral infections.

With other infectious microorganisms, however, the level of specific antibody circulating in the body does not alter significantly the final outcome of the infection. This is particularly true with the so-called "facultative intracellular parasites," e.g., *Salmonella, Mycobacterium, Brucella,* and *Listeria.* These organisms are not **obligate** intracellular parasites like the rickettsias or viruses, since they can be grown in test tubes containing various media. Once they initiate an infection, however, many of the leucocytes, as well as some somatic cells, will be found harboring these organisms. In this intracellular environment they are almost impervious to antibody, other bactericidal components of serum, and even antibiotics. Infections with these microorganisms often tend to be chronic, drawn-out affairs, but many individuals do eventually recover and then are immune to challenge with dosages of these organisms large enough to kill a normal animal. Thus, a different kind of immunity, termed **cellular immunity,** which is not dependent on circulating antibody, is recognized as being extremely important.

No one can visualize humans or animals existing for long in the normal environment, however, without both types of host defenses working hand-in-hand. Viral infections offer a good example. Often, the mere presence of circulating specific antibody is sufficient to protect an individual from a viral infection. Once the viral infection starts, however, and the viruses are intracellular, then convalescence depends largely upon the development of cellular rather than humoral immunity.

The precise mechanism of cellular immunity is still under investigation. We know that normal phagocytic cells, particularly with the aid of antibody and complement (see material on **complement** further on in this chapter), can ingest and kill a large variety of pathogens. But some pathogens, like the facultative intracellular parasites, are not killed after being engulfed. They not only survive but start to multiply intracellularly and eventually kill the phagocytic cell. In order to cope with this situation, the body must develop cellular immunity, which in turn rests on the production of a new (or altered) kind of phagocytic cells called **activated macrophages.**

Experimentation has shown that cellular immunity, unlike humoral immunity, cannot be transferred from one animal to another with serum but only with lymphocytes. This and other evidence have led investigators to believe that lymphocytes exposed to bacterial antigens become "sensitized." The sensitized lymphocytes then release one or more mediators (**lymphokines**) which, among other things, are believed to convert normal macrophages into activated macrophages. In this state, the macrophages have elevated enzyme levels, metabolism, etc., and are now able to engulf and kill almost all pathogens they encounter. Further details on cellular immunity will be found in the next chapter.

Opsonic Effect of Antibodies. If a small number of bacteria are mixed with a drop of freshly drawn blood, and the mixture is placed under a coverslip and examined with a microscope, the reaction of the phagocytic cells to the bacteria can be observed. When the phagocyte comes near to a bacterial cell, the former seems to sense the presence of the bacterium, moves directly toward it, and attempts to phagocytize it. The nature of this attraction is poorly understood (see discussions of macrophages and complement), but it is called **chemotaxis** (Gr. *chemo* = chemical; *taxis* = orientation toward). Certain encapsulated bacteria, on the other hand, often show what is called **negative chemotaxis.** Here, the phagocyte approaches the bacterial cell, suddenly stops, and then turns and moves away. Certain capsular substances thus seem to be offensive or toxic to the phagocytes.

Antibodies, when present, neutralize the toxic effects of capsules and greatly promote phagocytosis. This effect is known as **opsonization** (Gr. *opsonin* = to prepare food for). In addition, antibodies frequently cause bacteria to clump, and these irregular masses are easier for phagocytic cells to engulf than a single free-floating cell.

Surface Phagocytosis. Even without the clumping effect of opsonins, if the phagocytes can get the microorganisms against a surface from which the organisms cannot escape ("on the ropes," to use a prizefighter's term), the phagocytes can grasp the organisms much more effectively. The surfaces of adjacent tissue cells or of any uneven or rough surface, or strands of fibrin, serve the purpose very well. Phagocytosis is, therefore, not necessarily dependent on antibodies, nor even on surfaces.

Genetic and Constitutional Factors. In addition to those cellular and humoral factors just enumerated that serve as defenses against infection, there are several rather general, *nonspecific* physiological and anatomical factors that determine susceptibility to infection. Such factors are nonspecific in the sense that they are not directed at any particular organism and are generally present (at least at some level) even if there has been no past history of a confrontation between host and pathogen. These may be grouped as follows:

I. Genetic factors
II. Physiological factors
    A. State of general health and age
    B. Mechanical and chemical factors
    C. Inflammatory response

Genetic Factors. Frequently, a particular host will not be susceptible to a microorganism normally considered a pathogen. For example, horses do not contract measles, chickens do not get syphilis, and pathogens of mammals generally do not infect plants. The facts are obvious, and sometimes the explanations are also evident: "cold-blooded" animals (normal temperatures from 40 to 80F) and birds (normal temperatures around 104F) are not susceptible to microorganisms that grow only at mammalian temperatures (normal temperatures around 98F). However, as Pasteur proved in a dramatic demonstration, chickens will die of anthrax, to which they are ordinarily resistant because of their high body temperature, if they are infected and then cooled by partial immersion in ice water. In many instances complete explanations of species and racial resistance are lacking. Species resistance undoubtedly involves complex chemical factors in addition to differences in temperature. Cultured cells (Chapt. 16) of an animal may be very susceptible to infections to which the intact animal is wholly resistant. For example, embryo chicks succumb to many infections that do not affect the adult bird.

Physiological Factors. STATE OF GENERAL HEALTH AND AGE. Persons weakened by overwork, starvation, exposure, alcoholism, drug addiction, age, and disease are known to become more susceptible to various infections such as tuberculosis and pneumonia. Children often suffer little from diseases that are fatal to older persons, and vice versa. The mechanisms are still obscure.

MECHANICAL AND CHEMICAL FACTORS. The **outer skin,** especially of adults, is an obvious mechanical barrier to the entrance of many foreign agents. It is chemically aided against microorganisms by its oily secretion and by the acidity of perspiration. The acetic acid in perspiration is quite toxic to many bacteria. The skin may be bypassed, however, by hypodermic needles, by wounds, and by entrance of certain microorganisms into deeper tissues via hair follicles and sweat glands.

The **hairs in the ears and nose** mechanically entangle or enmesh particles of dust, insects, and bacteria. Secretions of **mucus** cause all of the respiratory surfaces, such as those in eyes, nose, and throat and others in contact with the exterior, to be sticky. Foreign bodies accumulate in the mucus. Removal from the respiratory tract to the exterior is then accomplished by sneezing, coughing, salivation, and tears. The **deeper air passages** are lined with **ciliated epithelial cells.** The cilia maintain a constant upward-waving movement that pushes mucus (with entrapped bacteria and dust) up to the larynx and throat where the mucus is either coughed up or swallowed. Further, these mucous surfaces are always "policed" by small numbers of leucocytes.

In the **gastrointestinal tract, acidity** of the stomach (pH, about 2.0) kills many organisms. The **upper intestine** is freed of microorganisms to a great extent by the **bile** and other **digestive juices.** The lower small intestine and large bowel contain great numbers of bacteria, many of which are highly pathogenic if they gain entrance to the blood or body tissues. They are normally held in check by the thick mucous membranes lining the intestines and by phagocytic and other mechanisms. A ruptured intestine, stomach or appendix is a source of serious infection which, if untreated, is usually fatal.

The adult **genitourinary tract** is protected against most bacteria mainly by thick mucous membranes, leucocytes and the flow of *urine,* which is normally acid. A protective acidity is maintained on the mucous membranes of the vagina by growth of harmless, acid-producing bacteria of the genus *Lactobacillus.*

THE INFLAMMATORY RESPONSE. Inflammation is a complex response of tissues to damage or irritating agents of any sort: chemicals, burns, excessive ultraviolet irradiation (sunburn), mechanical injury, bites and stings of insects and infections. Inflammation is characterized by five distinctive features: **calor** (heat),

rubor (redness), **turgor** (swelling), **dolor** (pain), and **infiltration** by phagocytic cells. Inflamed areas feel hot because of local dilatation of blood vessels and increased blood supply; redness also is caused by dilatation of the capillary blood vessels locally. Swelling is caused by the extra blood in the dilated capillary vessels and by extravasation (seepage outward), under pressure, of lymph, plasma and serum into the local lymph and tissue spaces. This accumulation of fluid in tissues is called **edema.** Pressure on and irritation of local sensory nerve endings causes pain. Nonspecific substances called **pyrogens** (Gr. *pyr* = heat) emanate from the site of injury and circulate in the blood. If they occur in sufficient amount they cause **fever.**

Within a few minutes after plasma seeps into injured tissues, fibrin threads begin to form a meshwork in the edema fluid. This tends to contain and prevent the spread of the invading microorganisms. Less evident, but just as important, is the attraction of both phagocytic and antibody-producing leucocytes to the area. They start their journey by adhering to the linings of the local capillaries. Then, by their ameboid motility, the phagocytic leucocytes leave the capillary blood vessels, passing between the cells of the capillary walls by a process called **diapedesis,** somewhat as ghosts are said (by some) to pass through keyholes. The leucocytes congregate in tissues and blood wherever any injury or infection exists. The cause of the attraction of the leucocytes is not fully known, but it appears to be a substance released by injured cells in the affected area.

Often many phagocytic leucocytes are killed by the poisons of the bacteria or other noxious agents that they ingest, and by crowding, acidity, and lack of oxygen in the region of concentration and activity. The white, creamy material in a boil or other infected lesion, or around a festering splinter, is made up largely of dead and living white corpuscles, dead and living bacteria, tissue debris, lymph, serum, and possibly fibrin. It is called **pus,** and the dead leucocytes in it are called **pus cells.**

On the other hand, leucocytes are sometimes unable to kill or suppress the growth of microorganisms that they ingest. Leucocytes containing live, virulent microorganisms may then be carried by the blood or lymph to other parts of the body, there to set up **secondary** or **metastatic** infections. This is particularly true of the facultative intracellular parasites described earlier.

Infectious agents of any sort contain substances (**antigens**) that stimulate antibody-forming cells (mainly large lymphocytes) to produce specific antibodies. The antibodies evoked by the infecting agent begin to appear within a few hours after infection. It is probably because of these and fibrin formation, the constant presence of pre-existing normal or nonspecific antibodies in the blood (see next section), and phagocytic cells that inflamed tissue tends to **localize** and hold the infective invaders. Histological studies reveal that the host defenses actually establish a wall of specialized cells around the injured tissue. Increased phagocytosis of antibody-covered (**opsonized**) microorganisms also occurs. If the inflammatory reaction in the local tissues, with its antibodies and phagocytes, can then hold and destroy the invader, the infection is suppressed. If not, then the victim may succumb.

During the inflammatory process, new tissue cells (**fibroblasts**) grow in and around the area, attempting to immure and contain the infection and to replace dead tissue with tough, new tissue, which eventually becomes **scar tissue.**

26.4
ANTIMICROBIAL SUBSTANCES PRESENT IN BLOOD AND TISSUE FLUIDS

In addition to phagocytic cells and specific antibody mentioned earlier, there are constantly present in the normal blood, as defensive mechanisms, soluble, **nonspecific,** antimicrobial factors which are enzyme-like or antibody-like. The list of nonspecific factors has grown considerably over the years, as shown in Table 26–1, and will undoubtedly continue to expand with further research. We shall not attempt to describe and explain the function of all these factors, but shall concentrate on those that have been more fully characterized. Natural antibodies are those that occur in various body fluids in the absence of any *known* infection or vaccination. A controversy of long standing concerns the question whether such antimicrobial substances are the result of unknown, specific, antigenic stimulation (e.g., unnoticed infection) or whether they can occur in the absence of any antigenic stimulus. "Normal" serum has long been known to be actively bactericidal. The antibodies are especially active against gram-negative bacteria. Some act like specific agglutinins, some like bacteriolysins, and some like antitoxins. The bactericidal or bacteriolytic natural antibodies, like specific

TABLE 26-1. ANTIMICROBIAL SUBSTANCES ASSOCIATED WITH BLOOD AND TISSUE FLUIDS°

Substance	Source	Chemical Nature	Antibacterial Specificity
Complement	Serum	Euglobulin with carbohydrate	Gram-negative
Properdin	Serum	Euglobulin	Gram-negative
Lysozyme	Ubiquitous	Small basic protein	Gram-positive†
β-Lysin	Serum	Protein	Gram-positive
Phagocytin	Leucocytes	Labile protein	Gram-negative
Leucin	Leucocytes	Basic peptides	Gram-positive
Tissue polypeptides	Lymphatics	Basic peptides	Gram-positive
Plakin	Blood platelets	Peptide	Gram-positive
Histone	Lymphatics	Basic peptides	Gram-positive
Hematin	Red blood cells	Iron porphyrins	Gram-positive
Spermine	Pancreas	Basic polyamines	Gram-positive
Interferon	Infected cells	Small protein	Viruses

°Modified from Carpenter: Microbiology, 3rd. ed., W. B. Saunders Co., 1972.
†Gram-negative cells affected in the presence of EDTA, serum, etc.

antibodies (Chapt. 27), require the presence of Ca^{2+} and Mg^{2+} and of another nonspecific serum component called **complement.**

Complement. Complement is actually a mixture of about 11 proteins which normally exist in serum in an inactive form. By itself, complement (abbreviated C) has little or no antimicrobial power, but it aids and **complements** (hence its name) the action of various types of antibodies. The current designations for the various components of complement are: C1q, C1r, C1s, C4, C2, C3, C5, C6, C7, C8, C9. When complement becomes activated, the various components function *in sequence*, proceeding from left to right through the list above. For some reactions, all components are activated, while for others, the activation may stop at C3, C5, etc. The cations Ca^{2+} and Mg^{2+} are necessary for almost all functions of complement. Heating at 56C for 30 minutes or holding at room temperature for several hours destroys complement activity. Freshly drawn guinea pig serum is one of the best sources of complement for serological work, but complement (at some level) is found in most vertebrates.

Direct evidence of how complement might act in vivo to aid the body in host defense has been difficult to obtain, but there seems to be little reason to doubt that it behaves much the same as it does in in vitro tests. Of the various reactions in which complement can participate, several seem to be of special significance.

Lytic Reactions. Whenever antibody attaches to antigen to form a complex, complement attaches to this complex. If the antigen happens to be the membrane of a red blood cell, or the cell membrane of a gram-negative bacterial cell, the result of combination of C with cell and antibody is lysis of the cell. The lysis of red blood cells (RBC's) by C forms the basis of a valuable serological test called the **complement-fixation reaction** (see page 388). Gram-positive bacteria and mycobacteria are not subject to lysis by complement, and the basis for their resistance is poorly understood but may have something to do with the chemical composition of the cell walls.

Immune Adherence. Complement can greatly increase the adherence of immune complexes (antibody-antigen) to various types of particles, e.g., RBC's, platelets, starch granules, etc. This could be of some importance if the antigen happens to be a virus or bacterial cell, since the efficiency of phagocytosis is increased (see discussion of surface phagocytosis).

Leucocyte Chemotaxis. Antigen-antibody complexes which have bound complement have been found to attract phagocytic cells (granulocytes). Thus, complement acts as a chemotactic factor.

Undesirable Complement Reactions. For the most part, complement is considered a valuable adjunct to host defense mechanisms. Some humans and several species and strains of experimental animals have been found to lack certain of the C factors and their sera have been shown to possess greatly reduced bactericidal and hemolytic power.

On the other hand, complement can sometimes have a deleterious effect. Certain humans suffer from a disease called hereditary angioneurotic edema in which an inhibitor of C1, present in normal serum, is almost entirely lacking. This allows C1 to reach high levels and this can lead to life-threatening accumulation of edema fluid. This is sometimes sufficient to block the respiratory passages. In addition, certain components of C can interact to produce a small polypeptide cleavage product called **anaphylatoxin.** This substance can cause the release of histamine from mast cells and the effects can mimic the severe allergic reaction called **anaphylaxis** (see Chapter 29).

The Properdin System. A group of serum components that work together, the **properdin system** (L. *pro* = to prepare; *perdo* = to destroy) apparently plays an important role in nonspecific resistance to infection. There are three components in the system: **properdin,** a serum protein; **magnesium** and **calcium** ions; and **complement.** The precise nature of properdin is still unknown, but recent work indicates that it is a distinct serum protein that has nonspecific antimicrobial powers. The amount of properdin activity in the blood seems to be directly related to the degree of nonspecific resistance of an animal to numerous types of infection: bacterial, protozoal, and viral. Injection of properdin-rich serum from animals having naturally high levels of properdin activity in their serum increases resistance to infection. Any agent (hemorrhage, shock, electromagnetic irradiations, cancer, and infection) which lowers the properdin activity of the blood also lowers nonspecific resistance to infection. The injection of zymosan, or of certain lipopolysaccharides (**endotoxins**) found in the cell walls of many gram-negative bacteria, results in a rapid lowering of properdin activity.

Lysozyme. A bacteriolytic enzyme, **lysozyme** is found in tears, saliva, white of egg, tissues, and leucocytes. As previously stated, it was discovered in 1922 by Fleming, the discoverer of penicillin. It acts (at least in vitro) to destroy the cell wall of several species of gram-positive (rarely gram-negative) bacteria. It produces, as does penicillin, protoplasts (**L bodies**) in hypertonic media. Though theoretically an important nonspecific defensive agent, its true significance in vivo is still obscure.

Other Nonspecific Factors. In general, other nonspecific factors are not well defined. Their action is not clearly understood and has been demonstrated chiefly by in vitro experiments. It is suggested by some that these substances are released from tissue cells (including leucocytes) only after injury and rupture of those cells, and that they do not exist free in the blood until such injury occurs. However this may be, the fact has been amply demonstrated that, in vitro, the fresh blood or serum of most animals exerts actively microbicidal action against numerous species. It seems reasonable to suppose that such mechanisms observed in vitro also occur in vivo.

β-**Lysins.** The *β*-lysins are a group of poorly defined lytic agents that occur in fresh serum and exert bactericidal action mainly against gram-positive bacteria. Their origin and the nature of their action are obscure.

Basic Polypeptides. Basic polypeptide structure is known to characterize certain antibiotics (e.g., polymyxins, subtilin, bacitracin). Many of these are surface tension reducers. As previously indicated, low surface tension is inimical to numerous species of microorganisms, destroying the cell walls. Like the polypeptide antibiotics, certain serum- and tissue-derived polypeptides are surface tension reducers active mainly against gram-positive bacteria.

CELL-DERIVED FACTORS. Several antimicrobial substances are liberated by the disintegration of phagocytes, one of which, **phagocytin,** is lethal for gram-negative bacteria. Others, **leucins** and **plakins,** derived from platelets, adversely affect gram-positive bacteria. More studies of these substances are needed.

Interferons, substances that interfere with replications of viruses, are released by virus-infected cells. (See Chapter 17.)

26.5
INTEGRATED LINES OF DEFENSE

We have now reviewed most of the factors used by the host in defense against infectious microorganisms. The various mechanisms which pathogens use to overcome these defenses were reviewed in Chapter 24. The host defense system normally excludes, or removes, all microorganisms from the blood stream and tissues proper. To the degree that this is not achieved, the individual is infected.

The oral cavity — intestinal tract — anus passage should be considered merely a tube through the body. Organisms in this tube, like those on the skin, are still outside of the body tissues with regard to host defenses. Infections, when they do occur, may be obvious or unnoticeable (subclinical), acute or chronic, mild or severe, life-threatening or a minor nuisance, depending on the balance of factors depicted in Figure 26–7.

In order to gain a foothold, the pathogen must penetrate what might be called the **physicochemical barriers** or **first line of defense.** This consists of the tough protective layer of skin with its fatty acids and low pH, the mucous membranes with their flushing actions and con-

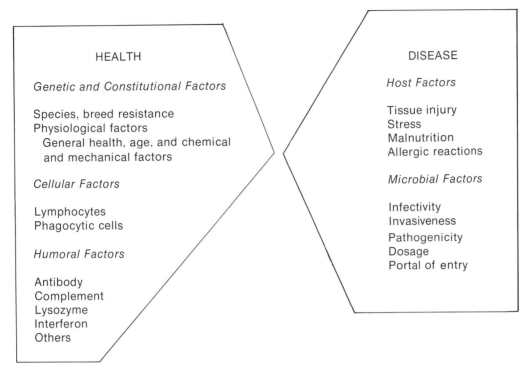

Figure 26–7. Various factors which interact to determine whether a particular host will appear healthy or diseased.

tent of lysozyme and other antimicrobial substances, the anatomical trap of the nasal passages with its sticky mucus and hairs which filter out any particle larger than 5 μm, the sticky mucus moved up the bronchioles by ciliated epithelial cells, and, finally, the blood itself with its full range of humoral factors (antibody, complement, properdin, etc.).

If all this fails to stop the invaders, the **cellular barrier** or **second line of defense** still has to be dealt with. This consists of the various types of phagocytic cells which have as their major task the job of engulfing, immobilizing, and destroying the pathogens.

The **third line of defense** is **specific immuno-globulins** or antibodies. These are produced by body cells after contact with the pathogens that have breached the physicochemical defenses.

In most encounters the phagocytic and humoral defenses will be greatly fortified by previous experience with the particular foreign agent, cellular or soluble. Such previous experience may consist of mild (subclinical) or severe infection, or injections of toxoids, vaccines, etc. Reinfection or restimulation by injections quickly **recalls** waning or vanished humoral and cellular immunity. The ability to recall an immune state is another of the remarkable host defenses. It will be discussed in the next two chapters.

CHAPTER 26
SUPPLEMENTARY READING

Barrett, J. T.: Textbook of Immunology. The C. V. Mosby Co., St. Louis. 1970.
Bellanti, J. A.: Immunology. W. B. Saunders Co., Philadelphia. 1971.
Colby, C., and Morgan, M. J.: Interferon induction and action. Ann. Rev. Microbiol., 25:333, 1971.
Hilleman, M. R., and Tytell, A. A.: The induction of interferon. Sci. Am., 225:26, 1971.
Pearsall, N. N. and Weiser, R. S.: The Macrophage. Lea & Febiger, Philadelphia. 1970.
Weiss, L.: The Cells and Tissues of the Immune System: Structure, Functions, Interactions. Prentice-Hall, Inc., Englewood Cliffs, N.J. 1972.

CHAPTER 27 • PRINCIPLES OF IMMUNOLOGY AND SEROLOGY

In the previous chapter we discussed various mechanisms of host defense. Both normal and specific antibodies were mentioned as protective serum proteins without trying to explain their origin, control, or precise interactions. In this chapter we will consider some of the principles of immunology and several of the tests and procedures which are of particular interest to microbiologists. Immunology is a large, rapidly expanding area of science, much of which has little to do with protection against infectious microorganisms but which does have as its major concern the overall health and protection of man and animals. The meaning of this statement will become even clearer in Chapter 29.

There is an enormous amount of research currently being performed in immunology, and as new facts are brought to light, it will become clear that certain statements made in this chapter will have to be modified. But this is true of any frontier in biology and should not deter a student from studying what must surely be one of the most fascinating adaptive responses we know of.

27.1
THE NATURE OF ANTIGENS

Specific antibodies and other immunological responses arise in vertebrates as a result of stimulation of certain tissue cells by **antigens.** An antigen (or **immunogen**), then, might be defined as any substance capable of provoking an immune response of any type. Antigens may be either man-made (synthetic) or, as is usually the case, some naturally occurring substance such as a large-molecular-weight protein or polysaccharide or a complex containing one or both of these substances. A bacterial cell, for example, would be expected to contain several dozen proteins or more which could act as antigens (see §27.7). Very few substances other than proteins, polysaccharides, or their complexes, are antigenic. Proteins, you will recall, are made up of "building blocks" called amino acids. Small numbers of linked amino acids within the protein may present an "unusual" surface configuration on the molecule and thus act as an immunologically specific **determinant group.** Most proteins which act as antigens generally contain a number of these determinant groups. The **specificity** of the immune reactions which we will be discussing depends both on the determinant groups of the antigen molecule involved and the elicited homologous antibody (or immune cells in the case of cell-mediated immunity).

Another feature of antigens (or immunogens) is that they are able to combine with the elicited antibodies in some observable or demonstrable way. This is generally shown by using one of the many serological tests available, e.g., precipitin test, complement-fixation test, immunodiffusion tests, etc. There are, however, small molecules called **haptens** which by themselves cannot evoke antibody synthesis, but which will combine with antibody once it is formed. How do you get the antibody if the hapten won't stimulate production? Simple! You chemically attach the hapten to a large **carrier** protein, e.g., albumin. The hapten then acts like one or more determinant groups stuck onto the protein. (See §27.7 for further details.)

In order to act as a good immunizing substance, each antigen should have the following properties:

1. A minimum molecular weight of 10,000.
2. A certain amount of molecular rigidity (aided by disulfide cross-linkages and aromatic amino acids).
3. Determinant groups that are exposed and not covered or blocked by folding of the molecule.
4. Exhibition of a fair degree of foreignness (not self, i.e., not from the host's own body).

Thus, antigenic substances range in character from components of infectious agents to usually harmless substances such as egg white. They may be viruses, blood, bits of tissue, whole cells, parts of cells, or soluble products of cells or enzymes. Antigens lose immunogenicity on being denatured,* digested, or hydrolyzed to residues of small molecular weight. Consequently, to retain their antigenicity they must gain access to the antibody-producing cells by a route other than via the digestive system. They may be introduced by hypodermic injection (as vaccines), by absorption through the skin or mucous membranes of the respiratory tract, eyes, or genitalia, or by infection. Such routes of entry, other than through the enteric or **enteral** tract, are said to be **parenteral**. In order to enter the body through the stomach and intestines, antigenic agents must be protected from the stomach acids and digestive enzymes by masses of food or by capsules.

27.2
SELF AND NONSELF

We may think of responses of the antibody-forming cells, to antigens, as resulting from ability of the cell (**self**) to recognize and react against anything that is **nonself**. Anything that is not derived from that cell, or from genetically identical cells, is nonself, i.e., a "foreign sub-

stance," and is reacted against or **rejected** or "resented." For example, if we inject horse serum (e.g., diphtheria or tetanus antitoxin derived from blood of horses) into a man, the man's tissue cells instantly recognize the equine proteins as "not self," and they respond by producing antibodies to destroy and reject it.

Iso-antigens. In an attempt to avoid such a rejection reaction against proteins from a different species (in this case equine) we might use antitoxin-containing serum taken from another person. But even this human protein (unless the donor were an identical [**monozygotic**] twin) would be considered "nonself" by the antibody-forming tissues of the recipient person and be rejected. However, because the human serum is more closely allied to the recipient than equine serum, the human protein is not rejected so rapidly or violently; it is said to be an **iso-antigen** (Gr. *isos* = the same). An iso-antigen is any substance from one individual that exhibits antigenic activity in another individual of the **same species**.

Important examples of human iso-antigens are the A, B, and Rh antigens of our erythrocytes that divide the human race into several immunological ("blood") groups: A, B, AB (AB = both iso-antigens present), or O (O = neither A nor B present), Rh+ (containing Rh antigen) or Rh− (containing no Rh antigen) (Table 27–1; Fig. 27–1). There are many such iso-antigens among humans, and they occur in other species of related vertebrates. The scheme given in Table 27–1 is actually a highly simplified one which makes no attempt to consider genotype as well as phenotype, but should suffice for now.

The Homograft Reaction. On the same basis, if we were "grafting" (transplanting) skin to cover a burned area on a person, we would not use horse or pig skin but human skin. But,

---

*An exception to this rule is the formation of toxoids by mild denaturation of bacterial toxins (see page 393).

TABLE 27–1. ISOHEMAGGLUTINATION: INTERNATIONAL SYSTEM OF BLOOD GROUPS

Sera from Persons of Group:	Agglutinate the Erythrocytes of Persons of Group:*			
	AB	B	A	O
AB (neither anti-A nor anti-B)	−	−	−	−
B (anti-A)	+	−	+	−
A (anti-B)	+	+	−	−
O (anti-A and anti-B)	+	+	+	−

*+ = Agglutination occurs; − = agglutination does not occur.

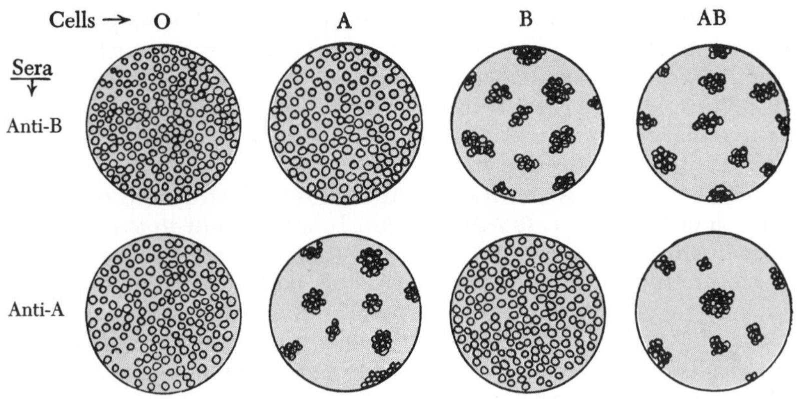

Cells → O   A   B   AB

Sera ↓

Anti-B

Anti-A

Figure 27–1. Diagram of hemagglutination in blood grouping. Anti-B serum and anti-A serum have been mixed on glass slides with erythrocytes of groups O, A, B, and AB. Anti-B serum agglutinates erythrocytes of persons of groups B and AB; anti-A serum agglutinates erythrocytes of persons of groups A and AB.

unless the donor were an identical twin, the recipient's body cells would reject even these human tissues as nonself, i.e., iso-antigens. Such a rejection is called a **homograft reaction.** Usually, skin grafts are successful only with the patient's own skin. Certain drugs have been found to suppress homograft reactions. Tolerance to "nonself," e.g., grafts, may also be induced as described below as **induced tolerance.**

Auto-antigens. Certain tissues of an individual's own body (eye lens protein, certain connective and nervous tissues, and some others) that would ordinarily be thought of as "self" may be attacked as nonself, possibly because during embryonic and fetal development and early infancy the antibody-forming tissues were too immature or too sequestered from those particular tissues. On maturation, the antibody-forming tissues may react, under certain conditions, to the lens, nerve, or connective tissues as nonself. Substances that, though naturally part of an individual's own body, nevertheless exhibit antigenic activity in the **same body** are said to be **auto-antigens.** (How are they different from iso-antigens?) Some types of arthritis, multiple sclerosis, and other "autoimmunization" diseases may have such reactions as a basic cause (see Chapter 29).

Acquired or Induced Tolerance. One of the problems of the "self" and "nonself" relationship is why "self" (except for the few special tissues that are auto-antigens) is not antigenic in its body of origin. It appears that in embryonic, fetal, and neonatal life the bursa-derived and thymus-derived cells (see next section) that react to antigens (i.e., that recognize foreign substances as "nonself") have not yet begun to function. The other tissues are not equipped to recognize nonself; to them, everything is "self."

Therefore, if antigenic substances are injected into the embryo or fetus, or during neo-natal life (the time varies in different animals and with different antigens), these antigenic substances are not recognized as nonself. The fully mature adult is therefore as tolerant to them as though they were self.* This type of early-life adaptation to antigenic substances is called **acquired** or **induced tolerance** or immunological paralysis. Practical use of such information is made in avoiding the use of immunizing "shots" for such diseases as polio, diphtheria, or tetanus in very young infants whose antibody-producing tissues still fail to recognize microbial antigens as nonself, i.e., whose tissues may be said to be "immunologically naïve" or to lack immunological competence. Currently, many studies are directed to means of avoiding homograft reactions, not only of bits of skin but of whole transplanted organs such as hearts and kidneys.

The surgery and postoperational care in such transplantations are mechanically and technically magnificent. Unfortunately, human efforts are apt to be frustrated by the rejection reaction in the patient. The immunological rejection response may be held in abeyance by matching the tissue antigens of the donor with those of the recipient as carefully as possible, i.e., checking for **histocompatibility.** Other procedures which greatly diminish the number of antibody-forming cells in the recipient are also used and may, under carefully controlled conditions, greatly extend the life of the transplant. These procedures include injecting the recipient with antilymphocyte serum (ALS) and using heavy doses of radiations, e.g., x-rays or cobalt-60. Large doses of analogues of purines, pyrimidines, etc., that interfere with DNA formation in the multiplying lymphocytes also suppress anti-

---

*The work of Sir MacFarlane Burnet, with Sir Peter Brian Medawar (Nobel Prize winners), led to this discovery.

body formation. The patient, however, in addition to sustaining various "side effects" of these treatments, may be deprived of necessary antibody defenses and may become highly vulnerable to infection by many pathogenic microorganisms and to some that are ordinarily harmless.

Tolerance to homografts and other foreign substances can also be induced in adult animals. The injection of massive doses of any antigen has long been known to produce "immunological paralysis." Large doses of antigen appear to saturate and overwhelm the antibody-producing reticuloendothelial and lymphocytic cells so that they fail to function. The tolerance induced in fetal or neonatal animals by injection of large doses of antigen may likewise depend on saturation, overloading, and "paralysis" of the embryonic or very immature antibody-producing systems.

27.3
THE IMMUNE RESPONSE

The ability to produce specific antibodies must be looked upon as a giant biological step forward in improving host defense mechanisms, albeit with a few drawbacks which we will discuss later. The combination of antibody with antigen is generally followed by the efficient destruction or removal of the antigen from the body. The ability to produce specific antibody appeared relatively late in the evolution of life on this planet, being limited to the vertebrates. The most primitive group of living vertebrates, the *Cyclostomata*, or jawless fishes (e.g., lamprey and hagfishes), are immunologically competent, but their immune responses are not as developed as those of the higher vertebrates.

The lymphocyte appears to be the key cell in the recognition of antigen and the subsequent immune response. It also appears that there are two populations of lymphocytes, the T-lymphocytes (T for thymus), which are primarily responsible for cellular immunity (see page 359) and the B-lymphocytes (B for bursa), which are primarily responsible for the production of circulating specific antibody, i.e., humoral immunity. Figure 27–2 summarizes how these two cell lines are derived from bone marrow. The stem cells from the bone marrow are released into the blood stream and eventually find their way to either of two primary lymphoid

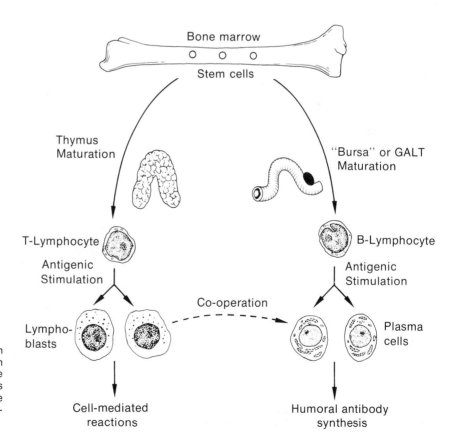

Figure 27–2. Bone marrow stem cells undergoing maturation in the thymus or the bursal tissue (gut-associated lymphoid tissues [GALT] in mammals) to become T- or B-lymphocytes, respectively.

tissues—the thymus and the bursal systems—which profoundly influence the role of each lymphocyte passing through them.

The thymus is a soft, fatty, cell-rich mass near the heart. It is minuscule in early fetal life but becomes large and active in late fetal and neonatal life and then shrinks to insignificant size at puberty. If the thymus is absent or removed early in life, the immunological processes of the adult are absent or defective.

The nature of the bursal system varies according to the species. In birds, for example, there is a distinct lymphoid organ called the **bursa of Fabricius** located near the hind gut. In man and other mammals there is no similar organ, but the same function is carried out by small collections of lymphoid tissue found in various parts of the body, e.g., tonsils, appendix, Peyer's patches of the gut. These tissues are sometimes referred to collectively as the **gut-associated lymphoid tissues** (GALT).

Those stem cells which mature in the thymus to become T-lymphocytes leave this primary organ, go into the circulation, and eventually colonize in a portion of the secondary lymphoid tissues, i.e., the spleen and the small lymph nodes located throughout the body. The T-lymphocytes are long-lived cells which, in the presence of antigens, are transformed into **lymphoblasts,** cells which do not secrete antibody but which are responsible for cellular immunity.

Stem cells maturing in bursal tissues (or GALT), on the other hand, leave these tissues as B-lymphocytes and also colonize in a specific portion of the spleen and lymph nodes (Fig. 27–3). The B-lymphocytes, upon contact with appropriate antigens, are transformed into **plasma cells** which are responsible for the production of antibody (immunoglobulins), i.e., humoral immunity.

We will discuss various aspects of both cellular and humoral immunity later, but for now it is important to understand that the mammalian host maintains two populations of lymphocytes which have as their major function the perpetuation of **immunological memory,** i.e., the ability to recognize foreign substances as nonself and to take affirmative action to rid the body of these substances. Depending on the nature of the antigen, dosage, and route of exposure, lymphocytes of either or both types may respond. Usually, both B and T cells are involved in the immunity which develops to an infectious microorganism. There is also good evidence to indicate that T cells may play the role of a "helper," perhaps by helping to concentrate antigen, in the response of B cells to antigens.

A full description of the precise mechanism of antibody induction, the types of cells involved, and the various biological mediators that have been identified is beyond the scope of this chapter. We can consider here only a few of the more interesting facets of immunology.

27.4
THE NATURE OF ANTIBODIES

Antibodies are proteins—immunoglobulins—found in the blood and lymph and also in various secretions of the body (e.g., mucus, colostrum). They are often referred to as immune γ-globulins. Several types are differentiated by their electrophoretic mobility. When a sample of serum is placed in an electrical field (see page 384 for technique) and the various components of serum are allowed to migrate

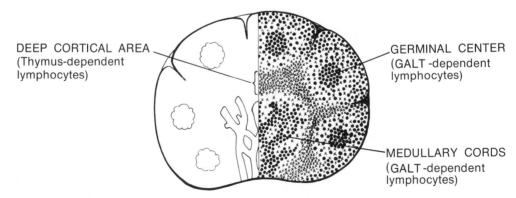

Figure 27–3. Diagrammatic sketch of a lymph node showing the areas colonized by T-lymphocytes (thymus-dependent) and B-lymphocytes (bursal or GALT-dependent).

according to their net electrical charge, a number of distinct bands or peaks are formed. Under the usual conditions, the albumin fraction migrates the most rapidly towards the anode and is followed by other fractions which were originally assigned Greek letters ($\alpha 1$, $\alpha 2$, $\beta 1$, $\beta 2$, and γ). Figure 27–4 shows the results of a typical electrophoretic separation of serum (see also discussion of immunoelectrophoresis on page 384). When these fractions were tested for antibody activity, all the fractions were either negative or feeble, except for the γ region. From that time on, the term **gamma globulin** (γ-globulin) became widely used for antibody.

We have now come to realize that there are various types or classes of antibodies (immunoglobulins), not all of which migrate exclusively in the γ region. They can be distinguished from one another by their physical and chemical properties (e.g., molecular weight, electrophoretic mobility, antigenic specificity) and also on the basis of the type of immunological phenomena with which they are associated. Table 27–2 lists the five classes of human immunoglobulins currently recognized by most authorities and compares their prominent features.

Properties of Immunoglobulin G. A clearer understanding of the distinguishing features of immunoglobulins will be gained after a

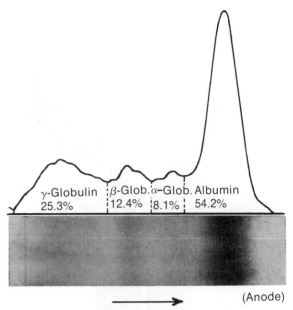

Figure 27–4. Electrophoretic separation of major serum proteins on a filter paper strip. The curve was plotted from points obtained with a photoelectric densitometer.

look at the structure of one of the most important immunoglobulins, called IgG or γG. The IgG fraction often accounts for 80 to 85 per cent of the antibody present in normal adult human

TABLE 27–2. PHYSICAL AND BIOLOGICAL PROPERTIES OF HUMAN IMMUNOGLOBULINS*

Property	IgG (γG)	IgA (γA)	IgM (γM)	IgD (γD)	IgE (γE)
Older terms	7S	B_2A	19S, B_2M	–	Reagin
Electrophoretic mobility	γ	β	β	γ	γ
Molecular weight	150,000	160,000 and polymers	900,000	185,000	200,000
Sedimentation value	6–7S	7–11S	19S	6–7S	8S
Heavy chains Molecular weight	γ 53,000	α 64,000	μ 70,000	δ ?	ϵ 75,000
Light chains Molecular weight	κ, λ 22,000	κ, λ 22,000	κ, λ 22,000	κ, λ –	κ, λ 22,000
Per cent total immunoglobulins	80	13	6	1	0.002
Serum concentration (mg/ml)	8–16	1–4	1	0.2	0.03
Half-life (days)	25–35	6–8	9–11	2–3	1–2
Placental transfer	Yes	No	No	No	No
Response to antigenic stimulation	Late	Intermediate	Early	–	–
Fix complement	Yes	–	Yes	–	–
Noted for activity in:	Precipitation, toxin neutralization	Presence in sero-mucous secretions	Agglutination, hemolysis, virus neutralization	Function unknown	Atopic allergy, (defense of respiratory tract?)

*Values approximate or estimates in some cases.

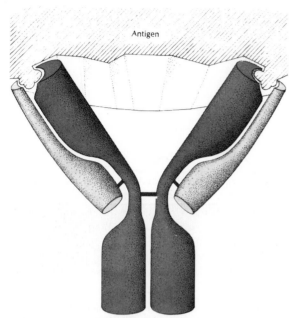

Figure 27–5. Diagrammatic sketch of a molecule of immunoglobulin (IgG) attached to two identical antigenic determinant groups. See text for description.

theria, measles, etc. (Chapt. 28). The maternal antibodies in the younger infants interfere with the antigenic effect of the vaccines.

Figure 27–5 shows a molecule of IgG as it might look attached to determinant groups on the surface of a red cell. Note that there are four pieces or chains of amino acids: two heavy (H) chains (the longer ones) and two light (L) chains, and that these are held together by disulfide bonds (represented by the dark connecting bands). The immunological specificity of the antibody molecule lies in the unique spatial configuration created at the distal ends of the molecule where the H and L chains come together. In the diagrammatic sketch this is illustrated as a contoured pouch or opening which just "fits" the determinant groups of the antigens (see discussion of specificity). Note also that IgG is divalent, i.e., contains two combining sites per molecule. This is important since it allows the antibody to connect two antigenic molecules, thus forming a complex. This in turn is the basis for many of the visible serological reactions we will be discussing.

It is only as a result of a great deal of research that we have arrived at our present concept of IgG, as shown in Figure 27–6. All other immunoglobulins, irrespective of class, are composed of similar basic structural units (see Figure 27–7). Each molecule has two identical L chains with a molecular weight each of about 22,000 and containing about 214 amino acids.

serum. In humans the IgG (but not the IgA or IgM) passes to the fetus and gives passive protection against many infectious agents for up to about six months of age. It is then customary to immunize infants against polio, tetanus, diph-

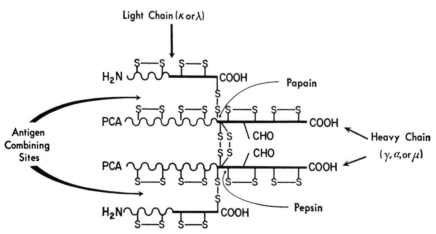

Figure 27–6. Diagram of the polypeptide chain structure of IgG; the structures of IgA and IgM are analogous. The two light chains ($H_2N \cdots COOH$, top and bottom) are of either κ or λ type; the two heavy chains (PCA $\cdots$ COOH, center) are of γ type in IgG, α type in IgA, and μ type in IgM. The chains are joined and their configurations are somewhat stabilized by disulfide bridges (—S—S—). Papain and pepsin cleave the heavy chains at the points indicated, and the intervening area, which includes —S—S— bridges linking the two chains, is a hinge region that provides flexibility to the entire molecule. The wavy portions are the regions of variable amino acid arrangement; this is almost exactly half of the light chains and is probably one-fourth to one-half of the heavy chains.

Each molecule also has two identical H chains with a molecular weight of about 53,000. The four polypeptide chains are held together and their configurations somewhat stabilized by disulfide (—S—S—) linkages. When one begins to make immunological comparisons of the IgG molecules from different individuals (i.e., to consider them as antigens), he finds that the molecules share some determinant groups but not others. A comparison of the other immunoglobulins (IgM, IgA, etc.) has shown similar variation. This has led to the conclusion that the immunoglobulins are quite heterogeneous with respect to their determinant groups.

Table 27–3 shows the subclasses and types of immunoglobulins now recognized for humans. Note that there are only two types of L chains, either κ or λ, regardless of the class of immunoglobin. The H chains, however, are more variable. Within each class, certain determinant groups are always present and can be used to differentiate those H chains from H chains of another class (i.e., the H chains of IgG always show γ specificity, whereas the H chains of IgM always show μ specificity). Within each class, the H chains also show some antigenic variation, so that in IgG, for example, we can have H chains showing $\gamma1$, $\gamma2$, $\gamma3$, and $\gamma4$ specificities.

But this is not the end of heterogeneity in immunoglobulins. You will note that Figure 27–6 shows both a constant and a variable region for both the light chains and the heavy chains. The constant portion of the molecule may play two roles: (1) It may determine the **antigenic** specificity of the molecules, leading to their classification as shown in Table 27–3. (2) It may determine some of the **biological** properties of the molecule, i.e., whether it will pass the placental barrier, be secreted, fix complement, bind to macrophages, etc.

The variable portion of the four polypeptide chains in each antibody molecule is believed to be responsible for the antibody specificity of the molecule; i.e., the sequence of amino acids in this portion of the molecule determines the spatial configuration which will just fit an antigenic determinant group of some kind.

Probably no single concept gives beginning students more trouble than when they're asked to remember that antibody (e.g., human immunoglobulin), being a protein, can also serve as an antigen when injected into another species of animal. Thus each molecule of immunoglobulin has antigenic specificity and, in most cases, also antibody specificity (that portion of the molecule which will combine specifically with an antigen).

Properties of Other Immunoglobulins. The overall structure of the various classes of immunoglobulins is compared in Figure 27–7. Note that IgG, IgD, and IgE are monomers, that IgM is a pentamer, and that IgA may be a monomer, dimer, or trimer.

IgA. IgA constitutes about 10 per cent of the immunoglobulins in human serum, but it is the principal immunoglobulin found in seromucous secretions (saliva, tears, colostrum). It can also be found in the lumen of the intestine, where it is known as **coproantibody.** In secretions, it is generally a dimer with an attached β-globulin of about 50,000 molecular weight (called T-piece, transport piece, or secretory piece). The secretory piece is attached to the IgA as the latter passes through the glandular cell producing the secretion. IgA is thought to be of special importance in the defense of membranes and exterior body cavities (mouth, vagina, etc.) against infections.

TABLE 27–3. THE VARIOUS CLASSES, SUBCLASSES, AND TYPES OF HUMAN IMMUNOGLOBULIN [*]

Immunoglobulin	H Chain Class	H Chain Subclass	L Chain Type	Molecular Formula
IgG	γ—	$\gamma1$	κ— λ—	$(\gamma1)_2\kappa_2$ $(\gamma1)_2\lambda_2$
		$\gamma2$	κ— λ—	$(\gamma2)_2\kappa_2$ $(\gamma2)_2\lambda_2$
		$\gamma3$	κ— λ—	$(\gamma3)_2\kappa_2$ $(\gamma3)_2\lambda_2$
		$\gamma4$	κ— λ—	$(\gamma4)_2\kappa_2$ $(\gamma4)_2\lambda_2$
IgA	α—	$\alpha1$	κ— λ—	$(\alpha1)_2\kappa_2$ $(\alpha1)_2\lambda_2$
		$\alpha2$	κ— λ—	$(\alpha2)_2\kappa_2$ $(\alpha2)_2\lambda_2$
IgM	μ—	$\mu1$	κ— λ—	$(\mu1)_2\kappa_2$ $(\mu1)_2\lambda_2$
		$\mu2$	κ— λ—	$(\mu2)_2\kappa_2$ $(\mu2)_2\lambda_2$
IgD	δ—	$(\delta$—$)$	κ— λ—	$\delta_2\kappa_2$ $\delta_2\lambda_2$
IgE	ϵ—	$(\epsilon$—$)$	κ— λ—	$\epsilon_2\kappa_2$ $\epsilon_2\lambda_2$

[*] Modified from Gordon and Ford: Essentials of Immunology, F. A. Davis Co., 1971.

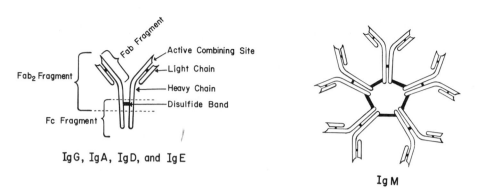

IgG, IgA, IgD, and IgE

IgM

IgA with Secretory Piece

Figure 27-7. Diagrammatic sketch of immunoglobulin structures. IgA may be a monomer, dimer, or trimer, and IgM is a pentamer.

IgM. IgM is a pentamer with increased binding capacity for antigen and thus is extremely efficient in agglutination and lytic reactions (see Table 27–2). IgM is usually one of the first types of specific antibody to appear in the blood during an infection or after immunization with most antigens. Thus it may have a special significance in combating a bacteremia or viremia. For some reason, the antibodies formed against lipopolysaccharides of gram-negative bacteria or against antigens of *Treponema* are of this class.

IgD. IgD immunoglobulins were first discovered in patients suffering from a disease of the lymphoid tissues known as multiple myeloma. Their main biological function is not known, but they are apparently present in certain intestinal fluids of normal individuals.

IgE. IgE immunoglobulins (also called **reagins**) are present in the serum in only very low concentrations. Their biological function is not clear. Serum levels rise during certain infections and they may, like IgA, be of some use in protecting mucous membranes. IgE, however, would appear to be more harmful than beneficial for many individuals who are allergic or hypersensitive to certain substances. This immunoglobulin, when transferred from a sensitized individual to a normal person, sensitizes the latter to the offending allergen (see Chapter 29).

27.5
THEORIES OF ANTIBODY FORMATION

We stated earlier that the B- and T-lymphocytes had the responsibility for recognizing foreign substances (antigens) and for taking remedial action. This, of course, does not explain how these cells know self from nonself or how they know what an antibody to bovine albumin, for example, ought to look like.

Many theories of antibody formation have been proposed. One current theory, called the **clonal selection hypothesis,** will be described. According to this theory, recognition of a substance as self or nonself depends on the following mechanism: During fetal development, the lymphoid cells in the primary lymphoid tissues differentiate so that many different cells are obtained, each having one (perhaps more) recognition sites on its surface. After a short time thousands of cells, each with one or more different recognition sites, are present in the cell population. As the cells migrate through the developing fetus, some of the cells encounter fetal tissue or soluble antigens which are complementary to their own recognition sites. The two combine, and the lymphoid cell dies. In this way, any lymphoid cells carrying recognition sites for "self" are destroyed during fetal life.

Cells which survive this purge can recognize and react only to nonself. They are free to locate in some lymphoid tissue and produce a **clone,** i.e., progeny like themselves. It has been postulated that by the time the individual becomes immunologically competent, large numbers of lymphoid cells with somewhere between 10,000 and 100,000 different kinds of recognition sites are present in the tissues. It is thought that this is sufficient to recognize almost any nonself antigen to which the host will be exposed.

Thus, when an antigen enters the body and makes contact with a B- or T-lymphocyte bearing the proper recognition site, the lymphocyte is stimulated to reproduce and form a clone. Its progeny (B or T) become either plasma cells (produce humoral antibody) or lymphoblasts (activate cellular immunity, i.e., phagocytic cells), depending on which cell line they represent. The more the antigenic stimulation, the greater the proliferation of the clones of cells possessing the homologous recognition sites.

The clonal selection hypothesis can also be used to explain the secondary or anamnestic response underlying the use of "booster" doses of vaccines, toxoids, etc. (see page 395).

27.6
SPECIFICITY

Specificity has been referred to as an attribute of antigens and antibodies that is a result of mutually corresponding molecular structures called determinants or recognition sites. The structural resemblance between molecules of an antigen and molecules of the corresponding antibody may be visualized as analogous to the relationship between a mold and its replica or casting. Each is **specific** for the other. Antigens with artificial determinants may be synthesized by combining protein with certain inorganic radicals having a given chemical structure (e.g., $NH_2C_6H_4 \cdot AsO_3H_2$ + protein). Such antigens will, upon injection into the body, engender antibodies in the serum that react only with that compound. For example, let us alter the antigen by substituting a $-SO_3H$ group in place of the $-AsO_3H_2$ group. Antibodies to the arsenilic antigen will not react significantly with this altered antigen. Almost any sort of chemical alteration in an antigen will alter its specificity. Too great an alteration will destroy its antigenicity (Table 27-4).

If the antigen molecule is small and contains only one combining site (i.e., is **univalent**), then only one antibody molecule can combine with each such antigen molecule (though two antigen molecules may combine with each antibody molecule since antibody molecules are typically bivalent). Such antigen-antibody combinations are small and soluble.

The more complex the form of the antigen molecule, the more determinant groups there are for specific combination with antibody, and

TABLE 27-4. THE EFFECTS OF SUBSTITUTION ON ANTIGENIC SPECIFICITY AS REVEALED BY THE ABILITY TO COMBINE WITH SPECIFIC ANTISERA*

Testing antiserum specific for azoproteins containing:	Test Antigens: Azoproteins Containing			
	ANILINE NH₂	p-AMINO-BENZOIC ACID NH₂ / COOH	p-AMINO-BENZENESUL-FONIC ACID NH₂ / SO₃H	p-AMINO-PHENYLARSONIC ACID NH₂ / AsO₃H₂
(1) Aniline	+++	−	−	−
(2) p-Amino-benzoic acid	−	+++±	−	−
(3) p-Aminoben-zenesulfonic acid	−	−	+++±	−
(4) p-Aminophenyl-arsonic acid	−	−	−	++++

*Reprinted by permission from Landsteiner: The Specificity of Serological Reactions, Revised ed., Harvard University Press, 1945.

the larger, more stable and insoluble the combination is. For example, most protein antigens have up to about five different determinant groups, each of which evokes an antibody molecule specific for that site. Thus, even though such a multivalent antigenic protein is a pure substance, the "antibody" produced in response to it is really a mixture of antibody molecules, each specific for a different determinant group on the antigen molecule. Saturation of all the determinant group sites with specific antibodies forms a large, stable, insoluble and **precipitable** colloidal complex. This is of basic importance and will be referred to again under Precipitins (see next section) and Lattice Formation (page 382).

Cross-Reactions. A slight reaction with antibodies to the arsenilate may occur if, instead of substituting a $^-SO_3H$ group, we introduce, say, a ^-Cl atom in place of an H atom. The antibodies produced in response to the original arsenilic antigen are said to **cross-react** with the chlorinated antigen. Such cross-reactions occur between many closely related antigens (Fig. 27–8).

27.7
MECHANISMS OF ANTIGEN-ANTIBODY REACTIONS

Both antigen molecules and antibodies (immunoglobulins) are substances of high molecular weight and are likewise colloidal in nature. Substances are in the colloidal state when they are in the form of ultramicroscopically minute particles stably suspended in a fluid (gas or liquid). For example, smoke is a colloidal suspension of minute particles of carbon, tars, and other substances in air; milk is a colloidal suspension of casein protein and fat globules in whey; enzymes are colloidal proteins.

Colloidal particles in aqueous suspensions generally have negative, mutually repellent, electric charges. These help to keep the particles suspended and prevent their coalescing and precipitating as floc, or from coagulating as in the souring of milk. Microorganisms, because of their minute size, generally have many of the properties of colloids.

Precipitin Reactions. Antigens and antibodies are both complexes of amino acids and also have positive and negative polar groups distributed over their surfaces in specific but reciprocal patterns. When antigen and corresponding antibody molecules are mixed, electrical attractions and repulsions, modified by van der Waals and ionic surface forces, result in an orientation of the corresponding antigen and antibody molecules with respect to their molecular forms and electrical charges, so that an absolute "fit" (mold and cast) is obtained. The compound colloidal particles formed as a result of the interaction may be very large. In the presence of electrolytes that, presumably, neutralize exterior colloidal charges, the particles become unstable and are thrown out of suspension. In a test-tube reaction they become visible as a cloudy precipitate (Fig. 27–9). This sort of reaction is commonly seen when specific anti-

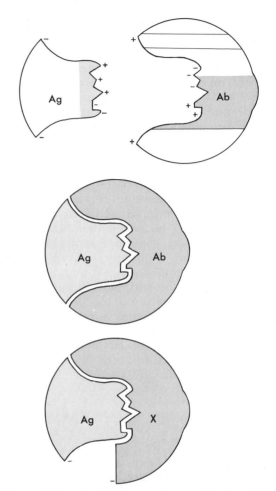

Figure 27–8. Specificity in antigen-antibody reactions. At top are shown, diagrammatically, antigen (Ag) and antibody (Ab), each with an indented margin, each margin complementary to the other. In the middle, antigen and antibody are shown forming a large colloidal complex by virtue of the perfect "fit" between Ag and Ab. At bottom is shown a colloidal complex formed by Ag with an antibody (X) of similar but not perfectly corresponding form. A visible precipitate may or may not be formed by such a combination.

body reacts with a **soluble protein** or polysaccharide antigen, i.e., not particulate antigen such as a whole cell. The reaction is called a **precipitin reaction.**

Stages of Antigen-Antibody Reactions. The first stage of antigen-antibody reaction is **adsorption** or complex physicochemical, sometimes reversible, combination between antibodies and antigens. This combination may proceed rapidly at temperatures near 37C. The bonds between antigen and antibody are not covalent: they are hydrogen bonds, ionic bonds, etc., and hence are not very strong. The second stage is usually a **visible reaction** (precipitation, cell lysis, or other effect; see §27.8). This stage often develops slowly and may be demonstrated best after 12 to 18 hours at 4 to 6C. As mentioned before, the presence of electrolytes (magnesium chloride, sodium chloride) and certain enzyme-like components of serum (**complement**) is necessary for certain types of reaction in the second stage.

Complex Antigens. Reactions of antibodies with antigens that are parts of entire cells instead of being free colloids are basically the same. However, since entire cells, instead of colloidal molecules, are involved as antigen, the result of the antigen-antibody combination is somewhat more complex and should not be called precipitation (which implies soluble components). Several cellular antigens are described in the following paragraphs.

In nature, antigens seldom occur in a pure state. This complicates the study of antigen-antibody reactions involving microorganisms. A bacterial cell, for example, may contain several antigens; e.g., the proteins of various enzymes, nucleoproteins, ribosomes, protein-polysaccharide complexes of the cell wall and capsule. The serum of a person or animal following immunization with or infection by such cells usually contains a mixture of antibodies specific for each separate antigen.

For illustrative purposes let us consider the surface antigens of a group of common rod-shaped, nonsporeforming bacteria, e.g., gram-negative eubacteria of the family Enterobacteriaceae. These organisms are more or less constant inhabitants of the intestinal tract of many animal species and are important bacteria. A few species are pathogenic, and most can act as opportunists (see Chapter 24). Their antigenic structure is representative of that of many species of bacteria.

Flagellar (H) Antigens. Various antigenic types of the protein **flagellin** are localized in the flagella of motile species of Enterobacteriaceae. Flagellar antigens are called **H antigens.*** H antigens are destroyed (denatured) by boiling, also by alcohol and dilute acids. In at least one genus, *Salmonella*, H antigens often exist in one of two different states of specificity, called **phases.** In the **specific phase** (phase I) they are specific for the species in which they occur. In the less specific or **group phase** (phase II) they resemble antigens in a group of closely related species or types (see heterogenetic antigens). They vary, often unpredictably, from one phase to the other, and this is spoken of as **phase variation.**

Thus each species or strain of flagellated bacteria has determinant groups located on the flagella capable of eliciting homologous antibodies with the ability to combine specifically with those determinant groups. The same determinant group may appear on the flagella of several strains or species and lead to cross-

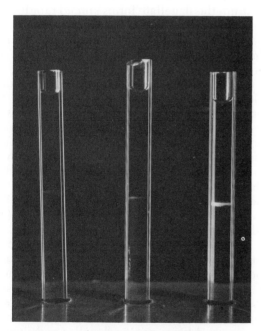

Figure 27–9. Precipitin test in narrow-bore glass tubes. The tubes were half filled with an antiserum, then carefully overlaid with three dilute antigen solutions, so as to create a sharp interface. A precipitate formed at the interface in the right-hand tube which contained the homologous reactants; i.e., the same antigen was used for the test as was used to elicit the antiserum. No precipitate developed with the unrelated antigen in the left-hand tube. A slight precipitate (cross-reaction) formed in the center tube, which contained an antigen related, but not identical, to the one on the right.

---

*The derivation of the terms H and O antigens is explained on page 378 under O antigens.

reactions in serological tests. Almost every strain or species, however, will have certain determinant groups which are unique for that type of organism and which will not cross-react.

Fimbrial Antigens. Fimbriae occur on many species of enterobacteria. They are strongly antigenic. They differ from flagellar antigens in being somewhat more resistant to heat and in resisting the effects of alcohol. They are not related to any other antigens of the cell surface.

Capsular Antigens. Most capsules are heteropolymers of various simple sugars with glucosamine and other sugar derivatives. As part of the cell they are conjugated with proteins. Some capsules, e.g., that of *Bacillus anthracis*, are composed of polypeptides. The possible number of different combinations of subunits in capsular heteropolymers of different species of bacteria is very large, and the number of determinant groups even larger. Capsular substance, when present, dominates the surface of the bacterial cell, its determinants depending on the species of organism. Since antibodies act chiefly at cell surfaces, the specificity of the capsular antigen of any species thus determines antigenic specificity of the entire cell, unless, of course, the cell is broken open, heated, or treated so as to destroy the capsule.

Among the Enterobacteriaceae, the capsule of *Klebsiella* serves as a typical example. It determines the antigenic specificity of the intact cell. The same situation is found among other, unrelated encapsulated bacteria, such as pneumococci and streptococci. Stripped of this **specific** surface antigen, the exposed antigens of the naked cells often cross-react with antibodies produced against related or even unrelated species. For example, the antigens of the naked cell walls of most pneumococci are immunologically alike. Antibodies for one react equally well with others. But the polysaccharide antigens of the capsules are not all alike. There are about 75 distinct types of pneumococcus capsular antigens. Each antigen represents a different antigenic or serological type of pneumococcus. Similar series of capsular types are found in *Klebsiella;* influenza bacilli (*Haemophilus influenzae*, types A, B, C, D, E); in meningococci (*Neisseria meningitidis*, types I, II, II alpha); and in numerous other species.

In addition to conferring antigenic specificity on the cell, capsules act as a protective coating against phagocytes, bactericidal components of serum, drying, and other unfavorable environmental influences.

K Antigens. The K (Ger. *Kapsel* = capsule) antigens resemble capsular antigens both in their location on the cell surface and in their specificity. Commonly they are not as readily demonstrable or as voluminous as most capsules. Their origin is not wholly clear; they may be derived by extrusion of material from inside the cell wall or by some modification of the exterior surface of the cell wall. Many workers prefer to call them envelope antigens rather than capsular or somatic antigens to indicate the uncertainty of their origin. Since they coat the surface, K antigens dominate the response of the cell to antibodies and confer antigenic specificity in the same manner as capsules. There are perhaps 45 to 50 different antigenic types of K antigens divided into three major subgroups (and designated, most inappropriately, L, A, and B). Like flagellar antigens (proteins), many of the K antigens are denatured by boiling.

O Antigens. The designations "H" and "O" for certain antigens are derived from early studies, by Weil and Felix, of a genus of Enterobacteriaceae called *Proteus*. Like many other enteric bacteria, *Proteus* species characteristically occur in two variant forms, motile and nonmotile. It was observed that on moist agar medium the flagellate forms spread rapidly over the entire surface of the agar in a thin, grey film that was described by the German observers as *Hauch* (German for fog or "cough in very cold air"). The nonflagellate variant formed the usual, discrete, round colonies *ohne Hauch* (German for "absence of the fog-like film"). The term **Hauch** (or **H**), associated with motility, soon became almost synonymous with the organs of motility, flagella (see preceding discussion of flagellar or H antigens). The **ohne Hauch,** or nonmotile form, was correspondingly called the **Ohne** or **O** form; O came to be associated with the bodies of the bacilli without flagella. O is now used to designate the body antigens (also called **somatic antigens** [Gr. *soma* = body]).

Further study of the family Enterobacteriaceae (Chapter 35) and other gram-negative rods has shown that O antigens are components of the lipopolysaccharide portion of the cell walls of most gram-negative organisms, in loose combination with somatic proteins. Unlike the species and type-specific flagellar and capsular antigens, O antigens of the Enterobacteriaceae, while specific, are less restricted in distribution. This is simply another way of stating that the determinant groups located on the cell wall are not always unique, but rather can be found distributed among several species in different groups of Enterobacteriaceae. For example, in the genus *Salmonella*, the O antigens designated as 9 and 12 (see Chapter 35) are common to some 75 species or serotypes that are included in the *Salmonella* Group D. Over 58 O antigens and dozens of groups are known in the genus *Salmonella* alone. Other O antigens, in addition to 9 and 12, are also present in some species of Group D. In such groups the individual species may be differentiated by their H antigens, but not by their O antigens. Several O antigens (i.e., various determinant groups) may occur in each individual species, though in different combinations. The antigenic structure of the Enterobacteriaceae is discussed more fully in Chapter 35.

O antigens are heat-stable (100C) and are not injured by alcohol or dilute acid. As previously indicated, O antigens may be completely covered and masked by capsules, a fact of great importance to diagnostic microbiologists.

Extracellular Antigens. Some of the extracellular metabolic products (sometimes called **metabolites**) of living cells are potent antigens. For example, diphtheria and tetanus toxins are antigenic, poisonous proteins released to the exterior of the cell by secretion or by lysis of the cell or both. Extracellular antigens like diphtheria toxin are called **exotoxins** (see Chapter 24). As protein antigens, all toxins, including enzyme proteins, stimulate the production of **antitoxins** or **antienzymes** (Fig. 27–10).

Heterogenetic (Shared, Common or Group) Antigens. As previously indicated, cells of two or more different species, whether closely or distantly related, may have certain antigens in common. (For example, pneumococci and chickens contain antigens like human blood-group A substances. Students of evolution may not be humiliated to learn that their blood contains antibodies related to proteins in certain fish.) The relation of cow to whale by immunological methods is perhaps an unexpected finding. Such antigens are said to be **heterogenetic.** **Antiserum** (serum containing antibodies) prepared by injecting cells of one species into an unrelated animal will react, to a greater or lesser extent, with different species of cells that contain the common antigen.

For example, three related species of dysentery bacilli (genus *Shigella*) may each contain four antigens (Fig. 27–11). Upon injecting species I into a rabbit, antibodies, a, b, c, and d will be engendered, corresponding to antigens A, B, C, and D. Upon injecting species II into another rabbit, antibodies c, d, e, and f will be called forth. Likewise, species III will stimulate production of antibodies e, f, g, and h. Now, the serum of rabbit I will react best of all with species I when these bacteria and the antiserum are brought into contact. Antiserum II will similarly react best with species II, and antiserum III with species III. However, since antiserum I contains antibodies c and d, it will cross-react to some extent with cells of species II, since the latter have these antigenic compounds in common with species I. There will be no cross-reaction between antiserum I and antigen III, but antiserum II will cross-react with species III. Such cross-reactions are common.

ANTIBODY ADSORPTION. If a volume of antiserum I (say, 5 ml) were mixed with a heavy suspension of cells of bacterial species II, then

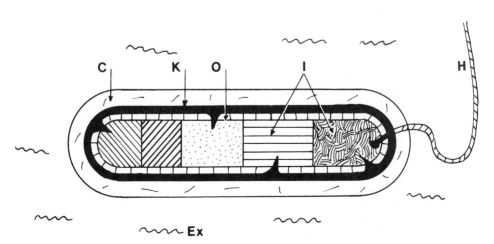

Figure 27–10. Location of various bacterial antigens. *C,* Capsular antigens; generally polysaccharides. *K,* Sheath or envelope antigens derived from O antigens. *O,* Lipopolysaccharide antigens of the cell wall. *I,* Internal antigens of the cell. *H,* Antigens of the flagella. *Ex,* Antigens excreted to the exterior like diphtheria toxin.

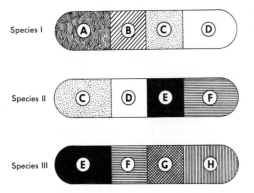

Figure 27–11. Sharing of antigens by different species. For explanation see text.

antibodies c and d and antigens C and D would combine, leaving antibodies a and b still free in the serum. By centrifuging the mixture, the bacteria of species II, with their attached antibodies c and d, can be removed, leaving serum I free from antibodies c and d and **specific** with regard to species I (i.e., the antiserum will no longer cross-react with species II; it will react only with bacteria containing antigens A and B). If we further adsorb antibody b by treatment of the serum with some species having only antigen B in common with species I, then we obtain a pure A antiserum (or anti-A serum) and the serum is said to be **monospecific.** Many such specific sera are thus prepared with a wide variety of both procaryotic and eucaryotic cells, and it has been possible to make extensive analyses of antigenic structures and to detect hitherto unsuspected antigenic and phylogenetic relationships.

Haptens. As previously pointed out (page 366), haptens are not immunogenic but can react like antigen. Haptens can be coupled chemically with a carrier protein and used to immunize an animal; antibody with specificity for the hapten portion of the molecule will be formed. Haptens may be complex substances of high molecular weight, such as partial proteins or protein derivatives, capsular polysaccharides (*without protein* attached), or they may be relatively simple compounds of low molecular weight, such as the arsenilate previously referred to, or certain drugs, cosmetics, and antibiotics.

The combination of antibody with fairly large haptens, called **complex haptens** (e.g., capsular polysaccharides), can result in the familiar, visible or demonstrable antigen-antibody reactions (i.e., precipitation or flocculation) previously described. Haptens of low molecular weight, sometimes called **simple haptens,** can combine with antibodies but they cannot cause any visible or demonstrable reactions such as precipitation. Combination of haptens with specific antibodies blocks any further combination of those antibodies with complete antigens. Haptens are therefore sometimes called **blocking, partial,** or **incomplete antigens.**

Complex haptens are well represented by the type-specific capsular polysaccharide ("SSS" or soluble specific substance) of Type II pneumococci. This (in some animals only) is not antigenic per se but, like complex haptens in general, precipitates readily with Type II antipneumococcus antibodies. When attached to its appropriate Type II pneumococcus protein it becomes completely antigenic, and engenders specific Type II antibodies. Many naturally occurring antigens are complex hapten combinations.

Simple haptens, such as certain drugs, cosmetics and antibiotics, may at times form complete antigens by combination with body proteins. Thus many persons who come into contact with such drugs or cosmetics, internally or externally, can become "sensitized" to them (see Chapter 29) and then experience allergy-like reactions with rashes, blotches, itching or gastrointestinal symptoms.

27.8
ASPECTS OF SEROLOGY

Serology is the study and use of various techniques employing serum. In most cases, the serum contains antibodies and thus, in reality, we are describing the interactions of specific immunoglobulins and various types of antigens. There are several manifestations of antigen-antibody reactions; some are visible in test tubes, some are demonstrable only by secondary tests. The different manifestations are spoken of as though many different antibodies were involved, but there are fewer types of antibodies than types of reaction. Thus, the same antibody might be involved in either precipitation or agglutination, depending on whether the determinant groups were attached to a soluble colloid or a particulate substance, such as a whole cell. Therefore, the kind of reaction seems to depend in great part on the kind and size of antigen and whether it consists of molecules or cells, the physical conditions of the suspending fluids, the presence of electrolytes, and other factors. For convenience, we shall speak of "types of antibody," as meaning "types of antigen-antibody reactions." Among the best understood "types" of antibody are **antitoxins** and **precipitins, agglutinins, cytolysins, immobilizing antibodies,** and **protective** and **neutralizing antibodies.**

Antitoxins and Precipitins. When bacteria gain a foothold in the body and liberate toxin into the blood, the toxin (a protein antigen) stimulates the production of antibody. As previously described, this antibody, called an **antitoxin,** combines specifically with the toxin and neutralizes it. The reaction, if in vitro, may be thought of as precipitin reaction in which the particles of combined toxin and antitoxin are too small to be seen. Under certain conditions in the laboratory in which the quantitative relations and electrolyte content are carefully adjusted, large, visible floccules are produced in the test tubes.

Quantitative Relations (The Flocculation Reaction). Note carefully that it is only when certain **proportions** of toxin and antitoxin are brought together in a test tube that a visible precipitate, or flocculation, occurs. The *total* amount of reactants in the tubes can vary over a wide range, but the amount of one reactant *relative* to the other is critical. This fact is of fundamental importance and is well illustrated by the following practical application for determining the concentrations or "strengths" of toxin or antitoxin. We set up a row of ten tubes. In the first we put serum containing a quantity of diphtheria (or any other) antitoxin, arbitrarily

spoken of as two units.* In the next tube we place four units, in the next six units, and so on. We then add to each tube a fixed amount, say 1 ml, of filtered broth culture of *Corynebacterium diphtheriae* that contains diphtheria toxin in an unknown amount. After a short time flocculation appears in one of the tubes, let us say the sixth tube (Fig. 27–12). Since this contained 12 units of antitoxin, we have a measure of the **precipitating** potency of the toxin broth. We say that it contains 12 **flocculation units** ($12L_f$) of toxin per milliliter. This is an arbitrary unit of potency and is used for convenience to define diphtheria, tetanus and other toxins. It shows antigenic combining power or flocculating potency but not necessarily toxicity. Partly deteriorated toxins or toxins which have been purposely denatured to convert them to **toxoids** will give the same undiminished L_f value though wholly devoid of **toxic power.**

Conversely, by using a series of tubes containing known, graded amounts of toxins (or toxoid), we may determine the number of l_f **units** of antitoxin in a serum of unknown potency.

Zone Phenomenon. The reaction just

---

*A unit (approximately) of diphtheria antitoxin is the least amount necessary to protect standardized (250 to 300 gm) guinea pigs against 100 minimal lethal doses (M.L.D.) of diphtheria toxin. An M.L.D. kills 50 per cent of such pigs in from four to five days (an LD_{50} dose for guinea pigs). Week-old chicks are advantageously substituted for guinea pigs in such determinations. Units of other toxins are determined by modified but analogous procedures.

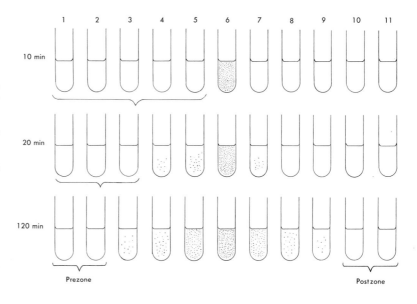

Figure 27–12. Determination of the relative optimal proportions of antibody and antigen. Tubes 1 to 11 contain increasing concentrations of antitoxin (2, 4, 6, etc., units) and fixed amounts of toxin-containing broth. Prompt antigen-antibody reaction occurs (10 min, tube 6, 12 units of antitoxin) in the presence of **optimal proportions** of antigen and antibody. Later (20 min, tubes 4, 5, and 7) reactions occur in the presence of suboptimal proportions of antigen and antibody. Still later (120 min), these reactions are extended to other tubes. In each series prezone (or prozone) and postzone are shown beyond the range of visible reactions.

described illustrates what is called a **zone phenomenon,** or prezone and postzone of inhibitions.

In the experiment just performed (Fig. 27–12) flocculation failed to occur *promptly* in a zone of five tubes before the sixth tube (**prezone**) and in a zone of four tubes after the sixth (**postzone**). Smaller amounts of floc appeared in some of the tubes near the sixth tube on long standing. This resulted from the fact that for **maximal** amounts of visible precipitate (or flocculation or any other antigen-antibody reaction) to occur most rapidly and extensively, **optimal relative proportions** (not necessarily actual **quantities**) of antigen and antibody must be present. If there is too great an excess of either, less precipitate, or none at all, will appear. Many factors are involved, among which are type of animal serum used, nature of antigen and antibody, pH, and electrolytes (Fig. 27–13).

Lattice Formation. Zone phenomena are related to the formation or nonformation of lattices. In any given reaction, if the antigen and antibody molecules are **univalent,** i.e., if each has only one combining site, they can form only pairs—small colloids usually not visible in test tube reactions. If, as is usually the case, an antigen molecule has several different combining sites (i.e., is **polyvalent**), it may combine with several antibody molecules. If, as is also usual, the antibody molecules are bivalent or multivalent, they can in turn combine with two or more antigen molecules.

A combination of polyvalent antigen with bivalent or multivalent antibody can thus form large complex masses or **lattices,** consisting of many antigen and antibody molecules joined together. The result is the precipitation of large, readily visible flocs as previously described (Fig. 27–14). This flocculation occurs, however, only if antigen and antibody molecules are present in approximately equal numbers or in **optimal relative** or **equivalent proportions.**

If excess antigen is present (relative to antibody), then all of the combining sites on the relatively few antibody molecules are saturated by antigen. Lattice (visible precipitate) formation is therefore minimal or absent. Conversely, if antibody is present in excess, antigen combining sites are saturated and lattice formation is likewise inhibited. The prezones (or **prozones**) and postzones previously described represent zones of antibody or antigen excess;* the midzone (of precipitation) represents the presence of antigen and antibody in optimal or equivalent proportions.

Lattices may be formed by reactions between antibodies and antigens that are in solution, as in the toxin-antitoxin precipitation reaction, or between antibodies and antigens that

_____

*Other contributing factors have been described. Consult any current textbook on immunology for details.

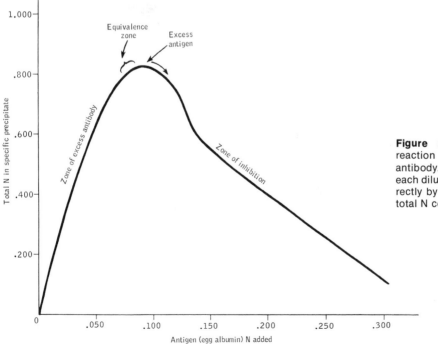

Figure 27–13. Quantitative precipitin reaction between egg albumin and its antibody. The amount of precipitate in each dilution and zone is determined indirectly by removing it and determining its total N content.

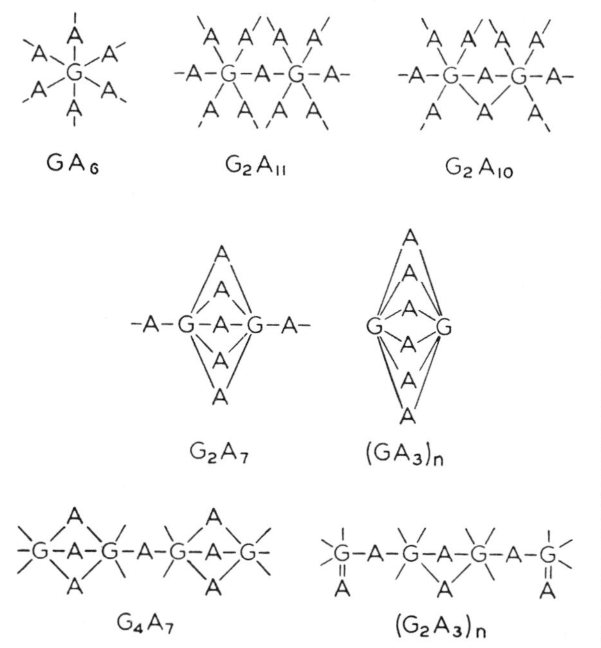

Figure 27–14. Lattice-like structures postulated by Heidelberger for antigen-antibody complexes with various proportions of G and A as indicated by the formulas G_2A_{11}, etc. G = antigen; A = antibody (IgG).

are attached to the surfaces of cells. Antibody-cell lattices consist of large, visible flocs of bacteria (or other cells) held together by antibodies linking their surface antigens. The cells are said to be **agglutinated,** a traditional term expressing the idea that the bacteria are "glued" together by special "sticky" antibodies called **agglutinins.** The terms agglutinin and agglutination are still used, but the underlying mechanism is now better understood.

Zone phenomena are important factors in determining the antibody content of serum, or the amount of antigen in various materials, by serial dilution methods. The series of dilutions should be made fairly extensive and the intervals between dilution steps not too great, otherwise the zone of optimal proportions either may not be reached or may be passed over. The same considerations hold for most other serological tests that involve antigen-antibody reactions.

Other Applications of the Precipitin Reaction. Because of the high degree of specificity that can be achieved by using adsorbed, monospecific sera in precipitin reactions, serological differentiation between soluble proteins of closely similar composition is easily made.

An interesting application is seen in the use of precipitin tests to determine the animal (**host**) from which a mosquito had its most recent blood meal. This illustrates the general method of using the precipitin test to identify "unknown" proteins.

In determining mosquito-host blood, the mosquito is crushed in 1 or 2 ml of saline solution and the blood is thus extracted. This fluid constitutes the antigen to be tested. The sera with which to test it are previously prepared in rabbits by injecting the rabbits with blood from various animal species. One rabbit receives bovine blood, another equine blood, and so on. The serum of each rabbit thus contains precipitins against a different species of animal. By mixing 0.25 ml of the mosquito extract (antigen) with a small quantity of each of the rabbit sera (specific antibody) in turn, one serum will usually be found which causes a definite precipitation. If that serum is from a rabbit immunized with bovine serum, then we may say that the mosquito probably got its blood meal from one of the nearby cattle. This information is of use in the control of mosquito-borne diseases. It guides efforts toward eradication of the mosquitoes that bite man.

In the above experiment, what would have to be done to check whether or not the mosquito's last blood meal was from a rabbit? Using an approach similar to that above, how could you tell whether hamburger had been adulterated with horse meat?

Precipitin Reactions in Gels; Immunodiffusion. An important method of demonstrating precipitin reactions makes use of gels of agar or starch. Precipitin reactions not demonstrable by other methods can readily be made visible by this method. For example, into a Petri dish one pours a warm, clear 1.25 to 1.7 per cent solution of agar. Prior to pouring the agar, 0.01 per cent merthiolate may be added as a preservative to hold contaminants in check. The addition of 0.2 per cent NaCl will aid in the development of sharp, easily seen precipitin lines. Additional salt and/or buffers may be required for certain sera and antigens. After the agar has hardened, a small paper disk soaked with the appropriate antiserum is placed in the center, or as is more frequently done, a cork borer is used to cut a **well** or cup in the agar to hold the antiserum. Several antigens to be tested are similarly applied, or placed in wells, around the central source of antiserum, leaving about 5 to 15 mm of agar between the discs or wells.

In many laboratories, a **micro**-version of the above is used, i.e., the agar is poured on a microscope slide instead of into a Petri dish and cut-off hypodermic needles (size 20) are used to cut the wells. Thus only tiny amounts of reactants (100 lambdas) are consumed. By using shorter distances, the reactions can sometimes be recorded in only six to eight hours rather than the usual 24 to 48 hours.

In both the macro- and micro-versions, the antigens and antibodies diffuse through the agar and encounter each other in series of diminishing concentrations of each. White lines of precipitate appear at the zones of optimal concentration between those antigens and antibodies having mutual reactivity (Fig. 27–15). This method and numerous variations (Fig. 27–16) are widely used in the study of many sorts of antigen-antibody relationships. The process is commonly called **immunodiffusion** or **Ouchterlony** plates (after the investigator who first described their use). The way in which precipitin lines intersect on a plate or slide can reveal whether the antigens in adjacent wells have any determinant groups in common (Fig. 27–15). The student should consult an immunology text for a full discussion of this technique, its pitfalls and variations. Here, we will simply point out that the method is very useful in following the purification of proteins, e.g., toxins, and for establishing the identity of various antigenic components.

Immunoelectrophoresis. This is a related procedure for first separating and then detecting individual antigens in a mixture, such as a bacterial extract or even serum (serum contains many antigenic proteins in addition to the immunoglobulins).

Again, there are macro- and micro-versions of the procedure, with the latter being the more

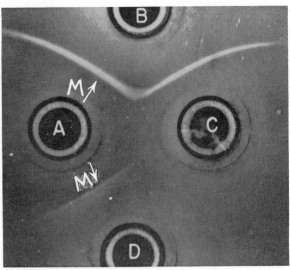

Figure 27–15. Immunodiffusion. Precipitin reaction in an agar gel in a Petri plate (actual size). From a cup in the agar at *A* a soluble protein antigen (M) has diffused outward into the surrounding agar in all directions. Serum containing antibodies (precipitins) against antigen M similarly diffuses outward from *B*. Along the line where the two advancing reagents meet in the agar, a precipitin reaction occurs, as shown by the broad white line of precipitate between *A* and *B*. *C* contains antigens related to those in *A*, which likewise react to some extent with the antibodies from *B*, as shown by the white line between *B* and *C*. *D* is a serum which contains only a very small amount of precipitins against the M proteins in *A*, as shown by the faint line between *A* and *D*.

popular because it conserves materials and can generally be completed more quickly. The number of antigen wells and antiserum troughs used on one slide can be varied also. A thin layer of buffered agar is poured onto a glass slide, and one or more antigen wells are cut.

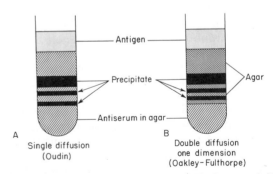

Figure 27–16. Two examples of qualitative precipitin tests. *A*, Oudin's single-diffusion method. Antigen solution diffuses into agar containing an antibody. Formation of a band of precipitate indicates a positive reaction. *B*, Oakley-Fulthorpe double-diffusion method. Antiserum is suspended in agar at the bottom of the test tube and the antigen is poured on top of a layer of agar. Both reagents diffuse through agar until they meet. A precipitate forms if they are homologous.

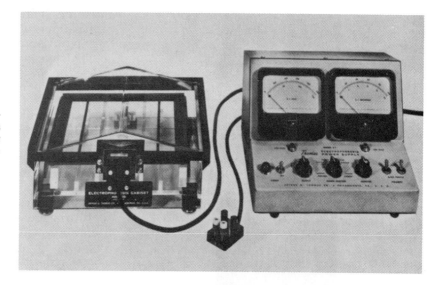

Figure 27–17. A form of apparatus used for electrophoresis on paper strips or in gels. (Courtesy of Arthur H. Thomas Co., Philadelphia, Pa.)

The slide is then placed in an apparatus which is hooked up to a DC power supply so that the gel will be placed in an electrical field, i.e., one end of the slide will act like the anode, the other end like the cathode (Fig. 27–17). After filling the antigen wells with appropriate samples, the power supply is turned on and the various proteins in each antigen well are allowed to migrate into the agar gel. This portion of the procedure is called **electrophoresis.**

The majority of naturally occurring proteins—immunoglobulins being a notable exception—at a pH of 7.2 to 8.0 will move toward the anode (+) end of the slide. But each protein can be expected to migrate at a slightly different speed, and thus they separate from one another. This is because the migration of each protein is governed by: (1) the size of the molecule (the larger the molecule, the more the resistance encountered in trying to move through the gel matrix) and (2) the net electrical charge on the molecule. Remember that proteins are amphoteric (they have both positive and negative charges), and that the relative number of minus to plus charges at a given pH will determine the net charge. The higher the net negative charge, the greater the "pull" toward the anode. After 25 to 40 minutes for the micro-slides, one to three hours for the macro-slides, the power must be turned off, or the fast-migrating proteins may leave the slide and go into the buffer chambers of the apparatus.

A trough is now cut adjacent to the antigen well, or between two wells, and is filled with antiserum. The slide is then left to incubate for 6 to 24 hours in a moist chamber. During this time, the reactants diffuse outward until they meet, each antigen forming a precipitin line or band where it meets a corresponding antibody (Fig. 27–18). The big advantage of immunoelectrophoresis over immunodiffusion is that the lines are spread out and thus easier to count, and the relationship between lines (nonidentity, etc.) is easily determined. Immunoelectrophoresis is widely employed for studying complex antigens and sera.

Agglutinins. In a foregoing discussion of lattice formation it was explained that multivalent antibodies can link antigen molecules together whether the antigen molecules are proteins in solution or are components of intact cells. In the latter case the linkages result in a gathering or clumping together of the cells, a reaction called **agglutination.** Both living and dead cells can be agglutinated provided that their surface antigens have not been altered by agents such as heat or chemicals. If the agglutinated cells are erythrocytes, the phenomenon is called **hemagglutination.**

As a defense mechanism, agglutination does not necessarily kill bacteria, but aids the leucocytes by gathering microorganisms into groups. A leucocyte or other phagocytic cell can engulf 50 agglutinated bacteria fifty times as easily as 50 separate ones, and in much less time (Figs. 27–19, 27–20). Furthermore, the agglutinins appear to **opsonize** the surfaces of the bacteria so that the phagocytes can grasp and engulf them more readily.

Diagnostic Use of Agglutinins. Agglutinins are widely used in the identification of bacteria and the diagnosis of disease. Let us assume that a patient has a febrile disease which has remained undiagnosed for a week or more.

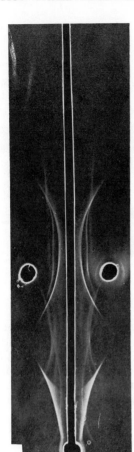

Figure 27–18. Immunoelectrophoretic pattern of sera of germfree mice before (left well) and 21 days after (right well) influenza A infection reacting with rabbit anti-mouse serum in center trough. During passage of the current from end to end (lengthwise) of the slide, various proteins from the "before" and "after" sera migrated to different zones in the gel. Anti-mouse antibodies then diffused laterally from the central trough and formed white lines of precipitate with the various proteins in their distinctive positions in the gel. Differences between the two samples of serum ("before" and "after") are hardly perceptible in this preparation.

antibodies. (There are also some other possible explanations which need not be discussed at this point.) Another test is generally made later.

This means of diagnosing typhoid fever is called the **Widal reaction** after Widal, who first published upon the subject. The term **Widal test** is sometimes (improperly) applied to *any* agglutination test.

Identification of Bacteria by the Agglutination Reaction. Conversely to the detection and identification of antibodies by means of the agglutination test just described, an unknown organism may be identified by using sera containing various *known* antibodies. The known sera are mixed with suspensions of the unknown bacterium, and agglutination is looked for. Suppose, for example, that we have a gram-negative rod that, by its cultural reactions, we know to belong to the typhoid-dysentery group. We may, as a preliminary test, set up two series of tubes: *A*, containing serial dilutions of serum of a typhoid-immune animal, and *B*, containing dilutions of serum of a dysentery-immune animal. A drop of our "unknown" bacterial suspension is added to each tube. If after several hours no change has occurred in the first series of tubes while the serum in the tubes of series *B* has caused the bacilli to agglutinate, we know that, since the serum in *B* contained only dysentery agglutinins, our unknown organism must be some species of *Shigella* (dysentery bacilli). Many other bacteria, saprophytic and parasitic, of agricultural, industrial and other special interests, may be identified in this way.

The Indirect (Passive) Hemagglutination Reaction. This procedure, widely used in immunology, illustrates very nicely the fact that antigens at the **surface** of a cell determine its

Presumably during this time demonstrable amounts of antibodies have accumulated in the blood. We draw a little blood from a vein and allow it to clot. We then remove the clear serum and mix it, suitably diluted (1:20; 1:40; 1:80; ... 1:2560), in a series of test tubes with, for example, a suspension of typhoid bacilli (*Salmonella typhi*). Cultures of these bacteria are commonly maintained in diagnostic laboratories for this purpose. If the patient has typhoid fever his serum will contain typhoid agglutinins and the bacilli in the test tube will be found in flocs or clumps (Fig. 27–21). If no agglutination occurs, either the patient does not have typhoid fever or he has not yet had time to develop

Figure 27–19. Agglutinated bacteria as seen with the microscope by darkfield illumination. Two different types of agglutination are seen here: H (flagellar or flocculent) agglutination, *left*; O (somatic or granular) agglutination, *right*. (Courtesy of Dr. Adrianus Pijper.)

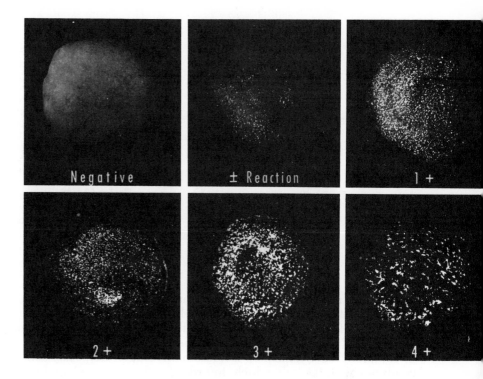

Figure 27-20. A slide agglutination test. Large drops (about 0.10 ml) of serial dilutions of a sample of serum containing antibodies specific for the bacterium under examination were placed in all squares except the upper left one. The serum in this square, marked "negative," was a "normal" control containing none of the specific antibodies. To each drop of diluted serum was added one drop of a heavy suspension of the specific bacteria and each mixture was gently stirred. A ± reaction (slight but definite agglutination) occurred in the highest ("weakest") dilution of serum, a 4+ reaction in the most concentrated serum. Intermediate reactions are indicated by 1+, 2+ and 3+. (Approximately actual size.) (Courtesy of Difco Laboratories, Inc., Detroit, Michigan.)

immunological specificity. Erythrocytes of a sheep or a cow are washed free from their serum. The surface of the cells is generally modified and the cells made more solid and stable in suspension by treating them with tannic acid or formaldehyde. The erythrocytes are again washed and then suspended in a saline solution containing any desired **soluble** antigen (viral or bacterial)—for example, an antigen extracted from "tuberculosis" bacilli. This antigen is adsorbed by the surfaces of the erythrocytes and covers them as a coating. The cells are removed from this suspension and excess antigen is removed.

A series of dilutions is now made with serum containing antibodies specific for the "tuberculosis" antigens with which the erythrocytes are coated. Into each serum dilution is introduced a drop of the suspension of antigen-coated erythrocytes. These behave as though

they were tubercle bacilli. Within a short time the coated sheep erythrocytes are agglutinated. "Tuberculosis" antibodies have no visible effect on the normal, untreated erythrocytes of a sheep or the uncoated tannic acid–treated erythrocytes (Fig. 27–22). It is interesting that totally inert particles of colloidal plastic, gum arabic, latex, and the like can be similarly coated with antigen and agglutinated by specific antibodies. Conversely, antigens may be detected by coating the cells or particles with antibodies.

Cytolysins and Complement. The somatic antigens of certain types of cells call forth antibodies that assist in the **lysis** of that cell. These antibodies are termed **cytolysins.** They are sometimes called **sensitizers** or, by an older term, **amboceptors.** The cytolytic antibody first combines with specific antigens on the surface of the foreign cell that called it into being. The cell may be a bacterium, an erythrocyte or a cell

Figure 27-21. Macroscopic agglutination test. Control tube (C) contains bacterial suspension only. Numbered tubes contain bacterial suspension plus the following dilutions of serum: 1:100, 1:200, 1:500, 1:1,000, 1:2,000, 1:5,000, 1:10,000, and 1:20,000. Agglutination is evident in dilutions 1:100 to 1:10,000 (tubes 1 through 7) but not in 1:20,000 (tube 8). The titer of the serum is therefore 1:10,000.

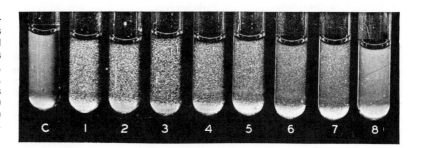

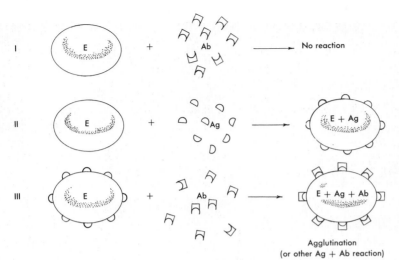

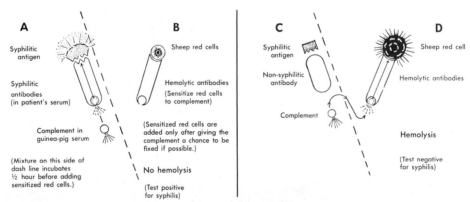

Figure 27-22. Indirect hemagglutination. *I,* Erythrocytes (E) with surfaces prepared by formaldehyde or tannic acid treatment are mixed with antibodies (Ab). No reaction occurs because there are no specific receptors for Ab on the surfaces of the erythrocytes. *II,* In another tube similarly (formaldehyde) prepared erythrocytes are suspended in a solution of antigen (Ag) specific for Ab. Ag is adsorbed to the surfaces of the erythrocytes (E + Ag). *III,* Suspended in a solution of specific Ab the erythrocytes, now specifically coated with Ag, combine with Ab and are agglutinated.

of other nature.* This simple combination is, however, not sufficient to destroy the cell. There is no visible reaction. A second substance, **complement** (previously described), is necessary to complete the lytic action. Complement, too, combines with the cell, which, in order for the complement to act, must already have been sensitized (hence the term **sensitizer**) by the specific cytolytic antibody. Lysis then results (Fig. 27–23). Among the complement-fixing immunoglobulins (sensitizers) IgM is most active, IgG much less so, and IgA, IgD, and IgE not at all.

———————————
* Not all types of cells are equally subject to cytolysis in this manner.

Complement cannot by itself destroy foreign cells; it must act through the intermediation of the sensitizer. The sensitizer is a specific antibody, but complement is nonspecific; it helps any sensitizer to complete its work.

Complement Fixation. After complement has combined with the sensitized cell, the complement is no longer active. It is said to be **fixed.** It is adsorbed onto the sensitized cells. The complement-fixation reaction, discovered in 1901 by Bordet, a famous Belgian scientist and Nobel Prize winner, is the basis of several valuable tests used in microbiology, diagnosis, and serology. As previously stated, complement is adsorbed (fixed) onto any finely divided material like chalk dust, clay or soot in aqueous

Figure 27-23. Diagram of the Wassermann test; a typical complement-fixation reaction. At far left (*A*) is seen a reaction in a test tube between syphilis antigen and syphilis antibodies (sensitizer) in a sample of patient's serum. Complement combines with the sensitizer to destroy the antigen. When sheep cells (*B*) that are already sensitized with sheep-cell-hemolytic antibodies are added, they are not destroyed by the complement because complement has all been used up (fixed) in the syphilitic antigen-antibody reaction. This constitutes a positive test for syphilis. Had there been no syphilitic antibody in the patient's serum (another, unrelated antibody is shown in *C*), no specific antigen-antibody combination could have occurred. Complement would not have been fixed and would have been left free to combine with and destroy (**hemolyze**) the sensitized sheep cells (shown exploding; *D*).

suspension. It is also fixed by precipitates formed by antigen-antibody combinations, but it is not a *necessary* component of such reactions as it is in cytolysis. The fixation of complement in the cytolytic reaction is evidence that cytolysis involves a type of antibody reaction on the cell surface.

The specificity of the complement-fixation reaction enables an investigator to identify unknown antigens or antibodies. For example, if antibody, antigen, and complement are mixed in a tube, we can determine whether antigen and antibody have combined by testing to see whether the complement has been fixed. If complement has been fixed, then we know that an immunologically specific antigen-antibody reaction has occurred. If we know the identity of one (antigen or antibody), we can identify the other. (But since complement is not visible, how can we know whether complement has been fixed? See Figure 27-23.)

Immobilizing Antibodies. The etiological (causative) agent of syphilis is a spirochete, *Treponema pallidum*. The organism is actively motile, rotating on its long axis and bending and flexing. In the serum of patients with syphilis *T. pallidum*–specific antibodies appear some days after initial infection. These antibodies **immobilize** and **kill** the spirochetes within a few hours when mixed with them in test tubes. This effect is readily seen by examining the mixture with a darkfield microscope. It is commonly spoken of as the TPI (**T. pallidum immobilization**) test. The immobilizing action does not take place unless complement is present. Curiously, little or no complement is **fixed** in this reaction. The role of antibody seems to be that of a sensitizer, but the action of the complement is not so obvious, as no lysis occurs.

Immobilization by specific antibodies also occurs in other microorganisms. This is readily seen in *Entamoeba histolytica* (the cause of amebic dysentery) when the active trophozoites of this protozoan are treated with the serum of *E. histolytica*–immune animals. Similarly, there are specific immobilizing antibodies for the ciliated larval state (**miracidium**) of a pathogenic worm (the fluke, *Schistosoma mansoni*) and for various motile bacteria other than *T. pallidum*.

Protective and Neutralizing Antibodies. All the immune reactions so far mentioned are demonstrable by in vitro methods. It was mentioned that immunity does not necessarily result solely from the presence of such antibodies. Indeed, it seems that, as previously indicated, most of them act principally by aiding in the process of phagocytosis. Some are clearly lytic, some antitoxic, some immobilizing.

The action of others is not demonstrable in vitro. The only reliable method of detecting and measuring such antibodies is to infect experimental animals (e.g., mice) and give them doses of the serum to be tested, before, after or simultaneously with the infection, to see whether the animals are thereby protected. This measures **protective** or **neutralizing power** directly, regardless of whether this power depends on agglutinins, cytolysins, or some still-undiscovered antibody. Such a test is known as a **protection test** and is widely used to measure the antigenic virtues of antigens (by measuring the immunological response) and the protective power of sera.

In dealing with viruses the term **neutralization test** is generally used. Instead of using live animals in neutralization tests one may use animal-tissue cells growing in tissue cultures. The serum, if effective, will prevent virus from producing cytopathic effects or from forming plaques in monolayer cell cultures. (Chaps. 15, 16, 17.)

Fluorescent* Antibody Staining. One of the most interesting and valuable advances in the field of microbiology is the fluorescent labeling of antibodies so that their combination with specific antigen can be detected visually and immediately. By means of this technique, microorganisms and their antigens, as well as antibodies, may be detected and identified within a few minutes, an improvement over the time-consuming and expensive methods of systematic study of cultures and antigenic tests and animal experiments previously described.

The fluorescent antibody–staining procedure is technically complex, although the principle is relatively simple. A first step is separation and concentration of the specific immune globulins from the bulk of the serum in which they occur. The concentrated immune globulins are then combined (conjugated) with a fluorescent dye, commonly fluorescein isothiocyanate. The antibodies are then said to be **labeled.** When illuminated with ultraviolet light they give off a brilliant yellow glow. We may use such a preparation of fluorescent antibody globulin to detect the corresponding specific antigen by four different methods: direct, indirect, inhibition, and indirect-complement. Only the

---

*By fluorescence is meant the property of emitting rays having a wavelength (color) different from that of the incident rays. Fluorescent objects are particularly brilliant in ultraviolet light.

direct and indirect methods need be described here.

Let us suppose that our fluorescent immunoglobulin is specific for an antigen on the surface of typhoid bacilli (*Salmonella typhi*). Let us suppose also that a single bacillus of this species is suspected to be present in a large section of gram-stained spleen tissue from a typhoid victim. It is impossible to find the one tiny organism by ordinary microscopic means, even though gram-stained, because of the relatively enormous mass of surrounding tissue, which takes the same stain as the bacillus. The bacillus is lost like the proverbial needle in a haystack.

Let us prepare another section of the same tissue, unstained. We flood it with fluorescein-labeled antibody specific for *S. typhi*. The fluorescent antibody attaches itself to the corresponding antigen in the bacillus. Then we wash out all of the unattached antibody. When the tissue section is illuminated with ultraviolet light and examined with the microscope the hidden bacillus reveals itself by its brilliant yellow fluorescent light (Figs. 27–24; 27–25, *I*).

A serious difficulty with this method is the necessity of preparing dozens of different fluorescent antibodies, each representing one of the dozens of different antigens which we might want to detect. The labeling procedure is difficult, time consuming and expensive. This difficulty is overcome in great part by using indirect methods, one of which is indicated in Figure 27–25, *II, IIa*. The fluorescent staining

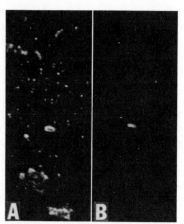

Figure 27–24. Use of fluorescent antibody to detect specific antigen in a mixture of materials. *A,* A sample of soil containing *Malleomyces mallei,* a very dangerous pathogenic bacterium. This is a fluorescent antibody-stained smear illuminated by ordinary light (darkfield preparation). It is impossible (without glancing at *B*) to distinguish the single cell of *M. mallei* from the many saprophytic soil bacteria and soil particles that are present in the sample. In *B,* illuminated by ultraviolet light, the single cell of *M. mallei* is brilliantly and exclusively evident.

method in various modifications is used to locate viruses in cells and tissues, identify organisms and many other procedures.

Ferritin Labeling. In an adaptation of the "labeling" principle of fluorescent antibody staining, antibody may be made visible in electron micrographs by labeling it with ferritin.

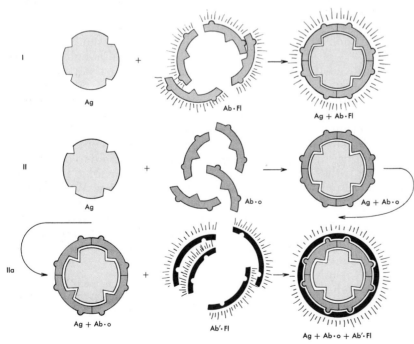

Figure 27–25. Direct and indirect fluorescent antibody staining. In the direct method (*I,* top row) antigen (Ag) is allowed to react with its specific antibody which has previously been conjugated with a fluorescent dye (Ab · Fl). When viewed in the microscope with ultraviolet illumination the antigen-antibody combination (Ag + Ab · Fl) glows brilliantly.

In the indirect method (*II,* second row) the antigen is allowed to combine with ordinary (nonfluorescent) specific antibody (Ab · o) in the usual manner. Viewed with ultraviolet light no fluorescence is seen. The invisible Ag + Ab · o combination is now treated (*IIa,* third row) with fluorescent antibody specific for gamma globulin (Ab' · Fl). Ab' · Fl therefore combines with the Ab · o (gamma globulin) on the surface of Ag, causing the particles of Ag + Ab · o to glow in ultraviolet light: (Ag + Ab · o + Ab' · Fl).

Ag   Ab · Fl   Ag + Ab · Fl

Ag   Ab · o   Ag + Ab · o

Ag + Ab · o   Ab' · Fl   Ag + Ab · o + Ab' · Fl

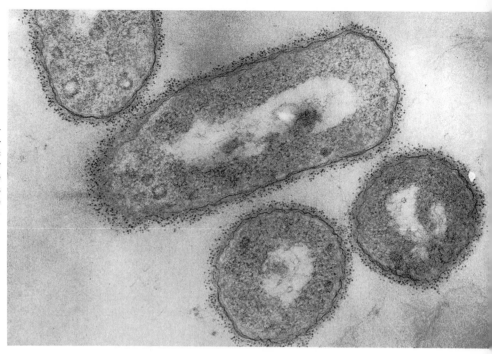

Figure 27–26. *Salmonella typhimurium* agglutinated by ferritin-conjugated antibody to somatic antigen. The antigen-antibody complexes, being electron-opaque due to ferritin, are revealed as myriads of black specks close around the bacterial cells. (60,-000×.)

Ferritin is an organic complex of ferric hydroxide and ferric phosphate associated with apoferritin, a protein of the liver and spleen that serves as a storage place for iron. Ferritin is **opaque to electrons.** Ferritin-conjugated antibody combines with specific antigen as does fluorescein-conjugated antibody. To visualize the electron-opaque ferritin-antibody, however, the electron microscope is used instead of ultraviolet light (Figs. 16–12, 27–26).

CHAPTER 27
SUPPLEMENTARY READING

Abramoff, P., and LaVia, M. F.: Biology of the Immune Response. McGraw-Hill Book Co., Inc., New York. 1970.

Bellanti, J. A.: Immunology. W. B. Saunders Co., Philadelphia. 1971.

Blood Group Antigens and Antibodies as Applied to Compatibility Testing. Ortho Diagnostics, Raritan, N.J. 1967.

Blood Group Antigens and Antibodies as Applied to the Hemolytic Disease of the Newborn. Ortho Diagnostics, Raritan, N.J. 1968.

Blood Group Antigens and Antibodies as Applied to the ABO and Rh Systems. Ortho Diagnostics, Raritan, N.J. 1969.

Carpenter, P. L.: Immunology and Serology, 3rd ed. W. B. Saunders Co., Philadelphia. 1972.

Clark, C. A.: The prevention of rhesus babies. Sci. Am., *219*:46, 1968.

Crowle, A. J.: Immunodiffusion. Academic Press, Inc., New York. 1961.

Eisen, H. N.: Immunology. Harper and Row, Publishers, Inc., Hagerstown, Maryland. 1973.

Good, R. A., and Fisher, D. W. (Eds.): Immunobiology. Sinauer Associates, Inc., Stamford, Conn. 1971.

Gordon, B. L., II, and Ford, D. K.: Essentials of Immunity. F. A. Davis Co., Philadelphia. 1971.

Leskowitz, S.: Immunologic tolerance. Bioscience, *18*:1030, 1968.

Levine, P.: Prevention and treatment of erythroblastosis fetalis. Ann. N.Y. Acad. Sci., *169* (Art. 1):234, 1970.

Roitt, I.: Essential Immunology. Blackwell Scientific Publications, Oxford. 1971.

Weiser, R. S., Myrvik, Q. N., and Pearsall, N. N.: Fundamentals of Immunology for Students of Medicine and Related Sciences. Lea & Febiger, Philadelphia. 1969.

ACTIVE AND PASSIVE IMMUNITY

It has been observed from antiquity that persons become immune to certain infectious diseases by surviving natural attacks of those maladies. It is now known that the body develops specific immunity, including specific antibodies, as an **active** response to a **natural** antigenic stimulus. Since this type of resistance is actively acquired in the course of natural events it is often spoken of as **active natural immunity**. In this chapter we will be primarily concerned with other ways in which immunity can be obtained. The nature of the immune response, various aspects of antibody formation, and a consideration of the types of microbial components which can serve as antigens were discussed in the two previous chapters.

28.1
ACTIVE NATURAL IMMUNITY

Following a naturally occurring infection, a period of specific immunity is frequently observed. By this we mean that either the individual is not susceptible to a repeat infection with the same microorganism, or, if a repeat infection does occur, the illness is terminated in a much shorter time than initially and usually with less severe symptoms. Immunity acquired from actual infection generally outlasts immunity acquired in any other way. The reason for this is that during infection (unless abruptly terminated by antibiotics), a considerable amount of

antigenic material is present, and this tends to elicit a maximum immune response. Also, the body may continue to harbor small numbers of the pathogen in deep tissues long after the convalescent period is over. This amounts to a small, continuous antigenic stimulus and is thought to keep specific immunity at a high level.

Ordinarily, in cases of any naturally occurring infection, the pathogens may be expected to be continuously or intermittently present in the environment and thus make repeated contacts with immune individuals (or with previously uninfected but resistant individuals) from time to time. Depending on the virulence of the microorganism, the dose, the route of inoculation, and the immune or resistant status of the individual, very mild, **inapparent** or **subclinical** infections frequently follow such encounters. With subclinical infections, the symptoms, if any, are so very minor that little or no attention is paid to them. Fortunately, the vast majority of infections are of this type. Repeated subclinical infections also provide renewed antigenic stimulus (**recall reactions**) and tend to maintain a high level of specific immunity (see §28.2). Thus, it is not difficult to understand why active naturally acquired immunity is often long-term, even lifelong. One does not expect to suffer from measles twice, for example.

Equally obvious, however, is that not all naturally occurring infections lead to long-term immunity. The reason is two-fold: first, there must be sufficient antigenic stimulus to elicit a pronounced immune response, i.e., appreciable antigenic material circulating in the blood and lymph. This criterion is often not met if the microorganisms remain localized in the superficial tissues. Secondly, the immune response shows considerable specificity, and host cells endowed with immunological memory (lymphocytes) will only recognize and respond to their specific antigenic counterparts. Thus immunity against a virus of one antigenic type often does not provide protection against a related, but antigenically different, virus. Many bacteria, e.g., *Salmonella,* and viruses, e.g., influenza virus, are known to show considerable antigenic variation and repeat infections with such organisms are commonplace.

While naturally occurring infections tend to be mild or subclinical for the most part, as previously mentioned, there are still a significant number of infections which are severe. They may be disabling and disfiguring or fatal, even though antibodies are formed. It would be much better if we could become immune by some means that we can control. Furthermore, we should like to become safely immune to infectious disease early in life and not have to wait for accidental natural infection, occurring perhaps at a very inconvenient time in adult life. It is often desirable to be able to produce immunity to certain diseases at certain definite times. For example, Americans traveling outside the North American continent were formerly required to have a valid certificate of vaccination against smallpox before they were allowed to re-enter the United States.* A person desiring to do laboratory research with yellow fever virus would like to be able to immunize himself before starting the work, since infection with the virus might otherwise prove fatal. So also, physicians and nurses or others working in plague areas or with poliomyelitis, diphtheria, or tuberculosis patients should be immunized safely and comfortably against these diseases in time to begin their work. All this, however, is too much to expect of natural processes.

In view of these needs, man has devised means of developing specific immunities artificially and safely. The methods involve natural processes, but are used under modified, carefully controlled conditions and are therefore called **artificial immunization.** Table 28–1 summarizes the relationship between the various routes of immunization.

28.2 ACTIVE ARTIFICIAL IMMUNITY

In active artificial immunity the patient's body is stimulated to develop resistance by being injected or infected with certain kinds of immunizing agents (Chapt. 27). These are of three general types: sterile, bacterial **exotoxins** or **toxoids;** sterile, microbial antigens consisting of **dead** microorganisms; and **living** infectious microorganisms, the virulence of which has been reduced or attenuated by various procedures so that no serious infection results (see Chapter 24). Active artificial immunity depends almost solely upon the induction of specific antibodies, the major exception being when certain live attenuated vaccines are used. In the latter case, both humoral and cellular types of immunity may be elicited.

Antigens Used in Active Artificial Immunization

Exotoxins. A culture of exotoxin-producing bacteria, such as *Clostridium tetani,* cause of tetanus or "lockjaw" (Chapt. 36), or *Corynebacterium diphtheriae,* cause of diphtheria (Chapt. 39), is made in broth. After sufficient growth of the bacteria, the culture is passed through very fine-pored filters which remove the organisms. The now-sterile **filtrate** (broth that has passed through the filter) contains the exotoxin liberated by the cells during their growth. This may be injected hypodermically (under the skin) into the persons to be immunized, in from one to five *very* minute doses, at weekly intervals. Eventually (usually after two to six weeks) their blood will be found to contain tetanus antitoxin (or diphtheria antitoxin) which protects them from tetanus toxin (or diphtheria toxin). There is danger from the toxin, however, and it is now *never* injected as such, even when mixed with antitoxin as "toxin-antitoxin" or T-A-T. Fatal accidents have occurred, caused by overdoses of toxin or by dissociation of toxin from the antitoxins in T-A-T.

Toxoids. In Germany in 1890, von Behring (Nobel Prize winner), Fränkel and Kitasato discovered that diphtheria and tetanus toxins

*The U.S. Public Health Service now requires smallpox vaccination only if an individual plans to travel to one of several parts of the world where smallpox is still an endemic disease. Travelers from such areas to the U.S. must also have vaccination certificates.

TABLE 28-1. A COMPARISON OF DIFFERENT TYPES OF SPECIFIC IMMUNITY

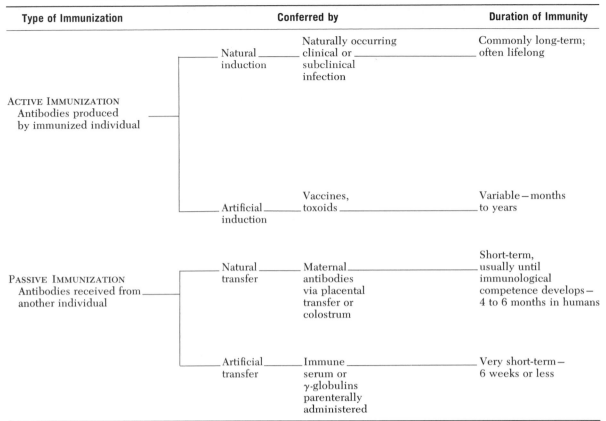

Type of Immunization		Conferred by	Duration of Immunity
ACTIVE IMMUNIZATION Antibodies produced by immunized individual	Natural induction	Naturally occurring clinical or subclinical infection	Commonly long-term; often lifelong
	Artificial induction	Vaccines, toxoids	Variable—months to years
PASSIVE IMMUNIZATION Antibodies received from another individual	Natural transfer	Maternal antibodies via placental transfer or colostrum	Short-term, usually until immunological competence develops— 4 to 6 months in humans
	Artificial transfer	Immune serum or γ-globulins parenterally administered	Very short-term— 6 weeks or less

that had been heated for one hour at 70C were no longer poisonous but could stimulate antibody production. The possibilities for the safe and effective prevention of tetanus and diphtheria were immediately recognized and used. In France in 1924, Ramon found that formaldehyde detoxifies the toxins of diphtheria and tetanus as well as heat. Formaldehyde-detoxified—but antigenic—exotoxin was called **toxoid.** Formaldehyde-treated toxoids soon came into general use for active immunization against diphtheria and tetanus.

In 1933 in the United States, Havens and Wheeler found that, on the addition of alum to broth containing toxoid, the toxoid is adsorbed onto the particles of alum, which then precipitates. The alum-toxoid may then be collected by sedimentation, concentrated and purified. Alum-precipitated (AP) toxoids are highly effective. Two or three injections at intervals of about one month commonly stimulate production of sufficient antibodies.

Adjuvants. When fluid (*not* AP) toxoid is injected under the skin, the soluble material is eliminated within a few hours. The antigenic stimulus is transitory. If the toxoid could be held in situ for several days, as occurs naturally in actual infections, the antigenic stimulus would be prolonged and continuous.

This is accomplished with AP toxoids since the alum precipitate remains undissolved where injected, releasing its adsorbed toxoid slowly, thus giving the patient a prolonged and continuous antigenic action which is highly effective. Substances that thus enhance the effectiveness of antigens are called **adjuvants** and are of several sorts, such as mineral oil, peanut oil, or bacterial lipids (see Chapter 29). The persistence in the body of any antigen, living or dead, maintains immunity by its continuous antigenic action.

This principle is not confined to diphtheria immunization but is of broad significance. For example, in a modified form it is used with antibiotics to maintain a high concentration of antibiotics in the blood over a long period without repeated injections.

The principle of adjuvants has also been adapted to bacterial vaccine against whooping cough. It is now common practice to mix diphtheria and tetanus toxoids as well as alum-precipitated bacterial antigens (especially whooping-cough bacilli). The combination of three antigens is even more effective with respect to each than any one of the antigens alone. In addition, the number of separate injections required for immunization is reduced.

28.3
PRIMARY AND SECONDARY STIMULUS

In the ordinary course of life in the United States one is constantly exposed to infectious organisms such as pneumococci, streptococci, tubercle bacilli, and influenza virus and polioviruses from healthy carriers and from ambulatory, mild, and inapparent cases of the diseases. In the nonimmunes after about four months of age, these exposures most commonly result in mild or unrecognized cases; much less commonly in perceptible or severe illness; least commonly in fatal illness. The survivors are actively, **naturally** immunized against the organism(s) that infected them. Renewed contact with each infectious agent has no outwardly perceptible effect, but inwardly and all unseen it restimulates the production of specific antibodies as previously mentioned. Thus, these **repeated antigenic stimuli** serve to keep one's immunity in a good state.

The beneficial and protective effect of **repeated** antigenic stimuli is exploited in artificial processes also. For example, suppose that a child is given a single, initial dose (or two or three weekly doses) of any antigen, say diphtheria toxoid. This initial dose or series of doses is called a **primary stimulus.** In about two weeks his blood, tested by appropriate methods, shows few if any antibodies. After four to six weeks, however, his blood is found to contain a satisfactory amount of diphtheria antitoxin (antibodies). The development of detectable amounts of antibody has been relatively slow (Fig. 28–1). A year later the concentrations (**titers**) of antibodies in his blood may be found to have declined to a very low level or to have disappeared entirely. This diminution of antibody titer from the serum is very common. However, **immunity** has not necessarily disappeared.

"Booster" Doses; The Anamnestic Reaction. Let us now give the child a second dose of diphtheria toxoid and test his blood for diphtheria antitoxin at short intervals. A surprisingly rapid and extensive response is now noted. After the first or **primary stimulus,** given a year before, response was slow. Response to this second dose (**secondary stimulus**) occurs at once. In a

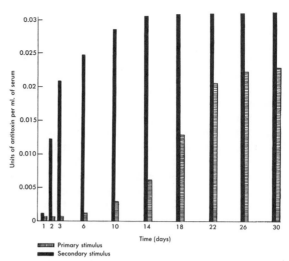

Figure 28–1. Quantities of antitoxin demonstrable in the blood following primary stimulus (light bars) and following secondary stimulus (dark bars). The rate of production by the stimulated cells is probably alike after both primary and secondary stimuli, but the **number of cells participating** is much greater after the secondary stimulus. These are representative data only, as individuals differ greatly in response to both primary and secondary stimuli.

few *hours* or a day or two the child may be found to have one or more units of diphtheria antitoxin per milliliter of blood. In practice, the secondary stimulus is often referred to as a **"booster" dose.** The underlying physiological response, as previously described, is called the **anamnestic** or **recall reaction** (Gr. *anamnesis* = to remember or recall; the reverse of Gr. *amnesia* = forgetfulness).

Clonal Selection Hypothesis and Booster Doses. A suggested explanation of the anamnestic reaction, and also of immunological specificity (see Chapter 27), is based on the **clonal selection hypothesis** of Burnet (Nobel Prize winner). In a very brief form this hypothesis postulates that each of the millions of developing lymphoid cells formed in early life undergoes a different mutation, so that the DNA in each and in its progeny is coded to recognize a different antigenic (determinant) group. An antigen gaining access to the B type lymphocytes for the first time (the primary stimulus) is thus practically certain to encounter a cell having a recognition site for its specific molecular structure. The two combine and the cell is stimulated to reproduce vigorously, and the progeny (cells of the clone) begin to make specific antibody. Under the stimulus of the antigen and after several days or weeks have passed to permit time for the cells to mature and multiply, this type of cell becomes very numerous. Its billions of progeny (selected clone), acting together, finally produce large enough amounts of the specific antibody to be detectable in the blood by appropriate tests (agglutination or precipitation).

After the subsidence of the infection, or removal of the antigen by other means, the proliferation of lymphocytes slows down drastically, the number of plasma cells producing the homologous antibody decreases, and the level of specific antibody in the circulatory system also decreases. It is doubtful, however, that the number of cells of one clonal type ever drops back to the level which existed at the time of the primary response, or at least not for a long time. Thus, in any subsequent encounters between host and antigen, a larger pool of lymphocytes capable of reacting to the antigen is available and will react to the secondary stimulus with great speed. These are called "memory cells."

Dead or Inactivated Microorganisms.
Many microorganisms pathogenic for man do not produce convenient exotoxins for the preparation of toxoids. The pathogenic properties of many are caused by **endotoxins**. For immunization against a number of these organisms it is common practice to use suspensions of killed, intact organisms with their entire complements of antigens. Such suspensions of killed or inactivated microorganisms are frequently referred to as vaccines. If prepared from bacteria, they are correctly termed **bacterins**. The term **vaccine** is properly restricted to smallpox vaccine, since smallpox vaccine is derived primarily from cows (L. *vacca* = cow). However, the term **vaccine** is widely used for any immunizing agent and is so used here.

Bacteria. Bacteria for the purpose of vaccination are usually cultivated on the surface of agar media, and the growth is removed with physiological (0.85 per cent) sodium chloride solution. The bacteria are **killed** by heating at 60 to 65C for 30 to 60 minutes. Such vaccines as typhoid, cholera, and whooping cough are prepared from gram-negative bacilli (see Chapter 35). Except for whooping-cough vaccine, vaccines made with gram-negative bacteria are less effective than many viral and rickettsial vaccines.

Viruses. In recent years many of the formerly used inactivated (killed) viral vaccines, such as the Salk polio vaccine, have been replaced by infectious but attenuated viral strains, e.g., Sabin oral polio vaccine. The attenuated strains produce a mild, limited infection which evokes a longer-lasting immunity than the killed vaccines. However, there are still some viruses for which a satisfactory live attenuated vaccine has not been developed. Influenza virus can be propagated in chick embryos and subsequently be inactivated by formalin. The vaccine is generally composed of several strains, particularly those which have been responsible for outbreaks of human influenza in various parts of the world. Since there are a great many strains which show antigenic variability, the pharmaceutical industry must constantly alter the composition of their vaccines so as to include those strains likely to be involved in forthcoming outbreaks or pandemics.

Rickettsias. Rickettsias are cultivated for rickettsial vaccines in live, chick-embryo yolk sacs. Vaccines against Rocky Mountain spotted fever are made from this material. The tissue and other extraneous matter are removed by filtration or centrifugation. The rickettsias are killed with formaldehyde or other substances. Rigid tests are made to insure that all of the microorganisms are dead. A minute amount (0.25 per cent) of phenol, tricresol, or some other antimicrobic agent is often added to insure sterility.

The principle of the secondary stimulus is widely used in connection with killed vaccines as well as with toxoids.

Attenuated, Living Infectious Agents.
A third method of active artificial immunization involves actual infection of the person or animal to be immunized. The immunizing organisms are so treated that their virulence is greatly **attenuated** or diminished.

There are at least four means of obtaining strains or clones of pathogenic organisms that can be safely used to induce active artificial immunity: passage of the organism in abnormal hosts or, virtually the equivalent, many successive passages of the pathogen in cultures of cells from some animal other than the usual host of the pathogen; prolonged search, by repeated trials, for clones of low virulence; treatment of the pathogen with unfavorable agents such as desiccation; cultivation of the pathogen under unfavorable conditions such as low surface tension or high temperature.

Passage in Abnormal Host. This is well illustrated by the development of a vaccine against fox distemper, a scourge to silver fox–fur farmers. The virus from a sick fox was injected into a ferret. Infectious fluid from the sick ferret was injected into another ferret. This ferret-to-ferret transfer (**animal passage**) was continued through a long series. The virus became adapted to, and enormously virulent for, ferrets. When fluid from the last ferret in a long series was tested in a fox, it had little or no virulence for the fox. However, the fox was afterward found to be completely immune to natural fox-distemper virus. The virus had become highly adapted to ferrets but **modified** or **attenuated** with respect to foxes. The method of animal passage, so valuable to fox farmers, has been much used with other viruses.

SMALLPOX VACCINE. The classic illustration of what was once thought to be the effect of animal passage on virulence is seen in the preparation of smallpox vaccine. It was originally thought that cows became infected with smallpox but that they developed only the relatively mild disease, **cowpox.** Contact with the cow (passage in an abnormal host) was supposed to have modified the virulence of the original smallpox virus. It was then called **vaccinia virus.**

The original supposition may be correct, but it may also be that the cowpox virus and smallpox virus are distinct "species" but so closely related that the one immunizes against the other. This sort of relationship is exemplified by the fact that one may immunize dogs to canine distemper by injecting them with tissue culture measles virus. These viruses are related but not identical.

The value of smallpox vaccination with cowpox virus was first demonstrated in 1774 in the home and family of Benjamin Jesty, a farmer of Somerset, England, and later by him publically in London in 1805. It was again demonstrated in 1798 by Edward Jenner, then a country doctor but later a famous British scientist.

Like Jesty, Jenner had observed that dairy workers associated with cows having cowpox, or **vaccinia,** did not succumb during epidemics of smallpox, which were then dreadful scourges. On the basis of this observation he again demonstrated experimentally that anyone could safely become immune to smallpox through infection with cowpox. The resistance to smallpox conferred by vaccination usually lasts at least three to seven years.

Vaccinia virus, as generally used today, is prepared by scratching the virus into the shaved and disinfected skin of a calf. When the pustules are fully developed the lymph is collected from them and put up in glass tubes ready for use (Fig. 28–2). Its potency and cleanliness are carefully controlled by the National Institutes of Health, in Bethesda, Maryland. Tissue culture virus preserved by lyophilization is also used, especially in tropical countries.

Revaccination should reveal either an allergic or an anamnestic response; both are commonly seen (Fig. 28–3).

Other examples of living vaccines prepared by the attenuation of virulence of viruses by animal passage or cell culture passage are rabies vaccines cultivated in cell cultures and in live avain embryos. Yellow fever vaccine is prepared in live chick embryos. Measles, German measles (rubella), mumps, and canine hepatitis vaccines are cultivated in live animal cells in tissue cultures.

Selection of Mutant Clones of Low Virulence; "Live" Polio Vaccine. Active (commonly called "live") tissue-cultured, attenuated polio vaccine has been available for public use since about 1958. It is administered orally and is called oral polio vaccine or Sabin vaccine. It has been administered to millions of persons of all age groups throughout the world. The immunizing effect against all three types of polio has been excellent, with antibodies appearing in one to two weeks. Attenuated poliovirus strains were developed only after years of devoted and grueling researches by many workers, notably Albert B. Sabin, Hilary Koprowski, and Herald R. Cox, each working independently.

Many methods of obtaining attenuated poliovirus were tried: induced mutation and recombinations, trial of naturally occurring strains of low virulence, selection and trial of clones of mutant virus from plaques in monolayer cell cultures, and so on. The work was triply arduous since it had to be done with each of three types of poliovirus.

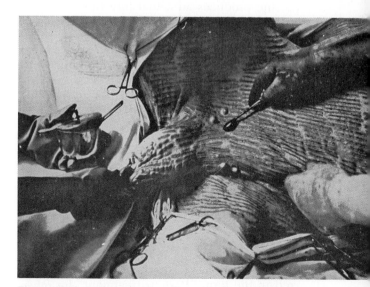

Figure 28–2. The collection of cowpox lymph from the skin of a calf. The skin has been shaved, cleaned, disinfected and inoculated in long parallel scratches (clearly seen in the picture) with cowpox lymph. Typical pustules have developed along the scratches, and the lymph from these is being collected with surgical cleanliness.

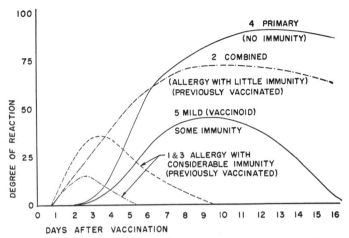

Figure 28–3. Types of reaction to smallpox vaccination. Curves *1* and *3* represent rapid, superficial reactions beginning during the first 24 hours, sometimes showing small vesicle formation, but terminating rather quickly after a mild course, without scar formation. These occur *only in previously vaccinated persons* (or in those convalescent from smallpox), and the quick reaction of short duration and mild degree is characteristic of immunity *with allergy* to the smallpox virus, living or dead. Allergy may exist when immunity has all but lapsed, as shown in curve *2*. This *begins* as an allergic reaction but goes on to a real "take" with papule, vesicle, pustule and eventual scar formation, much like the primary "take" in a wholly susceptible person without allergy, shown in curve *4*. Curve *5* is a milder reaction in a person without allergy but with a considerable degree of immunity. This is a "vaccinoid" reaction, usually without scar formation. Note that the reactions in the absence of allergy (curves *4* and *5*) do not *begin* until the second or third day. It is occasionally difficult to differentiate the types of reaction.

The clone of live polio-vaccine virus of each type finally selected is an attenuated mutant of that type obtained by picking and growing virus from single plaques (pure clones). Each of the attenuated strains is able to produce only a mild intestinal infection. Because oral vaccine produces an actual infection of the intestinal tract (see poliomyelitis), it is transmissible to other persons just as natural polio infection. This is a desirable result since these other persons, if they are not already immunized by natural means or by Salk inactivated vaccine, also become immunized by the attenuated strain spread by the vaccinated person.

The three types of virus may be administered separately, if desired, but this entails unnecessary bookkeeping and work. Most preparations are **trivalent,** i.e., they contain all three types. Booster doses are given at entry into school. In populations with low sanitary standards, other feces-transmitted viruses, such as Coxsackie, ECHO, "wild" (natural) polio, and adenoviruses, may be so prevalent as to interfere with the vaccine strains.

Desiccation. Pasteur's name is immortalized in the term "Pasteur treatment" for **rabies** or **hydrophobia.** Pasteur's process was originally a course of immunizing injections with the active virus of rabies attenuated by rabbit-to-rabbit passage and desiccation. In the modern Pasteur treatment of human beings, an attenuated virus cultivated in duck embryos and inactivated with β-propiolactone is widely used. An active attenuated strain, the HEP (high egg passage) Flury vaccine, is commonly used for **preventive** purposes in veterinary practice and may well become available for use in humans.

If the injections of Pasteur treatment are started soon enough after the bite of the infected animal, resistance or interference (?) develops from the injections before the virus from the infectious bite can cause disease.

Pasteur's first human immunization against rabies in a boy bitten by a rabid wolf, in 1885, was an extremely dramatic event and marked the beginning of an epoch in the war on disease. His experiments on rabies have been movingly described by Vallery-Radot, his grandson. (See Supplementary Reading.)

Cultivation in Special Media; Immunization Against Tuberculosis. A method for immunization against tuberculosis with living, attenuated tubercle bacilli has been known and widely and successfully used for years. The procedure is known as the **Calmette-Guerin process** and the attenuated cultures as BCG **(Bacillus Calmette-Guerin;** Calmette and Guerin were the French scientists who developed the method). The process consists of the injection of **live** tubercle bacilli, the virulence of which has been reduced by cultivation of the organisms on certain media containing bile. This method of cultivation results in the development of a stable variant of the tubercle bacillus having low virulence. This is the only important use, at present, of immunization of human beings with living bacteria.

Large-scale studies conducted in many countries for decades prove that BCG is a safe and valuable immunizing agent. In any procedure involving immunization of millions of human beings some unfortunate episodes are to be expected, and a few were reported in the days of pioneer use of BCG. Pioneering is often hazardous, but without the courage to undertake it there would be little progress. BCG is used especially for groups frequently exposed to the disease: nurses, doctors, and inhabitants of areas where tuberculosis is highly prevalent. It is discussed further under Tuberculosis (Chapt. 40). Vaccines of killed and disrupted BCG organisms are currently under investigation.

Cultivation at Unfavorable Temperatures. Although this method is not much used at present, its discovery and demonstration by Pasteur constitute an episode of great dramatic interest in the history of bacteriology. The principle was first used with *Pasteurella avicida*, the bacterial cause of fowl cholera. It was later developed in connection with studies of anthrax which, in the 19th century, decimated the sheep flocks of France. Following numerous preliminary experiments by the tireless Pasteur on variants of *Bacillus anthracis*, cultures of the organisms were prepared by incubation at unfavorably high temperatures of 39 to 43C (optimum, about 35C). This caused the bacilli to lose some of their virulence, and often their power of spore formation. The method has since been shown to vary in result, so that neither loss of spores nor cultivation at 42C should be assumed to have deprived these organisms of their dangerous properties.

As a result of animal and human experimentation, poliomyelitis, smallpox, rabies, tuberculosis, measles, yellow fever, malaria, rubella, whooping cough, diphtheria, tetanus, and many other scourges of the human race and of many animals are at the vanishing point.

28.4
PASSIVE NATURAL IMMUNITY

Some antibodies can pass from mother to infant through the placenta; consequently, they are commonly found in the blood of the infant at birth. They serve to protect the child for some months after birth. By three to six months after birth maternal antibodies have largely disappeared and the child becomes susceptible to many infectious diseases. However, the child is also immunocompetent by this time; i.e., the immune system has developed and matured to the point where the infant can develop its own immunity, including the formation of protective antibodies. It is therefore advisable to begin active immunization of the child early (second to fourth month) by injections of combined antigens of diphtheria, tetanus, and pertussis, followed by the oral administration of the trivalent oral polio vaccine at six months. If injections are begun too early, the maternal antibodies still in the child may interfere with the antigenic effect, or the immature antibody-forming mechanisms may be ineffective. Data from an interesting study of the interference of antibodies upon antigenic effect show that an Rh^- mother can be protected from the antigenic action of erythrocytes of an Rh^+ baby, live or aborted, by giving the mother antibodies (RhIG) to the Rh factor within 72 hours post-partum or post-abortion.

It is important to note that the types of antibodies found in the newborn infant reveal a good deal about the immune status of the mother. The expectant mother who does not have good antibody titers to diphtheria, tetanus, polio, rubella (German measles), pertussis, and possibly salmonellosis would do well to receive immunizations herself to confer passive immunity to these diseases on her child.

Maternal antibodies persist in the infant for one to several months after birth because the antibodies are of maternal origin; i.e., self or nearly self.

28.5
PASSIVE ARTIFICIAL IMMUNITY

In some cases it is necessary that a large supply of antibodies appear in the blood immediately in order to combat an overwhelming infection. This is especially well illustrated in such diseases as diphtheria and tetanus. The chief symptoms in these diseases are caused by the bacterial exotoxins in the body of the patient. The poisonous action is very rapid. When the patient is already ill there is no time to lose waiting for him to develop active immunity, natural *or* artificial; he must **passively** receive ready-made antibodies. Immunity resulting from injections of these ready-made antibodies is immediate and is called **passive artificial immunity.**

It is now possible to purchase at all well-stocked pharmacies and health departments syringes or ampules already filled with **antitoxic serum** prepared for just such emergencies. Such antibody-containing serum is obtained from animals, usually horses, that have received, weeks or months previously, repeated injections of the special antigen against which antibodies are desired.

Transitory Nature of Passive Immunity. If antibody-containing serum used for passive immunization or treatment is of *other than human* origin, the antibodies disappear from the body in two to three weeks (i.e., they are rejected as nonself). During their presence infection may be entirely prevented or, if infection occurs, the resulting disease is usually very mild.

Passive Immunity in the Prevention of Disease. Passive artificial immunity is used in the prevention (**prophylaxis** [Gr. *prophylassein* = to guard against]) as well as the treatment of disease. For example, if it is known or suspected that a person is likely to become exposed, or has very recently been exposed, to certain diseases, it is (under circumstances to be determined *only* by the physician) an excellent plan to inject a small quantity of serum, or some derivative of serum (e.g., purified gamma globulin) containing the appropriate antibodies, as a prophylactic measure.

As will be explained in the following chapter on allergy, a single injection of **foreign** (e.g., horse or rabbit) protein such as serum, whether or not it contains antibodies, not only can cause a serious, even fatal, allergic reaction but will usually make the recipient allergic for life. To avoid such effects γ-globulins of **human** origin are commonly used for preventive purposes; notably against German measles and infectious hepatitis. Diseases against which equine serum is occasionally used for prophylaxis are diphtheria and tetanus. Effective human sera are also available for prophylaxis against rabies and pertussis.

Passive Immunity and Serum Jaundice. It is mentioned elsewhere that a considerable number of apparently normal and healthy persons carry in their tissues and blood the viruses of **hepatitis**. This is a good example of latent viral infection. Unless suitable precautions are taken, the viruses are readily transmitted in the blood of donors, in γ-globulin and in other blood derivatives. Serum hepatitis is easily transmitted by improperly sterilized syringes, needles, razors, and other objects that can carry blood, serum, or lymph from one person to another.

CHAPTER 28
SUPPLEMENTARY READING

Bellanti, J. A.: Immunology, W. B. Saunders Co., Philadelphia. 1971.

Eichhorn, M. M.: Rubella: Will vaccination prevent birth defects? Science, *173*:710, 1971.

Evans, D. G. (Ed.): Immunization against infectious diseases. Brit. Med. Bull., *25*:121, 1969.

Raffel, S.: Immunity, 2nd ed. Appleton-Century-Crofts, New York. 1961.

Vallery-Radot, R.: The Life of Pasteur. Doubleday-Doran & Co., New York. 1926.

CHAPTER 29 · HYPERSENSITIVITY (ALLERGY)

29.1
THE HYPERSENSITIVE STATE

In the preceding three chapters, emphasis was placed on the protective role of the immune system. Unfortunately, not all immunological reactions are beneficial, and certain reactions are frequently observed to cause a variety of pathological effects. The adverse responses—called **allergy** (Gr. *allos* = changed; *ergon* = activity) or **hypersensitivity** (because the individual is, in a sense, overreacting to an offending substance)—can be divided into two major groups: (a) those that are mediated by humoral antibody (also referred to as immediate-type hypersensitivity because symptoms tend to appear rapidly), (b) those that are mediated by cells (delayed-type hypersensitivity). The various types of hypersensitivity reactions and their major characteristics are listed in Table 29–1,

TABLE 29–1. THE FIVE MAJOR TYPES OF HYPERSENSITIVITY (ALLERGIC) REACTIONS*

Antibody-Mediated ("Immediate"-Type)			Cell-Mediated (Delayed-Type)	
I ANAPHYLACTIC-TYPE REACTIONS	II IMMUNE COMPLEX REACTIONS	III CYTOTOXIC REACTIONS	IV ALLERGY OF INFECTION	V TRANSPLANT REJECTION
Immunoglobulin (IgE) binds to mast cells or circulating basophils. Shocking dose of antigen combines with bound antibody to trigger the release of various vaso-active amines such as histamine.	Humoral antibody combines with a fairly large shocking dose of antigen to produce Ag-Ab complexes which trigger the release of histamine and other highly active mediators. Blood vessels are damaged; inflammation and necrosis follow.	Antigen elicits the formation of antibody, which then combines with target cells. The attached Ab causes the cells to be destroyed by various means, e.g., phagocytosis, lysis by complement fixation, and destruction by lymphoid cells.	Antigens of the infectious organism sensitize the infected individual, causing a proliferation of specific lymphoid cells. These "sensitized" lymphoid cells react to the presence of the homologous antigens by releasing several soluble substances which mediate the hypersensitivity reaction. Humoral antibody not involved.	Tissue, e.g., skin, from a donor is grafted onto a recipient, e.g., a burn patient. A population of sensitized lymphoid cell arises and is instrumental in the destruction of the graft by mechanisms similar to those involved in allergy of infection. Antibody-mediated hypersensitivity can also be involved in graft rejection.
Example: Anaphylaxis, atopic allergy (e.g., hay fever).	Example: Arthus reaction, serum sickness.	Example: Isoimmune reactions, certain autoimmune reactions, drug reactions.	Example: Tuberculin reaction and similar manifestations.	Example: Allograft (homograft) rejection.

*See text for further details.

401

along with some clinical examples of each type. A fuller explanation of each type follows.

The administration of antigen (immunogen) for the purpose of eliciting an antibody response was referred to as **immunization** in the previous chapters. When hypersensitivity reactions are discussed, a slightly different terminology is frequently employed. Thus, the antigen or immunogen is called the **allergen** or sensitizing antigen and the immunization is referred to as **sensitization.** The immunized individual is called hypersensitive, allergic, or sensitized. **Reagin** is the older term for IgE immunoglobulin,* which participates in the anaphylactic type of immediate reactions. **Atopy** or atopic allergy (Gr. *atopia* = strangeness) simply means that the allergy has developed in response to an allergen which is rare or unknown, or one which does not elicit an allergic response in the vast majority of individuals.

Although the various hypersensitivity reactions vary considerably as to the mechanisms involved and the symptoms, they all have two features in common. First, the individual must be sensitized by exposure to the offending allergen and some time must lapse, during which an immune response develops. Second, the individual must be exposed to a shocking or eliciting dose of the same allergen in order to produce the hypersensitive reaction.

29.2
ANTIBODY-MEDIATED HYPERSENSITIVITY

Anaphylactic Reactions. The general characteristics of this type of reaction may be summarized as follows:

1. An induction period of one to several weeks must elapse between the sensitizing dose of antigen and the administration of the **shocking** or **eliciting** dose of antigen. This induction period allows time for antibody production.
2. A demonstrable, often visible, reaction of hypersensitive tissues begins *within seconds* or *minutes* after contact with the shocking dose of antigen and generally disappears within one to two hours. There is little or no infiltration of leucocytes.

3. The tissues affected are primarily **smooth muscle, blood vessels,** and **supporting tissues** such as cartilage and fibrous tissue, but other tissues may also be affected. Symptoms vary considerably, depending on the type of animal involved.
4. **Desensitization** can be achieved by administering very small doses of the offending antigen, keeping well below the shocking dose. This procedure can be dangerous.
5. The reaction is mediated by antibodies and can be passively transferred from animal to animal by serum.

One of the most studied and best-known, though still incompletely understood, manifestations of antibody-mediated allergy is **anaphylaxis.** The antibody involved appears to be IgE immunoglobulin. The anaphylactic reaction received its name because it was first thought to be a paradoxical immune reaction **against immunity** (Gr. *ana* = against; *phylassein* = to protect). It is not, however, a paradox but a sort of perversion or overreactivity of a truly protective mechanism. The sensitizing dose of allergen may be very small and the amount of antibody produced may also be small; the shocking dose of allergen must be relatively large if an acute reaction is to occur.

Anaphylaxis results apparently from antigen-antibody combinations in the shock tissues where antibodies concentrate. The eliciting dose of allergen can enter naturally, by the bites of venomous animals, but is generally brought about artificially by injection of antigen. Bee stings and penicillin injections have occasionally produced anaphylactic reactions.

Within about five minutes after introduction of the shocking dose of allergen, the laboratory animal (usually a guinea pig) becomes uneasy, scratches at its nose, coughs, and is evidently embarrassed for air. Gagging movements occur, and the animal gasps for breath. Urination and defecation take place, the animal falls on its side, and ceases to breathe. Death may occur within a few minutes. If the attack is not fatal, recovery is often abrupt and seemingly complete within an hour or two, and the animal will not exhibit hypersensitivity to the same antigen for some days or weeks afterward. The animal is said to be **desensitized.**

Histamine and several other substances— e.g., **serotonin, bradykinin,** and a "slow-reacting substance in allergy" (SRS-A)—are released by the cells involved in the antigen-antibody re-

---

*Some authorities have argued that it has not been proved that all reaginic antibody is necessarily IgE, but we need not be concerned with this minor point here.

action, especially by basophils, mast cells,* and platelets. Histamine and the related substances induce smooth-muscle contraction. Many of the symptoms of anaphylaxis result from histamine-induced contraction of smooth muscle fibers (i.e., smooth muscle is the most obvious shock organ in anaphylaxis); hence the value of anti-histaminic drugs such as diphenhydramine or chlorpheniramine maleate in immediate-type allergic reactions. Large amounts of smooth muscle are present in the lungs of guinea pigs. Thus, constriction of the air passages and consequent asphyxia is a marked feature of the anaphylactic reaction in them, but not necessarily in other species of animals. Epinephrine, because of its broncho-dilating action, will generally prevent death in guinea pigs undergoing anaphylaxis. In rabbits, dogs, and other animals the symptoms of anaphylaxis vary, partly owing to differences in anatomical location of smooth muscle. In pregnant animals abortion often occurs because the uterus consists largely of smooth muscle. In dogs and man there is much damage to the liver, which alters the clinical picture markedly.

During anaphylaxis there is damage to blood vessels resulting in dilatation and escape of fluid into the tissues, causing swelling (**edema**) and, if extensive, **shock.**

There is also decreased ability of the blood to coagulate (**hemophilia**). This appears to result from the removal of fibrin components from the blood by **heparin,** a normal anticoagulant that is produced in excessive amounts by mast and liver cells during anaphylaxis.

Atopic Allergy. Roughly ten per cent of the human population suffers to some extent from allergies which may be described as **localized** anaphylactic reactions. The offending allergens include such substances as mold spores, pollens, animal danders, house dust, etc. When the allergen contacts cell-bound IgE immunoglobulin present in the conjunctival tissues, nasal mucosa, or bronchial tissues, histamine and other mediators are released. The resulting symptoms, depending on which tissues are most afflicted, are generally those of hay fever (tears and redness of the eyes along with nasal secretion) or asthma (constriction of the bronchia). With atopic hypersensitivity, the sensitizing and shocking doses of allergen are generally acquired by some natural route of exposure rather than through injection (certain insect stings or bites being an exception).

Individuals suffering from atopic allergy are often skin-tested with various antigenic extracts in order to identify the offending allergen. Dilute solutions of the antigens are injected intradermally and the injection site observed for a typical **wheal** and **flare** reaction. This is a hard, white swelling surrounded by an irregularly shaped, and often quite extensive, zone of redness (flare). If the skin contains bound IgE which is specific for the allergen in question, histamine and other mediators will be released. This, in turn, causes a dilation of local blood capillaries (producing an erythema or flare) and the release of fluids into the local tissues, causing an increase in pressure. The pressure shuts off local vessels, which excludes blood and thereby produces the white central swelling (wheal).

Some individuals show hypersensitivity toward certain foods, e.g., strawberries. Soon after ingesting such food the individual breaks out in hives (generalized urticaria). Apparently, certain antigens from the offending food are absorbed from the gut and make their way to the skin via the blood stream. The reaction that follows appears to be closely related to the wheal and flare reaction just described. Again, antihistamines are generally quite effective in treating such cases.

Desensitization of hypersensitive individuals by the repeated injection of small amounts of the offending antigen is frequently attempted, with variable results. Some people respond favorably, others do not. The basis for desensitization, when it works, is not clearly understood. One theory is that repeated injections of the antigen elicit the formation of IgG. Circulating IgG would presumably act as a protective barrier by reacting with the allergen before the latter could react with cell-bound IgE.

Immune Complex Reactions. This type of reaction differs fundamentally from anaphylactic reactions, since it depends on the formation of complexes between the antigen and antibody (mostly IgG) rather than on cell-bound IgE. The two most thoroughly studied reactions of this type are the **Arthus reaction** and **serum sickness.**

Arthus Reaction. This reaction, named after its discoverer, Maurice Arthus, is frequently observed if an animal is repeatedly injected with an antigen to elicit a high antibody titer. Subsequent injection of the soluble anti-

---

*Cells that resemble basophilic granulocytes, found in connective and other tissues and that, in allergic reactions, release histamine and probably heparin. Heparin reduces coagulability of blood.

gen into the skin leads to the development of an edematous, hemorrhagic skin reaction within an hour, and it becomes progressively more severe, finally leading to a necrotic lesion or abscess. The underlying mechanism of this reaction is very complex, but in essence the antigen, as it diffuses into the tissues, combines locally with the homologous antibody to form an immune complex. As a result, complement is fixed and various biological mediators are released, including anaphylatoxin, which triggers the release of histamine (see page 402), and chemotactic factor, which attracts polymorphonuclear leucocytes. The phagocytic cells attempt to rid the area of the immune complexes, and in so doing, release various proteolytic enzymes which cause further damage to the local tissues. Various vasoactive amines in addition to histamine are also released and intensify the local reaction. The reaction is easily prevented by carefully controlling the concentration of antigen and the frequency of injection during the immunization period. In man, Arthus-like reactions have been observed when a sensitized individual has inhaled the offending allergen and developed severe respiratory difficulties some six to eight hours later.

Serum Sickness. At one time, antiserum obtained from animals, e.g., equine diphtheria antitoxin, was frequently employed for prophylactic or therapeutic purposes, and it was not uncommon to find patients showing allergic manifestations some 8 to 12 days after the injection. The symptoms included general swelling of lymph nodes, swollen eyelids, face, and ankles, itching, hives, fever, and painful joints. In this case, the horse serum (antitoxin) was acting as an antigen (since it was a foreign protein), and the individual responded by making antibody against it. By about the eighth day sufficient antibody had been made and had combined with the antigen (horse serum) to form a considerable number of soluble immune complexes. These complexes fixed complement and triggered the release of various vasoactive amines, including histamine. Localization of the immune complexes—for example, in the capillary bed of the glomeruli in the kidneys—leads to extensive damage (glomerulonephritis). Lesions can also be produced in vascular, cardiac, and cutaneous tissues.

In man, serum sickness, also called acute immune-complex disease, has been observed following exposure not only to foreign serum, but also to certain soluble bacterial or viral antigens and sometimes to antigens derived from the individual's own tissues (a type of autoimmune reaction). Repeated exposure to the offending antigen will result in a quicker, more severe response which may prove fatal.

Cytotoxic Reactions. Perhaps one of the best examples of this type of reaction is **erythroblastosis fetalis.** If a mother lacking the Rh blood group gives birth to an Rh⁺ baby, during the process she is exposed to the baby's erythrocytes carrying the Rh factor. Antibody is made against the Rh factor by the mother and is able to cross the placental barrier in a subsequent pregnancy. The antibody combines with the Rh antigen on the erythrocytes of the fetus (assuming the fetus is Rh⁺), complement is fixed, and the red cells lyse. This leads to anemia and jaundice in the fetus and causes the fetus to manufacture immature red blood cells (erythroblasts) at a high rate. A transfusion at birth coupled with the absence of further damage by maternal antibody allows the newborn infant to make a speedy recovery, provided the reaction has not been too severe. It is now customary to give an Rh⁻ mother, *within 72 hours* of first birth or abortion of an Rh⁺ offspring, an intramuscular injection of human anti-Rh immunoglobulins to suppress the formation of anti-Rh antibodies by the mother in response to the Rh⁺ cells from the fetus. (See also page 367.)

Reactions of the immune-complex type all have two points in common: (a) the antigen is attached to some type of cell and (b) the antibody (mostly IgG) is not fixed, as in the case of anaphylactic reactions. The attachment of antibody to the cells can actually trigger a variety of responses, e.g., lysis, adherence of phagocytic cells, phagocytosis, and attack by certain types of lymphoid cells. In addition to the isoimmune reactions involving the Rh factor and the ABO blood groups, this type of reaction accounts for damage done to the tissues in certain types of autoimmune diseases (e.g., Hashimoto's disease, where antibody is directed against the individual's own thyroid cells). Certain drugs, though not antigens by themselves, are apparently able to combine with cells of the body and then to act as haptens (see page 380). Antibody then reacts with the cells to which the drug is attached, and the cell is destroyed. Withdrawal of the drug usually leads to complete recovery from the allergic manifestation.

29.3
CELL-MEDIATED HYPERSENSITIVITY

Cell-mediated hypersensitivity or allergy differs from antibody-mediated allergic reactions, both as to mechanism and to manifestation:

1. A demonstrable or visible reaction of hypersensitivity begins only *hours* or *days* following the introduction of the shocking dose of antigen, hence the older term **delayed-type** hypersensitivity reaction.
2. The reaction generally reaches a maximum between 24 and 72 hours and then gradually subsides over a period of days.
3. Any vascular tissue may be affected; there is no specific "shock" organ or tissue like that seen in anaphylaxis.
4. Desensitization is difficult or impossible.
5. The hypersensitivity reaction can be transferred to a recipient only by lymphocytes (or in some cases, extracts made from these cells) and not by serum. Therefore, humoral (circulating) antibody does not appear to be involved in this reaction.
6. The lymphocytes mentioned above, most likely T-lymphocytes (see page 369) with some sort of fixed immunoglobulin-like recognition sites on their surfaces, react to the presence of the allergen by releasing several soluble substances which trigger various changes in cells of other types and thus produce the hypersensitivity reaction.

This form of hypersensitivity is frequently encountered following infection by certain bacteria, viruses, or fungi and, in fact, is sometimes referred to as **allergy of infection.** It is also encountered in some types of tissue rejection following transplantation and is frequently observed when adjuvants are used to enhance antibody response via immunization.

Adjuvants. Some proteins are not highly antigenic unless associated with lipids; e.g., the waxy substance in cells of dead tubercle bacilli or other species of *Mycobacterium*, lanolin, aqueous emulsions of vegetable oils, or mineral oil. So effective are such lipids in enhancing antigenicity that they are often used to promote antibody production and are called **adjuvants.** A well-known type, called Freund's adjuvant, is an aqueous emulsion of mineral oil mixed with heat-killed tubercle bacilli. Such lipid adjuvants also commonly promote cell-mediated (delayed) hypersensitivity. Since bacteria of the genus *Mycobacterium* contain more waxy components than other bacteria, infection by any species of *Mycobacterium* is particularly likely to induce delayed allergy. Indeed, infection by *M. tuberculosis* produces the classic model of delayed or infection allergy, readily demonstrable as the **tuberculin reaction.**

The Tuberculin Reaction. Hypersensitivity to the protein of tubercle bacilli resulting from an infection, either past or present, or induced by BCG vaccine, may be demonstrated by introducing into or onto (not under) the skin, tuberculo-protein in the form of dead tubercle bacilli, or the sterile filtrate of broth in which they have grown (**OT**, old tuberculin), or purified protein derivatives (**PPD**) extracted from the bacilli (Fig. 29–1). These tuberculo-proteins are called **tuberculin.** At the site of their introduction into the skin a red, **indurated** (firm and swollen) spot at least 5 mm in diameter appears after 24 to 48 hours. The red zone fades in a day or so, but the induration may be palpated for several days. The reaction is called a **tuberculin reaction,** and is representative of the delayed type of allergy. Analogous skin tests for delayed-type allery are used in several fungal infections: **histoplasmin** tests in histoplasmosis; **coccidioidin** tests in coccidioidomycosis (Chapt. 12). All are tests for hypersensitivity (allergy) to the respective organisms.

The tuberculin test is widely used in finding cases of tuberculosis. The tuberculin test,

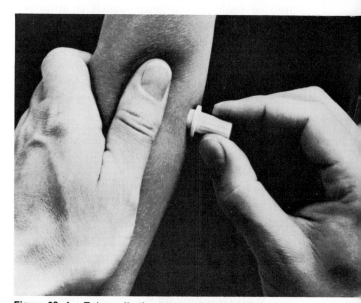

Figure 29–1. Tuberculin tine test employs a plastic holder with a stainless steel disk and four tines coated with Old Tuberculin and dried. The only preparation required is cleansing of the site of inoculation with alcohol. Methods such as the tine test, using disposable plastic units, offer distinct advantages in speed, accuracy, and economy. (Courtesy of Lederle Laboratories, a division of American Cyanamid Co., Pearl River, N.Y.)

modified for ocular application, is also widely used to detect tuberculosis in cattle.

Hypersensitivity and Tissue Destruction. Delayed hypersensitivity often results in necrosis of the tissues involved when the degree of hypersensitivity and dosage of allergen are sufficient. A tuberculin test in some persons may result in a large ulcer at the site of the intradermal injection.

Allergic necrosis of tissues occurs in numerous chronic infections. It is well exemplified by production of lung cavities in some tuberculous persons. Slowly, progressive destructive lesions of bone, arteries, and nerves are associated with syphilis, leprosy, and other slowly progressive diseases. In these allergies of infection, the hypersensitivity may be purely of the delayed type, or both immediate and delayed sensitivity may be present in varying degrees. These situations may give rise to varying symptoms and complex, little-understood reactions.

Massive tissue destruction is not inevitable in infection allergy and is the conspicuous exception rather than the rule. Allergy is basically a defensive mechanism, as will be explained shortly.

Transplant Rejection. There is a growing interest in the possibility of using healthy tissues and organs to replace diseased ones, and transplantation has now grown into a discrete area of biological science which is attracting considerable attention and talent. There are actually four major types of transplants:

1. Autograft: The tissue is simply removed from one location and grafted onto another in the same individual. Usually successful.
2. Syngraft (isograft): Donor and recipient are identical twins or of the same pureline strain (e.g., inbred mice). Usually successful.
3. Allograft (homograft): Donor and recipient are of the same species but different genetic constitution (e.g., man to man or mouse to mouse). Limited success.
4. Xenograft (heterograft): Donor and recipient are of different species (e.g., pig to man). Presently unsuccessful.

While it is beyond the scope of this chapter to cover the complexities of transplantation, the beginning student of microbiology should at least realize that grafts are usually rejected because they contain antigenic determinants which are **foreign** (nonself) to the recipient. The recipient frequently responds to these foreign antigens by developing both cell-mediated and antibody-mediated hypersensitivity reactions. The success of allograft transplants depends largely on how well the tissues of the donor and recipient can be matched (much the same as for blood transfusions). The fewer the number of foreign antigenic sites, the greater the possibility of success. New typing procedures have been developed to aid in this check for **histocompatibility.** Efforts are also being directed toward minimizing the immune response of the recipient toward those foreign antigenic groups that cannot be avoided, genetically identical twins being the only perfectly matched donor and recipient outside of autografts. In the next decade the incidence of success with allografts, and perhaps even xenografts, can be expected to increase markedly.

29.4
HYPERSENSITIVITY AS A DEFENSIVE MECHANISM

Binding Power of Immune Tissues. Once the body has been subjected to an antigenic stimulus, the tissues, especially the more superficial tissues (skin or mucous membranes), acquire a greatly enhanced power to **bind** and **localize** the specific antigen when later brought into contact with it. The effectiveness of this **tissue immunity** is greatly enhanced by hypersensitivity. Dermal hypersensitivity reactions generally are illustrations of this heightened reactivity of superficial tissues. The antigen thus bound in the tissue of initial contact, although perhaps causing necrosis where bound, cannot spread throughout the body and thus cause injury to deeper and more vital tissues.

Nonantitoxic Immunity to Toxin. An example of the effectiveness of the binding power of immune tissues is seen when rabbits and guinea pigs are made allergic by injections to somatic antigens of *Corynebacterium diphtheriae*. The animals are then able to survive *cutaneous* doses of live, **toxigenic** diphtheria bacilli which always kill normal animals. The toxin formed by the bacilli, instead of diffusing throughout the body as it does in the unimmunized animals, is held and bound in the skin of the allergic animals. Although it does severe damage locally, it cannot spread to the deep vital tissues, so the animals survive *in the complete absence of antitoxin.*

The Koch Phenomenon. Another example of the binding power of hypersensitive tissues is the **Koch phenomenon.** A guinea pig is inoculated in the right groin with virulent tubercle bacilli, the bacteria gain a foothold, form a local abscess, and then proceed almost unopposed from the abscess to the lymph nodes of the abdominal cavity, to the spleen, the liver, the lymph nodes of the thorax, and the lungs and kidneys; the pig dies of disseminated tuberculosis in six to eight weeks. Now, if on the second or third week of this progressive disease a second injection of tubercle bacilli is made into the left groin, there is a strong local allergic tissue reaction. The bacilli are *closely held* in the site where they are injected and *do not* progress farther, although they may cause a local abscess. The tissues are highly defensive because of the allergy due to the first infection.

Unfortunately, guinea pigs are so susceptible to tubercle bacilli that they eventually die in spite of the valiant defense put up by the tissues. Human beings are in general much more resistant. Many adult persons, especially dwellers in thickly populated districts, have had a mild, unrecognized, and long-since-healed infection with tubercle bacilli. Because of this they are much more resistant to tuberculosis than persons who have never had any contact with tubercle bacilli. The previously infected group are allergic (hence, resistant) to the bacilli, as shown by the fact that they react positively to tuberculin. The second group, those never previously infected and hence **tuberculin-negative,** may be made allergic (resistant) to tubercle bacilli by giving them a very mild infection, as is done in BCG vaccination. They become **tuberculin-positive.** Hypersensitivity or allergy is thus revealed as a potent defensive mechanism.

29.5
HARMFUL EFFECTS OF HYPERSENSITIVITY

Like many normal and beneficent physiological functions, hypersensitivity reactions may at times be so violent in some persons as to be harmful, "like an overzealous servant who, attempting to warm the living room with a good fire, burns the house down." While describing the various types of hypersensitivity reactions, we have listed a small number of specific examples of harmful effects. These few violent reactions, like violent acts, create more comment and attract more attention than the enormous number of normal and helpful reactions that go on unnoticed constantly. For example, certain persons appear to become excessively allergic to hemolytic streptococci. The heart and joints appear to be shock tissues in such allergy. There is believed to be a close relation between this allergy to hemolytic streptococci and rheumatic heart disease, one of the most important causes of disability and death in the United States. Many chronic, disabling conditions, especially forms of asthma and joint disease, such as rheumatoid arthritis, are thought to be related to allergic reactions of certain tissues to obscure microbial infection, such as chronic sinusitis. Many of the rashes and eruptions seen in bacterial, viral, and fungal diseases are allergic reactions of the delayed type. Reactions to poison ivy and poison oak are also allergic responses of the delayed type. Hypersensitivity, therefore, plays an important though "Jekyll and Hyde" part in infectious and other diseases.

Many allergens are entirely nonantigenic. Examples, among hundreds, are sulfonamides, certain cosmetics, certain dyes on fabrics, some plastics, penicillin, and some other antibiotics. Hypersensitivity to these substances, or certain molecular groups resulting from their disintegration, is commonplace.

Although these allergens are not antigens such as previously described, it is thought that, as haptens, they can become conjugated with proteins and other body components, especially lipids (adjuvants), so that they act like complete antigens.

CHAPTER 29
SUPPLEMENTARY READING

Becker, E. L., and Austen, K. F.: Anaphylaxis, *in*: Mueller-Eberhard, H., and Mischer, P. (Eds.): Immunopathology. Little, Brown & Co., Boston. 1967.

Dannenberg, A. M., Jr.: Cellular hypersensitivity and cellular immunity in the pathogenesis of tuberculosis: Specificity, systemic and local nature, and associated macrophage enzymes. Bact. Rev. *32*:85, 1968.

Flick, J.: Human reagins: Appraisal of the properties of the antibody of immediate-type hypersensitivity. Bact. Rev. *36*:311, 1972.

Gordon, B. L., II, and Ford, D. K.: Essentials of Immunology. F. A. Davis Co., Philadelphia. 1971.

Levine, B. B.: Immunochemical mechanisms of drug allergy. Ann. Rev. Med., *17*:23, 1966.

Roitt, I.: Essential Immunobiology. Blackwell Scientific Publications, Oxford. 1971.

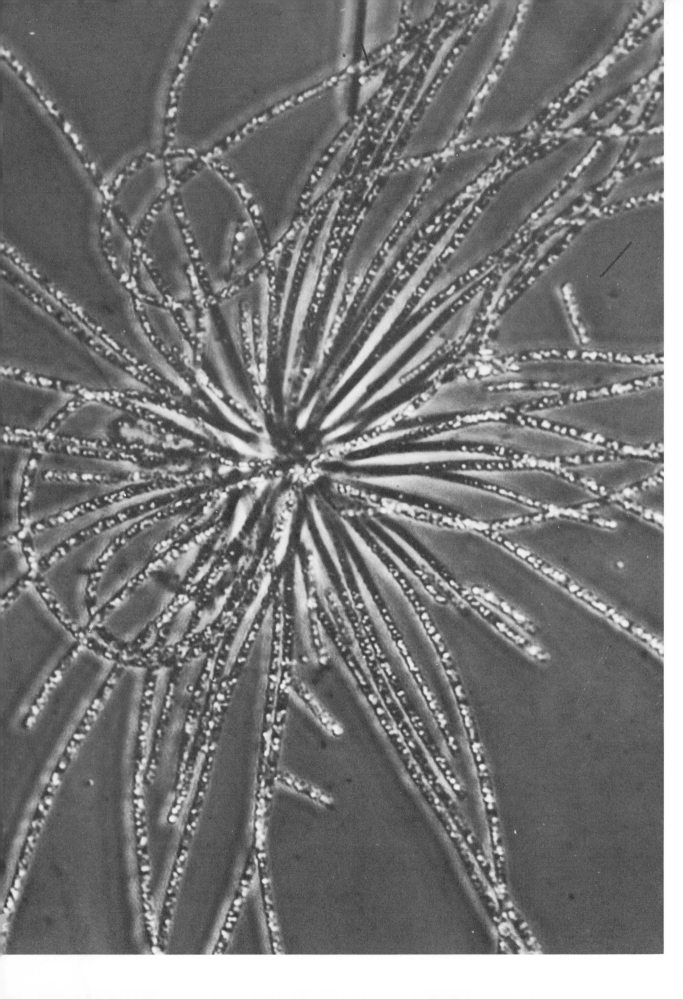

SECTION FOUR

INHABITANTS OF THE INVISIBLE KINGDOM

In preparing to journey in strange lands the prudent traveler first seeks whatever information is available about the physical, geological, geographical, and other significant characteristics of the places he is about to traverse: land and sea areas, climatic zones, means and routes of transportation, sources of food, and so on. Secondly, and of equal or greater importance, are data concerning the peoples to be met and dealt with: their various ethnic, political, religious, tribal, and other groupings; the bases of such groupings; distinguishing and identifying characteristics of each group; their attitudes toward, and influences upon, each other; and the most effective means of dealing with each, both as different segments of the population and as individuals, each with his own peculiarities.

As the voyager in foreign lands, so the tourist in the realms of the microbes: he obtains preliminary information. In the first three sections of this book he has acquired introductory, synoptic descriptions of the general field of microbiology, with guiding information about the general nature of micro-organisms, their minute size, and the curious environmental factors of their strange domiciles; he has also received some instruction in methods of study, and so on. Prepared with such teachings the wayfarer now personally meets, in this section, a major division of the fantastic inhabitants of the "Invisible Kingdom": the bacteria.

In this section a sort of panorama is laid out delineating the principal families and tribes, and the visitor is introduced to the "head men" (type species) of the many smaller groups (genera) that constitute the bacterial subkingdom as a whole. The distinguishing forms, habitats, colors, and physiological and anatomical peculiarities of each subdivision and of each main species are described, as well as their places and manner of living,

---

Facing page, Courtesy of F. E. Palmer and E. J. Ordal.

their foods, and their friendly or hostile attitude not only toward foreign investigators such as humans and microbiologists but also toward their own kind: antagonistic, mutually helpful, and so on.

The traveler returning from such an exploration into the unseen world can at least recount many strange sights and experiences to the untraveled; if so inclined he may use the information that he has gained in many profitable ways, some of which have been indicated in previous chapters, while others are outlined in the following section (Section Five). If he is imaginative, ingenious, and energetic he may even invent and patent new ways of exploiting microorganisms for himself.

CHAPTER 30 • THE BACTERIA— SYSTEMATIC STUDY

Isolation and Identification of Pure Culture

Any particular environment, no matter how closely it resembles another, may contain its own particular kinds of microorganisms that may differ markedly from the organisms of any other environment. This is because each environment provides conditions suitable for the growth of a particular kind of microorganism, and as a result a new and continually changing environment is being created. A new environment thus formed may make it possible for another group of microorganisms to flourish. In other words, any environment containing a viable population of assorted microorganisms will not remain the same for any extended period because microorganisms, however small, are extremely active and versatile, both individually and collectively.

If we are to understand the causes of some of the changes in an environment, we must at first understand the fundamental biological processes occurring in that environment. We may either study the entire array of organisms present, or isolate a particular kind of microorganism, to find the role played by each organism in relation to the total activities occurring in the environment. However, in assessing the physiological or biochemical capability of pure cul-

tures in nature, one must remember that the natural environment is quite unlike the conditions created in the laboratory; for example, these may be free from nutritional competition from other organisms, absence of antagonism, predation, synergism, and syntrophism. Nevertheless, to find a specific role played by a specific organism one must isolate the organism in **pure culture,** i.e., a culture containing only one kind of organism, or an **axenic** culture. The isolation of one kind of microorganism from a mixture of many different kinds is called "pure culture technique," and with this method Robert Koch (1870) was able to discover the etiological agents of various diseases and formulated what are now known as **Koch's postulates** (Chapt. 24).

Use of the pure culture technique is imperative if one desires to study morphological, cultural, biochemical, physiological, or serological properties, pathogenicity, susceptibility to specific bacteriophages, DNA base composition, DNA homology, or any other characteristics of the organisms. By understanding many phenotypic and genetic characteristics of an organism in detail, one is able to recognize similarities as well as differences among the

411

assorted organisms present in any given environment. Therefore, identification and classification of the organisms involves isolating pure cultures and investigating as many phenotypic and genetic characteristics as possible.

Ancient systems of biological classification were necessarily based almost entirely on characteristics discernible only with the naked eye. Today, however, identification and classification of all forms of life (including viruses) not only includes morphological study by high-powered electron microscopes but myriads of highly defined analyses of nutritional, biochemical, and physiological properties of organisms. For example, many of the growth factors, vitamins, and similar substrates used are pure compounds which enable us not only to detect precise nutrient requirements of the organisms, the kind of metabolic pathway by which the organisms obtain energy, or whether the kinds of enzymes produced are constitutional or induced, but also to identify end products produced from the metabolism of a certain substrate.

Some of these parameters as to bacteria are outlined in the chapter on classification (Chapt. 31). Classifications and identifying descriptions of viruses, chlamydias, rickettsias, and fungi are discussed in the chapters dealing with those organisms. (See also Tables 30–1 and 30–2 and Chapter 11.) Here we present a simple example of a type of procedure commonly used and available in most laboratories for the isolation

TABLE 30–1. DISTINCTIVE PROPERTIES OF PROCARYOTIC PROTISTS°

	Blue-Green Algae†	Bacteria‡
Photosynthesis	All species contain chlorophyll *a* only, with phycocyanin and β-carotene; free oxygen given off; typically aerobic.	Only species of order Rhodospirillales contain photosynthetic pigments bacteriochlorophyll or chlorobium chlorophyll; none contains chlorophyll *a* or phycocyanin; do not give off free oxygen; many anaerobic or facultative.
Cell differentiation	Slight; some "holdfasts"; some arthrospore-like resting cells: conidia, heterocysts, akinetes; no thermostable endospores; no sex cells known.	Very slight; conidiospores in mold-like species; extremely thermostable endospores in family Bacillaceae; arthrospore-like microcysts in most Myxobacterales and in some *Azotobacter*.
Multiplication	Transverse binary fission; asexual so far as known.	Generally transverse binary fission; asexual, although genetic recombination by a primitive type of conjugation occurs.
Cell groupings	Many form trichomes, chains or irregular masses within slimy or gelatinous matrix; some unicellular species resemble bacteria.	Generally free, undifferentiated, single cells; some occur in sheaths or in irregular masses within slimy or gelatinous matrix; some form trichomes, some form mold-like mycelia; some form chains.
Cell walls	Thick, rigid, morphologically distinct; supporting fibers of peptidoglycan; occasionally cellulose(?), capsules or sheaths of pectic materials usually prominent.	*Except* orders Myxobacterales, Spirochaetales, Mycoplasmatales, and probably the chlamydias, thick, rigid, morphologically distinct, supporting fibers partly or wholly of peptidoglycan; capsules or sheaths in some species, usually of polysaccharides, polypeptides.
Motility (when present)	Gliding, bending, oscillatory; no cytoplasmic streaming; no flagella or cilia.	*Except* orders Myxobacterales, Cytophagales, and Spirochaetales, which exhibit flexing, gliding, and rotation without flagella, actively progressive by means of flagella.
Size of cell	Generally smaller than eucaryotic cells; minimum commonly > 5μm; readily visible with good optical microscopes (50 to 600×).	Much smaller than Cyanophyceae; maximum diameters, commonly <5μm; minimum, >0.2μm; readily visible only with very good optical microscopes (400 to 1200×).
Cultivation and nutrition	Many species cultivable on inanimate media; typically autotrophic (photolithotrophic) though not necessarily obligately; osmotrophic.	*Except Rickettsia* and Chlamydiaceae, generally cultivable on inanimate media; osmotrophic; some photolithotrophic, some chemolithotrophic, some chemoorganotrophic, some photoorganotrophic; some obligately aerobic, some obligately anaerobic, some facultative.

°Except viruses.
†Sometimes called blue-green bacteria (Appl. Microbiol., 26:682, 1973).
‡Including rickettsias and chlamydias.

TABLE 30-2. GROUPS AND CHARACTERISTICS OF PROTISTS

I. Protozoa (Animal Kingdom):

Eucaryotic cell structure; no morphologically and chemically differentiated cell walls; cell cuticle or plasmalemma often contains chitin; typically without photosynthetic system; typically motile by ameboid, flagellate or ciliate action; nutrition may be osmotrophic or phagotrophic or both; typically organotrophic; diameters generally $>20\mu$m; sexual phenomena commonly evident; cells typically capable of asexual, independent multiplication; exhibit cytoplasmic streaming.

II. Protophyta (Plant Kingdom):

Eucaryotic or procaryotic cell structure; except mycoplasmas, have morphologically and chemically differentiated cell walls; *except* the Eumycotina, most bacteria, and all rickettsias and chlamydias, have photosynthetic system; motility may be by flagella or by gliding, flexing, oscillatory, or rotatory movements; nutrition typically osmotrophic; may be autotrophic or organotrophic; cells typically capable of independent multiplication.°

A. Algae (green):

Eucaryotic cell structure; cell walls contain cellulose; chloroplasts with chlorophyll *a* and generally *b* or *c* and α-, β-, and c-carotenes; do not contain phycocyanin (phycobilin); diameters generally $>20\mu$m; motility when present due to flagella, no cilia; in photosynthesis split H_2O and release **oxygen**; sexual phenomena commonly evident; aerobic; exhibit cytoplasmic streaming.

B. Eumycotina (Eumycetes, yeasts and molds):

Eucaryotic cell structure; cell walls commonly contain chitin; no photosynthetic systems or phycocyanin; motility generally restricted to flagellate gametes of aquatic species; nutrition osmotrophic, organotrophic; diameters of filaments (hyphae) generally $>15\mu$m; sexual phenomena commonly evident; obligately or preferentially aerobic; exhibit cytoplasmic streaming.

C. Schizophyceae (blue-green algae):†

Procaryotic cell structure; cell walls contain peptidoglycan; chromatophores contain *only* chlorophyll *a*; β-carotene and phycocyanin present; diameters generally $<20\mu$m; motility when present is of gliding, bending, and twisting type, slow, no flagella or cilia; in photosynthesis, split H_2O and release oxygen; no sexual phenomena described; aerobic; no cytoplasmic streaming.

D. Schizomycetes (bacteria):

Procaryotic cell structure; cell walls principally peptidoglycan (mucopolysaccharide and lipopolysaccharide complexes); may be photosynthetic or nonphotosynthetic; motility when present by flagella or alga-like bending, creeping, or twisting or (in Spirochaetales) rapid rotatory motion; nutrition osmotrophic; may be photo- or chemolithotrophic, photo- or chemoorganotrophic; diameters typically $<5\mu$m; sexual phenomena primitive, no true gametes or zygotes; aerobic, anaerobic, or facultative; no cytoplasmic streaming.

1. Order Rhodospirillales:

Photosynthetic bacteria; photosynthetic pigments in purple and brown species are bacteriochlorophyll with accessory carotenes; in green species, chlorobium-chlorophyll with carotenes; no phycobilins (phycocyanin); photosynthesis active in red and infrared; motility when present alga-like, no flagella; in photosynthesis split H_2S and release S; no sexual phenomena described; typically anaerobic.

III. Viruses:

Not cellular in structure; incapable of independent multiplication.

°Except rickettsias and chlamydias, which are obligate intracellular parasites.

†Sometimes called blue-green bacteria (Appl. Microbiol., 26:682, 1973).

---

and identification of a common species of bacterium.

In spite of the fact that, as previously explained, identifying characteristics of bacteria vary, it is rare that all characters vary simultaneously. Further, variations are usually recognized for what they are. Under uniform conditions of laboratory study, and in spite of variations, bacteria generally retain their fundamental distinguishing characteristics (Chapt. 31). It is necessary to know what these are and how to determine them in the laboratory.

30.1
ISOLATION OF PURE CULTURE

A pure culture is one that contains only one kind of microorganism. Several early microbiologists—A. de Bary, O. Brefeld, J. Lister, J. Schroeter, and R. Koch—were leading advocates of the use of pure culture techniques. The pure culture may, in fact, occasionally exist in nature within the confinement of a microenvironment, but for all practical purposes, the condition is routinely imposed in the laboratory.

The first step in an exact study of any bacterium is to separate it from other forms with which it might be mixed. These would introduce error into various biochemical experiments or tests performed in the course of the identification. This process is spoken of as isolation in pure culture. Many students (as well as veteran researchers) have fallen into difficulties by assuming the purity of a culture and neglecting this very important step. Even cultures of animal tissue cells have been found infected with protozoans, viruses, bacteria, and mycoplasmas derived from the animal source of the tissue cells. Such cultures also frequently become accidentally contaminated with yeasts, molds, and bacteria from extraneous sources. Microscopic examination of a smear stained by Gram's method may sometimes reveal the presence of contaminants and very often gives a valuable clue as to the genus or family of the organism, but it cannot be depended upon entirely, since many different bacteria, including contaminants, look and stain exactly alike. The culture must be purified.

Several techniques are available for isolating different species in pure culture from a mixture of many species. Selection of isolation media requires careful consideration to assure that the medium is appropriate to the kind of microorganisms to be isolated; for example, suitable inorganic media for isolation of photoautotrophs or chemolithotrophs; blood- or serum-enriched media for isolation of nutritionally fastidious organotrophic organisms; and living media such as tissue culture or growing chick embryo for isolation of obligate, intracellular bacteria or viruses; likewise for the conditions of incubation, i.e., aerobic or anaerobic; and for temperature, for example, isolation of strictly anaerobic, obligately thermophilic bacteria requires incubation under anaerobic conditions (Chapt. 36) and an incubation temperature of 55C or higher. However, for isolation of psychrophiles, the temperature may be as low as 0 to 10C to discourage the rapid growth of extraneous mesophilic or thermophilic organisms which may be present in the original sample.

30.2
PURE CULTURE TECHNIQUE

Serial Dilution Technique. A pure culture may be obtained by serially diluting the sample with sterile media to the point of extinction in numbers of cells (Chapt. 10). It is customary to consider that the growth occurring in the last tube of the series originated from a single cell; however, in practice the method is useful only if the organism being sought is present in large numbers in a mixed culture system. The dilution can be made with melted agar medium maintained at 45C. The culture so obtained is frequently of uncertain purity unless confirmed by streak, spread, or pour plate techniques.

Pour Plate Technique. This method involves plating of a suitably diluted sample mixed with melted agar medium. Since the number of bacterial cells in the original sample is not known, it is necessary to dilute the sample in suitable sterile broth to insure the development of well-isolated colonies. The agar medium is maintained in a liquid state (45C), and an aliquot of diluted inoculum is added to it. After thorough distribution of inoculum by mixing, the contents are poured into sterile Petri dishes, where they are allowed to solidify, and the plates are incubated. After the period of incubation, pour plates exhibit both surface and subsurface colonial growth. The method is also used for enumeration of a bacterial population as well as isolation of a pure culture from a mixed culture system.

Streak and Spread Plate Techniques. For the streak and spread plate techniques a small amount of specimen is transferred onto the surface of a suitable solid medium either by transfer needle or loop and the sample is streaked in such a way as to provide successive dilution and ultimately to obtain well-isolated colonies (Fig. 30–1).

Spread plate technique usually requires a serially diluted inoculum. An aliquot of diluted sample is transferred onto the agar surface, and it is spread uniformly with a sterile bent glass rod. Both streaking and spreading techniques are particularly useful in separating aggregates of cells in the sample; thus colonies developed may be considered as originating from a single cell, hence creating a pure culture. However, in practice, the colonies developed in the initial streak or spread plate are restreaked or respread several times until all the colonies developed are identical in colonial morphology and only then does one declare that the culture is pure. This repeated streaking or spreading is necessary, as there are many species of bacteria which may form clusters or aggregates by their genetic properties (diplococci, sarcinae, streptococci, and staphylococci) or by presence of heavy capsular or slime material to which cells of other species present may adhere to form a mixed population.

Enrichment; Selective and Differential Media. Isolation of a specific type of bacteria from a mixed population by the above method can be made more efficiently by either incorporating certain chemical agents in the media or modifying the physical conditions of incubation.

Enrichment Culture Technique (Beijerinck and Winogradsky). The method is often used if the type of bacteria to be isolated is present in relatively small numbers and its growth is slow in respect to other species present in the inoculum. For example, if we are to isolate from soil

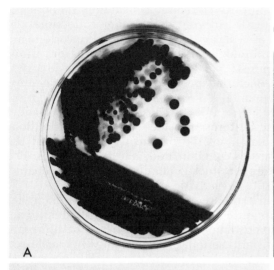

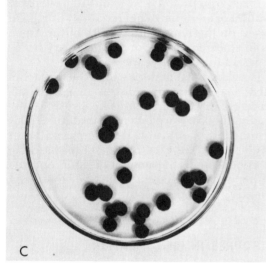

Figure 30–1. *A*, Three-phase discontinuous streak plate; note isolated colonies (*S. marcescens* on nutrient agar).

B, Turntable for spread plate technique. Bent glass spreading rod is immersed in 95 per cent alcohol and flame-sterilized. Inoculum is spread evenly over the surface of the agar.

C, Isolated colonies developed from spread plate (*S. marcescens* spread over nutrient agar).

chemolithotrophic bacteria capable of oxidizing ammonium to nitrite (nitrifying bacteria), the obvious substance to be incorporated into the medium is an ammonium salt in addition to other essential inorganic substances. Because chemolithotrophic bacteria gain energy from oxidation of inorganic elements or compounds and their growth is hindered by the presence of large amounts of organic substances, the medium should be devoid of organic nutrients; thus, most saprophytes present in the soil sample, requiring organic nutrients, are unable to grow, thereby favoring the growth of nitrifying bacteria.

Some enrichment media may contain highly complex organic substances. Blood, serum, extracts of plants or animals, or even certain antibiotics may be incorporated with standard nutrients such as nutrient agar to encourage the growth of nutritionally fastidious saprophytes or parasites.

Selective Media. Selective media contain specific chemicals which do not affect the growth of the organism being sought but which discourage the growth of other groups of bacteria. For example, incorporation of sodium azide at a specific concentration into culture media selectively isolates lactic acid bacteria because azide (N_3^-) binds tightly to the iron of the porphyrin ring of cytochrome, thus preventing the growth of organisms which possess the enzyme, cytochrome oxidase. Lactic acid bacteria, however, lack a cytochrome system and are unaffected by the presence of sodium azide.

There are many selective agents and among those more frequently used are dyes (such as crystal violet for isolation of brucellas), Cetrimide (*Pseudomonas aeruginosa*), tetrazolium, high concentration of NaCl (isolation of halo-philic bacteria), bile salts, antibiotics, specific sugars, to mention a few. Choice of these selective agents and concentration varies with the kind of microorganism to be isolated and kinds of extraneous organisms to be suppressed.

It should be mentioned here that it is also possible to treat the sample before application of the pure culture technique. For example, if sporeforming bacteria are to be isolated, the sample in aqueous suspension may be exposed to a temperature of 85C for three to five minutes before it is streaked or spread. Heating of the sample would destroy most of the nonspore-formers, and thus any colonies that developed would most likely be sporeforming species.

Differential Media. Differential media contain dyes, reagents, or chemicals which allow the observer to distinguish between types of bacterial colonies developed after incubation. For example, if raw sewage is streaked or spread on eosin–methylene blue (EMB) agar, some of the bacteria in the sample may produce colonies with brilliant green metallic sheen, while others may produce gummy, pink colonies with dark centers, which are called "fisheye" colonies. The former colonies may be considered as strains of *Escherichia coli* or related species, while the "fisheye" colonies may be strains of *Enterobacter (Aerobacter) aerogenes* or related species. EMB agar may be considered a differential as well as selective medium. The medium contains lactose as energy source, so that the medium allows the growth of organisms capable of producing the enzyme β-galactosidase, which breaks down lactose to glucose and galactose; hence, growth of organisms incapable of producing the enzyme may be suppressed. Similarly, the medium contains dyes (eosin Y and methylene blue) which suppress the growth of various gram-positive organisms.

30.3
MAINTENANCE OF STOCK CULTURES

It is extremely important to maintain isolated pure cultures of bacteria in viable condition for extended periods. Most well-trained bacteriologists usually maintain a large collection of such isolates as well as subcultures of authentic species (cultures deposited by microbiologists who proposed new or neotype species for experimental work) purchased from various culture collection centers. In the United States, for example, the American Type Culture Collection (ATCC) and Northern Regional Research Laboratory (NRRL) of the U.S. Department of

Agriculture maintain a considerable number of authentic cultures, and similar collection centers are established throughout the world to aid microbiologists in obtaining cultures for various purposes.

Bacterial species deposited in these centers are maintained in viable condition and are referred to as a "stock culture collection." Having these cultures in "stock" will prevent tedious reisolation of the organisms in the event the culture is contaminated, is lost, or mutates. The cultures in stock generally retain all the characteristics initially described.

There are several methods available for maintaining stock cultures. Most small bacteriology laboratories maintain stock cultures on agar slants, and they are periodically transferred to fresh media. (One may use sterile glycerol to cover the growth and agar slant to prevent dehydration of medium.) The frequency of transfer is based on the length of viability of the organisms in stock condition, and it may vary from a few weeks to several years (spores).

The larger laboratories and the collection centers may maintain cultures in lyophilized form (rapid drying of cultures in a frozen state), or maintained at very low temperature (−196C for liquid nitrogen) in well-sealed glass tubes. The choice of the type of storage is based on such conditions as the amount of labor required, time for which the culture must be stored, and availability of storage facilities.

30.4
PROCEDURES IN IDENTIFICATION

For purposes of discussion, let us proceed as though we have been given an unknown organism on an agar slant of unknown nutrient composition for systematic study and identification. The culture is presumed to be pure, but the source and habitat of the organism and how and what type of isolation medium was used are not known.

Initial Determination of Group. The first important step in the systematic study of an unknown microorganism is to determine to which of the major groups of microorganisms it belongs and whether other species are present as contaminants.

Much information can be obtained by examining the specimen under an oil-immersion lens. Assume the cells observed are spherical, having approximate dimensions of 0.7 to 1.0 μm in diameter. They show no evidence of yeast-like budding, no distinct eucaryotic-type nucleus, branching, or motility; therefore, the unknown is probably not a yeast, mold, or protozoan; since the unknown was grown on an agar slant, it is not an obligate intracellular parasite (including viruses); the cells are not pleomorphic and do not contain large bodies and elemental granules characteristic of L-forms or mycoplasmas. You can assume that in all probability it is a bacterium (see Tables 30–1 and 30–2).

Cultivation of Unknown Culture. Customarily it is advisable to start cultivation of an organism in one or more all-purpose media, that is, media known to support growth of a wide variety of organisms. If the organism fails to grow in such media, we must assume that the organism is either an autotroph or a nutritionally fastidious heterotroph, or the organism is thermophilic and requires a higher cultivation temperature. For the present purpose, let us inoculate the unknown culture onto three plates of blood-glucose-meat infusion agar, three meat- or yeast-extract agar plates, and three silica-gel plates. The blood agar and yeast-extract agar may support the growth of nutritionally fastidious saprophytes or parasitic bacteria, and inorganic synthetic silica-gel plates would support the growth of autotrophs, such as nitrifying bacteria. In addition to these, one may add three plates of an organic medium lacking a source of nitrogen for growth of nitrogen-fixing bacteria, such as *Azotobacter.* One plate of each kind of medium may now be incubated at 25C, one at 37C, and one at 55C for obligate thermophilic bacteria.

Initial Observation of Growth. After 24 to 48 hours, there may be no growth, in which case we may continue incubation for several more days. If there is no growth on the plates prepared for autotrophs or nitrogen-fixing organisms, but excellent growth (that of lemon-yellow colored colonies) is observed on the blood and yeast-extract agar media held at 25C, somewhat less growth at 37C, and sparse or no growth on media incubated at 55C, these data tell us that the unknown organism is a **mesotrophic** organotroph whose optimum temperature is near 25C. The organism is **neutrotrophic,** since it grew well in the blood and yeast-extract media, which have a pH near neutral. At this point our observations made in the initial determination of the group are further reinforced; that is, since it grows readily on organic, nonliving culture media with or without blood or serum, it is not one of the rickettsias, chlamydias, or viruses. Since it does not require sunlight for growth, it is neither an alga nor a photosynthetic bacterium. If, after prolonged incubation of the

silica-gel and nitrogen-free media, it fails to show growth, the organism is probably neither an autotroph nor a nitrogen-fixer.

An inspection of the growth on the agar gives an idea as to the size, shape, color, and consistency of the colonies. If all of the colonies on the plates are alike, we may assume that no contaminant is present. However, it should be noted that many different bacteria produce colonies of very similar appearance, and it may be necessary to examine stained smears made from selected colonies in order to detect the desired organism. Occasionally even this may yield no useful information because different species not only may produce colonies closely resembling each other but also can possess identical morphological and staining properties. In such cases investigators may select one or all of the distinct types and either identify all or discard some as the systematic study progresses.

For the present purpose, let us discuss one more situation; for example, if no growth occurred on any of the plates inoculated with the original material, we may assume that: (a) no living bacteria were present in the inoculum; or (b) the temperatures used were not suitable; or (c) some other medium, possibly with a different pH, is necessary; or (d) the bacteria may have been strict anaerobes. Suitable adjustments of conditions must then be made until growth is obtained.

30.5
SYSTEMATIC STUDY OF PURE CULTURE

The first important step in systematic study is to maintain a pure culture in stock. Well-buffered (to resist change in pH) agar slants (nutrient composition of media may be the same as that used in initial cultivation of unknown) may be inoculated by carefully transferring a portion of one of the selected colonies with a sterile needle. After proper incubation, one or more slants may be stored in a refrigerator and the remaining slant may be used as working stock for systematic study. Routine systematic study of a pure culture may include: morphology, cultural characteristics, nutritional requirements, biochemical characteristics, physiological properties, serology, and bacteriophage sensitivities.

Morphology

Microscopic examination of a hanging drop prepared with young broth culture is a useful means of determining size, shape, arrangement, and motility. The cells may be chains of cocci or bacilli, or cocci may occur in cubical packets (*Sarcina*), or occur in very irregular clusters resembling a bunch of grapes (*Micrococcus* and *Staphylococcus*), or simply in pairs or chains (*Diplococcus* or *Streptococcus*).

Motility. Motility, if present, is easily differentiated from Brownian movement (false motility) in a hanging-drop preparation. There are other means of determining motility of the organisms; for example, use of motility agar (semisolid: about 0.1 per cent agar and 3 per cent gelatin) with which motility is based on visible migration of the organism away from the line of inoculation (Fig. 30–2) or by flagella staining (Chapt. 14). Let us assume that our unknown is nonmotile and morphologically resembles *Micrococcus* or *Staphylococcus*.

Staining Reaction and Morphology. The bacterial cells may be stained readily with various aniline dyes (Chapt. 3). Among the more frequently used methods are Gram stain, flagella stain, metachromatic granules or volutin stain, capsule stain, negative stain, and sudanophilic granule stain (poly-β-hydroxybutyrate or PHB inclusion). Gram's stain is of great value because most bacteria may be classified as either gram-positive or gram-negative. It should be noted, however, that some organisms are gram-positive only when young or when cultivated on blood or serum media.

While gram-negative organisms generally retain gram-negativeness under all conditions, many gram-positive organisms may lose gram-positiveness with aging or in an acid environment; some are gram-variable under nearly all conditions. Let us assume that the organism under discussion is gram-positive.

Spores. Some bacteria may produce a unique, heat- and chemical-resistant body known as an **endospore** (Chapt. 36). However, spores may not readily be seen by ordinary staining, and their presence or absence may often have to be determined by staining an old (at least one week) agar slant culture with special spore-staining procedures (Chapt. 36). If spores are present and if the organism is an aerobic rod, it may belong to the genus *Bacillus* or *Sporolactobacillus*; if a coccus, it may be a species of *Sporosarcina*. If the organism is an anaerobic rod, it may belong to either genus *Clostridium* or genus *Desulfotomaculum* (based on the 8th edition of *Bergey's Manual*).

The cultural method may be used in reference to spore staining. Since most bacterial endospores are heat-resistant, a small mass of old cells may be suspended in sterile water and heated at 90C for 10 minutes. Afterward, inoculate some of the heated material into broth and incubate for several days. If growth occurs, it is practically certain that bacterial endospores were present, since most vegetative cells withstand 90C for 10 minutes. Some bacterial endospores, however, are considerably less heat-resistant. Let us say no evidence of endospore formation is demonstrated for the unknown either by a spore stain or by heat-treatment process.

Capsules. Capsules may or may not be visible, depending largely on the culture medium and whether we are dealing with an R or S form. Sometimes capsules are seen only on organisms occurring in pathological material or when growing in media containing serum or milk. They may be demonstrated by capsule stains, by negative stains with India ink or nigrosin, or by darkfield methods. Let us say our organism shows no demonstrable capsule under these particular conditions of examination.

Intracellular Granules. Some bacteria may accumulate intracellular reserve food re-

Figure 30–2. A test for motility. This meat-infusion medium contains 3 per cent gelatin and 0.1 per cent agar, with 0.2 per cent potassium nitrate (KNO_3) and an indicator of reduction, triphenyl tetrazolium chloride (TTC). The medium remains solid only at temperatures below 23C. Each tube is inoculated while firm by a stab as shown. The tubes are incubated for 16 hours at 35C, at which temperature the medium becomes liquid. Owing to the agar and gelatin the position of the stab inoculation is maintained. Migration of organisms into the liquid medium (at 35C) indicates motility (tubes 1, 2, 3, 4, 5) and is seen as clouding of the medium and by reduction (precipitation and darkening) of the TTC if the organisms reduce the KNO_3. Reduction may also be tested for as described in this chapter. Gelatin liquefaction is indicated in any tube if the medium remains liquid at 5C.

sembling starch-like or glycogen-like substances (iodophilic granules) or sudanophilic granules (PHB). However, it is generally observed that most bacteria accumulate either glycogen-like or sudanophilic substances, but not both. These substances are readily stained by either iodine solution or Sudan black dye. Let us say our unknown organism fails to accumulate such granules.

Morphology. The morphology of organisms may be studied with stained or **negatively stained** preparations (the background is stained but not the cells). Most bacteria produce cells of varying sizes. In the logarithmic phase of growth, cells are usually small but more uniform in size than in other phases of the growth cycle. It should be noted that if the bacteria are cocci, some of the individual cells may not be perfectly round; some may be oval or even lancet-like in shape; if bacilli, some may be long and thin, others short, oval, and thick; some may occur singly, others in pairs or chains or long filaments. However, the predominant form, size, and arrangement of the cells in pure culture are usually quite apparent.

Summarizing our knowledge of our "unknown" at this point, we may state that we are dealing with a nonmotile, gram-positive, unencapsulated coccus that grows well aerobically in irregular clusters on plain meat-extract medium at a pH of about 7.4, producing opaque, glistening, lemon-yellow colonies and preferring a temperature around 25C. There is still, however, a good deal to learn about our unknown before identification is complete.

Cultural Characteristics

Under appropriate cultural conditions, bacteria may show characteristic types of growth, and such observations frequently provide useful clues for identification. For example, colonies or growth formed on solid medium such as silica gel, gelatin, agar plates, and slants or stabs may yield several types of information about the unknown. The colonies may be observed in respect to size, margin, elevation, and chromogenesis (soluble or insoluble pigments); gelatin stab may show the type of growth as well as type of liquefaction (Fig. 30–3). However, all these properties are phenotypic characteristics and subject to modification by certain environmental conditions; for example, colonies on nutritionally poor media or in crowded areas may be smaller than those formed in rich media and uncrowded areas; *Serratia marcescens* may

fail to produce insoluble, red pigment when cultivated on medium lacking certain inorganic ions or when it is incubated at 37C or above (see Color Plate IV, *B*). Inability of the organism to produce pigment in certain media and at high temperature is not a genetic modification, since ability to produce pigment is readily restored by growing the culture in proper medium or incubating at room temperature (25C).

Nutritional Characteristics

All living organisms, whether autotrophs or heterotrophs, require water, a source of energy, carbon, nitrogen, sulfur, and phosphorus in either inorganic or organic form in addition to either small or trace amounts of several metallic elements. All living organisms require vitamins and vitamin-like compounds. Autotrophs are capable of synthesizing their entire requirements from simple inorganic compounds in the medium, but some heterotrophs fail to grow unless one or more of the preformed vitamins are provided in the medium. For example, *Bacillus anthracis* requires thiamine (B_1), *Lactobacillus* species may require cobalamine (B_{12}) for growth. Some bacteria, such as *Bifidobacterium*, require an unusual organic growth factor, N-acetylglucosamine, which is a biosynthetic precursor of N-acetylmuramic acid, a constituent of the cell wall material known as either murein or peptidoglycan.

Autotrophs obtain energy from the sun (photoautotrophs) or from oxidation of inorganic chemical compounds (chemoautotrophs). Their carbon, nitrogen, phosphorus, and sulfur are obtained from inorganic elements or compounds. However, all heterotrophs require organic carbon for both energy and synthesis of cell material. Some heterotrophs require organic nitrogen and sulfur, e.g., thiamine, while others may be able to use inorganic compounds such as $(NH_4)_2SO_4$ or KNO_3 as a source of nitrogen and $MgSO_4 \cdot 7H_2O$ as a source of sulfur. Further, some heterotrophs are able to grow in media devoid of nitrogen (nitrogen-fixing bacteria). Thus bacteria can be divided into many nutritional groups on the basis of their nutritional requirements.

Again, let us assume that our unknown is capable of utilizing inorganic ammonia as its sole source of nitrogen, whereas many species of *Staphylococcus* (which somewhat resemble the unknown) are unable to use inorganic nitrogen as in $(NH_4)_2HPO_4$. Therefore, the organism is more closely related to species of *Micrococcus* than *Staphylococcus*.

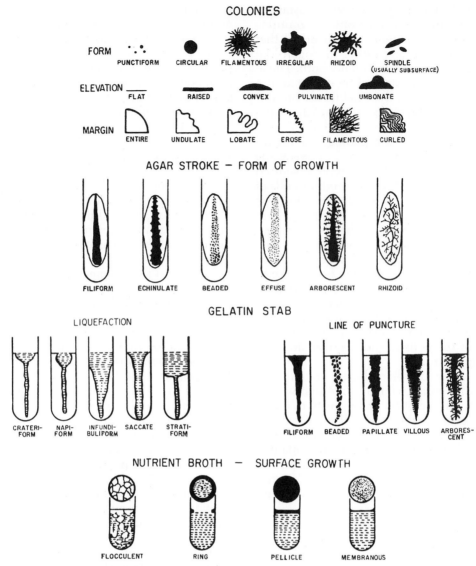

Figure 30–3. Cultural characteristics of bacteria.

Biochemical Tests

Since growth and multiplication of bacteria are consequences of nutrient metabolism, and the presence (or absence) of certain enzymes characterizes an organism's ability to utilize certain substrates, these characteristics are important in the systematic study leading to identification of organisms.

Routine biochemical tests may include fermentation and oxidation of carbohydrates; analysis of metabolic products formed from certain substrates; hydrolysis of polysaccharides, proteins, lipids, and related substances; reduction

of certain elements and compounds; and assay of various vitamin requirements.

In any of these tests it is important that the medium and conditions of incubation support good, vigorous growth of the organism. A negative result in the absence of good growth is obviously of no value. In order that uniform, standard, and authoritative results may be obtained, it is recommended that recognized, documented procedures such as those outlined in the *Manual of Clinical Microbiology* (published under the auspices of the American Society for Microbiology) or other authoritative methods be used throughout. It should be emphasized that results obtained in empirical media of the complex organic type, such as extract or infusion broth, can vary with changes in the quantity and quality of such ingredients as peptones, yeast extracts, and meat extracts; and that growth and enzymic activities are affected by presence or absence of certain kinds of hydrogen donors and acceptors, pH, temperature, and other factors. These variables are always standardized as far as possible and are carefully recorded for comparative purposes by competent workers. Here, only general principles and a few illustrative procedures of tests for **fermentation** of carbohydrates, **hydrolysis** of proteins and lipids, and **reduction** of various substrates will be outlined. A number of special test procedures are reserved to descriptions of particular organisms farther on in this book.

Fermentation Tests. As previously outlined, fermentation is the anaerobic, enzymic decomposition of organic compounds, typically carbohydrates, in which part of the substrate is oxidized, part reduced. Acids, alcohols, and carbon dioxide are common products of fermentation.

In the study of fermentative characters, tubes should contain a sufficient depth of broth containing the test substrate (e.g., glucose, lactose) to provide anaerobic conditions and should also contain a small inverted vial or other device (placed there *before* sterilization) to aid anaerobiosis and to catch any gas that may be formed as a result of the fermentation (Fig. 30–4). Gas might otherwise pass off into the atmosphere and not be detected. Two organisms, both of which ferment the same carbohydrate, may be sharply differentiated. For example, both *Enterobacter (Aerobacter) aerogenes* and *Lactobacillus arabinosus* ferment glucose, but *E. aerogenes* produces gases (CO_2, H_2) in addition to various other end products; *L. arabinosus* produces large amounts of lactic acid with no visible gases. The former are frequently termed "mixed acid fermenters," owing to the various end products produced, and the latter termed "homofermenters," since only one kind, i.e., lactic acid, is produced from fermentation of glucose.

When gas is produced from fermentable substrates by growing microorganisms, it is prima facie evidence of fermentation and is virtually always accompanied by acid formation since the gas is derived from formic or other acid resulting from fermentation. However, fermentation often occurs without gas production, and then acid formation is our only evidence that the organism has metabolized the

Figure 30–4. Durham fermentation tubes. Note that the inverted tube on the left is filled with liquid, whereas the one on the right is partially empty. What is the explanation for the difference between the two tubes?

substrate. Sometimes only alkaline substances are produced from substrates.

Acid or alkali production in the culture may be detected by adding to the medium an indicator or dye such as bromcresol purple, which changes from purple to yellow in the presence of acid. The ranges of color of various indicators at various pH's are shown in Table 30–3. Phenol red turns from yellow to red in the presence of alkali. The change in color of the indicator is our proof of fermentation or metabolic use of the substrate (see Fig. 35–9). However, some species can metabolize the acids produced by fermentation and cause a reversion from acid to alkaline. Others may produce ammonia from amino acids in sufficient quantity to mask acid production. In other instances, if too much buffer is used in the medium, it, too, will mask acid or alkali production. These and similar possibilities can be allowed for but must be kept in mind. Fresh serum added to the medium may vitiate the results because it often contains extraneous hydrolytic enzymes of animal origin and is also a potent buffer.

Observations of growth should be made every 24 hours in order that the culture may not revert to an alkaline reaction before acid formation has been noted.

Assay of Metabolic Products. Determination of intermediate or end products from carbo-

TABLE 30-3. SOME pH INDICATORS COMMONLY USED IN MICROBIOLOGY

Acid-Base Indicator°	pH range†											Color change‡ (Acid → Alkaline)
	0	1	2	3	4	5	6	7	8	9	10	
Thymol blue (acid range) (0.04)§		1.2———2.8										Red to yellow
Methyl orange (0.05)				3.1———4.4								Red to yellow
Bromphenol blue (0.04)				3.1———4.7								Yellow to blue
Bromcresol green (0.04)					3.8———5.4							Yellow to blue
Methyl red (0.02)					4.2———6.3							Red to yellow
Chlorophenol red (0.04)						5.1———6.7						Yellow to red
Bromcresol purple (0.04)						5.4———7.0						Yellow to purple
Bromthymol blue (0.04)							6.1———7.7					Yellow to blue
Phenol red (0.02)							6.9———8.5					Yellow to red
Cresol red (0.02)								7.4———9.0				Yellow to red
Thymol blue (alkaline range) (0.04)									8.0———9.6			Yellow to blue
Phenolphthalein (0.10)									8.3———10.0			Colorless to red

°Probably the most generally useful indicators in microbiology are methyl red, bromcresol purple, and phenol red.

†pH indicator ranges from Conn and Jennison (Eds.): Manual of Microbiological Methods, Society of American Bacteriologists, McGraw-Hill Book Co., Inc., 1957.

‡Color change for each indicator is given from maximum acid color to maximum alkaline color.

§Numbers in parentheses are recommended per cent concentrations of dye in solution.

hydrate, protein, and lipid metabolisms often yields valuable information in identifying unknown bacteria. For example, the name of the **IMViC** test stands for the following metabolic reactions: **I** refers to indole, derived from the amino acid tryptophan by certain species of bacteria (Fig. 30–5). **M** stands for the ability of an organism to produce a large amount of acid(s) from fermentation of glucose, which may be determined by the methyl red test; methyl red is an indicator (dye) whose color remains red if it is added to a culture medium having pH 4.2 or lower and becomes completely yellow at a pH of 6.3 (see Table 30–3). **Vi** stands for the Voges-Proskauer test, which is designed to detect acetyl-methyl-carbinol. **C** stands for the ability to utilize citrate as the sole source of carbon used in differentiation of typical strains of *Escherichia coli* from *Enterobacter aerogenes*. As an example, for *E. coli* the IMViC pattern is ++——, whereas for *E. aerogenes* it is ——++ (see Table 35–10).

Gases Produced by Bacteria. Two gases commonly given off by bacteria during fermentation are carbon dioxide and hydrogen in varying, more or less distinctive, proportions. One may determine the ratio of carbon dioxide to hydrogen by first marking the level of gas in a fermentation tube and then adding strong sodium hydroxide solution to the tube. This absorbs the carbon dioxide, leaving hydrogen. The difference in level of fluid in the gas tube is readily measurable.

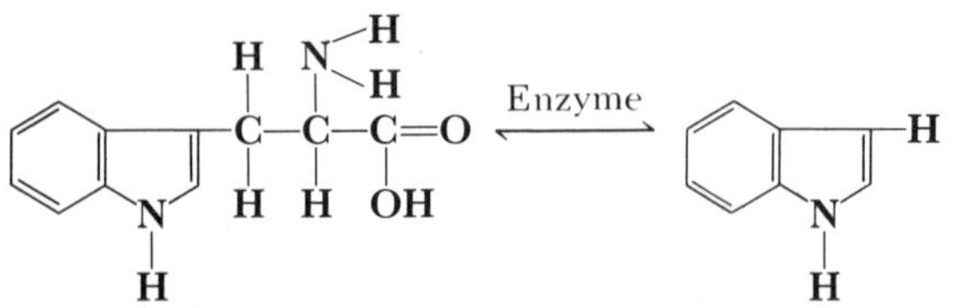

HYDROGEN SULFIDE. Many organisms in their metabolism of sulfur-containing organic compounds liberate hydrogen sulfide in considerable amounts. This gas is one of the most noticeable odors in putrefactive processes. Some organisms may be differentiated from others by their production of hydrogen sulfide. For example, *Salmonella paratyphi B* (a cause of gastroenteritis) produces it, while *Salmonella paratyphi A*, a closely related species, does not. This test therefore has diagnostic value.

Two methods for detecting the formation of hydrogen sulfide are the **lead-acetate-paper** method and the **stab** method. In the paper method hydrogen sulfide arising from the culture blackens lead acetate dried on a strip of filter paper suspended above the medium. In the stab method lead acetate already incorporated in the agar is blackened (Fig. 35–9).

METHANE. Methane is another gaseous product of bacterial metabolism. However, production of methane is rarely determined in routine procedures because the procedures are complex and require special apparatus. Carbon is a good hydrogen acceptor and is readily re-

tryptophan

indole
(red color with
p-dimethylaminobenzaldehyde)

Figure 30–5. Production of indole from tryptophan.

duced. In swampy places several species of strictly anaerobic bacteria attack the carbohydrates derived from dead vegetation. Cellulose is decomposed by the enzyme **cellulase**, yielding the disaccharide cellobiose which, in turn, is hydrolyzed to glucose. Carbon dioxide produced by fermentation of glucose can be reduced to methane by certain other bacteria (e.g., *Methanobacterium*). These bacteria also produce methane by the fermentation of acetic acid and of methyl alcohol and sometimes large amounts of the gas are given off. The bubbles seen rising during the summertime in woodland swamps are largely methane. Cellulase is also produced by a number of species of bacteria found mainly (or only) in the rumen of cattle. Microorganisms in sewage digestion tanks at sewage disposal plants produce such large quantities of methane (**sewer gas**) that it is profitable to collect it in tanks and use it as fuel. (Sewage contains a large amount of cellulose.) Accumulations of methane ("fire damp") in coal mines have sometimes caused disastrous explosions.

Ammonia and Nitrogen. Many bacteria, especially saprophytic species, produce ammonia and nitrogen from decomposition of proteins and other nitrogenous compounds. Some species can reduce sodium nitrate and sodium nitrite to nitrogen and ammonia. Like carbon, nitrogen is readily reduced. Ammonia and volatile amines may be detected by suspending a strip of paper impregnated with litmus or other indicator substance in the culture tube above the medium.

Assay of Extracellular Hydrolytic Enzymes. Many microorganisms produce exoenzymes (**hydrolases**) that hydrolyze many complex organic substrates, such as polysaccharides, proteins, and lipids. The ability to produce these enzymes is based on the organism's genetic composition; that is, some organisms may be able to break down cellulose (a polysaccharide) by producing cellulase, while other organisms are incapable of hydrolyzing cellulose because they lack the gene(s) for production of cellulase. For this reason, determination of production and activities of hydrolytic enzymes is frequently included in the systematic study of unknown bacteria and as an aid to identification. As stated in an earlier chapter, exoenzymes are secreted from the cell and catalyze reactions outside the cell. The function of the enzymes is the breakdown (**hydrolysis**) of large, complex, water-insoluble nutrients into simpler, soluble, and more immediately available foods.

The activities of exoenzymes, as of all enzymes, are very specific; that is, a certain enzyme can attack only a certain specific substrate, and the activity is usually confined to one step among the series of reactions needed to break down a large molecule into useful foods. Our discussion will be limited to those biochemical characteristics which are easily demonstrated and which are most commonly used in identification of unknown organisms.

A good method of testing the ability of many organisms to hydrolyze test substances like sugars, gelatin, fat, or starch is to mix the test substance with agar, pour into Petri plates, and, when solid, heavily inoculate the surface in streaks or spots. After good growth has occurred, a reagent reacting with the test substance to produce some distinctive appearance is flooded over the surface of the agar. Alternatively, the growth may be removed and the underlying surface treated with the reagent. If hydrolytic enzymes have been produced, the area of growth or the area surrounding the growth will show a distinctive reaction (or lack of reaction). For example, starch plates may be treated with Lugol's iodine solution. The starch-hydrolyzing colonies, or amylase-producing organisms, will be surrounded by colorless zones; the remainder of the plate will turn dark blue. In addition to starch, among the tests frequently used are hydrolysis of esculin, hippurates, chitin, and pectin.

Protein Hydrolysis (Proteolysis). The ability of a microorganism to hydrolyze the incomplete protein gelatin, so that it no longer solidifies in the cold or coagulates, is commonly taken as evidence that the organism hydrolyzes proteins in general. This is often true but there are numerous exceptions. However, the gelatin decomposition test is much used as a test for proteolytic power because of its convenience (Fig. 30–6). However, there are more dependable methods of detecting gelatin decomposition, and students are encouraged to read the suggested references presented at the end of this chapter.

Other proteinaceous materials used in determination of proteolysis are sterilized skim milk, a sodium-caseinate agar (milk protein, casein), and coagulated horse or beef serum. Hydrolysis of casein in skim milk often follows coagulation, and the milk then becomes brownish and translucent and the clot disappears. Similarly, if sodium caseinate agar is used, the hydrolyzed area becomes translucent and even more apparent as the plate is flooded with 1 per cent HCl, which will precipitate undigested casein.

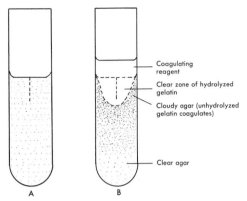

Coagulating reagent

Clear zone of hydrolyzed gelatin

Cloudy agar (unhydrolyzed gelatin coagulates)

Clear agar

A   B

Figure 30–6. Test for hydrolysis of gelatin. *A*, The tube of gelatin-agar just after having been inoculated by a stab at the top of the column. *B*, The tube after incubation and addition of the coagulating reagent. The gelatin-agar has become cloudy everywhere except in the zone of hydrolysis.

Lipolysis. Many microorganisms produce enzymes that hydrolyze one or more lipids or esters. One of the inconveniences in studying lipases is difficulty in bringing the lipid substrate into intimate contact with the organism. This contact is necessary since lipases often appear not to diffuse well into solid culture media. It may be that they are inducible enzymes and appear only in contact with their substrate. Another difficulty is in making the lipolytic effect evident. Lipolysis has not been studied as much as it should be because of such technical difficulties and it is not commonly mentioned in keys. Several methods are shown in Figure 30–7.

A reliable method is a recent improvement of the procedure devised by Eykman in 1901. A specially prepared mixture of mono- and diglycerides in palm oil is spread and allowed to cool on the inside bottom of a Petri plate. Two ml of a broth suspension of the organisms are then layered on the lipid and 6 to 7 ml of melted (45C) agar of appropriate composition are thoroughly mixed with the suspension by tilting and rotating the plate. Time and temperature of incubation are adapted to the organism under test. Lipolysis is indicated by the appearance of white, opaque spots of fatty acids or their calcium salts in and around the colonies. The method avoids the use of acid indicators that are toxic to some organisms and that introduce error if extraneous acids are formed from fermentable sugars in the medium.

Microbial Reductions. A distinctive property of many microorganisms is the power to reduce various substances such as tetrazolium, methylene blue, resazurin, selenium, litmus,

and nitrate. In the absence of free oxygen, many microorganisms use these substances as electron (hydrogen) acceptors.

Nitrate Reduction. Reduction of nitrate is a result of the use of potassium nitrate as an electron acceptor by facultative bacteria in the absence of oxygen. The microorganisms being investigated are incubated in broth containing about 0.1 per cent potassium nitrate (KNO_3). After several minutes of incubation, and at

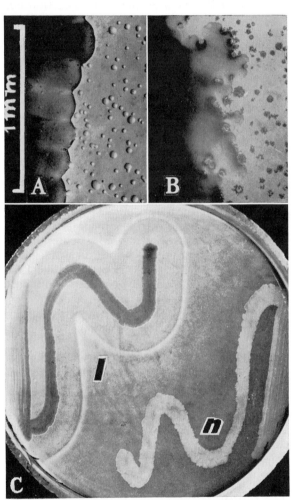

Figure 30–7. Lipase production and its detection. *A*, The margin of a colony of *Micrococcus* sp., on a plate of agar sprayed with a fine mist of olive oil. No lipase is evident since all the droplets of oil have remained unchanged, even in contact with the colony. *B*, The irregular margin of a colony of *Serratia* on agar sprayed with oil. Lipase activity has caused the oil droplets to become deformed and optically dense and refractive near, as well as at some distance from, the colony. *C*, A plate containing agar in which fat is emulsified and to which has been added a small amount of Nile blue sulfate (a dye which turns blue in the presence of lipolysis). On this plate are seen a strain of *Micrococcus* producing lipolysis (zone of color change and emulsion destruction, *l*), and a strain not producing lipolysis (no zone of color change or emulsion destruction, *n*).

longer intervals, a test is made for the presence of nitrites (KNO_2) by withdrawing a small sample of the culture from the bottom of the tube and immediately adding to that sample a drop of reagent to detect the presence of nitrite:

$$NO_3^- \xrightarrow{\text{nitrate reductase}} NO_2^-$$

The presence of free oxygen in nitrate broth may interfere with nitrate reduction, and to minimize this interference large amounts of actively growing cells are inoculated into the nitrate broth (1.0 ml) preheated in a water bath to 37C to drive out the dissolved oxygen in the medium. If the nitrite test is negative, a test for nitrate is made to insure that the organism is unable to utilize nitrate as the electron acceptor or that the nitrate is reduced completely. If both the nitrate and nitrite tests are negative, then we may consider that the organism reduced all of the nitrate as well as the nitrite to nitrogen or ammonia, which has dissipated. Unreduced nitrate may be tested for after the nitrate test by adding a little powdered zinc, which reduces KNO_3 to KNO_2.

Some organisms (denitrifying types) are able to reduce nitrite to either gaseous nitrogen or to ammonia. The organism's ability to reduce nitrite can be tested as in nitrate reduction, except that the quantities of nitrite used would be so small as to be just barely detectable (nitrites are toxic to many microorganisms, and the concentration of nitrite is usually limited to 0.002 to 0.004 per cent).

If the organism is capable of reducing nitrite, the culture will soon lose its nitrite because the organism will quickly reduce *all* of such a small amount of nitrite. A sterile control tube should be tested at the same time, since illuminating-gas fumes (as from Bunsen burners) often contain nitrous acid, which may be absorbed by the medium and give a slight reaction.

Reduction of Litmus. The ability of an organism to reduce other substances than nitrates and nitrites is often investigated. Some strictly anaerobic marine species and some found in petroleum (*Desulfovibrio* sp.) can reduce sulfates to hydrogen sulfide. Sulfur, like oxygen, carbon, and nitrogen, is readily reduced. Litmus serves to show whether the organism has strong reducing powers by becoming entirely decolorized when reduced. Litmus is used, like nitrates and nitrites, as an alternative electron acceptor by many facultative species. Just enough is added to the medium before sterilization to give a definite color.

The "Reductase" Test. Standardized solutions of methylene blue are often added to samples of market milk to estimate roughly whether a few, a moderate number, or enormous numbers of bacteria are present. When great numbers are present, the blue color disappears almost immediately. Several other oxidation-reduction dyes are now used in addition to methylene blue: tetrazolium salts, neutral red, and resazurin (see Chapter 47).

Assay of Specific Enzymes. With the advent of more sophisticated instrumentation and testing procedures, ability of an organism to produce certain enzymes is included in the routine systematic study of unknown bacteria for identification and classification.

Catalase is an enzyme which catalyzes the decomposition of hydrogen peroxide:

$$H_2O_2 \xrightarrow{\text{catalase}} H_2O + \tfrac{1}{2}\, O_2$$

It is produced by many cells, especially aerobic cells, rarely anaerobic cells. It is common in human saliva. Its presence or absence is often a valuable distinguishing character. Catalase activities of certain groups of bacteria are as follows:

BACTERIA	CATALASE
Micrococcus	+
Staphylococcus	+
Lactobacillus	−
Streptococcus	−
Propionibacteriaceae	+
Corynebacteriaceae	+
Bacillus	+
Actinomycetales	+

In a simple means of demonstrating the presence of catalase a previously inoculated and incubated plate of nutrient medium is flooded with a 3 per cent solution of hydrogen peroxide. Evolution of bubbles of oxygen is evidence of the presence of catalase. Special attention must be given to pH, which must be near 7.0.

Other enzymes frequently used in the systematic study of an unknown isolate are:

1. Oxidase, or indophenol (cytochrome) oxidase, which causes a successive change in color of a 1 per cent solution of tetramethyl-*p*-phenylenediamine.
2. Amino acid decarboxylase and deaminase—enzymes that liberate CO_2 from the carboxyl group of amino acids—and enzymes involved in removal of an

amino group from amino acid, with liberation of ammonia.

3. Phosphatase, an enzyme which splits phosphate from its organic compounds.
4. Coagulase, an enzyme causing coagulation of blood plasma.
5. Hyaluronidase, an enzyme that destroys hyaluronic acid, intercellular cementing substance of animal tissues.
6. Lecithinase, a lipolytic enzyme that destroys lecithins by separating the phosphate group attached to glycerol with two fatty acid groups.

Physiological Study of Unknown Bacteria

Physiological tests commonly used in the systematic study of unknown bacteria are: temperature range of growth, including minimum, maximum, and optimum temperatures (testing whether the organism is psychrophilic, mesophilic, or thermophilic); oxygen tolerance, that is, to determine whether the organism is strictly aerobic, strictly anaerobic, or facultatively anaerobic (Chapt. 36); pH range of growth, which determines whether the organisms are acidophilic, neutrophilic, or basophilic (alkalophilic); and maximum tolerance of the organism to hydrogen ion concentration of the medium. For example, some species of genus *Thiobacillus* may grow in a medium having a pH of 1 to 3, and it has been reported that *Thiobacillus thiooxidans* may be able to grow at pH 0, equivalent to the value of $1 N H_2SO_4$. At the extreme other end of the pH range are organisms capable of hydrolyzing urea, which results in considerable amounts of alkali, because 2 moles of ammonia are formed from each mole of urea, $CO(NH_2)_2$, hydrolyzed. For example, *Bacillus pasteurii* is unable to grow on all-purpose media at neutral pH in the absence of urea. Most basophiles and ureolytic organisms grow better at a pH of 8.5 or higher. Only a few species of any bacteria are able to grow at pH values of less than 2 or greater than 10.

Salt Tolerance. Because the effect of increased salt concentration on the growth of microorganisms varies with the species, the salt tolerance test is included in the routine systematic study of unknown bacteria. For example, obligate halophiles (NaCl is essential for growth) and extremely halophilic bacteria, such as species of genus *Halobacterium*, require a minimum of 15 per cent NaCl for growth; optimal growth may be obtained at about 25 per cent NaCl. Some microorganisms are facultative halophiles; that is, organisms which will tolerate media containing a high concentration of NaCl but which do not require it for growth. For example, *Staphylococcus aureus* is able to tolerate and grow in media containing 6.5 per cent NaCl; however, for most saprophytes, such as *Escherichia coli*, growth is inhibited by a much lower concentration of NaCl.

Bile solubility and tolerance or dye sensitivities (for example, sensitivities to dyes such as crystal violet) are frequently included in physiological testing of unknown bacteria. The principles involved and procedures of these tests are presented in detail in the reference material cited at the end of this chapter.

Serological Properties of the Unknown

As explained in Chapters 27 and 28, when whole cells or cell constituents, such as capsules, flagella, and cell wall preparations (antigens) are injected into an experimental animal, the animal produces antibodies specific to the injected antigens. If antigen preparations of an unknown culture react (agglutinate) with a known antibody (agglutinin) prepared from an authentic culture, the unknown culture may be either identical with the authentic culture from which the agglutinin was prepared, or at least the two organisms are very closely related in antigenic properties.

Serological typing of an unknown isolate is particularly useful when one narrows the identity of the isolate to the genus level. For example, organisms in either genus *Escherichia* or genus *Salmonella* can be differentiated not only as to species but also as to more specific serotypes. For this, group serum is followed by a series of specific sera (Chapt. 35). In recent years, fluorescent antibody techniques (fluorescence microscopy) have been extensively applied to serological reactions. As in other serological reactions, coupling of fluorescein-labeled antibody and bacterium is highly specific and sensitive. The methods have broad applications in medical as well as taxonomic identification of unknown organisms (Chapts. 25, 31).

Bacteriophage Typing of Isolates

Bacteriophage typing becomes a particularly useful tool as one narrows the identity of the isolate to the genus level. With phage typing it is possible to identify species within the same genus or even varieties within the same species. In fact, by use of specific bacteriophages one is able to distinguish organisms within the clone or serotype of the same species (see Chapters 16 and 35). Differentiations are established on the basis of differing susceptibilities to type-specific bacteriophages. However, one first usually uses multivalent phage which may attack organisms in one or more different genera, followed by phages of more specific types. The hyperspecific "phage types" to a certain extent parallel serotypes, since phage type susceptibility appears to be closely related to specific cell wall antigen content, and this, in turn, to DNA base sequence and content. However, phage typing often goes beyond serotyping. Certain strains of bacteria of great importance in industry and medicine are, at present, identifiable within species only by means of phage typing.

Genetic Characterization

Recently studies have established that similarity (or dissimilarity) among species may be determined on the basis of DNA homology or hybridization of DNA from one microorganism with that of another. The hybridization may be accomplished with living cells or with the extracted, purified DNA material. Genetic characterization such as DNA base per cent composition and hybridization (DNA:DNA or DNA:RNA) are more fully discussed in future chapters.

Pathogenicity

The term pathogenicity refers to the ability of a microorganism to enter a host and produce disease (physiological or anatomical changes). As indicated in earlier chapters, certain bacteria are pathogenic to plants or animals or both. For example, *Brucella abortus* may cause infection of cows, and a consequence of the disease may be abortion; *Erwinia atroseptica* may cause a disease of potatoes known as soft rot (or "blackleg"); and *Pseudomonas aeruginosa* may cause middle ear infection of man as well as soft rot of various plants. However, many pathogens attack only specific hosts and even specific loci of the host. Determination of pathogenic properties may be facilitated if the unknown isolate is obtained from an infected region of the plant or animal.

Pathogenicity may be determined if necessary by injecting 0.5 ml of a 24-hour broth culture intravenously or subcutaneously into a rabbit, guinea pig or other animal (Fig. 30–8).

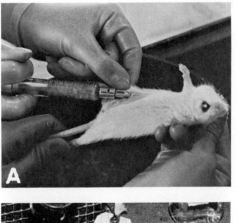
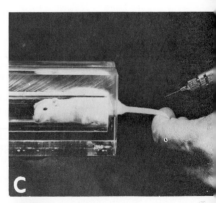
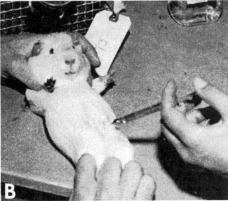
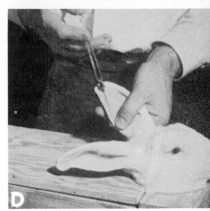

Figure 30–8. Tests for pathogenicity by injection of laboratory animals. *A* and *B*, Intraperitoneal injection of mouse and guinea pig, respectively. *C* and *D*, Intravenous injection of mouse (tail vein) and rabbit (marginal ear vein), respectively. (Courtesy of Pfizer Laboratories, Div. Chas. Pfizer & Co., Inc., New York.)

In all properly operated laboratories, animals, if required for experimental purposes, are maintained under most favorable conditions of nutrition, comfort, and sanitation, and anesthesia is used as it would be for humans. Strict regulations are published by the U.S. Department of Health, Education, and Welfare, Public Health Service.

If our unknown is truly pathogenic, abscesses will probably form or the animal may die.

30.6
RAPID MICROTECHNIQUES

By certain modifications of procedure many of the tests described above may be made more quickly and more economically by rapid techniques, which are based on the principles that small amounts of culture and of reagent may be used in small tubes; and since microorganisms rapidly multiplying in the logarithmic phase are most active in all enzymic functions, one may greatly speed up their effects by adding to **prewarmed medium** containing the test substrate a heavy suspension of young, actively growing organisms gently removed from a young agar-slant culture with a few drops of broth or saline solution. Alternatively, billions of active organisms may be concentrated by centrifugation of 10 ml of a young broth culture. They are resuspended in 0.5 ml of broth or saline solution. Of this suspension, 0.2 to 1.0 ml are added to 2 ml of the test medium previously warmed to the incubator temperature. The mixture is held in a water bath during the period of incubation. The ready-made population consisting of billions of young, active organisms sets to work immediately and brings about the desired characteristic changes (production of acid, hydrolysis of urea or gelatin, nitrate reduction, etc.) within a few minutes or hours (Fig. 30–9).

Disk and Tablet Methods. These methods simplify and expedite procedures by incorporating the culture medium or the test substrate and indicator in prepared, sterile, dried disks of filter paper or tablets of porous material. In one

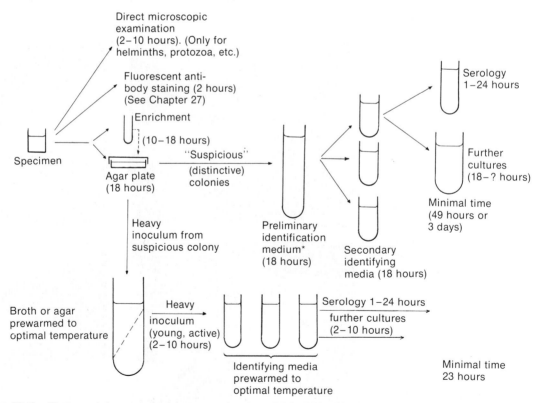

Figure 30–9. Steps and time required in conventional method of identifying a rapidly growing microorganism (e.g., typhoid bacilli) compared with steps and time required in "rapid" procedure. (Specimen is of feces.)

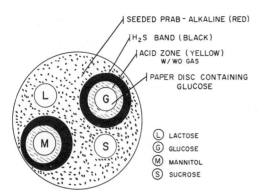

SEEDED PRAB - ALKALINE (RED)

H₂S BAND (BLACK)

ACID ZONE (YELLOW)
W/ WO GAS

PAPER DISC CONTAINING
GLUCOSE

Ⓛ LACTOSE
Ⓖ GLUCOSE
Ⓜ MANNITOL
Ⓢ SUCROSE

Figure 30–10. One form of disk method for identification of microorganisms. While still fluid at about 40C the base agar is heavily inoculated with young, active cells from a broth culture of the test organism. The agar contains an acid-base indicator and a source of sulfur (for H₂S production) with an iron salt to turn black in the presence of H₂S. Sterile paper disks containing lactose, sucrose, glucose, and mannitol, respectively, were placed on the surface of the agar before incubation. After overnight incubation the agar around the lactose and sucrose disks was colorless, showing that these sugars have not been fermented; agar around the glucose and mannitol disks showed a zone of yellow (indicator showed acid due to fermentation) and a black zone due to H₂S production from the sulfur source in the agar. If gas is produced it may appear as tiny bubbles in the depths of the agar.

procedure, 15 ml of melted nutrient agar at 45C in a tube are inoculated with a heavy suspension of young, active cells. The agar is then poured into a Petri dish and allowed to solidify. One then places, on the surface, small paper disks or tablets previously dried and sterilized after saturation with solutions of test substrate and suitable indicator. The test substance and indicator diffuse into the agar around the disk or tablet and are quickly acted upon by the bacteria in the agar (Fig. 30–10).

This procedure is easily modified for fluid media in tubes. Sterilized disks of filter paper are prepared by drying upon them the ingredients of various kinds of nutrient media plus the test substance (e.g., lactose) and indicator (e.g., bromcresol purple). With sterile forceps one merely adds the desired kind of sterile disk to 2 to 5 ml of sterile water in a tube. After a minute or two the ingredients in the disk dissolve and the tube is ready to be inoculated with a heavy suspension of active cells (Fig. 30–11). There are numerous ingenious modifications of these methods. In one, strips of filter paper are saturated with the test substance and indicator and then dried. Hundreds of these strips occupy little space and may be stored indefinitely.

A test is made by placing on the prepared strip a heavy suspension of the organisms scraped from agar (Fig. 30–12), or from the sediment pellet of a strongly centrifuged, luxuriant, broth culture. Readings can be made in 30 to 60 minutes. Results are satisfactory for routine purposes and can be checked by additional tests.

30.7
IDENTIFICATION OF THE UNKNOWN ORGANISM

After completion of the systematic study described throughout this chapter, the results of various observations and tests must be tabulated to facilitate identification of the unknown organism. The results may be divided into morphological, cultural, nutritional, biochemical, physiological, serological, genetic, and phage sensitivity tests. Let us assume the following observations are obtained from our unknown isolate, as shown in the table on the following page.

How are we now to determine the genus and species of the cocci which we have been studying?

Use of Keys. Although every experienced bacteriologist has at his finger tips, so to speak, all of the distinguishing cultural reactions and other identifying characters of the organisms

Insert sterile disk with desired reagent

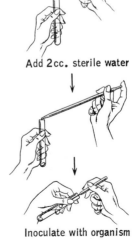

Add 2 cc. sterile water

Inoculate with organism

Figure 30–11. A rapid disk cultural method. For explanation see text. (Courtesy of Pennsylvania Biological Laboratories, Inc., Philadelphia.)

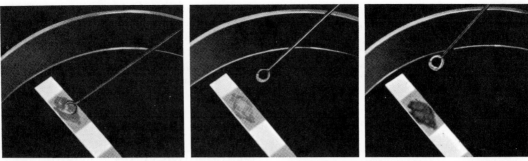

Figure 30–12. A rapid test for metabolic properties. A moist "paste" of pure culture of actively growing young cells is scraped from an agar surface with a loop and rubbed on the surface of a strip of test paper (in this case Patho-Tec) in the indicated zone. The zone contains test substrate (e.g., lysine) with an indicator that undergoes a distinctive change in color if the organisms decompose the substrate. The test in this case was for decarboxylation of the lysine (positive in case of *Salmonella*.) (Courtesy of General Diagnostics Division, Warner-Chilcott, Morris Plains, N.J.)

Morphological:
Shape:	Sphere
Size:	1.0 to 1.2 μm in diameter
Arrangement:	Occurring in clusters, pairs, and fours
Motility:	Nonmotile
Gram stain:	Gram-positive
Spore stain:	Spore absent

Cultural:
Gelatin colonies:	Lemon-yellow, raised, with undulate margin
Gelatin stab:	No liquefaction
Agar colonies:	Small, yellow, glistening, raised
Agar slant:	Lemon-yellow, smooth colonies
Broth:	Turbid with yellowish sediment, slight pellicle
Potato slant:	Thin, glistening, lemon-yellow colonies
etc.	

Biochemical tests:
Fermentation of:
Glucose:	Slight acid, no gas
Sucrose:	Slight acid, no gas
Lactose:	Not fermented
Mannitol:	Slight acid, no gas
etc.	

Hydrolysis of:
Starch:	Not hydrolyzed
Gelatin:	Not hydrolyzed
Fats:	Not hydrolyzed
Litmus milk:	Usually slightly acid, not coagulated
Casein:	Not hydrolyzed
Serum:	Not hydrolyzed
etc.	

Metabolic products:
Indole:	Not produced
Methyl red test:	Negative
Voges-Proskauer test:	Negative
Citrate:	Not utilized
Ammonia:	Produced from peptone

Reduction of:
Methylene blue:	Not tested
NO_3^-:	Not reduced
etc.	

Nutritional:
Accessory nutrients:	Not required
Nitrogen source:	Utilizes $NH_4H_2PO_4$ as sole source of nitrogen

Physiological:
Oxygen tolerance:	Aerobic
Optimum temperature:	25C
Salt tolerance:	Not tested
etc.	

Serological, genetic, phage sensitivity: Not tested

with which he is working, it is unusual, to say the least, to find one who knows *all* the characters of *all* the species. When an unknown organism is encountered that must be identified, the main morphological, physiological, and tinctorial features are determined in some such manner as just described, and then recourse is had to **keys** or other reference works. A much-used key for general bacteriological purposes in the United States is the 7th edition of *Bergey's Manual of Determinative Bacteriology*, 1957 (see page 438). As a valuable time-saver and as more up to date, one should also consult *A Guide to the Identification of the Genera of Bacteria*, 2nd edition, 1967, by V. D. B. Skerman.

The 8th edition of *Bergey's Manual* is in the process of publication. If we are to use the more established and presently available 7th edition of the *Manual*, we should first determine to which of the ten orders of the class Schizomycetes our culture belongs. On pages 33 and 34 of the 1957 edition of the *Manual* is to be found a brief synopsis of the characters used to differentiate the 10 orders. Obviously the species in question does not occur in trichomes (thread-like filaments); it does not multiply by budding, neither is it sheathed or stalked (Orders II, III, VI). It does not branch, nor is it acid-fast (Order V). It is not characterized by formation of sulfur granules or deposits, nor does it show gliding, jerky, or flexuous movements (Orders VII and VIII). It is neither spiral (Order IX) nor highly pleomorphic (Order X). We are thus left to consider Orders I (Pseudomonadales) and IV (Eubacteriales). All species of Pseudomonadales (Bergey, p. 35) are gram-negative, and they are also mostly rod-shaped and motile. Coccus-like forms are not common. Thus we may eliminate Order I and examine Order IV (Bergey, pp. 281–282). Our unknown clearly does not belong in Families I to VI, which are rod-shaped. We may eliminate Family VIII as gram-negative, and IX to XIII as rod-shaped.

Family VII (Micrococcaceae), therefore, would seem to be our objective, and we are referred to page 454. Here we find that the descriptions of Genera III to VI do not correspond with the organism in question, and that we must search in Genera I and II. The organisms of these genera resemble each other so closely that it is necessary to consider carefully the characters of the individual species.

A short study of the data we have already obtained by our cultural tests shows that our organism corresponds closely with the description of *Micrococcus luteus* (page 456), since it produces a lemon-yellow pigment, grows best at 25C, does not hydrolyze gelatin or ferment lactose, grows on potato with lemon-yellow pigment, does not reduce nitrates to nitrites, fails to produce indole, and is able to utilize $NH_4H_2PO_4$ as the sole source of nitrogen.

A further check upon the identity of the culture may be made by making other tests and comparing various characteristics of organisms closely resembling it, such as *Micrococcus flavus*.

A few repetitions of the tests (gelatin stab, litmus milk, oxidative or fermentative utilization of mannitol, lactose, etc.) usually serve to confirm the diagnosis or prove it to be in error, necessitating further study. Further details concerning micrococci will be given when these genera are taken up specifically.

Based on currently available information concerning the forthcoming 8th edition of *Bergey's Manual,* it will be divided into nineteen parts based primarily on the organisms' sources of energy, i.e., photoautotrophic (Part I), chemolithotrophic (Parts VII, VIII), or chemoorganotrophic bacteria; mode of locomotion; morphology of the cell; and Gram reaction. If we identify our unknown according to this new classification scheme, we will be able to eliminate Parts I through XIII and Parts XV through XIX because our unknown organism is not phototrophic and does not move by gliding; its cells are not sheathed, not prosthecate, and do not bud or possess appendages; nor is it spiral-shaped, curved, or filamentous. It is not a gram-negative or gram-positive rod, is not a strict anaerobe, does not form endospores, nor require living media for cultivation. Therefore, our choice is narrowed to Part XIV, which contains gram-positive cocci, which is divided into two parts, aerobic and/or facultative anaerobic organisms, and anaerobic organisms. Since our unknown is aerobic, this eliminates family Peptococcaceae, thus leaving two families: I, Micrococcaceae, and II, Streptococcaceae. However, Family II, Streptococcaceae, is composed of facultatively aerobic to microaerophilic organisms which produce a large amount of lactic acid from fermentation of various carbohydrates; therefore, this family may be eliminated. Family I, Micrococcaceae, contains three genera, *Micrococcus, Staphylococcus,* and *Planococcus*. After further investigation of biochemical and physiological properties of these three genera as described in the 8th edition of *Bergey's Manual* (and in the 7th edition), the student may arrive at genus *Micrococcus*, species *flavus*.

CHAPTER 30
SUPPLEMENTARY READING

Blair, J. E., Lennette, E. H., and Truant, J. P. (Eds.): Manual of Clinical Microbiology. American Society for Microbiology. The Williams & Wilkins Co., Baltimore. 1970.

Breed, R. S., Murray, E. G. D., and Smith, N. R. (Eds.): Bergey's Manual of Determinative Bacteriology, 7th ed. The Williams & Wilkins Co., Baltimore. 1957.

Gibbons, N. E. (Ed.): Bergey's Manual of Determinative Bacteriology, 8th ed. The Williams & Wilkins Co., Baltimore. 1974. (In press.)

Lockhart, W. R., and Liston, J. (Eds): Methods for Numerical Taxonomy. American Society for Microbiology, Bethesda, Md. 1970.

Skerman, V. B. D.: A Guide to the Identification of the Genera of Bacteria, 2nd ed. The Williams & Wilkins Co., Baltimore. 1967.

CHAPTER 31 · THE CLASSIFICATION OF BACTERIA

Biological classification, or **taxonomy,** is the systematic arrangement of organisms in groups or categories called **taxa** (singular, **taxon** [Gr. *taxis* = arrangement or grouping]) according to some definite scheme. Scientific names are given to organisms and taxa after adequate study, publication of the descriptions, and other prescribed formalities. Systematic naming constitutes the science of **nomenclature.** Scientific names are, by mutual agreement among scientists, abbreviated definitions or descriptions of the named organisms. The *Species Plantarum,* published by the famous Swedish botanist Carl von Linné (Carolus Linnaeus) in 1753, was the first biological classification to use the binomial system of nomenclature. The basic features of his scheme, **genus** and **species,** are now accepted as the units of classification in the field of taxonomy.

However, with bacteria, unlike higher plants or animals, the concept of species is exceedingly difficult to define, as there is no precise and universally acceptable definition of bacteria. The problem of defining bacterial species is further complicated by lack of apparent biological and evolutionary relationships among the bacteria and other living organisms. The extreme smallness and structural simplicity of bacteria offer only a few characters upon which to base classification. As a result, bacteriologists have always been forced to seek secondary characteristics, such as biochemical and physiological properties.

The difficulties involved in bacterial classification were eloquently expressed by Mueller (1786): "... the sure and definite determination (of species of bacteria) requires so much time, so much acumen of eye and judgement, so much of perseverance and patience that there is hardly anything else so difficult." More recently a similar sentiment was expressed by Leifson (Bact. Rev., *30*:257, 1966): "It would be naïve to expect universal agreement among bacteriologists on the definitive characteristics for all genera, not to mention species."

As for a definition of bacteria, the bacteriologist may not be accused of obscurantism or circular logic if he relies on five words and a letter: **bacteria are procaryons without chlorophyll a.** In giving this definition he must rely on the sophistication of his reader to realize that, although some bacteria are photosynthetic, they do not contain chlorophyll *a*, the type of chlorophyll that is found in all photosynthetic cells *except* bacteria; and that, except blue-green algae (which contain chlorophyll *a*), there are no other procaryons than bacteria (including as bacteria, mycoplasmas, rickettsias, and chlamydias). The viruses are excluded from this definition, though they may be included in the kingdom Protista "by courtesy."

Within the wide framework of this definition the bacteriologist draws up his systems of classification. Originally simple and limited in scope, the present systems have been built up during a century or more, little by little, being

433

modified and extended constantly as new organisms were discovered and as more and more data about them came to hand from microscopists, biochemists, biophysicists, geneticists, immunologists, and virologists. Because true genealogical relationships between bacteria have been generally obscure or moot, these classifications have evolved mainly as descriptive keys for identification of organisms. While artificial, these keys are, and have always been, indispensable for practical, systematic work.

31.1
DIFFICULTIES OF BACTERIAL CLASSIFICATION

Natural or Phylogenetic Classification. The zoologists and botanists have good guides to natural relationships among higher animals and plants based on demonstrable evolutionary and phylogenetic relationships. The structures and forms of higher organisms are usually patently related, and paleontology frequently gives "corroboratory" evidence. The arrangement of organisms systematically in an evolutionary series on the basis of inherited and stable structural and physiological resemblances or differences is called "natural" or phylogenetic classification.

Although there are reports that indicate discovery of bacterial fossils some 3.5 billion years old, the bacteriologists can only theorize as to the evolutionary and the phylogenetic history of these organisms. Their origin is merely the subject of shrewd speculation, their evolutionary history debatable, their properties "subject to change without notice."

In addition, new microbes are discovered each year, and many more remain to be discovered; new facts about existing microbes are published constantly. As a result, microbiologists do not yet know all there is to be known about even one species, much less all microorganisms. Therefore, when a microbiologist attempts to begin his classification of bacteria with a clear, concise definition of a bacterium, he immediately encounters difficulties. The first is at the border between the blue-green algae on the one hand and, on the other hand, photosynthetic bacteria and certain nonphotosynthetic microorganisms that, though classed as bacteria, closely resemble blue-green algae. These may really be forms of Cyanophyceae that have lost by mutation (or never gained?) the ability to form chlorophyll. He is equally puzzled when he reviews the mold-like properties of the Actinomycetales, the protozoa-like properties of the Spirochaetales and the bacteria-like (?) properties of the Eubacteriales and Pseudomonadaceae, to say nothing of the virus-like properties of the rickettsias, chlamydias, and the mycoplasmas.

Lack of Authentic Cultures and Standardized Procedures. Organisms discovered within the last decade or so generally have clearly described properties determined by specified and generally available methods. Sample cultures are available from the American Type Culture Collection, the International Collection of Phytopathogenic Bacteria at Davis, California, and other sources around the world for other workers to study. Rules for nomenclature are clear. Classification under such circumstances may not be too difficult a task. If, however, the person originally naming a supposedly newly discovered organism failed, as has frequently happened in the past, to keep cultures available for study; if he used undescribed or misleading methods to determine color, motility, or biochemical properties; and if in addition he was unacquainted with the **rules of nomenclature** and used incorrect names, or names already given to other organisms, it can be seen that confusion and disagreement are introduced. Many older descriptions are inadequate and might fit a dozen different species. On trying to duplicate them no exact descriptions of the methods used are to be found. "Definitions" of some genera and other groupings seem overinclusive. For example, in describing the members of a genus it suggests heterogeneity to say that *some* species are motile, *many* are gram-negative, *most* are encapsulated, *several* are pigmented, a *few* forms *tend to be* microaerophilic, and so on.

Alteration of a Gene. Even if a complete taxonomic system were finally agreed upon, a new difficulty would arise because microorganisms are variable and an organism having one set of characters today may have others tomorrow. For example, a culture of *Serratia marcescens,* which produces brilliant red pigment, may produce rare white colonies which appear after repeated streaking. The characteristics of the white colony are identical to red pigment–producing *S. marcescens* except it has lost the ability to produce red pigment. In some instances, white strains may revert to produce red-pigmented colonies among white colonies (reverse mutation). Such a sudden inheritable change in phenotype of an organism is known as mutation. It may occur naturally and spontaneously, or the change may be induced artificially by man; for example, transfer of R factors (antibiotic-resistant factors) or production of toxigenic strains of *Corynebacterium diphtheriae* from nonvirulent strains by prophage carrying *tox*+ determinant (transduction). Thus it is possible to transmute one "species" of microorganism into another!

31.2
EVOLUTION OF BACTERIAL CLASSIFICATION

Historical Development of Bacterial Taxonomy. Many systems of classification of microorganisms have been brought forward, but none has remained long without revision and enlargement. The first formal attempt to classify the microbes (in infusoria) was made by O. F. Mueller in 1773, and the generic name "*Vibrio*" was introduced for the first time. During the period between 1828 and 1838 the German zoologist C. G. Ehrenberg added two more generic names, "*Spirillum*" and "*Bacterium.*" With the advent of pure culture techniques (Pasteur, Koch, and Lister) and refined microscopes, increasing numbers of bacteria were described and named. One of the first schemes for bacteria was devised by Cohn, in 1872, based almost wholly on morphology. In 1897 Migula devised a scheme based not only on form but on color and some physiological characters, such as nitrogen fixation, and published the *System of the Bacteria.* During the same period (1896) Lehmann and Neumann started their *Atlas for Diagnostic Bacteriology.*

Active interest in classification of bacteria was not confined to European bacteriologists but rapidly spread among the American bacteriologists. D. F. Chester in 1899 and 1901 published the *Manual of Determinative Bacteriology,* which led to the formation of the Society of American Bacteriologists (SAB), now known as the American Society for Microbiology (ASM). Orla-Jensen, in 1909, made up a system based largely on physiological properties; this has served as a model for all later schemes. For example, in America, David Bergey, with assistance from the committee established by SAB, was given the task of revising Chester's *Manual,* and this grew into what is now known as *Bergey's Manual of Determinative Bacteriology,* which by 1957 had gone through seven editions.

31.3
THE UNIT OF CLASSIFICATION

Species and Genus. In bacteriology the terms **species** and **genus** are used, but the concept of these is somewhat vague. Until the 1950's there was little exact knowledge of genetics in relation to bacteria, and no established knowledge of evolutionary and phylogenetic relationships. Species of higher animals and plants are generally established on the basis of their demonstrable common ancestry and their ability to breed *only* within the species. **Breeding,** as the term is commonly used (i.e., fusion of sex cells), is doubtful in bacteria, except among a few kinds. However, recent experiments on interspecies **hybridization** among bacteria are revealing some surprising relationships.

In bacteriology a species is theoretically a single kind of bacterium, all individual cells of which are identical or nearly so. In actuality this identity of cells rarely exists. In any culture of a given species mutant cells may be found that, while having the outward form, staining properties, and other obvious characters of most of the cells in the group, differ in subtle and obscure ways. For example, they often possess different metabolic properties or different antigenic composition, and so on. Usually these differences are not extreme and may represent only temporary fluctuations from the principal type. It must be remembered that many ordinary 5-ml test tube cultures of a "species" consist of billions of individuals and represent a "population" many "generations" old even in 18 hours of growth. Many mutations can occur during that period. A **generation** (i.e., time required for one cell fission or division to occur) is as short as 20 minutes in many common species of bacteria. In terms of human generations of approximately 20 years, an 18-hour-old culture of bacteria would represent nearly 1,100 years or 54 human generations!

When two bacteria have one or more well-marked morphological differences, and exhibit important metabolic or other differences between them that are constant, the two may be regarded (by some!) as distinct species. But who is to determine what character or characters are "well-marked," constant, and of sufficient importance to be the basis of differentiation between species or genera? The same differences may be used as a basis of generic, tribal, or even familial distinction between some other kinds of bacteria. Bacterial species, therefore, are rather ill defined.

The concept of genera among bacteria is in many instances equally nebulous. A genus is theoretically and ideally a group of species all of which bear sufficient resemblance to one another to be considered closely related and easily distinguishable from members of other groups or genera. The boundaries of some genera are sharply defined by as few as three characteristics, as in the genus *Bacillus:* (1) aerobic, (2) endosporeforming, (3) rods. These (under certain defined conditions of growth!) are very definite, distinct, constant, and readily determined characters. The boundaries of other genera are sometimes more difficult to define, for example, the genera *Salmonella, Escherichia, Shigella,* and *Enterobacter.* All of these are nonsporeforming, gram-negative, facultative (capable of growing aerobically or anaerobically) rods of identical size and appearance, nonpigmented, and fermenting glucose. All occur more or less frequently in the intestinal tract. None forms pigment. Many possess certain O antigens (Chapt. 27) in common. All are motile except *Shigella.* But sometimes nonmotile variants of the others occur. In such a situation some experts prefer using the word **group** to avoid the restrictions imposed by rules of nomenclature and the definitions of **genus** and **species.** As stated by Taylor: ".... the situation is similar to the use of the terms 'boyfriend' and 'betrothed'; the former is carefree, the latter is a legal state with defined responsibilities." An organism of one genus may thus possess several of the important (?) characters of two or three or more other genera; its proper allocation to one of these is often difficult and must be decided on an arbitrary basis.

Strains. A term frequently used in microbiology is **strain.** A strain of microorganisms is a particular example, specimen, or culture of a given species. Strains may or may not show temporary or minor differences, which are referred to as strain differences.

Clones. A clone is a strain of microorganisms derived from a single cell and therefore asexually propagated. Clones of mammalian tissue cells are much used in tissue cultures for virology.

Type Species. There are certain central types of bacteria as, for example, *Streptococcus pyogenes, Bacillus subtilis, Clostridium butyricum,* and the like. Each of these is a well-known, thoroughly studied, easily identifiable species representative of a genus or a group of species. It is spoken of as the **type species** of that genus or group of organisms. Usually it is the first-described member of the species. Cultures of type species are usually maintained in various institutions (among them the American Type Culture Collection in Bethesda, Md.) for purposes of comparison.

To every experienced microbiologist the name of a type species conveys a very definite idea as to the characters of the group. However, ill-defined, partly studied organisms distinguishable from the type species or from each other only with the greatest difficulty, or not at all, are often included in such groups. On the other hand, organisms differing so markedly from the type species that the relationship seems very vague may also be included. Endless arguments often arise concerning such matters.

In certain new systems of classification (numerical or Adansonian classification) such groups are defined as "clusters," "taxospecies," "pleista," or "phenons," all equivalent terms meaning "cluster of strains with a considerable degree of relatedness."

31.4
BACTERIAL NOMENCLATURE

Following a tradition and accepted rules in biology, scientific names are in Latin because in the past Latin has been the common language of the learned of all nationalities (i.e., Latin transcended language barriers). When or if English becomes a universal language it may supplant Latin in scientific nomenclature.

The name of an organism is (or should be) a descriptive symbol. It should convey a definite idea of the organism named. This saves words, time and confusion. But it requires meticulous care to devise a name correctly, to avoid using the same name or previously used names for different organisms, and to describe the organism itself fully and accurately.

In naming a bacterium, certain official rules are followed. Each species is allowed a "first" and "last" name only. The two-name (binomial) scheme was originated in 1760 under the leadership of Linnaeus. The first name of a bacterium refers to the genus (pl., genera), and is usually a Latin or latinized word based on the morphology of the organism, the name of the discoverer, or some other distinguishing character, habitat, or the like. It is always used in the form of a Latin noun, singular. It is written with a capital letter. The second name is the species name and is usually an epithet descriptive of the noun, referring to its color, source, disease production, discoverer, or some other distinguishing point. The gender of the epithet must agree with that of the noun. It is not capitalized. Genus and species names are generally italicized. For example, in the name *Bacillus anthracis* the *Bacillus* indicates that the organism is a gram-positive, sporebearing, aerobic rod (properties of the genus *Bacillus*), while *anthracis* calls attention to the fact that this species of the genus *Bacillus* produces the disease anthrax. The name *Spirillum rubrum* tells us that this species is a heterotrophic bacterium, rigid, spiral in structure, nonsporeforming, motile with polar flagella, and gram-negative (all properties of the genus *Spirillum*), and that the species named is characterized by a red color (*rubrum*). The name *Clostridium novyi* indicates a heterotrophic, gram-positive, sporeforming, rod-shaped organism, saprophytic or parasitic, and restricted to growth in the total absence of free oxygen. These are properties of the genus *Clostridium*. This particular species bears the name *novyi* in honor of F. G. Novy of the University of Michigan, who discovered the organism and its relation to the disease gas gangrene. The practice of using personal names for newly discovered species of bacteria is obsolescent, although many generic names are derived from the discoverers or original students of the genus; for example the genus *Salmonella*, from an American microbiologist named *Salmon*; *Escherichia* from *Escherich*, a famous German scientist; *Shigella*, from a Japanese epidemiologist, *Shiga*. Sometimes an additional modifying term is added in the form of a varietal name; for example, *Streptococcus faecalis* var. *liquefaciens*, a gelatin-liquefying variety of a fecal streptococcus.

31.5
APPROACHES TO MICROBIAL CLASSIFICATION

Classical Approach. Any organism may be arranged into a taxon by measuring a variety of characteristics of different organisms. As a rule, structural and morphological characteristics, such as cell shape, stainability, motility, flagellation, spore formation, etc., are more stable and independent from external environment than are physiological and biochemical characteristics. For example, the shape of cells remains unaltered in a variety of environments because it is the product of numerous gene actions and is generally immune to drastic change as the result of a single gene mutation. In addition, these primary characteristics are readily determined, provided the size of the cells is within the limit of the microscope (0.5 μm or greater). The morphological and structural characteristics are frequently referred to as "qualitative unit characters." Such characteristics are most useful in classification because they are not quantitative or variable, as are size, rate of growth, the shade of color of a pigment, or its intensity. They are present or not present — *as a general thing*, that is!

Determination of biochemical, physiological, pathogenic, immunological, phage type, and other nonmorphological characteristics not only is useful but becomes imperative if the primary characteristics alone fail to separate a group of organisms into separate taxa. For example, it is impossible to separate certain *Micrococcus* from *Staphylococcus* on the basis of primary characteristics alone, but it is possible to separate one from the other if one employs a variety of nonmorphological characteristics. However, the student must be aware that some of these secondary characteristics may not be useful or applicable for separation of a particular group; he must decide how much information is necessary to effect sound classification. The chemical approach is neither natural nor rigidly systematic but an artificial system of classification which has proved very useful for the majority of microorganisms. The classical approach is discussed further in §31.6.

Adansonian Approach. This approach has been called numerical taxonomy and resembles the classical approach in that a large number of characteristics are determined for each organism of a group, but each phenotypic characteristic is given equal merit rather than placing more weight on some characteristics than on others. The taxonomic distance (affinity) between or among the organisms is determined by the number of phenotypic characteristics they share; distinct taxa are based on correlated characteristics. A detailed discussion concerning numerical taxonomy is presented in a later section (§31.7).

Genetic or Molecular Approach. Phenotypic characteristics of any given organism are the expression of a large number of genes that control enzymes; the expression of the genes is ultimately determined by DNA base sequence of the bacteria. The genotypic characteristics of most species are surprisingly constant, possibly because the segments of DNA that control them are too large to be affected by single point mutations. The fact that certain organisms may be classed as species results from this fact.

Genotypic characterization of the organism

is based on two different kinds of analyses. One is the analysis of base composition of DNA, to determine frequency of occurrence of guanine-cytosine base pairs among the various bacteria. This is commonly expressed as mole per cent of guanine + cytosine. The second test is hybridization between DNA of one organism with DNA of another, or between DNA and RNA. As you may recall (Chapt. 19), genetic recombination can occur only among closely related organisms. This is because gene recombination requires a close homology between the DNA molecules of the mating genomes. Detailed discussions on genetic classification are presented in §31.8.

31.6
CLASSICAL APPROACH—BERGEY'S MANUAL

A system of classification of Schizomycetes that is used by all bacteriologists and that has an international standing is *Bergey's Manual of Determinative Bacteriology* (commonly: *Bergey's Manual*). It represents the collaborative effort of over 100 of the best-qualified microbiologists at the time it was brought together and is a monumental work. The classification of the organisms is based on numerous properties: morphological, cultural, nutritive, biochemical, physiological, serologic, bacteriophage susceptibility, pathogenic, and genetic properties of the organisms. It is based on the *International Code of Nomenclature of the Bacteria and Viruses* established by the International Committee on Bacteriological Nomenclature in 1947. In 1959, a Committee on Terminology in Biological Nomenclature of the International Union of Biological Sciences met "to consider and report on possibilities of reconciliation of differing terminologies in the several International Codes (botanical, zoological, microbiological) of Biological Nomenclature."

So rapidly does microbiological science advance that *Bergey's Manual*, in its seventh edition, has needed revision since its appearance in 1957, and undoubtedly the same will prove true of the 8th edition of the *Manual*, soon to be published.

However, with the exception of some new names and changes that have come to be pretty generally accepted, bacteriologists continue to use *Bergey's Manual* simply because no other system has yet been put forward that is completed, officially sponsored, and as widely known. While publication of the 8th edition of the *Manual* is imminent at this writing, we believe a condensed version of the classification of the organisms as they appear in the 7th edition is worthy of presentation here, since it is still in general use.

All of the procaryons are included in Division I Protophyta, which is divided into three distinct classes.

Class I, Schizophyceae, includes all the procaryotic algae usually designated as "blue-green algae." The organisms possess chlorophyll *a* but lack an organized nuclear membrane–bound nucleus and chloroplasts. The cell wall contains some of the distinctive chemicals found in true bacteria, and some are attacked by cyanophages. They are to be classed as "blue-green bacteria" (cyanobacteria).

Class II Schizomycetes (the entire group of "true" bacteria) comprises some 1,500 species. These are divided into 10 orders differentiated from one another primarily on the basis of morphology (cell shape, arrangement, and type of locomotion, i.e., flagellar movement or gliding motility).

Class III Microtatobiotes includes usually obligate intracellular parasites (some are facultatively or exclusively extracellular parasites) comprising four families in the order Rickettsiales. Although these organisms at one time were thought of as large viruses, they are now regarded as true bacteria or modified or closely related to bacteria. Table 31–1 gives a résumé of primary characteristics used in distinguishing one order from another and these orders are described in detail.

Each order is divided into families, and these into tribes, genera, and species. Groups of similar species constitute genera; groups of similar genera constitute tribes; and so on. It is important to remember the endings attached to the names of taxa. These denote various grades:

Order..*ales*
(e.g., Pseudomonad*ales*)
Suborder ....*ineae*
(e.g., Rhodobacteri*ineae*)
Family ....*aceae*
(e.g., Thiorhod*aceae*)
Tribe...*eae*
(e.g., Escherichi*eae*)

These names are frequently used without other explanation and may be confusing unless clearly understood.

The names of genera and species have no such distinctive endings (see farther on). Some names of species and genera are changed with

TABLE 31-1. ORDERS OF SCHIZOMYCETES AND MICROTATOBIOTES*

Orders	Cell Morphology	Motility	Mode of Reproduction	Chapter Reference	Descriptive Features of the Orders
CLASS II: SCHIZOMYCETES					
I. Pseudomonadales	Cells rigid, spheroidal, or rod-like; straight, curved, or spiral	If motile, by means of polar flagella	Binary fission and by budding	32, 33	All are gram-negative; some are purple or green photosynthetic; chemolithotrophic and chemo-organotrophic; some groups form trichomes, some are prosthecated.
II. Chlamydobacteriales	Rod-like cells in trichomes often sheathed	Polar to subpolar flagella when present	Binary fission (produce gonidia); motile by polar flagella	34	All are gram-negative, frequently referred to as sheathed bacteria; may deposit iron or manganese hydroxide in sheath; some trichomes form rosettes; some deposit intracellular sulfur granules.
III. Hyphomicrobiales	Spheroidal or ovoid cells connected on stalk or thread	Some motile by polar flagella	Budding and longitudinal fission	34	All are gram-negative, frequently known as "budding" bacteria; swarm cells frequently form rosettes.
IV. Eubacteriales	Cocci, rods, curved, coryneform; some produce endospores	When motile, cells are peritrichously flagellate	Binary fission	35, 36, 37, 38, 39	Gram-negative or -positive, simple undifferentiated cells with rigid cell wall; all are chemoorganotrophs; some produce heat-resistant endospores.
V. Caryophanales	Short, disk-like cells in trichomes	Hormogonia; motile by means of peritrichous flagella	Binary fission, sporulation (hormogonia)	39	Gram-negative (now considered as gram-positive) filamentous bacteria; individual cells contain prominent nuclear bodies.
VI. Actinomycetales	Rod-like to filamentous; also spherical conidiospores	Mostly nonmotile; when motile, polar flagella	Binary fission; fragmentation of mycelia and sporulation	40	Gram-positive to variable, some are acid-fast; many species form filaments which tend to branch; some are frequently referred to as higher or mold-like bacteria.
VII. Beggiatoales	Rod-like or coccoid cells in trichomes	Gliding motility without flagella	Binary fission	34	Produce colorless alga-like trichomes; some filaments display gliding motility, while others show oscillatory or rolling motion; accumulate elemental sulfur; frequently referred to as alga-like bacteria.
VIII. Myxobacterales	Coccoid to rod-like or fusiform	Cells flexuous and gliding	Binary fission and microcyst formation	41	Some produce microcysts, while others produce prominent, visible, highly colored fruiting bodies; cells lack structural rigidity, are flexuous; frequently known as slime bacteria.
IX. Spirochaetales	Elongate spiral cells	Rotatory and flexing motion and translatory motility may be attributed to contraction of axial filaments	Binary fission	41	All are gram-negative, cell length ranges from a few μm to 500 μm; cells frequently possess axial filaments; cell wall not rigid.
X. Mycoplasmatales	Extremely pleomorphic; lacks cell wall	Nonmotile	Binary fission or by budding? Elemental body?	42	The organisms lack cell wall; frequently referred to as pleuropneumonia-like organisms (PPLO); some reproductive cells are filtrable.
CLASS III: MICROTATOBIOTES					
I. Rickettsiales	Very minute rods or cocci	Nonmotile	Binary fission	42	All are gram-negative, obligate intracellular parasites; some are nonfiltrable by ordinary bacteria-retaining filters, others are filtrable; some are transmitted by arthropods, others by aerosol, some unable to generate own ATP.

*Classification based on 7th edition of *Bergey's Manual*. Class Schizomycetes contains 10 orders, 49 families, and 190 genera.

successive revisions of the systems of classification, and different names are therefore sometimes used by different authors for the same organism, or two different organisms may be called by the same name. Thus there is some confusion in bacteriological literature. This is one of the signs of progress.

31.7
NUMERICAL TAXONOMY

The idea of numerical taxonomy was first described by Michel Adanson in 1757 and was first applied to bacteria by Sneath in 1956. It is the arithmetic method of classifying large numbers of bacterial strains on the basis of their overall similarity (affinity) to one another. The method requires the use of a large number of phenotypic (visible or demonstrable) characters which are either present (+) or absent (−), and each characteristic is given equal merit. However, the latter statement that "equal taxonomic merit or weight should be given to all characteristics of an organism" was not accepted with ease and enthusiasm. The mistrust of this principle is further augmented by difficulties which arise when some phenotypic properties are highly variable under different conditions of cultivation or when they vary as to degree of intensity of reaction: "+," "++," or "+++."

Selection of properties that are invariably present (+), or absent (−), simplifies analysis, but variable properties cannot always be ignored; personal judgment (frequently a source of error in any human endeavor) inevitably complicates the picture. However, the advent of computer science, electron microscopy, and more refined instrumental analysis of various phenotypic and genotypic characteristics has afforded large quantities of data for many hundreds of organisms. As a result, increasing numbers of publications rely heavily on computers for data processing, especially the problems of bacterial classifications. More and more frequently, separation and definition of taxa, established by the classical approach, are confirmed by numerical taxonomy. Although there are certain inherent shortcomings, the principles and usefulness of numerical classification are now widely accepted among microbial taxonomists.

Practical Application. The standard procedures generally used in numerical taxonomy are: collection of data, which includes selection of organisms, sample size, and selection of tests to determine the properties of the organisms. The organisms to be tested should be representative of the test set and ideally include cultures of historical importance, authentic cultures, or neotype cultures which have been used by other investigators as reference strains. The sample size is dependent on the availability of labor, facilities, and computer. One may be able to examine a set of 60 or more individual cultures of bacteria at a time. The tests to be used in characterization of organisms must be performed under standard conditions and should include as many characters as possible, preferably a minimum of 80. The characters may be phenotypic or genotypic or both, but the tests must provide admissible quantitative or qualitative information about an organism. Admissible information is derived from those tests which result in presence/absence or positive/negative observations. For suggested tests which may be used for numerical analysis, see Chapter 30.

Coding of Data. Once the distinctive (?) properties are selected, determined in the laboratory, and tabulated, they are coded for computer or numerical analysis (Table 31–2).

Tabulation of Coded Data and Evaluation of Results. Tabulations are made showing those organisms in which all properties agree (100 per cent similarity), those that differ in only one property, those that differ in two properties, and so on. Electronic computers are often used in such compilations. Percentage similarities (%S) of the organisms are calcu-

TABLE 31–2. EXAMPLES SHOWING PART OF CODED DATA*

Character States	Strains†			
	A	B	C	D
1	+	+	−	0
2	+	+	+	+
3	+	+	+	−
4	−	+	0	0
5	+	+	+	+
6	+	+	−	+
7	+	+	−	0
8	0	−	+	+
9	+	+	+	+
10	+	+	+	−
11	+	0	−	0
12	+	+	+	−

*From Ainsworth and Sneath (Eds.): Microbial Classification. Twelfth Symposium, Society for General Microbiology. Cambridge University Press, London. 1962.

†Symbols: + = character present; − = character absent; 0 = uncodable.

TABLE 31-3. PER CENT SIMILARITY COEFFICIENT (%S) ARRANGED IN AN ARBITRARY FASHION[*][†]

	1	2	3	4	5	6	7	8	9	10	11	12	13
1	100	50	48	56	84	49	39	51	52	51	53	48	61
2	50	100	50	85	51	53	43	51	93	49	48	94	50
3	48	50	100	52	50	34	32	86	52	93	89	50	52
4	56	85	52	100	57	54	39	58	93	54	52	89	50
5	84	51	50	57	100	51	41	52	53	52	56	59	64
6	49	53	34	54	51	100	89	80	55	82	87	53	47
7	39	43	32	39	41	89	100	29	43	29	39	44	39
8	51	51	86	58	52	80	29	100	80	88	78	51	52
9	52	93	52	93	53	55	43	80	100	51	50	89	43
10	51	49	93	54	52	82	29	88	51	100	81	49	52
11	53	48	89	52	56	87	39	78	50	81	100	48	53
12	48	94	50	89	59	53	44	51	89	49	48	100	50
13	61	50	52	50	64	47	39	52	43	52	53	50	100

[*]From Lockhart and Liston (Eds.): Methods for Numerical Taxonomy. American Society for Microbiology, Bethesda, Md. 1970.

[†]The completed matrix is cut by a principal diagonal of 100 per cent similarity, composed of values which represent comparison of each organism with itself. Only half of the remaining values need be considered, because the matrix is symmetrical about the diagonal, and the lower left portion is a mirror image of the upper right portion.

lated and are expressed as a **matching coefficient** or **%S value** for each organism:

$$\%S = \frac{Nsp}{Nsp + Nd} \times 100$$

or

$$\%S = \frac{Nsp + Nsn}{Nsp + Nsn + Nd} \times 100$$

Nsp = Number of similar positive matches (i.e., both positive); Nsn = Number of similar negative matches (i.e., both negative); Nd = Number of dissimilar matches (i.e., one positive, one negative).

A hypothetical example of the per cent sim-

ilarity coefficient (%S) between pairs of organisms is presented in Table 31-3.

These values (%S) may then be tabulated to show the relationship of each organism to all of the others in the group under investigation. For example, a similarity matrix (%S) table (Table 31-4) may be prepared by rearranging the organisms.

The values (%S) in Table 31-4 may be divided into equal intervals (e.g., 100-90 %S, 89-80 %S, etc.) and a different degree of shading assigned to each interval to reveal the relationship among the organisms; for example, the range between 100-90, solid black; 89-80, grey; etc. (Fig. 31-1).

The values may be arranged to form a sort of genealogical tree or **dendrogram** (Fig. 31-2).

TABLE 31-4. PER CENT SIMILARITY (%S) VALUES AFTER ARRANGING THE ORGANISMS INTO GROUPS[*]

2													
12	94												
4	85	89											
9	93	89	93										
3	50	50	52	52									
10	49	49	54	51	93								
11	48	48	52	50	89	81							
8	51	51	58	80	86	88	78						
6	53	53	54	55	34	82	87	80					
7	43	44	39	43	32	29	39	29	89				
1	50	48	56	52	48	51	53	51	49	39			
5	51	59	57	53	50	52	56	52	51	41	84		
13	50	50	50	47	52	52	53	52	47	39	61	64	
Organisms 2	12	4	9	3	10	11	8	6	7	1	5	13	

[*]From Lockhart and Liston (Eds.): Methods for Numerical Taxonomy. American Society for Microbiology, Bethesda, Md. 1970.

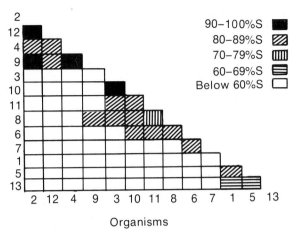

90–100%S ■
80–89%S ▨
70–79%S ▥
60–69%S ▤
Below 60%S ☐

Organisms

Figure 31–1. Differentially shaded similarity matrix.

However, the dendrogram of numerical taxonomy should not be thought of or implied as equivalent to a "phylogenetic tree"; it is simply a graphic way to show a cluster of phenotypically related organisms (phenon). Accordingly, whatever arrangement of the data is made, it is usually found that various groups of species appear in which the organisms have many similarities among themselves and dissimilarities from organisms of other clusters. Interpretation of these data reveals that the organisms clustering together or with certain reference organisms may be considered to be members of the same taxon. It should be remembered that computers can be programmed to perform most of the operations described thus far, including printing out a dendrogram. Finally, the student should also remember that excellence of computer analysis is dependent on: excellence of selection of phenotypic characters of the organisms, the procedures to be used for determination of phenotypic characters, and programming of these findings.

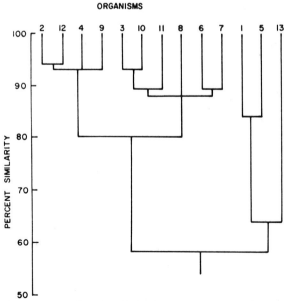

Figure 31–2. Completed dendrogram constructed from the data in Table 31–4.

31.8
GENETIC CLASSIFICATIONS

Attempts are at present being made to determine relationships between organisms not merely on the basis of relatively superficial phenotypic (e.g., enzymic) similarities but on the more fundamental property of likeness or **homology** between the genetic materials, i.e., the DNA's. One line of such investigations is concerned with the **DNA base per cent composition;** another, with degree of chemical hybridization between DNA and DNA, or between DNA and RNA of different organisms.

As discussed in earlier chapters, DNA contains four bases: adenine (A), thymine (T), guanine (G), and cytosine (C). For double-stranded DNA, a pairing always occurs between one purine base and one pyrimidine base, e.g., A=T and C≡G. Such pairs occur in random sequence and without restriction in respect to the molar ratio (A + T) : (C + G).

Recent studies of the molecular composition of DNA reveal that relative percentage of complementary pairs of guanine + cytosine:

$$\frac{G + C}{A + T + G + C} \times 100 = \text{mean DNA}$$

base per cent composition or % G + C

varies widely with different groups of bacteria. Variation became apparent even between microorganisms thought to be similar on the basis of tests of phenotypic properties. DNA base per cent composition of a species is presumably a fundamental and fixed property of each cell, dependent only on DNA base sequences and independent of age and all external influences except mutagenic agents.

DNA Base Per Cent Composition. Pure DNA can be extracted from cells that have been ruptured mechanically, as by osmotic shock; generally according to a method described by

Marmur in 1961. Because of the three H-bonds in the G≡C pairs, these pairs require a higher temperature to separate them; they have a higher "melting point" than A=T pairs. The per cent of G + C can therefore be calculated from the "melting point" (Tm), i.e., the mean temperature at which the DNA is denatured (thermal denaturation). Tm is determined by noting changes in optical density or light absorbance of a solution of DNA at a certain wavelength of monochromatic light (260 nm) during the heating period (Fig. 31–3): The mean Tm is directly related to per cent G + C:

$$\% \text{ G} + \text{C} = \text{Tm} - 63.54/0.47^*$$

DNA base per cent composition may also be determined by the relative rate of sedimentation (buoyant density) in cesium chloride (CsCl) solution, by hydrolyzing the DNA with formic acid and separating and measuring the resulting nucleotides by paper chromatography, and by other methods as well. It is worthy of mention that CsCl gradient method detects marked heterogeneity in the DNA of bacteria that possess episomes. Episomal DNA as a rule appears as a distinct satellite band in a gradient

*The formula listed for established species of Enterobacteriaceae and related organisms. For further discussion, see De Ley: J. Bact., *101*:738, 1970.

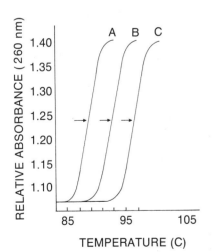

Figure 31–3. A typical melting point curve. In this diagram the mean melting temperature (*Tm*), as measured by increase in turbidity from 1.10 to about 1.24 (i.e., relative absorbency of light of wavelength 260 nm), is about 96C.

A and *B*, hypothetical cultures; *Tm* of culture *A* = 86.5C and *B* = 88.5C. *C* represents *Micrococcus* sp. L. A. 91. The midpoint of absorbancy increase (*arrows*) is the temperature (*Tm*) at which point approximately half of the hydrogen bonds between bases of DNA are broken. This temperature is directly related to GC content of DNA sample.

and this very observation eventually led to the discovery of competent DNA in mitochondria and chloroplasts. Such a discovery had a profound effect on theories regarding the evolution of eucaryotic organisms (endosymbiosis), and as a result the microorganismic nature of mitochondria and chloroplasts is being discussed.

Details of these complex and elegant methodologies are beyond the scope of this discussion (see Supplementary Reading list).

It is now clear that DNA base per cent composition is an extremely useful character which complements the phenotypic characterization of microorganisms. Organisms with similar phenotypes usually possess similar DNA base composition (within one or two per cent), as in certain groups of bacteria within the tribes Escherichieae, Klebsielleae, etc., or the genera *Pseudomonas* and *Rhizobium,* and many others. Similarly, it has also revealed that organisms once thought of as closely related (myxobacteria, 67 to 70 mole per cent GC, cytophagae, 33 to 47 mole per cent GC) have widely different DNA base composition values; closer observation of such groups indicates that these organisms are, indeed, not as closely related as had been supposed. Similarly, genera *Bacillus* (32 to 66), *Proteus* (36 to 53), *Neisseria* (40 to 52), and *Mycoplasma* (23 to 41 mole per cent GC) for many years have been considered as homogeneous groups, but DNA base composition studies show considerable heterogeneity among the species of the genera.

During the last decade, more than 600 bacterial species have been analyzed for DNA base composition, and it is now believed that bacteria possess wider ranges of G + C content than any other living organisms, with a low of 23 mole per cent GC to a high of 75 mole per cent GC. The information thus far accumulated indicates that the mole per cent GC ratio of all higher animals and plants ranges from 30 to 50, with vertebrates ranging from 35 to 45; vascular plants, 32 to 48 mole per cent GC. Eucaryotic microorganisms have somewhat wider ranges than either higher plants or animals; all algae, including blue-green algae, range from 35 to 71; protozoans range from 28 to 68; ciliates, from 22 to 35; all fungi, 28 to 62 mole per cent GC.

DNA Hybridization and Homology. Using a different procedure, estimates of genetic relatedness may be made by measuring the extent of **hybridization** or recombination between single strands of sheared (split) DNA helices of different organisms.

Hybridization is a term that was for long wholly foreign to bacteriology, since there was supposed to be no means of genetic recombination between bacteria. As explained in Chapter 19, it is now clear that there are several means of genetic recombination among bacteria.

Genetic Recombination. Nucleic acid analysis may permit us to recognize similarities between organisms but not necessarily the differences. However, occurrence of gene exchange through conjugation, transformation, or transduction may demonstrate that two organisms have certain similarities in genotypic characters, thereby providing useful information in classifying organisms. Mechanisms for genetic recombination appear to exist also among cells of rickettsias, between different viruses, and probably between certain cells of higher animals. It should be noted, however, that among bacteria such intercellular transfers of genetic material are limited to **genetically** related or **compatible** groups.

If recombination is to occur, there must be similarity in DNA base composition and also considerable agreement or homology between their DNA nucleotide sequences. For example, DNA base composition of certain strains of *Bacillus subtilis* and *B. stearothermophilus* is similar, but the DNA of *B. stearothermophilus* fails to transform the recipient, *B. subtilis.* Similarly, species within the genus *Neisseria* are transformed readily by interspecific DNA, except the saprophytic species, *N. catarrhalis.* While this species is able to transform its own DNA, it fails to serve as either recipient or donor to other species of *Neisseria.* This lack of transformation between species may be attributable to the wide differences in their DNA base composition. The mole per cent GC of *N. catarrhalis* is 40, whereas that of other *Neisseria* species is approximately 51 mole per cent GC. As a result, it has been proposed that *N. catarrhalis* be reclassified into the genus *Acinetobacter*, which has a DNA base composition ranging from 38 to 45 mole per cent GC.

Chemical Hybridization of Nucleic Acid. Hybridization is not a means of intercellular genetic transfer but a means of measuring the degree of similarity (homology) between the nucleotide base sequences in separated or sheared single strands of DNA from two different organisms. If the nucleotide sequences are absolutely identical, the two strands should theoretically combine or hybridize 100 per cent. Lesser degrees of agreement or homology would result in correspondingly lesser percentages of hybridization.

Degree of homology is determined by **reannealing** (mixing) the two kinds of single sheared strands of DNA and then measuring the extent to which they recombine (if at all). In the double-stranded DNA produced by this recombination or hybridization the two single strands involved may be differentiated if one is previously made radioactive by "feeding" the source organism with a pyrimidine base, say uracil, that has been artificially synthesized with ^{14}C.

For example, purified isotopic (with ^{14}C uracil) DNA is prepared from one organism and ordinary DNA is prepared from a closely related but different organism. The two kinds of DNA are mixed in a test tube containing pure 3 per cent agar and heated at a temperature above the Tm of either (105C for 10 minutes) to shear or break the H-bonds between the complementary strands in each helix of DNA. Four single strands result. The mixture is cooled to about 70C to permit whatever hybridization or recombination (**annealing**) might occur between the sheared strands. The mixture is then treated with an enzyme (phosphodiesterase) that hydrolyzes all the **single**-stranded (i.e., the **unhybridized** or unrecombined) DNA; the enzyme does *not* affect the newly hybridized or recombined **double** helices. Without going into the technical details, which (as is very evident) are complex and beyond the scope of the present description, it may be pointed out that more or less hybridization may occur and that the degree of hybridization is determined by measuring the amount of isotopic DNA of the one organism that combines (**per cent binding**) with ordinary DNA of the other organism.

Hybrids are formed only between single strands of DNA from closely related organisms having almost identical nucleotide sequences. For example, a single strand of DNA from *Pseudomonas fluorescens* (a common saprophyte of ditch and river water) will hybridize 100 per cent with isotopic single strands of DNA of its own species, and 83 per cent with strands from another species of the same genus; but *P. fluorescens* DNA will hybridize only 19 per cent with DNA of an organism of a different genus, even though of the same tribe or family, and not at all with a more distantly related and very different species. The degree of **genetic homology** (similarity of nucleotide sequences or **microhomology**) is high in one case, low or absent in the others (Table 31–5; also see Table 43–5). Strains that differ by more than about 2 per cent G + C composition cannot form DNA hybrids.

On the basis of similar genetic considerations, taxonomic groupings can also be made by means of transformation and transduction. These phenomena, like hybridization, are ge-

TABLE 31-5. THE DEGREE OF DNA HOMOLOGY OF SEVERAL FREE-LIVING NITROGEN-FIXING ORGANISMS WITH PSEUDOMONAS FLUORESCENS AND P. PUTIDA*

Organism	Strain Number	Hybridization with ^{14}C-DNA Fragments from P. fluorescens 488		Hybridization with ^{14}C-DNA Fragments from P. putida 520	
		LABELED DNA BOUND	%DNA BOUND RELATIVE TO P. FLUORESCENS	LABELED DNA BOUND	%DNA BOUND RELATIVE TO P. PUTIDA
Pseudomonas fluorescens	488	68.4	100	43.2	83
Pseudomonas putida	520	48.2	70.5	52.0	100
Derxia gummosa	III	8.2	12	8.3	16
Azotobacter vinelandii	C–1	27.4	40	26.0	50
Azotobacter beijerinckii	B–2	31.4	46	21.3	41
Azotobacter chroococcum	9125			25.5	49
Azomonas macrocytogenes	8200	30.8	45		
Azomonas macrocytogenes	9128	23.9	35	23.9	46
Azomonas macrocytogenes	9129	24.6	36	29.6	57
Azomonas insignis	9127	32.1	47	25.0	48
Beijerinckia derxii		19.2	28	15.1	29
Beijerinckia fluminensis				9.9	19
Azotococcus agilis	K	13.0	19	10.9	21
Azotococcus agilis	S	15.1	22	17.2	33
Azotococcus agilis	9	21.2	31	18.2	35
Pseudomonas azotogensis		2.7	4	3.6	7
agar		0.7	1	0.5	1

*From De Ley and Park: Ant. Leeuw., 32:6, 1966.

netic recombination in which DNA from one organism merges with that of another; only the means of transfer differ. Whatever the method of transfer, recombination occurs only between organisms of which the G + C base composition and nucleotide sequences are within certain, rather narrow, limits of similarity.

31.9
ANTIGENIC STRUCTURE

In attempts to introduce still greater accuracy into systems of classification, immunology has been extensively drawn upon. Striking antigenic differences between organisms formerly believed to be identical, and vice versa, have been found.

Immunological methods often reveal very subtle differences between individuals of a given species. For example, as previously described, all pneumococci (Streptococcus [Diplococcus] pneumoniae, cause of lobar pneumonia) were thought to be identical until it was found that the polysaccharides constituting their capsules fell into some 75 or more distinct antigenic groups because of differences in molecular structure of the capsular polysaccharides. The different sorts are called **types**: Type I, Type II, etc. Such types, based solely on anti-

genic differences, are often spoken of as **serotypes** because the antibodies used to determine the different types are found in serum. An analogous situation exists among hemolytic streptococci. We now speak of Group A or Group B hemolytic streptococci (Streptococcus pyogenes, cause of "strep throat"), depending on the chemical nature of their carbohydrate antigens. Some humans are of blood group A, some of group O, and so on.

31.10
BACTERIOPHAGE TYPES

As indicated in the section on bacteriophages (§16.3), within a species, and even within a clone or serotype, intraclonal or intraserotypic differentiations can be established on the basis of differing susceptibilities to type-specific bacteriophages. These hyperspecific **phage types** to a certain extent parallel serotypes, since phage type susceptibility appears to be closely related to specific cell wall antigen content and this, in turn, to DNA base sequence and content. However, phage typing often goes beyond serotyping. Certain strains of bacteria of great importance in industry and medicine are, at present, identifiable *only* by means of phage typing.

31.11
TOWARD UNIFIED BACTERIAL TAXONOMY

We have thus far discussed several approaches to more unified bacterial taxonomy. With the classical approach one closely examines morphological and cultural characteristics of organisms and groups them according to their similarities. After the establishment of grouping by primary characteristics, the DNA base composition of the group is analyzed and again the organisms are grouped according to the similarities in their mole per cent GC content. The organisms in a group or subgroup are subjected to as many biochemical and physiological tests as possible to determine, by Adansonian procedure, the similarity of their phenotypic characters. The groups of organisms so determined by these procedures may represent not only the concept of "species" but some implication of phylogenetic relationship with other groups.

The forthcoming 8th edition of *Bergey's Manual* includes many of these approaches and contains 19 different groups. The criterion for the first major division of the 19 groups is based on the mode of obtaining energy, that is, whether the organisms are phototrophic or chemotrophic. Chemotrophic organisms are further divided on the basis of their energy source. For example, organisms classified as chemolithotrophs are those which are capable of obtaining energy from oxidation of simple, inorganic elements or compounds, and chemoorganotrophs are those which are able to obtain energy from oxidation of simple or complex organic compounds. Both chemolithotrophic and chemoorganotrophic bacteria are further subdivided on the basis of primary characteristics such as mode of locomotion, cell morphology, Gram staining, and type of metabolism (aerobic or anaerobic). The main features of the new *Manual* are summarized in the following section:

Key to the 19 Parts

I. Phototrophic .... Part 1

Includes all the photosynthetic bacteria and contains one order, Rhodospirillales, and three families: Rhodospirillaceae (61 to 73 per cent GC), Chromatiaceae (46 to 67 per cent GC) and Chlorobiaceae (48 to 58 per cent GC). All are gram-negative. (See Chapter 32.)

II. Chemotrophic

    A. Chemolithotrophic

        1. Derive energy from the oxidation of nitrogen, sulfur, or iron compounds; do not produce methane from carbon dioxide.

            a. Cells glide .... Part 2

Includes members of the family Leucotrichaceae (*Leucothrix* and *Thiothrix*; 46 to 50 per cent GC) of the order Cytophagales. All are gram-negative. (See Chapter 34.)

          aa. Cells do not glide

             b. Cells ensheathed .... Part 3

Colorless, rod-shaped cells arranged in chains within a sheath; included are poorly defined genera such as *Crenothrix, Clonothrix, Phragmidiothrix*, etc. None of these organisms has been isolated in pure culture. (See Chapter 34.) Also see II.B.2.a., below.

            bb. Cells not ensheathed ....Part 12

Contains organisms oxidizing ammonia or nitrite, family Nitrobacteraceae (48 to 62 per cent GC), and includes such genera as *Nitrosomonas, Nitro-*

bacter, etc.; contains organisms metabolizing sulfur (*Thiobacillus* [58 to 68 per cent GC], *Sulfolobus,* etc.). Organisms depositing iron or manganese oxides are placed in the family Siderocapsaceae, which includes genera *Siderocapsa, Naumanniella, Ochrobium,* etc. (See Chapter 43.)

2. Do not oxidize nitrogen, sulfur, or iron compounds; produce methane from carbon dioxide...Part 13

Contains one family, Methanobacteriaceae. Members of the family have the ability to use CO_2 as electron acceptor and reduce it to methane (CH_4). All are strictly anaerobic, gram-negative. *Methanobacterium, Methanosarcina, Methanococcus* (38 to 52 per cent GC). (See Chapters 33 and 45.)

B. Chemoorganotrophic

1. Cells glide.. Part 2

Includes the order Myxobacterales (67 to 70 per cent GC), which consists of four families: Myxococcaceae, Archangiaceae, Cystobacteraceae, Polyangiaceae; and includes order Cytophagales, which includes four families: Cytophagaceae (33 to 48 per cent GC), Beggiatoaceae (*Vitreoscilla,* 44 to 45 per cent GC), Simonsiellaceae, and Leucotrichaceae (46 to 50 per cent GC). Also included are "Families and Genera of Uncertain Affiliation": Achromatiaceae and Pelonemataceae. All are gram-negative; life cycles involve aggregation, fruiting-body formation (*Myxobacter*) and microcyst formation (*Cytophaga*). (See Chapters 34 and 41.) Also see II.A.1.a., above.

2. Cells do not glide

a. Cells filamentous and ensheathed.. Part 3

Nonmotile, sheathed filaments or trichomes releasing polarly flagellated swarm cells, i.e., *Sphaerotilus, Leptothrix,* etc. (69 to 70 per cent GC). All are gram-negative. (See Chapter 34.)

aa. Cells not filamentous and ensheathed

b. Products of binary fission not equivalent (have appendages other than flagella and pili or reproduce by budding).................................... Part 4

The part includes budding and appendaged bacteria, *Hyphomicrobium, Caulobacter, Prosthecomicrobium,* etc. (60 to 69 per cent GC), *Pasteuria, Blastobacter,* etc., and those with nonliving bacterial appendages, *Gallionella* and *Nevskia,* etc. (See Chapter 34.)

bb. Not as above

c. Cells not rigidly bound

d. Cells spiral-shaped, have cell wall .... Part 5

Contains one order, Spirochaetales (34 to 46 per cent GC), and one family, Spirochaetaceae: *Spirochaeta, Treponema, Leptospira,* etc. Flexible, spiral, gram-negative organisms with one or more axial filaments. (See Chapter 41.)

dd. Cells not spiral-shaped, no cell wall...Part 19

One order, Mycoplasmatales (23 to 41 per cent GC). Contains two families. The species requiring cholesterol are grouped in Mycoplasmataceae and nonrequiring species into family Acholeplasmataceae. Also included are "Genera of Uncertain Affiliation." (See Chapter 42.)

cc. Cells rigidly bound

d. Gram-negative

e. Obligate intracellular parasites...Part 18

Contains two orders, Rickettsiales (30 to 46 per cent GC) and Chlamydiales. The family Rickettsiaceae is divided into three tribes: Rickettsieae, Ehrlichieae, and Wolbachieae. The order Chlamydiales contains one family, Chlamydiaceae, and one genus, *Chlamydia* (29 to 30 per cent GC; higher GC ratios, 40 to 45, have also been reported). The cells are rods, cocci to pleomorphic; contain both DNA and RNA. Some are filtrable; some are unable to generate their own ATP. (See Chapter 42.)

ee. Not as above

f. Curved rods... Part 6

Cells are spiral and curved but possess cell wall. The part includes Spirillaceae, which contains two genera, *Spirillum* (38 to 65 per cent GC) and *Campylobacter* (30 to 34 per cent GC), and "Genera of Uncertain Affiliation," which include parasites of bacteria, such *Bdellovibrio* (51 to 55 per cent GC), *Microcyclus*, etc. (See Chapter 33.)

ff. Not curved rods

g. Rods

h. Aerobic.. Part 7

Contains five families; all are considered strictly aerobic and are polarly flagellate. Some are nutritionally extremely versatile (Pseudomonadaceae, 58 to 69 per cent GC); some are able to fix free nitrogen nonsymbiotically (Azotobacteriaceae, 54 to 66 per cent GC); others fix nitrogen symbiotically (Rhizobiaceae, 58 to 65 per cent GC); some are able to oxidize methane, methyl alcohols, and related compounds (Methylomonadaceae, *Methylomonas*, 61 to 63 per cent GC); some require large amounts of salt for growth (*Halobacterium*, 55 to 68 per cent GC). Also included are "Genera of Uncertain Affiliation," which contains such genera as *Alcaligenes* (62 to 70 per cent GC), *Acetobacter* (55 to 65 per cent GC), *Brucella* (56 to 58 per cent GC), *Bordetella* (65 to 68 per cent GC), *Francisella* (34 to 36 per cent GC), etc. (See Chapter 33.)

    hh. Facultatively anaerobic.................................... Part 8

Contains two families, Enterobacteriaceae (36 to 63 per cent GC)—which includes *Escherichia* (50 to 52 per cent GC), *Salmonella* (50 to 54 per cent GC), *Shigella* (49 to 54 per cent GC), *Klebsiella* (52 to 59 per cent GC), *Serratia* (54 to 63 per cent GC), *Proteus* (36 to 53 per cent GC), *Yersinia* (46 to 48 per cent GC), *Erwinia* (50 to 57 per cent GC), etc.— and Vibrionaceae (43 to 63 per cent GC), composed of *Vibrio, Aeromonas, Photobacterium,* etc. Also "Genera of Uncertain Affiliation" such as *Zymomonas* (49 to 60 per cent GC), *Flavobacterium* (42 to 67 per cent GC), *Pasteurella* (35 to 40 per cent GC), etc. Organisms, if motile, peritrichously flagellate. Many are pathogenic to man and animals; some are plant pathogens. (See Chapter 35.)

    hhh. Anaerobic .... Part 9

Contains one family, Bacteroidaceae (32 to 43 per cent GC). All are strictly anaerobic rods to curved rods (vibrios); some are motile with peritrichous flagella (*Bacteroides,* 41 to 43 per cent GC). Also contains "Genera of Uncertain Affiliation," which includes such genera as *Desulfovibrio* (45 to 65 per cent GC), which are polarly flagellate and reduce sulfate to sulfides; *Butyrivibrio; Succinovibrio* (49 per cent GC); *Selenomonas* (54 to 60 per cent GC); etc. (See Chapter 35.)

  gg. Cocci or coccobacilli

    h. Aerobic..Parts 10, 7

Contains one family: Neisseriaceae (40 to 53 per cent GC). All are strict aerobes, some are arranged in pairs: *Neisseria* (47 to 53 per cent GC); some are coccoid: *Moraxella* and *Acinetobacter* (39 to 45 per cent GC). Also contains "Genera of Uncertain Affiliation": *Paracoccus* and *Lampropedia*. Two genera, *Methylococcus* (51 to 52 per cent GC) and *Halococcus* (57 to 66 per cent GC), from Part 7 are included in this section (gg.h.) The organisms are cocci to coccobacillus, rather than definite rods. *Methylococcus* is able to oxidize methane, meththanol, and related compounds. *Halococcus* requires a high concentration of salt for growth. (See Chapters 33 and 37.)

    hh. Anaerobic ....Part 11

All are strictly anaerobic, gram-negative cocci in pairs or clusters. Contains one family, Veillonellaceae, which contains three genera: *Veillonella* (36 to 38 per cent GC), *Acidaminococcus,* and *Megasphaera.* (See Chapter 37.)

dd. Gram-positive

e. Cocci

f. Endospores produced...Part 15

Family Bacillaceae. One genus, *Sporosarcina* (38 to 42 per cent GC) possesses spherical, motile vegetative cells. Able to decompose urea. (See Chapter 36.) For rod-shaped endosporeformers, see below (ee.f.).

ff. Endospores not produced...Part 14

Contains three families: Aerobic and/or facultatively anaerobic. Members include families Micrococcaceae (*Micrococcus*, 65 to 75 per cent GC); *Planococcus;* and *Staphylococcus* (30 to 40 per cent GC) and Streptococcaceae (31 to 44 per cent GC). Strictly anaerobic members are included in family Peptococcaceae (28 to 45 per cent GC). (See Chapters 37 and 38.)

ee. Rods or filaments

f. Endospores produced...Part 15

One family, Bacillaceae, which includes *Bacillus* (32 to 66 per cent GC), *Sporolactobacillus*, *Clostridium* (28 to 40 per cent GC), and *Desulfotomaculum* (42 to 50 per cent GC). Some are aerobic, while others are obligate anaerobes. Also included is "Genus of Uncertain Affiliation," *Oscillospira.* (See Chapter 36.)

ff. Endospores not produced

g. Straight rods...Part 16

Contains a family, Lactobacillaceae, with one genus, *Lactobacillus* (32 to 52 per cent GC). Aerobic to microaerophilic, may produce large amount of lactic acid. Also included are "Genera of Uncertain Affiliation" (*Listeria* (38 to 42 per cent GC), *Erysipelothrix*, and *Caryophanon*). Some are pathogenic to man and animals. *Caryophanon* forms trichomes, motile, peritrichously flagellate. (See Chapter 39.)

gg. Irregular rods (coryneform) or tend to form filaments or filamentous...Part 17

The part is divided into two major groups:

First, the coryneform group (Corynebacteriaceae) of bacteria, which includes genera *Corynebacterium* (48 to 75 per cent GC), *Arthrobacter* (60 to 64 per cent GC), *Cellulomonas*, and *Kurthia.* The *Corynebacterium* group is divided into three types: human and animal pathogens, plant pathogens, and nonpathogenic corynebacteria. The group also includes the family Propionibacteriaceae (58 to 70 per cent GC) and two "Genera of Uncertain Affiliation," *Brevibacterium* and *Microbacterium* (58 to 64 per cent GC).

The second major group includes the order Actinomycetales (43 to 74 per cent GC), which consists of eight families. Some tend to form filaments (Actinomycetaceae and Mycobacteriaceae, acid-fast); others are filamentous: Frankiaceae, Actinoplanaceae, Dermatophilaceae, Nocardiaceae, Streptomycetaceae, Micromonosporaceae. Some are strict aerobes; some produce antibiotics. (See Chapter 40.)

It should be noted that the above-described key to 19 parts is proposed for the 8th edition of *Bergey's Manual* and should be considered as provisional. Undoubtedly many of these classification schemes will be altered as more and more information accumulates for each and every organism thus far described. Perhaps bacteria may never be classified on the basis of their evolutionary affinity or phylogenetic hierarchical systems, but for the moment, the classification scheme used in the 8th edition of the *Manual* will serve excellently for determinative purposes which allow the ready assignment of newly isolated organisms to the correct genus.

CHAPTER 31
SUPPLEMENTARY READING

Bailie, N. E., Coles, E. H., and Weide, K. D.: Deoxyribonucleic acid characterization of a microorganism isolated from infectious thromboembolic meningoencephalomyelitis of cattle. Int. J. Sys. Bact., 23:231, 1973.

Johnson, J. L.: Use of nucleic-acid homologies in the taxonomy of anaerobic bacteria. Int. J. Sys. Bact., 23:308, 1973.

Jones, D., and Sneath, P. H. A.: Genetic transfer and bacterial taxonomy. Bact. Rev., 34:40, 1970.

Liston, J., Wiebe, W., and Colwell, R. R.: Quantitative approach to the study of bacterial species. J. Bact., 85:1061, 1963.

Lockhart, W. R.: Factors affecting reproducibility of numerical classifications. J. Bact., 94:826, 1967.

Lockhart, W. R., and Liston, J. (Eds.): Methods for Numerical Taxonomy. American Society for Microbiology, Bethesda, Md. 1970.

London, J., and Kline, K.: Aldolase of lactic acid bacteria: a case history in the use of an enzyme as an evolutionary marker. Bact. Rev., 37:453, 1973.

Marmur, J., Falkow, S., and Mandel, M.: New approaches to bacterial taxonomy. Ann. Rev. Microbiol., 17:329, 1963.

Skerman, V. B. D.: A Guide to the Identification of the Genera of Bacteria, 2nd ed. The Williams & Wilkins Co., Baltimore. 1967.

Sokol, R. R., and Sneath, P. H. A.: Principles of Numerical Taxonomy. W. H. Freeman & Co., San Francisco. 1963.

PHOTO-LITHOTROPHIC BACTERIA

The Order Rhodospirillales (Rhodobacteriineae) (Red, Purple, Green, and Carotenoid Pigmented Bacteria)

All photosynthetic bacteria are currently classified in the suborder Rhodobacteriineae of the order Pseudomonadales. The order Pseudomonadales (Gr. *pseudes* = imitating; *monas* = a monotrichous, flagellate protozoan) as arranged in the 1957 (7th) edition of *Bergey's Manual* is the largest and most heterogeneous of the 10 orders of the class Schizomycetes.

The order is divided into two suborders (Rhodobacteriineae and Pseudomonadineae) on the basis of presence or absence of photosynthetic pigments. The suborder Rhodobacteriineae includes Thiorhodaceae, Athiorhodaceae, and Chlorobacteriaceae, comprising 21 genera. Recently, more unified arrangements of the photosynthetic bacteria have been proposed for the 8th edition of *Bergey's Manual* (see Chapter 31).

The members of Rhodospirillales have red,

purple, brown, or green photosynthetic pigment and may or may not produce sulfur granules. The organisms in the order are alga-like, but their photosynthetic pigment can evolve S from H_2S, not O_2 from H_2O and lacks typical chlorophyll *a*, universally found in all oxygen-evolving photosynthetic plants. Based on this property, as mentioned elsewhere, bacteria might therefore be defined as procaryons without chlorophyll *a* (see Chapter 14). So far as is known, none of the Rhodospirillales is pathogenic for plants or animals.

The photosynthetic bacteria described in this chapter resemble the Cyanophyceae (blue-green algae) in that both utilize solar energy. The Rhodospirillales differ from the Cyanophyceae and, indeed, from all other photosynthetic cells in their method of using radiant energy. Before examining bacterial photosyn-

452

thesis let us review the usual type of photosynthesis that is familiar to all students of elementary biology.

32.1
PHOTOSYNTHESIS

Chemosynthetic and photosynthetic forms of metabolism, seemingly so different, are basically much alike. They differ only in the means by which energy is obtained. Once energy is made available in either type of cell, the energy is taken up in chemical bonds in the ADP-ATP system and in strongly reduced coenzymes, and is used in synthetic processes that are essentially alike in both forms of metabolism.

Photosynthesis is the biological process involved in the trapping of light energy by the chlorophyllous cells and its alteration into chemical energy, which is utilized for reduction and incorporation of CO_2 into cellular constituents (organic compounds). Hence the process in essence is the foundation of the pyramidal food chain that exists on earth.

Historical Development

Photosynthesis by green plants has been studied for many years, and present knowledge of the process is not the work of one individual. For example, the definitive demonstration of the central role of chlorophyll in the absorption of light is credited to the work of Engelman in 1880; Blackman's study (1905) showed that photosynthesis cannot be accelerated by increasing the intensity of illumination, and this discovery has been construed as evidence for a nonphotochemical reaction (dark reaction) as part of the photosynthetic process. Indeed, his findings led to the currently accepted view that CO_2 fixation occurs in the dark reaction and that the energy obtained from absorption of light by chlorophyll is made available for the reduction of the product of CO_2 fixation. The above observation that the photochemical reaction is basically a reduction process was augmented by Hill's discovery in 1937 that illumination of isolated chloroplasts is able to reduce ferric oxalate. The subsequent findings eventually led to "Hill's reaction" or the "chloroplast reaction" with the concomitant release of molecular oxygen

$$X + H_2O \xrightarrow[\text{chlorophyll}]{\text{light}} XH_2 + \tfrac{1}{2}\,O_2$$

Neither does the reaction involve CO_2 fixation, nor is carbohydrate involved.

From 1910 to 1940 the chemical structure of the chlorophyll molecule was elucidated by works of Willstätter and Fisher. Ruben (1941) provided definite evidence of "photolysis" of water in green plant photosynthesis. He showed, by the use of H_2O and CO_2 labeled with ^{18}O, that molecular oxygen, the product of photosynthesis, comes from H_2O, while the oxygen of CO_2 enters into the organic compounds.

Photosynthetic Pigments

As stated in the preceding paragraphs, photosynthetic activities are confined to organisms that possess chlorophyll. In the leaves of green plants, and even in procaryotic blue-green algae, chlorophylls exist in two forms, namely blue-green chlorophyll a ($C_{55}H_{72}N_4O_5Mg$) and yellow-green chlorophyll b ($C_{55}H_{70}N_4O_6Mg$). In both forms the basic molecular structure is porphyrin, as in the cytochromes, but chlorophylls contain an Mg atom instead of an Fe atom at the center of the porphyrin ring (Fig. 32–1). However, chlorophylls, unlike cytochromes, are not bound to protein but possess a lipid-soluble residue, **phytol** (a long-chain, optically active, aliphatic alcohol, $C_{20}H_{39}OH$), which is linked to the porphyrin by an ester bond (Fig. 32–1,E).

Chlorophyll b differs from chlorophyll a only in the presence of a CHO group instead of CH_3 at the boxed position in the upper right of Figure 32–1,C. Such a difference is readily detected by examining the spectral properties of the pigments. For example, ether extract of chlorophyll a shows strong absorption of red light at approximately 665 nm and blue light at 430 nm, whereas chlorophyll b absorbs red light at approximately 645 nm and blue light at 455 nm (Fig. 32–2).

There are other modified forms of chlorophyll, such as c, d, and e in certain algae, but chlorophylls b, c, d, and e transfer the radiant energy they absorb to chlorophyll a rather than use it directly. Such pigments are frequently termed **accessory pigments.** Nonchlorophyllous pigments, such as yellow **carotenoids** (and, in blue-green algae, phycocyanin), also serve as accessory pigments. The transfer of absorbed energy from chlorophyll a to chemically reactive sites in the cell and its fixation as chemical energy are discussed in the following sections.

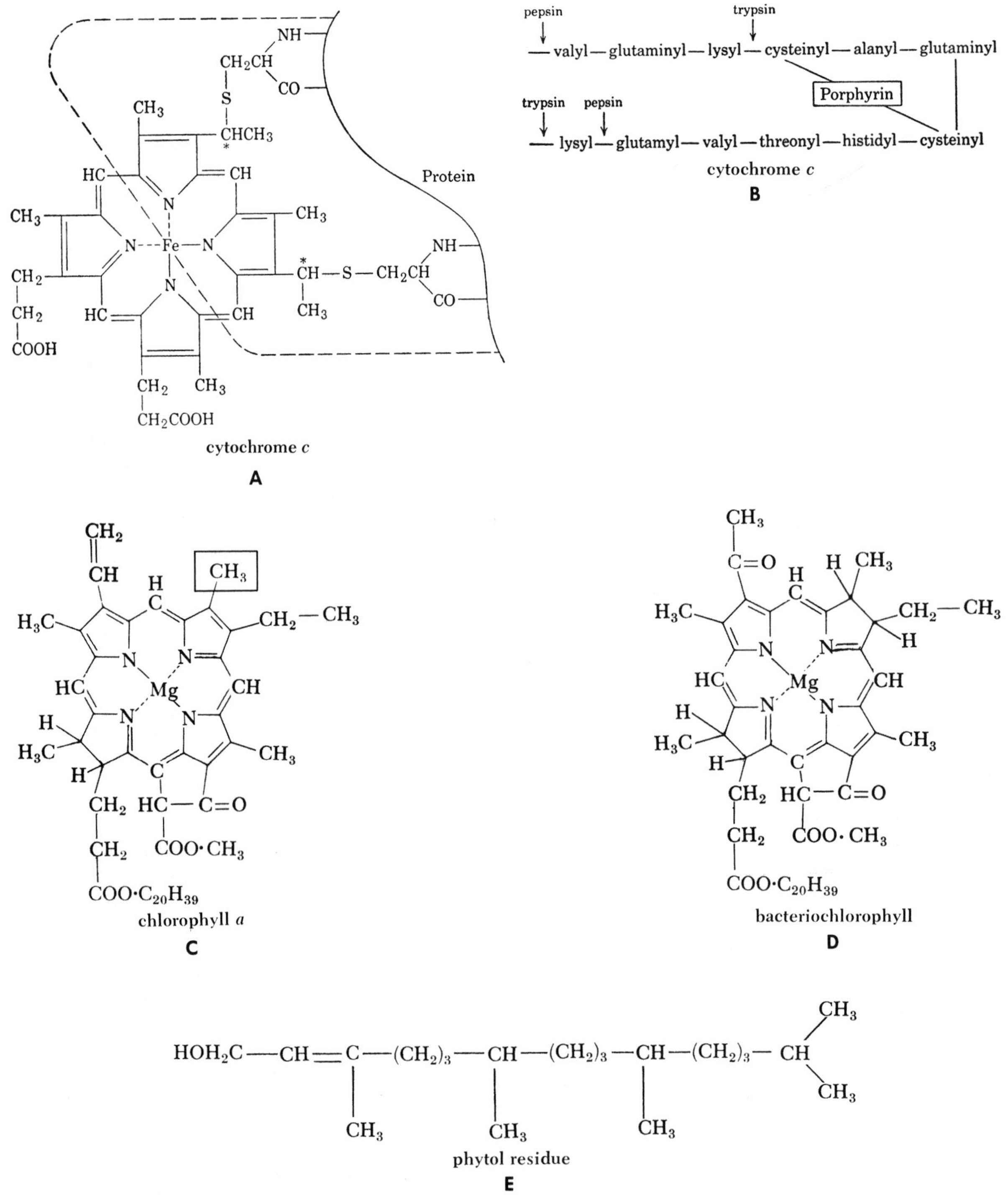

Figure 32–1. Chemical structure of cytochrome *c*, green plant chlorophyll *a*, and bacteriochlorophyll. *A*, Structure of cytochrome *c*. Note atom of iron chelated within porphyrin ring structure and bound to protein as shown in *B*; *B*, probable amino acid sequence in beef cytochrome *c*; *C*, green plant chlorophyll *a*; *D*, bacteriochlorophyll; *E*, phytol residue.

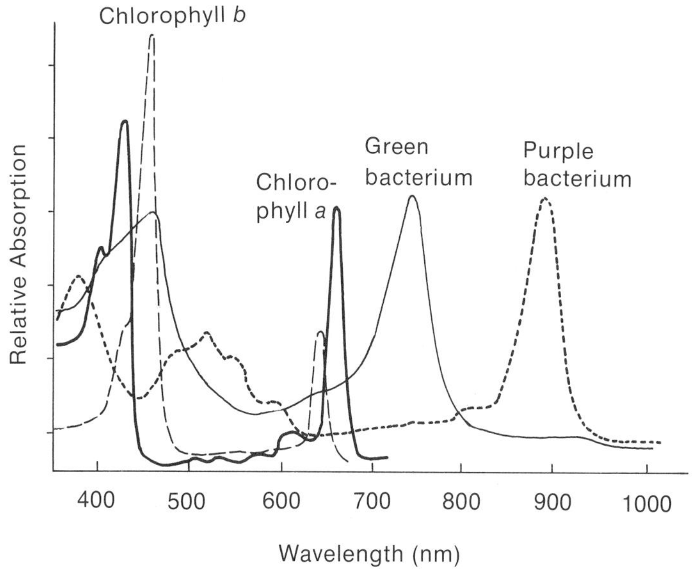

Figure 32–2. Absorption spectra of pigments of green bacterium and purple bacterium and purified chlorophylls *a* and *b* in ether. Note: there are two major peaks each for chlorophylls *a* and *b*: chlorophyll *a*, 430 and 665 nm; chlorophyll *b*, 460 and 640 nm.

Green Plant Photosynthesis

The process of photosynthesis is thus divided into two phases: (1) the **light phase,** in which radiant energy is trapped by chlorophyll and transferred into ATP with concomitant formation of NADH and NADPH; (2) the **dark phase,** in which the energy stored in ATP is used in the synthesis of organic compounds, as summarized in the well-known equation:

$$CO_2 + 2H_2O \xrightarrow[\text{energy}]{\text{light}} (CH_2O) + H_2O + O_2$$

or

$$6CO_2 + 12H_2O \xrightarrow[\text{(light and dark phases)}]{\text{photosynthesis}}$$

$$\underset{\text{glucose}}{C_6H_{12}O_6} + 6H_2O + 6O_2$$

The dark phase is independent of light and occurs alike in both photosynthetic and chemosynthetic cells when energy in the form of ATP is provided.

The Light Phase. When radiant energy is transferred to a photoreceptor (pigments such as chlorophyll *a*) in discrete units known as **quanta,** the energy is capable of altering the chemical nature of the photoreceptor. For example, the first effect of radiation of chlorophyll is to raise the energy level of certain of its electrons to such a pitch that one electron leaves each molecule of chlorophyll:

chlorophyll + quantum of light $\longrightarrow$ chl$^+$ + e$^-$

"Excited" electrons are potent reducing agents. Each is immediately taken up by an electron acceptor. The "extra" energy of the excited electrons is used in two processes: cyclic photophosphorylation and noncyclic photophosphorylation. Each process generates ATP. Noncyclic photophosphorylation also generates reducing potential in the form of NADPH (Fig. 32–3).

Cyclic Photophosphorylation. In this process electrons are the energy source, and the positively charged chlorophyll molecules serve as electron acceptors. As a quantum of energy strikes the chlorophyll, the "high-energy" electrons driven out of the chlorophyll are passed to ferredoxin and then to a photosynthetic electron-transfer system consisting in part of flavin nucleotides and cytochromes, analogous to the electron-transport systems in bio-oxidation. The electrons are transferred from coenzyme to coenzyme, yielding energy at each step. The energy of each electron generates two ATP from two ADP and two H$_3$PO$_4$. (H$_3$PO$_4$ is here indicated by P$_i$ = inorganic phosphoric acid.) The electron then returns to the chlorophyll, completing the cycle.

Noncyclic Photophosphorylation. In this reaction the electrons are taken up by ferredoxin and their energy is ultimately used for reduction of NADP in which the flow of electrons is in one direction only, hence noncyclic photophosphorylation. Ferredoxin is an iron-containing metalloprotein having enzyme-like properties and a uniquely high capacity for electrons; i.e., it thus becomes a potent reducing

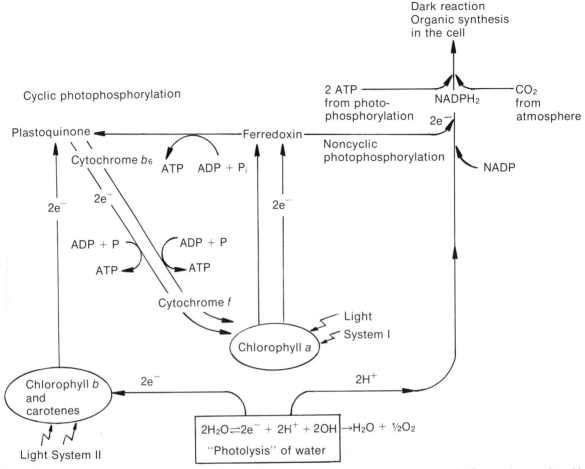

Figure 32–3. A simple diagram of cyclic and noncyclic photophosphorylation. Electrons lost by System I are replaced by electrons gained from photolysis of H_2O in System II. Electrons from ferredoxin may be passed from NADP to form NADPH or cycled back to plastoquinone, the cytochromes, and chlorophyll a to generate ATP.

agent. The electrons acquired by ferredoxin from irradiated chlorophyll are immediately transferred along two paths. Most of the energy is used to generate ATP from ADP and P_i. The remaining energy is used to "split" (oxidize) H_2O, ultimately yielding $2H^+$ and O^{2-}. It should be noted here that oxidation of water is unlike other bio-oxidations of inorganic compounds; for example, nitrifiers gain considerable energy from oxidation of ammonia to nitrate. On the other hand, the oxidation of water fails to yield energy but, on the contrary, requires a considerable amount of energy (endergonic reaction). The hydrogen ions are then reduced by electrons from ferredoxin and combine with NADP to form NADPH. Therefore, in noncyclic photophosphorylation positively charged chlorophyll must be reduced by electrons derived from an appropriate electron donor (H_2O, H_2S, etc.) to restore the energy level of chlorophyll as

in cyclic photophosphorylation. For example, the electrons left attached to the oxygen after water is split are immediately returned, at a very low energy level, to restore the chlorophyll that was robbed of its electrons by sunlight. The oxygen, minus its electrons, is given off as gaseous molecular oxygen. This terminates the light phase of photosynthesis.

Comparison of Green Plant and Bacterial Photosynthesis

Recent studies on the rate of photosynthesis by certain algae exposed to monochromatic light showed that green plants and algae contain two distinct types of photochemical reaction sites (light reaction centers I and II). Energy derived from reaction center I is enhanced by energy derived from reaction center II.

In light reaction system I, chlorophyll *a* is activated by the longer light waves. Energy thus derived from excited chlorophyll as described above is utilized in both photophosphorylation and/or is transferred to various intermediate reductants, including ferredoxin. It is then used in the reduction of NADP to NADPH (Fig. 32–4). However, investigations mentioned above show that energy produced by reaction system I alone (longer light waves with lesser energy) is insufficient to carry out *both* photophosphorylation *and* reduction of NADP with oxidation of water. The energy-supplementing accessory pigments of reaction system II, present in green plants, are needed for oxidation of water. It is believed that bacteria lack the energy-enhancing light reaction system II. The energy-harvesting system of bacteria is confined to light reaction system I and is insufficient to oxidize H_2O; hence there is no evolution of molecular oxygen in bacterial photosynthesis.

In reaction system II of green plants, chlorophylls are preferentially excited by blue light waves which are shorter and yield more energy than red light waves. Here electrons are derived from the oxidation of water. The electrons so derived are transferred via a photosynthetic electron-transporting system to the positively charged chlorophyll of light reaction system I. One molecule of ATP is formed by photophosphorylation.

Thus, in green plants the system I reaction carries out both cyclic photophosphorylation and reduction of NADP, while the system II reaction mediates noncyclic photophosphoryla-

tion. The reactions of the two systems result in synthesis of two ATP molecules, one NADPH, and molecular oxygen from oxidation of water.

32.2
BACTERIAL PHOTOSYNTHESIS

Photosynthetic bacteria do not contain any of the chlorophylls found in eucaryotic green plants or in the Cyanophyceae; the last contain only chlorophyll *a* plus accessory carotenoids and phycocyanin. Instead, the photosynthetic bacteria contain the structurally similar pigments **bacteriochlorophyll** *a*, *c*, and *d*. However, the structure of bacteriochlorophyll *b* is not yet elucidated. All of these differ somewhat from chlorophyll *a* of green plants (Fig. 32–1). Unlike eucaryotic photosynthetic cells, bacterial chlorophylls and the chlorophyll *a* of Cyanophyceae are not contained in complex, membrane-enclosed, highly differentiated, independently replicating chloroplasts, but in vesicular particles or disks or tubular structures in the cytoplasm called **chromatophores** (Gr. *chroma* = color; *phoros* = carrier) or, in the case of *Chlorobium* species, in ovoid or spindle-shaped bodies called "*Chlorobium* vesicles." In all groups the chromatophores appear to be associated with the cytoplasmic membrane and to contain oxidative and synthetic enzymes characteristic of cell membrane and photosynthesis. Procaryotic chromatophores are now seen to be very suggestive of the eucaryotic chloroplast. The structures differ in different families of bac-

Figure 32–4. Electron transporting system in green plant. There are two distinct types of photochemical reaction centers. System I activated by far-red light resembles photosynthesis of bacteria. By combined action of the two systems, NADP is reduced to NADPH by action of electrons derived from water and molecular oxygen is produced.

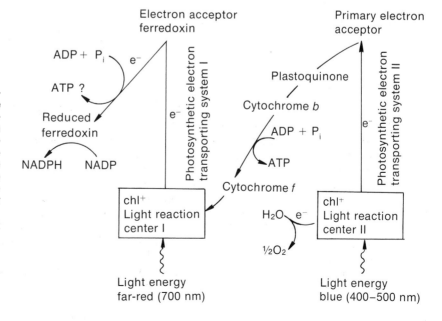

teria and are under active investigation. (See Figures 32–7, *A* to *D*, and 32–13, *A* and *B*.)

The chlorophylls (green chlorobium chlorophylls) of bacteria in Chlorobiaceae (Chlorobacteriaceae of the 7th edition of *Bergey's Manual*) absorb most light energy in the very far red spectra (720 to 750 nm). Invisible infrared energy (850 to 950 nm) is utilized by purple bacteriochlorophyll contained in organisms of the families Rhodospirillaceae and Chromatiaceae.

Light of these relatively long wavelengths may have been the very first to reach the primeval earth and permit development of primitive forms of photosynthesis. Since the energy of radiations is inversely proportional to their wavelength, it is evident that the chlorophylls of bacteria furnish less energy than the chlorophylls of green plants which use light of much shorter wavelengths. The energy trapped by bacteriochlorophylls is, in fact, insufficient to split H_2O, a process that requires high-level energy. Consequently, photosynthesis by bacteria differs from that of all eucaryotic plants and of Cyanophyceae. The photosynthetic bacteria cannot use water as a source of hydrogen to reduce CO_2 and therefore do not give off free oxygen as a result of photosynthesis. However, they can and do split H_2S, a chemically analogous substance $(2H^+ + S^{2-})$, or thiosulfates or certain organic compounds, as hydrogen donors in an equivalent manner; some can use molecular hydrogen directly. Those splitting H_2S evolve free S instead of O_2. These relationships have been summarized by Van Niel as analogous to eucaryotic plant photosynthesis in a generalized formula:

1. Green-plant photosynthesis:

$$CO_2 + 2H_2O \longrightarrow (CH_2O) + H_2O + O_2$$

2. Bacterial photosynthesis:

$$CO_2 + 2H_2A \longrightarrow (CH_2O) + H_2A + 2A$$

In reaction 2, H_2A serves as hydrogen donor to NADP at first, and NADPH reduces the CO_2 to sugar CH_2O. Hence, *A* may be sulfur or an organic residue, depending on species. Some sulfide-splitting species make themselves conspicuous by storing the sulfur as intracellular globules. Such species are called "sulfur bacteria."

The Dark Phase. The ATP, NADH, and NADPH resulting from the light phase reaction cooperate to introduce CO_2 into organic combination in the cell. This process can proceed in the dark as well as in the light. At least two processes appear to be involved. In one that has been most thoroughly studied by means of $^{14}CO_2$ by Calvin (Nobel Prize winner) and others, CO_2 is added directly to 1,5-ribulose diphosphate, a 5-carbon ketose. The 6-carbon compound thus formed is immediately split by the enzyme carboxydismutase into two molecules of 3-phosphoglyceric acid (PGA) (see Figure 32–5 and the Embden-Meyerhof scheme for anaerobic dissimilation of glucose, Chapter 8).

Using energy stored in ATP formed during the photophosphorylation reactions previously described, and apparently reversing the energy-yielding processes that occur during exothermic decomposition of glucose (Chapt. 8), the 3-phosphoglyceric acid molecules are reduced by NADPH and combined with phosphoric acid to form, first, triose phosphate, and from this, 6-carbon compounds such as fructose and glucose. These are polymerized as described elsewhere to produce sucrose, starch, or cellulose. PGA is also reduced to form fatty acids, which are the basis of lipids (Fig. 32–5).

In another, and possibly simultaneous, process CO_2 is reduced directly by ferredoxin in the presence of acetyl-CoA and an enzyme called **pyruvate synthetase**. The C_3 compound, pyruvic acid (pyruvate), is formed (see Chapter 8, discussion of the Krebs or citric acid cycle). As previously explained, pyruvate is a starting point for the synthesis of amino acids, the basis of proteins; fatty acids, the basis of lipids; and numerous other organic compounds that appear in the photosynthetic cell within a few seconds of exposure to light (Fig. 32–6). In Figure 32–6 the ferredoxin is reduced by activated electrons from irradiated chlorophyll. H_2O is split and ATP is formed, much as in Figure 32–3. CH_2O is merely a symbolic compound representing carbohydrates finally formed by the reduction of CO_2.

32.3
THE PHOTOSYNTHETIC BACTERIA

Order Rhodospirillales

As discussed in an earlier section, in the classification proposed in a forthcoming 8th edition of *Bergey's Manual* (see Chapter 31), all photosynthetic bacteria are placed in the new order Rhodospirillales (equivalent to suborder Rhodobacteriineae of the 7th edition of *Bergey's Manual*). Our discussion of photosynthetic bacteria will be based on the proposed classification.

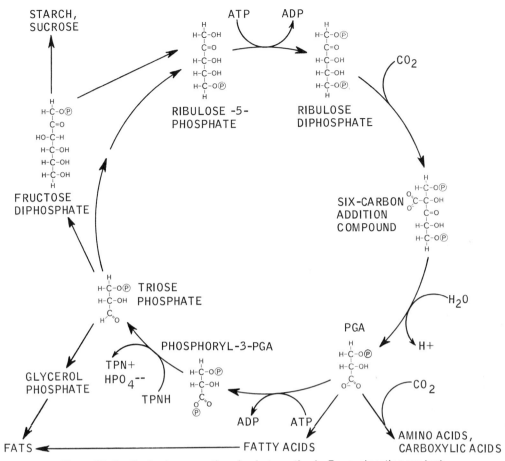

Figure 32–5. Dark-phase reactions in photosynthesis. For explanation see text.

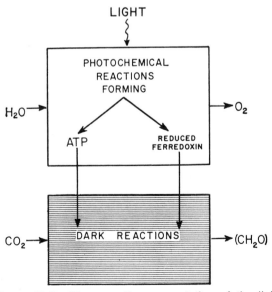

Figure 32–6. Diagrammatic representation of the light and dark reactions of photosynthesis in chloroplasts. For explanation see text.

Morphologically, the photosynthetic bacteria of the order Rhodospirillales are much like the bacteria that are their very common, numerous, and **nonphotosynthetic** cousins, the families Pseudomonadaceae and Spirillaceae. Rhodospirillales occur as unicellular, undifferentiated cocci, rods, vibrios, and spirilla (do not confuse with spirochete). Some Rhodospirillales secrete copious gelatinous material around themselves, in this respect resembling the Cyanophyceae.

All Rhodospirillales contain photosynthetic pigment. All can grow **anaerobically** (many are strict anaerobes) in the presence of light; some can also grow aerobically in light or dark. Obviously, photosynthesis cannot occur in the dark, so these aerobic species can grow in the dark like their nonphotosynthetic congeners; i.e., they are facultative phototrophs and facultative aerobes. Like all species of the family Pseudomonadaceae, all photosynthetic bacteria have **polar flagella** when motile. The flagella may occur at either or both poles, singly or in tufts (compare **peritrichously** flagellate Eubacteriales). All are gram-negative and none forms endospores or conidiospores.

As Van Niel states: "If, under certain conditions, one of the non-sulfur-storing photosynthetic bacteria should fail to produce its prominent pigment system, it would thereby become indistinguishable from a typical *Pseudomonas*, *Vibrio* or *Spirillum* species." It is interesting to note that nonphotosynthetic (pigment-free) mutants of photosynthetic bacteria have been discovered and that they are virtually indistinguishable from ordinary, nonphotosynthetic bacteria of the same morphological types. One may therefore imagine that photosynthetic bacteria evolved from corresponding nonphotosynthetic species by the acquisition (through mutations) of photosynthetic pigments. Backmutation (loss of pigment) has also been observed, and this produces the primitive, nonphotosynthetic forms again.

The Rhodospirillales are divided into three families—Rhodospirillaceae, Chromatiaceae, and Chlorobiaceae—each differentiated on the basis of cell morphology, major pigment system, photosynthetic electron donor, sulfur deposition, aerobiosis, and DNA base compositions (Table 32–1; Chromatophores, see Figure 32–7).

Family Rhodospirillaceae (Athiorhodaceae)

The family Rhodospirillaceae contains three genera: *Rhodospirillum*, *Rhodopseudomonas*, and *Rhodomicrobium*. DNA base composition of the members of the family ranges from 61 to 73 per cent GC. Most members of the family contain bacteriochlorophylls *a* and *b*, with maximum light absorption of the pigment around 820 and 1,025 nm, respectively. The family in the 7th edition of *Bergey's Manual* is termed Athiorhodaceae, and "athio" in the name implies lack of the ability to utilize H_2S as hydrogen donor in reduction of carbon dioxide. Therefore not only are the organisms unable to store sulfur intracellularly or excrete it extracellularly, but also the presence of H_2S is toxic to the organisms. For this reason these organisms are often called nonsulfur purple and brown bacteria. The bacteria are capable of utilizing hydrogen gas (H_2) as hydrogen donor in their photosynthesis:

$$CO_2 + 2H_2 \longrightarrow (CH_2O) + H_2O$$

However, some members of the family are not considered as strict photolithotrophs because they can grow aerobically in the dark, obtaining hydrogen and their carbon from organic carbon compounds, such as fatty acids.

Genus *Rhodospirillum*. As the name indicates, species of the genus are spiral in shape (Fig. 32–8), multiply by binary fission, and, if motile, are polarly flagellate.

Some species of the genus are able to grow photolithotrophically with molecular hydrogen, as well as photoorganotrophically. Growth occurs either anaerobically in light or under microaerophilic to aerobic conditions in the dark.

Most organisms with bacteriochlorophylls possess light absorption spectra of 375, 595, 805, 850, and 890 nm; with carotenoids, spectra of 465, 495, and 530 nm. However, it should be noted that these values vary somewhat from one species to another. Most species are widely distributed in aquatic environments, especially muddy freshwater ponds.

Genera *Rhodopseudomonas* and *Rhodomicrobium*. The members of the genus *Rhodopseudomonas* are generally curved rods to ovoid cells 1 to 1.5 μm wide and 2 to 5 μm long, and all motile strains possess polar flagella. Some strains reproduce by binary fission, others by budding (Fig. 32–9). However, morphology of budding cells is unlike that of organisms in the genus *Rhodomicrobium* in that they do not form tubes or filaments between the mother cells and buds (Fig. 32–10). The buds of *Rhodopseudomonas* are usually sessile on the mother cells and separate by constriction when the bud attains the size of the mother cell. As the buds separate, both mother and daughter cells bud at the newly formed pole.

The growth conditions of all species of both genera are similar to those of the Rhodospirillaceae described above. The photopigments consist of bacteriochlorophyll *a* and carotenoids of the spirilloxanthin series and others. All species except a few require complex organic substances for growth factors. Some species may accumulate intracellular granules of poly-β-hydroxybutyrate. DNA base composition of some members of the genera ranges from 62 to 67 mole per cent GC.

The organisms are readily isolated from stagnant bodies of water and mud.

Family Chromatiaceae (Thiorhodaceae)

The proposed family contains essentially the same genera as appear in the family Thiorhodaceae (*Bergey's Manual*, 1957 edition). The members of the family are motile or nonmotile rods, cocci, and pleomorphic; if motile they are polarly flagellate. The cells of some members (*Lamprocystis, Rhodothece,* and *Amoebobacter*) possess prominent gas vacuoles. As the name indicates, the organisms possess various colored pigments (purple, brown, red, orange, etc.) in addition to bacteriochlorophylls *a* and *b*; their absorption spectra are similar to those of Rhodospirillaceae. All are able to deposit sulfur granules intracellularly, except the 10th and last genus, *Ectothiorhodospira*, which deposits the sulfur extracellularly.

The Chromatiaceae (Thiorhodaceae) are generally called the sulfur purple bacteria.

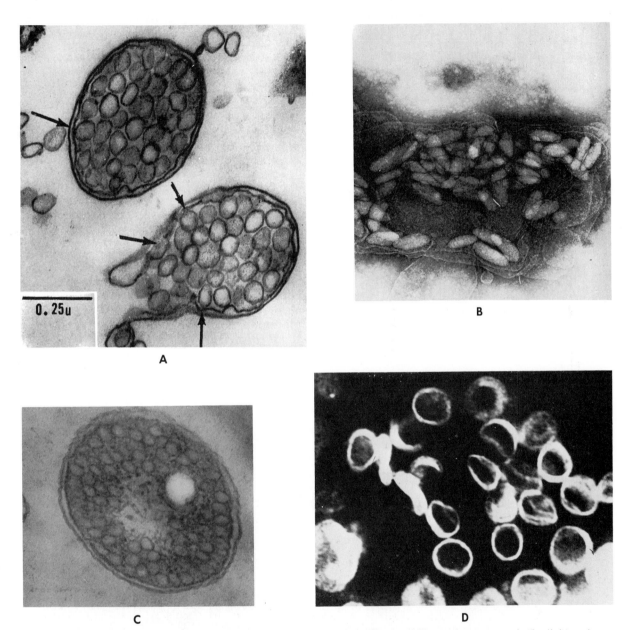

Figure 32–7. *A,* Transverse section of the photosynthetic bacterium *Rhodospirillum rubrum* grown in the light and ruptured by osmotic shock. The rupture of the cell wall is evident in one of the cells. The intracytoplasmic membranes (chromatophores) are still retained in the opened cell. Arrows indicate possible peripheral connections of the chromatophores to the peripheral membrane. (85,000×.)

B, Cells of the photosynthetic bacterium, *Chloropseudomonas ethylicum* (grown in the light and negatively stained with phosphotungstic acid) opened by mechanical (ballistic) disintegration. Note the distinct, ovoid vesicles. (88,400×.)

C, Thin section of fixed, anaerobically grown cells of *Rhodopseudomonas spheroides* poststained with lead hydroxide. Note chromatophores retained in the cell. (71,000×.) *D,* Purified chromatophores, negatively stained with phosphotungstate. (160,000×.) (*C* and *D,* Courtesy of Dr. Kenneth D. Gibson.)

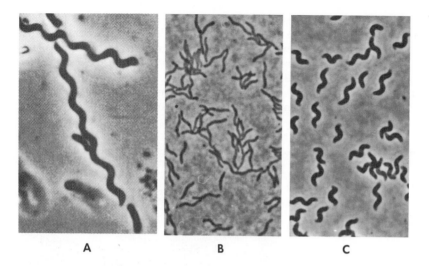

Figure 32–8. Cell morphology of three species of genus *Rhodospirillum.* *A,* Phase-contrast photomicrograph of motile *R. rubrum* S₁ cells grown anaerobically in the dark. (1,900×.) *B,* Phase-contrast photomicrograph of acetate-grown *R. tenue.* (1,700×.) *C,* Phase-contrast photomicrograph of acetate-grown *R. fulvum.* (1,700×.) (*B* and *C,* Courtesy of Dr. Norbert Pfennig.)

DNA base composition of the organisms ranges from 46 to 67 mole per cent GC. The reactions which follow are overall expressions of photosynthesis by these organisms:

$$2NADP + H_2S \longrightarrow 2NADPH + S$$
$$4NADPH + CO_2 \longrightarrow (CH_2O) + 4NADP + H_2O$$

In this process H_2S, thiosulfate, sulfite, and molecular hydrogen serve in place of water as an extraneous donor of hydrogen to reduce CO_2. The sulfur resulting from the oxidation of (re-

moval of H from) H_2S is stored as intracellular globules of elemental sulfur that is later oxidized as a source of energy to sulfuric acid (Fig. 32–11).

The members of the family are capable of utilizing thiosulfate as a hydrogen donor, and a few are able to use organic compounds as a hydrogen donor. Most species of the family do not require growth factors and grow photolithotrophically, but a few require vitamin B_{12}. All are considered as obligate anaerobes.

These photosynthetic bacteria are common

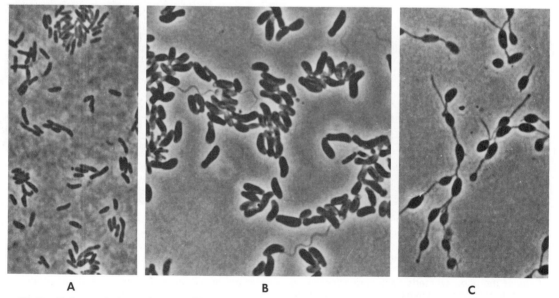

Figure 32–9. Cell morphology of genera *Rhodopseudomonas* and *Rhodomicrobium* species. All strains are succinate-grown and pictured by phase-contrast microscopy. (1,830×.) *A, Rhodopseudomonas palastris* strain 1850; *B, Rhodopseudomonas acidophila* type strain 7050; *C, Rhodomicrobium vannielii* strain 7255. (Courtesy of Dr. Norbert Pfennig.)

Figure 32–10. *Rhodopseudo-monas acidophila* strain 7050. Agar slide culture; pictures of the same cells taken at 30-min intervals to show the outgrowth of the sessile buds, the division by constriction, and the further outgrowth at the poles of the former division. (Phase-contrast, 1,500×.) (Courtesy of Dr. Norbert Pfennig; pictures by Heather M. Johnston.)

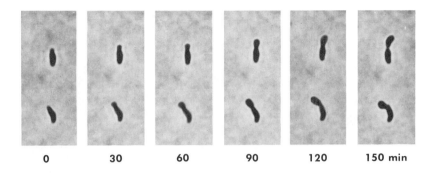

0   30   60   90   120   150 min

inhabitants of freshwater environments, and they thrive preferentially in anaerobic zones. The marine species, *Ectothiorhodospira mobilis,* is unlike all other species of the family in that it accumulates the sulfur extracellularly

and requires a salinity of 2 to 3 per cent NaCl. The organism is not identical with any of the species listed in *Bergey's Manual,* 1957 edition. It is spirilloid to vibrioid in shape, divides by binary fission, and is motile by means of a polar

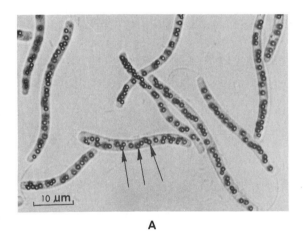

A

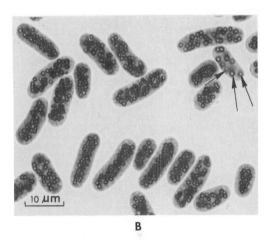

B

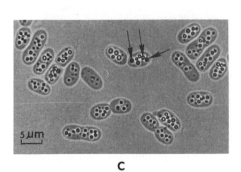

C

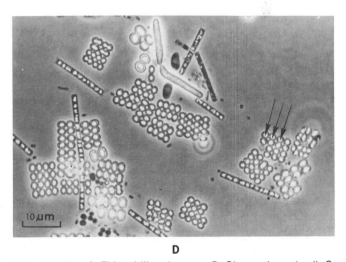

D

Figure 32–11. Chromatiaceae. The arrows indicate sulfur granules. *A, Thiospirillum jenense. B, Chromatium okenii. C, C. vinosum.* (*A, B,* and *C,* Courtesy of Dr. Norbert Pfennig.) *D, Thiopedia* sp. (the large cocci) in a mixed population in an enrichment culture from lake water. (*D,* Courtesy of Dr. Peter Hirsch.)

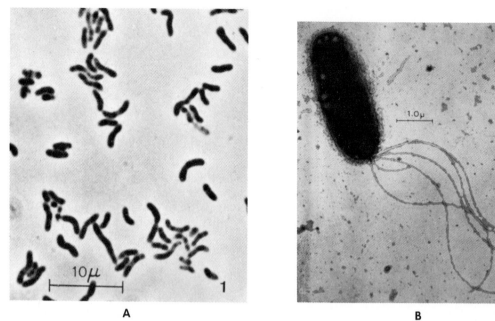

Figure 32–12. *A,* Phase-contrast photomicrograph of *Ectothiorhodospira mobilis* mounted in agar, showing dark inclusions. (2,000×.)

B, Flagellar tuft of *Ectothiorhodospira mobilis* strain 8113; young cell. Negative stain. (Courtesy of S. C. Holt, University of Massachusetts, Amherst.)

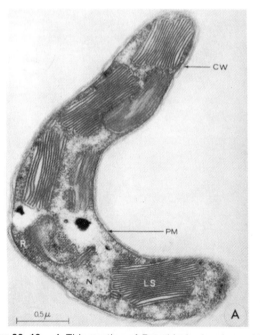

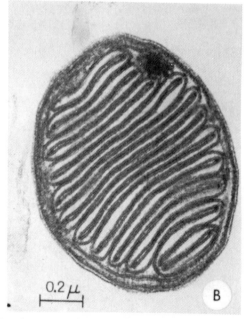

Figure 32–13. *A,* Thin section of *Ectothiorhodospira mobilis,* showing the general fine structure of the cell. A multilayered cell wall (*CW*), plasma membrane (*PM*), nucleoplasm (*N*), ribosomes (*R*), and six to seven lamellar stacks (*LS*) can be seen.

B, Thin section of *Ectothiorhodospira mobilis,* showing a common arrangement of the membranes with alternate and lateral infoldings. (58,500×.)

tuft of flagella (Fig. 32–12). It is obligately anaerobic and utilizes sulfide, sulfite, or thiosulfate, and a few organic compounds may be used as hydrogen donor, as with others in the family.

The photosynthetic apparatus of this organism appears to be more complex than that of other species of the family in that pigments are associated with a photosynthetic membrane laminated by alternate infolding (lamellar stacks, somewhat reminiscent of grana of chloroplasts) and originates from, and is attached to, the cell membrane (Fig. 32–13).

Although all bacterial photosynthetic membranes appear to be formed initially by invagination of the cell membrane and remain attached to it, in some species the invaginations may detach from the cell membrane to form vesicles. Both lamellar stack and vesicle types occur randomly throughout the cells of the organisms in the families Spirillaceae and Chromatiaceae, but only one species, *Ectothiorhodospira*, of the latter family possesses "lamellar stacks."

Family Chlorobiaceae (Chlorobacteriaceae)

The family Chlorobiaceae is provisionally created in place of Chlorobacteriaceae of *Bergey's Manual*, 1957 edition, and contains five genera: *Chlorobium, Prosthecochloris, Chloropseudomonas, Pelodictyon*, and *Clathrochloris*, hence the proposed family essentially retains

the genera designated in Chlorobacteriaceae. DNA base composition of the members of the family ranges from 48 to 58 mole per cent GC.

The members of the family are all green or yellowish green and fail to show the characteristic absorption band for bacteriochlorophyll at 590 nm. However, they contain green chlorobium chlorophylls which have characteristic infrared spectrum at approximately 750 nm.

The cells are motile or nonmotile rods, cocci, or pleomorphic, and some occur in strands or a loose irregular net. Motile cells possess polar flagella and the cells frequently contain gas vacuoles. Most members reproduce by binary fission and some may undergo ternary division. Some green bacteria, notably species of *Chlorochromatium*, are frequently found in association with larger, colorless bacteria; however, the nature of the ectosymbiotic relationship of these organisms with the others has not been clearly elucidated.

Like the Chromatiaceae, these bacteria (except *Clathrochloris*) metabolize H_2S and sulfites but deposit free sulfur extracellularly instead of intracellularly. The species of *Clathrochloris* deposit sulfur intracellularly. The S is often oxidized to sulfuric acid. They cannot grow without H_2S or other sulfides or sulfites. They are often called the **sulfur green bacteria.** They are strictly anaerobic, strictly photosynthetic and autotrophic.

Some of the distinguishing characteristics of photosynthetic bacteria are summarized in Table 32–1.

TABLE 32–1. SOME PROPERTIES OF FAMILIES IN THE ORDER RHODOSPIRILLALES (RHODOBACTERIINEAE)

| Family | DNA Base Composition (Mole Per Cent G + C) | Cell Shape | Reproduction | Pigments | | Photosynthetic H Donor | Sulfur Deposition | Photolithotrophic Growth | Organotrophic Growth |
				CAROTENOIDS	BACTERIO-CHLOROPHYLLS				
Rhodospirillaceae (Athiorhodaceae)	61–73	Spiral, rods, oval cells with filament; motile, polar flagella	Binary or budding	Red, purple, brown	B-chl *a* B-chl *b*	H_2, organic compounds; H_2S toxic	No	Photosynthetic growth anaerobic	In dark aerobic to microaerophilic
Chromatiaceae (Thiorhodaceae)	46–67	Rods, cocci, vibrioid short spiral; if motile, polar flagella	Binary	Red, purple	B-chl *a* B-chl *b*	H_2S, thiosulfate, H_2; some able to use organic compounds; some produce gas vacuoles	Most intracellularly; one species extracellularly (*E. mobilis*)	Obligate anaerobes; some halophilic	No growth in dark
Chlorobiaceae (Chlorobacteriaceae)	48–58	Rods, cocci, pleomorphic; if motile, polar flagella	Binary; some show ternary	Green	B-chl *c* B-chl *d*	H_2S, H_2; thiosulfate by some strains	Most extracellularly except *Clathrochloris*	Obligate anaerobe	No growth in dark

Habitat and Functions of Rhodospirillales

Species of photosynthetic bacteria are commonly found in sunlit fresh and marine waters, stagnant lakes, polluted bays, sea water, and brackish ditch water, where H_2S is plentifully produced by organic decomposition and where the O-R potential is very low because of chemosynthetic metabolism. Most photosynthetic bacteria are able to grow in aquatic environments where surface water is populated with algae. In such environments the photosynthetic bacteria usually flourish below the algal layer, where anoxic conditions exist. Though such an environment frequently lacks visible light, the ability of the photosynthetic bacteria to use far-red and infrared portions of the solar spectrum enables them to carry out photosynthesis. Thus, growth of these two organisms in terms of solar spectrum is complementary.

The Rhodospirillales are mainly lithotrophic. They are of no special industrial, agricultural, or medical interest but are of great importance as scavengers and deodorizers, since they utilize H_2S in photosynthesis. Photosynthetic bacteria are capable of fixing nitrogen under anaerobic conditions in the light; however, the quantities of reduced nitrogen formed are uncertain. They are much used by students of the enormously important process of photosynthesis.

While photosynthetic bacteria cannot be credited with contributing to the oxygen supply of our atmosphere, total photosynthetic activity by both land plants and marine algae will contribute to production of oxygen necessary for aerobic organisms. Without their photosynthesis, there might be very little oxygen in our atmosphere.

CHAPTER 32
SUPPLEMENTARY READING

Gest, H., San Pietro, A., and Vernon, I. (Eds.): Bacterial Photosynthesis. Antioch Press, Yellow Springs, Ohio. 1963.

McFadden, B. A.: Autotrophic CO_2 assimilation and the evolution of ribulose diphosphate carboxylase. Bact. Rev., 37:289, 1973.

Pfennig, N.: Photosynthetic bacteria. Ann. Rev. Microbiol. 21:285, 1967.

Pfennig, N.: *Rhodospirillum tenue*, a new species of the purple nonsulfur bacteria. J. Bact., 99:619, 1969.

Uffen, R. L., and Wolfe, R. S.: Anaerobic growth of purple nonsulfur bacteria under dark conditions. J. Bact., *104*:462, 1970.

Vernon, L. P.: Photochemical and electron transport reactions of bacterial photosynthesis. Bact. Rev., 32:243, 1968.

Walsby, A. E.: Structure and function of gas vacuoles. Bact. Rev., *36*:1, 1972.

GRAM-NEGATIVE, CHEMOORGANOTROPHIC BACTERIA:

Pseudomonadaceae, Vibrionaceae, and Spirillaceae

The suborder Pseudomonadineae of the order Pseudomonadales (*Bergey's Manual,* 1957) includes Nitrobacteraceae, Methanomonadaceae, Thiobacteriaceae, Pseudomonadaceae, Caulobacteraceae, Siderocapsaceae, and Spirillaceae. The members of these families lack photosynthetic pigment.

Members of the families Nitrobacteraceae, Thiobacteriaceae, and Siderocapsaceae are considered chemolithotrophic bacteria (autotrophic) in that they obtain their energy from oxidizing simple inorganic compounds and utilize CO_2 as their sole source of carbon. The remaining members of the suborder are considered chemoorganotrophs (heterotrophs). Many of the species are either parasites or opportunistic pathogens of man and animals. A few species of the genus *Pseudomonas* are unlike all other bacteria in that they are pathogenic for both plants and animals, including man, while members of the genus *Xanthomonas* are pathogenic for a variety of plants but not for animals.

Most organisms in the suborder are like those in the order Eubacteriales (*Bergey's Manual,* 1957) in that they are generally simple, undifferentiated, rod-shaped (a few are curved or spiral) cells with diameters of 0.7 μm and lengths of 8 μm, though some species approach 15 μm in diameter and lengths up to 100 μm. None forms spores and all are gram-negative. Motile forms have single or several **polar** flagella that may be mono- or lophotrichate. Unlike the Enterobacteriaceae of the order Eubacteriales, members of Pseudomonadineae are never peritrichate. A few representative characteristics of the families of the suborder Pseudomonadineae and order Eubacteriales are summarized in Table 33–1.

In this chapter we shall briefly describe a few representative members of the Pseudomonadaceae and related families.

Gram-Negative, Polarly Flagellate Chemoorganotrophic Bacteria

Bacteria conforming to the above description are frequently termed "aerobic pseudomonads" and have been placed in the family

TABLE 33-1. PROPERTIES OF AEROBIC PSEUDOMONADS AND SOME OTHER AEROBIC ORGANISMS (EUBACTERIALES)

Family	Gram Stain	DNA Base Composition (Mole Per Cent G + C)	Spores	Metabolism	Form	Flagella-tion (If Motile)	Relation to Oxygen	Pigment Formation
Pseudomonadaceae	Neg.	58–69	—	Oxidative	Straight rod	Polar	Aerobic	Common
Methylomonadaceae	Neg.	51–63	—	Oxidative	Rod and coccus	Polar	Aerobic	None
Spirillaceae	Neg.	30–65	—	Oxidative	Curved or spiral; rigid	Polar	Facultative	Few species
Enterobacteriaceae	Neg.	36–63	—	Fermentative	Straight rod	Peritrichous	Facultative	Few species
Achromobacteraceae	Neg.	55–65	—	Oxidative	Straight rod	Peritrichous	Facultative	Many species
Brucellaceae	Neg.	35–68	—	Oxidative	Straight rod	Peritrichous	Facultative	None
Azotobacteraceae	Neg.	53–66	—	Oxidative; fix N	Rods; pleo-morphic	Peritrichous	Strict aerobes	Brown
Rhizobiaceae	Neg.	58–65	—	Some fix N symbioti-cally in root nodules	Rods; pleo-morphic	Peritrichous	Aerobic	One species violet

° Based on representative organisms which appear in the 7th edition of *Bergey's Manual* and extreme low and extreme high DNA base composition certain organisms not included. Families Achromobacteraceae and Brucellaceae are eliminated in the 8th edition of the *Manual*.

Pseudomonadaceae. They are widely distributed in nature in soil, sewage, and water. Some aerobic pseudomonads are considered among the most metabolically versatile organisms known. Thus, these bacteria are responsible for the decomposition and mineralization of many insoluble and soluble compounds derived from the remains of plant and animal material in our ecosystem.

33.1
FAMILY PSEUDOMONADACEAE

The family Pseudomonadaceae (*Bergey's Manual*, 7th edition) is a very large group composed of some 150 species of the genus *Pseudomonas*, 75 of the genus *Xanthomonas*, and about 35 other species in 10 genera. Many species in the family are of marine origin, and many are very tolerant of high (3 to 30 per cent) salt concentrations, i.e., **halotolerant.** Some are parasitic on fish. Many species of the family produce blue, greenish, and yellow pigments that are water-soluble and diffusible and sometimes fluorescent. Some species form nondiffusible, carotenoid pigments.

Some others that exhibit remarkable properties are:

1. *Acetomonas* (polar flagellate *Acetobacter*—the "vinegar bacteria"), which oxidize alcohols to vinegar (Chapt. 49).
2. *Aeromonas*, parasitic on fish, which produce acid and gas from glucose and are often confused with *Enterobacter* (*Aerobacter*) species of Enterobacteriaceae although they do not produce gas from lactose.

3. *Azotomonas*, agriculturally important soil species that can synthesize their own proteins using the nitrogen of the air (nitrogen fixation).
4. *Photobacterium*, marine species that are associated with luminescence of light-organs of deep-sea fish and with luminosity of many other fish and decaying organic matter.
5. *Zymomonas*, which produce ethyl alcohol like yeasts.
6. *Halobacterium*, obligate halophils restricted to a life in nutrient solutions containing at least 12 per cent of salt, commonly found in the Dead Sea, Great Salt Lake (27 per cent salinity), salted fish, and pickling brines (30 per cent salinity) (Chapt. 44).

Several of these microbiological curiosities are discussed elsewhere in this book.

The aerobic pseudomonads have been investigated intensively with a view to additional knowledge and an improved taxonomic arrangement. An exhaustive study by Stanier et al. included *Pseudomonas, Xanthomonas, Methanomonas, Hydrogenomonas, Comamonas, Vibrio, Acetomonas, Alginomonas, Cellulomonas,* and *Cellvibrio.*

The groups eliminated from traditional aerobic pseudomonads are: Caulobacteraceae, Siderocapsaceae, Chlamydobacterales, *Aeromonas, Zymomonas, Photobacterium,* some *Vibrio, Desulfovibrio,* Rhodobacteriineae, Nitrobacteraceae, Thiobacteriaceae, and chemoheterotrophic spirilla. Hence it is clear from the foregoing discussion that the biological spectrum of the suborder Pseudomonadineae is wide and exceedingly complex, and the possible relationships between the constituent subgroups not only are arbitrary but are far from clear. A radical rearrangement of this order has been proposed, and some of the taxonomic treatments adopted in the following sections are provisional, as the scheme is based on the suggested outline of the forthcoming 8th edition of *Bergey's Manual*. Indeed, these classifications will be further modified as properties such as composition of DNA, nucleic acid hybridization, and episomal transfer, in addition to more unified stable biochemical and physiological properties of each organism, accumulate.

In the forthcoming edition of *Bergey's Manual*, the family Pseudomonadaceae contains four genera: *Pseudomonas, Xanthomonas, Zoogloea,* and *Gluconobacter*. The organisms are grouped together in Part VII, "Gram-Negative, Strictly Aerobic Rods and Cocci," with four other families: Azotobacteraceae (Chapt. 43), Rhizobiaceae (Chapt. 43), Methylomonadaceae, and Halobacteriaceae (Chapt. 44).

The organisms of the family Pseudomonadaceae as a group are characterized as aerobic, using O_2 as terminal electron acceptor; however, some may use NO_2^- or NO_3^- anaerobically. They are neither fermentative nor photosynthetic but are chemoorganotrophs utilizing many complex organic compounds, but a few species use H_2 (*Pseudomonas saccharophila*) or CH_4 (*P. methanica*; the organism is an "obligate methylotroph" and may be more appropriately called *Methylomonas methanica*) as energy source (see § 33.2). Many are pigmented and some are extremely metabolically versatile scavengers of the ecosystem while others are animal and plant pathogens. As a group, DNA base composition ranges from 58 to 69 mole per cent GC. The cells of these diverse genera are either straight or curved, and when motile they possess polar flagella, either mono- or lophotrichate.

The carbohydrate metabolism of aerobic pseudomonads and related organisms is worthy of detailed discussion, being unlike that of gram-positive bacteria (with a few exceptions). The pseudomonads metabolize glucose via the Entner-Doudoroff pathway (Fig. 33–1). The end product of this pathway is essentially the same as in the Embden-Meyerhof pathway (glycolytic cycle) since the triose phosphate is converted to pyruvate, thus yielding 2 moles of pyruvate from a mole of glucose, as in the glycolytic cycle.

However, the intermediary reactions are entirely different from those of the Embden-Meyerhof pathway described in Chapter 8. The 6-phosphogluconate formed by phosphorylation of gluconic acid by ATP is converted to 2-keto-3-deoxy-6-phosphogluconate by an enzyme (dehydrase) discovered by Entner and Doudoroff. This reaction is considered to be of major importance in aerobic metabolism of *Pseudomonas fluorescens* and requires the presence of Fe^{2+} and glutathione or cysteine. A second key enzyme that differentiates the two pathways is an aldolase which splits 2-keto-3-deoxy-6-phosphogluconate into pyruvic acid and glyceraldehyde-3-phosphate, which is later converted to pyruvic acid. Therefore, classification of aerobic pseudomonads not only includes source of energy (H_2, CH_4, and carbohydrates) but also includes complex nutritional and physiological characteristics in addition to DNA base composition and their DNA homology.

Genus Pseudomonas

Members of the genus *Pseudomonas* are among the most common and widely distributed

Figure 33–1. Abbreviated Entner-Doudoroff pathway of glucose metabolism.

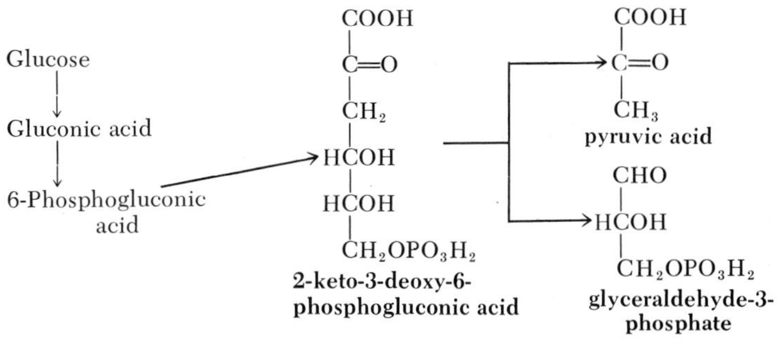

Glucose
↓
Gluconic acid
↓
6-Phosphogluconic acid
→
2-keto-3-deoxy-6-phosphogluconic acid

COOH | C=O | CH_2 | HCOH | HCOH | CH_2OPO_3H_2

pyruvic acid: COOH | C=O | CH_3

glyceraldehyde-3-phosphate: CHO | HCOH | CH_2OPO_3H_2

TABLE 33-2. DIFFERENTIAL CHARACTERS OF AEROBIC PSEUDOMONADS *

Group	DNA Base Composition (Mole Per Cent G + C)	PHB† as Reserve Food	Soluble Fluorescent	Phenazines Pigment	Glucose	Starch	Aromatic Compounds	Oxidase	Number of Polar Flagella
FLUORESCENT	58–68	–	+	+(–)‡	+	–	+	+	1 or >1
P. aeruginosa	64–68								
P. fluorescens	58–63								
P. putida	61–63								
P. syringae	58–60								
ACIDOVORANS	60–67	+	–	–	–	–	+	+	>1
P. acidovorans	66–67								
P. testosteroni	60–62								
ALCALIGENES	62–68	–	–	–	–(+)§	–	–	+	1 or >1
P. alcaligenes	66–67								
P. pseudoalcaligenes	62–63								
P. multivorans	66–68								
P. stuzeri	64–65								
P. maltophila	65–67								
PSEUDOMALLEI	67–69	+	–	v‖	+	+	+	+	>1 or nonmotile
P. pseudomallei	67–69								
P. mallei	67–69								

*Adapted and modified from Stanier, Palleroni, and Doudoroff: J. Gen. Microbiol., 43:159, 1966.
†PHB = poly-β-hydroxybutyric acid.
‡A few are negative.
§A few are positive.
‖Variable.

bacteria. As previously indicated, they are enzymically active, metabolizing as many as 100 different organic compounds for both carbon and energy sources. A single strain of *P. aeruginosa* can make use of a wide variety of proteins, fats, carbohydrates, and other organic compounds, including such aromatic compounds as phenol and naphthalene, and hydrocarbons. Thus they are excellent and ubiquitous scavengers. They are principally aerobic; a few are facultative. They are found in soil, fresh waters, and ocean waters and decomposing fish and other organic matter, including sewage. These organisms may be subdivided into four major subgroups on the basis of physiological characteristics as shown in Table 33-2. However, some investigators have divided the genus into five distinct groups on the basis of rRNA homologies (Palleroni et al., 1973).

Pseudomonas fluorescens. One of the commonest species in the genus is *Pseudomonas fluorescens*. Most strains of *Pseudomonas fluorescens* produce a greenish yellow, water-soluble, fluorescent pigment in cultures (Color Plate IV, A). *P. fluorescens* grows best at about 22C to 25C (room temperature). There are numerous similar species, widely distributed and extremely important as causing decomposition of petroleum fuels, asphalt, and many foods and other substances (Fig. 33-2).

Pseudomonas aeruginosa. *P. aeruginosa* is the type species of the genus. In addition to

the greenish yellow pigment characteristic of many *Pseudomonas*, it produces several water-soluble pigments: a turquoise-blue pigment, pyocyanin (Gr. *pyo* = pus; *cyanin* = blue), which may be extracted from broth culture with chloroform; fluorescent green pigment, pyoverdin; fluorescent reddish or ruby pigment, pyorubin (Color Plate IV, A); and water-soluble brown pigment, pyomelanin, which is produced in tyrosine- or phenylalanine-containing media.

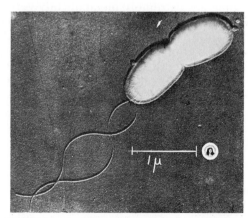

Figure 33-2. Electron micrograph of *Pseudomonas fluorescens,* a dividing cell. Note the polar flagella. The cytoplasm appears to have shrunken away from the cell wall, probably owing to drying. Note the line indicating 1 μ at this magnification and the small portrait of Leeuwenhoek, the symbol of the American Society for Microbiology. (Courtesy of The American Society for Microbiology and Drs. Houwink and van Iterson.)

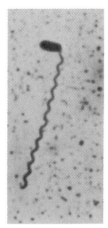

Figure 33–3. *Pseudomonas aeruginosa* KM 256 showing a long, polar flagellum, typical for this strain. Leifson flagella stain. (1,500×.)

Potentially pathogenic strains of *P. aeruginosa* grow poorly at 30C but well at 37 to 42C, whereas the plant pathogens (the fluorescent group) prefer the lower temperature.

Most strains of *P. aeruginosa* possess a single polar flagellum (Fig. 33–3), and some strains when cultivated on agar media typically produce phage-like plaques (autoplaques) of self-lysis (Fig. 33–4).

P. aeruginosa is commonly a saprophyte but is also an opportunistic pathogen able to establish infection and to invade when the natural resistance of an individual is severely lowered. The organism has a higher growth temperature range (41 to 42C) than saprophytic pseudomonads (20 to 30C) and frequently causes hospital infections associated with middle ear or urinary tract infections and wounds or ulcers that have not healed promptly. It is frequently found in patients with severe burns. The organism is naturally resistant to several antibiotics commonly used, and thus chemotherapy is very difficult.

Some outbreaks of diarrhea in adults and especially among newborn children are said to be caused by this organism. *P. aeruginosa* is also able to cause a leaf-rot disease in tobacco and lettuce and a fatal disease in poultry.

Several members of the genus, i.e., *P. aeruginosa* and *P. pseudomallei*, are denitrifiers, hence able to grow anaerobically by using nitrate as a terminal electron acceptor. As denitrifiers, the organisms play an important role in the nitrogen cycle in nature (Chapt. 43). The organisms may be isolated on agar media containing K_2HPO_4, $MgSO_4$, K_2SO_4, asparagine, proline, and alcohol.

Pseudomonas pseudomallei. *P. pseudo-*

mallei is pathogenic for man and the causal agent of melioidosis, a frequently fatal disease of man and animals. The disease is most common in Southeast Asia and is highly malignant, glanders-like, and relatively unaffected by antimicrobial therapy. The organism is commonly found in tropical soil, and the disease is generally acquired traumatically through contamination of wounds with soil or polluted water.

Pseudomonas mallei. The organism *P. mallei*, a true parasite of horses, mules, and asses, is the causal agent of glanders. The organism was originally known as *Actinobacillus mallei* because of its permanent immotility. However, recent studies indicate the organism is indistinguishable from *P. pseudomallei* in phenotypic and genetic properties such as identical DNA base composition and very high in vitro hybridization of their DNA (Table 33–2). Thus the organisms are now placed in the genus *Pseudomonas* and the only difference between the organisms is that *P. mallei* inhabits only specific animals and is nonmotile.

Glanders is characterized by nodules and eventual necrosis of the nasal mucous membrane, lymph glands, and skin. Human glanders is usually fatal, but occurrence of the disease is rare. Some species of the genus are harmless inhabitants of various animals, while some may

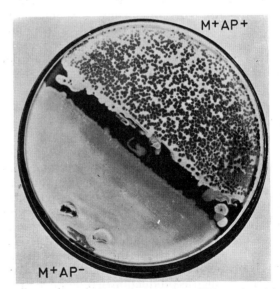

Figure 33–4. Phage-like autoplaques (AP) produced by spontaneous autolysis of a mucoid (M) strain of *Pseudomonas aeruginosa* (strain M⁺AP⁺) on agar medium in a Petri dish (upper portion of plate). In the lower portion is seen the growth of a variant that fails to produce AP (strain M⁺AP⁻). Some nonmucoid strains may also produce plaques (M⁻AP⁺). The phenomenon may be related to phage but the exact role of phage, if any, remains to be demonstrated.

cause human actinomycosis-like disease, either alone or in mixture with an assortment of gram-positive bacilli.

Pseudomonas saccharophila. P. saccharophila (DNA base composition, 69 mole per cent GC) is able to grow chemolithotrophically as well as chemoorganotrophically. It can obtain energy from oxidizing hydrogen (H_2). It is, thus, frequently considered as a facultative chemolithotroph, as are other hydrogen bacteria (*Hydrogenomonas*), which can also grow on a wide range of simple organic compounds. Other bacteria which can grow chemolithotrophically on H_2 include certain mycobacteria, actinomycetes, and some micrococci; not all of these bacteria are obligate chemolithotrophs, but they are rather common saprophytes ubiquitously found in soil.

P. saccharophila can grow in an atmosphere of $H_2 + O_2 + CO_2$ and gains energy from oxidative phosphorylation, with H_2 serving as electron donor to reduce NAD. This electron transfer reaction is activated by the enzyme hydrogenase. The NADH thus formed is oxidized via the electron transporting system with O_2 as terminal electron acceptor. Carbon dioxide incorporation is accomplished via the Calvin cycle (see Chapter 32, Figure 32–5).

It is believed that many facultative chemolithotrophic hydrogen-oxidizing bacteria are endowed with enzymic machinery similar to photolithotrophic organisms. Ecologically, hydrogen-oxidizing bacteria play a counter-role to certain members of the Enterobacteriaceae and clostridia which produce H_2 from the fermentation of carbohydrates.

Genus Xanthomonas

The genus *Xanthomonas* is placed as the second genus of the proposed family Pseudomonadaceae in the forthcoming 8th edition of *Bergey's Manual*. DNA base composition of the genus *Xanthomonas* ranges from 62 to 68 mole per cent GC, which is similar to that of genus *Pseudomonas* (58 to 69 mole per cent GC). The genus is separated from other members of the family on the basis of special physiological characteristics, such as production of protopectinase, pathogenicity to plants, production of yellow pigment, and others.

There are about 75 species of *Xanthomonas* (*Bergey's Manual*, 1957 edition), many of which are extremely important as pathogens of plants. They cause great losses to farmers, fruit-growers, horticulturists, lumber industries, and others

dependent on the plant kingdom. In most respects *Xanthomonas* is very much like *Pseudomonas;* several *Xanthomonas* were formerly classified as *Pseudomonas* and vice versa. The name *Xanthomonas* is derived from the Greek word, *xanthus*, for yellow, since *Xanthomonas* characteristically produces a **water-insoluble** yellow pigment on suitable media.

Xanthomonas and Plant Disease. Several species of *Xanthomonas* (and also several of molds, *Pseudomonas* and *Erwinia*) produce protopectin-hydrolyzing enzymes. Protopectins are a group of gum-like polysaccharides constituting the intracellular cement substance that binds plant cells together. When the protopectin is destroyed by protopectinase the plant cells fall together in a slimy mass; the plant is said to have a **soft rot** and often dies (Fig. 33–5). **Protopectinases** are, so far as plant pathology is concerned, analogous to the hyaluronidases produced by several bacterial pathogens of mammals: *Staphylococcus aureus, Streptococcus pyogenes,* and *Clostridium perfringens.* Hyaluronidases decompose **hyaluronic acid,** which is an intercellular cement substance of animals, analogous in function to protopectins of plants. Microorganisms that produce hyaluronidase are thus better able to penetrate between the tissue cells of animals and invade the host.

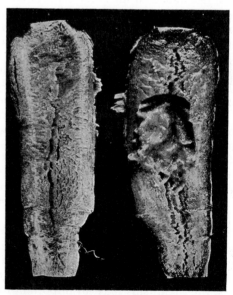

Figure 33–5. *Right,* soft rot of carrot due to inoculation with *Erwinia carotovora* and incubation for three days at 23C. The inoculum was taken from a rotting raw potato. The carrot was first washed, soaked in disinfectant and then cut with a cold, sterile knife. The uninoculated part (*left*) remained sound. (Courtesy of Erwin F. Smith.)

Figure 33–6. Bacterial wilt disease in a tomato plant due to *Pseudomonas solanacearum.* Note the wilted stems and leaves. (Courtesy of U.S. Department of Agriculture, Bureau of Plant Industry.)

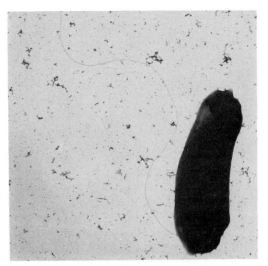

Figure 33–7. Electron micrograph showing a cell of *Zoogloea* species with subterminally inserted flagellum. (23,000×.) (Photograph by K. T. Crabtree.)

Numerous other species of *Xanthomonas,* which may or may not produce protopectinase, are the cause of destructive leaf spots, wilts (Fig. 33–6), and other diseases in many species of plants. Examples of some bacterial diseases of plants and the causative organisms are listed in Table 33–3. Some plant pathogens produce soluble toxic substances also.

Genus Zoogloea

Members of this genus were initially isolated from sewage and other polluted aquatic environments. The organisms, whose cells are motile by means of a single polar flagellum (Fig. 33–7), play an important role in the oxidation and stabilization of domestic sewage (Chapt. 45). They have a unique growth characteristic in liquid media in that they grow in

aggregates to form visible flocs (Fig. 33–8), and this is the basis of activated sludge in sewage treatment. The cells in flocs (submerged colonies) usually contain conspicuous deposits of poly-β-hydroxybutyrate, and these cells are frequently embedded in a gelatinous slime in which other bacteria may become entrapped. The organisms are nutritionally versatile, using many soluble carbohydrates and amino acids as both carbon and energy sources.

TABLE 33–3. SOME BACTERIAL PLANT PATHOGENS

Bacterium	Disease Produced
Erwinia tracheiphila	Wilt of cucumbers and melons
Erwinia carotovora	Soft rot of many species of plants
Pseudomonas solanacearum	Rot of potato and tomato
Pseudomonas savastanoi	Blight of olives
Xanthomonas hyacinthi	Yellow rot of hyacinth bulbs
Xanthomonas pruni	Blight of plums and peaches

Figure 33–8. Star-shaped "flocs" of *Z. ramigera* in broth culture. (23×.) (Photograph by K. T. Crabtree.)

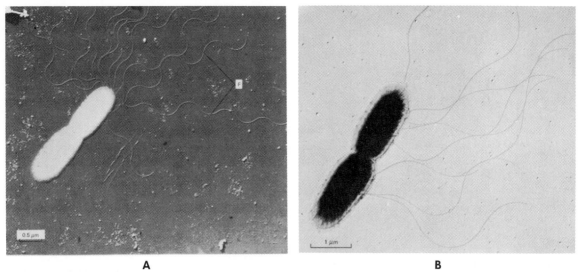

A　　　　　　　　　　　　　　　　**B**

Figure 33-9. Mode of flagellar insertion by two types of acetic acid bacteria. A, Acetomonas (Gluconobacter ?) with a sparse tuft of polar flagella. B, Acetobacter sp., a pseudomonad with polar and lateral flagella. (Courtesy of A. L. Houwink.)

Genus Gluconobacter

Gluconobacter is the fourth genus of the family Pseudomonadaceae (in the proposed 8th edition of *Bergey's Manual*). The genus contains only one species, *G. suboxydans*, formerly known as *Acetobacter suboxydans* because of its ability to form a relatively large amount of acetic acid from ethanol. Both for this reason and because of its being polarly flagellate (as are pseudomonads), the organism is sometimes referred to as genus *Acetomonas*. However, this organism, unlike other aerobic pseudomonads, fails to oxidize alcohol completely to CO_2 and possesses relatively high tolerance of acid conditions. *Gluconobacter* is differentiated from *Acetobacter* by ability to oxidize acetic acid, peritrichous flagellation (Fig. 33-9), and enzymes for the tricarboxylic acid (TCA) cycle (ability to oxidize acetate). However, the DNA base composition of *Acetobacter* and *Gluconobacter* are similar (Table 33-4), which indicates fairly close relationship between these two organisms. Systematic position of these organisms will be clarified by further investigation of DNA homology.

As its name implies, *Gluconobacter* "underoxidizes" higher alcohols and sugars; for example, glucose is first oxidized to gluconic acid, then to 5-ketogluconic acid; galactose to galactonic acid, mannitol to fructose, glycerol to dihydroxyacetone, and sorbitol to sorbose, which is used in the manufacture of ascorbic acid (vitamin C). The organism is apparently widely distributed in nature, such as on the surface of fresh and souring fruits and in wine and vinegar.

Genus Acetobacter

The genus *Acetobacter* shares many properties with *Gluconobacter*. However, the genus is placed in "Genera of Uncertain Affiliation" under Part VII, "Gram-Negative, Strictly Aerobic Rods" (proposed 8th edition of *Bergey's Manual*), along with genera *Alcaligenes*, *Thermus*, *Brucella*, *Bordetella*, and *Francisella*. The last three genera are discussed in Chapter 35.

One striking property of certain species of *Acetobacter* is their ability to synthesize cellu-

TABLE 33-4. COMPARISON OF GLUCONOBACTER WITH ACETIC ACID OXIDIZING ACETOBACTER

Genus	Mode of Flagellar Insertion	TCA Enzymes	DNA Base Composition (Mole Per Cent G + C)
Gluconobacter	Polar	−	58-63
Acetobacter	Peritrichous	+	55-65

lose from glucose or other sugars. An extracellular slime layer of cellulose in loose mesh or mat, mixed with bacterial cells, is formed in standing cultures. It is the thick, tough, leathery pellicle frequently referred to as "mother of vinegar" (Fig. 49–10).

Some species of *Acetobacter*, unlike *Gluconobacter*, are able to oxidize alcohols to acetic acid, which is further oxidized to CO_2. These species are frequently referred to as "overoxidizers," and are peritrichously flagellate. Industrial exploitation of acetic acid bacteria is presented in Chapter 49.

33.2
FAMILY METHYLOMONADACEAE

These organisms are related to the aerobic pseudomonads, but, based on ability to oxidize methane or methanol as sole source of carbon and energy, they are placed in the family Methylomonadaceae. The family was created to avoid nomenclatural problems which existed in the past. For example, the prefix "methano" has been used to denote both bacteria that oxidize methane and bacteria that produce methane. Thus, in the proposed nomenclature, the prefix "methylo" denotes bacteria which oxidize methane or methanol for energy and growth, as in Methylomonadaceae. The bacteria which have ability to produce methane are designated by use of the prefix "methano," as in *Methanobacterium*, *Methanococcus*, and *Methanosarcina*.

In the proposed classification (8th edition of *Bergey's Manual*), the family is placed with the family Pseudomonadaceae in Part VII, Gram-Negative, Strictly Aerobic Rods." The family contains two genera, *Methylomonas* and *Methylococcus*, and all the species in these genera may be considered as "obligate methylotrophs," since they obtain their energy exclusively from organic, one-carbon compounds such as methane, methanol, methylamine, and formate.

Genera Methylomonas and Methylococcus

The cells of *Methylomonas* are gram-negative rods, and the DNA base compositions range from 61 to 63 mole per cent GC. Cells of *Methylococcus* are aerobic, nonmotile, gram-negative, encapsulated cocci with distinctly diplococcoid arrangement, the DNA base compositions ranging from 51 to 52 mole per cent GC.

The process of cellular carbon synthesis by these obligate methylotrophs is somewhat reminiscent of lithotrophic organisms, although in a strict sense the organisms are chemoorganotrophs. For example, methane is first oxidized to methanol, then to formaldehyde, and finally to formate by a nonspecific primary alcohol dehydrogenase in the presence of ammonium ions. The carbon from either the formaldehyde or the formate unit combines with a pentose phosphate to form a rather special hexose known as allulose phosphate, which is the starting material for cellular synthesis. In addition, these organisms are capable of oxidizing certain primary alcohols and short-chain alkanes, but such oxidation is unable to support cellular growth (nongrowth oxidation).

Organisms able to use methane and methanol in addition to other organic compounds ("facultative methylotrophs") for both energy and growth are more widely distributed in nature than the Methylomonadaceae; for example, prosthecated, budding bacteria, such as species of genus *Hyphomicrobium* and vibrio bacterium *V. extorquens*. However, carbon incorporation by these organisms is unlike that of *Methylomonas* or *Methylococcus* in that methane is first oxidized to methanol, then to formaldehyde or formate, and these in turn react with the amino acid glycine to form serine. This, in turn, is converted to pyruvic acid, the "hub" of biochemical metabolism.

Both obligate and facultative aerobic methylotrophs are widely distributed in nature, especially in soil and aquatic environments, and they play a counterrole to the strictly anaerobic, chemolithotrophic methane-producers (*Methanobacterium*, *Methanococcus*, and *Methanosarcina*) in an ecosystem.

33.3
FAMILY VIBRIONACEAE

This family is newly created to accommodate organisms resembling members of either Pseudomonadaceae or Enterobacteriaceae. It is provisionally placed (in the forthcoming edition of *Bergey's Manual*) under Part VIII, "Gram-Negative, Facultatively Anaerobic Rods," which includes the family Enterobacteriaceae. The family Vibrionaceae includes five genera: *Vibrio*, *Aeromonas*, *Plesiomonas*, *Photobacterium*, and *Lucibacterium*. Most of these organisms were formerly placed in the family

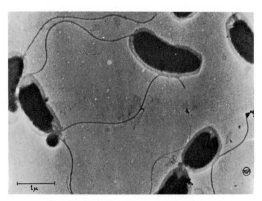

Figure 33–10. Electron micrograph of typical vibrios (*Vibrio cholerae*). Note curved cells and polar flagella. Compare with Figure 33–2. (Courtesy of American Society for Microbiology and Drs. Anderson and Pollitzer.)

Pseudomonadaceae and a few in the family Spirillaceae (*Bergey's Manual*, 1957 edition).

Although each genus of the family Vibrionaceae has a relatively distinct habitat and properties, as a family they are described as gram-negative, facultatively anaerobic, polarly flagellate rods with their DNA base composition ranging from 43 to 63 mole per cent GC.

Genus Vibrio

The name of the genus *Vibrio* is derived from the exceptionally rapid to-and-fro motility of these organisms (L. **vibrare** = to vibrate). Vibrios are short rods that are typically curved like a comma; each is a portion of a spiral turn (Fig. 33–10). This was the reason the genus has been included in the family Spirillaceae (*Bergey's Manual*, 1957 edition). Some are aerobes, while many are facultative anaerobes.

As a group they are considered more closely related to the Enterobacteriaceae than to Pseudomonadaceae or Spirillaceae on the basis of glucose fermentation. Many of the organisms carry out mixed acid fermentation without producing gas and share R factor (antibiotic resistance) with enteric bacilli. DNA base composition of species in the genus ranges from 40 to 49 mole per cent GC. Organisms with a base ratio less than this range (*V. fetus* and *V. bubulus*, DNA base composition 29 to 36 mole per cent GC) may be transferred to the genus *Campylobacter* (30 to 34 mole per cent GC). Similarly, those species of *Vibrio* having a DNA base composition ranging from 50 to 64 mole per cent GC (*V. percolans, V. cyclosites, V. neocistes*, and *V. alcaligenes*) may be transferred to

the genus *Comamonas* (DNA base composition, 62 to 64 mole per cent GC).

These proposed changes apparently stem from the failure of many vibrios to produce curved rods after in vitro cultivation or even after returning the culture to natural habitat. Hence, recent studies emphasize that the genus *Vibrio* be confined to those organisms which are motile by means of monopolar flagella, which ferment carbohydrates without gas production, and whose DNA base composition ranges from 40 to 49 mole per cent GC.

The cells of vibrios often remain attached end to end after fission, forming long spirals. The length of the individual vibrios seldom exceeds 10 μm; their diameter, 1.0 to 1.5 μm. They are principally saprophytic, and are worldwide in distribution in polluted rivers and lakes. Recently, a considerable number of marine vibrios have been isolated. Most grow readily in peptone media.

Many so-called "species" greatly resemble each other, having similar habitats and physiological properties and sharing several somatic and flagellar antigens. Many are named for the place where they were isolated: *V. danubicus, V. gindha,* and *V. massauah*. Some vibrios, like *V. metchnikovi* (named for its discoverer, Metchnikoff), are pathogenic for guinea pigs and pigeons, which is not true of *V. cholerae* unless special methods of injection are used.

V. cholerae. This organism, discovered in the feces of cholera patients by Koch in 1884, is the cause of Asian or classical cholera. It has been given various names such as *Spirillum cholerae asiaticae, S. cholerae,* and *V. comma* in the 7th edition of *Bergey's Manual*. However, the name *V. comma* has not gained acceptance. *V. cholerae* resembles other vibrios in living for long periods, possibly multiplying, in polluted river and lake waters. Various strains of *V. cholerae* are recognized: the Inaba, Ogawa, and Hikojima serotypes and others. These differ mainly in protein and carbohydrate structure (Linton groups) and in somatic antigen structure. The H or flagellar antigens are not highly specific; the O antigens are. Three major type-specific O antigens are recognized: A, B, C. The Inaba group is AC; the Ogawa group, AB; the third type, Hikojima AB(C), has been described but appears to be closely related to the Ogawa serotype. For identification of pathogenic vibrios, single O-group I antiserum containing Inaba and Ogawa factors is required. Identification is accomplished by use of commercially prepared polyvalent antiserum.

El Tor Biotype. The El Tor vibrios (isolated from pilgrims at a quarantine station at El Tor on the Sinai peninsula) are another cause of cholera or "paracholera," which has traditionally been endemic in Malaysia. In the 1960's and 1970's, cholera due to the El Tor vibrio (biotype) has spread pandemically throughout Southeast Asia, eventually encompassing India and Pakistan (formerly endemic foci of the classic version), the Middle East, and Africa. With rapid air transportation, the disease, though sporadically, has also been imported into such countries as Australia, the United Kingdom, and France, but Italy actually suffered an epidemic outbreak (1973). Further, the disease has occurred in the Western Hemisphere (Texas, 1973), which is a nonendemic area. El Tor vibrios (variously referred to as *Vibrio eltor* or *V. cholerae* biotype *eltor*) and classic *V. cholerae* and the diseases they cause are very similar. Some confusion has existed as to the distinction between El Tor vibrios and the classic *V. cholerae*. Most El Tor vibrios are strongly hemolytic when broth cultures are mixed with a suspension of sheep or goat erythrocytes and incubated for four to six hours (the Greig test). *V. cholerae* is not hemolytic (Greig-negative) under these conditions. However, nonhemolytic variants of El Tor vibrios are not uncommon. Other tests may be used (Table 33–5). In tests of hemolysis in blood-agar plates the El Tor vibrio usually produces hemolysis and no **hemodigestion;** *V. cholerae* causes destruction of the erythrocytes but by a process of hemodigestion rather than lysis.

Most World Health Organization officials now consider the El Tor vibrio to be a biotype. In addition to the above, *V. eltor* can be differentiated from *V. cholerae* by its resistance to polymyxin B; i.e., strains of *V. eltor* show no zone of inhibition around a disc, whereas *V. cholerae* is distinctly inhibited by polymyxin B. Similarly, most strains of *V. cholerae* are sensitive to the lytic action of cholera phage IV, whereas *V. eltor* is resistant (see Table 33–5).

Isolation. Most intestinal vibrios (and especially *V. cholerae*) grow at the surface of nutrient liquids in response to need for oxygen. They prefer a pH of 8 to 9. This pH retards the growth of many of the bacteria associated with vibrios in fecal material. These vibrios also metabolize peptone and hydrolyzed egg rapidly. Therefore, very rapid growth occurs at the surface of alkaline-egg-peptone solutions (Dieudonné medium) inoculated with feces of cholera patients. Transfers from the surface film of such cultures after six to eight hours of incubation often yield almost pure cultures of *V. cholerae*. This is a good example of **selective enrichment.**

The colonies on agar are small, colorless, and distinctively translucent. Most intestinal vibrios, including *V. cholerae*, are markedly proteolytic, liquefying gelatin and digesting casein.

Cholera. The vibrios are able to tolerate gastric acidity in sufficient numbers to set up a focus of infection in the small bowel. Cholera is characterized by intense vomiting, diarrhea, and prostration, due to production of the enzyme neuraminidase, which hydrolyzes an intercellular cement (N-acetylneuraminic acid), and a toxin which affects the sodium-potassium transporting system of the intestinal mucosa. These are intracellular substances that are liberated by mechanical disruption of the cells. Great damage is done to the patient by the action of the enzyme and the toxin, which causes profound electrolyte imbalance and enormous dehydration (loss up to 10 to 12 liters per day) through the rectum. The stools are thin, watery, and turbid with mucus, and are frequently described as **rice water stools.** They contain large numbers of microscopically demonstrable cholera vibrios.

For these reasons one of the most effective means of treating the cholera patient is to restore the fluid imbalance either by oral administration or intravenous feeding of properly prepared electrolyte solution. Antibiotic therapy is effective only in the very early stages of the disease and should be used as an adjunct to shorten

TABLE 33–5. DIFFERENTIATION OF EL TOR VIBRIO FROM V. CHOLERAE

		El Tor	V. cholerae
Hemolysis	Tube	+ (−)°	−
	Plate	+ (−) (no hemodigestion)	− (hemodigestion)
Susceptibility to phage specific for *V. cholerae* (phage type IV)		−	+
Slide agglutination of chicken erythrocytes		+	−
Virulence for chick embryos		+	−
Agglutination with Ogawa (O group I [A] serum)		+	+
Resistant to polymyxin antibiotic		+	−

°A few are negative.

the duration of the disease and prevent the carrier state. Cholera is transmitted primarily by feces and sewage-polluted drinking water and foods.

Less than a century ago, cholera was to be found in practically every large city in the world. It occurred often in places with a large transient population, or in centers for religious, military, or other concentrations of large numbers of people with no effective sanitary provisions with regard to sewage and pollution of water and food. In medieval Europe and later in America, cholera was an ever-present and often widespread and fatal scourge. It has played a sinister and strictly nonpartisan role in many disastrous military campaigns.

Until the onset of the seventh pandemic (1960–61), in this century cholera was found mainly in the Orient. One case was identified in the United States in the 1940's, and another recently in Texas (1973). By contrast, during an epidemic of cholera in 1958 there were nearly 50,000 (known) cases and over 20,000 (known) deaths in India, East Pakistan (now Bangladesh), Thailand, Cambodia, and Burma.

However, now even countries with sanitary engineering of sewage disposal and water supplies and constant vigilance by international health authorities as well as federal, state, and local sanitary administrations have experienced sporadic outbreaks of the disease (see El Tor Biotype, above). There is reason to fear that endemic foci may become established in Africa, and international authorities are concerned that the disease may be able to spread to other normally nonendemic areas, e.g., susceptible areas in Central and South America. This could be a serious problem in those underdeveloped countries without good sanitation; hence even the Western Hemisphere cannot be considered to be "immune" from the disease. Because vaccination does not prevent the healthy carrier state, travelers from any country are now allowed to enter the United States without a certificate of cholera vaccination (i.e., U.S. health authorities have recognized the limited value and duration of existing cholera vaccines).

Genus Aeromonas

Currently the genus Aeromonas is placed in the family Pseudomonadaceae (Bergey's Manual, 1957 edition) because members of the genus share a single morphological feature, polar flagella insertion, generally monotrichous.

In relation to O_2, the organisms are both respiratory and fermentative, and they break down carbohydrates by mixed fermentation with production of acids and gases (CO_2 and H_2) or acids only. For example, A. hydrophila carries out 2,3-butanediol fermentation accompanied by gas (CO_2 and H_2) production. Thus, this organism resembles the members of genus Enterobacter (Aerobacter). Likewise, the species A. formicans produces an enzyme, β-galactosidase, which is antigenically similar to that produced by strains of E. coli; thus Aeromonas species share some properties with members of Enterobacteriaceae as well as with species of Pseudomonas and Vibrio.

The DNA base composition of the genus ranges from 51 to 63 mole per cent GC, with the lower range resembling that of Enterobacteriaceae (tribe Eschericheae, 49 to 54 mole per cent GC) and the higher range resembling Pseudomonadaceae (58 to 69 mole per cent GC). Species of Aeromonas are widely distributed in nature: the habitat of A. hydrolytica and A. punctata is water and sewage, where these bac-

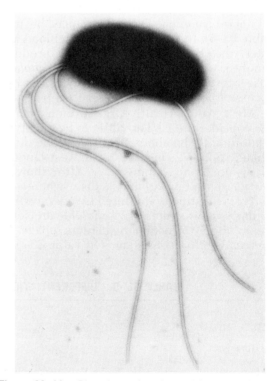

Figure 33–11. *Photobacterium fischeri* (strain 116) in exponential phase of growth in liquid medium. Marker indicates 1 μm. Negatively stained. (25,000×.) (Courtesy of R. D. Allen and P. Baumann.)

teria attain a concentration of 10^6 cells/ml. The *A. salmonicida* group is not found in surface water; they are strictly parasitic bacteria with a narrow host range. Some are known pathogens; for example, *A. hydrophila* causes red leg disease of frogs, and some are pathogenic for snakes. Strains of *A. salmonicida* are pathogenic to salmonids but rarely to other species of fish. The pathogenesis of *Aeromonas* in man is not elucidated, but cases of acute metastatic myositis caused by *A. hydrolytica* have been reported.

Genus Photobacterium

These organisms are polarly flagellate (Fig. 33–11), but unlike *Aeromonas* they emit light (luminescence). All are inhabitants of the sea, hence they are osmotically sensitive, requiring at least 1 per cent NaCl for growth. The DNA base composition of the genus ranges from 43 to 47 mole per cent GC. Many properties are shared with *Pseudomonas* and *Vibrio;* for example, the marine vibrio, *V. luminosus*, which gives off an eerie blue-green light when growing on seawater media or dead fish, is almost identical to species of *Photobacterium.*

While *Photobacterium* species are considered facultatively anaerobic, the property of luminescence is dependent on the presence of a small amount of free oxygen, and to some degree is controlled by nutritional environment. For example, the organisms when grown aerobically in synthetic medium supplemented with glucose as energy and carbon source may fail to luminesce; when such media are supplemented with various amino acids the luminescent property is restored.

Although the precise mechanism of each step is not yet elucidated, the mechanism of bacterial luminescence as well as other biological luminescence (fungi, protozoa, fireflies, etc.) has been studied intensively in the past. It is now believed that flavin mononucleotide (FMN) is the key starting compound which is responsible for *Photobacterium* luminescence. For example, energy (electron) transfer in bacterial luminescence occurs before the transfer of electrons to the cytochromes. Electron transfer is mediated by dehydrogenases such as FAD (flavin adenine dinucleotide), NAD, or NADP (nicotinamide adenine dinucleotide phosphate), an enzyme luciferase, aliphatic aldehyde, and O_2.

During normal electron transporting processes, dehydrogenases are reduced (NADH or NADPH) and thus become electron donors (see below):

$$
\begin{array}{ccc}
\text{Electron} & & \\
\text{transporting} & & \\
\text{system} & & \text{luciferase} \\
\downarrow & & + \\
\text{NADH} & & \text{aliphatic} \\
\downarrow & & \text{aldehyde} \\
O_2 \leftarrow \text{cytochrome} \leftarrow \text{FAD} \rightarrow \text{FMN} \xrightarrow[+\,O_2]{} \text{light}
\end{array}
$$

Electron transporting proceeds toward the right instead of normal electron transport (toward the left). For this reason, if the electron path via cytochromes is blocked, luminescence would be intensified. A similar phenomenon may be observed when a small amount of oxygen is introduced to the organism grown under anaerobic conditions because luminescence is oxygen-dependent. A complex chemical reaction occurs between reduced FMN and luciferase + aliphatic aldehyde. The reduced and modified FMN reacts with O_2, causing formation of peroxide in an excited state. When this compound releases energy (grounded), light is emitted, and again FMN is restored to the oxidized state.

In bacterial luminescence the energy available is not used for regeneration of ATP but is dissipated in the formation of light. Hence the physiological function of the luminescence is not clearly understood except when such bacteria exist symbiotically with deep-sea fish, where both may receive benefit. For further discussion of bioluminescence, students are referred to the Supplementary Reading list at the end of the chapter.

Genus Zymomonas

In the 7th edition of *Bergey's Manual*, this genus is classified in the family Pseudomonadaceae because *Zymomonas* is motile by means of polar flagella (lophotrichous). However, the organisms are unlike aerobic pseudomonads in that their metabolism is fermentative. For this reason, in the 8th edition of *Bergey's Manual*, the organisms have been transferred to Part VIII, "Gram-Negative, Facultatively Anaerobic Rods," and placed under "Genera of Uncertain Affiliation."

The organisms share certain properties with genus *Erwinia* of the family Enterobacteriaceae but differ in mode of glucose fermentation. For example, *Z. mobilis*, type species of the genus (DNA base composition, 49 mole per cent GC), ferments glucose with production of

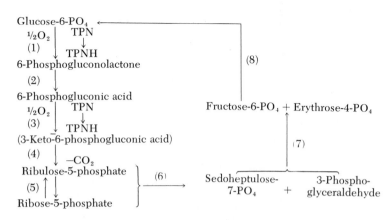

Figure 33–12. Modified hexosemonophosphate shunt.

Reaction	Enzyme
(1)	Glucose-6-phosphate dehydrogenase (Zwischenferment)
(2)	6-phosphogluconolactonase
(3)	6-phosphogluconic dehydrogenase
(4)	Spontaneous (?)
(5)	Phosphopentose isomerase
(6)	Transketolase
(7)	Transaldolase
(8)	Phosphohexose isomerase

an exceedingly large amount of CO_2 and lactic acid. However, the pathway of glucose fermentation to ethyl alcohol is unlike the typical yeast alcoholic fermentation (via the Embden-Meyerhof pathway) in that glucose is dissimilated via a hexose monophosphate shunt (Fig. 33–12).

33.4
FAMILY SPIRILLACEAE

Species of the family Spirillaceae are curved or spiral rods. Otherwise, most have the general properties and habitats of the family Pseudomonadaceae as described earlier in this chapter.

In the 7th edition of *Bergey's Manual*, the family contains 10 genera: *Vibrio, Desulfovibrio, Methanobacterium, Cellvibrio, Cellfalcicula, Microcyclus, Spirillum, Paraspirillum, Selenomonas,* and *Myconostoc.* The aggregation of this physiologically and biochemically diverse group into a single family is primarily based on morphology; that is, organisms are usually motile by means of a single flagellum or a tuft of polar flagella, in addition to the Gram reaction and cell shape. However, in the forthcoming edition of *Bergey's Manual*, most of the genera listed above are eliminated from the

family Spirillaceae, which is placed in Part VI, "Spiral and Curved Bacteria," and the family contains two genera, *Spirillum* and *Campylobacter*. Part VI also contains "Genera of Uncertain Affiliation," which includes *Microcyclus, Bdellovibrio, Pelosigma,* and *Brachyarcus.* All of these genera are chemoorganotrophs (some are parasitic, or bacteriovorous) and their DNA base composition ranges from 30 to 65 mole per cent GC.

Genus Spirillum

With the single exception of *Spirillum minus*, this genus contains only harmless saprophytes and scavengers living, along with vibrios, in stagnant or polluted water and putrefying materials.

Morphology. Most saprophytic spirilla are relatively large, ranging from 5 to 40 μm in length though only 0.5 to 3 μm in diameter. They are spirally twisted through one to five complete turns (Fig. 33–13), rigid, and motile by means of one or more flagella at one or both poles. The DNA base ratio of the genus *Spirillum* ranges from 38 to 65, with mean value at 55 mole per cent GC.

Cultivation. *Spirillum* species usually grow with difficulty on first isolation, are gen-

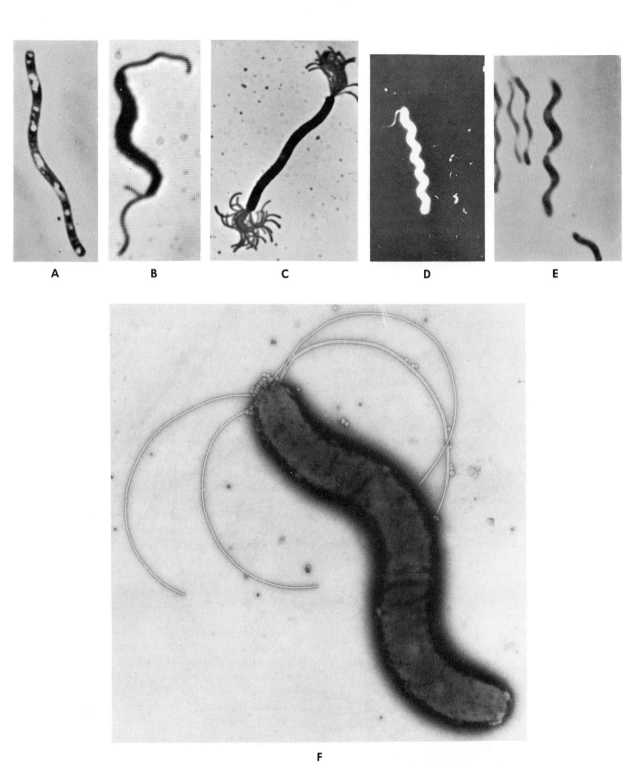

Figure 33–13. Various species of saprophytic spirilla. *A, S. serpens,* darkfield, live, flagella not visible. *B, S. serpens,* Leifson's flagella stain; compare with *A. C, S. sinuosum,* darkfield microphotograph. (600×.) *D, S. beijerinki,* Leifson's flagella stain. (1,250×.) *E, S. lunatus,* live, dark phase-contrast microphotograph. *F,* Electron micrograph showing normal cell of *S. itersonii.* Potassium phosphotungstate stain. (33,600×.) (*C,* Courtesy of A. Pijper, Institute for Pathology, Pretoria, South Africa. *F,* Courtesy of G. D. Clark-Walker.)

erally catalase-positive and require aerobic conditions and organic sources of carbon and energy. Some of them have been obtained in pure culture from infusions of stagnant water, dung, or sewage enriched with peptone, meat, or fish. After a few days at 25 to 30C the fluid usually swarms with these and other microorganisms. In one method for obtaining pure cultures, the surface fluid from such an infusion is sterilized and used as a medium by solidifying it with 2 per cent agar. It is poured into Petri dishes. Colonies of some species of *Spirillum* can be obtained by inoculation of the agar surface with the unsterilized infusion.

Spirillum volutans. *Spirillum volutans* (Ehrenburg, 1832) is of interest because of its active, tumbling volutions and its large metachromatic granules from which the term **volutin** (polymerized phosphates), in reference to the species, is derived. The rotatory and to-and-fro motility and the form of these spiral organisms are well demonstrated in hanging-drop and darkfield preparations. Negative staining or fluorescent-labeled preparations are useful in demonstrating their form and arrangement (not motility) (Fig. 33–14). DNA base composition of *S. volutans*, which is the type species, is 38 mole per cent GC.

Until recently this species could be cultivated in vitro only in association with other bacteria or their dialyzable products. In 1965 it was grown in a completely bacteria-free medium under special **microaerophilic** conditions. Similar species have been isolated by allowing them to grow through membrane filters lying on the surface of semisolid (0.1 per cent) agar medium. Some species grow in relatively simple synthetic or organic media. For example, strains of *S. volutans* and *S. itersonii* grow well in peptone-succinate-salts medium, and the latter organism is also able to utilize inorganic nitrogen compounds (ammonium salts) as sole source of nitrogen. This organism, unlike other species of *Spirillum*, is capable of growing anaerobically in the presence of nitrate as electron acceptor.

Spirillum minus and Rat-Bite Fever. *Spirillum minus*, one of the most minute species, causes a disease in man (rat-bite fever or **so-doku**) having several of the clinical features of a typical spirochetal disease (syphilis). *S. minus* has never been successfully cultivated. It is gram-negative.

S. minus occurs in the blood of rats, mice, and possibly other animals, and is transmitted from them to each other and to man by their bites. In man it causes intermittent fever and **spirillemia.**

Genus Campylobacter

The genus is newly created to include slightly or definitely curved organisms motile with polar mono- or multitrichous flagella and having DNA base composition ranging from 30 to 34 mole per cent GC. As a group, they do not produce acid from sugars, are nonproteolytic, and reduce nitrate to nitrite. The genus, therefore, contains a few species formerly placed in the genus *Vibrio* (V. fetus and V. bubulus).

C. fetus (V. fetus). The organism is a minute curved rod (S-shaped on initial isolation). Cells are motile with a single polar flagellum but S-shaped cells usually possess a single flagellum at each pole. DNA base composition ranges from 30 to 34 mole per cent GC and is significantly lower than that of species of genus *Vibrio* (40 to 49 mole per cent GC).

C. fetus is an important species because it causes abortion and considerable reduction in fertility in sheep, cattle, and horses and consequent serious economic losses among stock raisers in the United States. It can infect humans handling tissues of infected animals.

C. fetus is a rather highly adapted parasite. It apparently thrives only in the genital organs of male and female domestic mammals (possibly also in wild animals). The infection appears to be transmitted only by coitus. The organisms grow only on very moist organic media in small, translucent, colorless colonies. The slender, curved, individual cells are morphologically much like *V. cholerae*. However, unlike *Vibrio cholerae*, the organism neither ferments carbohydrates nor liquefies gelatin. The best means of diagnosis is by isolation of the organisms from the animals suspected. Chemically defined media for isolation and growth of *C. fetus* are now available.

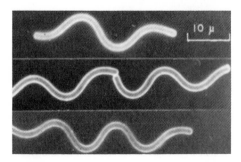

Figure 33–14. Living cells of *Spirillum volutans* labeled with fluorescent anti–*S. volutans* globulin and examined immediately.

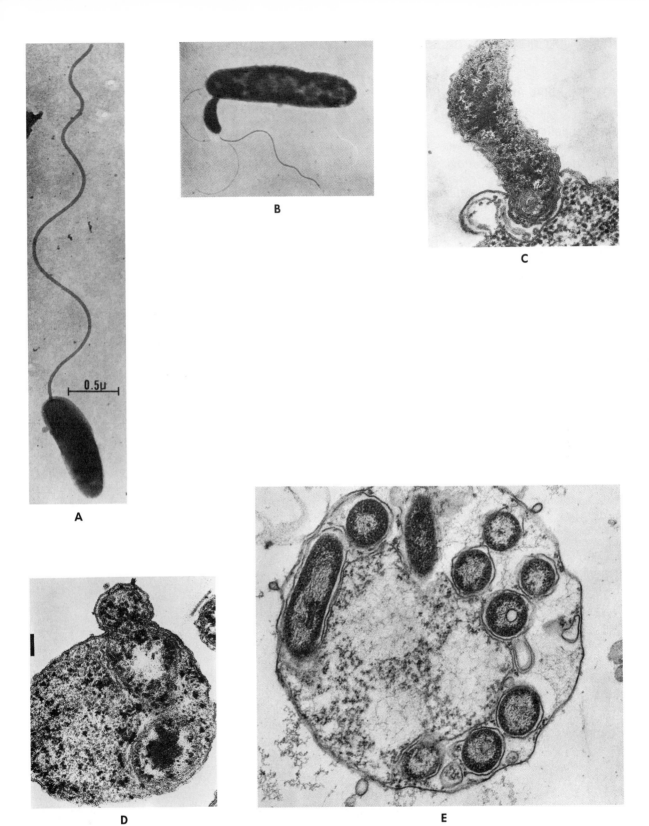

Figure 33–15. Electron micrographs of *Bdellovibrio bacteriovorus. A,* Parasitic *Bdellovibrio bacteriovorus* strain 109, grown with its host in NB/10 broth. Fixed with 1 per cent formaldehyde; stained with 0.5 per cent uranyl acetate. (43,000×.) Sequences of *Bdellovibrio* parasitizing the host cell: *B,* Electron micrograph of a shadowed whole-cell preparation showing *B. bacteriovorus* attacking *Pseudomonas.* (12,000×.) *C,* Electron micrograph of thin section of a bdellovibrio penetrating *E. coli.* (59,000×.) *D,* Late penetration. *E.* Complete invasion of the host and multiplication of the vibrios (numerous dark bodies) inside the *Pseudomonas* cell. (*A,* Courtesy of R. J. Seidler and M. P. Starr, *C* and *D,* Courtesy of J. C. Burnham, T. Hashimoto, and S. F. Conti.)

483

33.5
OTHER SPIRAL AND CURVED BACTERIA

Genus Bdellovibrio

Bdellovibrio bacteriovorus. One of the most interesting species of vibrios was first described in 1962. It is called *Bdellovibrio bacteriovorus* (Gr. *bdella* = leech; L. *vorare* = to eat). *B. bacteriovorus* is a minute (0.3 by 2.0 μm) comma-shaped rod with a single polar flagellum (Fig. 33–15). The DNA base composition of the genus ranges from 51 to 55 mole per cent GC. Although a few facultatively parasitic strains (*Bdellovibrio* species which can grow either endoparasitically in a bacterial host or saprophytically in a bacteria-free medium) have been isolated, the majority of bdellovibrios are unique among all known bacteria in being obligate, predatory parasites of bacteria. Cultivation of obligate *Bdellovibrio* species requires a specific host bacterium for growth and reproduction.

Species of *Bdellovibrio* are widely distributed in nature, in soil and water where large numbers of other gram-negative bacteria are found. The organisms have been found in activated sludge (Chapt. 45). They appear to attack preferentially members of either Pseudomonadaceae or Enterobacteriaceae and are most commonly associated with species of *Pseudomonas*

and a few other gram-negative bacteria, including *Escherichia* and *Salmonella*.

Unlike organisms that in nature require the mere presence of growth products of associated bacteria (e.g., *Spirillum volutans*), these minute vibrios actually absorb the contents of other bacteria like leeches, hence their name. The vibrios move at very high speeds. With phase microscopy they can be seen to collide violently with the host (perhaps victim is a better term) bacterial cell, become attached, and cause prompt lysis of the attacked cell. The attachments of the vibrios to their prey are sometimes reversible and repeatable and often multiple. The predators apparently live on the contents of the lysed cells; the vibrios continue to multiply long after the disappearance of all intact cells. The vibrios can also penetrate inside the host cells and metabolize the entire cell contents.

The mode of multiplication of obligate parasitic *Bdellovibrio* species is somewhat reminiscent of bacteriophage multiplication. If a culture of susceptible bacteria is mixed with a few cells of *Bdellovibrio* and the mixture then spread to grow as a "lawn" on an agar surface, the lawn will show plaques closely resembling plaques of phage (Fig. 33–16). The plaques are sites of the lytic action of the vibrios. That the plaques are not due to phage is clear from the facts that: plaques appear only after two days of incubation (*B. bacteriovorus* grows slowly) in-

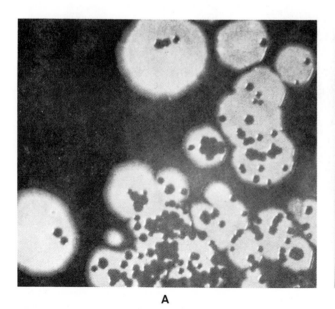
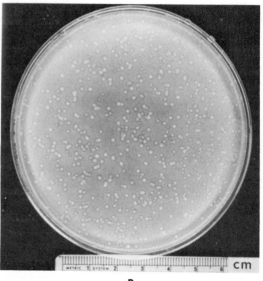

A   **B**

Figure 33–16. *A,* Lytic action of *Bdellovibrio bacteriovorus* on *Escherichia coli,* after streaking a mixture of parasites and host bacteria on nutrient agar. The colonies of *E. coli* are the larger light areas; colonies ("plaques") of the vibrios are small and dark. *B,* Four-day-old plaque colonies of *B. bacteriovorus* UKi2 on a lawn of *E. coli* B/r.

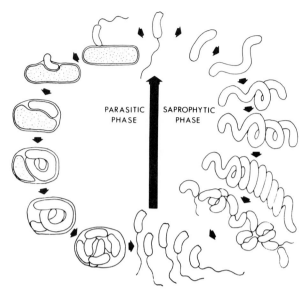

PARASITIC PHASE   SAPROPHYTIC PHASE

Figure 33–17. Growth cycle of *B. bacteriovorus* UKi2 as constructed on the basis of our light and electron microscopic observations. (Courtesy of J. C. Burnham, T. Hashimoto, and S. F. Conti.)

stead of within 18 hours or less as is true of phage plaques; the parasitic vibrios can be isolated from the plaques in greatly increased numbers; the plaque-forming units can easily be filtered or centrifuged out of the suspended growth so that the supernatant fluid or filtrate has no plaque-forming power. This is the reverse of phage. Unlike phage, the vibrios can also lyse heat-killed cells, if provided with some sterile lysate from another culture as a "starter" on which to grow before any of the new cells are lysed. Growth and cell multiplication under parasitic and saprophytic conditions are summarized in Figure 33–17. According to Burnham et al., the modes of growth, development, and ultrastructure of the organism grown either parasitically or saprophytically are essentially the same.

The ecological significance of bacteriovorous bacteria is not elucidated, but they may be profoundly involved in destruction of gram-negative bacterial populations in nature, as are bacteriophages.

CHAPTER 33
SUPPLEMENTARY READING

Barua, D., and Burrows, W. (Eds.): Cholera. W. B. Saunders Co., Philadelphia. 1974.

Burnham, J. C., Hashimoto, T., and Conti, S. F.: Ultrastructure and cell division of a facultatively parasitic strain of *Bdellovibrio bacteriovorus*. J. Bact., *101*:997, 1970.

Clark-Walker, G. C.: Association of microcyst formation in *Spirillum itersonii* with the spontaneous induction of a defective bacteriophage. J. Bact., 97:885, 1969.

Felsenfeld, O.: The Cholera Problem. Warren H. Green, St. Louis. 1967.

Hylemon, P. B., Wells, J. S., Jr., Krieg, N. R., and Jannasch, H. W.: The genus *Spirillum*: a taxonomic study. Int. J. Sys. Bact., *23*: 340, 1973.

McElroy, W. D., and Seliger, H. H.: Biological luminescence. Sci. Am. *207*:76, 1962.

Mandel, M.: Deoxyribonucleic acid base composition in the genus *Pseudomonas*. J. Gen. Microbiol., *43*:273, 1966.

Palleroni, N. J., Kunisawa, R., Contopoulou, R., and Doudoroff, M.: Nucleic acid homologies in the genus *Pseudomonas*. Int. J. Sys. Bact., *23*:333, 1973.

Pierce, N. F., Greenough, W. B., III, and Carpenter, C. J., Jr.: *Vibrio cholerae* enterotoxin and its mode of action. Bact. Rev., *35*:1, 1971.

Staley, T. E., and Colwell, R. R.: Deoxyribonucleic acid reassociation among members of the genus *Vibrio*. Int. J. Sys. Bact., *23*:316, 1973.

Stanier, R. Y., Palleroni, N. J., and Doudoroff, M.: The aerobic pseudomonads: A taxonomic study. J. Gen. Microbiol., *43*:159, 1966.

NONPHOTO-SYNTHETIC, CHEMOLITHO-TROPHIC, AND CHEMOORGANO-TROPHIC BACTERIA

• CHAPTER 34

(Gliding, Sheathed, Prosthecate, Budding, and Appendaged Bacteria)

Terms such as gliding, sheathed, prosthecate, budding, and appendaged are used to denote the taxonomic status of certain groups of bacteria. These morphologically and physiologically diverse groups are taxonomically quite widely separated in the 7th edition of *Bergey's Manual*. For example, many of the gliding bacteria are placed in the order Beggiatoales or Myxobacterales; the sheathed bacteria, in the order Chlamydobacteriales; and the budding bacteria, in the order Hyphomicrobiales. Some of these bacteria have certain more or less superficial similarities to blue-green algae; some are able to oxidize H_2S (sulfur bacteria), while others are able to obtain energy from oxidizing iron or manganese (iron bacteria); see Table 34–1.

All of the organisms discussed in this chapter are properly classed as bacteria, not algae, because: (1) their diameter is typically 1 to 5 μm and seldom exceeds 10 μm; (2) the chemical composition of their cell wall is distinctively bacterial (peptidoglycan); (3) their cellular structure is procaryotic; (4) they typically multiply only asexually by transverse, binary fission or by budding; (5) they are gram-negative; (6)

TABLE 34–1. BACTERIA SIMULATING OTHER CLASSES OF MICROORGANISMS *

Order	Suborder	Representative Family	Distinguishing Properties
PROTOZOAN-LIKE			
Spirochaetales		Spirochaetaceae	Thin, flexible cell wall or sheath; fibrillar structure; flexing, rotatory and translatory motility; no flagella
Myxobacterales		Myxococcaceae, Polyangiaceae	Thin, flexible cell wall; flexing and gliding motility; no flagella; form pseudoplasmodium and fruiting bodies
MOLD-LIKE			
Actinomycetales		Streptomycetaceae, Actinoplanaceae	Branching, mycelial growth; formation of conidia and/or motile sporangiospores
ALGA-LIKE			
Chlamydobacteriales		Chlamydobacteriaceae	Formation of free and sessile trichomes; aquatic habitat; sheath formation; motile swarm cells
Beggiatoales		Beggiatoaceae	Free and sessile trichomes; gliding motility; aquatic habitat
Pseudomonadales	Rhodobacteriineae	Thiorhodaceae	Cellular aggregates; aquatic habitat; photosynthesis
	Pseudomonadineae	Caulobacteraceae	Formation of branching stalks; aquatic habitat; free and sessile; accumulation of metallic oxides
		Siderocapsaceae	Aquatic habitat; accumulation of metallic oxide
Hyphomicrobiales		Hyphomicrobiaceae	Aquatic habitat; formation of stalks or filaments; free or sessile; photosynthesis in one species
Caryophanales		Caryophanaceae	Motile trichomes; distinctive nuclear structure; gonidia; aquatic habitat

* Grouping based on 7th edition of *Bergey's Manual.*

some are chemoorganotrophs, while a few are either mixotrophic (see §34.7) or chemolithotrophic.

The several curious nonphotosynthetic bacteria described in this chapter resemble the Cyanophyceae (blue-green algae) in the formation of trichomes, filaments, or branching stalks; in exhibiting gliding, nonflagellate motility; and in aquatic habitat. None resembles molds or "mold-like" bacteria (family Streptomycetaceae) in producing mycelia or conidiospores, and none resembles protozoans or "protozoan-like" bacteria (order Spirochaetales) in having thin, flexible cell walls or self-flexing movements. None produces thermostable endospores or **peritrichate** motile cells. All the members of Myxobacterales and Cytophagales (*Bergey's Manual,* 8th edition) resemble Cyanophyceae in having gliding, nonflagellate motility and in forming a slimy, enveloping matrix. This group is, however, discussed in this book with the protozoan-like bacteria because the organisms in it are characterized by thin, flexible, bacterium-like size and cell walls and by producing communal vegetative and "fruiting" groupings (see Chapter 41).

34.1
THE GLIDING BACTERIA

Myxobacterales and Cytophagales

There are several groups of nonphotosynthetic and nonflagellate bacteria capable of gliding across the surface of a solid substrate. While the mechanism of the motility is still obscure, these bacteria are definitely related to some of the blue-green algae in both morphology and mode of locomotion.

As stated earlier, the gliding bacteria are classified into several different orders and families (*Bergey's Manual,* 7th edition). However, in the proposed scheme (*Bergey's Manual,* 8th edition) all the gliding bacteria (both chemolithotrophs and chemoorganotrophs) are placed in Part II, "The Gliding Bacteria." Part II contains two orders, chemoorganotrophic Myxobacterales (Chapt. 41) and Cytophagales. The latter includes both chemolithotrophs (Leucotrichaceae) and chemoorganotrophs. Part II also includes "Families and Genera of Uncertain Affiliation," which contains family Achromatiaceae (*Achromatium*), family Pelonemataceae (*Pelonema, Achroonema, Peloploca,* and *Desmanthos*), and one taxonomically uncertain genus, *Toxothrix.*

The order Cytophagales is composed of four families: Cytophagaceae, Beggiatoaceae, Simonsiellaceae and Leucotrichaceae. Members of the family Cytophagaceae (genera *Cytophaga, Flexibacter, Herpetosiphon, Flexithrix, Saprospira, Sporocytophaga*) are unicellular to multicellular flexible rods (0.5 to 3 by 2 to 20 μm); the vegetative cells are motile by gliding or rotating on the solid substrate. Some species form microcysts, but unlike the members of Myxobacterales, they lack the fruiting body formation, and some species are capable of hydrolyzing cellulose (Chapt. 41). The DNA base compositions of the order are much smaller than those of the Myxobacterales (67 to 70 mole per cent GC) in that they range from a low of 33 to 38 mole per cent GC (*Cytophaga*) to a high of 35 to 48 mole per cent GC (*Saprospira*).

The Filamentous Gliding Bacteria

Currently the filamentous, nonphotosynthetic gliders are placed in the order Beggiatoales (7th edition, *Bergey's Manual*), which contains four families: Beggiatoaceae, Vitreoscillaceae, Leucotrichaceae, and Achromataceae. However, radical rearrangement of the order is proposed for the 8th edition of the *Manual.* According to the new scheme, the order Beggiatoales is replaced by Cytophagales, and this order includes nonfilamentous gliders (Cytophagaceae) and three families of filamentous gliders (Beggiatoaceae, Simonsiellaceae, and Leucotrichaceae). The separation of gliders into three families is based on their morphology, motility of filament, mode of reproduction, and DNA base composition, which ranges from 35 to 50 mole per cent GC for the members of these families, and on the average the range is slightly higher than that of the family Cytophagaceae. Some members of these families are frequently referred to as nonphotosynthetic "sulfur bacteria" because they are able to obtain energy from oxidation of inorganic sulfur.

Inorganic sulfur is available to bacteria in various stages of oxidation and reduction, ranging from the most reduced, H_2S, through elemental S, thiosulfates, and tetrathionates to the most oxidized form, sulfates. Any of these except sulfates may be oxidized as energy sources by certain species of bacteria; and any except H_2S may be reduced as electron acceptors by certain other species of bacteria.

The term "sulfur bacteria" is used merely as a convenience and includes a heterogeneous group of species that have little else in common. We have already discussed photosynthetic sulfur bacteria. The relations of these and nonphotosynthetic sulfur bacteria are summarized in Table 34–2.

The Family Beggiatoaceae

The family contains three genera: *Beggiatoa, Thioploca,* and *Vitreoscilla.* Species of the genera are colorless and are said to be **uniseriately multicellular;** that is, they form long, multicellular threads called trichomes. A **trichome** (Gr. *trichos* = hair) is a single, multicellular organism consisting of undifferentiated cells attached end-to-end like railway cars and clearly an entire multicellular structure. The term includes flagellate, nonflagellate, and gliding organisms. It does not include chains of obviously independent cells such as streptococci, individual cells of which have clung together accidentally after fission. The filaments of Beggiatoaceae are usually straight, cylindrical, and motile by means of gliding. None of the Beggiatoaceae is ensheathed and none branches. All species are structurally so very like the blue-green algae *Oscillatoria* that many authors regard the filamentous Beggiatoales as nonphotosynthetic variants of the *Oscillatoria.*

In addition to morphological, cultural, and habitat parallelisms, the two groups of organisms possess similar ranges of DNA base composition. For example, DNA base compositions of a member (*Vitrioscilla*) of the filamentous gliders in the family range from 44 to 45 mole per cent GC, which closely resembles the range of filamentous blue-green algae: Nostocaceae (38 to 46), Oscillatoriaceae (45 to 51), Scytonemataceae (42 to 48 mole per cent GC).

Only three families of Cytophagales are sulfur-storers: the filamentous Beggiatoaceae (*Beggiatoa* and *Thioploca*), Leucotrichaceae (*Leucothrix* and *Thiothrix*), and nonfilamentous Achromatiaceae, which belong to "Families of Uncertain Affiliation" under Part II.

Pure cultures of *Beggiatoa* and *Thioploca*

TABLE 34-2. RELATIONS OF THE SULFUR BACTERIA

Oxidize Sulfur and Its Inorganic Compounds				Reduce Sulfates	Produce H₂S from Organic Sulfur Compounds
INTRACELLULAR SULFUR GRANULES		EXTRACELLULAR SULFUR GRANULES			
Photosynthetic	*Nonphotosynthetic*	*Photosynthetic*	*Nonphotosynthetic*		
Chromatiaceae (Thiorhodaceae)° *Clathrochloris†*	Beggiatoaceae Leucotrichaceae Achromatiaceae *Sphaerotilus* Various chemolithotrophic bacteria: *Thiobacterium, Thiovulum, Micromonas, Macromonas,* etc.	Chlorobiaceae (Chlorobacteriaceae) *Ectothiorhodospira‡*	*Thiodendron Thiobacillus§*	*Desulfovibrio Desulfotomaculum Sporovibrio* (?)	Various pathogenic and saprophytic (putrefactive) species: *Proteus, Serratia, Clostridium*

°( ) Indicates family name according to the 7th edition of *Bergey's Manual.*
†Genus belonging to Chlorobiaceae.
‡Genus belonging to Chromatiaceae.
§The organism is chemolithotrophic and oxidizes H₂S, metal sulfides, and thiosulfate.

are able to use H_2S as an energy source, but in a strict sense these organisms are chemoorganotrophs. In pure culture studies, they fail to use CO_2 as their sole carbon source, and H_2S is used as an energy source when the growth medium is provided with simple organic compounds. In the presence of organic compounds such as acetate, the cells are able to oxidize H_2S and intracellularly deposit sulfur granules which may be further oxidized to sulfate. However, species of *Vitreoscilla* are chemoorganotrophs in that they oxidize organic compounds as an energy source. In nature, all frequently live together in the same habitats where H_2S is plentiful.

Oxidation of H_2S as source of energy may be expressed as follows:

$$2H_2S + O_2 \longrightarrow 2H_2O + 2S$$
$$2S + 3O_2 + 2H_2O \longrightarrow 2H_2SO_4$$

The acid combines with other substances to form sulfates. Sulfates are invaluable as the principal sulfur compounds, available to higher plants. However, these organisms, unlike species of *Thiobacillus*, sulfur-oxidizing chemolithotrophs of the soil (Chapt. 43), are unable to oxidize exogenously supplied reduced sulfur compounds, except H_2S.

Beggiatoa alba. *Beggiatoa alba* is a good representative of the sulfur-metabolizing filamentous species. It is common in all sewage and other polluted waters containing H_2S. Unlike most species of the nonphotosynthetic filamentous gliders, *B. alba* is microaerophilic (i.e., requires some free oxygen as indicated in the

above equations). Whether or not it is lithotrophic in natural habitats remains to be determined. Typical trichomes of *B. alba* range in diameter from 3 to 5 μm and up to several millimeters in length (Fig. 34-1). The cells are gram-negative. Reproduction is primarily by transverse binary fission of the individual cells constituting trichomes; secondarily by fragmentation of the trichomes. In contact with solid surfaces the trichomes show slow gliding and rotatory motility (Fig. 34-2). They also show slow, bending and waving movements. Flagella are absent. Cultivation of *Beggiatoa* is usually in enrichment cultures from sewage. Catalase has been found to favor growth and viability. When H_2S is abundant, like the sulfur purple bacteria (Chromatiaceae) *B. alba* stores col-

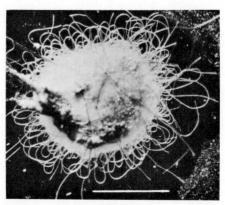

Figure 34-1. A dried colony of *Beggiatoa alba* showing loops of trichomes at edge of colony. The line indicates 1 mm. at this magnification.

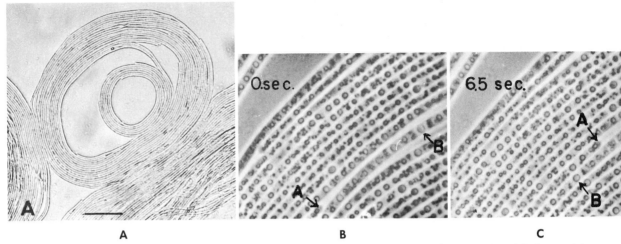

Figure 34-2. *Beggiatoa* growth on moist agar surface. *A,* Manner of coiling of the trichomes. *B* and *C,* Selected frames from a 16-mm motion picture film showing motion of gliding trichomes. In the course of 6.5 seconds point A passed point B. Note the sulfur granules in the trichomes. (Dark phase-contrast photomicrography.) The line in *A* indicates 50 μm; in *B* and *C* a similar line would indicate 10 μm.

loidal globules of sulfur inside its cells, giving the organisms a distinctive milky appearance, hence the name *alba,* or white. When H_2S is scanty, the stored H_2S is oxidized; when H_2S is absent the organisms grow poorly or die.

Families Simonsiellaceae and Leucotrichaceae

Heterotrophic members of the families Simonsiellaceae and Leucotrichaceae are distinguished from Beggiatoaceae by their distinct filament morphology. For example, multicellular filaments of *Simonsiella* and *Alysiella* are strongly compressed in ribbon-like units (Fig. 34–3) in which individual cells are closely apposed and flattened; thus the organisms of Simonsiellaceae are structurally reminiscent of the blue-green alga *Crinalium.* The DNA base compositions of the organisms range from 35 to 49 mole per cent GC. All are aerobic and found in saliva and oral cavities of man and other animals.

The habitat of Leucotrichaceae (*Leucothrix* and *Thiothrix*) is unlike that of Simonsiellaceae in that the former organisms live and reproduce primarily in an aquatic environment. Species of *Leucothrix* are found in nature only in marine

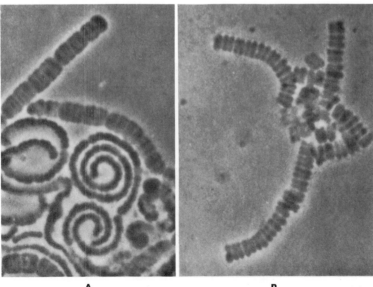

Figure 34–3. Photomicrographs of: *A,* compressed ribbon-like filaments of *Simonsiella* and *B, Alysiella.* Note terminal cells of *Simonsiella* are rounded while terminal cells of *Alysiella* are not. (1,800×.)

environments and grow abundantly in decaying material or as epiphytes on marine algae. However, the members of *Thiothrix* are considered obligate chemolithotrophs which obtain energy from oxidizing H_2S to S and to sulfate.

Large numbers of *Thiothrix* as single, gliding cells or rosettes of filaments surrounded by a delicate sheath are found in sulfur springs and other aquatic environments rich in H_2S. The rosettes are attached to solid surfaces by secreted "holdfasts" (Fig. 34–4,*A*). Thus the members of Leucotrichaceae are structurally similar to the blue-green alga *Calothrix*.

A

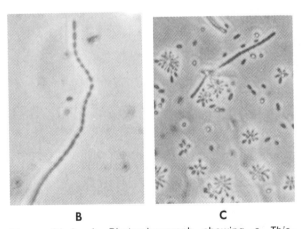

B   **C**

Figure 34–4. *A*, Photomicrograph showing a *Thiothrix* sp. rosette. The arrows indicate sulfur granules. *B*, Phase-contrast photomicrograph showing gonidia at the tip of *Leucothrix* filament. (309×.) *C*, Rosettes formed by aggregated *Leucothrix* motile cells. (279×.)

Although single cells of both *Leucothrix* and *Thiothrix* display gliding motility, their free filaments are unlike the members of Beggiatoaceae in that they are immotile. DNA base composition of Leucotrichaceae ranges from 46 to 50 mole per cent GC.

Reproduction of Leucotrichaceae is by transverse fission of free cells or segments of the filament. It is not restricted to the terminal cells but may occur throughout the length of the filament. Individual cells in the filament under unfavorable growth conditions may become round or ovoid structures called **gonidia**, which are frequently released from the tips of the filaments. Freed gonidia are motile and either aggregate to form rosettes which are held together by secreted holdfasts or divide successively to form nonmotile filaments, and the life cycle continues (Fig. 34–4, *B* and *C*). The organisms play an important role in the recycling of sulfur and mineralization of organic matter.

34.2
SHEATHED BACTERIA

There are numerous types of sheathed nonphotosynthetic, colorless, alga-like bacteria which occur in trichomes; many of these bacteria are currently classified in the order Chlamydobacteriales (Gr. *chlamydis* = cloak or covering) in the 7th edition of *Bergey's Manual*. However, their taxonomy is under active investigation, and recently organisms occurring in sheaths have been differentiated on the basis of unicellular or multicellular form, intracellular deposition of sulfur or iron, type of motility (if any), presence of gas vacuoles, type of sheath, and DNA base composition, in addition to morphology and nutritional and physiological characteristics.

In the provisional classification (*Bergey's Manual*, 8th edition), these organisms are placed in Part III, "Sheathed Bacteria," and the part includes seven genera: *Sphaerotilus, Leptothrix, Streptothrix, Lieskeella, Phragmidiothrix, Crenothrix, Clonothrix*. These organisms display very interesting morphological structures and reproductive cycles. However, with the exception of *Sphaerotilus*, very little information concerning their biochemical and physiological characteristics is available. Hence, some of the many "species" of these genera are undoubtedly variants of a few central species, and more unified classification of the group is anticipated as more information becomes available. In this chapter two main types, *Sphaerotilus* and *Leptothrix*, which have DNA base compositions ranging from 69 to 70 mole per cent GC, are discussed.

Sphaerotilus natans. *S. natans* is a common, much studied, and representative species.

It occurs worldwide in sewage and other polluted waters and becomes especially recognizable when those waters contain organic iron. The organism is constantly present in sewage treated by the activated sludge process (Chapt. 45) and is frequently used as an indicator of organic pollution of aquatic environments. It is obligately aerobic, organotrophic, and nonsporeforming.

The sheath of *S. natans* is flexible, looking and behaving much like a clear cellophane or paper tube, such as a drinking straw. Chemically, it resembles a modified cell wall because it is composed of the protein-polysaccharide-lipid complex found in the gram-negative cell wall, but the sheath lacks muramic acid, a distinctive component of all bacterial cell walls. It is, therefore, neither cell wall nor capsule; it is a unique structure (Fig. 34–5). It has been reported that the sheath is synthesized exclusively by the cells at the growing tip of the filament.

The individual cells inside the sheath are of the same structure and order of size as typical rod-shaped bacteria (1 μm by 2 to 10 μm), though the trichomes may be several milli-meters in length. The rods when motile possess one or more **polar** flagella. When the growth of *S. natans* is young, the filaments may resemble hyphae of coenocytic molds, though bacterial in diameter (1 to 3 μm). As the growth matures, the protoplast becomes more obviously divided into bacilli with lophotrichous flagella. These cells multiply by binary fission. The resulting motile bacilli, often called **swarm cells**, slip out at the ends of the sheaths or are liberated at the sides as the sheaths disintegrate. Sometimes the young cells cling to the outside of the sheath of origin and grow off at an angle. This is called **false branching.** In one variety, called *S. dichotomus* (possibly identical with *S. natans*), the false branching appears to be dichotomous.

S. natans and *S. dichotomus* usually do not accumulate much iron in the sheath except in matured filaments. Filaments encrusted with thick ferric and manganese oxides usually appear golden brown in color. The species of *Sphaerotilus* which deposit iron (or manganese) are frequently included in the ecological group known as "iron bacteria" and are incriminated as the cause of iron taste in drinking water. It is of interest to note that *S. natans* has also been described as a sulfur bacterium, depositing sulfur granules like *Beggiatoa alba*. *S. natans* also synthesizes prominent granules of poly-β-hydroxybutyric acid (Fig. 34–6).

S. natans is a tremendous nuisance when, because of its excessive growth, the tangled,

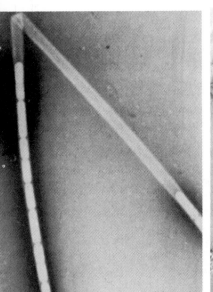

A

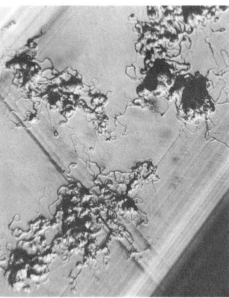

B

Figure 34–5. *Sphaerotilus natans. A,* Sheaths and bacillus-like cells within sheaths. Note also empty sheath. (About 2,500×.) *B,* Young colonies on nutrient agar after 24 hours at 28C. (About 50×.)

Figure 34-6. Globules of PHB in ensheathed cells of *S. natans*. (810×.)

filamentous masses cause blockage ("bulking") in the flow of sewage in disposal plants that use activated sludge. (See Chapter 45.)

Leptothrix. Species of *Leptothrix* (do not confuse with *Leucothrix*) are similar in several respects to *Sphaerotilus;* for example, the cells occur in chains within a uniformly thickened sheath, free reproductive cells are motile with a tuft of flagella, and filaments are not attached to any solid surface. However, unlike *Sphaerotilus,* the organisms in the genus *Leptothrix* actively oxidize iron and manganese and deposit the oxides within the sheath. The sheaths encrusted with iron and manganese oxides are completely dissolved in dilute hydrochloric acid, whereas sheaths of *Sphaerotilus* are insoluble.

Although the organisms have the ability to oxidize iron and manganese, there is no evidence to indicate that they derive energy from such oxidation. Likewise, the organisms have not been cultivated in the complete absence of organic compounds. Hence, the organisms are dependent on the oxidation of organic compounds for both energy and carbon and remain classified as chemoorganotrophs.

34.3
BUDDING, PROSTHECATE, AND APPENDAGED BACTERIA

Gram-negative, chemoorganotrophic, prosthecate (stalked) bacteria are currently classified in two orders, Pseudomonadales and Hyphomicrobiales, with most species placed in the genus *Caulobacter* of the family Caulobacteraceae (*Bergey's Manual,* 7th edition). In the 8th edition, *Rhodomicrobium,* a budding and prosthecate organism, has been transferred to the family Rhodospirillaceae (Athiorhodaceae) because the organism is photosynthetic.

The taxonomy of these alga-like organisms has been under intensive investigation and as a result the following genera are included in the group: *Hyphomicrobium, Hyphomonas, Pedomicrobium, Caulobacter, Asticcacaulis, Prosthecomicrobium, Thiodendron, Pasteuria, Blastobacter, Seliberia, Gallionella, Nevskia, Planctomyces, Metallogenium, Caulococcus,* and *Kusnezovia.* All members of these genera possess one distinct feature: they extend one or more stalks (composed of living cell parts or metabolic waste materials) in various forms and shapes. Some stalks have reproductive functions, while others may serve as an organ of attachment to a solid surface.

Caulobacters in which the stalk arises in line with the central, long axis of the cell are included in the genus *Caulobacter;* those in which the stalk arises excentrally and is not necessarily involved in the attachment process are grouped in a new genus, *Asticcacaulis* (Gr. *stichos* = alignment). Those bacteria with prosthecae extending in all directions from the cells are grouped in a new genus, *Prosthecomicrobium* (Gr. *prostheco* = appendage).

These organisms differ markedly in other characteristics; for example, some reproduce by binary fission, others by budding; many of their free cells are polarly flagellate and a few are immotile and possess gas vacuoles. These diverse characteristics are used in separating prosthecate bacteria into several subgroups. The DNA base composition of the prosthecate bacteria as a whole ranges from 55 to 67 mole per cent GC.

Caulobacter and Related Organisms

Caulobacter vibrioides and several similar species including *Asticcacaulis* of the stalk-formers are neither sulfur nor iron bacteria. The DNA base composition of the caulobacters range from 60 to 69 mole per cent GC and for *Asticcacaulis* about 55 mole per cent GC. Many of the species are currently placed in the family Caulobacteraceae of the order Pseudomonadales (*Bergey's Manual,* 7th edition). The cells are typically *Vibrio*-like: ellipsoidal, rod-like, fusiform, or banana-shaped. They resemble common species of *Pseudomonas* in size and procaryotic structure. They are chemoorganotrophic and multiply by transverse binary fission. They are nonsporeforming and have a single, unipolar flagellum when motile. Electron micrographs reveal such familiar structural details as cell wall, cell membrane, mesosomes, ribosomes, and nuclear area as seen in other bacteria (Fig. 34–7).

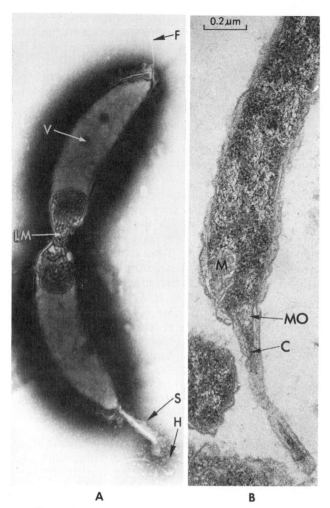

Figure 34–7. Electron micrograph of *Caulobacter crescentus* strain CB15. *A*, Large mesosomes at the equator of the organism symmetrically distributed between the two future daughter cells. Negatively stained with phosphotungstate. (3,600×.) *B*, Thin section of caulobacters from an oxygen-limited culture. (84,000×.) Abbreviations: *s*, stalk; *c*, core of stalk; *f*, flagellum; *h*, holdfast; *v*, volutin granule; *m*, mesosome; *lm*, large mesosome; *mo*, membranous organelle. (Courtesy of G. Cohen-Bazire, R. Kunisawa, and J. S. Poindexter.)

with, or possibly (as the cell matures) *including*, the polar flagellum. The stalk wall is a continuation of the cell wall itself; the core of the stalk is an extension of the cytoplasm, especially of its membranous parts (Fig. 34–7).

Formation of stalk and flagellum is related to what appears to be a primitive "life cycle" or cyclic type of cell division. In this cycle, a mature, stalked, **nonflagellate, vegetative** cell is attached by a strongly adhesive "holdfast" at the tip of its stalk, to some solid object. Adhesion of the stalks of caulobacters to other bacteria, especially gram-positive species, suggested at one time that the caulobacters became predatory by inserting the stalk into other cells as a "sucking proboscis." However, it now appears that attachment to other organisms is a purely mechanical process without ill effect on the organisms so invested by the caulobacters.

The stalked cell attached to a solid surface elongates and then divides transversely near the middle of the long cell. The *new*, distal cell is without a stalk but contains the accumulated stalk-forming material at its distal end. At that end an adhesive "holdfast" material and an active flagellum also appear. The new cell, now a motile "swarmer," separates from the parent cell and swims away. The older cell remains attached by its stalk and continues to repeat this vegetative process (Fig. 34–9).

The young swarmers soon attach to solid objects, usually other microorganisms such as protozoans, algae, fungi, other bacteria; most commonly, to each other. They thus form characteristic rosettes (Fig. 34–8, *B*). The new stalk, attached to the solid object by the holdfast, now develops from the cell, increasing the distance between the cell and the solid object. The stalked, sessile cell is now mature and begins the vegetative process described above. Swarmers do not undergo fission until they become sessile.

Recently a new genus, *Prosthecomicrobium*, has been added to the *Caulobacter-Asticcacaulis* group. The organism displays a bizarre shape, extending several prosthecae in all directions from the cell (Fig. 34–10). However, the prosthecae of the genus apparently lack the holdfast function. The cell divides by binary fission, and some progeny are motile with a single polar or subpolar flagellum, while others are immotile. The members of the genus are strictly aerobic and, unlike others in the group, are able to utilize numerous carbohydrates and sugar alcohols for both carbon and energy source and require vitamin B_{12}, thiamine, and biotin for growth. The DNA base

The caulobacters differ from all other known procaryotic organisms in the unique character of their dimorphic existence; i.e., two kinds of cells are usually found in a growing cell population. One type of cell (mother cell) is immotile, with a prostheca from which develops a motile daughter cell by binary fission; the cell possesses either a polar or subpolar single flagellum (Fig. 34–8).

Unlike the excreted stalks of *Gallionella*, the stalk of the caulobacters is not an excretion but a distinct *part of the cell itself*, a narrow, flexible tubular outgrowth closely associated

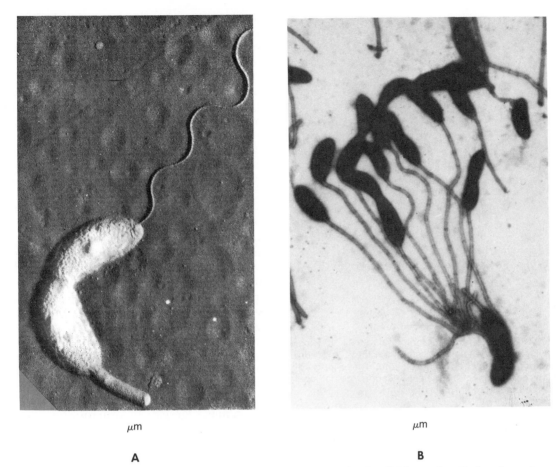

μm   μm

A   **B**

Figure 34–8. *A,* Vibrioid (*Caulobacter*) strain CB2. Electron micrograph of a dividing cell, with flagellum at one pole and stalk at the other. *B, Caulobacter crescentus* strain Sk1418. Stalked cell in rosette, cells adhering to the common mass of holdfast material by tips of stalks. (*B,* Courtesy of J. M. Schmidt and G. M. Samuelson.)

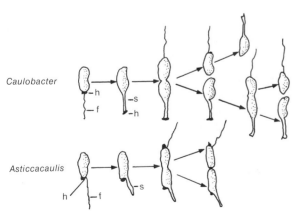

Caulobacter

Asticcacaulis

Figure 34–9. Diagrammatic representation of stages in the *Caulobacter* and *Asticcacaulis* life cycles. Abbreviations: *h,* holdfast; *f,* flagellum; *s,* stalk.

composition of the genus ranges from 67 to 71 mole per cent GC.

Hyphomicrobium and Other Prosthecate and Budding Bacteria

Organisms included in this group are unlike those of the previous group in that they multiply by budding rather than by binary fission. The most extensively studied organism of the group is in the genus *Hyphomicrobium,* currently placed in the family Hyphomicrobiaceae of the order Hyphomicrobiales (*Bergey's Manual,* 1957). The DNA base compositions of the group range from 60 to 68 mole per cent GC.

The organisms of this group are commonly

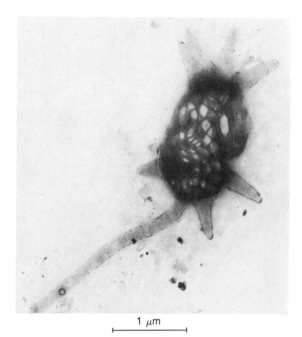

Figure 34–10. Electron micrograph of a cell of *Prosthecomicrobium pneumaticum*, strain 3a, that has an unusually long appendage (about 2.5 μm). These long appendages, less frequent than the shorter ones, do not taper toward the tip. (Courtesy of J. T. Staley.)

found in freshwater ponds, streams, and mud; also, in soil and sewage as well as in marine environments. The organisms at one time were considered to be chemolithotrophs because they grew in a medium apparently lacking a carbon source. However, subsequent investigations show that the organisms are capable of growing in media with carbon and energy sources derived from volatile compounds present in the atmosphere. It is now known that the organisms preferentially utilize one-carbon compounds such as methanol, formate, and cyanides.

The mode of reproduction is distinguished from the caulobacter group by the formation of yeast-like buds at the tip of a prostheca (stalk) extended from the mature mother cell (Fig. 34–11). Hence, the prosthecae of this group are involved in reproduction, whereas the sole function of prosthecae of caulobacters is adhesion to a solid surface. During the reproductive cycle, one pole of the motile, polarly flagellated cell may become attached to a solid substrate (frequently to other aquatic flora and fauna), and the flagellum is lost. As the cell becomes ovoid and matures (to a mother cell), it extends its hypha-like structure (prostheca), and the free end of the prostheca swells and becomes a bud. As the bud becomes mature, it forms a flagellum and is eventually liberated from the mother cell as a motile swarm cell which may aggregate to form rosettes. The cycle then repeats itself.

There are several variations in reproductive cycles; for example, in some strains buds may form directly from the mother cell, while other

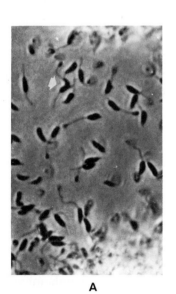

A

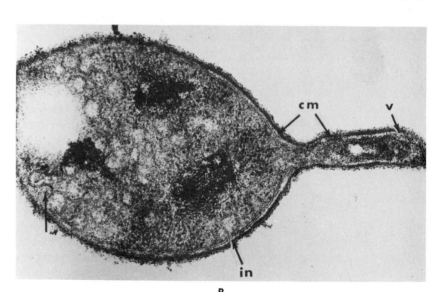

B

Figure 34–11. *A, Hyphomicrobium* T37 showing classic morphology and (*arrow*) a bizarre cell with regular buds. Phase contrast of living cells.

B, Hyphomicrobium strain H-526; longitudinal section showing the presence of numerous vesicles (*v*) within the cell and hypha. The interconnection of the vesicles (*arrow*), and invagination (*in*) of the cytoplasmic membrane (*cm*) can be readily observed. The relationship of the hypha to the rod portion of the cell is also apparent. (77,000×.) (*A*, Courtesy of P. A. Tyler and K. C. Marshall. *B*, Courtesy of S. F. Couti and P. Hirsch.)

strains may extend prosthecae from both poles of a mother cell to form two daughter cells. In some strains, the prosthecae may become branched or the motile stage may be absent; in such cases the daughter cell fails to separate from the mother cell to form a new prostheca from the free pole and continues to repeat the process, forming a microcolony. Nuclear division of the group is somewhat reminiscent of eucaryotic yeasts, in that one set of doubled nuclear material migrates through the prostheca to become a newly formed bud.

Recently, a new genus, *Ancalomicrobium* (Gr. *ancalo* = arm), has been added to this group. The species of the genus possess two to eight prosthecae extending from mature cells, but the prosthecae do not bear buds (Fig. 34–12). However, the mature cell reproduces by budding. As the bud matures, two or more prosthecae develop, and division occurs transversely when the mother and daughter cells have attained a similar size. The organism, unlike other hyphomicrobia, lacks both motile cell stage and holdfast and contains gas vacuoles. Also, the members of the genus are facultatively anaerobic and able to grow under strictly anaerobic conditions. The organisms are able to use numerous sugars and sugar alcohols for both carbon and energy source and require pantothenic acid for growth. DNA base composition of the genus ranges from 70 to 71 mole per cent GC.

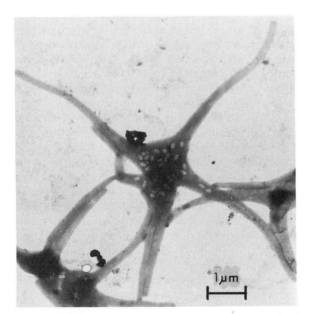

Figure 34–12. Electron micrograph of negatively stained cells of *Ancalomicrobium adetum*, strain 4a, containing gas vesicles. (Courtesy of J. T. Staley.)

Budding Bacteria

Recently, organisms of the family Pasteuriaceae (*Bergey's Manual*, 1957) and other budding and prosthecate bacteria have been unified into a taxonomic group called "budding bacteria," and the group has been placed in Part IV, "Prosthecate, Budding and Appendaged Bacteria" (8th edition, *Bergey's Manual*) thus far described. The group includes genera *Pasteuria, Blastobacter (Blastocaulis), Seliberia,* and *Planctomyces.* All are gram-negative chemoorganotrophs and share many properties with others described. For example, it has been reported that members of the genus *Pasteuria* multiply by longitudinal fission (like flagellate protozoans) or by budding of spherical or ovoid cells at the free end. Free, pear-shaped cells are nonmotile and frequently attach to each other or to a solid surface by a holdfast secreted at the narrow end of the cell.

The mode of reproduction and the life cycle of the organisms apparently resemble those of *Chamaesiphon,* a genus of blue-green algae. Some species of the group have been considered as epiphytes on various aquatic flora and fauna, while others may be parasitic on freshwater crustacea. The biochemical and physiological properties and reproductive cycles of this group need further investigation.

Appendaged Bacteria (The Iron Bacteria)

The organisms of this group are distinguished from other prosthecate bacteria by the nature of their "stalk" or "appendage." The stalk of this group is not composed of cell parts (extension of cell wall and membrane) but is formed by metabolic waste products containing ferric hydroxide excreted by the cell surface. It is considered nonessential in the reproductive cycle.

Most members of this group are currently classified in the genera *Gallionella, Siderophacus,* and *Nevskia* of the family Caulobacteraceae (*Bergey's Manual,* 1957). However, in the proposed classification, the group is included in Part IV, "Prosthecate, Budding and Appendaged Bacteria."

Gallionella ferruginea. *G. ferruginea* is a common and representative species of the stalk-forming family. Similar organisms are *Siderophacus* and *Nevskia.* Each cell of *Gallionella* forms a stalk which, as it matures, becomes encrusted with $Fe(OH)_3$. *Gallionella* does *not* form a sheath. It is said by some observers to be a true iron-oxidizing bacterium, but this is doubted by others. However, it grows only in waters bearing reduced iron (i.e., iron that can be oxidized as a source of energy). A curious metabolic feature is that all species require vitamin B_{12} (cyanocobalamin).

The cells of this organism are bean- or kidney-shaped and about 0.5 by 2 μm in size. Like other bacteria, they multiply by transverse binary fission. When motile, they resemble *Pseudomonas* and have polar flagella. From the concave side or end of each cell a flat, mucilaginous ribbon or stalk is excreted. This is attached by the distal end to some solid object. As each cell divides, dichotomy of the stalk occurs, so that complex tangles or rosettes of long stalks streaming from a common object are formed. The stalks are sometimes 0.2 to 0.3 mm in length (Fig. 34–13).

From the complex fibrillar structure of these stalks and their occasional independence of the cell, at one time it was inferred by some workers that these stalks or their fibers may possibly be living matter and play a role in the life cycle of the organism.

The stalks of *Gallionella* have the remarkable habit of twisting so that they resemble a loosely coiled rubber band. Large amounts of $Fe(OH)_3$ are later deposited in these stalks, giving them the appearance of a series of loops or strings of beads. The twisting habit renders identification of *Gallionella* easy, since no other organism of similar character is known to twist in just this way. As stated by Thimann: "... the gallionellas are more notable for their excreta than for themselves."

Gallionella is found in nature as widely distributed as *Sphaerotilus*. Like other iron-accumulating bacteria, *Gallionella* can multiply in water pipes and often causes extensive deposits and incrustations of iron which may eventually occlude the pipes. It is also responsible in part for the fouling of ship bottoms.

34.4
CHEMOLITHOTROPHIC IRON-DEPOSITORS

Gram-negative, nonphotosynthetic, chemolithotrophic organisms noted for deposition of iron in capsules or on cells are currently classified in the family Siderocapsaceae of the order Pseudomonadales. However, in the 8th edition of *Bergey's Manual* the members of the Siderocapsaceae are placed in Part XII, "Chemolithotrophic Bacteria," along with the family Nitrobacteraceae (see Chapter 31).

Representative genera of the Siderocapsaceae include *Siderocapsa, Siderococcus, Naumanniella,* and *Ochrobium.* All of these are found in iron- and manganese-rich, alkaline natural water in clusters of a few cells or as many as 60 or more, heavily encrusted with iron oxides. However, most of the members of the group have not been isolated in pure culture; thus many investigators consider the genera at best of doubtful significance.

Siderocapsa treubii. *S. treubii* is a coccobacillus about 0.5 by 2 μm in size. It is hardly alga-like but is representative of aquatic, iron-accumulating bacteria that grow in gummy masses, the Siderocapsaceae. *S. treubii* envelops itself in a thick, slimy or gelatinous extracellular substance and grows in fresh water in the form of a compact mass attached to some object such as a water plant or a stone. Sometimes these growths are extensive, and iron or manganese is deposited in large amounts around them. Like *Gallionella* and *Leptothrix,* they are important in the fouling of pipes, but their growth is not as extensive as that of *Gallionella* or *Leptothrix* or *Sphaerotilus.*

34.5
HABITAT AND ECOLOGICAL SIGNIFICANCE OF "SULFUR" AND "IRON" BACTERIA

Like the term "sulfur bacteria," "iron bacteria" is a term of convenience for bringing together some aquatic, alga-like bacteria that exhibit a curious relationship to iron in their metabolism but that are otherwise quite dissimilar. Nonphotosynthetic sulfur-utilizing bacteria, both sulfur-storing, oxidizing and reducing types, are common in sewage and other polluted waters, in decomposing organic matter and in swampy soils all over the world where putrefactive organisms are releasing H_2S from dead plants and animal wastes, or where sulfur-reducing species (*Desulfovibrio*) are reducing sulfates to H_2S. Some sulfur bacteria that are not at all like algae are found around free sulfur deposits, in oil wells or in sulfur springs. Some occur in acid coal mine waters, others in garden soil. (See discussion of sulfur bacteria, Chapter 43; also Table 34–2.) The strictly anaerobic photosynthetic species, of course, thrive in the sunlit situations where oxygen has been removed by chemosynthetic organisms and where H_2S occurs.

Reduced sulfur compounds such as H_2S are utilized by many different kinds of bacteria.

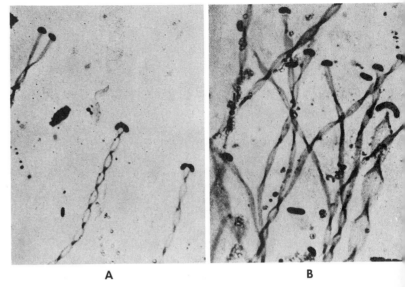

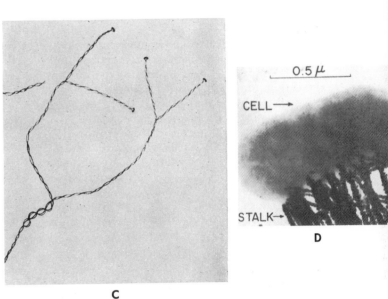

Figure 34–13. *A, Gallionella minor.* (1,100×.) *B, Gallionella major.* (1,100×.) *Gallionella ferruginea.* (650×.) *D,* Electron micrograph of one cell at the tip of ferric hydroxide stalk. (57,726×.)

For example, photosynthetic sulfur bacteria use H_2S as a source of reducing power and deposit the oxidized or free sulfur intracellularly or extracellularly; the deposited sulfur may be further oxidized to sulfate (SO_4^{2-}). Similarly, chemolithotrophic, "colorless, nonfilamentous, sulfur bacteria," *Thiobacillus, Thiobacterium, Thiospira,* and species of the family Thiobacteriaceae (*Bergey's Manual,* 7th edition) are able to utilize reduced sulfur compounds as a source of energy. Reduced sulfur compounds such as H_2S are first oxidized to free sulfur and deposited either intracellularly or extracellu-

larly, and additional energy may be obtained from oxidation of the sulfur to SO_4^{2-}.

However, the bioenergetics of H_2S oxidation by many of the filamentous gliding bacteria remain obscure, partly because these organisms are frequently on the border between autotrophic and heterotrophic life. For example, species of *Beggiatoa* fail to grow in the absence of organic compounds. However, the organisms grow well in medium provided with simple organic compounds, such as acetate, along with H_2S as energy source. This indicates that the organisms can use H_2S as an energy source although unable to fix CO_2. The group of bacteria able to use light or reduced inorganic compounds as energy source but unable to incorporate CO_2 as sole carbon source have frequently been called "mixotrophs." It should be emphasized that whether the organisms are photolithotrophic, chemolithotrophic, or mixotrophic sulfur bacteria, they are intimately involved in the sulfur cycle.

Nonphotosynthetic iron-oxidizing bacteria occur worldwide in fresh, sea, or brackish waters, especially waters rich in ferrous iron. Several species oxidize ferrous or manganous salts or organic compounds of iron or manganese as a source of energy and deposit $Fe(OH)_3$ or manganese oxides, not inside their cells but in **extracellular** structures. The yellow or reddish slime found on the mud and stones or water plants in iron-bearing waters is usually caused by ferric hydroxide in the sheaths or stalks or gum of iron-accumulating bacteria growing there.

The role of iron (or manganese) in the physiology of these organisms is interesting but is still under investigation. According to one view, the three groups of bacteria under discussion do not oxidize iron as a source of energy at all. They appear to utilize organic compounds containing iron but do not oxidize the iron itself.

For example, most sheathed bacteria (Chlamydobacteriales) and Siderocapsaceae can grow without gross amounts of iron or manganese, and then their sheaths or casings do not contain these metals. Young growths are free from iron or manganese. The deposition of the metals is a conspicuous and common feature of *mature* growths but clearly not an essential part of their physiology. The iron or manganese residue from the organism's metabolism remains as waste outside the cell in the sheath, stalk, or gum. According to one group of opinions, the metals are oxidized to ferric iron [or $Mn(OH)_3$] not by the cell but by free oxygen extraneous to the cell. The metals yield no energy to the cell. Such organisms are not true *iron bacteria* but might be called **iron-depositers.**

According to a logical view, only those organisms that oxidize **inorganic ferrous** iron compounds as a source of energy should be classed as true iron bacteria. A reaction often given to explain this process is:

$$4FeCO_3 + O_2 + 6H_2O \longrightarrow 4Fe(OH)_3 + 4CO_2$$

The iron is oxidized from the ferrous (Fe^{2+}) to the ferric (Fe^{3+}) state. There are several such species but they are not at all alga-like. They occur especially in acid drainage waters of iron and coal mines and are discussed elsewhere. (See discussion of *Ferrobacillus*, Chapter 43.)

These organisms are not pathogenic but are of great economic importance as scavengers because they decompose organic matter in water. They are also important economic nuisances because they grow in water-distributing pipe systems and create obstructions. Some large geologic iron deposits ("bog-iron") may represent the accumulation of iron over long periods by these microorganisms.

CHAPTER 34
SUPPLEMENTARY READING

Conti, S. F., and Hirsch, P.: Biology of budding bacteria. III. Fine structure of *Rhodomicrobium* and *Hyphomicrobium* spp. J. Bact., 89:503, 1965.

Hirsch, P.: Two identical genera of budding and stalked bacteria: *Planctomyces* Gimesi 1924 and *Blastocaulis* Henrici and Johnson 1935. Inst. J. Sys. Bact., 22:107, 1972.

Hirsch, P.: Re-evaluation of *Pasteuria ramosa* Metchnikoff 1888, a bacterium pathogenic for *Daphnia* species. Int. J. Sys. Bact., 22:112, 1972.

Jones, D., and Sneath, P. H. A.: Genetic transfer and bacterial taxonomy. Bact. Rev., 34:40, 1970.

Schmidt, J. M., and Samuelson, G. M.: Effects of cyclic nucleotides and nucleoside triphosphates on stalk formation in *Caulobacter crescentus*. J. Bact., 112:593, 1972.

Skerman, V. D. B.: A Guide to the Identification of the Genera of Bacteria, 2nd ed. The Williams & Wilkins Co., Baltimore. 1967.

Staley, J. T., and Mandel, M.: Deoxyribonucleic acid base composition of *Prosthecomicrobium* and *Ancalomicrobium* strains. Int. J. Sys. Bact., 23:271, 1973.

Tyler, P. A., and Marshall, K. C.: Pleomorphy in stalked, budding bacteria. J. Bact., 93:1132, 1967.

THE FAMILY ENTEROBAC-TERIACEAE AND OTHER GRAM-NEGATIVE BACTERIA

The order Eubacteriales, as classified in the 7th (1957) edition of *Bergey's Manual*, contains 69 genera and subgenera divided into 13 families, and these families include over 600 species. The order is not included as such in the new (8th) edition of *Bergey's Manual*.

Bacteria formerly in the order Eubacteriales occur in only two basic forms: (1) unicellular spheres or spheroids (**cocci;** singular, **coccus** [Gr. *kokkos* = grain; berry]); (2) simple, unicellular rods or rod-like forms (**bacilli;** singular, **bacillus** [L. *bacillum* = small rod]); various modifications and distortions of both forms are frequent.

At this point it may be well to emphasize that the term **bacillus** refers indiscriminately to any rod-shaped bacterium regardless of its other properties, whereas the term *Bacillus* (Chapt. 36) refers only to the genus of aerobic, rod-shaped bacteria that form heat-resistant endospores, i.e., the genus *Bacillus*. The word **bacterium** refers to any of the Schizomycetes regardless of form or properties.

Of the hundreds of species of Eubacteriales one or more is to be found almost anywhere on the surface of the earth: in the soil, thousands of feet in the air, in rivers and lakes and in the most profound marine depths (six miles or more), in mines, on mountain tops, in and on plants and animals, from pole to pole.

35.1
EUBACTERIALES

The Eubacteriales, as indicated by the name of this order, have what may well be called **typical** bacterial characteristics, being independently unicellular, undifferentiated cocci or rods. They are microscopic in size (about 0.5 to 2.0 μm in diameter; 5 to 20 μm in length if rod-shaped), enclosed in rigid, relatively thick, strong cell walls of or containing peptidoglycan, and are osmotrophic chemoorganotrophs in nutrition. They multiply by transverse binary fission.

All motile species are **peritrichously** flagellate. This is one of the major characters differentiating all Eubacteriales from all species of the other large group of typically unicellular bacteria of similar form and size, the order Pseudomonadales (now family Pseudomonadaceae and others). All Pseudomonadales, when motile, are **polarly** flagellate. (See Table 33–1.)

501

Many of the families listed in the 7th edition of *Bergey's Manual* are now divided primarily on the basis of Gram staining reactions, oxygen requirements, and cell shapes. Some of the families formerly in the order Eubacteriales have been eliminated or transferred to several other groups.

Gram-Negative, Rod-Shaped, Asporogenous Bacteria

Many of the species in the family Enterobacteriaceae and several other families of Eubacteriales listed in the 7th edition of *Bergey's Manual* that are gram-negative, nonsporeforming rods are extremely important in human activities such as medicine, agriculture, and industry. They exhibit a wide variety of properties. Some are adapted to very specialized situations such as sea bottoms at 4C and thousands of pounds pressure per square inch, others to sunwarmed soils or heaps of decaying matter at 35 to 40C. Many are adapted to very special ecological niches such as the intestinal tract of man and lower animals; some are highly specialized and fastidious parasites. Many species are motile; these species have **peritrichous** flagella. Nearly all are aerobic or facultative with respect to oxygen. An outstanding exception is the strictly anaerobic family Bacteroidaceae. The other families, as arranged in the 7th edition of *Bergey's Manual* (1957), are: (1) Azotobacteraceae, (2) Rhizobiaceae, (3) Achromobacteraceae, (4) Enterobacteriaceae, (5) Brucellaceae. Some of their properties are listed for comparison in Table 33–1.

The families Azotobacteraceae and Rhizobiaceae contain very highly specialized species that are important as causes of diseases of agricultural plants, e.g., *Agrobacterium,* and also the extremely valuable genus of *Rhizobium,* species of which are involved in the fixation of atmospheric nitrogen in symbiosis with leguminous plants. The family Azotobacteraceae is highly important in nonsymbiotic fixation of nitrogen in the soil. Both these families are more fully discussed in Chapter 43, dealing with microbiology of the soil.

The family Achromobacteraceae (7th edition, *Bergey's Manual*; this family has been eliminated in the 8th edition) contains five genera: *Alcaligenes, Achromobacter, Flavobacterium, Agarbacterium,* and *Beneckea,* comprising species that in general are small (0.5 by 1.0 to 4.0 μm) rods widely distributed, especially in aquatic and marine habitats, on seaweeds, in fish slime and soil. They are almost exclusively harmless saprophytes. Some are motile, others nonmotile. In spite of the family name, some produce red, yellow or orange carotenoid pigments. (For this reason and others, the genus *Achromobacter* has been eliminated from the 8th edition of *Bergey's Manual* and is now known as *Chromobacterium.*) Most grow at temperatures ranging from 4 to about 30C. All are aerobic or facultative. As a group they are not very active in fermentations, and rarely attack glucose or lactose anaerobically. One group (*Agarbacterium*) has the uncommon property of hydrolyzing the polysaccharide **agar,** a component of certain seaweeds (and of bacteriological culture media). Another group (*Beneckea*) actively hydrolyzes **chitin,** the skeletal material of crustacea and of fungal cell walls. Several groups, especially *Flavobacterium* and *Agarbacterium,* are actively proteolytic. Altogether, the family Achromobacteraceae constitutes a group of well diversified and important scavengers.

35.2
THE FAMILY ENTEROBACTERIACEAE

Most species of the family Enterobacteriaceae differ from the specialized, restricted and aerobic Rhizobiaceae and Azotobacteraceae in being able to thrive in a wide variety of habitats and under a wide range of environmental conditions. The Enterobacteriaceae are generally active fermenters of glucose and many other sugars and alcohols; many ferment lactose. Motile species are peritrichously flagellate. Many Enterobacteriaceae can grow in simple mineral solutions with a little glucose or peptone, and they are also readily cultivable in a variety of media at temperatures from 25 to 37C at pH of 6.5 to 8.0. In size, Enterobacteriaceae commonly range from 0.5 to 2.0 μm in diameter and lengths of from 1.0 to 10 μm.

Many species of the family possess pili (fimbriae) which must be distinguished from flagella. Their fine structure and the role played by the specialized F pili in chromosomal transfer during conjugation were discussed in an earlier chapter. Some organisms of the family Enterobacteriaceae, namely plant pathogenic species of the tribe Erwinieae, produce pectinase. Other pectinase producers are found in the family Pseudomonadaceae (Chapt. 33).

Whereas many species of Enterobacteriaceae are primarily environmental saprophytes and scavengers, all are sometimes (i.e., from rarely to invariably) found in the intestinal tract of man or lower animals, hence the family name. Some are dangerous primary pathogens, others are potential or secondary pathogens. Because of the relationship to environmental sanitation and to enteric diseases (typhoid fever, paratyphoid fevers, dysentery, and infant diarrhea) and to infections of the urinary tract and other organs, a major facet of interest in, and research concerning, the Enterobacteriaceae revolves around their taxonomy and the isolation and identification of the different species.

As arranged in *Bergey's Manual* (1957), there are some 60 species of Enterobacteriaceae. In the 8th edition they are arranged as shown in Table 35–1. These may be identified by groups on the basis of physiological and biochemical properties shown in Table 35–2. In addition, the kind, range, and proportion of the end products formed ("mixed acid" and butylene glycol fermentation) as the result of anaerobic fermentation of glucose are frequently used in separating the genera in the family.

The organisms of the family Enterobacteriaceae may be divided into two major groups on the basis of the fermentation end products of lactose; for example, mixed acid–producers, which include genera *Escherichia*, *Shigella*, *Salmonella*, *Erwinia*, and *Proteus*; butylene glycol–producers, which include three genera, *Klebsiella*, *Enterobacter* (*Aerobacter* or *Klebsiella*), and *Serratia*.

It is evident that no single biochemical test or physiological property can be used alone to identify any member of the family. It is to be noted also that the demonstrability of various physiological and biochemical properties is greatly influenced by the cultural conditions and methods used for such purposes. As stated by Edwards and Ewing: "A source of difficulty in work with enteric bacteria is the great lack of uniformity in methods employed . . . in different laboratories. . . ." This is true of other groups, also.

The DNA base composition of the family as a whole ranges from a low of 36 to 53 mole per cent GC (*Proteus* spp.) to a high of 54 to 63 mole per cent GC for *Serratia*. However, within the same genus or tribe the DNA base composition of various strains is very nearly the same; for example, 49 to 54 mole per cent GC for organisms in genera *Escherichia*, *Salmonella*, and *Shigella*.

TABLE 35–1. TRIBES IN THE FAMILY ENTEROBACTERIACEAE*

Tribe: Escherichieae
Genus   I. *Escherichia*
         II. *Edwardsiella*
       III. *Citrobacter*
       IV. *Salmonella*
        V. *Shigella*

Tribe: Klebsielleae
Genus   VI. *Klebsiella*
     VII. *Enterobacter*
   VIII. *Hafnia*
     IX. *Serratia*

Tribe: Proteeae
Genus   X. *Proteus*

Tribe: Yersinieae
Genus   XI. *Yersinia*

Tribe: Erwinieae
Genus XII. *Erwinia*

*Based on the 8th edition of *Bergey's Manual*. DNA base composition, mole per cent G + C of these organisms are presented in Table 35–7. Students should note that this book as well as other reference books uses various generic names (as shown in Table 35–2) to designate members of groups. For further classification by other generic names applied to tribe and genera of Enterobacteriaceae the student is referred to the reference material listed in the Supplementary Reading.

35.3
THE TRIBE ESCHERICHIEAE

The tribe contains five genera: *Escherichia*, *Edwardsiella*, *Citrobacter*, *Salmonella*, and *Shigella*. The organisms of these genera are described below.

Genus Escherichia. This genus is composed of *E. coli* and several biotypes. The organisms as a group are motile or nonmotile. Glucose and lactose may be fermented with the production of gases (CO_2, H_2 in the ratio 1:1), but some strains are anaerogenic. CO_2 and H_2 produced by *E. coli* are derived from formic acid through action of the enzyme formic hydrogenylase, and as a result equal moles of CO_2 and H_2 are produced:

$$HCOOH \xrightarrow[H_2O]{enzyme} H_2 + CO_2$$

Most strains are surrounded by microcapsules, and many strains possess fimbriae (pili). In some strains the sex (or F) fimbrial type may be detected by its affinity for special donor phage and by its antigenic properties (Fig. 35–1).

TABLE 35–2. SUBDIVISIONS OF THE ENTEROBACTERIACEAE *

Major Subdivisions	Differential Properties Indole[1]	Methyl Red	Acetylmethyl-Carbinol[1]	Utilizes Citrate	Splits Urea[2]	H₂S[1]	Phenylalanine Deaminase[1]	Groups or Genera	Lactose	β-Galactosidase (ONPG Test)	Glucose (with gas)	Sucrose	Motility	Mannitol	KCN[3]	Gelatin[7]	Urea[2]	Phenylalanine Deaminase	H₂S[1]	Decarboxylates L-Lysine	Deaminates L-Lysine	Extracellular DNase
Shigella-Escherichia	+(−)[8]	+	−	−	−	−	−	Shigella	−[4]	−	−[4]	−	±[5]	−	−	−	−	−	−	−	−(+)	−
								Escherichia	+	+	+	+	+							−(+)	−(+)	
Salmonella-Arizona-Citrobacter	−	+	−	+	−	+	−	Salmonella	−[4]	−	+[6]	−	+	+	−	−	−	−	+	+	−	−
								Arizona group	+[4]	−	+	−	+	+	−	+	−	−	+	+	−	−
								Citrobacter group	+[4]	+	+	±	+	+	+	−	−	−	+	−	−	−
Klebsiella-Enterobacter-Serratia	−	−	+	+	−	−	−	Klebsiella	+	+	+	+	−	+	+	−	−	−	−	+	−	−
								Enterobacter	+	+	+	+	+	+	+	+	−	−	−	−	−	−
								Hafnia	+	+	‥	+	‥	‥	‥	−	−	−	−	+	−	−
								Serratia	−	±	+	+	+	+	+	+	−	−	−	+	−	+
Proteus-Providence	+(−)	+	−	+(−)	+(−)	±	+(−)	Proteus group	−	±	±	+	±	+	±	+	+	+	±	+(−)	−(+)	±
								Providence group (Proteus inconstans)	−	+	−[4]	+	−	+	+	−	+	−	−	±	−	−

* It must be understood that in describing any group of bacteria allowance must be made for variation, modification, and mutation. Forms of any species not infrequently occur that are aberrant with respect to any physiological characteristic.

[1] Produced.
[2] *Rapidly* hydrolyzed.
[3] Grows in the presence of KCN.
[4] Some slowly positive.
[5] ± = variable.
[6] S. typhi acid but no gas.
[7] At 22C.
[8] + or − in parentheses = some variants.

In the study of genetics, strains of *E. coli* must be considered as the most extensively used among the myriads of bacteria. The three types of gene exchange mechanism—transduction, transformation, and conjugation—have been demonstrated largely among *E. coli* strains.

However, DNA homology studies indicate that recombination does not occur freely among all strains. Many strains of *E. coli* carry prophage; some strains produce colicins, enterotoxins, or hemolysins, some of which have been found to be genetically determined. Likewise, episomal transfer of antibiotic resistance factors (R) among different strains of *E. coli* has been demonstrated. (DNA base composition of 50 to 52 mole per cent GC for the genus has been reported.)

All strains are of particular interest to the sanitarian since they occur commonly in the normal intestinal tract of man and animals. Their presence in foods or drinking water, therefore, may indicate fecal pollution. *Escherichia coli* is the most distinctively fecal species and is *always* found in the normal intestinal tract. Certain strains of *E. coli* cause mild to severe diarrhea, especially in infants.

Pathogenic E. coli. If given a large enough dosage and sufficient opportunity, such as a very dirty wound or an old, slowly healing ulcer, *Escherichia coli* may act as a secondary invader, especially if the patient's general health and nonspecific resistance are low. *Escherichia coli* may cause more serious trouble by invading the bladder and pelvis of the kidney after surgery or instrumentation. It produces a stubborn and dangerous inflammation. In the bladder this is called **cystitis;** in the pelvis of the kidney, **pyelitis.** *Proteus* and *Pseudomonas* species are often also involved.

Extensive studies of bacteria in the feces of infants with diarrhea show that many cases of infantile diarrhea can be traced to certain particular kinds of *Escherichia coli.* These pathogenic strains can be distinguished from other strains of *E. coli* only by immunologic studies of their antigenic structure. Some of these strains of *E. coli* are designated as O26:K60(B6):H11;

O55:K59(B5):H6; O111a, 111b:K58(B4):HZ, and so on. The numbers and letters refer to O, K, and H antigens in the organisms. At least 13 such strains are known (Table 35–3).

These organisms and *Pseudomonas aeruginosa* are particular nuisances in children's institutions and nurseries. They also cause serious infections in adults. They are spread about by hands and fomites, as are other enteric pathogens, and at times are very difficult to eradicate.

Genus Edwardsiella. The members of this group closely resemble the members of the genera *Escherichia*, *Citrobacter*, *Proteus*, and *Salmonella*, and some of the biochemical differences and similarities are summarized in Tables 35–8 to 35–10. The organisms as a group are motile and produce H_2S from triple sugar-iron agar (TSI agar). Several serotypes have been described, but *E. tarda* is recognized as the type

TABLE 35–3. ESCHERICHIA COLI SEROTYPES[*] **REPORTED IN DIARRHEAL DISEASE**[†]

O Antigen	K Antigen	H Antigens (and Synonyms)
26	60 (B6)	NM (nonmotile) (E893), 11, 32
55a	59 (B5)	NM 6, 7
86a	61 (B7)	NM (E990), 11, 34
111a, 111b	58 (B4)	NM, 2 (D433), 4, 12, 21
112a, 112c	66 (B11)	NM (Guanabara)
119	69 (B14)	NM, 6 (Aberdeen 537-52)
124	72 (B17)	NM, 30
125a, 125b	70 (B15)	19 (Canioni)
125a, 125c	70	15, 21
126	71 (B16)	NM, 2 (E611), 27
127a	63 (B8)	NM (Holcomb)
128a, 128b	67 (B12)	2 (Cigleris), 7, 8, 9, 12
128a, 128c	67	NM, 12

[*]The simple numerical designations applied above to the K antigens are in accordance with the nomenclature recommended by Kauffmann et al.: Int. Bull. Bact. Nomen. Taxon., 6:63, 1956. For reference to descriptive literature regarding these serotypes, the reader is referred to Ewing, Davis, and Montague: Studies on the Occurrence of *Escherichia coli* Serotypes Associated with Diarrheal Disease. National Communicable Disease Center, Atlanta, Ga., 1963. Only the more prevalent serotypes are listed.

[†]From Blair, Lennette, and Truant (Eds.): Manual of Clinical Microbiology, American Society for Microbiology, The Williams & Wilkins Co., 1970.

species. The DNA base composition of *E. tarda* ranges from 51 to 52 mole per cent GC. The organisms normally inhabit the intestinal tracts of snakes. However, occasionally the organisms have been isolated from the stools of humans with diarrhea, from urine, and from the blood of man and animals.

Genus Citrobacter. This genus is created for the organisms conforming to the following general characteristics: motility (Fig. 35–2), fermentation of lactose with acid and gas ($CO_2:H_2$ in the ratio of 1:1), utilization of citrate as the sole carbon source, production of tri-

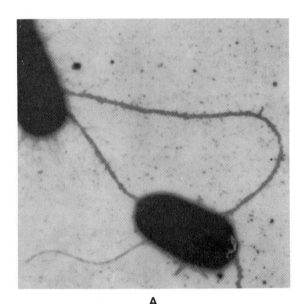

A

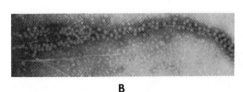

B

Figure 35–1. *A*, Electron micrograph of presumed specific pair between an Hfr cell (bottom) and an F⁻ cell. MS-2 was used to "stain" F pili. The specimen was prepared soon after mixing the Hfr and F⁻ cells. Two F pili have been used to make contact with a single F⁻ cell. (16,250×.)

B, Electron micrograph of two F pili from an Hfr cell with donor-specific RNA phage (MS-2) attached laterally. Numerous type I pili are recognized by their shorter length and their inability to adsorb MS-2. Negatively stained with phosphotungstic acid. (33,000×.)

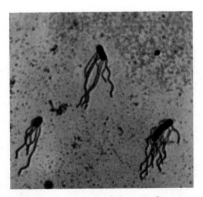

Figure 35–2. Peritrichous cells of *Citrobacter diversus* ATCC 27156. Twenty-four-hour broth culture at 26C. Staining by Leifson's method. (1,200×.)

methylene glycol from glycerol, and lack of growth inhibition by the presence of KCN. The organisms as a group have a DNA base composition of 50 to 53 mole per cent GC.

Two species, *C. freundii* and *C. intermedius*, are found in water, food, feces, and urine, but their pathogenicity is problematical. The organisms are thought to be normal intestinal inhabitants and are found constantly in healthy persons. Certain serotypes apparently cause sporadic infections of alimentary and urinary tracts, and infections of gallbladder, middle ear, and meninges have been reported.

Genus Salmonella. The genus *Salmonella* consists of rod-shaped cells, usually motile (except *S. gallinarum* and *S. pullorum*) (Fig. 35–3), and are able to use citrate as a carbon source. Most strains are aerogenic, but an important exception, *S. typhi*, never produces gas. The organisms as a group have a DNA base composition of 50 to 54 mole per cent GC, and the type species is *S. cholerae-suis*.

The nomenclature of the salmonellas does not follow the usual rules; for example, the first salmonellas were given names which indicated the disease and the animal from which they were isolated (*S. cholerae-suis*, *S. typhi-murium*, and *S. abortus-ovis*). However, such naming of the organisms was abandoned because of the implication that pathogenicity was limited to certain animal species; on the contrary, *S. typhimurium* and *S. bovis-morbificans* are frequently isolated from human infections. More recently, the name of the town, region, or country in which the first strain was isolated, such as *S. london*, *S. kentucky*, and *S. stanleyville*, has been used. As a result more than 1,200 bacterial species related to *S. typhi* have been isolated

and characterized. Scientifically, none of the present methods of nomenclature of *Salmonella* is satisfactory; however, it is generally accepted that the Kauffmann-White schema be used for diagnosis until a more satisfactory method is developed.

Some of the most common species of *Salmonella* recognized are:

Group A	Group C
S. paratyphi-A	S. paratyphi-C
S. senftenberg	S. cholerae-suis
Group B	S. thompson
S. paratyphi-B	Group D
S. typhi-murium	S. typhi
S. heidelberg	S. enteritidis
	S. sendai

Kauffmann-White Schema. With a collection of adsorbed, monospecific (often called monovalent) agglutinating sera (Chapt. 27), each representing a single different antigenic component of the various species of *Salmonella*, one may test any given organism for the presence of different H and O antigenic components, and assign to it a "formula" expressing the antigenic complex of which it is composed.

By use of such antigenic analyses, the salmonellas have been arranged in a series called the **Kauffmann-White Schema.** In this schema the O (somatic) antigens are given Arabic numbers. The flagellar (H) antigens have two series of numbers, depending on **phase variations:** small letters if in phase I, Arabic numbers if in phase II. The antigenic structure of any given species may therefore be expressed in terms of these numbers and letters. For example, *S. typhimurium* has the antigenic formula 1, 4, 5, 12: i;

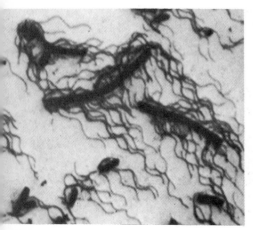

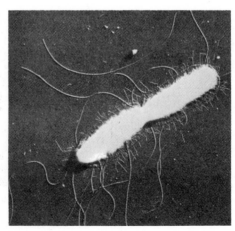

Figure 35–3. *A*, Bacterial flagella as seen in stained film of *S. typhi*.

B, *Salmonella typhi*; dividing bacilli with numerous fimbriae and a few flagella (the very long appendages). (12,500 ×.)

A   B

TABLE 35-4. ANTIGENIC FORMULAS OF SOME SALMONELLA SPECIES

			H Antigens	
Group	Species	O Antigen	PHASE I	PHASE II
A	S. paratyphi-A	1, 2, 12	a	—
B	S. schottmuelleri (S. paratyphi-B)	1, 4, 5, 12	b	1, 2
	S. typhi-murium	1, 4, 5, 12	i	1, 2
C₁	S. hirschfeldii (S. paratyphi-C)	Vi, 6, 7	c	1, 5
	S. cholerae-suis	6, 7	c	1, 5
D	S. enteritidis	1, 9, 12	g, m	—
	S. typhi	Vi, 9, 12	d	—
E₄	S. senftenberg	1, 3, 19	g, s, t	—

1, 2 (see Table 35–4). Species having one or more **somatic** antigens in common are placed in convenient groups: A, B, C, and so on.

The schema is essential in routine laboratory diagnosis of an organism suspected of being *S. typhi*. For example, if the reactions of the suspected culture are similar to those of *S. typhi* on TSI agar and tests for urease are negative, living and heat-killed cell suspensions of this isolate should be subjected to slide agglutination tests using *Salmonella* polyvalent antiserum, group D antiserum (9, 12) and polyvalent antiserum containing Vi antibody. Typhoid fever is caused by strains of *Salmonella* that possess the capsular Vi antigen, and strains which produce this antigen are given the designation of *S. typhi*. The type H (flagellar) antigen should be determined before a final report (including confirmatory biochemical tests, Table 35–5) is prepared.

Similar antigenic schemes are found in other groups of bacteria, such as *Clostridium, Shigella, Klebsiella, Corynebacterium,* and *Escherichia.*

VARIATION OF SPECIES. A confusing feature of such antigenic analysis is the fact that by various procedures, including transduction, one type may readily be changed into another! Undoubtedly such changes also occur in nature. New types are constantly being found.

Closely associated with the O antigens of Enterobacteriaceae is another antigen common to all species of these bacteria. It is called a "common antigen" (CA). CA differs from O antigen in being soluble in 85 per cent alcohol. It is highly antigenic when separated from O antigen, has marked opsonic activity, and can be identified by hemagglutination tests. Its presence is masked by the presence of overlying group-specific O antigens.

Salmonellosis. Salmonellosis may range in severity from almost imperceptible intestinal discomfort to fatal disease (notably in typhoid fever).

Of the more than 1,200 named salmonellas, *S. typhi* (the cause of typhoid fever) and the so-called "food-poisoning," or paratyphoid, group (*S. paratyphi-A, S. paratyphi-B, S. paratyphi-C, S. typhi-murium, S. enteritidis,* and *S. cholerae-suis*) are among the most important. Representative species of principal groups of *Salmonella* may be differentiated by the biochemical tests shown in Table 35–5. Further differentiations (serotypes) are based on antigenic structure (Table 35–4). The resulting diseases are characteristically gastrointestinal but may be septicemic (bloodborne) and completely generalized in the body.

S. gallinarum causes an infection of chickens known as fowl typhoid, and *S. pullorum* is the causative agent of chicken pullorum disease, both of which have a high fatality rate but rarely, if ever, infect man.

EPIDEMIOLOGY OF SALMONELLOSIS. The habitat of the organisms is mainly the intestinal

TABLE 35-5. DIFFERENTIAL PROPERTIES OF SOME SALMONELLAS °

Species	Xylose	Trehelose	Arabinose	H₂S	Citrate Agar †	Gas in Dextrose
S. typhi	v	+	—	+	—	—
S. paratyphi-A	—	⊕	⊕	—	—	+
S. schottmuelleri (S. paratyphi-B)	⊕	⊕	⊕	+	+	+
S. hirschfeldii (S. paratyphi-C)	⊕	⊕	⊕	+	+	+
S. enteritidis	⊕	⊕	⊕	+	+	+
S. typhi-murium	⊕	⊕	⊕	+	+	+
S. cholerae-suis	⊕	—	—	—	+	+

° Variants are not rare. Symbols: ⊕ = acid and gas; + = positive test or acid only; − = negative test; v = variable.
† Growth with citrate as sole source of carbon.

tract and tissues of infected animals, but the organisms can grow in feces-polluted foods and may survive in polluted or infected foods, in waters, and on fomites for periods of from a few hours to days. Hence, *Salmonella*, like all of the Enterobacteriaceae, is transmitted by feces or urine, or both, of patients or carriers. A common vector for all of them is feces-soiled hands. Another is food that has become infected and allowed to stand in a warm place after little or no cooking so that the organisms can grow. These organisms grow well at warm room temperatures (65 to 100 F). Infection of the food is sometimes by unwashed hands of a person with a very mild infection, or a human carrier.

Food **infection** (often incorrectly called "food poisoning") may also be caused by introduction of excrement of dogs, mice, rats, flies, or cockroaches that harbor, particularly, *S. typhimurium* as well as other salmonellas. So-called "meat poisoning" often results from eating or handling raw or improperly cooked flesh of cattle, swine, poultry, fish, or other animals suffering from infection with these organisms, especially *S. cholerae-suis* and *S. enteritidis*. These species are among the most commonly occurring in the United States. Even hens' eggs are often infected, before being laid, by maturation in an infected hen. The eating of raw egg or egg products is therefore not wise and has resulted in large outbreaks of salmonellosis caused by raw eggs in mayonnaise used in sandwiches.

Obviously, avoidance of these diseases means cleanliness in the kitchen; sanitary habits on the part of food handlers; care to see that food is properly cooked to kill all organisms, even those in the center of large masses; proper refrigeration of stored food; and avoidance of uncooked hand-prepared foods at club suppers or on picnics. Foods are often prepared during the morning or previous evening and then unwittingly incubated in the kitchen or in transit. *Salmonella* food infections are very common. Characteristically the onset is at least 10 (usually 18 to 24) hours after eating the infected food. The bacteria multiply during this incubation period.

Typhoid Vaccination. Typhoid fever outbreaks have been controlled most dramatically since 1900. Mortality rates due to typhoid and paratyphoid fevers in the United States from 1900 to 1970 have decreased from 37 per 100,000 to less than 1 per 100,000. This reduction is due principally to such public health measures as purification of water, treatment of sewage, pasteurization of milk, elimination of chronic carriers as food handlers, and general education

of the public through health officers and physicians. However, even today the fever is not completely eradicated, and breakdowns in milk pasteurization or in water-supply treatment have led to serious outbreaks of the disease.

Probably most students will have received typhoid "shots" at some time before studying microbiology, especially if they have seen military service. These injections are a good example of a method of active artificial immunization. The material injected is saline solution containing about one billion *Salmonella typhi* per milliliter killed by heating at about 65C for 30 minutes, or with formaldehyde. Sometimes included are killed *S. paratyphi-A* and *S. paratyphi-B* (TAB vaccine).

For initial immunization two doses (0.5 ml and 1.0 ml) at three- to six-week intervals are required. Revaccination with small doses every three or four years (0.1 ml intracutaneously or 0.5 ml subcutaneously) is recommended in order to maintain immunity at an effective level. This is a good example of the use of the secondary antigenic stimulus, or booster dose (Chapt. 28).

Chloramphenicol is used in the treatment of enteric fevers and *Salmonella* septicemia. However, the treatment of chronic carriers by antimicrobial drugs is seldom used; more frequently cholecystectomy (gallbladder removal) is performed, which terminates the carrier state in 9 out of 10 cases.

Genus Shigella. The genus *Shigella* is composed of nonmotile, unencapsulated rods (Fig. 35–4) which ferment glucose and other carbohydrates with production of acid but not gas (except certain species). The organisms are unable to utilize citrate as sole carbon source; growth is inhibited by bismuth sulfite; urease and H_2S are not produced. Growth is inhibited by KCN. The DNA base composition of the genus ranges from 49 to 54 mole per cent GC.

Figure 35–4. *Shigella flexneri;* dividing bacilli with numerous fimbriae surrounding the cells. (20,000×.)

The organism discovered by Shiga in 1896 during a frightful epidemic of dysentery in Japan with over 22,000 fatalities is now called *Shigella dysenteriae*. It is the type species of the genus. The principal distinguishing characteristics of the genus are shown in Table 35–6. After Shiga's discovery many other kinds ("species") of dysentery bacilli were discovered by Flexner, Boyd, Sonne, and others. Except *S. sonnei*, each represents several serotypes. The classification and differentiation of these species and serotypes present problems analogous to those related to classification of the salmonellas. H antigens are not involved. (Why?)

The four principal species, *S. dysenteriae*, *S. flexneri*, *S. boydii*, and *S. sonnei* are often referred to as subgroups A, B, C, and D. The organisms in subgroup A are separated from all others by their inability to ferment mannitol. The organisms in subgroup D (*S. sonnei*) are unlike all others in that they generally ferment lactose and sucrose after prolonged incubation. Organisms in subgroup B (*S. flexneri*) are differentiated on the basis of serological properties; i.e., all the members of the subgroup B are interrelated. However, members of subgroup C (*S. boydii*) are not related serologically to each other or to the other subgroups. The organisms in the subgroup formerly known as the *alkalescens-dispar* (A-D) are now excluded from *Shigella*, since they are more closely related to *Escherichia* both in biochemical and serological properties than to other members of Enterobacteriaceae.

Bacillary Dysentery. Any culture isolated from a suspected patient which proves nonmotile, produces an alkaline slant, an acidic butt, and no H_2S in slants of Kligler's or TSI agar, and is unable to hydrolyze urea rapidly should be considered as a possible *Shigella* strain and treated accordingly.

The shigellas cause intestinal disturbances ranging from very mild diarrhea to severe and sometimes fatal dysentery with intense inflammation and ulceration of the large bowel, often with scar formation and stricture of the bowel after recovery. Unlike *Salmonella typhi*, which always causes bacteremia, *Shigella* does not commonly invade the blood. In some epidemics of bacillary dysentery, especially those caused by *S. dysenteriae*, the fatality rate is high.

The pathogenicity of *S. dysenteriae* is dependent not only on production of endotoxin common to other shigellas but also upon production of a soluble protein exotoxin known as Shiga neurotoxin; apparently no other *Shigella* species forms exotoxins. Its toxicity can be completely inactivated by heat or neutralized with specific antitoxin.

The transmission and prevention of bacillary dysentery are similar to those aspects of salmonellosis, except that animals do not transmit dysentery.

Dysentery Vaccination. Vaccination against bacillary dysentery appears to be much less satisfactory than typhoid vaccination and is rarely practiced in the United States.

35.4
TRIBE KLEBSIELLEAE

The genus *Klebsiella* was placed in the tribe Escherichieae in the 1957 edition of *Bergey's Manual*. However, in the proposed 8th edition, the genus *Klebsiella* has been placed in the tribe Klebsielleae of the family Enterobacteriaceae. The tribe includes the genera *Klebsiella*, *Enterobacter* (*Aerobacter* or *Klebsiella*), *Hafnia* (*Paracolobactrum*), and *Serratia*. The DNA base composition of the group varies from 52 mole per cent GC in *Klebsiella* to 63 mole per cent GC in *Serratia*. Unlike organisms in the tribe Escherichieae, the organisms in this tribe are able to grow in broth containing KCN (i.e., they are KCN-tolerant).

Genus Klebsiella. The type species of this genus is *Klebsiella pneumoniae*. DNA base

TABLE 35–6. SOME CULTURAL REACTIONS OF THE SHIGELLAS °

Serological Group	Principal Species	Mannitol	Lactose	Sucrose	Indole Production
A	*Shigella dysenteriae*	−	−	−	−
	Shigella ambigua	−	−	−	+
B	*Shigella flexneri*	+	−	−	±
C	*Shigella boydii*	+	−	−	±
D	*Shigella sonnei*	+	Slowly +	Slowly +	−

°There are exceptions to most of the reactions, in aberrant strains.

composition of the members ranges from 52 to 59 mole per cent GC. The organisms are short rods, are nonmotile, are heavily capsulated, and produce mucoid colonies on agar media (Color Plate VI, *B*). Most members produce 2,3-butylene glycol as a major end product of glucose fermentation (some strains produce 2,3-butylene glycol either very slowly or in too small an amount to cause a sufficient pH change to give a positive methyl red (MR) reaction. With other strains, the 2,3-butylene glycol will disappear before the Voges-Proskauer (VP) test is made, which is evidenced by presence of acetoin; apparently these reactions are responsible for the seemingly paradoxical MR and VP tests (i.e., ++ and −−).

Capsule formation is a variable characteristic and noncapsulated avirulent forms are practically identical with *Enterobacter*. Hence, in the proposed 8th edition of *Bergey's Manual*, nonmotile bacteria able to grow at 37C and listed in the 7th edition as *A. aerogenes* are included as *K. pneumoniae*. All *Klebsiella* strains not belonging to *K. ozaenae* or *K. rhinoscleromatis* are considered to be *K. pneumoniae*.

Though they are not motile, most strains of *K. pneumoniae* possess fimbriae of either a thick (MS, 6.5 to 7.0 nm) type, which adhere to red cells of the guinea pig and other animals except ox, or a thin (MR, 4.7 to 4.8 nm) type, which adhere to tanned cells (ox cells most suitable). Adhesion of MS type fimbriae to red cells is inhibited by presence of D-mannose, but adhesion of MR fimbriae is not.

Species of *Klebsiella* may be classified on the basis of capsular (K, types 1 to 6) and somatic (O, types 1 to 11) antigens. There are more than 80 K types and 11 different O types.

Pathogenesis and Ecology. Virulent forms of *K. pneumoniae* produce diseases principally of the respiratory system, and the organism (sometimes called Friedländer's bacillus) causes a frequently fatal type of pneumonia. *K. pneumoniae* is widely distributed in nature in soil, water, and grain and is normally found in the intestinal tracts of healthy humans and other animals.

K. ozaenae causes chronic disease of the respiratory tract (**ozena,** a progressive, fetid atrophy of the nasal mucosa). *K. rhinoscleromatis* may cause nodular enlargement of the nose and respiratory tract (**rhinoscleroma,** a destructive granuloma of the nose and pharynx).

Genus Enterobacter. The genus *Enterobacter* is composed of motile, nonsporeforming rods (Fig. 35–5). The organisms grow on ordinary media, reduce nitrates to nitrites, give a

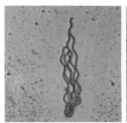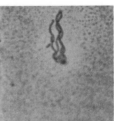

Figure 35–5. Peritrichous cells of *Enterobacter agglomerans* ATCC 27155. Twenty-four-hour broth culture at 26C. Staining by Leifson's method. (1,200×.)

negative oxidase reaction, and break down carbohydrates by a fermentative reaction that distinguishes them from other families in which the reaction is oxidative. The taxonomy of the genus is the most confusing among the Enterobacteriaceae in that in the past some members of the genus were classified as *Aerobacter* or *Cloacae* or *Klebsiella*, depending on the author's preference.

While the 7th edition of *Bergey's Manual* does not recognize this genus, the proposed 8th edition includes two species, *E. (Aerobacter) cloacae* and *E. (Aerobacter) aerogenes*. Biochemical properties of the genus are similar to other members of Enterobacteriaceae, i.e., glucose is fermented with production of acid and gas (CO_2:H_2 in ratio 2:1), the Voges-Proskauer reaction is usually positive, and citrate and acetate can be used as sole carbon source. DNA base composition of the genus ranges from 52 to 54 mole per cent GC, which is similar to other members of the tribe. The taxonomy of the genus requires further clarification.

Enterobacter is widely distributed in nature in soil, water, sewage, and dairy products and is occasionally found in urine, pus, and other pathologic materials from animals. *Enterobacter* is considered one of the indicator organisms of fecal pollution of aquatic environments as well as of food and potable water.

Genus Hafnia. This genus is included in the 8th edition of *Bergey's Manual* but not in the 7th edition. Taxonomic problems surrounding the genus are similar to those of *Enterobacter*. DNA base composition of the organisms varies from 52 to 57 mole per cent GC. The organisms have been isolated from feces of man and other animals, sewage, soil, water, and dairy products.

Genus Serratia. The organisms are morphologically similar to *Enterobacter*, i.e., motile rods, and some strains are capsulated and physiologically related to *Klebsiella* and *Enterobacter* but differ in certain biochemical properties. The type species, *S. marcescens,* is commonly found in soil and water. The DNA base composition of the organisms ranges from 54 to 63 mole per cent GC. The organisms produce DNase and a characteristic, red, pyrrole-containing, cellular pigment called **prodigiosin** when grown at

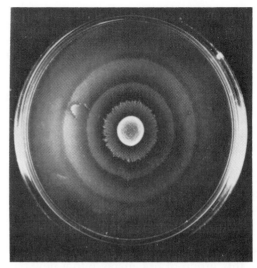

prodigiosin

approximately 25C on starchy agar media. Pigment production is variable and frequently influenced by temperature and mineral ions. Incubation temperatures higher than 38C appear to repress formation of the brilliant red pigment, and in some instances the ability to produce pigment is completely lost (Color Plate IV, *B*).

35.5
TRIBE PROTEEAE

The members of the tribe are motile rods. Most species produce phenylpyruvic acid by deaminating phenylalanine and produce indole. DNA base composition of the organisms ranges from 36 to 41 mole per cent GC for four species and 50 to 53 mole per cent GC for one species, *P. morganii*. Proteeae species are KCN-resistant.

Genus Proteus. *Proteus* species, although common in decaying matter, soil, and water, also are often found in the human intestine and in infections of the urinary or intestinal tract; they also cause diseases in lower animals. Most *Proteus* species are active in fermentation and, since they also actively decompose proteins, they are valuable scavengers. As shown in Table 35–2, they fail to ferment lactose and, being actively motile, are sometimes mistaken for *Salmonella* (paratyphoid) species. The power to *hydrolyze urea rapidly* is a distinctive character of *Proteus* that is used in differential diagnosis. Cultures of *Proteus* usually have a disagreeable, fetid odor.

The organisms may produce coccoid forms, irregular involution forms, filaments, and spheroplasts under certain conditions. Swimming motility is most pronounced at 20C. In addition, spontaneous swarming on the surface of solid media is considered a unique characteristic of most strains of *P. vulgaris* and *P. mirabilis*. The swarming cells usually spread

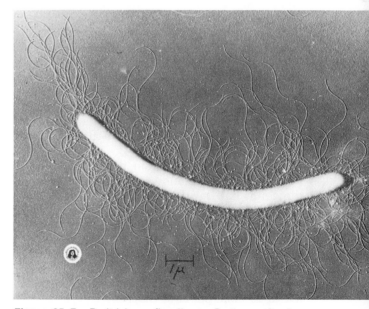

Figure 35–6. Swarming of *Proteus* on the surface of nutrient agar plate. The plate was inoculated in the center with a drop of a bacterial suspension, and was photographed after incubation at 37C for 20 hours.

rapidly, forming whorls and concentric bands (Fig. 35–6). The swarming cells are elongate and possess an enormous number of flagella (Fig. 35–7).

Bacteriocins are produced by all species except *P. rettgeri*, and virulent and temperate phages are known for all species.

Serological properties of *Proteus* are unique among the Enterobacteriaceae in that somatic (O) antigens 1, 2, and 3 of the schema react with

Figure 35–7. Peritrichous flagella on *Proteus vulgaris.* (Shadowed electron photomicrograph by C. F. Robinow and J. Hillier; A.S.M. LS-258.)

antibodies formed in man during some rickettsial infections, and are used in the **Weil-Felix reaction** for the diagnosis of typhus fever. Detailed descriptions of the Weil-Felix reactions with various rickettsial diseases are presented in Chapter 42.

35.6
TRIBE YERSINIEAE

In the past, organisms which caused plague in man and rodents or hemorrhagic septicemia in various other animals and in birds have been classified in the genus *Pasteurella* of the family Brucellaceae (*Bergey's Manual*, 7th edition). However, the 8th edition of the *Manual* segregates this group of organisms into a new tribe, Yersinieae, on the basis of DNA base composition (46 to 48 mole per cent GC) and physiological and biochemical properties. These organisms are more closely related to the enteric group than to other members of the genus *Pasteurella*, e.g., *P. multocida*. For example, certain species of the genus *Yersinia* are mixed acid–fermenters but without production of gases. They produce a powerful urease. Also, *Yersinia (Pasteurella) pestis* is sensitive to many of the bacteriophages that attack *E. coli*. The species *Y. pestis* and *Y. (Pasteurella) pseudotuberculosis* of the former genus *Pasteurella*— which produce an enzyme, β-galactosidase (as in other enterics), and carry out mixed acid fermentation—are combined into the genus *Yersinia*. Some strains of *Y. enterocolitica* isolated from human septicemia are included in the proposed 8th edition of *Bergey's Manual*. The organisms are also isolated from feces and lymph nodes of both sick and healthy animals and man, as well as material likely to be contaminated by feces, such as milk and ice cream. Some strains are reported to be β-galactosidase-negative, hence do not conform to the definition of the genus.

Genus Yersinia. From the standpoint of human disease, the most important member of the genus is *Y. pestis,* the cause of bubonic and pneumonic plague in man. Morphologically and culturally, *Y. pestis* resembles *P. multocida* but is slower and less vigorous in its growth and somewhat less active biochemically. The DNA base composition of *Y. pestis* is slightly higher (46 mole per cent GC) than that of *P. multocida* (37 to 40 mole per cent GC).

The morphology of *Yersinia* is distinctive. In pathologic material the organisms are short, oval rods about 0.5 by 3 μm, which tend to stain

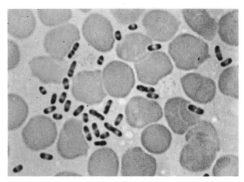

Figure 35–8. Stained smear of blood of mouse with septicemia due to *Yersinia* (*Pasteurella*) *pestis.* Note minute size (about 1 by 2 μm) and distinctive bipolar staining. (Courtesy of Naval Biological Laboratory.)

most heavily at the tips (**bipolar staining**). In cultures the bipolar appearance is often less definite (Fig. 35–8).

Bubonic Plague. This malady is the classic example of a bacterial disease transmitted by the bite of an insect. It is primarily a disease of rodents, a **zoonosis.** It is conveyed to human beings by the bite of infected fleas, commonly *Xenopsylla cheopis* and *Ceratophyllus fasciatus,* the rat fleas. The fleas usually derive the plague bacilli from the blood of infected rats, notably *Mus norvegicus* (sewer rat) and *Mus rattus* (house rat). When the rats are infected they show lesions similar to those found in man.

Rats and rat fleas maintain plague as an epizootic disease, much like hemorrhagic septicemia (caused by *Pasteurella multocida*), among themselves for long periods and act, therefore, as an **animal reservoir** of plague bacilli. When rats become excessively prevalent in any community, human plague is apt to occur because the opportunity for rat fleas to bite human beings greatly increases. Crowded populations living near dumps or in dirty, unsanitary conditions suffer most. Conditions following the devastation of war, with breakdown of disease-control systems, are ideal for the development of rats and therefore of rat-borne diseases. The pages of history are filled with disasters to armies and civil populations attacked by plague. Daniel Defoe's *Journal of the Plague Year* and Winsor's *Forever Amber* give dramatic descriptions of the plague in London in 1665. The rats often die in great numbers from the disease, and the fleas leave the cooling bodies by jumping on to the first warm animal which passes. Dead or dying rats are, therefore, potentially dangerous.

In 1900, plague was first found in human beings in this country. It has since been found in rats and in rodents other than rats, especially ground squirrels or prairie dogs. Many human cases have since been traced to contact with wild rodents. The disease in woodland- or wild-living animals is often spoken of as **sylvatic** (forest) or **campestral** (prairie) plague. The control of plague in wild rodents is a field problem of great importance.

Pseudotuberculosis. *Y. pseudotuberculosis* (48 mole per cent GC) is the causative agent of a disease known as pseudotuberculosis, which is characterized by diarrhea, wasting away, and death within a month. The organisms may enter the body via the intestinal tract, or through abrasions of the skin, or through bites of insects. The infection may be of septicemic form, and in such form death may occur in 1 or 2 days.

35.7
TRIBE ERWINIEAE

Genus Erwinia. This important group is named for one of America's outstanding pioneers in the field of plant microbiology and pathology, Erwin F. Smith. *Erwinia* was formerly included in the family Enterobacteriaceae because it resembles the coli-typhoid-dysentery organisms in some respects. In the 8th edition of *Bergey's Manual*, this group has attained the generic rank of a tribe in the family Enterobacteriaceae. The DNA base composition ranges from 50 to 57 mole per cent GC for the more well-established organisms, *E. amylova, E. carotovora,* and *E. atroseptica.* These organisms lack formic hydrogenylase (an enzyme responsible for breaking down formic acid to carbon dioxide and hydrogen); therefore, little or no demonstrable gas is produced.

Erwinias differ from most Enterobacteriaceae in that they are not commonly found in the intestine (although strains have been isolated from the intestinal tract of man and animals) but normally occur in the soil or as epiphytes on the surface of plants, or frequently associated with infected plants. They invade the tissues of living plants, producing various pathologic and destructive conditions. Representative species are *E. carotovora,* which, by means of a pectolytic enzyme common in *Pseudomonas* and *Xanthomonas* (Chapt. 33) but not found in Enterobacteriaceae, liquefies the tissues of carrots, cabbage, iris, and eggplant, causing the condition called **soft rot;** and *E. tracheiphila,* which grows so extensively in the sap channels of cucumber, melon, and related plants as to occlude the channels and, like *Pseudomonas,* cause the disease known as **wilt.**

Pathogenic and potentially pathogenic species in the family Enterobacteriaceae are summarized in Table 35–7.

35.8
ISOLATION OF MEMBERS OF ENTEROBACTERIACEAE

In practical diagnostic and sanitary bacteriology a time-honored (though imperfect) means of differentiation between the groups of Enterobacteriaceae has been the test for prompt (within 24 to 48 hours at 35C) fermentation of lactose. This property is the basis for selective media used for the isolation and preliminary identification of Enterobacteriaceae from feces, foods, water, or sewage. Scores of selective media, both fluid and solidified with agar, have been devised over nearly a century of work with these organisms. Basically they all consist of peptone or meat-extract or -infusion broth containing 0.5 to 1.0 per cent lactose; substances such as bismuth sulfite, deoxycholate, citrate, and eosin-methylene blue, or brilliant green, that inhibit extraneous organisms such as gram-positive cocci and sporeforming bacilli; and an acid-alkali indicator such as litmus, sulfite-reduced fuchsin or phenol red to detect acid formed by fermentation of the lactose. Most of the media are solidified with agar and are inoculated by streaking the agar in Petri dishes. They are then incubated 24 to 48 hours at 35C.

On most varieties of such media colonies of species that do not promptly ferment lactose are small, translucent and colorless whereas colonies of prompt lactose fermenters are usually deeply colored by the indicator. Colonies of desired types are fished to other media for further purification, Gram stain, verification of action on lactose, and species identification. More recently developed media for the isolation of the organisms are presented in both the *Difco Supplementary Manual* (1971) and the *Baltimore Biological Laboratory (BBL) Manual.*

Identification. The several tests necessarily used in identification have been ingeniously arranged in numerous schemes, each an improvement over previous ones because of newer knowledge. For example, one "screening" medium known as Kligler's iron agar, or triple sugar-iron agar (TSI), has been empirically devised to show the differential characteristics of these organisms in a single culture tube. For composition of TSI agar and other screening media, see reference material listed in the Suplementary Reading list at the end of the chapter.

TABLE 35–7. PATHOGENIC AND POTENTIALLY PATHOGENIC SPECIES IN THE FAMILY ENTEROBACTERIACEAE

Organism	DNA Base Composition (Mole Per Cent G + C)°	Diseases and Habitat
TRIBE: ESCHERICHIEAE Genus: *Escherichia* Species: *E. coli*	50 to 52	Members may show opportunistic pathogenicity. May cause enteritis, urinary tract infections in man, mastitis in cows, and infections in young pigs and may produce toxins genetically determined; also hemolytic strains are frequently found. Important as indicator of fecal pollution of water and food.
Genus: *Edwardsiella* Species: *E. tarda*	51 to 52	Occasionally isolated from stool of human with diarrhea; normal intestinal inhabitant of snake. May be isolated from blood of man and animals, and from urine of man.
Genus: *Citrobacter* Species: *C. freundii* (*E. freundii*) *C. intermedius* (*E. intermedia*)	50 to 53	Normally saprophytes in water, food, feces and urine. Their pathogenicity is problematical and certain serotypes may cause mass alimentary infection and infections of the urinary tract, gallbladder, middle ear, and meninges.
Genus: *Salmonella* Species: *S. cholerae-suis* *S. hirschfeldii* (*S. paratyphi*-C) *S. typhi* (*S. paratyphi*-A) *S. schottmuelleri* (*S. paratyphi*-B) *S. typhi-murium* *S. enteritidis* *S. arizona*, etc.	50 to 54	One species, *S. typhi*, causes typhoid fever, a form of salmonellosis. Others may cause enteritis in man and various other forms of salmonellosis, especially food infections in man. Species *S. arizona* may cause salmonellosis-like conditions in man.
Genus: *Shigella* Species: *S. dysenteriae* *S. flexneri* *S. boydii* *S. sonnei*	49 to 54	All species cause bacillary dysentery or shigellosis. Serotype 1 of *S. dysenteriae* is peculiar in that it produces a potent exotoxin (shiga toxin). Normal habitat of all species is the intestinal tract of higher primates and man.
TRIBE: KLEBSIELLEAE Genus: *Klebsiella* Species: *K. pneumoniae* *K. ozaenae*	52 to 59	Causes human pneumonia, enteritis, septicemia, urinary infections, etc. *K. ozaenae* causes ozaena and other chronic diseases of the respiratory tract. *K. rhinoscleromatis* is found constantly and exclu-

° Figures listed are average ranges of several strains and species.

(*Table 35–7 continued on opposite page.*)

A pure culture of the isolate is first stabbed, then streak-inoculated on a TSI agar slant. Following incubation for 24 to 48 hours at 35C, the medium reveals not only fermentation of glucose, sucrose, and lactose but also the production of hydrogen sulfide, which reacts with the ferrous ion in the medium to form black ferrous sulfide. In addition, the production of gases (H_2, CO_2) causes bubble formation in the agar.

The detection of the fermentation reaction of the sugar(s) is based on change in color of phenol red indicator. For example, a strain of *S. typhi* produces detectable amounts of acid in the butt of the medium but not on the slant surface, where alkaline products of peptone degradation sufficiently neutralize the less acid end products of respiration. Organisms such as *E. coli* and *P. vulgaris*, which are able to ferment both lactose and sucrose in addition to glucose, produce enough acid in the butt to diffuse throughout the medium, thus acidifying the entire slant (Fig. 35–9). The summary of TSI agar reactions of members of the family Enterobacteriaceae is shown in Table 35–8.

Recently, lysine-iron agar has been incorporated for refined screening of members of the

TABLE 35-7. PATHOGENIC AND POTENTIALLY PATHOGENIC SPECIES IN THE FAMILY ENTEROBACTERIACEAE
(Continued)

Organism	DNA Base Composition (Mole Per Cent G + C)°	Diseases and Habitat
		sively in patients with rhinoscleroma. Widely distributed in nature, in soil, water, grain and normally found in intestinal tract of man and other animals.
Genus: *Enterobacter* Species: *E. cloacae* (*A. cloacae*)	52 to 54	Occasionally produces urinary tract infections, septicemia, and frequently found in urine, pus and other pathological material from animals. The organisms inhabit the feces of man and other animals; found in water, sewage, soil and dairy products.
Genus: *Hafnia* Species: *H. alvei* (*E. alvei*)	52 to 57	Pathogenicity of the species is not established. Organisms inhabit feces of man and other animals, found in sewage, soil, water, and dairy products.
Genus: *Serratia* Species: *S. marcescens* *S. marcescens* subsp. *kiliensis*	54 to 63	Probably opportunistic pathogens. Presumably widely distributed. Occasionally found in pathological specimens from infections in man.
TRIBE: PROTEEAE Genus: *Proteus* Species: *P. vulgaris* *P. mirabilis* *P. rettgeri* *P. inconstans* *P. morganii*	38 to 41 50 to 53	*P. vulgaris* and *P. mirabilis* are considered a common cause of urinary tract infections; *P. inconstans* and *P. morganii* are a common cause of diarrhea, especially in infants. The organisms are most frequently found in the human digestive tract, feces and also in human clinical material.
TRIBE: YERSINIEAE Genus: *Yersinia* Species: *Y. (P.) pestis* *Y. (P.) pseudotuberculosis* *Y. enterocolitica*	46 to 48	Causes bubonic plague (formation of enlarged lymph glands) in man, rat and other rodents; transmitted from rat to rat and from rat to man by the rat flea. The organism produces potent endotoxins; highly invasive. If lungs are invaded, the agent may be transmissible from man to man by droplet infection (pneumonic plague). *Y. pseudotuberculosis* produces pseudotuberculous (false tuberculous) lesions generally in the mesenteric glands. The organism is known to produce an exotoxin. Pathogenicity of *Y. enterocolitica* is not well established.
TRIBE: ERWINIEAE Genus: *Erwinia* Species: *E. amylova* *E. carotovora*	50 to 57	Never pathogenic to warm-blooded animals. The organisms are plant pathogens. *E. amylova* causes "fire blight" of apple and pear. *E. carotovora* causes "soft rot" of various vegetables. Detailed discussion of the organisms is presented in Chapter 43.

°Figures listed are average ranges of several strains and species.

family Enterobacteriaceae. Use of both TSI and lysine-iron agar facilitates presumptive identification of the group. Reactions of members of the family Enterobacteriaceae in lysine-iron agar are presented in Table 35–9.

Further group and species identifications are made on the basis of additional tests for motility, gelatin liquefaction, sugar fermentations, production of various enzymes such as urease, lysine and ornithine decarboxylases, phenylalanine deaminase, and DNase (Table 35–1).

Recently, a commercially prepared combination of media, "Enterotube," which incorporates conventional media into an eight-compartmented, single, ready-to-use tube that allows simultaneous inoculation and gives 11 standard biochemical tests, has been made available (Fig. 35–10). Group differentiation and identification of gram-negative organisms (Enterobacteriaceae) using "Enterotube" is presented in Table 35–10. Several other excellent systems are also available, e.g., R/B system and API-20 tests.

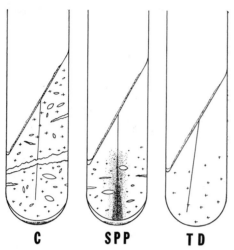

Figure 35-9. Cultures of Enterobacteriaceae in an agar medium designed to differentiate major groups. The medium is triple-sugar-iron (T-S-I) agar. It contains lactose, glucose and sucrose with acid-alkali indicator. $FeCL_3$ is added to detect the formation of H_2S. The slants are inoculated on the surface and by a stab of the needle into the depths of the butt of the agar. *C* shows the reaction of coliform organisms: gas bubbles and acid (crosses) throughout the agar. Note that the volume of gas formed has rent the agar slant at several places and pushed the butt of the agar slant away from the bottom of the tube, where a few drops of bacterial suspension have collected. *SPP* shows the reaction typical of *Salmonella* organisms and also of some strains of *Proteus* and Arizona group: acid and gas, with formation of H_2S (shading) in the butt; alkaline slant (upper portion). *TD* shows the reaction of *Salmonella typhi* and of the genus *Shigella* (typhoid-dysentery): acid butt, alkaline slant; no gas, no H_2S. Further differentiations are made on the basis of serological and additional biochemical tests.

Some species of Enterobacteriaceae do not ferment lactose rapidly but may do so after four to seven days of incubation at 35C in lactose media. This may result from their containing genetic units that permit induction of β-galactosidase (lactase) (Chapt. 7), or the selection of lactose-fermenting mutants, or the induced formation of appropriate permeases by cells genetically capable of fermenting lactose but not permeable to it. Among slow-lactose-fermenting Enterobacteriaceae are some strains of *Escherichia coli*, the Arizona group, *Hafnia*, *Citrobacter*, *Shigella*, and *Serratia*. By the preliminary selective plating methods that have been described, these types can easily be mistaken for true non-lactose-fermenters because their colonies on lactose-indicator plates after 24 to 48 hours at 35C are colorless like those of the true non-lactose-fermenters. Special consideration and tests must be used to detect and identify such organisms.

Of the species that do not ferment lactose

TABLE 35-8. SUMMARY OF TRIPLE SUGAR-IRON AGAR REACTIONS ON MEMBERS OF THE FAMILY ENTEROBACTERIACEAE° †

Organism	Slant	Butt	Gas	H₂S
Escherichia	A (K)	A	+(−)	−
Shigella	K	A	−	−
Salmonella typhi	K	A	−	+(−)
Other *Salmonella*	K	A	+	+++(−)
Arizona	K (A)	A	+	+++
Citrobacter	K (A)	A	+	+++
Edwardsiella	K	A	+	+++
Klebsiella	A	A	++	−
Enterobacter	A	A	++	−
E. *hafniae*	K	A	+	−
Serratia	K or A	A	−	−
Proteus vulgaris	A (K)	A	+	+++
P. *mirabilis*	K (A)	A	+	+++
P. *morganii*	K	A	−(+)	−
P. *rettgeri*	K	A	−	−
Providencia	K	A	+ or −	−

°From Blair, Lennette, and Truant (Eds.): Manual of Clinical Microbiology, American Society for Microbiology, The Williams & Wilkins Co., 1970; after Ewing: Lecture Outline Series II. Differential Reactions of Enterobacteriaceae, National Communicable Disease Center, Atlanta, Ga., 1966.

†Symbols: K = alkaline; A = acid.

TABLE 35-9. SUMMARY OF LYSINE-IRON AGAR REACTIONS ON MEMBERS OF THE FAMILY ENTEROBACTERIACEAE° †

Organism	Slant	Butt	Gas	H₂S
Escherichia	K	K or N	− or +	−
Shigella	K	A	−	−
Salmonella	K	K or N	−	+(−)
S. *typhi*	K	K	−	+ or −
S. *paratyphi-A*	K	A	+ or −	− or +
Arizona	K	K or N	−	+(−)
Citrobacter	K	A	− or +	+ or −
Edwardsiella	K	K	− or +	+
Klebsiella	K or N	K or N	+ or −	−
Enterobacter				
cloacae	K or N	A	+ or −	−
aerogenes	K	K or N	+(−)	−
hafniae	K	K or N	− or +	−
Serratia	K or N	K or N	−	−
Proteus				
vulgaris	R	A	−	−(+)
mirabilis	R	A	−	−(+)
morganii	K or R	A	−	−
rettgeri	R	A	−	−
Providencia	R	A	−	−

°From Blair, Lennette, and Truant (Eds.): Manual of Clinical Microbiology, American Society for Microbiology, The Williams & Wilkins Co., 1970; after Ewing: Lecture Outline Series II. Differential Reactions of Enterobacteriaceae, National Communicable Disease Center, Atlanta, Ga., 1966.

†Symbols: K = alkaline; N = neutral; A = acid; R = red (oxidative deamination).

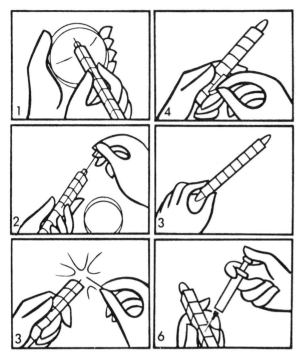

Figure 35–10. Multicompartmented ready-to-use Enterotube®. A ready-to-use tube permitting simultaneous inoculation and performance of 11 standard biochemical tests from a single bacterial colony. Tests:

Dextrose	Phenylalanine deaminase
Lysine decarboxylase	Dulcitol
Ornithine decarboxylase	Urease
H_2S	Simmons' citrate
Indole	Gas production
Lactose	

Detailed information and instructions may be obtained from Roche Diagnostics. (Courtesy of Roche Diagnostics, Division of Hoffman-La Roche Inc., Nutley, N.J.)

rapidly many are pathogenic: the typhoid bacillus (*Salmonella typhi*), paratyphoid bacilli (*Salmonella paratyphi-A*), other *Salmonella* species, and most dysentery bacilli (*Shigella* species). Slow-lactose-fermenters such as *Proteus*, Providence, and *Hafnia* species are also pathogenic but usually cause only secondary, though often very troublesome, infections. *Serratia* (*Aerobacter* C) is rare in the intestinal tract but has caused numerous serious infections of other tissues.

The rapid-lactose-fermenting types (*Escherichia, Klebsiella, Enterobacter* (*Aerobacter*), and *Citrobacter*) constitute the so-called "coliform" or "coli-aerogenes" groups, important as indicators of fecal pollution. They are generally nonpathogenic (except certain strains of *Escherichia coli* that cause infant diarrhea and other diseases).

Differentiation of Coliforms. The procedures for differentiating between the various coliforms are not technically difficult. Having obtained satisfactory pure cultures from either the standard or membrane filter method, tests are made of indole production, methyl red reaction, the Voges-Proskauer reaction, and ability to utilize sodium citrate as a sole source of carbon (Table 35–11). Technical details may be found in the Supplementary Reading list at the end of this chapter.

IMViC Formula. A mnemonic (memory-aiding) device, **IMViC**, is often used with plus and minus signs to express differences between coliform organisms by means of a "formula." **I** stands for indole reaction; **M**, for methyl red reaction; **V**, for the acetyl-methyl-carbinol test (originated by Voges-Proskauer); **i** is for the sake of euphony; and **C** stands for growth in mineral solution containing citrate as a sole source of carbon.

Thus, an organism of the coliform group designated as "IMViC + + − −" would be *E. coli,* since this gives positive indole and methyl red reactions but negative Voges-Proskauer and citrate reactions. These symbols are used in Table 35–11.

The Eijkman Reaction. A useful differential method, the **Eijkman reaction** is a good example of a temperature-dependent enzymic action. It is based on the fact that most strains of the fecal species of *E. coli* produce gas from lactose in a special, buffered broth when incubated at *exactly* 45.5C, whereas very few of the less frequently fecal strains of *Enterobacter* or *Klebsiella* and intermediates (e.g., *E. freundii*) do so.

A pure culture of the organism to be tested is inoculated into 0.3 per cent lactose-tryptose broth prewarmed to 45.5C and buffered at pH 7.4. The culture is held at 45.5C for 48 hours. The production of gas is a *positive* result (i.e., *Escherichia coli* is present). The reaction is extremely temperature-dependent, 0.2C downward fluctuation in temperature vitiating the test. The production of indole can be determined concomitantly if tryptophan-containing peptone is used.

Coliform Group – Index Organisms

Pollution of foods or water with fecal material, whether infected or not, is obviously undesirable, both from the standpoint of danger of infection and for purely esthetic reasons. The detection of fecal bacteria of any kind in food or domestic water is therefore of importance in determining its suitability for human consumption. For every typhoid bacillus or other pathogen (e.g., *Entamoeba histolytica,* or viruses of

TABLE 35-10. GROUP DIFFERENTIATION OF GRAM-NEGATIVE ORGANISMS (ENTEROBACTERIACEAE) USING ENTEROTUBE ® *

Groups[1]		Reactions †	Dextrose	H$_2$S	Indole	Phenylalanine Deaminase	Urease	Dulcitol	Lactose	Lysine (+ lactose) Decarboxylase	Simmons' Citrate
Escherichia-Shigella		Escherichia	+	−	+	−	−	d	+	−	−
		Shigella	+	−	∓	−	−	d	−[6]	−	−
Proteus-Providencia	Proteus	vulgaris	+	+	+	+	+	−	−	−	d
		mirabilis	+	+	−	+	+	−	−	−	d
		morganii	+	−	+	+	+	−	−	−	−
		rettgeri	+	−	+	+	+	−	−	−	+
		Providencia	+	−	+	+	−	−	−	−	+
Klebsiella-Enterobacter-Serratia	Enterobacter[2]	Klebsiella	+	−	∓	−	+	∓	+	−[7]	+
		cloacae	+	−	−	−	±	∓	+	−	+
		aerogenes	+	−	−	−	−	−	+	−[7]	+
		hafniae	+	−	−	−	−	−	∓	+	d
		liquefaciens	+	−	−	−	d	−	d	d	+
		Serratia	+	−	−	−	d	−	∓	+	+
Salmonella-Arizona-Citrobacter		Salmonella	+	+[4]	−	−	−	+[5]	−	+[8]	+[9]
		Arizona	+	+	−	−	−	−	d	d	+
		Citrobacter[3]	+	+	−	−	d	d	d	−	+
		Edwardsiella	+	+	+	−	−	−	−	+	−

*Courtesy of Roche Diagnostics, Division of Hoffman-La Roche, Inc., Nutley, N.Y.

†Reactions legend:
  + = Positive.
  − = Negative.
  ± = Majority positive.
  ∓ = Majority negative.
  d = Different biochemical types.

[1]Nomenclature according to Ewing, W. H.: Differentiation of Enterobacteriaceae by Biochemical Reactions, National Communicable Disease Center, Atlanta, Ga., 1968.

[2]Formerly Aerobacter.

[3]Includes Bethesda-Ballerup group.

[4]S. enteritidis (bioserotype Paratyphi A) and some rare biotypes may be H$_2$S-negative.

[5]Majority of Salmonella ferment dulcitol promptly; S. typhi, S. enteritidis (bioserotypes Paratyphi A and Pullorum) and S. cholerae-suis may give negative or delayed reactions.

[6]Shigella sonnei ferments lactose slowly.

[7]Positive reactions may occur after 24 hours.

[8]Bioserotype Paratyphi A is lysine-negative.

[9]S. typhi and bioserotype Paratyphi A are citrate-negative. S. cholerae-suis is delayed positive.

polio or hepatitis) in polluted water supplies or in any foods, there are usually millions of coliform organisms or fecal streptococci. Either or both may serve as an **index of fecal pollution.**

Methods for their detection and enumeration in water and foods are carefully prescribed by the American Public Health Association and affiliated societies, and are used daily in every health

TABLE 35-11. DISTINCTIVE PROPERTIES OF COLIFORM AND ASSOCIATED ORGANISMS°

Species	Lactose	Glucose	Sucrose	I	M	V	C	Gelatin	Motility
Escherichia coli	⊕	⊕	⊕	+	+	−	−	−	+
Citrobacter freundii	⊕	⊕	⊕	−	+	−	+	−	+
Klebsiella pneumoniae	⊕	⊕	⊕	−	−	+	+	−	−
Enterobacter cloacae	⊕	⊕	⊕	−	−	+	+	+	+
Proteus vulgaris	−	⊕	⊕	+	−	−	+	+	+
Providence group	−	⊕	⊕	+	+	−	+	−	+
Pseudomonas aeruginosa†	−	−	−	−	−	−		+	+

°Symbols: + = positive test; − = negative test; ⊕ = acid and gas formed.
†Produces two water-soluble pigments; blue **pyocyanin** and yellow **fluorescein**. (See Chapter 33, Order Pseudomonadales.)

department laboratory. A series of steps (tests for coliforms), the **presumptive test, confirmed test** and **completed test,** are carried out systematically. These tests are applicable to any properly prepared samples of foods, water, fruit juices, or dairy products. Detailed discussions on pollution and coliform and other indicator organisms are presented in Chapter 45.

Bacteriophage Typing

As indicated in the section on bacteriophages (Chapter 16), within a species, clone, or serotype, differentiation of the organisms can be made on the basis of differing susceptibilities to type-specific bacteriophages. Phage type susceptibility appears to be closely related to specific cell wall antigen content, and this, in turn, to genetic make-up of the organisms. Thus, bacteriophage typing has been used to differentiate and identify groups of bacteria, including *Salmonella typhi, Salmonella typhi-murium, Shigella sonnei, Klebsiella pneumoniae, Pseudomonas aeruginosa, Mycobacterium tuberculosis, Streptococcus pyogenes, Staphylococcus aureus,* and many others. Phages used in identification of various strains of pathogenic organisms are listed in the *Biological Reagent Catalogue* of the National Communicable Disease Center (NCDC).

Fluorescent Antibody Techniques

As with bacteriophage typing, fluorescent-labeled antibodies (Chapt. 27) have been used for identification of a variety of pathogenic as well as nonpathogenic organisms. Fluorescent antibody (FA) techniques for rapid "presumptive" identification of *E. coli* OB groups, group B and D *Shigella,* and *Salmonella typhi* O from

fecal specimens have been reported. However, identification of the members of the family Enterobacteriaceae by FA techniques still remain as "presumptive," and the technique is not a substitute for isolation and definitive serotyping of the microorganisms. Detailed discussions on fluorescent antibody techniques are presented in Chapter 27.

35.9 GRAM-NEGATIVE BACTERIA SIMILAR TO ENTERICS

The family Brucellaceae in the 1957 edition of *Bergey's Manual* includes the genera *Pasteurella, Brucella, Haemophilus, Actinobacillus, Calymmatobacterium, Moraxella,* and *Noguchia.* In the 8th edition this rank, family Brucellaceae, has been eliminated, and the genera have been distributed elsewhere. Here we shall describe several of the genera formerly in the families Brucellaceae and Bacteroidaceae.

Organisms of the family Brucellaceae, as described in the 7th edition of *Bergey's Manual,* are like the Enterobacteriaceae, i.e., the organisms are small, straight, asporogenous, gram-negative coccoid to rod-shaped cells which may become filamentous or pleomorphic (Table 33-1). Some species are motile at lower temperatures, while others may be motile at 37C. Unlike the ubiquitous Pseudomonadaceae, Enterobacteriaceae, and Achromobacteriaceae (the last no longer listed as such), these organisms represent highly evolved or specialized types of bacteria that require complex organic media and special nutrients for their optimum growth. They are not markedly fermentative or proteolytic. In general, unlike most Enterobacteriaceae, they lack the enzymic versatility and the ruggedness necessary for significant **growth** in the outer world, although they may **survive** for some time. No saprophytic species in this family is known. All appear to cause disease in man or lower mammals.

35.10
GRAM-NEGATIVE STRICTLY AEROBIC RODS

Genus Brucella. *Brucella melitensis* and two closely related species, *B. abortus* and *B. suis*, cause **Malta fever** or undulant fever in man, a disease common in most parts of the world. The generic name of the causative organisms is derived from the discoverer, Bruce, a British scientist. Bruce first (1887) found the organism now called *B. melitensis* on the island of Malta in the spleens of persons infected by the organisms in goats' milk. (Malta was called Melita by the ancients because of the fine honey [Gr. *meli* = honey] found there, hence **melitensis**). Brucellas are very small (0.3 tò 0.5 μm long), nonmotile, nonsporeforming rods without distinctive morphological features. DNA base composition of the genus ranges from 56 to 58 mole per cent GC.

B. abortus was first known as the cause of abortion in farm animals, especially cattle. It was discovered by a Danish worker, Bang, in 1895 and is still often called **Bang's bacillus,** and the disease in cattle, **Bang's disease.**

B. suis, commonly found in swine, was first observed by Traum in the United States in 1914.

The fact that the organisms discovered in 1887 in Malta, in 1895 in Denmark, and in 1914 in America are all closely similar species or variants of one type was revealed in 1918 by Evans in the United States.

Brucellosis. The diseases caused by members of the genus *Brucella* are collectively called brucellosis. The diseases may be transmitted in a variety of ways: by ingestion, via skin, by inhalation, and even by the venereal route. Hence the diseases are infectious as well as invasive. For example, in human brucellosis (from ingestion of contaminated milk), the organisms multiply in the mucous membranes of the gastrointestinal tract, and subsequently the organisms invade the lymphatics and blood stream, then localize in the reticuloendothelial system. The organisms generally display characteristic intracellular localization.

Disease symptoms include chills, fatigue, headache, and backache with fever that increases at night and drops in the daytime, hence the name "undulant fever." These disease symptoms are primarily attributed to the presence of an endotoxin. Sudden release of a large amount of endotoxin is observed following treatment of the disease with antibiotics, which causes bacterial lysis. The disease may be treated with prolonged use of either streptomycin or (preferably) tetracyclines.

Isolation. Brucellas grow rather slowly on first isolation from the blood, milk, or tissues of infected animals, or from the blood of man. They may be cultivated on slightly acid (pH 6.8) beef liver–infusion agar, or on tryptose or trypticase soy agar or broth. (See also under *Francisella tularensis*.) Exemplary selective media are used in isolating *Brucella* from contaminated material like feces. Selective agents may include polymyxin B, penicillin, Actidione, brilliant green, or crystal violet. These inhibit contaminants but not *Brucella*, *Yersinia*, or *Pasteurella*. *Brucella* can also be isolated from blood by injecting it into living chick embryos. *Brucella abortus* is aerobic but will grow at first *only* in an atmosphere containing about 7 to 10 per cent carbon dioxide. The three species of *Brucella* are closely similar but may be distinguished by special tests (Table 35–12 and Color Plate V, *C*). Specific phages (brucellaphages) have been found useful in the differentiation of these organisms.

Survival and Distribution. *Brucella* species can survive for considerable periods in dairy products, water, soil, dung, dust, and meats, and are transmitted by these agents. The blood of infected man and animals and tissues and fluids associated with aborted animals are highly infectious.

TABLE 35–12. USUAL DIFFERENCES BETWEEN BRUCELLA SPECIES°

	B. abortus	B. melitensis	B. suis
SENSITIVE TO:†			
Thionine	+	−	−
Fuchsin	−	−	+
Safranin O	−	−	+
Pyronin	−	−	+
Crystal violet	−	−	+
FERMENT:			
Inositol	+	−	−
Maltose	−	−	+
REQUIRES CO₂:‡	+	−	−
PRODUCE:			
Urease (4 hours)	+	−	+
H₂S§	2–4 days +	1–2 days +	4–8 days +
BRUCELLAPHAGE‖	+	−	+ #

°There are "unusual" types in *all* biological groupings.
†Cannot grow on agar containing the dye.
‡For initial isolation.
§Expressed as days of incubation during which H₂S is continuously produced from culture media.
‖Undiluted.
#Negative if highly diluted (critical dilution).

The three species, though originally associated with certain animals, are not restricted to those animals, but each may occur in any of the three species mentioned, as well as in man, dogs, horses, and possibly poultry.

Control of Brucellosis. Brucellosis may be considered an occupational disease in man because of the prevalence of the disease among meat packers, cattlemen, hog raisers, persons who drink uncertified or unpasteurized milk, bacteriologists, and veterinarians. Organisms are rarely transmitted from man to man, but when the organism is introduced into a herd of cattle, goats, or swine, it may spread rapidly and eventually infect the entire herd. Many animals recover completely but may continue to harbor virulent organisms and infect both healthy animals and man. Hence, control includes not only vaccination of the animals with a live attenuated vaccine but also segregation and slaughter of infected animals. No known safe vaccine is as yet available for human use.

Genus Bordetella. Species of the genus *Bordetella* are mammalian parasites and pathogens of the respiratory tract. The organisms produce dermonecrotic toxin but are weakly invasive and rarely invade the blood stream. The organisms possess a capsule-like sheath, but it fails to swell in antiserum. The genus includes *B. pertussis,* the cause of whooping cough, discovered by Belgian scientists Bordet (Nobel Prize winner) and Gengou in 1906. On initial isolation from patients these organisms are truly and rigidly hemophilic like *H. influenzae,* but on subculture soon become adapted to growth without either X or V factor. The **X factor** is the iron complex called **heme,** which is part of the red coloring matter of erythrocytes. The **V factor,** identified as coenzyme I or nicotinamide-adenine dinucleotide (NAD) or NADP, is found in the electron-transfer system of many types of cells: plants, yeasts, many bacteria. Bordetellas do, however, require nicotinic acid ("niacin"), a part of the NAD coenzyme molecule. On potato–glycerine–20 per cent blood agar the colonies are distinctive, being hemolytic and resembling minute pearls. A certain percentage of young, growing cells (early log phase) of *B. pertussis* may possess a large number of small vesicles on the cell walls, and it has been suggested that these vesicles may be associated with the release of toxins from the living cells (Fig. 35–11).

Two other species (*B. parapertussis* and *B. bronchiseptica*) are similar to *B. pertussis* antigenically and in causing pertussis-like disease. *B. bronchiseptica* differs markedly in be-

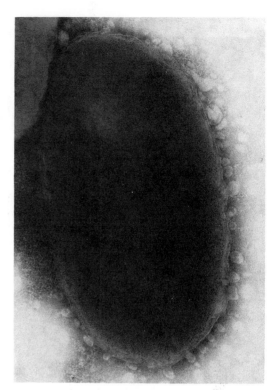

Figure 35–11. Electron micrograph of *B. pertussis* showing vesicles (cells stained by phosphotungstic acid). (100,000×.)

ing actively motile by means of lateral peritrichous flagella (Fig. 35–12) and being able to grow on simple organic media, such as peptone solution. The metabolism of members of the genus is strictly respiratory (aerobic), and the organisms produce catalase. DNA base composition of the genus ranges from 65 to 68 mole per cent GC. *B. parapertussis* differs from *B. pertussis* in formation of larger colonies, production of urease, and utilization of citrate as sole source of carbon. *B. bronchiseptica* has, at various times, been classified as a species of *Alcaligenes* (DNA base composition, 62 mole per cent GC), *Brucella* (DNA base composition, 56 to 58 mole per cent GC), *Bacillus, Bacterium,* and *Haemophilus* (DNA base composition, 38 to 42 mole per cent GC)–a sort of taxonomic orphan.

Recently, the genus has been classified on the basis of serology because the organisms have genus-specific, heat-stable O antigen and heat-labile dermonecrotic toxins, and each species has a specific agglutinogen.

The species of the genus cause whooping cough or whooping cough–like disease in man, and the agents are transmitted by droplets. Infants are usually vaccinated against whooping cough with killed, virulent (Phase I) *B. per-*

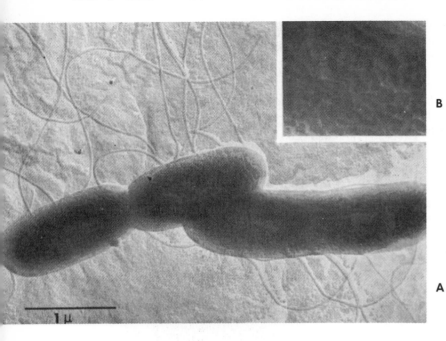

B

Figure 35–12. *A, Bordetella bronchiseptica* R-21 bacteria in this electron micrograph are coated with platinum-palladium. Note lobular and furrowed surface structure which is shown at higher magnification in the inset. *B,* Grooves measure about 10 μm. in width. (*A,* 25,000×; *B,* 75,000×.)

A

tussis, together with diphtheria and tetanus toxoids.

Genus Francisella. The organism *Francisella (Pasteurella) tularensis* much resembles *Y. (P.) pestis* and *P. multocida,* but it is somewhat more exacting in its nutritional requirements. DNA base composition of the genus ranges from 34 to 36 mole per cent GC. The organism had been classified among *Pasteurella* but recently a new genus, *Francisella,* has been designated (*Bergey's Manual,* 8th edition). As discovered by Francis in 1911 in Tulare County, California, the organism shows little or no growth unless cystine or (what amounts to the same thing) some compound containing the sulfhydryl (—SH) group, is added. Tryptose broth with thiamine, cysteine, glucose, ferric sulfate, potassium chloride, histidine, tris buffer, and agar can be used with excellent results. This medium is also good for *Y. pestis, Brucella* sp., and other fastidious bacteria. The organism does not possess β-galactosidase as does *Yersinia pestis.*

F. tularensis is found in much the same ecological relationship to rodents (rabbits, gophers, mice), biting insects (wood, dog, and rabbit ticks; rabbit lice, deer flies, horse flies), and man as are *Y. pestis,* fleas, and rodents. *F. tularensis* causes a disease called **tularemia.** (The name is derived from the Tulare swamps in California, where early observations were made on this disease.)

Tularemia ("Rabbit Fever"). This is a plague-like disease in many American rodents and other wild animals. As in bubonic plague, there are enlargements of regional lymph nodes, swelling of the spleen, and the appearance of tubercle-like nodules in spleen, liver, and elsewhere. The bacilli invade the blood from these foci. *F. tularensis,* the etiological agent, can also infect man. In man the disease is less fatal than bubonic plague but is often prolonged and debilitating.

One common means of transmission of tularemia to man is through the handling of infected wild rabbits, as in the marketing of these animals for food and pelts; hence **rabbit fever.** In some sections of the country the disease is known as **deer fly fever,** being transmitted largely by the deer fly, *Chrysops discalis.* In Arkansas and adjacent regions the disease is largely tickborne. It causes enormous losses among sheep if they graze in areas where there is tick-infested undergrowth. *F. tularensis* has been isolated from forest streams. The water is apparently infected by the carcasses of infected wild animals dying in the stream or on its nearby banks, another reason for disinfecting drinking water from questionable sources during field trips.

Genus Moraxella. The organisms in the genus *Moraxella* are small, short, rod-shaped cells which occur as diplobacilli (Fig. 35–13) and are sometimes described as diplococci. The organisms are nonmotile, are strictly aerobic, and do not require V or X factors for growth. It

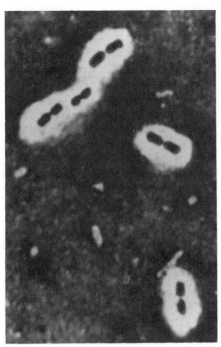

Figure 35-13. India ink preparation from 24-hour blood agar culture of *M. duplex* var. *nonliquefaciens,* demonstrating large capsules.

has been suggested that the genus should include all oxidase-positive strains of the *Herellea-Mima* group. DNA base composition of the genus ranges from 39 to 45 mole per cent GC.

The genus is named after Morax, a famous French ophthalmologist-bacteriologist, who first isolated the type species, *M. lacunata.* The organism causes subacute infectious conjunctivitis, or angular conjunctivitis. In the 8th edition of *Bergey's Manual,* the genus is placed in the family Neisseriaceae under Part X, "Gram-Negative, Strictly Aerobic Cocci and Cocci-Bacilli," which includes *Neisseria* (cocci, DNA base composition, 48 to 52 mole per cent GC), *Branhamella,* and *Acinetobacter* (rods; DNA base composition, 38 to 45 mole per cent GC). It has been suggested that the genus *Acinetobacter* should include all oxidase-negative strains of the *Herellea-Mima* group.

35.11
GRAM-NEGATIVE FACULTATIVELY ANAEROBIC RODS

Genus Pasteurella. The cells are small, ellipsoidal to elongate nonmotile rods. Specially stained cells may display characteristic bipolar staining (Fig. 35-8). In the 8th edition of *Bergey's Manual,* some species of the genus have been placed in genus *Yersinia* (*P. pestis* and *P. pseudotuberculosis*), and other species have been placed in genus *Francisella* (*P. tularensis*). The remaining species have been retained in the genus *Pasteurella.* DNA base composition of the genus ranges from 35 to 40 mole per cent GC.

The genus is named for Pasteur, who founded the science of immunology on his studies of vaccination against *Pasteurella avicida,* the cause of fowl cholera. *P. avicida* is now regarded as a variant of the type species of the genus, *P. multocida.* Other variants of *P. multocida* previously named for animals in which they were found as the cause of disease were: *P. bovicida* (cattle), *P. suilla* (swine), *P. muricida* (rats). They differ from one another only slightly in biochemical properties; therefore, these species were not recognized in the 1957 edition of *Bergey's Manual.*

Any infection with any species of *Pasteurella* is properly spoken of as pasteurellosis. *P. multocida* and its variants are highly pathogenic for most birds (causing fowl cholera). *Pasteurella multocida* can survive in, and is transmitted by, infectious dust, fomites, and animal secretions from stables, railroad cars, and stockyards. The organisms invade the lymphatic system and blood (**septicemia**), and may easily be cultivated on infusion media from all of the organs and body fluids of heavily infected animals. There are many small hemorrhages on various internal mucous surfaces, in the skin, and in the internal organs—hence the name **hemorrhagic** septicemia. In the animal disease there is much exudation of fluid from nose, mouth, and eyes. *P. multocida* rarely infects man.

Genus Haemophilus. All organisms in this genus are morphologically similar small (0.3 by 2 μm) rods, though they are pleomorphic and often vary from coccoid to long, filamentous or distorted forms (Fig. 35-14). None forms spores. They are among the most highly adapted and fragile parasitic and pathogenic bacteria, with fastidious nutrient requirements. Most grow well on **"chocolate agar"** (i.e., infusion agar, with about 10 per cent blood, heated to 90C for 10 minutes), though some require special nutrition.

The genus *Haemophilus* (Gr. *haemo* = blood; *philus* = requiring) is currently placed in "Genera of Uncertain Affiliation" under Part VIII, "Gram-Negative, Facultatively Anaerobic Rods" (8th edition of *Bergey's Manual*), and in-

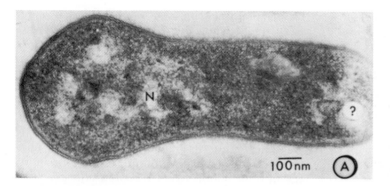

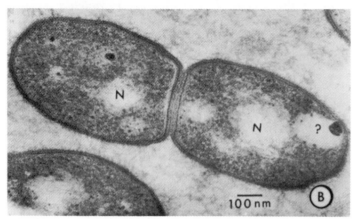

Figure 35–14. *A,* Longitudinal view of *Haemophilus vaginalis* 594 showing the pleomorphic bulbous enlargement of one end of the cell. Much of the nuclear material (*N*) appears to be contained within the bulbous area of the cell. (Question mark signifies unidentified organelle.)

B, Sections of *Haemophilus vaginalis* 594 showing distinct cross-wall formation during division. Nuclear regions (*N*) appear in both daughter cells.

cludes *H. influenzae* (formerly thought to cause influenza; influenza is now known to be caused by influenza virus) and several other pathogenic species, notably *H. ducreyi, H. parainfluenzae,* and *H. haemolyticus.* The organisms are non-motile, capsulated in young cultures, and frequently display very long filaments. DNA base composition of the genus ranges from 38 to 42 mole per cent GC.

The hemophils are excellent instances of the highly adapted and dependent bacterial parasite. For example, *H. influenzae* cannot live without certain blood components, X and V factors. *H. ducreyi* requires X but not V (Table 35–13). *H. influenzae* frequently inhabits healthy human respiratory tracts and may be isolated from acute respiratory infections, from conjunctivitis, and from purulent meningitis of children. *H. ducreyi* is the cause of the venereal disease chancroid, or "soft chancre."

Genus Actinobacillus. These organisms were formerly placed in the fifth genus of the

TABLE 35–13. SOME DIFFERENTIAL PROPERTIES OF HAEMOPHILUS AND BORDETELLA °

Organism	Usual Habitat	Production of		Growth Factors Required			Motility	Brown Color in Peptone Agar	Urease Produced	Colonies on Blood Agar	Fermentation of Glucose
		INDOLE	NaNO₂	X	V	NICOTINIC ACID					
H. influenzae	Resp. tract (man)	±	+	+	+	−	−	−	−	Small, "dewdrop" (some hemolytic)	+
H. ducreyi	Genitalia (man)			+	−	−	−	−	−	Small, gray, hemolytic	±
B. pertussis	Resp. tract (man)	−	−	+†	+†	+	−	−	−	Tiny, pearl-like, hemolytic	−
B. parapertussis	Resp. tract (man)	−	−	+†	+†	+	−	+	+	Resemble *B. pertussis*	−
B. bronchiseptica	Resp. tract (animal)	−	+	−	−	+	+	−	+	Large, white, like *E. coli*	−

°Adapted from Ripins: Medical Licensure Examinations, 11th ed., J. B. Lippincott Co., 1970.
†Requires neither after subcultivation.

family Brucellaceae, but in the 8th edition of *Bergey's Manual* they are placed in "Genera of Uncertain Affiliation" under Part VIII, "Gram-Negative, Facultatively Anaerobic Rods." The members of the genus have several properties in common with the members of the genera *Pasteurella, Brucella,* and *Haemophilus.* The DNA base composition of the genus ranges from 39 to 42 mole per cent GC. The species, *A. lignieresii,* the type species of the genus, is frequently isolated from actinobacillosis of cattle, a disease erroneously referred to as "actinomycosis" (see Chapter 40). Characteristic lesions are frequently found in lymph nodes and in muscles of the tongue (**wooden tongue**), where granulomatous tumors are formed. The tumor eventually breaks down to form abscesses in which the pus contains small, grayish white granules.

Genus Streptobacillus. The systematic position of this curious organism is not yet clear. Although much resembling *Mycoplasma,* it was included with the family Bacterioidaceae in 1957 mainly because of its *Bacteroides*-like filamentous character and pleomorphism, gram-negative stain, and heterotrophism. However, it is aerobic and facultatively anaerobic; *Bacteroides* is strictly anaerobic. *Streptobacillus* resembles *Bacteroides* also in growing best on blood or serum media. It is not motile. Because of the dissimilarities listed above, the genus *Streptobacillus* has been eliminated from the family Bacteroidaceae (in the 8th edition of *Bergey's Manual*) and reclassified under "Genera of Uncertain Affiliation" of Part VIII, which includes "Gram-Negative, Facultatively Anaerobic Rods" (genera *Haemophilus, Pasteurella,* and others).

Its most interesting characteristic is the fact that, when first studied by Klieneberger (later Klieneberger-Nobel) in 1935, there was associated with it what was thought to be a symbiotic species of PPLO. Klieneberger, working at the Lister Institute in London, called the supposed symbiont L_1, L being the initial of the Lister Institute named for Lord Lister, pioneer in aseptic surgery. The L_1 organism was later shown by Klieneberger and others to be a stable protoplast or L-form of *Streptobacillus moniliformis.* The two forms interchange under proper growth conditions. The letter L continues in bacteriological terminology for L-forms of bacteria in general. (See Chapter 42.) (Do not confuse L-forms of bacteria with L cells, a clone of animal tissue culture cells.)

Characteristics of each group of organisms are amply described in both the 7th and 8th editions of *Bergey's Manual.*

35.12
GRAM-NEGATIVE STRICTLY ANAEROBIC, ASPOROGENOUS RODS

Strictly anaerobic bacteria are found in several orders of the class Schizomycetes in the 7th edition of *Bergey's Manual.* Some of these (e.g., some of the suborder Rhodobacteriineae, genus *Actinomyces,* some of the family Micrococcaceae, and others) are discussed elsewhere. Here we shall give some attention to obligately anaerobic species other than genus *Clostridium* (Chapt. 36) of the order Eubacteriales.

Family Bacteroidaceae

The family Bacteroidaceae in the 8th edition of *Bergey's Manual* contains 10 genera, *Bacteroides, Fusobacterium, Leptotrichia, Desulfovibrio, Butyrivibrio, Succinivibrio, Succinomonas, Lachnospira, Selenomonas, Oscillospira.* These organisms are grouped together on the basis of two main characteristics: they are gram-negative and strictly anaerobic rods. They are small and very pleomorphic, sometimes branching or filamentous. Some species are motile with peritrichous flagella.

In this chapter only two genera pathogenic for man and other animals will be discussed.

Genus Bacteroides. The organisms in the genus are either nonmotile or motile with peritrichous flagella. Unlike other obligate anaerobes (*Clostridium* species), they usually do not produce large amounts of butyric acid from carbohydrate or nitrogenous compounds. The DNA base composition of the species in the genus ranges from 41 to 43 mole per cent GC, and the type species is *B. fragilis.* Many species of the genus are able to grow on complex laboratory media with initial oxidation-reduction potential below minus 100 mV; if above this potential, the culture may fail to grow even when large inocula are used. Addition of serum or ascitic fluid to such media promotes the growth of the organisms. Most species display some pleomorphic cells in broth or agar cultures, and cells may be more pleomorphic when grown in media that are not sufficiently reduced. Terminal or central swellings, vacuoles, or filaments are commonly found in cultures grown in laboratory media. It has been suggested the organism *B. corrodens* be transferred to a new genus (*Eikenella corrodens*) owing to its high DNA base composition (57 to 58 mole per cent GC).

Various species of *Bacteroides* are frequently found as apparent causative agents in lesions of the mucous membranes, in septicemia, in appendicitis, in abscesses of liver, lungs, and in necrotic lesions in other parts of the body. These pathogens are often overlooked in diagnostic microbiology because they grow only under **very** strictly anaerobic conditions on media containing blood or ascitic fluid. The colonies are small, inconspicuous, and colorless; the organisms are fragile, and it is difficult to maintain them alive. However, longwave ultraviolet light may be used as a diagnostic aid, as some species of *Bacteroides* produce brilliant red fluorescence (e.g., *B. melaninogenicus*).

Genus Fusobacterium. Organisms of the genus are much like *Bacteroides* in general properties, but, as the name implies, they have distinctly pointed ends. They are obligately anaerobic, nonmotile or motile with peritrichous flagella, and produce butyric acid as a major metabolic product of carbohydrate fermentation. DNA base composition of the species examined ranges from 28 to 34 mole per cent GC. The type species of the genus is *F. nucleatum* (DNA base composition, 28 mole per cent GC). *F. fusiforme* (sometimes known as *F. plautivincenti*) has a DNA base composition of 32 to 34 mole per cent GC. Separation from *F. nucleatum* is based on differences in G + C ratio in addition to gas formation by *F. fusiforme*. The organism is regularly found in the normal human mouth in association with *Borrelia vincentii*, a spirochete. The combination has been thought to cause ulcerative conditions of the buccal cavity, especially Vincent's angina, or "trench mouth," but there is doubt of the primary etiological status of the bacteria; they are secondary invaders.

Some human and animal diseases caused by gram-negative bacteria related to the enterics are summarized in Table 35–14.

TABLE 35–14. SOME HUMAN AND ANIMAL DISEASES CAUSED BY GRAM-NEGATIVE BACTERIA RELATED TO THE ENTERICS

Organism	Relation to O_2	DNA Base Composition (Mole Per Cent G + C)°	Disease	Epidemiology of Disease and Symptoms
Brucella melitensis *B. abortus* *B. suis*	Aerobic	56 to 58	Brucellosis	All three species are capable of producing disease (brucellosis) in a wide range of mammals, including man. Natural portal of entry is via the broken skin, the conjunctivae, or alimentary tract. Occasionally aerosol transmission may occur. Organisms at first invade local lymphatics, then spread to the regional lymph nodes followed by invasion of thoracic duct and blood stream, and finally localize in the spleen, liver, bone marrow, kidneys. In cattle and other animals, the organisms may accumulate in mammary glands, hence causing infection of the milk, and in the genital organs, i.e., pregnant uterus, often resulting in abortion. Onset of human brucellosis is usually insidious with malaise, fever, weakness, myalgia, and sweats with undulant fever (see text). Diagnosis of the disease includes cultivation of the organisms from biopsied materials (blood, bone lesions, etc.) and determination of antibody.
Bordetella pertussis *B. parapertussis* *B. bronchiseptica*	Aerobic	65 to 68	Whooping cough; canine distemper	*B. pertussis* causes "whooping" cough in children. Begins with benign nasopharyngeal symptoms followed by involvement of lower respiratory tract. Paroxysms of severe coughing with inspiratory "whooping" and often vomiting. Secondary bronchopneumonia is common and obstruction of the bronchi by mucous plug may cause anoxia resulting in convulsions. Encephalitis may occur as a complication. *B. parapertussis* causes mild form of whooping cough and *B. bronchiseptica* produces bronchopneumonia in dogs, guinea pigs, and other animals. These agents are transmitted by droplets, and active immunization of all children against pertussis at the age of three months is recommended.

°Figures listed are average ranges of several strains and species.

(*Table 35–14 continued on opposite page.*)

TABLE 35-14. SOME HUMAN AND ANIMAL DISEASES CAUSED BY GRAM-NEGATIVE BACTERIA RELATED TO THE ENTERICS *(Continued)*

Organism	Relation to O_2	DNA Base Composition (Mole Per Cent $G + C$)°	Disease	Epidemiology of Disease and Symptoms
Francisella tularensis	Aerobic	34 to 36	Tularemia (rabbit fever)	Transmission of the agent of human tularemia (rabbit fever) is usually by direct contact with tissues of infected rabbits via break in skin or through the mucous membranes of oropharynx or gastrointestinal tract. The agent may be transmitted by the bite of ticks and fleas or by the inhalation of infected aerosols. Development of ulcerating papules in the skin or mucous membranes is the first sign. The agent may be spread by the lymphatics to regional lymph nodes, which may suppurate. Transitory bacteremia may occur in some cases. If the infection occurs by droplet or aerosol, the disease is referred to as pneumonic tularemia; if the entry of agent is by gastrointestinal tract, the disease is called typhoidal tularemia. Diagnosis is by specific fluorescent antibody and cultivation of organism in special media.
Moraxella lacunata (Morax-Axenfeld bacillus or *Bacterium lacunatus*)	Aerobic	39 to 45	Infectious conjunctivitis or angular conjunctivitis	So far as is known, the agent is pathogenic only for the human eye. Experimental inoculation of the organism onto the conjunctival sac of man results in the development of blepharoconjunctivitis, either chronic or acute, and severe inflammation of the cornea may be produced.
Streptobacillus moniliformis	Aerobic		Rat-bite fever (Haverhill fever)	The organism is a causative agent of human rat-bite fever (another type of the disease is caused by *Spirillum minus*), and arthritis is common among infected persons. The organism may cause endocarditis. The infection may be acquired by infection of skin abrasion or by ingestion of contaminated foods; then it is frequently termed Haverhill fever.
Pasteurella multocida *P. septicaemiae* *P. haemolytica* *P. novicida*	Facultatively anaerobic	35 to 40	Hemorrhagic septicemia	The organisms are common inhabitants of the respiratory tracts of animals, and human infections are reported from time to time. Hemorrhagic septicemia is more common among rabbits, rats, horses, sheep, fowl, dogs, and swine.
Haemophilus influenzae *H. ducreyi* *H. aegyptius* *H. parainfluenzae*	Facultatively anaerobic	38 to 42	Acute bacterial meningitis; chancroid; conjunctivitis; bacterial endocarditis	*H. influenzae* infections are common in children but rare in adults. The common entry of the organism is the respiratory tract. The disease usually begins as a nasopharyngitis and may be followed by sinusitis or otitis media and even pneumonia, which is often complicated by empyema. In acute cases bacteremia may result in bacterial meningitis. Some strains of the organism may cause very serious disease known as obstructive laryngitis. *H. ducreyi* causes chancroid or soft chancre. The lesion produced by this venereal disease lacks the firm indurated margins of syphilitic ulcers. *H. aegyptius* is believed to cause purulent conjunctivitis, and *H. parainfluenzae* may occasionally cause bacterial endocarditis.

°Figures listed are average ranges of several strains and species.

(Table 35–14 continued on following page.)

TABLE 35-14. SOME HUMAN AND ANIMAL DISEASES CAUSED BY GRAM-NEGATIVE BACTERIA RELATED TO THE ENTERICS *(Continued)*

Organism	Relation to O_2	DNA Base Composition (Mole Per Cent G + C)°	Disease	Epidemiology of Disease and Symptoms
Actinobacillus lignieresii etc.	Facultatively anaerobic	39 to 42	Actinobacillosis (wooden tongue)	Several species of the genus are considered pathogenic to various domestic animals and human. The organisms may be involved in necrosis mucous membranes, lymphatics, and lymp glands and skin.
Bacteroides fragilis B. furcosus B. melaninogenicus B. nodosus B. oralis etc.	Strictly anaerobic	41 to 43	Assortment of nonspecific diseases	The role of *B. fragilis* and other species of the gen in disease production is not elucidated, but th organisms are frequently found in specimen from appendicitis, peritonitis, rectal abscesse pilonidal cysts, surgical wounds, and lesions the urogenital tract. *B. melaninogenicus* and i subspecies have been isolated from infections the mouth, soft tissues, and urogenital tract an are usually considered pathogenic but mainly association with other kinds of microorganism *B. nodosus* has been considered the causativ agent of foot rot in sheep and goats, and infecte hooves seem to be the only natural habitat. *oralis* has been isolated from gingival crevices man and from infections of oral cavities and uppe respiratory tracts, but significance of the organis in infection is unknown.
Fusobacterium nucleatum F. fusiforme F. mortiferum etc.	Strictly anaerobic	28 to 34	Assortment of nonspecific diseases	Some species within the genus are considere pathogenic and occur in various purulent gangrenous infections. The organisms are fr quently isolated from the oral cavity, infectio of the mouth, and upper respiratory and urogeni tracts, and occasionally from wounds. *F. mor ferum* has been isolated from necrotic abscesse of the human liver and rectum.

° Figures listed are average ranges of several strains and species.

CHAPTER 35
SUPPLEMENTARY READING

Blair, J. E., Lennette, E. H., and Truant, J. P. (Eds.): Manual of Clinical Microbiology. American Society for Microbiology. The Williams & Wilkins Co., Baltimore. 1970.

Brenner, D. J.: Deoxyribonucleic acid reassociation in the taxonomy of enteric bacteria. Int. J. Sys. Bact., 23:298, 1973.

Brenner, D. J., Steigerwalt, A. G., and Fanning, G. R.: Differentiation of *Enterobacter aerogenes* from Klebsiellae by deoxyribonucleic acid reassociation. Int. J. Sys. Bact., 22:193, 1972.

Edwards, P. R., and Ewing, W. H.: Identification of Enterobacteriaceae, 3rd ed. Burgess Publishing Co., Minneapolis. 1972.

Ewing, W. H., and Davis, B. R.: Biochemical characterization of *Citrobacter diversus* (Burkey) Werkman and Gillen and designation of the neotype strain. Int. J. Sys. Bact. 22:12, 1972.

Johnson, R., and Sneath, P. H. A.: Taxonomy of *Bordetella* and related organisms of the families Achromobacteraceae, Brucellaceae, and Neisseriaceae. Int. J. Sys. Bact., 23:381, 1973.

Prost, E., and Riemann, H.: Food-borne salmonellosis. Ann. Rev. Microbiol., 21:495, 1967.

Sneath, P. H. A., and Johnson, R.: Numerical taxonomy of *Haemophilus* and related bacteria. Int. J. Sys. Bact., 23:405, 1973.

Steele, J. H., and Galton, M. M.: Epidemiology of food-borne salmonellosis. Health Lab. Sci., 4:207, 1967.

CHAPTER 36 • ENDOSPORE-FORMING BACTERIA

The Family Bacillaceae and Anaerobiosis

Of all the families of Eubacteriales (7th edition, *Bergey's Manual*), one of the most interesting is the family Bacillaceae.

36.1
CLASSIFICATION

Two main genera of endosporeforming bacteria (so-called because spores are formed within the vegetative cells) are recognized, genus *Bacillus,* the species of which are aerobic or facultatively anaerobic, and genus *Clostridium,* which contains strictly anaerobic species. Most species of the family are gram-positive; a few are gram-negative or gram-variable (*Bergey's Manual,* 1957).

Members of the genus *Bacillus* produce the enzyme **catalase,** which decomposes hydrogen peroxide into water and oxygen, but the members of the genus *Clostridium* lack the genes for production of catalase, and for this reason they are obligately anaerobic (oxygen is toxic to the cells), and they have no way of disposing of the toxic H_2O_2 produced (see §36.6, Anaerobiosis).

Species of the genus *Bacillus* are among the most difficult to classify. For example, recent studies on nucleic acid homologies (by genetic transformation and hybridization) indicate that members of the genus *Bacillus* are extremely heterogeneous. Likewise, data on the DNA base compositions of the *Bacillus* species range as wide as 32 to 66 mole per cent GC with the majority of the species ranging between 33 and 56 mole per cent GC. Thus, on the basis of genetic material, these organisms can hardly be considered as closely related. However, similar studies on members of the genus *Clostridium* suggest less genetic heterogeneity. The DNA base compositions of *Clostridium* species vary from about 28 to 40 mole per cent GC.

While organisms in the genus *Bacillus* are not closely related genetically, physiological and morphological properties of various "species" overlap, and it requires the greatest care concerning composition of medium, its pH, temperature of incubation, age of culture, and numerous other factors to obtain reproducible results in the study of these species.

Partly for these reasons, primary groupings within the genus *Bacillus* are often based on the diameter and shape of the endospores (oval or spherical) and their location in the sporangium (see Table 36–1). However, separation of spe-

TABLE 36-1. SUBDIVISION OF THE FAMILY BACILLACEAE (GENERA BACILLUS AND CLOSTRIDIUM)

Endosporeforming, gram-positive to gram-variable rods			
Aerobic, produce catalase (32 to 66°)		Anaerobic, lack catalase (28 to 40°)	
Spore oval or cylindrical; facultative; casein and starch usually hydrolyzed.	Spore spherical; sporangia swollen; nonfermentative; casein and starch not hydrolyzed.	Fermentation of nitrogenous organic compounds.	Fermentation of carbohydrates.
Sporangia not swollen; spore wall thin.	Sporangia markedly swollen; spore oval; spore wall thick.	*B. sphaericus* 36 to 37° *B. pasteurii* AMINO ACID–FERMENTERS Fermentation of single or pairs of amino acids with formation of butyric or propionic acids, NH_3, CO_2, H_2.	Sugars, starch and pectin fermente to yield large amounts of butyric and acetic acids; some produce iso propanol, butanol, and CO_2, and H_2 may also be produced. Many are able to fix nitrogen.
MESOPHILES *B. cereus* 32 to 40° *B. megaterium* 37 to 38° *B. subtilis* 42 to 65°	MESOPHILES *B. polymyxa* 44 to 48° *B. macerans* 52 to 53° *B. circulans* 35°	PROTEOLYTIC *C. botulinum* *C. tetani* *C. sporogenes* NONPROTEOLYTIC *C. tetanomorphum*	*C. butyricum* 30 to 37° *C. perfringens* *C. pasteurianum* 31 to 32° *C. acetobutyricum* Fermentation of cellulose to acetic and succinic acids, and ethanol; CO_2 and H_2 may also be produced *C. cellulosolvens*
THERMOPHILES *B. coagulans*	THERMOPHILES *B. stearothermophilus* 47 to 56°	PURINE-FERMENTERS Purines and uric acid fermented with production of NH_3, CO_2 and acetic acid *C. acidiurici* 29 to 30°	Fermentation of mixtures of ethanol and acetic acid to higher fatty acids *C. kluyverii*
INSECT PATHOGENS *B. thuringiensis* 33 to 34°	INSECT PATHOGENS *B. larvae* *B. alvei* 32 to 33° *B. popilliae*		

° DNA base composition (mole per cent GC).

cies of the genus *Clostridium* is based primarily on energy source and energy-yielding mechanisms.

Electron microscopic studies of the distinctive surface ridges or sculpturings of spores (Fig. 36–1) and investigations of chemical constitution and enzymic content, are introducing taxonomic order among these confused species.

In the 8th edition of *Bergey's Manual*, all endosporeforming bacteria are combined in Part XV and this part is divided into two groups, aerobic and anaerobic sporeformers. Aerobic sporeformers, the family Bacillaceae, include the genera *Bacillus, Sporolactobacillus,* and *Sporosarcina;* the anaerobic genera consist of gram-positive *Clostridium* and gram-negative *Desulfotomaculum* (formerly known as *C. nigrificans*).

Regardless of the former or new system of classification, all species undergo changes in enzyme content as well as changes in cellular structure, which is frequently referred to as morphogenesis. One of the most remarkable examples of morphogenesis is that of endospore formation in bacteria within the family Bacillaceae. The vegetative cells of the family are able to form, within each cell, a more or less dormant, heat-resistant structure, the endospore.

Endospores must be distinguished carefully from conidiospores and sporangiospores of molds and Actinomycetales, from ascospores of yeasts and from conidia (often called spores) of the Streptomycetaceae. True bacterial endospores (family Bacillaceae) generally have a high degree of resistance to chemical disinfectants and to temperatures used in baking and sterilizing. None of the other spore types approaches this degree of resistance.

It was the heat-resistant spores of Bacillaceae that misled Needham and others to support the view that life began spontaneously in the infusions that they thought they had sterilized by heating. Even experienced bacteriologists are sometimes embarrassed by the appearance of sporeforming rods (*Bacillus* spp.) in supposedly sterile material or in pure cultures of bacteria. This is usually due to carelessness in the sterilizing room or to short-cuts in heating processes. Recall of heat-processed canned foods from supermarkets usually reflects these errors.

Because of the vast difference in thermostability between mature endospores and vegetative cells, presence or absence of these spores is most conclusively demonstrated by means of heat. Growth of a culture after exposure to 90C for 10 minutes in aqueous fluid (Chapt. 30) proves the presence of spores. Many spores are killed at lower temperatures or by shorter exposures to 90C. Most vegetative cells are killed by this exposure. Heat-resistant spores are also produced by a few gram-positive cocci (e.g.,

Sporosarcina ureae, DNA base composition ranging from 38 to 42 mole per cent GC), certain thermophilic actinomycetes, and sulfate-reducing, gram-negative bacilli, *Desulfotomaculum* spp.

Fine Structure of Mature Endospores

The fine structure of mature spores as observed with the electron microscope is vastly different from the structure of vegetative cells, and they also differ in both physiological and biochemical properties (Table 36–2). In general, the spore has a much more complex structure than that of the corresponding vegetative cell; for example, spores have many more defined enclosing layers (Fig. 36–2). The outermost layer, when present, is called the exosporium, and within this is a wall-like covering known as the spore coat. Underneath this spore coat is the cortex, which is composed of a multilaminated structure similar to the cell wall, the primary constituent being peptidoglycan. In addition, large amounts of dipicolinic acid (DPA) and a large number of calcium ions, probably in the form of calcium dipicolinate, are present. The chemical name of dipicolinic acid is 2,6-pyridine dicarboxylic acid:

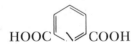

HOOC COOH

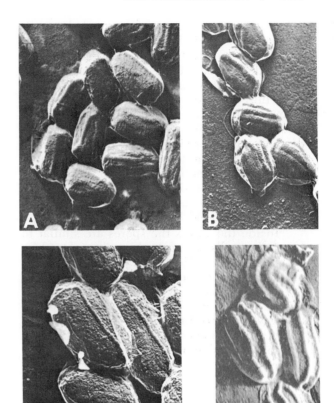

Figure 36–1. Distinctive sculpturing of spores of *Bacillus* species: *A, subtilis* (12,000×); *B, subtilis* (7,500×); *C, brevis* (9,000×); *D, polymyxa* (5,000×.)

TABLE 36–2. COMPARISON OF SPORES AND VEGETATIVE CELLS

Vegetative Cells	Spores
Vegetative cells are nonrefractile and young cells usually stain gram-positive; readily stained by ordinary stains such as methylene blue.	Very highly refractile, i.e., enveloped by layers of coverings, thick spore cortex, and coat; certain species may be enclosed by a loose outer covering, the exosporium. Impermeable to ordinary stains.
Sensitive to various physical and chemical agents such as heat, radiation, disinfectants and antibiotics; moisture content of cell 95 to 98 per cent.	Very high degree of resistance to heat, radiation, disinfectants and antibiotics. These properties may be associated with dehydration of protoplasm (5 to 10 per cent of dry weight).
Relatively low in protein, sulfur-bearing amino acids and calcium; high in polysaccharides. PHB present, DPA absent.	High in proteins, sulfur-bearing amino acids, calcium, and DPA. Low in polysaccharides; PHB absent.
Metabolic activities (O_2 uptake) and various enzymic activities are high; messenger RNA (mRNA) present, carries out synthesis of macromolecules; highly sensitive to the action of lysozyme.	O_2 uptake and enzymic activities low or practically absent. mRNA either low or absent; lack of synthesis of macromolecules; spores are resistant to the action of lysozyme.

Since vegetative cells contain much less calcium and are totally lacking in DPA, it was proposed that these substances play an important part in the heat resistance of the bacterial spore; considerable experimental evidence now supports this contention. DPA is liberated as calcium dipicolinate (CaDPA) in appreciable amounts during the germination process, and the cells which lose CaDPA become heat-sensitive. DPA was for some time thought to be a unique and distinctive component of bacterial endospores; it has since been found in considerable quantities as a waste product of the growth of the mold *Penicillium citreo-viride.*

Inside the cortex the spores are similar to vegetative cells, i.e., cytoplasm is enclosed by both cell wall and cell membrane. Thus, the major differences between spores and vegetative cells are primarily confined to the exterior of the spore cell wall. Detailed structural changes and derivation of each "spore covering" that occurs during conversion of vegetative cell to spore (sporulation) are discussed in the following section.

36.2
SPORULATION OF BACILLACEAE

Knowledge of the thermal resistance of spores, of their formation and germination, and of the factors affecting both of these is of great importance in various preserving industries,

manufacturing processes, medicine, and other human activities. In the strictly aerobic species of *Bacillus,* sporulation occurs *only* under aerobic conditions. In the strictly anaerobic *Clostridium* it occurs only anaerobically. In some of the aerobes that are facultative, sporulation may occur aerobically or anaerobically. However, in bacteria of both genera, sporulation does not occur during exponential growth but only when growth ceases as a result of exhaustion of one or more essential nutrients, or conditions become unfavorable for growth and multiplication of the organisms.

The process of **sporulation** in Bacillaceae has been admirably elucidated in great detail with the electron microscope. Sporulation is much alike in both *Bacillus* and *Clostridium.* Commonly the process begins by: (1) Elongation of the cell as though about to undergo fission. In *Bacillus* spp., which have been most used as models, many granules of poly-β-hydroxybutyric acid (PHB) now appear (Fig. 36–3). These are used as reserve food for energy required in the late and maturation phases of spore formation. By virtue of stored PHB, sporulation in *Bacillus* species can go to completion in the absence of extraneous food sources, e.g., sporulation can be completed in distilled water (**endotrophic sporulation**). In *Clostridium* spp., such internal food reserves have not been found, and these species require exogenous food sources during all stages of sporulation (**exotrophic sporulation**).

(2) In the elongated cell about to sporulate,

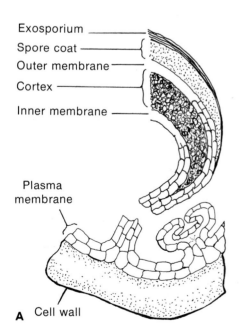

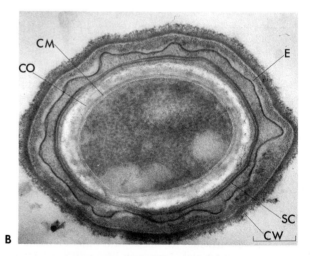

Figure 36–2. Ultrathin sections of sporulating cells of *B. cereus* showing developing spore. Untreated; Cell wall (*CW*) and developing cortex (*CO*); the cortical membrane (*CM*) is stained very intensely. No deposits are seen along the developing spore coat (*SC*) or exosporium (*E*). Marker represents 0.2 μm. (*B,* Courtesy of P. D. Walker and J. Short.)

the nuclear material is redistributed from end to end as an axial mass and is then divided so that one complete nucleus appears to be concentrated at one end of the cell (Fig. 36–4, *A*).

(3) The cell membrane now appears to grow inward, possibly from a **mesosome,** and to grow around the nucleus (Fig. 36–4, *B* and *C*), forming a complete, **two-layered septum** segregating the nucleus and a portion of the cytoplasm containing ribosomes, mesosomes, and enzymes, from the remainder of the cell. The material thus enclosed is now a **forespore** (Fig. 36–4, *D*), a new cell, and the septum is a **forespore membrane.** The forespore DNA becomes distinctively rearranged. Water may be withdrawn during forespore formation or at some other stage of the process; the mature spore is considerably but not wholly dehydrated.

(4) Between the outer and inner layers of the forespore membrane there now is laid down a thick shell or **cortex** (Fig. 36–4, *E*) that forms between the layers and, like a cell wall, encloses the spore protoplast. The spore cytoplasm, ribosomes, and nuclear material are enclosed within the cortex and the **inner layer** of the forespore membrane, which now appears to serve as, or to produce, a protoplast membrane.

(5) Around the cortex, and exteriorly to the **outer** forespore layer, there is formed a **spore coat.** This appears first in the form of segments, later as a complete, continuous, two-layered covering. Outside the spore coat a membrane is formed that is scalloped or rippled (**sculptured**) and that thickens and becomes multilaminate and often voluminous (Fig. 36–5). Some species may possess an additional layer, the **exosporium,** exterior to the spore coat.

Staining Spores. As noted previously, mature spores cannot be stained by ordinary methods such as Gram's. A rod with a spore inside, when stained by ordinary methods, may appear to have a hole in it (Fig. 36–6). However, the outer surface of the spore may readily be colored; in stained smears **free** spores may appear as tiny blue or red rings. There are special stains for spores. The Ziehl-Neelsen acid-fast stain often penetrates. Another method is outlined below:

Spore stain: Apply saturated malachite green (about 7.6 per cent) for 10 minutes. Flame until warm. Rinse with tap water for about 10 seconds.

Apply 0.25 per cent aqueous safranin for 15 seconds.

Rinse with tap water, blot dry.

Factors in Spore Formation. Endospores are formed most readily under good growth con-

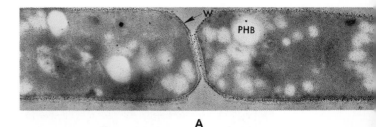

A

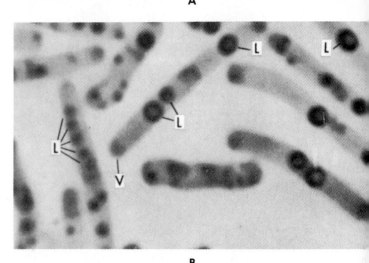

B

Figure 36–3. Ultrathin sections of *B. cereus* oxidized with periodic acid and stained with silver. *A*, Poly-β-hydroxybutyric acid granule formation. *B*, Lipid inclusion of *B. cereus* strain AC isolated after hypochlorite treatment.

ditions. Their formation appears to require an energy and carbon source, such as glucose. However, after the cessation of vegetative growth, if more energy source, such as glucose, is added to the culture medium, sporulation may be inhibited. Such a phenomenon is known as **catabolite repression,** i.e., certain nutrients inhibit the synthesis of essential enzymes involved in sporulation. Therefore, vegetative growth and sporulation are considered opposing processes. Nitrogenous food substances (as certain amino acids) are also required. In some species manganese ions are requisite; in others, Fe^{2+}, Cu^{2+}, Zn^{2+}, and Mo^{2+} greatly increase spore formation. At least part of the spore contents appear to be newly synthesized and are not merely old material of the vegetative cell. Energy for synthesis of spore materials is derived from endogenous sources, both protein and PHB (see above for catabolite repression). Depending on species and growth conditions, the process of sporulation is complete in about 5 to 13 hours from the time exponential growth ceases. It is generally accepted that sporulation proceeds when conditions are no longer favorable for growth.

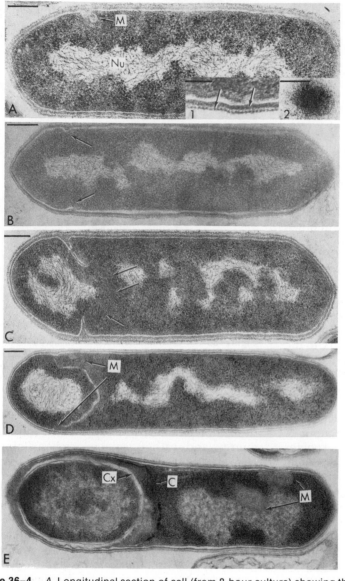

Cell — Wall, Membrane, Mesosome, Nuclear material

The cell membrane grows inward, possibly from mesosomes

Nuclear material redistributed from end to end as an axial mass

Membrane invaginates to form spore septum

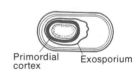

Spore septum grows around protoplast

Primordial cortex   Exosporium

Figure 36–4. *A,* Longitudinal section of cell (from 8-hour culture) showing the axial disposition of its nuclear material (*Nu*) during the initial stage of sporulation. Mesosome (*M*). Except where indicated otherwise, markers represent 0.2 μm. *1,* Enlarged longitudinal section showing the two layers (*arrows*) of the cell wall. Marker represents 0.1 μm. *2,* Oblique section showing the surface pattern on the outer cell wall.

B, Longitudinal section of cell (6-hour culture) showing slight invaginations (*arrows*) of the plasma membrane near one pole during the early part of spore septum formation.

C, Longitudinal section of cell (8-hour culture) with a completed spore septum (*arrows*).

D, Longitudinal section of cell (10-hour culture) during the early phase of forespore (*Fs*) engulfment. Mesosomes (*M*) are present at the periphery of the septum.

E, Longitudinal section of cell (10-hour culture) after completion of forespore engulfment. Substantial cortex (*Cx*) has formed, and several pieces of spore coat (*C*) are present at the mother-cell end of the spore. Mesosomes (*M*) are present in the nonsporulating portion of the cell. (Courtesy of L. M. Santo, H. R. Hohl, and H. A. Frank.)

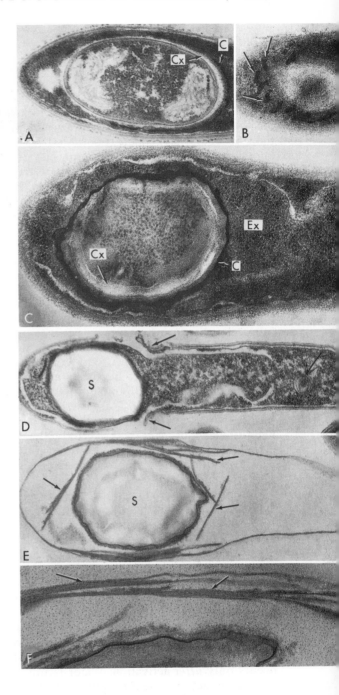

Figure 36–5. *A*, Longitudinal section of cell (16-hour culture) showing advanced stages in formation of spore coat (*C*) and cortex (*Cx*). Early exosporium formation is indicated by the thin longitudinal fissure, starting at the polar end of the cell, between the sporangial membrane and the cytoplasmic material around the spore protoplast.

B, Oblique section showing topside view of the discrete spore coat fragments (*arrows*).

C, Longitudinal section of cell (16-hour culture) showing encirclement of the cortex (*Cx*) following coalescence of spore coat (*C*) pieces. Development of the exosporium (*Ex*) is quite advanced at this stage.

D, Longitudinal section of cell (25-hour culture) before liberation of the spore (*S*). The circular inclusion (*arrow*) has formed by reaggregation of fragments resulting from partial lysis of the cell membrane.

E, Longitudinal section through a mature, free spore (*S*) (from 33-hour culture) showing exosporial layers (*arrows*) clustered around the spore.

F, Section showing exosporial multilayers (*arrows*) in greater detail. Marker represents 0.1 μm. (Courtesy of L. M. Santo, H. R. Hohl, and H. A. Frank.)

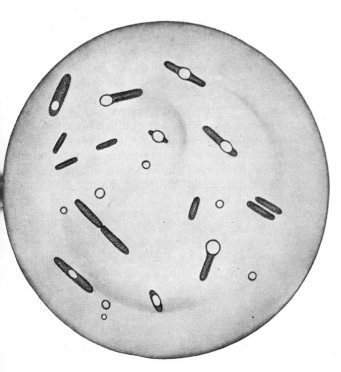

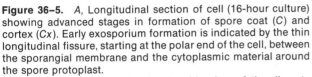

Figure 36–6. Various types of bacterial spores. Some of the spores have escaped from the sporangia. Stained with methylene blue, which does not penetrate inside the spore, only the outer surface of the spore is stained. (About 1,000×.)

36.3
SPORE GERMINATION

Spore Outgrowth

A spore is able to remain dormant for many years, and records of survival of spores for more than 50 years have been reported. If it were known how to make all spores promptly germinate and grow into vulnerable vegetative cells, methods of sterilization and preserving of foods and industrial materials could be much simplified at tremendous savings of time and money. The change from a dormant spore to an active vegetative cell involves at least three distinct stages: (1) **activation;** (2) **germination;** and (3) **outgrowth,** or **postgermination development.**

Activation. Under natural conditions, activation of endospores in aqueous media appears to occur only occasionally, slowly, and intermittently during hours or days at temperatures from 20 to 35C. Activation of all or nearly all the spores in a group occurs within a few minutes if the spores are heated in an aqueous fluid at about 65C for 15 to 60 minutes; higher temperatures (105 to 120C) are used for thermophilic spores. Spores so treated are said to be **heat-activated** or **heat-shocked.** Activation may also be brought about by certain chemicals, e.g., L-alanine, adenosine, glucose, and some reducing agents.

Activation is a reversible process. Unless agents are present that induce the activated spores to germinate after being activated, the spores revert to dormancy and may require reactivation. The mechanism of activation is not fully understood but may involve reversible alterations (rupture of S- and H-bonds?) in the tertiary structure of certain of the enzyme proteins in the spore.

Germination. Germination is an irreversible change from the activated but still inert state to the beginning of enzyme action. Germination occurs only *after* activation. Water is imbibed, metabolic activity begins, and nutrients are required: sources of P, N, S, and C and energy. Germination of heat-activated cells may be triggered by a variety of chemical agents: various hexoses, potassium nitrate, and some others but most notably, and specifically, L-alanine. The presence of electrolytes and of L-alanine is usually essential. D-Alanine, on the contrary, can suppress germination. An enzyme, referred to as a **germination enzyme,** that causes germination of *Bacillus* spores has been described.

Germination of all species of *Bacillus* can occur in the presence of air; many clostridia germinate only anaerobically in the presence of carbon dioxide. Water is essential in any case. Germination is not necessarily followed by outgrowth, and agents that trigger germination do not necessarily support outgrowth and vice versa.

The process of germination may be divided into two stages: (1) the **microlag,** extending from addition of triggering agent to the first visible change in the spore: *beginning* loss of refractility; (2) the **microgermination time,** which extends from first loss of refractility to *complete* loss of refractility. The spore has by now also lost its thermostability and it can also be stained with ordinary dyes. Swelling occurs, presumably caused by imbibition of water.

Dipicolinic Acid and Surfactants. It has been known for some time that dipicolinic acid and calcium form about 5 to 15 per cent and 1 to 3 per cent, respectively, of the dry weight of spores, and that these substances are released in considerable amounts, apparently from the cortex, when spores germinate and grow out. That DPA and calcium play a critical role is indicated by the observation that adding DPA, and chelates of calcium, magnesium or strontium to washed spores, aerobic or anaerobic, induces 100 per cent germination within about 20 minutes at pH 7.0 and temperatures of 25 to 35C in the absence of any activating agent. Equally fast or faster germination is induced, in the absence of heat-shocking or L-alanine, by some surface tension reducers, notably, *n*-dodecylamine at pH 7.0 and temperatures from 37 to 70C. The mechanisms of these actions are under intensive investigation.

Postgerminative Growth. As a seed may sprout in perfectly nonnutritious, sterile distilled water but fail to develop further, so germination of a spore does not necessarily imply postgerminative development into a bacterium. **Postgerminative** outgrowth of the spore occurs only if conditions for cell nutrition and growth are favorable. For example, amino acids, especially glutamic acid, with other energy sources such as glucose and minerals, are essential for the cells.

Assuming such favorable conditions, the postgerminative changes consist of swelling of the spore, increasing enzymic activity, splitting of the spore wall, elongation of the developing cell and, finally, emergence of the actively growing vegetative cell. The method of splitting of the spore wall may be distinctive of species (Fig. 36–7).

The DPA chelates mentioned above may

be added to a good nutrient broth, thus achieving activation, germination, and outgrowth in one operation.

36.4
GENUS BACILLUS

All species of *Bacillus* are more or less strictly aerobic, gram-positive or gram-variable, endosporeforming rods (Color Plate V, *D*). Most species are motile, and many contain numerous fat globules. Just prior to sporulation, they form many granules of PHB, conspicuous if properly stained; e.g., by fat stains. From the standpoint of cultivation in the laboratory, or growth in natural habitats, most species of *Bacillus* are not fastidious. They grow well in aqueous extracts of soil, in vegetable or yeast extracts, or in simple peptone media. The DNA base composition of the genus ranges from 32 to 66 mole per cent GC.

They are active and versatile producers of hydrolytic enzymes and consequently can utilize as food a wide variety of proteins, carbohydrates, lipids, glucosides, alcohols, and organic acids. They are thus seen to be of great importance as scavengers. Some of their enzymes are produced in quantity commercially and are used in industrial processes (leather, paper, silk, textiles, coffee). Several species are both famous and respected in the community because they produce valuable antibiotics: bacitracin and polymyxin, for instance. Others (sometimes the same) are both infamous and shunned because they grow in many sorts of valuable commodities (paper, various foods and drugs, wood, leather), producing spoilage and economic loss to human beings. One species, *B. anthracis*, is the cause of anthrax or "malignant pustule."

Important Species of Bacillus

Most species of *Bacillus* are harmless saprophytes of the soil. Their heat-resistant spores are ubiquitous and often plague the food-preserving industries, surgeons, microbiologists, and others who must work with sterile materials.

Bacillus anthracis. Only one species is definitely pathogenic for animals: *B. anthracis*, the cause of anthrax (Gr. *anthrax* = carbuncle). This organism is frequently considered as a variety of *B. cereus*; however, *B. anthracis* is not able to grow saprophytically in nature.

Anthrax. Not only was anthrax the first

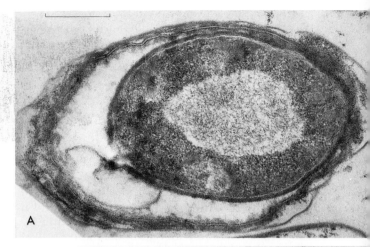

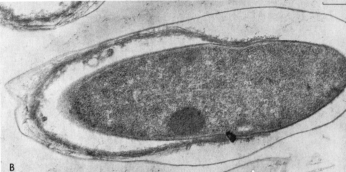

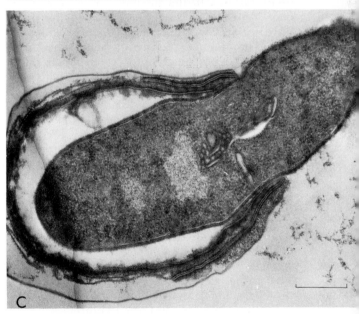

Figure 36–7. Electron micrographs of ultrathin section of germinating *C. pectinovorum*.

A, Young, developing cell after 0.5 hours of germination containing a large fibrillar nucleoplasm (*N*). Marker represents 0.5 μm; Glutaraldehyde fixation.

B, Elongate rod has ruptured the spore coat. Note the large, dense bodies in cell. Glutaraldehyde prefixation. Marker represents 1 μm.

C, Vegetative rods emerging from fractured, inflexible spore coats. R-K fixation. Marker represents 0.5 μm.

disease shown conclusively by Robert Koch to be caused by bacteria but his study of the disease provided one of the foundations (**Koch's postulates**) for the development of modern bacteriology as a science. Anthrax is primarily a disease of farm animals and many kinds of wild animals, including deer and birds. However, not all animals are equally susceptible to the agent of the disease.

Anthrax is transmissible to man, especially workers in hide, hair, and wool industries. In man, the spores most commonly gain entrance to the body from infected soil, dust, or animal hair or tissues, through a cut in the skin, and if untreated, may progress to a fatal systemic infection. The spores first germinate and grow at the point of entrance, forming a very rapidly progressive, angry, inflamed carbuncle or pustule (malignant pustule) which, when well developed, is covered with a black crust. This pustule teems with anthrax bacilli which are heavily encapsulated. The initial pustule frequently heals, but in other cases the bacilli invade the blood, multiply enormously, and are spread through all the organs of the body, where they tend to form local lesions which serve as further centers for dissemination. When growing in the body (i.e., in the absence of free O_2) or in cultures in the presence of atmospheres containing 50 per cent carbon dioxide, they produce no spores but develop large capsules of a polypeptide of D-glutamic acid. This may protect the organism from phagocytosis. The organism in pathologic material is highly invasive, growing well throughout the body, including the blood. Rapid sporulation can occur when the carcass is opened and exposed to the air.

The organisms also produce a group of potent, heat-labile protein exotoxins: I, II, and III. The toxins occur only in the blood of infected animals or in certain complex, semisynthetic media supplemented with large amounts of bicarbonates. The bicarbonates apparently stimulate the release of the toxins from the cells into the culture media. These toxins are responsible for most of the disease symptoms. The toxins may cause a physiologic shock syndrome, acute electrolyte imbalance, edema, hemoconcentration, and renal failure.

Another disease, known as "woolsorter's disease," is also caused by this organism, and, as the name indicates, the infection is prevalent among handlers of sheep's wool. The disease is a respiratory infection resulting from the inhalation of spores.

Treatment and Control. Anthrax in farm animals (specifically cattle and sheep) may be prevented (to a limited degree) by vaccinating the animals with *B. anthracis* **bacterins** (formaldehyde-killed bacilli). You may recall a famous and successful experimental vaccination of sheep performed by Pasteur. In fact, from this experiment Pasteur laid the foundation for the extensive vaccine research that ultimately produced live vaccines for a wide variety of infectious diseases. The animals may be protected from disease by injection of spore-vaccine made with living spores of graded, attenuated type, but no anthrax vaccine is known that is considered safe for man, even today. Therefore, control of the disease in animals is primarily through eradication measures by elimination of diseased animals from herds and by properly destroying carcasses. Animals dead of anthrax should be handled with care to avoid contaminating the premises and infecting the handlers and dust with the spores. The spores can survive many decades in dry soil, but in most agricultural soils spores do not persist more than a few years.

The disease may be effectively treated with penicillin, the tetracyclines, or erythromycin before the onset of bacteremia.

Bacillus Species Pathogenic for Arthropods

Pasteur was one of the first to make scientific studies of infectious diseases of arthropods, especially diseases of the valuable silkworms. Today, infection of arthropods is important from two opposite viewpoints:

First, as in Pasteur's day, it is essential to protect from infection such valuable arthropods as silkmoths and honeybees. Second, it is now known that specific infection is an important means of killing certain arthropod pests: (a) vectors of diseases and (b) arthropods that destroy crops. While there are numerous effective chemical insecticides, some have toxic effects on man and livestock, wild animals, and aquatic fauna, such as fish, some resist biodegradation and persist for a long period of time, as does DDT. Further, many noxious insects, notably malaria-transmitting mosquitoes and houseflies (*Musca domestica*), tend to become resistant to certain insecticides. To avoid some of these disadvantages, bacteria and viruses that are harmless to mammals and useful arthropods, but lethal to undesirable arthropods, are now being used more and more for pest control. The ideal microbial insecticide should be: (1) virulent and lethal at all times; (2) not sensitive to the environment where it is to act (sunshine, dryness, or dust); (3) persistent in the area of use, e.g., sporeforming; (4) fast-acting; (5) harmless to man, plants, and useful insects; (6) capable of mass production at low cost. There are few, if any, such agents; some have several of these desirable attributes.

There are several species of **entomogenous** bacteria (bacteria that live in insects) that cause disease of their insect hosts. Best known at present are species of *Bacillus* that cause (a) infection or (b) intoxication.

Infection. Infection is caused by the growth, in the larvae of the arthropod, of bacilli from ingested spores. At first, rapid growth occurs in the gut. This is followed by penetration of the gut wall and generalized invasion, with slow but massive growth of the bacilli in the hemolymph ("blood") of the larval stage. The invasion process is usually slow, requiring days or weeks to kill. Although some toxic substance may be produced by bacilli causing this type of disease, death appears primarily due to overwhelming invasiveness on the part of the bacilli.

Milky-White Disease. A "useful" disease, milky-white disease, is typical of the infection just described. It is caused by *B. popilliae* (Fig. 36–8) and *B. lentimorbus*. These are used to combat Japanese beetles (*Popillia japonica*). The bacilli grow to prodigious numbers in the "blood" of the larvae causing it to appear milky white. For use against the beetles, the infected larval juices, containing billions of spores, are dried, ground, and mixed with chalk dust or other powder. This is applied to the soil as a spray or dust. The beetles have disappeared almost entirely in areas where the spores have been applied.

Methods for production of large numbers of spores of the organisms by using artificial culture media have recently been developed. As a curious sidelight, these bacilli are confirmed drug addicts, i.e., they require barbiturates for vegetative growth in synthetic media.

Foulbrood. This illustrates an undesirable aspect of infection of arthropods. *Bacillus alvei* is one of several organisms that cause, or are associated with, a disease (**foulbrood**) of bees which results in great losses to beekeepers annually. There are several forms of the disease. American foulbrood is caused by a related organism, *B. larvae*, while *B. alvei* causes European foulbrood; certain streptococci (*S. apis*) also appear to cause the disease. The larvae of bees contain the infecting organisms in large numbers. This sort of disease appears to parallel the type of slowly progressive infection of Japanese beetle larvae just mentioned.

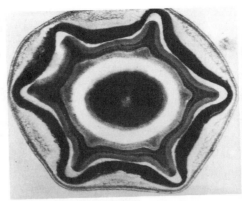 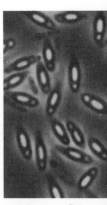

A **B**

Figure 36–8. *A*, Cross section of a *B. popilliae* spore, NRRL B-2309M, produced on solid medium. (35,000×.) *B*, Microscopic appearance of *B. popilliae* spores. Spores of B-2309 formed in larvae. (Phase-contrast, 2,375×.) (Courtesy of Eugene E. Sharpe; *A*, photomicrograph by S. M. Black.)

Intoxication. This is caused by a protein toxin produced by *B. thuringiensis*, *B. entomocidus*, and their several variants just prior to sporulation. The toxin (presumably a metabolic waste product) is formed in the **sporangium** of the bacilli as a cuboidal, triangular, or diamond-shaped crystal lying beside the spore. It is called a **parasporal body** (Fig. 36–9). Bacilli producing such crystals are said to be **crystalliferous.** Parasporal bodies are formed in the sporangia of many species of *B. cereus*-like bacilli, e.g., *B. popilliae*, but in *B. popilliae* the parasporal body does not appear to be very toxic, at least for Japanese beetles. Possibly it may be so for other species of insect, a good field for investigation in applied entomology.

Within a few minutes after ingestion by larvae, the toxin causes a paralysis of the gut with cessation of feeding and, within a few hours, total paralysis and death. Many species of larvae of *Lepidoptera* (moths and butterflies), including the valuable silkworms (*Bombyx mori*) and some species of flies, are susceptible to the toxin. An enzyme, phospholipase C, is also produced and appears to have a noxious effect. Unlike milky-white disease, infection and invasion ordinarily play little or no part in the lethal process.

The use of *B. thuringiensis* in the control of insects differs from the use of *B. popilliae* in that *B. thuringiensis* must be cultivated in large amounts to obtain sufficient toxic crystals. These, like any other insect poison, are sprayed repeatedly on crops. *B. popilliae*, on the contrary, is used in smaller amounts merely to start a self-propagating infection, among insect larvae, which maintains and spreads itself from

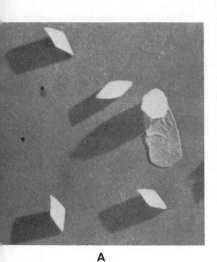
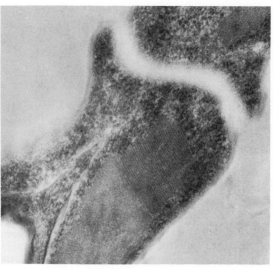

A   **B**

Figure 36-9. *A*, Electron micrograph of free crystals and a free spore surrounded by an exosporium of a crystal-forming *Bacillus*. (Metal-shadowed preparation; 7,500×.) *B*, Parasporal body of *B. thuringiensis* within a sporangium.

year to year. The production of *B. thuringiensis* toxin is now a considerable industry in some countries.

Other Important Bacillus Species

Bacillus subtilis is the type species of the genus and is found in dusty places everywhere. If hay is soaked in warm water for a day or two, the water will be found teeming with organisms of many kinds, including many beautiful protozoans; *B. subtilis* and other species of *Bacillus* will also be found.

Bacillus subtilis often forms long chains of bacilli sometimes called **streptobacilli**. Since the bacilli are motile, such chains swim with a writhing motion. Owing to avidity for oxygen, *B. subtilis* and many other species of *Bacillus* grow in a scum, or **pellicle**, at the surface of fluid media.

Because of its active attack on organic nitrogenous compounds, cultures of *B. subtilis* smell of ammonia. On slants of potato it grows luxuriantly, with a yellowish or pink color and a warty or vesiculated appearance.

B. subtilis is important as the source of the antibiotic, **subtilin. Bacitracin** is produced from a strain very like *B. subtilis*, often called *B. licheniformis*.

Bacillus coagulans is a species of importance as a cause of spoilage of canned foods. It is notable for its ability to grow in acid foods such as tomatoes. Since it produces no gas, spoilage (souring) is not discovered until the container is opened. *B. coagulans* is said to cause **flat sours,** so termed because the ends of

the can do not bulge as they would if gas were formed under pressure by the fermentation. Spores of *B. coagulans* are very heat-resistant and thus sometimes survive commercial processing. *B. coagulans* is either a facultative anaerobe or it can grow sufficiently in the small residuum of air enclosed in cans at the time of processing to produce its results.

B. stearothermophilus is also well known as a nuisance and a source of flat sours in the canning industry. Since it is exceptionally heat-resistant, its spores are often used to check the efficiency of heat sterilizing processes.

36.5
GENUS CLOSTRIDIUM

Clostridium is a large group, comprising nearly 100 "species." Many species are probably identical or are mere variants of each other. The group of clostridia is sometimes divided into physiological groups on the basis of enzymic properties, especially in relation to fermentation and proteolysis. DNA base composition of the genus ranges from 28 to 40 mole per cent GC. Some representative species and distinctive properties are shown in Tables 36-1 and 36-3.

All clostridia are obligately anaerobic, gram-positive, sporebearing rods. Unlike spores of genus *Bacillus* (which commonly exhibit ribbed or sculptured surface structure, Fig. 36-1), some species of genus *Clostridium* produce spores with prominent and elaborate protuberances or appendages. Such appendages

TABLE 36-3. REPRESENTATIVE SPECIES AND PROPERTIES OF CLOSTRIDIA

Species	Glucose	Lactose	Sucrose	Proteolysis	Motility	Capsules
C. tetani	−	−	−	−	+	−
C. histolyticum	−	−	−	+	+	−
C. novyi (A)	+	−	−	+	+	−
C. putrefaciens	+	−	−	+	−	−
C. botulinum	+	−	−	+	+	−
C. septicum	+	+	−	+	+	−
C. perfringens	+	+	+	−	−	+
C. butyricum	+	+	+	−	+	−

among different species are sufficiently prominent and varied to be considered taxonomically significant (Fig. 36–10). The chemical composition of the appendages is unlike cell wall or spore cortex in that it contains neither diaminopimelic acid (DAP) nor hydroxyproline. Similarly, it lacks muramic acid but is composed of glucosamine, phosphate, and some 17 common amino acids. Nearly all species of *Clostridium* are motile. They vary somewhat in size and shape in the manner, say, of cigars but average around 0.5 μm by 10 μm in dimensions. They require complex organic media such as cooked-meat medium, blood-glucose-infusion agar or broth, and the like. The group includes the organisms producing tetanus (lockjaw), gas gangrene, and botulism (food poisoning). The majority of clostridia are harmless and helpful saprophytes. Many of them produce enzymes, chemicals, and industrial fermentations of great value. All occur widely distributed in the soil. Some of them also live in the intestinal tract of man and animals. They are metabolically active and versatile.

Clostridium butyricum. *C. butyricum* is one of the earliest species of *Clostridium* to be studied (Prazmowski, 1880) and is the type species of the genus (DNA base composition, 37.4 mole per cent GC). It represents the group of industrially important clostridia. In general they are plump, actively motile rods having oval, excentric spores which swell the sporangium. They grow well in media made of dilute molasses or grain extracts, with starch and suitable nitrogenous and mineral (and sometimes vitamin) supplements. All ferment carbohydrates, with the production of one or more commercially valuable substances, e.g., **butyl, ethyl, amyl,** and **propyl** alcohols; **acetic, formic,** and **lactic** acids; **acetone; carbon dioxide;** and **hydrogen.** The products of fermentation depend on the species or variety of *Clostridium* used and the condition of the fermentation, i.e., pH, temperature and substrate. The industrial uses of these

species of *Clostridium* are more fully discussed in Chapter 49.

Anaerobic Nitrogen Fixation. An interesting property of some of these organisms (e.g., *C. pasteurianum*) is the power to fix atmospheric nitrogen. That is, they are not, like most other organisms, restricted to the use of nitrogen combined in the form of ammonia, nitrates, or amino acids, as are "higher" (more dependent!) plants and animals, but possess the power to cause free nitrogen of the air to combine in the synthesis of their cell substance. Several other microorganisms of the soil are even more important in nitrogen fixation. As will be seen later, without microbial nitrogen fixation there might be no human race at all. (See Nitrogen Cycle, Chapter 43.)

Some Pathogenic Clostridia

An important paradox is that although they are highly dangerous pathogenic organisms, *Clostridium botulinum* and *C. tetani* are not **parasites,** but strict **saprophytes.** They grow only in dead matter and cannot invade live tissue to a significant degree. They are commonly found in the soil and in human and other animal feces. The spores, consequently, are widespread in manured and sewage-polluted lands.

Clostridium tetani and Tetanus (Lockjaw). *Clostridium tetani* is one of the strictest anaerobes. Morphologically, the organism is distinguished by its spherical spore that occurs at the very tip end (**terminal**) of the rod. The round, terminal spore gives to the organisms what has been called a **drumstick** appearance (Fig. 36–11).

C. tetani gives off a potent exotoxin. Tetanus toxin is particularly active in the motor nerve centers, irritating them so that the muscles (most conspicuously those of the jaws) connected with them are thrown into a state of violent and continuous contraction (**tetanic con-**

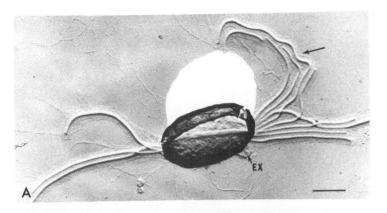

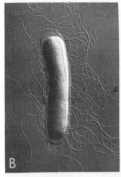

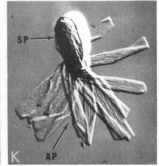

Figure 36–10. *A,* Replica of the free spore of a strain of *C. bifermentans* showing exosporium (*EX*) and appendages (*unlabeled arrows*). Strain 4407 spore has featherlike appendages. Bar indicates 0.5 μm.

B to *K,* Replica series, showing sequential steps in the development of the free spore of *Clostridium* sp. N1. Culture age varied from six days (*B* to *F*) to seven days (*G* to *K*). (Replicas, 7,000×.)

B, Vegetative cell with peritrichous flagella.

C, Early indication of spore (*SP*) formation (arrows).

D, Sporangium with maturing spore (*SP*); the spore body is rigid. This is indicated by the prominent shadow it casts compared with that cast by the vegetative portion of the sporangium. Spores at this stage are refractile under phase-contrast illumination.

E, Spore body (*SP*) is free and it has a rough surface; the vegetative (*V*) portion of the cell is undergoing lysis and disintegration.

F, Thin section showing appendage formation by *Clostridium* sp. N1. The appendage extends farther into vegetative cytoplasm.

G, First evidence of spirally arranged ribbon-like appendages (*AP*) in the disintegrating vegetative cell.

H, Spore body with exposed spiral of attached appendages (*AP, arrows*).

I, Free spore with some appendages (*AP*) still spirally arranged; last vestiges of vegetative cell have disappeared.

J, Free spore with appendages (*AP*) appearing now as a tuft of parallel ribbons.

K, Free spore with ribbon-like appendages (*AP*) now separated and flared. The connection between appendages and spore body (*SP*) is obscured. (*A,* Courtesy of L. J. Rode and L. D. Smith. *B* to *K,* Courtesy of L. J. Rode, M. A. Crawford, and M. G. Williams.)

vulsion, or **tetanus**). The use of antitoxin in the treatment and prevention of tetanus is a classical example of passive immunity and was the first to be discovered (von Behring and Fränkel, 1890).

Tetanus organisms gain entrance to the body with dirt or dirty objects when these are forced into the tissues, as in gunshot or shrapnel wounds, deep, extensive, and dirt-contaminated burns, or various other accidental means. Under

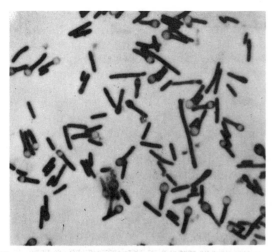

Figure 36–11. *Clostridium tetani,* a gram-stained smear showing terminal spores that swell the rods and produce the typical drumstick appearance. Cells of this species range from 0.3 to 0.8 by 2.0 to 5.0 μm. (Courtesy of General Biological Supply House, Chicago, Ill.)

phylactic antitoxin may be used. Even if serum has to be used, toxoid should also be given. (Why?) Antibiotics may be used to suppress growth of the bacteria in the wound.

Clostridium perfringens (C. welchii). *C. perfringens* is a rather short, thick rod with rounded ends. It usually grows singly, never in long chains or filaments. It forms oval, central, or subterminal spores which do not swell the cell. With perhaps a dozen uncommon or medically unimportant exceptions it is the only nonmotile species in the genus. It is rare also among clostridia in that it reduces sodium nitrate to sodium nitrite.

The constant presence of *C. perfringens* in feces has led at times to its consideration as an indicator of human fecal pollution when found in water (Chapt. 45). *C. perfringens* produces much hydrogen by fermentation and is often called the "gas bacillus" (Fig. 36–12).

Some strains of *Clostridium perfringens*

such circumstances some tissue is killed locally by the injury, and in deep wounds the low O-R potentials favor growth of anaerobic bacteria. *C. tetani* grows as a saprophyte on the necrotic (dead) tissue in the wound, liberating its deadly toxin, which is absorbed by the blood or nerves or both.

Tetanus toxin is one of the most potent poisons known. It requires only about 0.00025 gm of tetanus toxin to kill a man, while it requires twenty times as much cobra venom and about one hundred and fifty times as much strychnine.

Tetanus Immunization. Alum-precipitated fluid toxoids, in all respects analogous to diphtheria toxoids (Chapt. 27), are useful in producing active immunity to tetanus. Protection depends particularly on the action of a **primary** stimulus consisting of at least one, preferably two or three, doses of tetanus toxoid a month or so apart. This is routinely given on entering school or the armed forces. A booster dose is sometimes given about a year later.

To a **secondary** stimulus (such as entrance of tetanus toxin into the body as the result of infection of a wound) the toxoid-conditioned, antibody-producing cells respond quickly with the production of antitoxin. Similarly, a dose of toxoid is often used as prophylaxis in dealing with any fresh wound in previously immunized persons. Because of allergy, it is preferable to avoid the use of **serum** in dealing with wounds unless tetanus seems imminent. However, in case of severe, very dirty wounds, passive pro-

Figure 36–12. Tube of milk inoculated with *Clostridium perfringens* showing "stormy fermentation." For explanation see text. (Photograph courtesy of Communicable Disease Center, U.S. Public Health Service, Atlanta, Ga.)

have been implicated as causes of food poisoning in man and animals, especially by meat dishes in Great Britain. Typical symptoms are gastroenteritis with diarrhea. The particular strains involved in Britain are described as type A, distinguished by producing very heat-resistant spores and γ- or α-type colonies in blood agar (see discussion of streptococci in Chapter 38). In the United States, food-poisoning strains seem to be uncommon and to differ in several respects from British strains, especially in not producing highly thermostable spores regularly. The mechanism of the poisoning is still obscure.

Methods for isolation and enumeration of the organisms include 24 hours of anaerobic incubation of inoculated, highly selective agar medium containing sodium sulfite, polymyxin and sulfadiazine. Colonies blackened by reduction of the sulfite are transferred as pure cultures and tested for presence of gram-positive, nonmotile, sporeforming rods that reduce sodium nitrate to sodium nitrite. Animal tests for food-poisoning toxins in foods have not yet yielded useful results.

C. perfringens, common in the soil, always accompanies *C. tetani* in dirty wounds. Like *C. tetani*, *C. perfringens* grows as a saprophyte in deep, dead tissues and gives off several toxins: the enzyme **lecithinase** (α-toxin), which causes hemolysis, the "egg-yolk" reaction and tissue necrosis; κ-toxin, an enzyme (**collagenase**) that digests collagen protein fibers in tendons; μ-toxin or **hyaluronidase,** an enzyme destroying hyaluronic acid (see discussion of **spreading factor** in §24.4); and several others.

Gas Gangrene. In dirty wounds, in addition to *Clostridium perfringens* and *C. tetani*, there are nearly always one or more of about a score of similar species of clostridia of the soil, such as *C. novyi* and *C. histolyticum*. Some of these are able to digest dead tissue rapidly; these and others produce toxins and hemolysins. Some (*C. novyi, C. septicum*) can actually invade the blood. All of these bacteria are spoken of as gas-gangrene organisms.

The combined, unchecked growth of gas-gangrene organisms in dirty wounds such as crushed limbs, shell wounds, and nail punctures, where the lesion is deep and there is much dead tissue, produces pain, profound shock, toxemia, and the rapidly fatal condition known as **gas gangrene.** It used to be much feared by soldiers wounded on the battlefield but is now controlled (virtually eliminated) by prompt cleansing, surgery, antitoxins, and antibiotics.

A most important development in the treatment of gas gangrene has been the discovery and use of the fact that exposure of seriously ill gas-gangrene patients to an atmosphere of pure oxygen under 45 lb pressure per sq inch completely suppresses the formation of α-toxin and the growth of anaerobes in the infected tissues. Oxygen content of the plasma rises from a normal of 0.3 to 6.2 per cent, and the O-R potential in the tissues rises to a level that is inhibitory to the anaerobes. Cure of the gas gangrene is rapid and complete. Use of compressed oxygen can be very dangerous physiologically and explosively.

Clostridium botulinum. This is one of the two organisms (*Staphylococcus aureus* and *C. botulinum*) that are the usual causes of **food poisoning** (not **food infection**). *C. botulinum* is a strict anaerobe and forms large, oval spores in a subterminal position, often giving the sporangium a shape that is said to resemble a snowshoe (Fig. 36–13). These spores are very heat-resistant. The exotoxin that *C. botulinum* produces is the most poisonous biological substance known. It derives its name from the Latin word for sausage (*botulus*), which was given because the organism was first found in sausages that were the cause of an outbreak of fatal food poisoning (**botulism**) in persons who had eaten the sausages at a picnic. The interior of a sausage (or canned foods if not sterilized) obviously presents an ideal place for the growth of anaerobes, including *C. botulinum*.

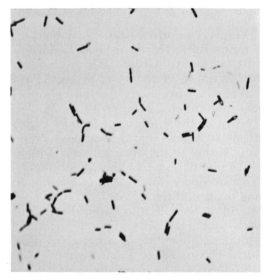

Figure 36–13. *Clostridium botulinum* type A from pure culture. Note the subterminal swollen spores and free unstained spores admixed with the vegetative cells. Fuchsin. (1,050×.)

Botulism. Botulism is not an infection but is **poisoning** by a bacterial exotoxin that is formed in foods under conditions of improper processing and storage in which strict anaerobes may find good pabulum and good anaerobiosis. Food poisoning due to botulinal toxin and staphylococcal enterotoxin must be distinguished from food infections due to *Salmonella* and other invasive pathogens. Both food poisoning and food infection are commonly called "food poisoning." The microorganisms are introduced into the containers when soiled foods are used. If the containers are not sufficiently heat-processed (autoclaved) and if storage (as is usual with canned goods) is at warehouse or household temperatures, anaerobic bacteria can grow and cause spoilage. This in itself is usually harmless. On the other hand, *Clostridium botulinum* may be present. The spores of this bacterium are often found in the soil and are likely to be present on any soil-contaminated food. Sometimes their growth is not sufficient to spoil the food noticeably and it may be eaten. This has often proved fatal.

Botulinal toxin is absorbed directly from the stomach and intestines. It affects the nerve-muscle complex, producing a flaccid paralysis, particularly of the face, eyes, and throat, and respiratory system. As in diphtheria and tetanus, after advanced symptoms appear antitoxin is of little value therapeutically.

There are at least six serological types of botulinal toxin: A, B, C, D, E, and F. (Strains of *C. botulinum* are frequently differentiated on the basis of toxigenic types: the proteolytic strains belong to toxigenic types A, B, and F; nonproteolytic strains belong to toxigenic types B, C, D, E, and F.) Toxigenic types C and D do not appear to cause botulism in man. Types A and B are the most common in the United States, though cases due to type E in fish have occurred. There is a specific antitoxin for differential diagnosis of each type, but **polyvalent** antitoxin is commonly used in treatment unless the type of toxin is known.

The four home-canned foods most commonly responsible for botulism are corn, beans, beets, and asparagus. Note that none is an acid food and that three of them are often contaminated with soil. Botulinal toxin is heat-labile and can be destroyed by heating to boiling for ten minutes (compare with staphylococcal enterotoxin). Botulism caused by commercially preserved foods is relatively uncommon in the United States due to perfection of processing by the trades concerned, but carelessness and errors sometimes occur.

C. botulinum ordinarily will not grow if foods are preserved in brines stronger than 10 per cent sodium chloride or that have an acidity greater than that represented by pH 4.5. Some foods that might be ruined by long processing are therefore acidified to a greater degree than pH 4.5 and processed for a much shorter period. The acidity makes the short heating more effective.

Some human diseases caused by aerobic and anaerobic endosporeforming organisms are summarized in Table 36–4.

36.6
ANAEROBIOSIS

The isolation of oxygen by Priestley in 1774, and subsequent observations by Lavoisier about 1775 on the role of oxygen in combustion and respiration, led to the conclusion that free oxygen (air) is necessary to all life. In 1861, however, Pasteur proved that certain yeasts and bacteria could multiply in the absence of air. He devised the term **anaerobiosis** to describe life without air. This was one of the epoch-making discoveries in biological science. Subsequent studies of the physiology of cells living in situations devoid of free oxygen revolutionized ideas of cell physiology and metabolism.

Since Pasteur's researches, many microorganisms capable of living without air have been discovered. These include many common species of bacteria.

Relation to Oxygen. Microorganisms are divided into several groups (orders, suborders, families) with respect to their relation to free oxygen, in addition to various other characteristics (Table 36–5). The 8th edition of *Bergey's Manual* has placed more emphasis on relation to free oxygen than did the 7th edition. For example, classification of organotrophic gram-negative bacteria is based on their cell morphology and their relation to free oxygen. All gram-negative bacteria, aerobic and anaerobic, are placed in Parts VII to XI. Part VII includes gram-negative strictly aerobic rods composed of four families (Pseudomonadaceae, Azotobacteraceae, Rhizobiaceae, Methylomonadaceae) containing a total of ten genera. Part VIII of the 8th edition includes gram-negative, facultatively anaerobic rods, including the family Enterobacteriaceae, which is divided into five tribes (Escherichieae, Klebsielleae, Proteeae, Yersinieae, and Erwinieae) with a total of twelve genera. Gram-negative, strictly anaerobic rods are placed in Part IX of the *Manual*, which is composed of a single family, Bacteroidaceae, which includes 10 genera. Gram-negative cocci are likewise separated into subgroups similar to the above groups in Parts X (strict aerobes) and XI (strict anaerobes).

Strictly Aerobic Species. Aerobic microorganisms cannot grow without free oxygen to act as final hydrogen acceptor. Their enzyme systems can transfer hydrogen to free oxygen only. They are said to have an aerobic or oxidative type of metabolism, and the substrate is usually completely oxidized to carbon dioxide and water or hydrogen peroxide. However, some aerobes are able to initiate growth under anaerobic conditions if the medium contains sufficient NO_3^- as electron acceptor. These organisms reduce NO_3^- beyond NO_2^-, namely to

TABLE 36–4. DISEASES PRODUCED BY AEROBIC° AND ANAEROBIC ENDOSPOREFORMING ORGANISMS

Etiological Agent	Disease	Symptoms	Habitat	Diagnosis and Epidemiology
B. anthracis	Anthrax	Primarily a disease of cattle and sheep. Portal of entry influences infection, i.e., subcutaneous inoculation most effective in induction of fatal septicemia.	In soil and hides of infected animals	Anthrax bacilli may be cultured from cutaneous lesions in the vesicular stage. In all types of anthrax disease, blood should be cultured. For cattle and sheep, the agent may be transmitted through the alimentary tract by swallowing spores while grazing in infected pastures. In human infections, spores enter through breaks in skin or through the alveoli. The bacilli are then transferred to the lymphatic system, producing septicemia. Agent may be transmitted via infected animals, their hides, exudates and excreta.
B. anthracis	Cutaneous anthrax (malignant pustules)	The most common form of anthrax in humans. Formation of localized boil or abscess – may lead to septicemic condition.		
B. anthracis	Pulmonary anthrax ("wool sorter's" disease)	Inhalation of the agent causes the most dangerous form of the disease in man. May lead to hemorrhagic mediastinitis and to meningitis. The organism also produces exotoxins but functions of the toxins are not well understood.		
C. tetani	Tetanus	The agent of the disease is a noninvasive pathogen. The tonic spasm usually begins near the site of infection followed by severe spasm of masseter muscles (lockjaw). Death usually attributed to muscular spasms affecting respiratory system. Produces one of the most potent neurotoxic biological poisons known. The toxin apparently acts on the anterior horn cells of the spinal cord and may act at inhibitory synapses where it interferes with function of the inhibitory transmitter which causes spasmic action.	Soil, feces of both wild and domestic animals	Usually depends on clinical findings alone. Culture may be isolated and characterized but usually considered too time-consuming, therefore therapeutic use of antitoxin is recommended as soon as possible. The spores frequently gain access to traumatic or surgical wounds and germinate to produce the potent neurotoxin that causes tetanus. Also produces a hemolysin known as tetanolysin.
Clostridium spp. C. perfringens (C. welchii) C. septicum C. novyi (C. oedematiens) C. histolyticum C. sporogenes C. bifermentans	Gas gangrene	Following incubation period of a few hours to 3 days, sudden onset of severe pain which consistently involves the muscles. Fever, tachycardia, tachypnea, and hypotension may accompany the muscle pain. Final stage may be accompanied by severe prostration, a peculiar apathy, and irreversible shock. Numerous exotoxins are produced by species listed at the left. C. perfringens, for example, produces α-toxins which are lethal, necrotizing, and hemolytic and considered as a type of lecithinase. It breaks up lecithin in the cell membrane, hence the toxin can cause damage throughout the body. Also, oxygen labile θ-toxin is produced. It is hemolytic, necrotizing and considered as cardiotoxic.	Soil	Clinical finding essential because most gas gangrene infections are mixed. The finding of gram-positive sporeforming bacilli from infected wound or from cervix and uterus are considered as tentative diagnosis of the disease. The wound often exudes a foul, serous discharge and gas (H_2) in the subcutaneous tissues and muscle may be noticed. The organisms are common inhabitants of the soil. The spores enter through deep wounds. Commonly gas gangrene occurs after attempted abortion or forceps injury after prolonged labor. Such infection is usually accompanied by bacteremia and delayed death may occur due to renal failure.
C. botulinum	Botulism	Not an infectious disease. It is an intoxication produced by a powerful exotoxin. For detailed information, see Table 48–2.		Intraperitoneal injection of mice with the patient's serum or an aqueous extract of the implicated food. If the toxins are present in large amount, the mice may succumb in a few hours to a few days. Isolation of bacilli also aids in diagnosis (see Table 48–2).

°Majority of aerobic. sporeforming bacilli (*Bacillus*) are harmless saprophytes, except illness may result from ingestion of large numbers of cells of *Bacillus* spp. (e.g., *B. cereus*); four species, *B. popilliae, B. alvei, B. paraalvei* and *B. larvae* cause various diseases of insects (see text). *B. subtilis* may occasionally cause mild human eye infection. *B. anthracis* is only member of this large group that is pathogenic for man.

TABLE 36-5. OXYGEN RELATIONSHIPS OF SOME BACTERIAL TYPES*

STRICT AEROBES
   Most species of the genus *Bacillus*
   Families Pseudomonadaceae, Azotobacteraceae, Rhizo-
      biaceae, Methylomonadaceae
   Genus *Brucella*
   Family Nitrobacteraceae
   Genus *Thiobacillus* (except *T. denitrificans:* facultative
      with $NaNO_3$ as H acceptor)
   Genus *Acetobacter*
   Genus *Mycobacterium*
   Bordetella pertussis
   Genus *Micrococcus*
   Family Streptomycetaceae
   Family Nocardiaceae
   Order Myxobacterales (except *Cytophaga fermentans*)

FACULTATIVE
   Family Enterobacteriaceae
   Family Streptococcaceae (a few are strict anaerobes; mostly
      indifferent)
   Family Spirillaceae
   Genus *Staphylococcus*
   Family Neisseriaceae
   Genus *Alcaligenes*
   Family Lactobacillaceae (mostly *indifferent*)

STRICT ANAEROBES
   Genus *Clostridium*
   Genus *Actinomyces*
   Family Bacteroidaceae
   Genus *Desulfovibrio*
   Family Chromatiaceae
   Family Chlorobiaceae

MICROAEROPHILS
   Genus *Leptospira*

*Based on proposed 8th edition of *Bergey's Manual.*

N_2, and such organisms are frequently referred to as denitrifiers (Chapt. 43).

Facultative Organisms. The facultatives grow either aerobically as above or, in the absence of free oxygen, they can use some other easily reducible substance (e.g., sulfur, carbon, or sodium nitrate) as hydrogen acceptor, i.e., they have the **faculty** of growing aerobically or anaerobically. This appears to be because they possess both aerobic and anaerobic enzyme systems. However, they generally grow better aerobically (the so-called **Pasteur effect**), i.e., complete oxidation yields more energy than incomplete oxidation. During anaerobic growth the facultatives exhibit the less efficient fermentative type of metabolism, i.e., the substrate is not completely oxidized (Chapt. 8).

Strictly Anaerobic Species. Obligately anaerobic organisms have two peculiarities: (a) oxygen is **toxic** to them, probably because certain of their enzymes can be blocked by oxygen;

(b) their enzyme systems cannot transfer hydrogen to free oxygen. With some possible minor exceptions they must use other hydrogen acceptors.

Hydrogen Peroxide and Anaerobiosis. Bio-oxidation in the presence of free oxygen commonly results in the formation of hydrogen peroxide, if free oxygen is used as hydrogen acceptor. H_2O_2 is very toxic. Thus, strict anaerobes, while possibly capable of some aerobic growth, immediately commit suicide by producing H_2O_2 when they attempt it!

"But," you say, "H_2O_2 is produced by many **aerobic** bacteria. Why do *they* not die?" Ah! But most of these produce **catalase,** an enzyme which immediately decomposes H_2O_2! And you (being a well-informed student) say, "True, but many vigorous aerobes do *not* produce catalase. Why does *their* H_2O_2 not kill *them?*" A valid question! We reply, "These are not sensitive to H_2O_2." You would like to ask, "Why aren't they sensitive?" and we would say, "Because they have enzyme systems not affected by H_2O_2." Unfortunately we do not yet know exactly why these enzyme systems are not affected by H_2O_2. Some organisms can use H_2O_2 as a hydrogen acceptor: $H_2O_2 + 2(H) \longrightarrow 2H_2O$.

Indifferent Organisms. With (as usual!) "a few exceptions," organisms of the genera *Streptococcus* and *Lactobacillus* are unique in their relation to free oxygen. Ordinarily they can grow in the presence of air but, like the strict anaerobes, do not contain the cytochrome system that would enable them to use free oxygen as a hydrogen acceptor; they neither require nor utilize oxygen. Consequently, when growing aerobically they do not produce H_2O_2 (at least in significant amounts). This is lucky for them because they do not produce catalase, either. Very few bacteria can grow in the presence of air unless they protect themselves from H_2O_2 by also producing catalase. Since the streptococci and lactobacilli neither need free oxygen nor are adversely affected by it, they are **indifferent** to it (they can "take it or leave it"); they do not exhibit a Pasteur effect. Note, however, that some substances, e.g., glycerol, are used by these organisms with the production of much H_2O_2. Unless the H_2O_2 is removed they die.

Microaerophilic Species. These require limited or lowered oxygen tension but not strict anaerobiosis. This peculiarity has not been explained fully but probably reflects a sensitivity to blockage of some of their enzymes by free oxygen, as in the case of strict anaerobes, but to a lesser degree.

36.7
CULTIVATION OF ANAEROBIC BACTERIA

Many types of anaerobic devices exist, but only one fundamental purpose is involved: the removal of free oxygen from the immediate environment of the bacteria, or the maintenance of a low oxidation-reduction (O-R) potential by adding a reducing or oxygen-absorbing agent to the medium itself.

Oxidation-Reduction Potentials. Energy metabolism always involves an oxidation-reduction reaction (e.g., energy source becomes oxidized while another substance becomes reduced). Although some oxidation-reduction reactions involve free oxygen, many can occur in the absence of oxygen. This is because the real basis of the reaction is electron transfer instead of oxygen transfer. Completion of oxidation-reduction (**redox**) always requires two reactants, one of which serves as the electron donor and the other the electron acceptor. The substance which donates or gives up electrons is said to be oxidized, and the substance which accepts electrons is said to be reduced.

$$MBH + X \xrightarrow[\text{H}_2\text{O}]{\text{enzyme}} MB + XH$$

The substrate MBH (reduced substrate) is the energy source and electron donor; generally the more reduced the material is, the more energy it contains. X (oxidized substance) is the electron acceptor. When MBH donates electron (H) to become MB, it is said to be **oxidized,** and when X (electron acceptor) becomes XH, it is said to be **reduced.** When an electron donor has been oxidized, it usually no longer serves as an energy source but conversely may serve as an electron acceptor.

Chemical substances (nutrients) vary in their ability to donate electrons and become oxidized. O-R potential of any given material, such as a bacterial culture or, what is actually the same thing, sewage, can readily be measured. An electrode is placed in it and it thus becomes a half-cell. The cell is then completed by means of a U-tube with salt agar between this half-cell and a **reference half-cell** of known potential. The potential of the complete cell thus formed is the tendency of electrons to flow from one half-cell to the other, the direction and force of flow (EMF) depending on their relative electron pressures, or O-R potentials. To measure the EMF of the culture, the EMF of this complete cell is balanced against that of a **known (Weston) standard cell** by means of a slide-wire resistance

and a potentiometer. Knowing the potential of the reference half-cell and of the standard cell, the potential of the half-cell consisting of culture or other material is easily calculated.

Oxidation-Reduction Requirement of Microorganisms. Anaerobic organisms require absence of oxygen or low O-R potentials (i.e., an electron-accepting or reducing environment) for growth. Some are much more sensitive to oxygen (high O-R potentials) than others. For example, a negative O-R potential of −0.2 V is optimum for the initiation of growth by most species of anaerobes; some will start to grow only at −0.4 V. Anaerobic spores usually require low O-R potentials for outgrowth. Once growth has started, the O-R potentials of cultures of all bacteria decline. This is especially marked in cultures of anaerobes, since these use every available hydrogen (electron) acceptor (tend to release electrons) in their respiratory processes.

Anaerobic Methods. Anaerobic conditions in culture media are brought about by numerous ingenious variations of two basic procedures: (1) use of substances that combine with free oxygen, i.e., reducing agents; and (2) mechanical exclusion of free oxygen. Most procedures involve combinations of (1) and (2), as shown in the following examples:

Cultures may be enclosed in an airtight vessel with sticks of phosphorus or with a freshly made mixture of potassium hydroxide and pyrogallol. These substances absorb large amounts of oxygen and leave mainly the inert gas, nitrogen, and a partial vacuum.

The **combustion of small amounts of alcohol** or the burning of a **small candle** in a closed jar containing the cultures will use up some of the free oxygen. Combustion ceases when the carbon dioxide content approximates 7 per cent. This method results in only partially anaerobic conditions. It is widely used to increase the carbon dioxide content of the atmosphere, a condition favorable to many organisms, both aerobic and facultative, since most require some carbon dioxide for cell synthesis. The reduction of oxygen tension favors aerobic and microaerophilic organisms rather than strict anaerobes.

Hydrogen Jars. A means of absolute **anaerobiosis** is to allow a fine stream of hydrogen to enter a closed vessel containing the cultures, impinging, as it enters, on a small mass of some catalytic agent such as finely divided platinum that causes it to combine with the free oxygen, forming water. The platinum catalyst acts rapidly only when heated. Heat is usually applied by means of an electric current. A drying agent is enclosed in the vessel to absorb the

1. EVACUATION

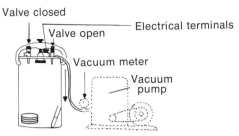

2. REPLACEMENT WITH HYDROGEN

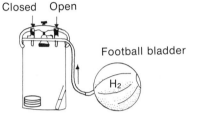

3. CATALYSIS
(Residual O_2 combines with H_2)

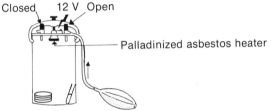

Figure 36-14. Hydrogen jar.

Figure 36-15. Simplified form of anaerobe jar utilizing hydrogen and carbon dioxide from the "Gaspak" envelope. Combination of hydrogen and oxygen in the jar is catalyzed at room temperature. (Courtesy of Baltimore Biological Laboratories, Division of Becton-Dickinson Laboratories, Baltimore, Md.)

water that is formed. There is no vacuum, the remaining gas being a mixture of hydrogen and nitrogen (Fig. 36-14).

In a modern, simplified form of jar, hydrogen and carbon dioxide are evolved from a mixture of dry chemicals in a plastic (Gaspak) envelope (Fig. 36-15). Ten ml of water is added to the contents of the bag and the jar is immediately closed and sealed. The evolved hydrogen combines catalytically at room temperature with all the free oxygen in the jar. A dye indicates, by a change of color, when all the free oxygen has been removed. There is no vacuum; the remaining atmosphere consists of nitrogen and carbon dioxide.

Thioglycollate. A widely used and effective method of anaerobiosis depends on chemical absorption of oxygen from air trapped by a specially shaped cover in a very thin layer over the surface of special agar medium in a Petri dish (Fig. 36-16). The oxygen in this air is absorbed by sodium thioglycollate, or some similar compound (e.g., cysteine) having an affinity for oxygen, incorporated in the agar. By this means the O-R potential of the medium is held very

low and even very sensitive anaerobes will grow on the agar surface. The method is especially useful in obtaining pure cultures by permitting formation of discrete colonies of strict anaerobes.

Fluid Reducing Media. Unless we are cultivating some of the strictly chemolithotrophic anaerobes, the addition of sodium thioglycollate (0.1 per cent), sodium formaldehyde

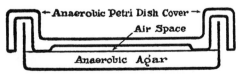

Figure 36-16. Cross section showing Brewer anaerobic Petri dish cover in use. The anaerobic agar contains the reducing agent, sodium thioglycollate. Note that, at the periphery of the agar surface, the Petri dish cover is in contact with the agar, thus sealing the air space. The thioglycollate absorbs the oxygen from the air space. (Courtesy of Baltimore Biological Laboratories, Division of Becton-Dickinson Laboratories, Baltimore, Md.)

sulfoxalate (0.1 per cent) or cysteine hydro-chloride (0.2 per cent) to glucose broth or other similar fluids adapts them to anaerobic requirements. These substances maintain a satisfactorily low O-R potential. (Why are such media unsuitable for some chemolithotrophs?) The further addition of 0.1 per cent agar creates a very slight viscosity which reduces aeration of the solution by convection currents from the air surface.

For chemoorganotrophic anaerobes, organic media such as milk, infusion broth, and infusion agar with blood, with the addition of reducing reagents, are recommended, since these organisms require not only low O-R potential but media rich in organic matter with a pH of about 7.2. Chopped brain, fish, or other tissues are also often used. The addition of glucose provides a readily available source of carbon and energy which promotes the growth of nearly all heterotrophic anaerobes.

Deep Media. Anaerobiosis in tubes of broth is satisfactory if the medium for heterotrophic species contains bits of chopped tissue: **cooked-meat medium.** The tissue acts as a reducing agent. The meat also serves as pabulum for the bacteria. Most heterotrophic anaerobic bacteria grow well in cooked-meat medium.

If the columns of medium in the tubes are 10 to 15 cm deep, all that is necessary is to heat the medium in boiling water for 10 minutes to drive off dissolved air and decompose "organic peroxides," cool rapidly, and inoculate in the depths.

Shake Tubes. Deep tubes of glucose-infusion agar are also used to cultivate anaerobes. Infusion agar in tubes 8 to 10 cm in depth is melted and cooled to about 40C. The inoculum

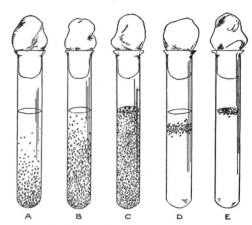

Figure 36–17. Deep tubes of agar inoculated with bacteria of various oxygen relationships. *A*, Fairly strict anaerobe, like *C. botulinum*; *B*, less strict anaerobe, like *C. perfringens*; *C*, Facultative aerobe-anaerobe, like *E. coli*; *D*, microaerophilic organism like *B. abortus*; *E*, strict aerobe, like *Pseudomonas fluorescens*.

is put in and mixed thoroughly. The agar is then made to solidify rapidly in cold water and is incubated. Strict anaerobes will grow only in the depths and will not appear at all within a centimeter or more of the surface. Less strict anaerobes will grow in the depths and will also grow somewhat nearer to the surface, while facultative anaerobes will grow on the surface as well as in the depths. Organisms having a narrow zone of tolerance to both oxygen and strict anaerobiosis (**microaerophils**) may grow in a narrow zone some distance below the surface (Fig. 36–17). Such preparations are often spoken of as **shake tubes** because shaking is used to mix the agar and the inoculum. Formation of foam is carefully avoided. (Why?)

CHAPTER 36
SUPPLEMENTARY READING

Bulla, L. A., Jr. (Ed.): Regulation of insect populations by microorganisms. Ann. N.Y. Acad. Sci., *217*, 1973.

Dunlop, W. F., and Robards, A. W.: Ultrastructural study of poly-β-hydroxybutyrate granules from *Bacillus cereus*. J. Bact., *114*: 1271, 1973.

Halvorson, H. O., Hanson, R., and Campbell, L. L.: Spores, I–V. American Society for Microbiology, Bethesda, Md. 1972.

Jones, D., and Sneath, P. H. A.: Genetic transfer and bacterial taxonomy. Bact. Rev., *34*:40, 1970.

Ohye, D. F., and Murrell, W. G.: Exosporium and spore coat formation in *Bacillus cereus* T. J. Bact., *115*:1179, 1973.

Rode, L. J., and Smith, L. D.: Taxonomic implications of spore fine structure in *Clostridium bifermentans*. J. Bact., *105*:349, 1971.

Santo, L. M., Hohl, H. R., and Frank, H. A.: Ultrastructure of putrefactive anaerobe 3679h during sporulation. J. Bact., *99*:824, 1969.

Santo, L. Y., and Doi, R. H.: Crystal formation by a ribonucleic acid polymerase mutant of *Bacillus subtilis*. J. Bact., *116*:479, 1973.

THE COCCI: MICRO- COCCACEAE AND NEISSERIACEAE

Although a few species of spherical or spheroidal bacteria occur scattered among several of the orders of Schizomycetes listed in the 7th edition of *Bergey's Manual*, the great majority of cocci were grouped together in the order Eubacteriales (families Micrococcaceae and Neisseriaceae and the tribe Streptococceae of the family Lactobacillaceae).

As described in Chapters 31 and 35, the order Eubacteriales has been discontinued in the 8th edition of the *Manual*, and we now find most of the cocci assigned to three groups in the new scheme: (1) Part X consists of the gram-negative strictly aerobic cocci or coccobacilli— family Neisseriaceae, with representative genera including *Neisseria* (40 to 53 mole per cent GC), *Moraxella* (39 to 45 mole per cent GC), *Acinetobacter* (38 to 45 mole per cent GC), and *Branhamella*. Also, the family includes taxonomically uncertain but morphologically interesting and physiologically versatile organisms, *Paracoccus* and *Lampropedia*. (2) Part XI consists of the gram-negative strictly anaerobic cocci (family Veillonellaceae). (3) The gram-positive, aerobic and/or facultatively anaerobic cocci—families Micrococcaceae (30 to 75 mole per cent GC), Streptococcaceae (31 to 44 mole per cent GC), and Peptococcaceae (28 to 45 mole per cent GC)—are placed in Part XIV.

While all of the organisms are morphologically similar, they show considerable heterogeneity in DNA base composition and in their physiological and biochemical properties. Certain of these families are, however, of greater interest for our present discussion than others, regardless of the classification system used. Thus, the streptococci have been singled out for separate consideration in the next chapter. This chapter will be devoted largely to the representative genera in the families Micrococcaceae and Neisseriaceae, both of which are of considerable medical importance.

37.1
THE FAMILY MICROCOCCACEAE

The Micrococcaceae are gram-positive, nonsporeforming, and mostly free-living saprophytes. The family includes genera *Micrococcus*, *Staphylococcus*, and *Planococcus*, and the DNA base composition ranges from 30 to 75 mole per cent GC, which indicates that the group is exceedingly heterogeneous. One curiosity, *Sporosarcina (Sarcina) ureae*, is a gram-positive, motile, endosporeformer but is morphologically a coccus. It is now considered by most to belong in the family Bacillaceae, which is discussed in detail in Chapter 36.

The Microccocaceae are usually spherical or nearly so and, unlike the Streptococceae, typically divide in two or three planes. If the divisions are in planes at irregular angles, irregular clusters or masses of cocci are formed (Fig. 37–1, F), resembling bunches of grapes (genera *Micrococcus* and *Staphylococcus*). If the divisions are in *two horizontal planes* at right angles, flat, square groups of four (**tetrads**) are formed (genus *Gaffkya**). Divisions in two horizontal planes and a perpendicular plane produce very distinctive cubical packets of eight or more cocci (genus *Sarcina* [L. *sarcina* = packet]). During rapid growth many single

---

*Gaffky was the German bacteriologist who first described these organisms. The generic status of the organism is uncertain, and some authors suggest that the genus may be included in *Aerococcus* or *Micrococcus*.

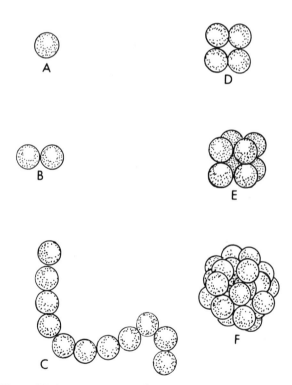

Figure 37–1. Arrangement of cocci showing: *A,* A single coccus, the basis of all other arrangements and frequently seen with most species. *B,* Fission in one plane, forming diplococci, e.g., *Streptococcus (Diplococcus) pneumoniae* or *Neisseria* sp. *C,* Fission in one plane with adherence, forming chains of streptococci, e.g., *Streptococcus pyogenes. D,* Fission in two planes at 90 degrees, forming tetrads, e.g., *Gaffkya tetragena. E,* Fission in three planes at 90 degrees, forming cubical packets of eight cells, e.g., *Sarcina* or *Micrococcus* sp. *F,* Fission in various planes, forming irregular (grape-like) clusters, e.g., *Staphylococcus* and *Micrococcus* sp.

cells occur, and cells of any genus in the process of division may often appear in pairs, temporarily simulating diplococci.

Actually, the ability to form tetrads or packets has turned out to be quite a variable property, and many organisms formerly placed in the genera *Gaffkya* or *Sarcina* have been shown to be identical to organisms in the genus *Micrococcus* with regard to their metabolism, cell wall chemistry, and DNA composition. Thus, classification of these organisms is confused at present. In all probability, *Gaffkya* will be dropped as a genus designation, and the anaerobic *Sarcina* will be included in the family Peptococcaceae in Part XIV, along with *Peptococcus, Peptostreptococcus,* and *Ruminococcus.* (DNA base composition of the group resembling *Micrococcus* has been reported to be 54 to 73 mole per cent GC, while that of organisms resembling *Peptococcus* has been reported to be 29 to 31 mole per cent GC.)

It is not clear at this time how many genera will be recognized as Micrococcaceae in the 8th Edition of *Bergey's Manual,* but three are likely, as shown in Table 37–1. For the time being we will continue to use the older terms and will concentrate on the two major genera: *Staphylococcus,* mainly parasitic, facultative anaerobic cocci producing acid from glucose under anaerobic conditions; and *Micrococcus,* mainly saprophytic, aerobic cocci producing acid from glucose aerobically but not anaerobically. Both genera give a positive catalase and negative oxidase test. A number of biotypes, representing species in some cases, can be distinguished in each genus (Table 37–2) by their physiology.

Genus *Methanococcus* (classified in Micrococcaceae in the 1957 edition of *Bergey's Manual*) is now placed in the family Methanobacteriaceae, Part XIII, "Methane-Producing Chemolithotrophic Bacteria" (8th edition, *Bergey's Manual*). The organisms are found in black mud in marshes, sewage, sludge and similar stagnant habitats. They are active producers of methane (marsh gas, or sewer gas). Carbon dioxide and hydrogen are often mixed with the methane in the bubbles rising from the bottom in such locations.

Saprophytic Micrococcaceae
Genus *Micrococcus*. The cocci of genus *Micrococcus* are enzymically versatile and are physiologically heterogeneous and able to thrive under a wide range of conditions, including in dairy products, soil, dust, sea water,

TABLE 37-1. SOME DIFFERENTIAL CHARACTERISTICS OF GENERA OF MICROCOCCACEAE°†

Characteristic	Staphylococcus	Micrococcus	Planococcus
Shape: Spherical	+	+	+
Arrangement			
Irregular clusters	+	+	−
Tetrads	−	v	+
Glucose fermentation‡	+	−	−
Cytochromes	+	+	+
Catalase	+	+	+
Hydrogen peroxide formation	−	−	−
Motility	−	−	+
Yellow-brown pigment	−	−	+
GC content of DNA (mole %)	30–40	30–75§	39–52

°Adapted from Baird-Parker, *in*: Cohen (Ed.): The Staphylococci, Wiley-Interscience, 1972.

†Symbols: + = most (90% or more) strains positive; − = most (90% or more) strains negative; v = characters inconstant and in one strain may sometimes be positive, sometimes negative; d = some strains positive.

‡Growth and acid anaerobically from glucose in the standard medium proposed by the ICSB Subcommittee on the Taxonomy of Staphylococci and Micrococci (Recommendations, 1965).

§Except for strains of *Micrococcus saprophyticus* and some *M. lactis*, the range is 57 to 75.

pickling brines, and many foods not too acid or alkaline (pH 6.5 to 8.0), at temperatures from about 22 to 38C. Most are killed by pasteurization, but some thermoduric species can survive. Of course, various species vary somewhat in these tolerances. All are organotrophic and grow well on ordinary peptone or meat-infusion media as commonly used in the laboratory. Within the genus, the DNA base composition ranges from 65 to 75 mole per cent GC, which is considerably higher than either *Staphylococcus* (30 to 40 mole per cent GC) or *Streptococcus* (33 to 44 mole per cent GC). They are strict aerobes and cannot utilize glucose or mannitol

anaerobically. They are distinguished from staphylococci by this, by their failure to be lysed by the enzyme-like antibiotic **lysostaphin** and by failure to produce coagulase. These four properties are of importance in diagnostic and other aspects of differential bacteriology.

On solid media most species of *Micrococcus* form opaque, butyrous colonies with white or yellow pigments; various shades of red and orange are also common, e.g., *M. roseus, M. agilis,* and *M. morrhuae.* Many species have a marked tolerance for sodium chloride and can be isolated from mixed cultures on selective media containing 5 to 8 per cent salt, a concen-

TABLE 37-2. SCHEME FOR CLASSIFYING STAPHYLOCOCCI AND MICROCOCCI°†

	Group I Staphylococcus (Rosenbach)						Group II Micrococcus (Cohn)							
SUBGROUP:‡	I	II	III	IV	V	VI	1	2	3	4	5	6	7	8
Pink pigment	−	−	−	−	−	−	−	−	−	−	−	−	−	+
Acid from glucose														
Aerobic	+	+	+	+	+	+	+	+	+	+	+	+	±	±
Anaerobic	+	+	+	+	+	+	−	−	−	−	−	−	−	−
Coagulase	+	−	−	−	−	−	−	−	−	−	−	−	−	−
Phosphatase	+	+	+	−	−	−	−	−	−	−	−	+	−	−
Acetoin	+	+	−	+	+	+	+	+	+	+	−	−	−	−
Acid from														
Arabinose	−	−	−	−	−	−	−	−	−	+	v	+	−	−
Lactose	+	+	v	−	+	v	−	+	v	+	+	+	−	−
Maltose	+	+	−	v	+	v	v	+	+	+	+	−	−	±
Mannitol	+	−	−	−	−	+	−	+	+	+	+	−	−	−

°From Baird-Parker, *in*: Gibbs and Skinner (Eds.): Identification Methods for Microbiologists, Part A, Academic Press –London, 1966.

†Symbols: ± = weak or negative; v = variable.

‡*Staphylococcus* subgroup I corresponds to *S. aureus* and subgroups II to VI to *S. epidermidis; Micrococcus* subgroup 7 corresponds to *M. luteus* and subgroup 8 to *M. roseus.*

tration that inhibits growth of many other organisms. Several species occur in sea water and salt lakes (e.g., *M. morrhuae*).

The micrococci, and the very similar cocci formerly designated as aerobic *Sarcina*, are of importance mainly as scavengers. Common in dust, they are familiar airborne contaminants in laboratory cultures. Many of them actively digest proteins such as gelatin and casein and attack various carbohydrates and numerous other organic substances. Some species of *Micrococcus (M. flavus, M. caseolyticus)* are of commercial importance in the ripening and flavoring of cheese, since they attack casein and lactose with the production of aromatic substances having pleasing flavors. They are said to be *acidoproteolytic*. On the other hand, some micrococci produce various undesirable slimy conditions such as **ropy milk** (*M. freudenreichii* and other slime producers, sometimes called *M. cremori-viscosi*).

Pathogenic Micrococcaceae

In this family the principal pathogen is *Staphylococcus aureus*.

Genus *Staphylococcus*. As previously stated, *Staphylococcus* is differentiated from *Micrococcus* by the ability of *Staphylococcus* to utilize glucose, mannitol, and pyruvate anaerobically and by its sensitivity to **lysostaphin.** Microscopically, species of the two are virtually identical, although cells of staphylococci are slightly smaller than those of micrococci. Staphylococci are usually to be found on the skin or mucous membranes of the animal body, especially of the nose and mouth, where they often occur in large numbers even under normal conditions. There are two principal species. *Staphylococcus aureus*, distinguished primarily by its ability to produce coagulase, is notorious as the cause of suppurative (**pyogenic** or **pus-**forming) conditions: **mastitis** of women and cows, **boils, carbuncles, infantile impetigo, internal abscesses,** and **food poisoning.** *S. epidermidis (S. albus)* is a lesser pathogen or commensal on the skin and mucous membranes. It is described farther on in this chapter.

Staphylococcus aureus. Staphylococci isolated from pathologic materials are generally *S. aureus*. These cocci typically: (1) **ferment** mannitol and lactose; (2) are **proteolytic;** (3) produce **coagulase** (an enzyme-like principle that causes citrated blood plasma to coagulate); (4) produce **golden pigment;** (5) produce **lipase;** (6) produce wide zones of **hemolysis** aerobically in blood agar plates (Color Plate V, *A*); (7) grow in media containing 10 per cent **sodium chloride.**

S. aureus may be isolated from contaminated material by first cultivating in a selective, protein-digest, broth medium containing 10 per cent NaCl. Growth in such a selective fluid may then be streaked on "Staphylococcus medium 110." This medium is specially designed to select *Staphylococcus aureus* and reveal some of its distinctive properties. It contains: protein digest (trypticase or tryptone), 1 per cent; yeast extract, 0.25 per cent; gelatin, 3 per cent; D-mannitol, 1.0 per cent; lactose, 0.2 per cent; NaCl, 7.5 per cent; K_2HPO_4, 0.5 per cent; agar, 1.5 per cent. Phenol red may be added as an indicator. Simple, 1.0 per cent mannitol–7.5 per cent NaCl agar may also serve. Distinctive colonies on these media may be picked and subcultured and then tested for coagulase production.

Isolates of *Staphylococcus aureus* often produce one or more hemolysins which have been assigned letters of the Greek alphabet and which lyse various types of mammalian erythrocytes. The hemolysins are best identified and distinguished from one another by using specific antisera and certain serological tests. However, the appearance of the zone of hemolysis surrounding a colony growing on a blood agar plate has frequently been used as an indication of the type of hemolysin produced. In general, α-hemolysin gives a wide zone of clear hemolysis on sheep or rabbit erythrocytes. δ-hemolysin gives a narrow zone of hemolysis under the same conditions. β-hemolysin does not lyse rabbit cells and gives only partial clearing on sheep cells. β-hemolysin has sometimes been called a hot-cold lysin, since the sheep cells will advance to complete lysis if the plates are stored overnight in the refrigerator after they are removed from the incubator. Still other staphylococcal hemolysins have been described but need not be discussed here. The student should note that the terminology developed for staphylococcal hemolysins bears no relationship to streptococcal hemolysins (in fact, the meaning of α and β are reversed!).

Hemolysins, by destroying erythrocytes and creating local anoxia, could conceivably aid the pathogen in establishing a foothold in the tissues. Other toxic properties have also been described for some of the hemolysins, such as α-hemolysin (also called α-toxin) described below. The production of hemolytic substances, however, does not always correlate with pathogenicity. For example, a number of harmless saprophytic species of *Bacillus* produce excellent zones of hemolysis on blood agar plates. Certain strains of *Staphylococcus* which have lost their ability to produce hemolysins have also been shown to retain their virulence.

Staphylococci from pathologic materials produce other noxious substances, notably *leucocidins* that destroy leucocytes. Under aerobic conditions they form α-**toxin,** which has three distinct effects that are *demonstrable in rabbits:* **hemolysis; rapid lethality;** localized dermal **necrosis** on subcutaneous injection. Most pathogenic strains also produce the enzyme **lysozyme** and a fibrin-digesting enzyme activator called **staphylokinase** (see also discussion of **streptokinase** in §24.4). Coagulase production is often, but not necessarily, associated with virulence of *S. aureus.*

Coagulase. Staphylococcal coagulase acts only in conjunction with a serum factor called **coagulase reacting factor** (CRF). Normal, physiological coagulation of blood results from a complex series of reactions, the last two of which are:

$$prothrombin + prothrombinase + CaCl_2 \longrightarrow$$
$$thrombin$$
$$thrombin + fibrinogen \longrightarrow fibrin (= clot)$$

The role of staphylococcal coagulase appears to be to displace both prothrombinase and prothrombin. Coagulase first combines with CRF, a substance much like prothrombin:

$$CRF + coagulase \longrightarrow CRF \cdot coagulase$$

This complex then replaces thrombin in the formation of fibrin:

$$CRF \cdot coagulase + fibrinogen \longrightarrow$$
$$fibrin (= clot)$$

A substance commonly called "bound coagulase" that causes *Staphylococcus* cells to clump together in the presence of fibrinogen during the coagulase test is more correctly called **clumping factor.** There is no relationship between clumping factor and coagulase.

DEMONSTRATION OF COAGULASE. Many methods for detecting coagulase have been described. A simple technique consists of mixing 0.5 ml of a vigorous, young, broth culture, or a heavy suspension of young growth on agar, with 0.5 ml of fresh or lyophilized citrated plasma (preferably human). The plasma must be known, by previous and simultaneous control tests, to be rich in CRF and fibrinogen (i.e., coagulable by known coagulase) and to be free from fibrinolytic (staphylokinase) properties and coagulase inhibitors. Pooled human plasmas suitable for such purposes are available commercially. Numerous organisms other than staphylococci will cause citrated plasma to clot, but this commonly results from their metabolic destruction of the citrate, which thus liberates the Ca^{2+} (essential in coagulation) from the citrate · $CaCl_2$ chelate. As a precaution against this possibility, the chelating agent EDTA can be added to the broth and will not be metabolized.

Antibiotic-Resistant Staphylococcus aureus. When penicillin first came into clinical use about 1944 during World War II, virtually every strain of *S. aureus* was highly sensitive to it. A few mutants were resistant, however, and it was found that these were usually vigorous producers of an enzyme, **penicillinase,** that very rapidly destroys penicillin. It was then discovered that penicillinase production is readily induced by contact with the drug. As a result, with more and more widespread use of antibiotics, especially penicillin, most strains of *S. aureus,* especially in hospitals, are now completely resistant to penicillin. The genetic mechanism for penicillinase production is transmissible, like an episome (Chapt. 17), but probably requires a transducing phage.

These penicillin-resistant strains became widespread in hospital personnel and in the environmental dust, on walls, floors, bedding and furniture. Infection with such strains has been a major problem in many institutions, especially in maternity wards and nurseries, where they often cause mastitis and impetigo or pemphigus neonatorum. Control requires the most assiduous attention to handwashing, exclusion of infected personnel and disinfection of premises, clothing and bedding. Fortunately the resistance of staphylococci to penicillin has been overcome to some extent by the development of semisynthetic penicillins (Chapt. 23) that are not vulnerable to penicillinase and that are fully as effective as natural penicillin. Still, physicians are careful these days to use antibiotics only when necessary, to use a sufficient dosage to achieve a high blood level, and to select a drug to which the pathogen is susceptible. Antibiotics should be used only under close medical supervision.

Staphylococcal Food Poisoning. Staphylococci are notorious as a cause of food poisoning. This is because some strains of staphylococci release a heat-stable exotoxin called **enterotoxin** when growing in many common foods: "tenderized" hams, milk, custards, "cream" fillings, soups, stews. Excessive growth of the organisms in an abnormal intestinal tract can cause similar symptoms.

When absorbed, staphylococcal enterotoxin causes nausea, vomiting, diarrhea and prostration. Staphylococcal food poisoning is seldom fatal but it has ruined many a dance date! The symptoms come on usually within 2 to 12 hours after eating the toxin. This permits differentiation from salmonellosis, which produces very similar symptoms, but only after a necessary incubation period, usually of 12 to 24 hours. Approximately 30 per cent of strains of S. aureus produce enterotoxin.

There are at least five distinguishable types of enterotoxin, A, B, C, D, and E, analogous to the several types of botulinal toxin. Unlike the thermolabile botulinal toxin, all staphylococcal enterotoxins are exceptionally thermostable proteins. They withstand boiling for at least 30 minutes. Distinguish carefully between staphylococcal **enterotoxins**, which are **exotoxins**, and the **endotoxins** of Enterobacteriaceae (Chapts. 24, 35).

Considerable growth of the cocci is necessary to produce enough enterotoxin to cause symptoms. Staphylococcal food poisoning therefore usually involves stale, moist, not-too-acid foods that have been contaminated from a human source and then held for several hours (incubated) at room temperature (22 to 40C) (Chapt. 25). There is only one dependable proof that a suspected food caused an outbreak of staphylococcal food poisoning; i.e., demonstration of enterotoxin in the food. Second-best evidence is demonstration of staphylococci in enormous numbers in the suspected food and proof that the isolated strain is capable of producing enterotoxin.

The most satisfactory method for demonstrating enterotoxin in suspected foods is by means of extraction, purification, and concentration followed by immunodiffusion techniques using specific-antitoxin serum. Older methods involve cat and monkey feeding tests that are difficult and not very reliable.

Most strains of enterotoxigenic staphylococci (usually, but not always, S. aureus) are distinguished by the production of DNase capable of resisting autoclaving (121C for five minutes). Demonstration of this enzyme in foods strongly indicates the growth of pathogenic staphylococci in the food.

Growth or enterotoxin production by staphylococci tends to be suppressed in foods in which large numbers of the various harmless saprophytes common in foods are actively mul-tiplying, especially at temperatures between 15 and 25C. Saprophytes whose growth is especially inhibitory to staphylococci include species of *Escherichia, Proteus, Klebsiella*, lactic streptococci, *Pseudomonas, Micrococcus*, and enterococci. Some species of *Bacillus* appear to stimulate staphylococci, others to inhibit them.

There is no wholly reliable, single laboratory test for identification of strains of staphylococci that are pathogenic for man. Coagulase production is commonly, but by no means invariably, associated with pathogenicity; further, its actual inimical effect, if any, remains obscure. No single enzyme, toxin, or product can be pointed to as the sole determinant of virulence in the staphylococci. Rather, it is more likely that various products act in concert to enhance the invasiveness and pathogenicity of the staphylococci. It should be noted, however, that normal, healthy individuals are seldom seriously infected with staphylococci even if they are nasal carriers of virulent strains. Most serious infections occur in debilitated hosts, i.e., burn patients, individuals undergoing cancer therapy, people showing allergic manifestations, etc.

Staphylococcus epidermidis. *S. epidermidis* typically does not ferment mannitol or produce coagulase, α-toxin or lipase. Other characteristics, including constituents of the cell wall, have been used to differentiate *S. epidermidis* from *S. aureus*, as summarized in Table 37–3. *S. epidermidis* frequently forms chalky white colonies on agar, whence its older name, *S. albus* (Color Plate V, *A*). Gelatin liquefaction is slow or absent; hemolysis is slight or absent. Pathogenicity is slight as a rule. Note

TABLE 37–3. DIFFERENTIAL CHARACTERISTICS OF SPECIES OF STAPHYLOCOCCUS °†

Characteristic	aureus	epidermidis
Coagulase	+	−
Mannitol		
Acid aerobically	+	d
Acid anaerobically	+	−
α-toxin	+	−
Heat-resistant endonucleases	+	−
Biotin for growth	−	+
Cell wall		
Ribitol teichoic acid	+	−
Glycerol teichoic acid	−	+
Protein A	+	−

*Adapted from Baird-Parker, *in:* Cohen (Ed.): The Staphylococci, Wiley-Interscience, 1972.

†Symbols: + = most (90% or more) strains positive; − = most (90% or more) strains negative; d = some strains positive, some negative.

that glucose is fermented anaerobically by both organisms, whereas mannitol is not fermented anaerobically by *S. epidermidis*.

"Genus Gaffkya." *Gaffkya tetragena* has been reported to be occasionally pathogenic for man, but most likely it infects only when resistance is low or absent. However, some species are very pathogenic for mice and some other forms of life. For example, *Gaffkya homari* (*Homarus* = genus of lobsters) is highly pathogenic for lobsters, producing a commercially costly disease with a high fatality rate.

As pointed out previously, the status of *Gaffkya* as a genus designation is uncertain. It now appears that *Gaffkya homari* is nearly identical to *Aerococcus viridans*, an organism widely distributed in air, meat brines and vegetables. In the future it is likely that *G. homari* will be called *A. viridans*, or a variety of this species, while other species formerly assigned to *Gaffkya* will be classified as species of *Micrococcus*.

37.2
THE FAMILY NEISSERIACEAE

This family derives its name from a German bacteriologist, Neisser, who in 1879 discovered and studied one of the most important species: *N. gonorrhoeae*, the cause of gonorrhea. The family Neisseriaceae, containing the genus *Neisseria*, is considered along with several other groups of related organisms (*Branhamella*, *Moraxella*, *Acinetobacter*, and others) in Part X of the 8th edition of *Bergey's Manual*. A new family, Veillonellaceae (Part XI), now includes genera *Veillonella* (DNA base composition, 35 to 36 mole per cent GC), *Acidaminococcus*, and *Megasphaera*. The *Veillonella* are small, gram-negative, anaerobic diplococci that occur chiefly in the mouth, respiratory, and alimentary tracts of mammals. They appear to have no significance as pathogens.

Genus Neisseria. Cocci of the genus *Neisseria* are rather small, gram-negative diplococci, each cell characteristically flattened where it is in contact with its mate, and each having somewhat the shape of a coffee bean (Fig. 37–2). The various species of *Neisseria* are indistinguishable morphologically. However, recent studies on DNA base composition indicate that two distinct groups are found in genus *Neisseria*: the *N. catarrhalis* group (which includes *N. catarrhalis*, *N. caviae*, and *N. ovis*, with 40 to 51 mole per cent GC) and the *N. meningitidis* group

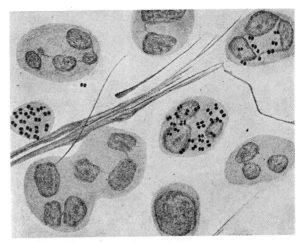

Figure 37–2. Gonococci in leukocytes in smear of pus from case of gonorrhea, stained with methylene blue. The long, fibrous objects are shreds of fibrin and mucus in the pus. Note that the gonococci are nearly all *within* the leukocytes. (1,000×.) (From Ford: Bacteriology.)

(which includes *N. meningitidis*, *N. gonorrhoeae*, and several other species, with 47 to 53 mole per cent GC). The *Neisseria* are aerobic, reluctantly facultative, and grow best at 35 to 37C. Unlike the ubiquitous and busy Micrococcaceae, all but one species of *Neisseria* are found only in the human upper respiratory tract. The exception is *N. gonorrhoeae*, which is associated mainly but not exclusively with the genital tract and gonorrhea.

Of the several species of *Neisseria*, two commonly cause disease: *N. gonorrhoeae* (the gonococcus) and *N. meningitidis* (the meningococcus). The latter causes epidemic meningitis, or cerebrospinal fever. *N. gonorrhoeae* and *N. meningitidis* are very much alike; one is probably a minor variant of the other.

These two species are classic examples of fragile microorganisms that have become highly adapted to an existence as parasites of man. In moist, warm material such as pus, urine, serum, or cultures in incubators, the meningococci and gonococci autolyze and die in a few hours. They are extremely fragile in the outer world. Unless special protective methods are used they do not survive for more than two to six hours outside the body, since drying is very deleterious to gonococci, and chilling, to meningococci. Their nutritive requirements are complex. They grow only at 35 to 37C, on **infusion** media containing blood heated to 90C ("**chocolate agar**") or other special mixtures of a similar nature. Some contain selective antibiotics and nutrient supplements such as yeast autolysate and vitamins, e.g., Thayer-Martin medium. Incubation must

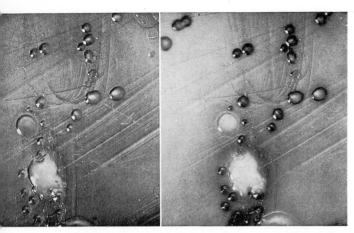

Figure 37–3. The oxidase test for the identification of meningococcus colonies. Mixed culture on blood agar. *Left,* colonies of meningococci and contaminants before the application of tetramethyl-*p*-phenylenediamine solution. *Right,* the same colonies after the application of the reagent. Note that the meningococcus colonies show the development of color first about the edges, and there is slight discoloration of the medium. (5×.) (From Burrows: Textbook of Microbiology, 20th ed., W. B. Saunders Co., 1973.)

be in a very humid atmosphere with 5 to 10 per cent carbon dioxide. The colonies are from 1 to 4 mm in diameter, clear, colorless, moist, and watery-looking (Fig. 37–3).

These cultural characteristics serve to differentiate the meningococci and gonococci from all other species of the genus *Neisseria*, e.g., *N. flava, N. catarrhalis,* and *N. sicca* (Table 37–4). These three species, and a few similar ones, grow well on blood-free media at temperatures as low as 25C, and are relatively resistant to drying, chilling, and light. Several species are pigmented. These species, while parasites in a broad sense, being restricted to a life on a mammalian host, are usually quite harmless.

However, as is true of many bacteria generally regarded as harmless, some of these respiratory *Neisseria* under certain conditions may cause meningitis, and some can cause nonvenereal infections of the genitalia in preadolescent girls, these infections being sometimes confused with gonorrhea, with tragic results. Therefore,

no diagnosis of gonorrhea, especially in females, can be said to be complete and accurate without a full bacteriological study of the organism involved. Institutional outbreaks of vulvovaginitis in female children are often caused by fomite-borne infections of *N. catarrhalis, N. flava,* or *N. sicca.* Many such outbreaks are also due to gonococci.

The Oxidase Test. All of the *Neisseria* produce an enzyme (**oxidase** or indophenoloxidase) that causes a 1 per cent solution of tetramethyl-*p*-phenylenediamine to turn, successively, pink, rose, magenta, and finally black. The oxidase test is applied by moistening a colony of the suspected organism with a drop of the dye solution. The changes in color begin in a few moments (Fig. 37–3). The same test for oxidase can be made on all sorts of microorganisms in addition to *Neisseria* and is a valuable differential method for general use.

The Catalase Test. The test for catalase is made by putting a drop of hydrogen peroxide on any suspected colony. If catalase is present, bubbles of oxygen will appear almost instantly. All *Neisseria* produce catalase. So do many other aerobic organisms: *Micrococcus, Staphylococcus,* various Myxobacterales, *Corynebacterium,* and *Bacillus.*

Gonorrhea. This is one of several diseases commonly spoken of as **venereal diseases,** deriving this appellation from the name of Venus, goddess of love. The inappropriateness of this term will become obvious in the discussion of the infections.

Gonorrhea is an acute inflammatory disease due to infection, by *N. gonorrhoeae,* of the mucous surfaces and adjacent glandular structures of the reproductive organs of men and women. Much pus forms and appears as a white discharge (**leukorrhea**) from the genitalia. Gonorrhea is an alarmingly prevalent disease, with well over a million cases under medical care in the United States in 1973 and a million or more unreported and untreated cases. Teenage venereal disease is common and is sometimes a sign of stupidity or ignorance, irresponsibility or social and psychological maladjustment.

TABLE 37–4. CHARACTERISTICS DIFFERENTIATING COMMONLY ENCOUNTERED SPECIES OF NEISSERIA

Species	Normal Habitat	Fermentation of Glucose	Sucrose	Maltose	Growth at 25C	Growth on Plain Agar
N. gonorrhoeae	Genital tract	+	−	−	−	−
N. meningitidis	Respiratory tract	+	−	+	−	−
N. catarrhalis	Same	−	−	−	+	+
N. flavescens	Same	−	−	−	+	+
N. sicca	Same	+	+	+	+	+
N. flava	Same	+	−	+	+	+

Infection of the genitalia with *N. gonorrhoeae* commonly results from sexual contact with an infected person; rarely, by other means of transmission. Gonorrhea is seldom fatal but is sometimes difficult to cure, especially in females. Patients often believe themselves cured only to find later that the disease has reappeared in a chronic form. The sulfonamide drugs promised for a time to eliminate gonorrhea, but drug-fast strains of gonococci rapidly developed, and indiscriminate use of the drugs by the medically ignorant has robbed such therapy of its effectiveness. Penicillin now offers the best hope for cure in all cases, although the incidence of penicillin-resistant strains is steadily increasing and much larger doses are being required for effective therapy.

It can be readily understood that stupid, careless, ignorant, or malicious people can spread gonorrhea widely. Prostitution is one of the chief means by which the disease is propagated; promiscuity is another. Adequate medical treatment with penicillin very close to the time of exposure will prevent many cases from developing. Inadequate treatment is in several respects worse than none. Drugs other than penicillin, e.g., spectinomycin, are now in use.

Gonorrhea, untreated, often results in sterility. The intense inflammation caused by the organism destroys the tissues lining the genitourinary tract, the tissue being replaced with scars. Such scars contract strongly and obstruct the **fallopian tubes** of the female and the **vas deferens** of the male (tubes through which the reproductive cells pass). Such scarred obstructions are called **strictures**. Stricture of the urethra in the male interferes with urination and may require surgical intervention. Gonococci sometimes invade the body, localizing in the joints and the heart valves. In the former case, a very painful and stubborn type of arthritis results, while in the latter case a very damaging disease of the heart occurs, with permanent injury and sometimes death.

Gonorrheal Ophthalmia. An intensely painful acute inflammatory infection of the eye (**ophthalmia**) results when gonococci are rubbed into the eye. Loss of sight usually results in a few days unless treatment is prompt. A gonorrheal mother may infect her child's eyes at birth. Infection of the eye of the newborn is called **ophthalmia neonatorum.** Much blindness has been caused by gonorrheal ophthalmia neonatorum. To prevent this, most cities, states, and countries require physicians, nurses, or midwives attending births, *regardless of any circumstances*, to instill into the eyes of the infant a few drops of weak (1 per cent) silver nitrate, penicillin, or other legally approved disinfectant solution. This rapidly destroys gonococci (and other organisms) before they can start an infection in the eye. Ophthalmia neonatorum may also be caused by pneumococci, streptococci, and staphylococci. Approved disinfectants are obtainable at any health department or drugstore, ready for use.

Laboratory Diagnosis of Gonorrhea. The diagnosis of **acute** gonorrheal infection in the adult male is usually based on microscopic examination of the pus stained by Gram's method. The gram-negative gonococci (see Color Plate III, *C*) appear within the leucocytes (Fig. 37–2). Such organisms found in adult males with acute urethritis are usually gonococci. In the female genitourinary tract many other organisms are present, and gonococci are frequently not discoverable with the microscope. By the use of a method employing antigens washed with saline solution from the surface of freshly isolated gonococci and deposited on the surface of tanned erythrocytes (see discussion of **indirect hemagglutination** in §27.8), hemagglutinating antibodies are readily detected in the body fluids of persons **chronically** infected by *N. gonorrhoeae*.

As noted previously, respiratory *Neisseria*, as well as gonococci, may cause gonorrhea-like infection of preadolescent girls, being transmitted by hands and towels soiled with oral or nasal secretions. Isolation and complete cultural and serological identification of the organism is therefore of especial importance under such circumstances.

Tribe Mimeae. Certain gram-negative, aerobic, pleomorphic rods or coccobacilli found during diagnostic studies of gonorrhea were originally classed as a new tribe, Mimeae. They are not found as such in either the 7th or the 8th edition of *Bergey's Manual*, and their systematic position, and even the existence of some of them as distinct species, is dubious. They grow vigorously on common, blood-free, laboratory media at approximately 37C, and some, because they frequently grow in the form of gram-negative diplococci, have at times been confused with some of the less fastidious species of *Neisseria*, although both rod and coccal forms often occur together. The name Mimeae is from the Latin word for mimic. None is motile or forms spores.

On the basis of physiological characters two genera are commonly recognized: *Mima polymorpha* and *Herellea vaginicola*. A variant, *M. polymorpha* var. *oxidans*, produces oxidase-positive colonies like *Neisseria*. Neither *M. polymorpha* nor *H. vaginicola* produces oxidase: all produce catalase. Colonies of all on agar are 1 to 3 mm in diameter, creamy, glossy, convex, smooth, and opaque and have regular margins. *H. vaginicola* differs from *Mima* species in oxidizing glucose and lactose; *Mima* attacks no carbohydrates. Neither reduces nitrate nor produces indole or H_2S.

Most reactions of the species are somewhat variable, and the organisms may at times bear one or more resemblances to a considerable variety of other gram-negative rods and cocci. For example, in the course of studies over some decades by numerous investigators, they have been named as one or more species of *Achromobacter, Acinetobacter, Bacterium, Cytophaga, Diplococcus, Moraxella, Neisseria,* and *Alcaligenes.*

They are widely distributed in and on man and other animals and in the environment. Members of the tribe Mimeae are frequently found associated with other bacteria in pathologic (also normal) materials but appear to have little or no primary pathogenicity per se; they appear to be mainly saprophytes or secondary invaders.

Meningitis and Neisseria meningitidis.

The term meningitis is drawn from pathology and means, simply, inflammation of the membranes (meninges) covering the brain and spinal cord. Meningitis may result from mechanical irritations or infection by viruses or many kinds of bacteria, both pathogens and saprophytes, which may localize in the meninges. The meningococcus is the only common cause of **epidemics** of meningitis. The organisms are transmitted, as are other respiratory microorganisms, in oral and nasal secretions.

Carriers of meningococci are common, but meningitis is not. There is evidence that the meningococcus often causes conditions like rhinitis, catarrh, or purulent colds, which heal and attract no particular attention because the etiological agent is unsuspected.

The meningococci gain entrance to the meninges from the upper respiratory tract. Diagnosis is based on clinical appearance of the patient and is confirmed by observing the meningococci in gram-stained smears of pus cells in the spinal fluid. Chemotherapy is instituted on this basis. Meningococcal meningitis is too rapidly progressive and fatal to await cultivation of the organisms, although this is done for confirmatory purposes.

Serological Types. The meningococci are separable into four main serological groups: A, B, C, and D, on the basis of agglutination reactions. The **quellung reaction** (see in next chapter) for grouping is available, analogous to that used in typing pneumococci, since freshly isolated strains of meningococci possess type-specific carbohydrate capsules.

CHAPTER 37
SUPPLEMENTARY READING

Baird-Parker, A. C.: Methods for classifying staphylococci and micrococci, *in*: Gibbs, B. M., and Skinner, F. A. (Eds.): Identification Methods for Microbiologists, Part A. Academic Press–London, 1966.

Baird-Parker, A. C.: The use of Baird-Parker's medium for the isolation and enumeration of *Staphylococcus aureus, in*: Shapton, D. A., and Gould, G. W. (Eds.): Isolation Methods for Microbiologists. Academic Press, Inc., New York. 1969.

Blair, J. E., Lennette, E. H., and Truant, J. P. (Eds.): Manual of Clinical Microbiology. American Society for Microbiology. The Williams & Wilkins Co., Baltimore. 1970. (See section dealing with genus in question.)

Burrows, W.: Textbook of Microbiology, 20th ed. W. B. Saunders Co., Philadelphia. 1973.

Cluff, L. E., and Johnson, J. E.: Clinical Concepts of Infectious Diseases. The Williams & Wilkins Co., Baltimore. 1972.

Cohen, J. O.: The Staphylococci. Wiley-Interscience, New York. 1972.

Collins, C. H., and Lyne, P. M.: Microbiological Methods, Chapters 24 and 26. Butterworth & Co. Ltd., London. 1970.

Gilbert, R. J.: Media for the isolation and enumeration of coagulase-positive staphylococci from foods, *in*: Shapton, D. A., and Gould, G. W. (Eds.): Isolation Methods for Microbiologists. Academic Press, Inc., New York. 1969.

Joklik, W. K., and Smith, D. T.: Zinsser, Microbiology, 15th ed. Appleton-Century-Crofts, New York. 1972.

Mudd, S. (Ed.): Infectious Agents and Host Reactions. W. B. Saunders Co., Philadelphia. 1970. (See Chapters 9 and 12.)

CHAPTER 38 • THE COCCI: STREPTO-COCCACEAE

As pointed out in the previous chapter, the classification scheme for the cocci has been rearranged in the 8th edition of *Bergey's Manual*. In the 7th edition, the tribe Streptococceae was considered a division of the large and important family Lactobacillaceae. The name of this family was taken from the Latin word *lactis*, for milk. Not only were many species of this family found in milk and dairy products, but all fermented carbohydrates such as glucose and lactose with the formation of considerable amounts of L-(+)-lactic acid. Such bacteria were often spoken of collectively as the **lactic-acid bacteria, or lactics.**

In spite of their common physiological traits, however, the family did encompass two morphologically distinct groups, the Streptococceae and the Lactobacilleae. The latter is now considered a distinct family rather than a tribe and is discussed in the next chapter. This chapter will be devoted to the streptococci, now placed in the family Streptococcaceae and covered in part XIV of the 8th edition of *Bergey's Manual*.

Members of the family Streptococcaceae divide in parallel vertical planes and tend to cling together after fission, thus forming chains of spheres like beads on a string. Chains may consist of only three or four cells or up to thousands, depending on species and growth conditions. Streptococci are often pleomorphic (Fig. 38–1); under some growth conditions several species of streptococci elongate and form chains of cells indistinguishable from lactobacilli and

diphtheroids (Chapt. 39). Streptococci are most strongly gram-positive in young cultures, especially in media containing blood or serum. They often appear gram-negative in old or acidified cultures. They have complex nutritive requirements, including specific amino acids, vitamins, and, for several pathogenic species, factors found in blood or serum.

The family is divided (at present) into five genera: *Streptococcus*, *Leuconostoc*, *Pediococcus*, *Aerococcus*, and *Gemella*. The first genus is characteristically parasitic in and on animals;

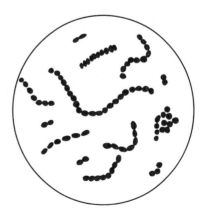

Figure 38–1. Representative forms and arrangements of various species of *Streptococcus*. Note variations in size, and that many cells are ovoid. Others are spherical and still others are flattened together. (Composite drawing of gram-stained cultures in serum broth.) (About 1,500×.)

561

some species are dangerous pathogens. However, other species of *Streptococcus* (e.g., *S. lactis* and *S. cremoris*) are common in dairy products and are harmless. The genera *Leuconostoc* and *Pediococcus* are probably of plant origin and are important in the dairy and brewing industries. DNA base composition of the members of the family ranges from 31 to 44 mole per cent GC.

All streptococci are typical eubacteria, nonsporeforming and nonmotile. Lactics are heterotrophs and generally aerobic or facultative. Some are microaerophilic. Although regarded as catalase-negative, several types have been shown to produce a distinctive type of hydrogen peroxide–decomposing enzyme, "pseudocatalase," which lacks the heme prosthetic group and is insensitive to treatment with cyanide or azide. The lack of a cytochrome system clearly differentiates the streptococci from other morphologically similar bacteria which possess a cytochrome system (hence iron-porphyrins). The latter thus give a positive benzidine test,* while the streptococci do not.

Because of their fermentative type of metabolism they do not fully utilize the energy of carbohydrates by decomposing them completely to carbon dioxide and water. Glucose is changed to pyruvate, and this is hydrogenated to form lactate. Most species convert about 90 per cent of lactose or glucose into lactic acid, with little or no production of CO_2 or acetic or other acids:

$$C_{12}H_{22}O_{11} + H_2O \longrightarrow 4CH_3 \cdot CHOH \cdot COOH$$

These species are therefore said to be **homofermentative**. Species of the genus *Leuconostoc* convert lactose or glucose into about 50 per cent lactic acid, about 25 per cent CO_2 and about 25 per cent acetic acid and ethyl alcohol. *Leuconostoc* species are therefore said to be **heterofermentative**. Differentiation between heterofermentative and homofermentative activities may depend somewhat on conditions of growth and nutrition. In fermenting **pentose** sugars instead of hexoses, for example, both types may produce lactic and acetic acids:

$$C_5H_{10}O_5 \longrightarrow$$
$$CH_3 \cdot CHOH \cdot COOH + CH_3COOH$$

38.1
GENUS STREPTOCOCCUS

This genus has traditionally been divided into four major groups by using the criteria pub-

*Benzidine dihydrochloride with hydrogen peroxide forms a blue color in the presence of cytochromes.

lished by Sherman in 1937, along with some additional tests to help differentiate the species, as shown in Table 38–1. The first group, called the **enterococcus** or **fecal** group, includes commensal species always found in the feces of normal animals and man; also, a few that are potentially pathogenic. The second group is the harmless **lactic** group, associated with the lactobacilli and other lactic-acid bacteria found in dairy products and in fermented foods such as sauerkraut and pickles. The third, or **viridans** group, is also mostly, but not entirely, nonpathogenic and is commonly found in milk and milk products and the nose and throat of man. These organisms are named after the type of hemolytic reaction they give with mammalian erythrocytes (see blood agar types). The last group, the **pyogenic** (pus-forming) streptococci are frequent pathogens in man and animals and can cause serious disease. DNA base ratios of the genus range from 33 to 44 mole per cent GC.

The problem with most classification systems is that eventually specimens are found which cut across divisional lines. This has proven true for the streptococci, and certain newly recognized species cannot be placed in any one of the four traditional groups listed in Table 38–1. While recognizing this problem, we have elected to retain the classic approach in describing the properties of these organisms. This should help the beginning student recognize the major differences between important species with regard to habitat, physiological characteristics, and pathogenicity.

Two of the more useful characteristics used to differentiate species are the types of **hemolysins** and **antigens** produced. Serological typing is described in §38.5. A description of hemolytic types and methodology follows.

Blood Agar Types. The terms **hemolytic, viridans,** and **nonhemolytic** streptococci refer to the action of colonies of the streptococci in plates of infusion agar containing about 5 per cent mammalian blood.

Blood Agar Plates. To determine the blood agar type, a tube containing about 15 ml of melted meat-infusion agar, cooled to 45C (still fluid, yet not hot enough to injure the microorganisms), is inoculated with a loopful of pus, milk, broth culture, or other material containing the desired streptococci. About 5 per cent sterile defibrinated blood is added aseptically and well mixed with the agar. The mixture is poured into plates and incubated **aerobically** for 24 hours at 25C for lactic streptococci and 37C for others. Students should be careful not to confuse the terminology used for strepto-

TABLE 38–1. PHYSIOLOGICAL AND BIOCHEMICAL TESTS FOR IDENTIFICATION OF STREPTOCOCCUS SPP.*†

Physiological Group	Serological Group	Hemolysis (Sherman)	Growth at 10C	Growth at 45C	Growth at pH 9.6	0.1% Methylene Blue	Survive 60C/30 min	NH₃ from Arginine	Streptococcus spp.	Usual Habitat or Source
Enterococcus	D	α, β, or γ	+	+	+	+	+	+	*faecalis*	Intestine of man and other warm-blooded animals
	D		+	+	+	+	+	+	*faecalis* var. *zymogenes*	
	D		+	+	+	+	+	+	*faecalis* var. *liquefaciens*	
	D		+	+	+	+	+	+	*faecium*	
	D		+	+	+	+	+	+	*faecium* var. *durans*	
	D	α or γ	−	+	−	−	−	−	*bovis*	Bovine intestine
	D		−	+	−	−	−	−	*equinus*	Equine intestine
Lactic	N	γ	+	−	±	+	±	±	*lactis*	Dairy utensils, milk and milk products, vegetable material
	N		+	−	±	+	±	±	*lactis* var. *diacetilactis*	
	N		+	−	−	+	±	−	*cremoris*	
Viridans	Ungrouped	α or γ	−	+	−	−	∓	−	*mitis*	Human mouth and throat
	Ungrouped		−	+	−	−	∓	−	*salivarius*	Human mouth and throat
	Ungrouped		−	+	−	−	∓	−	*thermophilus*	Pasteurized milk and cheese
Pyogenic	A	β	−	−	−	−	−	+	*pyogenes*	Human — pathogen
	B		−	−	−	−	−	+	*agalactiae*	Cattle — pathogen
	C		−	−	−	−	−	+	*equi*	Horse — pathogen
	C		−	−	−	−	−	+	*equisimilis*	Human — parasite
	C		−	−	−	−	−	+	*dysgalactiae*	Cattle and sheep — pathogen
	C		−	−	−	−	−	+	*zooepidemicus*	Animal — pathogen
	E		−	−	−	−	−	+	sp.	Cattle — parasite
	F		−	−	−	−	−	+	*anginosus*	Human — pathogen
	G		−	−	−	−	−	+	sp.	Human and animal — parasite
	H‡		−	−	−	−	−	±	*sanguis*	Human — parasite

Species Differentiation tests (column headings): β-HEMOLYSIS, Na HIPPURATE HYDROLYSIS, LACTOSE, TREHALOSE, SORBITOL, GELATIN LIQUEFACTION, 0.04% TELLURITE TOLERANCE, STARCH HYDROLYSIS, GLYCEROL (ANAEROBIC), TETRAZOLIUM REDUCTION, MANNITOL, GROWTH AT 50C, ARABINOSE, MELIBIOSE, MELEZITOSE, MALTOSE, GROWTH AT 40C, GROWTH AT pH 9.2, GROWTH IN 4% NaCl, NH₃ FROM ARGININE, ACETOIN PRODUCTION.

For the Pyogenic group: "Serological grouping and typing are the preferred methods of differentiation in this group."

Source annotations: Occasional pathogens (Enterococcus *faecalis* group); Mostly nonpathogenic (*bovis*, *equinus* and Lactic group).

*Adapted from Sharpe, Fryer, and Smith *in* Gibbs and Skinner (Eds.): Identification Methods for Microbiologists, Part A, Academic Press–London, 1966.

†Symbols: + = positive; − = negative; Sl = slight; ± = majority positive; ∓ = majority negative.

‡Serological groups K–T have also been designated.

coccal hemolysins with that employed for staphylococcal hemolysins (Chapt. 37).

α-Type Hemolytic Streptococci. Colonies of these streptococci are surrounded by a nearly colorless zone of hemolysis resulting from destruction of erythrocytes, and a zone of **discolored,** but **intact,** erythrocytes close in around the deep colonies. These erythrocytes have a green or brownish green color; hence the term **viridans streptococci** (L. *viridis* = green). Peripheral to the inner ring of discolored cells the outer zone of clear, almost colorless hemolysis may be of great or small extent, and may sometimes be so small as to coincide with the zone of green cells. It usually widens on refrigeration of the plate (Fig. 38–2, W and X).

Only the use of a microscope can be relied upon to make the distinction and *only colonies that are deep in the agar* are always thus char-

acterized, surface colonies sometimes producing deceptive appearances or growth spreading over and hiding hemolytic zones. Although the terms α-type hemolytic streptococci and "viridans" streptococci are sometimes used interchangeably, the latter should be reserved for those streptococci which give α-type hemolysis and do not belong to any serological group (see Table 38–1).

β-Type Hemolytic Streptococci. The hemolytic zones around deep colonies of streptococci of this type in blood agar plates are seen to be entirely clear, almost colorless, and *free from any intact erythrocytes* (Fig. 38–2, U and V). Such streptococci are loosely spoken of under the general term of "hemolytic strep."

Double-Zone β-Type Streptococci. Certain species, almost exclusively of bovine origin and not uncommon in dairy products, after producing a zone of hemolysis like that of other β-type streptococci and then allowed to cool, produce a second ring of hemolysis separated from the first by a ring of red erythrocytes.

γ-Type Streptococci. When deep colonies of streptococci in blood agar pour plates show no visible change in the blood cells surrounding the colony, they are said to be of the γ-, **indifferent,** or **nonhemolytic** type.

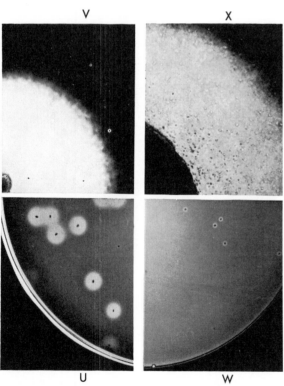

Figure 38–2. Colonies of hemolytic streptococci in blood agar. *U,* Clear zones of complete hemolysis around colonies of β-type (*Streptococcus pyogenes,* Lancefield group A), natural size. *V,* One β-type colony enlarged to show edge of colony at lower left and absence of erythrocytes in clear hemolyzed zone. *W,* Small hemolytic zones of α-type colony, *S. mitis,* natural size. *X,* One α-type colony enlarged to show edge of colony at lower left, with many intact erythrocytes in hemolyzed zone. (Preparations by Dr. Elaine L. Updyke. Photograph courtesy of Communicable Disease Center, U.S. Public Health Service, Atlanta, Ga.)

38.2
THE ENTEROCOCCUS (FECAL) GROUP

As the name implies, these are streptococci commonly found in feces of man and other animals. They also occur on plants. They are regularly found in the normal mouth, whence they probably get into the intestinal tract. These are all relatively hardy organisms capable of resisting growth conditions wholly inhibitory to the more fragile pyogenic group (Table 38–1).

Many of the enterococci have a tendency to produce short chains and pairs of plump, ovoid cocci, and are commonly found in clumps suggestive of micrococci; some were first described as micrococci.

Unlike the highly host-adapted pyogenic streptococci, enterococci are characterized by wide tolerance of heat and cold and other influences unfavorable to other streptococci: 60C; 6.5 per cent sodium chloride solution; bile (low surface tension); the presence of 0.1 per cent methylene blue; pH 9.6; and conditions of life in feces and the outer world. Enterococci are even more hardy than the lactic groups.

The enterococci are typified by *Streptococ-*

cus *faecalis.* This is most commonly found in the intestinal tract of man and lower vertebrates and on plants. A closely similar species, found primarily in the intestinal tract of lower vertebrates but also in man and on plants, is called *S. faecium.* Properties that distinguish between *S. faecalis, S. faecium,* and its variant, *S. faecium* var. *durans,* are shown in Table 38–1.

Varieties of *S. faecalis* are *liquefaciens* and *zymogenes,* the last differing from other enterococci in producing β-type zones in blood agar. Varieties *liquefaciens* and *zymogenes* hydrolyze proteins. As a result, they give rise to strong bitter flavors in cheese. They have been found occasionally in certain pathologic conditions. The present tendency is to include these variants in the species *S. faecalis.*

All contain Lancefield group D antigen (see §38.5). They grow readily at 22 to 40C on meat infusion and similar organic media commonly used in the laboratory.

S. bovis and *S. equinus* are found in large numbers in the intestinal tracts of cattle (and other ruminants) and horses, respectively. Thus they are fecal streptococci and as such are as resistant to bile as are the true enterococci. They are also heat-resistant and possess the group D antigen like the enterococci. However, because of some of their other physiological traits, certain workers are reluctant to call them enterococci. *S. bovis* is apparently a potential pathogen.

Fecal Streptococci in Water Supplies. Because fecal streptococci, especially *S. faecalis,* are constantly present in the human intestines in enormous numbers, survive for weeks in feces-polluted water supplies, and are easily cultivated and identified in the laboratory, their presence in water is commonly used as an indication of fecal or sewage pollution. Identifying tests are easily done. (See pages 563 and 707.)

38.3
THE LACTIC GROUP

This group contains: (a) *Streptococcus lactis,** the common, milk-souring streptococcus useful in the dairy industry; (b) its close relative, *S. diacetilactis,* valuable because of its ability to form large amounts of diacetyl; and (c) another close relative called *S. cremoris.* These three species are referred to collectively as the

group N streptococci. They form a homogeneous group, as indicated by their DNA base composition (40 to 44 mole per cent GC). These species are usually of the gamma type in blood agar.

S. lactis. *S. lactis,* a γ-type streptococcus, is always present in market milk, even of the best quality. It occurs in cow dung, dust, and soil, and on plants and utensils; its entrance into the milk is easily explained. Its persistence in such environments shows that it is a relatively hardy organism. It is quite harmless to man. *S. lactis* and *S. cremoris* (see below) are, together, responsible for pleasant flavors in dairy products such as butter and cottage cheese. The important role of these two species and of *S. thermophilus* (see next section) in the dairy and food industries is discussed in Chapters 48 and 49.

S. lactis sometimes forms long chains but chiefly occurs in short chains or pairs, and the cells are oval. It grows rapidly in milk at 28C (82F), causing souring of the milk. The acidity usually suppresses the development of other organisms, some of which might otherwise cause the milk to putrefy. Contributing to the suppression of undesirable gram-positive organisms in milk is the potent antibiotic **nisin** produced by some varieties of *S. lactis. S. lactis* can readily be cultivated on agar containing milk, whey, or tomato juice. It grows best in the presence of glucose or lactose.

S. cremoris. *S. cremoris* is one of numerous lactic-acid streptococci that are similar to *S. lactis. S. cremoris* tends to form very long chains. It may be differentiated from *S. lactis* by characters given in Table 38–1.

38.4
THE VIRIDANS GROUP

The α-type hemolytic streptococci must be distinguished from those of the enterococcus group, which are physiologically very different. As represented by *S. mitis,* the α-type streptococci are generally delicate, fragile, highly adapted parasites found mainly in the normal mammalian upper respiratory tract. Although usually regarded as having low pathogenicity, they are opportunists and sometimes produce serious infections, often of a chronic nature. For example, in man α-type streptococci are sometimes found in abscessed teeth, infected sinuses, and diseased tonsils. From teeth and tonsils they may be carried to the joints and produce rheumatic conditions (arthritis). They can also infect injured heart valves and cause a serious heart

---

* *S. lactis* represents a **cluster** of variants or closely similar species.

disease, **bacterial endocarditis.** S. *faecalis* (a viridans species of the fecal group) is also sometimes involved in such conditions.

Streptococcus salivarius, representative of the harmless "salivary" streptococci, forms thick, slimy, or gummy colonies if cultivated on media containing 5 per cent sucrose. The gumminess results from the synthesis of viscous polysaccharides (generally dextrans or levans) from the sucrose, and it is a good example of the ability of many bacterial species and all plants to synthesize extracellular polysaccharides. Several species of streptococci, S. *salivarius*, S. *mutans* (a variant of S. *salivarius*?), S. *mitis*, and S. *sanguis*, have been implicated in the production of dental caries because of their ability to form dental plaque on teeth and to produce enamel-destroying acid from fermentable sugars, particularly sucrose.

S. thermophilus. Common in dairy products, S. *thermophilus* is perfectly harmless. Although not truly thermophilic, it is thermoduric. It has the property of surviving temperatures as high as 63C (pasteurization) and may appear in large numbers in pasteurized milk, giving the false impression that the milk has not been properly pasteurized. It forms "pinpoint" colonies in agar plates used to determine the numbers of bacteria in milk. On the other hand, it is very useful in the ripening of cheeses made with high processing temperatures, such as Swiss cheese, because it grows well at about 45C.

38.5
THE PYOGENIC GROUP

Any agent, living or inanimate, that stimulates an increase in leucocytes locally or generally is said to be **pyogenic.** With reference to the streptococci, however, the term is generally understood to mean the β-type hemolytic streptococci. As previously mentioned, some α-type streptococci also occasionally produce pyogenic infections. In addition, some species of the fecal group produce β-type hemolysis and can produce pyogenic lesions but are not usually very important pathogens.

The α- and β-hemolytic streptococci differ markedly from the other streptococcal groups in being more highly dependent on a parasitic life in the mammalian body. They are less able to grow and survive under conditions of very low surface tension, high osmotic pressure, low pH, or high and low temperatures. Such influences as these are encountered in the intestinal tract,

ripening cheese, fermented and brined food products, and the outer world, which environ the fecal and lactic groups. (See Table 38–1.)

The fastidious and delicate pathogens of the pyogenic group are best cultivated at 37C in meat-infusion agar or broth containing serum or blood and 0.1 per cent glucose, in an atmosphere containing 5 to 10 per cent carbon dioxide. Their colonies on blood or serum agar are generally about 1 to 2 mm in diameter, colorless or greyish, watery, translucent, and inconspicuous. They may be mucoid, smooth or exhibit different degrees of roughness.

Lancefield Groups of β-Type Streptococci. Lancefield made extracts of cultures of β-type streptococci from different sources by means of hot, 0.2N HCl. These extracts contain specific, somatic carbohydrate antigen. The extracts are used as antigens in precipitin tests. The corresponding antibodies are obtained by injecting the antigens into animals.

By means of such precipitin tests several distinct groups of β-type hemolytic streptococci can be differentiated with respect to origin. Lancefield designated these groups by letters: A, B, C, D, E, F, G, and so on, according to source or other characters. They are accordingly known as **Lancefield groups.** There are nearly a score of such groups. Streptococci belonging to the enterococcus and lactic groups have also been shown to possess certain group antigens (Table 38–1). Only two of the Lancefield groups of the pyogenic streptococci need be described here.

Group A Streptococci. Although certain physiological characteristics may be helpful in distinguishing group A from others, serological tests are preferred and used routinely. S. *pyogenes* causes scarlet fever, septic sore throat, empyema, blood poisoning, puerperal sepsis, and many other serious, epidemic and acute pyogenic diseases in human beings. S. *pyogenes* can also infect the udders of cows. The unpasteurized milk of such cows then becomes a dangerous vector of streptococcal disease.

M TYPES OF GROUP A. Several colony forms of S. *pyogenes* have been described: a smooth, *glossy* form, a **mucoid** form, and a dull, granular form called *matte*, which has turned out to be simply a dehydrated mucoid form. Colonies giving either the matte or mucoid form represent cells producing a nonantigenic hyaluronic acid capsule. Previously it was thought that the M-protein, another somatic antigen made by these streptococci, determined colonial form, but we now know that this characteristic is determined by the capsule. The (somatic)

M protein antigens evoke **type-specific** precipitins. Over 40 serological types (**M types**) of *S. pyogenes* (all group A) have been found. Several physiological varieties of *S. pyogenes* were formerly given various names, such as *S. scarlatinae*, *S. erysipelatus*, and *S. epidemicus*.

The importance of determining the M type of *S. pyogenes* becomes clear when it is realized that resistance to *S. pyogenes* infection is M type–specific. For example, one may have resistance to a type 6 *S. pyogenes* yet succumb in a type 19 epidemic with what *clinically* appears to be the same disease. The importance of typing in the study of group A streptococcal disease can scarcely be overestimated. For example, the whole problem of the control of rheumatic heart disease appears to be bound up with the epidemiology of *S. pyogenes* infection of various M types. Other distinctive, precipitin-eliciting protein antigens in the cell walls of group A streptococci are designated T and R. Studies of T and R antigens are used in special diagnostic work.

STREPTOLYSINS. The hemolytic toxins (streptolysins) of group A streptococci are of two sorts: (1) **streptolysin S,** unstable in the presence of heat and acids, and (2) **streptolysin O,** unstable in oxygen. Streptolysin S produces β-type hemolysis in blood agar plates. It is not antigenic. Streptolysin O is an antigen. In persons infected with group A streptococci, antibodies to streptolysin O are generally found and easily measured. The measurement (titration) of antibodies against streptolysin O, especially when the titer rises sharply over a period of two weeks or so in connection with an illness such as sore throat, affords not only a valuable diagnostic test but also a measure of the reaction of the patient. Over 80 per cent of rheumatic fever patients have a considerable titer of antistreptolysin O.

The virulence of any given strain of group A β-hemolytic streptococci seems to depend largely on the ability to produce copious amounts of both hyaluronic acid capsule and M protein. Both of these surface components inhibit phagocytosis in the early stages of infection. Strains producing reduced amounts of either or both of these components are more quickly phagocytized and destroyed. These organisms also elaborate a number of toxins and enzymes which can cause extensive damage once an infection becomes established. Further details can be found in the Supplementary Reading list at the end of the chapter.

Group B Streptococci. Streptococci of group B differ from all other pyogenic streptococci (except enterococci) in hydrolyzing sodium hippurate and in producing double zones of β-hemolysis.

Hemolytic streptococci of group B are usually of bovine origin but are occasionally found in human infections. *S. agalactiae*, an important member of group B, is of particular interest to the farmer because it causes severe mastitis in cattle and stoppage of milk flow. Of minor pathogenic significance for man, *S. agalactiae* may be confused with *S. pyogenes* by workers seeking to eliminate *S. pyogenes* from milk supplies.

38.6
THE PNEUMOCOCCI

The many serological types of the organism formerly called *Diplococcus pneumoniae* are referred to collectively as the pneumococci. When it was found that transformation (see Chapter 19) could take place between the pneumococci and representatives of all four groups of streptococci, i.e., enterococcus, lactic, viridans, and pyogenic groups, the continuance of *Diplococcus* as a distinct genus seemed unwarranted. Hence, the pneumococci have been renamed *Streptococcus pneumoniae* in the 8th edition of *Bergey's Manual.*

This organism will be described separately because it is an important pathogen and differs from the other pyogenic streptococci in that it produces M proteins and somatic carbohydrate antigens which are distinct from the other streptococci. The various strains of this organism can be differentiated from each other by the antigenic specificity of their polysaccharide capsules. This organism is a cause of lobar pneumonia and, like *Streptococcus pyogenes*, group A, causes numerous other serious, acute pyogenic conditions: meningitis, septicemia, empyema and peritonitis. It was a frequent cause of death before the advent of chemotherapy and is still a dangerous killer.

The organism often forms short chains of ovoid diplococci and produces α-type colonies on blood agar plates. The chains are usually made up of from two to eight pairs of encapsulated cocci. The pneumococci are rarely spherical, commonly having the form of short artillery projectiles placed base to base. Methods of cultivation and study are like those used for other pyogenic streptococci. *S. pneumoniae* is found in the saliva and sputum of patients with lobar pneumonia and, like other viridans streptococci, also occurs frequently in the normal mouth and throat.

Diagnosis. Pneumococci are extremely pathogenic for white mice when freshly isolated from the body. Advantage is often taken of this fact to isolate pneumococci from sputum of patients for diagnosis. The sputum is injected into the mice intraperitoneally. After 6 to 24 hours the mice die or become very ill, and enormous numbers of pneumococci are found in the peritoneal cavity and heart blood. The cocci found on the peritoneum of the mouse may be identified by: (1) their morphology; (2) their **capsules** (Fig. 38–3); (3) their **solubility** in bile or in solutions of various other surface tension reducers; (4) ability to ferment inulin.

Serological Types of Pneumococci. Pneumococci may be divided into more than 80 serological types, which are designated by Roman numerals. Unlike the nonantigenic, and therefore nonspecific, hyaluronic acid capsules of *S. pyogenes,* serological type-specificity of pneumococci is conferred by their capsules. These organisms produce several other types of antigens, some type-specific and others species-specific, but these need not be described here. If encapsulated pneumococci or other organisms with type-specific capsular antigens are mixed with a type-specific immune serum, swelling of the capsules is seen (Fig. 38–3). This is spoken of as a quellung (German for swelling) reaction. This organism is one of the few pathogens believed to have a single determinant for virulence, namely the capsule. Nonencapsulated strains are completely avirulent.

Disease Transmission. Transmission of pneumonia is chiefly through droplets of infected saliva and nasal and pulmonary mucus, and by inhalation of infected dust. Kissing undoubtedly transmits the infection as well as other respiratory pathogens but, obviously, not every such infection results in disease. Romance has a powerful ally in natural resistance to respiratory disease.

38.7
GENUS LEUCONOSTOC

The name of genus *Leuconostoc* is derived from the name of a common blue-green alga, *Nostoc,* and the Greek *leukos,* meaning white or colorless. *Nostoc,* the alga, is characterized by tangled trichomes of spherical green (chlorophyll-containing) cells with a thick, firm, gelatinous outer coating. *Leuconostoc,* the bacterium, resembles *Nostoc* in forming spherical cells in tangled chains and in generally forming thick outer coatings of slime or gum (dextrans, *L. dextranicum* and *L. mesenteroides*) in sucrose media (Fig. 38–4). However, *Leuconostoc* cells are colorless and small (1.0 to 1.5 μm), whereas *Nostoc* cells are relatively enormous—25 to 200 μm—compared with *Leuconostoc* cells. *Leuconostoc* is much like *Lactobacillus* (see next chapter) in most respects, even in morphology; *Leuconostoc* is pleomorphic and often closely resembles lactobacilli. *Leuconostoc* is distinguished among the lactic acid bacteria in being heterofermentative and also in lacking aldolase, a key enzyme in glycolysis. There are various types of *Leuconostoc,* differentiated serologically and by colony form, the amount and molecular nature of gum formed, and by amount of carbon dioxide produced. They produce both levo- and dextrorotatory lactic acid. DNA base ratios of the organisms as a group are similar (39 to 42) to those of streptococci (31 to 44 mole per cent GC), thus indicating a close genetic relationship between these two groups of organisms.

L. mesenteroides and L. dextranicum. These two important species are widely distributed on growing plants and, like *S. lactis,* are commonly found in dairy products and vegetable products. They ferment actively and produce acid in such carbohydrate-rich plant materials as sauerkraut, ensilage, and in such plant juices as are used in making cane sugar and beet sugar. They constitute a great nuisance in the sugar refining industry because their slime and gum formations clog vats, pipes, and machines. Indeed, so much of the specific carbohydrate of *Leuconostoc* is left in commercial cane, beet, and other sugars that specific pre-

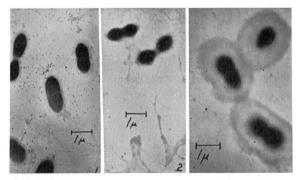

Figure 38–3. Electron micrographs of *Streptococcus pneumoniae.* The first picture shows the capsules in their normal state. The center picture shows the capsules virtually unaffected by serum of a heterologous type. The picture on the right shows the effect of homologous-type serum on the capsules—a well-marked quellung reaction.

cipitin tests can be obtained by mixing a little "sugar water" with immunologically specific serum.

L. citrovorum. This species produces little or none of the gum that is so characteristic of other species of *Leuconostoc*. It so closely resembles *S. lactis* that some workers have regarded it as a streptococcus (*S. citrovorum*). (It has also been classified as a species of *Pediococcus*, a sarcina-like lactic-acid coccus most commonly found in spoiled beer and beer wort.) A very closely related variety has often been called *L. paracitrovorum*. These organisms are closely allied to *S. cremoris*, are commonly found in its company, and have some of its principal characters and produce characteristic flavors. Precise generic designation of these organisms is not clear and requires further investigation.

Lactic-Acid Bacteria and Diacetyl. The important point about *L. citrovorum* and *S. cremoris* is that they decompose citric acid (usually present in small amounts in normal milk) with the formation of acetic acid, carbon dioxide, and, most important (like *S. lactis* var. *diacetilactis*), **diacetyl.** Diacetyl is responsible for the pleasant, buttery flavor of dairy products. Consequently, pure cultures of *L. citrovorum* and *S. cremoris* and/or *S. lactis* var. *diacetilactis* are commonly added to milk or cream that is to be made into butter or cheese. Citric acid also is often added to increase the amount of diacetyl formed. The overall reactions are shown in Figure 38–5.

Citric acid is first decomposed to yield acetyl-methyl-carbinol and other products. Acteyl-methyl-carbinol is then oxidized in the presence of acid to diacetyl. Since diacetyl is best produced in an acid medium, sometimes a little acetic or other harmless acid is added to ensure proper pH. Also, aeration is necessary or, as shown in the foregoing reaction, the acetyl-methyl-carbinol, instead of being **oxidized** to diacetyl, is **reduced** to 2:3 butylene glycol, which is tasteless. These facts are of vital importance to dairymen.

An important branch of the dairy industry is the selection, propagation, maintenance and sale of pure cultures of *S. lactis*, *S. lactis* var. *diacetilactis*, *L. cremoris*, *L. citrovorum*, various lactobacilli, and mixtures of these, for the manufacture of butter, fermented milks, yoghurt and cheeses. (See Chapter 47.) On the other hand, these organisms are nuisances in the citrus and other fruit juice industries, imparting "sour milk" and "buttermilk" flavors to the products.

Leuconostoc and Vitamin Assay. Like

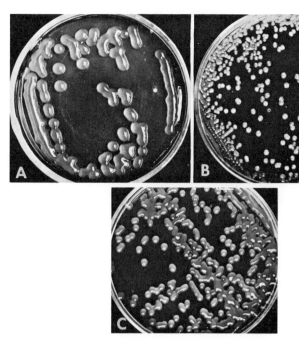

Figure 38–4. Colonies of *Leuconostoc* sp., growing on carbohydrate agar. Note the large amount of gummy, polysaccharide capsule-like material around the colonies in *A* and *C. B* is a species like *L. citrovorum* that produces little gum. (Photograph courtesy of American Society for Microbiology.)

many species of *Lactobacillus*, some *Leuconostoc* species are important because of synthetic enzyme deficiencies; for example, nicotinic acid is necessary for *L. mesentericus*, and the so-called citrovorum factor (CF) or leucovorin (the vitamin, folinic acid) is required by *L. citrovorum*. The organisms are commonly used in the assay of these vitamins. (See Table 39–3 and Chapter 49.)

38.8 GENUS PEDIOCOCCUS

Pediococcus (Gr. *pedion* = plane) is characterized by division in parallel vertical planes, forming short chains, or in two vertical planes at right angles, forming tetrads. Pediococci have often been called *Sarcina*, though the formation of cubical packets like *Sarcina* (see next section) probably does not occur. DNA base ratios of the organisms as a group are slightly higher (38 to 44 mole per cent GC) than typical *Streptococcus*. In some respects pediococci resemble *Leuconostoc citrovorum*, although pediococci are homofermentative. Pediococci have the general properties of other lactics.

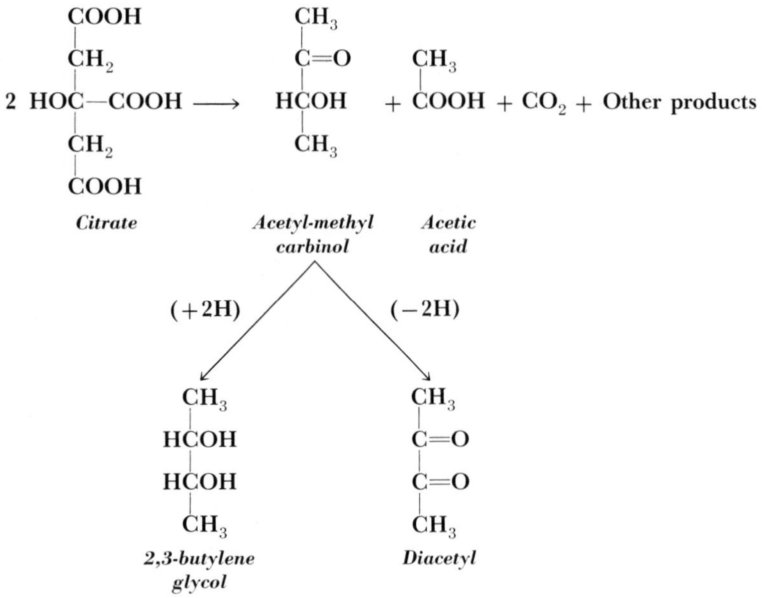

$$2 \ HOC\text{—}COOH \longrightarrow$$

Citrate   Acetyl-methyl carbinol   Acetic acid

$+ COOH + CO_2 +$ Other products

(+2H)   (−2H)

2,3-butylene glycol   Diacetyl

Figure 38–5. Overall reactions in production of diacetyl from citrate; for explanation see text. Diacetyl can also be produced from pyruvate. (2, 3-Butylene glycol is useful in some phases of rubber synthesis but is *no* good in butter!)

Pediococci are of considerable importance as nuisance saprophytes in fermenting vegetable products, notably sauerkraut, fruit juices, beer, and beer wort. They produce acidity, turbidity, and diacetyl and, consequently, buttery odors and flavors ("sarcina odor") in fruit juices and beer. Most true beer connoisseurs *despise* buttermilk-flavored beer!

38.9
FAMILY PEPTOCOCCACEAE

The family includes four genera, *Peptococcus, Peptostreptococcus, Ruminococcus,* and *Sarcina.* DNA base rations of the family range from 28 to 45 mole per cent GC. The members of the family are considered anaerobic and are frequently found in the rumen of herbivorous animals (sheep, goats, cattle). For this reason, members of the family, along with some methane-producing cocci (*Methanococcus*) and methane-utilizing cocci (*Methylococcus*), are frequently referred to as "rumen bacteria."

Peptostreptococci (Gr. *pepto* = to digest; i.e., proteolytic) occur mainly in pyogenic, putrefactive, and gangrenous lesions and wounds. Most are highly pathogenic. Physiologically the group is somewhat heterogeneous. Some are able to ferment cellobiose, the end product of cellulose decomposition, while others, such as *P. elsdeni*, are able to ferment lactate with the production of large amounts of acetate and small amounts of propionate and butyrate, characteristic end products of the anaerobic fermentation of carbohydrates. Except in being strongly proteolytic, peptostreptococci resemble streptococci of the pyogenic group; some are hemolytic, others not. Most species are distinguished by being, when first isolated from pathologic material, **rigidly anaerobic,** though they often later become aerotolerant.

CHAPTER 38
SUPPLEMENTARY READING

Blair, J. E., Lennette, E. H., and Truant, J. P. (Eds.): Manual of Clinical Microbiology. American Society for Microbiology. The Williams & Wilkins Co., Baltimore. (See section dealing with genus in question.)

Burrows, W.: Textbook of Microbiology, 20th ed. W. B. Saunders Co., Philadelphia. 1973.

Collins, C. H., and Lyne, P. M.: Microbiological Methods, p. 297. Butterworth & Co. Ltd., London. 1970.

Frazier, W. C.: Food Microbiology, 2nd ed. McGraw-Hill Book Co., Inc., New York. 1967.

Ginsburg, I.: Mechanisms of cell and tissue injury induced by group

A streptococci: Relation to post-streptococcal sequelae. J. Infect. Dis., *126*:294, 419, 1972.

Joklik, W. K., and Smith, D. T.: Zinsser, Microbiology, 15th ed. Appleton-Century-Crofts, New York. 1972.

Mudd, S. (Ed.): Infectious Agents and Host Reactions. W. B. Saunders Co., Philadelphia. 1970. (See Chapters 7 and 8.)

Sharpe, M. E., and Fryer, T. F.: Identification of the lactic acid bacteria, *in* Gibbs, B. M., and Skinner, F. A., (Eds.): Identification Methods for Microbiologists, Part A. Academic Press–London. 1966.

Wannamaker, L. W., and Matsen, J. M.: Streptococci and Streptococcal Diseases. Academic Press, Inc., New York. 1972.

CHEMOORGANO- • CHAPTER 39
TROPHIC, GRAM-POSITIVE, NONSPOREFORM- ING RODS AND FILAMENTOUS EUBACTERIA

Organisms of the families Brevibacteriaceae, Propionibacteriaceae, Corynebacteriaceae, and Lactobacillaceae, tribe Lactobacilleae, as outlined in the 1957 edition of *Bergey's Manual*, belong to the order Eubacteriales. All are gram-positive, nonsporeforming rods. Some are motile, some nonmotile; some are coryneform, and others tend to form filaments. The systematic position of many of the genera in these families is doubtful. Some are microaerophilic, while others are strict anaerobes, and they are distributed widely in nature in soil, grain, milk, and dairy products and in the intestinal tracts of man and other animals. They are physiologically rather heterogeneous and contain both nonpathogenic and pathogenic species. The tribe Lactobacilleae and some other gram-positive, nonsporeforming rods are compared in Table 39–1. The name of the tribe is derived from the fact that all species ferment lactose, with production of lactic acid.

A radical rearrangement of the tribe Lactobacilleae and family Corynebacteriaceae has been proposed for the 8th edition of *Bergey's Manual*. For example, the tribe Lactobacilleae

itself and most of the genera in the tribe have been either eliminated or transferred to other parts for the purpose of establishing more unified taxonomic groupings. The family Lactobacillaceae, which contains one genus, *Lactobacillus*, is placed in Part XVI, "Gram-Positive, Nonsporeforming Rod-Shaped Bacteria," together with "Genera of Uncertain Affiliation," which includes *Listeria, Erysipelothrix,* and *Caryophanon*. These organisms currently (*Bergey's Manual*, 1957) belong to three different families encompassing two orders, Eubacteriales and Caryophanales.

Here we shall consider only representative genera of the groups described above.

39.1
GENUS LACTOBACILLUS

The species of genus *Lactobacillus* are among the most important and most studied of the Lactobacilleae. They, with certain streptococci important in dairying, are sometimes spoken of collectively as "lactics."

572

TABLE 39-1. COMPARISON OF SOME GRAM-POSITIVE NONSPOREFORMING RODS*

Characteristics	Lactobacilleae	Propionibacteriaceae	Brevibacteriaceae	Corynebacteriaceae
Morphological peculiarities	Long, slender rods; pleomorphic	Short rods; occasional branching; pleomorphic	Short rods; red, yellow, and orange pigments; pleomorphic	Short rods; pleomorphic; metachromatic granules; club forms
Distinctive metabolic products or characters	Lactic acid	Propionic or butyric acids; ferment lactic acid	Lactose not fermented	As Propionibacteriaceae
Proteolysis	Little or none	As Lactobacilleae	Active	Various
Catalase production	−†	+	+	+
Relation to oxygen	Indifferent or microaerophilic	Anaerobic to microaerophilic	Aerobic and facultative	Strictly aerobic to microaerophilic or facultative
Special nutritive requirements	Numerous vitamins and amino acids	As Lactobacilleae		Few
Habitat (saprophytic species)	Feces, soil, dairy products, fermenting vegetable matter	As Lactobacilleae	As Lactobacilleae	Soil, plants, dairy products
Pathogenesis	Some pathogenic on mammals	None	None	Several species on animals and plants

*Characteristics according to 7th edition, *Bergey's Manual*; there are some exceptions to nearly all of the properties listed.
†Some catalase activity has been described in some species.

Lactobacilli are usually long, slender, nonmotile rods, often forming filaments of cells (Fig. 39–1). Several species that are motile with peritrichous flagella have been described. Lactobacilli are pleomorphic; under some growth conditions they form chains of cells that resemble streptococci or *Leuconostoc* (a distinctive species of dairy streptococci having some properties and habitat in common with lactobacilli). Encapsulated strains of lactobacilli have been described. Lactobacilli occur in soil, in dust, on plants, in the vertebrate intestinal tract, and in foods and dairy products.

Lactobacillus species are rarely of pathogenic significance, with the possible exception of a few oral species; e.g., *L. casei*, which, with acid-forming oral streptococci, appear to be involved in the production of dental caries. The principal role of lactobacilli is in the dairy industry. For example, *L. bulgaricus* and related species are used in the preparation of yogurt; *L. delbrueckii* and others are used in the manufacture of lactic acid; other species are used in production of silage, pickles, and sauerkraut. Detailed discussions concerning use of lactobacilli in food manufacturing and for industrial use are presented in Chapters 48 and 49. In many ways lactobacilli resemble *Leuconostoc* and are often found associated with *Leuconostoc* in dairy products and fermenting plant materials. At least one species of lactobacilli, *L. casei*, produces a specific capsular polysaccharide analogous to but less conspicuous than that of *Leuconostoc* (see Chapter 38).

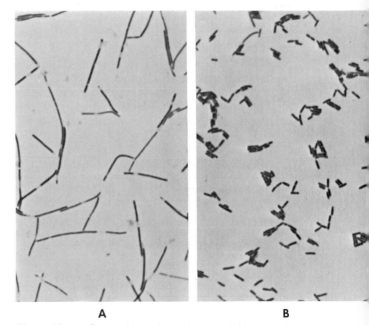

A   B

Figure 39–1. Comparison of morphology of *Lactobacillus leichmannii* grown in media of different compositions. *A,* Cells from complex medium containing thymidine; *B,* cells from complex medium containing vitamin B_{12}.

Taxonomy of Lactobacilli. Traditionally, members of the genus *Lactobacillus* have been subdivided into two major fermentation (biochemical) types: **homofermenters**, which produce virtually a single fermentation product: two moles of lactic acid from a mole of glucose; and **heterofermenters**, which produce numerous organic acids and alcohols in addition to lactic acid. These two fermentation types are further divided into three subgenera: all heterofermentative organisms comprise the **betabacterium** group, while homofermenters are divided into two groups on the basis of optimum growth temperature. The homofermenters which grow well at 15C but poorly or not at all at 45C are called **streptobacterium**; those others which grow well at 45C or higher but show no growth at 15C are the **thermobacterium** group. Representative species of each of the subgenera are presented in Table 39–2.

Recently, DNA base compositions (mole per cent G + C) of various genera of lactics have been added as a taxonomic tool in unifying the group. Comparison of DNA base compositions with the classification of lactics by more traditional methods reveals some very interesting relationships. For example, DNA base composition of the homofermentative streptococci ranges from 38 to 40 mole per cent GC; similarly, heterofermentative *Leuconostoc*, 39 to 42 mole per cent GC; *Pediococcus* (cells in tetrads), 38 to 44 mole per cent GC. Thus the studies not only show very little variation from strain to strain but also show that DNA base composition studies support the classification based on phenotypic characters.

DNA base composition of the members of the genus *Lactobacillus* ranges from 32 to 52 mole per cent GC, which indicates that membership of the genus is heterogeneous. However, such heterogeneity is not unique for the lactobacilli but occurs within genera *Proteus* (36 to 53), *Neisseria* (40 to 53), *Bacillus* (32 to 66), and *Mycoplasma* (23 to 41 mole per cent GC).

Closer observation of Table 39–2 shows that the genus *Lactobacillus* may be further divided into genetic species on the basis of DNA base ratio. Group 1 includes species with GC content from 35 to 38 mole per cent; Group 2, 42 to 48 mole per cent GC; and Group 3, 49 to 52 mole per cent GC. Hence, the "Betabacterium" group contains two separate groups (2 and 3); "Streptobacterium" consists of a single group (2); and "Thermobacterium" contains possibly two entirely separate groups (1 and 3). These findings suggest that further studies (including serological investigations) of the genus are necessary; undoubtedly many species of *Lactobacillus* may become members of separate genera. For example, one member of the lactics sometimes classified as *Lactobacillus bifidus* has been transferred into the family Actinomycetaceae of the order Actinomycetales primarily on the basis of DNA base composition, which ranges from 57 to 63 mole per cent GC (*Bifidobacterium bifidus*). The organism has a unique habitat and is more fully discussed in Chapter 40.

Nutrition of Lactobacillus. Several species of *Lactobacillus* and of *Leuconostoc* are important because of curious deficiencies in their synthetic-enzyme equipment. In the laboratory they generally do not grow well on the usual peptone and meat-extract media. Instead they require vitamin-containing vegetable or fruit juices, yeast extracts, whey or milk, carbon dioxide for carbon, and carbohydrates as sources of carbon and energy. While some can use ammonium salts as a source of nitrogen, and from them synthesize all necessary amino acids and proteins, others are **absolutely dependent** on

TABLE 39–2. REPRESENTATIVE LACTOBACILLUS SPECIES FOR EACH OF THE SUBGENUS GROUPS

	Heterofermentative		Homofermentative		
"BETABACTERIUM"	DNA BASE COMPOSITION (MOLE PER CENT G + C)	"STREPTOBACTERIUM"	DNA BASE COMPOSITION (MOLE PER CENT G + C)	"THERMOBACTERIUM"	DNA BASE COMPOSITION (MOLE PER CENT G + C)
L. brevis	45 (2)°	L. plantarum	45 (2)	L. salivarius	35 (1)
L. buchneri	45 (2)	L. casei	47 (2)	L. acidophilus	37 (1)
L. pastorianus	45 (2)			L. jugurti	38 (1)
L. fermenti	52 (3)			L. helveticus	38 (1)
L. cellobiosus	52 (3)			L. bulgaricus	38 (1)
				L. leichmannii	50 (3)
				L. lactis	50 (3)
				L. delbrueckii	50 (3)

°Numbers in parentheses ( ) indicate taxonomic group based on DNA base composition (mole per cent G + C); value of mole per cent may differ slightly from strain to strain and from one investigator to another.

certain ready-formed amino acids that they cannot synthesize. Others are similarly dependent on certain specific vitamins or other specific compounds (purines and fatty acids), the specific requirements depending on species (see Table 39–3). Nutritive conditions often determine specific requirements. For example, some species do not require the amino acid L-serine if the vitamin folic acid is present. It would seem that if they have folic acid they can synthesize their own L-serine.

Because of these specific nutritive requirements, various species of lactic-acid bacteria are of immense value and are widely used in **assaying** (measuring) the vitamin and amino acid content of foods and drugs for industrial and pharmaceutical purposes. More details of many industrial aspects of lactic-acid bacteria are given in Chapter 49.

Isolation of Lactobacilli. Many species such as *L. fermenti, L. bulgaricus, L. brevis, L. delbrueckii, L. plantarum,* and *L. lactis* are quite acid-resistant and thermoduric. As a result they can grow in warm and acid situations such as curing ensilage and sauerkraut; fermenting beer, wine, and whiskey mashes; and fruits, vegetables, and their juices as prepared for canning. Several species are resistant to high osmotic pressures and grow well in salt-pickling vats and meat-corning and -curing brines, in sugar-refining vats and salty cheeses.

The members of the genus may be selectively isolated by use of complex media (yeast extract or various amino acids) containing carbohydrates and acetic acid and having initial pH of approximately 4.5. The addition of acetic acid and the low pH of the medium usually discourage the growth of practically all other bacteria except a few "over-oxidizing" strains of acetic acid bacteria. However, unlike species

of lactobacilli, acetic acid bacteria are not only gram-negative but also obligately aerobic and grow best at lower temperatures. Their growth is readily inhibited by incubation conditions.

39.2 GENUS LISTERIA

The organisms of genus *Listeria,* named for Joseph, Lord Lister, British scientist and founder of aseptic surgery, are in most respects much like diphtheroids (see §39.5, Family Corynebacteriaceae), except that they are more regular in form and are motile (with peritrichous flagella) in a tumbling pattern and occur in pairs resembling diplococci, or form filaments. *Listeria* grows readily on blood and serum media at 37C and is facultative with respect to oxygen. The genus is currently placed in the Corynebacteriaceae (*Bergey's Manual,* 1957) but in the 8th edition of the *Manual* the genus is transferred and grouped together with *Lactobacillus.*

The organism has been isolated from decaying vegetable matter, from soil, and from feces of healthy animals and healthy human carriers. Distribution of the organism is worldwide as evidenced by listeric infections occurring in a wide geographical area.

Listeriosis. *Listeria* appears to be regularly pathogenic for man and a great variety of lower animals, causing a febrile disease **(listeriosis)** characterized especially by swollen lymph nodes and necrotic lesions in various organs (glandular fever) and the appearance, in the blood, of the organisms and large numbers of white cells, called **monocytes** — hence the type species name, *Listeria monocytogenes* (38 mole per cent GC). The organism may be transmitted to man and domestic animals from wild animals, but the epidemiology of listeriosis remains uncertain.

Identification of cultures may be determined by intravenous inoculation into a rabbit, which results in increase in monocytes (up to 40 per cent), or by the Anton test, which shows marked purulent conjunctivitis developing within 24 to 36 hours after instillation of *L. monocytogenes* into the conjunctival sacs of young rabbits or guinea pigs. These tests should be considered presumptive tests.

39.3 GENUS ERYSIPELOTHRIX

The name of genus *Erysipelothrix* (Gr. *erysipelo* = erysipelas; *-thrix* = thread) refers

TABLE 39–3. SOME ESSENTIAL GROWTH REQUIREMENTS OF CERTAIN LACTOBACILLACEAE*

Specific Growth Requirement	Species
p-Aminobenzoic acid	*Lactobacillus plantarum*
Folic acid	*Lactobacillus casei* *Streptococcus faecalis*
Folinic acid	*Leuconostoc citrovorum*
Nicotinic acid (niacin)	*Leuconostoc mesenteroides* *Lactobacillus plantarum*
Riboflavin (vitamin B_2)	*Streptococcus lactis*
Thiamine (vitamin B_1)	*Lactobacillus fermenti*

*Characteristics according to 7th edition, *Bergey's Manual.*

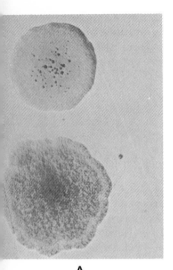
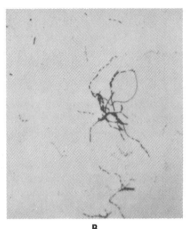

A   **B**

Figure 39–2. *A,* Smooth (*above*) and rough (*below*) colonial forms of a single strain of *Erysipelothrix rhusiopathiae* (48 hours 30×.)

B, Typical morphology of *Erysipelothrix rhusiopathiae* from rough colonies. (900×.)

to the inflammation produced and the often-filamentous appearance of the organism. A farmer who has raised many hogs or poultry for market probably knows about swine erysipelas. Only the fortunate farmer has escaped the infection himself (erysipeloid of man). *Erysipelothrix rhusiopathiae* (*E. insidiosa* in the 7th edition of *Bergey's Manual*), the causative organism, is widely distributed in soil, dung, dust, and sewage, and can infect sheep, birds, rodents, and fish. It is a common cause of infection in commercial-fish handlers.

E. rhusiopathiae in many respects resembles the diphtheroids. However, it is often pleomorphic and filamentous, like lactobacilli, especially in the R phase (Fig. 39–2). It is nonmotile. Because of its tendency to filament formation it has sometimes been classified in the Actinomycetales. It is facultatively aerobic. It is quite resistant to preservatives like smoking, pickling, drying, and outdoor conditions generally and therefore can persist stubbornly in barns, vats, dust, and dirt of animal pens, buildings, and vehicles where infected animals have been kept. Wild rodents may harbor the organisms and act as reservoirs for infection in swine. The organism probably lives in soil as a saprophyte. It is transmitted by inhalation and ingestion of infected dirt and animal products, and by way of cuts and scratches.

Swine Erysipelas. Swine erysipelas, caused by *E. rhusiopathiae*, is usually slowly progressive, though the infection at times is highly and rapidly fatal in swine herds (and

very costly to stock raisers). It can be isolated on blood agar from the dung, urine, and lesions of infected animals. Because the reddish skin lesions are often roughly diamond-shaped, swine erysipelas is sometimes called **diamond disease.**

39.4
GENUS CARYOPHANON

The members of the genus *Caryophanon* (Gr. *caryum* = kernel; *phanus* = conspicuous) are currently classified as gram-negative, unsheathed, long (4 × 40 μm) alga-like trichomes, typically with rounded ends (*Bergey's Manual*, 1957). However, recent studies show that the organisms not only are aerobic but are gram-positive. For these reasons the order Caryophanales and the family Caryophanaceae have been eliminated in the 8th edition of *Bergey's Manual*, and two species, *C. latum* and *C. tenue*, are included in Part XVI, "Gram-Positive, Nonsporeforming Rod-Shaped Bacteria," along with the organisms thus far discussed.

C. latum is one of the largest bacteria (3 × 20 to 30 μm). Grown in pure culture on agar medium, *C. latum* trichomes may form long, unsheathed filaments which contain bacillus-like cells. The individual cells are cylindrical or discoid, and each contains a distinctive, **conspicuous,** disk-shaped or ring-like body that resembles a nucleus (Fig. 39–3). The trichomes in some species are motile with peritrichous flagella. Multiplication is by transverse division of the trichomes and by bacillus-like swarm cells (hormogonia or gonidia) released from the trichomes. None is photosynthetic. They are generally nonpathogenic saprophytes and occur

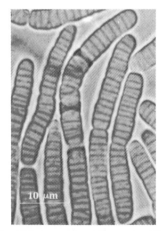

Figure 39–3. Living unstained filaments of *Caryophanon.*

in dung, the oral mucous membranes of mammals, and decaying organic matter.

39.5
FAMILY CORYNEBACTERIACEAE

The name of the family Corynebacteriaceae is derived from the Greek word **koryne** for club, and refers to the bizarre club-shaped cells often formed by some species, notably *Corynebacterium diphtheriae.*

Wide use of the terms "diphtheroid" or "coryneform" bacteria has caused considerable confusion in classification of the organisms conforming to these morphological descriptions. For example, typical coryneform bacteria may belong to one of several groups of bacteria bearing the generic names *Mycobacterium, Nocardia, Jensenia, Listeria, Erysipelothrix, Kurthia, Brevibacterium, Arthrobacter, Cellulomonas,* and *Microbacterium,* as these bacteria display "coryneform" shape during some stage in their growth cycle on artificial media. As a result, the members of the family Corynebacteriaceae display great heterogeneity of cultural, morphological, biochemical, and physiological characteristics: some are motile rods, others nonmotile; some are catalase-positive, others catalase-negative; some are aerobic, others facultatively anaerobic; some are animal pathogens and others are either plant pathogens or common saprophytes distributed in a wide variety of habitats.

According to the 7th edition of *Bergey's Manual,* the family contains six genera: *Corynebacterium,* of which *C. diphtheriae,* the cause of diphtheria, is type species; *Listeria,* of which *L. monocytogenes,* the cause of listeriosis in man and animals, is the only species; and *Erysipelothrix,* of which the cause of swine erysipelas, *E. rhusiopathiae (E. insidiosa),* is the only species. Both *L. monocytogenes* and *E. rhusiopathiae* have been described in previous sections because these organisms have been transferred from the proposed family Corynebacteriaceae. Three other genera, *Microbacterium, Cellulomonas,* and *Arthrobacter* are saprophytes of dairy products, the soil, dust, and the skin, and are common as accidental contaminants of laboratory cultures.

In the 8th edition of the *Manual* radical changes are also made with organisms belonging to the family Corynebacteriaceae and Brevibacteriaceae. These irregular rods (coryneform) which tend to form filaments are grouped into two families, Corynebacteriaceae and Propionibacteriaceae. These two families and

Genera of Uncertain Affiliation are placed in Part XVII, Actinomycetes and Related Organisms. The proposed Family Corynebacteriaceae includes genera *Corynebacterium, Cellulomonas, Arthrobacter (Brevibacterium* and *Microbacterium)* and *Kurthia.* Hence the family Brevibacteriaceae (1957 edition, *Bergey's Manual)* will be eliminated in the 8th edition of the *Manual.*

Members of the family Corynebacteriaceae may be described as gram-positive (or gram-variable), nonsporeforming, generally short, nonmotile, plump rods ranging around 1.0 by 8.0 μm in dimensions and exhibiting various degrees of pleomorphism. All are heterotrophic. Most species grow best on infusion media at 25 to 40C (pH 7.0 to 8.0) particularly if serum (or blood) and glucose are added. Most species are aerobic or facultative and catalase-positive. DNA base compositions of the family extend over a rather wide range, from about 50 to 70 mole per cent GC. This wide per cent GC range may be narrowed considerably with further investigation of the genera in the family.

Genus Corynebacterium. Generic descriptions given here are based on the better-described species which are parasites on, and pathogens of, man and animals, and excludes plant pathogenic species because some or all of the plant pathogens may be improperly classified as members of *Corynebacterium.* Organisms of the genus are chemoorganotrophic, gram-positive, straight to slightly curved rods which frequently show club-shaped swellings with irregularly stained segments. They are nonmotile, are aerobic to facultatively anaerobic, are catalase-positive, and metabolize carbohydrates fermentatively as well as by respiratory means.

Recently, the cell wall composition has been included as one of the major criteria for distinguishing species of *Corynebacterium* from others. The cell walls of the genus contain mesodiaminopimelic acid as the diamino acid of the peptidoglycan, along with a polysaccharide containing arabinose and galactose and often mannose. Some pathogenic species produce powerful exotoxins. The DNA base compositions of most species range from 57 to 60 mole per cent GC. However, DNA base ratios of the plant pathogens appear to be distinctively higher (70 mole per cent GC, or higher) than those of animal and human pathogens. As a result, the species in genus *Corynebacterium* frequently have been divided into three groups: I, human and animal pathogens; II, plant pathogens; III, nonpathogens. The type species representing animal parasites or pathogens is *C. diphtheriae* (see Table 39–5).

Corynebacterium diphtheriae. From the standpoint of human health the two most important species are *C. diphtheriae*, cause of diphtheria, and the anaerobic *C. acnes*, implicated as a partial cause of acne, so often a temporary thorn in the flesh of the young and therefore beautiful.

C. diphtheriae is usually distinguishable by: (1) great variation in length of the cells, from coccoid to spindles or clubs 10 to 15 μm in length; (2) great variation in shape, from club-shaped to sperm-like, needle-shaped or boomerang-shaped forms; (3) irregularity of arrangement often described as "Chinese-character configuration"; (4) conspicuous intracellular granules, bars, and masses of **volutin**, which consists of polymetaphosphate granules (Fig. 39–4) having marked affinity for **methylene blue**. The result of this affinity is that the granules, bars, or the entire cell (depending on the distribution of the material in the cell) stain a *very intense blue or* **metachromatic** (red) color.

Diphtheroids. Species of corynebacteria having more regular length, form, and arrangement, properties that distinguish them from *C. diphtheriae*, are spoken of collectively as diphtheroids (Fig. 39–5). They characteristically arrange themselves side by side in distinctively

Figure 39–5. *Corynebacterium xerosis*, a typical diphtheroid common in the normal throat of man. Compare with *C. diphtheriae* (Fig. 39–4). Note the comparative uniformity in size and shape and the occasional *C. diphtheriae*-like beaded cell. (About 2,000×.) (Drawing of a methylene blue–stained smear.)

regular parallel rows spoken of as "palisade configuration." This is thought to result from their method of fission, in which incomplete breakage of the cell coatings leaves the two cells attached by a fragment that acts as a hinge, like the bark of a partly broken green twig, on which the two cells swing around by a "snapping movement" to lie side by side. Diphtheroids are generally nonpathogenic for man. Two common species found in humans are *C. xerosis* and *C. pseudodiphtheriticum*. Most species of *Corynebacterium* have similar **diphtheroidal** morphology.

The colonies of corynebacteria on solid media such as blood-infusion agar vary considerably in size and appearance, depending on type (gravis, mitis, or intermedius; see below), but are generally white or yellowish, opaque, and round and range in diameter from about 1 mm to 3 or 4 mm. Colonies of pathogenic types may show a narrow band of hemolysis but soluble hemolysin is not produced. The colonies are usually soft and butyrous, but some species form irregular and brittle (R) colonies. Many species, especially the saprophytic species, form brilliant pigments.

Diphtheria. Diphtheria is a specific disease caused by *Corynebacterium diphtheriae*. The organisms are transmitted in the same manner as others causing respiratory disease, i.e., by secretions of the upper respiratory tract. Healthy carriers are not uncommon and are doubtless sources of infection. The bacteria establish themselves on the mucous mem-

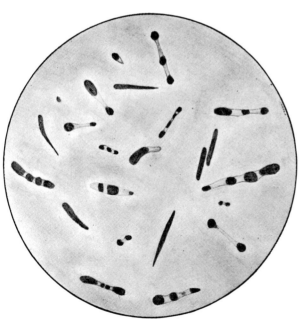

Figure 39–4. *Corynebacterium diphtheriae.* These have been stained with Loeffler's alkaline methylene blue solution. Note the great variation in length, the pleomorphism, and the polymetaphosphate (volutin) arranged as bars and granules and sometimes filling the entire cell. (2,500×.)

TABLE 39–4. CULTURAL CHARACTERS DIFFERENTIATING PRINCIPAL TYPES OF CORYNEBACTERIUM DIPHTHERIAE

Characteristics	Gravis	Intermedius	Mitis
Morphology	Short, irregular rods; few metachromatic granules	Long, barred forms; few granules	Long, curved pleomorphic rods; many granules
Colonies on blood tellurite medium	Low, rough; grey or black center, pale periphery; 2 to 3 mm	Low, rough or smooth; greyish black, 0.5 to 1 mm	Domed, smooth; dark grey to black; 1 to 2 mm
Hemolysis in broth°	−	+	+
Formation of pellicle in broth	+	−	−
Reversal of pH†	Early (7.5 to 8.4)	Late (6.5 to 7.4)	pH does not revert (6.5 to 7.4)
Fermentation of			
Starch	+	−	−
Glycogen	+	−	−
Glucose	+	+	+
Sucrose‡	−	−	−

°Mix 0.5 ml 48-hour broth culture with 0.5 ml 2 per cent washed human erythrocytes.
†In absence of glucose, after 5 days at 37C.
‡A few sucrose-fermenting strains have been reported.

branes, most commonly of the tonsils or pharynx and larynx and nares. In a susceptible person the bacilli set up an intense local inflammation with swelling and fibrinous, purulent exudate that forms a spreading **pseudomembrane**. As the bacteria grow they release diphtherial exotoxin.

Diphtherial Exotoxin. An organism's ability to produce toxins is determined by the presence of prophage carrying a specific determinant called *tox+*. A nontoxigenic strain (mitis variety) may be converted to a toxigenic strain by lysogenizing the cell with the temperate phage carrying *tox+* determinant.

The toxin of *C. diphtheriae* is one of the most powerful biological poisons known. When the organisms infect the mucous membranes, this poison is absorbed by the blood and damages heart, kidneys, adrenals, and nerves, and may cause death unless antitoxin is (a) already present as a result of previous infection or artificial immunization or (b) is developed rapidly by the cells of the patient as an anamnestic or secondary response or (c) is injected into him in the form of antitoxic serum from some outside source (passive immunity). The toxin is also very poisonous to rabbits, mice, and guinea pigs and to chicks and other birds.

Corynebacterium diphtheriae is a classical illustration of virulence depending almost entirely on toxigenicity. This organism, although often growing extensively on the surfaces of the respiratory and other mucous mem-

branes, has little ability to invade the tissues beyond the mucous membranes.

Cultural Types of C. diphtheriae. Three distinct cultural types, **gravis, intermedius,** and **mitis**, are recognized in accordance with the clinical severity of the cases from which the different types are most frequently isolated (Table 39–4). One type formerly known as **minimus** is now eliminated on the basis that the strains of the type are identical with intermedius. The gravis types were thought to cause more severe and fatal diphtheria than the other types, but extensive studies in the United States have failed to support that view, since no constant correlation between cultural type and clinical severity of diphtheria has been found. All produce the same toxin.

Immunity to Diphtheria. Some persons possess **natural, active,** specific immunity to diphtheria, resulting from **subclinical** attacks during early childhood. They retain their immunity throughout life, probably as a result of repeated reinfection. Their blood usually, but not always, contains a small amount of antitoxin, which helps combat ordinary infection.

Active Artificial Immunization. Because today there are few cases and carriers of diphtheria, **natural** active immunization and repeated reinfections are much diminished. It is therefore desirable, and sometimes required by law for community protection, to immunize **artificially.** The antigens used consist of harmless toxoids and are generally mixed with antigens against whooping cough and tetanus.

Any physician or health department will do this on request. The process requires only two (preferably three) injections of toxoid, alum-precipitated, given four to six weeks apart. Since this immunity wears away in time, it is customary to reimmunize with "booster doses" one year later, and then every two to three years through 10 to 12 years of age. There are various modifications of this regimen.

Passive Immunization. A child exposed to the disease by proximity to a diphtheria patient may have immediate need of antibodies to combat the disease or ward off infection. Endangered persons may, under special circumstances to be judged by the physician, receive immediate protection through injections of serum that contain large quantities of antitoxin. Injections of serum for any purpose are always to be avoided if possible because of dangers of allergy (Chapt. 29).

Other Parasitic Species of Corynebacte- **rium.** Species of genus *Corynebacterium* that are of importance to the farmer and veterinarian are: *C. pyogenes* (58 to 59 mole per cent GC), which is common in purulent lesions of cattle, swine, and sheep; *C. equi,* causing pneumonia in foals; and *C. renale,* which causes a necrotic disease of the urinary tract in domestic livestock.

Currently, the taxonomic position of *C. pyogenes* (Color Plate III, *B*) is questioned because the organisms differ from the type species in a number of important characteristics. For example, *C. pyogenes* is catalase-negative, has a markedly different cell wall composition, produces soluble hemolysin, and apparently shares cell wall polysaccharide antigen with streptococci of Lancefield group G. Thus the organism, while well recognized as the cause of pyogenic infections in domestic animals, has little similarity to other pathogenic corynebacteria. Some distinctive features of species of *Corynebacterium* parasitic on man and animals are summarized in Table 39–5.

Currently the genus *Corynebacterium* in-

TABLE 39–5. SOME DISTINCTIVE FEATURES OF SPECIES OF CORYNEBACTERIUM PARASITIC ON MAN AND ANIMALS

Species	DNA Base Composition (Mole Per Cent G + C)	Hemolysis	Fermentation of Sucrose	Reduction of Nitrate	Gelatinase	Urease	Other Features
C. diphtheriae	52–60	+	−	+	−	−	Pathogenic: specific exotoxin; specific bacteriophages. See text for detailed discussion on pathogenesis.
C. pseudotuberculosis	67.5	+	d°	d°	d°	d°	Pathogenic: specific exotoxin. Causes ulcerative lymphangitis, abscesses, and other purulent infections in sheep, goats, horses, and man.
C. xerosis	55–59	−	+	+	−	−	Not pathogenic. Barred morphology often predominant.
C. renale	53–57	−	−	−	−	+	Pathogenic: causes pyelitis and cystitis in experimental animals and cattle. Zone of clearing on milk agar; alkaline reaction and burgundy-color in litmus milk.
C. kutscheri	57–60†	−	+	−	−	+	Specific parasite of rats and mice.
C. pseudodiphtheriticum	57–60†	−	−	+	−	+	Short, regular rods; strongly gram-positive; no acid produced from any carbohydrate.
C. equi	57–60†	−	−	+	−	−	Oval or coccoid forms on solid media; capsulated; salmon-pink pigment; no acid produced from any carbohydrate; bronchopneumonia in foals, isolated from aborted equine fetuses.
C. bovis	57–60†	−	−	−	−	+	Lipolytic for butterfat; will grow in media containing 9 per cent NaCl; requires long-chain fatty acids; may be cause of bovine mastitis.

°Variable results reported in the literature.
†Probably the majority fall into the rather narrow range of 57 to 60 mole per cent G + C.

cludes a number of important plant pathogens, such as *C. michiganese*, the cause of tomato wilt and canker, and *C. insidiosum*, cause of a destructive disease of alfalfa. As stated earlier, most plant-pathogenic corynebacteria conform morphologically to the genus description, but they differ significantly in other properties. For example, some are motile, most species show differences in cell wall composition, and the DNA base ratio of the plant pathogens ranges from 68 to 75 mole per cent GC, which is distinctly higher than that of the animal and human pathogens (57 to 60 mole per cent GC).

39.6
OTHER GENERA OF CORYNEBACTERIACEAE

Genus Arthrobacter. *Arthrobacter* spp. are widespread in soil and water and nutritionally are extremely versatile, decomposing various and unusual organic compounds, including some herbicides. The organisms exhibit a characteristic life cycle in complex growth media. The cells display morphological alteration from spheres to rods when inoculated into fresh medium and subsequently revert to coccoid form at the end of exponential growth. In some stages the rods seem to exhibit hinge arrangements like corynebacteria. The morphological variation may also be influenced by composition of the culture medium, as shown in Figure 39–6.

Cell wall compositions of the organisms are similar to *Corynebacterium* species which contain mesodiaminopimelic acid as the diamino acid of the peptidoglycan. The morphological change (rods to spherical cells) occurring at the end of the exponential growth phase is apparently attributed to modification in cross-linking of the cell wall peptidoglycan. The spherical forms, like *Azotobacter* cysts, are more resistant to drying than are rod-shaped cells. Some strains of *Arthrobacter* are apparently able to carry out heterotrophic nitrification (oxidizing ammonium to hydroxylamine, and to a bound hydroxylamine substance presumed to be a primary nitro compound, nitrite or nitrate). The DNA base compositions of arthrobacters range from 60 to 64 mole per cent GC.

Genus Cellulomonas. The members of the genus *Cellulomonas* are similar to either brevibacteria or kurthiae in many respects. Some are motile with peritrichous flagella, and some nonmotile, but all differ from other coryneform bacteria in their ability to decom-

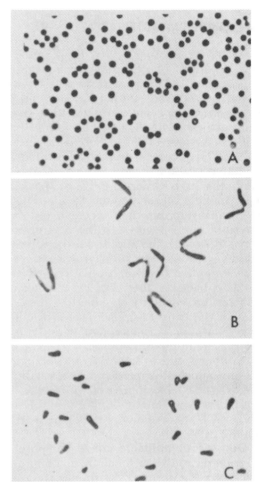

Figure 39–6. Phase-contrast photomicrographs of *A. crystallopoietes,* showing various cell shapes. All are taken at mid-logarithmic phase but grown in different media. (1,600×.)

pose cellulose. The organisms are common inhabitants of soil and play a role similar to cellulose-hydrolyzing species of *Cytophaga* in our ecosystem.

Genus Kurthia. The species of *Kurthia* are motile with peritrichous flagella and are aerobic to facultatively anaerobic. The members of the genus are unable to ferment any carbohydrates. One species, *K. catenaforma,* is able to produce a large amount of the amino acid proline from either aspartic or glutamic acid; commercial-scale production of proline by this organism is being investigated. The organism has been isolated from human feces and poultry manure, and the organisms are believed to be widely distributed in putrid organic matter.

39.7
FAMILY PROPIONIBACTERIACEAE

Members of the family Propionibacteri-aceae are much like *Lactobacillus:* gram-positive, nonmotile, nonsporeforming. They are generally short rods with rounded ends but may assume very pleomorphic, club-shaped, and branched forms. Unlike the family Lacto-bacillaceae, most of these organisms produce catalase. They are nutritionally fastidious, are facultative anaerobes, and are heterofermenta-tive. In the 1957 edition of *Bergey's Manual* the family included three genera: *Propionibacteri-um, Butyribacterium* (one species), and *Zymo-bacterium* (one species). In the 8th edition of *Bergey's Manual*, the family will be altered to exclude two genera, *Butyribacterium* and *Zymobacterium*, because, unlike typical propi-onic acid bacteria, they are catalase-negative. *Eubacterium*, currently classified as the second genus of the tribe Lactobacilleae (7th edition, *Bergey's Manual*), is included with Propioni-bacteriaceae in the 8th edition.

Although these bacteria morphologically and physiologically resemble lactobacilli, the DNA base compositions of propionic acid bacteria are considerably higher (58 to 70 mole per cent GC) than of any of the lactic acid bac-teria (32 to 52 mole per cent GC).

Genus Propionibacterium. The propionic acid bacteria, unlike lactobacilli, do not produce lactic acid but ferment it, producing carbon dioxide, acetic acid, and propionic acid. The ability to ferment carbohydrates with produc-tion of propionic acid is considered a dis-tinguishing characteristic of the genus, along with DNA base composition (58 to 68 mole per cent GC).

Species of *Propionibacterium* are common in dairy products. *P. freudenreichii* (type species), *P. shermanii*, and some others are com-monly found in various hard cheeses. During the ripening of these, *Propionibacterium* con-tributes flavors with its fermentation products, among them propionic acid. For example, the distinctive sweet and bitter flavor of Swiss cheese is thought to be due, in part at least, to propionates. The "eyes" of Swiss cheese are largely caused by carbon dioxide. Pure cultures of *Propionibacterium* are added to milk in the manufacture of Swiss, Münster, and similar cheeses.

Genus Eubacterium. Members of the genus are nonmotile, straight or curved rods occurring in very short chains; the cells neither form filaments nor branch, and they ferment carbohydrate with production of formic, acetic, propionic, butyric, and lactic acids. Many species grow in peptone water, produce gas, and liberate volatile, offensively odorous amines and ammonia. Some species are hemo-lytic and pathogenic to small animals and have been isolated from a variety of gangrenous and suppurative infections in various parts of the body. Currently, twenty species are recognized. Separation of species is based on production of CO_2 from culture media and whether the organism produces a fetid odor. Taxonomic maintenance of these 20 species is questioned, as differences among species are superficial.

Other Coryneform Bacteria. In the 8th edition of *Bergey's Manual*, genera *Brevibac-terium* and *Microbacterium* are placed in "Gen-era of Uncertain Affiliation" in Part XVII, along with families Corynebacteriaceae and Pro-pionibacteriaceae.

Genus Microbacterium. Members of this genus are related to *Arthrobacter* or to plant-pathogenic corynebacteria. The taxonomic posi-tion of the genus is quite nebulous, and it has been proposed that most members of the genus, except *M. thermosphactum* (36.1 mole per cent GC), which resembles streptococcus (33 to 44 mole per cent GC), should be transferred to either *Arthrobacter* or *Corynebacterium* on the basis of DNA base composition (58 to 64 mole per cent GC).

The organisms have been frequently iso-lated from raw and pasteurized milk and from milk products, cheese, and butter-making utensils. Some of the organisms are thermoduric and survive exposure to temperatures between 80 and 85C for 10 minutes; hence, they are among the most heat-tolerant of the nonspore-forming organisms. The role of the organisms in nature is not clearly understood. They are a nuisance in pasteurization control because they survive proper pasteurization. However, they are not pathogenic.

Genus Brevibacterium. The brevibacteria are similar to *Propionibacterium* in being short, pleomorphic, gram-positive, nonsporeforming rods. A few species are motile with peritrichous flagella. Unlike typical lactic-acid and pro-pionic-acid bacteria, brevibacteria commonly produce red, yellow, or brown pigments, and do *not* ferment lactose. Like *Lactobacillus* and *Propionibacterium*, species of *Brevibacterium* are widely distributed in nature and occur in dairy products, where they are important mem-bers of the cheese-making community. Unlike anaerobic or microaerophilic *Lactobacillus* and *Propionibacterium*, brevibacteria are aerobic and grow well on outer surfaces. For example, *Brevibacterium linens* (type species) grows with a yellow-orange pigment on the outside surfaces of ripening cheeses such as Limburger and Camembert. Since the organism is proteolytic, it contributes to the softening and flavoring of such cheeses.

CHAPTER 39
SUPPLEMENTARY READING

Blair, J. E., Lennette, E. H., and Truant, J. P. (Eds.): Manual of Clinical Microbiology. American Society for Microbiology. The Williams & Wilkins Co., Baltimore. 1970

Johnson, J. L., and Cummins, C. S.: Cell wall composition and deoxyribonucleic acid similarities among the anaerobic coryneforms, classical propionibacteria, and strains of *Arachnia propionica*. J. Bact., *109*:1047, 1972.

Miller, A., III, Sandine, W. E., and Elliker, P. R.: Deoxyribonucleic acid base composition of lactobacilli determined by thermal denaturation. J. Bact., *102*:278, 1970.

Pappenheimer, A. M., and Gill, D. M.: Diphtheria. Recent studies have clarified the molecular mechanisms involved in its pathogenesis. Science, *182*:353, 1973.

Peppler, H. J. (Ed.): Microbial Technology. Reinhold Book Co., New York. 1967.

Rogosa, M., Franklin, J. G., and Perry, K. D.: Correlation of the vitamin requirements with cultural and biochemical characters of *Lactobacillus*. J. Gen. Microbiol., 25:473, 1961.

ORDER ACTINOMYCETALES (MOLD-LIKE BACTERIA)

• **CHAPTER 40**

The name of the order Actinomycetales is derived from the term **Actinomyces** (Gr. *actino* = radial emanations, e.g., sunlight; *mykes* = fungus; hence "ray-fungus"). The organisms are sometimes referred to collectively as **actinomycetes.** At one time the actinomycetes were thought to be fungi because of their mycelial structure and formation of conidia on aerial branches, both of which are very similar to filamentous fungi. However, it has been demonstrated not only that they are procaryons but also that many aspects of cell structure and chemical composition are more closely related to unicellular bacteria than to true fungi. The term *Actinomyces* was first used by Harz in 1878 to describe an organism found in the pus of cattle suffering from the disease now called **actinomycosis** or "lumpy jaw." Harz used the term *Actinomyces* as descriptive of the **radial** arrangement of the branching, mold-like threads of the organism when growing in infected tissues. The term has since been adopted as the name of one genus of the order Actinomycetales.

Bergey's Manual, 7th edition, includes four families: Mycobacteriaceae, Actinomycetaceae, Streptomycetaceae, and Actinoplanaceae (Table 40–1). In the 8th edition, the order Actinomycetales has been placed in Part XVII, "Actinomycetes and Similar Organisms," which also includes families Corynebacteriaceae and

TABLE 40-1. THE ORDER ACTINOMYCETALES (TRUE-BRANCHING BACTERIA)*

Family: Actinomycetaceae (limited mycelia that fragment readily)
    Genus: *Actinomyces* (anaerobic; cause actinomycosis) ⎫
    Genus: *Nocardia* (aerobic; some cause nocardiosis) ⎬ (slightly acid-fast)
 	
Family: Mycobacteriaceae (mycelial character limited to occasional branching cells: T, Y)
    Genus: *Mycobacterium* (strongly acid-fast; tuberculosis, leprosy)
    Genus: *Mycococcus* (not acid-fast)
	No
	Conidia
	or
	Spores

Family: Streptomycetaceae (filament and mycelium formers; do not readily fragment)
    Genus: *Streptomyces* (well-developed aerial mycelia; curled chains of conidia; many produce
        antibiotics)
    Genus: *Micromonospora* (single conidia; subsurface mycelia)
    Genus: *Thermoactinomyces* (50 to 65C)
Family: Actinoplanaceae (spores formed in sporangia)
    Genus: *Actinoplanes* (no aerial mycelium; spores motile)
    Genus: *Streptosporangium* (much aerial mycelium; spores nonmotile)
	Conidia
	or
	Spores
	Present

*Classification based on 7th edition of *Bergey's Manual.*

584

TABLE 40–2. THE ORDER ACTINOMYCETALES (TRUE-BRANCHING BACTERIA)

Family I.	Actinomycetaceae	Produce a well-defined mycelium which later fragments into simple branched or unbranched rods, diphtheroid and coccoid forms; nonmotile, aerobic, anaerobic to microaerophilic; some species cause actinomycosis; some species are thermophilic; composed of 5 genera.
Family II.	Mycobacteriaceae	Cells may be unbranched or have rudimentary branching; nonmotile; acid-fast (hydrophobic—some species do not stain readily with water-soluble dyes); aerobic; some species cause tuberculosis, leprosy; composed of a single genus.
Family III.	Frankiaceae	See 8th edition, *Bergey's Manual.*
Family IV.	Actinoplanaceae	Filamentous, branching mycelium retained even with age; conidia produced in coiled chains or irregularly arranged within sporangia; conidia are motile for some species and germinate and produce branched mycelium; aerobic; composed of 10 genera.
Family V.	Dermatophilaceae	Branching mycelium produced from coccoid bodies, undergoes initial subdivision by transverse septa 5 to 30 μm behind tip of the hypha as growth proceeds, subsequently progressive division of each cell by further transverse septa until individual cells are 0.3 to 0.5 μm long; each cell enlarges to form sarcina-like packet of coccoid cells (some are motile cocci, zoospores); some cause dermatitis; contains 2 genera.
Family VI.	Nocardiaceae	Produce well-defined mycelium which later fragments completely into branched or unbranched bacillary elements; aerobic; some are motile with flagella; acid fastness occurs in some species; some species cause nocardiosis; composed of 2 genera.
Family VII.	Streptomycetaceae	Filamentous mycelium usually remains intact with age; chains of conidia are produced from aerial hyphae; sporangia are not produced; aerobic; many produce antibiotics; composed of 4 genera.
Family VIII.	Micromonosporaceae	Filamentous, branching substrate mycelium only produced in young culture which does not disintegrate with age; conidia are produced singly, doubly, or more; some are thermophilic; some conidiophores are dichotomously branched; composed of 6 genera.

*While families Corynebacteriaceae and Propionibacteriaceae are placed in Part XVII, Actinomycetes and Similar Organisms, in the 8th edition of *Bergey's Manual*, they are not included in this table.

Propionibacteriaceae (see Chapter 39). The order Actinomycetales has been considerably enlarged and contains families Actinomycetaceae, Mycobacteriaceae, Frankiaceae, Actinoplanaceae, Dermatophilaceae, Nocardiaceae, Streptomycetaceae, and Micromonosporaceae (Table 40–2).

The Actinomycetales differ from most other bacteria in that **true branching**, a distinctly mold-like character, is normally found in all of its species. None of the Actinomycetales is sheathed, stalked, or photosynthetic; none appears to accumulate sulfur, iron, or other free elements in or on the cells. All of the Actinomycetales are chemoorganotrophic, although many will grow on simple mineral media with an organic carbon source. The organisms in the order include aerobic, facultatively anaerobic, and strictly anaerobic species; some possess catalase, while others completely lack this enzyme.

All are gram-positive. Only a few species are motile. All but a few species are harmless saprophytes, widely distributed in decomposing organic matter in soil, dung, and marine and fresh waters. None forms endospores, although some species form mold-like conidia (or conidiospores) borne on **conidiophores** (or **sporophores**) (Table 40–1).

Molds and Actinomyces. Striking and constant differences between the mold-like Actinomycetales and true molds (Eumycotina) are: (1) the minuteness of the filaments of Actinomycetales (1 to 5 μm in diameter and seldom more than a few millimeters in length; true molds range around 10 to 20 μm in diameter and their mycelia are often several inches in length; (2) the procaryotic structure of Actinomycetales (true molds are eucaryotic); (3) the cell walls of Actinomycetales contain peptidoglycan, like bacteria, and are not chitinous as in true molds and contain both muramic and diaminopimelic acids, found only in bacteria; (4) sexual phenomena such as occur in many molds are absent.

40.1
FAMILY ACTINOMYCETACEAE

In the 7th edition of *Bergey's Manual*, the family Actinomycetaceae includes two genera of mycelium-formers: the aerobic genus *Nocardia* and the strictly anaerobic genus *Actinomyces*. There have been changes since 1957; for example, in the 8th edition of *Bergey's Manual* the family includes five genera: *Actinomyces, Arachnia, Bifidobacterium, Bacterionema,* and *Rothia.* The DNA base composition of the family (in part) ranges from 43 to 70 mole per cent GC. All of these organisms fragment into simple branched or unbranched filaments. The fragmented cells are often diphtheroid or coccoid in form. The organisms of this family are nonmotile, are non-acid-fast, and lack aerial mycelia and spores. They usually grow as facultative anaerobes, but some are anaerobic while others are aerobic; hence, ability to form catalase may or may not be present.

Genus Actinomyces. The filaments of the genus *Actinomyces* produce true though limited branching, and the branched rods may form V, Y, and T forms (Fig. 40–1). The organisms as a group are considered facultative anaerobes, but most grow preferentially as anaerobes, with a few species being aerobic. Hence organisms are frequently isolated in an atmosphere of 80 per cent N_2, 10 per cent H_2, and 10 per cent CO_2; or 95 per cent N_2 and 5 per cent CO_2. Unlike organisms in Eubacteriales (7th edition, *Bergey's Manual*), the cell walls of all species currently recognized as *Actinomyces* lack diaminopimelic acid (DAP) and the majority of organisms lack arabinose.

The 8th edition of *Bergey's Manual* recognizes five species: *A. bovis, A. odontolyticus, A. israelii, A. naeslundii,* and *A. viscosus.* The first four organisms are facultative anaerobes and catalase-negative; the last, facultatively aerobic and catalase-positive. All are chemoorganotrophs requiring a complex medium for growth, and certain species are pathogenic for man and other animals. The DNA base composition of genus *Actinomyces* ranges from 43 to 60 mole per cent GC.

Actinomyces bovis, the cause of **lumpy jaw** or **actinomycosis** in cattle, is one of the few Actinomycetales rarely if ever found free in the environment. It occurs as a normal inhabitant of the oral cavity in cattle and other animals and man. However, infection in man has not been established. *A. bovis* may be introduced into the flesh of the jaws or tongue of cattle by thorns, splinters, and the like. Swellings are produced by the growing *Actinomyces,* and the surrounding tissues become hard and indurated; hence the confusion with **wooden tongue** caused by *Actinobacillus lignieresii* (see Chapter 35).

The hard texture of the lesion may be due to precipitation of calcium phosphate by the organisms. Eventually the infected tissue becomes riddled with abscess-like cavities which are filled with pus.

Sulfur Granules. The organisms, in contact with the defensive substances in the tissues, grow in compact, granular masses that acquire a bright yellow color. They are commonly called **sulfur granules.** Crushed, they are seen with the microscope to consist of masses of mycelium held in a hard matrix of a polysaccharide-protein complex that is rich in calcium phosphate. The embedded mycelia tend to send out filaments radially; hence the origin of the term "ray-fungus." The tips of the filaments appear at the surface of the granules. Coated with the cement-like calcified substance, they appear club-shaped.

A. israelii is the causal agent of human actinomycosis and occasionally of infections in cattle. "Sulfur granules" are produced in both human and animal infections.

Genus Arachnia. The genus *Arachnia* was newly created for the organisms resembling those in genus *Actinomyces* but differing markedly by having diaminopimelic acid in their cell walls and fermenting glucose anaerobically, with propionic acid as the major product. Certain species of the genus are pathogenic for man, causing lacrimal canaliculitis and typical actinomycosis.

Genus Bifidobacterium. Previously this organism was classified as *Lactobacillus bifidus,* which was apparently a composite description of *B. bifidum* and *L. acidophilus.* The organism has been removed from the genus *Lactobacillus* and placed as a genus in the family Actinomycetaceae. *Bifidobacterium,* like *Actinomyces,* is catalase-negative, is a microaerotolerant anaerobe, and ferments sugars, with the formation of lactic acid as the end product. The characteristics of this genus are that the cells are swollen, irregular, forked or branched (bifurcated Y and V forms; Fig. 40–2), and frequently described as coryneform. The morphology is readily influenced by nutritional conditions. DNA base compositions of both homofermentative and heterofermentative lactobacilli are similar, with both types having 32 to 52 mole per cent GC; *Bifidobacterium* species range from 57 to 63 mole per cent GC, a range which is considerably higher than that of the organisms of the genus *Lactobacillus.*

Bifidobacterium is a predominant inhabitant of the intestinal tract in breast-fed human infants, but it is quickly displaced by other in-

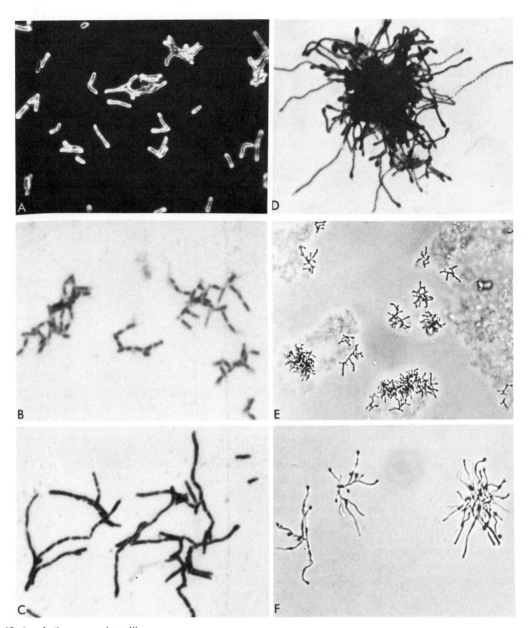

Figure 40–1. *Actinomyces israelii.*

A, Darkfield from 72-hour growth in Thioglycollate Broth, showing V and Y forms and an indication of granular cytoplasm. (1,260×.)

B, Gram stain from 72-hour growth in Thioglycollate Broth, showing irregularly stained rods of variable length and angular arrangements. (1,580×.)

C, Gram stain from 72-hour growth in Thioglycollate Broth, showing elongated filaments, branching, and some irregular staining. (1,890×.)

D, Gram stain from 72-hour growth in Thioglycollate Broth, showing a mass of branching intertwined filaments. Some filaments irregularly stained and some with small bulbous ends. (1,240×.)

E, Unstained 24-hour microcolonies of *A. israelii* growing on BHI, showing filaments with multiple short angular branches. Resembles *A. naeslundii* microcolonies. (410×.)

F, Unstained 24-hour microcolonies growing on BHI, showing branching, filamentous or "spider" colonies with no distinct center. Similar colonies may be formed by *A. propionicus,* rough *A. odontolyticus,* and rough *A. bovis.* (410×.) (Courtesy of J. M. Slack, S. Landfried, and M. A. Gerencser.)

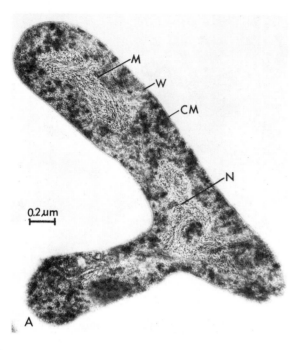

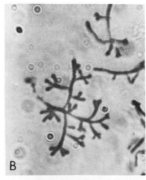

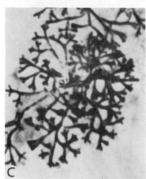

Figure 40–2. *A*, A section of the bifid form of *Bifidobacterium (Lactobacillus) bifidus.* No cross-wall can be seen. *N*, nucleoid; *W*, cell wall; *CM*, cell membrane; *M*, mesosome. (38,-000×.) Development of cellular branching induced by NaCl at 18 hours (*B*) and 22 hours (*C*) after the beginning of incubation. (1,500×.) (Courtesy of M. Kojima, S. Suda, S. Hotta, K. Hamada, and [*A*] A. Suganuma.)

testinal flora after infants are weaned. The organism requires a very specific growth factor, N-acetylglucosamine, which is present in human milk but not in cows' milk. The student may recall that N-acetylglucosamine is the precursor of cell wall peptidoglycan, and, apparently, **bifid** or coryneform morphology of this organism results from an insufficient supply of the growth factor. When the organism is grown in a medium containing an excess of N-acetylglucosamine, the cell structure becomes much more regular.

Genera Bacterionema and Rothia. These organisms possess morphological and physiological similarities to other genera of the family Actinomycetaceae. Morphologically, species of *Bacterionema* are characterized by attachment of the bacillus to a filament ("whip handle"), and reproduction by filament septation and fragmentation to form bacilli which germinate to produce one to four filaments. The organisms include facultative aerobes and obligate anaerobes. Their cell walls contain both diaminopimelic acid and arabinose. However, *Rothia* grows best aerobically; the major product of glucose fermentation is lactic acid. The cell walls of *Rothia* species, like those of *Actinomyces*, do not contain diaminopimelic acid or arabinose. The DNA base composition ranges from 65 to 70 mole per cent GC. The habitat of organisms in both genera is the oral cavity of

man, and they may be readily isolated by inoculating human dental plaque or crushed calculus into various complex organic media.

40.2
FAMILY MYCOBACTERIACEAE

This family contains only the bacterium-like genus *Mycobacterium*, which includes soil-living species, some of which are minor pathogens (all often grouped as "atypical acid-fasts"), and two highly pathogenic species: *Mycobacterium tuberculosis*, cause of tuberculosis, and the closely related *Mycobacterium leprae*, cause of leprosy. The mycobacteria rarely form true mycelia but sometimes exhibit limited branching. Any short filaments that are formed tend to break up at once into bacillus-like fragments. The organisms are nonmotile, nonsporulating, aerobic rods. They are gram-positive. The DNA base composition of the family ranges from 60 to 65 mole per cent GC.

Species of *Mycobacterium* are distinguished from all other bacteria by a peculiar staining property called **acid fastness** (Chapt. 3). When cells are stained with such dyes as basic fuchsin, they resist decolorization with either dilute acid or acid alcohol, while other bacteria are readily decolorized. This property is found only in the mycobacteria and a few related species of the order Actinomycetales, especially some nocardias. Even the staining of the nocardias is not like the definite, strong and constant acid fastness of *Mycobacterium*. Acid fastness is conferred by the presence of large amounts of waxy substances (on the cells) with which the dye molecules couple tightly.

Morphologically mycobacteria are much like ordinary bacteria except for being somewhat curved and spindle-shaped, and sometimes beaded or granular, the beads and granules probably being artifacts resulting from staining processes. None forms conidia or aerial hyphae. Most species are harmless saprophytes living in the soil, where they are very important as scavengers that bring about decomposition of a wide variety of complex organic compounds. Saprophytic mycobacteria may also be found in acid-fast–stained smears of material from preputial secretions (*Mycobacterium smegmatis*) or from folds of the skin, as in the buttocks or axillae. Other saprophytic acid-fast bacilli (*M. phlei*; see Figure 10–4) are found in dust, soil, butter, and manure or on hay.

Cultivation. Mycobacteria are commonly cultivated on rich, solid organic media (Petroff's, Petragnani's, Löwenstein-Jensen) made with eggs, milk, and potatoes and containing selectively inhibitory dyes. Any species may also be cultivated in simple, aqueous, mineral solutions containing an organic source of carbon or nitrogen, e.g., ammonium citrate plus glucose and bovine albumin. The surface tension of the fluid must be lowered with a surfactant substance such as Tween 80 (a surface tension–reducer, a polyoxyethylene derivative of sorbitan monooleate, a sort of liquid soap) or WR1339, so that the fluid **wets** the waxy bacilli. Most species, both saprophytes and pathogens, grow rapidly in such fluids with diffuse turbidity. Virulent tubercle bacilli distinctively form long, tangled **cords** of growth on the surface of fluid or solid media (Fig. 40–3). They also bind the dye **neutral red.** This is of ancillary value in diagnosis.

Mycobacterium tuberculosis. The organism causing tuberculosis, *M. tuberculosis* (DNA base composition, 60 to 65 mole per cent GC), is typical of the genus *Mycobacterium*. It is the most strongly acid-fast species of the genus. It was first isolated and shown to be the cause of tuberculosis by R. Koch in 1882.

Tubercle bacilli may remain alive for long periods outside the human or animal body. In dried sputum in dark corners they may live

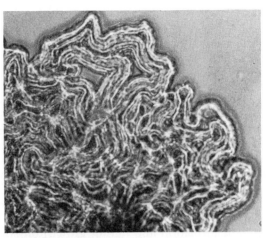

Figure 40–3. *Mycobacterium tuberculosis,* human strain at 21 days on WR 1339 medium, showing thick cords formed. (100×.)

six to eight months; in particles of dried and powdered sputum which can float through the air as dust, they can remain alive for days and may be inhaled. Exposure to sunlight for a few hours kills them; so does pasteurization. This is of importance since they can infect cows' udders and hence their milk, thus transmitting the disease from the cattle to persons who drink unpasteurized or improperly supervised (uncertified) raw milk.

Types of Tubercle Bacilli. There are several kinds of tubercle bacilli, varying according to the animal infected. For example, there is the human type, *Mycobacterium tuberculosis* (or *M. tuberculosis* var. *hominis*) and the bovine type (Color Plate VI, *D*), *M. bovis* (or *M. tuberculosis* var. *bovis*). The BCG bacillus used for vaccination (Chapt. 28) is a modified bovine strain. A third mammalian type, called the **vole bacillus** (*M. microti*), was discovered in 1937. This organism is highly virulent for voles but produces only a mild **immunizing** infection in man and has been used in place of BCG. (See Chapter 28.) There is also a bird or avian type (*M. avium*), which grows well at about 40C and fish (*M. marinum*, DNA base composition 63 to 64 mole per cent GC) and other cold-blooded animal types that grow well at lower temperatures (18 to 30C). All look alike microscopically and may be cultivated on similar media. The cold-blooded animal types do not as a rule infect the warm-blooded animals or birds, and vice versa, a good example of **species** (not of **specific**) resistance to infection.

Atypical Acid-Fasts. Some species of

mycobacteria (e.g., *M. fortuitum* [Color Plate VII], *M. intracellularis, M. kansasii* [DNA base composition 60 to 65 mole per cent GC]) are found associated with lesions in the lung resembling tuberculosis and with nodular lesions in cattle and other animals. These mycobacteria are spoken of by medical microbiologists as **atypical, unclassified,** or **Runyon groups** of **acid-fasts.** Their etiological relationship to such lesions has long been in question, but organisms of Groups I and possibly III of the Runyon classification appear to be definitely pathogenic for man.

The atypical acid-fasts have been divided into several groups, primarily on the basis of pigment formation and rate of growth on solid media.

Certain species of the genus *Mycobacterium* have been reported to produce soluble pigment on various media. For example, strains derived from *M. fortuitum* have been reported to produce pigment varying from straw-colored to black without change in other morphological and physiological characteristics of the parent strain. Dark pigment of the strain is considered an unstable property of the strain (Color Plate VII).

Differentiation between Mycobacteria. In the control of tuberculosis, both human and bovine, it is clearly necessary to differentiate between the pathogens, possible pathogens such as the atypical acid-fasts, and harmless saprophytes. Several tests used in such differentiations are summarized in Table 40–3. Space does not permit a detailed description of these tests but data may be found in the literature cited. It is important to note that, as in many microbiological differentiations, no *single* test is wholly reliable by itself.

The atypical acid-fast bacilli produce a tuberculin type of allergy detectable with tuberculin-like antigens prepared from the appropriate strains, but often cross-reacting with tuberculin from *M. tuberculosis.*

TABLE 40–3. DISTINCTIVE PROPERTIES OF VARIOUS MYCOBACTERIA*†

Species or Subgroup	Niacin Produced	Nitrate Reduction		Catalase Produced (mm)			Tween Hydrolysis (Days)		Tellurite Reduction in 3 Days	Pigment Formation		Growth on 5 Per Cent NaCl	Growth in Less Than 7 Days at 37C	Produce Arylsulfatase Enzyme +3 Days	Growth on MacConkey Agar
		>1+	>3	>40	>50	68+	+5	+10		DARK	LIGHT				
M. tuberculosis•	++	++	++	++	−	−	∓	∓	−	−	−	−	−	−	−
M. bovis	−	−	−	++	−	−	−	−	−	−	−	−	−	−	−
Runyon Group I (Photochromogenic) (lemon-yellow pigment)															
M. kansasii•	−	++	++	++	++	++	++	++	−	−	++	−	−	−	−
M. marinum•	−	−	−	±	∓	−	++	++	−	−	++	−	∓‡	−	−
Runyon Group II (Scotochromogenic) (yellow-orange to dark red)															
M. sp. *scrofulaceum*	−	∓	−	++	++	++	−	−	−	++	++	−	−	−	−
M. sp. *aquae*	−	−	−	++	++	++	+	++	−	++	++	−	−	−	−
M. flavescens	−	++	+	++	++	++	++	++	−	++	++	±	−	−	−
Runyon Group III (no pigments in light)															
M. avium§•	−	−	−	++	−	++	−	−	±	−	∓	−	−‡	−	−
M. intracellulare§•	−	−	−	++	−	++	−	−	+	−	−	−	−‡	−	∓
M. xenopei	−	−	−	++	−	++	−	−	−	∓‖	∓‖	−	−	±	−
M. gastri	−	−	−	++	−	−	++	++	−	−	−	−	−	−	−
M. terrae complex	−	++	+	++	++	++	++	++	−	−	−	−	−	−	−
"V" (*M. triviale*)	−	++	+	++	++	++	+	++	−	−	−	++	−	∓	−
Runyon Group IV (rapid growers) (some scotochromogens)															
M. fortuitum	−	±	±	++	++	++	−	∓	++	−	−	++	++	++	++
M. smegmatis	−	++	∓	++	++	++	++	++	++	−	−	++	++	−	−
M. phlei§	−	++	∓	++	++	++	++	++	++	++	++	++	++	−	−
M. vaccae§	−	++	+	++	++	++	±	++	+	++	++	++	++	−	−
M. borstelense	V	−	−	++	++	++	−	−	−	−	−	−	++	++	±
(*M.*) *rhodochrous*	−	++	±	++	++	++	−	++	±	++	++	++	++	−	±

*Modified after Current Item No. 165, Laboratory Program, 1968. National Communicable Disease Center. Courtesy of George Kubica, former Chief, Mycobacteriology Unit, NCDC, Atlanta.

†Key to percentage of strains reacting as indicated: ++ = 85 per cent or more; + = 75–84 per cent; ± = 50–74 per cent; ∓ = 15–49 per cent; − = <15 per cent; V = variable.

‡*M. ulcerans* and *M. marinum* grow best at about 32C; *M. avium* and *M. intracellulare* at 41C.

§With tests listed the pairs of organisms so indicated cannot be separated; colonial morphology on 7H-10 may be helpful in the case of *M. phlei*-*M. vaccae.*

‖Pigment increases with age.

•Mycobacteria that are of clinical significance. The others are rarely or never significant.

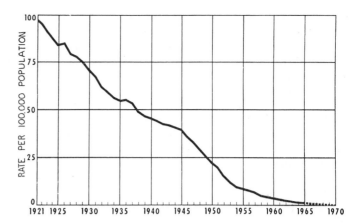

Figure 40–4. Decline in tuberculosis mortality in the United States from 1921 to 1970. Note that this applies to **deaths**, not to new or existing cases.

Tuberculosis

Tuberculosis is much more common than is generally supposed. While **deaths** in the United States have declined greatly in four decades (Fig. 40–4) because of improved methods of finding cases, diagnosis, and treatment, the number of **newly reported cases** each year remains appallingly high: over 35,000 in 1971, with nearly 7,000 deaths. Undoubtedly, many more infections (actual and potential cases), possibly 250,000 are **unreported.**

Tubercles and Tuberculosis. When tubercle bacilli gain a foothold in an uninfected but susceptible animal or person, the tissues in which the bacilli settle immediately react against the organisms in a very characteristic way. Numbers of tissue cells begin to grow around the bacilli in an attempt to incarcerate them, or wall them in. A tiny, pearly gray mass of cells is thus formed with tubercle bacilli at the center. This lesion is called a **tubercle.** Such infections occur very commonly. In a large majority of people, in good health and environments, resistance is high. Under such circumstances these infections heal and confer immunity. The tuberculin test becomes positive (see Chapter 29).

If the resistance of the host is low, because of malnutrition, overwork, or other debilitating factors, the tissues are unable to arrest the bacilli. They continue to grow, killing the surrounding cells and destroying the fibrous walls. Numbers of adjacent tubercles may thus coalesce. The dead tissue at the center of such masses of tubercles becomes cheesy and yellowish and is said to be **caseated.** It may rapidly involve the major part of the infected organ. If the tubercle is in a lung, the necrosis (death of tissue) may extend till it invades and breaks through the wall of a bronchus. The caseous

material, which may contain millions of living tubercle bacilli, is discharged with the sputum by coughing. A cavity is left behind (**cavitation**).

Sometimes the caseous process breaks through the wall of an artery and then hemorrhage of the lung occurs, which may be fatal. In the vast majority of human beings these tuberculous processes tend to heal and to form lifelong scars whether the patient recovers or not. Sometimes they remain latent only to start up again later in life under stress of illness and other weakening influences.

Hansen's Disease (Leprosy)

While the number of cases of leprosy in the world is not accurately known, it has been estimated that the number may vary from two to seven million, mainly in the tropics: Central Africa, India, China, Burma, Asia, Brazil. The number of active cases in the United States is estimated to be between 500 and 1,000. Many of the patients are confined in the National Leprosarium in Carville, Louisiana. The World Health Organization conducts worldwide studies of the disease and its control and cure.

Etiology. *Mycobacterium leprae* was discovered in 1874 by the Norwegian physician Hansen. The disease and the bacillus are often referred to as **Hansen's disease** and **Hansen's bacillus.** Morphologically and in degree of acid fastness, the bacillus is almost exactly like the tubercle bacillus, but the first differs from it in being an **obligate intracellular parasite.** It has never been artificially cultivated in a **virulent form.** (Compare Reiter treponeme in relation to syphilis in Chapter 41.)

Mycobacterium leprae (Color Plate VI, *C* is always present in leprous tissue and never occurs in normal tissue. For many years this has been the only one of Koch's postulates to be demonstrated in connection with the etiology of leprosy. Demonstration of this organism in histological sections or **biopsy** material (live tissue removed during life) is the most useful and conclusive diagnostic procedure. Within the last decade the organisms have been transmitted experimentally to the footpads of mice. A very similar organism, *M. lepraemurium,* causes an enzootic disease of rats and some other animals.

There are several strains of so-called *M. leprae,* isolated from lepers, which may be cultivated easily on ordinary culture media. They are not virulent and have the properties of saprophytic species of *Mycobacterium.* Several reports of successful inoculation of human beings with cultures of *M. leprae* and even with leprous material have appeared. None has ever been confirmed. It is interesting to note that leprous lesions occur principally in the outer parts of the body that are relatively **cool**: skin, peripheral nerves, and nasal passages.

The method of transmission of leprosy under natural conditions is obscure. Contrary to a notion as old as history, Hansen's disease is *not* highly contagious. The rate of conjugal infection is low. Many children who become infected heal spontaneously, as in tuberculosis. Also as in tuberculosis, the outcome of infection depends greatly on the resistance of the individual. Lesions and nasal discharges contain the bacilli in large numbers and probably transmit the disease. However, prolonged (for months or years) close contact appears to be necessary. Persons under 25 years of age are more likely to contract the disease. The incubation period appears to range from a few months to as long as 20 years.

Allergy in Hansen's Disease. As in tuberculosis, allergy of the delayed type develops in Hansen's disease. A skin test, the **lepromin test,** in some ways analogous to the tuberculin test, is used in diagnostic work. The antigen may be made from the bacilli themselves (**bacillary lepromin**) or, preferably, from leprous tissue (**integral** or **Mitsuda-Hayashi lepromin**). These antigens act not only to detect allergy but also, it seems, to induce some degree of immunity. BCG can induce lepromin positivity and there is encouraging evidence that BCG vaccination may be a valuable weapon in immunizing against this ancient enemy of mankind.

History. Hansen's disease is well known to students of the Bible and medieval history. Fear and horror of lepers have been a human tradition since antiquity. In former days many disfiguring diseases were confused with leprosy: various fungal infections, yaws, protozoal infections, and so on. The loathing of leprosy arises in great part from the disfigurement of the body and destruction of tissue, with scar formation, which accompanies it (Fig. 40–5). Modern surgery is doing much to combat this.

In ancient times lepers were excluded from all public contacts and left to die of exposure and starvation. In some areas they still are excluded, but provisions for their comfort and well-being are being greatly improved in developed areas. There are now drugs (sulfones, antibiotics) that are of great assistance, but much remains to be done.

40.3
FAMILY DERMATOPHILACEAE

The family contains two genera, *Dermatophilus* and *Geodermatophilus.* The organisms in both genera have very complex life cycles. In *Dermatophilus,* spores or coccoid bodies

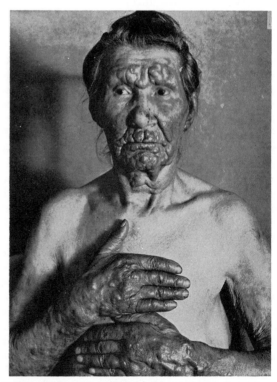

Figure 40–5. An advanced lepromatous case with leonine face, both furrowed and nodulate, and with marked involvement of the forearm and hands, less of the upper arms and still less of the body.

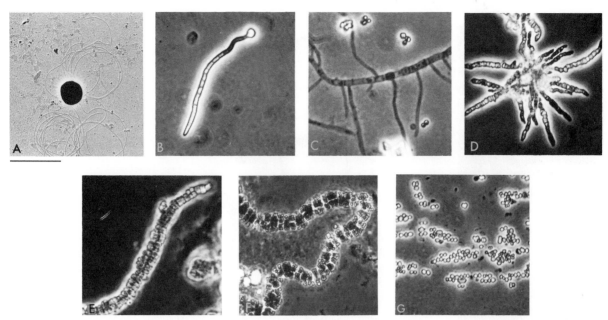

Figure 40–6. Life Cycle of *Dermatophilus congolensis* (ATCC14638). *A,* Shadow-cast electron micrograph of the zoospore; marker represents 2 μm; *B,* germinating zoospore; *C,* filaments with secondary branching perpendicular to axis of the main filaments; *D–E,* cuboidal cells resulting from transverse and longitudinal septation; *F,* filaments composed of cuboid or coccoid cells forming sarcina-like packets; *G,* liberated cuboid or coccoid cells which become zoospores. (Courtesy of G. M. Luedemann.)

germinate to produce filaments resembling true mycelia; long, tapered or relatively uniformly thickened filaments are septate and branch perpendicularly to the axis of the main filament. The thallus appears to undergo subdivision by formation of transverse septa, usually behind the tip of the thallus. After repeated division by septa formation, the outer wall of the thallus disintegrates, and the coccoid or spherical spores enlarge to form cuboidal, sarcina-like packets which are often retained in a gelatinous matrix. The coccoid bodies become spherical, motile (flagellate) zoospores. After a period of motility, zoospores germinate, and the germinating spores produce a long, relatively rigid, uniform filament within which transverse septation begins (Fig. 40–6).

Some species of *Dermatophilus* cause dermatitis in wild and domestic animals and man. However, the organisms belonging to genus *Geodermatophilus* are apparently nonpathogenic for experimental animals (rabbits and guinea pigs). In *Geodermatophilus* species, unlike *Dermatophilus* species, filaments are usually not produced from coccoid bodies or spores, and when present, they are rudimentary and resemble pseudomycelia. The secondary branching occurs in a manner similar to budding and is not confined to a plane per-

pendicular to the main axis of the filament. Zoospores are more elliptical to lanceolate than spherical, and division appears to occur by budding. The germinating cell may divide directly to produce a short germination tube or a longer flexible filament which may branch at various angles and produce septa simultaneously.

The habitat of the organism is soil, and cultures frequently resemble some of the black, sooty molds. However, the cultures are apparently insensitive to antifungal agents such as nystatin, griseofulvin, and tolnaftate but sensitive to antibacterial agents such as streptomycin, neomycin, chlortetracycline, and others. The ecological role of this organism in nature is not clearly understood.

40.4
FAMILY NOCARDIACEAE

Genus Nocardia. Nocardias differ from *Streptomyces* in that: (1) the filaments, at first coenocytic, later become septate and tend to fragment readily into bacillus-like and coccus-like segments (Fig. 40–7); (2) no conidia or spores are formed. According to the 8th edition of *Bergey's Manual,* the organisms are assem-

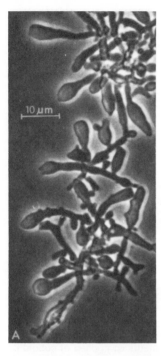

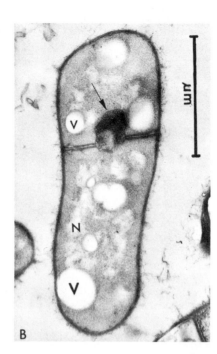

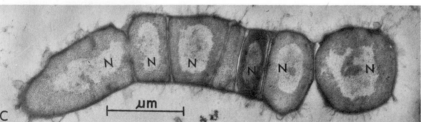

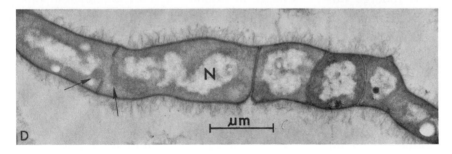

Figure 40–7. *A,* Phase-contrast micrograph of *Nocardia* 721-A grown on BHIA for 12 hours at 30C.

B, Electron micrograph of thin section of *Nocardia* 721-A grown on BHIA for 36 hours at 30C. Complex vesicular mesosomes were frequently observed associated with transverse septum formation (*arrow*).

C, Thin section of *Nocardia* 721-A grown on BHIA for 48 hours. Note that the filament has divided into coccoid units possessing a distinct and separate nuclear region.

D, Thin section of *Nocardia* 721-A grown on nutrient agar for 24 hours. Note complex mesosomes associated with the newly forming septum (*arrow*). (Courtesy of B. L. Beaman and D. M. Shankel.)

bled into Nocardiaceae, the sixth family of the order Actinomycetales, and the family contains two genera, *Nocardia* and *Pseudonocardia.* DNA base composition of the family ranges from 62 to 72 mole per cent GC.

Differentiation between some species of *Nocardia* and *Streptomyces* is often difficult, but may be made on the basis of differing forms of diaminopimelic acid present in their cell walls.

Most of the nocardias are saprophytes that live in the soil as scavengers, decomposing complex organic substances of a great variety,

such as cellulose, proteins, polysaccharides, lipids, paraffins and even carbolic acid (phenol), naphthalene, rubber, and cresol, as sources of energy and carbon. Except for these curious sources of carbon and energy, their minimal food requirements are simple minerals. Some species are able to grow autotrophically at the expense of H_2.

Like the Actinoplanaceae, nocardias tend to grow on, and in, the surface of solid media such as agar, rarely producing much aerial mycelium.

Several *Nocardia* species are parasitic

in man or animals, causing tuberculosis-like diseases or ulcerative lesions (**nocardiosis**). Saprophytic species of *Nocardia* are frequently found as contaminants in laboratory cultures. The mycelial fragments are often rod-shaped with some branching. Misshapen, clubbed, and knobbed forms are easily mistaken for *Corynebacterium diphtheriae*, *Lactobacillus*, and other gram-positive rods. A few species are somewhat acid-fast, especially in pathologic exudates (see §40.2, Family Mycobacteriaceae).

Many species form brilliant pigments, as do *Streptomyces*. The colonies of *Nocardia* on agar, however, are usually not mycelial but butyrous like those of true bacteria, ranging in diameter from 1 to 10 mm.

40.5
FAMILY STREPTOMYCETACEAE

As presented in the 8th edition of *Bergey's Manual* this family includes genera *Streptomyces*, *Streptoverticillium*, *Sporichthya*, and *Microellobosporia*.

Genus *Streptomyces*. Members of this genus form long, much-branched, aerial mycelia consisting of mold-like, nonfragmenting, very fine filaments. The DNA base composition of *Streptomyces*, in part, ranges from 67 to 74 mole per cent GC. These organisms are typically procaryotic, with extensive internal membranous bodies. Their branching vegetative hyphae are embedded in the substrate, and aerial hyphae, by extensive septation, produce spores (Fig. 40–8). Their cell walls contain some of the components of chitin, a major constituent of cell walls of true fungi, but true chitin is not present. For example, the cell walls of *Streptomyces*, *Sporichthya*, and *Microellobosporia* contain L,L-DAP which is cross-linked with glycine residue (see Chapter 14). Multiplication is by conidia produced asexually at the tips of **conidiophores** or **sporophores**. The conidia form long, straight, curved, or coiled chains, giving a curious appearance to the mycelium as a whole (Fig. 40–9). The directions and forms of the coils are fairly constant for any given species under standard cultural conditions and are important in the classification of *Streptomyces*.

The conidia of *Streptomyces*, when examined with the electron microscope at magnifications of 8,000×, reveal striking surface configurations, "sculpturing," or "ornamentations." They are of at least four types: spiny, hairy, warty, and smooth (Fig. 40–10). Under standardized cultural conditions these characters are

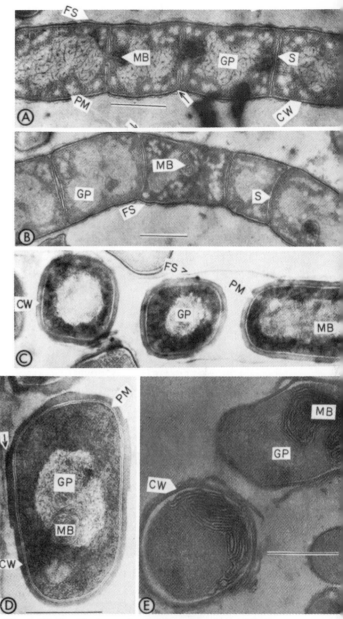

Figure 40–8. Electron micrographs of sections of *Streptomyces venezuelae*. Marker bars denote 0.5 μm. Abbreviations *CW*, cell wall; *FS*, fibrous sheath; *GP*, germ plasm; *MB*, membranous body; *PM*, plasma membrane; *S*, septum.

A–C, Sections of aerial hyphae stained with uranyl acetate. Arrows indicate the points where the inner layer of the cell wall turns in to form septa.

D, Section of normal spore, stained with uranyl acetate. Arrow indicates attached strip of the outer layer of the cell wall of the aerial hypha.

E, Section of germinating spore, stained with lead acetate.

fairly constant as to species and can serve as an aid in classification, usually a frustrating process when based only on the notoriously

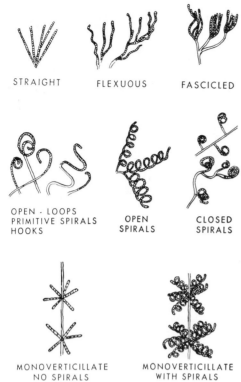

STRAIGHT   FLEXUOUS   FASCICLED

OPEN - LOOPS
PRIMITIVE SPIRALS   OPEN   CLOSED
HOOKS   SPIRALS   SPIRALS

MONOVERTICILLATE   MONOVERTICILLATE
NO SPIRALS   WITH SPIRALS

Figure 40–9. Distinctive arrangements of conidial filaments of various types of *Streptomyces.*

variable biochemical characters of *Streptomyces.*

The conidia of *Streptomyces* and other Actinomycetales are not as heat-resistant as the endospores of true bacteria, being killed by 10- to 30-minute exposures at 65C, a temperature only slightly higher than that required to kill the vegetative mycelium.

The colonies of *Streptomyces* are usually tough, dense-textured, and often very adherent to the medium owing to vegetative (subsurface) mycelia. They have a woolly or velvety appearance because of the mycelial structure. The growth of many species is brilliantly colored: red, orange, yellow (Color Plate VI, *A*). Colonies usually range in diameter from less than 1 mm to several millimeters, definitely smaller than the huge colonies of true molds.

The colonies on infusion or extract agar are often papillate, and frequently the surface is thrown into radial folds or ridges. They give off a distinctive, musty odor characteristic of damp cellars and newly turned soil. This has been ascribed to a neutral volatile oil called **geosmin**.

Most species of *Streptomyces* are saprophytes and are active in decomposition of a wide range of organic materials. A few species are pathogens in animals or plants. *Strepto-*

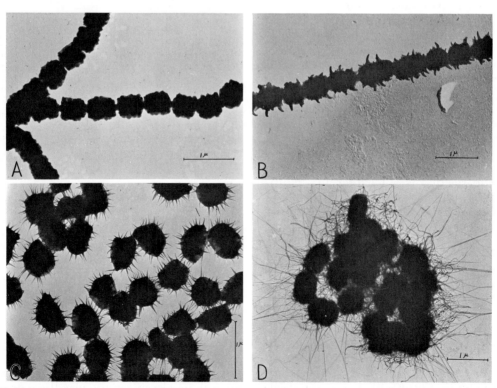

Figure 40–10. Examples of distinctive "sculpturing" on the surfaces of conidia of various species of *Streptomyces. A, S. olivaceus; B, S. purpurascens; C, S. diastatochromogenes; D, S. albogriseolus.*

myces scabies, for example, produces a troublesome disease ("scab") of potatoes.

Growth is usually best at temperatures about 25C. Unlike most true molds, optimal growth occurs at pH 8 or 9, and is greatly depressed by reactions near pH 5.

It is obvious from the optimal pH that liming of acid soils will encourage growth of the saprophytic actinomycetes of all families. Their growth increases fertility since they actively decompose complex organic materials so that other bacteria and farm crops can make use of them. The various species causing **scab** of potatoes and other root crops are also encouraged by liming of soil so that one must consider both the nature of the crop to be raised and soil pH before indiscriminate liming. As the Streptomycetaceae and other actinomycetes of the soil are aerobic, it is apparent why draining swamp lands in addition to liming increases the fertility of such soils.

Streptomyces and Antibiotics. The genus *Streptomyces* is one of the most extensively investigated groups of bacteria, because various species produce antibiotics of immense value in human or veterinary medicine, in the control of plant diseases and industrial spoilage by molds and bacteria, and in scientific research (see Chapters 23, 43, and 49).

40.6
FAMILY MICROMONOSPORACEAE

This family is newly created in the 8th edition of *Bergey's Manual* and contains six genera: *Micromonospora, Thermoactinomyces, Actinobifida, Thermomonospora, Microbispora,* and *Micropolyspora.* The organisms in the family are similar to *Streptomyces* in many respects but devoid of aerial mycelia. As a rule, mycelia grow well into the substrate, do not disintegrate with age, and form compact, smooth, frequently brightly colored colonies. The organisms multiply by means of conidia which are produced singly (*Micromonospora;* Fig. 40–11), in pairs *(Microbispora),* or in larger groups *(Micropolyspora)* at the ends of special conidiophores on the surface of substrate mycelia. In some, conidiophores are dichotomously branched (*Actinobifida;* Fig. 40–12). Most members of the family are aerobic, mesophilic saprophytes of soil, lake mud, and similar environments; some are anaerobic and thermophilic (able to grow at temperatures between 45 and 55C, *Thermomonospora;* and others grow at temperatures as high as 55 to 65C,

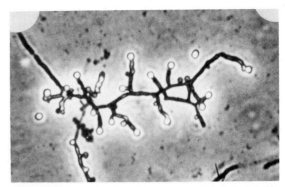

Figure 40–11. *Micromonospora chalcea,* showing spherical conidiospores borne singly at the tips of hyphae. (Phase contrast, 1,965×.) (Courtesy of G. M. Luedemann.)

Thermoactinomyces). The organisms as a whole are proteolytic and diastatic, and some are cellulose-fermenters; hence they play an important role in decomposition of organic matter and the cycle of elements. The DNA base composition of the family (in part) ranges from 69 to 73 mole per cent GC: *Micromonspora,* 72 to 73; *Microbispora,* 70 to 73; *Thermomonospora,* 69 mole per cent GC.

40.7
FAMILY ACTINOPLANACEAE

This family has been expanded in the 8th edition of *Bergey's Manual* to include ten genera: *Actinoplanes, Spirillospora, Ampullariella, Streptosporangium, Amorphosporangium, Pilimelia, Planomonospora, Planobispora, Dactylosporangium,* and *Kitasatoa.* The members of the family are aerobic mesophiles with simple nutritional requirements. These genera are distinguished from one another not only by physiological and growth characteristics but by the shape of the sporangium and the structure of the sporangiospores as well as by motility of the spores.

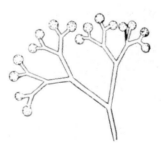

Figure 40–12. *Actinobifida* sp.

Some of the species of family Actinoplanaceae are among the most mold-like of bacteria. They are commonly found in soil and in aquatic habitats or growing as saprophytes on dead leaves. They form a much-branched, occasionally septate, inconspicuous but widespreading **vegetative** mycelium similar to *Streptomyces* and a more or less conspicuous aerial mycelium.

In many respects the organisms of the family resemble aquatic Phycomycetes. For example, swelling occurs at the tip of hyphae in either aerial or substrate mycelia, and swollen portions are then cut off by transverse septa, and a portion becomes a sporangium (Fig. 40–13). The sporangium may contain a few to several sporangiospores, and the spores are eventually released from the sporangium by rupture of the sporangial wall.

While some spores borne in sporangia are nonmotile (for example, in genera *Streptosporangium* and *Amorphosporangium*), spores in the genera *Actinoplanes*, *Ampullariella*, and *Spirillospora* are actively motile with polar flagella when released from the sporangium (Fig. 40–14). This is suggestive of the aquatic mold *Saprolegnia*, which also forms motile sporangiospores (**zoospores**). However, the major constituents of the cell walls of the organisms in the family Actinoplanaceae are peptidoglycans (directly cross-linked *m*-diaminopimelic acid); also, electron micrographs of *Actinoplanes* show it to be procaryotic, not mold-like (eucaryotic), in structure. The DNA base composition of the family (in part) ranges from 70 to 73 mole per cent GC: *Actinoplanes*, 72 to 73; *Spirillospora*, 71 to 72; *Ampullariella*, 73; *Planomonospora*, 72; *Planobispora*, 70 to 71; *Dactylosporangium*, 71 to 73 mole per cent GC.

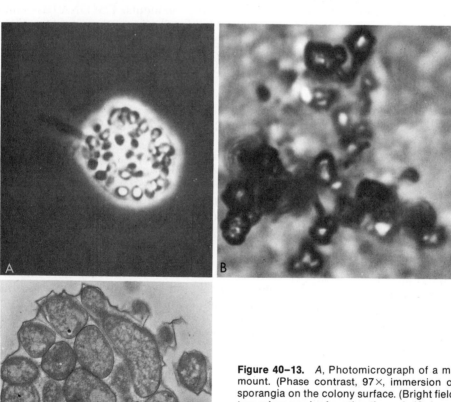

Figure 40–13. *A,* Photomicrograph of a mature sporangium in a water mount. (Phase contrast, 97×, immersion objective.) *B,* Several mature sporangia on the colony surface. (Bright field, 40 × dry objective.) *C,* Electron micrograph of section through 40-day-old sporangium; note that the sporangium wall is continuation of the outer sheath of the sporangiophore. (Courtesy of H. A. Lechevalier and P. E. Holbert.)

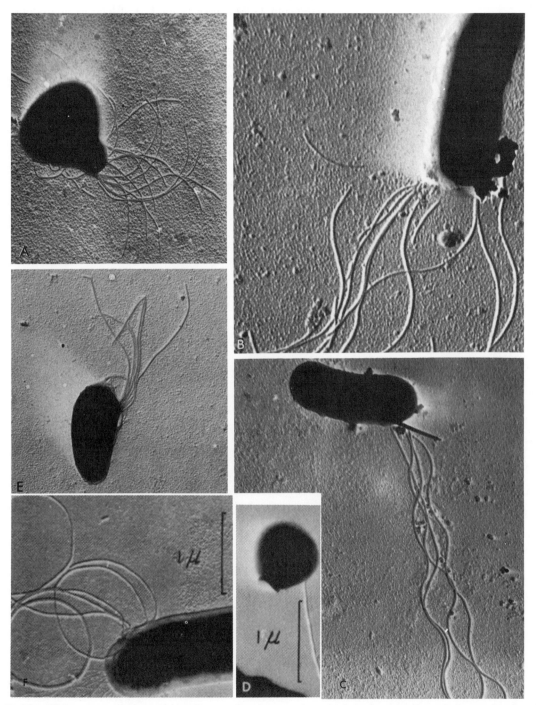

Figure 40–14. All figures are electron photomicrographs of germanium-shadowed preparations. *A,* Spore of *Actinoplanes* sp. L.G.W.; note hernia; *B,* spore of *A. campanulata; C,* spore of *Spirillospora* sp.; *D,* spore of *Sporichthya polymorpha;* note collar-like scar; *E,* spore of *Dactylosporangium* sp.; *F,* spore of strain 8–32 (C₄ group). (Courtesy of M. L. Higgins, M. P. Lechevalier, and H. A. Lechevalier.)

CHAPTER 40

SUPPLEMENTARY READING

Blair, J. E., Lennette, E. H., and Truant, J. P. (Eds.): Manual of Clinical Microbiology. American Society for Microbiology. The Williams & Wilkins Co., Baltimore. 1970.

Luedemann, G. M., and Casmer, C. J.: Electron microscope study of whole mounts and thin sections of *Micromonospora chalcea* ATCC 12452. Int. J. Sys. Bact., 23:243, 1973.

Poupard, J. A., Husain, I., and Norris, R. F.: Biology of the bifidobacteria. Bact. Rev., 37:136, 1973.

Schleifer, K. H., and Kandler, O.: Peptidoglycan types of bacterial cell walls and their taxonomic implications. Bact. Rev., 36:407, 1972.

Shirling, E. B., and Gottlieb, D.: Cooperative description of type strains of *Streptomyces*. Int. J. Sys. Bact., 22:265, 1972.

Skerman, V. D. B.: A Guide to the Genera of Bacteria, 2nd ed. The Williams & Wilkins Co., Baltimore. 1967.

Waksman, S.: The Actinomycetes, Vols. 1 and 2. The Williams & Wilkins Co., Baltimore. 1971.

CHAPTER 41 • ORDERS MYXO-BACTERALES AND SPIROCHAETALES (GLIDING, PROTOZOAN-LIKE BACTERIA)

The terms "mold-like," "alga-like," and "protozoan-like" have little taxonomic significance and are used here merely for convenience in describing the organisms with such characteristics. There are a variety of bacteria which show gliding motility; among them, however, characteristics that are suggestive of protozoans (thin, flexible cell wall with either gliding or active rotatory as well as translatory motility) distinguish two orders: Myxobacterales and Spirochaetales.

The Myxobacterales have thin, flexible cell walls and definite, translatory motility without flagella. They are distinguished from Spirochaetales by their alga-like, gliding motility in contact with solid objects. They are distinguished from all other bacteria by their relatively complex life cycle that superficially but strikingly resembles the life cycle of certain protozoans, the ameba-like "slime molds" or myxamebas of the order Acrasiales (see §41.2, Myxomycetes). The Spirochaetales are characterized by relatively complex helicoidal forms; thin, flexible cell walls; a multifibrillar, flexible, **axial filament** inside the cell wall; and active rotatory as well as translatory motility.

They do not exhibit flagellar or gliding motility (Table 34–1). The cell walls of spirochetes resemble those of other gram-negative bacteria, the only difference being that m-DAP of the peptidoglycan is replaced by L-ornithine. On the other hand, the cell walls of myxobacteria are composed of directly cross-linked m-DAP–containing peptidoglycan (see Chapter 14).

The Spirochaetales and Myxobacterales are nonchlorophyllous (not photosynthetic) and do not form trichomes, branching filaments, mycelia, flagella, conidia, or endospores. They are not sheathed or stalked, nor do they accumulate granules of sulfur, iron, or other elements. With a few exceptions they are bacterial in dimensions and, so far as is known, procaryotic in structure. All multiply by transverse binary fission and are chemoorganotrophic. True sexuality has not been observed. Not all have been cultivated, but several representative species have been grown on artificial media at mesophilic temperatures and a pH of about 7.2. Most species are harmless saprophytes of the soil or of marine or fresh aquatic muds and are active as scavengers in decomposing organic matter. A few species of Spirochaetales are

601

among the most dangerous and widespread pathogens; a few species of Myxobacterales are pathogenic on fish or plants.

41.1
ORDER MYXOBACTERALES

The Myxobacterales (myxobacters) (Gr. *myxo* = slime) are gram-negative coccus-like or often fusiform rods, varying in length from 3 to 12 μm (Fig. 41–1, *A* to *C*), and are especially characterized, in addition to the aforementioned peculiarities, by the production of distinctive masses of slime, in which they live communally. These communal structures consist of flat, spreading, slimy colonies with lobular extensions, and are usually called a **swarm stage** or **pseudoplasmodium**. The latter term is derived from their resemblance to a similar structure called a **plasmodium** formed by true eucaryotic protozoa, Mycetozoa (or myxomycetes, slime molds) (see farther on). The pseudoplasmodium is one stage of the life cycle of myxobacters. Some of the myxobacters are distinguished by the production of "fruiting bodies."

Classification of Myxobacteria. In the 8th edition of *Bergey's Manual*, myxobacteria or "slime bacteria" are placed in Part II, "The Gliding Bacteria," which contains two orders, Myxobacterales and Cytophagales. Establishment of the new order, Cytophagales (formerly the family Cytophagaceae of order Myxobacterales), is based on the lack of "fruiting bodies" and resting cells, i.e., physiological characteristics, in addition to unrelatedness of the DNA base compositions.

In the new *Manual*, the Myxobacterales contains four families: Myxococcaceae, Archangiaceae, Cystobacteraceae, and Polyangiaceae. The former family Sorangiaceae is incorporated into the family Polyangiaceae in the 8th edition of the *Manual* because the differences between the organisms in these two families are considered superficial, namely shape of cysts and vegetative cells. Transfer of the family Cytophagaceae and related organisms, i.e., genus *Sporocytophaga* from former Myxobacterales into a new order is based not only on the lack of fruiting body or resting cells but on DNA base composition. For example, fruiting myxobacteria have a DNA base composition with a range of 67 to 70 mole per cent GC, but *Cytophaga* and *Sporocytophaga*, which were formerly placed in the family Myxococcaceae, form myxospore-like structures but not fruiting bodies. These two groups of organisms have a DNA base composition ranging from 31 to 42

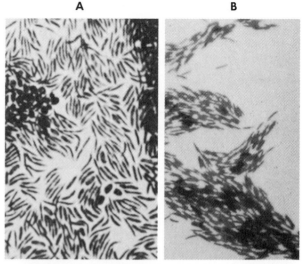

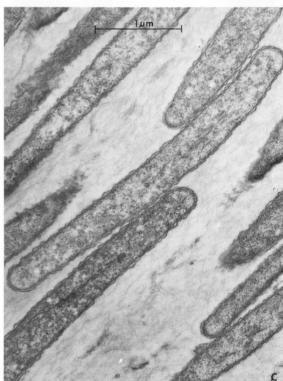

Figure 41–1. *A* and *B, Myxococcus,* a myxobacterium. Long, thin, tapering, rod-shaped vegetative cells.

C, Stigmatella aurantica. Swarm cells from the base of a young fruiting body of strain aur W1 C1, apparently all oriented in one direction. Note the fibrous material in the space between the cells which may be condensed slime that was secreted by the cells. Fixation with osmium tetroxide vapors followed by picric acid formaldehyde. (18,500×.) (*C,* Courtesy of H. Voelz and H. Reichenbach.)

mole per cent GC, which is considerably less than that of organisms in the new order Myxobacterales. The DNA base composition of other nonchlorophyllous gliding bacteria is similar to the organisms of genera *Cytophaga* and *Sporocytophaga*. For example, genera *Flexibacter* and *Saprospira* of the family Cytophagaceae have a DNA base composition of 30 to 47 mole per cent and 35 to 48 mole per cent GC, respectively. Similarly, genus *Vitreoscilla* of Beggiatoaceae and genus *Leucothrix* of Leucotrichaceae have a DNA base composition of 44 to 45 and 46 to 50 mole per cent GC, respectively. All the organisms mentioned above are now placed under Order II, Cytophagales, of Part II, "The Gliding Bacteria," in the 8th edition, and some of the organisms are discussed more fully in Chapter 32.

Life Cycle. All the genera of the four families in the order Myxobacterales (8th edition, *Bergey's Manual*) produce resting cells and fruiting bodies in their life cycles, and hence the organisms are frequently referred to as "fruiting myxobacteria." The vegetative cells are gram-negative, nonflagellate rods, and the cells in some stage of the life cycle undergo alternation of form: the swarm stage and the encystment or fruiting stage. Therefore, the life cycle of the organisms is suggestive of alternation of generation in eucaryotes.

Swarm Stage. This growth may occur on any decaying organic matter. In this stage the bacteria multiply by binary fission and secrete a slimy matrix in which they all live together. The slime appears to be formed within the cell wall and extruded to the exterior. This slimy material is ". . . a distinct, firm, hyaline, gelatinous base, secreted by the colony as it extends itself, over which the individuals may move or in which they may become imbedded, and is so coherent a structure that whole colonies may be stripped intact by means of it, from the surface of nutrient agar, for example" (Thaxter) (Fig. 41–2). The swarm stage lasts for periods varying from a day to a week. Under favorable growth conditions the swarm stage continues and the growth may cover an area of several square centimeters. Encystment (fructification) then begins, apparently as a result of increased nutritional deficiencies (Fig. 41–3).

Encystment or Fruiting Stage. Under appropriate conditions, a swarm of vegetative cells gathers together (possibly a chemotactic response) at different points in the slimy matrix, and they heap themselves up to form "fruiting bodies." The swarm stage continues as long as there is adequate nutrient for vegetative growth, but upon exhaustion of one or more essential amino acids, namely tyrosine and phenylalanine, the cells are induced to form the fruiting stage. The heaps may become distinctly raised above the substratum, or slightly raised. For example, the fruiting body of the genus *Myxococcus* is considered most primitive among myxobacteria, and never develops beyond a low, round, microscopic mound or heap (Fig. 41–4, *A*). However, the fruiting bodies of species of *Stigmatella* (Family III, Cystobacteraceae) and *Chondromyces* (Family IV, Polyangiaceae) consist of a thick stalk supporting a number of individual cysts (Fig. 41–4, *B*). The stalk is apparently composed of debris of swarm cells and nonliving slime or discontinuous tubules, suggesting empty hulls of vegetative cells, which lie parallel to the longitudinal axis of the stalk (Fig. 41–5). Fruiting bodies are colored bright yellow (*Myxococcus virescens*) to reddish brown (*Stigmatella* and *Chondromyces*), with carotenoids as dominant pigments. The (fruiting body) pigment formation is enhanced in the light, and the function of carotenoid pigment formation may be thought of as photoprotection.

Within a fruiting body or cyst the rod cells undergo differentiation into shortened, rounded, or dormant resting cells known as **myxospores** (which are frequently referred to as **cysts** or **microcysts**). The shortened, thick-walled cells in the cyst (Fig. 41–6) are usually enveloped by large capsules or gelatinous slime. As

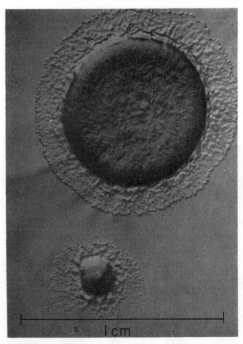

Figure 41–2. Vegetative colonies of *Myxococcus xanthus*, strain FB.

Figure 41–3. Successive stages in the development of the complex, branched fruiting body of *Chondromyces*. (From Bonner: Morphogenesis, 1952. Reprinted by permission of Princeton University Press.)

these extracellular materials dry, the myxospores become more resistant than the vegetative cells to various physical agents, such as desiccation, UV radiation, sonic vibration, and heat. Myxospores within dried fruiting bodies may remain viable for several years. However, the degree of resistance of myxospores to heat or other physical agents is considerably less than that of the bacterial endospores discussed in Chapter 36.

Isolation and Cultivation of Myxobacteria. All myxobacteria are chemoheterotrophs (saprophytes) and are strictly aerobic. Their natural habitat is soil. While it has been reported that the vegetative cell growth is prevalent in the soil, fructification of the organism in soil is considered either rare or not yet observed. The fruiting bodies on decaying wood or leaves, and more frequently on the surfaces of dung pellets of rabbit and deer, are observed with a dissecting microscope.

The fruiting myxobacteria may be readily isolated by one or more methods. For example, they can be isolated on water agar (1.5 per cent agar) seeded with a suspension of any gram-negative or gram-positive bacterium, such as *E. coli* or *Micrococcus* cells, as a source of food for the myxobacteria. One usually places a small amount of soil or other decaying material in the center of the plate as inoculum. The plates are kept moistened for the duration of the incubation period to promote swarming of vegetative cells. Many of the species in the order Myxobacterales are able to lyse bacterial cells and use their liberated products as nutrients.

Within a few days to a week of incubation, vegetative cells of myxobacteria emerge from the inoculum and begin to swarm across the seeded plate and form rather flat colonies. Pure cultures may be obtained from cells of the fruiting head (cyst) or from the peripheral cells of swarming microcolonies. The pure culture so

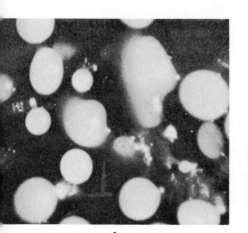

A

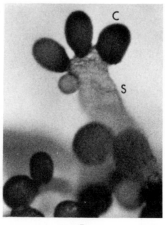

B

Figure 41–4. *A*, Fruiting bodies of the mature organism *Myxococcus*, which is composed of spherical microcysts. (30×.)

B, Side view of a fruiting body of *Stigmatella aurantiaca*, consisting of a stalk (*S*) and cysts (*C*). (300×.) (*B*, Courtesy of H. Voelz and H. Reichenbach.)

For many species of myxobacters, one enrichment method is to place pellets of sterilized rabbit feces close together on the surface of fresh soil in covered dishes, keeping the whole quite moist for a week or two at about 35C. The organisms grow well at 10 to 20C. Pure cultures of some species may be obtained on rabbit dung agar or on infusion agar.

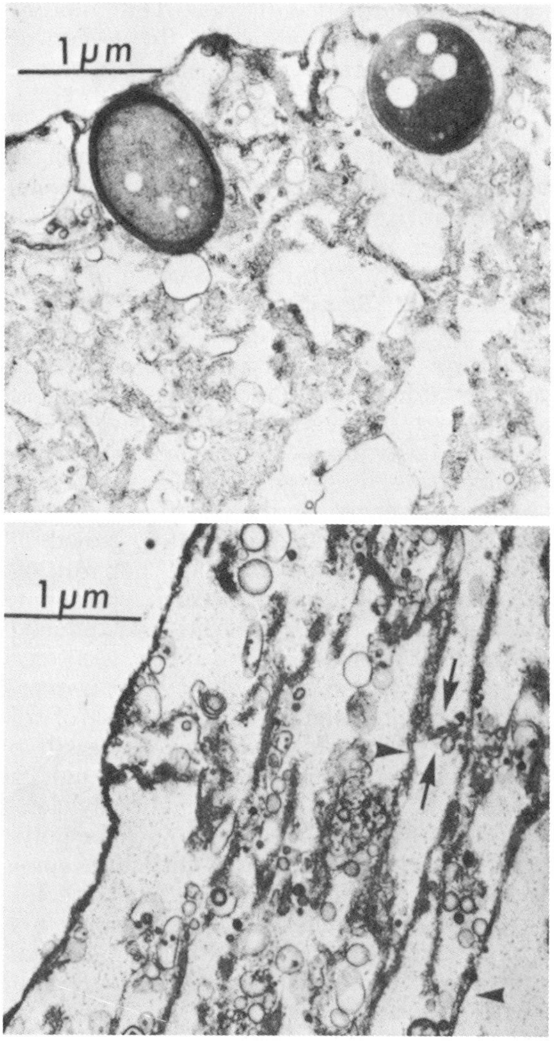

Most fruiting myxobacteria, unlike organisms in the order Cytophagales, are unable to decompose cellulose, but some members of Archangiaceae and Polyangiaceae have been

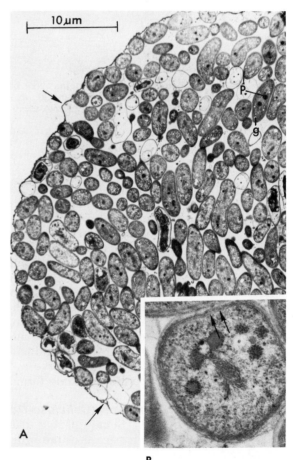

Figure 41–5. *A,* Transverse section through the slime stalk of a fruiting body of strain cro C1 1. Empty spaces appear to be channels or empty hulls of vegetative cells which are embedded in fibrous material. A few more or less mature myxospores are also present.

B, Longitudinal section of a slime stalk of strain cro C1 1. It appears to be constructed of discontinuous tubules. At the short arrow, a double track is shown. The tubules appear to be wedged into each other but are also aligned front to rear (*opposing arrows*). Glutaraldehyde fixation with osmium tetroxide postfixation. (Courtesy of H. Voelz and H. Reichenbach.)

Figure 41–6. *A,* Section through the central portion of a cyst of strain aur W1 C1, containing mature, randomly oriented myxospores with dense granules (*p*) and electron-translucent inclusions (*g*). The space between the myxospores is filled with cell debris and empty hulls. *Arrows,* cyst wall. Glutaraldehyde fixation with osmium tetroxide post-fixation.

B, Enlarged section through a cyst of strain aur W1 C1, similar to that in *A* but fixed differently. Myxospores appear slightly deformed. They are enclosed by a fibrous capsule. Electron-translucent inclusions (*g*) seem to be surrounded by ribosome-like particles. The cell wall is often wavy (*arrows*) apposing the fairly smooth cytoplasmic membrane. Fixation with osmium tetroxide vapors, followed by picric acid–formaldehyde. (32,000×.) (Courtesy of H. Voelz and H. Reichenbach.)

obtained may be inoculated onto solid organic media composed of peptone or casein hydrolysate, rich in amino acids or small peptides. On such organic media the cells spread rapidly and radially with gliding motion. The cells or aggregates of cells continue to break away from the edge of the microcolony, and as they move the cell or cells usually leave a characteristic slime behind on their trail (Fig. 41–7).

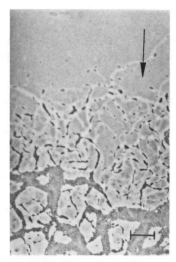

Figure 41–7. Peripheral spots of *M. xanthus* FB$_t$ on fruiting agar after eight days of incubation. Bar = 50 μm. A trail left by gliding FB$_t$ cells is indicated by arrow. (Courtesy of Robert P. Burchard.)

shown to utilize cellulose in addition to other sugars. Hence these organisms may be cultivated in defined mineral basal media supplemented with powdered cellulose or filter paper. While precise nutritional requirements of myxobacteria are not well established, some members of *Myxococcus* have been cultivated in aerated liquid medium containing a mixture of amino acids. Fructification proceeds as the medium is depleted of certain aromatic amino acids, namely tyrosine and phenylalanine, and myxospore formation within the cyst is induced by addition of glycerol (which may serve as a possible precursor compound for slime formation).

Isolation of Cytophagae (Cellulose-Decomposing Organisms). Some gliding bacteria of the family Cytophagaceae of the order Cytophagales, like organisms in Myxobacterales, are able to decompose cellulose (some decompose chitin and other sugars). For example, organisms of genera *Cytophaga* and *Sporocytophaga* are able to destroy the structure of cellulose fibers (i.e, filter paper). The organisms produce the hydrolytic enzyme cellulase, but apparently the enzyme is closely associated with the cell surface and is unable to diffuse away from it; hence, the cells must be directly in contact with cellulose fibers for decomposition.

One species *(Cytophaga columnaris)*, of importance because it is a pathogen of fish of commercial value, has been cultivated on media containing peptone (tryptone), yeast and meat extracts, and sodium acetate. Such media may be fluid or solidified with agar. The colonies of these organisms are beautifully stellate or arborescent, especially when floating in a fluid medium.

Sometimes they swing by one end from a fixed surface and oscillate like a pendulum. It is of interest that a bacteriophage **(myxophage)** active in the lysis of *C. columnaris* has been demonstrated, the first to be observed in any species resembling fruiting myxobacteria.

Ecological Significance. Many of the fruiting myxobacteria are considered to be predators of other procaryotic and eucaryotic organisms. For example, they lyse the cells of true bacteria, blue-green algae (Fig. 41–8), and some higher fungi. They utilize as food the substances thus liberated. Practical use of this property is described in the isolation and cultivation section. Curiously, vegetative growth of the myxobacters is not abundant, but fruiting bodies are readily formed on such nutriment, further evidence that fructification is induced by nutritional deficiencies.

This lytic action must be of great importance in the interrelationship (ecology) of all organisms in the soil, where myxobacters live in contact with other microorganisms, though its full significance has not yet been entirely clarified. The lytic action is reminiscent of the polypeptide cell wall–destroying antibiotics produced by *Streptomyces* and *Bacillus* species from the soil, and suggests that the myxobacters may be a rich source of valuable antimicrobial products. In some species the lytic agent resembles lysozyme.

Except a few parasitic species, myxobacters and cytophagae are of great importance to humanity as scavengers. As a group they are very active in the decomposition of **insoluble** organic materials such as dead vegetable matter (cellulose, agar) and the exoskeletal material (chitin) of insects and crustacea, as well as a wide variety of other animal and vegetable matter, thus not only removing waste matters but obligingly transforming them into predigested soluble foods for plants.

41.2
MYXOMYCETES (MYCETOZOA)

(True Slime Mold)

The mycetozoa, acellular slime molds, are eucaryotic organisms that have been studied by

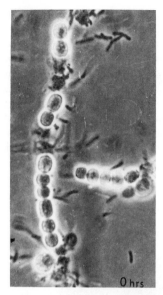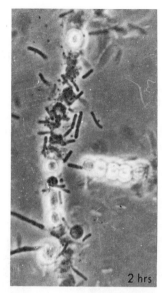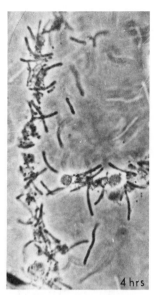

Figure 41–8. Sequence of lysis (in hours) of Nostoc filament with high multiplicity of the myxobacters obtained by the thin-agar technique. Zeiss phase-contrast microscope. (3,150×.) (Courtesy of Dr. Miriam Shilo.)

both protozoologists and mycologists. Recently, the organisms have been included in the phylum Myxomycotina (slime fungi), which includes the classes Myxomycetes and Acrasiomycetes, because the organisms produce not only structures resembling fruiting bodies of fungi but also spores with cellulose-containing cell walls. However, regardless of their taxonomic position, they resemble protozoans as much as fungi. The organisms undergo several morphological changes in their life cycles: **spore stage,** flagellate **swarm cell, cyst, ameba, plasmodium, sclerotium,** and **fruiting body,** in which spores are borne.

The spores of acellular slime molds may germinate to form amebas, and the amebas may form either swarm cells with whiplash flagella or cysts, depending on environmental conditions. The amebas or swarm cells may fuse together to form diploid amebas, which in turn undergo cell division to form a mass of living protoplasm (a true **plasmodium,** not a **pseudoplasmodium** consisting of inert slime (see Figures 12–13 and 12–14). These masses of multinucleate protoplasm are capable of ameboid motion and, like true amebas, can ingest solid particles of food, including bacteria. These are distinctly **animal** characters. The mycetozoa live on rotten logs and the like in much the same situations as myxobacters and move about in the moisture and shade like amebas.

After several days of plasmodial existence they cease to move, and fructification begins. The protoplasm sends up **stalks,** on the tips of which **sporangia** (spore-bearing cysts) are formed, in a great variety of the most graceful and delicate forms and of the most brilliant colors. Each sporangium contains many spores; each spore contains a haploid nucleus and is surrounded by a cellulose wall. The spores are dispersed by the wind. In water, each germinates, forming a separate, naked, flagellate or ameba-like swarm cell. These forms multiply rapidly by cell division. They later conjugate. Eventually they reaggregate to form the **multinucleate swarm stage** again.

Generally, cultivation of these organisms in both vegetative and fruiting stages is possible only if living bacterial cells (or, to a much lesser extent, dead cells or extracts of bacteria) are provided as pabulum. Without bacteria, development of the vegetative stage was not possible until 1962, when one species was cultivated through both stages on an axenic medium containing embryo extract, serum albumin, tryptose, glucose, certain vitamins, and necessary minerals.

The resemblance between the myxobacters and mycetozoa is so close, yet so superficial, as to suggest the idea that a comparative study in morphology and function might have been in progress, an example of **convergent evolution.**

41.3
ORDER SPIROCHAETALES

(Spiral, Flexible Bacteria)

The term spirochete is often used in a general sense to include all species of the order Spirochaetales. It will be so used here. The spirochetes are chemoorganotrophic, gram-negative, unicellular organisms with distinct morphology and mechanisms of locomotion. The length of cells ranges from 6 μm (*Treponema*) to 500 μm (*Spirochaeta*), and the width of the cells ranges from 0.1 μm (*Leptospira*) to 3.0 μm for *Cristispira* species. All spirochetes are either coarsely or tightly coiled and possess a delicate, flexible wall. The primary structure of the spirochete peptidoglycan is similar to that of other gram-negative bacteria, with the only difference being that *m*-DAP is replaced by

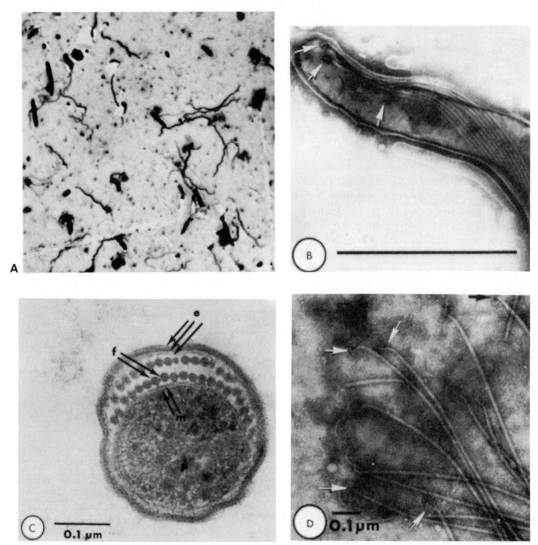

Figure 41–9. *A*, Smear from Vincent's angina showing the characteristic mixture of spirochetes (*Borrelia vincentii*) and fusiform bacilli. (Fuchsin, 1,250×.) (Courtesy of P. E. Harrison.)

B, Extremity of spirochete in gingival scrapings. Note terminal bend in axial fibrils leading to attachment discs in protoplasmic cylinder (*arrows*). Negative contrast. (50,000×.)

C, Cross section through large spirochete in gingival lesion of acute necrotizing ulcerative gingivitis. Note substructure of axial fibrils (*f*) suggestive of the presence of a dense central core. Fibrils are located between three dense layers of outer envelope (*e*) and two dense layers of membrane (*m*) covering protoplasmic cylinder (*p*). Kellenberger fixation.

D, Axial fibrils inserted into a crushed protoplasmic cylinder by means of attachment discs (*arrows*). Negative contrast.

L-ornithine. All agents except *Borrelia* (not yet demonstrated) possess one or more fibrillar structures known as the axial filaments or **axostyle** (Fig. 41–9). The organisms are actively motile in liquid medium, and their motility may be attributed to the axial filament, which is thought to be responsible not only for spiral configuration of the cells but also for causing the cells to shorten or lengthen by contractions without changing the highly coiled structure. All spirochetes are flexible *and* spiral. Differentiate the term spirochete from the name of the genus *Spirillum*, family Spirillaceae. *Spirillum*, though spiral, is not flexible. On the other hand, Myxobacterales, while flexible, are not spiral. The DNA base composition of the order Spirochaetales ranges from 34 to 46 mole per cent GC.

Classification. In the 8th edition of *Bergey's Manual*, the order Spirochaetales is placed in Part V, "Spirochaetes," and contains only one family, Spirochaetaceae. The family contains five genera: *Spirochaeta, Cristispira, Treponema, Leptospira,* and *Borrelia.* Thus the new classification eliminates the former family Treponemataceae, the organisms of which are now placed in genus *Treponema.* Likewise, the former genus *Saprospira* of Spirochaetaceae is eliminated on the basis that these organisms, while spiral-shaped, show definite gliding motility and lack the axial filaments. *Saprospira* is placed in the family Cytophagaceae.

Spirochetes are widely distributed in nature, especially in aquatic environments and bodies of warm-blooded animals. Many of these are harmless saprophytes, while others cause diseases of animals and man, namely syphilis and other treponematoses, borreliosis, and leptospirosis, of which the first is the most important.

Family Spirochaetaceae

Genus *Spirochaeta*. The name of the genus is derived from the term "Spirochaeta," by which name Ehrenberg, in 1838, designated a very long (100 to 500 μm) slender (0.5 to 1.0 μm) spiral and flexible organism that he found free-living in stagnant water. This organism, now the type species of the genus *Spirochaeta*, is called *Spirochaeta plicatilis.* The ends are blunt. Large intracellular granules of "volutin" or inorganic polyphosphates and fat are present.

The cells are in a tight spiral, and axial filaments are easily visible (Fig. 41–10). Motility is achieved by creeping and twisting movements. The organism may be found in sewage

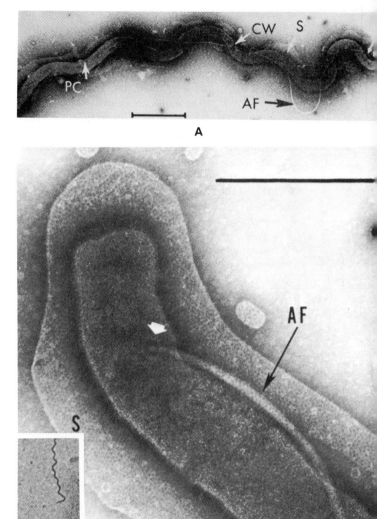

Figure 41–10. *A,* Electron micrograph of *S. stenostrepta* Z1. The subterminally inserted (*arrows*) axial filament (*AF*) traverses the cell and appears to lie over the protoplasmic cylinder (*PC*). The protoplasmic cylinder is enclosed by a cell wall (*CW*) which is surrounded by a sheath (*S*). Negative contrast. The bar indicates 1 μm.

B, S. aurantia strain J1. Electron micrograph of the terminal portion of a cell. The disc-like insertion structure (*white arrow*) of an axial fibril (*AF*) is visible. Note the polygonal substructure of the sheath (*S*). Negative stain. The marker bar represents 0.5 μm.

C, Phase-contrast photomicrograph of *S. aurantia* strain J1 (wet-mount preparations). The bar indicates 10 μm.

and stagnant and brackish water. Pure cultures have been made on medium containing extracts of leaves, and on 1.5 per cent agar with red blood cells. Growth occurs as a thin film on the agar, rarely forming ordinary colonies. *Spirochaeta* is microaerophilic and grows in a pH range of 6 to 9 at about 26C, and is entirely saprophytic.

Genus Cristispira. Members of the genus *Cristispira* also are entirely saprophytic organisms resembling the foregoing in habitat and overall morphology. They are somewhat shorter and thicker than *Spirochaeta* (40 to 120 by 0.5 to 3.0 μm). The cells are coarsely spiral, and the axial filaments are easily visible by phase-contrast microscopy. The habitat of some *Cristispira* is considered unique because they are found in the esophagi (crystalline style) of oysters and related mollusks. Cristispiras are distinguished by a **crista**, a sort of keel or end-to-end membrane, apparently constructed of flagella-like fibrils, which winds spirally about the organism from one end to the other, one edge free, the other attached to the cell. This structure is suggestive of the undulating membrane seen in one genus of pathogenic Protozoa, the trypanosomes. However, there is no evidence that the organisms are harmful to the host. Physiological and biochemical characteristics of the organisms are not established because the organisms have never been cultivated on artificial medium. Therefore, the ecological relationship between this organism and the host is not known, except that the organisms are more frequently found in healthy mollusks than in diseased ones.

Pathogenic Spirochetes

Members of genera *Treponema, Leptospira,* and *Borrelia* are the smallest and most tenuous of all the spirochetes. The family Treponemataceae of Bergey's 7th edition has been eliminated in the 8th edition of *Bergey's Manual*, but all the genera in the family are retained. All three genera contain dangerous and widely distributed pathogens of man and lower animals.

Structure. Because of the minute diameters of these organisms, the details of their cell structures have had to await the development of very high-powered electron microscopes. Studies of various species show that treponemes, leptospires, and borrelias, while differing in size and form, have analogous structures. All appear to consist of three principal parts: (1) An **outer envelope** or sheath or cell wall that is trilaminate like a "unit membrane" in some species. In other species the sheath is thin, vague, fragile, and easily removed by washing or agitation, more like a capsule than a cell wall (Fig. 41–11). Differences in appearance may result from differences in preliminary treatment of the cells. (2) The cell membrane is very thin (Fig. 41–9, *B*). Like the outer structure, this also appears to be multilayered or of the "unit membrane" type; in some species it appears double. In *Treponema* it is easily destroyed by surfactants (emulsifying agents) and may be rich in lipids. (3) All species except *Borrelia* (not yet demonstrated) have an **axial filament** that may consist of from 2 to 20 fibrils, depending on species and perhaps on extraneous factors. (See Figures 41–9 to 41–11.)

These fibrils appear to arise, like flagella, from discs or basal granules or groups of granules close to one end of the protoplast and to be wound spirally, close around the outer surface of the protoplast, inside the cell wall, from one end of the cell to the other (Fig. 41–9, *C* and *D*). When the cells divide, extensions of these fibrils sometimes appear beyond the tips of the cells as "terminal fibrils," once regarded as flagella. The fibrils from opposite ends of a pair of dividing cells are separate but appear at times to overlap on the protoplasmic cylinders. Fibrils may consist of many subfibrils that sometimes become frayed and resemble flagella, or the fibrils may be solid or have a distinct core. These appearances seem to be variable. Some spirochetes appear to have many large axial fibrils that seem to be tough and elastic. In all species motility appears to be caused by the contractility and elasticity of the fibrils and the rotation of the helical cell.

The internal structure of the protoplasmic cylinder appears granular or homogeneous in some species; markedly laminate or membranous in others. The nuclear material seems to be of the procaryotic type.

41.4
GENUS TREPONEMA

Genus *Treponema* contains the organisms that cause venereal syphilis, nonvenereal syphilis (bejel), the widespread tropical scourge yaws, and some other diseases **(treponematoses)** of man and other animals.

Treponemes are slender (0.25 to 0.4 μm in diameter) and seldom exceed a length of about 15 μm. Their size is therefore comparable with that of true bacteria. The organisms have neither crista nor septa. The 8 to 14 spirals found in *Treponema* are close and regular unless the contractions that characterize these organisms change them. The ends of the organisms are drawn out to extremely fine fibrils. These terminal fibrils have no function in the motility of the organism (Fig. 41–12). They are

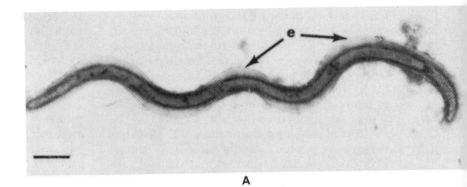

A

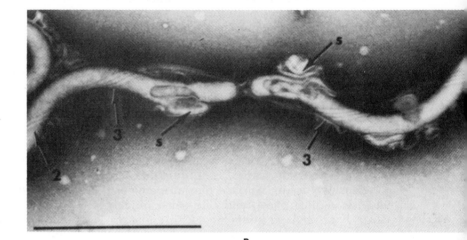

B

Figure 41–11. *A,* Large spirochete in gingival debris demonstrating well-pre-served outer envelope. (*e*). (Negative contrast. 9,300×.)

B, Dividing *Treponema microdentium.* Note three fibrils on either side of division site and two fibrils farther away. Also note localized disruptions of proto-plasmic cylinder (*s*) on either side of point of division. Negative contrast. (46,000×.)

probably extensions of fibrils of the axial fila-ments, as previously mentioned.

Treponema is not easily stained or seen in moist, unstained preparations by ordinary illumination. Indeed, the first *Treponema* to be described, that causing syphilis (Schaudinn, 1905), was named *Treponema pallidum* because of its pale appearance. Other methods are there-

Figure 41–12. *A,* Scrapings from a syphilitic chancre as seen in the microscope with dark-field illumination. The spirochetes are *Treponema pallidum.* The eight small rounded objects are erythrocytes; the two larger, rounded objects are tissue cells or pus cells; the smallest irregular objects are salt crys-tals, cell detritus, and other bacteria.

B, Electron micrograph of starting sam-ple of *T. pallidum* extract for zonal centrifuge run. (6,250×.) (*A,* Courtesy of Chas. Pfizer & Co., Inc., Brooklyn, N.Y. *B,* Courtesy of Dr. Myrtle L. Thomas.)

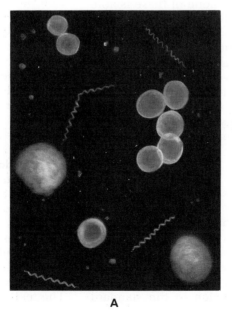

A

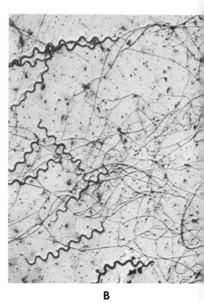

B

fore used to demonstrate treponemes microscopically. One is the method of negative staining; another, widely used to examine exudate from lesions of syphilis for diagnostic purposes, is the darkfield apparatus. A third, used mainly by the pathologist to demonstrate spirochetes in infected tissues, is termed **silver impregnation** (Fontana's method or Levaditi's method).

Resistance and Cultivation. *T. pallidum* is a relatively fragile, highly parasitic organism. It has never been cultivated in a virulent form in artificial media, in chick embryos, or in tissue cultures, although it may be maintained alive and virulent, without multiplication, for several days in certain complex artificial media; an important factor in the diagnosis of syphilis.

T. pallidum, under ordinary circumstances, can survive for only very short periods outside the tissues of man or experimentally infected animals; hence, nonvenereal infection of man is rare. However, when quickly frozen and maintained at −76C by means of solid carbon dioxide, syphilis spirochetes, as well as many other bacteria and viruses, remain viable and fully infectious for years. *T. pallidum* does not long survive drying. Surface tension–reducers, such as ordinary soap and bile salts, quickly cause lysis of *T. pallidum*, suggesting that the membrane of the protoplasmic tube may be rich in lipids. The organism is quickly killed by ordinary disinfectants. In citrated blood stored in blood banks the spirochetes quickly die out.

Reiter Treponeme. Some similar spirochetes, possibly mutants of *T. pallidum* (notably the Reiter treponeme), have been cultivated and grow vigorously, but none of these is able to cause infection. The Reiter treponeme is of especial importance because it appears to be antigenically almost identical with *T. pallidum*. It is therefore of great value in the diagnosis of syphilis by means of specific serological tests. Media used for cultivating the Reiter treponeme are complex organic mixtures containing numerous amino acids, fatty acids, vitamins, salts and serum. The organism is strictly anaerobic.

The Treponematoses

Infection of humans or other animals by any species of treponeme is properly spoken of as **treponematosis**. The most important of these diseases in humans are syphilis and bejel (*T. pallidum*), yaws (*T. pertenue*), and pinta or carate (*T. carateum*). The treponemes causing them are identical or very closely related, and all cause a positive reaction to the standard serological tests for syphilis, as well as very extensive destruction of body tissues. There is considerable cross-immunity between syphilis and yaws. With the possible exception of pinta all may be transmitted by contact with open (infectious) lesions: syphilis, venereally; yaws and bejel, nonvenereally; the last two are probably also transmitted by biting flies. All are extensively distributed: syphilis worldwide, yaws in tropics around the world, bejel in southeast Mediterranean areas, and pinta, a discoloration of the skin, in the tropics of the Americas. A summary of venereal and nonvenereal treponematoses is presented in Table 41–1.

Syphilis. Syphilis is usually transmitted by sexual intercourse. When so transmitted, and when it develops typically, it begins as a small ulceration on the mucosal surface of the genitalia within two to eight weeks after exposure. The spirochetes rapidly migrate from there to the deeper tissues of the body.

The ulcer increases in size, becoming rather hard **(indurated)** and flat. Upon removal of the crust, serous fluid oozes from the surface. This, upon examination with a darkfield apparatus, is found swarming with *Treponema pallidum* (Fig. 41–12, *A* and *B*). When syphilis is acquired through kissing, the ulcer may appear on the lip. The **primary ulcer**, oral or genital, is spoken of as a **hard chancre** (pronounced "shank'er"). It tends to heal spontaneously because of the development of antibodies after two to six weeks. The victim may believe himself cured. Attempts are made by the regional lymph nodes to arrest the migrating spirochetes. The nodes become swollen and firm and are sometimes called **buboes**. Their efforts, unfortunately, are futile.

What really happens is that by this time a certain degree of immunity has developed. The treponemes have long since migrated from the primary lesion, probably within less than an hour after exposure, and have been carried all over the body. They localize in various organs, particularly the liver, spleen, walls of arteries, heart, brain, skin, and mucosal surfaces, setting up **secondary lesions**. These begin to manifest themselves after two to four months. When situated on the skin these often appear as red blotches or an extensive rash and may be very infectious if open or moist, since they contain the treponemes. White patches may also appear in the mouth and on the genitalia. In such conditions of the mouth, kissing of other persons results in infection of the lips, tongue, or gums. The teeth may loosen and come out, as well as

TABLE 41–1. VENEREAL AND NONVENEREAL TREPONEMATOSES

Etiological Agent (Organisms)	Disease and Symptoms	Distribution	Laboratory Diagnosis	Epidemiology
T. pallidum	Syphilis. Formation of a characteristic primary inflammatory lesion known as a chancre followed by a generalized systemic skin rash (secondary lesion). Untreated patients may recover from secondary lesions but often enter a tertiary stage in which symptoms of the disease are difficult to recognize. Tertiary lesions may lead to general paresis or tabes dorsalis (see text for detailed information).	Worldwide.	Darkfield or fluorescent microscopy of the specimen obtained from primary or secondary lesions. Flocculation or complement-fixation tests for Wassermann antibody. If STS (serological test for syphilis) is negative and TPI (treponema immobilization test) is positive, FTA (fluorescent antibody) test should be made.	Ordinarily transmitted by sexual contact. The organism may be present in either external lesions or in discharge from deeper genitourinary tract. Organism most commonly penetrates mucous membranes but may be introduced through breaks in skin.
T. pertenue	Yaws (frambesia). Nonvenereal tropical syphilis resembling *T. pallidum* syphilis except for the character of lesion produced. Mother yaws (primary lesion) appears in 3 to 4 weeks as painless red papule surrounded by a zone of erythema (resembling a raspberry, framboise). Secondary lesions are similar to primary but appear several weeks later. The lesion on the soles of feet may become "crab yaws," hyperkeratotic lesions. Disfigurement of the nose and face are common.	Equatorial Africa and other tropical regions of heavy rainfall.	Since it shows positive Wassermann reaction and other serological reactions, diagnosis to differentiate this disease from venereal syphilis is based on character of chancre, mode of transmission, and frequency of visceral and tertiary lesions.	Common mode of transmission is contact between individuals. Biting flies and other flies which feed on serous fluid containing large numbers of organisms may also transmit the disease.
T. carateum	Pinta (mal de pinto [Mexico] or carate [Colombia]). Nonvenereal type and clinically a skin disease. Skin may undergo dyschromic changes (gray, bluish gray, or pinkish, but eventually becomes white).	Tropical regions; common in South America.	Spirochetes may be demonstrated in serous fluid from lesions or from fissures in plantar keratosis. Negative for TPI test, but serological reaction of Kahn is identical with that of syphilis. Wassermann reaction negative for primary stage but positive for secondary stage.	Serous fluid from early lesion is highly infectious when it comes in contact with lacerated skin. Usually transmitted by person-to-person contact. Flies have been implicated as possible vectors.

Mode of Transmission

Treponematosis $\longrightarrow$ Infected person

(rare) $\longrightarrow$ Nonsexual transmission from uninfected person through break in skin $\longrightarrow$ Infected person

$\longrightarrow$ Sexual contact with uninfected person $\longrightarrow$ Infected person

$\longrightarrow$ Placental transmission $\longrightarrow$ Infant with congenital syphilis

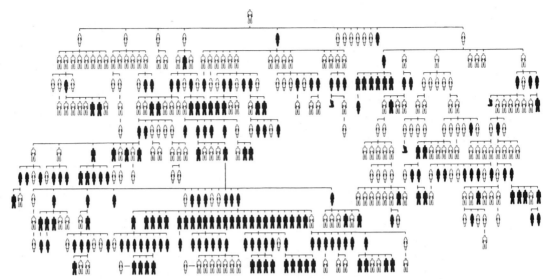

Figure 41–13. Persons involved in a typical syphilis epidemic. White figures represent adults; black figures represent persons under 20; small black figures represent infants. (Courtesy of U.S. Public Health Service Task Force on Eradication of Syphilis, L. Baumgartner, M.D., Chairman.)

the hair. Probably these and later lesions represent a reaction of delayed allergy. After a time, weeks or months, these outwardly visible secondary lesions slowly disappear in great part and the patient may again believe himself cured. Spontaneous cure actually may occur but this is not the usual outcome.

In untreated syphilis the treponemes usually slowly cause extensive **tertiary lesions**, called **gummata**, in various internal organs and on the skin. These tend to heal and form scars as the process continues. The liver becomes damaged and scarred (syphilitic **cirrhosis** of the liver), and bulges appear in the aorta where lesions in the layers of the vessel have weakened it. These bulges are called **aneurysms**. When they burst, death from hemorrhage often ensues. Gummata also occur in many of the bones.

The treponemes also damage the brain and spinal cord. Various nerve centers are slowly destroyed and characteristic forms of insanity and paralysis result. Death follows, sometimes after a period of many years. Mothers recently infected practically always have miscarriages, stillbirths, or sickly, syphilitic children.

Prevalence and Control. Syphilis is not pleasant and, indeed, the disease is one of the most insidious and dangerous. It is an epidemic disease of major importance (Fig. 41–13). Many more persons die annually of this disease in the United States than of poliomyelitis, diphtheria, mumps, measles, scarlet fever, ty-

phoid fever, and malaria combined. Over 25,000 new cases are reported each year and many more are never recorded. Venereal disease is *not* beaten. It is on the increase, especially among young people and teenagers, because of complacency, ignorance, and general laxity, as well as homosexuality and drug addiction.

As in the case of gonorrhea, prostitution and sexual promiscuity are the chief means by which syphilis is spread. In spite of sustained efforts by federal, state, and local authorities to educate the public to the dangers of syphilis and to enlist the aid of legislatures, of medical and civic authorities, and of *the people themselves* who are endangered, it increases steadily.

As one great physician has said, "The greatest obstacles to public health are the ignorance and indifference of the public!"

Syphilis Serology. The diagnosis of syphilis after the disappearance of the primary lesion in which the spirochetes are easily demonstrable microscopically is made by means of serological tests.

The Precipitin Test Applied to Syphilis. In complement-fixation tests, including the **Wassermann test** for syphilis, the complement is fixed because it is adsorbed onto finely divided particles of antigen-antibody precipitate. In the Wassermann test the precipitate is not visible. It would be a great advantage if we could see the precipitate directly, as in other precipitin reactions. This can be arranged.

A specially prepared and very concentrated

PLATE V

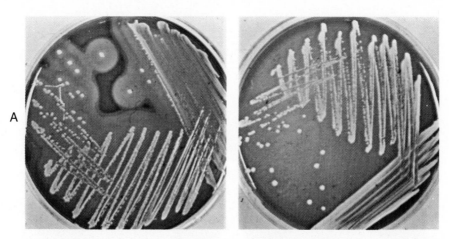

A, Staphylococci on blood agar. Note the wide area of hemolysis around *S. aureus* (left view) and the lack of hemolysis around the nonhemolytic *S. albus* (right view).

B, *Staphylococcus* colonies surrounded by satellite colonies of *Haemophilus* species.

C, A demonstration of the sensitivity of *Brucella suis* to four different dyes. Growth is not inhibited by thionin but is inhibited by pyronine, basic fuchsin, and crystal violet. Other *Brucella* species show different dye sensitivities, and these tests can therefore be used to differentiate among them. (See also Table 35–12.)

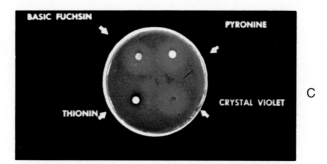

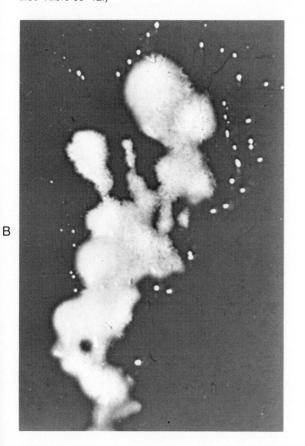

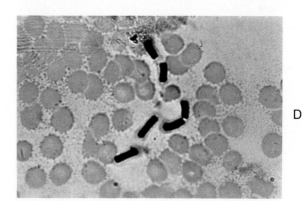

D, A blood smear showing infection by *Bacillus anthracis*. Note the characteristic rod-like shape of the bacilli. Giemsa stain.

PLATE VI

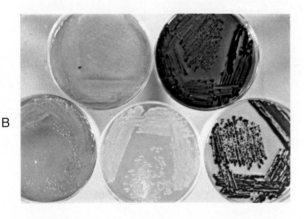

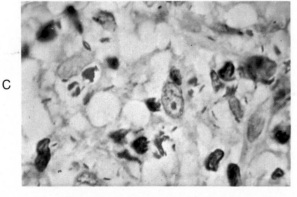

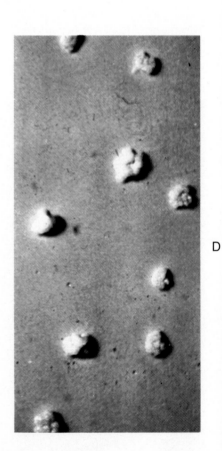

A, Color characterization of *Streptomyces lomondensis* sp. n., which produces the antibiotic Lomofungin. There are six types of agar culture media, and the appearance of the culture surface and underside are shown routinely on Ektachrome film to establish either the identity or the novelty of an antibiotic. (Courtesy of A. Dietz.)

B, Growth of the enteric bacterium *Klebsiella* on five selective media. The selective media help differentiate *Klebsiella* from the closely related *Enterobacter (Aerobacter)* and *Serratia.* Note the characteristic mucoid colonies on the media where growth has taken place.

C, A characteristic skin lesion in Hansen's disease (leprosy), a disease caused by the organism *Mycobacterium leprae. M. leprae* is always present in leprous tissue but apparently never occurs in normal conditions. This section utilizes a Ziehl-Neelsen stain; original magnification, 400×. (See also page 591.)

D, Characteristic colonies of *Mycobacterium bovis* on Löwenstein-Jensen medium. Note the dry, wrinkled appearance of the colonies.

alcoholic antigen is used in which the reactive substances are present but in the form of large, unstable colloidal complexes. These are brought by proper dilution with saline solution to a state where, in contact with syphilitic serum, they precipitate in a visible form. Generally, no precipitation occurs in the presence of normal serum *under proper circumstances.* Tests based on this principle are the Kahn test, Eagle test, Hinton test, Mazzini, and V.D.R.L.

It is worth noting that the antigen used in these tests is not specific. Commercial antigens used in the tests are derived from normal beef heart and are called **cardiolipins.** They can give false-positive reactions, with always distressing, sometimes tragic, results. The advantages of tests using **specific antigens** are therefore obvious.

Three specific tests are now available for the exact diagnostic study of syphilis, great steps forward in syphilology. All are dependent on **specific antibodies** against *T. pallidum.* One is the **T.P.I.** (*Treponema pallidum* immobilization)

test; another, the **immune adherence phenomenon;** the third, a **complement-fixation test,** all differing fundamentally from the Wassermann test in using extracts of *T. pallidum* as a specific antigen. A near-specific test is the complement-fixation test using purified antigens extracted from the Reiter treponeme. These are available commercially. Reiter treponemes are also used as antigen in detecting syphilis by the fluorescent antibody–staining technique (Chapt. 27).

41.5
GENUS BORRELIA

Borrelias resemble treponemes in many respects. However, *Borrelia,* unlike treponemes, can be stained readily by Gram's method or by means of polychrome stains (Jenner's stain, Wright's stain) used for Protozoa. Borrelias are gram-negative. They often have a less definite spiral form, being more wavy and open than treponemes, especially in microscopic, stained preparations (Fig. 41–14, *A* and *B*). In death

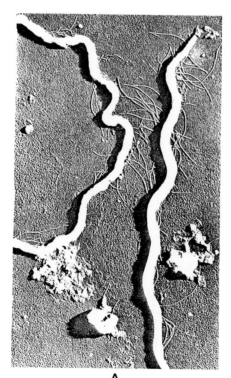

A

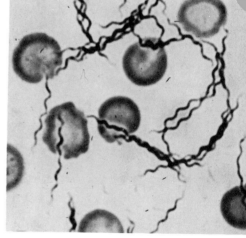

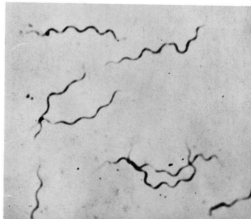

B

Figure 41–14. *A, Borrelia recurrentis* (relapsing fever). The flagella-like objects are presumably parts of intracellular fibrils of the spirochetes extruded as a result of mechanical damage to the cells. (50,000×.)

B, Relapsing fever spirochetes. *Top: Borrelia duttonii* of Central African relapsing fever; *bottom: Borrelia kochii* of East African relapsing fever. (2,000×.) (Courtesy of Dr. Kral.)

TABLE 41–2. BORRELIOSES

Etiological Agent (Organisms)	Disease and Symptoms	Distribution	Laboratory Diagnosis	Epidemiology
B. recurrentis and other Borrelia species	Relapsing fever. Disease characterized by sudden onset of fever after incubation of 3 to 4 days; initial period lasts for about 4 days followed by afebrile (relapse) period of a few days. The agent may be found in the blood and urine but not during the afebrile period. The disease becomes progressively less severe.	Worldwide but to lesser extent in Australia.	Direct microscopic examination of blood sample obtained during a febrile attack. Animal inoculation (young white rat) may aid in diagnosis.	The agent is transmitted by both ticks and lice; wild rodents are natural reservoirs.
B. vincentii* T. microdentium*	Vincent's angina, or trench mouth; an acute ulcerating disease of the oropharynx.	Worldwide.	Darkfield microscopic examination of specimen taken from ulcerative condition of mucous membrane.	B. vincentii is regularly found in association with Bacteroides fusiformis, gram-negative anaerobic bacillus. Borrelia as etiological agent of the disease remains open to question.

Mode of Transmission

Borreliosis   Infected rodents ⟶ Uninfected ticks or lice become infected ⟶ Infected ticks or lice bite uninfected man ⟶ Infected man   ⟶

*Precise agent is not known.

they seem to relax and lose their regular, coiled form. They are also somewhat thicker and coarser-looking than *Treponema*. Cultivation of some has been accomplished in media with sterile tissue, although it is not very satisfactory. They grow well in living chick embryos.

Commensal species of *Borrelia* occur, often in large numbers, in the normal mouth *(Borrelia buccalis)* and on the external genitalia *(B. refringens)*. Some of these so closely resemble *Treponema pallidum* in appearance as to create confusion at times in the diagnosis of syphilis by microscopic methods. They appear to be harmless. They are sometimes designated as *Spirillum*, *Spirochaeta*, or *Treponema*. Leeuwenhoek probably was the first to observe these.

Borreliosis. Most species of pathogenic *Borrelia*, like *B. recurrentis* and *B. novyi*, are blood parasites causing fever of a recurrent nature (**relapsing fever**). Many species infect lower mammals and birds. The spirochetes occur in large numbers in the blood during the numerous febrile relapses characteristic of the disease. In North America they are transmitted mainly by certain ticks; in Asia, Africa, and South America, by body lice (thus they are good examples of **arthropod-borne zoonoses**).

A summary of the more common borrelioses is presented in Table 41–2.

Trench Mouth. Some anaerobic species of *Borrelia* (notably *B. vincentii*) are found associated with ulcerative conditions (**trench mouth,** or **Vincent's angina**) in the mouth. They may be seen readily in gram-stained smears from such conditions, mixed with gram-negative fusiform bacilli *(Fusiformis fusiforme)*. The name **trench mouth** originated from the frequent occurrence of outbreaks of the disease in soldiers in trenches during World War I. The disease was thought to be transmitted by unclean eating utensils and other articles that carry saliva directly from mouth to mouth. It seems more correctly to be associated with dietary deficiencies, or the microorganisms may act merely as secondary invaders of lesions due to other causes, such as caries or herpes.

41.6
GENUS LEPTOSPIRA

Leptospira is the smallest (5 to 10 by 0.1 to 0.2 μm) of the spirochetes. The spirals of leptospires are so fine and so closely wound that when observed in the darkfield with optical

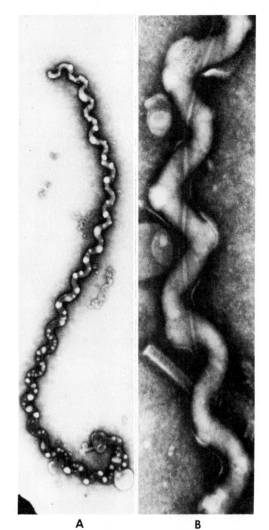

A **B**

Figure 41–15. Electron micrograph of complement-treated leptospire serotype canicola cells with adhering pieces of sheath, showing positions of axial filaments and intracellular dense bodies. Rectangular sheath fragments are evident. (*A*, 13,000×; *B*, 65,000×.) (Courtesy of D. L. Anderson and R. C. Johnson.)

microscopes only the outer curves of the spirals are seen, and the organisms appear like strings of minute, illuminated beads. The leptospires are further characterized by being bent into a hook at one or both ends (Fig. 41–15, *A*). Their motion consists of a writhing and flexing movement and a rapid rotation around the long axis. Their progression from place to place is rapid and can be readily explained on the basis of their screw-like form and their rotation (Fig. 41–15, *B*).

The structure of *Leptospira* is much like that of other Spirochaetaceae but is distinctive in that the long, cylindrical cell is wound as a helix around an extracellular, apparently uni-fibrillar, axial filament. The arrangement has been likened to a segment of plastic hose wound around a thinner segment of rather rigid wire. The DNA base composition of pathogenic leptospires ranges from 36 to 39 mole per cent GC.

Species and Serotypes. Numerous species of leptospires have been named in the past, but differentiation between many of them is difficult or impossible with methods presently available. As a matter of convenience two species are recognized: one, representing the harmless, free-living saprophytic types, *Leptospira biflexa*; a second, representing the pathogenic types, *Leptospira icterohaemorrhagiae*. *L. biflexa* and others are not sensitive to the purine analogue 8-azaguanine; pathogenic species are highly sensitive. *L. icterohaemorrhagiae* is divided into several serological (antigenic) groups and serotypes. Over 100 serotypes or subserotypes are recognized. Many of these were formerly given species names (see Table 41-3).

Distribution. *L. biflexa* (and related species or variants) is widely distributed in river and lake waters, mine drainage, bilge, stagnant ponds, sewage, and the like as a normal habitat where it appears to multiply. *L. icterohaemorrhagiae* probably does not normally grow significantly in such situations but is often found there as a result of pollution by the urine or dead bodies of infected mammals. *L. icterohaemorrhagiae* can live for months in polluted water and can infect humans and animals drinking or bathing in that water.

Cultivation. Leptospires may readily be cultivated in simple mineral solutions with peptone and serum—for example, Noguchi's or Korthof's medium, which contains NaCl, $NaHCO_3$, KCl, $CaCl_2$, KH_2PO_4, Na_2HPO_4, and 1 per cent peptone. All such media also require 10 per cent rabbit serum containing some hemoglobin. Leptospires may also be cultivated in colony form on the surface of 1 per cent agar containing tryptose phosphate and serum. The colonies give a positive oxidase reaction (Chapt. 37).

As a differential character, the saprophytes will also grow in media such as hay infusion, dilute feces, and egg water, without serum and at 10 to 15C. Some can be cultivated with minerals, vitamins, and tryptose phosphate and no serum. The pathogenic species require serum, prefer temperatures around 30C, and will not grow in mixed cultures such as feces or hay infusion. All are morphologically identical. Aside from the above cultural differentiations the different types are distinguishable only by special serological procedures.

Filtrability. Certain of the spirochetes (e.g., *Leptospira* and *Borrelia*) are readily filtrable through porcelain or Seitz-type filters which hold back ordinary bacteria. It is thought by some that the filtrability of these organisms is a result of the formation of minute granules which may represent phases in a life cycle. Material (e.g., lake water) that has been filtered and found on microscopic examination to contain no visible spirochetes has later been shown to contain spirochetes by cultural methods or animal inoculation. However, this does not prove conclusively that *only* invisible granules were present, since microscopic examination of a drop or two of fluid might easily fail to detect the presence of a few spirochetes in the greater, unexamined portion.

Small buds or granules appear to form at the end or on the side of some species, possibly as part of a reproductive cycle. Some of these may be merely blebs in the spirochetal sheath. Such granules may form in *Leptospira* and *Borrelia* and, if truly viable, account for their filtrability.

Leptospirosis. Infection with any species or serotype of *Leptospira* is properly called **leptospirosis** (see Table 41-3). Leptospirosis in man is fairly common, though often wrongly diagnosed. The more common leptospiroses are summarized in Table 41-4.

The various forms of leptospirosis are basically alike, although the symptoms vary. In general, the infection is common among numerous species of mammals. It is most commonly transmitted from animal to man through

TABLE 41-3. SOME COMMON LEPTOSPIROSES

Serotype	Usual Animal Host	Disease*
icterohaemorrhagiae	Rats	Weil's disease
canicola	Dogs	Canicola fever
pomona	Cattle, swine, horses	Swineherd's disease
autumnalis	Rats	Autumnal fever
hebdomadis	Field mice, other rodents	Seven-day fever
grippotyphosa	Field mice, other rodents	Swamp fever

*Terms commonly used for various leptospiroses.

TABLE 41-4. LEPTOSPIROSES

Etiological Agent (Organism)	Disease and Symptoms	Distribution	Laboratory Diagnosis	Epidemiology
L. ictero-haemorrhagiae (*Spirochaeta icterogenes*)	Weil's disease (infectious jaundice). Incubation period of 6 to 12 days. Onset abrupt, chills, followed by high fever, headache, photophobia, severe muscular pain. Conjunctivitis is the most apparent physical sign.	Worldwide.	Demonstration of the organism in the blood, and rise in titer of leptospiral antibody. Agglutination test, complement-fixation test may be used.	The disease is a zoonosis; wild rodents and domestic animals serve as principal reservoirs. The agent is present in urine of infected rats and is transmitted to man and other animals via stagnant water contaminated with rat urine.
L. autumnalis	Pretibial fever (swamp fever). The disease is characterized by an erythematous rash concentrated over the shins and other parts of the body.	Worldwide.	Isolation and microscopic demonstration of the organism from the blood of patients; rise in specific agglutinin antibody titer.	Similar to the above but also includes skunks, muskrats, raccoons, and others as reservoirs.
L. pomona	Swineherd's or Bouchet-Gsell disease; an acute illness which ordinarily lasts 3 to 10 days. Human infection from this disease is considered an occupational disease. In U.S., considerable numbers of cattle and wild animals have been affected by the disease.	Worldwide.	Laboratory isolation of the agent from blood and serological tests, such as agglutination and complement-fixation would help in identifying the organisms.	Domestic animals, especially swine and cattle (also dogs) serve as source of human infection. The agent may be disseminated among animals by contaminated water.

Mode of Transmission

Leptospiroses   Infected rodents, other ⟶ Feces, urine, etc., of infected ⟶ Consumption of, or contact with, contaminated water ⟶ Infected man
animals   animals contaminate water   by uninfected man or animal   or animal

the urine or bodies of infected animals and by anything so polluted (water, food) coming into contact with the skin or being ingested. Leptospirosis is thus another typical zoonosis. Doubt-less the same mode of transmission occurs among animals. Many wild and domestic animals carry leptospires in their urine. Transmission from man to man is uncommon.

CHAPTER 41
SUPPLEMENTARY READING

Benenson, A. S. (Ed.): Control of Communicable Diseases in Man, 11th ed. American Public Health Association, New York. 1970.

Breznak, J. A., and Canale-Parola, E.: *Spirochaeta aurantia*, a pigmented, facultatively anaerobic spirochete. J. Bact., 97:386, 1969.

Cockburn, A.: The Evolution and Eradication of Infectious Diseases. The Johns Hopkins Press, Baltimore. 1970.

Blair, J. E., Lennette, E. H., and Truant, J. P. (Eds.): Manual of Clinical Microbiology. American Society for Microbiology. The Williams & Wilkins Co., Baltimore. 1970.

McNeil, K. E., and Skerman, V. B. D.: Examination of myxobacteria by scanning electron microscopy. Int. J. Sys. Bact., 22:243, 1972.

ORDERS MYCOPLASMA-TALES (PPLO) RICKETTSIALES CHLAMYDIALES

42.1

PLEUROPNEUMONIA-LIKE ORGANISMS (PPLO)

In 1898 the French scientists Nocard and Roux (see Figure 3–10), studying pleural fluids of cattle suffering from a disease called pleuropneumonia, discovered organisms that were unlike any other microorganisms then known. The organisms were aerobic and were cultivable only on rich organic media containing about 20 per cent of animal serum. When cultivated on plates of such specially prepared agar these organisms were sometimes found to be spheroid in form, but they also produced a bewildering variety of minute granules, thin, branching filaments, stellate or asteroid structures, and many other highly irregular forms (Fig. 42–1). **Pleomorphism** (Gr. *pleo* = many; *morphe* = forms) is now recognized as one of the outstanding properties of these organisms. The species discovered by Nocard and Roux was called *Asterococcus mycoides* (now known as *Mycoplasma mycoides*), meaning, "rounded and stellate forms with radial, mold-like filaments."

After the original description of the bovine pleuropneumonia organisms, similar organisms were found in various other animals: sheep, goats, dogs, rats, mice, human beings. They are associated with various pathologic conditions, especially rheumatic or arthritic diseases, infections of the mammary glands, respiratory tract, and adjacent tissues, and inflammations of the genitourinary system. Similar organisms were found growing as saprophytes in decaying organic matter. All of these organisms have been referred to as **pleuropneumonia-like organisms** or PPLO, and are now commonly called **mycoplasmas**. The cells of these organisms lack a cell wall and fail to revert to walled organisms, hence the cells are delicate and plastic; apparently all the mycoplasmas are nonmotile.

Classification of Mycoplasmas

The group of long, branching, filamentous, and coccoid (pleomorphic, PPLO) organisms lacking cell walls was given the taxonomic status of order Mycoplasmatales under the class Schizomycetes in the 7th edition of *Bergey's Manual*. The order contains the family Mycoplasmataceae and genus *Mycoplasma*. However, in the 8th edition of the *Manual*, the order Mycoplasmatales contains two families, the division based on sterol requirement for growth: those mycoplasmas requiring sterol are grouped

621

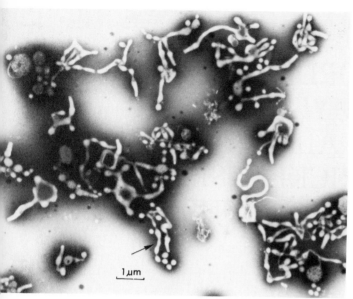

Figure 42-1. *M. pneumoniae*, negatively stained. The electron micrograph shows numerous forms of variable morphology, some of which are clearly ring-shaped with lobes and some which have beaded filaments (*arrows*). (Courtesy of E. S. Boatman.)

in Family I, Mycoplasmataceae, and the sterol-nonrequiring group is placed in Family II, Acholeplasmataceae, with each family having one genus (*Mycoplasma* and *Acholeplasma*, respectively). The genus *Thermoplasma* has been placed under the family Acholeplasmataceae, but its taxonomic position is still under discussion. The DNA base composition of order Mycoplasmatales spans the range from 23 to 41 mole per cent GC, with *Acholeplasma* species ranging from 30 to 36 and *Mycoplasma* species, except for *M. pneumoniae*, ranging from 23 to 36 mole per cent GC; *M. pneumoniae* ranges from 39 to 41 mole per cent GC.

In 1967, and again in 1972, with greatly increased knowledge of the organisms, the Subcommittee of the International Committee on Nomenclature of Bacteria published *Recommendations on Nomenclature of the Order Mycoplasmatales* and *Proposal for Minimum Standards for Description of New Species of the Order Mycoplasmatales*. These include: the classification of an organism as a member of the order should be based on such criteria as lack of a cell wall, typical colonial appearance, filtrability through a 450-nm membrane filter, and absence of reversion to a bacterium with cell walls under appropriate conditions; classification of the family should be based on sterol requirements.

Differentiation of the species of *Mycoplasma* may employ such systematic studies as fermentation of glucose and other sugars; hydrolysis of arginine and urea; production of carotenoids; serological tests; complement-fixation tests or double immunodiffusion techniques; phosphatase activity; tetrazolium reduction; proteolytic activity, including ability to digest coagulated horse serum, casein, and gelatin; hemolysis of red blood cells and adsorption of erythrocytes. Determinations of GC content of DNA, as well as nucleic acid homology studies, have also been recommended.

The proposal also includes guidelines for establishment of new species of Mycoplasmatales, and these will be presented in the sections below.

Cultivation of Mycoplasmas

Mycoplasma cells are able to grow on non-living substrates and, since the cells contain both DNA and RNA, it is clear they are not viruses. The coccoid or elementary bodies are the minimum reproductive units and have the smallest amount of DNA per cell of any known free-living microorganism, yet there is sufficient genetic information present to be able to synthesize all the necessary protein.

The organism to be isolated should be cloned by filtering the culture through a membrane filter having a pore diameter of 200 to 450 nm; the isolated colonies developed from the filtrate on a solid medium are picked and purified by repeated filtration. It has been reported that *M. laidlawii* and *M. granularum* strains passed through 220-nm filters (although some investigators believe 330 nm to be the limit of the smallest viable mycoplasmas) but were completely retained by filters having an average pore diameter of 100 nm. The purified cultures of mycoplasmas are best cultivated on solid (agar) medium, although they will grow, usually reluctantly, in liquid media. In general, media for their growth must be fairly concentrated and have a considerable osmotic pressure. A rich meat-infusion agar, pH 7.8, containing about 1 per cent peptone and 25 per cent serum, is satisfactory. Yeast extract, about 25 per cent, is often added, especially for forms parasitic on the animal body. Incubation is at about 37C and may be aerobic, although an increased (5 per cent) content of carbon dioxide is usually helpful. Some species are anaerobic.

To determine to which of the two families

the culture belongs, it is necessary to investigate its dependence on sterol for growth. If the organism belongs to the family Acholeplasmataceae, it will grow in serum-free fluid growth medium, or the same medium with or without Tween 80 and albumin, or the same medium supplemented with cholesterol plus albumin and Tween 80. However, if the culture belongs to the family Mycoplasmataceae, it will grow *only* in the above serum medium supplemented with cholesterol, but it should be tested with and without Tween 80 because growth of some nutritionally fastidious members of the family may be inhibited by Tween 80. Likewise, members of the family will exhibit a growth response to increasing cholesterol concentration in the above medium.

Mycoplasmas are also readily cultivated in living chick embryos and, as previously mentioned, in tissue cell cultures (Chapt. 16).

Subculture. Because, as will be explained, colonies of mycoplasmas adhere strongly to agar media, subculture of mycoplasmas is most convenient if a small agar block containing the desired colony is carefully cut out of an agar plate. It may be used to inoculate an appropriate fluid medium by macerating it in the fluid. However, it is usually more satisfactory to inoculate

agar by sliding the piece of colony-containing agar, inverted, over the surface of the fresh medium.

Colony Form. In addition to extreme pleomorphism a very distinctive property of mycoplasmas is the form of their colonies on agar medium. Growth is commonly initiated just beneath the surface of the agar. Expanding, it erupts onto the surface and spreads outward, the filaments tending to penetrate into the agar (Fig. 42–2). Thus the colony appears densest at its central point and, when mature, can be removed from the agar only with difficulty. Unlike colonies of other organisms cultivable on agar, individual cells of mycoplasmas are not visible in their colonies even with high-powered optical microscopes. The mature colony reveals a dark center and, over the nearby agar, many rounded or pleomorphic bodies ranging in size from barely visible to large, vacuolated or apparently empty sacs. These are regarded by some as stages in a reproductive cycle, by others as degenerate, nonviable, involution forms. The whole colony is said to have a "fried egg" or lacy appearance and is highly distinctive of mycoplasmas (Fig. 42–3). Colonies are always extremely minute, ranging from about 10 to 600 μm and usually

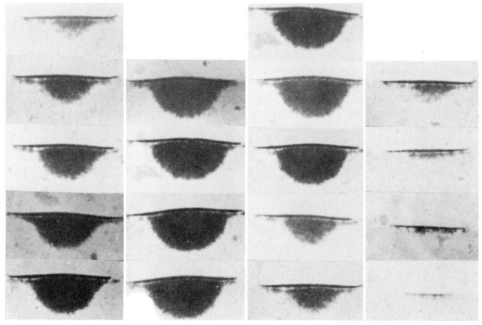

Figure 42–2. Photomicrographs of serial, longitudinal, 4-μm sections through a characteristic "fried-egg" colony of *M. pneumoniae,* suggesting distinct surface and agar growths. (832×.) (Courtesy of D. L. Knudson and R. MacLeod.)

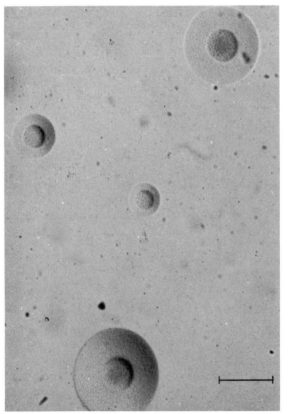

Figure 42-3. Colonial morphology of *Mycoplasma* S-743 strain on horse serum medium after five days of incubation. Marker represents 100 μm. (105×.) (Courtesy of J. G. Tully and S. Razin.)

resembling fine dust on the agar surface. They must be differentiated from **pseudocolonies:** minute crystals or other specks and irregularities that are common on agar surfaces. One group of mycoplasmas called **T strains** is distinguished by its **tiny** colonies. The DNA base composition of this group ranges from 27 to 28 mole per cent GC.

Morphology and Structure

Staining. Mycoplasmas can be seen much more readily if they are stained with an aniline dye. A good method of staining is that of Dienes, one of the first investigators of mycoplasmas and related organisms. Cut out a block of agar on which the colonies are growing and place it upon a slide. Then invert upon it a coverslip on which is dried methylene blue azure (deposited from alcoholic solution, Dienes' stain). The microscope is focused on the inverted

coverslip. Most species of living bacteria decolorize the stain (why?); mycoplasmas do *not.*

Mycoplasma cell preparations, under phase-contrast or darkfield microscopy, should show the typical pleomorphic morphology, characterized by very small coccoid bodies, ring forms, and fine filaments, some of which exhibit branching. However, the organism should be examined by electron microscope to confirm the characteristics listed above (Fig. 42–4, *A* to *D*). Mycoplasmas, like animal cells, have no demonstrable cell walls. Their sole retaining structure is the cytoplasmic membrane, which, like most other cell membranes, is of the three-layered or "unit" type (Fig. 42–5). The membrane of mycoplasmas is extremely thin, "limp," plastic, and elastic, hence the extreme pleomorphism of mycoplasmas. Because of the fragility of this membrane, mycoplasmas are very liable to mechanical distortion of form (as, for example, in laboratory manipulations) and also to osmotic rupture, unless the surrounding fluid has a considerable protective osmotic tension. (See §14.4, The Cell Wall.)

It is noticeable that lobule and filament formations by mycoplasmas are enhanced by "feeding" the organisms increased amounts of unsaturated fatty acids. The resulting increased lipid content of the membrane seems to make the membrane more elastic and thus more easily distorted but less likely to rupture. Low temperatures and polyvalent cations also seem to strengthen the mycoplasma membrane. Because of the large amount of lipid in the cell membrane, many mycoplasmas are markedly sensitive to lipid solvents like alcohol, or to lipid-emulsifying agents such as bile salts and detergents.

Internal Structure. The internal structure of mycoplasmas, as revealed by the electron microscope, is typical of procaryons generally and closely resembles that of other bacteria. Nuclear structures are present but are less evident in mycoplasmas than in typical bacteria. Ribosomes are clearly visible within the cells (Fig. 42–6). Mesosomes are conspicuously absent. This may result from the elasticity of the cell membrane (Fig. 42–7), which "stretches" so that invaginations of the membrane, such as mesosomes, cannot occur (Fig. 42–5).

Enzyme systems and metabolic characteristics of mycoplasmas differ somewhat from those of familiar bacteria but, in general, the differences are not extreme and, like many properties of mycoplasmas, can be ascribed to the physiological adaptation of mycoplasmas to life without a cell wall.

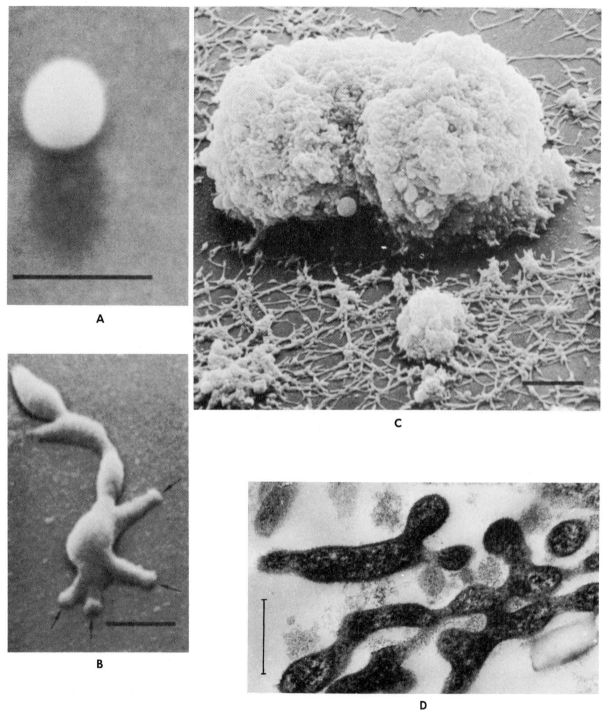

Figure 42–4. *A* and *B,* Scanning electron micrographs of *Mycoplasma pneumoniae* from cultures inoculated with filtered organisms. The bars represent 0.5 μm. *A,* Spherical organism from a culture incubated for 2 hours. *B,* An irregular, filamentous organism with several protrusions and constrictions. Note the knob-like swellings at the endings (*arrows*) of the protrusions; six-day-old culture.

C, Scanning electron micrograph of a six-day-old culture inoculated with unfiltered *Mycoplasma pneumoniae* organisms. Note the dense network of filamentous forms growing on the surface and the rounded appearance of the organisms in the colonies. The bar represents 10 μm.

D, Mycoplasma orale, type 1. Filamentous mycoplasma fixed with 0.25 per cent glutaraldehyde at 270 mOsm/kg. (*D,* Courtesy of Ruth M. Lemcke.)

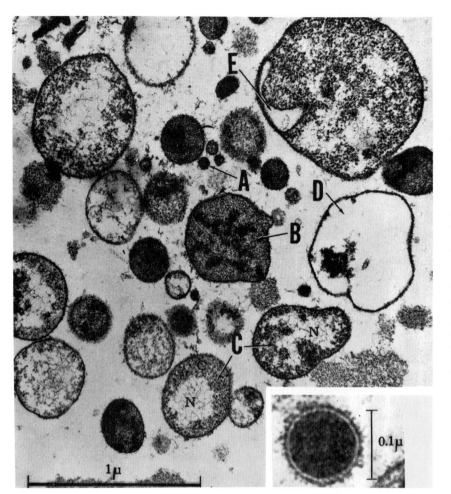

Figure 42–5. Culture of *Mycoplasma hominis* showing a variety of forms. At (*A*) is a small dense form ("elementary body"). One type of the large forms (*B*) has a finely granular protoplasm divided into light and dark areas. A second major type (*C*) has its protoplasm divided into a central nuclear area (*N*) of net-like strands and a cytoplasm (*c*) containing ribosome-like granules. The internal material in several of the large forms has a watery appearance, and sometimes only an empty plasma membrane is seen (*D*). One of the organisms in this field has a membrane-bound vacuole (*E*) at its periphery. (53,000×.) *Inset* shows body similar to *A* at a higher magnification. (200,000×.)

Reproduction of Mycoplasmas

The dimensions of mature cells of mycoplasmas are of the same order as those of typical bacteria. However, cells of mycoplasmas divide unevenly. Very minute bodies, called **elementary bodies** or **minimal reproductive units,** are commonly formed inside the **large bodies** or mature cells. These elementary bodies range in size from about 330 nm to around 450 nm. The smaller sizes of these can pass through bacteria-retaining filters (450 nm) like viruses but are viable on ordinary media. The elementary bodies have bacteria-like structures. These minute, living bodies of mycoplasmas are often cited as the smallest independently living entities (see insert, Figure 42–5) (viruses are not independently living).

These minute bodies represent a stage in the life cycle of mycoplasmas. They enlarge to form long filaments and mycelia and chains of minute spheres like conidia but much smaller. It is thought by some that these conidia-like bodies are liberated and that each increases in size to become a large body several micrometers in diameter, inside which new elementary bodies are formed. These are released by rupture of the membrane of the large body (Fig. 42–5). It has been reported that the growth rate of mycoplasmas is moderately rapid, with a generation time of one to three hours, depending upon the species and nutrient conditions.

Relation to Antibiotics

Mycoplasmas, like typical viruses and animal cells, are completely resistant to penicillin. Since the principal action of penicillin is to stop synthesis of the peptidoglycan complex of the procaryotic cell wall, it is understandable that it would have no effect on organisms without any cell wall. That mycoplasmas are not resistant to all antibiotics in general is shown by their sensitivity to the tetracyclines and to other antibiotics that do not depend for

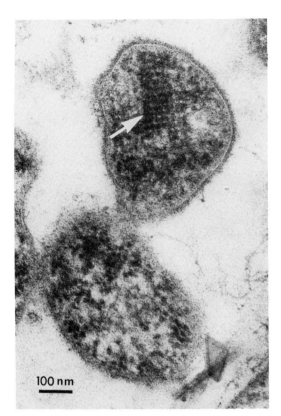

Figure 42–6. Thin section of *T-mycoplasma* (strain Cook), 12-hour culture. Ribosomes partly arranged in a regular geometrical pattern (*arrow*). Scale marker = 100 nm. (Courtesy of Finn T. Black.)

their effect on elimination of cell walls, e.g., kanamycin. (See also Color Plate I, *A*.) It has also been reported that certain species are attacked by mycoplasma phages.

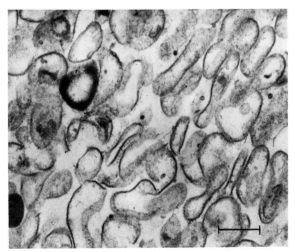

Figure 42–7. Thin sections of *Mycoplasma hominis* membranes isolated by digitonin. (24,000×.) Scale marker = 500 nm. (Courtesy of S. Rottem and S. Razin.)

Pathogenic Mycoplasmas

Mycoplasmas have usually been found in pathologic conditions of lower animals. A few harmless saprophytic species are known, a common one being *Mycoplasma laidlawii* (named for discoverer, Laidlaw). A number of species have also been found associated with a variety of diseases of various organs in man, but their etiological relationship, while probable, remains to be established in most cases. One species, *M. hominis*, is especially common in such situations. Some of the more common diseases caused by mycoplasmas are summarized in Table 42–1.

The first undoubted human pathogen of the mycoplasma group is the agent causing primary atypical pneumonia (PAP). Many viruses (e.g., influenza) cause pneumonias of this type, and it was long thought that only viruses were involved. About 1950 it was discovered by Eaton that many cases of PAP were caused by an agent that, like viruses, was filtrable, that was (at first) not visible with optical microscopes, and that, like viruses, could infect cells in tissue cultures. However, it was completely unlike true viruses in that, although insusceptible to penicillin, it was fully susceptible to other antibiotics. It was called "Eaton's agent." In 1962 it was shown to be cultivable on inanimate media and to be, in fact, a mycoplasma. It has since been called *Mycoplasma pneumoniae*. This brief history shows the importance of looking for mycoplasmas in conditions in which other microorganisms cannot be demonstrated. In a recent study, for example, mycoplasmas have been isolated from, and observed in, benign human tumors. The observation is stimulating but the exact etiological role of the mycoplasma remains to be fully investigated.

Other mycoplasmas have since been found as important and insidious contaminants in many tissue cultures, including some used in the propagation of viruses for human vaccines. These tissue cultures had formerly been thought to be "pure" cell lines. (Compare the difficulties with contamination experienced by Koch, Pasteur, and others a century ago.) In contaminated tissue cultures the mycoplasmas apparently live intracellularly, unrecognized although some produce definite CPE.* The commonest sources of these contaminations (mainly *M. hominis* and a species called *M. orale*) appear to be (a) the oropharynges of

---

*Cytopathic effect.

TABLE 42–1. PATHOGENIC MYCOPLASMAS*

Etiological Agent (Organisms)	Disease and Symptoms	Distribution	Laboratory Diagnosis	Epidemiology
M. pneumoniae (39 to 41)†	Primary atypical pneumonia (PAP); the disease is characterized by fever, nonproductive cough, headache.	Worldwide.	Complement-fixation test; hemagglutination test; immunofluorescence microscopy; demethylchlortetracycline has been used with moderate success. PAP patients may produce cold hemagglutinins in the serum.	Microorganisms are usually found in the pharyngeal exudate, serous exudate, or pleural fluid. Spread by infected droplets from the respiratory tract. No vaccine is yet available.
M. hominis (type I) (29 to 30)	Nongonococcal urethritis—Reiter's syndrome.	Worldwide.	See above.	See above.
M. mycoides (25 to 28)	Bovine pleuropneumonia; extensive consolidation and subpleural effusion in lungs.	Worldwide.	See above.	See above.
M. agalactiae (28 to 32)	Contagious agalactia; atrophy of lactating glands, eyes, and joints.	Most common in the Mediterranean region.	See above.	See above.

*While some species of mycoplasmas have been studied extensively (i.e., *M. mycoides, M. pneumoniae*), many aspects of disease production and epidemiology remain uncertain.
†DNA base composition (mole per cent GC) is in parentheses. For DNA base composition of other *Mycoplasma* spp., see Williams et al.: J. Bact., 99:341, 1969.

persons working with the cell cultures and (b) the calf serum usually added to the culture medium. Only the most assiduous aseptic technique can eliminate these contaminants. Contaminated cultures may be "cured" by using mycoplasma-immune serum (see Chapter 27) or antibiotics to which the mycoplasmas are sensitive. Contamination may be detected by several means, including culture methods on agar of appropriate composition; also by microscopic examination of tissue cultures stained with fluorescent antibodies **specific** for the mycoplasma involved (see discussion of fluorescent antibody staining, Chapter 27).

L-Forms

As explained in Chapter 14, protoplasts or spheroplasts may be prepared from various bacteria either by removing the cell wall or by metabolically blocking synthesis of the cell wall. An enzyme (such as lysozyme) which dissolves the cell wall and an antibiotic (such as penicillin) which inhibits cell wall synthesis are most frequently used in preparation of the protoplasts. Such protoplasts in hypertonic or isotonic fluid are often able to grow and multiply on agar media. The colonies developed from such protoplasts resemble the colonies of myco-

plasmas (compare Figures 42–3 and 42–10). These cells without cell walls have been designated as stable L-forms. These observations, about 1953, brought into question the status of *Mycoplasma* as a separate and distinct group of microorganisms. Recently, we have been able to differentiate some stable L-forms frequently found in animals and man from true mycoplasmas. For example, DNA base composition and nucleic acid homology or hybridization studies have shown that many of the stable L-forms are unrelated to mycoplasmas. DNA base composition of *Mycoplasma* species ranges from 23 to 36 mole per cent GC, whereas most other bacteria and the L-forms usually have higher per cent GC content. For example, DNA base composition of L-forms of *Corynebacterium* species was reported to be 63 to 64 mole per cent GC, which is within the range of plant pathogenic species (60 to 74 mole per cent GC).

Some bacteria undergo spontaneous mutation to loss of cell walls. For example, some bacterial mutant auxotrophs (Chapt. 18) require diaminopimelic acid (DAP) among their nutrients in order to form their cell walls. DAP is peculiar to bacterial cell walls. If DAP is withheld, *and* the medium is **osmotically protective**, the bacteria grow *without* a cell wall; i.e., they are L-forms of the bacteria from which they were derived (Figs. 42–8, 42–9). If DAP

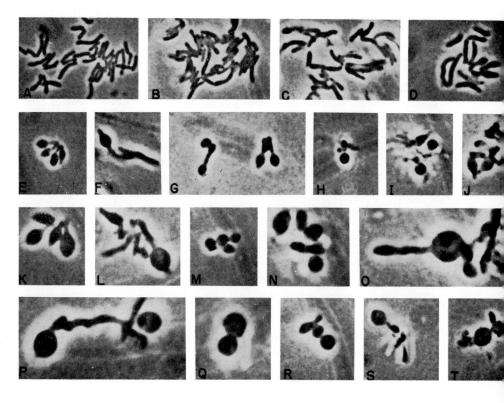

Figure 42–8. Formation of stabilized spheroplasts of *Mycobacterium tuberculosis* by cultivation in a low-surface-tension fluid medium containing lysozyme and ethylenediaminetetraacetate (EDTA) with Tris buffer, Mg^{2+} and osmotically protective (0.34 M) sucrose. Changes in morphology from normal bacillary form (*A, B, C*) to swollen translucent forms at 6 days (*E*) and 10 days (*F, L*) and to globular and spheroidal (spheroplast) forms after incubation for 14 days (P-T) are clearly shown.

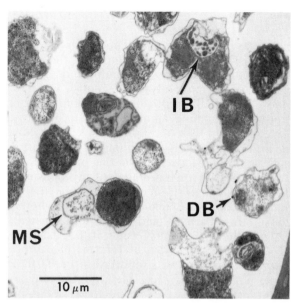

Figure 42-9. *B. abortus* L-forms from a 96-hour broth culture with penicillin added initially and at 48 hours, showing variations in size, shape, and cytoplasmic density of the cells; small internal bodies (*IB*); internal membranous structures (*MS*); and small dense bodies (*DB*) on the surface of some cells. (Courtesy of Betty A. Hatten.)

is fed to them they revert to their original form. Like mycoplasmas, none forms colonies on ordinary media, but colonies are formed on media **osmotically protective;** i.e., media with about 3 per cent sucrose or other nontoxic osmotically active substance. L-forms possess virtually all of the distinguishing features of mycoplasmas (Fig. 42–10). Eucaryotic fungi may also grow as L-forms (Fig. 42–11).

Stability of L-Forms. Usually, when separated from the influence of the agent that removed (or prevented formation of) their cell walls (penicillin is commonly used), L-forms more or less rapidly revert to their original status as bacteria, identical in virtually all respects with the bacteria from which they originated. Such forms are said to be "transition" or "unstable" L-forms. However, under certain conditions of cultivation in the laboratory some may retain the wall-less state tenaciously for considerable periods without reversion. They become fairly well stabilized as L-forms even when cultivated in ordinary media. For example, it has been reported that the L-forms of *Proteus* and *Salmonella* strains which have been transferred for 15 to 20 years have failed to revert even after repeated growth in media in the absence of penicillin. Should such well-stabilized L-forms, if not known to have origi-

nated previously from bacteria, be called mycoplasmas? *No undoubted mycoplasma has ever been demonstrated to originate from, or revert to, a bacterium.* However, several organisms long supposed to be mycoplasmas have lost their status as mycoplasmas by reverting to common types of bacteria and back again. Thus some workers, understandably, regard L-forms and PPLO or mycoplasmas as identical, differing only in origin and stability of the wall-less state and in other relatively minor properties. The differentiation of these two groups should not depend solely on the knowledge of origin and lack of reversion but should also include DNA base composition and nucleic acid hybridization studies.

Colony Types. In the bacterial species *Proteus mirabilis*, at least two types of L-forms are differentiated by their colonies: 3A and 3B. Type 3A L-forms induced by penicillin require serum for growth and initially produce in agar very small subsurface colonies containing minute, deeply staining granules. The subsurface growth appears on the surface later.

Type 3B L-forms produce larger surface colonies that are definite in outline and slightly raised, having a dark, vacuolated or granular center, with a central density caused by growth beneath the surface. The peripheral surface growth may appear granular in the center, lacy and vacuolated at the periphery (Fig. 42–10, *C*). Type 3B L-forms do not require serum, and they readily revert to the parent bacterial form when grown in the absence of penicillin.

Type 3A L-forms are so slow and reluctant to revert to the parent bacterial forms that they could be regarded as **stable** or permanent L-forms (mycoplasmas?) if their origin were not known and if DNA base composition were considerably less than 36 to 41 mole per cent GC, which represents typical *Proteus* species.

Just how mycoplasma species evolved from normal walled bacteria is not known, but it is conceivable that the organisms may have evolved from special habitats, such as lysozyme- and penicillin-rich environments.

42.2
RICKETTSIALES

Rickettsiales are very small, gram-negative coccoid cells or short rods (0.3 to 1.0 μm) and are obligate intracellular parasites in a certain group of blood-sucking arthropods, such as fleas, mites, lice, and ticks. Some rickettsias apparently cause no injury to the arthropods,

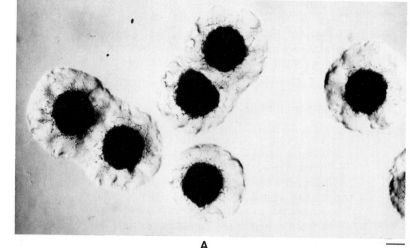

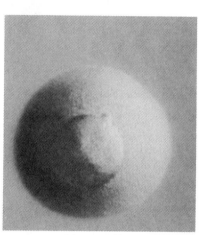
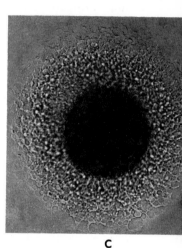

A

B

C

Figure 42-10. *A,* Comparison of a mature L colony (*Streptococcus faecium*) as photographed after growth on a Millipore filter (*A*) with mature L colonies as developed on the agar surface surrounding the Millipore filter. Bar represents 200 μm.

B, Meningococcal L-forms on agar surface. (140×.)

C, Unstained L colony induced from a penicillin G resistant strain of *Staphylococcus aureus* in a medium containing 500 μg/ml of methicillin. (115×.) (*A,* Courtesy of Harry Gooder. *B,* Courtesy of D. N. Wright.)

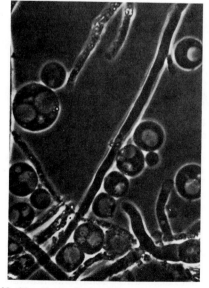

Figure 42-11. Protoplasts of *Fusarium culmorum.* Note the variation in size of the protoplasts and the large vacuoles in some of them. (3,000×.)

but they are often pathogenic to man and other animals (vertebrate hosts). As a rule, rickettsias are transmitted to the vertebrate host by the bites of the arthropod vectors (ticks and mites) or via their feces (lice and fleas). Some rickettsial diseases are severe and often fatal; for example, epidemic typhus. The causative agent of the disease is *Rickettsia prowazeki,* and it is transmitted to man by body lice. The disease has long been associated with war, famine, human misery, and suffering through the history of mankind.

Other important human rickettsial diseases of man and other vertebrates are tick-borne Rocky Mountain spotted fever (*R. rickettsi*), louse-borne trench fever (*R. quintana*), flea-borne endemic typhus (*R. mooseri*), mite-borne scrub typhus (*R. tsutsugamushi*), commonly known in Asia as "tsutsugamushi disease," and Q fever. The latter disease is caused by *Coxiella burneti* and is associated with cattle, goats, and sheep. The agent is transmitted by "fomites," inanimate vectors such as dust from infected animals, or improperly pasteurized milk (see Chapter 47). Some ticks harbor, and may transmit, *C. burneti*.

Classification of Rickettsias

The precise relationship of the rickettsias to other bacteria is not known, but in the past they have been thought to occupy a taxonomic position between true bacteria and viruses. The organisms were considered to be the product of "degenerative evolution"; that is, evolved from other bacteria by slow but progressive loss of true bacterial function as a result of mutation and obligate intracellular adaptation. Much attention has been directed to cultivate them or to maintain them in a viable state outside the host cell, but thus far they have not been unequivocally cultured in nonliving media. For this reason, it has been difficult to gather information about their mode of reproduction, metabolic activities, and biochemical pathways. However, with the advent of arthropod and mammalian tissue cultures, with the discovery that the organisms can be readily cultivated in the yolk sacs of chick embryos, and with the use of the electron microscope, much of the needed information concerning classification of the rickettsias has been elucidated.

The organisms multiply by binary transverse fission within the host cell; they are gram-negative, are coccoid or rod-shaped cells (0.3 to 1.0 μm), and are generally considered nonfiltrable. Electron micrographs of thin sections of the organisms show distinct cell walls and cell membranes which are much like normal bacterial morphology (Fig. 42–12).

In addition, the cell walls of the rickettsias contain peptidoglycan, and the cells contain both DNA and RNA as do the bacteria (viruses contain either RNA or DNA but never both). The DNA extracted from rickettsias is of double-stranded form; the base composition of *R. prowazeki* is 30 to 32 mole per cent GC and of *C. burneti* is 44 to 45 mole per cent GC. The organisms appear to be able to generate their own ATP (chlamydias and viruses are unable to generate ATP), as evidenced by oxidation of amino acid and glutamate, by cell suspensions. DNA base compositions of other rickettsias are listed in Table 42–2.

In the 8th edition of *Bergey's Manual*, rickettsias are grouped in Part XVIII, which consists of two orders: Order I, Rickettsiales,

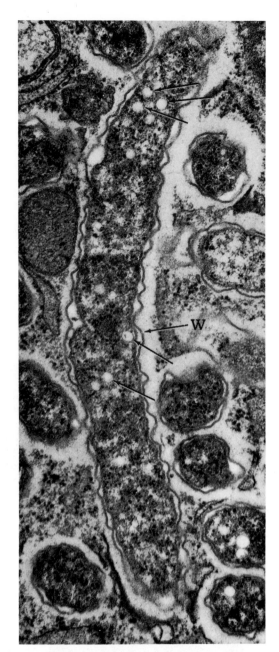

Figure 42–12. Longitudinal section through *Rickettsia prowazeki* which is 4 μm long. Several cross sections are also included in the field. The rippled cell wall (*W*) is obvious, and the underlying plasma membrane which bounds the rickettsial cytoplasm is easily seen in several places. The internal material consists of ribosomes and DNA strands, and this species also has electron-lucent spherical structures, some of which are indicated with arrows. (45,000×.)

and Order II, Chlamydiales. The order Rickettsiales contains three families, Rickettsiaceae, Bartonellaceae, and Anaplasmataceae. The family Rickettsiaceae is further divided into three tribes: Rickettsieae, Ehrlichieae, and

Wolbachieae. The family known as Chlamydiaceae in the 1957 edition of the *Manual* is now considered an order and is composed of the family Chlamydiaceae and a genus, *Chlamydia* (Chapt. 31).

Only two genera of the tribe Rickettsieae (genera *Rickettsia* and *Coxiella*) will be described here, since they are fairly representative of the entire order. Formerly the tribe Rickettsieae included two genera, *Rickettsia* (comprising four subgenera, A to D) and *Coxiella*. The new *Manual* (8th edition) lists three genera, *Rickettsia*, *Rochalimaea*, and *Coxiella*, with the organisms belonging to the first genus considered not filtrable and the second filtrable under certain conditions, while *Coxiella* species are filtrable through bacteria-retaining filters.

Genus Rickettsia

Discovery. Howard Taylor Ricketts, an American medical scientist, while studying Rocky Mountain spotted fever in 1909, described as the causative agents of that disease a group of microorganisms that differed from any that had been previously described. A year later he discovered similar organisms as the cause of typhus fever (not typhoid fever caused by *Salmonella typhi*) while working in Mexico. During the latter studies he contracted the disease and died.

In 1916, H. da Rocha-Lima, a Brazilian scientist, made further observations of the organisms described by Ricketts and named the genus *Rickettsia*, in honor of its discoverer. He also gave the species name *prowazeki* to the rickettsias associated with louse-borne or classical typhus fever, in honor of another scientist, Stanislaus von Prowazek, of Hamburg, who had also lost his life in the study of that disease. The causative agent of louse-borne typhus fever is therefore called *Rickettsia prowazeki*. It is the type species of the genus *Rickettsia*.

Properties of Rickettsias

Morphology and Staining. The rickettsias are much smaller than typical bacteria, having diameters of about 0.5 to 1.0 μm and lengths seldom exceeding 5 μm and often much less than this. However, they are larger than any of the viruses and can be seen readily with ordinary microscopes at magnifications around 1,000×. They multiply by transverse binary fission, like other bacteria. They are variously shaped: cigar- or rod-shaped, spherical or ovoid (Fig. 42–13). They may cling in pairs or chains after fission. Sometimes relatively long filaments are formed. No spores are produced. Rickettsias are not motile.

Unlike most bacteria, it is difficult to stain rickettsias with ordinary basic aniline dyes. They can, however, be colored very well with Giemsa's (special stains used for blood and protozoans) or Machiavelli's stain. If stained by Gram's method they are gram-negative. The rod forms tend to stain more intensely at the tips, often giving short rods the appearance of a pair of minute spheres or cocci (bipolar staining).

Electron microscope studies of ultrathin cross sections of *Rickettsia* (magnifications around 100,000×) indicate that rickettsias possess a "unit" type (three-layered) cell membrane and a two- or three-layered cell wall. Biochemical studies show that, like typical bacteria, the rickettsial cell walls contain muramic acid, a component of peptidoglycan that occurs only in bacterial cell walls. The internal structure as seen in ultrathin sections closely resembles that of familiar bacteria, and includes ribosomes and nucleoids of the procaryotic type (Fig. 42–12). Invaginations of the cell membrane suggest mesosomes or a primitive endoplasmic reticulum.

Cultivation. In the matter of growth, rickettsias resemble the viruses and chlamydias in that they are not cultivable on nonliving material. However, rickettsias, like many bacteria, are easily cultivated in **live chick embryos.** They will also grow to some extent in **living** cells of tissue cultures, such as those used to cultivate viruses (see Chapter 16). They appear to grow best in the live cells lining the egg-yolk sac. Yolk-grown rickettsias are widely used as antigens in procedures for the serological diagnosis of rickettsial diseases and for the preparation of **rickettsial vaccines.**

Metabolism and Antibiotics. Rickettsias have been shown to have autonomous, though incomplete, metabolic activity. Difficulties have been encountered in elucidating some of the biochemical properties of the organisms because not only do such studies require large populations but the organisms must be pure or free from contaminating host tissues which may confuse biochemical and metabolic studies. The difficulties have been compounded by the fact that rickettsial cells freed from host tissues rapidly lose their viability. Loss of viability may be retarded by addition of some key

TABLE 42–2. RICKETTSIAL DISEASES

Etiological Agent (Organisms)	Disease and Symptoms	Distribution	Laboratory Diagnosis	Epidemiology
R. prowazeki (30 to 32)°	Epidemic typhus fever or European typhus characterized by abrupt, severe aches, fever, and chills, followed by rash on back and chest.	Balkans, Ukraine, Asia, Mexico, and South America.	High agglutinins (Weil-Felix reaction) for OX-19; low for OX-2; negative for OX-K.	Transmission via body lice to man and animals. Infected man may serve as reservoir along with lice.
R. rickettsi (32 to 32.5)	Rocky Mountain spotted fever. Abrupt chills, fever, and prostration. Rash occurs peripherally on the ankles, wrists, and forehead and spreads to trunk (reverse of typhus fever).	Eastern and western United States.	Complement-fixation test; Weil-Felix reaction positive for OX-19, OX-2.	The bite of wood tick (*Dermacentor andersoni*) is responsible for transmission of Rocky Mountain spotted fever in western U.S., and *D. variabilis* is usually responsible in the eastern U.S. Agents may be transmitted from tick to tick, or tick to rodent, or dog to tick to human.
R. mooseri (30 to 31)	Murine typhus (flea-borne) or endemic typhus. Latent illness in both rat and rat flea hosts. Relatively mild form of typhus occurs endemically in man.	Worldwide.	Weil-Felix agglutination reaction positive for OX-19 and complement-fixation tests.	The agent is transmitted from rat to rat and rat to man by bite of rat flea (*Xenopsylla cheopis*). The agent is shed by rat flea in feces.

R. tsutsugamushi (30 to 33)	Tsutsugamushi disease or scrub typhus; abrupt chills, fever, and intense headaches. Within a few days rash and pneumonitis become evident. Stupor and prostration are often pronounced. Site of mite bite may develop an eschar—regional lymph nodes enlarged.	Southeast Asia and Pacific region.	Patient develops elevated titer of OX-K, but negative for others; isolation of the organism from patient's blood during acute illness and examination with microscope.	Transmitted by mite (bite) to man. Mite acts as both reservoir and vector. Wild rodents may serve as reservoir. (Man is an incidental host.)
R. akari (32 to 33)	Rickettsialpox—vesicular rickettsiosis. Disease characterized by a distinctive red cutaneous papule and enlarged regional lymph nodes; the rash is vesicular and simulates chickenpox (varicella).	United States, Soviet Union.	Etiological agent does not elicit antibodies to *Proteus* OX strains.	Disease is transmitted by bite of blood-sucking mite; mite may serve as reservoir and vector. The house mouse may become reservoir.
C. burneti (44 to 45)	Q fever in man is pneumonitis without rash; onset of the disease is abrupt with chills and fever; pneumonitis is considered "atypical pneumonia."	Worldwide.	Isolation of organisms from sputum, urine, or blood during acute illness. Specific complement-fixation test.	Disease may be acquired by inhalation of contaminated aerosol shed by infected cattle and other domestic animals. Ticks, body lice, wild rodents, cattle, sheep, and goats may become reservoirs.

*DNA base composition (mole per cent GC) is in parentheses. DNA base composition of other rickettsial organisms: *R. typhi* (30 to 31), *R. conori* (30), *R. canada* (30), *Rochalimaea quintana* (38 to 39). All DNA base compositions from Tyeryar, F. J., Jr., et al.: Science, *180*:415–417 (April 27), 1973. Copyright 1973 by the American Association for the Advancement of Science.

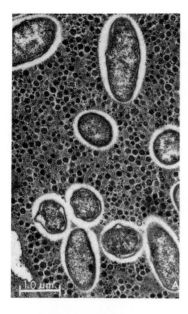

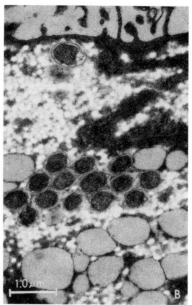

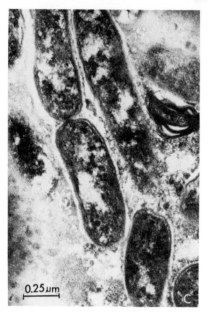

Figure 42–13. *A, R. canada* in Malpighian tubule cells of male *D. andersoni*. Each organism has a well-defined "halo" surrounding it. Cytoplasm of tubule cell contains numerous glycogen-like particles.

B, R. canada in Malpighian tubule cell of unfed female one month after molting. Rickettsias are clustered adjacent to microvillar lined luminal surface of cell.

C, R. canada in the cytoplasm of a hypodermal cell, adjacent to the basement membrane and integument. These atypical type II growth forms, including division form, have a "coagulated" appearing cytoplasm and apparently are devoid of ribosomes. (Courtesy of L. P. Brinton and W. Burgdorfer.)

coenzymes, such as NAD and CoA, which are believed to be supplied by the host cell during the intracellular multiplication. Hence it has been suggested that the rickettsias possess a more permeable membrane than do free-living bacteria, as evidenced by incorporation of large coenzyme molecules which do not readily penetrate membranes of free-living bacteria. Therefore, the rapid death of the cells outside of the host tissue is attributed to leakage of intracellular constituents. This phenomenon may explain the rigid requirement of a vector (arthropod) for mediating animal-to-animal

transmission of the agents, except the agent of Q fever, *Coxiella burneti*. This ability to survive between hosts in the free state may explain why the cells of *C. burneti* are less easily damaged during purification, and for this reason many biochemical studies have been performed with this organism.

It has been demonstrated that rickettsias do possess a complete cytochrome system and are able to carry out oxidative phosphorylation. However, the organisms are unable to oxidize amino acids, except glutamate, through the reactions of the citric acid cycle; i.e., they are able to utilize pyruvate and succinate using NADH derived from the oxidation of glutamate as electron donor. Hence it is implied that, unlike chlamydias (see §42.3), they are able to generate their own ATP. Though they appear to have lost (or never to have had?) their own independent systems of digestive and synthetic enzymes (degenerative evolution?), they have apparently retained (or gained?) at least some energy-mediating enzymes, since they are markedly susceptible to antibiotics that interfere with energy-mediating enzymes.

Habitat. Rickettsias seem to be primarily parasites of insects, and to appear only secondarily in man and other animals. The microorganisms characteristically inhabit the cells lining the intestines and other tissues of arthropods. Arthropods are usually not affected by the parasites and thus serve as a natural reservoir of infection. Rickettsias not pathogenic for man have been found in ticks, fleas, lice, bedbugs, spiders, and mosquitoes. Human-pathogenic species of rickettsias primarily inhabit arthropods that bite man or animals or both.

Mode of Transmission and Pathogenesis

Some species of pathogenic rickettsias are found in the salivary glands of biting arthropods, whence they may be transmitted to man. Other species of *Rickettsia* occur in the intestinal contents of sanguisugent (L. *sanguis* = blood; *sugere* = to suck) arthropods and appear in the feces. Transmission of rickettsias to animal hosts may thus be obtained by bites or by rubbing or scratching the fecal deposits of arthropod vectors into the skin. Some rickettsias pathogenic for man are also pathogenic to their insect hosts; e.g., *R. prowazeki* kills body lice.

The onset of rickettsial disease begins in the vascular system following the bite of an infected arthropod. The organisms first multiply primarily in the endothelial cells. Subsequently the agent is widely disseminated by way of the blood and frequently obstructs small blood vessels and as a result forms small thrombi. As a rule, the incubation period lasts one to four weeks, followed by headache, chills, fever, and hemorrhagic rash, and the disease may lead to shock in severe cases. The organisms may produce toxins which can be neutralized by homologous group-specific immune serum. The organisms also produce a hemolysin which lyses red blood cells of rabbits and sheep. However, the roles of toxins and hemolysins

in human rickettsial disease have not been fully elucidated.

Diagnosis of the disease is usually made by analysis of antibodies produced by the patient during the course of the illness. Two kinds of antigens, soluble "group antigens" and the insoluble "type-specific" antigens (which are obtained from rickettsial cells cultivated in yolk sacs of chick embryos) are used as antigens to react with the patient's serum. Positive complement-fixation reaction (Chapt. 26) of the patient's serum with group antigens reveals the relationship of the etiological agent to one major rickettsial group (i.e., epidemic typhus, Rocky Mountain spotted fever, scrub typhus, Q fever, trench fever, etc.), while a reaction with the type-specific antigen identifies the etiological agent as of a certain species, or strain within a species.

The Weil-Felix reaction is widely used in the diagnosis of rickettsial disease. It is not rickettsial antigens that are involved in this reaction but the polysaccharide O antigen of a certain strain of *Proteus*, called the X strain, and the antigen is referred to as **Proteus OX antigen.** The reaction is based on the fact that antibodies from patients with rickettsial disease react with the *Proteus* OX antigens. The Weil-Felix reaction utilizes three *Proteus* OX strains: OX-2, OX-19 and OX-K. The agglutination reactions between serum from the suspected diseased patient and all three strains help to differentiate among the various rickettsial diseases. Results of the Weil-Felix test may be misleading because patients infected only by *Proteus* species may also give positive reactions.

Genus Coxiella

There is only one species in this genus, *Coxiella burneti.* The genus is named for Herald L. Cox, codiscoverer of the organism in the United States; the species is named for F. M. Burnet (Nobel Prize winner), who first studied the organism in Australia. *Coxiella burneti* causes an influenza-like disease of the respiratory tract. The disease is called **Q fever,** Q standing for **Query.** The word was applied to the unknown diagnosis of the first cases to be observed by Derrick in Queensland, Australia. Q fever was first thought to be confined to Australia, but later it was found that the disease is worldwide. In some areas of the United States the prevalence of infection among cattle may reach more than 50 per cent of some herds. Infected cows shed *Coxiella* in milk, but the organisms are destroyed by modern pasteurization temperatures (145F for 30 minutes or 162F for 15 seconds); hence there has been no major epidemic originating from ingestion of milk obtained from infected cows.

Q fever may be acquired by inhaling contaminated dust and aerosols. The infected animals also shed the organisms in their nasal discharge and salivary secretions. Unlike other rickettsial organisms, *Coxiella* may remain viable for prolonged periods in dried exudates, secretions, milk, and also water. Man-to-man infection is rare, but authenticated cases have been reported.

The disease is rarely fatal in man, is usually characterized by pneumonitis without rash, and may be severe enough to require a prolonged convalescence. The disease is apparently common among slaughterhouse employees and farm workers who handle sheep, or livestock handlers in general.

The cells of *Coxiella burneti* may display various shapes and sizes. The organism is pleomorphic, and some forms will pass through filters that retain other rickettsias (Fig. 42–14).

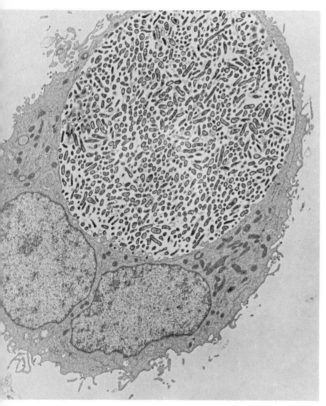

A

B

Figure 42–14. *A*, Mouse L cell in vitro infected with *C. burneti*. In this cell, all the rickettsias appear to be confined to a single large vacuole; calculations indicate that at least 3,600 rickettsias occupy the vacuole. (3,600×.)

B, Dividing *C. burneti* cell in mouse L cell in vitro 64 hours post-inoculation. (164,000×.) (Courtesy of P. R. Burton.)

TABLE 42–3. NATURAL CYCLE OF RICKETTSIAL ORGANISMS

EPIDEMIC TYPHUS FEVER

Infected man as reservoir → Uninfected body louse bites infected man and becomes infected → Infected louse bites man; fecal transmission via break in skin → Infected man → Ad infinitum

ROCKY MOUNTAIN SPOTTED FEVER

Infected tick → Ovarian transmission → Infected tick bites wild rodent or domestic animal (dog, etc.) → Uninfected tick bites infected rodent or dog and tick becomes infected → Infected tick bites uninfected man and man becomes infected

MURINE TYPHUS

Infected flea bites uninfected rat → Infected rat bitten by uninfected flea, which becomes infected → Infected flea bites uninfected man; man becomes infected

SCRUB TYPHUS

Infected mite → Ovarian transmission → Infected mite bites uninfected field mouse; mouse becomes infected → Uninfected mite bites infected field mouse, becomes infected → Infected mite bites man; man becomes infected

Q FEVER

Infected tick bites small uninfected mammal; which becomes infected → Infected mammal bitten by uninfected tick, which becomes infected → Infected tick bites cattle → Infected cattle shed contaminated aerosol or pass organism in milk → Contaminated aerosol inhaled by man, or raw milk ingested by man, who becomes infected

In summary, *Coxiella burneti* differs from *Rickettsia* species as follows:

1. Some morphological forms are filtrable.
2. Stable outside host cells.
3. Transmitted by inanimate vectors (inhalation of contaminated aerosols derived from infected animal nasal and salivary secretions, wet or dry) as well as by bite of an infected arthropod.
4. Fails to elicit antibodies for OX strain of *Proteus*.
5. The disease is characterized by pneumonitis without rash.
6. More resistant to heat, drying, and chemical disinfectants such as formaldehyde and phenol.

Some of the more common rickettsioses are summarized in Table 42–2, and the natural cycles of rickettsial organisms are presented in Table 42–3.

42.3
CHLAMYDIALES

According to the 7th edition of *Bergey's Manual of Determinative Bacteriology*, the family Chlamydiaceae of the order Rickettsiales contains five genera: *Chlamydia, Colesiota, Ricolesia, Collettsia,* and *Miyagawanella.* In the 8th edition of the *Manual* only one genus, *Chlamydia*, is retained under the family Chlamydiaceae of the order Chlamydiales. The order was named for its supposed relation to a group of viruses known as cloak or mantle viruses.

Genus Chlamydia

The genus *Chlamydia* (Gr. *chlamys* = a cloak) of the family Chlamydiaceae was originally recognized (1930) as the cause of a type of pneumonia now called **psittacosis** and known for many years to be contracted by humans from infected birds of the psittacine type (L. *psittacus* = parrot), popularly called "parrot fever." The causative agent was positively identified in 1932 by Bedson and Bland, and the organisms are sometimes called Bedsonias. Since the original identification, similar organisms have been found as the cause of psittacosis-like pneumonias (**ornithosis**) in many nonpsittacine birds such as pigeons, and in man and other mammals. Closely related organisms also have been found to cause diseases of the eyes of man: **trachoma** and **inclusion conjunctivitis** or **blennorrhea**, and a venereal disease, **lymphogranuloma venereum.**

All of the chlamydias are very much like the psittacosis or ornithosis organism in morphology, structure, developmental cycle, and protein (**group** antigen) composition. The various species causing the different diseases named can be distinguished only by minor differences in protein (**species** antigen) composition and by their host specificity and distinctive pathologic effects. In fact, all appear to be minor modifications of a single ancestral type.

The organisms of this group were for many years regarded as "large" viruses. The cells are more or less spherical and slightly smaller than those of rickettsias, being 200 to 700 nm in diameter (Fig. 42–15). Thus, many chlamydias are smaller than some of the true viruses, such as smallpox (200 to 300 nm).

The chlamydias are no longer regarded as viruses but, like rickettsias, as derivatives from a bacterial type modified and adapted to a life of obligate, intracellular parasitism. They must live intracellularly and cannot grow in inanimate media. They multiply by binary fission (Fig. 42–15, *B*). Their great reduction in size, like that of rickettsias, would be a logical result of their loss of many of the physiological mechanisms necessary to an extracellular existence. Therefore, the organisms may be considered as a further stage in degenerative evolution from the rickettsias. The chlamydias' closer relationship with bacteria than with viruses was first suspected when it was discovered that the organisms were susceptible to penicillin and other antibiotics whose actions are confined to bacteria.

Like all bacterial cells and totally unlike viruses, the chlamydias possess cell walls containing muramic acid and D-alanine, and for this reason they are sensitive to the action of penicillin. Chlamydias purified from infected cells are gram-negative, and the metabolic properties of such cell preparations reveal that they have retained some of the autonomous enzyme systems found in rickettsias. For example, they possess the enzyme for oxidation of glucose to pentose, but they are unable to generate energy from this oxidation. Similarly, they are able to synthesize D-alanine, lysine (cell wall constituents), growth factor, folic acid, and other macromolecules. Unlike the viruses, chlamydias contain both RNA and DNA; the DNA base composition ranges from 29 to 30 mole per cent GC (higher values,

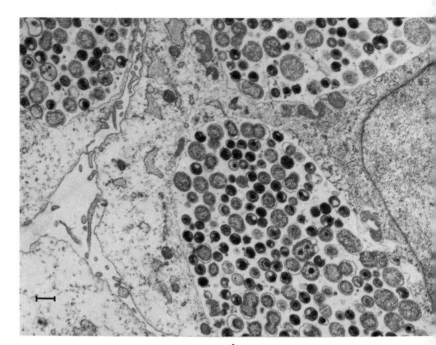

A

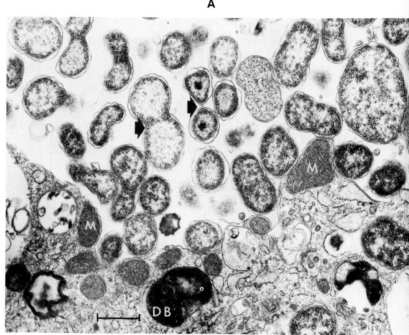

B

Figure 42–15. *A,* Two cells containing terminal populations of *Chlamydia psittaci* are shown 30 hours after infection.

B, Development of chlamydias after 20 hours in the host phagocytic vacuole. Numerous parasites dividing by binary fission are present: a typical dividing reticulate body and a transitional form or intermediate form are indicated by arrows. Marker on each micrograph represents 500 nm.

40 to 45 mole per cent GC, have also been reported), and the molecular weight of DNA is approximately one-tenth that of *Escherichia coli* but twice that of vaccinia virus. Chlamydias also contain about 40 per cent lipids and 35 per cent proteins. Nuclei are like bacterial nucleoids, as would be expected of such procaryons. Ribosomes also are evident; antibiotics, such as tetracycline, that inhibit the ribosomes of rickettsias and bacteria also inhibit chlamydias. Chlamydias are not affected by low temperature but are easily destroyed by standard pasteurization temperature and disinfectants such as ether, phenol, and formalin.

Electron micrographs of ruptured chlamydias show cytoplasmic membranes and struc-

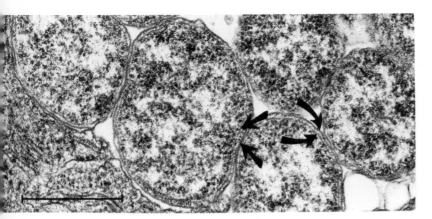

Figure 42–16. Several reticulate bodies present in a cell infected for 24 hours are shown. The chlamydial cell walls, comprising a trilaminar membrane and a diffuse zone, or space, are indicated by arrows. Parasite plasma membranes are also marked with arrows. Marker represents 500 nm. (Courtesy of Robert R. Friis.)

tures closely resembling bacterial cell walls but much less rigid; the chlamydia cell wall comprising trilaminar membrane is shown in Figures 42–16 and 42–17; ruptured chlamydia cells, "resembling discarded grape skins" (Moulton), are shown in Figure 42–17.

A virus-like aspect of infection of susceptible cells and animals with chlamydias is the establishment of **latency** (do not confuse with latent or lytic period of viral growth cycle). In latent infections the organisms often remain for long periods apparently wholly inactive, though

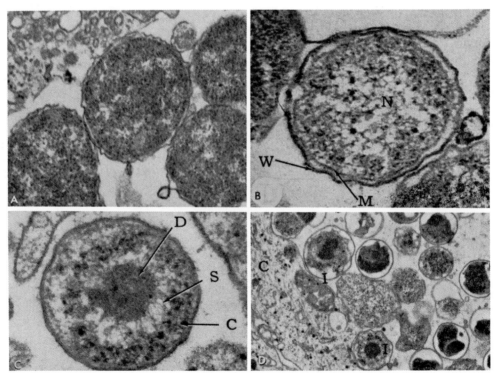

Figure 42–17. *A,* Initial bodies of chlamydias in ornithosis. They are confined to a membrane-lined space in the host-cell cytoplasm in which are seen, at upper left, a Golgi apparatus and a mitochondrion. (45,000×.) *B,* Higher magnification of an initial body of organism of ornithosis. Note the cell wall (*W*) and the cell membrane (*M*). In this initial body, the DNA strands are concentrated centrally to form a nuclear area (*N*) while the cytoplasmic components have migrated peripherally. (90,000×.) *C,* Intermediate form of organism of ornithosis. Homogeneous dense material (*D*) has formed within a central area of DNA strands (*S*), and the peripheral cytoplasmic components (*C*) include prominent ribosomes. (90,000×.) *D,* Portion of a vacuole containing organism of ornithosis in late stages of development. Many elementary bodies are seen, along with a few intermediate forms (*I*) and initial bodies. The membrane lining the vacuole is obvious in several places. The thin rim of host-cell cytoplasm (*C*) contains several mitochondria and other organelles. (15,000×.)

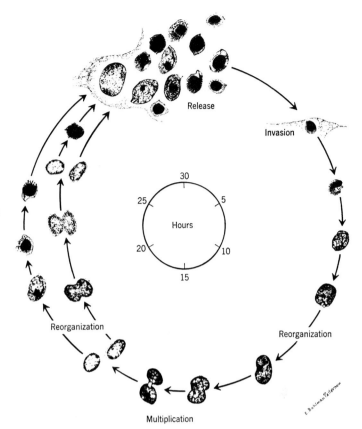

Figure 42–18. Idealized representation of the growth cycle of the psittacosis group.

they are fully viable inside infected cells. This relationship appears not to be analogous to lysogeny of phages; the mechanism of latency of *Chlamydia* infections remains to be fully elucidated.

As stated above, unlike all other cells, including rickettsias, but like viruses, chlamydias apparently depend, at least to a large extent, on the host cell for their **energy**, i.e., they cannot synthesize their own ATP. For this

TABLE 42–4. COMPARISON OF SEVERAL TYPES OF PROCARYONS

	Viruses	Chlamydias	Rickettsias	Schizomycetes	Blue-Green Algae
Depend on host NA for multiplication	+	−	−	−	−
Visible with optical microscope (1500×)	−	+	+	+	+
Cultivable in inanimate media	−	−	−	+°	+
Obligate intracellular parasites	+	+	+	−†	−
Synthesize proteins by own enzymes	−	+	+	+	+
Energy metabolism, e.g., synthesize ATP	−	−	+	+	+
Contain both DNA and RNA	−	+	+	+	+
Binary fission	−	+	+	+	+
Cell wall peptidoglycan	−	+	+	+	+‡

°A few possible exceptions, e.g., *Treponema pallidum.*
†A few possible exceptions, e.g., *Mycobacterium leprae.*
‡Multilayered.

TABLE 42-5. PATHOGENIC CHLAMYDIAS

Etiological Agent (Organisms)	Disease and Symptoms	Distribution	Laboratory Diagnosis	Epidemiology
Chlamydia psittaci spp.	Psittacosis-ornithosis (parrot fever) causes respiratory infection; illness begins abruptly or insidiously; chills, fever, headache, and may lead to migratory pneumonitis. Infected birds may appear sick with shivering, apathy, and diarrhea, or they may seem healthy. Many birds become healthy carriers. Obligate intracellular parasites.	Worldwide.	X-ray findings; complement-fixation with strain-specific antigen or agglutination. Skin test is used in screening turkey flocks.	The agent is present in droppings of infected birds, and dried fecal material is major source of infection for man and other mammals. Main sources of the agent are parrots and parakeets. Also considered as occupational disease of farmer or turkey processing plant workers.
Chlamydia species	Lymphogranuloma venereum or venereal bubo. Venereal disease of man. Manifestation of the disease appears as a herpetiform vesicle on the genitals followed by enlarged or suppurated regional lymph nodes. Edema of genital skin apparent.	Worldwide.	Allergic response to an intra-dermal injection of killed organisms (Frei antigen); response is delayed reaction; complement-fixation reaction.	Under natural conditions only man is infected. Host range more restricted than that of psittacosis-ornithosis group. It is transmitted by sexual intercourse, hence considered as direct contact disease.
Chlamydia species	Trachoma. The disease may cause inflammatory conjunctivitis with hyperplastic nodules on the conjunctiva, particularly of the upper lids. Vascularization and scarring of the cornea occur, leading to blindness. Extremely restricted host and tissue range.	Egypt, North Africa, Near East.	Cytological demonstration of the agent as inclusion bodies in epithelial cells scraped from the lesions. Skin test with Frei antigen is negative.	The agent is spread by direct contact. The agent is more prevalent in areas with poor personal hygiene.

| *Chlamydia* species | Inclusion conjunctivitis. Disease is restricted to epithelial cells of the conjunctiva; transitional epithelium of the external cervical os and also male urethra. The agents of trachoma and inclusion conjunctivitis together are referred to as TRIC agents. | Worldwide. | Cytological examination of conjunctival epithelial cells. Trachoma apparently never develops in newborn; inclusion conjunctivitis does not involve cornea. | Newborn may acquire this agent from mother's genital tract during birth. Rarely transmitted from eye to eye. Venereal transmission may maintain the organisms in the human population. Swimming pool can be source of infection. |

Mode of Transmission

PSITTACOSIS-ORNITHOSIS
Infected bird droppings ———→ Inhalation of contaminated aerosol originating from dried droppings ———→ Infected man or birds ———→

LYMPHOGRANULOMA VENEREUM
Infected man ———→ Uninfected man through direct contact, as in sexual intercourse, becomes infected ———→ Infected man ———→

TRACHOMA
Infected man ———→ Transmission of agent to uninfected man by direct contact with infected man or fly ———→ Infected man ———→

INCLUSION CONJUNCTIVITIS
Infected man (mother) ———→ Transmission of agent via genital tract during birth ———→ Infected child
Infected man ———→ Uninfected man in contact with infected man or contaminated swimming pool becomes infected ———→ Infected man ———→

reason the organisms are frequently referred to as "energy parasites."

Multiplication

Like rickettsias and other cells, chlamydias appear to multiply, in at least one phase of their developmental cycle, by a modified type of binary fission (Fig. 42–15, *B*). They also exhibit a yeast-like budding process. Additional phases of their multiplicative processes involve "small bodies," "large bodies," and intermediate (developmental?) sizes somewhat suggestive of the developmental cycles of mycoplasmas and L-forms (Fig. 42–17). Intracellular binary fission, budding, and variously sized particles are clearly visible in electron micrographs. The growth cycle, as seen in electron micrographs, suggests intracellular, virus-like eclipse and latent (lytic) periods and a long lag period (Fig. 42–18). The large particles appear early in the presumed "eclipse" period and are of low infectivity; the small particles appear late in the latent period and seem to be of high infectivity. Unlike infection by viruses, the nucleus of the chlamydia-infected cell is not necessarily affected. It is worth noting that chlamydias can complete their growth cycle in cells deprived of their nuclei, indicating that, unlike viruses, chlamydias do not depend on host DNA for their synthesis.

For comparative purposes (Table 42–4) we may think of viruses as intracellular parasites that depend almost wholly on the host for materials for synthesis and can utilize only the host's energy sources and mechanisms (utilization of host's ADP-ATP system). Chlamydias seem to have much more effective cell-synthesizing mechanisms, including RNA, DNA, and proteins, but they cannot synthesize their own ATP and therefore depend on the host for their energy. Rickettsias can synthesize themselves and can form their own ATP but require certain micronutrients that can be found only inside living cells. Bacteria are, in general, "self-supporting," though some, as the leprosy bacillus and the syphilis spirochete, appear to require intracellular micronutrients if they are to multiply in an **infective** state.

Pathogenesis

Three groups of the most common chlamydias are: (a) psittacosis-ornithosis (parrot fever);

(b) lymphogranuloma venereum; and (c) trachoma and inclusion conjunctivitis (sometimes referred to as TRIC agents). As a rule, agents that are transmitted by inhalation cause atypical pneumonia, and to a lesser extent an infectious septicemia. Psittacosis falls into this category, and it was once thought that the disease was transmitted exclusively by parrots. However, transmission of the agents to man is mediated not only by parrots and a great variety of other birds but also by aerosol via the respiratory tract. The disease is diagnosed by means of complement-fixation and agglutination tests. The disease may be promptly arrested by use of tetracyclines.

Lymphogranuloma venereum is a dangerous venereal disease; the organism causes lesions primarily in the genital skin (vaginal fistulas) and corresponding lymph nodes, and it may also cause rectal stricture or abscesses. Diagnosis is by isolation of the causative agent confirmed by complement-fixation test. A presumptive diagnosis of the disease may be made by the **Frei test**, which involves an intradermal injection of killed organisms. The allergic reaction is delayed-type hypersensitivity and is positive if reddening and slight swelling of the skin at the injection site appear in 24 to 48 hours. The disease has been effectively treated with tetracyclines. Table 42–5 lists various human diseases caused by chlamydias.

The TRIC agents may be transmitted by direct contact with patients or fomites. The diseases trachoma (TR) and inclusion conjunctivitis (IC) not only have an extremely narrow host range but also limited locus. For example, under natural conditions, the agents of the diseases invade epithelial cells of the conjunctiva of man; however, experimentally the organisms have been transmitted to a few other higher primates, and certain strains have been adapted to grow in embryonated eggs. The agent of IC may infect the transitional epithelium of the external cervical os. The disease may be transmitted to a newborn infant from the mother's genital tract during birth. Diagnosis is usually made by demonstrating inclusion bodies in scrapings of epithelial cells from the infected person. Differential diagnosis between TR and IC is based on the fact that the trachoma agent apparently never infects newborn infants and the agent of inclusion conjunctivitis does not invade the cornea. Treatment with tetracyclines and sulfonamide are highly effective in curing the diseases.

PLATE VII

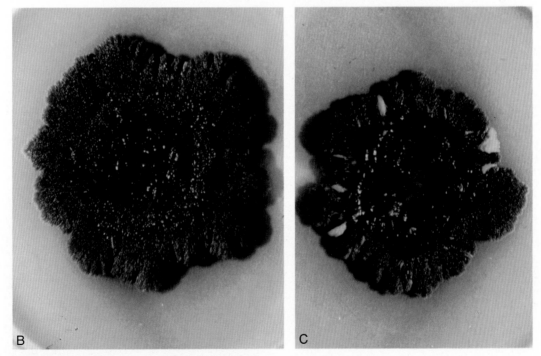

A, A nonpigmented colony of *Mycobacterium fortuitum* Cruz, strain 1000.

B, A dark colony of the same organism.

C, Sectors of the original whitish growth appear in some cultures of the dark strains. (Courtesy of Dr. R. E. Gordon.)

PLATE VIII

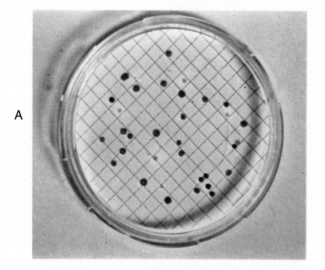

A

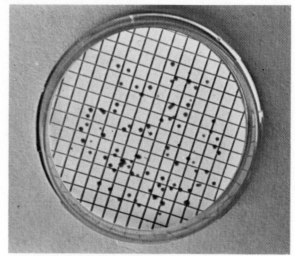

B

C

A, An indicator of polluted water, a growth of fecal coli-form colonies, is seen on this membrane filter. Incubation was in m FC broth (Difco) for 24 hours at 44.5C. (Compare with Color Plate IV, C.) (Courtesy of Millipore Filter Corporation, Bedford, Mass.)

B, Fecal streptococci in a water supply can be shown, as seen here, on a membrane filter. Incubation was in KF Streptococcus broth (Difco) for 48 hours at 35C. (Courtesy of Millipore Filter Corporation, Bedford, Mass.)

C, A more visible indicator of water pollution—algae growing in the recreational area of a lake in Tennessee. (Courtesy of Dr. Benjamin Wilson.)

CHAPTER 42
SUPPLEMENTARY READING

Black, F. T., Birch-Andersen, A., and Freundt, E. A.: Morphology and ultrastructure of human T-mycoplasmas. J. Bact., *111*:254, 1972.

Blair, J. E., Lennette, E. H., and Truant, J. P. (Eds.): Manual of Clinical Microbiology. American Society for Microbiology. The Williams & Wilkins Co., Baltimore. 1970.

Dienes, L.: Permanent alterations of the L-forms of *Proteus* and *Salmonella* under various conditions. J. Bact., *104*:1369, 1970.

Krass, C. J., and Gardner, M. W.: Etymology of the term mycoplasma. Int. J. Sys. Bact., *23*:62, 1973.

Manilof, J., and Morowitz, H. J.: Cell biology of the mycoplasmas. Bact. Rev., *36*:263, 1972.

Maramorosch, K. (Ed.): Mycoplasma and mycoplasma-like agents of human, animal, and plant diseases. Ann. N.Y. Acad. Sci., *225*, 1973.

Razin, S.: Structure and function in *Mycoplasma* 1531. Ann. Rev. Microbiol., *23*:317, 1969.

Tyeryar, F. J., Jr., Weiss, E., Millar, D. B., Bozeman, F. M., and Ormsbee, R. A.: DNA base composition of rickettsiae. Science, *180*:415, 1973.

Weiss, E.: Growth and physiology of rickettsias. Bact. Rev., *37*:259, 1973.

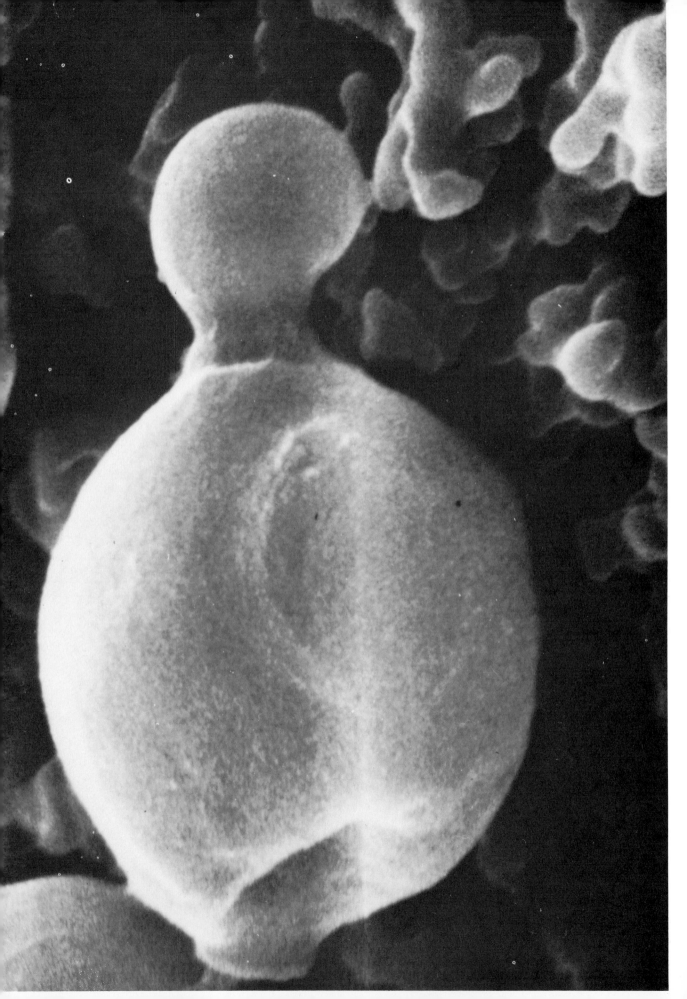

SECTION FIVE

MICROORGANISMS IN INDUSTRY AND THE ENVIRONMENT

In the course of foregoing discussions it has been mentioned frequently that there are thousands of species of harmless microbes that are ignored by all except a few prying microbiologists. They (the microorganisms) thrive by the billions, "born to blush unseen," in what are to us the most unutterably dismal and uninhabitable milieus, from the highest mountain tops to the sludge at the bottom of the deepest oceans. Whole chapters have been devoted to such biological curiosities. They are all very interesting no doubt, but the practical question arises, as of mosquitoes, of what use or value are they?

Section Five presents an answer to this question, and a justification of the subtitle of Chapter 1 of this book. In Section Five it is explained how, were it not for certain specialized microbes of the soil and oceans, there would, in all probability, be no students (pro or con), no cats, dogs, flies, worms, dwellers in high-rise condominiums, or any other living thing; the planet earth would, at this moment, be a scene of barren rocks and lifeless seas— sterile, magnificent desolation. It is shown in this section also how, thanks solely to harmless microorganisms, the picture is enlivened by forests and fields, by the sight of farmers using their knowledge of microorganisms of the soil to increase their crops, and by that of citizens of "developed" nations using their knowledge of both pathogenic and harmless species to combat microbial diseases, to clean and disinfect their polluted water supplies (are there any that aren't polluted?), to purify the air that they must breathe, and to decontaminate the potentially infectious or poisonous foods they must eat. To view the situation in a more cheerful and congratulatory

Facing page, Courtesy of R. Albrecht and A. MacKenzie.

mood, *Homo sapiens* is obviously making effective use of his knowledge of all kinds of microorganisms, both harmful and innocuous.

The final chapter of this section shows how the student could conceivably progress from rags to billionaire (possibly only to millionaire) through exploitation of the—to us—bizarre dietary, digestive, and metabolic eccentricities of various "remote and dismal" microbes. Certain species of these organisms, properly coddled and persuaded, daily produce for their exploiters a long list of foods, drugs, household goods, agricultural and pharmaceutical products, and so on. Better health and larger dividends reward the efforts of any citizen in any situation who takes the trouble to make himself microbiologically knowledgeable. For example, it has recently been proposed that, in view of the energy and fuel crisis, some of the many species that can produce energy-yielding fuels like industrial alcohol and other flammable liquids and gases from garbage, sewage, and other wastes be put to work immediately; of course, under the guidance of competent microbiologists!

CHAPTER 43 · MICROBIOLOGY OF THE SOIL

Microorganisms in Cycles of Biogeochemical Activity

The cycles of nutrient elements within the biosphere involve myriads of microorganisms, and their activities (decomposition and transformation of compounds) are essential for the continuance of life itself. It should be noted that the rarity of undecomposed organic matter in the biosphere is evidence of the metabolic versatility of microorganisms; in fact, all organic matter formed naturally is eventually decomposed by some microorganism. However, certain compounds contrived by man have proven recalcitrant. For example, some plastics and pesticides remain unaffected by microorganisms for a considerable length of time and pollute the environment.

Soil, either of mineral or organic origin, is the product of a series of biogeochemical events, not of the metabolic versatility of a single organism but of the versatility of the microbial world as a whole. For example, mineral soil is the product of physical, chemical, and biological processes on rock. Photosynthetic organisms, algae, mosses, and lichens (symbiotic plants composed of ascomycetes and blue-green algae) grow on rocks and convert atmospheric carbon dioxide to complex cellular organic matter. As these plants die, they are decomposed by various heterotrophs, bacteria, and fungi, and they release not only carbon dioxide but also various organic acids. These metabolic products are involved in the dissolution of rocks. Microcrevices created on the rocks are enlarged by alternate freezing and thawing of water in the crevices, and ultimately raw soil in which higher plants can develop is formed.

The raw soil is further enriched by addition of humus, a relatively stable organic residue produced by microbial degradation of plant and animal remains. Within the soil the cycles of elements continue—photosynthetic microorganisms fix carbon dioxide and release oxygen into the atmosphere, and this oxygen is responsible, through cycling, for replenishing atmospheric carbon dioxide. Carbon dioxide may react to form $CaCO_3$ under alkaline conditions, and calcareous rocks so formed are again solubilized by a change in hydrogen ion concentration by microorganisms, freeing the Ca^{2+} ion and CO_2. Some microorganisms fix free nitrogen (N_2); some reduce NO_3^- back to free nitrogen; still others are capable of reducing inorganic compounds of N, S, and other elements; and sequestration of the nutrient elements to the biosphere continues.

Microorganisms are both a panacea and a curse, contributing to environmental pollution as well as to its control and abatement. As already stated, microorganisms play the role of biochemical incinerators that convert pollutants in soil and water to harmless products, such as CO_2 and water. However, the transformation of elements such as phosphorus frequently contributes significantly to eutrophication of lakes and streams. Transformation and mislocation of H_2S, NO_2^-, and NO_3^- may damage health and property. Some microorganisms chelate (chemically bind) and effectively remove toxic matter from the environment, but others activate metals that are hazardous to health. Recently it has been observed that certain bacteria are capable of solubilizing mercury by a process known as methylation. Methyl mercury enters the aquatic food chain and thereby becomes a public health concern. In fact, the activities of microorganisms in the biosphere are so varied and vast that many colleges and universities now offer courses in microbial ecology.

43.1
THE SOIL AS AN ENVIRONMENT

In Chapter 10 and again in Chapter 44 it is pointed out that many substances useful as food for microorganisms tend to be adsorbed upon surfaces immersed in fluid media. A mass of tiny particles, such as sand or charcoal in a fluid culture medium, furnishes multitudes of tiny, protected niduses and extensive surfaces where foods become abundant and digestive enzymes tend to be concentrated. A comparable relationship occurs in soil. In moist, fertile loam, for example, each particle of soil has its film of moisture and its swarm of microorganisms on its surface. On a rainy day, in angles and depressions in the soil, tiny pools or puddles may develop and persist for a day or so. In this fluid myriads of microorganisms grow in warm weather.

The topsoil is indeed an entire universe

TABLE 43-1. MICROORGANISMS IN FERTILE LOAM SOIL AT VARIOUS DEPTHS

Depth (inches)	Bacteria (per gram[*])
1	4,000,000,000
4	3,000,000,000
8	2,000,000,000
12	1,000,000,000
20	500,000,000
30	1,000,000
72	100

[*]These figures are rounded to the nearest billion or hundred thousand. They represent only viable cells capable of growth on the particular medium provided, found in the particular sample of soil examined. Wide variations occur, depending on time of year, moisture, recent manuring, cropping, etc.

where billions of minute organisms—algae, bacteria, viruses, protozoans, nematodes, fungi, and others—for millions of years have lived their pigmy lives, multiplied in their minute-long generation times, struggled together for space, food, and survival, and have finally died, only to be replaced by others. Nevertheless, like each human being, however obscure, each microorganism leaves its effect on the universe.

Most soil microorganisms occur in the upper few inches or feet of soil. In a fertile loam in central New Jersey, for example, few microorganisms are found below 3 feet (Table 43–1) and none at 5 feet, unless they are accompanying plant roots or other extensions from the surface. Their numbers and species vary. Good, fertile, moist loam may contain from 100,000 to half a billion or more live microorganisms of all classes per gram of soil, depending on moisture and food. Each type of soil is a special study in itself. It is impossible to give an "average" number for each kind of microorganism in soil, but the ranges for different soils are summarized in Table 43–2.

Composition of Soils. Mineral soil (more predominant in most areas) consists primarily of inorganic particles derived from disintegrating rocks (mainly complex aluminum silicates)

TABLE 43-2. NUMBERS OF MICROORGANISMS IN SOILS (NUMBERS PER GRAM OF SOIL)[*]

Microorganism	Lower Limit	"Usual" Range	Higher Limit
True bacteria	1,000–10,000	1,000,000–10,000,000	1,000,000,000–10,000,000,000
Actinomycetes	100–1,000	100,000–1,000,000	5,000,000–10,000,000
Protozoa	none–100	10,000–100,000	500,000–1,000,000
Algae	none–100	1,000–100,000	200,000–500,000
Molds	1–100	1,000–100,000	200,000–500,000

[*]From Sarles et al.: Microbiology: General and Applied, 2nd ed., Harper & Row, 1956.

ranging in size from large boulders through gravel and sand to microscopic specks, mixed in varying proportions, all more or less compacted together, but having interstices between them as a result of their irregularity in shape. These interstices contain more or less water and air, carbon dioxide, hydrogen sulfide, ammonia, and other gases in small amounts, the proportions of each depending on rainfall, drainage, barometric pressure, winds, temperature, atmospheric humidity, microbial activity, and other factors (Fig. 43–1).

Soil as a Culture Medium. Soil, like air, contains free water (not bound by colloidal material). The water exists in soil interstices as a thin film of very irregular shape. On a volume basis, well-drained soil may contain 40 per cent water (plus 10 per cent air and 50 per cent soil particles), which is available to soil microorganisms for respiration and growth.

The water in good agricultural soil may be considered a dilute nutrient broth. It contains, in solution, ions like K^+, Na^+, Mg^{2+}, Ca^{2+}, Fe^{3+}, NO_3^-, SO_4^{2-}, CO_3^{2-}, PO_4^{3-}, and others. In addition, the hydrogen ion concentration (pH) of good top soil ranges from 6.0 to 8.0, at which most kinds of soil microorganisms grow best; however, pH ranges from 1 to 10 have been reported.

The temperature of soil varies from day to day, controlled by climate. Soil, as long as it is not frozen, can support the growth of microorganisms. Most soil microorganisms grow best at temperatures between 15 and 45C (i.e., are mesophiles), while psychrophiles are able to grow at lower temperatures and thermophiles at higher temperatures.

Availability of free oxygen, availability of hydrogen donor and acceptor compounds (oxidation-reduction systems), and physicochemical conditions of soil play important roles in microbial degradation of organic matter. Most soil bacteria are aerobic to facultative, and practically all actinomycetes and molds require free oxygen for their respiration and growth. Loose, open-textured soils or relatively dry soils maintain aerobic conditions. Water and air usually compete for the interstitial space between the soil particles, and water usually drives off the air. As a result, water-logged soils are completely devoid of free oxygen (anaerobic). Such soils usually have very low redox potentials, therefore supporting the growth of anaerobes. These have a profound influence on soil fertility resulting from loss of NO_3^- and SO_4^{2-} into the atmosphere when reduced to NH_3 or N_2 and H_2S.

In addition, soil nutrients may depend on the decomposition of the original rocks, on farm cropping, on manuring and fertilizing practices, on the microscopic and macroscopic flora and fauna, and on other factors. Various ions, it will be seen, represent elements essential in culture media for all forms of life. In a fertile soil these elements, in mineral form, are supplemented by a variety of organic compounds derived from the decomposition of animal and plant residues and from the synthetic activities of microorganisms, the latter producing carbohydrates ranging in complexity from glucose to starches, cellulose, lignins, and polysaccharide gums; nitrogenous compounds ranging from urea to amino acids, peptones, and complex proteins; fats; waxes; organic acids such as acetic; pyrimidines; numerous vitamins; and so on. Thus, the ground water in a fertile soil (aqueous extract of soil) is actually an excellent culture medium for many microorganisms. Lignins are a class of substances that bind together the cellulose fibers in wood. They are complex heteropolymers of *p*-hydroxycinnamyl alcohol residues distantly related (chemically) to plastics such as Bakelite and Formica. Lignins are much harder, tougher, and more resistant to hydrolysis than is cellulose, and they occur especially in the roots and hearts of trees and woody plants.

Variations in Soil. As stated, the soil en-

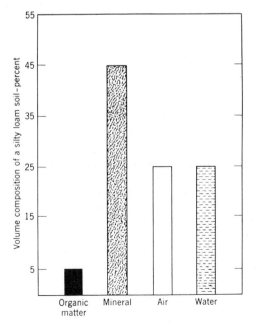

Figure 43–1. Percentage distribution (by volume) of important classes of constituents of a representative silty, fertile loam soil.

vironment is a highly variable one. Obvious variables are daily and seasonal temperatures and water content. It is clear also that if a heavy crop of clover and timothy grass is plowed under, the soil is aerated, some moisture is lost, and an enormous amount of readily assimilable soluble substances in the plant juices is introduced. These soluble substances (proteins, carbohydrates, lipids, minerals, and vitamins) quickly undergo hydrolysis and other complex changes of metabolism. There is a resulting great increase in internal temperature of the soil; in its acidity, resulting from fermentation; in content of CO_2, NH_3, and ammonium salts; and in relatively simple organic food substances: various fatty and amino acids, peptones, and alcohols. These support a tremendous upsurge in numbers of all heterotrophic forms as well as of facultative autotrophs capable of thriving in such an actively fermenting, acid, partly aerobic, partly anaerobic environment.

Oxidative microorganisms soon use up all the immediately available free oxygen. *Clostridium* spores can germinate, and strict, as well as facultative, anaerobes can thrive. If all air is excluded, as in swamps, heavy clays, and compacted soils, then fermentative processes predominate and the soil becomes acid. Drainage, aeration, and liming aid in *sweetening* such soils. Later, the acids are metabolized or neutralized, carbonates are formed, and the initial acidity reverts to alkalinity, especially if the soil is well aerated by tillage and drainage.

The tremendous growth of microorganisms that occurs after plowing under manures and green crops temporarily depletes the soil water of nutrient compounds that are in solution, especially those of nitrogen, phosphorus, potassium, and sulfur. They are removed from solution by combination as new microbial substance. As a result, for two weeks or more most newly planted crop plants find the soil a rather unfavorable medium. However, many of the microorganisms soon die, especially nonsporing species, and they release the elements and nutrient compounds for crop use. Finally, a more or less complete equilibrium of dormancy or low-level activity is re-established, awaiting the next change, perhaps the planting of a corn crop with liberal application of lime or commercial fertilizer, to stir things up once again.

Prominent in this residual community are the various saprophytic Myxobacterales, Actinomycetales, and Eumycetes, all of which are active in decomposing many resistant substances, including cellulose, chitin, keratin, lignin, and even paraffin and vulcanized rubber!

The numerous protozoans in soil convert much organic matter into "protoplasm." A principal item of their diet is bacteria; thus protozoans are the basis of an important ecological control relationship in soil, as they are in sewage. Another control mechanism active in soil, as also in water and sewage, consists of bacteriophages, antibiotics, and probably bacteriocins, all produced by growing members of the ecosystem.

Many worms, ranging from microscopic nematodes to large earthworms (night crawlers, the delight of fishing enthusiasts) eat organic matter. They digest it with the aid of their own enzymes and intestinal bacteria and return part of it to the soil in the form of simpler, more soluble substances as food for plants. Similarly, burrowing animals and the larvae of insects such as Japanese beetles help to transform organic matter in the soil.

43.2
SYNTROPHISM IN THE SOIL

Syntrophism (Gr. *syn = mutual* or together; *trophe* = nourishment) is that ecological relationship in which organisms provide nourishment for each other. In so varied a community as a fertile loam soil the nutritional relationships can be exceedingly complex.

For example, **cellulose**, a principal component (with lignin) of wood, is a complex polymer of glucose. Numerous species of soil bacteria (some species of Pseudomonadaceae, certain species of Cytophagaceae and of Spirillaceae, and Actinomycetales), with numerous species of Eumycetes, hydrolyze the cellulose molecule into molecules of cellobiose, a disaccharide resembling maltose. **Starch** likewise is hydrolyzed into maltose and dextrins.

Some organisms can utilize cellobiose, maltose, and dextrins, as such. Cellobiose, maltose, and dextrins may also be further decomposed by other organisms to glucose. This is an almost universal source of energy, and under anaerobic conditions (fermentation) it is decomposed by microorganisms into a great variety of still smaller molecules that can be used as food by one organism or another. At each stage in the decomposition of cellulose or starch a new group of microorganisms is found, capable of metabolizing the products of decomposition produced by other organisms. In the microscopic world, the waste of one is the indispensable food of another (Fig. 43–2).

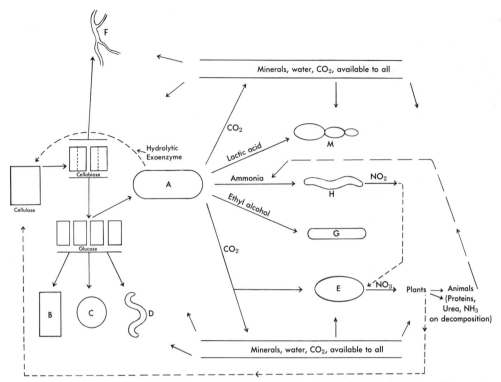

Figure 43–2. Some of the many complex syntrophic relationships occurring in the soil. Heterotrophic organism *A* (left of center) secretes a cellulolytic enzyme (*curved dash line*) which hydrolyzes cellulose to cellobiose. Cellobiose is used directly by some organisms (represented by mold *F*). Cellobiose is also hydrolyzed to glucose by organism *A* as well as by various other species not shown. Organisms *B*, *C*, and *D* as well as *A* then ferment the glucose with the production of various waste products (CO_2, etc.) (center of diagram). These serve as foods for still other species represented by *M* and *G*. Ammonia (center) may be produced from protein if organism *A* or any of the other species attacks proteins or amino acids. Ammonia so produced is oxidized to NO_2 as a source of energy by organism *H*. This NO_2 is oxidized to NO_3 by autotroph *E*. The NO_3 is then used by plants (lower right) which nourish animals; the plants and animals producing more cellulose and more protein for organism *A*, thus renewing a cycle of foods. Syntrophic relationships in the soil are so complex that to place in one diagram all that we know about would result in an almost solid mass of tangled lines.

Proteins are built up of amino acid units much as cellulose is built up of glucose units. Like cellulose, proteins are hydrolyzable and are metabolically decomposed to peptones, polypeptides, and amino acid molecules. Each product may be used by many species as sources of energy, carbon, and nitrogen.

Satellitism. Syntrophism is not confined to microorganisms in the soil, but is found wherever complex mixtures of organisms live together. Syntrophism in a simplified form is well illustrated by the phenomenon often called **satellitism** (a form of **commensalism;** a relationship between different microorganisms living in proximity without harm to either, and in which one or both organisms may benefit from the association). For example, inoculate, with organism A, the whole surface of a plate of agar medium that lacks an essential metabolite (let us say, NAD) of organism A. Then **spot inoculate** the plate with organism B, which is

known to produce NAD. After incubation, colonies of A appear as satellites *only* around the spot of growth of B, *nowhere else*. B is obviously a syntrophe of A (Fig. 43–3). (See also Color Plate V, *B*.)

Many other such interrelationships are seen in nature. **Lichens** (Color Plate II, *C* and *D*) are examples in which a fungus and an alga live together and in some instances will not live separately. Various animals are dependent on several B vitamins and essential amino acids synthesized by bacteria of the gastrointestinal tract; this is notably true of equines and in that very complex ecological system, the bovine **rumen.** Termites depend on cellulose-decomposing protozoans in their guts to digest the wood ingested as food. Man depends on intestinal bacteria for vitamin K (unless he buys it in "7-a-week" capsules).

Formation of Humus. The least digestible parts of cellular components of soil micro-

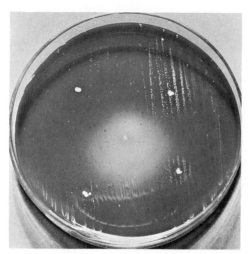

Figure 43-3. Minute colonies of *Haemophilus influenzae* growing as satellites around colonies of *Staphylococcus aureus* on agar in a Petri plate. Colonies of *H. influenzae* are barely seen except in areas around the two larger *S. aureus* colonies at upper and lower right that liberate into the agar an essential metabolite (NAD) absolutely required by *H. influenzae*. Organisms (not *S. aureus*) in the large colonies at upper and lower left liberate little or no NAD. (Courtesy of Drs. H. E. Morton and E. E. Long; collection of American Society for Microbiology.)

organisms, of plant tissues (e.g., lignin, resins), and of animal carcasses (e.g., waxes, hair, horn, and bone) undergo slow decomposition. The mixture of slowly decaying remains makes up a soft, spongy, brownish residual material called **humus.** It improves the texture of the soil, making it more friable; it holds moisture like a sponge; it also serves as a reserve store of slowly released food for microorganisms and crop plants.

Compost is complex organic matter used for fertilizing as well as improving the soil texture. Piled leaves and other plant materials are decomposed by soil microorganisms, and the resultant product is the humus described above. During decomposition of compost material, a succession of different kinds of microorganisms dominate. This change in microbial flora is primarily attributed to the rise in temperature (60 to 70C) of compost resulting from retention of the heat generated by microbial oxidation of organic matter. Many thermophilic bacteria, actinomycetes, and fungi flourish. These organisms are able to decompose cellulose, and the products formed are made available to other soil microorganisms. Therefore, the principle of syntrophism discussed in an earlier section is well demonstrated in the compost decomposition process.

43.3
MICROBIOLOGICAL EXAMINATION OF SOIL

Fertile soil contains such a wide variety of microorganisms that no single method can be given for cultivating or enumerating soil microorganisms in general.

Plating Methods. Plating methods were described in Chapter 10. They are applicable to the enumeration and isolation of bacteria, yeasts, and molds in any substance, including soil. Suitable modifications are made to meet the cultural requirements of the kinds of microorganisms that it is desired to enumerate.

Selective methods are often used to cultivate or isolate some particular species from the soil. For example, to enumerate soil microorganisms capable of metabolizing cellulose, a weighed sample of soil is placed in water and well shaken. One-tenth milliliter serial dilutions of this water are spread over agar medium containing cellulose as the only source of carbon. Hence only cellulose-metabolizers can grow. One may estimate the number of cellulose-digesters present in a gram of soil by counting the colonies surrounded by a clear zone resulting from hydrolysis of cellulose.

There are several commercially available selective and differential media for isolating particular groups of bacteria. For example, isolation of the coliform group may readily be accomplished by use of either eosin methylene blue (EMB) or Endo agar media; similarly, species of fecal streptococci by use of m Enterococcus agar, or particular pseudomonads (e.g., *P. aeruginosa*) by use of cetrimide agar.

One may follow (and adapt to other uses as the student's ingenuity may suggest) the clever scheme of Winogradsky to select and enumerate autotrophic organisms that obtain energy by oxidizing ammonia to nitrite. An inorganic medium is prepared with $(NH_4)_2SO_4$ as the sole source of energy. Since these organisms are strict autotrophs and will not grow (in the laboratory) in contact with organic matter like agar or gelatin, the medium is solidified with silica gel. The surface is coated with powdered chalk, giving it a white, opaque appearance. As the NH_3 [$(NH_4)_2 SO_4$] is oxidized to HNO_2 by the growing colonies of ammonia-oxidizers, the $CaCO_3$ is destroyed and a clear zone appears around each colony.

Phototrophic bacteria (Chapt. 32) may be isolated by use of the membrane filter method described in Chapter 45.

Microscopic Examination. By making stained smears of soil and examining them with the microscope, we may count various morphological types of bacteria and other microorganisms, especially filaments of fungi. Since dead as well as living cells are counted, microscopic counts are higher than plate-culture counts. Error may arise in microscopic counts from difficulty of staining some species, and confusion of bacteria with soil particles. These sources of error may be largely eliminated, and the organisms specifically identified and observed in their natural relationships by staining soil preparations with specific, fluorescent antibodies.

43.4
THE RHIZOSPHERE

The rhizosphere is a zone of increased microbial growth and activity in the soil around the roots of plants and may extend several inches into the soil around the roots. Sometimes the microorganisms form a sort of living mantle close around the roots. Numbers and kinds of bacteria and their metabolic activities are generally much greater in the rhizosphere than in the root-free zone. Therefore, one finds practically all the ecological interactions—**syntrophism, satellitism, mutualism, commensalism,** including **antagonism** between plants and microorganisms and among different microorganisms, and **predation.** Some are favorable to plants, some indispensable, some unfavorable, and others lethal.

We know, for example, that some bacteria or fungi make nitrogen available to plants as nitrates or in organic form. Sulfur-oxidizers make sulfur available as sulfates. Heterotrophic metabolism makes carbon available as carbon dioxide for photosynthesis. Production of acids by microbial action makes rock or bone phosphorus available as soluble phosphates. Some bacteria synthesize auxins or phytohormones (e.g., indole-acetic acid), which greatly stimulate root growth, and certain fungi (*Gibberella* species) synthesize the growth auxin **gibberellic acid** (Fig. 43–4).

Plant roots reciprocate in kind. The roots of leguminous plants secrete soluble, organic nitrogenous compounds (e.g., amino acids) into the soil around them to be used by microorganisms and other plants. In fact, most amino acid–requiring bacteria are more concentrated in the rhizosphere than in any other part of the soil. Many plant roots appear also to give off simple soluble carbon compounds (foods for bacteria) such as malic acid, pentoses, and phosphatides.

The sloughing off of bark and root coverings, as well as the death of roots, provides a rich source of carbohydrates and derivatives to support a luxuriant flora of nitrogen-fixers and other helpful forms close around the plant roots. The cellulose-digesters, amylolytic, and other hydrolytic forms transform plant material into humus, glucose, and other valuable foods for plants and microorganisms. A good heavy growth of microorganisms absorbs nitrogen, sulfur, phosphorus, potassium, and other elements in soluble forms which might otherwise be removed **(leached)** from the soil by rain and drainage. While the organisms withhold these elements temporarily from plant use (some-

Figure 43–4. Effect of gibberellin on plant growth. The first (left) lima bean seedling was grown from seeds dusted with plain talc. The talc used to dust the seeds of seedlings 2, 3, and 4 contained, respectively, 10, 20, and 40 gm of gibberellic acid. (Courtesy of Boyce Thompson Institute for Plant Research Inc.)

times with damage to the plants), the elements are eventually released on death of the microorganisms. Thus, the higher plants act as a food manufacturer and storehouse for microorganisms of the soil and rhizosphere, while microorganisms act as collectors, processors, and treasurers of foods for the higher plants.

Thus, the rhizosphere may be considered a most active factory for the transformation of essential elements into living constituents as well as for returning the essential elements to the soil upon the death and decomposition of the various organisms.

43.5
CYCLES OF THE ELEMENTS

All the elements that are essential components of protoplasm undergo cyclical alternations between an inorganic state, free in nature, and a combined state in living organisms. This repeated transformation of elements from living protoplasm to the free state in nature constitutes the cycle of elements or matter. Among the essential elements undergoing biological transformation are carbon, oxygen, hydrogen, nitrogen, sulfur, and phosphorus. When the above elements (except phosphorus) are incorporated into protoplasm, there is usually a change in the oxidation state. In protoplasm, many of these elements are in a reduced state, but, as they are returned, they frequently are in an oxidized state. Therefore, these elements not only serve as protoplasm constituents but also as sources of energy (for chemolithotrophs) by oxidation; others may serve as electron (H) acceptors in oxidation-reduction reactions.

43.6
THE CARBON CYCLE

Transformation of carbon occurs constantly and ubiquitously. Carbon is introduced into the organic system from its most oxidized state, carbon dioxide, and is reduced primarily by photosynthesis. In plant photosynthesis, water serves as hydrogen donor and reduces carbon dioxide to sugar; in bacterial photosynthesis, hydrogen sulfide and other reduced compounds serve as hydrogen donor to reduce carbon dioxide. A lesser amount of atmospheric carbon dioxide is utilized by certain chemolithotrophs, as it is more readily supplied by CO_2 derived from dissolution of carbonates and bicarbonates. Carbon dioxide used by green plants is transformed into various forms; first, sugar, which

will be converted to CO_2 via energy metabolism, or to become part of the cell components as organic carbon (protein, amino acids, nucleic acids, etc.). In both events the carbon is temporarily immobilized until decomposition of the cells and tissues proceeds and the metabolic products such as organic acids, alcohols, and so on are used as sources of energy or cell substance by other organisms.

In the biosphere the carbon cycle is the beginning of the food chain; primary producers through their photosynthesis establish the food chain upon which the remaining heterotrophs, including man, depend. (It has been estimated that nearly 70 per cent of free atmospheric CO_2 is used by photosynthetic organisms.) Organic materials thus formed are utilized by primary, secondary, and tertiary consumers, either to become their cell material or to be released to the atmosphere through their respiration. All organisms eventually die and decay, and through the aerobic decomposition process CO_2 and H_2O are once again recycled into the atmosphere or temporarily immobilized in the form of living matter, carbonates, limestone, and related rocks. The carbon in carbonates may be returned to the atmosphere by acids (H_2SO_4, HNO_3) produced by microorganisms.

Anaerobic decomposition of organic carbon differs considerably from the aerobic system. For example, anaerobic decomposition of organic materials may yield end products such as CH_4, H_2, and CO_2 in addition to various organic acids and alcohols. Some of the organisms capable of producing CH_4 from oxidation of hydrogen and reduction of CO_2 (both of which are derived from organic carbon) are species of *Methanobacterium*, *Methanococcus*, *Methanosarcina*, and some species of *Clostridium*.

If all existing supplies of CO_2 in the atmosphere or dissolved in the waters of the earth were to be continuously removed from the atmosphere and combined in organic matter or in carbonate rocks, life on the earth would cease in a generation or so. But carbon is continuously reoxidized and returned to the atmosphere, and thence to the seas, as CO_2 in a variety of familiar ways: mainly by combustion of coal and organic fuels and biooxidations, and also by volcanic activities, all of which liberate CO_2. Biological activities include not only fermentations that yield CO_2, but metabolism by certain rare bacteria that oxidize methane as a source of energy, e.g. species of *Pseudomonas* or *Methylomonas*. Some of these are *wholly dependent* on the **methyl** group as in methane or methanol, e.g., *Methylococcus capsulatus*.

Carbon monoxide is a relatively rare gas

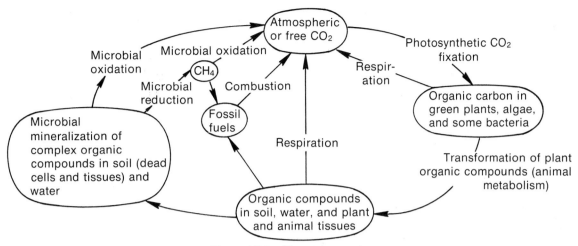

Figure 43–5. The carbon cycle.

under ordinary conditions and results commonly from partial combustion (an estimated 70 million tons of carbon monoxide are produced per year by incomplete combustion of fossil fuels). Exceedingly poisonous for most aerobic organisms, including man, it is relished as a source of energy and carbon by at least one autotrophic bacterial species, *Carboxydomonas oligocarbophilia* (*Bergey's Manual*, 1957) that oxidizes CO to CO$_2$. So the carbon "goes 'round and 'round," alternating between organic and inorganic, reduced and oxidized, like sulfur and nitrogen (see §43.10 and §43.11) (Fig. 43–5).

43.7
THE OXYGEN CYCLE

Oxygen production is limited to photosynthetic organisms which utilize water to reduce carbon dioxide, and thus oxygen is produced only by green plants which possess chlorophyll *a*. It is believed that photosynthetic bacteria do not participate in production of oxygen through reduction of carbon dioxide with water as hydrogen donor. • ·

All aerobic organisms, regardless of complexity, require oxygen for their energy metabolism; microaerophilic organisms must have small amounts of oxygen present if they are to grow and reproduce. Only anaerobes lack the requirement, and the presence of oxygen may actually be inhibitory to their growth and reproduction.

Oxygen is necessary in the combustion of fossil fuels, as well as alcohols; wood, etc.; and such oxygen is usually combined with carbon to form CO$_2$, which is recycled as described in the carbon cycle. A simple oxygen cycle is presented in Figure 43–6.

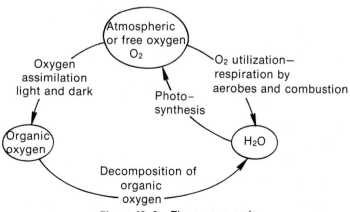

Figure 43–6. The oxygen cycle.

43.8
THE PHOSPHORUS CYCLE

This cycle involves an alternation in the form of phosphorus between inorganic and organic, and soluble and insoluble forms. Most microorganisms and plants utilize soluble inorganic phosphate for synthesizing their nucleotides and nucleic acids (along with phosphorylation of hexoses in the glycolytic cycle with which you are familiar). Assimilated phosphate is incorporated into organic compounds by esterification ($R—O—PO_3^{2-}$), and after the death of the cell it is again liberated as soluble organic compounds of phosphorus, such as DNA, RNA, ADP, ATP, and various other compounds. Fragments of DNA or RNA may be incorporated directly into cells as in transformation, but most organisms hydrolyze these to soluble inorganic phosphates before they can incorporate them into their cellular constituents. Therefore, phosphorus, unlike other elements, is neither reduced nor oxidized during the transformation. No organisms are known that reduce phosphates, or that oxidize phosphorus as a source of energy.

In the natural environment, occurrence of soluble phosphate is considered rare because most soluble inorganic phosphates readily form insoluble salts of calcium, magnesium, and iron. Phosphorus enters the soil in relatively insoluble, inorganic forms as phosphates in the rock from which the soil is derived. It is added to agricultural soils as $Ca_3(PO_4)_2$ in the form of **bone meal** and in commercial fertilizers as **rock phosphates.**

Insoluble phosphates may be solubilized by various acids produced by microorganisms. For example, nitrous and nitric acids produced by nitrification organisms in soil, or sulfuric acid produced by sulfur-oxidizing bacteria, can react with insoluble calcium phosphate to form soluble phosphates.

$$Ca_3(PO_4)_2 + 2HNO_3 \rightarrow 2CaHPO_4 + Ca(NO_3)_2$$
$$Ca_3(PO_4)_2 + 4HNO_3 \rightarrow Ca(H_2PO_4)_2 + 2Ca(NO_3)_2$$
$$Ca_3(PO_4)_2 + H_2SO_4 \rightarrow 2CaHPO_4 + CaSO_4$$

In many environments, phosphorus is considered a limiting factor in the growth of living organisms, yet our nation's waterways are plagued with phosphate pollution. The presence of excessive phosphates (0.01 mg/1 or more) in lakes and streams stimulates the pestiferous water hyacinth (*Eichornia crassipes*) and enhances the growth of algae to the point of bloom. The major portion of phosphates entering the nation's lakes and streams originates from household phosphate detergents via sewage and from phosphate fertilizer runoff from farm areas. The student should note that phosphate is the easiest to control among the nutrients required for algal bloom. Nitrogen can be obtained from atmosphere through nitrogen fixation, and CO_2 may be obtained from the atmosphere as well as from carbonates. All other essential trace elements are usually present in natural water or added to lakes and streams by rain and surface runoff water.

43.9
THE IRON CYCLE

Iron in nature exists in two forms, the ferrous (Fe^{2+}) and ferric (Fe^{3+}) states, and these two forms are readily convertible under the influence of pH and redox potential of the environment. Ferrous iron under neutral pH is spontaneously oxidized to the ferric state, forming highly insoluble ferric hydroxide. Frequently water from deep wells may contain large amounts of ferrous iron, causing the clogging of sand filters in potable water treatment. However, this very reaction is used in flocculation of particulate matter in the treatment of both potable water and sewage.

Many microorganisms require iron in trace amounts, as it is the key element in the heme group of the cytochrome system. Some bacteria are capable of reducing ferric iron to ferrous iron, which lowers the redox potential of the environment. However, organisms such as *Ferrobacillus ferroxidans* and *Thiobacillus ferroxidans* (autotrophic iron and sulfur bacteria) can oxidize ferrous iron to ferric hydroxide. Through this process these two strict autotrophs obtain energy for fixation of CO_2. The amount of energy gained by the oxidation is relatively small in comparison with, for example, oxidation of nitrite to nitrate; consequently, these bacteria must oxidize large amounts of ferrous to ferric iron. Ferric hydroxide so formed usually accumulates outside the cell, in gelatinous coats around the cell. These iron-oxidizing bacteria may be responsible for the iron precipitate found in either acidic springs or acid mine drainage.

43.10
THE NITROGEN CYCLE

Almost all microorganisms, higher plants, and animals require combined nitrogen, either

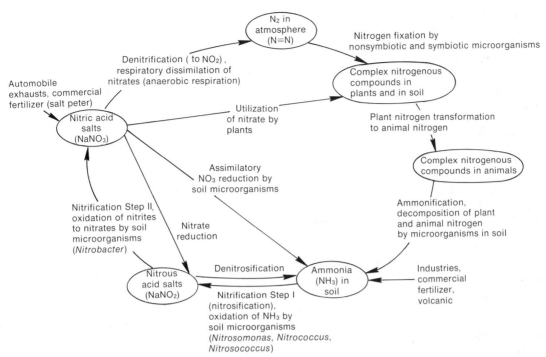

Figure 43-7. The nitrogen cycle. At the top of the cycle are shown the means by which atmospheric nitrogen is converted directly into living matter by soil microorganisms (see Table 43-3 for the kinds of microorganisms which might be involved in nitrogen fixation). Proceeding in a clockwise manner, after nitrogen is at last incorporated in plants as protoplasm, etc., it is converted into animal tissues. When plants and animals die, and their wastes decay, saprophytic microorganisms in soil convert the nitrogen back to the form of ammonia and other nitrogenous compounds. This process, digestion of plant protein to proteoses to ammonia, is referred to as ammonification. At this point, agricultural soil may receive artificially fixed nitrogen (commercial fertilizer) in form of aqueous ammonia or ammonium salt. Oxidation of ammonia to nitrous acid salts (nitrosification or nitrification Step I) is carried out by various chemoautotrophs, such as species of *Nitrosomonas*. Nitrites so formed are further oxidized (nitrification Step II) to nitrates by other chemoautotrophic soil microorganisms *(Nitrobacter)* and nitrification is completed. Nitrites and nitrates are also produced from atmospheric O and N by lightning flashes and photooxidation of exhaust from internal combustion engines of automobiles, etc., and brought to the soil by rain water. Nitrates are now available to plants as well as to facultative and anaerobic bacteria of the soil. Some bacteria may reverse the process by using nitrate as a source of cellular nitrogen by reduction (mechanism is not clearly understood) and is generally referred to as assimilatory nitrate reduction. Nitrates or nitrites may be dissimilated by various soil organisms as a source of electron acceptor. The process of reducing nitrate to molecular nitrogen is termed denitrification (anaerobic respiration), but if nitrate is reduced only to nitrite, the process is called nitrate reduction. If nitrites are reduced to ammonia, it is called denitrosification. Loss of molecular nitrogen to the atmosphere by anaerobic respiration of soil microorganisms recommences the cycle.

with H (as in ammonia or amines) or O (as in nitrate). These forms of nitrogen, as well as organic nitrogen (protein, amino acids and sugars, nucleic acids, and others), are relatively scarce in soil and water, and their concentration frequently becomes a limiting factor in crop nutrition. For this reason the transformation of nitrogen has attracted considerable attention from soil microbiologists. The main features of the nitrogen cycle are fixation of gaseous nitrogen (N_2); ammonification of cellular nitrogen; nitrification and denitrification (Fig. 43-7). Each of these processes involves either reduction or oxidation of nitrogen, and each is mediated by a complex series of specific enzymes. Many of these steps have been clearly elucidated, while some steps are not clearly understood.

Nitrogen Fixation

Nitrogen fixation is the process of causing free nitrogen gas to combine chemically with other elements. In the atmosphere covering one acre of soil it is estimated that there are some 35,000 tons of free nitrogen. Yet, though absolutely essential to life, not a molecule of it can be used as such by higher plants, animals, or man (in his primitive state; since becoming "civilized," he has learned to fix atmospheric nitrogen by artificial means) without the intervention of the nitrogen-fixing microorganisms.

A number of soil and aquatic microorganisms are capable of utilizing molecular nitrogen in the atmosphere as their source of nitrogen. These organisms are divided into two groups according to their mode of nitrogen fixation (Table 43–3). **Nonsymbiotic nitrogen-fixers** are those capable of converting molecular nitrogen to cellular nitrogen independently of other living organisms; **symbiotic nitrogen-fixers** are those which fix nitrogen by living in the roots of legumes and other plants.

Nitrogen-fixers enzymically combine atmospheric nitrogen with other elements to form organic compounds in living cells. In organic combinations nitrogen is more reduced than when it is free. From these organic compounds, upon their decomposition, the nitrogen is liberated in a **fixed** form, available to farm crops either directly or through further microbial action. It is of interest that some of these nitrogen-fixing enzymes are inducible or adaptive in contact with nitrogen.

Nonbiological nitrogen fixation may occur in the atmosphere through lightning discharge, and nitrogen oxides (exhaust fumes) formed by internal combustion engines such as in automobiles may undergo photochemical reactions. The combined nitrogen thus formed is brought to earth in rainwater.

Since the development of the Haber process in the early 1900's, man has chemically synthesized large quantities of ammonia from molecular nitrogen. It should be noted here that ammonia contains more concentrated combined nitrogen (82 per cent by weight) available to plants than any other substance. An aqueous form of this ammonia is applied to cropland, because it can be readily oxidized by nitrifying bacteria in soil and, in addition, ammonia, unlike nitrate fertilizer, resists leaching from the soil. Combined nitrogen also occurs in nature as deposits in the semiarid regions of Chile (Chilean saltpeter, KNO_3), but its use as fertilizer has diminished in value since the development of chemical fixation of atmospheric nitrogen.

TABLE 43–3. NITROGEN-FIXING ORGANISMS*

Symbiotic	Free-Living		
	BACTERIA	BLUE-GREEN ALGAE	YEASTS
Rhizobium species with legume plants	Nonphotosynthetic species of:	Species of:	Species of:
Klebsiella species with *Psychotria*	*Azotobacter*	*Anabaena*	*Pullularia*
Actinomycetes and/or fungi with species of:	*Azotomonas*	*Calothrix*	*Rhodotorula*
Alnus	*Bacillus*	*Chlorogloea*	
Casuarina	*Beijerinckia*	*Cylindrospermum*	
Ceanothus	*Chromobacterium*	*Fischerella*	
Ceratozamia	*Clostridium*	*Mastigocladus*	
Cercocarpus	*Derxia*	*Nostoc*	
Comptonia	*Desulfovibrio*	*Scytonema*	
Coriaria	*Enterobacter*	*Stigonema*	
Cycas	*Nocardia*	*Tolypothrix*	
Discaria	*Pseudomonas*		
Dryas	*Spirillum*		
Elaeagnus			
Encephalartos	Photosynthetic species of:		
Hippophae	*Chlorobium*		
Macrozamia	*Chromatium*		
Myrica	*Methanobacterium*		
Podocarpus	*Rhodomicrobium*		
Purschia	*Rhodopseudomonas*		
Stangeria	*Rhodospirillum*		
Shepherdia			

*From Carpenter: Microbiology, 3rd ed., W. B. Saunders Co., 1972.

Nonsymbiotic Nitrogen Fixation. The first microorganism discovered (by Winogradsky, 1895) to possess the property of fixing atmospheric nitrogen without symbiotic aid was the anaerobic species *Clostridium pasteurianum*, common in boggy soils. Aerobic nonsymbiotic nitrogen-fixing bacteria (*Azotobacter* and *Beijerinckia*) were discovered in the soil by Beijerinck in 1901. Since those discoveries the phenomenon of nonsymbiotic nitrogen fixation has been observed in numerous other *Clostridium* species; also, in most blue-green algae, in many photosynthetic bacteria like *Rhodospirillum*, in *Desulfovibrio*, and in many other bacteria. One of the most interesting is Beijerinck's *Azotobacter*.

Genus Azotobacter. *Azotobacter* thrives in all well-aerated, neutral or slightly alkaline (pH about 7.5) arable soils. It is pleomorphic, strictly aerobic, normally encapsulated, nonsporing, and motile with peritrichous flagella. It has been reported that the organism has the highest rate of oxygen uptake of any living organism. An especially distinctive feature is the formation of thick-walled, spherical, dormant **cysts.** These have some properties suggestive of primitive spores but, although very resistant to ultraviolet and γ-radiations and sonic vibrations as well as drying, they are not highly thermostable as are true bacterial endospores.

The complex *Azotobacter* cyst is shown in Figure 43–8. The cyst, which structurally resembles the bacterial endospore, is a spherical body that contains one or more central bodies resembling a vegetative cell in overall structure. The central body is surrounded by an inner **intine** and an inner **exine,** which are sometimes surrounded by other layers, the outer intine and outer exine. DNA base composition of *Azotobacter* ranges from 54 to 66 mole per cent GC.

Cysts may contain a large amount of reserve food (PHB) but carry out little or negligible endogenous respiration. Since the central body is not resistant to various agents, the cyst coat (inner and outer intines and exines) may be responsible for the cyst's ability to survive adverse environments. Germination of the cyst is analogous to that of the endospore in that it can be halted but not reversed. It is in a medium favorable for growth that germination occurs, i.e., enlargement of the central body and disappearance of PHB and inner intine and exine, followed by formation of a break in outer intine and exine, and young vegetative cell(s) emerging from the disintegrating cyst structure.

In the soil *Azotobacter* species grow almost autotrophically; however, organic substances are needed as energy source. These energy sources are probably derived from the decomposition of cellulose, starches, and the like by other microorganisms of the soil. Carbohydrates added to the soil in a form such as molasses or starch wastes stimulate the accumulation of nitrogen in the soil through the growth of the *Azotobacter* and other nonsymbiotic, nitrogen-fixing organisms. The nitrogen combined in their structures is taken from the atmosphere and is released in organic form as secretions and on the death of the nitrogen-fixing organisms.

Azotobacter grows readily in such nitrogen-free solutions as the following:

H_2O	1,000.0	ml
Mannitol (or other organic source of energy and carbon)	15.0	gm
K_2HPO_4	0.2	gm
$MgSO_4 \cdot 7H_2O$	0.2	gm
$CaCl_2$	0.02	gm
$FeCl_3$ (10 per cent aqueous solution)	0.05	ml
Molybdenum salt	Trace	

Adjust to pH 7.2; for solid medium add 15 gm agar or silica gel before adjusting the pH.

In the absence of molybdenum, nitrogen fixation will not occur; the metal ions appear to activate an enzyme (**nitrogenase**) essential in the fixation process. Molybdenum can be replaced only by vanadium.

Algae. Blue-green algae, *Nostoc* and *Anaboena*, can also fix nitrogen, both in soil and in aquatic environments (Table 43–3). These two organisms are frequently used as indicators of lake and stream eutrophication (Color Plate VIII, *C*). They are able to bloom in an aquatic environment with little or no combined nitrogen as long as adequate amounts of soluble phosphates are present. No satisfactory estimates have been made of total nitrogen turnover by all the nonsymbiotic nitrogen-fixers, partly because spontaneous loss of combined nitrogen by denitrification occurs. However, some authors have reported that, on the average, 6 to 10 lb of combined nitrogen is added by these organisms in an acre of soil annually in temperate areas.

Symbiotic Nitrogen Fixation. In this system, the fixation of molecular nitrogen results from a mutualistic association between leguminous plants (and others) and a bacterium, neither of which is able to fix nitrogen alone. The association between nitrogen-fixing bacteria (e.g., species of *Rhizobium*) and legu-

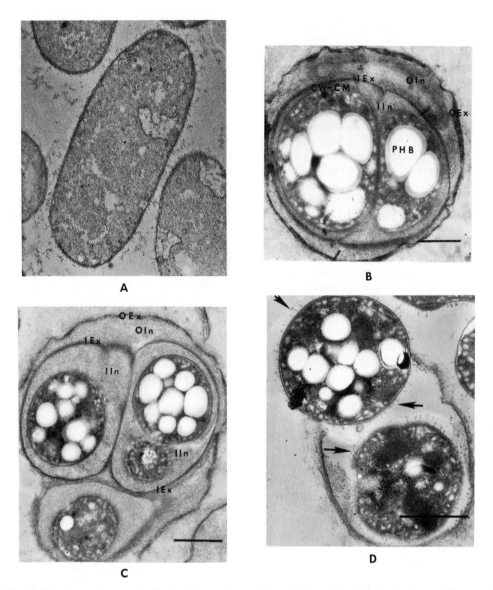

Figure 43–8. *A*, Electron micrograph of ultrathin sections of vegetative cells of *Azotobacter* sp. The vegetative cells show the internal structure commonly seen in vegetative bacteria.

B, Thin-section of a cyst of *A. vinelandii* with two central bodies completely enclosed in a second cyst coat. Morphological structures of the cyst include the vegetative body cell wall-cell membrane *(CW-CM)*, dispersed poly-β-hydroxybutyric acid granules *(PHB)*, an inner intine *(IIn)*, an inner exine *(IEx)*, an outer intine *(OIn)*, and an outer exine *(OEx)*. Bar represents 1 μm.

C, Electron micrograph of a complex structure of a cyst with a double coat in which the outer cyst coat *(OEx-OIn)* has been ruptured and three intact cysts, enclosed by a second coat *(IEx-IIn)*, are emerging. Bar represents 1 μm.

D, Thin-section of normal cyst in the process of germination. Intine vesicles are indicated *(arrows)*. Bar represents 1 μm. (*B* to *D*, Courtesy of G. C. Cagle and G. R. Vela.)

minous plants (e.g., clover, lupine, peas, beans, alfalfa) is more specifically called **endosymbiosis** as compared to the relationship known as **ectosymbiosis.** In endosymbiosis, the bacteria or any other microorganisms grow in the host's cells and tissues. In ectosymbiosis the microorganisms remain outside of the cells or body and tissues (e.g., on the surface of the body or body cavity), as exemplified by attachment of

certain blue-green algae on the body of a protozoan.

Genus Rhizobium. (Gr. *rhizo* = root; *bios* = life; hence, "living in roots"). When young and actively growing, these bacteria are organotrophic, aerobic, nonsporing, pleomorphic, gram-negative rods. They are usually motile with variably placed flagella. Enzymically they are restricted and feeble. Their pleomorphism is regarded by some as evidence of a somewhat complex life cycle. They grow on ordinary organic laboratory media, especially if made with yeast extracts, at 20C and pH 7.2. DNA base compositions have been determined on several species of *Rhizobium*, and they range from 59 to 65 mole per cent GC, which indicates close genetic relationship among the different species of the genus. Similarly, nucleic acid homologies as revealed by hybridization studies show a close relationship, not only among *Rhizobium* species but also with *Agrobacterium*, whose DNA base composition ranges from 58 to 65 mole per cent GC.

Their most characteristic activity and form are seen when they grow in the tissues and within the cells of leguminous plant roots. In the plant cells they are morphologically distinctive as **bacteroids.** In stained smears made from crushed **nodules** from the roots of leguminous plants they are often seen as large, oddly angular, stellate, budding or L, Y, V, T, and X forms. These contain metachromatic masses and bands which are thought by some to represent special reproductive mechanisms. Others regard these swollen forms as degenerative: the end of the supposed life cycle (Fig. 43–9).

Nodule Formation. Although they invade plant roots, *Rhizobium* species are unable to hydrolyze the cellulose of plant cell walls. However, they appear to find noncellulosic points at the tips of root hairs of legumes. Through these, entrance into the root tissue is made.

INFECTION THREADS. The bacteria utilize various carbohydrates found in the soil or juices of the host plant to synthesize gummy coatings around themselves. These coverings help the bacteria to invade plant roots. The rods advance endwise, several abreast in their gummy coating, into the plant tissue cells. A long, gummy thread is formed. This is surrounded by a tube of cellulose produced by the plant cells. The whole constitutes what is called an **infection thread.** These threads penetrate well into roots (Fig. 43–10) and into the plant cells. The infection thread inside a cell bulges and then ruptures, and the bacteria are liberated intracellularly. Each bacterium becomes a **bacteroid,** and undergoes fission while encysted within a double-layered membrane produced by the plant cell (Fig. 43–11). The presence of the bacteria stimulates multiplication of the plant cells around the localization, with resulting formation of a protective **nodule** of tissue (Fig. 43–12). Nutrients from the plant juices nourish the bacteria.

Process of Symbiotic Nitrogen Fixation. The rhizobia, when thus growing in leguminous plant nodules, take nitrogen directly from the air. By combined action of plant cells and bacterial cells, this is built into nitrogenous compounds such as amino acids and polypeptides that are found in the plants, the bacteria, and the surrounding soil. Neither bacteria nor legumes can alone fix nitrogen. If, in the absence of rhizobia, combined nitrogen is not available in the soil, legumes die. In contrast, if proper species of *Rhizobium* are present, legumes not only thrive in nitrogen-deficient soil but enrich it with fixed nitrogen while doing so.

Soil Inoculation. Because of the value of nitrogen fixation by legumes with the bacteria, it is customary to inoculate virgin soils, or soils not known to support good growth of legumes, with the proper species of *Rhizobium* preparatory to planting such crops as alfalfa or soybeans for the first time. Once introduced, the bacteria continue to live in the soil.

SPECIES SPECIFICITY. There are species or varieties of the genus *Rhizobium.* They exhibit considerable specificity as to the species of legume that they can infect. For example, *Rhizobium japonicum* produces nodules only on the soybean, whereas *Rhizobium meliloti* will not do so.

A number of other microorganisms that have been reported to fix atmospheric nitrogen more or less efficiently are listed in Table 43–3.

Value of Nitrogen Fixation. A well-nodulated crop such as red clover may intro-

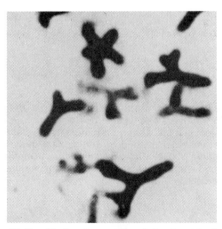

Figure 43–9. Photomicrograph of bacteroids from a nodule on a leguminous plant. These oddly shaped rhizobia are found only in the nodule and will appear as small uniform bacilli when grown on laboratory media. (Enlarged about 2,000×.)

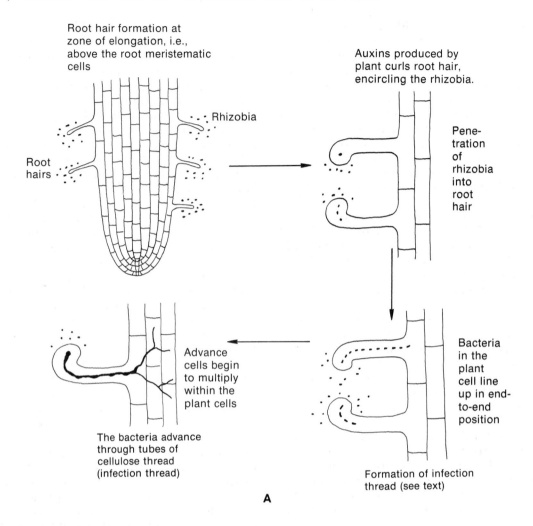

Root hair formation at zone of elongation, i.e., above the root meristematic cells

Rhizobia

Root hairs

Auxins produced by plant curls root hair, encircling the rhizobia.

Penetration of rhizobia into root hair

Advance cells begin to multiply within the plant cells

The bacteria advance through tubes of cellulose thread (infection thread)

Bacteria in the plant cell line up in end-to-end position

Formation of infection thread (see text)

A

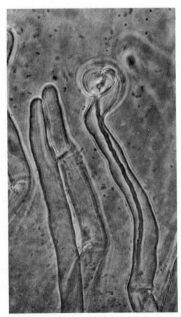

B

Figure 43–10. *A*, Mode of invasion of root hair by rhizobia. *B*, Infection threads formed in root hairs of a legume by *Rhizobium trifolii* about two weeks after inoculation. The right-hand thread clearly shows the matrix of the thread surrounded by a thin, plant-derived, cellulose wall, and the column of bacteria (dark line) inside the matrix. Many bacteria have escaped into the cell cytoplasm from the matured hairs. Phase-contrast microscopy. (About 600×.)

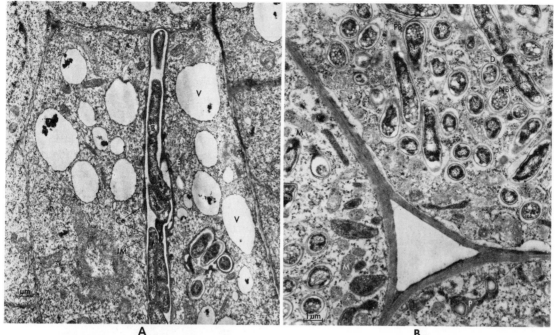

Figure 43-11. *A,* Infection thread crossing a central tissue cell of a nodule aged one to two days. The bulge *(Bu)* was seen in other sections of this cell to be connected in another plane to the group of bacteria *(C)* which is partially enclosed by cellulose and partially by a cellulose-free membrane. *IM,* Infection thread membrane; *Ce,* infection thread cellulose; *V,* vacuole. *B,* bacterium.

    B, Later stage (five to seven days). The host-cell cytoplasm is filled with bacteria, each within its own membrane envelope. Division of bacteria within the envelopes is just commencing *(D). M,* Mitochondria; *P,* proplastid; *PB,* electron-translucent granules, probably poly-β-hydroxybutyric acid; *NB,* bacterial nuclear filaments. (Courtesy of D. J. Goodchild and F. J. Bergerson.)

Figure 43-12. Different types of root nodules formed by *Rhizobium* sp., on leguminous plants.

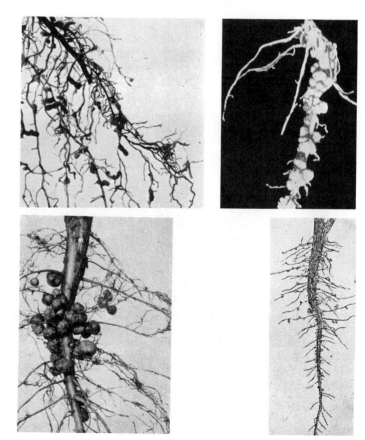

duce as much as 100 pounds of organically combined (fixed) nitrogen per acre per season. In the form of a commercial fertilizer this nitrogen costs in the neighborhood of six hundred dollars (in 1973).

Inoculation of swampy, acid soils is money wasted because *Rhizobium* will survive and grow only in fertile, well-drained, aerated, and nearly neutral soils. Crops of nonlegumes grown in association with legumes (as vetch and rye, or clover and corn) have been known for centuries to be superior.

Mycorrhiza. The symbiotic relationship between a fungus and rootlets of plants is termed **mycorrhiza;** the word literally means root fungus (Fig. 43–13). There are two types of association: **endotrophic,** in which the fungus mycelium grows in root tissues; and **ectotrophic,** in which the fungus mycelium ensheathes the root or forms tubercles around rootlets with limited penetration of hyphae into root tissues. The mycorrhizal fungi have not been isolated from nature except in relationship with roots, and they can therefore be considered obligate symbionts. From this relationship, the fungus receives simple carbohydrates, a few vitamins, and other nutrients from the root secretions. The plant, likewise, receives benefit from the mycorrhizal fungus in that it supplies mineral nutrients. Some of the observations

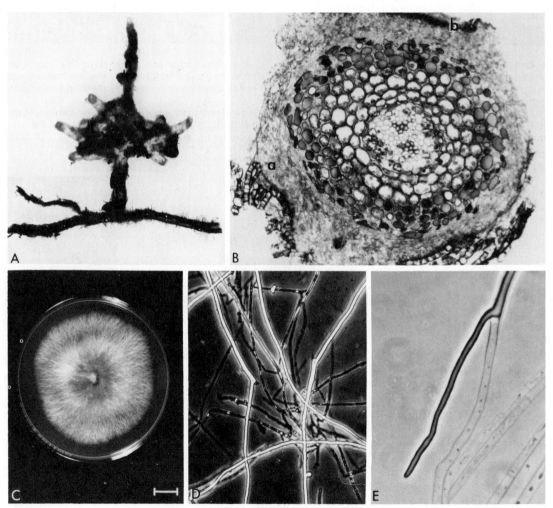

Figure 43–13. Pure culture synthesis of mycorrhizae with Douglas-fir seedlings and *Rhizopogon vinicolor*. *A,* Pinnate mycorrhiza formed after 4 months. Tubercle did not develop but surfaces irregularly covered by patches of dark brown mycelium. (3.7×.) *B,* Cross section of mycorrhizal element. Mantle *(a)* is thick and irregular; dark, superficial mycelial patch *(b)* is composed of same amber aseptate hyphae as tubercle rind. (100×.) *C,* Mat of *Rhizopogon vinicolor* on potato dextrose agar medium. *D,* Teased-apart mycelium including aseptate hyphae with characteristic crook like those of tubercle rind. Note remaining stubs of thin-walled septate hyphae. *E,* Aseptate hypha with crook, formed from attached thin-walled septate hypha. (Courtesy of Dr. Bratislav Zak.)

indicate that mycorrhizal plants are more efficient in absorbing nutrients than nonmycorrhizal plants, especially when they are planted in poor soil side by side. There is no evidence that nitrogen fixation occurs in either endotrophic or ectotrophic mycorrhizal plants.

Ammonification

Were all fixed nitrogen to remain inextricably bound up as organic matter, then the agricultural use of manures, animal carcasses, fish, and fertilizer would be of no avail. Dead animals would not decay, manure would not rot, and dead fish would remain dead fish. The only forms of combined nitrogen available for living organisms would be the rare nitrogen compounds produced by lightning. All of the non-nitrogen-fixing forms of life would have to await the slow activities of the nitrifying and nitrogen-fixing microorganisms in order to obtain properly combined nitrogen. Such, however, is not the case. As soon as any organism ceases to live, and as soon as any organic waste matter returns to the soil, it begins to undergo biological decomposition; its fixed nitrogen is released.

Protein, nucleic acids, and some other organic nitrogenous compounds are hydrolyzed to amino acids and similar compounds, and these are broken down to other, simpler compounds when they are metabolized by microorganisms. The amino groups ($—NH_2$) are split off to form ammonia (NH_3); this series of enzymic reactions resulting in release of ammonia from complex organic nitrogenous compounds is collectively termed "ammonification." Ammonification usually occurs under aerobic conditions, but if protein decomposition proceeds under anaerobic conditions, the process is called **putrefaction** because some of the amino acids are converted to offensive odor-producing amines and related products. In the putrefaction process, the release of ammonia is not immediate, but in the presence of air the amines are oxidized with liberation of ammonia. These reactions are usually brought about by anaerobic sporeformers such as species of the genus *Clostridium*.

Urea ($NH_2 \cdot CO \cdot NH_2$), a waste product found in the urine of man and other animals, is also decomposed by numerous microorganisms (e.g., *Proteus* species and *Micrococcus ureae*), with liberation of ammonia. An ammoniacal odor is an outstanding impression in an uncleaned stable or in the infant's not-promptly-changed diaper wet with urine, owing to rapid decomposition of urea.

Nitrification

In soil, ammonia liberated during the ammonification of organic nitrogenous matter is rapidly oxidized to nitrates by two highly specialized groups of strictly aerobic chemolithotrophic bacteria. This oxidation of ammonia to nitrate by the microorganisms is called **nitrification** and occurs in two steps. In the first step, ammonia is oxidized to nitrite, and then secondly, nitrite is oxidized to nitrate. In both oxidation steps the involved organisms (in the family Nitrobacteraceae) gain energy.

Family Nitrobacteraceae. These bacteria are common in fertile soils. They are so called because they are concerned in **nitrification.** Nitrates are the most useful and most expensive (and for many crops, the *only*) form of nitrogen for crop plants, though some plants can use ammonia and/or nitrites if necessary.

The Nitrobacteraceae are strictly autotrophic. None forms spores. Some are simple rods, motile with polar flagella, others are spirals or cocci; some are gram-positive, others gram-negative.

DNA base compositions of the family Nitrobacteraceae range from 48 mole per cent GC (*Nitrosomonas* group, ammonia-oxidizers) to 62 mole per cent GC (*Nitrobacter* group, nitrite-oxidizers).

The oxidation of ammonia to nitrates in the soil by Nitrobacteraceae involves two distinct stages, each stage carried out by different genera. The first stage, the oxidation of ammonia to nitrites, is sometimes called **nitrosification:**

$$2NH_3 + 3O_2 \rightarrow 2HNO_2 + 2H_2O + 79{,}000 \text{ cal}$$

Oxidation of Ammonia to Nitrite. *Nitrosomonas*, *Nitrosolobus*, and *Nitrosospira* (Fig. 43–14) are very small oval rods, each with a single, polar flagellum. They are strictly aerobic and are very sensitive to acidity. Since oxidation of ammonia, and especially of ammonium sulfate, creates acidity caused by HNO_2 and H_2SO_4, *Nitrosomonas* soon ceases growth unless a soil is well limed or otherwise buffered. The optimum pH is around 8.6. It should be noted here that no other microorganisms are able to obtain their energy from the oxidation either of ammonia to nitrites, or of nitrites to nitrates.

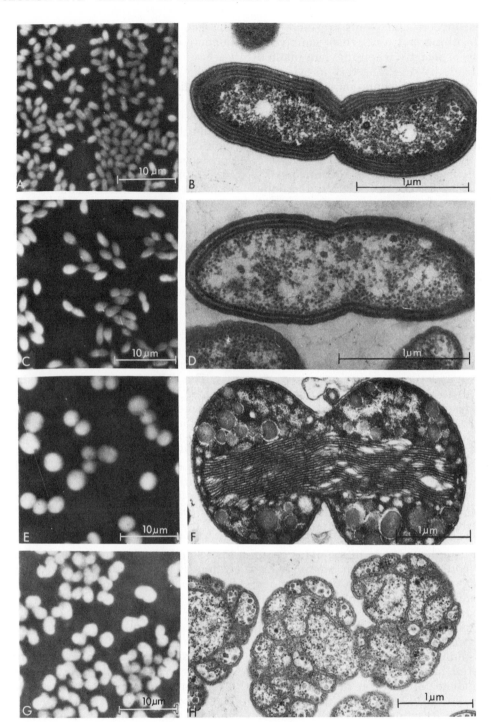

Figure 43–14. Phase-contrast and electron micrographs of *Nitrosomonas, Nitrosocystis, Nitrosolobus, Nitrosospira, Nitrobacter, Nitrococcus,* and *Nitrospina* species. *A–J,* organisms which oxidize ammonium to nitrite; *K–P,* organisms which oxidize nitrite to nitrate.

    A to *H,* Phase-contrast and electron micrographs of *Nitrosomonas, Nitrosocystis,* and *Nitrosolobus* species. *A,* Phase-contrast photomicrograph of *Nitrosomonas europaea* (C-31). (2,500×.) *B,* Electron micrograph of a marine *Nitrosomonas* species (C-56) showing peripheral cytomembranes. (37,600×.) *C,* Phase-contrast photomicrograph of *Nitrosomonas* species (C-91) isolated from the Chicago sewage disposal plant; note connections between cells. (2,500×.) *D,* Electron micrograph of *Nitrosomonas* strain shown in *C*; note peripheral membranes. (45,000×.) *E,* Phase-contrast photomicrograph of *Nitrosocystis oceanus* (C-107). (2,500×.) *F,* Electron micrograph of *Nitrosocystis oceanus* (C-107) showing flattened vesicles in the central region of the cell. (27,100×.) *G,* Phase-contrast photomicrograph of *Nitrosolobus multiformis* (C-71) showing lobular shape of cells. (2,500×.) *H,* Electron micrograph of *Nitrosolobus multiformis* (C-71) showing lobular shape of cells and partial compartmentalization of cytoplasm by cytomembranes. (25,000×.)

Figure 43–14 continued on opposite page.

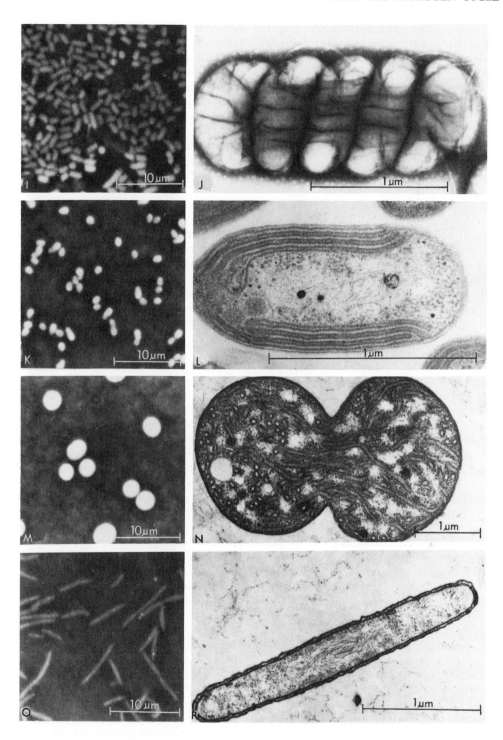

Figure 43–14. *Continued.* *I* to *J*, Phase-contrast and electron micrographs of *Nitrosospira, Nitrobacter, Nitrococcus,* and *Nitrospina* species. *I*, Phase-contrast photomicrograph showing *Nitrosospira briensis* (C-76); note that the spiral nature of the cells is not immediately apparent. (2,500×.) *J*, Electron micrograph showing negatively stained cell of *Nitrosospira briensis* (C-76); note spiral nature of cells. (40,100×.) *K*, Phase-contrast photomicrograph of *Nitrobacter winogradskyi* (NB-255) showing wedge- to pear-shaped cells. (2,500×.) *L*, Electron micrograph of *Nitrobacter winogradskyi* (NB-255) showing polar cap of cytomembranes. (71,800×.) *M*, Phase-contrast photomicrograph of *Nitrococcus mobilis* (NB-231). (2,500×.) *N*, Electron micrograph of *Nitrococcus mobilis* (NB-231) showing tubular cytomembranes. (23,900×.) *O*, Phase-contrast photomicrograph of *Nitrospina gracilis* (NB-211). (2,500×.) *P*, Electron micrograph of *Nitrospina gracilis* (NB-211) showing cells lacking cytomembranes. (42,600×.) (Courtesy of Stanley W. Watson and M. Mandel.)

These species are chemolithotrophic and can be cultivated in a solution of minerals such as the following:

INGREDIENT	PER CENT
(NH$_4$)$_2$SO$_4$ (source of energy and nitrogen)	0.20
K$_2$HPO$_4$ (buffer)	0.10
MgSO$_4$	0.05
FeSO$_4$	0.04
NaCl	0.04
CaCO$_3$	0.10
MgCO$_3$	0.10

This medium may be solidified with silica gel but not with agar, since all Nitrobacteraceae are strict autotrophs.

Some of these bacteria, notably *Nitrosouva*, which reproduce by budding with highly irregular cell shapes, and *Nitrosocystis oceanus*, a marine species, exhibit very complex intracellular and pericellular membranous structures or organelles that are suggestive of the photosynthetic lamellae or thylakoids in blue-green algae (Fig. 43–14). Most nitrifying bacteria possess large amounts of respiratory enzyme (cytochrome), as well as characteristic internal membranes bearing electron transporting systems.

Oxidation of Nitrite to Nitrate. This process is called nitrification. Both nitrosification and nitrification are sometimes spoken of together as nitrification. Most higher plants cannot utilize nitrites as their source of nitrogen. In fact, nitrites are toxic to many plants and animals. The most immediately useful form of nitrogen for agricultural purposes is nitrate. Since nitrate does not commonly occur spontaneously in soil, its development is dependent on the presence of the genera *Nitrobacter*, *Nitrococcus*, and *Nitrospina* (Fig. 43–14), which oxidize nitrites to nitrates:

$$HNO_2 + \tfrac{1}{2}O_2 \rightarrow HNO_3 + 21,600 \text{ cal}$$

A difficulty with nitrates as fertilizers is that they are very soluble and are quickly leached from the soil.

Nitrobacters are nonmotile rods. They occur in soil, rivers, and streams, and are worldwide in distribution. Under laboratory conditions they grow well only in the entire absence of organic matter. *Nitrobacter* may be cultivated in solutions such as the preceding by substituting sodium nitrite for ammonium sulfate as a source of energy. However, it should be remembered that these organisms, even under optimal conditions, grow very slowly, especially in laboratory medium, and isolation of pure cultures is, therefore, a time-consuming and tedious process.

Other Nitrogen-Oxidizers. In addition to the Nitrobacteraceae (chemolithotrophs), certain heterotrophic bacteria have been shown to oxidize ammonia to nitrite (e.g., *Streptomyces* and *Nocardia* species). However, several species of eucaryotic fungi (*Aspergillus flavus*, *Penicillium* sp., *Cephalosporium* sp.) carry out both steps, oxidizing **organic** nitrogen (possibly first forming ammonia from it?) to nitrite and nitrate.

Nitrate Reduction and Denitrification

Many aerobic bacteria are able to reduce nitrate to nitrite under anaerobic conditions. The reduction is mediated by **nitrate reductase;** nitrate serves as the electron acceptor. The end product of the reaction below:

$$NO_3^- + 2e^- + 2H^+ \longrightarrow NO_2^- + H_2O$$

nitrite, is highly toxic, and only certain organisms, such as nitrosification bacteria, can tolerate it in large amounts.

In denitrification, nitrates are reduced to nitrite and subsequently to gaseous nitrogen or to ammonia. Nitrate may be reduced to ammonia by a variety of microorganisms (Table 43–4). From such reduction, these organisms are able to obtain cellular nitrogen, and the process is frequently referred to as **assimilatory nitrate reduction.** The extent of nitrate reduction to gaseous nitrogen depends on the species involved, the amount of H donor compounds, and the availability of free oxygen. Denitrification processes are likewise mediated by special enzymes which enable the electrons to be transferred to nitrate, and nitrate is ultimately reduced to molecular nitrogen (the biochemical steps between nitrite and N$_2$ are not yet clearly understood). Various facultative and anaerobic soil organisms are able to carry out denitrification processes, but only a few aerobic bacteria (species of *Pseudomonas* and *Bacillus*) can reduce nitrate beyond the level of nitrite. For these organisms nitrate serves as an effective electron acceptor, as shown by the reaction below:

$$2NO_3^- + 10e^- + 12H^+ \rightarrow N_2 + 6H_2O$$

The end product of denitrification, unlike nitrate reduction, is nontoxic, nonreactive molecular nitrogen. For this reason the denitrification process, unlike the nitrification process, is not

TABLE 43–4. GENERA OF CHEMOSYNTHETIC MICROORGANISMS CONTAINING SPECIES REPORTED TO ASSIMILATE NITRATE NITROGEN °

$$\text{?}$$
$$(NO_3^- \rightarrow NO_2^- \rightarrow [X] \rightarrow NH_2OH \rightarrow NH_3 \rightarrow \text{amino acids})$$

Bacteria	Filamentous Fungi†	Yeasts‡
Aeromonas	*Actinomucor*	*Brettanomyces*
Agrobacterium	*Alternaria*	*Bullera*
Arthrobacter	*Aspergillus*	*Candida*
Azotobacter	*Cladochytrium*	*Citeromyces*
Bacillus	*Coprinus*	*Cryptococcus*
Clostridium	*Fusarium*	*Kekkera*
Cytophaga	*Helminthosporium*	*Endomycopsis*
Edwardsiella	*Neurospora*	*Hansenula*
Enterobacter	*Nowokowskiella*	*Leucosporidium*
(*Aerobacter* or *Klebsiella*)	*Penicillium*	*Pachysolen*
Escherichia	*Phymatotrichium*	*Rhodosporidium*
Hafnia	*Phyctochytrium*	*Rhodotorula*
Hyphomicrobium	*Phytophthora*	*Sporobolomyces*
Micrococcus	*Rhizophlyctis*	*Sporidiobolus*
Nocardia	*Scopulariopsis*	*Sterigmatomyces*
Pasteurella		*Trichospora*
Providencia		
Pseudomonas		
Rhizobium		
Sporocytophaga		
Thiobacillus		
Vibrio		
Yersinia		

°From Payne: Bact. Rev., 37:409, 1973.
†Very likely an incomplete listing.
‡Based on taxonomy from Lodder (Ed.): The Yeasts: A Taxonomic Study, 2nd ed., North Holland Publishing Co., Amsterdam, 1970. *Debaromyces* and *Pichia* species assimilate nitrite but not nitrate nitrogen.
§Evidence for assimilatory nitrate reduction by members of the genera identified in Pichinoty: Arch. Mikrobiol., 76:83, 1971, and Pichinoty et al.: Ann. Inst. Pasteur, 116:27, 1969, is based on their production of enzyme B.

TABLE 43–5. GENERA OF CHEMOSYNTHETIC BACTERIA CONTAINING SPECIES REPORTED TO REDUCE NITRATE DISSIMILATIVELY AND TO DENITRIFY °

Nitrate Respiring ($NO_3^- \rightarrow NO_2^-$)		Denitrifying ($NO_3^- \rightarrow NO_2^- \rightarrow NO \rightarrow N_2O \rightarrow N_2$)
Achromobacter	*Halobacterium*	*Achromobacter*
Actinobacillus	*Leptothrix*	*Alcaligenes*†
Aeromonas	*Micrococcus*	*Bacillus*
Agarbacterium	*Micromonospora*	*Chromobacterium*
Agrobacterium	*Mycobacterium*	*Corynebacterium*‡
Alginomonas	*Nocardia*	*Halobacterium*
Arizona	*Pasteurella*	*Hyphomicrobium*
Arthrobacter	*Propionibacterium*	*Micrococcus*
Bacillus	*Proteus*	*Moraxella*
Beneckea	*Providencia*	*Nitrosomonas*‡
Brevibacterium	*Pseudomonas*	(Not known if observed nitrite reduction to
Cellulomonas	*Rettgerella*	nitric oxide and nitrous oxide serves respi-
Chromobacterium	*Rhizobium*	ratory function)
Citrobacter	*Salmonella*	*Propionibacterium*
Corynebacterium	*Sarcina*	*Pseudomonas*‡
Cytophaga	*Selenomonas*	*Spirillum*
Enterobacter	*Serratia*	*Thiobacillus*
(*Aerobacter* or *Klebsiella*)	*Shigella*	*Xanthomonas*
Erwinia	*Spirillum*	
Escherichia	*Staphylococcus*	
Eubacterium	*Streptomyces*	
Flavobacterium	*Vibrio*	
Haemophilus	*Xanthomonas*	

°From Payne: Bact. Rev., 37:409, 1973.
†One species reduces nitrite to nitrogen but does not reduce nitrate (Chatelain: Ann. Inst. Pasteur, 116:498, 1969).
‡Certain species and strains yield nitrous oxide as a terminal product of reduction.

agreeable to the farmer! Several microorganisms are able to carry out the above reactions (Table 43–5), and the process which uses nitrate as a source of electron acceptors is frequently referred to as **dissimilatory nitrate reduction.**

The reduction of nitrates accounts in part for the lack of fertility of constantly wet soils that support growth of nitrate-reducing anaerobic species. Some of these species are *Thiobacillus denitrificans* and various species of *Clostridium.*

As mentioned previously, part of the fixed nitrogen represented by ammonia, whether the ammonia is derived from decomposing organic matter or by denitrification, escapes into the atmosphere. More would be lost from the living cycle, were it not for its immediate combination in the soil as ammonium salts, and for the nitrifying microorganisms (see previous section) that oxidize ammonia to nitrites and nitrates (Fig. 43–7). In the last form it is again available for plants; thus it re-enters the organic cycle.

The principal microorganisms and processes involved in the nitrogen cycle are summarized in Table 43–6.

TABLE 43–6. PRINCIPAL MICROORGANISMS AND PROCESSES INVOLVED IN THE NITROGEN CYCLE

I. NITROGEN FIXATION:
  A. Nonsymbiotic (independently living bacteria)
    1. Azotobacteraceae
      Azotobacter
      Beijerinckia
    2. Miscellaneous others
      Certain species of *Enterobacter (Aerobacter),* *Nocardia,* Rhodospirillales, *Clostridium,* certain fungi, most blue-green algae
  B. Symbiotic (bacteria living symbiotically with leguminous plants)
    1. *Rhizobium meliloti, R. trifolii,* and so forth
II. NITROGEN OXIDATION (production of nitrites and nitrates):
  A. Nitrobacteraceae
    1. NH_3 to NO_2
      Nitrosomonas
      Nitrosococcus
      Nitrosocystis
      Nitrosolobus
    2. NO_2 to NO_3
      Nitrobacter
      Nitrospira
      Nitrococcus
  B. Miscellaneous others (NO_2 and NO_3)
      Nocardias, *Streptomyces*
      Aspergillus sp., and other molds and higher fungi
III. NITROGEN REDUCTION (ammonia production and denitrification):
      Various microorganisms causing ammonification by producing NH_3 from decomposition of proteins, and denitrification by use of NO_3 and NO_2 as hydrogen acceptors.

43.11
THE SULFUR CYCLE

In many respects the sulfur cycle resembles the nitrogen cycle. Sulfur is essential to all living organisms, as it is part of protein molecules in the form of sulfur-bearing amino acids (methionine, cystine, and cysteine). If sulfur is to be assimilated into organic compounds, the sulfur atom as sulfate must become reduced, because in living organisms sulfur occurs primarily in reduced form as —SH or —S—S— groups.

Sulfur undergoes the familiar alternations between organic and elemental states and between oxidation and reduction. Like nitrogen also, sulfur is most available to green plants in its most oxidized form, i.e., as sulfates. Sulfur is commonly added to agricultural soils as gypsum or as ammonium sulfate. In nature sulfur is often found in the elemental state or in volcanic ("medicinal") waters as hydrogen sulfide (H_2S) and other sulfides. It is released from organic compounds (e.g., proteins) by anaerobic decomposition (putrefaction) in its most reduced state, H_2S, analogous to ammonia (NH_3). Sulfates are also reduced to H_2S by certain bacteria. Like nitrates, fully oxidized sulfur (sulfate) is expensive and quickly leached (dissolved) from soil by rains.

Oxidation of Sulfur and Sulfur Compounds

Many colorless and photosynthetic bacteria are able to oxidize various forms of sulfur, especially hydrogen sulfide, to sulfate. Regardless of the oxidation state of sulfur (except sulfates), when the organisms oxidize sulfur or sulfur in compounds to the next higher oxidation state, the cells gain energy.

Many photosynthetic organisms belonging to the family Chromatiaceae of the order Rhodospirillales are sulfur-oxidizers. Nonphotosynthetic, chemolithotrophic organisms such as *Thiobacillus, Thiobacterium,* and *Thiospira,* are also able to oxidize sulfur and are single, independent, gram-negative, cocci, straight or curved rods, or spirals, generally about 0.5 μm by 10.0 μm in dimensions. Motile species have polar flagella. Many are strict or facultative chemolithotrophs.

Genus Thiobacillus. Thiobacilli may be found in environments which contain large amounts of sulfur; for example, mud, sea water, sewage, boggy places, coal-mine drainage, sulfur springs or areas with low oxidation-reduction potential, such as the benthic zone of lakes, stagnant, polluted aquatic environments.

Thiobacilli oxidize sulfur or its reduced inorganic compounds as energy sources in a variety of ways, depending on species:

1. $5Na_2S_2O_3 + H_2O + 4O_2 \longrightarrow$
$$5Na_2SO_4 + H_2SO_4 + 4S$$

2. $2Na_2S_2O_3 + \frac{1}{2}O_2 + H_2O \longrightarrow$
$$Na_2S_4O_6 + 2NaOH$$

The sulfur in equation 1 above may be further oxidized by other thiobacilli to sulfuric acid:

3. $2S + 3O_2 + 2H_2O \longrightarrow 2H_2SO_4$

All thiobacilli are strict autotrophs. Aqueous solutions such as the following meet all of their nutritive requirements.

INGREDIENT	PER CENT
S	1.000
$Na_2S_2O_3$	0.500
$(NH_4)_2SO_4$	0.030
KH_2PO_4	0.025
$CaCl_2$	0.050
$FeSO_4$	0.001
KCl	0.050
$MgSO_4$	0.020
$Ca(NO_3)_2$	0.050

Note the absence of carbon source. This diet and metabolism are truly marvelous when compared with the complex organic requirements of heterotrophic bacteria or man. Instead of lipids, carbohydrates, proteins, and their derivatives as sources of energy and cell substance, thiobacilli use a few minerals. Instead of complex organic wastes in urine and feces, these organisms excrete corrosive H_2SO_4!

The metabolism of *T. denitrificans* is of special interest, since this represents one of the factors responsible for losses of fertility in certain anaerobic (swampy) soils (**denitrification**, or reduction of nitrates):

$$5S + 6HNO_3 + 2H_2O \rightarrow 5H_2SO_4 + 3N_2 \ (+ \ \epsilon)$$
(In this equation, ϵ = energy.)

Thiobacillus thiooxidans oxidizes sulfur and thiosulfates to sulfuric acid **aerobically**. As sulfuric acid is formed in considerable amounts, it might be thought that the organisms would quickly inhibit their own further growth. This species, however, is of interest in having a great resistance to acid. It is "distinctive in that it is able not only to tolerate but to produce higher concentrations of acid than any other living organism yet known" (Starkey). Some growth is said to occur at a pH of 1, and it grows readily at pH 3. Another species, *T. intermedius*, requires both organic and reduced inorganic sulfur for best growth (Fig. 43–15). The DNA base composition of the genus ranges from 58 to 68 mole per cent GC.

An interesting physiological question arises, and remains unanswered, as to how sulfur particles, water-insoluble, pass through the bacterial cell wall and membrane. In spite of their strange properties these organisms have the same general structures as familiar, heterotrophic, gram-negative bacteria. Could pinocytosis operate in a cell coated by a cell wall?

An important aspect of acid formation by any microorganism lies, on the debit side, in the corrosive and destructive properties of the acids on industrial steel, pipes, and other acid-sensitive products. On the credit side is the very desirable solvent action of acids on phosphate rocks that contain the indispensable element phosphorus in otherwise insoluble forms. (See The Phosphorus Cycle, §43.8.)

Thiobacillus ferrooxidans, a species closely similar to *T. thiooxidans*, is found in acid drainage waters of iron and bituminous coal mines. *T. ferrooxidans* can oxidize ferrous iron salts as well as sulfur:

$$4FeSO_4 + 2H_2SO_4 + O_2 \rightarrow 2Fe_2(SO_4)_3 + 2H_2O$$

$$Fe_2(SO_4)_3 + 6H_2O \longrightarrow 2Fe(OH)_3 + 3H_2SO_4$$

Similar species called *Ferrobacillus ferrooxidans* and *F. sulfooxidans* have been described. These are all true "iron bacteria," i.e., they oxidize iron as a source of energy (Fig. 43–16). The above two organisms have been thought to be variants of *Thiobacillus ferroxidans*.

Bacterial Reduction of Sulfur

As in the nitrogen cycle, sulfate produced by sulfur-oxidizing bacteria may be reduced to hydrogen sulfide (oxidation state of $^-2$) by a few anaerobic species. Sulfate-reducing organisms are distributed widely in nature where anoxic conditions prevail. For example, the organisms have been found in sewage, polluted water, sediment of lakes, sea and marine muds, oil wells, and the rumen of bovine animals. Two groups of organisms are able to reduce sulfate to hydrogen sulfide.

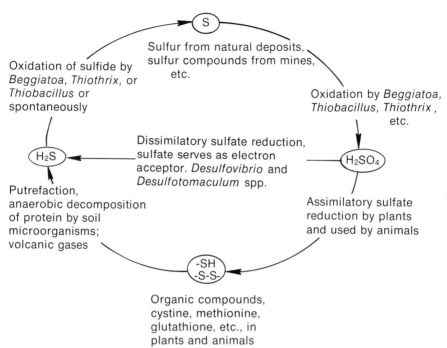

Figure 43–15. The sulfur cycle. At the top, sulfur either enters the cycle from oxidation of H_2S or enters from inorganic sources. Clockwise, sulfur is oxidized by microorganisms to H_2SO_4, which may enter organic structure in plants and animals *(at bottom)* or undergo dissimilatory reduction to H_2S by sulfate reducers.

(1) *Desulfovibrio desulfuricans*, the best-known species of the reducers, is a gram-negative, pleomorphic, curved rod (vibrio-like), motile with polar flagellum (Fig. 43–17). The organism is anaerobic, though it has a cytochrome system like oxidative organisms. The DNA base composition of the genus ranges from 45 to 65 mole per cent GC. Like all typical anaerobes it requires low O-R potentials and must have iron for its cytochrome. Organic materials are dehydrogenated, and the hydrogen is transferred to sulfites, sulfates, and thiosulfates, which are reduced to H_2S:

$$2CH_3 \cdot CHOH \cdot COONa + H_2SO_4 \longrightarrow$$
sodium lactate

$$2CH_3 \cdot COONa + H_2S + 2CO_2 + 2H_2O$$
sodium acetate

Some sulfate-reducers can use molecular hydrogen in the reduction of sulfate:

$$4H_2 + H_2SO_4 \longrightarrow H_2S + 4H_2O$$

(2) The other group of sulfate-reducing organisms are obligate anaerobic sporeforming rods belonging to the genus *Clostridium* (*Bergey's Manual*, 7th edition; however, in the 8th edition of the *Manual* the organism is placed under genus *Desulfotomaculum*, gram-negative, sporeforming rods). In sulfate reduction, sulfite (SO_3^{2-}) is the only intermediate whose role is clearly elucidated, but conversion of sulfite to hydrogen sulfide remains unclear. DNA base composition of *Desulfotomaculum* ranges from 42 to 50 mole per cent GC.

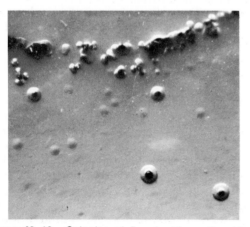

Figure 43–16. Colonies of *Ferrobacillus sulfooxidans*, an autotrophic, sulfur- and iron-oxidizing bacterium. The colonies are on a wholly inorganic nutrient agar (pH, 4) containing $FeSO_4 \cdot 7H_2O$ as the sole source of energy. Note the red (dark) central areas of oxidized iron in the larger colonies. (50×.)

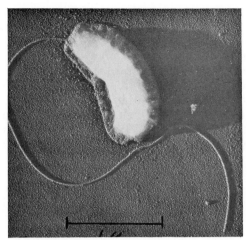

Figure 43–17. Shadowed electron micrograph of *Desulfovibrio desulfuricans.* (18,540×.)

43.12
PLANT DISEASES

Among the unfavorable relationships between higher plants and soil organisms are (a) **parasitism** of plants by pathogenic microorganisms, such as many species of *Xanthomonas* and *Erwinia* (rots, wilts, blights, and spots), *Agrobacterium* (galls, hairy root), eucaryotic fungi (rusts, rots, wilts), and viruses (mosaics, curly top); and (b) **predation** by insects, rodents, nematodes, and the like.

The organisms causing plant diseases (**phytopathogens**) live in the soil, often as saprophytes. They possess protopectinolytic enzymes and other properties enabling them to live in or upon plant tissues, causing disease. Protopectin is a plant gum or cementing substance that holds the plant structures in place.

Genus *Agrobacterium*. Interesting and important bacteria, *Agrobacterium* species are much like *Rhizobium.* DNA hybridization studies indicate that the two genera are closely related (Table 43–7), and, further, these two organisms bear from one to four flagella per cell, generally attached at a slight distance away from the pole (degenerately peritrichous flagellation). Likewise, they both live in, or closely associated with, plant tissues. DNA base composition of *Agrobacterium* ranges from 58 to 65 mole per cent GC. The type species, *A. tumefaciens,* is well known to the floral and horticultural industries as the cause of crown galls and tumors on plants such as the Paris daisy and many other families. Growth of *A. tumefaciens* in the plant tissues stimulates local overgrowth (tumors) of the tissues much as *Rhizobium* stimulates nodule growth on roots. Unfortunately, *Agrobacterium* species do not fix atmospheric nitrogen. Studies of the tumorigenic effects have given some interesting leads in research on human neoplasms (Fig. 43–18).

A related species, *Agrobacterium rhizogenes,* stimulates abnormal root growth, prob-

TABLE 43-7. RELATEDNESS OF VARIOUS DNA SAMPLES TO DNA FROM R. LEGUMINOSARUM 321 AS DETERMINED BY THE SPECTROPHOTOMETRIC METHOD, AND BY THE DNA/DNA AND DNA/RNA MEMBRANE FILTER TECHNIQUES[*]

Source of DNA	Relatedness to R. leguminosarum 321 (±S.E.) by the Spectrophotometric Method[†] (per cent)	Relatedness to R. leguminosarum 321 by DNA/DNA Hybridization[‡] (per cent)	Relatedness to R. leguminosarum 321 by DNA/RNA Hybridization[§] (per cent)
*R. leguminosarum*111	81.6 ± 2.85	79.0	87.3
R. trifolii TA1	70.6 ± 2.50	59.3	74.5
A. tumefaciens 371	37.9 ± 2.33 ⎤	27.4 ⎤	37.9 ⎤
R. meliloti 118	31.1 ± 2.14 ⎬	28.4 ⎰	33.3 ⎱
R. lupini WU8	30.7 ± 2.26 ⎪	25.8 ⎱	30.6 ⎰
R. phaseoli 468	26.1 ± 2.97 ⎪	24.2 ⎰	29.8 ⎱
R. japonicum 372	23.8 ± 2.08 ⎦	18.2	23.8 ⎦
S. marcescens 378	7.3 ± 1.49	5.3	9.6

[*]From Gibbins and Gregory: J. Bact., *111*:129, 1972.

[†]Relatedness is expressed as a percentage of the homologous reaction and is based on triplicate determinations. The coefficient of variation and overall standard deviation were 10.62 and 4.11 per cent, respectively. Means that are joined by braces are not significantly different at the 5 per cent level, according to Duncan's multiple range test.

[‡]Relatedness is expressed as relative hybridization values (per cent). The coefficient of variation and overall standard deviation were 2.6 and 0.87 per cent, respectively (based upon relative hybridization values).

[§]Relatedness is expressed as relative hybridization values (per cent). The coefficient of variation and overall standard deviation were 2.53 and 1.03 per cent, respectively (based upon relative hybridization values).

Figure 43–18. Tumor (crown gall) on a species of chrysanthemum inoculated seven months previously with *Agrobacterium tumefaciens*. (About one-fourth natural size.) (Courtesy of Erwin F. Smith.)

and Myxomycotina (slime molds) contain several species that are pathogenic to plants. For example, *Spongospora* of the class Myxomycetes causes powdery scab of potato tubers; *Phytophthora* and *Rhizopus* species of the class Phycomycetes cause late blight of potato and soft rot of fruits and vegetables, respectively; species of *Claviceps* (class Ascomycetes) are known to produce the disease known as ergot of rye; *Ustilago* and *Puccinia* species (class Basidiomycetes) produce smut of corn and wheat rust, respectively.

The most common and useful method of combatting plant diseases, regardless of viral, bacterial, or fungal origin, lies in breeding disease-resistant plants, augmented with the practices of crop rotation, inspection, and rigid quarantine.

ably by the synthesis of a hormone-like factor (**auxin** or **phytohormone**). It causes **hairy root** of pomaceous plants (apples and pears).

Other Phytopathogens. Other phytopathogenic bacteria of soil origin are: certain species of *Pseudomonas*, which cause spot and stripe of leaf, wilt, and similar diseases; the genera *Corynebacterium* and *Pseudomonas* include animal as well as plant parasites and pathogens. Some species of corynebacteria cause a vascular disease of alfalfa, and ring rot in tomatoes and potatoes. *Erwinia* species produce the enzyme pectinase, which enables them to invade tissues of living plants, causing soft rot as well as wet or dry necrosis and galls in economically important crops such as carrots, potatoes, and cucumbers. One of the best-known organisms in this group is *E. amylovora* (sometimes known as "fireblight" bacteria), which causes cankers in apples and pears (Fig. 43–19). Some *Streptomyces* species are responsible for potato scab.

The plant pathogens of soil are usually disseminated by wind, water, soil rodents, insect vectors, or even infected farm equipment.

Soil Fungi and Plant Diseases. Infectious diseases caused by fungi are extremely important because many of the diseases occur in epiphytotic proportions and spread rapidly, often resulting in complete crop failure. The classic example of a fungus epiphytotic is that of potato blight in Ireland and to some extent wheat rust caused by species of *Puccinia*. All plants are susceptible to some fungus infection. In fact, all classes of Eumycotina (true molds)

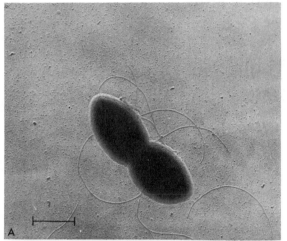

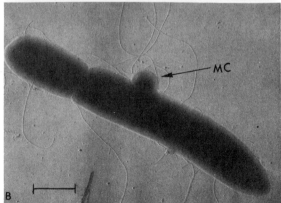

Figure 43–19. Minicells appear to possess both cell wall and cytoplasmic membrane, but do not appear to contain nuclear material. *A,* Dividing normal rod-shaped cell of *E. amylovora* and *B,* filamentous cell with minicell *(MC).* (14,000×.) The markers represent 1 μm. (Courtesy of P. Huang and R. N. Goodman.)

43.13
ANTAGONISMS

In a natural habitat, one microorganism may injure, inhibit the growth of, or even kill, a neighboring organism of another species, and this relationship is referred to as **microbial antagonism.**

There are many antagonisms among soil microorganisms that benefit the plant grower (Fig. 43–20). For example, soil infested with *Phytophthora parasitica*, the fungal cause of **damping-off** of tomato seedlings, may be virtually rid of the pest by inoculation with *Penicillium patulum*, which is antagonistic to the parasitic fungus. It has been observed that *Staphylococcus aureus* and strains of *Pseudomonas aeruginosa* are antagonistic toward the noxious fungus *Aspergillus terreus*. Both *S. aureus* and *P. aeruginosa* produce a diffusible antifungal substance that either causes hyphal swelling or prevents germination of the fungus species. A genus of eucaryotic fungi, *Trichoderma*, produces a substance that greatly reduces infectivity of tobacco mosaic virus. Many similar examples are found in the literature.

Antibiotics and Plant Diseases. Many plant pathogens are quite susceptible to antibiotics, including some antibiotics that are used for treating infections in higher animals (streptomycin, griseofulvin, cycloheximide, the tetracyclines). We know that many of these antibiotics are produced by soil microorganisms: *Streptomyces, Penicillium, Bacillus,* and numerous others. Other antibiotics not suitable for use in human or veterinary medicine are excellent for control of various plant pathogens when used as sprays, dusts, or dips. It is scarcely to be doubted that the antibiotic-producers of the soil produce their antagonistic agents in their natural habitat and that they exert a tremendous influence on the soil microflora. They undoubtedly control plant pathogens to a great degree.

Antibiotics added to the soil or water in which plants or cuttings are growing are soon taken up in the plant and distributed to all parts. Obviously, if a heavy growth of organisms that produce penicillin or streptomycin or polymyxin is present in the rhizosphere, not only is it likely to prevent growth of pathogens in the soil but it may also prevent growth of desirable species.

Other Types of Microbial Interaction. The soil is the natural habitat of countless microorganisms, displaying several types of interaction between or among the many different kinds. If the populations of different species of microorganisms are capable of living in confinement without affecting each other, the relationship is frequently referred to as **neutralism.** Such existence is possible because the two organisms have dissimilar ecological niches. However, if two different organisms have the same ecological niche, they would most likely compete for the same but limited nutrients, and this phenomenon is termed **competition. Synergism** or **mutualism** (a form of **symbiotic** relationship) is found in various natural habitats, including soil. For example, some essential growth metabolic products which neither organism produces alone are produced when these two organisms grow together in the same environment. Also, putrescine production from arginine is accomplished if *E. coli* and *S. faecalis* are grown together, but neither alone can produce this ptomaine. Many similar examples have been reported.

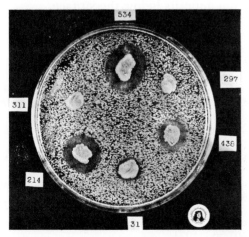

Figure 43–20. Microbial antagonism. The entire surface of the plate was inoculated with *Shigella paradysenteriae*. "Spot" inoculations were then made with various cultures of *Escherichia coli*. After incubation, growth of the *Shigella* appeared as grey "pebbling" except in zones around certain antagonistic cultures: 534, 214, 31, and 438. Cultures 311 and 297 showed no antagonism. (Compare with Figures 17–9 and 43–3.) (Courtesy of Dr. S. P. Halbert and the American Society for Microbiology.)

43.14
MICROBIOLOGY AND PETROLEUM

No final conclusions as to the mode of origin of petroleum may be reached on the basis of present knowledge. However, it is generally held that it originated from living organisms and that microorganisms had a part in it. Crude petroleum contains many hydrocarbons (e.g., paraffin, kerosene), as well as compounds of nitrogen, **reduced** sulfur, phosphorus, and other elements in proportions and relations suggestive of derivation from organic matter. Studies of the subject strongly indicate that: (a) the temperature of petroleum formation was within a range compatible with microbial life (30 to 80C); (b) pressures up to 100,000 pounds per square inch or more are within the limit of microbial viability; (c) petroleum was formed in or near its present locations that at the time were probably sea bottom; (d) conditions were highly anaerobic; (e) salinities were probably elevated (5 to 10 per cent?) but not excessive. The only hydrocarbon known to be produced by bacteria is methane. A question is whether any known microorganisms could produce any of the higher homologues in the hydrocarbon series. While some experimental evidence suggests that this could occur, no conclusive demonstrations on the point have been made.

All higher plants synthesize fats and carbohydrates. Huge vegetable deposits like those that formed coal, when decomposed by certain microorganisms, could conceivably liberate large amounts of the hydrocarbons found in petroleum, but the exact mechanism is not clear.

Destruction of Petroleum. While microorganisms may or may not have produced petroleum, there is a large group of organisms that actively attack and destroy petroleum hydrocarbons. We have already noted some species that oxidize methane. Others, common in the soil near petroleum wells, vigorously oxidize ethane (*Mycobacterium* species and *Pseudomonas* species). Others, as *Desulfovibrio*, oxidize higher homologues, such as petroleum oils and paraffin.

Many microorganisms can decompose the hydrocarbons in gasoline and are of considerable importance in the petroleum industry as causes of spoilage. Among these are *Pseudomonas* and *Chromobacterium* species; also *Alcaligenes*, *Mycobacterium*, *Aspergillus*, *Monilia*, and *Sarcina*. Several species of microorganisms capable of metabolizing petroleum hydrocarbons cause pitting and erosion of tanks, including fuel tanks of aircraft, because of acid formation.

Prospecting for Petroleum. Microorganisms that utilize ethane and higher hydrocarbon vapors as their carbon and energy source are sometimes used to find hidden sources of petroleum. Culture mixtures, complete in all respects *except carbon source*, are placed in flasks and inoculated with an appropriate species of organism able to utilize only petroleum vapors as carbon source. On being lowered into suspected oil-bearing strata and left for some days, growth will occur if petroleum vapors are present. Patents have been issued for some processes of this kind. Error can arise from the fact that methane produced by anaerobic microorganisms of the surrounding soil, e.g., *Methanobacterium*, can confusingly support growth of some hydrocarbon-users quite as well as hydrocarbon vapors from deep oil deposits.

The finding of large numbers of hydrocarbon-oxidizing microorganisms in soil also suggests the presence of hydrocarbons from petroleum deposits below the surface.

CHAPTER 43
SUPPLEMENTARY READING

Baker, K. F., and Snyder, W. C. (Eds.): Ecology of Soil-Borne Plant Pathogens. Prelude to Biological Control. University of California Press, Berkeley. 1970.

Gibbons, A. M., and Gregory, K. F.: Relatedness among *Rhizobium* and *Agrobacterium* species determined by three methods of nucleic acid hybridization. J. Bact., *111*:129, 1972.

Jobson, A., Cook, F. D., and Westlake, D. W. S.: Microbial utilization of crude oil. Appl. Microbiol., 23:1082, 1972.

Kellogg, W. W., Cadle, R. D., Allen, E. R., Lazrus, A. L., and Martell, E. A.: The sulfur cycle. Science, 175:587, 1972.

Payne, W. J.: Reduction of nitrogenous oxides by microorganisms. Bact. Rev., 37:409, 1973.

Watson, S. W., and Mandel, M.: Comparison of the morphology and deoxyribonucleic acid composition of 27 strains of nitrifying bacteria. J. Bact., *107*:563, 1971.

Zak, B.: Characterization and classification of mycorrhizae of Douglas fir. II. *Pseudotsuga menziesii* + *Rhizopogon vinicolor*. Can. J. Bot., *49*:1079, 1971.

CHAPTER 44 • MICROBIOLOGY OF NATURAL WATERS

Approximately three-fourths of the earth's surface is occupied by water, and nearly 24,000 cubic miles of water falls on land annually. Yet today many hydrobiologists are suggesting an imminent and general water crisis throughout the United States.

Available water within the realm of the hydrologic cycle (Fig. 44–1) is defined as total precipitation on land minus evaporation and transpiration. It has been estimated that the available water supply in the United States is approximately 315 to 400 billion gallons per day, with an estimated intake of about 360 bil-

lion gallons per day in 1965; nearly 450 billion gallons per day has been predicted for 1975, or an increase in water intake of about 2.5 per cent per year. From these estimates the student may realize that intensive reclamation and recycling of water will become imperative as domestic and industrial consumptive uses (i.e., uses that make water unfit for potable or other use without further treatment) increase.

Water is a universal solvent and all life depends on it: for green plants and algae (primary producers), water serves as an electron or hydrogen donor in photosynthesis; all life-support-

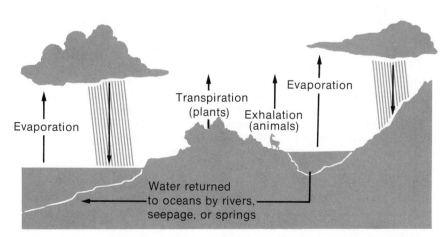

Figure 44–1. The water cycle. From this diagram it is possible to see a number of ways in which water, once it reaches the surface of the earth, is returned to the atmosphere. The constant recycling of water is essential to the maintenance of life in an ecosystem.

ing enzymic activities depend on the presence of water. Most of our potable water supplies are from surface waters which include streams, rivers, lakes, and oceans (by desalination or desalting processes), and these waters are likely to be polluted with domestic and industrial wastes, including agricultural run-off. As the population increases, pollution problems become more serious (viruses, bacteria, hydrocarbons, pesticides, and insecticides, including heavy metals), and such waters can endanger the health and life of humans and other animals. Regardless of the source, surface or ground waters when polluted by fecal material become potential carriers of pathogenic organisms, such as causative agents of typhoid and paratyphoid fevers, dysentery, melioidosis, cholera, and even viral diseases, for example, hepatitis. It will be an enormous task for sanitary or environmental microbiologists to maintain quality water for years to come.

44.1
FACTORS AFFECTING NUMBERS AND KINDS OF MICROORGANISMS IN NATURAL WATERS

The numbers and kinds of microorganisms in natural waters depend largely on the available nutrient supply, the environment, other organisms present, and physical factors.

Available Nutrients. Of primary importance are the available nutrient substances in the water (e.g., ferrous iron for certain "iron bacteria," H_2S for sulfur-oxidizing bacteria; CH_4 for methane-oxidizers; decaying vegetation or animal matter or sewage for organotrophic saprophytes; and so on). Most natural waters contain all of the minerals necessary for all microscopic life.

As pointed out in Chapter 10, nutrient substances dissolved in water tend to accumulate at solid surfaces and in minute interstices and niduses (microenvironments) in porous materials. Hydrolytic and other enzymes secreted by microorganisms, which would otherwise be lost by dilution and removed by water currents, tend to remain more concentrated and effective in such microenvironments. Consequently, the largest numbers and greatest variety of microscopic plants and animals are found at the very surface of the film of standing water (ponds and lakes), where 10 to 1,000 times more bacteria have been reported in the topmost millimeter than in the immediately underlying layers of the water. Likewise, large populations of bacteria

are found near the bottoms (**benthic zone** [Gr. *benthos* = depth of the sea]) and banks (**littoral zone** [L. *litus* = the seashore]) of lakes and rivers.

In the sea, sources of nutrients and populations of microorganisms are most numerous generally within a few miles of shores (**neritic province**) and river outlets (**estuary zone** [L. *aestus* = the tide]). The numbers of microorganisms (especially bacteria) in the open sea and far out in great inland lakes are usually small. The species in mid-ocean are usually indigenous (i.e., specifically adapted to this environment). Bacteria from such waters are often difficult to cultivate in the laboratory unless sterilized aged sea water is included in the medium.

Presence of Toxic Substances. Some waters contain substances unfavorable to certain microorganisms. For example, sea water or water from the Dead Sea is too saline for many species, whereas hydrogen sulfide produced from organic matter by putrefactive microorganisms is toxic to algae and some other microorganisms. Photosynthetic bacteria, however, may use H_2S as electron or hydrogen donor compounds for reduction of CO_2.

Acidic and metallic industrial wastes kill not only microorganisms but tons of larger aquatic and marine plants and fish. Many thousands of miles of streams, which receive billions of gallons of acid mine drainage, are virtually devoid of life except for a few acid-tolerant sulfur and iron bacteria. Acids, such as H_2SO_4 from certain sulfur-oxidizing bacteria, or organic acids, alcohols, and other products of fermentations occurring in muds or stagnant ooze, are unfavorable to many microorganisms as well as macroorganisms.

Predation and Antagonism. The microscopic flora and fauna of waters also depend to some extent on the kinds and numbers of other living things. For example, most plankton organisms feed on bacteria, algae, and one another. Among themselves they maintain an equilibrium between life and death. **Protozoans** and **bacteriophages**, and possibly bacterial predators like *Bdellovibrio*, in water destroy billions of bacteria. Probably some species of marine and aquatic bacteria and other microorganisms produce **antibiotic substances** that destroy other species. Several bacteriophages specific for certain marine bacteria have been found. Cyanophages, likewise, may destroy various species of blue-green algae. The role of **bacteriocins** may be of importance but is still to be evaluated.

Physical Factors. Physical factors are of

critical importance. The kinds and numbers of microorganisms in waters are inevitably determined to a great extent by temperature, pH, osmotic pressure (salinity favoring halophilic bacteria), hydrostatic pressure (depth predisposing toward barophilic bacteria), aeration, and penetration of sunlight.

The microflora in any one aquatic environment may be divided into two types: indigenous and transient populations. Sulfur springs may maintain several species of "sulfur bacteria" or sulfur-oxidizing bacteria in addition to transient populations introduced into the spring via rainwater or surface run-off. Likewise, within the same pond, vertical and horizontal distribution of microorganisms may differ markedly. At the surface, growth of aerobic organisms and algal populations may dominate, while the bottom zone, which may lack dissolved oxygen, is more likely to support the growth of facultative and anaerobic microorganisms.

Thus, the microbiology of natural waters is an exceedingly complex subject, and much remains to be learned about it. However, it is obvious that the microscopic flora and fauna of a shallow, sun-warmed river below the outfall of a large sewage-disposal plant will differ from that of water from the middle of the Atlantic Ocean, water from the Great Salt Lake of Utah, icy water from the Mendenhall glacier in Alaska, or water from the hot sulfur springs of Yellowstone Park.

44.2 AQUEOUS ENVIRONMENTS

Freshwater Environments

The ultimate source of water in various freshwater environments is precipitation water (rain, snow, hail) which falls on land and evaporates, is transpired, or runs off directly into streams, rivers, and lakes, or percolates through soil and eventually forms subterranean water. Therefore, within the water cycle, surface water at one time may be subterranean water at another, and again may be freed by activities of man or nature.

Among the more prominent freshwater environments are lakes, ponds, marshes, bogs, rivers, and springs. The study of these habitats is known as **limnology** and encompasses all branches of science, including aquatic microbiology. While these subjects appear to be beyond the scope of beginning microbiologists, the current emphasis on ecology makes it imperative that the student become familiar with the terms presented below.

Lakes and Ponds. Any depressions of land retaining water from rain, rivers, or springs are called lakes or ponds and are either formed naturally (usually depressions made by glacier scarring or volcanic action) or are man-made. As a group, the waters in them differ markedly in chemical composition and are frequently classified on the basis of nutrient content or degree of biological productivity. For example, relatively infertile lakes low in nutrients, which are primarily of **autochthonous** origin (Gr. *auto* = self, *chthon* = earth, land; i.e., nutrients recycled within), are said to be **oligotrophic** (Gr. *oligos* = small; *trephein* = to nourish). However, biologically productive lakes receiving large amounts of **allochthonous** (Gr. *allo* = other + *chthon*; i.e., externally added) nutrients in addition to autochthonous nutrients are said to be of **eutrophic** (Gr. *eu* = well + *trephein*) type. Limnologists frequently call the lakes which fall between these two extremes **mesotrophic** (Gr. *meso* = middle) lakes.

All lakes eventually mature and become extinct, either through slow, natural succession processes or accelerated eutrophication (Figs. 44-2, 44-3). Some of the major causes of accelerated eutrophication processes are discharge of improperly treated effluents from sewage plants and septic tanks, and farm run-off. The large amounts of soluble fertilizer now applied to farm land frequently are leached into the water table or carried off in flood waters. Nitrate and phosphate pollution of lakes by these sources frequently disturbs the balance of population in the aquatic ecosystem. Some of the undesirable consequences of accelerated eutrophication are increase in turbidity, promotion of toxic algal bloom, accumulation of organic matter, and depletion of dissolved oxygen. Bacteria as decomposers in the ecosystem deplete the dissolved oxygen through aerobic decomposition of dead algae, weeds, etc. Oxygen is further depleted by oxidation of reduced compounds such as H_2S and ferrous iron formed in the bottom sediment by anaerobic microorganisms, and as a result there is a reduction in zooplankton followed by a reduction in desirable fishes.

Lakes and ponds have some characteristics in common. When both environments are sufficiently deep, they display temperature stratification. The region of rapid temperature drop with depth is known as the **thermocline** (Gr. *therme* = heat; *klinein* = to slope), and the

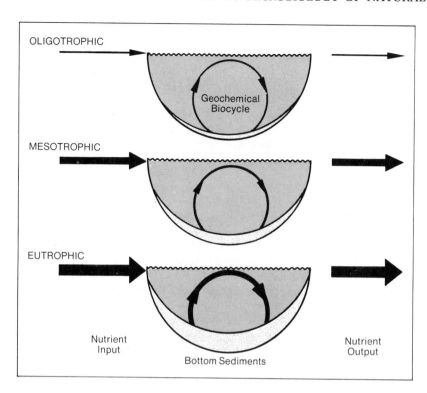

Figure 44–2. The natural process of ecological evolution of a lake, or eutrophication. This leads through a series of successional stages, from oligotrophic through mesotrophic to eutrophic in nature, in which the amount of nutrients recycled and the biomass which accumulates in bottom sediments gradually increase, eventually completely filling in the lake.

body of water above the thermocline is called the **epilimnion** (Gr. *epi* = upon; *limne* = sea); the body of water below is called the **hypolimnion** (Gr. *hypo* = under) (Fig. 44–4). The thermocline layer prevents mixing of epi- and hypolimnion waters during summer; but in spring and fall when surface water is cooled to 4C (the temperature at which water is most dense), it sinks to the bottom, replacing warmer water and causing extensive mixing of both layers of water.

Lakes and larger ponds usually have a surrounding **littoral** (L. *litus* = the seashore) zone (along the shore), where nutrient accumulation is highest; a **limnetic** or **photic** zone, where sufficient light is available for photosynthetic activities by plant life; and a **profundal** (L. *fundus* = bottom) zone, where photosynthetic activi-

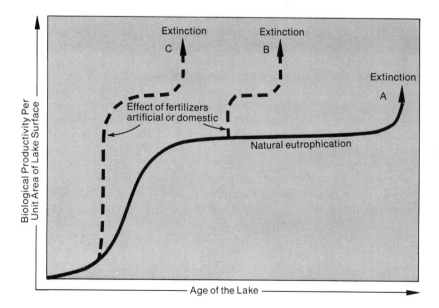

Figure 44–3. Accelerated eutrophication in a lake. The curve *A* represents the slow process of ecological succession from an oligotrophic to a eutrophic condition illustrated in Figure 44–2. Curve *B* shows the effect when the increase in net primary productivity is advanced by the addition of nitrates and phosphates into the biochemical cycling processes of the natural ecosystem. Curve *C* represents even heavier dosages of such additions.

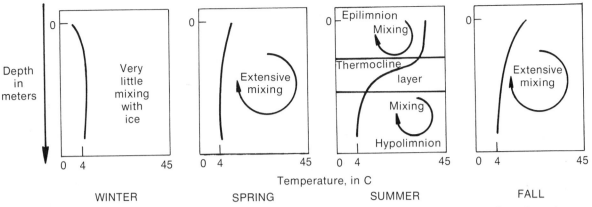

Figure 44-4. Four graphs representing the general temperature structure of most lakes throughout the year.

ties cease as a result of decreased light penetration. Between the **photic** (Gr. *photos* = light) and profundal zones there is a "compensation level," or arbitrary zone or level separating two zones, i.e., photic above and **benthic** below (Fig. 44–5). The benthic zone, or bottom of a lake, is usually composed of soft mud or remains of decaying organic material. The thickness of the benthic zone increases as eutrophication of the lake progresses or the **geobiochemical cycle** (decomposition and mineraliza-

tion) is unable to maintain an equilibrium as an ever-increasing amount of organic matter is introduced into the lake by cultural eutrophication.

Qualitatively and quantitatively microbial populations differ markedly among the various zones: littoral and limnetic zones constitute the most productive regions and contain the greatest varieties (bacteria, algae) of biotypes. The littoral region tends to contain more nutrients because most of the allochthonous nutrients

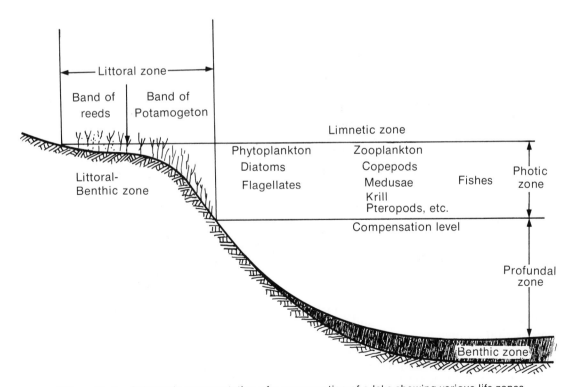

Figure 44-5. Schematic representation of a cross section of a lake showing various life zones.

as well as transient microbial populations are introduced to this area. In addition, nutrients tend to concentrate on the surfaces of solid objects (rocks, various decaying debris, etc.), which are readily utilized by various **epiphytes** (Gr. *epi* = upon; *phyton* = plant), slime-producers, or zoogloea-producing microorganisms. As a rule, especially in eutrophic lakes, facultative heterotrophs dominate the profundal zone, but the nutrient-rich benthic zone harbors facultative as well as anaerobic organotrophic decomposers, e.g., species of *Clostridium.*

Massive winter fish kills, very common in excessively eutrophic lakes, are attributed not to fish pathogens but primarily to lack of dissolved oxygen in the water. Depletion of dissolved oxygen, as stated earlier, is due to aerobic bacterial decomposition of organic material in the lake.

Marshes. The marsh environment is essentially that of a shallow or aging lake characterized by extensive growth of vascular plants (cattails) and so forth, and it is high not only in nutrients but also in the varieties of microorganisms that are present. In many respects, the marsh is an enriched littoral zone with an extensive benthic layer.

Bogs. Bogs are usually formed as a result of aging or ecological succession processes of a lake. There are several ecological successional stages in bog classification; some are semi-aquatic, others may be more nearly terrestrial in characteristics. The nature of the habitat is often revealed by distinctive biotypes, for example, sundew and pitcher plant (carnivorous plants), as well as acid-tolerant vegetation (pH 3 to 5). If the environment is low in nutrients, especially nitrate, nitrogen used by plants is characteristically taken from rainwater or ammonium ions derived from decaying proteinaceous material (aquatic vegetation). The water of bogs, as a rule, is stagnant and contains little or no dissolved oxygen, and the environment as a whole is considered to have very low oxidation-reduction potential. Anaerobes, together with some of the more acid-tolerant microorganisms, are able to flourish, and sulfate-reducing organisms, such as species of *Desulfovibrio*, have been isolated.

Streams and Rivers. Streams and rivers are flowing water fed by surface or seepage water. Therefore, both inorganic and organic nutrients as well as microorganisms are derived primarily from the terrestrial environment (forests, grasslands, rural and urban run-off), including effluents from domestic and industrial sewage. Microbial populations differ markedly according to the location of sampling sites. For example, water samples taken from a river near the sewage outfall from a large municipality contain more microorganisms per unit volume (including the frequent presence of human pathogens, such as species of *Salmonella*) than samples taken from a river in the middle of grassland, where microbial populations may be limited to a mere few hundred per milliliter.

Springs. Springs are a source of water issuing from the ground. The physical and chemical characteristics of the water are generally constant for any given spring, but considerable differences are found among the different springs of various locales. For example, the temperature of hot springs found in either active or inactive volcanic areas varies considerably. Some spring water may reach near the boiling point (e.g., 98C, Yellowstone National Park, Wyoming, or the Hakone area near Fujisan in Japan). Certain sulfur bacteria of Boulder Springs (Yellowstone National Park) are able to function at 90 to 93C. These bacteria possess a cell envelope structure quite different from either mesophilic or thermophilic bacteria. While they possess no morphologically distinct peptidoglycan cell wall layers, the walls are quite thick and diffused (Fig. 44–6).

It has been observed that certain species of blue-green algae not only survive but reproduce in temperatures as high as 50 to 56C, and thermophilic bacteria may thrive at temperatures 10 to 15 degrees higher. The explanation of survival of organisms in such adverse environment is that the organisms' enzymes and other proteins, as well as cell membranes, are much more heat-resistant than are those of psychrophiles or mesophiles. In general, procaryotic organisms (both nonphotosynthetic bacteria and blue-green algae) are more heat-tolerant than are eucaryotic microorganisms such as fungi, green algae and others, and protozoans. However, a few species of thermophilic fungi are able to thrive at temperatures as high as 60C.

Qualitative differences in microflora in a spring are frequently governed by the types of nutrients present. Springs rich in reduced iron may support several species of iron-oxidizing bacteria, such as species of *Gallionella* and *Sphaerotilus*. Springs high in reduced sulfur (H_2S) frequently contain sulfur-oxidizers, such as species of *Thiobacillus*, and photosynthetic bacteria, such as *Chlorobium* and

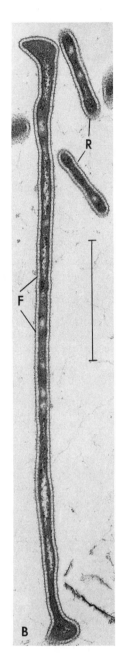

Figure 44–6. *A,* General view of Boulder Spring. The upper source is erupting and the lower source is to the left behind a large boulder. *B,* Overall view of rod-shaped *(R)* and filamentous *(F)* bacteria from Boulder Spring. Bar indicates 1 μm. *C,* Portion of a filament. *CW,* cell wall with inner and outer light regions separated by a dense layer *(arrow); PM,* plasma membrane. Bar indicates 0.15 μm. *D,* Enlarged view of a rod-shaped bacterium. *PM,* plasma membrane; *CW,* cell wall with inner and outer regions and a dense middle layer *(arrow); R,* ribosomes. Bar indicates 0.25 μm. (Courtesy of T. D. Brock, M. L. Brock, T. L. Bott, and M. R. Edwards.)

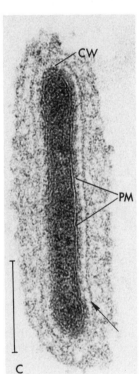

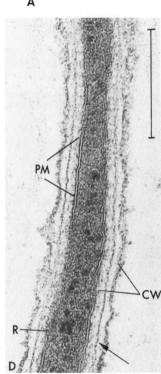

Chromatium. (Iron- and sulfur-oxidizing bacteria are more fully discussed in Chapters 34 and 43.)

Marine Environments

More than 70 per cent of the earth's surface is covered by the sea (salt water), and this environment as a whole has received considerable attention among scientists of various disciplines. The study of the sea is known as oceanography and encompasses all branches of science, including marine microbiology. Perhaps the most striking difference between the sea and freshwater lakes, streams, and rivers is the high concentration of salt and mineral ions in the sea. The most prevalent ions in sea water are chlorine (19.4 gm/kg), sodium (10.7 gm/kg), sulfur (2.7 gm/kg), and magnesium (1.3 gm/kg). In addition, all the naturally

TABLE 44-1. ELEMENTS IN SEA WATER°†

Element	Tons Per Cubic Mile	Element	Tons Per Cubic Mile	Element	Tons Per Cubic Mile	Element	Tons Per Cubic Mile
Cl	89,500,000	Li	800	Ni	9	Ne	0.5
Na	49,500,000	Rb	570	Va	9	Cd	0.5
Mg	6,400,000	P	330	Mn	9	W	0.5
S	4,200,000	I	280	Ti	5	Xe	0.5
Ca	1,900,000	Ba	140	An	2	Ge	0.3
K	1,800,000	In	94	Co	2	Cr	0.2
Br	306,000	Zn	47	Cs	2	Th	0.2
C	132,000	Fe	47	Ce	2	Sc	0.2
Sr	38,000	Al	47	Y	1	Pb	0.1
B	23,000	Mo	47	Ag	1	Hg	0.1
Si	14,000	Se	19	La	1	Ga	0.1
F	6,100	Sn	14	Kr	1	Bi	0.1
A	2,800	Cu	14			Nb	0.05
N	2,400	As	14			Tl	0.05
		U	14			He	0.03
						Au	0.02

°From The Physical Resources of the Ocean, by E. Wenk, Jr. Copyright © 1969 by Scientific American, Inc. All rights reserved.

†Concentration of 57 elements in sea water is given in this table. Only sodium chloride (common salt), magnesium, and bromine are now being extracted in significant amounts.

Sea water contains an average of 35,000 parts per million of dissolved solids. In a cubic mile of sea water, weighing 4.7 billion tons, there are therefore about 165 million tons of dissolved matter, mostly chlorine and sodium. The volume of the ocean is about 350 million cubic miles, giving a theoretical mineral reserve of about 60 quadrillion tons.

occurring elements known to man are found in the sea (Table 44–1). The concentration of these elements in ionic form is frequently expressed as degree of **salinity.**

In general, organisms living in the estuarine habitat must be able to tolerate extreme fluctuations in salinity. Salinity is defined as total inorganic salt concentration (in grams) in one kilogram of sea water. Salinity of the ocean averages about 35 parts per thousand (gm/kg), or approximately 3.5 per cent by weight.

The ocean is now considered a major source of food for the human population, yet this environment, like others, is not free from pollution caused by human activities. For centuries man has used the sea as a most convenient dumping ground for liquid and solid wastes (the United States dumped about 112 million tons of oil, toxic chemicals, heavy metals, and pesticides into the sea in 1971). Why should students of microbiology be concerned with the flora and fauna of the sea? Because the sea and other aquatic environments harbor multitudes of microorganisms collectively known as "plankton" (Gr. *planktos* = wandering). Plankton may be further divided into zooplankton (protozoans and other microscopic animals) and phytoplankton (photosynthetic microorganisms, marine algae). The latter are considered the more important plankton organisms because they are the primary producers of organic mat-

ter in the pyramidic food chain (Fig. 44–7). The marine environment, as does land, contains algae, protozoans, molds, yeasts, bacteria, and bacteriophages, as well as viruses from animals that live in the sea. However, the field of marine microbiology remains in an embryonic stage compared to soil, food, or dairy microbiology. Consequently, very little information about the species composition of microflora is available even today. Most of the available information has been obtained from coastal and photic zones. The marine (saltwater) environment has frequently been classified into **estuary, coastal, intertidal, pelagic,** and **abyssal** zones.

Estuary Zone. The estuary zone is the area or passage, as the mouth of a river or lake, where the tide meets the river current, or simply an arm of the sea at the lower end of a river. Microbial populations in the environment include semihalophilic organisms indigenous to saline water and organisms from salt and fresh water. (In general, organisms living in the estuarine habitat must be able to tolerate extreme fluctuations in salinity.) The area is richly supplied by nutrients from feedwater which may contain considerable amounts of both nitrates and phosphates leached from the land mass. These ions contribute to the fertility and productivity of the estuarine zone as well as to coastal and pelagic zones.

Coastal and Intertidal Zones. The coastal

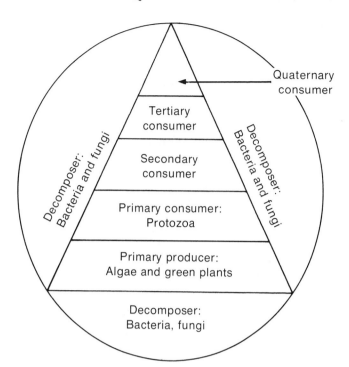

Figure 44-7. Principal trophic levels in the ecosystem.

environment includes the water of bays and inlets as well as of the sea along the coast. The intertidal zone is the area between high and low tide, and the micro- and macroflora and -fauna that live in the intertidal zones are subject to repeated covering and uncovering by the cyclical tides. The tidal zone, as a rule, is subject to extreme fluctuations in salinity, temperature, and aeration, but the area is provided with adequate nutrients both from the estuary region and from the benthic zone through upwelling of ocean currents or by wind action. Biological productivity of coastal and

tidal regions is slightly greater than that of dry agricultural land (Fig. 44-8).

Pelagic and Abyssal Zones. The pelagic zone (open sea) is the area beyond the outer border of the littoral zone, frequently called the **euphotic** zone, to which light penetrates. The area is low in nutrients with somewhat variable salinity, especially in tropical regions as a result of greater evaporation. The abyssal zone is the deep sea or bottom water of the deep sea. Phytoplankton and other plant life are absent because of lack of light penetration. The animals which thrive in these areas are

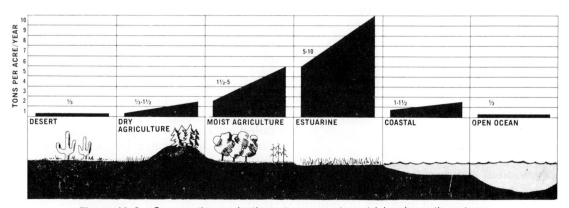

Figure 44-8. Comparative production rates among terrestrial and aquatic systems.

carnivorous; some are blind and others are luminescent. Some luminescence of fishes is caused by bacteria which grow in symbiosis with fish. The mechanism of psychrophilic bacterial biochemical luminescence is poorly understood.

Abyssal and benthic zones may support the growth of psychrophilic and halophilic types, as the temperature of the area may range from 5 to —2C and in euphotic zones from 9 to 12C. Temperatures of water in polar regions may reach as low as —2C (sea water freezes at about —2C), while in equatorial areas temperatures may reach as high as 30 to 40C. The primary function of bacteria in these areas is decomposition of settled organic matter (dead plants, fishes, and other animals) and ultimate mineralization. Nutrients so produced are brought to the surface by the upwelling action of tides and wind and ocean currents, and again converted to organic form by action of photosynthetic phytoplankton. Thus, life in the sea is a vast, complex interaction among micro- and macroflora and -fauna. Bacteria play an essential role in the ecology of the ocean, and without their activities the pyramid of the food chain would cease to exist because they are closely linked in the cycle of elements, as shown in the illustration below.

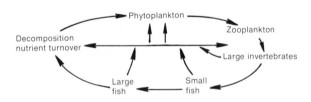

The nutrient turnover by bacteria is continuous, and it is delicately balanced. The cycle must not be upset by massive pollution. It is in the sea that the major portion of **oxygen** needed for survival is produced, in addition to the **foods** needed by all living organisms.

44.3
MICROORGANISMS IN AQUATIC ENVIRONMENTS

Freshwater Environments

Unpolluted Waters. In lakes and rivers free from sewage pollution (if any such now exist!) the concentration of nutrients in solution is usually much lower than in polluted streams like the Hudson, Danube, or Ganges rivers.

Consider a placid woodland pool fed by surface run-off and springs. The water is clear and looks "pure."

The number of bacteria floating free in the water away from zones of nutrition at the bottom and shores is often quite limited, perhaps only a dozen or so per milliliter. These may include various species of soil saprophytes that can grow to some extent in the small amount of organic and mineral substances in solution in the water: species of *Micrococcus, Flavobacterium, Chromobacterium, Bacillus, Proteus, Pseudomonas, Leptospira,* and others.

Aquatic environments, like soil, harbor considerable numbers and kinds of psychrophilic bacteria. For example, psychrophilic strains belonging to genera *Chromobacterium, Aeromonas, Alcaligenes, Arthrobacter, Corynebacterium, Escherichia, Flavobacterium, Klebsiella, Micrococcus, Proteus, Pseudomonas, Rhodomicrobium, Streptococcus, Vibrio, Clostridium,* and *Bacillus* have been reported. Hence, the organisms may be rods, cocci, vibrios, and sporeformers, and they may be gram-negative or gram-positive, strictly aerobic, anaerobic, or facultative. Psychrophilic bacteria are found in many environments in greater numbers than mesophiles, and often in greater numbers than thermophiles; thus their role in various cycles of matter cannot be disregarded. As a rule these organisms not only are able to grow but grow well at 0C or even at subzero temperatures (Figs. 44–9, 44–10).

If there is much decaying organic matter at the bottom, species of *Clostridium* and other anaerobes, strict and facultative, are often found, including sulfur bacteria and gram-negative, strictly anaerobic sulfate-reducing species (*Desulfovibrio* and endosporeforming *Desulfotomaculum*).

Among the myriads of bacteria which thrive in freshwater environments, some of the most interesting are the prosthecate (Gr. *prostheke* = appendage) bacteria known as *Ancalomicrobium* and *Prosthecomicrobium*. The organisms of the first genus possess several long appendages and reproduce by budding; the organisms in the second genus possess several short appendages tapering toward a blunt tip and reproduce by binary fission (Fig. 44–11) (a detailed discussion of prosthecate bacteria is presented in Chapter 34). Other prosthecate bacteria, such as *Caulobacter* and *Hyphomicrobium* (budding bacteria), and sheathed bacteria, such as *Leptothrix, Crenothrix, Clonothrix, Lieskeela,* and *Sphaerotilus,* and other "alga-like" forms may be found growing on the surfaces of rocks and logs near the shore.

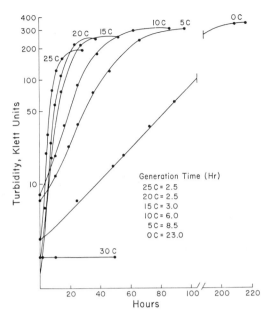

Figure 44-9. Effect of temperature on the growth of psychrophilic *Bacillus (B. globisporus)* isolate W25 in trypticase soy broth.

If hydrogen sulfide is being produced by anaerobic decomposition of organic matter at the bottom and if the pool is not too shaded, species of phototrophic bacteria of the family Rhodospirillaceae may be present.

When during a summer windstorm a large tree falls into the water and stirs up the bottom sediment of a pool, the whole flora changes almost instantly. The organic matter stirred up from the bottom furnishes a rich and varied pabulum. Cellulose-digesters and fermentative types thrive. Numerous species of saprophytes, previously present in small numbers, multiply enormously and some, previously numerous, are suppressed by newly multiplying, antagonistic species. Total numbers of microorganisms per milliliter may rise to 100,000 or more until an equilibrium is again reached.

In a high mountain stream derived from melting snow the numbers and variety of microorganisms to be found are ordinarily small. Unless the stream runs over polluted soil or soil rich in decaying vegetable matter, the water is likely to be almost sterile. It may contain a few spores of *Bacillus*, Streptomycetaceae, molds, or yeasts, but they will not be very actively germinating because of the low temperature. They have probably been caught from the air by the falling rain or snow. A few other microorganisms such as micrococci, corynebacteria (diphtheroids), or gram-negative rods (mostly from dust of the air or from soil), caught by snow or rain, might be found. However, they would not be multiplying either, because of cold and lack of dissolved nutrient substances. Psychrophilic species would predominate in the indigenous flora.

Polluted Waters. The lower Hudson River (among others!) has for decades had a flora representative of sewage pollution. One may assume that *Escherichia coli* and other Enterobacteriaceae, as well as fecal streptococci and various species of intestinal *Clostridium*, are present in large numbers. Many soil saprophytes, such as *Spirillum*, *Vibrio*, *Sarcina*, *Micrococcus*, *Mycobacterium*, *Bacillus*, yeasts,

Figure 44-10. Flagella *(A)* and spores *(B)* of *Bacillus globisporus* strain W25. Flagella stains are from 24-hour-old cultures on trypticase soy agar, and spore strains are from four-day-old cultures on nutrient agar. Incubation of all cultures was at 20C. (2,400×.)

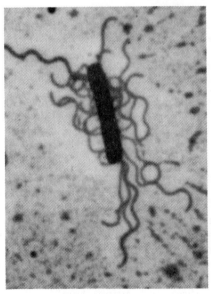

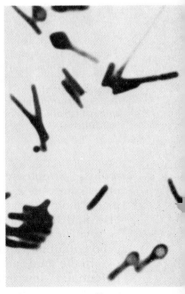

A   B

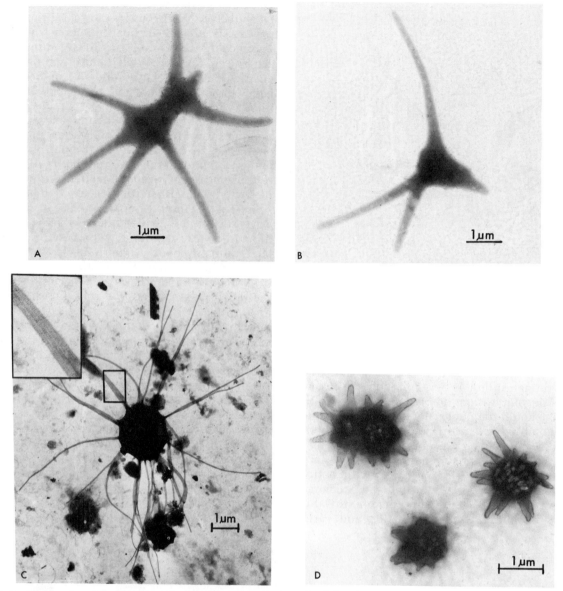

Figure 44–11. *A* and *B*, Electron micrographs of negatively stained *Ancalomicrobium adetum* (type strain). *A* shows a cell prior to division, whereas *B* shows a cell after division.

C, Electron micrograph of a microorganism with many fine (diameter ca. 0.1 μm), slightly tapering appendages having transverse striations.

D, Electron micrograph of negatively stained cells of *Prosthecomicrobium pneumaticum* (type strain) from the original isolation colony. Note the electron-transparent inclusions interpreted as the vesicles of gas vacuoles. (Courtesy of J. T. Staley.)

molds, Streptomycetaceae, *Leptospira* and other spirochetes, *Beggiatoa*, *Sphaerotilus*, and many mold-like and alga-like species, would also find the organic matter in raw or treated sewage or garbage found in such rivers to be good pabulum (Color Plate VIII, *A* to *C*).

In the mud and ooze at the bottom, the O-R potential is low and anaerobic species exist: *Clostridium*, *Desulfovibrio*, and various facul-

tative bacteria, the species depending on the physicochemical nature of the sediment.

In the more aerated surface layers, strict anaerobes do not thrive, and the odors and tastes of putrefaction and fermentation are not so perceptible. The total numbers of microorganisms may reach into the millions per milliliter of water.

There is nothing constant or necessarily

predictable about the flora of a moving body of water, such as a river or tidal water, except within wide limits. For example, conditions change hourly, and the flora changes in response to tide and pollution. The temporary pollution from large passenger ships is a case in point.

In any body of water, saprophytic organisms serve the purpose of scavengers. They decompose organic wastes and make them available as food for other organisms in the water: algae, higher plants, protozoans, and worms. These in turn support fish and other commercially useful marine or aquatic life, and so contribute to human welfare. The suppression of aerobic (i.e., nonputrefactive and non-fermentative) saprophytic microorganisms and other aquatic life by excessive sewage pollution, with its demands on every molecule of dissolved oxygen (BOD; i.e., Biological Oxygen Demand), and by microbicidal industrial waste is one of the major problems of the progress of civilization.

Bacteria in Marine Environments

Bacteria in a marine environment may be divided into two general types: (1) those indigenous to the sea (natural habitat being saline water) and not growing on media without sea water, and (2) transient organisms whose natural habitat is terrestrial and which are able to grow in media without sea water yet able to tolerate varying degrees of salinity. For general studies of many species of marine bacteria a representative medium contains: sea water (aged) 1,000 ml; peptone, 5.0 gm; soluble starch, 2.0 gm; KNO_3, 1.0 gm; $FePO_4$, 0.1 gm; agar, 15 gm. Adjust to pH 7.9 and sterilize. Marine bacteria appear to have a distinctive, specific requirement for Na^+ and other ions in sea water. Many seem to be otherwise identical with familiar terrestrial species but to be, in comparison, osmotically fragile.

Bacteria in Estuarine Environments. In polluted areas of estuarine regions (rich in organic nutrients), organisms such as *Beggiatoa*, *Thiothrix*, *Thiovolum*, and various species of *Thiobacillus* may be predominant; likewise, many transient heterotrophic bacteria, such as species of *Bacillus, Corynebacterium, Actinomyces, Sarcina*, and gram-negative, vibrio-like organisms (Table 44–2). Recently, several strains of terminal-spored anaerobes were isolated from marine sediments. These organisms, unlike typical *Clostridium* species described in Chapter 36, often form cells with two terminal endospores (Fig. 44–12). The organism with two endospores apparently does not belong to any recognized species; hence it has been designated as a new species, *C. oceanicum*.

Photosynthetic purple sulfur bacteria usually occur below algal mats. Such environments are usually anaerobic, which favors growth of either green or purple sulfur bacteria. This is probably because of the wavelengths of light

TABLE 44–2. OCCURRENCE OF BACTERIA IN VARIOUS ESTUARINE ENVIRONMENTS (PERCENTAGE OF SPECIES) *

Species	Water			Sea-Grass Community
	BOTTOM	1 m FROM BOTTOM	SURFACE	
Bacillus subtilis	45	39.5	22	10–25
B. megaterium	18	7	5.5	0
B. sphaericus	0	7.5	0	0
Corynebacterium globiforme	0	7.0	0	0
C. flavum	0	0	0	10
C. miltimon	0	0	0	5
Actinomyces spp.	18	0	5.5	10–25
Staphylococcus candidus	0	8	8	0
S. roseus	0	8	0	0
Mycoplana dimorpha	19	23	54	40
M. citrea	0	0	0	5
Sarcina lutea	0	0	5.5	0
Pigmented strains	20	38	27.5	50
Ratio gram-positive to gram-negative strains	1.9	2.0	0.7	0.9

*From Marine Microbial Ecology, by E. J. F. Wood. © 1965 by Litton Educational Publishing, Inc. Reprinted by permission of Van Nostrand Reinhold Company.

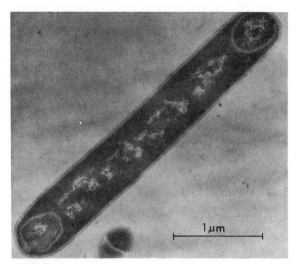

Figure 44–12. Electron photomicrograph of a sporulating cell of *C. oceanicum.* (Courtesy of L. D. Smith.)

required by the bacteria for photosynthesis; these are complementary to those absorbed by algal pigments. Some estuarine sediments with low redox potential and high sulfide concentration may contain green sulfur bacteria (a detailed discussion of photosynthetic bacteria is presented in Chapter 32).

Bacteria in the Ocean. Nearly all marine bacteria adhere to the surfaces of particles, their distribution on particles is more uniform than in sediments, and the numbers of viable bacteria increase with increase in numbers of particles. However, unlike standing freshwater ecosystems, vertical distribution of bacterial populations increases near the 20 m depth

(Fig. 44–13). Lesser numbers of bacteria per milliliter of sample above this depth are partly due to "grazing" of bacteria by the planktonic population.

The most important single group of bacteria in the sea are the pleomorphic, gram-negative, usually motile, psychrophiles resembling species of either genus *Vibrio* or genus *Mycoplasma* in their natural habitat. *V. marinus*, for example, is gram-negative with typical membrane and cell wall, and it forms coccoid or round bodies. Such bodies may contain at least one, and often three or four, cell units. The multicell-unit round bodies are apparently produced from constrictive cell division of the organism (Fig. 44–14). DNA base composition of *V. marinus* is found to be 41 to 43 mole per cent GC. These organisms, however, apparently assume bacillus form in stock cultures and are frequently considered as species of *Pseudomonas* (Fig. 44–15) or species of either *Chromobacterium* or *Flavobacterium*. In general, in marine environments pleomorphic gram-negative rods are more frequent than in fresh water and soil.

Marine Halophiles. Most truly marine bacteria are sensitive to small changes in salinity above or below that of sea water (i.e., they are **stenohaline**). Some bacteria peculiar to salt lakes (e.g., the Dead Sea) *require* salinities of 13 per cent or more (i.e., are **halophilic** [Gr. *halo* = the sea, salt; *philus* = loving]).

Many species of halophilic *Spirillum* and *Vibrio* may be found in tidal bays. Some are able to pass through very fine filters (0.45 μm but not 0.22 μm). Some marine species appear to exist as protoplast-like forms. The salinity

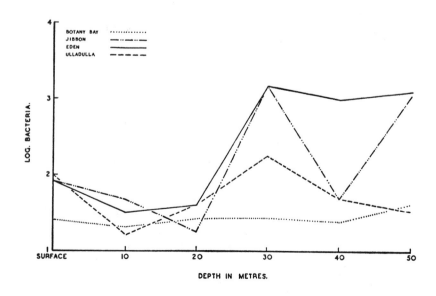

Figure 44–13. Viable counts of bacteria 20 miles off the Australian coast; i.e., east of the continental shelf, except for the Botany Bay station, which is on the shelf. The other three stations show a minimum between 10 and 20 m and maxima at or below 30 m.

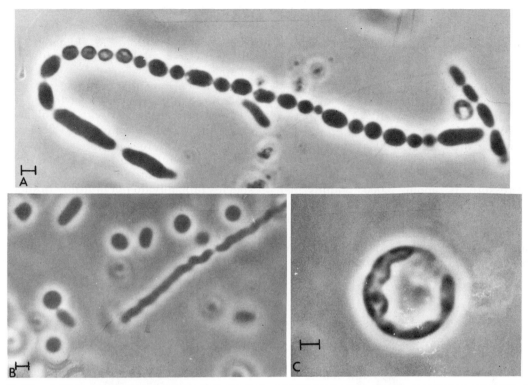

Figure 44–14. *A*, Phase-contrast micrograph of *V. marinus* showing long chain of cells of varied morphology; 48-hour culture. (3,000×.) *B*, Culture at 48 hours showing numerous round bodies. (3,000×.) *C*, Round body showing internal structure. (5,000×.) (Courtesy of R. A. Felter, R. R. Colwell, and G. B. Chapman.)

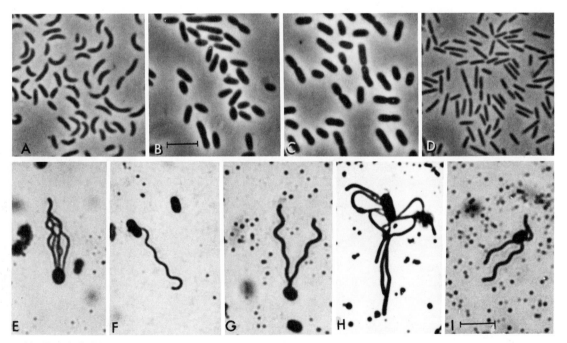

Figure 44–15. *A–D*, Phase-contrast micrographs of cells in exponential phase of growth. (2,000×.) Marker in *B* represents 5 μm, *Alteromonas communis*, strain 8; *B, A. macleodii*, strain 107; *C, P. marina*, strain 140; *D, P. nautica*, strain 179.
E–I. Variation in Leifson flagella stains characteristic of strains in groups B–1, B–2, I–2, and *P. marina* (group F–2). Photomicrographs illustrating single polar flagella have not been included. (2,500×.) Marker in *I* represents 5 μm; *E, P. marina*, strain 143; *F*, group B–1, strain 52; *G*, group I–2, strain 211; *H, P. marina*, strain 140; and *I, P. marina*, strain 143. (Courtesy of L. Baumann, P. Baumann, M. Mandel, and R. D. Allen.)

of sea water would cause the water to be osmotically protective.

Barophilic Bacteria. At the floor of the profoundest depths, pressure becomes very high. In marine environments, unlike those on land, organisms must be able to tolerate a considerable amount of hydrostatic pressure. The rate of pressure increase is 1 atmosphere (1 atm; 760 mm Hg) every 10 m; therefore, an animal which thrives at 2,000 m depth in an ocean trench may experience 200 atm all around. At a depth of 1 mile (1,600 m) the hydrostatic pressure is about 2,000 pounds per square inch. Yet many bacteria thrive in the bottom muds at 6 miles and deeper, where the temperatures range about 3C. These bacteria seem to be very well adapted to these conditions. For example, in the laboratory, many grow only at 3C and 15,000 pounds of pressure per square inch. Those requiring such high pressures are said to be **obligately barophilic** (Gr. *baros* = weight; *philus* = loving). (See Chapter 20.)

It is curious that obligate thermophiles, thought of as growing only at temperatures of 50C or over, are found both in polar seas and in samples of mud from the sea floor at depths of over 30,000 feet, where the temperature ranges constantly *below 10C*. Numerous strictly aerobic bacteria are also found. Some of these species probably do not multiply in such waters and may have reached the ocean floor from above as spores carried by ocean currents.

Photogenic (Luminous) Bacteria. There are several species of photogenic bacteria indigenous to the sea. They may be cultivated

A

B

C

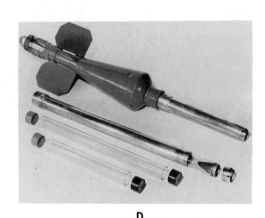

D

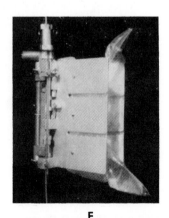

E

Figure 44–16. Water and sediment samplers for bacteriological study. *A*, Gemware-JZ bacteriological sampler; *B*, Birge-Ekman dredge; *C*, Orange Peel dredge; *D*, Phleger corer; *E*, Niskin sampler. (*A to D*, Courtesy of Kahl Scientific Instrument Corp., El Cajon, Calif.)

upon seawater agar with peptone. *Photobacterium phosphoreum* from various marine fish, and various species of *Vibrio* (e.g., *V. pierantonii*) are isolated from luminous marine fish. Many of the photogenic species luminesce only in waters with salinity equivalent to that of sea water (about 3 per cent).

Marine Bacteria and Petroleum. Very interesting observations have been made concerning the possible role of marine microorganisms in the formation of petroleum. Marine bacteria as a group are enzymically active, like most soil and sewage bacteria. The various organisms inhabiting deep ocean beds and marine sediments can, as a group, attack almost any sort of organic matter. There is some evidence, although it is not conclusive, that these marine anaerobic microorganisms can transform certain organic substances into petroleum-like matter. (See also Chapter 43.)

44.4
SAMPLING METHODS IN AQUATIC MICROBIOLOGY

The first difficulty encountered in investigation of microbial ecology in any aquatic system, particularly in the marine environment, is to obtain representative samples. In limnetic areas and in estuaries, the microorganisms vary greatly in quality and quantity from day to day, from week to week, and frequently from year to year. One can hardly consider a sample of a few hundred milliliters obtained from a lake or ocean as a representative sample, especially when one considers such factors as wind, precipitation, currents (both surface and subsurface), and difficulty in determining sampling sites, both location and depth. In limnetic zones and even in estuaries these problems are reduced because distances are less, requiring smaller vessels and permitting ready acquisition of supplementary samples when needed.

Bacteriological Sampling. There are several bacteriological samplers on the market today. For example Gemware-JZ and Niskin samplers (Fig. 44–16, *A* and *E*) can be attached on a bathythermograph, and samples may be obtained at depths down to 10,000 m without possible contamination. Similarly, sediment samples (grab samples) may be obtained by such devices as the Birge-Ekman dredge and Orange Peel dredge (*B* and *C*). While grab samples obtained with such samplers are adequate for qualitative studies of microbial ecology, for quantitative studies one must obtain samples with corers, such as the Phleger corer (Fig. 44–16, *D*). In addition, the study of the microbiology of water is hindered by available microbiological methods and techniques.

CHAPTER 44
SUPPLEMENTARY READING

Baumann, L., Baumann, P., Mandel, M., and Allen, R. D.: Taxonomy of aerobic marine eubacteria. J. Bact., *110*:402, 1972.

Brock, T. D., Brock, M. L., Bott, T. L., and Edwards, M. R.: Microbial life at 90 C: the sulfur bacteria of Boulder Spring. J. Bact. *107*:303, 1971.

Ford, J. N., and Monroe, J. E.: Living Systems. Principles and Relationships. Canfield Press, San Francisco. 1971.

Hutchinson, G. E.: A Treatise on Limnology. Vol. I. Geography, Physics, and Chemistry. John Wiley & Sons, Inc., New York. 1957.

Reid, G. K.: Ecology of Inland Waters and Estuaries. Reinhold Publishing Corp., New York. 1961.

Wenk, E., Jr.: The physical resources of the ocean. Sci. Am., *221*: 167, 1969.

Wood, E. J. F.: Marine Microbial Ecology. Reinhold Book Corp., New York. 1964.

MICROBIOLOGY OF POTABLE WATER AND SEWAGE

• CHAPTER 45

Potable water, or drinking water, is defined as water which is free from pathogenic microorganisms and chemicals that are deleterious to human health. However, other factors, such as taste, odor, and color must be absent if the water is to be palatable. Most of our urban population is served by surface waters (rivers, streams, and lakes) which are cleaned, "polished," and generally disinfected. The raw waters obtained from these sources are frequently contaminated with domestic or industrial sewage, or both, because any body of water is a convenient place for the disposal of sewage and refuse. As our population grows, the recirculation and reclamation of this essential raw material increases, and water is no longer considered to be an unlimited natural resource. It should be pointed out here that the perfectly safe, clean, and clear water you drink may be derived from surface water, of which a greater portion was constituted of domestic or industrial sewage at an earlier stage. The term "sewage" as used in this chapter refers to used water supplies of homes, communities, or industries, or polluted waters. The recycling of water increases as the body of water moves downstream. For example, the water supplies of the cities of Minneapolis and St. Paul have been less reused than that of St. Louis, which also obtains its drinking water supply from the Mississippi River.

We have pointed out in previous chapters that many enteric pathogens (viruses, bacteria, protozoans) are transmitted via water used for domestic purposes. Therefore, water for domestic use in many cities is treated by elaborate means (deep-drilled well water requires less treatment). Surface waters laden with biochemical pollutants are frequently collected and stored in reservoirs, and during this storage many oxidizable organic materials are biochemically stabilized and discrete particles settle. This is usually followed by filtration (in some cases, water conditioning) and disinfection before the water is ready for use.

In the United States, the treated-water budget per capita ranges between 100 and 200 gallons per day, with average domestic consumption of 150 gallons per capita per day. Then we must assume that each one of us produces an average of 150 gallons of sewage per day. In the second half of this chapter we will present sewage treatment procedures, including biological, physical, and chemical means of removing potentially harmful bacteria and chemicals, which may otherwise fertilize the receiving body of water.

45.1
MICROBIOLOGY OF POTABLE WATER SUPPLIES

Most impounded waters, regardless of source (surface or subsurface) contain sufficient nutrients to support growth of various organ-

698

isms; for example, algae, which require only mineral nutrients from the water and energy from sunlight. There are many bacteria, photolithotrophic or chemolithotrophic, which are able to grow in environments with exceedingly dilute nutrients, such as impounded waters. Once these autotrophs flourish, a succession of heterotrophs emerges as they decompose organic material of dead autotrophic cells or organic matter introduced by wind, rain, or surface run-off from the surrounding soil. Some may participate in transformation of iron, mineralization of organic matter, increase in CO_2, or change in pH, which results in corrosion of ironworks; others foul the potable water supplies.

Tastes and odors of microbial origin are probably the most complex of any that create problems in the treatment of a water supply. Among the wide array of microorganisms incriminated as producers of tastes or odors are various species of algae, protozoans, and bacteria commonly known as "iron bacteria." These organisms cause unpleasant tastes and odors that are associated with the growth, death, and subsequent decomposition of the various types by other saprophytes. In addition, some produce color and slime, which causes clogging of water filters and water pipes. The iron bacteria, either ensheathed or stalked, are typical water organisms, and they are aerobic and widely distributed in nature, especially in stagnant waters such as reservoirs for potable water supplies. The more prominent bacteria which have been identified as taste- and odor-producers and filter-clogging bacteria are *Siderocapsa*, *Sphaerotilus*, *Clonothrix*, *Leptothrix*, *Crenothrix*, *Caulobacter*, and *Gallionella* (Fig. 45–1). Growth of some of these filamentous iron bacteria has been referred to as "water calamities" or "red water" by operators of water treatment facilities. The organisms are able to accumulate a considerable amount of ferric hydrate around their cells, and some are able to oxidize manganese compounds. The precipitate frequently is deposited in gelatinous material surrounding the cells. Other bacteria which may participate in fouling of water supplies are sulfur bacteria and sulfate-reducing bacteria. The latter reduce sulfate to H_2S and are responsible for various transformations of iron, both inorganic and organic forms, indigenous to impounded surface waters.

Sanitation of Water For Domestic Use

Preventive Treatment. The most effective and common means of control of taste- and odor-producing microorganisms is to prevent the growth of the causative algae, which are primary producers. Copper sulfate is the most frequently used toxicant for algae control. The amount of copper sulfate should always be determined after microscopic examination of the water sample. While many water-consuming animals may tolerate copper sulfate concentrations up to 12 ppm or more, 0.3 ppm or 2.5 pounds per million gallons (or lower) will kill most of the taste- and odor-producing organisms.

Other methods, such as chlorination by gaseous chlorine or hypochlorous acid–yielding compounds like sodium and calcium chlorite, have been tried successfully. Recently, ozone has been successful in controlling both taste- and odor-producing organisms, as well as oxidation of problem-causing organic chemicals. Ozone has disinfecting power similar to chlorine but is more expensive than chlorine; however, it has a decided advantage over chlorine in that no aftertaste is produced. Physical adsorption processes, such as activated carbon have been tried for removal of taste, odor, and color with considerable success. These methods, however, are not designed to remove all of the impurities present in raw water, and further treatment is necessary before water is considered palatable.

Filter Plants. The main function of water treatment processes is to remove water impurities (either in suspension or solution) which are detrimental to both the appearance and the aesthetic appeal of the water. It is imperative to remove or render harmless any impurities, such as bacteria, which may affect the safety and well-being of the consuming public.

In purifying and disinfecting drinking water for municipalities, three major operations— (1) sedimentation, (2) flocculation and filtration, and (3) disinfection—are generally carried out.

Sedimentation. All waters that have been polluted or contaminated by the introduction of wastes (soil, domestic sewage, or industrial wastes) usually undergo some degree of purification during storage in ponds or reservoirs. The rate at which this purification proceeds depends upon the kinds and amounts of pollutants as well as physical, chemical, and biological conditions of the reservoir and the stored water. Adequate storage time is essential; other factors are temperature of air and water, amount of insolation, and velocity of flow. For example, quiescent sedimentation in a reservoir for a period of 30 to 60 days may result in purification equivalent to that of filtration. However, bacteria or viruses introduced with sewage may persist long after the visible evidence of pollu-

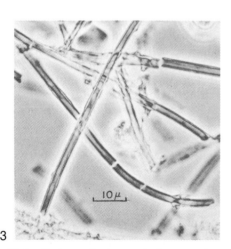

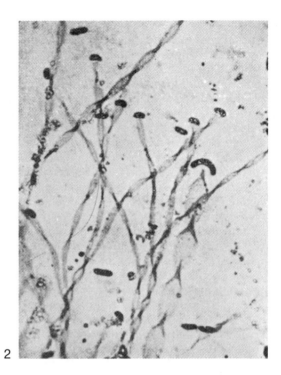

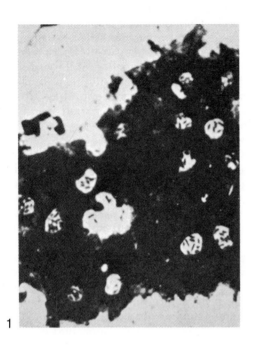

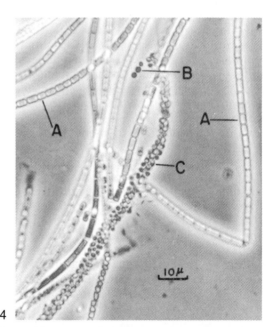

Figure 45–1. Taste- and odor-producing organisms prevalent in potable water. *1, Siderocapsa treubii.* Multiple colonies surrounded by ferric hydrate. (About 500×.)

2, Gallionella major. Curved cells at the ends of excretion bands. (About 1,120×.)

3, Fragments of *Leptothrix ochracea*, phase-contrast photomicrograph. (Taken with the assistance of Dr. J. C. Ensign).

4, Phase contrast photomicrograph of *Crenothrix polyspora.* (A), typical bacillus cells within a sheath similar to that of *Sphaerotilus natans;* (B) and (C) non-motile, spherical cells (conidia) unique to *Crenothrix.*

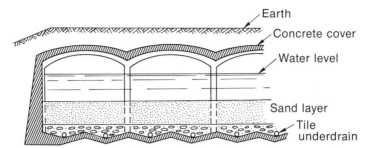

Figure 45–2. Slow sand filter. A slow sand filtration plant consists essentially of a covered concrete basin about 10 or 12 feet deep. Open-jointed tile drains are placed on about 6 foot centers and lead to a central connecting pipe or main drain. These tile drains are covered with about 12 to 18 inches of graded gravel with the largest sizes on the bottom, which is in turn covered with about 3 feet of sand. The cover of the structure should be at least 6 feet above the surface of the sand in order to provide for an adequate depth of water over the sand and sufficient head room during the cleaning operation. The cover usually consists of a concrete slab supported by columns with several feet of soil on top to prevent freezing. Figure 45–3 is a typical cross section of such a filter.

tion has disappeared; consequently, further treatment is generally necessary for production of potable water.

Filtration. After sedimentation, the clarified water is passed through sand filters, of which there are several types.

THE SLOW SAND FILTER. While the trend is toward the construction of rapid sand filtration plants (because of demand for more water), there are many slow sand filtration plants in existence which are effective in providing a safe, potable water supply. Large sand and gravel beds an acre or more in area are built up over drain pipes, starting with coarse gravel (5 cm in diameter) at the bottom and graduating in size to fine sand (0.25 to 0.35 mm in diameter) at the top (Fig. 45–2). The water is led onto the sand and allowed to trickle slowly through (Fig. 45–3). The area of the slow sand filter is necessarily large because the water passes slowly through it.

As filtration proceeds, day after day, there accumulates, around each grain of sand and in the interstices, especially in the upper three or four inches of sand, a slimy, gelatinous film called a **schmutzdecke** (German for "dirt layer") composed of millions of bacteria, protozoans, and other microorganisms. This slowly closes up the pores between the sand grains and makes the filter bed more and more effective. At best, slow sand filters yield about five million gallons of filtered water per acre per day.

Through the action of enzymes, biological oxidation and reduction processes, and the in-

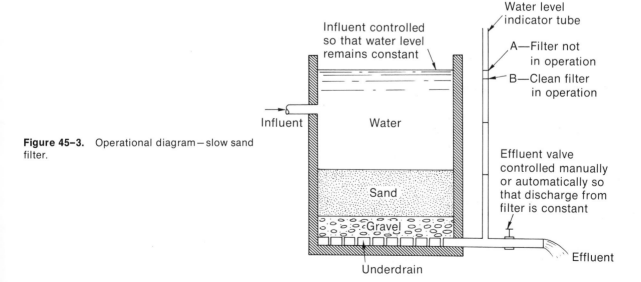

Figure 45–3. Operational diagram—slow sand filter.

gestion of bacteria by myriads of protozoans inhabiting the slimy film, the bacterial and chemical content of the water is greatly reduced. When the gelatinous film finally becomes too thick, the filter is taken out of service and the schmutzdecke is removed by cleaning machines.

The effectiveness of the filtration is constantly tested by bacteriologists in the plant, who determine the numbers and kinds of bacteria present in the water during different stages of the filtration process, as well as in the finished product. Procedures are described in Chapter 35. The filters can remove 99 per cent of the bacteria present in the raw water.

Coagulation-Flocculation. The effective and maximum permissible filtration rate by slow sand filters is at best five million gallons per acre per day; larger demand for water production with slow sand filters would therefore require an extremely large filter area, which is already scarce in many large cities. In addition, slow sand filters do not tolerate turbid water because of rapid clogging of the bed. However, if suspended solids and bacteria are first coagulated and the flocculate settled, it is possible to use filter beds with coarse-grained sand and operate them at a much faster rate (30 to 40 times the volume of slow sand filters), and this has led to general use of rapid sand filters.

Colored materials and turbidity in raw waters may be removed by addition of coagulant and flocculent chemicals, such as filter alum $(Al_2(SO_4)_3 \cdot 18H_2O)$, copperas $(FeSO_4)$, $FeCl_3$, $Fe(SO_4)_3$, sodium aluminate (activated alum which contains silica), and black alum (which contains activated carbon), to raw water. The chemicals react with the basic substances in water to form a white or pale yellow precipitate. Gentle mixing of the coagulated precipitate causes bacteria, suspended solids, and other impurities to adhere to each other, forming larger flocculent masses or "flocs."

Coagulation and flocculation are usually complete within a half-hour after mixing. The coagulant and flocs are allowed to settle in a sedimentation tank before the effluent is filtered through the slow or rapid sand filter, primarily to reduce the load on the sand filter. The physicochemical mechanism of flocculation is very complex and beyond the scope of this discussion. Nevertheless, effluent from the sedimentation tank is clear and practically free of suspended solids and bacteria (99 per cent removal).

THE RAPID SAND FILTER. The principle is similar to the slow sand filter, except that its area is much less and it does not depend on the growth of a schmutzdecke. The rapid sand filter is capable of producing as much water as 200 million gallons per acre per day. The plant consists of a clean bed of fairly coarse sand to remove previously coagulated solids remaining after sedimentation. The effective size of the sand is larger (0.35 to 0.55 mm as compared to 0.25 to 0.35 mm for the slow sand filter). The plant is designed above a special filter bottom with an underdrainage system capable of uniformly collecting and distributing a large flow of filtered water. The arrangement is such that the filters may be effectively "backwashed" by reversing the flow of cleaned water and bubbling air through them. The wash water is "wasted" (the filter inflow valve is shut and the filter wastewater valve is open) (see Figure 45–4). The advantages of rapid sand filters are that they require not only less space for more water filtration but also less maintenance expense.

DIATOMITE FILTERS. A more compact unit with a less elaborate filtration system is the diatomite filter, which consists of central cores or tubes with many small openings, on which a thin layer of diatomaceous earth (siliceous fossil remains of microscopic marine diatoms mined as a filter aid) is applied. The flow of water keeps the filter aid in place. The filter requires less space than do the slow or rapid sand filters and may be made in portable units. The diatomite filter plants generally do not include coagulation of the water; therefore, the treatment is limited to relatively clear water, as with the slow sand filter. The efficiency of bacteria and turbidity removal is comparable to that obtained by conventional sand filters, however. By utilizing various types of filter material, such as activated carbon, the system will find increasing use in years to come, especially for removal of taste, odor, and color (Fig. 45–5).

MEMBRANE FILTER REVERSE OSMOSIS. Until a decade ago, distillation was about the only process of choice to recover large quantities of water from dilute solutions of minerals or pollutants. However, increased public attention has focused on the recovery of potable water from the sea or brackish water, and recently on the reclamation of domestic or industrial sewage for potable water supplies. The increased demand for reclamation of wastewater led to development of processes other than distillation. Reverse osmosis is one of the most exciting and promising processes because it requires low energy and capital expenditure and is highly selective in removal of dissolved matter.

Reverse osmosis is a new term in the tech-

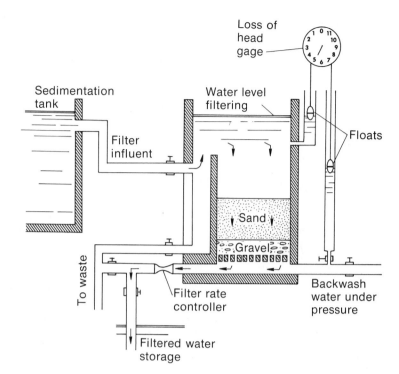

Figure 45–4. Schematic diagram of a cross section of a rapid sand filter. The terms "rapid sand filters" and "mechanical sand filters" are synonymous. The first expression is based on the fact that the rate of filtration is usually about forty times the rate of filtration through slow sand filters, whereas the latter expression comes from the fact that mechanical washing equipment is used to clean the beds.

nology of water purification, while the process of osmosis has been known for many centuries. Osmosis is the process of selective transport of aqueous solutions through a semipermeable membrane, and its principal mechanisms are presented in Figures 45–6 and 45–7.

The principle of desalination by reverse osmosis is essentially the same as that of the osmotic process, except that the process is reversed. The water is forced through a semipermeable membrane (cellulose acetate membrane) under high pressure up to or greater than 1,500 psi (usually 40 psi for tap water and 1,500 psi for sea water). The membrane allows the water to pass through but excludes the majority of dissolved solids. For example, the process may exclude 95 to 99 per cent of the sodium chloride in salt water, while most other salts are

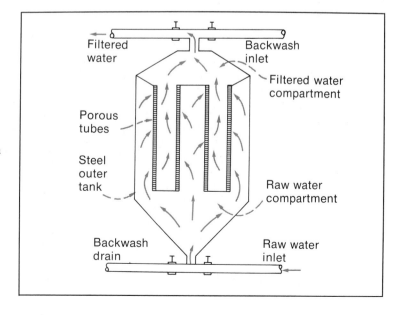

Figure 45–5. Diagrammatic section of a diatomaceous earth filter.

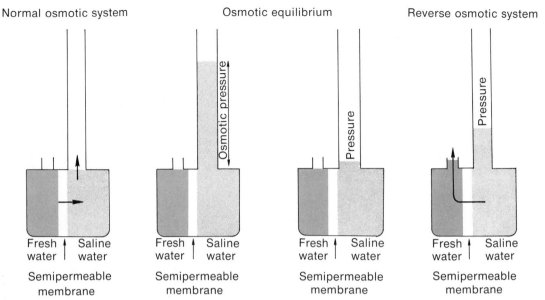

Figure 45–6. In a normal osmotic system, water moves from the more dilute solution through the semipermeable membrane to the less dilute solution, the movement ceasing when both solutions are equally dilute or when sufficient pressure builds to create an osmotic equilibrium. Reverse osmosis, the flow of water from the less dilute solution through the semipermeable membrane to the more dilute solution, can be brought about by exerting pressure greater than the osmotic pressure on the less dilute solution, causing water to diffuse in reverse of normal osmotic flow.

excluded to a greater extent (99 to 99.7 per cent for $CaSO_4$). Discrete and particulate organic material, proteins, bacteria, and viruses are excluded to an even greater extent than typical inorganic salts indigenous to polluted or saline water. Currently, the potential application of the process to treatment of water not amenable to conventional biological treatment is being considered, and the process may be used in place of oxidation ponds now used in tertiary treatment of sewage effluent.

Disinfection of Potable Water. Disinfec-

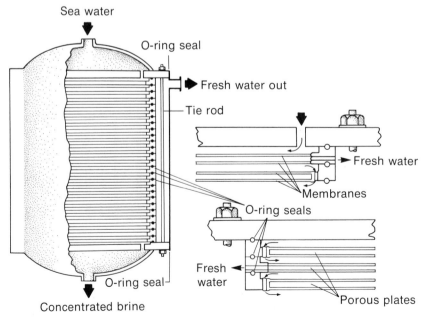

Figure 45–7. Multiple plate type desalination cell.

tion of public water supplies by chlorination or ozonation represents the most important process used in the production of water of safe, sanitary quality. Disinfection refers to reduction of the bacterial population to a safe level, as contrasted with sterilization, which refers to the total destruction of the microbial and viral populations.

Historically, disinfection of public water supplies began in Jersey City in 1908 with use of a solution of hypochlorite. In 1912 for the first time commercial equipment was developed for the application of gaseous chlorine to a pub-

lic water supply, and today more than two-thirds of the nation's population, or more than 17,200 communities, are served by disinfected public water supplies.

As a result of disinfected water supplies and improved methods of sewage disposal, typhoid fever, dysentery, both bacterial and amebic, cholera, and other waterborne enteric diseases remain at a low level or absent in cities with properly operated water filtration plants. Some common bacterial diseases transmitted by fecally contaminated water supplies are summarized in Table 45–1.

TABLE 45–1. SOME HUMAN BACTERIAL DISEASES TRANSMITTED BY FECALLY CONTAMINATED WATER

Etiological Agent (Organisms)	Disease	Pathogenesis
Salmonella typhi	Typhoid fever	Acquired also by ingestion of contaminated food. Incubation period of 7 to 14 days. The organism multiplies in the gastrointestinal tract and may enter intestinal lymphatics and migrate through thoracic duct to the blood stream, then to entire body. The organism may be excreted in urine and in feces. During this period malaise, anorexia, headache, and fever may be experienced; characteristic "rose spots" may appear on the trunk and may last for a few days. In severe cases, there may be intestinal hemorrhages or perforation of the bowel, causing peritonitis. The organism grows luxuriantly in the biliary tract and is thus able to grow in media containing bile salts or deoxycholate.
S. paratyphi (*S. paratyphi-A*), *S. schottmuelleri* (*S. paratyphi-B*), *S. hirschfeldii* (*S. paratyphi-C*)	Paratyphoid fever (enteric fever)	Enteric fever caused by these organisms is usually milder; short incubation period (1 to 10 days). Onset of bacteremia occurs early, and fever usually lasts for one to three weeks, but appearance of rose spots is rare. *S. schottmuelleri* is considered the most common causative agent of enteric fever in the U.S.
S. cholerae-suis	Salmonella septicemias	Causes prolonged septicemia, and the disease is characterized by high remittent fever. Bacteremia may occur without apparent involvement of gastrointestinal tract.
S. typhi-murium and others	Salmonella gastroenteritis	Short incubation period of 8 to 48 hours followed by sudden headache, chills, abdominal pains, vomiting, and diarrhea. Bacteremia is rare, and the disease is usually confined to the gastrointestinal tract.
Shigella dysenteriae	Shigellosis (bacillary dysentery)	Produces not only endotoxin common to all shigellas but also a soluble protein exotoxin known as shiga neurotoxin. The disease is characterized by sudden onset of abdominal cramps, diarrhea, and fever, and appearance of both mucus and blood in the stool are common. Acute diarrhea may cause electrolyte imbalance in infants and young children.
Vibrio cholerae (*V. comma*)	Cholera	The organism multiplies in the small intestine. Sudden onset of nausea, vomiting, diarrhea, and abdominal cramps after incubation period of two to four days. The liquid stool ("rice-water") usually contains mucus, epithelial cells, and large numbers of the organisms. Loss of liquid stool may reach 10 to 12 liters per day, with electrolyte imbalance (loss of K^+, HCO_3^-) resulting in hemoconcentration, acidosis, and hypokalemia (paralysis of the sodium-potassium pump, which controls electrolyte gradient between intracellular and extracellular fluid).

MECHANISM OF DISINFECTION. Chlorine or any of its derivatives (i.e., salts of hypochlorite, chloramines) reacts with water to form hypochlorous acid:

$$Cl_2 + H_2O \longrightarrow HCl + HOCl$$

Hypochlorous acid

Hypochlorous acid is a very unstable compound and decomposes quickly by releasing singlet oxygen:

$$HOCl \longrightarrow HCl + O$$

This oxygen (**nascent oxygen**) is a strong oxidizing agent, and its actions on cellular components are indiscriminate; it oxidizes protein and irreversibly denatures essential cellular enzymes. Destruction of cells by direct combination of chlorine with cell membrane or other protein is considered minor in chlorination of water. Another gas, ozone, behaves in a similar manner as it releases singlet oxygen:

$$O_3 \longrightarrow O_2 + O$$

Some of the factors which influence the disinfectant action of chlorination are: (1) presence of suspended solids and organic matter that may either shield bacteria from the disinfectant action of chlorine or react so that the disinfectant power is lost, and (2) presence of ammonia ions that may reduce disinfecting power by forming chloramine. The higher the pH and temperature of water, the more chlorine necessary per unit volume of water to be disinfected. Presence of reduced ions or compounds, such as nitrite, H_2S, manganese, or iron, may reduce disinfecting properties of chlorine because nascent oxygen will act as an oxidizing agent of these ions rather than on bacteria. Ideally, enough chlorine must be added to leave a **residual** of 0.2 to 1.0 mg per liter of free chlorine after all microorganisms and extraneous organic matter have been saturated with chlorine (**break-point** chlorination) (Fig. 45–8). The chlorinated water is commonly stored in underground cisterns.

ULTRAVIOLET DISINFECTION OF POTABLE WATER. Surface disinfection of inanimate objects by germicidal ultraviolet rays is not a new process, but currently the use of ultraviolet disinfection of potable water is gaining acceptance among certain industries. For example, Sanitron® ultraviolet water purifiers are capable of disinfecting as few as 75 and as many as 20,000 gallons per hour. The purifiers are capable of emitting a dosage in excess of 30,000 microwatt seconds per square centimeter (μW sec/cm²). Figure 45–9 shows overall arrangement of the liquid purifiers. Table 45–2 indicates the ultraviolet energy at 254 nm required for complete destruction of various microorganisms. The method is relatively cheap and safe and requires a minimum of operational maintenance as compared with either chlorination or ozonation. The mechanism of killing the bacteria is similar to that discussed in Chapter 22.

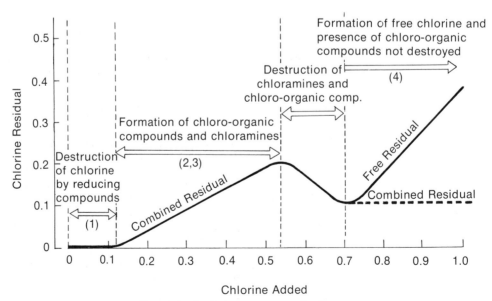

Figure 45–8. Reactions of chlorine in water.

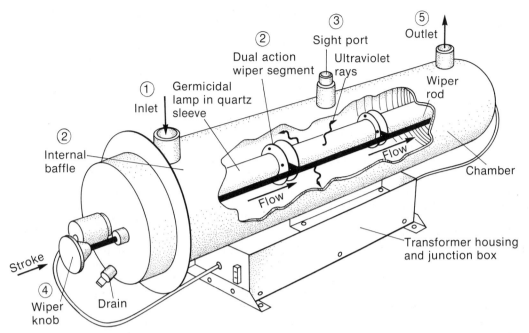

Figure 45–9. Ultraviolet water disinfectant (Sanitron®). *1,* The water enters the purifier and flows into the annular space between the quartz sleeve and the outside chamber wall. *2,* The internal baffle and wiper segments induce turbulence in the flowing liquid to insure uniform exposure of suspended microorganisms to the lethal ultraviolet rays. *3,* The sight port enables visual observation of lamp operation. *4,* The wiper assembly facilitates periodic cleaning of the quartz sleeve without any disassembly or interruption of purifier operation. *5,* Water leaving the purifier is instantly ready for use.

45.2
BACTERIOLOGICAL EXAMINATION OF DRINKING WATER

Periodic bacteriological examination is performed to detect sewage pollution of the water and thus eliminate the possibility that disease may be transmitted by its use. As a rule, water may contain many types of relatively harmless saprophytes whose normal habitats are soil, water, and air. However, water may be contaminated by domestic sewage which may contain human pathogens, such as agents of typhoid or paratyphoid fevers, dysentery, hepatitis, and cholera. The causative agents of these diseases are considered to be of fecal origin and are of primary importance in water examination. However, contrary to the usual belief, no examinations are made for specific pathogenic microorganisms. In fact, the routine bacteriological examination of water is based on the approximate determination of the total numbers of bacteria present and the presence or absence of more common organisms of intestinal or sewage origin.

Index Organisms of Fecal Pollution

Coliform Group. Among the commonest organisms of the intestine or sewage (or feces) are the bacteria of the coliform group (see Chapter 35). The group includes all the aerobic and facultatively anaerobic, gram-negative, nonsporeforming bacilli that produce acid and gas from the fermentation of lactose within 48 hours when incubated at 35C. The most prevalent species of this group are various

TABLE 45–2. ULTRAVIOLET ENERGY AT 254 nm REQUIRED FOR COMPLETE DESTRUCTION OF SOME MICROORGANISMS*

Organism	μW sec/cm²
Bacillus anthracis	8,700
Bacillus subtilis	11,000
Corynebacterium diphtheriae	6,500
Shigella dysenteriae	4,200
Salmonella typhi	4,100
Escherichia coli	7,040
Influenza virus	6,600
Staphylococcus aureus	6,600

*From Sanitron Form #200, Atlantic Ultraviolet Corp., Long Island City, N.Y., 1972.

strains of *Escherichia coli*, followed by *Enterobacter (Aerobacter) aerogenes*. The advantage of testing for these organisms rather than specific pathogens is that the coliform organisms are constantly present in both healthy and diseased human intestines in large numbers, and billions of these organisms are excreted daily by an average person. It is estimated that for every typhoid bacillus or other pathogen (e.g., *Entamoeba histolytica* or viruses of polio or hepatitis) in polluted water supplies, there are usually millions of coliform organisms, especially *E. coli*. Another advantage is that organisms of the coliform group survive longer (Fig. 45–10) in an aquatic environment than do most intestinal pathogens, thus it is possible to detect recent as well as earlier pollution. In addition, the presence of the organisms is easily detected in a short period of time, in contrast to the more tedious, time-consuming identification of a specific pathogen. Consequently, the potential consumer of fecally polluted water may be warned within 24 hours, thus preventing further exposure to possible infection.

Other Index Organisms. Three other groups of bacteria invariably present in human (and animal) feces are: (a) fecal streptococci, especially *S. faecalis*; (b) *Clostridium*, especially *C. perfringens*; and (c) certain species of anaerobic bacteria, *Bifidobacterium (Lactobacillus) bifidus*. These species are easily isolated from water, foods, or dairy products by the use of relatively simple methods of selective cultivation and are readily identified. The first two especially are used frequently as indices of fecal pollution in both waters and foods.

Streptococci. The enterococci are readily isolated by the use of selective media such as sodium azide glucose broth. Sodium azide inhibits gram-negative bacteria especially. Tubes of such broth are often inoculated as a **presumptive test**. Streptococci growing in this medium are transferred to ethyl violet azide glucose broth. Ethyl violet and azide inhibit virtually all bacteria except fecal streptococci. This serves as a **completed test**. Streptococci growing in ethyl violet azide broth are almost certainly of human or fecal origin. *S. faecalis* are rarely found in any but human feces, while *S. bovis* and *S. equinus* generally occur only in bovine, equine, or other animal feces (see Table 38–1).

In the past, the terms "fecal streptococcus" and "enterococcus" have frequently been used interchangeably, but a more recent concept restricts the use of fecal streptococci to denote organisms belonging to Lancefield Group D, which includes *S. faecalis*, *S. faecalis* var. *liquefaciens*, *S. faecalis* var. *zymogenes*, *S. durans*, *S. faecium*, *S. bovis*, and *S. equinus* and their biotypes. Detection of these organisms indicates fecal pollution. Enumeration of fecal streptococci by membrane filter using KF

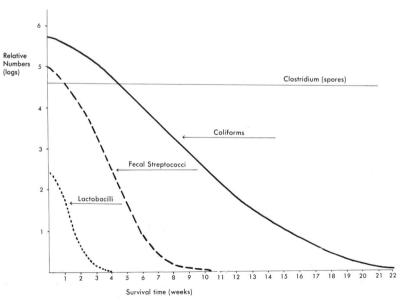

Figure 45–10. Approximate curves of survival of various organisms used as indices of sewage pollution of water. These are merely hypothetical curves, based on data from various situations and types of pollution, and so are subject to considerable variation.

Streptococcus agar (Difco) is shown in Color Plate VIII, *B*.

Significance of Index Organisms. The survival time of these indicator organisms in water is of significance (Fig. 45–10). The **fecal streptococci** seem unable to multiply significantly in open water and do not survive long. *B. bifidus* are quite unable to multiply, and survive only a short time. Their presence in considerable numbers, therefore, suggests relatively recent pollution—a few hours or days. The **coliforms** generally outnumber the streptococci and, possibly being able to multiply to some extent in open polluted waters, may survive for weeks or months, depending on conditions in the water. *C. perfringens*, because of its resistant spores, can survive indefinitely. The presence of this organism in the absence of the others, therefore, suggests pollution that may have existed for a considerable time if it occurred at all. Clostridia grow naturally in many soils and polluted waters.

None of these organisms is a perfect index of human pollution because all occur in the soil or in animal dung, or both, as well as in human excrement. Examination of any water or food for these index species must be supplemented by a sanitary survey-examination of the terrain for their proximity to sources of pollution. It is essential to know the probability of human fecal pollution before final conclusions can be drawn or legal restrictions placed on the use of a water supply or food.

Routine Bacteriological Analysis

Methods for routine detection and enumeration of coliforms and other indicator organisms are carefully prescribed by the American Public Health Association and affiliated societies and are used daily in every health department laboratory.

1. **Standard plate count.** This test is designed to enumerate "total" viable population and not to detect either coliform or pathogenic microorganisms that may be present. The procedure is used as a guide in determining the efficiency of the treatment (e.g., sedimentation, flocculation, filtration, and disinfection). It is generally understood that properly treated water should yield less than 100 colonies per ml.
2. **Test for coliforms.** A series of tests—the presumptive test, confirmed test, and completed test—is carried out in systematic order according to the results of each step. The stepwise standard method of water analysis to detect the presence of coliforms is summarized in Figure 45–11.
3. **Enumeration of coliform organisms.** The presumptive test procedure may be used in estimating the number of coliform organisms present in a given sample.

Enumeration by Most Probable Number (MPN) Method. Multiple tubes of lauryl tryptose broth are inoculated with different volumes of sample (e.g., 5 tubes with 10 ml, 5 tubes with 1.0 ml, and 5 tubes with 0.1 ml). If the number of positive and negative tubes in each dilution is known, it is possible to calculate the probable number of coliforms present in a given volume of water. A combination of the numbers of positive and negative tubes provides an index of pollution which is usually expressed as the "most probable number (MPN) of coliform bacteria per 100 ml of sample." The number of organisms present is not absolute, but is a statistical estimate, i.e., the index represents the number of coliforms which, more frequently than any other number, would give the observed result (Table 45–3). A detailed discussion of the subject is presented in *Standard Methods for the Examination of Water and Wastewater.*

The Membrane Filter Method. Some of the difficulties in determining the presence of coliform organisms by the procedure shown in Figure 45–11 lie in the facts that: (a) only relatively small samples can be examined at one time; (b) several days are required for the incubations of the successive cultures; (c) the test is largely qualitative; only relatively rough estimates of numbers of coliform organisms being feasible; (d) it requires considerable amounts of expensive medium and equipment; (e) it is not readily done in the field on the spot, but necessitates transportation of samples to the distant laboratory, with accompanying risk of deceptive changes in coliform content of the samples.

By means of membrane filters (see Chapter 21), it is possible to filter rapidly, on the spot, and immediately after collection, large samples of water or other fluid samples to be tested for coliforms (Color Plate IV, *C*), fecal coliforms (Color Plate VIII, *A*), fecal streptococci (Color Plate VIII, *B*), or, indeed, any and all organisms that will grow under these particular conditions.

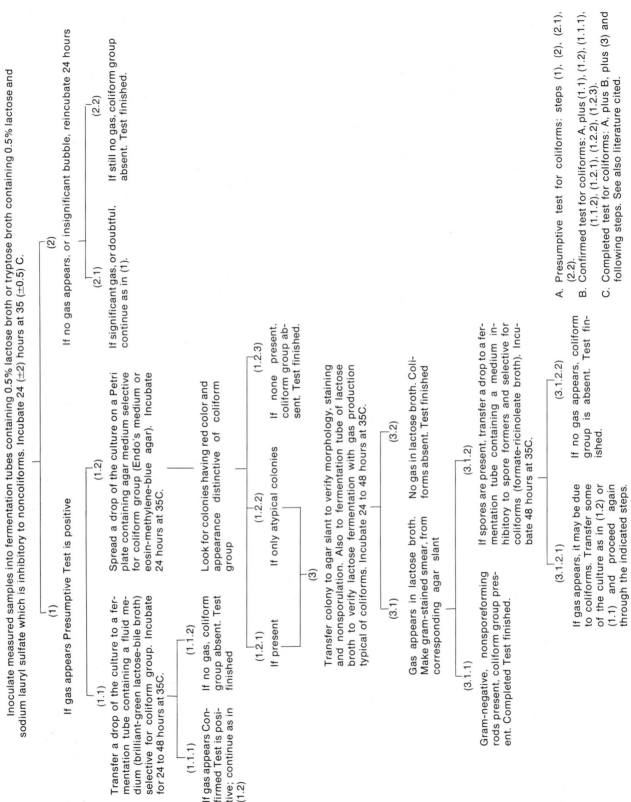

Figure 45–11. Outline of official completed test for coliform organisms in fluids.

TABLE 45–3. MPN INDEX AND 95 PER CENT CONFIDENCE LIMITS FOR VARIOUS COMBINATIONS OF POSITIVE AND NEGATIVE RESULTS WHEN FIVE 10-ML PORTIONS, FIVE 1-ML PORTIONS, AND FIVE 0.1-ML PORTIONS ARE USED *

Number of Tubes Giving Positive Reaction out of			MPN Index per 100 ml	95 Per Cent Confidence Limits		Number of Tubes Giving Positive Reaction out of			MPN Index per 100 ml	95 Per Cent Confidence Limits	
5 OF 10 ML EACH	5 OF 1 ML EACH	5 OF 0.1 ML EACH		LOWER	UPPER	5 OF 10 ML EACH	5 OF 1 ML EACH	5 OF 0.1 ML EACH		LOWER	UPPER
0	0	0	<2								
0	0	1	2	<0.5	7	4	2	1	26	9	78
0	1	0	2	<0.5	7	4	3	0	27	9	80
0	2	0	4	<0.5	11	4	3	1	33	11	93
						4	4	0	34	12	93
1	0	0	2	<0.5	7						
1	0	1	4	<0.5	11	5	0	0	23	7	70
1	1	0	4	<0.5	11	5	0	1	31	11	89
1	1	1	6	<0.5	15	5	0	2	43	15	110
1	2	0	6	<0.5	15	5	1	0	33	11	93
						5	1	1	46	16	120
2	0	0	5	<0.5	13	5	1	2	63	21	150
2	0	1	7	1	17						
2	1	0	7	1	17	5	2	0	49	17	130
2	1	1	9	2	21	5	2	1	70	23	170
2	2	0	9	2	21	5	2	2	94	28	220
2	3	0	12	3	28	5	3	0	79	25	190
						5	3	1	110	31	250
3	0	0	8	1	19	5	3	2	140	37	340
3	0	1	11	2	25						
3	1	0	11	2	25	5	3	3	180	44	500
3	1	1	14	4	34	5	4	0	130	35	300
3	2	0	14	4	34	5	4	1	170	43	490
3	2	1	17	5	46	5	4	2	220	57	700
3	3	0	17	5	46	5	4	3	280	90	850
						5	4	4	350	120	1,000
4	0	0	13	3	31	5	5	0	240	68	750
4	0	1	17	5	46	5	5	1	350	120	1,000
4	1	0	17	5	46	5	5	2	540	180	1,400
4	1	1	21	7	63	5	5	3	920	300	3,200
4	1	2	26	9	78	5	5	4	1600	640	5,800
4	2	0	22	7	67	5	5	5	2400		

* From Standard Methods for the Examination of Water and Wastewater, 13th ed., American Public Health Association, 1971.

The bacteria in the sample are held on the surface of the filter membrane (Fig. 45–12). By methods, adopted as standard by the American Public Health Association in 1965, it is possible, within 24 hours at 35C (total coliform) or 24 hours at 44.5 ± 0.2C (fecal coliform) not only to enumerate the coliform or fecal coliform colonies by their distinctive color (see color plates cited above) but to finish the procedures involved in the confirmed test in the procedure outlined in Figure 45–11.

Filter membranes cannot be used with water or other fluids (e.g., whole blood, fruit juice) containing any considerable amount of algae, cells, silt, or sediment likely to clog the filter, or with samples heavily contaminated with noncoliforms. The standard lactose-broth test and the standard membrane filter method do not measure the same flora, and neither gives results that, on the basis of present knowledge, may be expressed accurately in terms of the other. However, each is a recognized and reliable method.

45.3
SEWAGE AND ITS DISPOSAL

Wastewater from a home, community, or industry is collectively called sewage. The quantity and chemical composition of domestic or industrial sewage varies from hour to hour and from day to day, and even greater differences occur among the different industries. For example, packing plant sewage may be high in nitrogenous organic matter such as manure, blood, fresh grease, and hair, and heavily laden with many different types of microorganisms, as compared with sulfite liquors from wood-pulp plants, high in cellulose, lignin, and bisulfite. The wastes are very difficult to treat because of antiseptic chemicals and high oxygen demand. Almost all industries (dairy, tannery, cannery, distillery, oil refinery, textile, coal and coke, synthetic rubber, steel, etc.) produce their own characteristic sewage; some are readily treated, while others are practically unamenable to biological treatment.

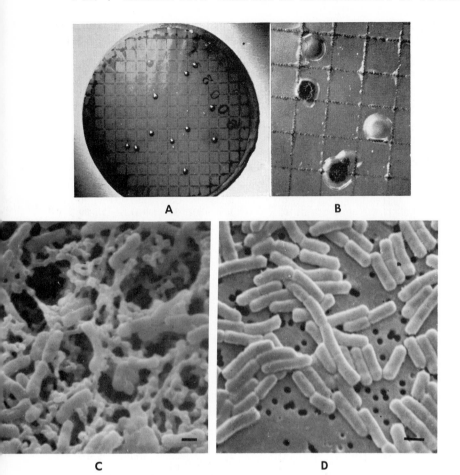

Figure 45–12. *A,* Membrane or ultra-filter disk with colonies of *Escherichia coli.* The disk is on a circular pad saturated with Endo medium, which is selective for coliform organisms. *E. coli* colonies are seen here as dark and glistening. (About actual size.) They are actually deep magenta in color and have a green-violet metallic iridescence. *B,* Section of membrane filter disk with colonies of coliform organisms. *Escherichia coli* colonies appear here as whitish, opaque, and slightly mounded. *Enterobacter aerogenes* are larger, more mucoid, with dark centers. (4×.) (Courtesy of Environmental Health Center, U.S. Public Health Service, Cincinnati, Ohio.) See also Color Plates IV, *C,* and VIII, *A.*

C, Scanning electron micrograph of *Bacillus subtilis* on a 0.45-μm Millipore membrane filter. (5,000×.) *D,* Scanning electron micrograph of *Bacillus subtilis* on a 0.4-μm Nuclepore membrane filter. (5,000×.)

Much of our population is unaware that sewage must be treated before discharge into a receiving body of water for protection of public health, recreational, economic, and aesthetic values. Likewise, most city dwellers are unable to conceive that each individual may produce more than 100 gallons of sewage per day. Domestic sewage simply will not disappear underground. Sewage is transported to a treatment plant either by sanitary sewers or by combined sewers, which may carry both surface run-off (storm sewer) and domestic sewage in a single sewerage system.

Composition of Sewage. Pollution of natural water is largely attributable to municipal sewage and organic industrial wastes. The characteristics of these wastes which are significant in pollution are the amounts of suspended solids (SS), oxygen demand of the organic matter, and microorganisms of intestinal origin. The bulk of municipal sewage consists of water (99.8 to 99.9 per cent), but there is sufficient inorganic and organic matter in suspended and soluble forms to provide enough food for many kinds of microorganisms. As a rule, fresh urban sewage contains dilute excrement and paper, along with other city wastes, such as comminuted garbage, laundry water, and the like. The major constituents of suspended solids (SS) are: lignocellulose, cellulose, protein, fats, and various kinds of inorganic particulate matter; these may exist in the colloidal state, as in undigested starch. Dissolved materials include various sugars, fatty acids, alcohols, practically all the amino acids, and countless inorganic ions.

The organic content of sewage is measured in terms of its oxygen equivalence by means of the **BOD** (biochemical or biological oxygen demand) test. BOD may be defined as that quantity of oxygen required during the stabilization of decomposable organic matter and oxidizable inorganic matter by aerobic (oxidative) biological action. Efficiency of mechanical, chemical, and biological treatment is based on the amount of BOD reduction. Effluent having zero BOD when discharged into receiving water would not deplete dissolved oxygen in the recipient water, since the effluent is completely stabilized. Suspended solids and BOD concentrations of some typical wastes are presented in Table 45–4.

The pH of sewage ranges from 6.8 to 8.5, with an average of 7.0. The temperature varies seasonally (5 to 25C), which favors growth of mesophiles, but sewage contains both psychrophiles and thermophiles, and some of these are pathogenic for man. It should be noted that sewage contains the flora and fauna of the human intestinal tract, as well as many soil and water species.

Microorganisms in Sewage. These include many aerobes, strict anaerobes, and facultative anaerobes, mostly saprophytic chemoorganotrophs. Common types of sewage bacteria derived from soil and intestine are: Enterobacteriaceae, fecal streptococci, *Clostridium, Bacteroides, Cytophaga, Micrococcus,* Pseudomonadaceae, spirochetes, Lactobacillaceae, *Chromobacterium, Aeromonas, Comamonas,* yeasts, molds, and others.

Sphaerotilus, Crenothrix, Beggiatoa, and filamentous Rhodospirillales (if sunlight is present) characteristically form slimy growths on the sides and bottoms of sewage-containing ditches, pipes, and tanks. They are often called **sewage fungi.** Certain real fungi (Phycomycetes), *Saprolegnia* and *Leptomitus,* are often found among them. All the organisms aid in decomposition of organic matter in the sewage. In addition to these organisms, sewage may contain several methanogenic bacteria, which convert H_2 and CO_2 to CH_4 (methane). This **sewer gas** is frequently entrapped and used as a power energy source by treatment plants.

At least 60 types of enteric viruses have been found in sewage, mainly polioviruses, ECHO, and Coxsackie viruses, as well as hepatitis and adenoviruses. Unless very heavy chlorination is used, some of these can appear in the final effluent.

45.4
WASTEWATER DISPOSAL PLANTS

If raw domestic or "sanitary" sewage is discharged without previous biological or chemical treatment directly into lakes or streams, the water becomes a vector of disease; the available oxygen and other hydrogen acceptors in the water are soon used by the microorganisms living in the sewage; and foul-smelling, anaerobic processes develop and create a nuisance. Also, growth of fish and other forms of higher aquatic life is impossible, and considerable economic loss results. Further, the water that receives the sewage is ruined for drinking or recreational purposes. Sewage disposal plants are therefore operated to accomplish several ends as follows.

Physical or Mechanical Treatment

Screening. It is necessary to remove bulky foreign matter such as bottles, paper, wooden boxes, and other extraneous refuse as well as to allow grit and gravel to settle out in **grit chambers** or **preliminary settling** basins. This is largely a matter of mechanics and does not concern us.

Separation. Various kinds of **preliminary settling** or sedimentation tanks are employed, all designed to allow the solid matter to settle out as much as possible. The sewage is held in these tanks usually from 2 to 10 hours. This is a modification of the natural process of sedimentation that goes on constantly in rivers, lakes and the ocean. **Flotation** of fats, wood, and the like may also occur, and surface-skimming is often an important part of preliminary settling. From 40 to 60 per cent of the solid matter of sewage settles out of suspension as primary sludge in three hours in these tanks. Treatment at this point is frequently referred to as sedimentation or primary treatment, and treatment efficiency ranges from 30 to 40 per cent BOD removal. A small septic tank of this nature is shown in Figure 45–13.

Two-Story Tanks. Some tanks for sedimentation purposes are made in two compartments, an upper and a lower. The upper portion is like a long, double, V-shaped trough and serves to introduce fresh sewage (Fig. 45–14). The solid matter settles out through slots at the

TABLE 45–4. CHARACTERISTICS OF SOME TYPICAL ORGANIC WASTES*

	Suspended Solids (ppm)	BOD$_5$ (ppm)	pH
Sewage	100–300	100–300	–
Pulp and paper	75–300	–	7.6–9.5
Dairy	525–550	800–1,500	5.3–7.8
Cannery	20–3,500	240–6,000	6.2–7.6
Packing house	650–930	900–2,200	–
Laundry	400–1,000	300–1,000	–
Textile	300–2,000	200–10,000	–
Brewery	245–650	420–1,200	5.5–7.4

*Reprinted with permission from Eckenfelder and O'Connor: Biological Waste Treatment, Pergamon Press, Inc., 1961.

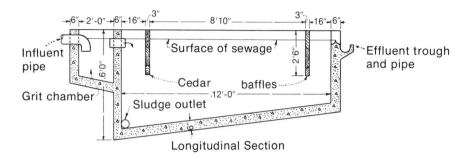

Figure 45-13. Typical septic tank for school or factory.

bottom of the V's into a deep **sludge-sump** as the fluid flows along the trough. The solid matter, after settling to the lower compartment, is held there for some time to permit digestion and decomposition. The sludge is periodically removed to other digestion tanks. The best known device of this form is the **Imhoff** tank.

Biological Stabilization of Sewage

All biological waste treatment depends upon microbial degradation activities (hydrolysis, oxidation-reduction) on decomposable nutrients in sewage. Therefore, the treatment process is considered a very specialized ecosystem, and within this system each microorganism's growth and reproduction is governed by its ecological niche (lack of O_2 or essential growth factors or nutrients, presence or absence of toxic material, predators, etc.) within the process. For example, the waste products of hydrolysis and metabolism produced by one organism are excellent pabulum for still another species, and so on. This is a form of **syntrophism** or mutual nutrition. The original molecules of wood, fat, meat, or starch in sewage are finally changed, usually through the combined actions of several species, into soluble, relatively simple substances available to algae and other plants.

Decomposition and stabilization of wastes may be accomplished either anaerobically or aerobically, depending on the choice of the engineer, and both methods depend on characteristics and amount of waste being treated. However, regardless of whether aerobic or anaerobic, waste stabilization occurs in a series of more or less discrete stages. These stages are frequently grouped under the headings of **primary, secondary,** and **tertiary,** or advanced, **treatment,** and the amount of BOD removed is roughly correlated with the number of stages in the treatment process.

Anaerobic Processes. **Sludge Digestion.** As we have seen, many of the common saprophytes of soil, water, and the intestinal tract are anaerobic or facultative and, acting together, possess marked powers of hydrolysis of proteins, fats, and related compounds. Nearly all forms of saprophytes can hydrolyze one or more carbohydrates. The woody materials in sewage are decomposed by numerous species capable of hydrolyzing cellulose. A piece of linen fabric or a thick sheet of cellulose filter paper will be digested and disappear com-

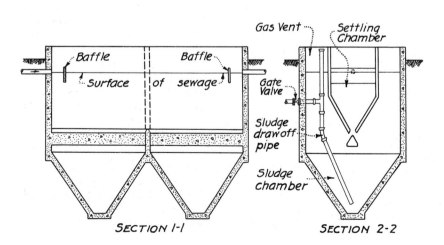

Figure 45-14. A typical Imhoff tank. For explanation see text.

pletely in active sewage in five to seven days; faster in warm weather or in warmed sewage. Several species of microorganisms in sewage together hydrolyze all manner of organic matter: phenol, rubber, paraffin, and so on. Some of these organisms have been mentioned. Others, especially certain gram-negative, anaerobic, cellulose-digesters of sludge, are still imperfectly known. Treatment processes, such as with septic or Imhoff tanks, are anaerobic (as in tank-digested sludge), and reduced compounds are formed: e.g., hydrogen sulfide, nitrogen, ammonia, and methane. These are volatile. Other products of decomposition consist of microbial substance and reduced nonliving materials rich in nitrogen, sulfur, and phosphorus.

Aerobic Processes. Under aerobic conditions, as in shallow sewage oxidation lagoons, aerating (trickling) filters, and activated sludge, all of the complex, putrescible, and organic matters in sewage are eventually changed largely into oxidized, inorganic materials: sulfates, phosphates, nitrates, carbon dioxide, and water: i.e., they are **mineralized.** If one desires to compact the time of mineralization of organic matter, aeration is the key note in sewage disposal. The fluid part of the sewage, after passage through the sedimentation tanks or above the sludge compartment of the Imhoff tank, still contains much putrescible organic matter subject to oxidation. It may be recirculated through the system. The clarified fluid is subjected to various aeration treatments.

Sewage Oxidation Lagoons or Ponds. The oxidation lagoon is nothing more than a well-sealed shallow hole in the ground used for storage, equalization, evaporation, and sedi-

mentation, along with aerobic or anaerobic stabilization of wastes. The lagoon treatment process is usually recommended only for small communities in rural areas, where large acreages of land are available. The lagoon may be a single unit, or two ponds in series, or multiple ponds; the choice is dependent on the degree of treatment desired (Fig. 45–15). Detention time of wastewater in a lagoon varies considerably (20 days to no discharge), according to the climate of the area and holding capacity of the lagoons. The depth of the oxidation ponds should not exceed 5 ft if they are to maintain aerobic biological oxidation. Oxygenation of the wastewater is accomplished by natural means (e.g., wind action and photosynthetic activities of algae) or by mechanical aeration. Efficient and properly operated lagoons are capable of reducing BOD by 75 to 95 per cent, or greater. Because of long detention time, the oxidation ponds are very effective in reducing numbers of both bacteria and viruses. After detention of 45 to 60 days, reduction of coliforms may reach 99.7 to 99.9 per cent (Fig. 45–16). Similarly, reductions of polioviruses, adenoviruses, Coxsackie virus B-3, ECHO virus, and reovirus are noted in raw sewage and primary effluent treated in a lagoon or pond.

Currently, the oxidation pond system is receiving considerable attention, especially in many arid areas, partly as a result of recognition that ground water must be conserved for domestic, industrial, irrigation, and recreational purposes. For example, a certain state in the western United States has proposed converting an existing pond to an aquatic park, featuring three lakes, with combined surface area of 19 acres. At present, the water in the oxidation

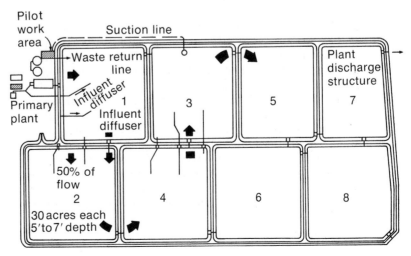
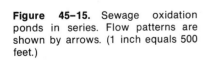

Figure 45–15. Sewage oxidation ponds in series. Flow patterns are shown by arrows. (1 inch equals 500 feet.)

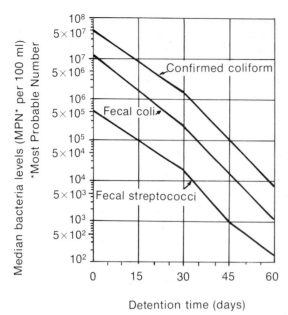

Figure 45–16. Median bacteria counts through oxidation ponds.

pond is generally unsuitable for use in the lake without further treatment. However, with the addition of extensive primary treatment (chemical and biological) and an effective disinfection system, the effluent may be used not only for recreational activities such as swimming, boating, and fishing, but also as a potential source of potable water. In the future, there will be closed systems; i.e., the same water may be used again and again.

Aerating Filters. The fluid from any form of settling tank **(primary treatment)** may be sprayed onto, and allowed to trickle and splash continuously or intermittently through, artificial beds of broken stone or coke. The process is considered aerobic, and such **secondary treatment** is frequently referred to as a **trickling filter** or **bacteria bed.** These beds may be single-stage (1 filter), 2-stage, or even multiple-bed, depending upon the degree of treatment and amount of sewage to be treated. The filtrate from such filters or beds is collected by underdrains, much as in slow-sand filtration of water. The pieces of stone become coated with a living film of aerobic, strongly oxidative microorganisms, much like the schmutzdecke in a slow-sand water filter or the floc in activated sludge (see next section). These organisms feed upon and oxidize the organic matter of the sewage, the result being a much less offensive liquid (Fig. 45–17).

The filtrate or effluent from the trickling filter may be recirculated (part of the treated effluent is returned and mixed with primary effluent) to decrease further the BOD, and the remaining effluent is fed into a secondary settling tank. Here many bacterial mats may be sloughed from stones, or discrete solids may settle and form what is known as humus sludge, which will be treated along with sludge in the primary sludge digester (Imhoff tank). The stabilized sludge from the digester will be dried on sand beds and may be used as soil builder or for fill. The effluent from the secondary settling tank is disinfected prior to discharge into the receiving body of water. BOD reduction of properly operated trickling filters ranges from 80 to 90 per cent, or higher.

Activated Sludge. The aerobic sewage treatment in which flocculated biological growth is continuously circulated and in contact with organic waste in the presence of oxygen is called the **activated sludge process.**

If large volumes of compressed air are forced through sewage in a tank, aerobic conditions are maintained throughout the liquid (Fig. 45–18). Particles of suspended matter flocculate, after a time, into small, gelatinous masses swarming with aerobic microscopic life and capable of oxidizing organic matter readily. These gelatinous masses are called **activated sludge.** As the process continues, the volume of the floc, or activated sludge, increases as more and more sewage and air are passed through the tank.

ACTIVATED SLUDGE ORGANISMS. The particles of floc in activated sludge consist of mixed species of bacteria which embed themselves in a mass of polysaccharide gum called **zoogloea** (Gr. *zoos* = living; *gloea* = glue; hence, "living glue"). One of the principal zoogloea-forming organisms is the *Pseudomonas*-like species called *Zoogloea ramigera*. This and related organisms have complex nutritional requirements. They oxidize sewage materials very rapidly and are very active in forming floc. Numerous other familiar microorganisms may also form (or are found in) zoogloeal masses under the conditions of activated sludge: *Escherichia,* various species of *Pseudomonas, Alcaligenes,* and *Bacillus,* and filamentous organisms like *Sphaerotilus* as well as numerous types of protozoans (Fig. 45–19). The sticky zoogloeal material formed by these species gathers up, by adhesion and by adsorption, much of the colloidal material, bacteria, color, and odors of the sewage fluid.

Within the activated sludge system, different bacterial flora may act in syntrophism or

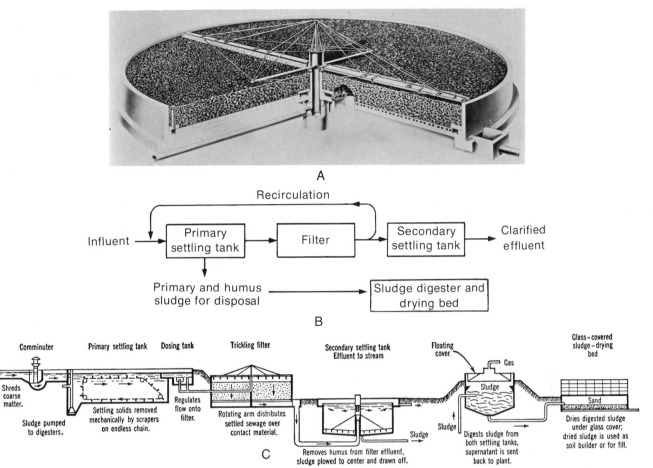

Figure 45–17. Sewage flow diagram of trickling filter. *A,* A typical trickling filter equipped with Dorrco rotary distributor (sparger). *B,* Flow diagram of aerating filter. *C,* The solids are comminuted mechanically and collected mechanically from a primary settling tank. The aerating filter is of the sparger type. A secondary settling tank removes residual sludge or humus to a digestion tank which has a "floating" cover to trap sewer gas for use as fuel. The sludge is finally removed to a covered sludge drier.

antagonism, and those bacteria most adapted to the environment become dominant. Later, as the bacterial population increases, free-swimming ciliates and flagellates prey on bacteria, and an increase in these protozoans is usually accompanied by an increase in rotifers and nematodes. The end result is mineralization of organic carbon, nitrogen, and phosphorus. However, the mineralization of organic matter by this process has been criticized in recent years, because effluent from the process, while clear and no longer containing oxygen-demanding substances, does contain relatively large amounts of inorganic phosphates and nitrates (produced by oxidation of ammonia by nitrifying bacteria in the activated sludge tank), both of which stimulate the growth of algae in receiving streams and lakes.

Rapid clarification, enzymic decomposition, oxidation, and decrease in bacterial content

are achieved, provided aeration continues and enough activated sludge is intimately mixed with the flowing sewage. Complete mixing and vigorous aeration is, therefore, essential in efficient oxidation-reduction of organic wastes. The oxygen uptake rate of a gram of activated sludge ranges from 10 to 20 mg/hr during active respiration; similarly, for dairy waste, 40 to 45; cannery refuse, 35; pulp and paper, 10 to 15; and certain pharmaceutical wastes, 76 mg/hr have been recorded. Oxygen uptake of bacteria and protozoans is listed in Table 45–5. The basic principle of the process is similar to that of aerating filters, the microbe-laden stones of sewage filters being replaced by the living particles and air in the activated sludge.

The fluid part of the activated sludge–sewage mixture is at length passed into a final settling tank. Part of the activated sludge is

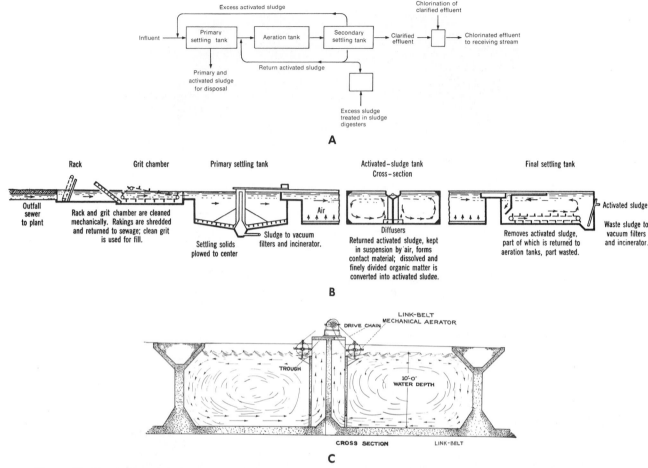

Figure 45–18. Activated sludge process. *A,* Flow diagram showing conventional activated sludge process. *B,* The partly clarified fluid is aerated and treated by the activated sludge process. *C,* Enlarged diagram of cross section of activated sludge tank. Link belt mechanical top aeration agitators used in some activated sludge tanks.

retained in the activated sludge tank as "seed," to reinoculate new sewage with the floc-forming bacteria, and the excess is removed.

A properly operated activated sludge process is capable of obtaining BOD reduction of 85 to 95 per cent, or greater. Table 45–6 shows BOD reduction of various treatment processes.

SLUDGE DIGESTION AND DISPOSAL. Sludge produced from primary and secondary treatment processes usually undergoes biochemical change even during settling. However, in most municipal sewage treatment processes the sludge is piped into a sludge digester for further treatment. The digester is designed to promote anaerobic decomposition. Both facultative and anaerobic organisms digest organic substances, including bacterial cells, into soluble substances, such as various organic

acids and gases (H_2S, N_2, H_2, CO_2, CH_4). Among these, CH_4 comprises nearly 70 to 75 per cent and CO_2 20 to 25 per cent. The gas mixture has a relatively high BTU value and has been used as fuel for generating electricity for aerators as well as steam to heat sludge digesters (30 to 40C) to accelerate metabolic activity of thermophilic bacteria. However, even with ideal pH (7.0) and temperature, complete digestion requires 30 to 40 days.

The sludge remaining after digestion usually requires dewatering prior to final disposal. Methods in common use today include air drying on sand filter sludge beds, vacuum filtration, centrifugation, and wet air oxidation (Zimpro process; the concentrated sludge is heated under high pressure of air to carry out autooxidation). Wet air oxidation not only reduces the volume of sludge up to 90 per cent,

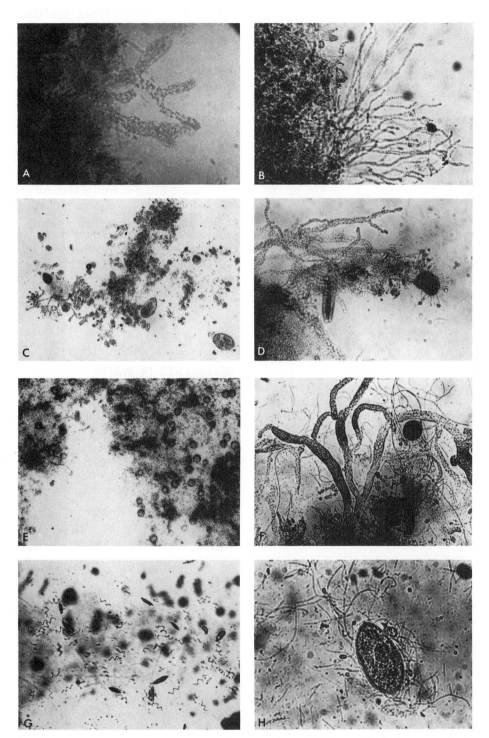

Figure 45–19. *A, Zoogloea ramigera* from laboratory model activated sludge process. (64×.) *B*, Fingered *Zoogloea* colonies from untreated raw sewage. (64×.) *C*, Fingered *Zoogloea* colonies with protozoans. (32×.) *D*, Fingered branching *Zoogloea* colonies with *Podophyra fixa*. (64×.) *E*, Fingered *Zoogloea* colonies with *Vorticella microstoma* and other protozoans. (26×.) *F*, Fingered branch-bearing *Zoogloea* colonies with the filamentous *Sphaerotilus natans*. (64×.) *G*, Spirilla with flagellated protozoans. (64×.) *H*, Long rods with *Paramecium* species. (64×.)

TABLE 45-5. SPECIFIC OXYGEN UPTAKE RATES (k_r) FOR VARIOUS ORGANISMS*

Organism	Temperature (C)	k_r (mgO$_2$/hr/gm)
Luminous bacteria	20	11.2
Azotobacter chroococcum	10	300
	30	13.1–250
Escherichia coli	30	12.8
Streptomyces griseus	27	16–48
Paramecium sp.	20	1.1
Euglena gracilis	25	3.2
Chlorella pyrenoida		
endogenous	25	1.4
exogenous	25	11.2
Neurospora sp.		
spores, dormant	?	0.2
spores, germinating	?	12.8
Yeast	30	152
Yeast	25	43
Planaria agilis (worms)	20	0.24

*From Biological Treatment of Sewage and Industrial Wastes, Vol. 1, edited by McCabe and Eckenfelder. © 1956 by Litton Educational Publishing, Inc. Reprinted by permission of Van Nostrand Reinhold Company.

but also the sludge so processed is considered sterile and has very good dewatering characteristics. Filtrate from the sludge may be piped into aerators or discharged into the receiving stream after proper chlorination. The sludge as conditioned (dewatered) is pulverized, incinerated, or sold as fertilizer. The heating kills all pathogens. **Milorganite** is a good example of such fertilizer. Important constituents are various soluble compounds of nitrogen, phosphorus, and potassium—essentials of plant growth.

45.5
CHLORINATION AND FINAL DISPOSAL

Unless chlorinated and poured into some convenient body of water, the final effluent may be disposed of, where a porous and dry soil is available, by surface ditches or by a subsurface irrigation system of tile pipes. These can furnish excellent fertilizer for farm crops raised on the land. There is little danger of infection, as typhoid, dysentery, and cholera organisms are largely killed by the antagonistic action of protozoans or bacteria in the settling tanks or soil, or are filtered out by the soil and die. However, tubercle bacilli, *Salmonella,* and poliovirus have been isolated from Imhoff-tank effluent as well as from secondary effluents. Promiscuous use of undisinfected sewage effluents is therefore not entirely safe. Epidemic (infectious) hepatitis virus is common in untreated water (Fig. 45–20).

45.6
DETERGENTS IN WATER

One phase of water pollution involves the use of synthetic detergents in household and industry. These can cause foaming of water in the house, in natural bodies of water, and in sewage plants. The problem in respect to drinking waters is almost entirely one of esthetics; in respect to sewage disposal it is one of me-

TABLE 45-6. EXTENT OF PURIFICATION OF RAW SEWAGE BY VARIOUS TREATMENT PROCESSES*

Process	Approximate Percentage of Reduction		
	5-DAY BOD	SUSPENDED SOLIDS	BACTERIA
Plain sedimentation	30–40	40–75	25–75
Septic tank	25–65	40–75	40–75
Chemical precipitation	60–75	70–90	40–80
Sedimentation + contact beds	50–75	70–80	50–80
Sedimentation + trickling filters (low-rate)	80–90†	80–90†	90–95
Sedimentation + activated sludge	85–95†	85–95†	90–98
Sedimentation + intermittent slow sand filters	90–95†	85–95†	95–98
Oxidation ponds	75–95‡	§	>99.9

*From Warren: Biology and Water Pollution Control, W. B. Saunders Co., 1971; after Klein.
†Even higher reductions are sometimes attained.
‡BOD after filtration.
§Suspended solids may be high because of presence of algae.

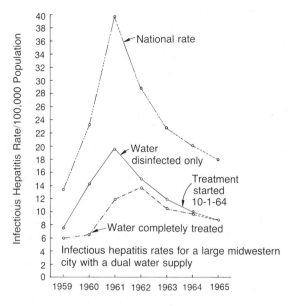

Figure 45–20. Correlation between frequency of infection by epidemic (infectious) hepatitis virus and type of treatment of municipal water supply over a six-year interval including an epidemic period. Disinfection of the water, and the combination of disinfection with other procedures of purification in this city markedly lowered the local rate below the national rate even during the epidemic year.

chanics. The actual amounts of detergents, even in sewage-polluted water, are relatively minute. Detergents as ordinarily used are not harmful, although some persons are allergic to certain detergents ("enzyme presoaks") and highly alkaline detergents. Fundamentally the problem of disposal is a bacteriological one and revolves around "biodegradation" of the detergents, i.e., their decomposition by the microorganisms in sewage disposal plants or soil.

The effective agents in most household synthetic detergents are anionic surface tension reducers (surfactants). They may be mixed with perfumes, colors, complex phosphates, and whiteners. None is toxic in ordinary use. Until recently they were of the class of polypropylene (alkyl) benzyl sulfonates (ABS); long, **branched-chain** molecules (tetrapropylene derivatives). There is some evidence the detergent is toxic to certain fish populations. Most states have discontinued the use of ABS "nondegradable" and excessively foaming detergents.

The long, branching, alkyl-benzene chains either resist direct enzymic attack by microorganisms or are too large to permeate into the microbial cells. Hence they are not readily de-

composed by microorganisms in sewage. They are said to be "hard" (difficult) detergents. Bacteriological studies show that some of these can be destroyed by certain specially adapted species of bacteria. Straight-chain detergents, i.e., **linear** alkyl sulfonates (LAS), were found to be much more easily decomposed ("biodegradable") than the branched-chain compounds; LAS are called "soft" detergents. As a result, progressive detergent manufacturers are using more of the LAS, such as sodium lauryl sulfate. These are up to 84 per cent biodegradable in ordinary disposal plants as against only about 50 per cent or less of ABS.

Recently, high-phosphate (up to 60 per cent by weight) detergents have received considerable attention. The phosphate in household detergents is believed to be a leading cause of eutrophication of streams and lakes. As a rule, household sewage may contain phosphate in amounts ranging from 20 to 60 mg/l. As mentioned in the previous chapter, the two substances most critical are nitrates and phosphates, both of which stimulate and accelerate eutrophication of lakes and streams. Phosphate is not easily removed by biological processes and must depend on chemical procedures, at least in part. While nitrates may be found in natural ecosystems, phosphate is the limiting factor in most bodies of water. Inhibition of algal growth probably requires removal of phosphate to a concentration well below 0.05 mg/l in the discharged effluent, and this would require **tertiary treatment**, along with chemical precipitation. The choice of treating sewage to this level is entirely the choice of the population, based on their ability to pay for it. However, many localities in the United States now have discontinued the use of phosphate-based detergents to reduce the eutrophication of aquatic environments not yet plagued with algal bloom.

45.7
PREVENTION OF WATERBORNE DISEASE IN THE ABSENCE OF FILTRATION

Disease transmission by water in the absence of elaborate filtration and chlorination systems (as on camping trips) is easily prevented. With the single possible exception of hepatitis virus, all of the common intestinal pathogens are readily killed by boiling for five minutes as well as by contact for at least two hours with 2 to 5 ppm of chlorine, iodine, or other suitable disinfectants.

Ice made from polluted water is as dangerous as the water itself. Water for camp use, provided it is clear and clean, may be treated with chlorine or iodine by using one of the numerous disinfectants for the purpose available on the market: **HTH, B-K, Wescodyne,** or **Globaline.** A teaspoonful of fresh chloride of lime (hypochlorite) in 25 to 50 gallons of water will also prove an adequate safeguard, unless the water is very heavily polluted or dirty. At least an hour must be allowed for the chlorine to act. Ordinary laundry bleaches (5 per cent sodium hypochlorite) will also serve. Directions are found on the labels of the bottles.

To persons contemplating a camping trip, or travel in countries where water and sanitation are poor, publications of the Public Health Service of the Department of Health, Education, and Welfare can be of great assistance.

45.8
WILL THERE BE ENOUGH WATER?

Water Reclamation. As the human population grows, and more and more water is demanded for both industrial and domestic uses, the natural sources of supply threaten to become inadequate. For example, even today nearly 15 to 20 per cent of the municipalities discharge untreated sewage into bodies of water, 20 to 25 per cent discharge effluent from primary treatment (sedimentation only), and less than 50 per cent of the total municipal waste receives some form of secondary treatment. Tertiary treatment of sewage is confined to a

few municipalities, and some facilities are designed for demonstration rather than as practical reclamation of used water.

Accompanying the increasing population is the ever-growing need for more clean water, and the concern over the water problem is certainly justified. For example, the available U.S. water supply (Table 45–7) was slightly more than 315 billion gallons per day in 1965, but the estimated water intake for the same year was 356 billion gallons per day. Since use exceeded available water supply, one must assume some portion of the water was recycled. Many of our industries now reclaim their water for manufacture (Table 45–8), and it is very clear that by 1975 the estimated gross intake of water will reach or exceed the available supply (Table 45–9). We must reclaim water, as water is an essential natural resource for sustenance of life; otherwise, "a permanent water shortage affecting our standard of living will occur before the year 2000."*

---

*Bradley: Science, *138*:480–91, 1962.

TABLE 45-8. U.S. INTAKE AND GROSS USE OF WATER FOR MANUFACTURING*

| | Billion gal/yr | | |
Year	TOTAL INTAKE	GROSS USE	Recirculation Ratio
1954	11,570	21,042	1.82
1959	12,131	26,257	2.16
1964	14,045	30,599	2.18

*Reprinted with permission from Dykes, Bry, and Kline: Env. Sci. Tech., *1*:780, 1967; after Census of Manufacturers. Copyright by the American Chemical Society.

TABLE 45-7. AVAILABLE U.S. WATER SUPPLY*

	Billion gal/d
Precipitation	4,300
Evaporation and transpiration	−3,040
Net stream flow	1,260
Unusable floodflow	−630
Dependable minimum flow	630
Navigation, fish, wildlife	−315
AVAILABLE SUPPLY	315
ESTIMATED INTAKE, 1965	356
NET CONSUMPTION, 1965	138

*Reprinted with permission from Dykes, Bry, and Kline: Env. Sci. Tech., *1*:780, 1967. Copyright by the American Chemical Society.

TABLE 45-9. ESTIMATED GROWTH IN INTAKE OF WATER IN U.S.*

| | Billion gal/d | | Average Annual Increase (Per Cent) |
	1965	1975	
Irrigation	142	170	1.8
Steam electric	111	145	2.7
Industrial	74	98	2.8
Public water	23	30	2.7
Rural domestic	6	7	1.5
TOTAL	356	450	2.4

*Reprinted with permission from Dykes, Bry, and Kline: Env. Sci. Tech. *1*:780, 1967. Copyright by the American Chemical Society.

CHAPTER 45
SUPPLEMENTARY READING

Eckenfelder, W. W., and O'Connor, D. J.: Biological Waste Treatment. Pergamon Press, Inc., New York. 1961.

Hawkes, H. A.: The Ecology of Wastewater Treatment. Pergamon Press Ltd., Oxford. 1963.

Keup, L. E., Ingram, W. M., and Mackenthun, K. M. (Eds.): Biology of Water Pollution: A collection of selected papers on stream pollution, wastewater and water treatment. Federal Water Pollution Control Administration. U.S. Department of the Interior. 1967.

Standard Methods for the Examination of Water and Wastewater, 13th ed. American Public Health Association, New York. 1971.

Warren, C. E.: Biology and Water Pollution Control. W. B. Saunders Co., Philadelphia. 1971.

MICROBIOLOGY • CHAPTER 46
OF THE
ATMOSPHERE

No organisms are indigenous to the atmosphere. Microorganisms of the air within 300 to 1,000 or more feet of the earth's surface are merely organisms of soil that have become attached to fragments of dried leaves, straw, or dust particles light enough to be blown about by the wind. Live microorganisms are more numerous in air in dry weather than just after a rain, because rain washes them out of the air and settles the microbe-laden dust. Species vary considerably in their sensitivity to any given degree of relative humidity, temperature, and exposure to sunlight.

Kinds and numbers of microorganisms found in the atmosphere depend on where the air samples are collected, weather, speed and direction of air currents, especially strong updrafts related to thunderstorm activities, and population. For example, on the Atlantic coast in a howling nor'easter, marine bacteria, plankton, and seaweed as well as sand and sea water are to be found in the atmosphere well inland.

46.1
DISTRIBUTION OF MICROORGANISMS IN AIR

More microorganisms are found in air over land masses than far at sea, although Darwin, during his famous voyage on the *Beagle*, found various microbial spores in the air a thousand miles at sea west of Africa. Spores of fungi, especially *Alternaria, Hormodendrum, Penicillium,* and *Aspergillus,* are more numerous than other forms over the sea within about 400 miles of land in both polar and tropical air masses at all altitudes up to about 10,000 feet (Table 46–1).

TABLE 46–1. NUMBER OF BACTERIA AND MOLDS WHICH DEVELOPED ON PLATES OF SEAWATER AND FRESHWATER MEDIA EXPOSED FOR 1 HR AT DIFFERENT DISTANCES FROM LAND

Distance from Land (Nautical Miles)	Seawater Medium		Freshwater Medium		Ratio SW:FW	
	BACTERIA	MOLDS	BACTERIA	MOLDS	BACTERIA	MOLDS
0–10	45	115	20	200	2.25	0.57
10–150	48	79	13	69	3.69	1.14
150–400	71	20	39	36	1.82	0.56

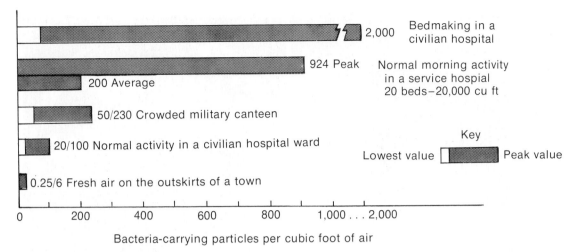

Figure 46-1. Bacterial content of air in civilian and military establishments, as measured with the slit sampler.

Total numbers of aerial organisms at such altitudes may range from less than one per cubic foot of air over oceans to several hundred per cubic foot over land. Much of the microflora of the lower air strata tends to settle out as land air moves out over the sea, leaving relatively more microorganisms in the upper levels.

Microorganisms found in air over populated land areas, below altitudes of about 500 feet, in clear weather with moderate breeze include spores of *Bacillus* and *Clostridium,* ascospores of yeasts, fragments of mycelium and conidia of molds and Streptomycetaceae, pollen, cysts of protozoans, unicellular algae, and some of the more resistant nonsporeformers, such as *Micrococcus luteus,* nonpathogenic species of *Corynebacterium* (diphtheroids), and some few gram-negative rods, such as some coliform species, *Chromobacterium,* and so forth. In fact, almost any of the microorganisms discussed in this book except certain of the more fragile parasitic and aquatic species may at times be found in the atmosphere.

In the dust and air of theaters, schools, and hospital wards (Fig. 46–1) or the rooms of persons suffering from infectious diseases, such organisms as tubercle bacilli, streptococci, pneumococci, and staphylococci have been demonstrated. These respiratory bacteria are dispersed through the air in the droplets of saliva and mucus that are always produced by coughing, sneezing, talking, and laughing. Viruses of the respiratory tract and probably (to some extent, at least) those of the enteric tract are also transmitted by dust and air.

The spores of many pathogenic fungi causing costly crop diseases, plant pollens and seeds,

and probably animal pathogens can be transmitted from continent to continent by air currents. Many larger objects of biological importance are also carried long distances by high winds and air currents: fragments of soil with seeds and plant pathogens; parts of plants, often diseased; soil nematodes; virus- and bacteria-infected insects; birds; and even parts of small rodents and fish. Consequently, transportation of pathogenic microorganisms via air currents is of prime concern to epidemiologists.

46.2
COLLECTION AND ENUMERATION OF AERIAL MICROORGANISMS

Microorganisms of the upper air have been collected by means of airplanes or other aerial devices. For the collection of microorganisms in the upper layers of the atmosphere, special apparatus must be used that excludes contamination of the sample by dust from the aircraft or its occupants, that prevents damage to organisms by impact at the high speed of today's planes, and that operates efficiently at all altitudes and atmospheric temperatures.

Microorganisms in air at low levels may be collected by the simple method of allowing dust to settle on an open Petri dish containing nutrient agar (Figs. 46–2, 46–3). This is useful in enclosed spaces. Dust and microorganisms may also be collected by drawing air through a tube containing a filter of wet sand or cotton (Fig. 46–4). The cotton or sand is then shaken in broth. This is one of the oldest and simplest devices.

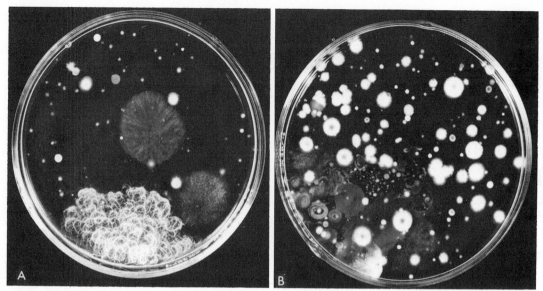

Figure 46–2. Colonies developed on nutrient agar. Petri plates have been opened for 5 min (*A*) and 20 min (*B*) in a crowded school cafeteria. (Photograph by K. T. Crabtree.)

There are many other types of devices for collecting microorganisms in the air. One **impingement device** consists of an agar-covered cylinder rotating slowly around its axis vertically. An air stream, carrying dust and microorganisms, impinges on the sticky agar surface as the drum rotates. There are also **bubbling devices, atomizing devices,** and **electrostatic devices.** In addition, the **membrane filter** (Chapt. 21) is adaptable to direct collection by filtration of air. Some representative devices are illustrated here (Figs. 46–5 to 46–7). None of the devices collects and counts all the microorganisms in the air sample tested. Some microbial cells pass entirely through, or are destroyed in, all the processes.

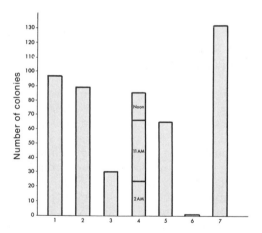

Figure 46–3. Approximate numbers of colonies of microorganisms expected to appear on Petri plates containing "standard methods agar" incubated 18 hours at 35C after remaining open for 1 hour under the conditions indicated: *1,* broom sweeping, busy railway station, 10 A.M.; *2,* crowded downtown lunch room, noon; *3,* large secretarial room, 10 A.M.; *4,* main corridor, large city hall, at hours indicated; *5,* busy downtown street, summer, humid; *6,* open country after snowstorm; *7,* Broadway subway car, winter, 5 P.M.

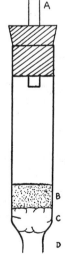

Figure 46–4. Tube for collecting dust from the air for bacteriological analysis. Air enters at *A*, and deposits its dust on the sand (*B*), which is supported by a cotton plug (*C*). The air leaves at *D*. The sand is later washed with broth, from which a plate count is made.

Figure 46–5. *A*, Sieve device (a form of impingement device), with box and cover, containing a standard Petri dish with agar on which most of the particles of dust impinge. Arrows show the course of air through the perforations, along the agar surface, and out beneath the Petri dish. *B*, Sieve plate with wings. *C*, Agar plate culture of room air obtained with sieve device. (*C*, Courtesy of Environmental Services Branch, National Institutes of Health, Public Health Service.)

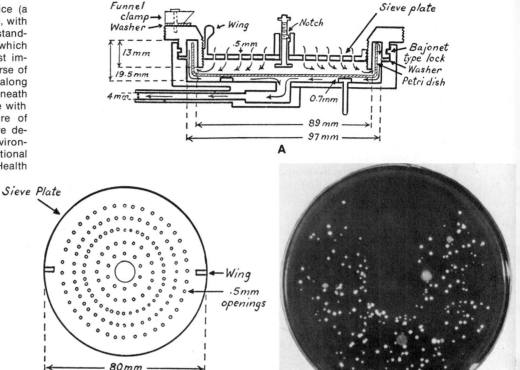

Figure 46–6. Bead-bubbler device, consisting of a 250-ml suction flask containing broth, glass beads 5 mm in diameter, and a glass bubbler kept in place by a rubber stopper. A plate count is made from the broth after sufficient air has been sampled.

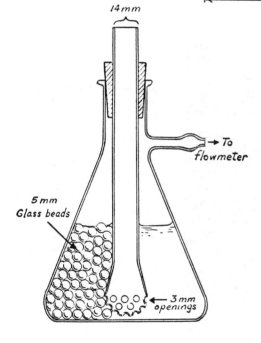

Figure 46–7. A multistage, liquid impingement device. *B* is a sectional drawing at 90 degrees to *A*, *B* being rotated counterclockwise to reveal *A*; about one-half actual size. Three chambers (stages) (*1, 2, 3,*) are arranged vertically. Access to each is through rubber bungs *13, 14, 15*. Air entering stages *1* and *2* flows over the surface of disks of sintered glass (*9, 10*) sticky with collecting fluid, the upper surface of the fluid being below the surface of the disks. The air then passes through the nozzle (*7*) and across a shallow layer of fluid in stage *3*. Under prescribed conditions of operation particles of graded sizes are caught on the two wetted disks and in the fluid in stage *3*.

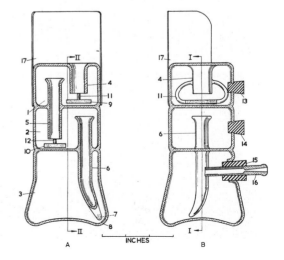

46.3
DUST, DROPLETS, AND DROPLET NUCLEI

Dust in and around urban dwellings usually arises from airborne sand, ash and soot, soil, and lint from bedding, clothing, and carpets. Most dust particles are laden with a variety of microorganisms and are relatively large (i.e., 10 to 100 μm) and tend to settle rapidly. Pathogens in dust are primarily derived from objects contaminated with infectious secretions that, after drying, become **infectious dust.** The ability to withstand drying is one factor affecting the ability of any organism to be a successful respiratory parasite.

Droplets are usually formed by sneezing, coughing, and talking (Fig. 46–8). Each **droplet** consists of saliva and mucus, and each may con-

tain thousands of microorganisms. It has been estimated that the number of bacteria in a single sneeze may be between 10,000 and 100,000. Most droplets are relatively large (about 100 μm) and, like dust, tend to settle rapidly in quiet air. Inhaled, they are trapped in the defensive hairs and mucus of the upper respiratory passages of persons inhaling them.

Small droplets in a warm, dry atmosphere are dry before they reach the floor and thus quickly become **droplet nuclei.** These are small (i.e., 2 to 5 μm) and light and may float about for many minutes or even hours. The microorganisms in them are protected by the dried mucus which coats them. Being very small, droplet nuclei tend to pass the mechanical traps of the upper respiratory tract and enter the lungs. All nuclei so entrapped have an opportunity to set-

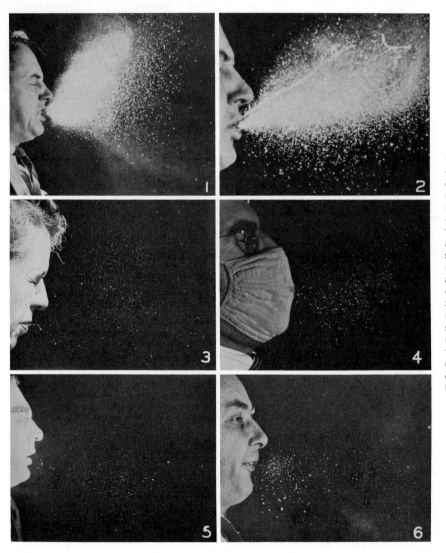

Figure 46–8. The atomization of mouth and nose secretions demonstrated by high speed photography. *1,* A violent sneeze in a normal subject; note the close approximation of the teeth, resulting in effective atomization. *2,* Head cold sneeze; note ·the strings of mucus and the less effective atomization of the viscous secretions. *3,* A stifled sneeze. *4,* Sneeze through a dense face mask. *5,* Cough; note the smaller discharge than in the uninhibited sneeze. *6,* Enunciation of the letter "f." (Courtesy of M. W. Jennison, Department of Plant Sciences, Syracuse University.)

tle on the alveolar tissue. Transmission of pulmonary pathogens such as *Streptococcus (Diplococcus) pneumoniae* (also *Klebsiella pneumoniae*) and *Mycobacterium tuberculosis* occurs primarily by such droplet nuclei.

The size, weight, moisture content, and opacity to ultraviolet light of airborne particles are of importance in considerations of methods of sampling air for bacteriological examination and of methods for disinfecting air. For example, with respect to sampling, impingement methods will tend to catch the larger and heavier particles but not very small and light ones. With respect to disinfection, ultraviolet rays or sunlight may reach and kill organisms in droplets or nuclei if they are very small. Ultraviolet-opaque coatings and opaque dust particles will protect microorganisms from ultraviolet irradiation which, as we have seen (Chapt. 20), has little power to penetrate. Humid atmospheres (about 50 per cent) have been shown to be more lethal than very dry or very moist atmospheres (20 or 80 per cent), though both bacteria and viruses differ in this respect.

46.4
AIRBORNE DISEASES

Most airborne diseases are transmitted by either infectious dust or droplet nuclei via the respiratory route. When a person with a respiratory infection sneezes and coughs, he is likely to exhale millions of microdroplets of saliva which contain varying numbers of infectious microorganisms. The moisture surrounding such droplets quickly evaporates, leaving in the air the mucus-coated living bacteria, which contaminate every person who inhales them. However, in some cases, especially in crowded areas, the pathogens may be transmitted in droplets (without becoming nuclei) directly from person to person, and this mode of transmission is frequently called "droplet infection." Some of the common but important human diseases transmitted by exhalation droplets via respiratory route are summarized in Table 46–2.

Control of Airborne Infection. Thorough **dilution** of contaminated air by **ventilation** is a very effective means of controlling airborne diseases indoors. However, it is sometimes expensive because of costs of heating or the installation of air ducts and blowers.

Under certain conditions disinfection or, more rarely, sterilization of air is desirable. Three general methods in addition to ventilation are available for the control of microorganisms in the air of rooms and buildings. This does not take into consideration such procedures as passing air through mechanical (e.g. fiberglass) filters or spray-washing devices. These reduce the numbers of organisms in air but do not necessarily disinfect or sterilize it.

Radiation. Irradiation with ultraviolet light, as pointed out in Chapter 20, is lethal to numerous microorganisms. A radiation wave length 254 nm is generally used, as this is sufficiently microbicidal and at the same time not excessively irritating (Fig. 46–9). Ultraviolet-producing electric lamps are attached to walls, overhead, or at other strategic points. Deflectors are used to prevent direct exposure of persons to the rays, which can cause serious "sunburn," and to protect the eyes, which may be seriously and permanently injured by more than very limited direct observation of an ultraviolet source.

A difficulty arises from the necessity that in occupied rooms the microorganisms must circulate in air well above the heads of the people in the room in order to come within the range of action of the lethal rays. Only the lighter particles do this. Also, dust is little affected, and only places directly exposed are disinfected. Actually bactericidal, ultraviolet irradiation is difficult to apply with general effectiveness except under very special situations where the rooms are not constantly occupied, as in storage facilities, hospitals, industry, and research. If air is to be **recirculated**, it may be first filtered and then passed through a tube, where it is irradiated by powerful ultraviolet sources. These devices appear to be effective (Figs. 46–10, 46–11).

Bactericidal Vapors. Many substances are lethal to microorganisms in the vapor phase (formaldehyde, ethylene oxide, β-propiolactone). Probably the most effective for disinfection of air are propylene glycol and triethylene glycol. These are odorless, tasteless, nonirritating, nontoxic, and not explosive or corrosive. They are highly effective in killing bacteria in the air in the form of **vapor**, although (curiously enough) relatively ineffective in the form of concentrated aqueous solutions in vitro. It is the vapor molecules and not droplets of fine mist of these substances that are the effective disinfecting agent. As little as 0.5 mg of propylene glycol vapor per liter of air can virtually sterilize heavily contaminated air in 15 seconds. Triethylene glycol is almost 100 times as effective.

Difficulties in the use of these vapors are chiefly of an engineering nature. Air conditioning appears inevitable to their effective use. The temperature and humidity of the air are important factors. If the air is cold and dry or ex-

TABLE 46–2. SOME AIRBORNE DISEASES TRANSMITTED BY EXHALATION DROPLETS*

Etiological Agent (Organisms)	Disease	Pathogenesis
Streptococcus pyogenes β-hemolytic type	Streptococcal infections or "strep throat"	The organism first lodges in the upper respiratory tract or in the pharynx. The organism produces erythrogenic toxins, which may cause an outbreak of red skin, and the disease is then called scarlet fever. An infection may, if untreated, cause complications such as mastoiditis, erysipelas, puerperal sepsis, and peritonitis. Acute infections may lead to rheumatic fever and may irreversibly damage heart tissue.
Corynebacterium diphtheriae	Diphtheria	Organisms usually remain localized in the upper respiratory tract. Powerful neurotoxin (an exotoxin) is produced and may cause toxemia. Organisms frequently cause formation of fibrinous "pseudomembranes," which often may block the air passage.
Mycobacterium tuberculosis	Tuberculosis	The bacterium lodges in alveolar tissue and subsequently multiplies within lesions of the lung to form tubercles. The cells within the tubercles may be sloughed out and spread to other parts of the body or be exhaled by a violent cough. The organisms are acid-fast and may be detected in saliva or gastric washings.
Streptococcus (Pneumococcus) pneumoniae (gram-positive) and *Klebsiella pneumoniae* (gram-negative)	Pneumococcal pneumonia (lobar pneumonia); atypical pneumonia	Pneumococcal pneumonia causes congestion in the lobes of the lung. Disease may cause complications such as pericarditis, meningitis, etc. Organisms produce heavy capsules, and this property is used in diagnosis of the disease (quellung reaction—swelling of capsules in the presence of type-specific immune serum). *K. pneumoniae* is commonly called Friedländer's bacillus. Lesions due to this organism may occur in any part of the body; the most frequent manifestation is pneumonia, with a high fatality rate. It is commonly found in sinusitis, pharyngitis, peritonitis, and other conditions. The organism appears to resist the action of penicillin.
Neisseria meningitidis	Meningococcal meningitis	The organism is usually carried harmlessly in the nasopharynx but occasionally enters the blood stream and localizes in the membranes surrounding the brain and spinal cord (meninges).
Yersinia (Pasteurella) pestis	Bubonic plague or pneumonic plague	The organism is usually transmitted by *Xenopsylla cheopis* (rat flea), and the most common manifestation is bubonic plague, which is characterized by chills, fever, nausea, and enlargement, ulceration, and suppuration of lymph nodes, called "buboes." The agent of primary pneumonic plague is disseminated via air, and the disease is characterized by the lungs' becoming congested and filled with exudate and blood. Large numbers of the bacilli are found in the alveoli and sputum.
Bordetella pertussis	Whooping cough	The organism is sometimes called the Bordet-Gengou bacillus (named after the discoverers). The disease is characterized by a typical paroxysmal cough that results in inspiratory crowing sound, or whoop. The organism rarely penetrates the mucosa of the respiratory tract—it is usually confined to the upper respiratory tract.
Haemophilus influenzae	Children's influenza	The name "influenza" has no etiological significance, since influenza is caused by viruses. The organism causes respiratory infection in children and is considered the most common cause of bacterial meningitis in children.

*Other common airborne infections are caused by viruses (such as smallpox, chickenpox, mumps, and epidemic influenza [Asiatic or Hong Kong flu]), as well as by fungi (for example, histoplasmosis).

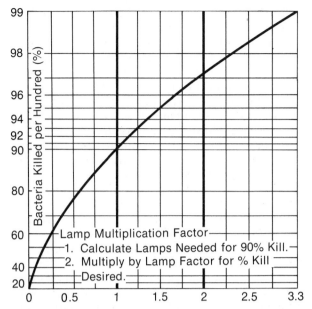

Lamp Multiplication Factor
1. Calculate Lamps Needed for 90% Kill.
2. Multiply by Lamp Factor for % Kill Desired.

Figure 46–9. In the average commercial installation, a 90 per cent kill of airborne microorganisms is recommended. This rate of kill is used most to protect personnel from cross-infection. However, a higher kill rate, at least 98 per cent, is more desirable in hospital and pharmaceutical applications, where absolute sterility is often necessary. This graph indicates the number of lamps required to increase the kill rate to the desired level.

cessively humid, the effectiveness of aerosols is reduced. Relative humidities of about 40 per cent at about 76F are favorable.

Other agents such as orthophenylphenol and related compounds have been recommended for **surface** application to supplement aerosols, especially as the phenylphenols cling to surfaces on which dust settles and render the surfaces bactericidal for prolonged periods under ordinary atmospheric conditions. Creation of persistent films of phenolic disinfectants in contact with the skin can be poisonous and destructive, though films of such disinfectant on floors and other surfaces is undoubtedly of importance in control of dustborne disease (page 309).

Dust Control. In Chapter 25 it was pointed out that dust control is important in preventing disease transmission, and the fact is re-emphasized here. Methods were described in the earlier discussion (page 729).

Effectiveness of Methods. Each of the methods just mentioned has been shown beyond doubt to be effective in reducing the number of aerial bacteria, both in experimentally contaminated laboratory atmospheres under various controlled conditions and in such places as barracks, schoolrooms, and hospital wards. When it comes to reduction of **disease,** however, data are less conclusive.

A great difficulty is that a person may spend his daytime hours in a protected environment such as an air-conditioned building with sterile air, but as he goes home in crowded subways or buses, all of the expensive protection is nullified. This is well supported by published experience.

46.5
AIR POLLUTION

Many microorganisms, to say nothing of us humans, are unfavorably affected by the horrible exhalations of some industries that now pollute parts of our atmosphere. Some of the

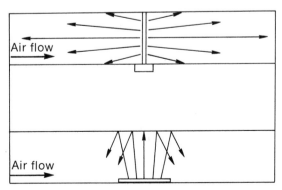

Figure 46–10. For full irradiation effect (along length of duct) Sterile Conditioners (American Ultraviolet Corp.) may be installed at right angles to air flow. When bank of lamps is used, plane of lamp bank is installed across the air flow.

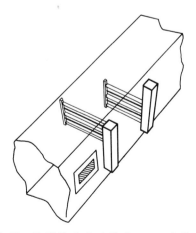

Figure 46–11. D-36 Units installed across air flow of air conditioning system.

consequences of industrial air pollution are irritation of skin, eyes, and respiratory tract. The consequences of irritation frequently result in loss of normal protective barriers, and as a result, microorganisms otherwise considered harmless may become established and cause injury to the irritated area. Death of individuals from pneumonia during heavy industrial air pollution is no longer an isolated incident in the large, industrialized cities of the world.

We may, however, consider the contribution of the individual human being to the all-enveloping pall of smog, perceptible or imperceptible, which now surrounds him in this day of advanced civilization, "from cradle to grave."

It is clear that the main source of infection-producing, microbial air pollution, so far as normal, peacetime human sources are concerned, is the mixture of saliva and mucus from their upper respiratory tracts. For example, at one time pollution of the air by penicillin-resistant strains of staphylococci reached near-epidemic proportions in hospitals and created a serious problem in controlling dispersion of contaminated air.

Aside from methods of air sampling already discussed, two important questions in studies of

pollution of indoor atmospheres by human beings are: (1) what organism shall serve as an index of pollution by saliva and respiratory mucus? (compare use of *Escherichia coli* and *Streptococcus faecalis* as indices of sewage pollution, Chapter 45); and (2) what shall be used as a standard culture method to isolate and identify the organisms from the air? Final answers to these questions have not been formulated. As a test organism *Streptococcus salivarius*, one of the usually harmless oral, viridans-type streptococci (Chapt. 38), has been proposed. Objections to it are that it may occur elsewhere, and that it is a fragile organism which soon disappears from the air. In other words, does its presence truly indicate salivary (and *only* salivary) pollution; does its absence prove the reverse?

Various media have been proposed for enumeration of airborne bacteria of sanitary significance. For general purposes the media used in milk or water examination are often used. As in all such enumerations, only those microorganisms capable of growth under the cultural conditions provided will be counted. However, the problem has not been fully explored and offers a good field for the ingenious and well-informed student.

CHAPTER 46
SUPPLEMENTARY READING

Adams, A. P., and Spendlove, J. C.: Coliform aerosols emitted by sewage treatment plants. Science, *169*:1218, 1970.

Dimmick, R. L., and Akers, A. B. (Eds.): An Introduction to Experimental Aerobiology. Wiley-Interscience, New York. 1969.

Gregory, P. H., and Monteith, J. L. (Eds.): Airborne Microbes. Seventeenth Symposium, Society for General Microbiology, Cambridge University Press, London. 1967.

Runkle, R. S., and Phillips, G. B. (Eds.): Microbial Contamination Control Facilities. Van Nostrand Reinhold, New York. 1969.

MICROBIOLOGY OF DAIRY PRODUCTS

Nutritionally, milk is considered one of the most complete foods for man and microorganisms. A "typical" chemical composition of milk is nearly as difficult to describe as a "typical" industrial waste. Butterfat, protein, calcium and other ionic contents, etc., differ not only among the different breeds and herds of cows but also vary in composition from day to day and from month to month within individual cows. However, the composition of an "average" sample may be:

Water	87.5 per cent
Sugar (lactose)	5.0 per cent
Butterfat	3.6 per cent
Protein (casein)	2.5 per cent
Albumin and globulin	0.7 per cent
Minerals (ash)	0.7 per cent
pH	6.7 to 6.9

Milk also contains vitamins A, B complex, C, and D in varying amounts. Therefore, unlike natural water, milk provides essential nutrients for excellent growth of many bacterial pathogens; hence contamination of milk by any pathogens becomes a very serious problem.

Many milkborne epidemics of human diseases in the past and present have been caused by contamination of milk through unsanitary handling by milkers or processors of milk. Undesirable microorganisms are introduced into milk not only by diseased cows but more frequently by dirty hands of dairymen; unclean, contaminated utensils; flies; unsanitary condi-

tions of cows, barn, and milk house; and even polluted water supplies.

Except milk itself, alone or in mixtures like ice cream, all dairy products, e.g., buttermilk, yogurt, butter, and all cheeses, are manufactured from raw or pasteurized milk through desirable chemical changes caused by certain microorganisms. For these reasons dairy bacteriologists and health and environmental bacteriologists closely inspect not only the nutritional quality of milk but also the sanitary quality and the conditions of milk production (*Milk Ordinance and Code*, U.S.P.H.S., 1967).

47.1
NORMAL FLORA OF MILK

Milk secreted into the udders of healthy cows is sterile. It has a pH of about 6.8. Some saprophytic bacteria of the outside environment, such as species of Micrococcaceae, Bacillaceae, Escherichieae, Corynebacteriaceae and Lactobacillaceae, are able to grow a short way up into the milk duct of the teat, so that the first milk drawn usually contains from hundreds to thousands of bacteria per milliliter. As a rule this milk is discarded. Except in cases in which extra precautions are taken at the time of milking, the milk receives contributions of organisms from the pail or mechanical milker and other dairy utensils, from soil and dust in the air, from the flanks, tails, and udder of the cow, and from the hands of milkers. Yeasts, molds, and numerous other saprophytes find their way into the milk. These constitute the **normal flora** of market milk.

The presence of these nonpathogenic bacteria in milk is usually not a serious matter, but if they are allowed to multiply they can and will cause the milk to sour quickly, putrefy, or develop undesirable flavors or conditions like bitter milk (*Streptococcus cremoris*), blue milk (*Pseudomonas syncyanea*), red milk (*Serratia marcescens*), or ropy (slimy) milk (*Alcaligenes viscolactis, Enterobacter aerogenes*, and others). Their presence in very large numbers shows the milk to be stale or dirty. Entrance of these organisms into milk in large numbers can be prevented only by clean handling and routine **effective** sanitization of milk-handling equipment, as recommended by the "milk ordinance."

Pasteurization. The general principle of pasteurization is selective destruction of the heat-sensitive microbial population in milk and foods through application of mild heat. *Pasteurization* consists in holding the milk in tanks at 145F (63C) for 30 minutes ("low-temperature holding," or LTH) and immediately refrigerating. **Disinfection** of the milk is accomplished, *not* sterilization. In many dairies the same result is achieved by heating the milk rapidly in a tube or in thin layers between metal plates, (a heat exchanger) to 161F (71.7C) and holding at that temperature for 15 to 30 seconds, then rapidly cooling. This latter method of pasteurization is called the high-temperature short-time (HTST or flash) method. This method not only saves time and money and is effective, but as far as the sanitation of milk is concerned, it is favored because milk can be processed on a continuous flow basis with little or no change in flavor. However, for both methods (LTH or HTST), pasteurization equipment must be designed so that every particle of milk or milk product is heated to the specified temperature for the specified time.

Coxiella burneti. Prior to the discovery of *Coxiella burneti*, the causative agent of Q fever, pasteurization temperature was based on the complete destruction of *Mycobacterium tuberculosis*. At one time *M. tuberculosis* was considered the most heat-resistant pathogen likely to be found in milk. The organism is destroyed when heat-treated at a temperature of 140F for 10 minutes. Thus the "milk ordinance and code" was set at 143F for 30 minutes (slightly higher temperature and longer period of time) to insure complete destruction of the organism. However,

C. burneti is able to survive in milk pasteurized at this temperature and time, and for this reason it became necessary to raise the temperature 2F to 145F to insure complete destruction of *C. burneti*.

Infection by this organism is prevalent among cattle; in certain areas of the United States it can reach as high as 65 per cent of some herds. Since the disease can be transmitted to man via milk, the modification of the pasteurization recommendations in the milk code became necessary. No major epidemics from ingestion of contaminated milk have been reported. The disease is most common among slaughterhouse employees and farm workers who handle sheep. The pathogenesis and epidemiology of the organism are presented in Chapter 42.

The Phosphatase Test. The presence of excessive numbers of coliform bacteria in pasteurized milk suggests staleness or (since coliforms are killed by prescribed pasteurization) improper pasteurization or illegal adulteration with raw milk. Concerning the adequacy of pasteurization, definite and accurate information can rapidly be obtained by means of the **phosphatase test.**

This test is based on the destruction by pasteurization of the heat-sensitive enzyme, **phosphatase,** normally present in fresh milk. The test for presence or absence of phosphatase is based on the power of phosphatase to liberate phenol from phosphoric-phenyl ester added to a sample of the milk.

In the **Scharer Rapid Phosphatase test,** 0.5 ml of milk sample is added to 5.0 ml of **buffered substrate** (disodium phenyl phosphate buffered with $NaHCO_3 \cdot Na_2CO_3 \cdot 2H_2O$) and held at 40C for 15 minutes. If the milk is unpasteurized or insufficiently pasteurized, the phosphatase enzyme normally present in the milk will be active and will decompose the added phenyl phosphate, liberating phenol. The tell-tale phenol turns blue if 2,6-dichloroquinonechloroimide (CQC), with $CuSO_4$ as catalyst, is added to the mixture. The appearance of blue color (indophenol blue) indicates the presence of free phenol liberated by undestroyed phosphatase in the milk. The indophenol blue is extracted by shaking with neutral *n*-butyl alcohol. The color content of the tested sample of milk is then compared with the color of standards containing known amounts of phenol and treated with the same reagents. When milk is pasteurized at 143F for 30 minutes, 96 per cent of the enzyme is destroyed. Only a trace of blue color should appear. Heating at 145F or above for 30 minutes insures complete inactivation of the phos-

$$\text{disodium phenyl phosphate} + \text{phosphatase} \rightarrow \text{phosphate} + \text{phenol} + \text{CQC}$$

substrate   enzyme from raw   blue color
or improperly
pasteurized milk

---

phatase. No blue color should develop. The reaction may be summarized in the above equation.

When milk has been underheated (in respect to either temperature or time) or when there is an admixture of raw milk afterward, the phosphatase will be present in larger amounts than when the milk is properly processed, and a definite blue color appears in the phosphatase test. The phosphatase test can be made quantitative by comparing color with known standards.

This test will detect 0.5 per cent raw milk mixed with pasteurized milk, or one degree below standard temperature, or five minutes of underheating during pasteurization. Color values (in the Scharer Rapid Method) greater than 0.5 μg of phenol per ml of milk indicate progressive degrees of improper handling of milk. The Grade "A" pasteurized milk ordinance (1967) recommends that the amount of phosphatase in Grade A pasteurized milk and milk products (except cultured products) should be "less than 1 μg/ml by Scharer Rapid Method or equivalent by other means."

Sanitary Significance of the Phosphatase Test. In pasteurization, *Coxiella burneti* and *Mycobacterium tuberculosis* (the most resistant of the nonsporeforming pathogens commonly found in milk) are destroyed more quickly than phosphatase. Therefore, a heat treatment adequate to inactivate completely the enzyme likewise kills these organisms and all other common pathogenic microorganisms (Fig. 47-1). A sample of milk that does not have more phosphatase present than the standard allows can be regarded as both safely pasteurized and free from subsequent contamination with raw milk.

Sources of Error. In some cases the phosphatase seems to become reactivated after proper pasteurization. This is because certain harmless bacteria can give falsely positive results in properly pasteurized milk. They produce a thermostable phosphatase *before* pasteurization. This remains active even *after* proper pasteurization.

47.2
CHANGES IN THE FLORA OF MILK

Since milk is an excellent medium for bacterial growth, the numbers of bacteria in it increase steadily the longer it stands, even if pasteurized and refrigerated (Fig. 47-2). Even if milk is refrigerated so that growth of thermophiles and heat-resistant sporeformers is retarded, psychrophilic species will grow. Under storage at 3C some of these can soon cause discoloring, ropiness (Fig. 47-3), off-flavors and other undesirable conditions. Common dairy psychrophiles are *Pseudomonas, Chromabacterium, Alcaligenes, Flavobacterium,* and *Micrococcus* species. Not all are killed by pasteurization, and the survivors can multiply rapidly.

It is important that refrigeration be very near 0C. Much commercial refrigeration is at

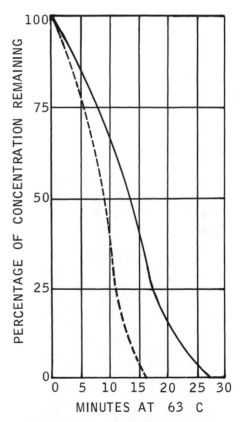

Figure 47–1. Representative time-concentration curves for destruction of pathogenic bacteria, here represented by *Coxiella burneti* (–––––), and phosphatase (———) in milk during pasteurization at 63C. Ordinates show approximate percentages of original concentrations or numbers remaining at 63C after minutes numbered on abscissa. Note that at all times destruction of the bacteria is more rapid and extensive than destruction of the phosphatase.

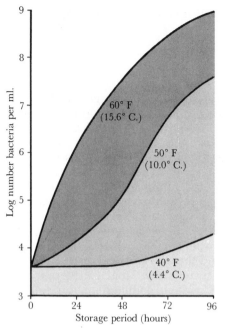

Figure 47–2. Effect of storage temperature on bacterial multiplication in raw milk.

about 10C, a temperature ideal for many psychrophiles. The temperature of the average household refrigerator (not freezer) is about 5C. At best, such refrigeration is effective for not much over 24 hours.

Figure 47–3. A bottle of ropy milk. The growth of *Alcaligenes viscosus* has made the milk so viscous that it can be picked up with a forceps.

Raw Milk. If allowed to stand at about 25 to 28C, the flora in raw market milk rapidly undergoes a series of changes. Numbers of bacteria increase to almost astronomical figures within 24 hours. The first organisms to increase are Enterobacteriaceae, lactic streptococci (e.g., *S. lactis*), *Micrococcus,* some sporeformers such as *Bacillus polymyxa*; and other saprophytes that thrive at a pH near neutrality. They grow rapidly and dominate the picture at first. The lactose is fermented. As acidity increases, these species are inhibited. The aciduric lactic organisms, commonly known as "lactics," especially *Lactobacillus* and species of *Leuconostoc,* then gain the ascendancy. These include two distinct biochemical types, homo- and heterofermentative (e.g., streptococci, lactobacilli, and others). In **homofermentation,** lactic acid is the major or only product of lactose fermentation; no gas is formed. **Heterofermentative** organisms, however, produce lactic, acetic, propionic, and some other acids and some alcohols and, in addition, gases such as CO_2 and H_2. Further discussion of lactics is presented in Chapter 34. Many *Clostridium* species will also ferment the lactose with acid and gas formation, but the presence of large numbers of *Clostridium* spores in fresh milk is unusual and indicates excessive contamination of the milk with soil or dung.

When the acidity reaches a pH of about 4.7, curdling occurs. The curd shrinks and settles out. Eventually, organisms capable of attacking lactic acid develop, especially aciduric yeasts and molds growing on the surface. These diminish the acidity of the milk by metabolizing the lactic and other acids and by producing alkaline products of protein decomposition: amines, ammonia, and the like. Since the carbohydrates (lactose) have been decomposed by this time, fermentation does not reoccur.

Organisms capable of hydrolyzing the fat and casein now thrive: eucaryotic fungi, Bacillaceae, both aerobic and anaerobic, Pseudomonadaceae, and many other lipolytic and proteolytic saprophytes. As the oxidation-reduction potential of the milk is reduced, species of *Clostridium* and other anaerobes, both obligate and facultative, gain the ascendancy, and the odors (ammonia, odoriferous amines, mercaptans, hydrogen sulfide, rancid odors) and effects of putrefaction become evident. The casein is hydrolyzed; the milk is darkened (Figs. 47–4, 47–5). After the situation has somewhat stabilized, a more prolonged decomposition continues, mainly by eucaryotic fungi and various

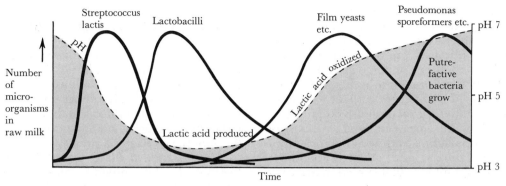

Figure 47–4. Changes in numbers and types of microorganisms in raw milk held at summer temperature. For explanation see text. Note the initial decline in pH (increase in acidity) during fermentation of the lactose, with later rise in pH as the resulting lactic acid is metabolized and alkaline products of casein putrefaction accumulate.

microbial enzymes. Thus raw milk (like soil) is considered as one of the best substrates for investigation of microbial ecology. Students can observe the qualitative and quantitative successions of microbial populations which are rigidly controlled by one another's ecological niches. There is first a dominance of lactics, followed by other acidophiles, then proteolytics and lipolytics; likewise, with respect to oxygen relationship, a dominance of facultative, then microaerophilic, then anaerobic, and ultimately aerobic microorganisms, with each group creating an environment suitable for another's existence.

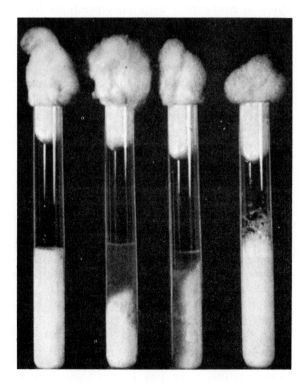

Figure 47–5. Changes produced in milk by some common bacteria. A, *Streptococcus lactis* produces lactic acid to form a solid curd with no gas or proteolysis. B, An aerobic spore former, *Bacillus cereus*, produces a soft curd due to bacterial rennin followed by rapid proteolysis. C, *Streptococcus liquefaciens* produces acid and rennin enough to form a shrunken curd, and then rapidly digests the curd with a caseinase. D, *Enterobacter (Aerobacter) aerogenes* produces much gas from lactose and little or no curd.

Pasteurized Milk. Pasteurized milk does not promptly undergo souring because many of the lactose-fermenting species, being nonspore-formers, are killed by the heat of the process. The milk may then undergo **sweet curdling** caused by rennet formation by bacteria, especially proteolytic streptococci and aerobic spore-formers (*Bacillus*). Often pasteurized milk does not sour, but the casein undergoes digestion and, later, putrefaction by the proteolytic entero-cocci, sporeformers, and other thermoduric, proteolytic saprophytes that survive pasteurization.

Significance of Coliform Organisms in Milk. Coliform organisms are always present in market milk *before* pasteurization. They are derived from hay, soil, dust, dung, and utensils. They do not necessarily indicate human fecal contamination. However, these organisms do not survive in milk in significant numbers if pasteurization is properly carried out. It is possible to detect coliforms in milk by methods similar to those used for determining coliforms in water (Chapt. 45). Plating the milk in deoxy-cholate lactose agar or violet red bile agar (both selective media) is especially recommended. Red colonies may be counted as coliforms and transferred for **completed test** if desired. The details for all of the important laboratory procedures in connection with dairy products are to be found in *Standard Methods for the Examination of Dairy Products* (see the Supplementary Reading list).

Unless almost-surgical precautions are used, a few coliform organisms gain entrance to milk *after* pasteurization, during cooling or bottling. Small numbers are of little significance and are not inimical to health, but their numbers give a good indication as to postpasteurization cleanliness and refrigeration or staleness. Coliform organisms are particularly undesirable in milk to be used for cheese, since they cause rapid souring, gassy fermentation, and undesirable odors and flavors in cheese and other products made from the milk.

If considerable numbers of coliforms (more than about 1 to 5 per milliliter) are found in pasteurized milk it is evident that: (a) the milk has not been properly pasteurized; or (b) it has been excessively contaminated after pasteurization by unclean conditions, possibly by sewage, feces, or dung; or (c) the milk has been held unduly long above about 15C after pasteurization; or (d) raw milk was mixed with it after pasteurization. (See The Phosphatase Test in §47.1.)

47.3
ENUMERATION OF BACTERIA IN MILK

In order to have some measure of the conditions under which milk has been produced and handled and to have a legal control over its sanitary quality, health departments and dairymen have set up various standards by which to judge milk. Important among these standards is the number of bacteria present.

The Standard Plate Count (SPC). Numbers may be determined by one or more of several methods, legally recognized procedures for which are detailed in literature cited. One of these, the plate (colony) count, is closely analogous to the plate-count procedure described in Chapter 10.

For counting bacteria in milk the plates may be incubated at 32 or 35C for 48 hours for routine work, at 7C for 7 to 10 days to enumerate psychrophiles, or at 55C for 48 hours to enumerate thermophiles. The plating medium officially recognized for all of these, as well as for enumerating the bacteria in drinking water and widely used for counting bacteria in other materials (e.g., foods), contains: agar, 1.5 per cent; yeast extract, 2.5 per cent; protein digest, 5.0 per cent; glucose, 1 per cent. This medium is generally called **Standard Methods Agar** and must conform in bacteriological quality to that of a standard lot of medium specified by the American Public Health Association in 1971. The SPC method is the most widely used procedure for estimating the numbers of bacteria in both raw and pasteurized milk. Some of its advantages over the direct count are: it counts viable organisms only and is applicable to milk having a low bacterial count. Some of the disadvantages or limitations are that the method detects only those organisms capable of growing in the particular nutrients (Standard Methods Agar), and colonies developed may be derived from either single cells or clumps of cells.

Direct Count. Bacteria may also be enumerated by the direct microscopic examination of milk in a smear. The smear is prepared by spreading exactly 0.01 ml of the sample on a slide over an area of exactly one square centimeter (100 mm²). After staining with specially prepared methylene blue solutions designed to remove fat globules, the smear is examined by means of a microscope calibrated with a stage micrometer so that the field is exactly 0.206 (or 0.146) mm in diameter. Such a field represents approximately 1/300,000 (or 1/600,000) ml of milk. The number of bacteria seen in it must therefore be multiplied by the reciprocal of this fraction (the **microscopic factor** or *MF*) to determine the numbers of bacteria per milliliter:

$$MF = \frac{10,000}{\pi r^2}$$

The advantages attributed to this method are: it is simple and less time-consuming, counts are made of all the cells regardless of viability or nutritional type, the different morphological types may be distinguished, and the slide may be kept as a permanent record. The disadvantages are that it cannot be applied to pasteurized milk because dead cells may be stained and counted; the microbial population must be equal to, or greater than, MF to reveal one or more bacteria in each field.

Numerous large **clumps** of bacteria indicate unclean utensils. Many **pus cells** indicate infected udders. Streptococci and staphylococci indicate mastitis. It is common practice to report results in terms of **clump counts** and, in reporting, to translate the microscopic observations directly into such terms as "good," "bad," and "mastitis."

Numerical Relationship between Counts. The **total** direct microscopic count is usually five to ten or more times as high as the plate count. This is because the total direct count enumerates individual cells, even those in clumps, and also dead bacteria. As in the plate count applied to water and soil, the plate count applied to milk enumerates only live bacteria capable of developing in the medium and under the environmental conditions used. In the plate count each clump, even though it may contain scores of live cells, counts only as a single bacterium since each clump forms only one colony. The plate count is usually closer to the clump count than to the total direct count.

47.4
QUALITY AND OXIDATION-REDUCTION POTENTIAL

The Reduction Test. Most actively growing bacteria cause a lowered oxidation-reduction (O-R) potential in their medium. The presence in milk of large enough numbers of growing bacteria to produce a significantly lowered O-R potential can be detected by the use of various dyes which undergo color change when reduced or oxidized. For example, methylene blue is colored when oxidized, but the dye becomes colorless when the oxidation-reduction

potential of the medium is lowered to about 0.01 V. It should be remembered that some redox reactions involve oxygen, but many do not. Many redox reactions (energy metabolisms) involve electron transfer. One reactant must serve as the electron donor and become oxidized, and the other reactant must serve as the electron acceptor and become reduced. Therefore, the rate at which methylene blue (electron or H acceptor) is reduced can be interpreted as the rate at which the substrate (milk) is oxidized by living organisms.

The reduction (so-called "**reductase**") test is used principally with raw milk, and furnishes a rough but useful approximation of the number and kinds of living bacteria present. In performing the test, 10 ml of milk sample are pipetted into a sterile tube, and 1 ml of a standard (certified) methylene blue solution (final concentration about 1:250,000) is added. The tube is closed with a rubber stopper and slowly inverted three times to mix. It is placed at 36.0C in the water bath immediately. At the end of each hour during the test the tube is inverted once. Observations are made after 30 minutes, 1 hour, and later.

MBRT. The **methylene blue reduction time** (MBRT) is the interval between the placing of the tubes in the water bath at 36.0C and the disappearance of the blue color from the milk.

The shorter the MBRT, the greater the number of active bacteria in the milk and the lower its bacteriological quality. However, MBRT does not give an accurate **count** of bacteria present; only an overall measure of its bacteriological quality. Milk with an MBRT of 30 minutes is of very poor quality; an MBRT of six hours is very good.

Resazurin. A dye related to methylene blue, resazurin undergoes a **series** of color changes (slate blue to pink to colorless, with many other colors in between), depending on O-R potential changes, whereas methylene blue changes only from blue to colorless. Resazurin, therefore, permits readings of **degrees** of reduction at shorter intervals than does methylene blue. In one procedure (the **one-hour test**), color of the milk, initially blue, is compared after one hour at 37C with several exactly described (**Munsell**) color standards designated as 5PB7/4 (pink-blue or mauve), 10PB7/5.5, 5P7/4, and 10P7/8 (pink). Milk showing no greater change during the hour than from blue to 5PB7/4 is Grade 1; from 10PB7/5.5 to 10P7/8, Grade 2, and so on to grade 4, complete decolorization. In the **triple reading** or **three-hour** test, three successive readings are made at one-hour intervals to see how long it takes to reach the color 5P7/4. High-grade (acceptable) milk requires at least three hours (resazurin reduction time, or RRT, three hours).

47.5
FACTORS AFFECTING BACTERIOLOGICAL QUALITY

Bacteriophages in Milk. A factor of importance, especially to manufacturers of dairy products that depend on early, rapid, and luxuriant growth of certain bacteria such as *Streptococcus lactis*, is the presence of bacteriophage lytic for that species. These phages are common in dust in and around dairies. They are known to interfere with many sorts of dairy work (e.g., cheese-making) that are dependent on bacterial growth. Very rigid aseptic technique in preparing and handling the pure **starter** cultures is necessary to eliminate the phages. Steel filters with triple layers of fiberglass have been successfully used to remove phages from air of dairy laboratories and work rooms.

Antibiotics and Disinfectants. Another important factor in bacteriological studies of milk is the possibility that the results have been influenced by (1) preservatives illegally added; (2) residues of disinfectants used to sanitize the dairy equipment; and (3) antibiotics or other drugs used to control udder infections (frequently supplemented in feed) or otherwise administered to cattle. Antibiotics, disinfectants, and preservatives can interfere with growth of the bacteria used in the manufacture of cheese, butter, and cultured milks (e.g., yogurt), and in such bacteriological controls as reduction tests and plate counts. In most manufacturing procedures, uninhibited growth of *S. cremoris, Leuconostoc citrovorum, Streptococcus lactis, Lactobacillus* species, and other bacteria is essential.

Penicillin is especially undesirable in milk because it can produce severe allergic reactions in persons hypersensitive to it. The Grade A pasteurized milk ordinance (1967) of the U.S. Public Health Service recommends that antibiotic content should be less than 0.05 units/ml by *Bacillus subtilis* method or equivalent, and this recommendation is applied to Grade A raw milk.

47.6
GRADES OF MILK

The actual numbers of bacteria permissible in milks of various grades vary in different cities. A good guide is the standard milk ordinance and code of the Public Health Service of the U.S. Department of Health, Education, and Welfare. Various localities may have somewhat different standards.

Grade A Raw Milk for Pasteurization. Grade A raw milk for pasteurization is raw milk from properly supervised producer dairies conforming to standards of sanitation of workers, cattle, premises, and equipment as prescribed in the ordinance. Cattle and personnel must be free from diseases transmissible in milk. The bacterial plate count or the microscopic clump count of the milk, as delivered to the pasteurizing plant for pasteurization, shall not exceed 200,000 organisms per milliliter, as determined by standard methods of the American Public Health Association or must not have an MBRT of less than 5.5 hours or RRT of less than 2.75 hours.

Grade A Pasteurized Milk. In all cases Grade A pasteurized milk shall show efficient pasteurization, as evidenced by satisfactory phosphatase test ("less than 1 μg/ml by Scharer Rapid Method or equivalent by other means"), and at no time after pasteurization and before delivery shall the milk have a bacterial plate count exceeding 20,000 per milliliter, or a coliform count exceeding 10 per milliliter, as determined by Standard Methods of the American Public Health Association (Table 47–1).

Grade B Pasteurized Milk. Grade B pasteurized milk is pasteurized milk that does not meet the bacterial-count standard for grade A pasteurized milk, and certain other sanitary requirements. Such milk may be used in some commercial processes.

Most communities now permit the sale only of grade A milk, pasteurized.

Criteria of Good Milk. Quality tests such as bacterial enumerations and reduction tests, as well as others, have a distinct value and usefulness from the standpoint of **cleanliness** but not necessarily with regard to **infection**. The present status of microbiology of milk was well summarized by Robertson, who said:

None of the routine laboratory procedures for estimating the number of bacteria in milk will determine whether or not infectious bacteria are present. The best assurance of freedom from infectious bacteria is that provided by proper pasteurization of the milk. The best assurance of pasteurization is that demonstrated by a satisfactory phosphatase test on the bottled pasteurized milk. The best assurance of freedom from recontamination in freshly bottled milk after pasteurization is a satisfactory coliform test in 1.0 ml portions of the bottled product.

TABLE 47-1. CHEMICAL, BACTERIOLOGICAL, AND TEMPERATURE STANDARDS FOR GRADE A MILK AND MILK PRODUCTS

Grade A raw milk for pasteurization	Temperature:	Cooled to 50F or less and maintained thereat until processed.
	Bacterial limits:	Individual producer milk not to exceed 100,000 per ml prior to commingling with other producer milk.
	Antibiotics:	Less than 0.05 units/ml by the *Bacillus subtilis* method or equivalent.
Grade A pasteurized milk and milk products (except cultured products)	Temperature:	Cooled to 45F or less and maintained thereat.
	Bacterial limits:	Milk and milk products — 20,000 per ml.
	Coliform limit:	Not exceeding 10 per ml.
	Phosphatase:	Less than 1 μg per ml by Scharer Rapid Method (or equivalent by other means).
Grade A pasteurized cultured products	Temperature:	Same as above.
	Coliform limit:	Same as above.
	Phosphatase:	Same as above.
	Bacterial limits:	Exempt.

*From: Grade "A" Pasteurized Milk, 1965 Recommendation of the U.S. Public Health Service, 1967.

Certified Milk. If milk is to be offered for sale unpasteurized, it is often required that it be produced only under very carefully supervised conditions. The American Association of Medical Milk Commissions has established rules and regulations concerning veterinary inspection of cows, especially for tuberculosis, brucellosis, and infectious mastitis, and sanitation of barns and utensils. These very rigid regulations are often used by health departments and milk dealers in certifying qualified farms to produce such milk. It is usually called **Certified milk** or **baby milk.** The use of certified milk has much to recommend it, especially its cleanliness. The coliform standard for Certified milk–raw, is 10 per milliliter; for Certified-pasteurized, 1 per milliliter. Certified milk is said to contain a larger proportion of certain vitamins essential for infants than milk which has been heated.

Most cities and states, as well as the A.A.M.M.C., require that all persons occupied in preparing Certified milk, or, indeed, any food for the public, be examined periodically for typhoid, paratyphoid, and dysentery bacilli. Examinations for presence of organisms that cause diphtheria, tuberculosis, scarlet fever, and other transmissible diseases are also required for Certified milk handlers.

47.7
PRESERVATION OF MILK AND MILK PRODUCTS

Once microorganisms gain entrance into milk, they cannot be removed effectively; therefore, preventing contamination of milk is the first and most important factor in maintaining quality milk. In addition to maintenance of healthy herds, the practice of asepsis in all phases of production is essential. The milk may be preserved by refrigeration and pasteurization for a limited time, as well as by dehydrating it to various degrees for a longer "shelf life."

Milks from which part or all of the water has been withdrawn are termed **concentrated** or **dried.** Assuming that approved standards of cleanliness, freshness, sanitation, and chemical content (fat, solids, and so on) have been met in selecting the milk to be dehydrated, the microbiological quality of the finished product is determined largely by: (a) temperatures and time of storage (if any) prior to processing; (b) times and temperatures of processing; (c) cleanliness of the apparatus and final containers; and (d) time and temperature of final processing of canned milks. Dairy products, such as fermented milks (various buttermilks; see next section) and cheeses are preserved partly by lactic acid produced by "starter" and to a lesser extent by other organisms involved in manufacture of the products. The curing or ripening process of cheeses contributes but little if anything to the keeping quality of the products, except that the cheeses may dehydrate further and form a protective rind.

Evaporated Milk. The raw milk is first cooled and may be clarified. Fat and solids contents are "adjusted" to meet required standards. The milk is then heated to boiling or nearly so (94 to 100C) for about 20 minutes. This kills all but the most heat-resistant microorganisms. Water is driven off in vacuo at about 55C, a temperature which favors development of undesirable thermophiles, e.g., *Bacillus stearo-thermophilus* and which therefore requires careful bacteriological control.

After homogenization, the product is cooled, canned, and sterilized. Spoilage problems can result from inadequate heating and subsequent storage at unduly high temperatures that favor growth of spores of thermoduric molds and Bacillaceae.

Sweetened Condensed Milk. Milk can be doubly preserved by the addition of about 20 per cent of sucrose or glucose, or both, to whole milk, and by subsequent heating at temperatures near boiling and by partial dehydration, as with evaporated milk.

Nonfat Dried Milk. Nonfat dried milk is prepared by preliminary steps similar to those used for evaporated milk. The fat is removed by high-speed cream separators. The water is partly removed by preliminary heating (about 85C) and then by: (a) spraying the milk as a mist into a current of hot (about 120C) air in a closed chamber; or (b) by spreading the milk as a thin film on hot rollers or drums—at about 145C if no vacuum is used, or at less than 100C in a vacuum chamber.

The dried **flakes** from hot drums or the **powder** from sprayed milk are packaged to prevent access of moisture. Dried milks should be kept dry and cool at all times or they may spoil since they are not sterile. As previously pointed out, dry heat is an inefficient sterilant. Microorganisms surviving the process are mainly Bacillaceae and thermoduric streptococci, lactobacilli, micrococci, and species of *Microbacterium*. Nonsporing pathogens are eliminated. The presence of coliform bacteria or of pathogenic streptococci or other heat-sensitive microorganisms is of the same sinister significance that it is in pasteurized whole milk.

Reconstitution. In reconstituting dried or evaporated milk it is clearly desirable to use clean, cool, and hygienically acceptable water and utensils. The reconstituted milk should be kept and handled under the same conditions of sanitation and refrigeration as are recommended for whole milk.

Standards of Quality. Quality standards for dewatered milks are similar to those for whole milk, taking into account the changes in flora caused by heating. The standards are established by the U.S. Department of Agriculture, The American Dry Milk Institute, and the Evaporated Milk Association (Evaporated Milk Industry Sanitary Standards Code, Chicago, Ill.). Standards for various grades of dried milk, for example, are 50,000 per gram for "extra" quality and 100,000 per gram for "standard" quality. Counting procedures are the same as for fresh milk after the dried milk is reconstituted with sterile water.

47.8
SOME MANUFACTURED DAIRY PRODUCTS

Market milk contains numerous species of microorganisms in varying numbers whose uncontrolled action is too unreliable to serve as a basis for commercial operations requiring uniformity of product. Pure-culture inocula (called **mother** or **stock cultures**) of constant properties are essential to continued success in this highly competitive field. For these purposes **lyophilized** ("freeze-dried"), or other pure, stock cultures of desired organisms may be maintained in the dairy if a competent bacteriologist and adequate laboratory facilities are available. Otherwise it is best to obtain stock cultures from dairy-supply houses. Cultures of species of lactic streptococci, *Leuconostoc*, and *Lactobacillus* are used especially.

Starter Cultures. In practice, mother cultures in about 2 per cent volume are added as nearly aseptically as possible to about 600 ml of sterile or very-low-count milk (previously heated 30 minutes at about 88C and cooled to 21C), and incubated. The lactic organisms soon outgrow other species, if any are present. This culture is called a **starter**. It may be used to inoculate a tank-size batch of milk or cream for butter or cheese, or to inoculate a still larger lot of starter.

Fresh, high-quality (low-count) pasteurized milk is brought *quickly* to the desired incubating temperature. A large, *pre-emptive*, virtually pure, starter inoculum of vigorously growing young cells of the desired lactic organism is added and thoroughly mixed with the milk or cream. Before the other bacteria in the milk have time to recover from their previous refrigeration or pasteurization and overcome their lag phase, the acidity quickly produced by the actively metabolizing added lactic starter suppresses them.

Butter. Butter is generally made by churning cream that has been soured by lactic acid bacteria.

Two species of bacteria, each with a distinct function, are added to the cream simultaneously. *Leuconostoc citrovorum* is depended on for flavor, *Streptococcus cremoris* or *S. lactis* is selected primarily for *rapid*, initial lactic acid production. If high acidity (pH 4.3) is not produced promptly, numerous undesirable contaminants may grow excessively. When the pH reaches about 4.3, *Leuconostoc* ceases growth, but its enzymes attack the citrates in the milk and produce diacetyl. This substance gives butter and similar products their characteristic buttery flavor and aroma. Neither *S. cremoris* nor *Leuconostoc* alone can produce the desired result in commercial practice.

Cheese. The manufacture of all cheeses depends upon the activities of selected microorganisms in the milk. The milk is curdled by addition of a "starter culture," lactic acid bacteria (*S. lactis, S. cremoris, L. helveticus, L. bulgaricus, S. thermophilus*, etc.). The enzyme **rennin** (casein coagulase) obtained from calves' stomachs is also added to coagulate the casein (milk protein). The combined action of acid produced by starter culture, and of rennin, causes milk to curdle or coagulate, and the whey is separated by various methods, depending on the amount of moisture desired by various types of cheeses.

Whey, the watery liquid extracted from curdled milk, is rich in carbohydrates (lactose), minerals, and vitamins (93 per cent water and 7 per cent suspended or dissolved solids) and is the major by-product of the cheese industry. Because of its highly nutritive characteristics, much attention has been given to recovery of the nutrients, such as lactic acid and its derivatives. Currently, whey may be dehydrated and sold as supplement for domestic animals in powdered form, or used in lactic and other acid fermentations (Chapt. 49).

Cheeses may be divided into three general types: (a) soft- or cottage-type cheese, and cream cheese (these are eaten in a fresh or unripened state); (b) hard- or rennet-curd cheese, including Roquefort, American Cheddar-type ("rat-bait"), Edam, and Swiss (these are **ripened** by the enzymes and slow growth of bacteria or molds or both, which cause some, but not extensive, proteolysis); (c) soft or semisoft rennet-curd cheese, of which Camembert, Limburger, and Liederkranz are types (these are ripened by proteolytic and lipolytic organisms which soften the curd and give it flavors). The hardness of cheese depends to some extent on moisture and fat content as well as on heating and acidity of the curd, draining, salting, and conditions of storage. A list of common cheeses is given in Table 47–2, and various cheeses on the market are shown in Figure 47–6.

Soft, Acid-Curd Cheese. In making cottage cheese, starters containing mixtures of *Leuconostoc citrovorum, L. dextranicum, S.*

TABLE 47–2. TYPES OF NATURAL* CHEESE

Representative Cheeses	Distinctive Organisms in Ripening Flora
Soft:	
Cottage	
Cream	Not ripened
Liederkranz	*Streptococcus liquefaciens, Brevibacterium*
Camembert	*Penicillium camemberti, Brevibacterium*
Semisoft:	
Blue (or Bleu)	
Roquefort	*Penicillium* strains such as *P. roqueforti*
Gorgonzola	
Hard:	
Swiss	*Propionibacterium* species
Cheddar	Lactic group, *Geotrichum*
Parmesan	Lactic group (brine-cured)

* Not processed artificially. (Pasteurized and processed cheeses and cheese spreads are not included, since they are made almost entirely from the natural cheeses such as those listed above.)

lactis, and the like are added to pasteurized milk. These ferment the lactose, *Leuconostoc* adding flavor and aroma. The lactic acid coagulates the casein. Rennet may be added to hasten the coagulation and make the curd firmer. The curd is cut into small cubes. To firm the curd and separate it from the whey, the mass is heated slowly to about 50C and held so for 30 minutes. Water is added; the curd settles. The water, with the whey, is drained off and the curd is pushed into heaps to drain. It is washed a second time with water and drained. About 0.5 per cent salt is added. Just before packaging many manufacturers add a little cream.

Hard-Curd Cheese. In the preliminary stages, nearly all natural (i.e., not "processed") cheeses are much alike. Differences result from different methods of treating the curd: degree of acidity, addition of different amounts of salt, special ripening microorganisms, moisture, temperature, and humidity of ripening and other factors (Figs. 47–7, 47–8).

For yellow cheeses of the Cheddar type, color is added. After a slight acidity has developed, rennet is added (Fig. 47–9) to make an elastic, rubbery curd which is later cut into pieces about 1 inch in diameter (Fig. 47–10) and warmed to about 39C. The curd becomes firmer, and the whey separates and is drained off and may be used for stock feed. The clumped masses of firm curd are chopped (**milled**) again (Fig. 47–11) and piled up to press out whey.

Figure 47–6. Various cheeses sold on the market. (Courtesy of the National Dairy Council, Chicago, Ill.).

This is called **cheddaring** in Cheddar cheese making. The curd is again milled, and then is salted (Fig. 47–12), drained, and pressed in hoops to cure (Fig. 47–13). Curing of Cheddar cheese proceeds at about 15C. It becomes "sharper" with aging.

CURING OR RIPENING OF HARD CHEESES. During the curing process of Cheddar and other hard cheeses, various microorganisms, the varieties depending on the kind of cheese, continue a slow fermentative, lipolytic, and proteolytic action, the products of these processes

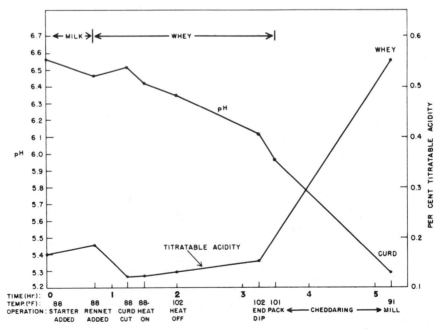

Figure 47–7. Cheddar cheese: changes in temperature and in acidity as measured by determination of pH and by titration of total acid, during successive early phases, from addition of starter (low left) to milling (lower right); about five hours in all. At the time starter is added at 88F. the acidity is low (pH 6.58; titrable acidity 0.18 per cent). As lactose is fermented acidity quickly increases. At pH about 6.46, rennet is added and the curd forms. It is soon cut and heated, and the whey is drawn off (dipped). Cheddaring and milling then proceed. Further increases in acidity occur during the ripening process.

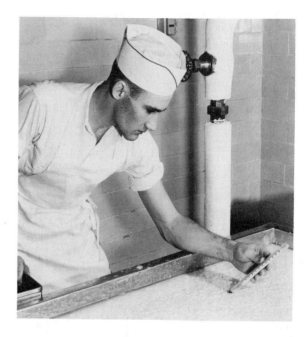

Figure 47-8. Every step in cheese-making process is carefully controlled. (Courtesy of the National Dairy Council, Chicago, Ill.)

Figure 47-9. Making cheddar-type cheese. Rennet added to hasten coagulation and to make curd firmer. (Courtesy of the National Dairy Council, Chicago, Ill.)

Figure 47-10. After a curd has formed, it is cut with stainless steel wire knives into small cubes, to help expel some of the whey. Following this, automatic paddles will stir the contents of the vat while it is heated, to shrink and firm the curd and expel more whey. (Courtesy of the National Dairy Council, Chicago, Ill.)

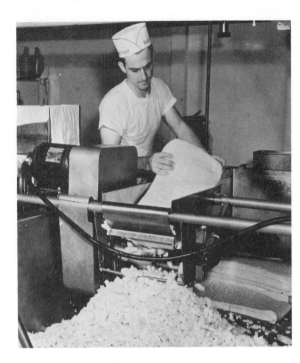

Figure 47–11. Milling the curd. An electrically driven machine cuts the slabs of curd into small pieces. (Courtesy of the National Dairy Council, Chicago, Ill.)

Figure 47–12. Salt is added to the milled curd. (Courtesy of the National Dairy Council, Chicago, Ill.)

Figure 47–13. Cheese curing room, Michigan State University. (Courtesy of the National Dairy Council, Chicago, Ill.)

yielding the characteristic flavors, textures, and aromas of various cheeses. Prominent among these flavors are diacetyl, lactic, butyric, caproic, and acetic acids, and various amines, as well as various esters such as those that give flavors to ripe fruit juices. In addition, since many of these organisms synthesize vitamins, especially nicotinic acid and vitamins of the B complex, the nutritive value of the cured (or ripened) cheeses is increased.

Gas-formers, such as species of *Clostridium* and *Escherichia*, are undesirable because they produce gassy cheeses (Fig. 47–14) and off-flavors; they may be especially active in the early stages. They generally occur in milk of poor quality.

Swiss cheese is heated to 50C after cutting the curd. Starters therefore usually contain the *thermoduric* lactics: *Streptococcus thermophilus* and *Lactobacillus bulgaricus* or *L. lactis*, as well as *Propionibacterium shermanii* and *P. freudenreichii.*

The cheese is soaked in 23 per cent brine for some days at 13C. Propionibacteria are then favored by incubation at 22C. Later the cheese is ripened for months at about 13C.

The "eyes" in Swiss cheese are a result of the production of carbon dioxide by species of *Propionibacterium*, while its bitter-sweet flavor is caused in part by the formation of glycerol and propionic and succinic acids by *Propionibacterium* species while ripening.

Semisoft cheese, such as Roquefort, Gorgonzola, or Blue (or **Bleu**), contains much fat and as high as 5 per cent salt and relatively little moisture. The high salt content prevents continued growth of most bacteria, as does the low ripening temperature (7 to 8C) and humidity (60 per cent). A species of mold (*Penicillium roqueforti*) is introduced by the inoculation of spores into the milk or into the curd as it is put into hoops for ripening. The mold grows in the interior, producing the masses of blue-green conidia and the sharp flavor so characteristic of this type of cheese. As the mold is aerobic, perforations are made in the cheese to aerate the interior.

Soft Cheeses. Limburger, Liederkranz, and Camembert cheeses are cured mainly by the growth of organisms in a red-orange, slimy coating on the outer surface. Numbers of microorganisms in this slime sometimes exceed 10 billion per gram. In Limburger cheese, yeasts (*Geotrichum* species) begin to grow on the surface after subsidence of the initial acidity caused by lactic organisms. They persist in the

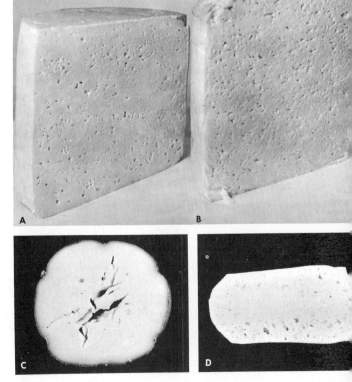

Figure 47–14. Defects in cheeses due to gas-forming (aerogenic) microorganisms. *A* and *B,* Gas formed by *Enterobacter aerogenes* in early ripening of Cheddar cheeses (*A*, yellow; *B*, white). (Courtesy of The Borden Co.) *C,* Gas formed late in ripening of Provolone, an Italian cheese similar to Cheddar. Compare with "stormy fermentation," Figure 36–13. *D,* Gas formed by a lactose-fermenting yeast in early ripening of Brick cheese.

surface slime for about a week. *Brevibacterium linens* and *B. erythrogenes* then grow all over the surface, forming a reddish brown coating, or "**smear**," commonly seen on soft and semisoft cheese (brick, Camembert). Camembert cheese is sometimes inoculated on the outer surface (or before curd formation) with a pure culture of the mold, *P. camemberti*. The enzymes of the various micrococci, yeasts, or molds in the slime penetrate into the interior of the cheese, producing the flavors, softening, and the famous aroma of Limburger and similar cheese.

47.9
FERMENTED MILK BEVERAGES

In certain countries lactobacilli have been used for centuries in combination with certain yeasts and streptococci to produce foods of fermented milk. The **yogurt** of eastern central Europe (now available in all grocery stores), the **busa** of Turkestan, the **kefir** of the Cossacks, the **koumiss** of central Asia, and the **leben** of Egypt are examples of these. Formerly, the microbial nature of these processes was unknown. In all of these fermented milks lactobacilli act in company with other microorganisms: yeasts, lactic streptococci, and various rods. For example, kefir, made from milk of various domestic animals (e.g., cows, goats, mares), is prepared by putting **kefir grains** (small, cauliflower-like masses) into the milk. These grains consist of dried masses of lactobacilli (*L. brevis*), yeasts (*Saccharomyces delbrueckii*), *Streptococcus lactis,* and probably other lactic organisms held together in a matrix of coagulated casein and bacterial polysaccharide gum. The kefir grains increase in size and break apart as the fermentation proceeds. The combined growth of the

TABLE 47-3. COMMON TYPES OF SPOILAGE OF MILK AND MILK PRODUCTS

Dairy Products	Types of Spoilage	Microorganisms Involved	Apparent Signs of Spoilage
Raw and pasteurized milk	Souring or acid production	*Streptococcus lactis*	Milk becomes sour; some caused by enterocci, coliforms, microbacteria, and lactobacilli.
	Gas production	Coliform bacteria and *Clostridium* spp.	Milk first curdles, then curd is ripped apart by gas production; "stormy fermentation" by *Clostridium* spp.
	Proteolysis	*Micrococcus* spp., *S. faecalis* var. *liquefaciens*, *Pseudomonas* spp., *Chromobacterium* spp., *Flavobacterium* spp.	Milk usually becomes alkaline in time, "sweet curdles" the milk (curdles with rennin) before digesting protein.
	Ropiness	*Alcaligenes viscolactis*, *Klebsiella pneumoniae, Enterobacter (Aerobacter) aerogenes*, and some lactics	Stringiness or sliminess caused by slime or polysaccharide capsules from the cells during low-temperature storage.
Fermented milk and milk products	Gassiness	Coliform bacteria (*Enterobacter aerogenes*), *Leuconostoc* spp., *K. pneumoniae*	Causes gas holes (eyes) in the curd, produces off-flavor in various types of cheeses.
	"Late gas"	Lactate-fermenting *Clostridium* and *Propionibacterium* spp.	Spoilage occurs during ripening, causing alteration of texture and flavor; excessive eye formation in various cheeses.
	"Grey rot"	*Clostridium* spp.	"Foul-smelling" due to putrefaction of casein; dark color due to sulfides.
	"Rusty spot"	*Lactobacillus* and *Propionibacterium* spp. *Brevibacterium* spp.	Reddish to grayish brown discoloration by pigment of bacterial cells; discoloration mostly on the surface.
	"Dairy mold"	*Geotrichum, Cladosporium,* and *Penicillium* spp.	Spoilage of cured cheeses; various color formations (color due to growth of the organisms on surface or in crevices); gives off-flavor. Frequently bump of growth becomes filled with white chalky mass.

mixed flora yields a characteristically flavored, soured milk containing small amounts of alcohol. The kefir grains are found in the bottom of the vessels of fermented milk.

Yogurt. In the United States this is made of pasteurized milk thickened by rennet or addition of dried milk, and soured by species of *Lactobacillus*.

Acidophilus Milk. Milk soured by *Lactobacillus acidophilus* is thought by some to have medicinal properties in the intestinal tract. **Bulgarian buttermilk**, a similar beverage, is prepared with pure-culture starters of *Lactobacillus bulgaricus*.

Buttermilk. Much of the product commonly sold in the United States as buttermilk is in reality pasteurized skim or whole milk soured mainly by *Streptococcus cremoris* with *Leuconostoc citrovorum* and then beaten so as to produce a smooth, creamy beverage. It is a pleasant, nourishing drink. Addition of 0.15 per cent of citric acid to the milk results in formation of increased flavor because diacetyl is produced by *Leuconostoc citrovorum* from citrate.

47.10
SPOILAGE OF MILK AND MILK PRODUCTS

Most milk and milk products, with the exception of powdered milk and condensed milks, have a limited "shelf life" because preservation processes applied either kill only part of the microorganisms present, as in pasteurization, or inhibit the growth of the remainder by refrigeration or addition of lactic acid by the group of lactics, as in fermented dairy products. Therefore, many of the products spoil readily if the methods of storage are inadequate. Milk, raw or pasteurized, may undergo spoilage of several types: souring, gas production, proteolysis, ropiness, etc.; a few of the more common types of spoilage of milk and its products are summarized in Table 47–3.

Spoilage of cheeses may occur during the manufacturing process, during the ripening process, and in the finished product during storage. Sporeforming gas-producers, such as species of *Clostridium*, can cause spoilage in raw and pasteurized milk, as well as in ripening cheeses (Fig. 47–14). Unripened cheeses, such as cottage cheese, are subject to spoilage during manufacture or storage. Cottage cheese may undergo deterioration, such as proteolysis, gas production, sliminess, and off-flavor, usually caused by soil and water bacteria; for example, species of *Pseudomonas*, *Alcaligenes*, *Bacillus*, and even coliform bacteria, especially *Enterobacter (Aerobacter) aerogenes*. Spoilage of cured cheeses increases with their moisture content; for example, Limburger and Brie (soft cheeses) are more readily subject to spoilage than hard cheeses like Cheddar and Swiss.

CHAPTER 47
SUPPLEMENTARY READING

Foster, E. M., Nelson, F. E., Speck, M. L., Doetsch, R. N., and Olson, J. C.: Dairy Microbiology. Prentice-Hall, Inc., Englewood Cliffs, N.J. 1957.

Henderson, J. L.: The Fluid-Milk Industry, 3rd ed. AVI Publishing Co., Inc., Westport, Conn. 1971.

Methods and Standards for the Production of Certified Milk. American Association of Medical Milk Commissions, New York. 1962.

Milk Ordinance and Code. U.S. Public Health Service Publication No. 229, Washington, D.C. 1967.

Standard Methods for the Examination of Dairy Products, 12th ed. American Public Health Association, New York, 1967.

MICROBIOLOGY OF FOODS · CHAPTER 48

The modern human diet includes a wide variety of substances from many sources. However, most of the food consumed by man may be classified into one of eight main divisions: vegetables, fruits, milk and dairy products, meat and poultry, eggs, seafood, sugar and sugar products, and cereal and cereal products. Microorganisms produce both desirable (flavoring, aroma, and preservative) and undesirable (unpleasant appearance, odor, taste, and color) changes in food. Undesirable changes in food quality are generally referred to as food spoilage.

Foods are subject to natural contamination by many different kinds of microorganisms, including some that are dangerous pathogens, and they are modified with many additives (antimicrobial preservatives, hygroscopic agents, colors, flavors) and by various preparative processes involving many species of microorganisms; consequently, the microbiology of foods is an exceedingly complex subject.

48.1
CLASSES OF FOODS

Since water and dairy products are dealt with elsewhere, they will not be included here. Other common foods may be grouped as follows:

1. Fresh foods (e.g., meats, vegetables, fruits, and fish)
2. Foods preserved by:
   a. Drying
   b. Canning
   c. Pickling, brining, salting, or fermenting
   d. Low temperatures, especially rapid freezing
   e. Antimicrobial substances
3. Breads
4. Eggs

For purposes of this discussion, fresh foods may be defined as those recently harvested or prepared that are in their natural or original state, not affected by any means of artificial preservation excepting refrigeration (not freezing) for limited periods, and unchanged by effects of holding for sale or use beyond *slight* wilting or drying. Foods still edible but held so long that perceptibly undesirable changes in volume, weight, color, flavor, odor, appearance, or other properties of fresh products have occurred may be classed as stale. An *accurate* definition of **fresh** and **stale** or **spoiled** is very difficult indeed, as are chemical or bacteriological determinations of these conditions. We shall use the terms in their commonly accepted meanings.

Foods may also be classified on the basis of stability:

1. **Perishable** foods such as meat and fish
2. **Semiperishable** foods such as potatoes
3. **Stable** foods such as cereals, flour, and sugar.

Of course, any stable or semistable foods that become over-moist or water-soaked are no longer stable.

750

We may divide microorganisms of foods into three general groups on a functional basis:

1. Those causing spoilage or **undesirable** changes in the food
2. Those producing **desirable** changes
3. Those producing **disease**

Attempts to combat undesirable microorganisms or to encourage desirable microorganisms in food are the basis of industries and research in this country involving billions of dollars annually.

Autolytic Enzymes. In any discussion of the stability of foods, autolysis must be considered. When cells die, certain incompletely understood intracellular enzymic processes act within a few hours to disintegrate the cells. Such self-disintegration is called **autolysis.** Autolysis proceeds best under conditions optimal for enzymic action: pH, temperature and so on. Several bacteria contain N-acetylmuramidase, which hydrolyzes certain bonds in peptidoglycans of the cell walls.

Many meats undergo autolysis. Venison and poultry are often "hung" until tender (i.e., until some autolysis has taken place). Beef is more tender after a ripening period in refrigerators, partly due to autolysis. Autolysis may cause the uninitiated housewife to wonder why the pound of liver she bought in the butchershop seemed to melt away to a half pound as she drove home on a warm summer day. Since conditions for autolysis and microbial growth are parallel to a great extent, autolysis of any fresh, cellular food under market conditions is usually accompanied by microbial action unless foods are sterile. Autolysis will occur even in sterile foods if the autolytic enzymes have not been destroyed by heat.

Leafy vegetables (lettuce, spinach, endive), bananas, and other fruits are made very soft during autolysis, especially in warm weather. Loss of weight, color, flavor, and nutritive value result during excessive autolysis. Decomposition by yeasts, molds, and bacteria can advance rapidly when over-autolysis has prepared the way. Hence in preserving foods, efforts are made to (1) prevent autolysis beyond a certain desirable **ripening** point, and (2) to prevent decomposition by microorganisms. Usually the same preservative measures are effective against both. For example, in preparation of vegetables for preservation by freezing, autolytic enzymes, as well as some of the superficial microorganisms, are partly destroyed by **blanching** or scalding for a few minutes before packing.

48.2
FRESH FOODS AND MICROORGANISMS

Meat

When an animal dies of natural causes (disease or age), before death there is a short period (the agonal or moribund state), during which there is a collapse of the defensive mechanisms which normally prevent invasion of the blood (and thence of the tissues) by microorganisms in the gastrointestinal tract, respiratory tract, skin, and other body surfaces. Examination of muscle, liver, and other tissues of such an animal (and especially lymph nodes which arrest bacteria coming from such sources) immediately after death reveals the presence, often in considerable numbers, of microorganisms characteristic of the intestine, respiratory tract, and so on. These include species of *Clostridium*, Enterobacteriaceae, *Micrococcus*, fecal and respiratory streptococci, *Proteus*, *Pseudomonas*, and the like. Meat from such an animal will spoil quickly.

If a healthy animal is *suddenly* killed, as in the abattoir by a blow on the head, relatively little postmortem invasion of the blood and tissues occurs. Cutting a large vessel with a sharp, clean knife, as in killing hogs, probably introduces very few microorganisms. If the animal is then immediately dismembered in a cleanly manner, as is done in well-conducted abattoirs, relatively few organisms are to be found in the depths of solid tissues; spoilage must proceed mainly on the surfaces.

Surface Flora of Meat. Meat becomes contaminated immediately upon exposure in the abattoir. Dust from hides and hair, bacteria on gloves, hands, and cutting and handling instruments, all contribute to this. The principal damage done by such organisms is to cause decomposition of organic substances in the surface tissues. Preventive and preservative measures include prompt hanging in refrigerators and exposure to ultraviolet light.

The microorganisms involved in surface decomposition and spoilage of meat and other protein foods include saprophytes of the genera *Pseudomonas*, *Bacillus*, and *Micrococcus*; various Actinomycetales, yeasts, and molds; *Chromobacterium*, *Proteus*, enterococci, *Clostridium*, *Corynebacterium*, *Escherichia*, and *Enterobacter* (*Aerobacter*). All of these are ubiquitous inhabitants of the common environment. Some psychrophilic microorganisms cause unsightly but harmless blackening, greening, or other discoloration during refrigeration and may produce undesirable tastes and odors, such as rancidity owing to decomposition of fats.

Ground Meat. The flora of freshly ground meat is much richer in numbers and types of microorganisms than that of large pieces of meat such as roasts, because in the ordinary processes of preparing ground meat, the meat is cut into small pieces, scraps are used, and the surface contamination is thoroughly mixed with the meat as it is ground. It is also usually warmed somewhat. The microorganisms present in the product are those found initially on the surface and, in addition, molds, yeasts, sporeformers, Actinomycetales, and others from the grinding machine, hands, implements, and dust. Unless effectively refrigerated, the microorganisms grow throughout the whole mass and the meat spoils rapidly.

Bacteria in Other Comminuted Foods

The basic principles that apply to ground meats hold true also for products such as separated, fresh crab meat; "shucked" (removed from the shell) shellfish; flaked, fresh fish; ground horseradish; chopped spinach; salad mixtures; coleslaw, and the like. The more foods are handled and the more they are chopped, ground, or grated (**comminuted**), the more the surface microorganisms are mixed with the interior and with richly nutritive sap or juice, and the heavier the inoculation. Such foods must be kept refrigerated at all times, and never sold if more than 24 to 48 hours old, the time depending on temperature, pH, salinity, and so on.

Poultry, Fish, and Shellfish

Poultry. If the birds are fresh-killed, and there has been rigorous application of standards of hygiene (asepsis), bacterial contamination of the product is minimized. Of the bacteria most frequently found in poultry, organisms of genera *Pseudomonas*, *Proteus*, and *Chromobacterium* are especially undesirable. Bacteria of the coliform group are frequently introduced by food handlers, and, infrequently, organisms such as *Salmonella* and other related pathogens may also be introduced. The sources of contamination of food products will be discussed in a later section.

Fish. Bacteria of the genera *Pseudomonas*, *Chromobacterium*, *Micrococcus*, *Flavobacter-*

ium, *Corynebacterium*, *Sarcina*, and *Serratia* have been found in association with the slimy coats of fish. Some of these organisms are psychrophilic, and the first two above-mentioned organisms are the most destructive of freshly caught fish. Intestinal flora may include species of genera *Bacillus*, *Escherichia*, and *Clostridium*. It is common practice on commercial fish boats or landing places to gut fish immediately and pack them in ice. During the cleaning, fishermen soon become heavily contaminated with bacteria and transfer them to the flesh of the fish. Autolysis is slow in fish muscle, and unopened fish have better keeping quality than opened fish because contamination of the body cavity is avoided. Chlorinated water, heavy salting with sodium chloride, ice with calcium chloride, and sodium nitrite have been used to preserve fresh fish. (Use of sodium nitrite has been questioned because it is poisonous and may cause fatal methemoglobinemia, loss of ability of hemoglobin to carry and release oxygen to the body.)

Shellfish. Under market conditions, shellfish are often rather heavily contaminated with dirt from the shells and from benches at which **shucking** (removal of shells) is done. The flora includes marine species similar to those found on fish, organisms from soil, and those from equipment used in handling. If the water in which they have grown has a rich bacterial flora, the shell liquor may contain considerable numbers of various harmless bacteria.

Oysters, mussels, and some other shellfish fatten on sewage, and have long been known as vectors of typhoid fever if taken from polluted waters. Typhoid bacilli may live for two weeks or longer in live and in shucked oysters. Formerly common, typhoid due to shellfish is now very rare. Uncooked shellfish are vectors of enteric viruses, especially that of epidemic hepatitis, and they should be regarded with suspicion in respect to polio and other viruses. In the United States, state and government supervision of the shellfish beds has virtually removed infected shellfish from the market. It has been shown that fluids inside the shells of steamed clams may not always reach temperatures high enough to inactivate the heat-stable virus of epidemic hepatitis.

The bacteria that may be present in shellfish are similar to those in fish, i.e., species of *Bacillus*, *Alcaligenes*, *Proteus*, and the coliform group. These are not harmful per se but represent potentially dangerous sewage pollution. Tests for the coliform group of bacteria in shell liquor, or in the liquid around shucked shellfish, are made in much the same manner as tests are made for this group in drinking water. The results are expressed in a similar manner. However, coliforms in shellfish may be entirely nonsewage in origin if the shellfish are taken from clean beds. Details of procedure may be found in *Recommended Procedures for the Bacteriological Examination of Sea Water and Shellfish*, published (1970) by the American Public Health Association.

Oysters to be offered for sale raw are generally **floated** (allowed to remain for some hours or days) in clean chlorinated water. In the process they pass a large volume of the clean water through and around their bodies and thus greatly reduce their bacterial content.

Fruits and Vegetables

The principles underlying the microbiology of fresh meat and fish products apply equally to fresh vegetables and fruits. That is to say, the internal tissues of whole, healthy plants and fruits contain very few bacteria, but contamination of the exterior surfaces by microorganisms from the soil, hands, and packages occurs. Vegetables growing in the soil, such as root crops, have adhering to them soil saprophytes such as species of *Bacillus, Pseudomonas*, and many others, depending on the nature of the soil.

Soaking and washing by agitation may tend to distribute spoilage organisms from damaged tissues to the entire crop. Recirculated or sewage-contaminated waters are likely to add more organisms, including potential pathogens. Washing with mild detergents or germicidal solutions may tend to reduce the number of microorganisms on the products (Table 48–1).

TABLE 48–1. AVERAGE NUMBERS OF BACTERIA ON PEAS DURING PROCESSING IN THIRTEEN PEA-FREEZING PLANTS*

Point of Sampling	Bacteria on Peas (Nos./gm)
Platform	11,346,000
After washing	1,090,000
After blanching	10,000
End of flume	239,000
End of inspection belt	410,000
Entrance to freezer	736,000
After freezing	560,000

*From Western Regional Laboratory, USDA, Albany, Cal., 1944.

Intact vegetable skins free from superficial water resist invasion of microorganisms for considerable periods, especially if kept in a cool, dry place. Nonsucculent vegetables with whole skins, such as turnips and potatoes, will stand storage better than soft, succulent, nonacid vegetables and fruits like lettuce, asparagus, spinach, and ripe peaches, the juices of which offer good pabulum to microorganisms. Such soft vegetables soon autolyze and then decay. Invasion and decomposition are retarded by gentle handling and storage under cool, dry conditions. Yams autolyze even at refrigerator temperatures (about 5C).

48.3 ANTIBIOTICS IN FRESH FOOD

During the 1950's, the Food and Drug Administration (FDA) approved certain antibiotics for use in preservation of fresh foods, particularly meat, poultry, and fish. However, in 1967 the FDA revoked the regulation granting such permission. Prior to 1967, the broad-spectrum antibiotics, such as chloramphenicol and the tetracycline group, had been injected intravenously into cattle before killing. The method has been useful in the tropics and in situations where refrigeration is not economically feasible. The advent of treatment of products with antibiotics considerably extended the "shelf life"; however, such practice frequently resulted in relaxation of the rigid application of standards of hygiene and cleanliness. The use of penicillin is no longer permitted because it is highly allergenic. It should be noted that antibiotics in any material can interfere greatly with bacterial examination.

Antibiotic-Resistant Organisms in Foods. A serious problem in the use of antibiotics for preservation of foods of any sort concerns the development of antibiotic-resistant species. Antibiotic-resistant pathogens (e.g., *Salmonella*) capable of infecting man as well as animals are developed by the continuous use of antibiotics; the various spoilage saprophytes also become resistant.

48.4 EGGS AND EGG PRODUCTS

These may be considered under two general headings: (a) shell eggs (fresh or storage); and (b) egg products such as liquid eggs; eggs that are frozen, dried or whole; egg whites and yolks; and eggs that are salted, plain, sugared, desugared, or fermented.

Microorganisms of Shell Eggs. As soon as an egg is laid, the outer surface becomes contaminated. If, as is the practice in many large-scale poultry plants, the eggs are laid by clean, healthy hens on a clean wire frame and automatically collected immediately, contamination is at a minimum. The sanitary and keeping qualities of such eggs (and of commercial egg products derived from them) are clearly superior to those of eggs laid in damp, mud-and-feces-fouled straw nests and collected perhaps once a day (or every two or three days!).

Surface microorganisms are prevented from entering the egg for some days largely by the dried, mucilaginous surface coat, a sort of natural varnish.

Molds and some bacteria can grow on the outer mucinous coating if eggs are stored in humid atmospheres (above 70 per cent saturation) at ordinary climatic temperatures. The microorganisms eventually penetrate the shell and contaminate the interior. These give the eggs "off" odors and tastes and unsightly appearances. Eventually, the mucinous protective film is entirely decomposed, and the interior of the eggs is overwhelmingly invaded, with consequent decay.

The flora of stale or bad eggs is largely of fecal and soil origin and includes the various saprophytes listed elsewhere.

Pseudomonas and Pyoverdin. Early spoilage of eggs stored at 15C is often due to *Pseudomonas*, especially *P. fluorescens* and *P. ovalis* from soil or dirty wash water. These bacteria impart "musty" tastes and odors to so-called **fresh** eggs long before overt, late-stage decomposition develops. These bacteria produce a fluorescent pigment called **pyoverdin** (see Color Plate IV, *A*) that can readily be detected by the ingenious expedient of examining eggs, shell or liquid, or extracts of liquid products, in ultraviolet light. Higher than normal fluorescence indicates heavy growth of *Pseudomonas* and usually other bacteria, incipient spoilage, and low commercial quality.

Infection of Eggs. Eggs may contain pathogenic bacteria when laid, even though the hen *seems* healthy. *Salmonella* species are commonly found. These bacteria have at times caused serious outbreaks of gastroenteritis. *Salmonella* may be detected by means of selective cultivation and fluorescent antibody staining.

Microorganisms of Egg Products. In commerce, the best (freshest) eggs are sold in the shell, while second or lower grade eggs are used for liquid, frozen, or dried products. Plate or direct microscopic counts of such egg products may run from a million to a billion or more per gram. This does not necessarily make the eggs unfit for use in cooking, any more than high bacterial counts destroy the commercial value of lower grades of milk. The problem is one of profit and loss. Egg products are a picnic for microorganisms, just as are milk, crab meat, and ground meat. If the egg products are not promptly frozen, refrigerated, or dried, the heavy initial contamination quickly ruins them.

Liquid Eggs. Egg whites are often sold separately. To improve their keeping quality, they (and some other food products) are often heavily inoculated with the nonproteolytic *Escherichia coli* or, better, with bakers' yeast, *Saccharomyces cerevisiae*. The whites are held at about 28C for 24 to 48 hours. Glucose in the whites, main source of energy for putrefactive organisms, is thus promptly removed by fermentation with nonputrefactive *E. coli* or *yeast*. The whites must be initially clean or they will spoil during the intentional fermentation.

48.5
BREAD

The better grades of bakers' bread are produced by allowing bakers' yeast (varieties of *Saccharomyces cerevisiae*) to ferment sugars (glucose, maltose, sucrose) in a mixture **(dough)** consisting mainly of flour and water (or milk), with some salt. In commercial baking softening and hygroscopic agents, and flavoring, raisins, caraway seeds, vitamins, and shortening are commonly added. Small amounts of cane sugar are often included in home bread making, both for flavor and to stimulate fermentation.

The dough is a soggy, plastic mass at first. *S. cerevisiae* does not attack the starch in flour. However, the necessary amylases and proteinases from the grains are present in the flour. These enzymes hydrolyze the starch and proteins of the dough. Sometimes **malt** (see under Beer in §49.6) is added to aid this process. The products of the hydrolysis support growth of the yeast. Early growth of lactobacilli, derived from the grains, gives the dough an initial acidity which helps suppress undesirable organisms and favors yeast. The lactobacilli and yeast also contribute to the flavor and aroma of the bread.

The yeast, or **leaven** (L. *levare* = to raise), produces carbon dioxide, water, and ethanol. The gas causes the bread to rise and gives it its foamy texture. Baking drives off the alcohol and partly dries and **sets** (firms) the bread.

Leavens. An important phase of the work of the microbiologist in some bakeries is the preparation of leavens. Many bakeries maintain their own leavens as trade secrets. The leaven may consist of a pure culture of a selected strain of yeast cultivated massively in aerated wort made of malted grains (Chapt. 49) or other starch derivatives. Many tons of yeast in dried, compressed cakes are prepared in commerce daily for use by brewers, by bakers, and in the home.

Leaven may also consist of a mixture of pure cultures of **aerogenic** (gas-producing) bacteria: *Enterobacter (Aerobacter) cloacae*, the heterofermentative *Lactobacillus brevis*, and *Leuconostoc*, as well as yeasts. *Clostridium* species are undesirable in leaven because they often produce rancid flavors. Bacterial leavens are used to prepare bread made of sour dough (sauerteig), and "self-rising" or salt-rising breads. Such leavens often consist of previous lots of sour dough and are fortuitous mixtures: the standby of the old-time prospector, or "sourdough," of Alaska.

Microorganisms in Bread. Most organisms in bread, aside from those added as leaven, appear to come originally from the flour or meal and include the familiar list of the environmental saprophytes. Bacterial counts of flour range from a few hundred thousands to several millions per gram. These organisms, especially molds, can cause spoilage in the stored product, especially under humid conditions.

During ordinary baking most of the vegetative forms of molds, yeasts, and bacteria are killed. Heat-resistant bacterial spores and conidia of molds may survive. Temperatures inside the loaf of baking bread rarely rise much above 80C.

Ropy Bread. If bread is made with ingredients containing large numbers of spores of slime-forming species of *Bacillus* (*B. polymyxa* and *B. pumilis*), some of the spores can survive baking and may grow in the bread, producing a mucinous slime. When the bread is broken apart this slime is drawn into long threads, resulting in a defective product called **ropy bread.** These difficulties with bacterial spores may be largely eliminated by cleanliness and modern methods of inducing germination of spores before baking (Chapt. 36).

Red or "Bloody" Bread. This product is caused by *Serratia marcescens*, and the red color is pigment produced by the organism. In ancient times, growth of this organism was thought to be the mysterious appearance of an apparent drop of blood and was considered sacred or miraculous.

Moldy Bread. Moldy bread is usually the result of extraneous contamination of cut surfaces or crust by hands, dust, and knives, followed by holding under humid conditions at household temperatures. Molds primarily involved in spoilage of bread are *Rhizopus nigricans* ("bread mold") and species of *Penicillium* and *Aspergillus*. Other fungi involved are species of *Monilia (Neurospora)*, *Mucor*, or *Geotrichum;* other genera of molds may develop occasionally. A long cooling period and excessive exposure of the surface by slicing enhance the growth of molds. Certain species of the genus *Aspergillus* are known to produce poisonous substances (mycotoxins) called **aflatoxins** when grown in certain media. When ingested by animals aflatoxins have caused illness and sometimes death. Their role in human disease remains to be determined. Slight surface growth of mold gives bread a musty odor and taste.

Bread in the Kitchen. Since bread is not sterile, it should be cooled promptly after baking. The wise housewife keeps bread in the refrigerator in warm, humid weather. Wrapping bread in waxed paper or a plastic bag helps keep the bread clean and prevents drying out, but conserves a humid atmosphere and favors growth of molds and bacteria, even if refrigerated.

Nonsporeforming pathogenic bacteria cannot survive proper baking of bread but may be introduced by unsanitary handling after baking.

48.6
SOME FERMENTED FOODS

As noted elsewhere, soil and plants harbor, among many other microorganisms, numerous species of lactic-acid bacteria, such as *Lactobacillus*, *Leuconostoc*, and *Pediococcus*. These can ferment the carbohydrates and metabolize the other nutrients in the tissues, sap, and juices of green plants. This is the basis of the making of several kinds of fermented plant foods: ensilage for cattle, sauerkraut and pickles for humans. The methods of preparing each kind of food differ somewhat in form, but the principles and microorganisms are identical, or very similar, in all three processes. The preparation of ensilage for cattle is representative.

Ensilage. Finely chopped, partly mature plants such as corn stalks or alfalfa are tightly packed in tall cylindrical tanks (**silos**). Microorganisms of many kinds start to grow in the plant juices and ferment the carbohydrates. As fermentation proceeds, the material becomes warm and acid. The heat can be reduced if the rate of oxidation is decreased (exclusion of free oxygen) by tight packing. Oxygen is used up rapidly, so that molds and other strict aerobes soon cease growth. Only facultative and strict anaerobes continue.

In the first stages probably the Enterobacteriaceae and other rapidly growing nonaciduric but fermentative microorganisms predominate. These are undesirable, since many produce gas and unpleasant flavors. As acidity increases they subside, and the more aciduric homofermentative lactic-acid bacteria predominate. These produce more lactic acid with small amounts of other products of fermentation (diacetyl and other volatile substances) which give an aroma and flavor to ensilage that is relished by cattle. The last stages of the fermentation and final increase in acidity are caused by the very acidophilic and aciduric lactobacilli, e.g., *L. delbrueckii* and *L. plantarum.*

After three or four weeks, the process slows and the fermented mass gradually cools. Carbon dioxide and oxides of nitrogen derived from reduction of nitrate accumulated in the plant tissues are produced during the fermentation process and often settle in the lower part of silos, so that a person ignorant of this may die if he stays in the depths of a poorly ventilated silo.

It has been suggested that a desirable type of fermentation in silos may be facilitated by introducing cultures of various fermentative bacteria, such as *Streptococcus lactis* or *Lactobacillus* sp., as the material is packed. Under natural conditions various other organisms are doubtless involved, including the bacteria of the soil. In some sections of the country molasses is added to promote fermentation by the acid-formers, and to improve palatability.

If too much soil is introduced with the fodder, undesirable and excessive putrefactive processes may spoil the product. For example, butyric-acid organisms such as *Clostridium butyricum* get in and ruin the silage by producing butyric acid, which makes it rancid. The action of such organisms constitutes a "disease" of silage. *C. botulinum*, a soil anaerobe forming a deadly exotoxin, has also at times caused much damage to livestock by its growth in silos.

Sauerkraut. In the production of this savory delicacy pure cultures of lactobacilli are sometimes used to aid the process. Commonly, however, the fermentation is allowed to proceed naturally. Salt is placed between layers of shredded cabbage as it is packed in large crocks or barrels. The salt inhibits undesirable bacteria and draws out the juices of the cabbage. Wooden frames are placed on the cabbage to keep it submerged (anaerobic). Except for the salt, sauerkraut is analogous to silage. Only facultative, anaerobic, aciduric and acidophilic and thermoduric forms not sensitive to salinities of two per cent or slightly higher can grow. During the first two to five days species of *Leuconostoc*, especially *L. mesenteroides*, become increasingly common. Acids, esters, and diacetyl give pleasant aromas and flavors. Temperatures near 70F favor the best fermentations. Fermentation is complete in two to three months.

Pickles. Fermentation by mixtures of organisms normally present, in a manner analogous to the manufacture of ensilage and sauerkraut, is part of the processing of pickles, ripe olives, and the like. The process is made selective by progressively increasing salt concentrations from an initial 8 to 10 per cent (or lower) up to 16 per cent.

The actions of the lactic acid, brine, and microbial metabolism change the color, consistency, and flavor of the cucumbers, olives, etc. In making pickles, after 8 to 10 weeks the vat is emptied and the pickles, now called **salt stock**, are "freshened" by soaking in water. Then they are packed in fluid containing vinegar, sugar, and various flavorings, spices, dill, and so on.

Slimeforming organisms such as *Leuconostoc mesenteroides;* sporeforming bacilli; pectinase-formers that destroy vegetable tissues; organisms that destroy lactic acid; molds; and other hydrolytic species may ruin the pickles if temperatures remain long near 20C or if salinity is too low (below about 8 per cent). Stock to be used for dill pickles (low in acid and salt) is particularly liable to such spoilage. Methods of prevention of spoilage with organic acids are discussed on page 761.

48.7
PRESERVATION OF FOODS

Modern home or commercial canners have to consider not only the killing of bacteria likely to cause spoilage or disease but also the effect of the processing on palatability and appearance of the food. It has been stated in the introduction of this chapter that today our diet consists very largely of readily perishable foods; i.e., food that may be deteriorated by growth of microorganisms, especially bacteria, yeasts, and molds. Methods of food preservation include one or more of the following principles:

1. **Aseptic** methods prevent the entrance, or effect the removal, of undesirable microorganisms. Vigorous application of standards of hygiene and cleanliness to processes, utensils, and the surrounding environment greatly reduces the premature spoilage of food.
2. **Bacteriostatic** methods include: drying, freezing, refrigeration at temperatures slightly above freezing, various types of pickling, brining, salting, smoking, the use of antibiotics as already described, and other preservatives such as various organic acids.
3. **Bactericidal** or **sterilizing** methods involve canning and such processes as making jams, jellies, and preserves (i.e., heating processes). The use of highly penetrating, microbicidal, ionizing radiations will also probably take its place among these methods in the future.

Bacteriostatic Preservation. Preservation by low temperature employs the principle of storing perishable foods in a household refrigerator to control the growth and activities of microorganisms and reduce the action of autolytic enzymes inherent in plant or animal tissues. While most modern refrigerators maintain temperatures from slightly above 0C (32F) to 5C (41F), it must be borne in mind that meat, milk, vegetables, fruit, and other common foods stored in the refrigerator provide an excellent habitat for the growth of psychrophilic or cryophilic microorganisms; growth of most mesophilic saprophytes and pathogens will be retarded at these temperatures. Most psychrophilic organisms can initiate slow growth at the temperature range of −10C (freezing point depressed by increase in solute content) to 0C without danger to health. However, the common food-poisoning organisms—i.e., species of *Staphylococcus* and *Clostridium*—usually cease to grow at temperatures below 3C, but growth may be initiated at higher temperatures, up to 10C, and more rapid growth occurs at temperatures up to 37C (98F). (See Figures 48–1 and 47–2.) Some of the more common food poisonings caused by contaminated food are summarized in Table 48–2.

It should be noted that the growth of bacteria and fungi in food at low temperatures can lead to alteration in food quality and eventually to spoilage, as seen in spoilage of wieners, butter, bacon, and green, mold-laden oranges! As a general rule, the lower the temperature the slower the spoilage; only when food is solidly frozen does microbial growth cease.

Preservation by Freezing

Frozen foods represent one of the greatest money-exchange items in the American dietary. These foods may be classified for microbiological purposes as (1) uncooked and (2) precooked.

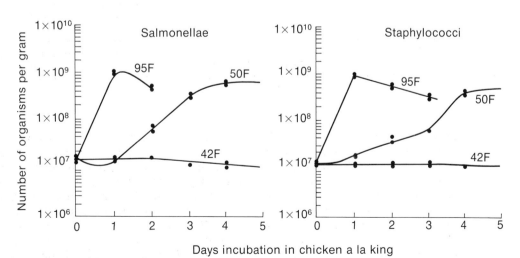

Days incubation in chicken a la king

Figure 48–1. Salmonellas and staphylococci multiply rapidly in chicken a la king incubated at room temperature. Curves also show growth at other temperatures.

TABLE 48-2. SOME FOOD POISONINGS AND DISEASES CAUSED BY CONTAMINATED FOODS *

Etiological Agent (Organisms)	Disease	Foods Usually Involved and Pathogenesis
Clostridium botulinum	Botulism	Usually involves ingestion of improperly smoked or uncooked meat or fish, improperly canned vegetables, or low-acid foods. The disease is not an infectious disease but an intoxication caused by ingestion of foods in which the organism has grown and produced one of the most potent exotoxins known to man. There are six types of toxins (A to F); types A, B, and E cause human food poisoning, with type A most severe and most prevalent. The toxins are usually destroyed by boiling. The toxin is absorbed via the gastrointestinal tract into the blood stream and attacks susceptible neurons (blocking release of acetylcholine from demyelinated ends of cholinergenic motor nerves). Symptoms include: double vision, uncoordination of eye muscles, difficulty in swallowing and speech, and bulbar paralysis. Death may be due to respiratory failure or to cardiac arrest.
Staphylococcus aureus	"Staph" food poisoning or staphylococcal enterotoxemia	Potato salad, ham, cream-filled bakery goods, and dry skim milk have been incriminated as common sources of food poisoning. The food poisoning is due to the action of enterotoxins (type B, heat-stable) released by the organisms that grew in the food before it was eaten. The disease is characterized by sudden nausea, vomiting, diarrhea, and often shock occurring within 6 to 12 hours after ingestion of contaminated food.
Streptococcus faecalis and other species	Enterococcus food poisoning	Caused by inadequately refrigerated food contaminated with organisms which are often referred to as "enterococci" because they are frequently found in the human gastrointestinal tract. The disease is characterized by nausea, frequently vomiting, colicky pain, and diarrhea. These symptoms usually appear within 8 to 12 hours after ingestion of contaminated food.
Clostridium perfringens	*C. perfringens* food poisoning	Improperly or unrefrigerated cooked meat products, especially those left at room temperature environment for several hours, are usually involved. While the organism is a normal inhabitant of the intestinal tracts of animals, in man it is able to cause mild illness with diarrhea and abdominal pain, usually without vomiting. The patient recovers in a few hours without treatment. The organism is more important as a cause of gas gangrene, which consistently involves the muscles. While the mechanism of food poisoning by this organism is not known, various specific exotoxins (α, β, γ, δ, etc.) that cause gas gangrene are well documented.
Bacillus cereus	*B. cereus* food poisoning	Common source of food poisoning is inadequately refrigerated starchy foods. The disease symptoms, such as nausea, sometimes vomiting, colicky pains, and diarrhea, may occur 8 to 12 hours after the ingestion of contaminated foods.

*Commonly includes *Salmonella* infection (salmonellosis), which may be transmitted by inadequately cooked egg and poultry products. The disease is an infection rather than a food poisoning and is characterized by abdominal pain, diarrhea, and frequently vomiting, accompanied by chills, fever, and prostration typically beginning 10 to 24 hours after ingestion. (See Chapter 35.)

Uncooked Frozen Foods. These are fresh foods to be eaten raw (e.g., frozen berries) or to be cooked (e.g., frozen steaks). At the time of freezing, fresh foods have on or in them the original microbial flora plus any bacteria that are added during handling, packaging, or processing. Factors affecting the microbial content of uncooked frozen foods include: freshness, condition at time of freezing, pH, preliminary processing, packaging, time and temperature of frozen storage, and kind of food (protein, starch, comminuted). If foods are dirty or have been held long at room temperatures prior to freezing, they may be on the verge of spoilage at the time of freezing and the microbial counts may be very high. However, since freezing and storage at −25C soon destroys many bacteria, low (deceptively favorable) plate counts may be obtained with long-frozen food that was initially of low quality. In general, storage at higher temperatures (0C to −10C) is more bactericidal, especially in acid foods such as orange juice, than storage at lower temperatures: −25C (the approximate temperature of the home freezing locker) in freezer lockers to −75C in carbon dioxide ice.

Vacuum drying after nearly instantaneous freezing ("freeze-drying") (analogous to **lyophilization** used to preserve viruses, bacteria, etc.; see Chapter 20) appears to offer practical usefulness in the preservation of foods.

As with milk, a direct microscopic count on frozen foods reveals much concerning initial numbers of now-dead bacteria. Microorganisms that will not grow in ordinary plate-count media are revealed. The same is applicable to foods preserved by any of the methods described farther on. High microscopic count and low plate count suggests food of initially poor bacterial quality.

Precooked Frozen Foods. Some precooked frozen foods, such as certain "TV dinners," are mixtures of cooked or partly cooked components that are not further cooked before packaging and freezing. Some of these foods are meant to be oven-baked before serving. Others, such as meat pies, are said to be ready to eat after mere warming. Bacterial counts on cleanly produced, promptly frozen, cooked foods are relatively low. Excessively high bacterial counts, especially high direct microscopic counts, indicate staleness or dirtiness of ingredients or processing.

Baking meat pies (a representative precooked food) does not necessarily sterilize them. Commercial-size meat pies held at 425F for 40 minutes have very low plate counts, but the same temperature for 20 minutes is wholly inadequate and may, by merely warming the interior, actually incubate the pathogens in the pies.

Sanitation in Frozen Foods. Virtually all frozen foods are subject to some sort of contamination before freezing. The usual environmental saprophytes are found. Since many foods contain animal products, and are handled by people, fecal streptococci, coliforms, and staphylococci are frequently found. The full sanitary significance of these indicator organisms in frozen foods is not yet fully clarified, but they are generally regarded with suspicion.

As indicators, enterococci tend to persist in contaminated frozen foods, while the coliforms tend to diminish in numbers and disappear with continued frozen storage.

Thawing of Frozen Foods. Freezing during preparation, and thawing for use, should be completed within two to four hours under ordinary conditions. Most frozen foods remain wholesome so long as they remain frozen. On thawing, frozen foods undergo rapid deterioration because the natural resistance of fully active tissues is reduced or absent. Autolysis occurs and microbial decomposition can set in with little delay at room temperature. Pathogens can grow at summer temperatures (70 to 100F).

Interruptions to electric current supplying frozen-food lockers may cause serious spoilage if long enough for frozen foods to thaw. Indeed, food refrozen after being thawed without the consumer's knowledge may be a very dangerous product. Large masses of ice in the bottom of the package suggest that thawing and refreezing have occurred.

General Rule. A sure means of making any food, canned, frozen, or otherwise, safe from infectious organisms and from botulinal (but *not* staphylococcal) toxin is to heat all of it to 100C for at least 15 minutes just before eating.

As a matter of experience of over three decades, food poisoning (botulinal or staphylococcal) and food infection (salmonellosis, shigellosis) due to **commercially** prepared foods are relatively uncommon (but by no means unknown!) in the United States.

Preservation by Heating

Preservation of foods by use of high temperature (with exclusion of pasteurization as described in the preceding chapter) involves killing all microorganisms (both vegetative cells and spores) and destruction of autolytic enzymes therein. As stated elsewhere, heat destroys microorganisms by coagulation of proteins and irreversibly destroys the enzymes necessary for metabolism.

Various factors affect the complete destruction of vegetative cells and spores of microorganisms, especially those of bacteria. If canned foods are processed long enough, they can be absolutely sterilized. This may require such prolonged heating in steam under pressure (autoclaving) (especially of nonacid foods like corn, beans, or asparagus) that the foods become mushy and discolored and have bitter flavors. Prolonged heating also adds to the cost of canning. The aim of the canner, then, is to heat as little as possible, consistent with safety from food poisoning and loss from spoilage. For this reason food microbiologists are well acquainted with the **thermal death time** of organisms most likely to alter food quality. Thermal death time

TABLE 48-3. THERMAL DEATH TIMES OF BACTERIAL CELLS

Bacterium	Time (min)	Temperature (C)
Neisseria gonorrhaeae	2–3	50
Salmonella typhi	4.3	60
Staphylococcus aureus	18.8	60
Escherichia coli	20–30	57.3
Streptococcus thermophilus	15	70–75
Lactobacillus bulgaricus	30	71

*From Frazier: Food Microbiology, McGraw-Hill Book Co., Inc., 1958.

is the time required at a certain temperature to kill stated types of vegetative cells (Table 48–3) or spores (Table 48–4) under specifically described conditions.

From the standpoint of food poisoning, the only organism likely to resist routine autoclaving is *Clostridium botulinum.* Even the most resistant spores of this species are killed by heating at 121C for 20 minutes at pH 7.0.

Acidity greatly reduces the time and temperature necessary to preserve foods by heat, even though they may contain resistant spores. Canned tomatoes, rhubarb, and acid fruits (pH 3.5 to 4.5), for example, require only a few minutes at 100C to preserve them. On the contrary, nearly neutral materials, such as meats, corn, spinach, and beans, require much longer periods, depending on solidity and size of packs, and preheating. The cans are often preheated to 100C and sealed in a vacuum (Table 48–5).

Commercially Sterile Foods. Canned foods may contain spores of organisms that are viable but fail to grow under ordinary conditions of storage. These foods may be said to be **virtually sterile** or **preserved by heating** or **commercially sterile.** For example, spores of nonpathogenic *Bacillus stearothermophilus* (designated in the trade as No. 1518) and *Clostridium ther-*

TABLE 48-4. THERMAL DEATH TIMES OF BACTERIAL SPORES

Spores of	Time to Kill at 100C (min)
Bacillus anthracis	1.7
Bacillus subtilis	15–20
Clostridium botulinum	100–330
Clostridium calidotolerans	520
Flat sour bacteria	Over 1,030

*From Frazier: Food Microbiology, McGraw-Hill Book Co., Inc., 1958.

mosaccharolyticum (No. 3814) generally withstand ordinary processing. The spores of these species are among the most heat-resistant of all. They are often used to test the efficacy of heating. Both can cause sour spoilage of canned goods. However, being obligate thermophiles, they do not grow unless the food is stored in a *very* warm place (55 to 70C). Under ordinary conditions of storage the food is, from a practical standpoint, sterile. In any case, care must be taken to eliminate the spores of *Clostridium botulinum,* cause of deadly botulism (Chapt. 36).

The effectiveness of expert canning is well illustrated by the history of several cans of meat prepared in London several years after the process of canning was devised by Appert, in 1805 in Paris, at the request of Napoleon Bonaparte for a means of preserving foods for his troops. A can of roast veal prepared in London and dated 1823 was carried the next year on the expedition of H.M.S *Fury* (Sir Edward Parry, R.N., Captain) in search of a Northwest Passage. The *Fury* was crushed in the ice and was abandoned, with stores including the can in question, for four years in the arctic. The *Fury* was visited by another ship in 1829 and the stores were brought back to London. When the can was opened, in 1958, after 125 years, the meat was found in good, unspoiled condition although the fat had emulsified somewhat. Also in 1958, a can of plum pudding prepared in London in 1900 and kept in South Africa was found on opening to be in excellent, unspoiled condition.

TABLE 48-5. ACIDITY (pH) RANGE OF SOME REPRESENTATIVE FOODS

Range of Acidity	pH	Examples in Order of Acidity
Very acid	2.5 to 3.5	Lemons Cranberries Rhubarb Grapes Pineapples
Moderately acid	3.6 to 4.5	Oranges Apples Tomatoes Pears
Slightly acid	4.6 to 6.8	Carrots Squash Spinach Fish Beef Poultry and eggs Corn Shellfish

Chemical Preservatives

Several **organic acids** and their salts—acetic, propionic, lactic, benzoic, salicylic, butyric, caproic, sorbic, citric, and others—have marked microbiostatic and microbicidal action. Some of these acids may be developed during fermentation of the food. Lactic acid forms during fermentation of milk (see preceding chapter), and some others form during fermentation of vegetables, such as pickles and sauerkraut, or certain meat products, e.g., salami sausage.

Artificially introduced food preservatives usually undergo extensive toxicity tests before use and must pass very stringent requirements of the Food and Drug Administration. Chemicals, while they may be effective microbiostatic or microbicidal agents, must not cause chronic toxicity, accumulative effects or be carcinogenic over extended periods of use. Ideally, preservatives should not only be water soluble but also be odorless, colorless, and tasteless in addition to being microbiostatic. As previously indicated, the effectiveness of organic acids like lactic, sorbic, etc., is dependent mainly on the toxic action of the **undissociated** acid or salt. Two factors that diminish dissociation of these acids or their salts are salinity and low pH. The higher the salinity and the lower the pH, the more effective these organic acids are as preservatives. Undesirable **yeasts** and **molds** in brined or fermented pickles, in sauerkraut, and in unfermented acid products, such as fresh apple juice and tomato catsup, are well controlled by the addition of 0.05 per cent to 0.10 per cent sodium benzoate or sorbic acid.

Products containing high concentrations of alcohol, as in distillery products, have self-preserving properties. Ethyl alcohol acts as a cell protein coagulant and is most germicidal between 50 and 70 per cent; higher and lower concentrations are considered to have antiseptic properties. However, at low concentrations (30 per cent or less), products such as wine may be subject to spoilage by acetic acid–producing bacteria.

The most commonly used inorganic acids and salts are sodium chloride, sulfites, sulfurous acid, and sodium salts of nitrite and nitrate.

Sodium nitrate or nitrite, or mixtures of these, are commonly added to sodium chloride in mixtures for curing meats (corned beef, ham). Whether the nitrates or nitrites have any bactericidal or bacteriostatic action per se seems to be undecided, but meat packers find less spoilage when these salts are used. They appear to have an adjuvant action in acid solutions (most cured meats are acid, pH near 5.8) and on the preservative effect of sodium chloride. Nitrites and nitrates are particularly desired because they give good red color to meat. Recently, however, the generous use of nitrate and nitrite in meat and fish products for purposes of preventing formation of putrefactive odors has been questioned. Both nitrite and nitrate (reduced to nitrite in the intestinal tract by bacteria) are poisonous if present in potable water or food products in more than minimal amounts. Nitrate poisoning by both drinking water and treated fish has been reported in recent years.

Sulfur dioxide has a bleaching effect desired in some fruits and also suppresses growth of yeasts and molds. It is applied as a gas to treat drying fruits and is also used in molasses.

Other Methods of Preservation

High Osmotic Pressures. The salting and brining of fish, corning of beef, and sugar-curing of hams and brining of green olives are examples of the use of solutions of high osmotic pressure for food preservation. Salt concentrations of 20 to 30 per cent are commonly used in brines; 50 to 70 per cent of sugar, in sugar syrups. Their preservative action depends almost entirely on their withdrawal of water from microorganisms. Other uses of sugar are in the making of jams and jellies. Heating of jams and jellies greatly reduces the initial count of microorganisms and enhances the effectiveness of sugar. Dry salt- and sugar-cured products will spoil if allowed to stand in very humid atmospheres, since they are not sterile and since sugar and salt are strongly hygroscopic, thus tending toward "self-dilution." Spoilage of salted products, e.g., salt fish, is sometimes caused by halophilic microorganisms in the brines or salt. Clean ingredients and equipment help prevent losses due to spoilage. Molds and yeasts are causes of early spoilage of sugared and salted products, since these fungi grow well in solutions of high osmotic pressure.

Drying. Drying of foods is another time-tested means of withdrawing water from spoilage microorganisms and their environment. Drying is largely microbiostatic in effect; therefore, dried foods are not sterile and will decompose promptly if kept warm in humid atmospheres.

Smoking. Meats and fish to be smoked are usually first salt-cured to prevent rapid deterioration, since smoke curing is slow. Modern "streamlined" injection processes preserve and flavor more rapidly than salt or smoke curing. The preservative factors in wood smoke are various cresols, a mixture of formic, acetic, and other organic acids, alcohol ("pyroligneous" acid), and formaldehyde, all of which are bactericidal or bacteriostatic, as well as effective against fungi and viruses. However, it should be noted that direct application of formic acid or formaldehyde to food is not allowed except as a minor constituent of wood smoke. Such chemicals are gradually absorbed in small quantities by the tissues exposed to the smoke. The meat is thus cured and rendered impervious to the action of most microorganisms, as well as given an agreeable aromatic flavor. Kept dry, the meat is preserved almost indefinitely. In warm, humid weather it may become moldy and rancid, especially on the surface.

Radiations. As pointed out elsewhere, streams of electrons (cathode rays) under sufficient potential (1 million volts or more) are potentially lethal ionizing radiations. β-Rays and γ-rays from radioactive cobalt and other wastes of atomic research have similar properties, and these have been tried on an experimental scale for the preservation of certain foods and food products.

X-rays, cathode rays, and rays from radioactive substances, particularly γ-rays, or high-energy particles, cause ionization of molecules of absorbing material, and as a result, microorganisms associated with food may be killed.

Fresh perishable foods (as well as culture media, surgical supplies, and drugs) in packages, e.g., plastics, that can be readily penetrated by such rays can be sterilized *without heat* (frequently referred to as **cold sterilization**). Such products may be kept in their moist, "fresh" condition with *no refrigeration*. The possibility of keeping foods fresh without refrigeration, whether by antibiotics or irradiation or other means, obviously offers enormous advantages. During 1963, radiation sterilization of bacon was approved by the FDA, but later (1969), the procedure was revoked for reasons similar to the discontinuation of use of antibiotic additives to foods for preservation purposes.

The necessary amounts of radiation are expensive and produce undesirable changes in flavor, color, consistency, and possibly in vitamin and other nutrient content in many foods. However, nonsterilizing amounts of radiation can be used economically as a preliminary treatment. Spores are sensitized to heat by radiation. The use of irradiation for pasteurizing (not sterilizing) as a means of reducing spoilage is still in the experimental stage.

48.8
MICROBIOLOGICAL EXAMINATION OF FOODS

Quantitative Microbiology of Food. Estimating the total numbers of bacteria in milk, water, ground meat, vegetables, frozen foods, and the like involves, for all, basically the same procedures. The main differences lie in methods of collecting, preparing, and measuring the sample. These steps are guided by the nature of the material. Fluids are diluted and plated directly, as are water, juices, and milk. In the examination of solid foods such as meat or vegetables, washings, swabbings, or scrapings from measured surfaces may be collected and shaken in measured amounts of diluting fluid. Deep samples of solid foods are cut from within after sterilization of the surface with a hot spatula.

Colony Counts. Any solid material is first weighed and then comminuted in a Waring blender or similar device to get the microorganisms of the sample of food into a **fluid suspension**. This can then be diluted and put into Petri plates with Standard Methods Agar for colony counting. The diluted sample must not be too cloudy or contain visible particles, since particles confuse colony enumeration. For inoculation into tubes of broth, clarity of inoculum is not so essential. Highly acid products should be neutralized. Incubation may be at 15C, 21C, or 32C, or all three, depending on the types of food.

Bacterial counts of some food products are extremely difficult to interpret. For example, bacteriological standards for ground meat are difficult to establish and to evaluate. According to some workers, using a medium like that used for enumerating bacteria in milk, incubated at 32C, an aerobic plate count of 10 million organisms per gram of meat is a reasonable maximum. Occasional counts in ground meat of market quality may run into the billions per gram, and this is exclusive of anaerobes, molds, and other organisms that do not grow under the cultural conditions provided.

Direct Microscopic Count. The direct microscopic examination is very useful, both quantitatively and qualitatively.

It is made in much the same manner as the Breed or direct count for milk. High microscopic counts and low plate counts on a food generally indicate that an initially poor product has been treated in some microbicidal manner or that the organisms seen will not grow in the medium provided for the plate count. Numbers of yeasts and molds and other special morphological types that do not grow on the routine media may also be estimated in the direct count.

Qualitative Microbiology of Food. Methods of examining water or milk for the coliform group or for fecal streptococci typify **qualitative** procedures in examining foods for these groups. Tubes of selective broths (for determining MPN) or of selective plating medium (for colony counts) may be used. The same sorts of media and procedures are used, with minor modifications, as for milk.

In several dried and frozen foods, fecal streptococci appear to be more dependable indicators of fecal contamination than coliforms. The coliform organisms are so ubiquitous that their presence in many foods does not accurately indicate fecal pollution at all. Further, they do not survive in preserved foods, especially acid foods, as do the enterococci.

In adaptation of microbiological procedures to the examination of foods, the nature of the medium used, the temperature (thermophilic, mesophilic, or psychrophilic), the conditions of incubation (aerobic or anaerobic), and other factors (acidity, salinity) must be modified to suit the flora. For example, if a plate count of yeasts and molds is desired, plate with acidified potato-dextrose agar (pH 3.5) or with malt agar (pH 3.5). For lactic acid bacteria, use orange-juice (serum) agar (pH 5.5). Other special media for coliforms and staphylococci are suggested in the literature cited. Numbers of bacterial spores may be estimated (not accurately counted) by heating the material to 90C for 10 minutes before plating. Other selective procedures will occur to the ingenious microbiologist and can be adapted to various problems.

48.9
FOOD SPOILAGE: DOMESTIC

The two principal aims of the domestic kitchen director are much the same as those of the commercial food handler: (a) prevention of spoilage; and (b) prevention of food infection and food poisoning.

Some of the more common types of non-canned food spoilage occurring in home and restaurant kitchens, along with the causative agents, are summarized in Table 48–6. As the table indicates, acidic and high-carbohydrate foods are spoiled not only by bacteria but also by yeasts and molds. However, high-protein foods, such as meat, fish, and eggs, are spoiled primarily by oxidative, proteolytic organisms belonging to genera *Pseudomonas* and *Chromobacterium.*

The causative organisms of spoilage may be divided into three major metabolic groups on the basis of the major chemical constituents of the foods. The approximate organic composition of various types of foods is presented in Table 48–7.

Spoilage of relatively high-carbohydrate foods is caused by carbohydrate-fermenting or **saccharolytic** organisms; spoilage of high-protein foods is likely to be caused by **proteolytic** organisms whose end products may include malodorous amines and hydrogen sulfide; high-fat foods, such as butter and bacon, are damaged by growth of **lipolytic** organisms.

Factors influencing the spoilage of foods in addition to the hydrogen ion concentration (pH) and the chemical composition are: the amount of available water, the presence of free oxygen, and the temperature of the foods. Water is absolutely necessary for the growth of all organisms, and the presence of free oxygen stimulates both aerobic and facultative saprophytes. Refrigeration temperatures (0 to 4C) tend to discourage the growth of pathogens, but many organisms intimately involved in the spoilage of foods may be able to initiate growth, provided all other environmental and nutritional requirements are present.

Prevention of Spoilage. The housewife relies mainly on low temperatures, on osmotic pressure (e.g., in jams or jellies), and on heating for preservation of foods. As a domestic economist she is interested in saving from spoilage those foods bought in quantities at lower-than-usual prices, or bought only during weekly or monthly trips to food stores. The culinary expert is also interested in "leftovers." The surest mark of high achievement in the art and economics of the kitchen is the skillful and appetizing use of foods that are kept from one meal to another—the pot of good soup (excellent bacteriological medium!) that is saved for tomorrow's dinner, the leftover porridge or bread for pudding, and so on. All are valuable assets to the resourceful domestic economist. But they are also subject to spoilage by microorganisms.

The household refrigerators or freezers offer the readiest means for saving foods. When these places are overcrowded or unavailable, spoilage can be retarded by reheating contaminated foods in covered vessels and *not opening* the vessels until more of the food is needed. For example, the pot of soup or other nonacid food may be brought to a boil for a few minutes in a vessel with a good dust-tight cover. The heat kills any vegetative forms of microorganisms and the spores of wild yeasts and molds that might have gotten in while it was cool and uncovered. A good many bacterial spores will be killed also,

TABLE 48–6. COMMON TYPES OF PERISHABLE FOOD SPOILAGE

Representative Foods	Types of Spoilage	Causative Agent(s) of Spoilage	
		BACTERIA	FUNGI
Fresh vegetables and fruits	Bacterial soft rot	Erwinia carotovora	
	Rhizopus soft rot		
	Black mold rot		R. nigricans
			Aspergillus niger
	Sliminess or souring	Saprophytic bacteria	
	Blue mold rot		Penicillium spp.
Fermented vegetables	Slimy or ropy kraut	Lactobacillus plantarum	
	Pink kraut		
			Rhodotorula (asporogenous yeast)
	Soft pickles	Bacillus spp.	
	Black pickles	Bacillus nigrificans°	
Sugar products, honey, syrups, molasses	Yeasty honey		Zygosaccharomyces, Torula
	Ropy or stringy syrup	Enterobacter (Aerobacter) aerogenes	
	Green syrup	Pseudomonas fluorescens	
	Gassy or frothy molasses	Clostridium	Zygosaccharomyces
Fruit and vegetable juice	Off-flavor	Lactobacillus spp.	
	Souring	Lactobacillus	
		Acetobacter	
Bread	Ropy bread	Bacillus spp.	
	Moldy bread		Species of Rhizopus, Penicillium, Aspergillus
	Red bread	Serratia marcescens	
Fresh meat	Souring (aerobic)	Chromobacterium spp., Pseudomonas spp., Lactobacillus spp.	
	"Putrefaction"	Clostridium, Pseudomonas, Chromobacterium, Proteus	
Cured meat	"Souring"	Chromobacterium, Bacillus, Pseudomonas	
	Greening	Lactobacillus	
	Slimy	Leuconostoc	
	Moldy		Penicillium, Aspergillus, Rhizopus spp.
Poultry	Odor (tainted or sour), slime	Chromobacterium spp. Pseudomonas spp. Alcaligenes spp.	
Eggs	Green rot	Pseudomonas fluorescens	
	Colorless rot	Pseudomonas, Chromobacterium, coliforms	
	Black rot	Proteus	
	Moldiness (fungal rotting)		Penicillium, Cladosporium spp.
Fish	Discoloration	Pseudomonas spp.	
	Putrefaction ("putrid")	Chromobacterium, Flavobacterium	

°According to Frazier: Food Microbiology, 2nd ed., McGraw-Hill Book Co., Inc., 1967.

TABLE 48–7. APPROXIMATE COMPOSITION OF VARIOUS TYPES OF FOODS°

Type of Food	Per Cent of Organic Matter		
	PROTEIN	CARBOHYDRATE	FAT
Fruits	2–8	85–97	0–3
Vegetables	15–30	50–85	0–5
Fish	70–95	0	5–30
Poultry	50–70	0	30–50
Eggs	51	3	46
Meats	35–50	0	50–65
Milk	30	40	30

°From Carpenter: Microbiology, 3rd ed., W. B. Saunders Co., 1972.

but some will survive. However, these probably will not grow sufficiently to cause spoilage for 18 hours or more. If the soup should be brought to a boil the next day, the cover not having been removed in the meanwhile, virtual **fractional sterilization** has been accomplished, and the food should keep for some time.

With solid foods not immersed in water, such as a roast of meat, spoilage tends to occur mainly on the surfaces. Much of this is introduced by utensils. Clean utensils introduce less contamination than soiled ones. If the cooled roast is reheated in a very hot oven for a few minutes after the meal, so as to cook only the exterior, and is left in the oven without opening the door, thus avoiding recontamination from the air, surface growth is definitely checked and the meat may remain good for some days, depending on the extent and depth to which initial contamination was introduced. In reheating food in this manner, it is desirable to use a self-basting roaster or other covered dish to prevent drying out of the food.

If food is not reheated, it should be promptly placed in an efficient refrigerator. It is always poor practice to allow perishable food, including milk, to stand in the warm kitchen overnight— incubating microorganisms into the millions, possibly forming large amounts of toxin—and then put it in the refrigerator next morning. This is locking the barn after the horse is stolen. When feasible, cooking of moist foods that have stood more than about 2 hours at temperatures above 15 to 20C is desirable.

Failure or delay of refrigeration is one of the commonest causes of spoilage of any nonsterile food. Another is overlong holding at refrigerator temperatures several degrees *above* zero instead of *at* or *near* 0C. Bacterial spoilage of many refrigerated foods may occur twice as rapidly at 5C as at −1C. At temperatures above 5C psychrophiles grow well. Many microorganisms not definitely classed as psychrophiles can also grow to a significant extent at low temperatures. Maximum numbers of live bacterial cells accumulate *slowly* at the lower temperature. When the refrigerated products are afterward gradually warmed for use, these enormous, active populations (along with their accumulated enzymes, and other growth products) act very rapidly indeed to effect spoilage.

Prevention of Infection. With respect to infection of foods with enteric pathogens, the "four F's" are easily born in mind and are particularly pertinent:

$$\text{Feces} \begin{array}{c} \nearrow \text{Fingers} \searrow \\ \\ \searrow \text{Flies} \nearrow \end{array} \text{Foods}$$

With respect to respiratory pathogens, one may remember the "four S's":

$$\text{Saliva} \begin{array}{c} \nearrow \text{Staph.} \searrow \\ \\ \searrow \text{Strep.} \nearrow \end{array} \text{Sickness}$$

Proper attention to handwashing after toilet and after nose-blowing is an easy and effective prophylactic against respiratory and enteric organisms. Most bacterial pathogens of the enteric

tract and of the respiratory tract can grow luxuriantly in common nonacid foods at temperatures between 23 and 38C.

48.10
INDUSTRIAL FOOD SPOILAGE

Industrial spoilage of foods may be classified as of the general type or of the specific type.

General Spoilage. This is a result of growth of a heterogeneous mixture of various types. Contamination with mixtures of organisms found in soil, dust, decks of fishing boats, floors of slaughter houses, canneries, and dairies is really the most difficult to control. It cannot be dealt with by means of a single, specific measure designed especially to eliminate one particular source of difficulty such as a single contaminated piece of machinery. It can be eradicated only by generalized and vigorous application of asepsis. This may involve expensive manpower, steam and hot water, general disinfection on a very wide scale, and costly measures against continuous reintroduction of dirt.

Specific Spoilage. This often presents very challenging detective problems in diagnosis of "diseases" of food products analogous to Pasteur's "diseases" of wine and beer.

Sometimes conditions in food manufacturing or preserving processes become especially favorable to the luxuriant growth of a single species of organism. This can then dominate the flora of a given food, or of a preserving or pickling fluid, and may produce some striking alteration of taste, odor, pigment, sliminess, or the like. A good example is ropy bread caused by the use of flour heavily contaminated with spores of a slime-producing *Bacillus* species. Butter sometimes acquires a rancid, fishy or oily taste or odor because of the growth in it of large numbers of some particular organism, such as *Pseudomonas fluorescens*, which hydrolyzes fat or produces other bad-smelling and bad-tasting compounds. Perhaps a single dirty churn may have heavily seeded the whole lot of butter. Sometimes a red, blue, or yellow or other color is imparted to milk or fish by heavy growth in it of pigment-formers: a *Serratia* species, a *Pseudomonas*, or a *Flavobacterium.* Orange juice sometimes acquires a buttery flavor because of certain diacetyl-producing species of *Lactobacillus.* Various spottings, ammoniacal decompositions, and peculiar types of fermentation are produced under special circumstances and by specific organisms.

These spoilages are usually owing to the

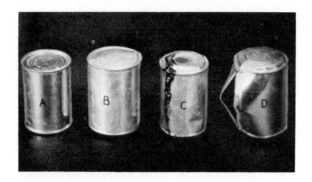

Figure 48–2. A normal can of food and cans that have swelled because of gas produced by anaerobic sporeforming bacteria. *A,* The normal can has concave ends because of the partial vacuum inside. *B,* A swelled can with ends bulged out. *C,* A swelled can with leaks where the seams are breaking. *D,* A burst can.

use of a stale product to begin with, or to a dirty tank, an implement, or a particular lot of a preservative heavily contaminated with some organism that can grow in the particular food or circumstances involved. Usually the control of such specific types of spoilage is not too difficult once the organism is isolated and its growth peculiarities and source studied.

Spoilage of Heat-Processed Foods.

Spoilage of heat-processed, commercially canned foods is confined (almost entirely) to the action of thermophilic, heat-resistant, sporeforming, anaerobic bacteria. Two major causes of canned food spoilage are (1) survival of the organisms resulting from underprocessing and (2) defective containers which permit the entrance of organisms after the heat process.

Low or medium acid food spoilage by thermophilic sporeforming bacteria may be divided into three major types:

1. **Flat sour.** This is caused by species of *Bacillus* and occurs primarily in low-acid foods, such as peas and corn. Souring of foods is caused by lactic and/or other acids produced by the organism. The ends of the can remain flat; i.e., there is no internal pressure due to gas.
2. **Thermophilic anaerobe ("T.A.") spoilage.**

The bacterium responsible for this type of spoilage is most commonly *Clostridium thermosaccharolyticum,* an obligate thermophilic, sporeforming bacillus. The organism produces acid and gases (CO_2, H_2) in low- and medium-acid foods, but hydrogen sulfide is not produced. Spoilage of canned food by this organism may be detected by change in external appearance; i.e., the can may swell to various degrees or it may explode from internal gas pressure (Fig. 48–2).

3. **Sulfide or "stinker" spoilage.** This is caused by *Desulfotomaculum desulficans (Clostridium nigrificans),* which produces hydrogen sulfide. The spoilage is characterized by the odor of hydrogen sulfide and by blackening of both food and can from sulfide of iron. Canned corn and peas are most frequently affected.

Spoilage of low- to medium-acid canned food may be caused by mesophilic putrefactive sporeforming anaerobes. Among this group, *Clostridium botulinum,* causative agent of acute food poisoning (botulism), is best known and has been discussed in Chapter 36.

CHAPTER 48
SUPPLEMENTARY READING

Frazier, W. C.: Food Microbiology, 2nd ed. McGraw-Hill Book Co., Inc., New York. 1967.

Jay, J. M.: Modern Food Microbiology. Van Nostrand-Reinhold Co., New York. 1970.

Joslyn, M. A.: Methods in Food Analysis. Physical, Chemical and Instrumental Methods of Analysis, 2nd ed. Academic Press, Inc., New York. 1970.

Riemann, H. (Ed.): Food-borne Infections and Intoxications. Academic Press, Inc., New York. 1969.

Shapton, D. A., and Board, R. G. (Eds.): Safety in Microbiology. Academic Press, Inc., New York. 1972.

Thatcher, F. S., and Clark, D. S. (Eds.): Microorganisms in Foods: Their Significance and Methods of Enumeration. University of Toronto Press, Toronto. 1968.

CHAPTER 49 · INDUSTRIAL MICROBIOLOGY

Industrial microbiology is a field of microbiological science involved with utilization of microorganisms in industrial processes. The term "microorganisms" includes both procaryotic (many kinds of bacteria) and eucaryotic (yeasts and molds, etc.) organisms. The useful products produced are the result of metabolic (enzymic) activities of microorganisms upon certain raw materials. The raw materials must be cheap and readily available, and the processes must be economically practical. The products formed must be able to compete with the product manufactured by other means. Some of the processes are the uncontrolled actions of a mixture of microorganisms, while other processes utilize pure cultures of a specific strain with carefully controlled conditions.

Today, microorganisms and their enzymes are the basis of industries grossing billions of dollars annually. For example, in 1969 the antibiotic industry alone produced more than 13 million pounds of bulk antibiotics valued at 155.5 million dollars. The magnitude of economic importance is readily conceived by the amounts of penicillins (2,462 trillion units) and tetracyclines (2.1 billion grams) produced in the same year. Some industrial products of microbial origin are listed in Tables 49–1 and 49–2.

TABLE 49-1. SOME COMMERCIALLY VALUABLE PRODUCTS PRODUCED BY MICROORGANISMS

Commercially Valuable Products	Microorganisms	Raw Materials	Nature of the Process
ACIDS, ALCOHOLS AND SOLVENTS*:			
Lactic acid	*Lactobacillus delbrueckii*	Various wastes containing glucose	Fermentation
Acetic acid	*Acetobacter* spp.	Alcoholic liquors	Aerobic; oxidation
Citric acid	*Aspergillus niger*	Carbohydrate mashes (molasses)	Aerobic dissimilation; aerated tanks; media deficient in Fe^{3+}, Zn^{2+}, Mn^{2+}
Dihydroxyacetone, 5-ketogluconic acid (sorbose)	*Acetobacter suboxidans*	Yeast extract with glucose, sorbitol, glycerol	Aerobic; metabolic wastes

*Products manufactured for use as industrial solvents, chemicals, chemical intermediates, etc.

(*Table 49-1 continued on the following page.*)

TABLE 49-1. SOME COMMERCIALLY VALUABLE PRODUCTS PRODUCED BY MICROORGANISMS (Continued)

Commercially Valuable Products	Microorganisms	Raw Materials	Nature of the Process
Butanol, acetone, ethanol	*Clostridium acetobutylicum*	Carbohydrate mashes; molasses, corn steep liquor	Fermentation
2,3-butanediol	*Bacillus polymyxa*	Carbohydrate mashes	Aerobic; metabolic wastes
Glycerol	*Saccharomyces cerevisiae*	Molasses, corn steep liquor	Fermentation; sulfite added for good yield
ANTIBIOTICS (SEE CHAPTER 23):			
Penicillin	*Penicillium chrysogenum*	Corn steep liquor + sugar	Aerobic
Streptomycin	*Streptomyces griseus*	Soybean meal extract	Aerobic
Tetracyclines	*Streptomyces* spp.	Corn steep liquor; peanut meal extract	Aerobic
Erythromycin	*Streptomyces erythreus*	Soybean meal extract; corn steep liquor + fatty acids	Aerobic
Bacitracin	*Bacillus licheniformis*	Complex organic media + Mn^{2+}	Aerobic
Chloromycetin	*Streptomyces venezuelae*	Soybean meal extract	Aerobic
AMINO ACIDS, VITAMINS, GROWTH FACTORS, HORMONES:			
Glutamic acid	*Micrococcus glutamicus*	Molasses	Aerobic
Lysine	*Micrococcus glutamicus*	Molasses and corn steep liquor	Aerobic
Valine	*E. coli* (ATTC 13,005) (mutant strain)	Molasses, corn steep liquor	Aerobic

(Table 49–1 continued on the opposite page.)

49.1
GENERAL TYPES OF INDUSTRIAL PROCESSES

Industrial processes based on microbial action are of several general types. Microorganisms may be cultivated:

1. In food products (e.g., fermented vegetable products and dairy products) for the purpose of producing certain **flavors, consistencies,** and **nutritive values** in the products (Chapt. 48).
2. In media where they decompose (ferment) various substrates (usually carbohydrates), the **products of fermentation** (various alcohols, organic solvents, lactic, acetic, and citric acids) being recovered, purified, and sold.
3. In flavored nutritive solutions, notably fruit juices and extracts of grains that are fermented, the entire culture fluid (after clarification and processing) then becoming **beverages** such as **beer** and **wine.**
4. In contact with a specific substance such as a sex hormone (the chemical group of **steroids**) so that a single, specific enzyme of the microorganism(s) brings about a **specific transformation** of the **substrate molecule** into a molecule of another desired substance (bioconversion or biotransformation).
5. In media so that the **enzymes** or other substances that the organisms **synthesize** (amylase, protease, antibiotic) may be collected, purified, and sold for commercial or medical use.

In addition to these uses, organisms *themselves* (principally yeasts, fungi, some algae) are sometimes cultivated for use as **food.** Such food products are generally used as feed for poultry and livestock. This is not as yet a large

TABLE 49–1. SOME COMMERCIALLY VALUABLE PRODUCTS PRODUCED BY MICROORGANISMS *(Continued)*

Commercially Valuable Products	Microorganisms	Raw Materials	Nature of the Process
Vitamin B$_2$	*Ashbya gossiypi*	Corn steep liquor	Aerobic
Vitamin B$_{12}$	*Streptomyces olivaceus*	Malt extract, corn steep liquor + Co^{2+}	Aerobic
Gibberellin	*Fusarium moniliforme* or *Gibberella fujikuroi*	Sugar-containing substrate	Aerobic
11-α-hydroxy-progesterone	*Rhizopus nigricans*	Corn steep liquor and progesterone	Aerobic hydroxylation
Corticosterone	*Curvularia lunata*	Corn steep liquor and progesterone	Aerobic dihydroxylation
FOODS (ADDITIVES AND SUBSTITUTES), BEVERAGES, AND SPIRITS:			
Diacetyl	*Leuconostoc citrovorum*	Dairy wastes with citrate	Fermentation
Dextran	*Leuconostoc mesenteroides*	Wastes, carbohydrate mashes	Fermentation (used as food stabilizer as well as plasma substitute)
Beer	*Saccharomyces cerevisiae* (special strain)	Malt extract, mashes of grains	Fermentation
Wine	*Saccharomyces cerevisiae* (special strain)	Grape and other juices	Fermentation
Spirits	*Saccharomyces cerevisiae* (special strain)	Various mashes of grains, potatoes, molasses	Fermentation
Bakers' yeast	*Saccharomyces cerevisiae* (special strain)	High sugar substrate, molasses	Aerobic
Mushrooms	*Agricus bisporus*	Bed of soil + horse manure or other decaying material	Aerobic, cool temperature

industry in the United States, where other foods are plentiful. However, in view of enormously increasing populations, the possibilities are being seriously considered and studied experi- mentally. Cultivation of certain eucaryotic algae in waste materials as a source of oxygen and food in space and prolonged submarine travel is also under experimentation.

TABLE 49–2. SOME INDUSTRIAL ENZYMES; THEIR SOURCES AND APPLICATIONS

Types of Enzymes	Source Microorganisms	Typical Industrial Applications or Products
Amylases (starch hydrolysis)	Malt, *Aspergillus* sp., *Bacillus* sp.	Bread; beer; whiskey; textile fibers; preparation of glucose syrups
Proteases (protein hydrolysis)	*Aspergillus* sp., *Bacillus* sp.	Clarifying (chillproofing) beer; whiskey making, leather, baking bread and crackers, meat tenderizers
Pectinases (hydrolysis of pectin)	*Aspergillus* sp.	Clarification of fruit juices
Glucose oxidase	*Penicillium notatum*	(1) Removal of glucose from eggs to be dried, to prevent fermentation; (2) Removal of oxygen from canned fruits, dried milk, and other products subject to oxygen spoilage
Invertase	*Saccharomyces cerevisiae*	Preparation of soft-center candies (cordial cherries)

Microorganisms are sometimes used for **special industrial purposes,** such as removal of certain sulfur compounds from petroleum, and for vitamin assay. These have highly specialized applications.

The **uncontrolled actions** of fortuitous **mixtures of microorganisms** in processes such as retting of flax, preparing hides for leather, and coffee-bean hulling are time-honored practices. They are now being replaced in more advanced industries by processes in which purified cultures or enzymes of the **effective** species in the mixtures are used under carefully controlled conditions. Since the older processes are not scientifically designed, they are not discussed further here.

49.2
DEVELOPING AN INDUSTRIAL PROCESS

In developing an industrial process based upon the action of microorganisms, many details must be given consideration. Important among these are the type of culture needed, the cultural conditions, and the adaptability of the organism to large-scale production.

Purity and Nature of Cultures. It must be ascertained whether absolutely pure cultures must be used, or whether the mere predominance of one organism is sufficient. This may be a deciding factor, as the cost of preparing and maintaining pure cultures and sterile apparatus throughout a process is relatively high.

Cultural Conditions. The organism must be able to grow well in the medium to be used and under the conditions of the process. This necessitates exact studies and careful control of **optimum** conditions of aerobiosis or anaerobiosis, temperature, nutrition, and pH. Appropriate adjustments of the process and apparatus must be made to provide the necessary conditions.

Productive Mutants. The organisms selected must be such as will produce the desired substance(s) or results in the medium under the conditions furnished, in amounts sufficient to yield a profit. Some firms have "pet" strains of microorganisms that so excel in producing certain products, such as butyl alcohol, certain antibiotics, or itaconic acid, that they have "developed" (selected mutants) for these purposes. It has been found possible to induce industrially valuable mutations in microorganisms by ultraviolet radiations (Chapt. 18). Where sexual processes are known to occur, the breeding of yeasts and molds for similar purposes is analogous to breeding of farm animals for special purposes.

Medium or Raw Material. The substrate or medium should support luxuriant growth of the organism to be used, and it must be available constantly at costs compatible with profit. Expensive handling machinery may be needed for some substrates.

An important item is the possible necessity of a preliminary treatment such as liming of very acid yeast slops, distillery wastes, molasses, and whey. Some substrates, such as sawdust or fiber, may need preliminary "digestion" with hot acid or alkali to hydrolyze them to fermentable substances. This all adds to the expense and time. Some of the more common substrates (raw materials) used in various industrial processes are included in Table 49–1.

Nature of the Process. The more complicated and exacting the system of cultural details and preliminary heatings, dilutings, and digestions, as well as the type of machinery (cracking stills, tanks, pumps) to handle the end and by-products and the final wastes, the greater will be the cost and therefore the less the commercial practicability of any process. Any time-consuming aging or ripening process eats into profits. Sometimes, very desirable end or by-products may be found in commercial fermentations, yet the cost of their recovery may be prohibitive.

Preliminary Experimentation. The microbiologist working with 10-ml test tube cultures may find many valuable things. When attempts are made to reproduce the test tube experiments on a 100,000-gallon factory scale, however, the laboratory discoveries often fail to yield the promised result. Any process developed in the experimental laboratory must next prove its worth in the factory. A small-sized model, or pilot plant, is usually tried after the preliminary laboratory work. All may depend on such a seemingly far-removed detail as international relations. These may affect the cost of importation of some raw product essential to the process under investigation. Then the industrialist turns to home resources, goes to Washington, or *employs a resourceful microbiologist!*

The whole matter is a complex of microbiology, chemistry, engineering, and economics. Only the microbiology can be discussed here. Many chemical and microbiological processes in use at present are patented and secret, and specific strains of bacteria, yeasts, and molds, which are zealously guarded, are often carefully developed in the laboratories of manufacturing concerns. As a result of continuous and intensive research by industrial microbiologists, methods change or are superseded frequently.

49.3
TYPES OF FERMENTATION PROCESSES

Industrial fermentation processes may be divided into two main types: (1) batch fermentation and (2) continuous process. There are various combinations and modifications of these.

Batch Fermentations. A tank or fermenter is filled with the prepared **mash** (material to be fermented, e.g., diluted molasses, comminuted potatoes, "digested" corn cobs). The proper adjustments of pH, temperature, nutritive supplements, and so on are made. In a pure culture process, the mash is steam-sterilized, the entire fermentation tank sometimes being the autoclave. The inoculum, a **pure culture,** is added from a separate pure culture apparatus. The fermentation proceeds. Some pressure may be maintained within the tank to prevent inward leakage of contamination and sometimes to maintain increased tension of special gases. After the proper time, the contents of the fer-

menter are drawn off for further processing, the fermenter is cleaned, and the process begins over again. Each fermentation is a discontinuous process divided into **batches** (Fig. 49–1).

The Continuous-Growth Process. In continuous-growth processes, the substrate is fed into a container continuously at a fixed rate. The cells grow (or enzymes act) continuously as the material passes through the apparatus. The organisms and process are said to be in a **steady state** or condition of **homeostasis.** The mechanical device which maintains steady exponential growth phase is commonly called a **chemostat** (Figs. 49–2, 49–3). The product or fully fermented mash is drawn off continuously. The engineering arrangements may be complex, permitting aeration, cooling or heating, adjustment of pH, or addition of nutrients continuously during the process. There must also be means of controlling rate of growth, phase of growth curve, and removal of dead organisms. The culture must remain pure and must not undergo any variation.

One may conceive of such a process as taking place in a long pipe (actually it may be a rotating conical tank or series of connected tanks). At one end the prepared mash enters. It at once encounters the growing organisms.

Preparation of inocula

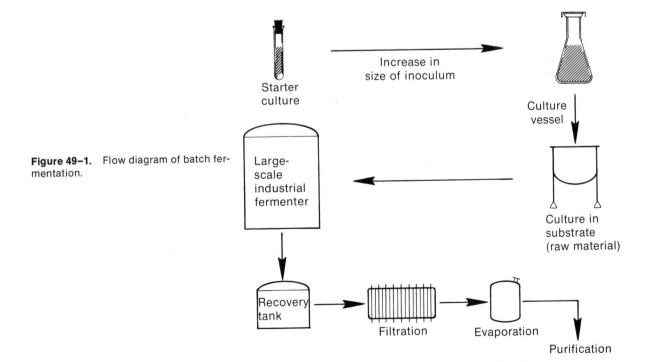

Figure 49–1. Flow diagram of batch fermentation.

Starter culture

Increase in size of inoculum

Culture vessel

Culture in substrate (raw material)

Large-scale industrial fermenter

Recovery tank

Filtration

Evaporation

Purification

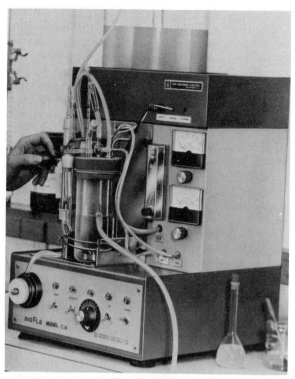

Figure 49–2. Bio Flo—the bench-top chemostat for continuous culture of microorganisms. (Courtesy of New Brunswick Scientific Co., Inc., New Brunswick, N.J.)

vessels for further processing (e.g., distillation). The animal alimentary tract may be thought of as a natural, continuous-growth process.

Submerged Aerobic Cultures. Many industrial processes, casually called "fermentations," are carried on by strictly aerobic microorganisms: for example, production of penicillin by *Penicillium notatum*, a strictly aerobic mold. In older aerobic processes it was necessary to furnish large surfaces of culture media exposed to air. The limitations of space, difficulties from contamination, and expense of hand labor can well be imagined, though little expense for power equipment was necessary. Now it is common commercial practice to carry on such "fermentations" in closed tanks with **submerged cultures.** Aerobic conditions are maintained by constant agitation of the contents of the tank with an **impeller** and constant **aeration** by forcing sterilized air through a porous **diffuser.** The flow of air through the tank removes gases such as ammonia and carbon dioxide. In each sort of process very careful adjustments of O-R potentials, mechanical agitation, ratio of dissolved oxygen to other ingredients in the medium, and pH are necessary. This is one of the many fascinating and potentially very lucrative fields for research in industrial microbiology.

These act on the substrate as it flows through the system. At the stage at which the valuable product of the fermentation is at its maximum concentration, the fluid is drawn into receiving

49.4
ANTIBIOTICS

Industrial production of antibiotics is presented in Chapter 23.

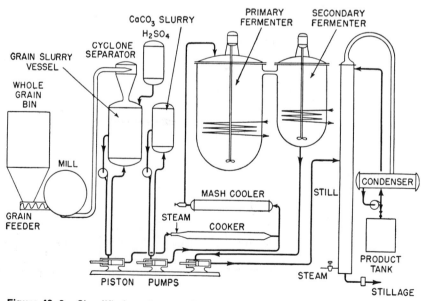

Figure 49–3. Simplified continuous alcoholic fermentation process flow diagram.

49.5
INDUSTRIAL ETHYL ALCOHOL MANUFACTURE

Much industrial ethyl alcohol is now made from by-products of **cracking** petroleum to make gasoline. However, the manufacture of ethyl alcohol from fermentation by yeasts is still an important industry. It serves to illustrate industrial fermentation processes in general. Crude molasses is often used as mash. It generally requires only to be diluted and the pH adjusted (usually with sulfuric acid) to 4.5. This pH is favorable to the yeast and unfavorable to many bacteria. A source of nitrogen such as ammonium sulfate or ammonium phosphate is usually added. The final solution is a richly nutrient carbohydrate culture medium or **mash.**

This is rather heavily inoculated with an aciduric and alcohol-resistant strain of yeast, the variety depending on the conditions under which the fermentation is to proceed and the exact end products desired. A good strain of *Saccharomyces cerevisiae* is commonly used. The inoculum comes from a large tank of carefully maintained pure culture, previously inoculated from a smaller seed tank, and the latter from a flask or tube of culture in the laboratory. At present stainless steel continuous-culture apparatus is available for maintaining constantly large amounts (many gallons) of **pure** cultures of inoculum.

The maintenance of purity of the inoculum is a responsibility of the microbiologist, and woe betide him if some sporeformer, *Lactobacillus*, wild yeast, or bacteriophage gets in and ruins 100,000 gallons of mash! The mash and all of the machinery are generally sterilized before the inoculation and then cooled. The microbiologist is kept busy at every stage of the process, making cultural and microscopic examinations of the water, mash, and apparatus to detect and eliminate contamination.

In the batch process, much used for this purpose, fermentation in enormous tanks (Fig. 49–4) is allowed to continue for about 48 hours at a carefully controlled temperature of about 25C until the yeast stops growing because of the concentration of alcohol and other products. Aeration with filtered air is used at first to promote rapid growth, but anaerobiosis is soon established to promote fermentation and alcohol accumulation, and to prevent its oxidation to carbon dioxide and water.

The fermented mash contains the crude alcohol or **high wine,** as it is called. This is usually a mixture of ethyl alcohol and a small amount of glycerol with **fusel oil.** The last contains amyl, isoamyl, propyl, butyl, and other alcohols with acetic, butyric, and other acids, as well as various esters. The high wine is driven off from the mash or **beer** by heat, and further purified by fractional distillation, which is a problem in chemical engineering.

The chemical reactions involved in the fermentation are complex; the principal stages follow the Embden-Meyerhof scheme. The overall reaction in the production of alcohol from glucose is:

$$C_6H_{12}O_6 \xrightarrow{\text{yeast}} 2C_2H_5OH + 2CO_2$$

The chief constituents of fusel oil are probably

Figure 49–4. Lower level of 50,000-gallon fermentation tanks. (Courtesy of Commercial Solvents Corp., Terre Haute, Ind.)

derived from the action of the yeasts on amino acids in the mash. The large amounts of carbon dioxide evolved are purified and compressed in tanks or made into solid carbon dioxide. Part of this may be used for cooling the fermentation vats.

49.6
ALCOHOLIC BEVERAGE INDUSTRIES

Whiskey

In principle, the production of alcoholic distilled beverages is similar to the production of industrial ethyl alcohol. Refinements are introduced in beverage production with respect to flavor, aroma, color, and sanitation that are not necessary in the making of industrial alcohol.

There are four general types of distilled liquor: brandy, from fermented fruit juices; rum, from fermented molasses; neutral spirits, from fermented mash of mixed grains; and whiskey, from fermented mashes made with single or predominant (51 per cent) types of grains: rye, straight or rye whiskey; corn, bourbon whiskey. Scotch whiskey is traditionally made from malted barley. To make whiskey and neutral spirits, the grain, mixed with water, is autoclaved, cooled, and diluted, and 1 per cent barley **malt** (aqueous extract of sprouted barley; see next section on beer) is added to hydrolyze the starch and proteins of the grain. The "mashing," or hydrolysis, proceeds in a special tank at about 65C for about 30 minutes. The mash is then pumped to the fermentation tanks. Here, as in beer-making, it is heavily inoculated with a starter of selected yeast, which has been cultivated in a mash previously made somewhat acid (pH 4.0) with lactobacilli. Fermentation is complete in about 72 hours, as in industrial alcohol production. The mash is then removed to the distillery, and the ethanol, with various by-products, is recovered.

Beer

This time-honored and popular beverage is one of the class of malt liquors: stout, porter, ale, and others. To prepare beer, grain, usually barley, is kept moist for two or three days to induce sprouting, or **malting**. Amylase enzymes that are released in the malt grains during the sprouting process hydrolyze the starches of the grains to simple sugars, mainly maltose and dextrins. Malting (Ger. *malz* = to soften) is necessary, since brewer's yeast does not produce amylase and therefore cannot directly attack the starch of the grain. At the same time, proteases in the malt grains convert proteins in the grains and flour to soluble nitrogenous foods.

The sprouts are removed mechanically, and the **malt grains** are dried. They are later crushed and soaked, or **mashed**, in warm water. The aqueous extract of these malt grains and flour, prepared at just the time when there are maximum amounts of maltose, dextrins, and protein derivatives, constitutes a rich nutrient medium. It is called **beer wort**. The beer wort is now drained off and heated to kill contaminating microorganisms. Hops are added for additional antibacterial (stabilizing) effect, color, flavor, and aroma. (The hop vine, *Humulus lupulus*, is cultivated for the papery scales of the female flower, which are dried and powdered for use.)

After cooling, a large inoculum of pure culture of *Saccharomyces cerevisiae* (brewers' yeast or **barm)** is added to the wort. This is called **pitching**. Rival brewers maintain very special strains of yeast for this process. Some are "top growers," some "bottom growers." The inoculated wort is aerated at first to stimulate **rapid growth** of yeast; anaerobic conditions prevail later on to favor **fermentation**, when carbon dioxide and 3 to 6 per cent ethanol are produced. After fermentation is complete, the beer is clarified (**chill-proofed**) (see Table 49–2), and pasteurized and otherwise processed and aged (**lagered** [Ger. *lager* = to be stored, i.e., to age]). (See Figure 49–5.)

Unless scrupulous care is taken, many contaminants (*Pediococcus, Lactobacillus*) will grow vigorously in beer wort, producing buttery flavor, turbidity, and off flavors. It was in the study of such spoilages, or "diseases," of beer and wines that Pasteur first became famous and developed **pasteurization** to prevent them. He was one of the first industrial microbiologists.

Wine

The term wine is broadly used to include any properly fermented juice of ripe fruits (traditionally grapes), or extracts of certain vegetable products, such as dandelions and palm shoots. The juices or extracts contain glucose and fructose in concentrations of from 12 to 30 per cent. Fermentation of these sugars by various species of yeasts produces carbon dioxide and ethanol up to concentrations of 7 to 15 per

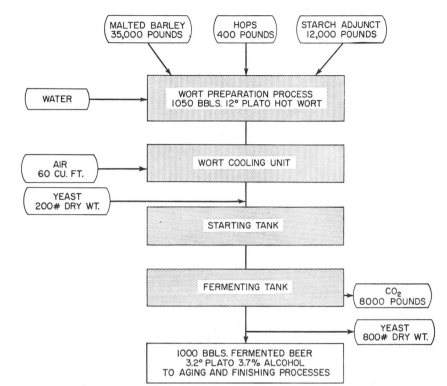

Figure 49–5. Flow diagram for the fermentation of 1,000 barrels of beer.

cent, the alcoholic content depending on the kind of juice and yeast involved and the conditions of fermentation. In Europe the fermentation is produced mainly by **wild** yeasts, i.e., those brought to the fruits (largely by insects) from soil or other fruits. Yeasts similar to the species called *Saccharomyces ellipsoideus* are common in such wines.

Although other organisms are usually present, the yeasts soon predominate in the fermenting juice under suitable conditions. Tartaric, malic, and other acids, as well as tannin and other substances, including added sulfur dioxide in commercial wine, tend to inhibit growth of many undesirable organisms in the juices.

Even though practices may differ in different wineries, basically they are similar. Commonly in modern American commercial practice, sterilized fruit juices are inoculated in vats with a pure culture of a desirable species of yeast. The preparation and maintenance of the yeast inocula are the special tasks of the microbiologist.

The inoculated juice is, as in beer-making, at first aerated to promote active and pre-emptive growth of yeast. Were this to continue, only carbon dioxide and water and massive growth

of yeast cells would result. As soon as a good growth of yeast has occurred, the aeration is stopped and the fermentation proceeds anaerobically, so that ethanol, in concentrations of from 7 to 15 per cent (by volume) is produced. The new wine is placed in large casks to settle, clarify, and age.

After alcohol production begins, spoilage by alcohol-oxidizing species of *Acetobacter*, molds, and other **aerobic** microorganisms may occur if conditions are not anaerobic and the reaction not sufficiently acid.

49.7
OTHER INDUSTRIAL PROCESSES

Production of Butanol

The production of butanol is outlined as an example of an industrial fermentation based on a species of bacterium.

As is true of industrial ethanol production, much butanol is now derived as a by-product of petroleum "cracking." However, the biological process is still used to some extent and illustrates important microbiological principles.

There are numerous species of *Clostridium* that ferment carbohydrates, with the production of butyl alcohol and other materials of value in drugs, paints, synthetic rubber, explosives, and plastics. Some species produce isopropyl alcohol and acetone as well. Important among these organisms are *C. acetobutylicum* and *C. felsineum*. The name of another species suggests its potentialities as an industrial agent: *Clostridium amylosaccharobutylpropylicum!*

Many wastes are rich in fermentable carbohydrates, e.g., fruit cannery refuse. Complete sterilization of all apparatus is essential. Conditions cannot be kept as acid (and antibacterial) as they are in yeast fermentations because *Clostridium* has its optimum pH near 7.2. Particularly troublesome contaminants are species of *Lactobacillus*. An organism once incorrectly called *B. volutans*, a gram-positive, nonspore-forming rod (possibly a species of *Lactobacillus?*), is also especially dangerous.

Fermentation proceeds anaerobically for about three days. Normally butyl alcohol, acetone, and ethyl alcohol, with carbon dioxide and hydrogen in large amounts, predominate when *C. acetobutylicum* acts in a mash rich in glucose. Other substances may occur in smaller amounts. Riboflavin (a vitamin of the B complex) is a valuable constituent of the residue after distillation of the fermented mash. Butyl and isopropyl (rubbing) alcohols are important among the volatile fermentation products of a related species, *C. butylicum*. The successive reactions in the production of butanol and various side-products from glucose are given in Figure 49–6.

Production of Gluconic Acid

Gluconic acid and its derivatives may be used in pharmaceutical, food, feed, textile, and other industries. The products are manufactured by conversion of a glucose solution containing nutrient salts and sodium carbonate, using a strain of *Aspergillus niger* or species of *Penicillium*. The industrial process is aerobic, and glucose may be converted by *A. niger* either by surface growth in shallow pans or by the submerged-culture technique, which is preferred. The oxidation of glucose to gluconic acid requires adequate oxygen, as indicated by the reaction below:

$$C_6H_{12}O_6 + \frac{1}{2}O_2 \longrightarrow C_6H_{12}O_7$$

glucose   gluconic acid

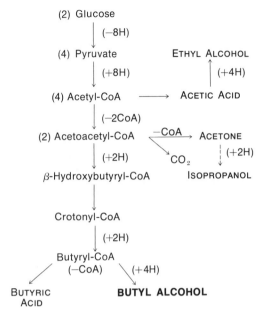

Figure 49–6. Reactions in production (by *Clostridium* spp.) of butanol and side-products from glucose.

It should be noted that gluconic acid can be chemically synthesized by oxidation of glucose with hypochlorite (hypochlorous acid, which releases a powerful oxidizing agent, singlet oxygen: $HOCl \longrightarrow \frac{1}{2}O_2 + HCl$) or by the electrolysis of sugar in the presence of bromine.

Production of Lactic Acid

Lactic acid is commonly produced from acid-hydrolyzed corn and potato starches, molasses, and whey or from lactose or glucose by species of homofermentative lactic acid bacteria, notably *Lactobacillus delbrueckii* or *Lactobacillus bulgaricus*.

$$\underset{\text{lactose}}{C_{12}H_{22}O_{11}} + H_2O \longrightarrow \underset{\substack{\text{glucose} \\ \text{and} \\ \text{galactose}}}{2C_6H_{12}O_6} \longrightarrow$$

$$\underset{\text{pyruvate}}{4C_3H_4O_3} \xrightarrow{(+4H_2)} \underset{\text{lactic acid}}{4CH_3 \cdot CHOH \cdot COOH}$$

Whey, the watery part of milk separated from the curd during cheese-making, has been used extensively for manufacture of lactic acid, as well as for protein and carbohydrate supplement for animal feed. Whey contains a relatively large amount of lactose (Chapt. 47) and proteinaceous substances, minerals, and some essential vitamins. Therefore, it is considered

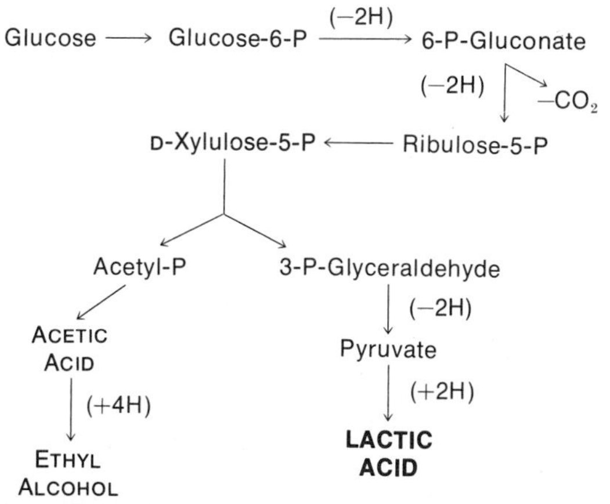

Glucose $\longrightarrow$ Glucose-6-P $\xrightarrow{\text{(−2H)}}$ 6-P-Gluconate

$\text{(−2H)} \searrow -CO_2$

D-Xylulose-5-P $\longleftarrow$ Ribulose-5-P

Acetyl-P   3-P-Glyceraldehyde

$\downarrow$ (−2H)

ACETIC
ACID   Pyruvate

$\downarrow$ (+4H) $\downarrow$ (+2H)

ETHYL   **LACTIC
ALCOHOL   ACID**

Figure 49–7. Reactions in production (by *Lactobacillus* spp.) of lactic acid and by-products from glucose.

an excellent medium for growth of many bacteria. The homofermentative lactics, such as *L. bulgaricus*, are able to grow and ferment lactose to lactic acid. Heterofermentative species of lactobacilli like *L. buchneri* produce lactic acid and several by-products (Fig. 49–7).

The fermentations in Figure 49–7 are carried out at a temperature of 43C (110F) to discourage the growth of undesirable organisms. Calcium lime, $Ca(OH)_2$, is usually added to neutralize the acid and to promote a good yield of calcium lactate. Calcium lactate may be sold as food supplement for treatment of calcium deficiency or converted to lactic acid and its derivatives (sodium lactate, *n*-butyl lactate) for various industries. Lactic acid fermentation from whey not only produces industrially important products but also aids in abatement of pollution of our environment. Millions of gallons of whey are produced annually, and if untreated whey is disposed of in our waterways, it results in disastrous consequences, because whey is very high in BOD (800 to 1,500 mg/l; also see Chapters 46 and 47). Thus, in the future more attention should be directed to development of valuable products from otherwise useless waste products just to maintain a clean environment.

Production of Citric Acid

Many industrially important acids are produced by molds, and citric acid production by *Aspergillus niger* or *A. wentii* is a good example. Citric acid ($COOH \cdot CH_2 \cdot C(OH) \cdot COOH \cdot CH_2 \cdot COOH$) is now produced by submerged culture method, and it has made the United States self-sufficient for production of citric acid. Various carbohydrate-rich raw materials—cornstarch, ground corn, beets, blackstrap molasses, and commercial glucose—are aerobically converted to citric acid. Good yields of citric acid from various carbohydrates are influenced by various factors. For example, metallic ions such as Fe^{3+}, Zn^{2+}, and Mn^{2+} would give a low yield of citric acid, but addition of methyl, ethyl, and isopropyl alcohols in concentrations of up to 2 per cent of the raw materials may increase the tolerance for metal ions and increase the production of citric acid. The overall reaction involving production of citric acid by *Aspergillus* spp. is given in Figure 49–8.

Production of Vinegar

Acetic acid is almost entirely responsible for the sour taste of vinegar. Indeed, a slightly

Figure 49–8. Reaction in production (by *Aspergillus* spp.) of citric acid from glucose.

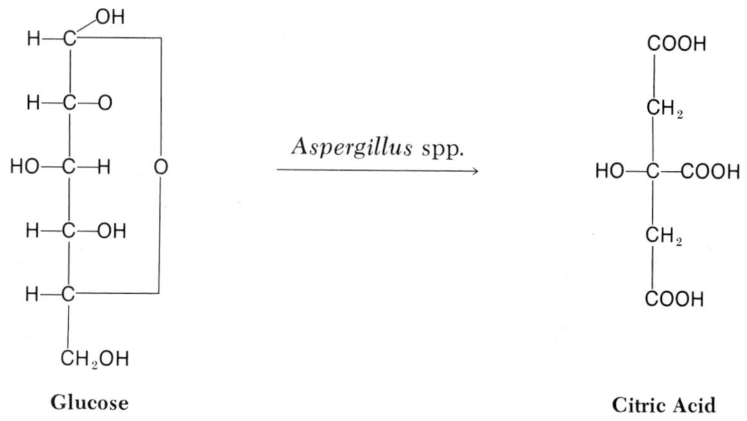

$Aspergillus$ spp.

Glucose

Citric Acid

sweetened, 3 per cent, aqueous solution of acetic acid makes a reasonable substitute for vinegar.

The occurrence of vinegar in fermented fruit juices was known to the ancients, although they had no knowledge of its cause. The bacteria involved were called *Mycoderma aceti* in 1862 by Pasteur.

The acid of natural vinegar is derived from alcohol by the oxidative action of bacteria of the family Pseudomonadaceae (genus *Acetobacter*). However, in the 8th edition of *Bergey's Manual* the organism is placed under "Genera of Uncertain Affiliation" in Part VII, "Gram-Negative, Strictly Aerobic Rods." Pleasant flavors of natural vinegar are given by traces of various esters like ethyl acetate, and by alcohol, sugars, glycerin, and volatile oils produced in small amounts by microbial action. Flavors are also derived from the fermented fruit juice, malt, or other alcoholic liquor (wine, beer, hard cider) from which the vinegar was made.

In commercial vinegar-making by biological methods, preliminary fermentation of fruit juices to produce the necessary alcohol is often carried out by means of *Saccharomyces cerevisiae* (brewers' yeast). The *Acetobacter* then utilize the alcohol as a source of energy, oxidizing it to acetic acid in the presence of air. They utilize other substances in the fermented liquor as foods.

The alcoholic fluid is aerated as much as possible by various devices. In vinegar "generators" the alcoholic liquor trickles over the surface of aerated shavings, coke, gravel, or other finely divided material (Fig. 49–9) inoculated with *Acetobacter*. Such an arrangement is called a **two-phase continuous process:** one phase is the down-trickling alcoholic liquor; the other phase is the column of coke or other material covered with growth of *Acetobacter* (see also Aerating Filters, Chapter 45).

Genus *Acetobacter*. These are nonspore-forming, polar or peritrichous flagellate, gram-negative rods about 0.5 by 8.0 μm, although species vary in size. Branching involution forms and large swollen cells frequently occur, especially in **mother-of-vinegar,** the gummy or slimy growth-phase of the organisms sometimes seen in natural vinegar or sour cider (Fig. 49–10). Various species are found in souring fruits and vegetables. A species of historical interest is *Acetobacter (Mycoderma) aceti*, originally used by Pasteur to demonstrate the biological nature of vinegar formation. In practice, several species of *Acetobacter* usually

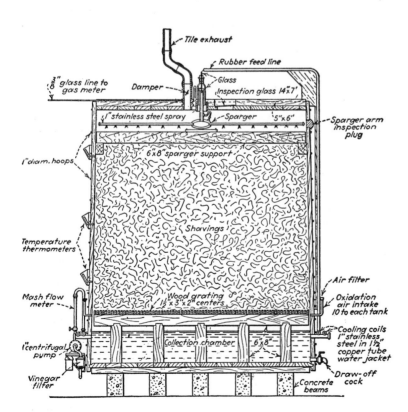

Figure 49–9. Cross section of the Frings vinegar generator. The alcoholic liquor is sprayed over the shavings by the rotating stainless steel spray (sparger) near the top. Note the thermometers, cooling coils, and air intakes.

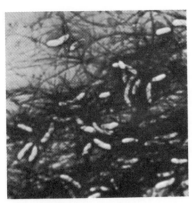

Figure 49–10. Extracellular formation of cellulose by *Acetobacter xylinum*. (1,213×.) The bacterial cells are entangled in a mesh of cellulose fibrils.

act jointly. The alcoholic and acidic nature of the process suppresses most contaminants. The overall reactions probably are as follows:

$$C_2H_5OH + {}^{1}\!/_2 O_2 \longrightarrow CH_3CHO + H_2O$$
alcohol   acetaldehyde

$$CH_3CHO + {}^{1}\!/_2 O_2 \longrightarrow CH_3COOH$$
acetic acid

In a generator such as that shown in Figure 49–9, the rapid oxidation of alcohol by the organisms produces so much heat that careful control of the internal temperature by cooling coils is necessary.

49.8 FOODS FROM WASTES

Whey from the dairy industry is utilized for the production of lactic acid, or whey may be dried and used as feed supplement for domestic animals. Similarly, feed supplement in the form of torula yeast is produced from pulp and paper wastes. For example, wood chips are cooked for 6 to 18 hours at 60C in solutions of calcium bisulfite with free sulfur dioxide. The waste **sulfite liquor,** after the cooking process and removal of the wood fibers for paper, contains much valuable wood sugar (largely **xylose**) and other extractives. These form a good nutrient for asporogenous yeast or *Torula*.

The nutrients in such a medium may be turned into masses of the yeast by adjustment of pH to about 5.0, removal of SO_2, aeration, addition of nitrogen and phosphorus as $(NH_4)_2HPO_4$ and NH_4OH, and inoculation with *Torulopsis utilis*. **Aerobic** growth is induced in aerated vats so that alcohol is not produced. The separation, drying and pressing of the resulting yeast growth are mechanical details. Yields of up to 50 per cent of the total reducing sugar consumed, in terms of dry torula, are obtainable. (At present, attempts are being made to cultivate yeasts on petroleum wastes.) The yeast cells are rich in proteins, fats, and vitamins. These are fed to stock or poultry and thus turned into meat and dairy products. Surely the transformation of a knotty old pine slab into a succulent pork chop or a fried egg is modern magic!

Other by-products from spent sulfite liquor include vanillin, food flavorings, and various organic acids. Manufacture of even more by-products from these liquors will require new market development first. Another industrial process for use of waste products is the potential commercial-scale recovery of vitamin B_{12} from domestic activated sludge. It has been found that this sludge may contain B_{12} concentrations ranging from 1 to 10 mg/kg dry weight of sludge.

Amino Acid Production. Not only may yeasts serve as foods themselves, but some species, especially *Saccharomyces cerevisiae* and *Torula utilis*, while growing can synthesize large amounts of various amino acids. With decreases in availability of nitrogen-rich foods like meats and fish, means of producing nitrogenous nutrients like amino acids (components of proteins) assume increasing importance. L-Lysine production by microbial action is a good example. This is an expensive amino acid widely used to "fortify" many familiar foodstuffs. It is requisite that the growth medium for the yeast contain L-adipic acid or its derivatives. The yeasts use the adipic acid derivatives as **precursors** (i.e., as the molecular raw material) for their synthesis of L-lysine. Commercial lysine synthesis by bacteria involves a two-step process utilizing two different species of bacteria. The first step is carried out by special strains of *E. coli* in which raw materials — glycerol, corn steep liquor, and $(NH_4)_2HPO_4$ — are converted to diaminopimelic acid. This acid is converted to lysine by an enzyme, diaminopimelic acid decarboxylase, produced by a strain of *Enterobacter (Aerobacter) aerogenes*, which is particularly useful because the organism lacks lysine decarboxylase (Fig. 49–11).

Another amino acid, glutamic acid, has been commercially produced by bacteria and molds. Species of *Micrococcus, Arthrobacter,* and others are capable of converting α-keto-

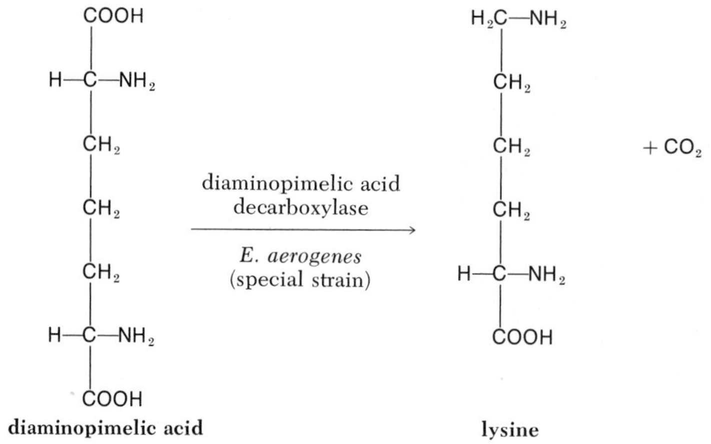

COOH H_2C—NH_2

H—C—NH_2 CH_2

CH_2 CH_2 + CO_2

CH_2 CH_2

    diaminopimelic acid

    decarboxylase

CH_2

    E. aerogenes

    (special strain)   H—C—NH_2

H—C—NH_2 COOH

COOH

diaminopimelic acid   lysine

Figure 49–11. Conversion of diaminopimelic acid to lysine by a strain of *Enterobacter aerogenes.*

glutaric acid to glutamic acid (in a simple nutrient mixture containing biotin) with an enzyme, glutamic dehydrogenase. Millions of pounds are produced annually and sold primarily as a flavor-enhancing agent in the form of monosodium glutamate.

The production of glutamic acid is a two-stage process: α-ketoglutaric acid (precursor of glutamic acid) is produced by a strain of *Micrococcus glutamicus* and species of *Arthrobacter* and *Brevibacterium* and others via the Krebs cycle. The raw materials include glucose (also molasses and occasionally acetic and succinic acids as carbon sources), organic nitrogenous material (soybean cake, fish meal), or inorganic nitrogenous compounds—NH_4NO_3, $(NH_4)_2SO_4$, $(NH_4)_2HPO_4$—for conversion of α-ketoglutaric acid to glutamic acid. The reaction is accomplished by glutamic dehydrogenase in the presence of reduced nicotinamide-adenine dinucleotide ($NADH_2$) and am-

monia (Fig. 49–12), or by the transamination reaction in which other amino acids act as donors of an amino group to α-ketoglutaric acid, and the reaction is catalyzed by an enzyme, transaminase, with the aid of a coenzyme (Fig. 49–13).

It is apparent that many commercial processes can be developed for other amino acids, especially methionine, threonine, and tryptophan, all of which are important dietary amino acids. It is well known that microorganisms are capable of synthesizing amino acids in excess of the amount they normally need if certain physical and nutritional conditions are met.

Hydrocarbons for Protein. Various hydrocarbon (petroleum) wastes are metabolizable by certain yeasts, eucaryotic fungi, and also some bacteria, e.g., *Bacillus* spp., especially thermophilic species. One difficulty (or possibly a source of profit?) is the production of large amounts of heat by biooxidation of hydrocar-

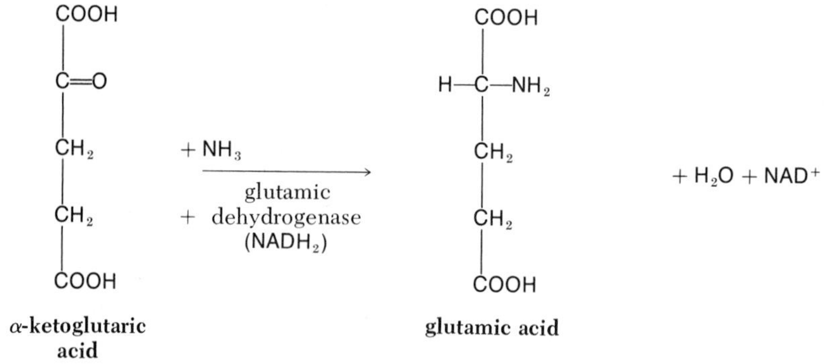

COOH   COOH

C=O   H—C—NH_2

CH_2 + NH_3 CH_2

    glutamic   + H_2O + NAD^+

CH_2 + dehydrogenase CH_2

    ($NADH_2$)

COOH   COOH

α-ketoglutaric   glutamic acid

acid

Figure 49–12. Conversion of α-ketoglutaric acid to glutamic acid.

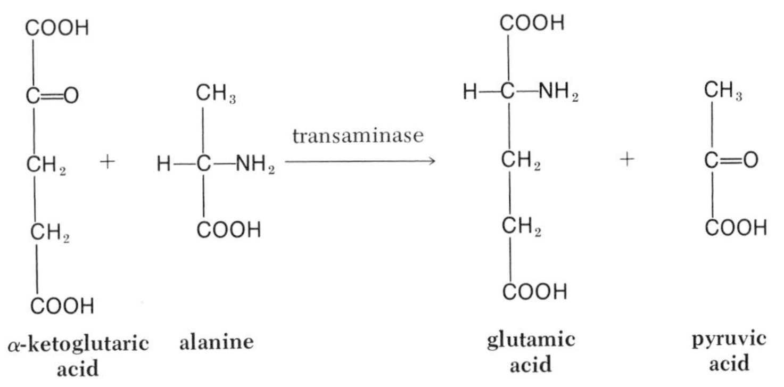

Figure 49–13. Transamination reaction for conversion of α-ketoglutaric acid to glutamic acid.

bons. The proteins produced in the process are of high nutritive value, and the method is potentially profitable. Pure cultures may be used, producing "single-cell protein," i.e., protein produced by a single species.

49.9
STEROID TRANSFORMATIONS (BIOCONVERSION)

An outstanding example of the use of microorganisms to change one substance into another is the transformation of the steroids. Steroids are physiologically active compounds of complex structure (hormones) (Fig. 49–14) and are represented by cholesterol, ergosterol or vitamin D, sex hormones such as testosterone and progesterone, and the adrenal steroids such as corticosterone and its derivative, cortisone. Microbial transformations of these compounds differ basically from industrial fermentations previously described. In industrial fermentation the alcohol or other product results from the action of numerous enzyme systems in the overall metabolism of a substrate, such as the sugar in molasses. The same product might be made with any of several different microorganisms that ferment saccharose or glucose. In steroid transformation one particular form of molecule is changed into another by the action of a single, specific enzyme. The requisite enzyme may be present in only a single species of microorganism. Many steroid derivatives thus obtained are of immense value in the treatment of various disease conditions or in the development of other hormones. On a commercial scale, many are at present available *only* through the action of certain specific microorganisms.

A single example will illustrate the type and importance of these transformations. Corticosterone, an important hormone from the cortex of the adrenal gland of mammals, was originally obtainable only from animals. It had a wide use in the treatment of shock and other prostrating conditions. A still more valuable derivative was made by chemical, and later by microbiological, transformation of the corticosterone molecule. One of these alterations was the introduction by microbial action of an —OH group into the 11 position in the corticosterone molecule (Fig. 49–14). The resulting compound is the now-familiar cortisone, widely used in treating arthritis and many other inflammatory conditions.

The sex hormones, testosterone, estradiol, and progesterone, are closely related in molecular structure to corticosterone and cortisone. They differ only in the nature and location of attached side groups, especially —OH and —CO · CH$_2$OH. These groups may be added or withdrawn or shifted about on a practical scale only through the action of certain specific microorganisms. Some of these transformations are indicated in Figure 49–14. The resulting compounds are often of much greater value than the natural hormone or steroid from which they are derived.

Similar transformations can be brought about in other kinds of molecules, e.g., various alkaloids. The microbiological chemist has an open field in the development of such processes.

49.10
ENZYMES OF MICROORGANISMS IN INDUSTRY

Knowledge that many of the essential chemical changes that occur in microbiological processes are entirely enzymic led to attempts to separate and concentrate the purified enzymes themselves on a commercial scale. The production of microbial enzymes for industrial use is now a considerable industry in itself. The enzymes are derived mainly from molds,

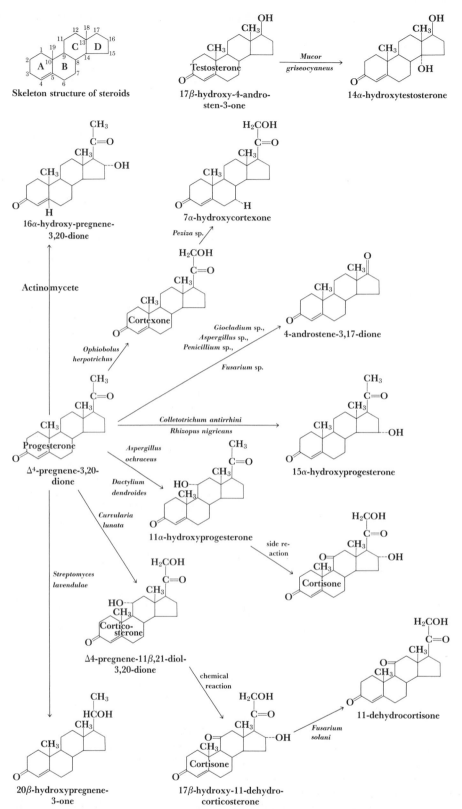

Figure 49–14. Some examples of steroid transformation by microorganisms. Note that the structural formulas of all are basically alike, differing only in the presence or absence, or molecular position, of OH, H, CH₃, O, etc. Organisms that effect the transformations shown are named. An example of stepwise transformation is given: from progesterone, with *Curvularia lunata,* to **corticosterone**; from **corticosterone** to **cortisone** by chemical alteration. The numbers 11, 15, etc., refer to positions in the molecule shown diagrammatically at the upper left of the figure. Note that several fungi, among them the familiar molds *Penicillium, Rhizopus,* and *Aspergillus,* as well as certain mold-like bacteria, *Streptomyces,* and an actinomycete, perform steroid transformations.

yeasts, and bacteria already familiar to us. A few of the more widely used organisms, their enzymes, and their uses are listed in Table 49–2.

Mold-Bran Process. For obtaining enzymes from *Aspergillus* and *Penicillium*, the **mold-bran** process is often used. To provide an extensive aerated surface, flaky or fibrous material, commonly wheat bran, is moistened with a nutrient medium of composition and pH appropriate to the mold being cultivated and the enzyme desired. The nutrient bran is sterilized, spread out in shallow trays and inoculated with the mold conidia. The trays are incubated in carefully air-conditioned cabinets. After sufficient growth the moldy bran is thoroughly extracted with water or other solvent to remove the enzyme. This fluid may be filtered, centrifuged, and concentrated, and the enzyme precipitated and dried for sale.

For bacterial enzymes the desired species of *Bacillus* is generally cultivated on the surface of broad, shallow layers of liquid medium. This is often prepared from inexpensive cannery or dairy refuse (e.g., whey) rich in organic matter. After incubation the bacteria are removed by filtration or centrifugation, and the enzyme is extracted and processed as indicated above.

The submerged culture process may be used in enzyme production by molds or bacteria, in much the same manner as in antibiotic production.

Gibberellin (Gibberellic Acid). This sensationally effective plant-growth stimulant, now available in every garden-supply house, was discovered by Kurosawa, in Japan in 1926. Analogous to penicillin, it is a waste product of a mold, *Fusarium moniliforme*, or *Gibberella fujikuroi*, and is a mixture of gibberellins. In nature the mold grows in young rice plants and causes the "overgrowth disease" **bakanae.** A pest in rice paddies, the mold and its **gibberellin** are welcomed by agriculturalists and gardeners. Gibberellin is produced on a commercial scale by submerged aerated growth in media and by methods similar to those used in producing penicillin (Chapt. 23).

49.11
MICROBIOLOGICAL ASSAY

Microbiological assay is a highly specialized application of the fact that certain organisms lack certain specific synthetic powers, i.e., are auxotrophs. *Lactobacillus plantarum*, for example, is unable to synthesize nicotinic acid ("niacin"). We may furnish the organism with a medium that is complete and satisfactory in all other respects, but if niacin is lacking, absolutely no growth occurs. (Humans are no better off; without niacin they die of pellagra.) If a minute amount (say, 0.01 μg) per milliliter of niacin is added to the medium for *L. plantarum*, some growth will occur. More growth will occur in the presence of more of the missing factor. Up to the point of satiation or acidification, growth bears a linear relationship to the amount of the specific growth factor added.

For example, to assay the nicotinic acid content of fresh green beans, we prepare a medium for *L. plantarum* that is complete in all respects except niacin. This we omit. We now prepare two series (A and B) of 10 sterile tubes each. Each tube receives 10 ml of the niacin-deficient medium. To each tube in series A we add known and graded amounts of pure niacin. To each tube in series B we add graded amounts of bean extract, niacin content unknown. All tubes are now inoculated with carefully washed (niacin-*free!*) cells of *Lactobacillus plantarum*. Accurate, photometric measurements are then made of the growths (turbidities) obtained in the cultures (Fig. 49–15). If the medium contains glucose, titrations of acidity instead of turbidity may be used as a measure of growth (Fig. 49–16). By comparison of growths in series A and B, it is possible to estimate closely the concentration of niacin in the green beans. This method of estimation of growth factors is spoken of as **microbiological assay.**

Although the basic principle of all microbiological assays is the same, there are other methods of measuring the growth (or other physiological) response. These affect the cultural methods used. A commonly used procedure is the measurement of carbon dioxide produced by fermentation of sugar in the test medium. Yeast is routinely used in the microbiological assay of **thiamine** (vitamin B_1) by this method. In **pyridoxine** (vitamin B_6) assays, the mold *Neurospora* is the test organism. After sufficient incubation, the culture is steamed, and the entire mycelium of *Neurospora* is removed from the culture medium, dried, and weighed. Dry weight is directly proportional to concentration of pyridoxine in the sample of material being assayed. Another assay procedure depends on the spheroplast-producing power of the assayed substance.

Certain organisms lend themselves very well to such assay procedures. *Lactobacillus casei* and *L. arabinosus* are easy to cultivate, relatively hardy, harmless, and wholly dependent upon several growth-factors, including

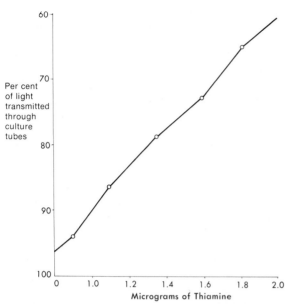

Figure 49–15. Representative curve obtained with a series of "standard" tubes for microbiological assay of thiamine with *Lactobacillus casei*. A similar form of curve would be obtained in any assay by this method, such as that described for niacin (*L. plantarum*) in the text. Slight deviations from the theoretical straight line are due to slight technical errors. Growth was measured photoelectrically: increasing growth (turbidity) reduced the amount of light transmitted through the culture tubes.

various amino acids, riboflavin, biotin, pantothenic acid and nicotinic acid. Other organisms may be used for assay of other substances, for example, *Streptococcus lactis* for folic acid. Ultraviolet-induced, synthetically deficient auxotrophs of molds, yeasts and bacteria are extremely valuable in assay work.

Even though the basic principle of microbiological assay is easily understood, the technological details are often exceedingly complex and filled with pitfalls. Many obscure factors affect the test organisms, and they may also undergo mutation and other changes without notice. Mutational and other injuries may be held to a minimum by storage of the stock cultures in containers with liquid nitrogen at −196C (−321F). Temperature, pH, and presence or absence of air may be of critical importance. For example, under **aerobic** conditions, *Lactobacillus lactis* will die before it will grow without vitamin B_{12}! **Anaerobically,** it sneers at vitamin B_{12}! There are many other examples. We may smile, but knowledge of this and many other peculiarities is essential to successful assay procedures.

49.12
INDUSTRIAL SPOILAGE

In contrast with the useful activities of bacteria, a word may be said of their destructive action. Several causes of industrial spoilage (e.g., "diseases" of fermentations) have been mentioned in this chapter and in the chapters on soil, food, and water bacteria. Species of *Micrococcus, Alcaligenes, Flavobacterium, Serratia, Clostridium,* coliform organisms, yeasts, and molds are common causes of spoilage.

Each type of product is attacked by certain species of microorganisms that can metabolize the substance especially well. For example, spoilage of cellulosic products such as lumber; telephone poles; paper; sisal, jute, and flax fibers; and tobacco and cotton is brought about by cellulose-decomposers like molds, various species of *Clostridium, Cellulomonas, Cytophaga,* and many other such organisms of the soil. Fermentable substances such as syrups and beverages are attacked by yeasts, lacto-

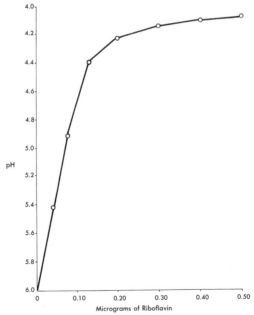

Figure 49–16. Representative curve obtained in the assay of a food substance for riboflavin content by measurement of pH. This is the "standard" curve used as the basis for measurement of the "unknowns." Note the linear relationship between growth of the test organism (*Lactobacillus arabinosus*) as measured by pH, and amount of riboflavin in the first part of the curve. In the later parts of the curve acidity inhibits unlimited response of the organism to the larger amounts of riboflavin.

bacilli, organisms of the coli-aerogenes group, and various environmental bacteria, including the genus *Clostridium*. Spoilage of proteins such as meats, fish, milk, and so on results from the action of proteolytic species such as *Pseudomonas, Bacillus, Proteus, Micrococcus, Clostridium*, and many others. Petroleum hydrocarbons are attacked by certain soil bacteria, as already mentioned, and rubber insulation of vital communication wires is attacked by certain bacteria and eucaryotic fungi.

Lactobacilli and *Leuconostoc* species have already been noted as particular villains in the acid-food, fermentation, and distillery industries. Species of both can ruin fruit or vegetable juice or various industrial mashes (beer and wine) during processing. They produce a buttermilk flavor. Pasteur found *Lactobacillus* and *Leuconostoc* causing "diseases" of beers and wines. They are just as active today. Lactobacilli also discolor meats, especially producing greenish discoloration (oxidized porphyrins) of cured hams and sausages.

The slimy dextran- or levan-forming species, such as *Leuconostoc mesenteroides* and *L. dextranicum* and some lactobacilli and micrococci, produce slimy and ropy conditions in a great variety of human endeavors: sugar refineries, pickle brines, dairy products, ham-curing cellars, and the like. These organisms prefer acidified products such as partly fermented foods, mashes, and citrus juice. Examples of several of these types of spoilage have been given in discussions of the various products.

Development of undesirable flavors in fatty products such as butter, especially rancidity, is due in great part to the formation of butyric acid as a result of lipolysis. It is caused by species of *Aspergillus* and other molds, *Pseudomonas* species, and streptococci related to *S. liquefaciens*. These difficulties do not arise when clean equipment, clean milk, and proper precautions to avoid contamination are used.

Proteolytic organisms such as *S. liquefa-*

ciens are responsible for undesirable bitter flavors and early spoilage of cheeses and other protein products. Gas production is usually caused by coliforms and *Clostridium;* putrefaction or digestion by *Clostridium, Pseudomonas*, and *Bacillus*. Such conditions result mainly from dirty milk or other food equipment, or careless handling.

Included among other sabotage activities of bacteria are corrosion of the inside of structural aluminum alloys used in fuel tanks of jet aircraft. Pitting and scaling of the metal occurs beneath heavy, slimy growths of hydrocarbon-utilizing bacteria such as species of *Pseudomonas* and *Desulfovibrio* (see Bacterial Reduction of Sulfur under §43.11). Some species of molds are also involved. Bacteria of the genera *Mycobacterium* and *Nocardia*, among others, are also implicated in the deterioration of bituminous products, including asphalt highways and asphalt coatings and pipe-linings. The microorganisms seem to utilize the high-viscosity hydrocarbons and resins in asphalt.

Prevention of Spoilage. This, in each instance, is a problem that can be solved only by careful examination of the process involved to find: (a) the nature of the organism(s) involved, (b) where the contamination is getting in, and (c) then devising means of excluding it. It is impossible to lay down a blanket rule for industrial spoilage in general. Everything depends on maintaining conditions unfavorable to, or excluding by asepsis, organisms that can grow on or in the particular product involved. This may involve complete steam sterilization of fermentation equipment (tanks, pipes, pumps); drying; refrigeration; aeration; the use of inhibitory salt, sugar, or acid concentrations; radiation with ultraviolet light; exposure to sunlight; and treatment with substances such as creosote, sodium benzoate, and the like. In some processes specific antibiotics may be used, as penicillin and tetracyclines in alcoholic fermentation of molasses.

CHAPTER 49
SUPPLEMENTARY READING

Corum, C. J. (Ed.): Developments in Industrial Microbiology, Vol. 10, American Institute of Biological Sciences. 1970.

Kaplan, A. M.: Industrial microbiology: concepts, challenges and motivation. Bioscience, *21*:468, 1971.

Prescott, S. C., and Dunn, C. G.: Industrial Microbiology, 3rd ed. McGraw-Hill Book Co., Inc., New York. 1959.

Umbreit, W. W. (Ed.): Advances in Applied Microbiology, Vol. 10. Academic Press, Inc., New York. 1968.

glossary

abscess. A localized accumulation of pus.

acid-fast. Capable of retaining the initial stain despite washing with dilute acid; e.g., *Mycobacterium* species.

actinomycete. Filamentous, mold-like bacterium found in the order Actinomycetales.

active immunity. Immunity acquired by an individual as a result of his own reactions to pathogenic microorganisms or their products either through infection or immunization.

active transport. Passage of a substance into the cell through the cell membrane against a concentration gradient; energy must be supplied.

adaptive enzyme. See **inducible (adaptive) enzyme.**

adenine. A purine base found in nucleosides, nucleotides, and nucleic acids.

adenosine. A mononucleoside consisting of adenine and a pentose sugar.

adenosine diphosphate (ADP). Contains one less high-energy phosphate than adenosine triphosphate. See **adenosine triphosphate (ATP).**

adenosine triphosphate (ATP). A compound containing a purine (adenine), a pentose (ribose), and three phosphate groups; a major carrier of phosphate and energy in biologic systems.

aerobe. A microorganism capable of growing or metabolizing in the presence of free oxygen.

aflatoxin. Poisonous substance produced by certain species of mold (*Aspergillus* spp.).

agar (agar-agar). A polysaccharide obtained from *Gelidium* (red alga) and other seaweeds; used as a solidifying agent in culture media.

agglutinin. Term used for any antibody capable of causing the clumping or agglutination of bacteria or other cells.

allergy. See **hypersensitivity.**

amensalism. See under **symbiosis.**

ammonification. Series of enzymic reactions resulting in the release of ammonia from complex organic nitrogenous compounds.

amphitrichous. Describing a bacterial cell having at least one flagellum at each end.

amylase. An enzyme that attacks glycosidic bonds and hydrolyzes starch into simple sugars.

anaerobe. A microorganism capable of growing or metabolizing in the absence of free oxygen; may be facultative or obligate (will perish in the presence of free oxygen).

anaphylaxis. A type of hypersensitivity reaction that can be fatal; evoked by the injection of a specific antigen into a sensitized individual.

antagonism. The inhibition, injury or killing of one species of microorganism by another.

787

antibiotic. A substance of microbial origin that kills or inhibits certain other microorganisms.

antibody. A general term for the several types of immunoglobulins found in immune serum and capable of reacting specifically (in vitro or in vivo) with the antigen or hapten which elicited the immune response.

anticodon. A triplet of nucleotide bases on a molecule of transfer ribonucleic acid (tRNA) which is complementary to a particular codon (triplet of bases) on messenger ribonucleic acid (mRNA).

antigen. A substance (usually a protein, polysaccharide, or complex) that evokes a specific immunologic response in an animal.

antimicrobial. A chemical or biological agent capable of killing or inhibiting microorganisms.

antiseptic. A chemical which opposes sepsis, i.e., kills or inhibits microorganisms, especially in contact with the body.

antiserum. Blood serum containing antibodies; serum from an individual that has been immunized against a specific antigen.

antitoxin. An antibody capable of uniting with and neutralizing a specific toxin, generally a microbial exotoxin.

apoenzyme. Protein component of an enzyme molecule.

arthropod. An animal of the phylum Arthropoda; bilaterally symmetrical, joint-legged invertebrate (e.g., insect or crustacean). Many species of arthropods can act as vectors of infectious diseases.

arthrospore. An asexual fungal spore formed by the segmentation of the hyphae.

ascospore. A sexual spore of Ascomycetes; spore contained within a sac-like structure (ascus) following the formation of a zygote.

asepsis. Absence of unwanted microorganisms; absence of infectious microorganisms in living tissues.

aseptic. Free of microorganisms capable of causing contamination or infection.

assimilation. The transformation of nutrients into protoplasmic constituents.

attenuated. Describing a mutant microorganism (including virus) having reduced virulence for one or more hosts.

autoclave. An apparatus for effecting sterilization by steam under pressure.

autolysis. Digestion and rupture of cells by action of their own enzymes.

autotroph. A microorganism that requires only CO_2 and other inorganic nutrients for growth.

auxotroph. A bacterial mutant that requires one or more growth factors that the wild-type strain (prototroph) can synthesize.

axenic. Not contaminated by or associated with any foreign organisms. Used in reference to pure cultures of microorganisms or germ-free animals.

bacillus. May refer to any rod-shaped bacterium or, more specifically, to the genus *Bacillus*.

bacteremia. A condition in which infectious bacteria are present in the blood stream.

bactericidal. Capable of killing bacteria.

bactericide. Any substance or agent that kills bacteria, generally within a specified time (one hour or less).

bacterin. A suspension of killed or attenuated bacteria used for artificial immunization of man or animals.

bacteriocin. Bactericidal, proteinaceous substances produced by strains of a number of bacteria which are active against only a few closely related organisms.

bacteriophage (phage). A bacterial virus, i.e., a virus that infects bacteria.

bacteriostatic. Inhibiting the growth of bacteria, generally without killing them.

bacterium. A procaryon without chlorophyll *a*.

bacteroid. Irregularly shaped bacterial forms found under special conditions (e.g., *Rhizobium* in root nodules of legumes).

basidiospore. A sexual fungal spore produced following the union of nuclei in a specialized clublike structure known as the basidium.

basidium. The specialized, club-like spore-producing cell found in the Basidomycetes.

BCG vaccine. Attenuated strain of *Mycobacterium tuberculosis* (Bacillus Calmette-Guerin) used to immunize against tuberculosis.

benthic zone. The bottom region of a lake or other body of water.

bioluminescence. Emission of light by living organisms.

biosphere. That part of earth occupied by living matter, generally the lower atmosphere and upper layers of soil and water.

blastospore. An asexual fungal spore formed by budding from a hypha or single cell.

BOD (biochemical oxygen demand). The amount of dissolved oxygen required by aerobic and facultative microorganisms to stabilize organic matter in sewage or water.

botulism. Food poisoning due to ingestion of *Clostridium botulinum* exotoxin.

brownian movement. The purposeless, undirected movement of suspended minute particles caused by molecular bombardment.

buffer. Any substance in solution that tends to maintain a constant pH despite the addition of moderate amounts of acid or alkali.

calorie. The amount of heat necessary to warm 1 gm water from 15 to 16C; 1,000 cal = 1 kcal (kilocalorie).

capsid. The protein coat of a virion (virus).

capsomer. A morphologically distinct structural unit of the viral capsid.

capsule. A gelatinous or slimy layer exterior to the cell wall of certain microorganisms.

carrier. An individual who harbors and may disseminate an infectious agent but who does not display clinical symptoms of disease.

catabolism. The dissimilation or breakdown of complex organic molecules, generally for the purpose of obtaining energy or simple compounds needed for synthesis of other organic matter.

catalase. An enzyme that catalyzes the decomposition of hydrogen peroxide (H_2O_2) into water and oxygen.

cellulase. An enzyme secreted by certain bacteria and fungi that hydrolyzes cellulose to cellobiose.

cellulose. A complex polymer of glucose, i.e., a polysaccharide; major structural component of plant cell walls.

chemolithotroph. Organism that obtains energy through chemical oxidation and uses inorganic or "mineral" substrates as electron donors.

chemostat. A device for maintaining a bacterial culture in the logarithmic phase of growth.

chemotaxis. The movement of an organism toward (positive) or away from (negative) a substance or object because of a chemical concentration gradient; particularly noticeable with certain types of phagocytic cells and motile bacteria.

chemotherapy. The treatment of infectious disease by chemical agents that cause little or no injury to the patient when properly used.

chemotroph. Organism that obtains energy through chemical oxidations as opposed to photosynthesis.

chitin. A polysaccharide yielding acetylglucosamine upon hydrolysis; present in the cell walls of many fungi, the exoskeletons of certain protozoans, insects, and crustaceans.

chlamydospore. An asexual fungal spore produced through fragmentation of a hypha; typically thick-walled and resistant to unfavorable conditions.

chloroplast. An intracellular, membrane-bound organelle containing chlorophyll in eucaryotic cells.

chromatophore. In microbiology, generally the vesicular particle or tubular structure which contains bacterial chlorophyll.

cilium (pl., cilia). A minute hair-like appendage on certain eucaryotic cells used for locomotion or feeding.

cistron. A subunit of a gene that can complement another subunit of the same gene; also used synonymously with gene.

citric acid cycle. See **Krebs cycle.**

clone. In microbiology, the asexual progeny of a single cell.

coagulase. An enzyme characteristically produced by *Staphylococcus aureus* and capable of clotting citrated blood or plasma.

coccus. A spherical bacterium.

codon. A triplet of nucleotide bases in messenger ribonucleic acid (mRNA) constituting the code for a molecule of transfer ribonucleic acid (tRNA) carrying a particular amino acid or for the start or finish of polypeptide synthesis.

coenocytic. Term generally applied to a cell or an aseptate hypha containing numerous nuclei.

coenzyme. A small nonprotein organic component of an enzyme.

cofactor. A small nonprotein inorganic component of an enzyme, frequently a metallic ion such as magnesium, zinc, copper, or iron.

coliform bacteria. All the aerobic and facultatively anaerobic, gram-negative, nonsporeforming bacilli that produce acid and gas from the fermentation of lactose.

colony. A macroscopically visible growth of microorganisms on a solid culture medium, generally circular and frequently representing a clone.

commensalism. See under **symbiosis.**

communicable. Capable of being transmitted from one individual to another.

complement. A thermolabile complex of at least 11 proteins in the sera of most warm-blooded animals. It reacts with a variety of antigen-antibody complexes and often exerts a lytic effect on the cell membranes of blood cells and certain bacteria.

conidiophore. A mold hypha bearing conidia on its tip.

conidium (pl., conidia). An asexual fungal spore borne on the tip of a fertile hypha called a conidiophore.

conjugation. The act of joining together; in microbiology, usually the mating of cells, with genetic material being exchanged or transferred from one cell to another. Common in some bacteria and protozoans.

constitutive enzyme. An enzyme produced by a cell, whether or not its substrate is present.

contagious. Capable of being transmitted from one individual to another.

coryneform. Straight or curved rods which stain irregularly (metachromatic granules) and frequently show club-shaped swelling; "snapping" division forms angular and palisade arrangement of cells.

culture. Any growth, population, or cultivation of microorganisms.

cutaneous. Pertaining to the skin.

cytochrome. Any of a group of iron-bearing proteins found in aerobic cells and distinguished from one another by their absorption spectra. Active in the transfer of electrons along an oxidative pathway.

cytopathic. Pertaining to pathologic changes in cells; in microbiology, frequently refers to the death or morphologic changes in cells in tissue culture infected with a virus.

cytoplasm. The protoplasm of a cell exclusive of the nucleus.

dalton. A unit of molecular weight equal to the weight of one hydrogen atom. Molecular weights of macromolecules are frequently expressed as multiples of daltons.

deaminase. An enzyme that removes an amino group from a molecule, liberating ammonia.

deamination. Removal of an amino group, usually from an amino acid.

decarboxylase. An enzyme that liberates carbon dioxide from the carboxyl group of a molecule, e.g., an amino acid.

decarboxylation. Removal of a carboxyl group (—COOH).

dehydrogenase. An enzyme that catalyzes the transfer of hydrogen and electrons from a substrate, often to a hydrogen carrier.

dehydrogenation. Removal of hydrogen from a molecule.

denitrification. Microbial reduction of nitrates to free nitrogen, commonly observed with certain types of organisms utilizing anaerobic respiration.

deoxyribonucleic acid (DNA). A single- or double-stranded macromolecular chain of nucleotides (each consisting of phosphoric acid, deoxyribose, and either a purine or pyrimidine base), the sequence of which determines the genetic code.

deoxyribose. A pentose (five-carbon sugar) found in DNA and having one oxygen atom less than ribose.

dermatophyte. Any one of a group of fungi causing superficial infections of the skin, hair, or nails, e.g., *Microsporum* or *Trichophyton*.

desiccation. The act of drying, e.g., the preparation of dry microorganisms, cell fractions, enzymes, etc., usually by the use of vacuum, desiccants, or mild heat.

dextran. A water-soluble glucose polymer (polysaccharide) produced by certain bacteria as a capsule or slime layer; used as a blood plasma substitute.

dialysis. The separation of solute (soluble salts, low molecular sugars, etc.) from colloidal material through the use of a semipermeable membrane.

diaminopimelic acid (DAP). Amino acid found in cell walls of most species of bacteria.

dimorphic. Having two forms or morphologies; also referred to as biphasic.

diphtheroid. Cells of irregular morphology; frequently arranged as palisades or having a V, Y, or T shape.

dipicolinic acid (DPA). Substance found in large amount in all bacterial spores but undetectable in vegetative cells.

diplococci. Cocci occurring in pairs.

diploid. Having a pair of each chromosome; twice the haploid number.

disaccharide. A sugar composed of two monosaccharides.

disinfectant. An agent that destroys the vegetative cells of infectious microorganisms.

disinfection. Destruction (or removal) of organisms capable of causing infectious disease.

DNA. See **deoxyribonucleic acid (DNA).**

DNA base ratio. See **GC content.**

dysentery. A lower intestinal infection caused by protozoans, bacteria, or viruses, and associated with cramps and diarrhea. Stools may contain blood and mucus.

eclipse period. Early part of the latent period of a viral infection cycle, prior to maturation of the virion and during which no complete infective virions are contained within the infected host cell.

ecology. The study of interrelationships of living organisms and their environment.

ecosystem. The fundamental unit in ecology; organisms in relation to their biotic and abiotic environment in a certain specified area.

ectosymbiosis. See under **symbiosis.**

edema. An abnormal accumulation of fluid in the intercellular spaces of tissue, often recognized by swelling or pain, or in the lungs, by difficult breathing.

endemic disease. Present in a community or area at all times, but with a low or normal incidence.

endergonic. Characterized by the absorption of energy; requires free energy.

endoenzyme. An intracellular enzyme that acts only within the cell.

endogenous. Originating from within.

endospore. A thick-walled spore formed within bacteria; particularly characteristic of the family Bacillaceae.

endosymbiosis. See under **symbiosis**.

endotoxin. A highly toxic polysaccharide-lipoprotein complex released from the cell walls of certain gram-negative bacteria, particularly following autolysis or cell disintegration.

enteric bacteria. Bacteria indigenous to the intestines, mostly gram-negative rods. Although not generally pathogenic, some may act as opportunists.

enteritis. Inflammation of the intestine.

enterotoxin. A microbial toxin specific for cells of the intestine and causing "food poisoning" when ingested.

enzymes. Specialized proteins that act as organic catalysts; often composed of a protein part, the apoenzyme, and a nonprotein part of low molecular weight, the coenzyme or cofactor. The combined components constitute the holoenzyme.

epidemic. The sudden occurrence in a community of numerous cases of a specific disease.

epidemiology. The study of factors that influence the frequency and distribution of infectious diseases in man and animals.

episome. A segment of DNA capable of existing in two alternate forms: one replicating autonomously in the cytoplasm, the other replicating as part of the bacterial chromosome.

esterase. One of a group of enzymes that hydrolyze esters.

etiological agent. The specific cause of a disease.

etiology. The theory or study of causes of disease.

eucaryotic cell. A type of cell which has a true nucleus, i.e., a well-defined nuclear membrane. These cells also have other distinctive features which are absent in procaryotic cells.

exergonic. Characterized by the release of energy; energy-yielding (reaction).

exoenzyme. An extracellular enzyme catalyzing chemical reactions outside of the cell.

exogenous. Originating from without.

exotoxin. Toxin produced by microorganisms and excreted into the surrounding medium or tissues during the growth phase. Generally inactivated by heat and easily neutralized by specific antibody.

exudate. The viscous fluid containing blood cells and debris which accumulates at the site of inflammation or lesion.

facultative (anaerobe). Microorganism that grows under either aerobic or anaerobic conditions.

fastidious organism. An organism having complex nutritional requirements; generally difficult to isolate or cultivate using ordinary culture procedures.

fecal coliforms. Coliform bacteria that produce gas from lactose in a special, buffered broth incubated at exactly 45.5C. Used to differentiate *Escherichia coli* of fecal origin from coliforms of other habitats.

fermentation. Enzymic, anaerobic oxidation of carbohydrates and related compounds by microorganisms. An organic compound serves as the final electron (hydrogen) acceptor. Alcohol production by yeasts is a classic example.

filamentous. Composed of long, thread-like structures.

filter, bacterial. Any one of a group of special filters through which bacterial cells cannot pass.

filtrate. Liquid that has passed through a filter.

fimbria. See **pilus.**

fission, binary. An asexual, single nuclear division followed by division of the cytoplasm to form two daughter cells of approximately equal size.

flagellar antigen. Antigenic material present on bacterial flagella; gives very characteristic serological reactions in vitro with its homologous antibody. Also known as H antigen.

flagellum. A filamentous appendage on cells used as an organ of location.

floccule. An aggregate of microorganisms or other materials floating in or on a liquid.

flocculent. A chemical which enhances floc formation in the treatment of sewage or water.

flora. The group or types of microorganisms present in a given locality, e.g., skin, intestinal tract, nares, etc. They may be either transient or resident (i.e., they will re-establish themselves if temporarily displaced or disturbed).

fluorescence. Generally, the emission of visible light from a substance which has absorbed radiation from another source (e.g., ultraviolet light).

fluorescence microscopy. Microscopy in which the microorganisms are stained with some form of fluorescent dye and observed by illumination with ultraviolet light.

fomite. Any inanimate object or substance that serves to convey infectious microorganisms from one host to another.

food infection. Generally, gastroenteritis resulting from the ingestion of salmonella-contaminated food, often erroneously called food poisoning.

food intoxication. See **food poisoning.**

food poisoning. A poorly defined term applied to all stomach or intestinal disturbances due to food contaminated with certain microorganisms, their toxins, chemicals, or poisonous plant material.

fungi. Multinucleate, sometimes multicellular, eucaryotic organisms having an absorptive mode of nutrition and rigid cell walls, but lacking chlorophyll (e.g., molds, yeasts, toadstools). Some authorities would place them in a separate kingdom.

fusiform. Spindle-shaped, tapered at the ends.

gamete. Reproductive cell which fuses with another reproductive cell to form a zygote; a sex cell.

GC content. The molar percentage of guanine + cytosine of the total base content in DNA. Can also be expressed as a molar ratio, i.e., $(A+T):(C+G)$.

gelatin. A product obtained by partial hydrolysis of collagen derived from the skin, connective tissues, and bones of animals; used in culture media for the determination of a specific proteolytic activity of microorganisms.

gelatinase. An exoenzyme that liquefies gelatin.

gene. The biologic unit of heredity; a cistron; a genetically functional group of nucleotides, existing as a sequence of bases in either RNA or DNA. Each gene specifies a single polypeptide chain.

genome. The entire genetic complement of a cell.

genotype. The types of particular genes contained by a cell, whether or not all of the genes are expressed.

genus. Ideally, a group of species all of which bear sufficient resemblance to one another to be considered closely related and easily distinguished from members of other groups; a group of related species.

germ. A pathogenic microorganism.

germicide. Generally, any chemical agent that kills the vegetative cells of pathogenic microorganisms.

glycogen. A polysaccharide which yields glucose upon complete hydrolysis; the chief carbohydrate storage material in animals.

glycolysis. Anaerobic process of glucose dissimilation to pyruvic acid by the Embden-Meyerhof pathway (also known as the glycolytic and hexose diphosphate pathway).

glycopeptide. See **peptidoglycan**.

Gram stain. A differential stain for bacteria developed by a Danish physician, Hans Christian Gram. The stain is based on the inability of certain bacteria to be readily decolorized with alcohol once they have been stained with a combination of crystal violet and iodine.

gram-negative. Describing bacteria which are readily decolorized with alcohol in Gram's method of staining and will then pick up the color of the counterstain (generally the red dye safranin).

gram-positive. Describing bacteria which are not readily decolorized with alcohol in Gram's method, thus retaining the violet to blue-black color of the crystal violet–iodine dye complex.

H antigen. See **flagellar antigen**.

halophile. A microorganism that requires a high salt concentration for growth; salt-loving.

haploid. Having a single set of unpaired chromosomes, as normally carried by a gamete.

hemagglutination. Agglutination (clumping) or erythrocytes in the presence of specific antibody or certain types of virus.

hemolysin. A substance released by certain bacteria that causes hemolysis of erythrocytes; hemolysins may also act as exotoxins.

hemolysis. The lysis of erythrocytes with the release of hemoglobin.

heterofermentation. Fermentation which results in a mixture of end products, i.e., various acids and gases.

heterologous. Different with respect to type or species.

heterotroph. A microorganism that requires its carbon in organic form, unlike the autotroph, which can utilize CO_2 as its sole carbon source.

holoenzyme. The entire enzyme; synonymous with enzyme.

homofermentation. Fermentation which results in a single end product, i.e., lactic acid.

homologous. The same with respect to type or species; corresponding in origin or structure, e.g., the specific antibody elicited by a particular antigen.

humus. A dark mass of decayed animal and plant matter that gives soil a loose texture and black or brown color.

hyaluronic acid. Essential intercellular cementing substance of animal cells; capsular substance of Group A beta hemolytic streptococci (*S. pyogenes*).

hybridization. Reannealing of single-stranded DNA prepared from two different organisms forming double-stranded (complementary) DNA molecules; used in study of genetic relatedness of the organisms.

hydrolysis. The cleavage of a compound by the addition of a water molecule, the hydroxyl group being incorporated in one fragment and the hydrogen atom in the other.

hypersensitivity. A type of immune response, frequently harmful, that may occur when a sensitized individual is exposed to the offending antigen, e.g., anaphylaxis.

hypha. A single filament of a fungal mycelium.

immunity. Resistance of the body to the effects of pathogenic microorganisms or their toxins. Immunity results from the presence of specifically reactive cells and/or immunoglobulins (antibodies) capable of neutralizing, opsonizing bacteria, etc.

immunoglobulin. A blood serum globulin (protein) having antibody activity and evoked as a result of antigenic stimulation. Term frequently used interchangeably with antibody and immune gamma globulins.

IMViC test. A series of four physiological tests (indole, methyl red, Voges-Proskauer, and citrate) helpful in differentiating between various coliform microorganisms.

inducible (adaptive) enzyme. An enzyme produced by a microorganism in response to the presence of its substrate (or a related substance) in the environment.

induction. In virology, the process whereby prophage becomes excised from the bacterial DNA and enters the vegetative state.

infection. Generally, a diseased state due to the invasion of a host by a pathogenic or opportunistic organism (bacterium, virus, protozoan, fungus, etc).

infectious. Capable of being communicated to susceptible hosts by infection.

inflammation. The response of tissues to injury; characterized by pain, heat, redness, swelling, and, sometimes, loss of function.

inoculation. The introduction of microorganisms, infective material, or serum into culture media or living animal or plant tissues.

inoculum. Material (microorganisms, viruses, etc.) used in inoculation.

intercellular. Between cells.

interferon. A protein released by mammalian cells infected by certain viruses, or in response to the injection of foreign nucleic acid or certain other agents, which when applied to uninfected cells of the same species will inhibit virus replication.

intracellular. Within a cell.

in vitro. Within a glass container, e.g., culture flask, test tube, etc., as opposed to in vivo.

in vivo. Within the living body.

Kauffmann-White schema. A means of identifying and classifying various serotypes ("species") of *Salmonella* by antigenic analysis of their O, H, and Vi antigens.

Krebs cycle. Also known as the citric acid or tricarboxylic acid cycle. A metabolic sequence which in aerobic organisms converts pyruvic acid to carbon dioxide with the release of energy and also provides many intermediates necessary for the synthesis of cell components.

lactose. A disaccharide consisting of glucose and galactose; also known as milk sugar.

latency. A state of apparent inactivity; e.g., individuals may harbor infectious microorganisms for long periods of time and not manifest any clinical symptoms.

latent period. In virology, the time during the intracellular replicative cycle of a virus prior to the release of progeny virions.

LD$_{50}$ (lethal dose 50). The number of pathogenic microorganisms required to cause death in 50 per cent of the inoculated animals in a test.

lesion. Any injured or diseased area or pathologic change in a tissue.

leucocyte. General term for any of the white cells of the blood, e.g., polymorphonuclear leucocytes, lymphocytes, and monocytes.

lichen. In microbiology, a plant-like growth composed of a symbiotic alga and fungus.

lipase. An enzyme capable of splitting lipid or fat-like molecules.

lipid. Generally, any of a group of naturally occurring substances containing fatty acids which are insoluble in water, but soluble in fat solvents such as chloroform, ether, alcohol, etc.

lithotroph. See **autotroph.**

litmus. An extracted lichen pigment useful as an indicator for pH and oxidation or reduction.

lophotrichous. Describing a bacterial cell having a tuft of flagella at one end.

lyophilization. Also known as freeze-drying. The creation of a stable preparation of a biologic substance (microorganisms, serum, toxin, etc.) by rapid freezing, followed by dehydration under high vacuum.

lysis. The disruption of cells (e.g., bacteria or erythrocytes) by hemolysins or by the action of specific antibodies plus complement.

lysogeny. The genetic property of a bacterial cell enabling it or its progeny to produce a lytic phage without external infection; bacteria carrying prophage are lysogenic.

lysozyme. An enzyme present in tears, saliva, egg white, and many animal fluids which has the ability to lyse certain types of bacteria, especially gram-positive cells, by splitting the peptidoglycan contained in the cell wall.

maltase. An enzyme that hydrolyzes maltose to produce glucose.

maltose. A disaccharide formed when starch is hydrolyzed by amylase.

meiosis. A type of nuclear division occurring in the production of gametes, whereby each daughter cell receives half the number of chromosomes found in the parent cell (i.e., it becomes haploid).

mesophile. A microorganism whose optimum temperature for growth ranges from about 30 to 45C.

mesosome. A structure observed in certain bacterial cells that appears to be a highly convoluted invagination of the cytoplasmic membrane. Its function is not clear.

metabolism. The sum of all chemical and physical processes by which living organized substance is produced and maintained.

metachromatic granules. Cytoplasmic granules in the cells of certain bacteria that stain intensely with basic dyes, such as methylene blue, but appear red (or a different) color.

microaerophile. A microorganism whose optimum growth occurs in the presence of small amounts of atmospheric oxygen.

mitosis. A type of nuclear division whereby each daughter cell has the same number of chromosomes as the parent cell.

mold. A fungus characterized by a filamentous or cottony type of mycelium.

Monera. New kingdom proposed for the unicellular procaryons, namely bacteria and blue-green algae.

mononucleotide. The basic building block of nucleic acids (DNA and RNA); consists of phosphoric acid, a pentose, and either a purine or pyrimidine base.

monosaccharide. A simple sugar (e.g., a hexose or pentose) which cannot be decomposed further by hydrolysis.

monotrichous. Describing a bacterial cell having only one flagellum at one end of the cell.

most probable number (MPN). Statistical estimate of a bacterial population through the use of dilution and multiple tube inoculations.

mucocomplex. See **peptidoglycan**.

mucopeptide. See **peptidoglycan**.

muramic acid. N-acetylmuramic acid is one of the two important amino-sugars forming the disaccharide backbone of the unique peptidoglycan found in the cell walls of procaryons.

murein. See **peptidoglycan**.

mutagenic agent. A chemical or physical agent capable of inducing mutations.

mutant. An organism with a changed physiological trait, morphology, or characteristic which will breed true; an offspring or daughter cell with an altered genotype.

mutation. Change of the sequence of nucleotides within a gene to a new sequence, resulting in change of the polypeptide specified by the gene (caused by insertion, deletion, or alteration of a base or portion of the nucleic acid chain).

muton. Minimal genetic unit capable of mutation.

mutualism. See under **symbiosis.**

mycelium. The vegetative body of a fungus, consisting of a mass of hyphae.

mycoplasmas. A group of bacteria characterized by highly pleomorphic cells lacking cell walls.

mycorrhiza. See under **symbiosis.**

mycosis. Any disease caused by a fungus.

negative staining. A technique whereby the background rather than the cells is stained, e.g., a nigrosin stain for demonstrating capsules surrounding bacterial cells.

nitrate reduction. The reduction of nitrates to nitrites or ammonia.

nitrification. The transformation of ammonia to nitrites and nitrates.

nitrogen fixation. The formation of nitrogen compounds from free atmospheric nitrogen.

nucleic acid. Polynucleotide macromolecules, the principal types being DNA and RNA. See also **deoxyribonucleic acid (DNA)** and **ribonucleic acid (RNA).**

nucleocapsid. The nucleic acid and the surrounding capsid of a virion. Synonymous with the complete virion or virus where an envelope or a tail is not involved.

nucleoid. The central nuclear area in a procaryon containing the DNA of the cell, but not limited by a nuclear membrane, as in a eucaryon.

nucleotide. See **mononucleotide.**

nucleus. The central, membrane-bound, DNA-containing structure of eucaryotic cells.

O antigen. See **somatic antigen.**

obligate. Necessary or required; e.g., obligate aerobes grow only in the presence of free oxygen; obligate intracellular parasites grow only within living cells.

operon. A group of genes that works under the control of a single operator gene to synthesize related proteins.

opportunists (opportunistic microorganisms). Organisms which are not normally regarded as pathogens but which may cause infection under certain circumstances, particularly if the host is debilitated (e.g., patients suffering from burns, malnutrition, etc.) or if the normal competitive resident flora is absent.

organelle. Cell substructure that performs a specific function, e.g., mitochondrion, chloroplast, etc.

organic compound. A compound containing carbon, as distinguished from an inorganic compound, which lacks carbon. Most workers do not consider CO_2 or carbonates as organic compounds when discussing microbial physiology.

organotroph. See **heterotroph.**

oxidation. A chemical reaction in which a substance loses electrons or a hydrogen atom or gains an oxygen atom.

oxidation-reduction (O-R) potential. The electron-yielding or electron-accepting potential of any given material under specified conditions.

oxidative phosphorylation. The conversion of inorganic phosphate to the high-energy phosphate of ATP by reactions associated with the electron transport (cytochrome) system.

pandemic. A widespread, often worldwide, epidemic, e.g., as seen with certain strains of influenza virus.

parasitism. See under **symbiosis.**

parenteral. Administered to the body by some means other than the intestinal tract.

passive immunity. Immunity acquired by an individual after receiving antibody either through injection or by natural means (colostrum, placental transfer).

pasteurization. A mild heat treatment, commonly of milk and other beverages, designed to kill any pathogenic microorganisms present and to reduce the load of spoilage organisms without materially affecting taste and other characteristics of the products. Named after Pasteur, who developed the technique.

pathogenic. Capable of producing disease.

pentose. A simple sugar with five carbon atoms, e.g., ribose.

peptidase. An enzyme that catalyzes the hydrolysis of peptide bonds in polypeptides and proteins.

peptide. A compound of low molecular weight consisting of two or more amino acids.

peptidoglycan. A unique polymer found in the cell walls of procaryons and consisting largely of alternating units of N-acetylglucosamine and N-acetylmuramic acid, to which short peptide chains are attached to form a coarse molecular meshwork. Also known as mucopeptide, glycopeptide, mucocomplex, murein, etc.

peptone. An intermediate product formed in the digestion of protein.

peritrichous. Describing a bacterial cell having flagella protruding around the entire surface of the cell.

permease. A stereospecific membrane transport system which requires energy input and which can establish a concentration gradient, i.e., the concentration of a substance within the cell will exceed the concentration in the environment.

pH. The measure of the acidity or alkalinity of a solution; defined as the negative logarithm of the H ion activity (7.0 is neutral, less than 7.0 is acid, and more than 7.0 is alkaline).

phagocyte. Any of the leucocytes, e.g., neutrophils (PMN) or macrophages, capable of ingesting and destroying microorganisms and other particulate matter.

phagocytosis. Ingestion of particulate matter by fixed or wandering leucocytes.

phenotype. The outward detectable expression of the hereditary constitution, or the function of a gene, under a given set of conditions.

phosphorylation. The addition of phosphate onto an organic molecule.

photolithotroph (photoautotroph). Organism that obtains energy through photosynthesis and can utilize CO_2 as its major source of carbon.

photoorganotroph (photoheterotroph). Organism that obtains energy through photosynthesis but cannot utilize CO_2 as its sole source of carbon, e.g., certain types of photosynthetic bacteria.

photophosphorylation. The conversion of light energy into that of high-energy phosphate bonds.

photosynthesis. Synthesis of carbohydrates from carbon dioxide and water using the energy of light secured with the aid of chlorophyll.

phototroph. Organism that obtains energy through photosynthesis.

pilus (or fimbria; pl., pili and fimbriae). Short, minute, hair-like projections found on the surface of certain bacteria. Function is not entirely clear, but it is not related to motility.

plankton. Minute, freely floating plant or animal organisms in large, natural bodies of water.

plaque. In virology, a clearing (indicating lysis) in a lawn of virus infected cells seeded upon some type of solid culture medium.

pleomorphic. Irregular or variant morphologic form.

pleuropneumonia-like organisms (PPLO). See **mycoplasmas.**

polypeptide. A polymer formed by joining many amino acids together. A dipeptide would consist of two amino acids, a pentapeptide of five, etc.

polysaccharide. A polymer formed by the joining of many mono- and disaccharides, e.g., starch, glycogen.

polysome (polyribosome). A chain of ribosomes, held together by a messenger ribonucleic acid (mRNA) molecule, that together with transfer ribonucleic acid (tRNA) joins amino acids in sequence according to the code in the mRNA.

potable. Fit to drink; free of harmful chemicals, bacteria, and toxins.

precipitation. In immunology, the aggregation of soluble antigen molecules by antibody until a visible precipitate forms and settles out.

procaryotic cell. A type of cell lacking a nuclear membrane and having other features which distinguish it from eucaryotic cells; e.g., bacteria and blue-green algae.

prophage. Phage DNA integrated into bacterial DNA and replicated simultaneously so that each daughter cell receives prophage.

prosthecate bacteria. Cells having appendages other than pili, flagella, or stalks.

protein. A macromolecular polymer of amino acids. All enzymes and most structural components of the cells are made up largely of protein.

proteolytic. Capable of cleaving peptide bonds and thereby degrading protein.

Protista. A third kingdom proposed for the relatively simple, predominantly unicellular organisms (bacteria, fungi, algae, protozoans) as an addition to the traditional plant and animal kingdoms.

protoplast. In bacteriology, an actively metabolizing bacterial cell devoid of any cell wall and suspended in hypertonic medium to prevent lysis.

psychrophile. Generally, a microorganism capable of growth at 0C or lower.

pus. The viscous fluid from lesions or sites of inflammation and consisting of leucocytes, necrotic tissue debris partially liquefied by enzymes, blood proteins, and other microbial products that are characteristically produced in certain bacterial infections.

putrefaction. Microbial decomposition of animal or plant matter, especially proteins, producing disagreeable odors.

pyogenic. Pus-producing.

recombination. The introduction of a new segment of DNA into an existing segment which results in an altered genotype.

recon. The minimal genetic unit transmissible from one chromosome to another.

reduction. A chemical reaction in which a substance gains electrons or a hydrogen atom or loses an oxygen.

rennin. Enzyme from the gastric juice of a calf that transforms the soluble casein of milk into insoluble paracasein.

replicon. A segment of DNA that replicates as a unit.

respiration. In microbiology, a type of metabolism in which oxygen serves as the final electron and hydrogen acceptor in an electron transport system that conserves energy in the form of ATP.

rhizoid. Root-like "holdfasts" of certain fungi and plants.

rhizosphere. The zone of increased microbial activity and growth in the soil immediately surrounding the roots of plants.

ribonucleic acid (RNA). A single- or double-stranded macromolecular chain of nucleotides (each consisting of phosphoric acid, ribose, and either a purine or pyrimidine base), the sequence of which can specify the order of amino acids in polypeptide synthesis. There are three types: ribosomal (rRNA), messenger (mRNA), and transfer (tRNA).

ribose. A pentose (five-carbon sugar) found in ribonucleic acid (RNA).

ribosome. A submicroscopic particle consisting mostly of RNA and protein which is the site of protein synthesis in both eucaryotic and procaryotic cells.

rickettsias. Small bacteria which are obligate intracellular parasites and responsible for a number of important diseases of man and animals, sometimes in epidemic proportions; organisms of the genus *Rickettsia*.

RNA. See **ribonucleic acid (RNA).**

saccharolytic. Capable of degrading or splitting sugar compounds.

sanitize. Generally, to clean and reduce the microbial load on inanimate objects to a level regarded as safe by public health authorities; also, to disinfect.

saprophyte. An organism living on dead and decaying organic matter.

sepsis. The presence of infectious microorganisms in living tissues; the presence of unwanted microorganisms.

septate. Possessing cross-walls, e.g., septate hyphae.

septicemia. The presence of pathogenic microorganisms, sometimes with their toxins, in the blood stream of an infected individual.

septum. A cross-wall dividing a cell into two or more compartments, depending on the number of septa.

serology. The study and use of serum, particularly as it relates to tests involving antibody and antigen.

serum. Broadly, any clear, watery animal fluid except urine; more specifically, the clear liquid exuded from clotted blood or obtained from plasma by removal of the fibrin. Serum is rich in antibody.

sewage. The used water supply of a home, community, or industry, and often including storm run-off water. Sewerage is the system of sewers and pipes carrying sewage.

slime layer. A slimy layer exterior to the cell walls of certain microorganisms, but generally more easily washed away than a true capsule.

sludge. The semisolid part of sewage that has sedimented, or the floc which has formed and settled out during sewage processing.

somatic antigen. Antigen derived from the body of the cell (as opposed to appendages such as flagella), particularly lipopolysaccharide antigens derived from the outer layers of the cell wall. Often used synonymously with O antigen.

species. A taxonomic category subordinate to a genus and superior to a variety or strain; ideally, a group of organisms all of which bear sufficient resemblance to one another to be considered related and also easily distinguished from members of other species.

spheroplast. A spherical bacterial cell retaining at least some remnants of the cell wall and kept in hypertonic medium to prevent lysis.

spirillum. A rigid spiral or corkscrew-shaped bacterium; organism of the genus *Spirillum*.

spirochete. A flexible spiral form of bacterium; e.g., organism of the genus *Spirochaeta*.

sporangiophore. A specialized fungal hypha bearing a sporangium (containing spores) at its tip.

sporangium. A sac-like structure, containing asexual spores, at the end of a sporangiophore.

spore. A special type of resting cell characterized by its resistance to unfavorable environmental conditions; a reproductive body of a protist (fungus, protozoan, etc) produced by either asexual or sexual means. Endospores are formed within a cell, exospores are born exogenously.

sporulation. Production of spores.

staphylococci. Spherical bacteria (cocci) occurring in irregular clusters, sometimes resembling a bunch of grapes, and typical of members of the genus *Staphylococcus*.

sterile. Free from all living organisms (also viruses). Sterility is produced by the use of lethal agents or by the removal of the organisms.

sterilization. The process of destroying or removing all living organisms (also viruses).

stock cultures. Species or strains of microorganisms (pure cultures), known or as yet unidentified, maintained in the laboratory for various tests and study.

strain. A group of organisms within a species or variety, characterized by some particular quality (e.g., toxin production, colonial morphology, etc.); often used interchangeably with the terms biotype and serotype.

streptococci. Spherical cells (cocci) occurring in long or short chains; bacteria of the genus *Streptococcus*.

substrate. The substance acted upon by an enzyme.

supernate. The fluid remaining after centrifugation and removal of suspended matter; the clear liquid over a precipitate or sediment.

suppuration. The formation of pus.

symbiosis. The living together of two or more dissimilar organisms or populations. The term is most frequently used to describe symbionts where the association is advantageous, or often necessary, to one or both. In a broader sense, the association may be beneficial to both (mutualism), beneficial to one without effect on the other (commensalism), beneficial to one and harmful to the other (parasitism), detrimental to one without effect on the other (amensalism), or detrimental to both (synnecrosis). One organism may grow inside the tissue of the host (endosymbiosis) or around the host with little penetration into the host tissue (ectosymbiosis). Fungi may also be associated with the roots of a higher plant (mycorrhiza).

synergism. In microbiology, the ability of two or more strains to bring about changes, typically chemical, that neither could accomplish alone.

synnecrosis. See under **symbiosis.**

syntrophism. Stimulation of growth of a microorganism resulting from the presence of another species or strain.

systemic. Pertaining to or affecting the body as a whole.

taxon (pl., taxa) A taxonomic group, e.g., class, order, family, genus, or species.

taxonomy. The classification or systematic arrangement of organisms into groups or categories called taxa.

temperate bacteriophage. One whose nucleic acid is able to be integrated into the bacterial cell's DNA molecule, rendering that cell lysogenic.

thermal death time. The time required to kill all the bacteria of a certain species in a given substance at a stated temperature.

thermoduric. Nonsporeforming microorganisms capable of surviving at high temperature.

thermophile. An organism that grows at temperatures of 45 to 50C. Thermophilic species that thrive only at temperatures above 50C are called obligate thermophiles.

thylakoid. The lamellar structure observed in the cytoplasm of blue-green algae and containing the photosynthetic pigments.

toxin. Any of a number of poisonous substances elaborated by certain microorganisms, plants, and animals; typically, they are proteins and antigenic.

toxoid. A toxin treated with heat or chemical agent (e.g., formaldehyde) to destroy its toxicity without affecting its antigenic properties.

transduction. A virus-mediated transfer of bacterial DNA from one bacterium to another.

transformation. In bacteriology, a genetic change in a recipient bacterium resulting from the absorption of DNA released from another cell. Note that the DNA is unprotected or free as opposed to the transfer in either conjugation or transduction.

tribe. A taxonomic division containing a group of related genera within a family.

tricarboxylic acid (TCA) cycle. See **Krebs cycle.**

trichomes. Filamentous structures; the long multicellular threads formed by uniseriately multicellular organisms such as *Beggiatoa.*

tubercle. A nodule, especially the specific lesion caused by *Mycobacterium tuberculosis.*

tuberculin. A sterile liquid containing the growth products of, or specific substances extracted from, the tubercle bacillus. It is of great value in skin tests for tuberculosis.

urease. An enzyme that catalyzes the hydrolysis of urea to ammonia and carbon dioxide.

vaccination. Artificial immunization against smallpox by inoculation with coxpox virus; frequently used synonymously with the term immunization.

vaccine. Lymph containing cowpox virus and used for vaccination against smallpox; a suspension of killed or attenuated microorganisms administered for the prevention or treatment of infectious diseases.

vacuole. A cavity or space in the cytoplasm of a cell, e.g., contractile vacuoles found in many protozoans.

variant. An organism having one or more characteristics that differ from the parent strain.

vegetative stage. Active growth as opposed to sporulation or the resting state.

viable. Capable of living, provided a suitable environment is furnished.

viremia. The presence of viruses in the blood stream of an infected individual.

virion. A complete virus particle.

virulence. The capacity of a given strain or pure culture to produce disease.

Voges-Proskauer reaction (VP test). A test for the presence of acetylmethylcarbinol following glucose fermentation; useful in distinguishing between species of the coliform group.

yeast. A type of fungus that grows primarily as single cells, reproducing by budding.

zoonosis. An animal disease that may be transmitted to man.

zygote. A cell produced by the union of two gametes.

illustration credits

The following list of credits includes original published sources of figures used in this text. (The courtesy lines in the figure legends indicate photographers and authors who generously loaned original prints for enhanced reproduction quality.)

Figure 1–1   From Illingworth, Rose, and Beckett: J. Bact., *113*:373, 1973.
Figure 1–2   From Parson, Warner, Anderson, and Snustad: J. Virol., *11*:807, 1973.
Figure 1–4   From Hardy: The Open Sea, Vol. I, William Collins & Sons, Ltd., 1966.
Figure 1–5   From Fleming: Brit. J. Exp. Pathol., *10*:228, 1929.
Figure 1–8   From Bolduan: Public Health and Hygiene, W. B. Saunders Co., 1929.

Figure 2–4   From Dobell: Antony van Leeuwenhoek and His "Little Animals," Dover Publications, Inc., 1960.
Figure 2–6   Advertisement from *The Microscope*, Vol. II, No. 6, p. 6, 1883, edited and published by C. H. Stowell and L. R. Stowell.
Figure 2–7   From Schulze: Ann. Physik Chemie, *xxix*:487, 1836.
Figure 2–8   From Schwann: Ann. Physik Chemie, *xli*:184, 1837.
Figure 2–9   From Bulloch: History of Bacteriology, *in* Medical Research Council: A System of Bacteriology in Relation to Medicine, Vol. 1, 1930. By permission of the Controller of Her Britannic Majesty's Stationery Office, London.
Figure 2–11 From Pasteur: Ann. Sci. Nat., *xvi*:5, 1861.
Figure 2–12 From Tyndall, Floating Matter of the Air in Relation to Putrefaction and Infection, London, 1881.
Figure 2–13 From Cohn (Dolley, Tr.): Bull. Hist. Med., *7*(1):49, 1939.
Figure 2–14 From Cohn (Dolley, Tr.): Bull. Hist. Med., *7*(1):49, 1939.
Figure 2–16 From Kummel: History of the Earth, W. H. Freeman & Co., 1961.
Figure 2–17 From Beutner: Life's Beginning on the Earth, The Williams & Wilkins Co., 1938.
Figure 2–18 From Fox: Ann N.Y. Acad. Sci., *194*:71, 1972.
Figure 2–19 From Bulloch: History of Bacteriology, Oxford University Press, 1938. Reproduced by permission of Athlone Press of the University of London.

Figure 3–8   From Thompson: Microbiology and Epidemiology, 5th ed., W. B. Saunders Co.
Figure 3–9   From Koch, Beitr. Biol. Pflanzen, Bd. II, Heft *3*:399, 1877.
Figure 3–11 From Lillie et al.: H. J. Conn's Biological Stains, 8th ed., The Williams & Wilkins Co., 1969.
Figure 3–12 From Bartholomew and Finkelstein: J. Bact., vol. 75.
Figure 3–27 Adapted from Everhart and Hayes: Sci. Am., *226*:55, 1972. (© W. B. Saunders Co.)

Figure 6–7   From Mazur and Harrow: Textbook of Biochemistry, 10th ed., W. B. Saunders Co., 1971.

Figure 6–8   After Feughelman et al., in J. R. Oncley (Ed.): Biophysical Science—A Study Program, John Wiley & Sons, 1959.

Figure 6–9   From Calvin and Calvin: Am. Sci., 52:163, 1964.

Figure 6–10   From The Bacterial Chromosome, by John Cairns. Copyright © 1966 by Scientific American, Inc. All rights reserved.

Figure 6–11   From The Bacterial Chromosome, by John Cairns. Copyright © 1966 by Scientific American, Inc. All rights reserved.

Figure 6–12   From Smyth et al.: J. Biol. Chem., 238:227, 1963.

Figure 6–13   From Pauling and Corey: Proc. Nat. Acad. Sci., 37:729, 1951.

Figure 6–14   From Cantarow and Schepartz: Biochemistry, 4th ed., W. B. Saunders Co., 1967.

Figure 6–15   From Villee: Biology, 6th ed., W. B. Saunders Co., 1972.

Figure 7–2   From Mazur and Harrow: Textbook of Biochemistry, 10th ed., W. B. Saunders Co., 1971.

Figure 7–3   From Koshland, D. E., Jr.: Science, 142:1533–1541 (December 20), 1963. Copyright 1963 by the American Association for the Advancement of Science.

Figure 9–3   From Snow: Bact. Rev., 34:99, 1970.

Figure 10–1   From ZoBell: J. Bact., vol. 46.

Figure 10–5   From Carpenter: Microbiology, 3rd ed., W. B. Saunders Co., 1972.

Figure 10–8   From Daneo-Moore and Higgins: J. Bact., 109:1210, 1972.

Figure 10–9   From Higgins and Daneo-Moore: J. Bact., 109:1221, 1972.

Figure 10–21   From Schultz and Gerhardt: Bact. Rev., 33:1, 1969.

Section Two Lefthand Page. From Kleinschmidt, Lang, Jacherts, and Zahn: Biochim. Biophys. Acta, 61:857, 1962.

Figure 11–1   Adapted from Margulis: Evolution, 25:242, 1971; and Whittaker: Science, 163:150, 1969.

Figure 11–2   A, From The Living Cell, by Jean Bachet. Copyright © 1961 by Scientific American, Inc. All rights reserved. B, From Villee: Biology, 6th Ed., W. B. Saunders Co., 1972.

Figure 11–3   A, From Brock: Biology of Microorganisms, Prentice-Hall, Inc., 1970. B, From Dodge: An Atlas of Biological Ultrastructure, Edward Arnold (Publishers) Ltd., 1968.

Figure 11–4   From Silva and Sousa: Appl. Microbiol., 24:471, 1972.

Figure 12–2   B, From Carpenter: Microbiology, 3rd ed., W. B. Saunders Co., 1972.

Figure 12–3   From Dodge: An Atlas of Biological Ultrastructure, Edward Arnold (Publishers) Ltd., 1968.

Figure 12–4   From Stanier et al.: Microbial World, 3rd ed., Prentice-Hall, Inc., 1970.

Figure 12–5   Adapted from Alexopoulos: Introductory Mycology, 2nd ed., John Wiley & Sons, Inc., 1962.

Figure 12–6   From Alexopoulos: Introductory Mycology, 2nd ed., John Wiley & Sons, Inc., 1962.

Figure 12–7   1, From the collection of the American Society for Microbiology; Agar and Douglas: J. Bact., vol. 70.

Figure 12–8   From Carpenter: Microbiology, 3rd ed., W. B. Saunders Co., 1972.

Figure 12–9   Adapted from Villee: Biology, 6th ed., W. B. Saunders Co., 1972.

Figure 12–10   From Poindexter: Microbiology: An Introduction to Protists, © Copyright, The Macmillan Co., 1971.

Figure 12–11   From Predatory Fungi, by Joseph J. Maio. Copyright © 1958 by Scientific American, Inc. All rights reserved.

Figure 12–12   From Conant et al.: Manual of Clinical Mycology, 3rd ed., W. B. Saunders Co., 1971.

Figure 12–13   From Lechevalier and Pramer: The Microbes, J. B. Lippincott Co., 1971.

Figure 12–14   Line drawing from Lechevalier and Pramer: The Microbes, J. B. Lippincott Co., 1971.

Figure 13–1   From Villee: Biology, 6th ed., W. B. Saunders Co., 1972.
Figure 13–2   From Villee: Biology, 6th ed., W. B. Saunders Co., 1972.
Figure 13–3   From Villee: Biology, 6th ed., W. B. Saunders Co., 1972.
Figure 13–4   From Villee: Biology, 6th ed., W. B. Saunders Co., 1972.
Figure 13–5   From Villee: Biology, 6th ed., W. B. Saunders Co., 1972.
Figure 13–6   From Leedale: Euglenoid Flagellates, Prentice-Hall, Inc., 1967.
Figure 13–7   From Small, E. B., and Marszalek, D. S.: Science, *163*:1064–1065 (March 7), 1969. Copyright 1969 by the American Association for the Advancement of Science.
Figure 13–8   From Wessenberg and Antipa: J. Protozool., *17*:250, 1970.
Figure 13–9   *A*, From MacDougall and Hegner: Biology, McGraw-Hill Book Co., Inc. *B*, From Beutner: Life's Beginning on the Earth, The Williams & Wilkins Co., 1938.
Figure 13–10   From Villee: Biology, 6th ed., W. B. Saunders Co., 1972.
Figure 13–13   From Villee: Biology, 6th ed., W. B. Saunders Co., 1972.

Figure 14–2   From Fawcett: The Cell: Its Organelles and Inclusions, W. B. Saunders Co., 1966.
Figure 14–4   From Kerridge et al.: J. Mol. Biol., *4*:227, 1962.
Figure 14–5   *A*, From Cohen-Bazire and London: J. Bact., *94*:458, 1967. *B*, Adapted from De Pamphilis and Adler: J. Bact., *105*:376, 1971.
Figure 14–8   From Burdon and Williams: Microbiology, 6th ed. © Copyright, The Macmillan Co., 1968.
Figure 14–9   From Taylor and Juni: J. Bact., *81*:688, 1961.
Figure 14–12   From DePetris: J. Ultrastruct. Res., *19*:45, 1967.
Figure 14–13   Adapted from DePetris: J. Ultrastruct. Res., *19*:45, 1967.
Figure 14–14   From Kawata, Asaki, and Takagi: J. Bact., *81*:160, 1961.
Figure 14–15   From Silva and Sousa: Appl. Microbiol., *24*:463, 1972.
Figure 14–16   *A*, From Silva and Sousa: Appl. Microbiol., *24*:463, 1972. *B*, From van Iterson: Bact. Rev., *29*:299, 1965.
Figure 14–17   From Ryter: Curr. Top. Microbiol. Immunol., *49*:151, 1969.
Figure 14–18   From Huxley and Zubay: J. Mol. Biol., *2*:10, 1960.
Figure 14–19   From Boatman: J. Cell. Biol., *20*:297, 1964.
Figure 14–20   From Harold: Bact. Rev., *30*:772, 1966.
Figure 14–21   From Villee: Biology, 6th ed., W. B. Saunders Co., 1972.
Figure 14–22   From Wolk: Bact. Rev., *37*:32, 1973.

Figure 15–1   From Mayor, D. H., and Melnick, J. L.: Science, *137*:613–615 (August 24), 1962. Copyright 1962 by the American Association for the Advancement of Science.
Figure 15–2   From Strandberg and Carmichael: J. Bact., *90*:1790, 1965.
Figure 15–3   Redrawn from Caspar: Adv. Protein Chem., *18*:37, 1963.
Figure 15–4   From Reginster: J. Gen. Microbiol., *42*:323, 1966.
Figure 15–5   From Smith, Gehle, and Trousdale: J. Bact., *90*:254, 1965.
Figure 15–6   From Mayor, D. H., and Melnick, J. L.: Science, *137*:613–615 (August 24), 1962. Copyright 1962 by the American Association for the Advancement of Science.
Figure 15–7   From Dales, S.: Penetration of animal viruses into cells. Progr. Med. Virol., *7*:1–43 (Karger, Basel/New York 1965).
Figure 15–8   From Horne and Waterson: J. Mol. Biol., *2*:75, 1960.
Figure 15–9   *A*, From Hummeler, Koprowski, and Wiktor: J. Virol., *1*:152, 1967. *B*, From Hitchborn, Hill, and Hull: Virology, *28*:768, 1966. *C*, From Howatson and Whitmore: Virology, *16*:466, 1962.
Figure 15–10   From Horne and Wildy: Virology, *15*:348, 1961.
Figure 15–11   Sketch and sizes based on Lechevalier and Pramer: The Microbes, J. B. Lippincott Co., 1971.

Figure 16–1   From Burrows: Textbook of Microbiology, 19th ed., W. B. Saunders Co.
Figure 16–3   From Trent and Scott: J. Bact., *88*:702, 1964.
Figure 16–4   From Lerner, Takemoto, and Shelokov: Proc. Soc. Exp. Biol. Med., vol. 95.
Figure 16–5   From Froman, Will, and Bogen: Am. J. Pub. Health, vol. 44.

Figure 16–6   1, From Valentine, Chen, Colwell, and Chapman: J. Bact., 91:819, 1966.
2, From Dawson, Smillie, and Norris: J. Gen. Microbiol., 28:517, 1962.
3 and 4, From Bradley: J. Gen. Microbiol., 31:435, 1963.
Figure 16–7   From Hyde and Randall: J. Bact., 91:1363, 1966.
Figure 16–8   From Margaretten, Morgan, Rosenkranz, and Rose: J. Bact., 91:823, 1966.
Figure 16–10   From Cummings, Chapman, deLong, and Mondale: J. Virol., 1:193, 1967.
Figure 16–12   From Howe, Morgan, St. Cyr, deVaux, Hsu, and Rose: J. Virol., 1:215, 1967.

Figure 17–2   Redrawn from Cowdry and Kitchen: Am. J. Hyg., 11:227, 1930.
Figure 17–3   From McAllister, Landing, and Goodheart: Lab. Invest., 13:894, 1964.
Figure 17–4   From McAllister, Wright, and Tasem: J. Pediat., 64:278, 1964.
Figure 17–5   From Goodheart: An Introduction to Virology, W. B. Saunders Co., 1969.
Figure 17–6   From Burrows: Textbook of Microbiology, 20th ed., W. B. Saunders Co., 1973; after Schwerdt.
Figure 17–7   From Jawetz, Melnick, and Adelberg: Review of Medical Microbiology, 6th ed., Lange Medical Publications, 1964.
Figure 17–8   From Braun: Bacterial Genetics, 2nd ed., W. B. Saunders Co., 1965.
Figure 17–9   From Ikari, Robbins, and Parr: Proc. Soc. Exp. Biol. Med., vol. 98.

Figure 18–4   From Avery, MacLeod, Colin, and McCarty: J. Exp. Med., vol. 79.
Figure 18–5   From Barber: J. Gen. Microbiol., vol. 13.
Figure 18–6   From Shimwell and Carr: Ant. Leeuw., 26:169, 1960.
Figure 18–7   Adapted from Szybalski: Science, vol. 116.

Figure 19–1   From Adelberg and Pittard: Bact. Rev., 29:161, 1965.
Figure 19–2   From Hayes: The Genetics of Bacteria and Their Viruses, John Wiley & Sons, Inc., 1964.
Figure 19–3   From Jacob, F.: Science, 152:1470–1478 (June 10), 1966. Copyright 1966 by the American Association for the Advancement of Science.

Figure 20–7   Modified from The Westinghouse Sterilamp and the Rentschler-James Process of Sterilization, Westinghouse Electric & Manufacturing Co., Inc.
Figure 20–10   From Marr and Cota-Robles: J. Bact., vol. 74.

Figure 21–3   Modified from Walker and Harmon: Appl. Microbiol., 14:584, 1966.

Figure 22–3   From Perkins: Principles and Methods of Sterilization, Charles C Thomas, 1960.

Figure 23–8   From Fairbrother and Rao: J. Clin. Pathol., vol. 7.

Figure 25–1   From Domonkos: Andrews' Diseases of the Skin, 6th ed., W. B. Saunders Co., 1971.
Figure 25–3   From Witten: Microbiology with Application to Nursing, 2nd ed., McGraw-Hill Book Co.
Figure 25–5   From Wilson and Podas: Mod. Sanit., May, 1950.
Figure 25–6   From Barton: Appl. Microbiol., vol. 2.
Figure 25–7   From Stitt, Clough, and Branham: Practical Bacteriology, Hematology, and Parasitology, McGraw Hill Book Co.
Figure 25–8   Adapted from Williams, Zumbro, and MacDonald, in: Animal Diseases, The 1956 Year Book of Agriculture, U.S. Department of Agriculture.

Figure 26–4   From Smith and Wood: J. Exp. Med., vol. 86.

Figure 27–2   Modified from Roitt: Essential Immunology, Blackwell Scientific Publications, 1971.
Figure 27–3   Modified from Alexander and Good: Immunobiology for Surgeons, W. B. Saunders Co., 1970.
Figure 27–4   From Carpenter: Immunology and Serology, 2nd ed., W. B. Saunders Co., 1965.

Figure 33–9   From Poindexter: Microbiology: An Introduction to Protists, © Copyright, The Macmillan Co., 1971.
Figure 33–11   From Allen and Baumann: J. Bact., *107*:295, 1971.
Figure 33–13. *A, B, D,* and *E,* From Williams and Rittenberg: Int. Bull. Bact. Nomen. Taxon., vol. 7. *F,* From Clark-Walker: J. Bact., *97*:885, 1969.
Figure 33–14   From McElroy, Wells, and Krieg: J. Bact., *93*:499, 1967.
Figure 33–15   *A,* From Seidler and Starr: J. Bact., *95*:1952, 1968. *B,* From Stolp, H., und Petzold, H.: Untersuchungen über einen obligat parasitischen Mikroorganismus mit lytischer Aktivität für Pseudomonas-Bakterien. Phytopathol. Z., *45*:364–390, 1962. *C* and *D,* From Burnham, Hashimoto, and Conti: J. Bact., *96*:1366, 1968. *E,* From Starr and Baigent: J. Bact., *91*:2006, 1966.
Figure 33–16   *A,* From Stolp and Starr: Ant. Leeuw., *29*:217, 1963. *B,* From Diedrich: J. Bact., *101*:989, 1970.
Figure 33–17   From Burnham, Hashimoto, and Conti: J. Bact., *101*:997, 1970.

Figure 34–1   From Faust and Wolfe: J. Bact., vol. 81.
Figure 34–2   From Faust and Wolfe: J. Bact., vol. 81.
Figure 34–3   From Skerman: A Guide to the Identification of the Genera of Bacteria, 2nd ed., The Williams & Wilkins Co., 1967.
Figure 34–4   *A,* From Poindexter: Microbiology: An Introduction to Protists, © Copyright, The Macmillan Co., 1971, *B* and *C,* From Harold and Stanier: Bact. Rev., *19*:49, 1955.
Figure 34–5   *A,* From Stokes: J. Bact., *67*:278, 1954. *B,* From Dondero, Phillips, and Heukelekian: Appl. Microbiol., vol. 9.
Figure 34–6   From Heukelekian and Dondero (Eds.): Principles and Applications in Aquatic Microbiology, John Wiley & Sons, Inc., 1964.
Figure 34–7   From Cohen-Bazire, Kunisawa, and Poindexter: J. Gen. Microbiol., *42*:301, 1966.
Figure 34–8   *A,* From Poindexter: Bact. Rev., *28*:231, 1964. *B,* From Schmidt and Samuelson: J. Bact., *112*:593, 1972.
Figure 34–9   Adapted from Schmidt and Stanier: J. Cell Biol., *28*:423, 1966.
Figure 34–10   From Staley: J. Bact., *95*:1921, 1968.
Figure 34–11   *A,* From Tyler and Marshall: J. Bact., *93*:1132, 1967. *B,* From Conti and Hirsch: J. Bact., *89*:503, 1965.
Figure 34–12   From Staley: J. Bact., *95*:1921, 1968.
Figure 34–13   *A* to *C,* From N. Cholodny, 1929, and reproduced by Starkey, *in* Thimann: The Life of Bacteria, 2nd ed. © Copyright, The Macmillan Co., 1963. *D,* From Heukelekian and Dondero (Eds.): Principles and Applications in Aquatic Microbiology, John Wiley & Sons, Inc., 1964.

Figure 35–1   From Curtiss, Carol, Allison, and Stallions: J. Bact., *100*:1091, 1969.
Figure 35–2   From Ewing and Davis: Int. J. Sys. Bact., *22*:12, 1972.
Figure 35–3   Courtesy of E. Leifson and the Journal of Bacteriology.
Figure 35–4   Reprinted by permission from J. P. Duguid and J. F. Wilkinson (Eds.): Eleventh Symposium, Society for General Microbiology, Cambridge University Press, 1961.
Figure 35–5   From Ewing and Fife: Int. J. Sys. Bact., *22*:4, 1972.
Figure 35–6   From Jones and Park: J. Gen. Microbiol., *47*:369, 1967.
Figure 35–7   From Carpenter: Microbiology, 3rd ed., W. B. Saunders Co., 1972; ASM LS-258.
Figure 35–11   From Lane: Appl. Microbiol., *16*:1400, 1968.
Figure 35–12   From Richter and Kress: J. Bact., *94*:1216, 1967.
Figure 35–13   From Bottone and Allerhand: Appl. Microbiol., *16*:315, 1968.
Figure 35–14   From Criswell, Stenback, Black, and Gardner: *109*:930, 1972.

Figure 36–1   From Bradley and Williams: J. Gen. Microbiol., vol. 17.
Figure 36–2   *A,* Redrawn from Young and Fitz-James: J. Cell Biol., *12*:115, 1962. *B,* From Walker and Short: J. Bact., *98*:1342, 1969.
Figure 36–3   *A,* From Walker and Short: J. Bact., *98*:1342. *B,* From Williamson and Wilkinson: J. Gen. Microbiol., *19*:198, 1958.
Figure 36–4   From Santo, Hohl, and Frank: J. Bact., *99*:824, 1969. Drawings adapted from Murrell, *in* Rose and Wilkinson (Eds.): Advances in Microbial Physiology, vol. 1, Academic Press-London, 1967.

Figure 36–5   From Santo, Hohl, and Frank: J. Bact., 99:824, 1969.

Figure 36–7   From Hoeniger and Headley: J. Bact., 96:1835, 1968.

Figure 36–8   From Sharpe et al.: Appl. Microbiol., 19:681, 1970.

Figure 36–9   A, From Hannay and FitzJames: Can. J. Microbiol., 1:694, 1955. B, From Norris and Proctor: J. Bact., 98:824, 1969.

Figure 36–10   A, From Rode and Smith: J. Bact., 105:349, 1971. B to K, From Rode, Crawford, and Williams: J. Bact., 93:1160, 1967.

Figure 36–13   From Burrows: Textbook of Microbiology, 20th ed., W. B. Saunders Co., 1973.

Figure 36–14   Redrawn from Gillies and Dodd: Bacteriology Illustrated, The Williams & Wilkins Co., 1965.

Figure 38–3   From Mudd, Heinmets, and Anderson: J. Exp. Med., vol. 78.

Figure 39–1   From Deibel et al.: J. Bact., vol. 71.

Figure 39–2   From Blair, Lennette, and Truant (Eds.): Manual of Clinical Microbiology, American Society for Microbiology, The Williams & Wilkins Co., 1970.

Figure 39–3   From Pringsheim and Robinow: J. Gen. Microbiol., 1:278, 1947.

Figure 39–6   From Krulwich, Ensign, Tipper, and Strominger: J. Bact., 94:734, 1967.

Figure 40–1   From Slack, Landfried, and Gerencser: J. Bact., 97:873, 1969.

Figure 40–2   A, From Kojima, Suda, Hotta, Hamada, and Suganuma: J. Bact., 104:1010, 1970. B and C, From Kojima, Suda, Hotta, and Hamada: J. Bact., 95:710, 1968.

Figure 40–3   From Lorian: Appl. Microbiol., 14:603, 1966.

Figure 40–4   From Dublin: Am. J. Pub. Health, vol. 48, and Statistical Bureau, Metropolitan Life Insurance Co.

Figure 40–5   From Hunter, Frye, and Swartzwelder: Manual of Tropical Medicine, 4th ed., W. B. Saunders Co., 1966.

Figure 40–6   From Luedemann: J. Bact., 96:1848, 1968.

Figure 40–7   From Beaman and Shankel: J. Bact., 99:876, 1969.

Figure 40–8   From Bradley and Ritzi: J. Bact., 95:2358, 1968.

Figure 40–9   Adapted from Pridham, Hesseltine, and Benedict: Appl. Microbiol., vol. 6.

Figure 40–10   From Tresner, Davies, and Backus: J. Bact., vol. 81.

Figure 40–11   From Luedemann: Adv. Appl. Microbiol., 11:101, 1969.

Figure 40–12   From Krassilnikov: Mikrobiologica, 33:935, 1964.

Figure 40–13   From Lechevalier and Holbert: J. Bact., 89:217, 1965.

Figure 40–14   From Higgins, Lechevalier, and Lechevalier: J. Bact., 93:1446, 1967.

Figure 41–1   A, © E. J. Ordal. From Henrici and Ordal: Biology of Bacteria, 3rd ed., D. C. Heath & Co., 1948. B, From Stanier et al.: Microbial World, 3rd ed., Prentice-Hall, Inc., 1970. C, From Voelz and Reichenbach: J. Bact., 99:856, 1969.

Figure 41–2   From Dworkin: J. Bact., 86:67, 1963.

Figure 41–3   From Bonner: Morphogenesis, 1952. Reprinted by permission of Princeton University Press.

Figure 41–4   A, From Stanier et al.: Microbial World, 3rd ed., Prentice-Hall, Inc., 1970. B, From Voelz and Reichenbach: J. Bact., 99:856, 1969.

Figure 41–5   From Voelz and Reichenbach: J. Bact., 99:856, 1969.

Figure 41–6   From Voelz and Reichenbach: J. Bact., 99:856, 1969.

Figure 41–7   From Burchard: J. Bact., 104:940, 1970.

Figure 41–8   From Shilo: J. Bact., 104:453, 1970.

Figure 41–9   A, From Burrows: Textbook of Microbiology, 20th ed., W. B. Saunders Co., 1973. B to D, From Listgarten and Socransky: J. Bact., 88:1087, 1964.

Figure 41–10   A, From Holt and Canale-Parola: J. Bact., 96:822, 1968. B and C, From Breznak and Canale-Parola: J. Bact., 97:386, 1969.

Figure 41–11   From Listgarten and Socransky: J. Bact., 88:1087, 1964.

Figure 41–12   B, From Thomas et al.: Appl. Microbiol., 23:714, 1972.

Figure 41–14   A, From Martin: Thirteen Steps to the Atom, Franklin Watts, Inc. B, From Burrows: Textbook of Microbiology, 20th ed., W. B. Saunders Co., 1973.

Figure 41–15 From Anderson and Johnson: J. Bact., *95*:2293, 1968.

Figure 42–1 From Boatman: J. Bact., *106*:1005, 1971.
Figure 42–2 From Knudson and MacLeod: J. Bact., *101*:609, 1970.
Figure 42–3 From Tully and Razin: J. Bact., *98*:970, 1969.
Figure 42–4 *A* to *C*, From Biberfeld and Biberfeld: J. Bact., *102*:856, 1970. *D*, From Lemcke: J. Bact., *110*:1154, 1972.
Figure 42–5 From Anderson and Barile: J. Bact., *90*:180, 1965.
Figure 42–6 From Black, Birch-Andersen, and Freundt: J. Bact., *111*:254, 1972.
Figure 42–7 From Rottem and Razin: J. Bact., *110*:699, 1972.
Figure 42–8 From Thacore and Willet: Proc. Soc. Exp. Biol. Med., *114*:43, 1963.
Figure 42–9 From Hatten et al.: J. Bact., *99*:611, 1969.
Figure 42–10 *A*, From Gooder and Wyrick: J. Bact., *105*:646, 1971. *B*, From Wright and Stewart: J. Bact., *99*:899, 1969. *C*, From Kagan, Molander, and Weinberger: J. Bact., vol. 83.
Figure 42–11 From Rodriguez Aguirre, Garcia Acha, and Villanueva: Ant. Leeuw., *30*:30, 1964.
Figure 42–12 From Anderson et al.: J. Bact., *90*:1387, 1965.
Figure 42–13 From Brinton and Burgdorfer: J. Bact., *105*:1149, 1971.
Figure 42–15 From Friis: J. Bact., *110*:706, 1972.
Figure 42–16 From Friis: J. Bact., *110*:706, 1972.
Figure 42–17 From Anderson et al.: J. Bact., *90*:1387, 1965.
Figure 42–18 From Moulder: The Psittacosis Group as Bacteria, John Wiley & Sons, Inc., 1964.

Figure 43–1 From Waksman and Starkey: Soil Microbiology, John Wiley & Sons, Inc., 1952.
Figure 43–8 *A*, From Scolofsky and Wyss: J. Bact., vol. 81. *B* to *D*, From Cagle and Vela: J. Bact., *112*:615, 1972.
Figure 43–9 From Sarles, Frazier, Wilson, and Knight: Microbiology: General and Applied, Harper & Bros., 1951.
Figure 43–10 *B*, From Fahraeus: J. Gen. Microbiol., vol. 16.
Figure 43–11 From Goodchild and Bergersen: J. Bact., *92*:204, 1966.
Figure 43–12 From Waksman and Starkey: Soil Microbiology, John Wiley & Sons, Inc., 1952.
Figure 43–13 From Zak: Can. J. Bot., *49*:1079, 1971.
Figure 43–14 From Watson and Mandel: J. Bact., *107*:563, 1971.
Figure 43–16 From Kinsel: J. Bact., vol. 80.
Figure 43–17 From Campbell, Frank, and Hall: J. Bact., vol. 73.
Figure 43–19 From Huang and Goodman: J. Bact.: *102*:862, 1970.

Figure 44–1 Redrawn from Adams, Baker, and Allen: The Study of Botany, Addison-Wesley Publishing Company, 1970.
Figure 44–2 From Sawyer: J. Water Poll. Contr., *38*:737, 1966.
Figure 44–3 From Hasler: Ecology, *28*:383, 1947.
Figure 44–4 Redrawn from Ford and Monroe: Living Systems: Principles and Relationships, Canfield Press, 1971.
Figure 44–6 From Brock, Brock, Bott, and Edwards: J. Bact., *107*:303, 1971.
Figure 44–8 From Inger, Hasler, Bormann, and Blair (Eds.): Man in the Living Environment, University of Wisconsin Press, 1972. © Board of Regents of the University of Wisconsin. Redrawn from Teal and Teal: Life and Death of the Salt Marsh, Little, Brown & Co., 1969.
Figure 44–9 From Larkin and Stokes: J. Bact., *91*:1667, 1966.
Figure 44–10 From Larkin and Stokes: J. Bact., *94*:889, 1967.
Figure 44–11 From Staley: J. Bact., *95*:1921, 1968.
Figure 44–12 From Smith: J. Bact., *103*:811, 1970.
Figure 44–13 From Marine Microbial Ecology, by E. J. F. Wood. © 1965 by Litton Educational Publishing, Inc. Reprinted by permission of Van Nostrand Reinhold Company.
Figure 44–14 From Felter, Colwell, and Chapman: J. Bact., *99*:326, 1969.
Figure 44–15 From Baumann, Baumann, Mandel, and Allen: J. Bact., *110*:402, 1972.
Figure 44–16 *E*, From Marine Microbial Ecology, by E. J. F. Wood. © 1965 by Litton Educational Publishing, Inc. Reprinted by permission of Van Nostrand Reinhold Company.

Figure 45–1   1 and 2, From Keup, Ingram, and Mackenthun: Biology of Water Pollution, Federal Water Pollution Control Administration, U.S. Department of Interior, 1967. 3 and 4, From Wolfe, in Heukelekian and Dondero (Eds.): Principles and Applications in Aquatic Microbiology, John Wiley & Sons, Inc., 1964.

Figure 45–2   Redrawn from Manual of Instruction for Water Treatment Plant Operators, New York State Department of Health, Albany, N.Y.

Figure 45–3   Redrawn from Manual of Instruction for Water Treatment Plant Operators, New York State Department of Health, Albany, N.Y.

Figure 45–4   Redrawn from Manual of Instruction for Water Treatment Plant Operators, New York State Department of Health, Albany, N.Y.

Figure 45–5   Redrawn from Manual of Instruction for Water Treatment Plant Operators, New York State Department of Health, Albany, N.Y.

Figure 45–6   Redrawn with permission from Gentry: Env. Sci. Tech., 1:124, 1967. Copyright by the American Chemical Society.

Figure 45–7   Redrawn with permission from Gentry: Env. Sci. Tech., 1:124, 1967. Copyright by the American Chemical Society.

Figure 45–8   Redrawn from Manual of Instruction for Water Treatment Plant Operators, New York State Department of Health, Albany, N.Y.

Figure 45–9   Redrawn from Sanitron Form #200, Atlantic Ultraviolet Corp., Long Island City, N.Y., 1972.

Figure 45–11 Adapted from Standard Methods for the Examination of Water and Wastewater, 13th ed., American Public Health Association, 1971.

Figure 45–12 C and D, From Todd and Kerr: Appl. Microbiol., 23:1160, 1972.

Figure 45–13 From Bulletin No. 16, Engineering Experiment Station, University of Washington.

Figure 45–14 From Keefer: Sewage Treatment—How It Is Accomplished, Smithsonian Institution Publication No. 4281.

Figure 45–15 Redrawn with permission from Dryden: Env. Sci. Tech., 2:269, 1968. Copyright by the American Chemical Society.

Figure 45–16 Redrawn with permission from Dryden: Env. Sci. Tech., 2:269, 1968. Copyright by the American Chemical Society.

Figure 45–17 A, From Keefer: Sewage Treatment—How It Is Accomplished, Smithsonian Institution Publication No. 4281. C, From Fair and Geyer: Water Supply and Wastewater Disposal, John Wiley & Sons, Inc., 1954.

Figure 45–18 B, From Fair and Geyer: Water Supply and Wastewater Disposal, John Wiley & Sons, Inc., 1954. C, From Keefer: Sewage Treatment—How It Is Accomplished, Smithsonian Institution Publication No. 4281.

Figure 45–19 From Amin and Ganapati: Appl. Microbiol., 15:17, 1967.

Figure 45–20 From Taylor et al.: Am. J. Pub. Health, 56:2093, 1966.

Figure 46–1   Reproduced in part from Ellis and Raymond, in Medical Research Council: Studies in Air Hygiene, Special Report Series 262, 1948. By permission of the Controller of Her Britannic Majesty's Stationery Office, London.

Figure 46–5   A and B, From duBuy, Hollaender, and Lackey: Supplement No. 184, U.S. Public Health Service.

Figure 46–6   Reproduced from Wheeler et al.: Science, vol. 94.

Figure 46–7   From May: Bact. Rev., 30:599, 1966.

Figure 46–8   From Burrows: Textbook of Microbiology, 20th ed., W. B. Saunders Co., 1973.

Figure 46–9   Redrawn from Form SC 090 168 R-70, American Ultraviolet Company, Chatham, N.J.

Figure 46–10 Redrawn from Form SC 090 168 R-70, American Ultraviolet Company, Chatham, N.J.

Figure 46–11 Redrawn from Form SC 090 168 R-70, American Ultraviolet Company, Chatham, N.J.

Figure 47–2   Data of Ayers et al.: U.S.D.A. Bull. 642, 1918.

Figure 47–3   From Sarles et al.: Microbiology: General and Applied, 2nd ed., Harper & Row, 1956.

Figure 47–4   From Carpenter: Microbiology, 3rd ed., W. B. Saunders Co., 1972.

Figure 47–5   From Sarles et al.: Microbiology: General and Applied, 2nd ed., Harper & Row, 1956.

Figure 47–7   From Brown and Price: J. Dairy Sci., vol. 17.
Figure 47–14   *C* and *D*, From Foster et al.: Dairy Microbiology, Prentice-Hall, Inc.

Figure 48–1   Redrawn from Angelotti, Foter, and Lewis: Am. J. Pub. Health, *51*:76, 1961.
Figure 48–2   From Sarles et al.: Microbiology: General and Applied, 2nd ed., Harper & Row, 1956.

Figure 49–3   From Altsheler, Mollet, Brown, Stark, and Smith: Chem. Eng. Progr. (Trans. Sec.), *43*(9):467, 1947.
Figure 49–5   Reprinted with permission from Schneider: J. Agr. Food Chem., 1:241, 1953. Copyright by the American Chemical Society.
Figure 49–6   After Jawetz, Melnick, and Adelberg: Review of Medical Microbiology, 6th ed., Lange Medical Publications, 1964.
Figure 49–7   After Jawetz, Melnick, and Adelberg: Review of Medical Microbiology, 6th ed., Lange Medical Publications, 1964.
Figure 49–9   From Hansen: Food Indust., vol. 7.
Figure 49–10   From Frateur: La Cellule, *53*:3, 1950.

Color Plate III   *B*, From Kramer: Comparative Pathogenic Bacteriology: A Filmstrip Presentation, W. B. Saunders Co., 1972. *C*, From Lynch and Raphael: Medical Laboratory Technology and Clinical Pathology: A Filmstrip Presentation, W. B. Saunders Co., 1970.
Color Plate V   From Kramer: Comparative Pathogenic Bacteriology: A Filmstrip Presentation, W. B. Saunders Co., 1972.
Color Plate VI   *A*, From Johnson and Dietz: Appl. Microbiol., *17*:755, 1969. *B* and *D*, From Kramer: Comparative Pathogenic Bacteriology: A Filmstrip Presentation, W. B. Saunders Co., 1972. *C*, From Lynch and Raphael: Medical Laboratory Technology and Clinical Pathology: A Filmstrip Presentation, W. B. Saunders Co., 1970.
Color Plate VII   From Gordon and Pang: Appl. Microbiol., *19*:862, 1970.
Color Plate VIII   *C*, From Jones, Netterville, Johnston, and Wood: Chemistry, Man and Society, W. B. Saunders Co., 1972.

index

Note: "g" refers to glossary entry.

ammonia, detection of, 423
oxidation of, 669
to nitrite, 669
ammonification, g, 669
ammonium halides, 310
Amoeba proteus, 184
amphitrichous, g
amphoteric substances, 70, 282
amylase, g, 92
amylose, 73
Ancalomicrobium, gas vacuoles in, 497
anaerobe(s), g, 98, 547
cultivation of, 548
facultative, 98
obligate, 118
strict, 104, 107
anaerobic bacteria. See *anaerobe(s)*.
anaerobic methods, 548
anaerobic respiration, 105, 118
anaerobiosis, 89, 545
hydrogen peroxide and, 547
thioglycollate in, 549
analogues, base, 254
metabolite, 97, 317
anamnestic response, 395, 397
anaphylactic-type reactions, 401
anaphylatoxin, 364
anaphylaxis, g, 402
anemia, pernicious, 122
angular aperture, 37, 38
animal(s), axenic, 334
germ-free, 334
gnotobiotic, 334
animal bites, as disease vectors, 350
animal kingdom, origin of, 176
animal passage, 336, 396
animal reservoir, 512
Animalia, 150, 152
anion, 60
anionics, 311
annealing, DNA, 444
anode, 60
Anopheles mosquitoes, 188–190
antagonism(s), g, 332
chemical, 295
reversal of, 295
metabolite, 97, 317
reversal of, 317
in natural waters, 682
ionic, 283
antheridium, 163
anthrax, 537
control of, 538
toxins of, 538
treatment of, 538
anthrax bacterins, 538
anthrax vaccine, 398
antibiosis, 332
antibiotic(s), g, 319–331
and plant diseases, 679
assay of, 330
broad-spectrum, 328
from *Bacillus*, 329
from molds, 319
from *Streptomyces*, 327, 597
in milk, 740
macrolide, 329
nonmedical uses of, 329
polypeptide, 329
producers of, 10

antibiotic(s) *(continued)*
resistance to, 317
sensitivity testing of, 330
standardization of, 330
surfactant, action of, 329
therapy with, presurgical, 333
antibody(ies), g
adsorption of, 379
and mycoplasmas, 626
and phagocytes, 359
formation of, theories of, 374
immobilizing, 389
labeled, 389
maternal, 399
molecular structure of, 372
natural, 362
nature of, 370–374
neutralizing, 389
nonspecific, 362
opsonic effect of, 360
protective, 389
Rh, 399
specific, 32, 359
antibody staining, fluorescent, 389
anticodon, g, 85
antigen(s), g, 362
auto-, 368
blocking, 380
capsular, 378
complex, 377
determinant groups, 366
envelope, 378
extracellular, 379
fimbrial, 378
flagellar, 377, 378
group, 379
H, 377, 378
heterogenetic, 379
immunogenicity of, 367
immunoglobulins as, 373
inactivated organisms as, 396
incomplete, 380
iso-, 367
K, 378
mixed, 379
nature of, 366–369
O, 378
of chlamydias, 640
of rickettsias, 637
partial, 380
Proteus OX, 637
route of entry, enteral, 367
parenteral, 367
shared, 378, 379
somatic, 378
synthetic, 375
tumor, 251
antigen-antibody combinations, 375–376
antigen-antibody reactions, stages of, 377
antigenic stimulus, primary and secondary, 395
antigenic structure, of bacteria, 445
antilymphocyte serum (ALS), 368
antimetabolite, 317
antimicrobial, g
antimicrobial drugs. See *drugs*.
antimicrobial factors, nonspecific, 362
anti-Rh immunoglobulins, 404
antiseptic, g, 290
antiserum, g